Springer Monographs in Mathematics

This series publishes advanced monographs giving well-written presentations of the "state-of-the-art" in fields of mathematical research that have acquired the maturity needed for such a treatment. They are sufficiently self-contained to be accessible to more than just the intimate specialists of the subject, and sufficiently comprehensive to remain valuable references for many years. Besides the current state of knowledge in its field, an SMM volume should ideally describe its relevance to and interaction with neighbouring fields of mathematics, and give pointers to future directions of research.

More information about this series at http://www.springer.com/series/3733

V.I. Bogachev · O.G. Smolyanov

Topological Vector Spaces and Their Applications

Springer

V.I. Bogachev
Department of Mechanics
and Mathematics
Moscow State University
Moscow
Russia

and

Higher School of Economics
National Research University
Moscow
Russia

O.G. Smolyanov
Department of Mechanics
and Mathematics
Moscow State University
Moscow
Russia

ISSN 1439-7382 ISSN 2196-9922 (electronic)
Springer Monographs in Mathematics
ISBN 978-3-319-57116-4 ISBN 978-3-319-57117-1 (eBook)
DOI 10.1007/978-3-319-57117-1

Library of Congress Control Number: 2017939903

Mathematics Subject Classification: 46A03, 58C20, 28C20

The book is an expanded and revised version of the Russian edition under the same title, published in Regular and Chaotic Dynamics, Moscow-Izhevsk, 2012.

Printed on acid-free paper

This Springer imprint is published by Springer Nature
The registered company is Springer International Publishing AG
The registered company address is: Gewerbestrasse 11, 6330 Cham, Switzerland

Contents

Preface

This book gives a concise exposition of the fundamentals of the theory of topological vector spaces, complemented by a survey of the most important results of a more subtle nature, which cannot be qualified as basic, but knowledge of which is useful for applications, and, finally, some of such applications connected with differential calculus in infinite-dimensional spaces and measure theory. Almost half of the book is devoted to these applications, which makes it very different from the whole series of known texts on topological vector spaces. Another notable difference between this book and known treatises like Bourbaki [**87**], Edwards [**150**], Grothendieck [**207**], Jarchow [**237**], Kelley, Namioka [**270**], Köthe [**292**], Narici, Beckenstein [**365**], Pérez Carreras, Bonet [**385**], Robertson, Robertson [**420**], Schaefer [**436**], Trèves [**530**], and Wilansky [**567**] is that we decided to include also some results without proofs (this does not concern the fundamentals, of course) with references instead, which enables us to inform the reader about many relatively recent achievements; some of them are disguised as exercises (with references to the literature), such exercises should not be confused with usual exercises marked by the symbol °. Thus, with respect to the presented information, our book is not covered by any other book on this subject (though, we cannot claim that it covers any such book).

Chapter 1 contains the fundamentals of the theory, including a large list of concrete examples, some general concepts (convex sets, seminorms, linear mappings) and a number of facts, the most important of which is the Hahn–Banach theorem on extensions of functionals in its diverse versions.

The main material of Chapter 2 is connected with projective and inductive limits (including strict inductive limits and inductive limits with compact embeddings, which is not sufficiently discussed in the existing literature), and also Grothendieck's method of constructing Banach spaces embedded into locally convex spaces.

Chapter 3 contains the classical material related to the so-called duality theory, i.e., introduction of different locally convex topologies on a given space giving the same set of continuous linear functionals. The central topics here are the Mackey–Arens theorem on topologies compatible with duality, the results on weak compactness, including the Eberlein–Šmulian and Krein–Šmulian theorems, and also some concepts and facts connected with completeness of locally convex spaces.

Chapter 4 is devoted to the fundamentals of the differentiation theory in locally convex spaces. It presents a general scheme of differentiability with respect to a system of sets (partial cases of which are Gâteaux, Hadamard and Fréchet differentiabilities) and a thorough discussion of important for applications differentiabilities with respect to systems of bounded and compact sets.

Chapter 5 gives a concise introduction to measure theory on locally convex spaces. Here we discuss extensions of cylindrical measures, the Fourier transform and conditions for the countable additivity in its terms (in particular, the Minlos and Sazonov theorems and their generalizations), covariance operators, measurable linear functionals and operators, measurable polynomials, and some important classes of measures (such as Gaussian, stable, and convex).

Each chapter opens with a brief synopsis of its content. All chapters contain many additional subsections with some more specialized information related to the main themes of the chapter, and also many exercises are given (more difficult ones are provided with hints or references). The book ends with the historic-bibliographic comments, the list of references (with indication of page numbers of citing the included works), and the author and subject indices.

The prerequisites for the first chapter of this book are just a grasp knowledge of calculus and linear algebra and some experience with basic concepts of topology, but for a thorough study it is advisable to be acquainted with a university course of functional analysis (following any text, e.g., Kolmogorov, Fomin [**284**] or Rudin [**425**]).

We are very grateful to T.O. Banakh, E.D. Kosov, I. Marshall, S.N. Popova, A.V. Shaposhnikov, A.S. Tregubov and E.V. Yurova for useful remarks and corrections.

Our work on this book began 25 years ago by the initiative of Vladimir Ivanovich Sobolev (1913–1995), the author of a series of widely known texts on functional analysis (including one of the first Russian texts, published as early as in 1951), and its completion is a tribute to the memory of this remarkable scientist and teacher.

CHAPTER 1

Introduction to the theory of topological vector spaces

In this chapter we present basic concepts and examples connected with topological vector spaces.

1.1. Linear spaces and topology

A topological vector space is a linear (or vector) space equipped with a topology which agrees with the linear structure. So we first recall separately basic concepts related to linear spaces and topological spaces. Let $\mathbb{K}$ be an algebraic field (throughout we deal with the field $\mathbb{R}$ of real numbers or with the field $\mathbb{C}$ of complex numbers; so a reader not acquainted with the general notion of an algebraic field can safely ignore it). A set E is called a *linear* (or *vector*) space over the field $\mathbb{K}$ if the elements of E (called vectors) can be added and multiplied by the elements of $\mathbb{K}$, i.e., we are given mappings

$$E\times E \to E,\ (u,v)\mapsto u+v, \quad \mathbb{K}\times E \to E,\ (\lambda,v)\mapsto \lambda v$$

satisfying the following conditions:

(i) $u+v=v+u$ for all $u,v\in E$,

(ii) there is a unique element $0\in E$ (the zero element) for which $v+0=v$ for all $v\in E$,

(iii) for every $v\in E$ there is a unique element $-v$ for which $v+(-v)=0$,

(iv) $\lambda(u+v)=\lambda u+\lambda v$, $\lambda(\mu v)=(\lambda\mu)v$, $(\lambda+\mu)u=\lambda u+\mu u$ and $0v=\lambda 0=0$ for all $u,v\in E$ and $\lambda,\mu\in\mathbb{K}$.

Below we often omit the explicit mentioning the field $\mathbb{K}$ and its elements will be called scalars, and in case of $\mathbb{K}\subset\mathbb{C}$ they will be called numbers. About general fields, see Kurosh [**306**].

1.1.1. Example. Let $\mathbb{K}=\mathbb{R}$ and let T be a nonempty set. Let $\mathbb{R}^T$ be the set of all real functions on T, where the linear operations are defined pointwise:

$$(f+g)(t):=f(t)+g(t), \quad (\lambda f)(t):=\lambda f(t).$$

Then $\mathbb{R}^T$ is a linear space; it is called the product of T copies of the real line or a *power of the real line*.

V.I. Bogachev and O.G. Smolyanov, *Topological Vector Spaces and Their Applications*, Springer Monographs in Mathematics,
DOI 10.1007/978-3-319-57117-1_1

Throughout, if it is not explicitly stated otherwise, we assume that $\mathbb{K}$ is a nondiscrete normed field. A norm on the field $\mathbb{K}$ is a mapping $\mathbb{K} \to [0, +\infty)$ (its value on the element $x \in \mathbb{K}$ is denoted by $|x|$) such that the following conditions are fulfilled: $|x| > 0$ for $x \in \mathbb{K} \setminus \{0\}$ (nondegeneracy), $|0| = 0$, $|xy| = |x|\,|y|$ (multiplicativity) and $|x + y| \leqslant |x| + |y|$ (the triangle inequality) for all $x, y \in \mathbb{K}$. A field equipped with a norm is called a *normed field.* For example, the field $\mathbb{C}$ of complex numbers is normed if $|a|$ is the usual absolute value of the number $a \in \mathbb{C}$. A field is *nondiscrete* if it has an element $k \neq 0$ with $|k| \neq 1$.

A *seminorm* on a vector space E is a function $p\colon\ E \to [0, \infty)$ such that

(1) $p(kx) = |k|p(x)\ \ \forall\, k \in \mathbb{K},\ x \in E$;

(2) $p(x_1 + x_2) \leqslant p(x_1) + p(x_2)\ \ \forall\, x_1 \in E,\ x_2 \in E$.

A seminorm p is called a *norm* if $p(x) > 0$ whenever $x \neq 0$. For example, $\|x\| = (x, x)^{1/2}$ is a norm on a *Euclidean* space E with an inner product $(\,\cdot\,,\,\cdot\,)$, i.e., $x \mapsto (x, y)$ is linear, $(x, y) \equiv (y, x)$ for real E and $(x, y) \equiv \overline{(y, x)}$ for complex E, $(x, x) \geqslant 0$ and $(x, x) = 0$ only for $x = 0$.

Two norms p and q are called *equivalent* if for some numbers $c_1, c_2 > 0$ and all x the inequality $c_1 p(x) \leqslant q(x) \leqslant c_2 p(x)$ holds.

A collection of vectors in a linear space is called *linearly independent* if the equality $\lambda_1 v_1 + \cdots + \lambda_n v_n = 0$, where v_i are distinct vectors in this collection and λ_i are scalars, implies that $\lambda_i = 0$ for all $i = 1, \ldots, n$; otherwise this collection is *linearly dependent.*

A linearly independent collection of vectors v_α is called an *algebraic basis* (*Hamel's basis*) in the space X if every vector in X is a finite linear combination of the vectors v_α. In the zero space the zero element is a basis by definition. Below we prove the existence of a Hamel basis in every linear space; moreover, different Hamel bases have the same cardinality. The cardinality of a Hamel basis is called the *dimension* of the space.

Let E and F be two vector spaces over the same field. A mapping $A\colon\ E \to F$ is called *linear* (or a *linear operator*) if

$$A(\lambda u + \mu v) = \lambda A(u) + \mu A(v)$$

for all vectors $u, v \in E$ and all scalars λ, μ. A linear mapping with values in the field of scalars is called a *linear functional.*

The set $\operatorname{Ker} A := A^{-1}(0)$ is called the *kernel* of the linear mapping A and the set $\operatorname{Ran} A := A(E)$ is called the *range* of A.

For every vector space E, the symbol E^* denotes the vector space of all linear functions on E; it is called the *algebraic dual* to E. The algebraic dual space should not be confused with the *topological dual* considered below and consisting of all continuous linear functions. For the general theory and applications, the topological dual spaces are most important, but the algebraic dual is useful for some examples and constructions.

The quotient (or the factorspace) E/E_1 of a vector space E by its its subspace E_1 is defined as follows: the elements of E/E_1 are equivalence classes of the set E, where $x \sim z \Longleftrightarrow x - z \in E_1$. Thus, if $Z \in E/E_1$, then there exists (non unique) $z \in E$ such that $Z = z + E_1$. The linear operations in E/E_1 are defined as follows: suppose that $X = x + E_1$, $Z = z + E_1$, $\lambda \in \mathbb{K}$; then $X + Z = (x + z) + E_1$, $\lambda X = \lambda x + E_1$.

The dimension of the space E/E_1 is called the *codimension* of the subspace E_1 in the space E. A *hypersubspace* in a vector space E is its subspace G with $\dim E/G = 1$, i.e., there exists a nonzero vector v such that every vector in E is a linear combination of v and some vector in G. In this case, we say that the codimension of G in E is one.

A subset Γ in a vector space E is called a *hyperplane* if E contains a hypersubspace G and an element a such that $a + G = \Gamma$ (then $a \in \Gamma$). In other words, a subset Γ in vector space E is called a hyperplane precisely when for some (hence for any) element $b \in \Gamma$ the set $\Gamma - b$ is a hypersubspace (a hypersubspace is a hyperplane passing through the origin).

A subset A in a vector space E is called an *affine subspace* or a *linear manifold* if it is nonempty and for all $a, b \in A$ and every $t \in \mathbb{K}$ one has the inclusion $ta + (1 - t)b \in A$. The set $\{ta + (1 - t)b\colon\ t \in \mathbb{K}\}$ is (for $a \neq b$) the straight line passing through a and b. In other words, the set A is an affine subspace if it has the form $a + L$ for some vector subspace L and some (then for any) element $a \in A$.

The *linear span* of A is the smallest linear subspace containing A.

1.1.2. Definition. *A set V in a real or complex vector space is called convex if $tu + (1 - t)v \in V$ for all $u, v \in V$ and $t \in [0, 1]$.*

In other words, a set is convex if along with every two its points it contains the interval joining them. The interval $[a, b]$ with the endpoints a and b is defined by the equality

$$[a, b] := \{x\colon\ x = ta + (1 - t)b, t \in [0, 1]\}.$$

Set also

$$(a, b) := [a, b]\backslash\{a, b\}, \quad [a, b) := [a, b]\backslash\{b\}, \quad (a, b] := [a, b]\backslash\{a\}.$$

The *convex hull* (or *convex envelope*) of a nonempty set A in a real or complex vector space E is the intersection $\operatorname{conv} A$ of all convex sets containing A.

Thus, the convex hull of the set A is the smallest convex sets containing A. It is readily verified that it consists of all possible sums of the form $t_1a_1 + \cdots + t_na_n$, where $a_i \in A$, $t_i \geqslant 0$, $t_1 + \cdots + t_n = 1$.

1.1.3. Definition. *A set M is called circled or balanced if $\lambda x \in M$ for all $x \in M$ and $|\lambda| \leqslant 1$.*

A convex circled set is also called absolutely convex.

The *circled* and *convex circled* (or *absolutely convex*) hulls of a set A in a linear space are, respectively, the smallest circled set and the smallest convex circled set $\operatorname{abs\,conv} A$ containing A.

1.1.4. Definition. *If A and B are sets in a linear space E, then we say that A absorbs B (or that the set B is absorbed by the set A) if there exists a number $r > 0$ such that $kB \subset A$ whenever $|k| < r$, $k \in \mathbb{K}$.*

A set in E is called absorbing (or absorbent) if it absorbs every singleton (and then every finite set) in E.

A simple example of a set which does not absorb itself is $\mathbb{K}\setminus\{0\}$; every balanced set absorbs itself (take $r=1$). If a normed field $\mathbb{K}$ is discrete, then the property to absorb, although formally meaningful, has no useful content since then $\{0\}$ absorbs every set with $r=1$.

For nonempty sets A and B in a vector space and a scalar λ, we set

$$A+B:=\{a+b\colon\ a\in A, b\in B\},\quad \lambda A:=\{\lambda a\colon\ a\in A\};$$

$A+B$ is the *algebraic* (*vector*) *sum* of sets. Further,

$$A-B=A+(-B)=\{a-b\colon\ a\in A, b\in B\}.$$

Throughout we also use two set-theoretic concepts: the equivalence relation and the partial ordering relation.

Suppose that a certain set $\mathcal{R}$ if pairs of elements in the set X is given, i.e., a subset $\mathcal{R}\subset X\times X$. We say that $\mathcal{R}$ defines an *equivalence relation* on the set X and write $x\sim y$ for $(x,y)\in\mathcal{R}$ if

(i) $x\sim x$ for all $x\in X$,

(ii) if $x\sim y$, then $y\sim x$,

(iii) if $x\sim y$ and $y\sim z$, then $x\sim z$.

The reader can easily see by simple examples that these three conditions are independent.

The equivalence relation splits X into disjoint *equivalence classes* consisting of pairwise equivalent elements. For example, if $x\sim y$ only when $x=y$, then every class consists precisely of a single element; if, in the opposite direction, all elements are equivalent, then we obtain only one equivalence class. Yet another example: let $x\sim y$ for $x,y\in\mathbb{R}^1$, if $x-y\in\mathbb{Q}$. Then the equivalence classes are countable. If is often useful to be able to choose a representative in every equivalence class. It turns out that for realizing this, at the first glance absolutely innocent desire, a special axiom is required.

The axiom of choice. *If we are given a collection of nonempty pairwise disjoint sets, then there exists a set containing precisely one element from each of these sets.*

The use of this axiom is essential for many issues in functional analysis, and without this axiom at least for countable collections very little remains from the continuous mathematics at all. Nevertheless, it is useful to remember that this is indeed an axiom that does not follow from the so-called naive set theory.

We say that a set X is equipped with a *partial order* or *partial ordering* if a certain collection $\mathcal{P}$ of pairs $(x,y)\in X\times X$ is distinguished, for which we write $x\leqslant y$, such that (i) $x\leqslant x$, (ii) if $x\leqslant y$ and $y\leqslant z$, then $x\leqslant z$. If $x\leqslant y$, then we also write $y\geqslant x$. Note that we do not include the equality $x=y$ in the case where $x\leqslant y$ and $y\leqslant x$, unlike some other authors (though, one can pass to this case by identifying such elements). For example, our definition applies to the relation $f\leqslant g$ almost everywhere for measurable functions on an interval.

It is not required that all elements be pairwise comparable. For example, the plane $\mathbb{R}^2$ can be partially ordered in the following way:

$x = (x_1, x_2) \leqslant y = (y_1, y_2)$ if $x_1 \leqslant y_1$ and $x_2 \leqslant y_2$.

If all elements of X are pairwise comparable, then X is called *linearly ordered.*

For example, the real line with its usual ordering is linearly ordered, and the aforementioned coordinate-wise ordering of the plane is not linear. However, the plane can be naturally linearly ordered: take the so-called *lexicographic order*, when $x \leqslant y$ if either $x_1 < y_1$ or $x_1 = y_1$ and $x_2 \leqslant y_2$.

In a partially ordered set some parts can be linearly ordered. Such parts are called *chains*. For example, the real line as a part of the plane with the coordinate-wise order is a chain.

If X is a partially ordered set and $M \subset X$, then an element $\mu \in X$ is called an *upper bound* of the set M if $m \leqslant \mu$ for all $m \in M$. If m is an upper bound of M such that $m \leqslant \widehat{m}$ for every other upper bound $\widehat{m}$ of the set M, then m is called the *least upper bound* M. An element $m \in X$ is called *maximal* if there is no element $m' \in X$ such that $m \leqslant m'$. It is not required that all elements of X be smaller than m. For example, if $x \leqslant y$ only when $x = y$, then every element is maximal. Similarly one defines a *lower bound, the greatest lower bound* and a *minimal* element.

A linearly ordered set X is called *completely ordered* if every nonempty part of X has a minimal element.

For example, the set of natural numbers with its natural order is completely ordered, but the sets of rational and real numbers are not.

The axiom of choice is equivalent to the following assertion (if we accept it as an axiom, then the axiom of choice becomes a theorem); for a proof, see Kolmogorov, Fomin [**284**], Kurosh [**306**].

The Zermelo theorem. *Every nonempty set can be completely ordered.*

Let us give yet another corollary of the axiom of choice (equivalent to it).

The Zorn (or Kuratowski–Zorn) lemma. *If every chain in a partially ordered set has a majorant, then this set contains a maximal element.*

We recall that a maximal element need not be unique. Let us give an example of how Zorn's lemma works.

1.1.5. Proposition. *Every real or complex linear space possesses an algebraic basis. Moreover, any two such bases have the same cardinality. In addition, any algebraic basis in a linear subspace can be complemented to an algebraic basis of the whole space.*

PROOF. We assume that our space X contains nonzero vectors. Then X contains systems of algebraically independent vectors. Let Λ denote the collection of all such systems and let us introduce in Λ the following relation: $\lambda_1 \leqslant \lambda_2$ if $\lambda_1 \subset \lambda_2$. Clearly, we obtain a partial order. We have to show that the set Λ has a maximal element, i.e., a system λ of algebraically independent vectors that is not a proper subset of any other system of independent vectors. Such a maximal system will be a basis, since the existence of a vector v not representable as a linear combination of vectors in λ would mean that the system $\lambda \cup v$ is also independent, which contradicts the maximality of λ. The existence of a maximal element follows from Zorn's lemma, but in order to apply it we have to show that every chain Λ_0 in Λ has an upper bound. In other words, having a set Λ_0 of independent collections of vectors such that every two collections are comparable (i.e., one of the two is contained in the other), we have to find an independent system of vectors containing all systems from Λ_0. For such a system we simply take the union Λ_1 of all systems from Λ_0. The fact that the obtained system is independent is clear from the following. If vectors $v_1, \ldots, v_n$ belong to Λ_1, then there exist systems $\lambda_1, \ldots, \lambda_n \in \Lambda_0$ such that $v_i \in \lambda_i$ for $i = 1, \ldots, n$. Since the systems λ_i are pairwise comparable, among them there is the largest one λ_{i_0}. Then all v_i belong to λ_{i_0} and hence are linearly independent.

A minor modification of this reasoning enables us to complement algebraic bases in a subspace to bases of the whole space: it suffices to take for elements Λ independent systems containing a fixed basis in the given subspace. By the way, this reasoning is true for every field.

Finally, the assertion about the equicardinality of algebraic bases in the space X in the case of a finite-dimensional space is known from linear algebra. If X is infinite-dimensional and γ_1 and γ_2 are two its algebraic bases, then the cardinality of γ_2 does not exceed the cardinality of γ_1. Indeed, every element $v \in \gamma_2$ corresponds to a finite set of elements $S \subset \gamma_1$ through which it is linearly expressed. This finite set S is associated to at most finitely many elements in γ_2 (not more than the cardinality of S, since by using k vectors we cannot linearly express more than k linearly independent vectors). Hence the cardinality of γ_2 does not exceed the cardinality of the set of finite subsets of γ_1, which has the same cardinality as γ_1 (see Hrbacek, Jech [**229**, p. 136]). Thus, the cardinality of γ_2 does not exceed the cardinality of γ_1, and the opposite inequality is true. □

With the aid of this result any linear mapping T defined on a linear subspace E_0 of a vector space E and taking values in a vector space F can be extended to a linear mapping of the whole space E to F. It suffices to complement an algebraic basis in E_0 to a basis of the whole space E, define T by zero on the additional elements of the basis and then define by linearity on all vectors.

Let us now proceed to necessary topological concepts. For more details, see Arkhangel'skiĭ, Ponomarev [**22**], Bogachev, Smolyanov [**72**], Edwards [**150**], Engelking [**154**], and Kelley [**268**].

A *topology* on a set X is a family τ of subsets of this set possessing the following properties:

(i) $X, \varnothing \in \tau$;

(ii) if $V_1, V_2 \in \tau$, then $V_1 \cap V_2 \in \tau$;

(iii) the union of every collection of sets from τ belongs to τ.

Hence the minimal topology is $(X, \varnothing)$ and the maximal topology is 2^X, the class of all subsets of X.

A *topological space* is a pair (X, τ), where X is a set, called the set of elements of this topological space, and τ is a topology on X. The elements of τ are called *open subsets* of the topological space X.

A subset of a topological space is called *closed* if its complement is open. The topology in X can be also defined by introducing the class $\mathcal{F}$ of all closed sets, which must satisfy the following conditions:

(i) $X, \varnothing \in \mathcal{F}$;

(ii) if $F_1, F_2 \in \mathcal{F}$, then $F_1 \cup F_2 \in \mathcal{F}$;

(iii) the intersection of every collection of sets from $\mathcal{F}$ belongs to $\mathcal{F}$.

An important subclass of the class of topological spaces is formed by metric spaces. Although a grasp knowledge of them is assumed, we recall that a *metric space* (M, d) is a pair, where M is a set and $d\colon\ M\times M \to [0, +\infty)$ is a function, called a *metric*, that satisfies the following conditions:

(i) $d(a, b) = d(b, a)$, in addition, $d(a, b) = 0$ if and only if $a = b$,

(ii) $d(a, c) \leqslant d(a, b) + d(b, c)$ (*the triangle inequality*).

A linear space with a norm $\|\cdot\|$ (a *normed space*) is a metric space with the metric $d(x, y) = \|x - y\|$.

Let $a \in M$ and $r > 0$. The set

$$\{x \in M\colon\ d(x, a) < r\}$$

is called the *open ball* with the center a and radius r. If we define open sets in M as arbitrary unions of open balls (with arbitrary centers and radii) and the empty set, then we obtain a topological space (a simple verification is left as an exercise). Here any open ball will be an open set (which is easily verified with the aid of the triangle inequality). The *closed ball* with the center a and radius r is defined as the set

$$\{x \in M\colon\ d(x, a) \leqslant r\}$$

The topological space is called *metrizable* if its topology is obtained in the indicated way from some metric on it. Different metrics can generate the same topology. For example, the usual metric on the real line generates the same topology as the bounded metric $d(x, y) = \min(1, |x - y|)$. Below we encounter many examples of nonmetrizable spaces, so we do not give artificial examples of this sort. The *discrete topology* on X is $\tau = 2^X$.

We assume that the concept of a complete metric space is known (anyway, we recall it in § 1.7).

A *pseudometric* on a set M is a function $\varrho\colon\ M \times M \to [0, \infty)$ with the following properties:

(1) $\varrho(x, x) = 0$;

(2) $\varrho(x, y) = \varrho(y, x)$;

(3) $\varrho(x, y) \leqslant \varrho(x, z) + \varrho(z, y)$.

If the triangle inequality (3) is written as

(3′) $\varrho(x, y) \leqslant \varrho(x, z) + \varrho(y, z)$,

then conditions (2) and (3) together will be equivalent to the pair of conditions (2) and (3′), but (2) will follow from (1) and (3′) by replacing z in (3′) with x.

The pseudometric ϱ on a nonempty set M generates a topology on this set in the same way as a metric: a set $V \subset M$ is called open in the topology generated by the pseudometric ϱ if for each $x \in V$ there is $\varepsilon > 0$ such that $\{z\colon \varrho(z,x) < \varepsilon\}$ is contained in V. In addition, any pseudometric generates a metric on the set of equivalence classes if we set $x \sim y$ when $d(x,y) = 0$.

Every subset X_0 of a topological space X becomes itself a topological space if we define open sets in X_0 to be the sets of the form $U \cap X_0$, where U is open in X. Certainly, such sets need not be open in X (if X_0 itself was not open in X). This topology on X_0 is called the *induced* topology.

An *open neighborhood of a point* x is any open set containing x. Sometimes it is useful to employ a broader concept of a *neighborhood* of a point (not necessarily open!) as an arbitrary set containing some open neighborhood of this point. For example, it becomes possible to speak of closed neighborhoods in this sense.

A *base of the topology* (*topology base*) is any collection of open sets with the property that all possible unions of the elements of this collection give all nonempty open sets.

A *base of the topology at a point* x or a *fundamental system of neighborhoods of the point* x is any collection of open neighborhoods of the point x with the property that every neighborhood of x contains an element of this collection. Sometimes, similarly to neighborhoods, bases of not necessarily open neighborhoods are used. A *prebase* of neighborhoods of a point in a topological space is a family of neighborhoods of this point finite intersections of elements of which form a base of its neighborhoods.

A point x in a topological space X is called a *limit point* of a set $A \subset X$ (or an *accumulation point* of A) if every neighborhood of x contains a point of A distinct from x. If every neighborhood of x intersects A, then x is called a *cluster point* of A. The closure $\overline{A}$ of a set A (the intersection of all closed sets containing A) is exactly the set of all its cluster points. The points of A that are not limit are called *isolated.*

If $X = \overline{A}$, then A is called *everywhere dense* in X. If X contains an at most countable everywhere dense set, then X is called *separable.*

If we are given a collection of nonempty topological spaces X_t, where $t \in T$, then the product $\prod_{t\in T} X_t$ is equipped with the *Tychonoff* product topology, in which open sets are all possible unions of the sets of the form $\prod_{t\in T} U_t$, where every U_t is open in X_t, but only for finitely many indices t the set U_t differs from X_t. See Exercise 2.10.26 about the competing box topology.

A mapping $f\colon X \to Y$ between topological spaces is called *continuous* if, for every open set V in the space Y, the set $f^{-1}(V)$ is open in X. The continuity at a point $x_0 \in X$ is defined as follows: for every open set V containing $f(x_0)$, there exists an open set U containing x_0 such that $f(U) \subset V$.

The continuity of f is equivalent to the continuity at every point. Indeed, if f is continuous and V is an open set containing $f(x_0)$, then $U = f^{-1}(V)$ is open, $x_0 \in U$ and $f(U) \subset V$. Conversely, let f be continuous at every point x and let $V \subset Y$ be open. For every point $x \in U := f^{-1}(V)$ there exists an open set $U_x \ni x$ such that $f(U_x) \subset V$. Then the set $W := \bigcup_{x\in U} U_x$ is open. Since $f(W) \subset V$ and $U \subset W$, we have $W = U$.

If X and Y are topological spaces, then a mapping $F\colon X \to Y$ is called a *homeomorphism* if it is one-to-one, $F(X) = Y$ and both mappings F and F^{-1} are continuous; in this case X and Y are called *homeomorphic*.

Let us introduce certain separation properties. A topological space (X, τ) is called a *Kolmogorov* space or a T_0-space if, for every two its different points, there exists an open set containing precisely one of them; (X, τ) is called a T_1-space if, for every two different points a, b in X, there are open sets $A, B \in \tau$ such that $a \in A \setminus B$ and $b \in B \setminus A$; (X, τ) is called a *Hausdorff* or *separated* space (or a T_2-space) if, for every two different points $a, b \in X$, there are open sets $A, B \in \tau$ such that $A \cap B = \varnothing$, $a \in A$, $b \in B$; a *regular* space (or a T_3-space) is a T_1-space every point of which possesses a base of closed neighborhoods. In a Hausdorff space every point is closed. A *completely regular* space is a space with the following property: for every closed set $F \subset X$ and every point $x \notin F$, there is a continuous function $g\colon X \to [0, 1]$ such that $g(x) = 0$ and $g = 1$ on F. A *Tychonoff* space (or a $T_{3\frac{1}{2}}$-space) is a Hausdorff completely regular space. If a pseudometric is not a metric, then the topology generated by it is not Hausdorff. We shall see below that topological vector spaces are completely regular.

A *cover* of a set is any collection of sets whose union contains this set.

1.1.6. Definition. *A subset of a topological space X is called compact if every cover of this set by open sets contains a finite subcover. If this is true for the whole space X, then X is called a compact or a compact space.*

A topological space is called *locally compact* if every point in it possesses a fundamental system of neighborhoods consisting of compact sets.

A subset of a topological space is called *relatively compact* if its closure is compact. This is equivalent to the following: this subset is contained in a compact set.

A topological space is called *connected* if it cannot be represented as a union of two disjoint nonempty open sets, or, which is the same, it cannot be represented as a union of two disjoint nonempty closed sets.

Useful tools for working with topological spaces are the concepts of a net and a filter. For the reader's convenience, we briefly explain these concepts, which are sometimes used below.

1.1.7. Definition. *A partially ordered set T is called directed if, for every two elements $t, s \in T$, there exists an element $\tau \in T$ such that $t \leqslant \tau$ and $s \leqslant \tau$.*

A net in a given set X is a family $\{x_t\}_{t\in T}$ of its elements indexed by a directed set T.

For example, the plane with the lexicographic order and the set of all neighborhoods of a given point in a topological space partially ordered by the inverse inclusion are directed sets. The set of all nonempty open subsets of the real line, partially ordered by the inverse inclusion, is not directed (two disjoint open sets have no common upper bound).

In a somewhat peculiar way (as compared to sequences) one introduces the notion of a subnet $\{y_s\}_{s\in S}$ of a net $\{x_t\}_{t\in T}$: it is required that there is a mapping $F\colon S\to T$ such that $y_s = x_{F(s)}$ for all $s\in S$ and, for each $t\in T$, there is $s_t\in S$ such that $F(s) > t$ whenever $s > s_t$ (certainly, a subsequence in a sequence satisfies this condition).

For example, in the countable net $\mathbb{Z}$ of integer numbers (indexed by the same set $\mathbb{Z}$ with its usual order) the subset of negative numbers is not a subnet, but the subset of natural numbers is a subnet. A subnet of a countable sequence can be uncountable.

Not every countable net is isomorphic to a sequence indexed by natural numbers with its usual order (say, it can occur in a countable net that for every index there are infinitely many smaller indices).

1.1.8. Definition. *A net $\{x_t\}_{t\in T}$ in a topological space X converges to a point x if, for every neighborhood U of the point x, there exists an index $\tau\in T$ such that $x_t\in U$ whenever $t\geqslant\tau$.*

Note that the set of indices $t\in T$ such that $x_t\notin U$ can be infinite. Hence even for countable sets T convergence of nets does not reduce to convergence of sequences. For example, if on $\mathbb{N}$ we introduce the order such that all odd numbers are smaller than 2 and on even and odd numbers separately the usual order is kept, then we obtain a directed countable set; the net $\{x_n\}$ in which $x_n = 0$ for even n and $x_n = 1$ for odd n converges to zero. One can give an example of a convergent countable net in a topological space that does not contain convergent subsequences (Exercise 1.12.27).

If b is an accumulation point of the set A, then there exists a net $\{a_t\}$ of elements of A converging to b. Indeed, let T be the collection of all neighborhoods of b, partially ordered by the inverse inclusion. Every such neighborhood of t by condition contains a point $a_t\in A$. The obtained net converges to b, since for every fixed neighborhood of $\tau\in T$ we have $a_t\in t\subset\tau$ if $t\geqslant\tau$.

1.1.9. Definition. *A filter on a set X is any nonempty set Φ of nonempty subsets of X satisfying the following conditions:*

(i) *if $A, B\in\Phi$, then $A\cap B\in\Phi$,*

(ii) *if $B\in\Phi$ and $B\subset C$, then $C\in\Phi$.*

A filter basis (base) on a set X is any nonempty set $\mathcal{B}$ of nonempty subsets of X such that the following condition is fulfilled: for every pair of sets $B_1, B_2\in\mathcal{B}$, there exists $B_3\in\mathcal{B}$ with $B_3\subset B_1\cap B_2$.

A filter Φ is majorized by a filter Ψ if $\Phi\subset\Psi$.

A filter basis is an integral term (not assuming any filter).

Among all filters containing a given filter basis $\mathcal{B}$ there exists the (unique) minimal filter $\Phi_{\mathcal{B}}$, which is called the filter generated by the filter basis $\mathcal{B}$. Then $\mathcal{B}$ is called a basis of the filter $\Phi_{\mathcal{B}}$. The minimal filter is the intersection of all filters containing $\mathcal{B}$ (such filters exist, for example, the class of all sets containing at least one set from $\mathcal{B}$). If τ is a topology on the set X and $x\in X$, then the set of all (not necessarily open) neighborhoods of the point x is a filter on X, called the filter of neighborhoods of this point with respect to τ and denoted by the symbol Φ_τ^x. Thus, a fundamental system of neighborhoods of zero is a basis of the filter of all neighborhoods of zero.

1.1.10. Definition. *A filter in X is called converging to a point x in the topology τ if it majorizes the filter of neighborhoods of this point.*

Maximal elements of the system of all filters on the set X, partially ordered by the relation of majorizing by inclusion, are called *ultrafilters* on X. One can deduce from the axiom of choice that every filter on X is majorized by some ultrafilter on X. A filter Ψ on X is an ultrafilter precisely when the conditions $A \cup B = X$ and $A \notin \Psi$ imply that $B \in \Psi$. As a simple example of application of filters we mention the following assertions, the proof of which we leave as an exercise.

1.1.11. Proposition. *A mapping f of topological spaces is continuous at a point x precisely when, for every filter Ψ converging to x, the filter generated by the filter basis $f(\Psi)$ converges to $f(x)$.*

Note that the image of a filter need not be a filter, but is always a filter basis. The proof is left as an exercise.

1.1.12. Proposition. *A subset of a topological space is compact if and only if every ultrafilter containing it converges to some element of this subset.*

1.2. Basic definitions

Here we give basic definitions connected with topological vector spaces and prove some simplest facts, but examples will be considered in the next section. Although in our discussion the field $\mathbb{K}$ is usually $\mathbb{R}$ or $\mathbb{C}$ (and occasionally nondiscrete normable), we give a general definition.

1.2.1. Definition. *A topological vector space over a topological field $\mathbb{K}$ is a vector space E over $\mathbb{K}$ equipped with a topology with respect to which the following two mappings are continuous, where $E\times E$ and $\mathbb{K}\times E$ are equipped with the products of the corresponding topologies*: 1) $(x_1, x_2) \mapsto x_1 + x_2$, $E\times E \to E$ (*addition of vectors*), 2) $(k, x) \mapsto kx$, $\mathbb{K}\times E \to E$ (*multiplication of vectors by scalars*).

Such a topology on E is called *compatible with the vector space structure* (or we say that it *agrees with the vector space structure*). A topological vector space E with the topology τ is denoted by the symbol (E, τ). We observe that in the definition of a topological field one requires the same conditions with $\mathbb{K}$ in place of E and the continuity of $k \mapsto k^{-1}$ outside of zero.

Two topological vector spaces over the same field are called *isomorphic* if there exists a continuous linear one-to-one mapping of one of the two spaces onto the other such that the inverse mapping is also continuous (i.e., a linear homeomorphism). The *dimension of a topological vector space* (E, τ) is the dimension of the vector space E.

The continuity of the mapping 1) implies that the topology of any topological vector space (E, τ) is invariant with respect to translations (i.e., for every $a \in E$ the mapping $x \mapsto x + a$ is a homeomorphism of E); hence the topology of a topological vector space can be reconstructed if we know a fundamental system of neighborhoods of zero.

If $\mathcal{U}$ is a base of neighborhoods of zero and $a \in E$, then the collection of sets of the form $a + V$, where $V \in \mathcal{U}$, is a base of neighborhoods of the point a. Thus, for defining a topology of a topological vector space it suffices to define a base of neighborhoods of zero; this is usually done in most of applications of the theory of topological vector spaces. However, not every system of subsets of a vector space can serve as a base of neighborhoods of zero of a topology compatible with the vector space structure; conditions sufficient for this are indicated in Proposition 1.2.7.

Before we proceed to that proposition, it is useful to prove the following result, according to which among fundamental systems of neighborhoods of zero in a topological vector space there are systems with particularly nice properties.

1.2.2. Proposition. (a) *Every base of neighborhoods of zero $\mathcal{U}$ in a topological vector space has the following properties*:

(1) *for every $V \in \mathcal{U}$ there exists a set $W \in \mathcal{U}$ such that $W + W \subset V$;*

(2) *every $V \in \mathcal{U}$ is an absorbent set.*

(b) *In every topological vector space there exists a base of neighborhoods of zero $\mathcal{U}_0$ having also the following properties*:

(3) *every $V \in \mathcal{U}_0$ is a circled closed set*;

(4) *if $V \in \mathcal{U}_0$, then $kV \in \mathcal{U}_0$ for every $k \in \mathbb{K}$, $k \neq 0$.*

PROOF. Let $\mathcal{U}$ be a base of neighborhoods of zero in a topological vector space E. Since the mapping $(x_1, x_2) \mapsto x_1 + x_2$, $E \times E \to E$ is continuous at the point $(0, 0)$ by Axiom 1, we obtain that $\mathcal{U}$ has property (1). Further, by Axiom 2 for every $a \in E$ the mapping $k \mapsto ka$, $\mathbb{K} \to E$ is continuous at the point $0 \in \mathbb{K}$; so if V is a neighborhood of zero in E and $x \in E$, then there exists $\varepsilon > 0$ such that $kx \in V$ whenever $|k| < \varepsilon$, so that an arbitrary neighborhood of zero V in E is an absorbent set. This means that $\mathcal{U}$ possesses property (2). Thus, part (a) of the proposition is proven.

For the proof of part (b) it suffices to show that the set $\mathcal{U}_0$ of all closed circled neighborhoods of zero in E is a base of neighborhoods of zero in E, since properties (3) and (4) are easily verified. Indeed, it is clear from the definition of the set $\mathcal{U}_0$ that it possesses property (3). By Axiom 2, for every fixed nonzero $k \in \mathbb{K}$, the mapping $x \mapsto kx$ is a linear homeomorphism of E. It follows that if V is a neighborhood of zero in E, then kV, where $k \in \mathbb{K}$, $k \neq 0$, is a neighborhood of zero, moreover, if the set V is closed and balanced, then kV has both properties, so that $\mathcal{U}_0$ possesses property (4). In order to verify that $\mathcal{U}_0$ is a base of neighborhoods of zero in E, we show that every neighborhood of zero in E contains a closed circled neighborhood of zero. Let W be an arbitrary neighborhood of zero in E. By the continuity at zero of the mapping $(x_1, x_2) \mapsto x_1 - x_2$, $E \times E \to E$, which follows from Axioms 1 and 2, there is a neighborhood of zero W_1 such that

$W_1 - W_1 \subset W$. We show that $\overline{W_1} \subset W$. To this end, we verify that if $x \notin W$, then $x \notin \overline{W_1}$. The set $x + W_1$ is a neighborhood of the point x not intersecting W_1 (if $z \in W_1 \cap (x + W_1)$, then $z = x + y$, $y \in W_1$ and $x = z - y \in W_1 - W_1 \subset W$, but $x \notin W$). The existence of such neighborhood means that $x \notin \overline{W_1}$. Further, by the continuity of the mapping $(k, x) \mapsto kx$, $\mathbb{K} \times E \to E$ at the point $(0, 0)$ there exist $\varepsilon > 0$ and a neighborhood of zero W_2 in E such that if $|k| < \varepsilon$ and $x \in W_2$, then $kx \in W_1$. Hence the set $W_3 = \bigcup_{|k|<\varepsilon} kW_2$ is a circled neighborhood of zero in E contained in W_1 (the fact that W_3 is a neighborhood of zero follows from the condition that $\mathbb{K}$ is not discrete, so there exists $k \neq 0$ with $|k| < \varepsilon$). The closure of a circled set is a circled set. Indeed, if $|k| \leqslant 1$ and G is a circled set, then $kG \subset G$, hence $k\overline{G} \subset \overline{kG} \subset \overline{G}$ (if $k \neq 0$, then $k\overline{G} = \overline{kG}$). Hence $\overline{W_3}$ is a closed circled neighborhood of zero and $\overline{W_3} \subset \overline{W_1} \subset W$. □

1.2.3. Remark. (i) In the proof we have actually shown that the family of closures of all possible sets in a base of neighborhoods of zero in a topological vector space is again a base of neighborhoods of zero (in fact, this is true for any topological group).

(ii) It has been also shown that every neighborhood of zero in a topological vector space is an absorbent set; this fact will be used all the time below.

(iii) Proposition 1.2.2 remains true if in its formulation in part (b) the word "closed" is replaced by the word "open": in other words, in every topological vector space there exists a base of neighborhoods of zero $\mathcal{U}$ having properties (1) and (4) and the following property (3′): every $V \in \mathcal{U}$ is a circled absorbent open set. The proof essentially coincides with the proof of Proposition 1.2.2, but is even simpler. As above, we verify that E has a base consisting of open circled sets. The existence of such a base follows from the fact that for every neighborhood of zero $W \subset E$ there exist $\varepsilon > 0$ and an open neighborhood of zero W_1 such that if $k \in \mathbb{K}$, $|k| < \varepsilon$ and $x \in W_1$, then $kx \in W$. Hence the set $W_2 = \bigcup_{|k|<\varepsilon} kW_1$ is an open circled neighborhood of zero contained in W.

Let us give some corollaries of the proven proposition.

1.2.4. Corollary. *Each point in a topological vector space possesses a base of neighborhoods consisting of closed sets (i.e., every topological vector space is a regular topological space, as well as any topological group).*

PROOF. Indeed, if $\mathcal{U}$ is a base of closed neighborhoods of zero, then $a + \mathcal{U}$ is a base of closed neighborhoods of the point a for every a. □

1.2.5. Corollary. *A topological vector space is a T_3-space (hence Hausdorff) if and only if it is a T_0-space.*

PROOF. By the previous corollary and the definition of a T_3-space it suffices to show that in a given space E the axiom T_1 is fulfilled. Suppose that $a_1, a_2 \in E$. Since the axiom T_0 holds by assumption, for one of these points — let it be a_1 — there exists a neighborhood of zero W such that $a_1 + W \not\ni a_2$. Then $a_2 - W \not\ni a_1$, since otherwise for some $z \in W$ we have $a_2 = z + a_1$, i.e., $a_2 \in a_1 + W$. Thus, $a_2 - W$ is a neighborhood of the point a_2 not containing a_1. □

Actually, more is true: a Hausdorff topological vector space is completely regular, which will be established in § 1.6.

1.2.6. Corollary. *A topological vector space is Hausdorff if and only if the intersection of all its neighborhoods of zero is the zero element of this space.*

PROOF. The sufficiency follows from the previous corollary; the necessity is obvious. □

The next result can be used for constructing vector topologies.

1.2.7. Proposition. *Let $\mathcal{B}$ be a filter basis in a vector space E consisting of circled sets and having the following properties* (*analogous to properties* 1, 2, 4 *from Proposition* 1.2.2):

(1)′ *for every $V \in \mathcal{B}$ there is $W \in \mathcal{B}$ such that $W + W \subset V$;*

(2)′ *every $V \in \mathcal{B}$ is an absorbent set;*

(4)′ *if $V \in \mathcal{B}$, then $kV \in \mathcal{B}$ for every $k \in \mathbb{K}$, $k \neq 0$.*

Then, there exists a unique topology in E compatible with the vector space structure such that $\mathcal{B}$ is a base of neighborhoods of zero (*not necessarily closed or open*).

PROOF. Let τ be the family of subsets of E defined as follows: $V \in \tau$ precisely when for every $a \in V$ there exists a set W in $\mathcal{B}$ such that $a + W \subset V$. Then τ is a topology in E. Indeed, the inclusions $\varnothing \in \tau$ and $E \in \tau$ and the stability of τ with respect to formation of arbitrary unions follow directly from the definition of τ. We show that τ is stable with respect to finite intersections. Let $V_1, V_2 \in \tau$; we have to show that $V_1 \cap V_2 \in \tau$. Let $a \in V_1 \cap V_2$. Hence there exist sets $W_1, W_2 \in \mathcal{B}$ such that $a + W_i \subset V_i$, $i = 1, 2$. Then $a + (W_1 \cap W_2) \subset V_1 \cap V_2$. Therefore, if $W_3 \subset W_1 \cap W_2$, $W_3 \in \mathcal{B}$ (such W_3 exists, since $\mathcal{B}$ is a filter basis), then $a + W_3 \subset V_1 \cap V_2$.

We show that the topology τ agrees with the vector space structure in E. We first show that $\mathcal{B}$ is a base of neighborhoods of zero in the topology τ. By the definition of the topology τ, if V is an open neighborhood of zero in τ, there is a set $W \in \mathcal{B}$ such that $W \subset V$. Hence it suffices to show that every set that is an element of $\mathcal{B}$ turns out to be a neighborhood of zero in the topology τ. Thus, let $W \in \mathcal{B}$. Let W^0 denote the set defined as follows: $x \in W^0$ if and only if there exists a set $W_1 \in \mathcal{B}$ such that $x + W_1 \subset W$. Since the zero element of the space E is contained in every set from the system $\mathcal{B}$ (these sets are circled), we have $0 \in W^0 \subset W$.

We now show that W^0 is open in the topology τ; this will mean that W is a neighborhood of zero in this topology. It suffices to be able, for every given $a \in W^0$, to find a set $W_2 \in \mathcal{B}$ such that $a + W_2 \subset W^0$. Let $a \in W^0$. Then by the definition of W^0 there exists a set $W_1 \in \mathcal{B}$ such that $a + W_1 \in W$; by property (1) there exists $W_2 \in \mathcal{B}$ such that $W_2 + W_2 \subset W_1$, i.e., $(a + W_2) + W_2 \subset W$. This means that $a + W_2 \subset W^0$.

Since by the very definition the topology τ is translation invariant, for every vector $a \in E$ the family of the sets of the form $a + V$, where $V \in \mathcal{B}$, is a base of neighborhoods of the point a. Hence for the proof of continuity in τ of the operation of addition (i.e., verification of Axiom 1 in Definition 1.2.1) it suffices to show that if $a = x_1 + x_2$ and $W \in \mathcal{B}$, then there exists a set $W_1 \in \mathcal{B}$ such that

$$(x_1 + W_1) + (x_2 + W_1) \subset a + W.$$

By Axiom 1′ there exists W_1 such that $W_1 + W_1 \subset W$; this W_1 satisfies the necessary relation.

Let us proceed to the proof of continuity of the operation of multiplication (i.e., verification of Axiom 2 from Definition 1.2.1). Let $a \in E$, $k \in \mathbb{K}$ and $W \in \mathcal{B}$ be given. We have to prove the existence of $W_1 \in \mathcal{B}$ and $\varepsilon > 0$ such that if $a_1 \in a + W_1$ and $|k_1 - k| < \varepsilon$, then $k_1 a_1 \in ka + W$. Since the addition operation is continuous, as we have just proved, there exists a set $W_2 \in \mathcal{B}$ such that $W_2 + W_2 + W_2 \subset W$. Since

$$k_1 a_1 - ka = (k_1 - k)a + (k_1 - k)(a_1 - a) + k(a_1 - a),$$

the required properties hold for the set W_1 and the number $\varepsilon > 0$ such that the inclusion $a_1 \in a + W_1$ and the inequality $|k_1 - k| < \varepsilon$ imply that

$$(k_1 - k)a \in W_2, \ (k_1 - k)(a_1 - a) \in W_2, \ k(a_1 - a) \in W_2.$$

Since the set $W_2 \in \mathcal{B}$ is circled, the relations $|k_1 - k| < 1$, $a_1 - a \in W_2$ yield that $(k_1 - k)(a_1 - a) \in W_2$; since the set W_2 is absorbing, there exists $\varepsilon_1 \in (0, 1)$ such that $(k_1 - k)a \in W_2$ if $|k_1 - k| < \varepsilon_1$. Finally, if $k = 0$, then we can take $W_1 = W_2$; if $k \neq 0$, then we find a neighborhood $W_1 \in \mathcal{B}$ such that $W_1 \subset W_2 \cap k^{-1}W_2$. Thus, in both cases $W_1 \in \mathcal{B}$, and at the same time the inclusion $a_1 - a \in W_1$ yields that $k(a_1 - a) \in W_2$. Therefore, we can set $\varepsilon = \varepsilon_1$.

Let us verify the uniqueness of the indicated topology. Let t be yet another topology in E compatible with the vector space structure for which $\mathcal{B}$ is a base of neighborhoods of zero. Then all sets of the form $x + W$, where $x \in E$ and $W \in \mathcal{B}$, give a base of both topologies, whence $t = \tau$. □

1.2.8. Remark. The last but one paragraph of this proof yields that condition $(4)'$ of the proven proposition can be replaced with the following weaker condition: for every $s \in \mathbb{K} \setminus \{0\}$ and every $V \in \mathcal{B}$ there exists a set $V_1 \in \mathcal{B}$ such that $V_1 \subset sV$. However, in the case where $\mathbb{Q} \subset \mathbb{K} \subset \mathbb{C}$, this last condition is a corollary of Axiom 1 along with the fact that all sets in $\mathcal{B}$ are circled. Indeed, Axiom 1 implies that for any natural number n and a set $W \in \mathcal{B}$, there exists $V \in \mathcal{B}$ such that

$$2^n V \subset \underbrace{V + V + \cdots + V}_{2^n \text{times}} \subset W,$$

i.e., that $V \subset 2^{-n}W$. Hence if we are given $\varepsilon > 0$ and $W \in \mathcal{B}$ and a number $n \in \mathbb{N}$ is such that $2^{-n} < \varepsilon$, there exists $V \in \mathcal{B}$ such that $V \subset 2^{-n}W \subset \varepsilon W$ (the last inclusion follows from the fact that the set W is circled). Thus, if $\mathbb{K} \subset \mathbb{C}$, then for the validity of the conclusion of the proposition above it suffices to require that $\mathcal{B}$ be a filter basis in E possessing properties (1) and (2) and consisting of circled sets.

1.2.9. Corollary. *Let E be a vector space over a field $\mathbb{K}$ and let τ be a translation invariant topology in E possessing a basis of neighborhoods of zero $\mathcal{B}$ consisting of circled sets and having properties* (1)′, (2)′, (4)′ *from Proposition* 1.2.7 *(and in case where $\mathbb{K} \subset \mathbb{C}$, just properties* (1) *and* (2)*). Then τ is compatible with the vector structure in the space E.*

PROOF. By Proposition 1.2.7 and the previous remark in this case in E there exists a topology τ_1 compatible with the vector structure such that $\mathcal{B}$ is a base of neighborhoods of zero. Since τ_1 is translation invariant, we have $\tau = \tau_1$. □

Among topological vector spaces over the fields of real and complex numbers, the most important for applications class is formed by locally convex spaces, the definition of which will be now given.

Note that the closure $\overline{A}$ of a convex subset A in a topological vector space is convex, since by the continuity of the vector operations we have

$$t\overline{A} + (1-t)\overline{A} \subset \overline{tA} + \overline{(1-t)A} \subset \overline{tA + (1-t)A} \subset \overline{A}.$$

Further, the convex hull $\operatorname{conv} W$ of an open set W is again open: this follows from the fact that $\operatorname{conv} W$ is the union of all possible sets of the form $\sum_{k=1}^{n} \alpha_k W$, where $n \in \mathbb{N}$, $\alpha_k \geqslant 0$, $\sum_{k=1}^{n} \alpha_k = 1$, each of which is open by the continuity of the operations of addition and multiplication by scalars.

In addition, the interior $\breve{A}$ of any convex subset A in a topological vector space is convex. Indeed, if $a, b \in \breve{A}$, then $\breve{A}$ is a neighborhood of the points a and b, and the set $t\breve{A} + (1-t)\breve{A}$ is an open neighborhood of the point $ta + (1-t)b$ contained in A for every $t \in [0, 1]$ (see also Proposition 1.4.2).

1.2.10. Definition. *A locally convex topological vector space is a topological vector space over $\mathbb{R}$ or $\mathbb{C}$ possessing a base of convex neighborhoods of zero.*

In place of the term a "locally convex topological vector space" it is customary to use the term a *locally convex space* or the abbreviation *LCS*. A topology τ in a vector space E (over $\mathbb{R}$ or $\mathbb{C}$) is called *locally convex* if the space (E, τ) is locally convex. In the definition of a locally convex space the Hausdorff separation property is often included, but we do not do this, although in most of the results in this book we shall consider separated spaces.

1.2.11. Proposition. *Let $\mathbb{K} = \mathbb{R}$ or $\mathbb{K} = \mathbb{C}$.*

(i) *Every locally convex space E over $\mathbb{K}$ has a base of neighborhoods of zero consisting of closed circled convex absorbent sets and invariant with respect to multiplication by nonzero numbers from $\mathbb{K}$.*

(ii) *If τ is a topology in a vector space E over $\mathbb{K}$ invariant with respect to translations and possessing a base of neighborhoods of zero that consists of circled convex absorbent sets and contains along with each set V the set $2^{-1}V$, then E is a locally convex space.*

(iii) *If $\mathcal{B}$ is a filter basis in a vector space over $\mathbb{K}$ consisting of circled convex absorbent sets and containing along with each set V the set $2^{-1}V$, then in E there is a unique topology making E a locally convex space such that $\mathcal{B}$ is a base of neighborhoods of zero.*

PROOF. If $\mathcal{B}$ is some base of neighborhoods of zero in E consisting of convex sets, then every set of the form $W_V = V \cap (-V)$, where $V \in \mathcal{B}$, is a convex circled neighborhood of zero in E in case of $\mathbb{K} = \mathbb{R}$; in case of $\mathbb{K} = \mathbb{C}$ for a convex circled neighborhood $W_V \subset V$ (for $V \in \mathcal{B}$) we take $W_V = \bigcap_{|z|=1} zV$. This is indeed a neighborhood (not necessarily open), since there are a neighborhood U and $\varepsilon > 0$ such that $kU \subset V$ if $|k| \leqslant \varepsilon$, whence $\varepsilon U \subset W_V$. In both cases the collection of all sets W_V is a base of neighborhoods of zero in E; by Remark 1.2.3 the same is true for the collection $\mathcal{B}$ of their closures, which are again convex and circled. Hence the family of all sets of the form kV, where $V \in \mathcal{B}$, $k \in \mathbb{K}$, $k \neq 0$, is a base of neighborhoods of zero in E, the existence of which is asserted in (i) (as already noted, every neighborhood of zero is an absorbent set).

The remaining assertions (ii) and (iii) follow from Proposition 1.2.7 and Corollary 1.2.9. It suffices to verify that the sets $\mathcal{B}$ mentioned in these assertions have property (1) from Proposition 1.2.7. Let $V \in \mathcal{B}$. Then $2^{-1}V \in \mathcal{B}$. By the convexity of V we have $2^{-1}V + 2^{-1}V = V$. □

1.2.12. Remark. Similarly one can prove that every locally convex space has a base of neighborhoods of zero consisting of open circled convex absorbent sets and invariant with respect to multiplication by nonzero numbers in $\mathbb{K}$. Indeed, let $\mathrm{int}A$ be the interior of A.

If V is a convex neighborhood of zero and $W \subset V$ is an open neighborhood of zero, then its convex hull $\mathrm{conv}\, W$ is open and is contained in V by the convexity of V. Since $W \subset \mathrm{conv}\, W$, the set $\mathrm{conv}\, W$ is an open convex neighborhood of zero contained in V and the set $W_0 = \mathrm{conv}\, W \cap (-\mathrm{conv}\, W)$ in the real case and the set $W_0 = \mathrm{int} \bigcap_{|z|=1} \mathrm{conv}\,(zW)$ in the complex case is an open convex circled neighborhood of zero (observe that there are $\varepsilon > 0$ and an open neighborhood of zero W_1 such that if $k \in \mathbb{C}$ and $|k| \leqslant \varepsilon$, then $kW_1 \subset W$; so $\varepsilon W_1 \subset W_0$), and we also have $W_0 \subset V$. Hence the family $\mathcal{U}_0$ of all such neighborhoods of zero is a base of neighborhoods of zero. The same is true for the family of all sets of the form kV, where $k \in \mathbb{K}$, $k \neq 0$, $V \in \mathcal{U}_0$.

1.3. Examples

Here we present a rich collection of model examples.

1.3.1. Example. Every algebraic field $\mathbb{K}$ is a one-dimensional vector space over $\mathbb{K}$ with respect to the operations of addition and multiplication in $\mathbb{K}$; this one-dimensional vector space over $\mathbb{K}$ is denoted by $\mathbb{K}^1$. If $\mathbb{K}$ is a topological field with respect to a topology τ, then $\mathbb{K}^1$ is a one-dimensional topological vector space over $\mathbb{K}$ with respect to the same topology; it is denoted again by $\mathbb{K}^1$ or $\mathbb{K}$.

1.3.2. Example. Let $\mathbb{K}$ be an arbitrary topological field, let T be a nonempty set, and let $\mathbb{K}^T$ be the vector space over $\mathbb{K}$ that is the product of T copies of $\mathbb{K}$ equipped with the product topology; so $\mathbb{K}^T$ is the set of all functions $x\colon T \to \mathbb{K}$ with the topology of pointwise convergence whose base consists of the sets

$$U_{x_0,t_1,\ldots,t_n,V} = \{x\colon\ x(t_i) - x_0(t_i) \in V, i = 1, \ldots, n\},$$

where $x_0 \in \mathbb{K}^T$, $t_i \in T$ and V is a neighborhood of zero in $\mathbb{K}$. Then $\mathbb{K}^T$ is a topological vector space. More generally, the product of any family of topological vector spaces over the field $\mathbb{K}$ is again a topological vector space over $\mathbb{K}$ with respect to the product topology of the factors.

For $T = \mathbb{N}$ and $\mathbb{K} = \mathbb{R}$ we obtain $\mathbb{R}^\infty$, the *space of all real sequences* with the topology of coordinate-wise convergence; it can be defined by the metric

$$d(x,y) = \sum_{n=1}^{\infty} 2^{-n} \min(|x_n - y_n|, 1), \quad \text{where } x = (x_n),\ y = (y_n).$$

1.3.3. Example. If the topology of a topological field $\mathbb{K}$ is discrete, then every vector space E over $\mathbb{K}$ equipped with a topology compatible with the structure of its additive group (this means the continuity of the mapping $(x_1, x_2) \mapsto x_1 - x_2$, $E \times E \to E$) and invariant with respect to the operation of multiplication by nonzero elements of $\mathbb{K}$ is a topological vector space over $\mathbb{K}$ (in particular, this condition is fulfilled for the discrete topology on E). Topological vector spaces over fields with the discrete topology are called *topological vector groups*.

Throughout we assume that the field $\mathbb{K}$ is not discrete. In most of the examples $\mathbb{K} = \mathbb{R}$ or $\mathbb{K} = \mathbb{C}$.

1.3.4. Example. Let E be a vector space over a nondiscrete normed field $\mathbb{K}$ and let $\mathcal{P}$ be a family of seminorms on E. The *open ball of radius $r > 0$ with the center at zero with respect to a seminorm p on E* is the set $\{x \in E\colon\ p(x) < r\}$. The collection of the intersections of all possible finite families of open balls of positive radii with respect to seminorms from the family $\mathcal{P}$ is a base of neighborhoods of zero for some topology $\tau_{\mathcal{P}}$ in E that agrees with the vector structure; it is customary to say that this *topology is given* (*or defined*) *by the family of seminorms $\mathcal{P}$*. Thus, the collection of open balls of all possible positive radii (for all given seminorms) is a pre-base of neighborhoods of zero in the topology $\tau_{\mathcal{P}}$. Note that all seminorms in $\mathcal{P}$ are continuous in this topology. If $\mathbb{K} = \mathbb{R}$ or $\mathbb{K} = \mathbb{C}$, the topology $\tau_{\mathcal{P}}$ is locally convex, since the sets $\{x\colon\ p(x) < r\}$ are convex; it is shown in § 1.4 that the topology of every locally convex space (over $\mathbb{R}$ or $\mathbb{C}$) is defined by some family of seminorms. A topological vector space is called *normable* if its topology can be given by a single norm. A *Banach* space is a normed space that is complete with respect to the metric generated by the norm (the concept of completeness is recalled in § 1.7). A *Hilbert* space is a complete Euclidean space. A criterion of normability of a topological vector space over $\mathbb{R}$ or $\mathbb{C}$ (discovered by A.N. Kolmogorov) will be given in § 1.5.

1.3.5. Example. Let $n \in \mathbb{N}$. The topology in $\mathbb{K}^n$ is generated by the norm given by the equality $\|(x_1, \ldots, x_n)\| = \max_{i=1,\ldots,n} |x_i|$, where the symbol $|\cdot|$ denotes the norm in $\mathbb{K}$. We could also take here $\sum_{i=1}^n |x_i|$ or $\bigl(\sum_{i=1}^n |x_i|^2\bigr)^{1/2}$, but below we need the fact that the set of values of the norm $\max_{i=1,\ldots,n} |x_i|$ coincides with the set of values of the norm $k \mapsto |k|$. It will be shown in § 1.5 that if the field $\mathbb{K}$ is complete, then every n-dimensional separated topological vector space over $\mathbb{K}$ is isomorphic to the space $\mathbb{K}^n$ (for $n = 1$ this is also true without the assumption of completeness of $\mathbb{K}$), and if the field $\mathbb{K}$ is locally compact, then a separated topological vector space over $\mathbb{K}$ is finite-dimensional precisely when it possesses a precompact neighborhood of zero. The first part of the last sentence means that in a n-dimensional real or complex topological vector space there exists precisely one separated topology compatible with the vector structure; this topology will be further called *standard*.

1.3.6. Example. Let $\mathbb{Q}$ be the field of rational numbers (with its usual topology defined by the norm equal to the absolute value of a number) and let α be an irrational real number. The set $\{\alpha q_1 + q_2 \colon q_1, q_2 \in \mathbb{Q}\}$ in $\mathbb{R}$ with the topology induced by the usual topology of the real line is a two-dimensional topological vector space over $\mathbb{Q}$ that is not isomorphic to the topological vector space $\mathbb{Q}^2$ (Exercise 1.12.26).

1.3.7. Definition. *A pseudonorm on a vector space E is a nonnegative function $p\colon E \to [0,\infty)$ such that*

(1) $p(0) = 0$, (2) $p(-x) = p(x)$, (3) $p(x_1 + x_2) \leqslant p(x_1) + p(x_2)$.

The pseudonorm p is called nondegenerate if $p(x) = 0$ only for $x = 0$.

Note that our definition differs from the one given in the book Schaefer [**436**], where the nondegeneracy is required along with the estimate $p(\lambda x) \leqslant p(x)$ whenever $|\lambda| \leqslant 1$. Though, for defining vector topologies this distinction is not important (see the end of the proof of Theorem 1.6.1).

Thus, a *seminorm* is a pseudonorm q having the following property, which is more restrictive than (2) and (1) together:

(2′) $q(\alpha x) = |\alpha| q(x) \quad \forall \alpha \in \mathbb{K}$.

Unlike a norm, a seminorm can be zero on nonzero elements. For example, the identically zero function is a seminorm.

If p is a pseudonorm on a vector space E, then the equality

$$\varrho(x_1, x_2) = p(x_1 - x_2)$$

defines a pseudometric invariant with respect to translations; this pseudometric, in turn, generates a topology on E compatible with the structure of an additive group of the space E; the pseudometric ϱ becomes a metric precisely when $p(x) = 0$ only for $x = 0$.

If E is a topological vector space the topology τ of which is metrizable, then on E there is a pseudonorm generating this topology in the described way (this will be proved in § 1.6). A criterion of metrizability of a topological vector space will be also given in § 1.6.

Note also that if p is an arbitrary pseudonorm on a vector space E, then the topology generated by p need not be compatible with the vector structure (give an example); in order it be compatible with the vector structure, it suffices (and is obviously necessary) that the pseudonorm p have the following additional properties:

(4) if $x_n \in E$, $t \in \mathbb{K}$, $p(x_n) \to 0$, then $p(tx_n) \to 0$;

(5) if $x \in E$, $t_n \in \mathbb{K}$, $t_n \to 0$, then $p(t_n x) \to 0$;

(6) if $x_n \in E$, $t_n \in \mathbb{K}$, $t_n \to 0$, $p(x_n) \to 0$, then $p(t_n x_n) \to 0$.

Property (6), as one can show, follows from properties (4) and (5); we leave the proof to the reader; these properties hold if and only if the operation of multiplication by scalars is continuous with respect to the topology generated by the pseudonorm p.

1.3.8. Definition. *A quasi-norm is a pseudonorm possessing properties* (4) *and* (5) (*hence also property* (6)). *Thus, a pseudonorm* p *defining the topology of a metrizable topological vector space is automatically a quasi-norm* (*having the property* $p(x) \neq 0$ *for all* $x \neq 0$).

1.3.9. Example. Let E be a vector space and let $\mathcal{P}$ be a family of quasi-norms on E. The *open ball of radius* $r > 0$ *with the center at zero with respect to the quasi-norm* $p \in \mathcal{P}$ is the set $\{x \in E\colon\ p(x) < r\}$; the collection of all open balls of all possible positive radii with respect to the quasi-norms in $\mathcal{P}$ is a prebase of neighborhoods of zero of some topology in E compatible with the vector structure; this topology is called the topology generated by the family $\mathcal{P}$ of quasi-norms. It will be shown in § 1.6 that the topology of every topological vector space can generated by a suitable family of quasi-norms. Note that all quasi-norms of a family defining the topology are continuous in this topology.

1.3.10. Example. Let (E, τ) be a topological vector space, let $E_1 \subset E$ be a vector subspace, and let τ_1 be the topology induced in E_1 by the topology τ. The topology τ_1 agrees with the vector structure. The topological vector space (E_1, τ_1) is called a *topological vector subspace* of the topological vector space E. If $\mathcal{U}$ is a base (or prebase) of neighborhoods of zero in (E, τ), then the family $\{V \cap E_1\colon\ V \in \mathcal{U}\}$ is a base (respectively, a prebase) of neighborhoods of zero in the space (E_1, τ_1). If (E, τ) is Hausdorff (or metrizable, or locally convex), then (E_1, τ_1) has the respective property. If the topology τ is given by some set of seminorms (or pseudonorms), then the topology τ_1 is defined by their restrictions to the subspace E_1.

The following sufficient condition for the closedness of E_1 as a subset in the topological vector space E is useful.

1.3.11. Lemma. *Let a vector subspace F in a Hausdorff topological vector space E be complete with respect to some metric defining the topology of this subspace. Then F is closed in E.*

PROOF. A very short proof of a generalization of this lemma can be given by means of the concept of a Cauchy filter (see Proposition 1.7.8); here we give a direct justification, which will be used also in the first proof of Theorem 1.5.1. We show that every point y in the closure $\overline{F}$ of the subspace F in E actually belongs to F. Let $\{V_j\colon\ j \in \mathbb{N}\}$ be a base of neighborhoods of zero in the metric topology of the subspace F. For every $j \in \mathbb{N}$, let W_j and W_j' be neighborhoods of zero in E such that $V_j = W_j \cap F$ and $W_j' - W_j' \subset W_j$ and also $W_{j+1}' \subset W_j'$. Then, for every $j \in \mathbb{N}$, we obviously have

$$((y + W_j') \cap F) - ((y + W_j') \cap F) \subset W_j \cap F = V_j.$$

Hence arbitrarily picked points $x_j \in (y+W'_j)\cap F$ form a Cauchy sequence in F converging to some point $x \in F$ by the completeness of F. It remains to verify that the same sequence converges also to y.

Let U be a neighborhood of zero in E; first we choose a neighborhood of zero U' in E such that $U'+U' \subset U$, then we take a number $k=k(U)$ such that $V_k \subset U'\cap F$, hence $V_k \subset U'$. Finally, we pick an element z in the nonempty set $(y+U')\cap\bigl((y+W'_k)\cap F\bigr)$. Then, for every $j \geqslant k$, we have

$$\begin{aligned} x_j - z \in \bigl((y+W'_j)\cap F\bigr) - \bigl((y+W'_k)\cap F\bigr) \subset (W'_j - W'_k)\cap F \subset \\ \subset (W'_k - W'_k)\cap F \subset V_k \subset U'. \end{aligned}$$

Therefore,

$$x_j = y + (x_j - y) = y + (x_j - z) + (z - y) \in y + U' + U' \subset y + U,$$

so that for every $j \geqslant k(U)$ we have $x_j \in y+U$, which proves convergence of $\{x_j\}$ to y. Since E is Hausdorff, we obtain $y = x \in F$. □

In the case of locally convex spaces the topology induced in a subspace possesses the following property.

1.3.12. Lemma. *Let E be a locally convex space, let E_1 be its vector subspace with the induced topology, and let U be an absolutely convex neighborhood of zero in E_1. Then, there exists an absolutely convex neighborhood of zero V in E such that $V\cap E_1 = U$.*

If $x_0 \in E\backslash\overline{E_1}$, then V can be taken such that $x_0 \notin V$.

PROOF. By the definition of the induced topology there is an absolutely convex neighborhood of zero $W \subset E$ for which $W\cap E_1 \subset U$. Let V be the absolutely convex hull of $W\cup U$. Then $U \subset V\cap E_1$. If $v \in V\cap E_1$, then $v = tw + su$, where $w \in W$, $u \in U$, $|t|+|s| \leqslant 1$. In addition, $tw = v - su \in E_1$. If $t=0$, then we obtain at once that $v = su \in U$, since U is absolutely convex. If $t\neq 0$, then $w \in E_1$, whence we obtain $w \in E_1\cap W \subset U$ and hence $v \in U$. Thus, $V\cap E_1 \subset U$, i.e., $V\cap E_1 = U$.

If $x_0 \in E\backslash\overline{E_1}$, then we take W such that $(x_0+W)\cap E_1 = \varnothing$. Therefore, $x_0 \notin V$, since otherwise, as above, we have $x_0 = tw + su$, whence we obtain $x_0 - tw \in E_1$ and $x_0 - tw \in x_0 + W$, which is impossible. □

1.3.13. Example. Let (E,τ) and E_1 be the same as in Example 1.3.10 and let E/E_1 be the quotient of the vector space E by its subspace E_1. The topology τ_2 in the space E/E_1, called the *quotient topology*, is defined as follows: a set $V \subset E/E_1$ is open in τ_2 precisely when its preimage under the canonical mapping $E \to E/E_1$ is open in τ.

Moreover, the canonical mapping $E \to E/E_1$ is open, i.e., it takes every open set to an open set. The topological vector space $(E/E_1,\tau_2)$ is called the *topological vector quotient* of the space (E,τ).

PROOF. We show that, given a base $\mathcal{U}$ of neighborhoods of zero in (E,τ), the family of images of its elements under the canonical mapping $E \to E/E_1$ is a base of neighborhoods of zero in the space $(E/E_1,\tau_2)$ (for prebases this is not true in the general case). First we show that the canonical mapping $f\colon E \to E/E_1$ is open, i.e., takes open sets into open sets. Let $V \subset E$ and let V be open in τ. Then the set

$$V + E_1 = \bigcup_{a \in E_1} (V + a)$$

is open in the topology τ as a union of open sets (obtained from the open set V by translations). We have $f^{-1}\bigl(f(V)\bigr) = V + E_1$, so that the set $f(V)$ is open by the definition of the topology τ_2.

If now $\mathcal{U}$ is a base of neighborhoods of zero in E, then, according to what we have proved, all sets $f(V)$ are neighborhoods of zero; in addition, if W is an arbitrary neighborhood of zero in $(E/E_1, \tau_2)$, then $W = f\bigl(f^{-1}(W)\bigr)$, moreover, $f^{-1}(W)$ is a neighborhood of zero in (E, τ) (since by the definition of the topology τ_2 the mapping f is continuous). Hence there exists a set V in $\mathcal{U}$ contained in $f^{-1}(W)$, i.e., $f(V) \subset W$. The fact that $f(\mathcal{U})$ is a base of neighborhoods of zero in the quotient topology yields that the quotient topology agrees with the vector structure in E/E_1. □

The quotient is not always Hausdorff.

1.3.14. Lemma. *The topological vector space $(E/E_1, \tau_2)$ is Hausdorff if and only if the subspace E_1 is closed in the space (E, τ).*

PROOF. The singletons in any Hausdorff topological space are closed. Hence our assumption that $(E/E_1, \tau_2)$ is Hausdorff implies that zero is closed. Hence its preimage $E_1 = f^{-1}(0)$ under the canonical mapping f is closed, since the latter is continuous.

Suppose now that E_1 is closed in (E, τ) and show that the topological vector quotient $(E/E_1, \tau_2)$ is Hausdorff (even if E itself is not). It suffices to show that E/E_1 is a T_0-space. To this end, by the invariance of the topology with respect to translations it suffices to show that every nonzero element a in E/E_1 possesses a neighborhood not containing zero. Let $a \in E/E_1$, $a \neq 0$ and $b \in f^{-1}(a)$. Then $b \notin E_1$. By the closedness of E_1 there exists an open neighborhood V of the point b such that $V \cap E_1 = \varnothing$. Hence $f(V)$ is an open neighborhood of the point $a = f(b)$ not containing zero of the space E/E_1. The fact that $0 \notin f(V)$ follows from the equality

$$f^{-1}\bigl(f(V)\bigr) \cap f^{-1}(0) = (V + E_1) \cap E_1 = \varnothing,$$

which is a corollary of the equality $V \cap E_1 = \varnothing$. The set $f(V)$ is open $(E/E_1, \tau_2)$, since V is open. □

If E is an arbitrary (possibly, not Hausdorff) topological vector space, then the closure $E_0 = \overline{\{0\}}$ of the singleton $\{0\}$ is a vector subspace in E; the Hausdorff topological vector quotient E/E_0 is called the *Hausdorff topological vector space associated with E*. Certainly, if E itself is Hausdorff, then the associated Hausdorff topological vector space is isomorphic to it.

The factorpace of an arbitrary locally convex space E by its arbitrary vector subspace E_1 is locally convex; this follows from the fact that the canonical mapping $f\colon E \to E/E_1$ takes any base of convex neighborhoods of zero in E to a base of convex neighborhoods of zero in E/E_1.

Note also that if the space E is metrizable and the subspace E_1 is closed, then E/E_1 is also metrizable (this follows from the following criterion of metrizability of a Hausdorff topological vector space proven below: the existence of a countable base of neighborhoods of zero). If E is normable, then the quotient is also normable.

1.3.15. Example. If a topological vector space E is the algebraic sum of its vector subspaces E_1 and E_2, i.e., $E_1 \cap E_2 = 0$ and $E_1 + E_2 = E$, one can consider the natural algebraic projections $p_1\colon E \to E_1$, $p_2\colon E \to E_2$; the subspaces E_1 and E_2 are called *algebraic complements* of each other. If the projections p_1 and p_2 are continuous, then E_1 and E_2 are called *topological complements* of each other. We shall see in § 3.9 that in a number of important cases algebraic complements are automatically topological ones; for example, this is the case for closed subspaces of complete metrizable topological vector spaces. In the general case the projections can be discontinuous.

For example, this happens if E_1 is the kernel of a discontinuous linear function and E_2 is an algebraically complementing one-dimensional subspace. There are also examples of closed algebraically mutually complementary subspaces E_1 and E_2 in an incomplete normed space with discontinuous projections.

1.3.16. Example. Let $\mathcal{L}^0 = \mathcal{L}^0(\lambda)$ be the vector space of all Lebesgue measurable real functions everywhere defined on $[0,1]$, let λ be Lebesgue measure on this interval, and let

$$V_n = \bigl\{f \in \mathcal{L}^0\colon\ \lambda\bigl(t \in [0,1]\colon\ |f(t)| > 1/n\bigr) < 1/n\bigr\}, \quad n \in \mathbb{N}.$$

The family $\mathcal{U} = \{V_n\}$ is a base of neighborhoods of zero of some (non-Hausdorff) topology τ in $\mathcal{L}^0$ that agrees with the vector structure. Let L^0 be the Hausdorff topological vector space associated with $(\mathcal{L}^0, \tau)$. It is metrizable, but is not locally convex (Exercise 1.12.29). It can be identified (as a vector space) with the space of the classes of λ-equivalent λ-measurable real functions on $[0,1]$; convergence of sequences in $(\mathcal{L}^0, \tau)$ and in L^0 is convergence in measure (of individual functions or equivalence classes of measurable functions). For a general bounded nonnegative measure μ on a measurable space $(\Omega, \mathcal{B})$ the spaces $\mathcal{L}^0(\mu)$ and $L^0(\mu)$ are defined similarly. Convergence in measure can be defined by the metric

$$d(f,g) = \int_\Omega \min(|f(\omega) - g(\omega)|, 1)\,\mu(d\omega).$$

In the remaining examples $\mathbb{K} = \mathbb{R}$ or $\mathbb{K} = \mathbb{C}$.

1.3.17. Example. Let n be a natural number and let $\mathcal{K}(\mathbb{R}^n)$ be the vector space of all continuous functions $f\colon \mathbb{R}^n \to \mathbb{K}$ with compact support (a function on $\mathbb{R}^n$ is with *compact support* if it vanishes outside of some bounded set); another notation: $C_0(\mathbb{R}^n)$. Let $\mathcal{F}_n$ denote the set of all continuous positive functions

$f\colon \mathbb{R}^n \to (0,+\infty)$ (certainly, $\mathcal{F}_n$ is not a vector space and $\mathcal{F}_n \cap \mathcal{K}(\mathbb{R}^n) = \varnothing$). For any $f \in \mathcal{F}_n$, consider the set

$$V_f = \bigl\{\varphi \in \mathcal{K}(\mathbb{R}^n)\colon\ |\varphi(x)| < f(x)\ \forall\, x \in \mathbb{R}^n\bigr\}.$$

The family of all sets of such a form is a base of neighborhoods of zero for some Hausdorff locally convex topology $\tau_{\mathcal{K}}$ in $\mathcal{K}(\mathbb{R}^n)$; below we assume, if it is not stated otherwise, that $\mathcal{K}(\mathbb{R}^n)$ is equipped with this topology. Let us also describe a family of norms defining this topology. For every function $f \in \mathcal{F}_n$ let p_f denote the norm on the space $\mathcal{K}(\mathbb{R}^n)$ defined by the equality $p_f(\varphi) = \max_x |f(x)\varphi(x)|$ and set $\mathcal{P}_{\mathcal{K}} = \{p_f\colon\ f \in \mathcal{F}_n\}$. The family of norms $\mathcal{P}_{\mathcal{K}}$ defines the introduced topology $\tau_{\mathcal{K}}$. The space $\bigl(\mathcal{K}(\mathbb{R}^n), \tau_{\mathcal{K}}\bigr)$ is not metrizable. A sequence $\{\varphi_i\}$ of elements of the space $\mathcal{K}(\mathbb{R}^n)$ converges to 0 if and only if the following two conditions are fulfilled:

(i) $\max_{x\in\mathbb{R}^n} |\varphi_i(x)| \to 0$,

(ii) there exists a bounded set B in $\mathbb{R}^n$ outside of which all the functions φ_i vanish.

Indeed, if $x_i \in \mathbb{R}^n$, $|x_i| \to \infty$, and $C_i \to \infty$, then there is $f \in \mathcal{F}_n$ with $f(x_i) = C_i$. There is no metric on $\mathcal{K}(\mathbb{R}^n)$ with respect to which convergence of sequences coincides with convergence of sequences in the topology $\tau_{\mathcal{K}}$ (this assertion, the subject of Exercise 1.12.30, is stronger than the assertion that $\mathcal{K}(\mathbb{R}^n)$ is not metrizable).

1.3.18. Example. Let E be a vector space over $\mathbb{K}$, let $\mathcal{P}_0$ be the family of all seminorms on E, and let $\mathcal{P}_1$ be the family of all quasi-norms on E. The topology in E defined by the family of seminorms $\mathcal{P}_0$ is the strongest among all locally convex topologies in E (it is called *the strongest locally convex topology* in E); the topology in E defined by the family of quasi-norms $\mathcal{P}_1$ is the strongest among all topologies in E that agree with the vector structure. One can prove (Exercise 1.12.32) that if the algebraic dimension of the space E over $\mathbb{K}$ is at most countable, then these two topologies coincide; otherwise they are distinct.

Note that in the strongest locally convex topology all seminorms on E are continuous, hence are continuous all linear functions and all linear mappings to any locally convex spaces.

It is instructive to find smaller classes of seminorms defining the strongest locally convex topology. For example, taking a Hamel basis $\{e_\alpha\}$ in E and a positive function φ on the set of indices α we can introduce seminorms of the form

$$p_\varphi(x) = \sum_\alpha \varphi(\alpha)|x_\alpha|, \quad \text{where} \quad x = \sum_\alpha x_\alpha e_\alpha.$$

The family of all seminorms of such a form also defines the strongest locally convex topology in E, since for every seminorm p on the space E we have the obvious estimate $p(x) \leqslant \sum_\alpha |x_\alpha| p(e_\alpha) \leqslant p_\varphi(x)$, where $\varphi(\alpha) = p(e_\alpha) + 1$.

1.3.19. Example. Let Σ be the space of rapidly decreasing sequences, i.e., real sequences $x = (x_n)$ with finite norms

$$p_k(x) = \sup_n n^k |x_n|, \quad k \in \mathbb{N}.$$

The same topology is defined by Euclidean norms q_k defined by

$$q_k(x)^2 = \sum_{n=1}^{\infty} n^{4k}|x_n|^2.$$

1.3.20. Example. Let $\mathcal{S}(\mathbb{R}^n)$ be the Schwartz space of all infinitely differentiable $\mathbb{K}$-valued functions φ (here $\mathbb{K} = \mathbb{R}$ or $\mathbb{C}$) defined on $\mathbb{R}^n$ and satisfying the following condition (further we set $t = (t_1, \ldots, t_n)$):

$$p_{rk}(\varphi) = \max_t (1 + |t|^r)\|\varphi^{(k)}(t)\| < \infty \quad \text{for all } k, r \in \mathbb{N},$$

where $|t| = \left(\sum_{k=1}^n |t_k|^2\right)^{1/2}$, $\|\varphi^{(0)}(t)\| = \max_{t\in\mathbb{R}^n} |\varphi(t)|$,

$$\|\varphi^{(k)}(t)\| = \max\left\{\left|\frac{\partial^k \varphi(t)}{\partial t_1^{k_1} \ldots \partial t_n^{k_n}}\right| : \ k_1 + \cdots + k_n = k, \ k_i \geqslant 0\right\}.$$

Each of the functions p_{rk} is a norm on $\mathcal{S}(\mathbb{R}^n)$; let $\mathcal{P}_S$ denote the set all such norms and let τ_S be the topology in $\mathcal{S}(\mathbb{R}^n)$ defined by this family of norms.

The topology in $\mathcal{S}$ can be defined also by Euclidean norms. For example, for $n = 1$ we can take Euclidean norms q_{rk} defined by

$$q_{rk}^2(\varphi) = \int_{-\infty}^{+\infty} (1 + t^2)^r |\varphi^{(k)}(t)|^2 \, dt.$$

Everywhere below we assume, if it is not stated otherwise, that the space $\mathcal{S}(\mathbb{R}^n)$ is equipped with the topology τ_S. The space $\mathcal{S}(\mathbb{R}^n)$ turns out to be a complete metrizable locally convex space (verify completeness!); such a space is called a *Fréchet space*. The space $\mathcal{S}(\mathbb{R}^n)$ plays an important role in the theory of distributions (generalized functions). It is worth noting that on $\mathcal{S}(\mathbb{R}^n)$ there is no norm making it a Banach space in the topology of which every function $\Psi_t\colon \varphi \mapsto \varphi(t)$, $t \in \mathbb{R}^n$, is continuous. Since in the topology τ_S all such functions are continuous, this yields that the topology τ_S is not normable (though, this property of τ_S follows immediately from Kolmogorov's criterion of normability of a topological vector space proved below). In applications of the space $\mathcal{S}(\mathbb{R}^n)$ in analysis one usually needs the continuity of the mappings Ψ_t, so the absence on $\mathcal{S}(\mathbb{R}^n)$ of a Banach norm with respect to which they are continuous is an example demonstrating insufficiency of the theory of Banach spaces for the purposes of analysis.

1.3.21. Example. Let $\mathcal{D}(\mathbb{R}^n)$ be the space of all infinitely differentiable functions $\varphi\colon \mathbb{R}^n \to \mathbb{K}$ with compact support and let $\mathcal{F}_n = \{g \in C(\mathbb{R}^n)\colon \ g > 0\}$. For every pair of functions f, r in $\mathcal{F}_n$, let $q_{f,r}$ denote the norm on $\mathcal{D}(\mathbb{R}^n)$ defined as follows:

$$q_{f,r}(\varphi) = \max_t \left\| f(t)\varphi^{([r(t)])}(t) \right\|,$$

where $[a]$ is the integer part of a number a (the notation $\|\varphi^{(k)}(t)\|$ is introduced in the previous example). Below we assume that $\mathcal{D}(\mathbb{R}^n)$ is equipped with the topology τ_D, defined by the family of norms $\mathcal{P}_D = \{q_{f,r}\colon \ f, r \in \mathcal{F}_n\}$. Then

$\mathcal{D}(\mathbb{R}^n)$ is a nonmetrizable locally convex space (Exercise 1.12.31). For $n = 1$ this topology is defined by the family of seminorms

$$p_{\{\alpha_k\},\{r_k\}}(\varphi) = \sum_{k=-\infty}^{\infty} \alpha_k \max_{t\in[k,k+1]} |\varphi^{(r_k)}(t)|, \ \alpha_k \in \mathbb{N}, r_k \in \mathbb{N} \cup \{0\}.$$

1.3.22. Example. Let $\mathcal{E}(\mathbb{R}^n)$ be the space of all infinitely differentiable functions on $\mathbb{R}^n$ taking values in $\mathbb{K}$ and, for every $n \in \mathbb{N}$, let us define a seminorm p_m on $\mathcal{E}(\mathbb{R}^n)$ by the equality

$$p_m(\varphi) = \max\{\|\varphi^{(k)}(t)\|: \ k = 0, 1, \ldots, m; \|t\| \leqslant m\}.$$

Below we assume, if it is not explicitly stated otherwise, that the space $\mathcal{E}(\mathbb{R}^n)$ is equipped with the topology $\tau_{\mathcal{E}}$ defined by the seminorms $\mathcal{P}_{\mathcal{E}} = \{p_m\colon m \in \mathbb{N}\}$. Then $\mathcal{E}(\mathbb{R}^n)$ is a complete metrizable locally convex space (verify this!). Note that the topology $\tau_{\mathcal{E}}$ cannot be defined by means of norms; moreover, on $\mathcal{E}(\mathbb{R}^n)$ there is no any continuous norm (in fact, these two assertions are equivalent).

Below, when we discuss properties of the spaces $\mathcal{K}(\mathbb{R}^n)$, $\mathcal{S}(\mathbb{R}^n)$, $\mathcal{D}(\mathbb{R}^n)$, and $\mathcal{E}(\mathbb{R}^n)$ independent of the dimension n of the space $\mathbb{R}^n$, symbols like $\mathcal{D}(\mathbb{R}^n)$, $\mathcal{S}(\mathbb{R}^n)$ will be frequently abbreviated to $\mathcal{D}$, $\mathcal{S}$, etc. As a rule, the considered properties of these spaces will not depend on our choice of the real or complex field; this is why in the previous examples the spaces of complex functions (over the field $\mathbb{C}$) and real functions (over the field $\mathbb{R}$) were denoted by the same symbols. Below in our discussion of the properties of the spaces $\mathcal{K}$, $\mathcal{S}$, etc. we do not mention the field of scalars at all if the discussed properties do not depend on it. In other cases we shall use expressions like "the complex space $\mathcal{S}$" or "the real space $\mathcal{S}$". Similarly we shall act dealing with the topological duals of these spaces $\mathcal{D}'(\mathbb{R}^n)$, $\mathcal{S}'(\mathbb{R}^n)$, etc. These spaces of linear functionals play a very important role in applications and are called "spaces of distributions" or "spaces of generalized functions" (the spaces $\mathcal{D}$, $\mathcal{S}$, etc., are often called "spaces of test functions"). The terminology is explained by the fact that "usual" functions define distributions by means of integration. For example, every locally integrable function f on $\mathbb{R}^n$ defines an element of $\mathcal{D}'$ the action of which on $\varphi \in \mathcal{D}$ is the integral of φf over $\mathbb{R}^n$. If f estimated (in the absolute value) by a polynomial, then f defines also a generalized function of class $\mathcal{S}'$.

1.3.23. Example. This example can be considered as an introduction to the duality theory. Let E be an arbitrary vector space over $\mathbb{K}$ (always non-discrete) and let G be some vector space of $\mathbb{K}$-linear functionals on E, i.e., $\mathbb{K}$-linear mappings from E to $\mathbb{K}$. Then, on E there exists a topology τ possessing the following properties:

(a) it agrees with the vector structure;

(b) every functional g belonging to G is continuous as a mapping from (E, τ) to $\mathbb{K}$;

(c) every $\mathbb{K}$-linear continuous functional on (E, τ) is an element of G.

Moreover, among all topologies in E possessing property (b) there exists the weakest one and it automatically possesses also properties (a) and (c); this topology, called the *weak topology in E defined by the elements of G*, will be now defined. Convergence in it is called *weak convergence*.

For every $g \in G$, let p_g denote the seminorm on E defined by the formula

$$p_g(\varphi) = |g(\varphi)|.$$

We now prove that the topology in E defined by the whole family of seminorms $\mathcal{P}_G = \{p_g\colon\ g \in G\}$ is the desired one. This topology will be denoted by the symbol $\sigma(E,G)$. Clearly, the topology $\sigma(E,G)$ possesses property (a).

If $g \in G$, then $|g(\varphi)| < \varepsilon$ whenever $p_g(\varphi) < \varepsilon$; this means that the functional g is continuous at zero; since it is linear, we obtain its continuity at every point, so that $\tau = \sigma(E,G)$ possesses property (b). Let now f be an arbitrary continuous $\mathbb{K}$-linear functional on $\big(E,\sigma(E,G)\big)$. We have to show that $f \in G$. By the continuity of f at zero in the space $\big(E,\sigma(E,G)\big)$, there exist elements $g_1,\ldots,g_n$ in the space G such that if $|g_1(a)| < \varepsilon,\ldots,|g_n(a)| < \varepsilon$, then $|f(a)| < 1$. It follows that

$$\operatorname{Ker} f \supset \bigcap_{k=1}^{n} \operatorname{Ker} g_k(\varphi). \tag{1.3.1}$$

Indeed, if this is not true and there exists an element $a \in E$ such that $f(a) \neq 0$, but $g_1(a) = \cdots = g_n(a) = 0$, then, using that the field $\mathbb{K}$ is not discrete, we choose an element $k \in \mathbb{K}$ such that $|k| > 1$. We have the relation $\big|f\big(ka/f(a)\big)\big| = |k| > 1$, although

$$\big|g_j\big(ka/f(a)\big)\big| = \big|kg_j(a)/f(a)\big| = 0 < \varepsilon,\ j = 1,2,\ldots,n.$$

However, inclusion (1.3.1) yields (this will be proved in the next lemma, sometimes called the lemma on three homomorphisms) that f is a linear combination of the functionals g_j. Since G is a vector space, this means that $f \in G$. Certainly, one could also prove directly that f is a linear combination of $g_1,\ldots,g_n$. If $n = 1$ and there is v with $g_1(v) = 1$, then $f = f(v)g_1$. For $n > 1$ we argue by induction and represent f as $c_1g_1 + \cdots + c_{n-1}g_{n-1}$ on $\operatorname{Ker} g_n$. Taking v with $g_n(v) = 1$, we find the final representation with $c_n = f(v) - c_1g_1(v) - \cdots - c_{n-1}g_{n-1}(v)$.

1.3.24. Lemma. *Let E_1, E_2, and E_3 be vector spaces and let*

$$f_{12}\colon\ E_1 \to E_2 \quad \textit{and} \quad f_{13}\colon\ E_1 \to E_3$$

be linear mappings such that $\operatorname{Ker} f_{13} \supset \operatorname{Ker} f_{12}$. *Then, there exists a linear mapping $f_{23}\colon\ E_2 \to E_3$ such that $f_{13} = f_{23}\circ f_{12}$.*

PROOF. The mapping f_{23} is first defined on the subspace $f_{12}(E_1)$ in E_2 by the equality $f_{23}(x) = f_{13}\big(f_{12}^{-1}(x)\big)$; the fact that f_{23} is well-defined and linear follows from the linearity of the mappings f_{12} and f_{13} and the inclusion $\operatorname{Ker} f_{13} \supset \operatorname{Ker} f_{12}$. Next the mapping f_{23} is extended in an arbitrary way as a linear mapping on the whole space E_2 (which can be easily done by using a Hamel basis, see § 1.1). This extension is the desired mapping. □

Now, returning to Example 1.3.23, in order to establish the relation

$$f = \sum_{j=1}^{n} k_j g_j,$$

it suffices to take $E_1 = E$, $E_3 = \mathbb{K}$, $E_2 = \mathbb{K}^n$, $f_{13} = f$ and define the mapping f_{12} as follows: $f_{12}(x) = \bigl(g_1(x), \ldots, g_n(x)\bigr)$. By the lemma above there exists a $\mathbb{K}$-linear functional $f_{23}\colon \mathbb{K}^n \to \mathbb{K}$ such that $f = f_{23} \circ f_{12}$. Every linear functional on $\mathbb{K}^n$ is defined by a collection $(k_1, \ldots, k_n)$ of n elements of the field $\mathbb{K}$ such that $f_{23}(h_1, \ldots, h_n) = \sum_{j=1}^n k_j h_j$. Hence we obtain $f(x) = \sum_{j=1}^n k_j g_j(x)$, i.e., we have $f = \sum_{j=1}^n k_j g_j$.

Thus, the topology $\sigma(E, G)$ possesses property (c). We now show that $\sigma(E, G)$ is the weakest topology with property (b). If τ is an arbitrary topology with this property, then the sets of the form

$$\bigl\{x \in E\colon\ |g_k(x - x_k)| < \varepsilon_k,\ k = 1, 2, \ldots, n\bigr\},\ n \in \mathbb{N},\ x_k \in E,\ g_k \in G,$$

are open in it. Since these sets form a base of the topology $\sigma(E, G)$, the topology $\sigma(E, G)$ is majorized by the topology τ (if S and T are two systems of sets such that $S \subset T$, we say that the system T majorizes the system S and that S is majorized by the system T). Justification of Example 1.3.23 is complete.

It is the right place to introduce the dual to a topological vector space. More precisely, we introduce even two duals: the algebraic and topological, the latter being particularly important.

1.3.25. Definition. *Let E a topological vector space over the field $\mathbb{K}$. The space of all continuous linear functions on E with values in $\mathbb{K}$ is called the dual (or topological dual) space to E and is denoted by the symbol E'.*

We emphasize that usually E' is much more narrow that the *algebraic dual* to E consisting of all linear functions and denoted in this book by E^*. Note that in many books (including our [**72**]) the notation is the opposite; we have decided to keep here the notation that is more traditional in the literature on topological vector spaces. It may happen that $E' = \{0\}$, but on Hausdorff locally convex spaces, as we shall see from the Hahn–Banach theorem, the topological dual E' separates the points of the space E.

If E is a locally convex space and $G = E'$, then the topology $\sigma(E, E')$ is called the *weak* or the *weakened topology in* E; note that $\bigl(E, \sigma(E, E')\bigr)' = E'$. The described situation can be regarded as symmetric with respect to E and G. Namely: every element $x \in E$ can be identified with some linear functional F_x on the space G defined as follows: $F_x(g) = g(x)$, $x \in E$, $g \in G$. Then the whole space E is identified with some vector space of linear functionals on G. If E is locally convex, the *weak topology* (or *weak-$*$*) in E' is the topology $\sigma(E', E)$.

1.3.26. Example. This is a generalization (actually, only formal) of the previous example. We shall say that two vector spaces E and G *are in duality* (or that they form a *dual pair*) if a bilinear function $b\colon\ E \times G \to \mathbb{K}$ (or a "bilinear form"; we say that it establishes a duality between the spaces E and G) is given with the following properties:

(1) if $x \in E$, $x \neq 0$, then there is $g \in G$ with $b(x, g) \neq 0$;

(2) if $g \in G$, $g \neq 0$, then there is $x \in E$ with $b(x, g) \neq 0$.

Suppose that, as in the previous example, E is a vector space, G is some vector space of linear functionals on E such that if $x \in E$, $x \neq 0$, then there is $g \in G$ with $g(x) \neq 0$, i.e., as we shall say, the set G *separates the points* in E. The bilinear form b on $E \times G$ defined by the equality $b(x, g) = g(x)$ possesses properties (1) and (2), hence establishes a duality between the spaces E and G (this bilinear form is called *canonical*).

Actually, the general case of two vector spaces in duality coincides with this special one. Namely, let E and G be two vector spaces in duality defined by a bilinear form b. Then every vector $x \in E$ defines a linear functional $g \mapsto b(x, g)$ on the space G and every element $g \in G$ defines a linear functional $x \mapsto b(x, g)$ on E. Thus, we have defined linear mappings $E \to G^*$, $x \mapsto [g \mapsto b(x, g)]$ and $G \to E^*$, $g \mapsto [x \mapsto b(x, g)]$.

It is readily verified (with the aid of properties (1) and (2) of the mapping b) that both mappings are linear and injective, i.e., are linear isomorphisms onto their ranges. Hence the space E can be identified with its image under the first of these isomorphisms, i.e., with some space of linear functionals on G, and the space G can be identified with its image under the second of these isomorphisms, i.e., with some space of linear functionals on E. In this way we shall act (even without special notice) when considering pairs of spaces in duality. In particular, elements of E will be called (linear) functionals on G and elements of G will be called (linear) functionals on E. Therefore, the construction from the previous example yields two "weak topologies defined by the duality between E and G": the topology $\sigma(E, G)$ on E and the topology $\sigma(G, E)$ on G.

One more remark about notation. Throughout the bilinear form establishing a duality between the spaces E and G will be usually denoted by the symbol $\langle \cdot, \cdot \rangle$; in particular, even in the case where, say, the vector space G is from the very beginning defined as a space of linear functionals on E, we shall often replace the symbol $g(x)$, where $g \in G, x \in E$, by the symbol $\langle g, x \rangle$ (or $\langle x, g \rangle$). The symbol $\langle \cdot, \cdot \rangle$ should not be confused with an inner product.

1.4. Convex sets

In this section we assume that the field of scalars is $\mathbb{R}$. The *closed convex hull* or *closed convex envelope* of a set A in a topological vector space is defined as the intersection of all closed convex sets containing A. The notation: $\overline{\mathrm{conv}}\, A$. Similarly, we define the *closed absolutely convex hull* $\overline{\mathrm{abs\,conv}}\, A$ of the set A as the intersection of all closed absolutely convex sets containing A.

1.4.1. Proposition. *Let E be a topological vector space, let $V \subset E$ be a convex set, and let $\check{V}$ and $\overline{V}$ be the interior and closure of V, respectively. Then, for any points $a \in \check{V}$ and $b \in \overline{V}$, the set $\check{V}$ contains the set*

$$[a, b) := \{ta + (1-t)b \colon\ t \in (0, 1]\}.$$

PROOF. We first consider the case where $b \in V$; as assume that $b = 0$ (in order to get this, it suffices to replace V with $V - b$; since the translation is a homeomorphism, it takes the interior of the set into the interior of its image, the closure into the closure, etc.). Let $x \in [a, 0)$, i.e., $x = \lambda a$, $\lambda \in (0, 1]$. The set $\breve{V}$ is an open neighborhood of the point a belonging to V. Then the set $\lambda\breve{V}$ is an open neighborhood of the point x and $\lambda\breve{V} \subset V$ by the convexity of V.

Let us consider the general case. Now, unlike the previous case, we assume that $x = 0$ (and not that $b = 0$), where x is a point in the interval

$$(a, b) := \{ta + (1 - t)b\colon\ t \in (0, 1)\};$$

if $x = a$, then there is nothing to prove, since $a \in \breve{V}$ by assumption. Thus, $b = \nu a$, where $\nu < 0$. The set $\nu\breve{V}$ is a neighborhood of the point b. Moreover, since $b \in \overline{V}$, there exists $z \in V \cap (\nu\breve{V})$, i.e., $z/\nu \in \breve{V}$. Therefore, the relation $x = 0 \in [z/\nu, z)$ yields that $x \in \breve{V}$ according to what is already proved. □

1.4.2. Proposition. *Let V be a convex set in a topological vector space E. Then the sets $\breve{V}$ and $\overline{V}$ are also convex. If $\breve{V} \neq \varnothing$, then $\overline{\breve{V}} = \overline{V}$ and $\breve{\overline{V}} = \breve{V}$.*

PROOF. The convexity of $\breve{V}$ and the equality $\overline{\breve{V}} = \overline{V}$ (under the assumption that $\breve{V}$ is not empty) are direct corollaries of the previous proposition. The convexity of $\overline{V}$ follows from the continuity of the mapping $\Phi_t\colon\ E \times E \to E$, $(x, z) \mapsto tx + (1 - t)z$ for every $t \in [0, 1]$. Indeed, by the continuity of Φ_t we have $\Phi_t(\overline{V}\times\overline{V}) = \Phi_t(\overline{V\times V}) \subset \overline{V}$; these inclusions mean the convexity of $\overline{V}$.

Let us verify that $\breve{\overline{V}} = \breve{V}$. Since $\breve{V} \subset \breve{\overline{V}}$, it suffices to verify that also the opposite inclusion is valid. Let $a \in \breve{\overline{V}}$ and $b \in \breve{V}(\neq \varnothing)$. We shall assume that $b = 0$ (as already noted, this does not restrict the generality). Then by the continuity of the mapping $\varphi\colon\ t \mapsto ta$ the equality $\varphi(1) = a$ and the openness of $\breve{\overline{V}}$ yield the existence of $\varepsilon > 0$ such that $\varphi(1 + \varepsilon) = (1 + \varepsilon)a \in \breve{\overline{V}} \subset \overline{V}$. Thus, $a \in [0, (1 + \varepsilon)a)$, $0 \in \breve{V}$ and $(1 + \varepsilon)a \in \overline{V}$. Hence by the previous proposition we have $a \in \breve{V}$. □

1.4.3. Remark. (i) For further references we note the following obvious fact: if E is a vector space, V and W are its convex subsets, and $\alpha, \beta \in \mathbb{K}$, then the set $\alpha V + \beta W$ is also convex and $(\alpha + \beta)V \subset \alpha V + \beta V$, and if $\alpha\beta > 0$, then this inclusion becomes the equality $(\alpha + \beta)V = \alpha V + \beta V$.

(ii) The assumption that $\breve{V}$ is not empty is essential for the validity of the proven proposition. For example, let E be a topological vector space on which there exists a discontinuous linear functional (this property holds, in particular, for all infinite-dimensional normed spaces and even all infinite-dimensional metrizable space, see Example 1.9.10). Then the kernel V of this functional is an everywhere dense vector subspace (hence a convex set) with the empty interior, so that $\overline{V} = E$, $\breve{\overline{V}} = E \neq \breve{V} = \varnothing$.

Let us discuss connections between convex sets and sublinear functions.

Let E be a vector space over the field of real numbers. A function (functional) $p\colon\ E \to \mathbb{R}^1 \cup \{+\infty\}$ is called *sublinear* or *homogenously convex* if

(1) $p(x + y) \leqslant p(x) + p(y)$,
(2) $p(0) = 0$, $p(\lambda x) = \lambda p(x)$, $\lambda > 0$.

If p does not assume the value $+\infty$, the latter equality holds also for $\lambda = 0$.

A sublinear function is convex. We recall that a function f on a convex set V in a linear space is called *convex* if

$$f(\lambda u + (1-\lambda)v) \leqslant \lambda f(u) + (1-\lambda)f(v) \quad \forall\, u, v \in V,\ \lambda \in [0,1].$$

A function g is called*concave* if $-g$ is convex.

Note that every linear functional f on E is sublinear; the functional $|f|$, where $|f|(x) = |f(x)|$, $x \in E$, is also sublinear; every seminorm on a vector space is a sublinear functional.

Our nearest goal is to describe the connection between nonnegative sublinear functionals and convex subsets in E containing zero.

1.4.4. Definition. *Let E be a vector space and $A \subset E$. The Minkowski functional or the gauge function of the set A is the function*

$$p_A\colon\ E \to \mathbb{R}^+ \cup \{+\infty\}$$

defined by the equality

$$p_A(x) = \inf\{\lambda > 0\colon\ x \in \lambda A\},$$

where $p_A(x) = +\infty$ if there is no such λ.

1.4.5. Definition. *A subset in a topological vector space is called bounded if it is absorbed by every neighborhood of zero.*

One should bear in mind that even if the topology of a topological vector space is defined by a metric, the boundedness in the sense of the previous definition does not reduce to the boundedness with respect to the metric. For example, the standard topology of the real line can be defined by a bounded metric. On the other hand, a bounded set in a topological vector space need not be bounded with respect to the metric defining the topology (see Exercise 1.12.36).

1.4.6. Proposition. *A set B in a topological vector space E is bounded precisely when for every sequence $\{a_n\} \subset B$ and every sequence of scalars $\{t_n\}$ converging to zero, the sequence $\{t_n a_n\}$ converges to zero in E.*

PROOF. If B is bounded, $a_n \in B$, scalars t_n tend to zero, V is a neighborhood of zero in E, and $\lambda > 0$ is such that $tB \subset V$ whenever $|t| < \lambda$, then $t_n a_n \in V$ for n so large that $|t_n| < \lambda$. Hence $t_n a_n \to 0$ in E.

Conversely, let the condition in this criteria of boundedness be fulfilled for the set B. If B is not bounded, then there exists a neighborhood of zero $V \subset E$ such that for every $n \in \mathbb{N}$ there is a number t_n such that $|t_n| < 1/n$ and $(t_n B)\backslash V \neq \varnothing$. Hence there exist points $a_n \in B$ with $t_n a_n \in (t_n B \backslash V)$. Clearly, the vectors $t_n a_n$ do not converge to zero, although $t_n \to 0$. □

Let us study the relations between the function p_A and the set A.

1.4.7. Proposition. (i) *If $x \in E$ and there is $\lambda \in (0, \infty)$ such that $x \in \lambda A$, then $p_A(x) < \infty$. Hence if A is an absorbent set, then $p_A(x) < \infty$ for every $x \in E$. If the set A is convex and the function p_A is finite, then A is an absorbent set.*

(ii) *If* $x \neq 0$ *and the set* $A_x = A \cap \{z \in E\colon\ z = \lambda x,\ \lambda \geqslant 0\}$ *is bounded in the standard topology of the one-dimensional subspace generated by the element* x, *then* $p_A(x) > 0$. *In particular, if* A *is bounded in some Hausdorff topology in the space* E *compatible with the vector structure, then* $p_A(x) > 0$ *for all* $x \neq 0$.

(iii) *If* $c > 0$, *then* $p_A(cx) = cp_A(x)$; *if* $0 \in A$ *and* $p_A(x) < \infty$, *then we have* $p_A(cx) = cp_A(x)$ *for all* $c \geqslant 0$.

(iv) *If the set* A *is convex, then*

$$p_A(x_1 + x_2) \leqslant p_A(x_1) + p_A(x_2) \quad \forall\, x_1,\, x_2 \in E.$$

(v) *If* A *is circled, then* $p_A(cx) = |c| p_A(x)$ *for all* $c \in \mathbb{R}$.

PROOF. All assertions are obvious except for (iv), in the justification of which we can assume that $p_A(x_1)$ and $p_A(x_2)$ are finite, since otherwise there is nothing to prove. Let $a_1 > p_A(x_1)$, $a_2 > p_A(x_2)$; then there exist numbers $a_i' \in \bigl(p_A(x_i), a_i\bigr)$ such that $x_1 + x_2 \in a_1'A + a_2'A \subset (a_1' + a_2')A$. This means that

$$p_A(x_1 + x_2) \leqslant a_1' + a_2' \leqslant a_1 + a_2.$$

Since the numbers a_1 and a_2 can be made as close to $p_A(x_i)$ as we wish, we have $p_A(x_1 + x_2) \leqslant p_A(x_1) + p_A(x_2)$. □

1.4.8. Proposition. (i) *If* A *is a convex set and* $0 \in A$, *then we have the inclusion* $\{x \in A\colon\ p_A(x) < 1\} \subset A$.

(ii) *We have* $A \subset \{x\colon\ p_A(x) \leqslant 1\}$.

(iii) *If the intersection of* A *with every one-dimensional subspace of the space* E *is closed (in the standard topology of the one-dimensional space) and* A *is convex and absorbing, then* $A = \{x\colon\ p_A(x) \leqslant 1\}$.

(iv) *If* A *is convex,* $0 \in A$ *and the intersection of* A *with every one-dimensional subspace in* E *is open (in the standard topology of the one-dimensional space), then* $\{x \in E\colon\ p_A(x) < 1\} = A$.

(v) *Let* p *be a nonnegative sublinear function on the space* E. *Then the sets* $A_1 = \{x\colon\ p(x) < 1\}$ *and* $A_2 = \{x\colon\ p(x) \leqslant 1\}$ *are convex and contain zero, in addition,* $p_{A_1}(x) = p_{A_2}(x) = p(x)$ *for all* $x \in E$.

(vi) *Under the assumptions in* (v), *if the function* p *is everywhere finite, then* A_1 *is open and* A_2 *is closed in the strongest locally convex topology of the space* E *(see Example* 1.3.18).

PROOF. (i) If $p_A(x) < 1$, then there is $\lambda \in (0, 1)$ such that $x \in \lambda A$, i.e., $x/\lambda \in A$. Since $0 \in A$ and A is convex, we have $\lambda x/\lambda + (1-\lambda)0 = x \in A$. Assertion (ii) is obvious.

For the proof of assertion (iii) it suffices to show that we have the inclusion $\{x\colon\ p_A(x) = 1\} \subset A$, since under our assumptions about A we have

$$\{x\colon\ p_A(x) < 1\} \subset A \subset \{x\colon\ p_A(x) \leqslant 1\}.$$

Let $p_A(x) = 1$, i.e., for every natural number n there exists a number $\lambda_n \geqslant 1$ such that $x/\lambda_n \in A$ and $\lambda_n - 1 < 1/n$. This means that $x_n = x/\lambda_n \to x$ in the one-dimensional space generated by the element x, moreover, we have $x_n \in A \cap \{\lambda x\colon\ x \in \mathbb{R}\}$ for each n, and the latter set is closed by assumption.

(iv) Since the set in the left-hand side of this equality is contained in the set in the right-hand side according to what has been proved above, it suffices to show that if $x \in A$, then $p_A(x) < 1$. So, let $x \in A$; we have to show that $p_A(x) < 1$. If $x = 0$, then this is true. Let $x \neq 0$ and let M be the one-dimensional subspace in E generated by x. Then $M \cap A$ is open in M. The inclusion $x \in A$, which is true by assumption, yields that $x/\lambda \in A$ for some number $\lambda \in (0, 1)$. This means that $p_A(x) < 1$.

(v) Clearly, $0 \in A_1 \subset A_2$. If $z = tx_1 + (1-t)x_2$, where $x_1, x_2 \in A_1$, $t \in [0, 1]$, then

$$p(z) = p(tx_1 + (1-t)x_2) \leqslant tp(x_1) + (1-t)p(x_2) < t + (1-t) = 1,$$

so that $z \in A_1$. Similarly we verify the convexity of A_2. Let us prove the equality $p_{A_1}(x) = p(x)$. Let $p_{A_1}(x_0) = \alpha$, $p(x_0) = \beta$. We consider two cases: $\alpha < \beta$ and $\beta < \alpha$. In the first case, there exists $\varepsilon > 0$ such that $\alpha + \varepsilon < \beta$. Then, on the one hand, $p_{A_1}\bigl(x_0/(\alpha+\varepsilon)\bigr) = \alpha/(\alpha+\varepsilon) < 1$, so that $x_0/(\alpha+\varepsilon) \in A_1$. On the other hand, $p\bigl(x_0/(\alpha+\varepsilon)\bigr) = \beta/(\alpha+\varepsilon) > 1$, so that $x_0/(\alpha+\varepsilon) \notin A_1$. In the second case ($\alpha > \beta$), there exists $\delta > 0$ such that $\beta + \delta < \alpha$. Then $p\bigl(x_0/(\beta+\delta)\bigr) < 1$, so that $x_0/(\beta+\delta) \in A_1$, and $p_{A_1}\bigl(x_0/(\beta+\delta)\bigr) > 1$, i.e., $x_0/(\beta+\delta) \notin A_1$, which contradicts the previous inclusion. Similarly we prove the equality $p(x) = p_{A_2}(x)$.

Assertion (vi) follows from the fact that every finite sublinear function on a vector space is continuous in this topology. Let us prove this. Let g be a sublinear function on E. Then the function $f\colon x \mapsto \max\{|g(x)|, |g(-x)|\}$ is, as one can readily verify, a seminorm on E, so that the set $\{x\colon f(x) \leqslant 1\}$ turns out to be convex, absorbent and circled, i.e., a neighborhood of zero in the strongest locally convex topology on E. This means that the seminorm f is continuous. By the inequality $|g(x_1) - g(x_2)| \leqslant f(x_1 - x_2)$ also the sublinear function g is continuous. □

The next result is often taken as a definition of a locally convex space.

1.4.9. Theorem. *The topology of any locally convex space over the field* $\mathbb{R}$ *or* $\mathbb{C}$ *can be defined by a family of seminorms.*

PROOF. Let E be a locally convex space and let $\mathcal{B}$ be a base of its neighborhoods of zero consisting of convex circled sets (such a base of neighborhoods of zero exists by Proposition 1.2.11). For every set V in the base $\mathcal{B}$, let p_V denote its Minkowski functional and let τ be the topology defined by the family of seminorms $\{p_V\colon V \in \mathcal{B}\}$. The topology τ coincides with the original topology of the space E. Indeed, on the one hand, every set of the form $\{x \in E\colon p_V(x) < \varepsilon\} = \varepsilon V$, where $\varepsilon \in (0, \infty)$, $V \in \mathcal{B}$, is a neighborhood of zero in the original topology; these sets form a prebase (see Example 1.3.9), which in this case is a base of neighborhoods of zero in τ. On the other hand, if W is an arbitrary neighborhood of zero in the original topology and V_0 is a set in the base $\mathcal{B}$ such that $V_0 \subset W$, then we have $\{x\colon p_{V_0}(x) < 1\} = V_0 \subset W$. □

It is worth noting that the topology of a locally convex space can be defined by a family of seminorms of the minimal possible cardinality (Exercise 1.12.25).

1.4.10. Remark. (i) Even in the case where our locally convex space is Hausdorff, its topology is not always defined by a family of norms. Indeed, every norm in such a collection is continuous in the topology defined by this family, but there exist Hausdorff locally convex spaces on which there are no continuous norms. For example, any infinite power of the real line with the usual product topology and the space $\mathcal{E}(\mathbb{R}^n)$ have this property (Example 1.3.22). Normability is discussed in § 1.5.

(ii) It is readily seen that the property to be Hausdorff for a locally convex space is equivalent to the following property: for every $x \neq 0$ there is a continuous seminorm p such that $p(x) \neq 0$.

(iii) The locally convex topology defined by a family of seminorms $\mathcal{Q}$ is stronger than the locally convex topology defined by the family of seminorms $\mathcal{P}$ precisely when for for every seminorm $p \in \mathcal{P}$ there exist seminorms $q_1, \ldots, q_n$ in $\mathcal{Q}$ and a number $C > 0$ for which $p(x) \leqslant C[q_1(x) + \cdots + q_n(x)]$ for all x.

Indeed, if this condition is fulfilled, then the topology generated by $\mathcal{Q}$ is stronger than the one generated by $\mathcal{P}$. On the other hand, if every neighborhood of zero in the second topology contains a neighborhood of zero in the first topology, the set $\{x\colon\ p(x) < 1\}$ must contain a set of the form $\{x\colon\ q_i(x) < r, i = 1, \ldots, n\}$ for some $q_1, \ldots, q_n \in \mathcal{Q}$ and $r > 0$. Then $p \leqslant r^{-1}(q_1 + \cdots + q_n)$, since otherwise by homogeneity of both sides there is an element x for which $q_1(x) + \cdots + q_n(x) < r$, but $p(x) > 1$.

Therefore, two collections of seminorms $\mathcal{P}$ and $\mathcal{Q}$ on a given space define the same locally convex topology precisely when in addition to the indicated condition also the symmetric condition is fulfilled: for every seminorm q in $\mathcal{Q}$ there exist seminorms $p_1, \ldots, p_k \in \mathcal{P}$ and $M > 0$ for which $q(x) \leqslant M[p_1(x) + \cdots + p_k(x)]$.

In locally convex spaces, there is the following simple description of boundedness.

1.4.11. Proposition. *A set B in a locally convex space E is bounded if and only if every seminorm from some system of seminorms defining the topology of E is bounded on B. Hence every continuous seminorm on E is bounded on every bounded set.*

PROOF. If p is a continuous seminorm on E, then $V_p = \{x \in E\colon\ p(x) \leqslant 1\}$ is a neighborhood of zero in E and hence for sufficiently large (in absolute value) scalars t we have $t^{-1}B \subset V$ if B is bounded. This means that $p(x) \leqslant 1$ if $x \in t^{-1}B$, i.e., $p(z) \leqslant |t|$ if $z \in B$. Thus, the boundedness on B of every continuous seminorm is proved.

Let now $\mathcal{P}$ be some system of seminorms on E defining the topology. Let us set $\alpha_p = \sup_{x \in B} p(x) < \infty$ for every seminorm $p \in \mathcal{P}$; thus, if $|t| > \alpha_p$, then $B \subset tV_p$. Since the sets V_p form a prebase of neighborhoods of zero, every neighborhood of zero in E absorbs B, i.e., the set B is bounded. Another proof can be easily extracted from Proposition 1.4.6. □

1.4.12. Proposition. *The closure and the circled hull of every bounded subset in a topological vector space are bounded. The convex hull of every bounded subset in a locally convex space is also bounded.*

PROOF. Our reasoning is based on the fact that any topological vector space possesses a base of neighborhoods of zero consisting of circled closed sets, and every locally convex space has a base of neighborhoods of zero consisting of convex sets. Hence, in order to verify that a set is bounded, it suffices, in the case of an arbitrary topological vector space, to verify that it is absorbed by every circled closed neighborhood of zero, and in the case of a locally convex space to verify that it is absorbed by every convex neighborhood of zero.

Thus, let B be a bounded subset of a topological vector space E, let $\overline{B}$ be its closure, let ΓB be its circled hull, and let $\operatorname{conv} B$ be its convex hull. If V is a closed circled neighborhood of zero in E and for some number t we have the inclusion $B \subset tV$, then also the inclusions $\overline{B} \subset tV$ and $\Gamma B \subset tV$ hold (since, e.g., $\overline{B}$ is the intersection of all closed sets containing B and tV is one of such sets). If E is a locally convex space and V is a convex neighborhood of zero in E and again $B \subset tV$, then $\operatorname{conv} B \subset tV$, since $\operatorname{conv} B$ is the intersection of all convex set containing B and tV is one of them. □

The convex hull of a bounded set in a topological vector space that is not locally convex is not always bounded (an example: the ball in the space from Example 1.3.16 or in the spaces $L^{1/2}$ and $l^{1/2}$ described before Corollary 1.11.14).

Note that this proposition immediately yields that the circled closed hull of a bounded subset in a topological vector space is also bounded; the same is true for the absolutely convex closed hull of a bounded subset of a locally convex space. The situation is more complicated with the preservation of compactness under such operations, which is discussed in § 1.8.

1.5. Finite-dimensional and normable spaces

In this section we prove the criteria of the membership of a topological vector space in the classes of spaces listed in the title; in addition, we prove that for every natural n, up to an isomorphism, there exists precisely one Hausdorff topological vector space of dimension n over $\mathbb{K}$ if $\mathbb{K}$ is a complete normed field.

A family of subsets of an arbitrary set is called *centered* if it is not empty and the intersection of every finite collection of its elements is also nonempty.

1.5.1. Theorem. *Every Hausdorff topological vector space of finite dimension n over a complete nondiscrete normed field $\mathbb{K}$ is isomorphic to the space $\mathbb{K}^n$ (the product of n copies of the field $\mathbb{K}$ regarded as a one-dimensional topological vector space over $\mathbb{K}$).*

PROOF. We give three proofs. The first one will work in the general case; the other two — in the case of a locally compact field $\mathbb{K}$. Since every two vector spaces of the same dimension over the same field are isomorphic as vector spaces, it suffices to prove that an arbitrary Hausdorff topology τ in the space $\mathbb{K}^n$ making it a topological vector space coincides with the product topology, which we denote by τ_0. To this end, we show the continuity at the point $(0, \ldots, 0)$ (hence everywhere) of the identity mappings

$$f_1^n\colon (\mathbb{K}^n, \tau_0) \to (\mathbb{K}^n, \tau) \quad \text{and} \quad f_2^n\colon (\mathbb{K}^n, \tau) \to (\mathbb{K}^n, \tau_0).$$

We verify that the continuity of the mapping f_1^n follows from the definition of the product topology and the axioms of a topological vector space. Indeed, let W be a neighborhood of zero in τ and let W_1 be a neighborhood of zero in τ such that $W_1 + \cdots + W_1$ (n terms) is contained in W.

Denote by e_j the element $(0,\ldots,0,1,0,\ldots 0) \in \mathbb{K}^n$ (the unit of the field $\mathbb{K}$ stands at the jth place). Let α_j be positive numbers such that $k_j e_j \in W_1$ whenever $|k_j| < \alpha_j$ (such numbers exist since every neighborhood of zero in a topological vector space is an absorbent set). For $\varepsilon > 0$ let $V_\varepsilon = \{x \in \mathbb{K}^n \colon \|x\| < \varepsilon\}$, where $\|x\| = \max_{j=1,\ldots,n} |x_j|$ is the norm introduced in Example 1.3.5, so that V_ε is a neighborhood of zero in the space $(\mathbb{K}^n, \tau_0)$. Let also $\alpha = \min\{\alpha_1, \ldots, \alpha_n\}$. Then, whenever $x = (x_1, \ldots, x_n) \in V_\alpha$ (which means that $|x_j| < \alpha_j$ for each index $j \in \{1,2,\ldots,n\}$) we have $x \in W_1 + \cdots + W_1 \subset W$. Thus, the continuity of the mapping f_1^n at the point $(0,\ldots,0)$ is proved.

We now prove the continuity of the mapping f_2^n at the same point. Since the sets $\{V_\varepsilon \colon \varepsilon > 0\}$ form a base of neighborhoods of zero in τ_0, it suffices to show that for every $\varepsilon > 0$ there exists a neighborhood of zero W in τ such that $W \subset V_\varepsilon$. We argue by induction. Suppose first that $n = 1$. Let $\varepsilon > 0$, $k \in V_\varepsilon$ (i.e., $|k| < \varepsilon$), $k \neq 0$; such an element k exists, because the field $\mathbb{K}$ is not discrete. The space $(\mathbb{K}^1, \tau)$ is Hausdorff, which yields the existence of a circled neighborhood of zero W not containing the element k. We have $W \subset V_\varepsilon$. Indeed, otherwise there is an element $k_1 \in W \backslash V_\varepsilon$. Then $|k_1| \geqslant \varepsilon$, $|k/k_1| < 1$. Since W is circled, we must have $k = \frac{k}{k_1} k_1 \in W$, but this is not true. Thus, for $n = 1$ the theorem is completely proved (without using the completeness of the field $\mathbb{K}$).

We now prove that if the theorem is valid for some $n \in \mathbb{N}$, then it will be valid also for $n+1$. It suffices to show that (under this assumption) the mapping f_2^{n+1} is continuous. We observe that every space $(\mathbb{K}^n, \tau_0)$ is complete due to the completeness of the field $\mathbb{K}$; the completeness of the product of complete topological vector spaces is proved below in Proposition 1.7.10, but in the case under consideration we are concerned with the product of complete metric (normed) spaces, the completeness of this product is a standard fact in elementary functional analysis. Hence every n-dimensional topological vector subspace in the space $(\mathbb{K}^{n+1}, \tau)$ is complete, since by the inductive assumption it is isomorphic to the space $(\mathbb{K}^n, \tau_0)$. By Lemma 1.3.11 we obtain that in $(\mathbb{K}^{n+1}, \tau)$ every n-dimensional vector subspace is closed.

For $\varepsilon > 0$ and $j \in \{1,2,\ldots,n+1\}$, the set $\{(k_1,\ldots,k_{n+1}) \in \mathbb{K}^{n+1} \colon |k_j| < \varepsilon\}$ will be denoted by V_ε^j. Since the sets V_ε^j form a prebase of neighborhoods of zero in $(\mathbb{K}^{n+1}, \tau_0)$, for the proof of the continuity of the mapping f_2^{n+1} (at the point $(0,\ldots,0)$, hence everywhere) it suffices to show that for any $\varepsilon > 0$ and $j \in \{1,2,\ldots,n+1\}$ there exists a neighborhood of zero W in the topology τ such that $W \subset V_\varepsilon^j$. The condition that the field $\mathbb{K}$ is not discrete yields the existence of $k \in \mathbb{K}$, $k \neq 0$ such that $ke_j \in V_\varepsilon^j$ (i.e., $|k| < \varepsilon$). By the closedness in τ of the subspace $G_j = \{(k_1,\ldots,k_{n+1}) \in \mathbb{K}^{n+1} \colon |k_j| = 0\}$ there is a circled neighborhood of zero W_0 in the topology τ such that $(ke_j + W_0) \cap G_j = \varnothing$. Therefore,

$$ke_j \notin G_j + W_0. \tag{1.5.1}$$

We derive from this that W_0 is a neighborhood of zero in τ contained in V_ε^j. This is proved with the aid of a reasoning similar to the one used above in the proof of the inclusion $W \subset V_\varepsilon$. Let $a = (k_1, \ldots, k_{n+1}) \in W_0$, $a \notin V_\varepsilon^j$. Since W_0 is circled, we have $ka/k_j \in W_0$ (because $|k/k_j| \leqslant 1$ due to the fact that $|k| < \varepsilon$, $|k_j| \geqslant \varepsilon$); but this contradicts (1.5.1), since $ka/k_j \in ke_j + G_j$. The first proof of the theorem is complete.

Two other proofs differ from the given one by their nontrivial parts concerned with the verification of the continuity of the mapping f_2^n. However, these proofs apply only to the case where the topological field $\mathbb{K}$ is locally compact. The local compactness of the field $\mathbb{K}$ yields the local compactness of the space $(\mathbb{K}^n, \tau_0)$, a corollary of which is the compactness in the topology τ_0 of every closed ball $W_\varepsilon = \{x \in \mathbb{K}^n\colon\ \|x\| \leqslant \varepsilon\}$, which will be used below. The first of these different proofs is based on the concept of a filter, due to this it applies not only to topological vector spaces, but also to pseudotopological vector spaces (the definition of a pseudotopological space can be found in § 4.10(vi) or in the book Smolyanov [**489**]).

Thus, suppose that the mapping f_2^n is not continuous (by its linearity this is equivalent to the property that f_2^n is not continuous at zero) and obtain a contradiction. Since f_2^n is discontinuous at zero, in $\mathbb{K}^n$ there exists a filter Φ converging to zero in the topology τ, but not convergent in the topology τ_0. The latter, in turn, means that there exists $\varepsilon_0 > 0$ such that for every $\varphi \in \Phi$ we have

$$\varphi \cap (\mathbb{K}^n \setminus V_{\varepsilon_0}) \neq \varnothing, \tag{1.5.2}$$

where $V_\varepsilon := \{x \in \mathbb{K}^n\colon\ \|x\| < \varepsilon\}$, $\varepsilon > 0$.

We now consider two cases. Suppose first that there exists $c > \varepsilon_0$ such that for all $\varphi \in \Phi$ we have $(W_c \setminus V_{\varepsilon_0}) \cap \varphi \neq \varnothing$. Then $\{(W_c \setminus V_{\varepsilon_0}) \cap \varphi\colon\ \varphi \in \Phi\}$ is a filter basis in $\mathbb{K}^n$; if Φ_1 is some ultrafilter majorizing the filter Φ_c generated by this basis, then Φ_1 majorizes the filter Φ. In addition, since the set $W_c \setminus V_{\varepsilon_0}$ is compact in $(\mathbb{K}^n, \tau_0)$, the ultrafilter Φ_1 converges in the topology τ_0 to some element a of the set $W_c \setminus V_{\varepsilon_0}$. Since we earlier proved that the mapping f_1^n is continuous, the ultrafilter Φ_1 converges to a also in the topology τ. This contradicts the assumption that τ is Hausdorff, since Φ_1 must converge in the topology τ to zero, because Φ_1 majorizes the filter Φ converging to zero. The obtained contradiction means that for every $c > \varepsilon_0$ there exists $\varphi \in \Phi$ such that $(W_c \setminus V_{\varepsilon_0}) \cap \varphi = \varnothing$ (and (1.5.2) holds). In particular, for every $c > 0$ and every $\varphi \in \Phi$ we have $(\mathbb{K}^n \setminus W_c) \cap \varphi \neq \varnothing$ (verify this); in other words, all elements of the filter are unbounded in the norm.

Next, for $\eta > 0$ we set $W_\eta^1 = \{k \in \mathbb{K}\colon\ |k| \leqslant \eta\}$. Let us show that if $\eta > 0$, $\varphi \in \Phi$ and a real number $c_0 > \varepsilon_0 > 0$ is such that there exists an element $k_0 \in \mathbb{K}$ with $|k_0| = c_0$ (such c_0 exists, since the normed field $\mathbb{K}$ is not discrete), then

$$(W_\eta^1 \cdot \varphi) \cap (W_{c_0} \setminus V_{\varepsilon_0}) \neq \varnothing.$$

Indeed, by the unboundedness of the set φ there is a point $x_0 \in \varphi$ such that $\|x_0\| \geqslant c_0/\eta_0 > 0$, where $0 < \eta_0 \leqslant \eta$, and there exists an element $k_\eta \in \mathbb{K}$ for which $|k_\eta| = \eta_0$, moreover, there exists also an element $k' \in \mathbb{K}$ such that

$$|k'| = \|k_\eta \cdot x_0\| = \eta_0\|x_0\| \geqslant c_0 = \|k_0\| > 0.$$

Hence

$$|k_0 k_\eta / k'| = (|k_0|/|k'|) \cdot |k_\eta| = (|k_0|/|k'|)\eta_0 \leqslant 1 \cdot \eta = \eta,$$

whence we obtain that $k_0 k_\eta / k' \in W^1_\eta$, $(k_0 k_\eta / k')x_0 \in W^1_\eta \cdot \varphi$ and

$$\|(k_0 k_\eta / k')x_0\| = c_0 \cdot (1/|k'|) \cdot \|k_\eta x_0\| = c_0.$$

Hence the family of sets

$$\{(W^1_\eta \cdot \varphi) \cap (W_{\delta_0} \setminus V_{\varepsilon_0})\colon\ \eta \in \mathrm{I\!R},\ \eta > 0, \varphi \in \Phi\}$$

forms a filter basis in $\mathbb{K}^n$, which by the continuity of multiplication also converges to zero in the topology τ. The filter generated by this basis possesses all those properties of the filter Φ which we used in the previous part of the proof. As we have seen, they yield that the topology τ is not Hausdorff. The second proof of the continuity of the mapping f_2^n, hence also of the theorem, in the case of a locally compact field is complete.

The third proof of the continuity of the mapping f_2^n, which we now give, at the level of ideas is very close to the second one, but it does not use the concept of a filter. We give it in the case where $\mathbb{K} = \mathbb{C}$ or $\mathbb{K} = \mathrm{I\!R}$, although it can be easily adapted to the general case of a locally compact field (we leave this to the reader). Suppose again that the mapping f_2^n is discontinuous. This means that there exists $\varepsilon > 0$ such that the set W_ε contains no neighborhood of zero in the topology τ. In particular, for any closed circled neighborhood of zero V in this topology we have $V \not\subset W_\varepsilon$. It follows that V has a nonempty intersection with the set $S_\varepsilon = W_\varepsilon \setminus V_\varepsilon$. Indeed, S_ε is a sphere in $\mathbb{K}^n$ with respect to the norm $\|\cdot\|$. Since there exists an element $a \in V$ such that $a \notin W_\varepsilon$, we have $\|a\| = \alpha > \varepsilon$. Using that V is circled and $a \in V$, we obtain $\varepsilon a/\alpha \in V$, but $\|\varepsilon a/\alpha\| = \varepsilon$, i.e., $\varepsilon a/\alpha \in S_\varepsilon$ and $V \cap S_\varepsilon \neq \varnothing$. The set W_ε is compact in τ_0 by the local compactness of the field $\mathbb{K}$. Hence the set S_ε is compact in τ_0. Since the mapping f_1^n is continuous (which has been proved earlier), the set S_ε is compact in the topology τ. On the other hand, the intersection of two neighborhoods of zero is again a neighborhood of zero. Hence the sets of the form $V \cap S_\varepsilon$, where V is a closed circled neighborhood of zero in τ, form a centered family of closed subsets of the set S_ε, which is compact in the topology τ. Hence the intersection of all these sets is nonempty; let a be its element. Then a belongs to every neighborhood of zero in the topology τ, which contradicts that it is Hausdorff. The third proof of the theorem is also complete. □

1.5.2. Corollary. *Every finite-dimensional subspace in a Hausdorff topological vector space over a complete field is closed.*

PROOF. In this case $\mathbb{K}^n$ is complete. □

1.5.3. Corollary. *Let F be a closed vector subspace of finite codimension in a Hausdorff topological vector space E. Then every algebraic complement G of the subspace F in E is also a topological complement* (*see Example* 1.3.15).

PROOF. The quotient space E/F is Hausdorff and finite-dimensional and the natural projection $\pi\colon E \to E/F$ is continuous. Its restriction $\pi|_G$ to G is an algebraic, hence by the proven theorem also a topological isomorphism between G and E/F. The projection $p_G\colon E \to G$ has the form $p_G = (\pi|_G)^{-1} \circ \pi$, hence is continuous. □

1.5.4. Definition. *A subset A of a topological vector space E is called precompact or totally bounded if, for every neighborhood of zero V in E, one can find a finite set $\{a_1, \ldots, a_n\}$ in E such that $A \subset \bigcup_{k=1}^{n}(a_k + V)$.*

The set $\{a_1, \ldots, a_n\}$ is called a finite V-net (or an ε-net if V is a ball of radius ε in a metric space). It is easily seen that every compact set in a topological vector space is precompact. We observe (this fact is not needed now, so it will be proven in § 1.8 after we discuss completions of topological vector spaces) that a subset of a topological vector space is precompact precisely when its closure in the completion of this topological vector space is compact. However, the closure of a precompact set in an incomplete space may fail to be compact.

1.5.5. Lemma. *Every precompact subset of a topological vector space is bounded.*

PROOF. Let A be a precompact subset of a topological vector space E and let V be a neighborhood of zero in E; we have to prove that there exists $\nu > 0$ such that $A \subset tV$ if $|t| > \nu$. Let W be a circled neighborhood of zero such that $W + W \subset V$ and let $a_1, \ldots, a_n$ be elements of E for which $A \subset \bigcup_{k=1}^{n}(a_k + W)$. Let $\nu > 1$ be such that $\{a_1, \ldots, a_n\} \subset tW$ if $|t| > \nu$. Then for such numbers t we have

$$A \subset \bigcup_{k=1}^{n}(a_k + W) \subset tW + W \subset tW + tW = t(W + W) \subset tV,$$

which shows that A is bounded. □

1.5.6. Theorem. *A Hausdorff topological vector space E over the field $\mathbb{R}$ or $\mathbb{C}$ is finite-dimensional if and only if it possess a compact neighborhood of zero. Moreover, it is sufficient that it possess a precompact neighborhood of zero.*

PROOF. The necessity is clear from Theorem 1.5.1, since any Hausdorff topological vector space (over a nondiscrete complete normed field $\mathbb{K}$) of finite dimension n is isomorphic (as a topological vector space) to the space $\mathbb{K}^n$, and if S is a compact neighborhood of zero in $\mathbb{K}$, then the product of n copies of S is a compact neighborhood of zero in $\mathbb{K}^n$.

Let us prove the sufficiency (the given proof is due to Gleason). Let V be a precompact neighborhood of zero in E and let $a_1, \ldots, a_n \in E$ be elements such that

$$V \subset \bigcup_{k=1}^{n}\Bigl(a_k + \frac{1}{2}V\Bigr). \tag{1.5.3}$$

We show that the linear span of the set $A = \{a_1, \ldots, a_n\}$ coincides with the whole space, i.e., that every element in E is a linear combination of elements of the set A. Thus, let $b \in E$ and let $t \in \mathbb{R}\backslash 0$ be such that $tb \in V$. By (1.5.3) there exists an element $a_{k_1} \in A$ such that $tb - a_{k_1} \in V/2$. If for some natural number r we have already proved that there exist $a_{k_1}, \ldots, a_{k_r} \in A$ such that $tb - \sum_{j=1}^{r} 2^{1-j} a_{k_j} \in 2^{-r}V$ (for $r = 1$, as we have just noted, this is true), then by the relation

$$2^{-r}V \subset \bigcup_{k=1}^{n} (2^{-r}a_k + 2^{-r-1}V),$$

which follows from (1.5.3), there exists an element $a_{k_{r+1}} \in A$ such that

$$tb - \sum_{j=1}^{r+1} 2^{1-j} a_{k_j} \in 2^{-r-1}V.$$

Since the precompact set V is bounded, it follows that the sum of the series of $2^{1-j}a_{k_j}$ is tb. Indeed, let W be an arbitrary neighborhood of zero in E and let a number $r_0 > 0$ be such that $2^{-r}V \subset W$ if $r > r_0$. Then for any $r > r_0$ we have

$$tb - \sum_{j=1}^{r} 2^{1-j} a_{k_j} \in 2^{-r}V \subset W.$$

On the other hand, $\sum_{j=1}^{r} 2^{1-j} a_{k_j} = \sum_{i=1}^{n} \nu_i^r a_i$ for each natural number r, where $\nu_i^r = \sum_{k \in B_{ir}} 2^{1-k}$, $B_{ir} = \{j\colon\ k_j = i,\ j \leqslant r\}$. Hence there exist numbers $\nu_i \in \mathbb{K}$, $i = 1, \ldots, n$, such that $\nu_i^r \to \nu_i$ as $r \to \infty$ for every $i \in \{1, \ldots, n\}$ (every number ν_i is the sum of a series the elements of which form a subsequence of the sequence 2^{-n}). The axioms of a topological vector space yield that $\sum_{j=1}^{r} 2^{1-j} a_{k_j}$ converges to $\sum_{i=1}^{n} \nu_i a_i$ as $r \to \infty$. In any Hausdorff topological vector space a sequence can converge to a unique limit, hence we obtain that $tb = \sum_{i=1}^{n} \nu_i a_i$. $\square$

A.N. Kolmogorov [**279**] discovered the following important fact.

1.5.7. Theorem. *A real or complex topological vector space is normable if and only if it is Hausdorff and possesses a convex bounded neighborhood of zero.*

PROOF. If p is a norm on a topological vector space E generating its topology, then (by the definition of the topology generated by a norm) the collection of all sets $\{x \in E\colon\ p(x) < \varepsilon\}$, each of which is convex, forms a base of neighborhoods of zero in E; since each of them is obtained from every other one by multiplying by a scalar, these sets are bounded in E. The topology in E is Hausdorff: the norm vanishes only at zero, hence if $x \in E$, $p(x) \neq 0$, then $x \notin \{z\colon\ p(z) < p(x)\}$. Thus, the necessity of the indicated conditions is proven.

Let us prove the sufficiency. Let V be a convex bounded neighborhood of zero in E. By Proposition 1.2.2 there exists a circled neighborhood of zero $W_0 \subset V$. Its convex hull W is a convex circled neighborhood of zero; it is contained in V and hence is bounded. The Minkowski functional p_W of the set W is a norm on E generating the original topology. The fact that p_W is a norm follows from the properties of W: $p_W(x)$ is finite for every $x \in E$, since W is absorbing. The convexity of W gives the semiadditivity of p_W. The equality $p(tx) = |t|p(x)$ for all $t \in \mathbb{K}$ and $x \in E$ holds, since W is circled. The boundedness of W implies the positivity of p_W on nonzero elements. Finally, the most important property that this norm generates the original topology follows from the fact that the topology generated by the norm p_W possesses a base of neighborhoods of zero consisting of all possible sets of the form $\{x \in E\colon p_W(x) < \varepsilon\}$, $\varepsilon \in (0, \infty)$. Indeed, on the one hand, each of these sets is a neighborhood of zero in the original topology (this follows from the inclusion $\varepsilon W/2 \subset \{x \in E\colon p_W(x) < \varepsilon\}$ for every ε); on the other hand, an arbitrary neighborhood of zero V_0 in the original topology contains one of these sets, since by the boundedness of the set W we have the inclusion $\{x \in E\colon p_W(x) < \delta\} \subset \delta W \subset V_0$ for a sufficiently small number $\delta > 0$. □

The convexity of the neighborhood of zero in the formulation is important: the nonnormable space $l^{1/2}$ (see p. 79) has bounded balls.

1.6. Metrizability

Here we obtain a condition of metrizability of a topological vector space, but before that we establish a connection with metrics for every vector topology.

1.6.1. Theorem. *The topology of every topological vector space can defined by a family of quasi-norms.*

PROOF. First we shall show that if (E, τ) is a topological vector space and $\mathcal{B} = \{V_n\}$ is a sequence of circled neighborhoods of zero such that

$$V_{n+1} + V_{n+1} \subset V_n \quad \forall n \in \mathbb{N}, \tag{1.6.1}$$

then the vector space E can be equipped with a quasi-norm q such that

$$\Bigl\{x \in E\colon q(x) \leqslant \frac{1}{2^{n+1}}\Bigr\} \subset V_n \subset \Bigl\{x \in E\colon q(x) \leqslant \frac{1}{2^n}\Bigr\}. \tag{1.6.2}$$

Let $k(\mathbb{N})$ denote the set of finite subsets $\mathbb{N}$ and let $\mathbb{Q}_1$ be the set of binary-rational numbers in $[0, 1)$, i.e., the numbers $r \in \mathbb{Q}_1$ have the form

$$r = \sum_{k\in\Phi(r)} 2^{-k}, \quad \text{where } \Phi(r) \in k(\mathbb{N});$$

if $r = 0$, then $\Phi(r) = \varnothing$. We define a mapping W of the set $\mathbb{Q}_1$ to the set of subsets in the space E as follows:

$$W(r) = \sum_{n\in\Phi(r)} V_n.$$

In particular, for $r = 2^{-n}$ we have $W(r) = V_n$; $W(0) = \{0\}$. Thus, $W(r)$ is a circled neighborhood of zero in E if $0 \neq r \in \mathbb{Q}_1$.

For $x \in E$ set

$$q(x) := \inf\bigl\{r \in \mathbb{Q}_1\colon x \in W(r)\bigr\}$$

if $x \in \bigcup_r W(r)$, $q(x) = 1$ if $x \notin \bigcup_r W(r)$. We show that q is a quasi-norm on E. We first verify that q is a pseudonorm. It follows directly by the definition of q that $q(0) = 0$ and $q(x) \in [0, 1]$ for all $x \in E$. Since the sets $W(r)$ are circled, it follows that $q(x) = q(-x)$ for all $x \in E$. We now show that

$$q(x_1 + x_2) \leqslant q(x_1) + q(x_2) \quad \forall x_1, x_2 \in E. \tag{1.6.3}$$

If $q(x_1) + q(x_2) \geqslant 1$, then (1.6.3) holds, since $q(z) \leqslant 1$. Let $q(x_1) + q(x_2) < 1$. For the proof of inequality (1.6.3) in this case we show that

$$q(x_1 + x_2) \leqslant q(x_1) + q(x_2) + 2\varepsilon \quad \forall \varepsilon > 0. \tag{1.6.4}$$

Thus, let $\varepsilon > 0$ and $q(x_1)+q(x_2)+2\varepsilon < 1$. Then there exist r_1, $r_2 \in \mathbb{Q}_1$ such that $r_i < q(x_i) + \varepsilon$, $x_i \in W(r_i)$, $i = 1, 2$ (without loss of generality we can assume that $r_2 \leqslant r_1$, hence $r_2 < 1/2$). For the proof of (1.6.4) it suffices to verify the inclusion

$$x_1 + x_2 \in W(r_1 + r_2),$$

because then $q(x_1 + x_2) \leqslant r_1 + r_2 < q(x_1) + q(x_2) + 2\varepsilon$. For this, in turn, it suffices to prove the inclusion

$$W(r_1 + r_2) \supset W(r_1) + W(r_2). \tag{1.6.5}$$

Note that directly from the definition of W (and also from (1.6.1)) we obtain (1.6.5) in the case where $r_2 = 0$ and $r_1 = r_2$; in turn, (1.6.5) yields the monotonicity of the function $W(\cdot)$, i.e., $W(r_1) \subset W(r_1 + r_2)$ if $r_i \in \mathbb{Q}_1$.

We now proceed to the general case. Set

$$r = r_1 + r_2, \ r_i^1 = r_i, \ i = 1, 2$$

and construct by induction a finite sequence of some length $s_0 \in \mathbb{N}$ the elements of which are pairs (r_1^s, r_2^s) of numbers in $\mathbb{Q}_1$ such that for every $s \in \{1, \ldots, s_0\}$ we have

$$r_1^s + r_2^s = r, \tag{1.6.6}$$

$$W(r_1^s) + W(r_2^s) \supset W(r_1) + W(r_2), \tag{1.6.7}$$

where $r_2^{s_0} = 0$ and hence $r_1^{s_0} = r$ (the value of the last number s_0 in this sequence will be defined at the last step of our construction and the very fact that it exists will be proved after we describe the inductive process).

The construction of such a sequence will give the proof of inclusion (1.6.5), since for $s = s_0$ inclusion (1.6.7) coincides with it. Suppose that $j \in \mathbb{N}$, $j \geqslant 2$ and for all $s \in \mathbb{N}$, where $s < j$, we have already constructed pairs of numbers (r_1^s, r_2^s) from $\mathbb{Q}_1$ for which (1.6.6) and (1.6.7) hold. We define r_1^j and r_2^j by the relations

$$\Phi(r_1^j) = \bigl(\Phi(r_1^{j-1}) \cup \Phi(r_2^{j-1})\bigr) \setminus \bigl(\Phi(r_1^{j-1}) \cap \Phi(r_2^{j-1})\bigr),$$

$$\Phi(r_2^j) = \bigl(\Phi(r_1^{j-1}) \cap \Phi(r_2^{j-1})\bigr) - 1 = \bigl\{n \in \mathbb{N}\colon\ n + 1 \in \Phi(r_1^{j-1}) \cap \Phi(r_2^{j-1})\bigr\}.$$

Let us show that also for $s = j$ relations (1.6.6) are (1.6.7) are true. Equality (1.6.6) follows from the similar equality for $s = j-1$ and the definition of the sets $\Phi(r_i^j)$. Note that if $s \in \mathbb{Q}_1$, $s < 1/2$, then $\Phi(s) - 1 = \Phi(2s)$. In the case under consideration, this and the relation $r_1^{j-1} + r_2^{j-1} = r < 1$ yield that $r_2^j = 2s^{j-1}$, where the number s^{j-1} is defined by the equality $\Phi(s^{j-1}) = \Phi(r_1^{j-1}) \cap \Phi(r_2^{j-1})$

(so that $s^{j-1} < 1/2$). Inclusion (1.6.7) follows from the next chain of relations:

$$\begin{aligned}
W(r_1)+W(r_2) &\subset W(r_1^{j-1})+W(r_2^{j-1}) \\
&= \sum_{n\in\Phi(r_1^{j-1})} W(2^{-n}) + \sum_{n\in\Phi(r_2^{j-1})} W(2^{-n}) = \\
&= \sum_{n\in\Phi(r_1^{j-1})\backslash\Phi(r_2^{j-1})} W(2^{-n}) + \sum_{n\in\Phi(r_1^{j-1})\cap\Phi(r_2^{j-1})} W(2^{-n}) \\
&+ \sum_{n\in\Phi(r_1^{j-1})\cap\Phi(r_2^{j-1})} W(2^{-n}) + \sum_{n\in\Phi(r_2^{j-1})\backslash\Phi(r_1^{j-1})} W(2^{-n}) \\
&\subset \sum_{n\in\Phi(r_1^{j})} W(2^{-n}) + \sum_{n\in\Phi(r_2^{j})} W(2^{-n}) = W(r_1^j)+W(r_2^j),
\end{aligned}$$

where the last inclusion follows from (1.6.1). If the set $\Phi(r_2^j)$ is nonempty, i.e., $r_2^j \neq 0$, then $\Phi(r_2^{j-1})$ is also nonempty and the maximal value of the elements in the set $\Phi(r_2^j)$ is strictly smaller than the maximal value of the elements in the set $\Phi(r_2^{j-1})$. Hence there exists $j \in \mathbb{N}$ such that $\Phi(r_2^j) = \varnothing$ (hence $\Phi(r_2^{j+k}) = \varnothing$ for all integers $k \geqslant 0$). We define s_0 to be the smallest of such numbers j; this brings our construction of the required sequence to the end. Thus, the proof of the fact that q is a pseudonorm is complete.

Let us verify that for q we have inclusions (1.6.2). The second one follows from the definition of q. The first one follows from the definition of q and (1.6.1). Indeed, if $q(x) \leqslant 2^{-n-1}$, then there exists $r \in \mathbb{Q}_1$ such that $x \in W(r)$ and $r < 2^{-n}$. This means that all numbers in the set $\Phi(r)$ are larger than n, whence on account of (1.6.1) we obtain that

$$x \in W(r) = \sum_{k\in\Phi(r)} W(2^{-k}) \subset V_n.$$

We now prove that the pseudonorm q is a quasi-norm. Property (5) in the definition of a quasi-norm (Example 1.3.6) follows from the fact that all sets $W(r)$ ($r \neq 0$) are absorbing, and property (6) is seen from the fact that they are circled.

Next, we show that q has property (4) from Example 1.3.6. Let $t \in \mathbb{R}$ and $\{x_n\} \subset E$ be a sequence such that $q(x_n) \to 0$; we have to prove that $q(tx_n) \to 0$, and for this it suffices to verify that for every fixed $j \in \mathbb{N}$ the inequality $q(tx_n) < 2^{-j}$ holds for sufficiently large $n \in \mathbb{N}$. Since the sets $W(r)$ used in the definition of q are circled, we can assume that $t > 0$. Let n_t denote an (arbitrary) integer number greater than t, and let $j \in \mathbb{N}$, $j_0 = j + n_t$. Then by (1.6.1) we have

$$V_{j_0} + V_{j_0} + \cdots + V_{j_0} \subset V_j$$

if the number of terms in the left-hand side is 2^{n_t}. Let $i_0 \in \mathbb{N}$ be such that $q(x_i) < 2^{-j_0-1}$ whenever $i > i_0$. Then, for such i, by (1.6.2) we have $x_i \in V_{j_0}$, whence

$$tx_i = \frac{t}{n_t} n_t x_i \in \frac{t}{n_t} n_t V_{j_0} \subset \frac{t}{n_t} 2^{n_t} V_{j_0} \subset \frac{t}{n_t} V_j \subset V_j,$$

so that $q(tx_i) < 2^{-j}$. Thus, q is a quasi-norm; we observe that property (6) follows, as noted above, by properties (4) and (5), but for the constructed function q this property is very easy to verify directly, which has been done.

We continue the proof of the theorem. Let $\mathcal{U}$ be an arbitrary base of circled neighborhoods of zero in E. For any $V \in \mathcal{U}$, let $\mathcal{B}_V$ denote the sequence $\{V_n\}$ of elements $\mathcal{U}$ with property (1.6.1) for which $V_1 = V$, and let q_V be the quasi-norm q on E constructed for this sequence as we have just described. Then, by (1.6.2), the family of quasi-norms $\{q_V : V \in \mathcal{U}\}$ defines the original topology. We finally observe that the constructed quasi-norms possess the property that $q(\lambda x) \leqslant |\lambda| q(x)$ whenever $|\lambda| \leqslant 1$. □

1.6.2. Remark. The proven theorem asserts that on the space E there exists a family $\mathcal{P}$ of quasi-norms such that the balls centered at zero with respect to the quasi-norms in $\mathcal{P}$ form a prebase of neighborhoods of zero in E. Actually we have proved that such balls form even a base of neighborhoods of zero in E.

1.6.3. Theorem. *If a topological vector space E possesses a countable base of neighborhoods of zero, then there a exists a quasi-norm on E generating its topology.*

PROOF. Let $\mathcal{U} = \{V_n^1\}$ be a countable base of circled neighborhoods of zero in E. Let $V_1 = V_1^1$; if the set V_n is already defined, then let V_{n+1} be an element of $\mathcal{U}$ such that $V_{n+1} + V_{n+1} \subset V_n$ and $V_{n+1} \subset \bigcap_{i=1}^{n+1} V_i^1$. Let q be the quasi-norm defined by the sequence $\mathcal{B} = \{V_n\}$ in the same way as it has been done in the proof of the previous theorem. Then q defines the original topology in the space E. □

We now obtain a simple criterion of metrizability.

1.6.4. Corollary. *The topology of a topological vector space E can be defined by a metric precisely when E is Hausdorff and has a countable base of neighborhoods of zero. Moreover, the corresponding metric can be taken translation invariant. If the space E is locally convex, then its topology is defined by a countable family of seminorms* $\{p_n\}$ *and for a metric generating the topology one can take* $d(x, y) = \sum_{n=1}^{\infty} 2^{-n} \min\bigl(p_n(x-y), 1\bigr)$.

PROOF. The necessity of the stated conditions is obvious. Let us prove the sufficiency. If the space E has a countable base of neighborhoods of zero, then by the above theorem its topology is defined by a single quasi-norm q. This means that the topology of E is generated by the pseudometric $\varrho(x_1, x_2) = q(x_1 - x_2)$. This pseudometric is a metric, since E is Hausdorff. It is clear that ϱ is translation invariant. If E is locally convex, then its topology is defined by a family of seminorms (Theorem 1.4.9), among which we can choose a countable part due to the existence of a countable base of neighborhoods of zero. Taking such a part $\{p_n\}$, it is easy to verify that d is a metric that defines the same topology as $\{p_n\}$. □

1.6.5. Theorem. *Every topological vector space is completely regular and every Hausdorff topological vector space is Tychonoff.*

PROOF. Let F be be a closed set in a topological vector space E, let $x_0 \notin F$, and let V be an open neighborhood of the point x_0 disjoint with F. Let us take a quasi-norm q on E such that $\{x\colon q(x) < 1\} \subset V - x_0$ (its existence follows from Remark 1.6.2). The function $f(x) = \min\bigl(q(x - x_0), 1\bigr)$ is continuous on E and possesses the following properties: $0 \leqslant q \leqslant 1$, $q(x_0) = 0$, $q(x) = 1$ if $x \in F$. □

1.7. Completeness and completions

We recall that a sequence $\{a_n\}$ in a metric space (M, ϱ) is called *fundamental* or *a Cauchy sequence* if for every $\varepsilon > 0$ there exists $n_0 \in \mathbb{N}$ such that $\varrho(a_k, a_n) < \varepsilon$ for all $k, n > n_0$. It is obvious that every convergent sequence is Cauchy. If A is a subset of (M, ϱ) and every Cauchy sequence in A converges to an element of A, then A is called *complete.* If this is true for $A = M$, then M itself is called a complete space.

For arbitrary topological spaces, the concept of a Cauchy sequence, hence also the given definition of completeness is not meaningful. Moreover, it can happen that on the same set E one can introduce two metrics ϱ_1 and ϱ_2 generating the same topology such that under one of these metrics E is a complete metric space, but under the other one is not. For example, $E = \mathbb{R}^1$ with respect to the usual metric $\varrho_1(x, y) = |x - y|$ is complete, but with respect to the metric $\varrho_2(x, y) = \operatorname{arctg}|x - y|$ is not, although the topologies induced by both metrics coincide.

At the same time there exist "geometric" objects close to topological spaces, but possessing many features common with metric spaces: the so-called uniform spaces introduced by A. Weil in 1938. Their special cases are topological groups and topological vector spaces. More precisely, a topological group or a topological vector space can be equipped (but not in a unique way) with a uniformity generating the original topology (under the additional requirement of the translation invariance this uniformity is unique). We do not consider arbitrary uniform spaces (see Bourbaki [**88**], Kelley [**268**], and Engelking [**154**] and a brief discussion in § 1.12(i)); all necessary results about topological vector spaces that are actually special cases of theorems about uniform spaces will be proved directly (though, the proofs of such theorems for arbitrary uniform spaces differ very little from the proofs of their specializations for topological vector spaces). Throughout this section (E, τ) is a topological vector space over the field of real or complex numbers.

1.7.1. Definition. *A sequence $\{a_n\} \subset E$ is called a Cauchy sequence (or a fundamental sequence) if for every neighborhood of zero V in E there exists $n_0 \in \mathbb{N}$ such that $a_n - a_k \in V$ whenever $n, k \geqslant n_0$.*

A subset $A \subset E$ is called sequentially complete if every Cauchy sequence consisting of its elements converges to some element of A.

A net $\{a_t\}_{t\in T} \subset E$ is called a Cauchy net (or a fundamental net) if for every neighborhood of zero V in E there exists an index $t_0 \in T$ such that $a_t - a_s \in V$ whenever $t, s \geqslant t_0$.

A net $\{a_t\}_{t\in T}$ in a locally convex space is Cauchy if and only if for every seminorm p from a family defining the topology and every $\varepsilon > 0$ there is $t_0 \in T$ such that $p(a_t - a_s) < \varepsilon$ whenever $t, s \geqslant t_0$.

A filter Φ of subsets of E is called a *Cauchy filter* (or a fundamental filter) if for every neighborhood of zero V in E there exists a set $F \in \Phi$ such that $F - F \in V$ (a set F for which the latter is true is called small of order V).

In particular, a sequence $\{a_n\} \subset E$ is Cauchy if the corresponding elementary filter (whose basis consists of all sets $F_n = \{a_k\colon\ k \geqslant n\}$) is a Cauchy filter. Other examples of Cauchy filters are arbitrary convergent filters, in particular, the filter of all neighborhoods of an arbitrary point. Note also that a filter Φ of subsets of E is a Cauchy filter precisely when the filter $\Phi - \Phi$ (this is the filter in E generated by the basis $\{A - B\colon\ A, B \in \Phi\}$) converges to zero. Hence any filter majorizing a Cauchy filter is a Cauchy filter as well.

1.7.2. Definition. *A subset $A \subset E$ is called complete if every Cauchy filter in E containing A as an element converges to some element of A.*

If every Cauchy filter in E converges, then E is called a complete space.

A space E is called quasi-complete if every bounded subset of E is contained in a complete set.

A complete metrizable locally convex space is called a Fréchet space.

Below we prove that the completeness of A is equivalent to convergence of all Cauchy nets in A. Hence any complete topological vector space is sequentially complete (but not conversely). Furthermore, for a subset of a metrizable topological vector space E the sequential completeness and completeness are equivalent (this is seen from the existence of a countable base of zero); it is easy to verify that if ϱ is a translation invariant metric on E defining the topology (which exists by Corollary 1.6.4), then the completeness of E as a topological vector space and the completeness of the metric space (E, ϱ) are also equivalent properties. It is readily seen that any closed part of a complete set is complete. Hence the quasi-completeness of E is equivalent to the completeness of all closed bounded sets in E.

1.7.3. Proposition. *If Φ and Ψ are two filters in a topological vector space E, $\Phi \subset \Psi$, Φ is a Cauchy filter and Ψ converges to an element $x \in E$, then Φ converges to x as well.*

PROOF. Let V be a neighborhood of zero in E. We show that $F \subset V + x$ for some $F \in \Phi$ (then $V + x \in \Phi$). Let W be a neighborhood of zero in E such that $W + W \subset V$. Since Φ is a Cauchy filter, there exists $F_1 \in \Phi$ such that $F_1 - F_1 \subset W$, and since Ψ converges to x, there exists $F_2 \in \Psi$ such that $F_2 \subset x + W$, hence $F_2 - x \subset W$. Let $F_3 = F_1 \cap F_2$. Then we have $F_3 \in \Psi$ (and hence $F_3 \neq \varnothing$). We have $F_1 - x \subset F_1 - F_3 + F_3 - x \subset W + W \subset V$, so that we can take $F = F_1$. $\square$

Note that this proposition is a generalization of the following fact: every Cauchy sequence of elements of a topological vector space or a metric space containing a convergent subsequence converges itself to the same limit.

1.7.4. Corollary. *A set A in a topological vector space E is complete precisely when every Cauchy ultrafilter in A converges.*

PROOF. The necessity is obvious; on the other hand, if Φ is an arbitrary Cauchy filter in A, then by the previous proposition its convergence follows from convergence of a majorizing ultrafilter, which is automatically a Cauchy filter. □

We now describe the completeness of topological vector spaces in terms of nets (which may be intuitively more clear due to some analogy with sequences).

1.7.5. Corollary. *A set A in a topological vector space is complete precisely when in A every Cauchy net is converging.*

PROOF. Let A be complete and let $\{x_t\}_{t\in T}$ be a Cauchy net in A. Let us consider the filter Φ whose base are the set A and all possible sets $F_s := \{x_t\colon\ t \geqslant s\}$, where $s \in T$ is a fixed element. It is clear from the definition that this filter is fundamental. By the completeness of A it converges to some point $x \in A$. Let V be a neighborhood of x. Then, by the definition of convergence of a filter, we have $V \in \Phi$. This means that V contains some of the sets F_t, which proves convergence of $\{x_t\}$ to x.

Conversely, suppose that in A every Cauchy net is converging. We verify that every Cauchy ultrafilter Φ in A converges. We use it to construct a Cauchy net in A. Let us take Φ for the index set equipped with the natural partial order by the inverse inclusion, i.e., $\varphi \leqslant \psi$ if $\psi \subset \varphi$. Since $\varphi \cap \psi \in \Phi$ for all Φ, we obtained a directed set. In each set $\varphi \in \Phi$ we choose an element x_φ. Let us show that the net $\{x_\varphi\}$ is Cauchy. Let V be a neighborhood of zero in E. Since Φ is fundamental, there exists a set $\varphi \in \Phi$ with $\varphi - \varphi \subset V$. Then, whenever $t, s \geqslant \varphi$, we have $x_t \in t \subset \varphi$, $x_s \in s \subset \varphi$, whence $x_t - x_s \in V$. Hence $\{x_\varphi\}$ is fundamental and converges to some $x \in A$. Then also the ultrafilter Φ converges to x. Indeed, let W be a neighborhood of the point x. We have to show that $W \cap A \in \Phi$. If this is false, then $A\backslash W \in \Phi$, as explained in § 1.1. By convergence of $\{x_\varphi\}$ to x there is φ_1 such that $x_\varphi \in W$ for all $\varphi \geqslant \varphi_1$, i.e., whenever $\varphi \subset \varphi_1$. In particular, for all $\psi = \varphi_1 \cap (A\backslash W)$ we must have $x_\psi \in W$, which is impossible, since $x_\psi \in A\backslash W$. □

Suppose that E is a metrizable locally convex space and let its topology be defined by a countable family of seminorms $\{p_n\}$. It follows from what has been said above that E is a Fréchet space precisely when it is complete with respect to the metric $d(x, y) = \sum_{n=1}^\infty 2^{-n} \min\bigl(p_n(x-y), 1\bigr)$.

1.7.6. Proposition. *Let A be a complete subset of a locally convex space E. Then A is complete in every locally convex topology on E that is stronger than the original one and possesses a base of neighborhoods of zero consisting of sets closed in the original topology.*

PROOF. Let τ_0 be the original topology and let τ be a stronger topology with the indicated properties. If a net $\{a_t\} \subset A$ is Cauchy in the topology τ, then it is Cauchy and hence converges to some $a \in A$ also in the topology τ_0. We show that $a_t \to a$ in τ. Let V be a τ_0-closed neighborhood of zero in τ. By condition there exists an index t_1 such that $a_t - a_s \in V$ for all $t, s \geqslant t_1$. Since $a_s \to a$ in τ_0 and A is closed in τ_0, we have $a_t - a \in V$ for all $t \geqslant t_1$, i.e., $a_t \to a$ in τ. □

1.7.7. Corollary. *In any locally convex space every set complete in the weak topology is complete.*

The properties to be closed and complete are connected in the following way.

1.7.8. Proposition. *A subset A of a complete Hausdorff topological vector space E is closed if and only if it is complete.*

PROOF. Indeed, if A is closed and Φ is a Cauchy filter in E such that $A \in \Phi$, then Φ converges in E by the completeness of E and the limit belongs to A, because A is closed. On the other hand, if A is complete, a is a cluster point of the set A and $\mathcal{U}$ is the filter of neighborhoods of zero in E, then $(a + \mathcal{U}) \cap A$ is a basis of the Cauchy filter Φ in E containing the set A as an element. Since the set A is complete, this filter must converge to some element $b \in A$. At the same time it converges to a; since E is Hausdorff, we have $a = b$.

Another proof can be given in terms of nets. Namely, if we have a Cauchy net in A, then it converges in E due to completeness, and the limit belongs to A because A is closed. In the opposite direction, if A is complete and $x \in E$ is its cluster point, then there is a net in A converging to x. By completeness of A this net must converge to a point of A, so $x \in A$, since E is Hausdorff. □

1.7.9. Proposition. *A subset of a sequentially complete Hausdorff topological vector space E is sequentially closed (i.e., contains the limits of all convergent sequences of its points) if and only if it is sequentially complete.*

PROOF. The justification of this proposition is similar to the previous one; the only difference is that in place of filters (or nets in the second proof) one has to consider sequences. □

In relation to the following proposition on products of topological vector spaces we observe that the product of an arbitrary family of topological vector spaces equipped with the product topology (Tychonoff's topology) is a topological vector space (in more detail products of topological vector spaces are discussed in Chapter 2).

1.7.10. Proposition. *The product of a nonempty family of topological vector spaces is complete if and only if every space in this family is complete. In particular, every power $\mathbb{R}^T$ of the real line is complete.*

PROOF. In order to prove that the completeness of the spaces E_α yields the completeness of their product E it suffices to observe that if Φ is a Cauchy filter in E, then its projection to each of the spaces E_α is again a Cauchy filter and that a filter in E converges precisely when all filters obtained as its projections converge. On the other hand, if one of the spaces in the family $\{E_\alpha\}$, say, E_{α_0}, is not complete, then let Φ_{α_0} be a Cauchy filter in it that has no limit and, for every $\alpha \neq \alpha_0$, let Φ_α be the filter of all neighborhoods of zero in E_α. Then the product-filter is a Cauchy filter in E without limit. An alternative justification employs nets: if we have a Cauchy net in the product, then the projects form Cauchy nets in the factors, hence converge, which implies convergence of the original net in the product topology. □

1.7.11. Example. Every infinite-dimensional normed space B with the weak topology $\sigma(B, B')$ is not complete.

PROOF. We show that for every linear functional $F \in (B')^*$ there exists a net $x_\alpha \in B$ such that $l(x_\alpha) \to F(l)$ for all $l \in B'$. Then, taking a discontinuous functional F on B' (existing on every infinite-dimensional normed space), we obtain a net in B that is Cauchy in the topology $\sigma(B, B')$ and has no limit in B. For constructing $\{x_\alpha\}$ it suffices to verify that for every finite collection $l_1, \ldots, l_n \in B'$ there exists a vector $z \in B$ with $F(l_i) = l_i(z)$. The existence of such a vector z follows from the Hahn–Banach theorem proved in § 1.11, which we apply to the space $\bigl(B', \sigma(B', B)\bigr)$, taking into account that the space of continuous linear functionals on $\bigl(B', \sigma(B', B)\bigr)$ is exactly B (see Example 1.3.23). □

Let us give a rather exotic example of a complete locally convex space.

1.7.12. Example. Any locally convex space E is complete in its strongest locally convex topology (see Example 1.3.18).

PROOF. We recall that, given a Hamel basis $\{e_\alpha\}_{\alpha\in A}$ in E, the strongest locally convex topology can be defined by the family of seminorms of the form

$$p_\varphi(x) = \sum_\alpha \varphi(\alpha)|x_\alpha|, \quad \text{where} \quad x = \sum_\alpha x_\alpha e_\alpha,$$

with all possible positive functions φ on the set of indices α. Suppose that a net $\{v_t\}_{t\in T}$ is Cauchy in every such seminorm. We verify that it converges. Let $v_t = \sum_\alpha v_{t,\alpha} e_\alpha$. For every fixed α we obtain that the coordinates $\{v_{t,\alpha}\}$ are fundamental. Hence this net converges to some scalar c_α. We observe that only finitely many scalars c_α are nonzero. Indeed, if there is an infinite sequence $c_{\alpha_n} \neq 0$, then we take the function φ for which $\varphi(\alpha_n) = n|c_{\alpha_n}|^{-1}$ and $\varphi(\alpha) = 0$ for the remaining indices. Since $\{v_t\}$ is Cauchy in the seminorm p_φ, there exists an index t_0 such that $p_\varphi(v_t - v_{t_0}) \leqslant 1$ for all $t \geqslant t_0$. Hence for all $t \geqslant t_0$ we have $\sum_{n=1}^\infty n|c_{\alpha_n}|^{-1}|v_{t,\alpha_n} - v_{t_0,\alpha_n}| \leqslant 1$. There are finitely many nonzero numbers v_{t_0,α_n}, so for some $m > 1$ we have $v_{t_0,\alpha_n} = 0$ for all $n \geqslant m$. Thus, $\sum_{n=m}^\infty n|c_{\alpha_n}|^{-1}|v_{t,\alpha_n}| \leqslant 1$ for all $t \geqslant t_0$. In particular, $m|c_{\alpha_m}|^{-1}|v_{t,\alpha_m}| \leqslant 1$ for all $t \geqslant t_0$, whence in the limit we obtain the false estimate $m\, c_{\alpha_m}|^{-1}|c_{\alpha_m}| \leqslant 1$. Thus, we can define the vector $v = \sum_\alpha c_\alpha e_\alpha$. We have $p_\varphi(v_t - v) \to 0$ for every seminorm p_φ of the indicated form. Indeed, given $\varepsilon > 0$, taking t_0 such that $p_\varphi(v_t - v_s) < \varepsilon$ for all $t, s \geqslant t_0$, we obtain that $p(v_t - v) \leqslant \varepsilon$ for all $t \geqslant t_0$. Indeed, if $t \geqslant t_0$ is fixed, taking all nonzero coordinates $v_{t,\alpha}$ and c_α of the vectors v_t and v (the corresponding finite set of indices will be denoted by M), we find that $p(v_t - v) = \sum_{\alpha\in M} \varphi(\alpha)|v_{t,\alpha} - c_\alpha| \leqslant \varepsilon$, since in the inequality with finitely many terms $\sum_{\alpha\in M} \varphi(\alpha)|v_{t,\alpha} - v_{\alpha,s}| < \varepsilon$ for $s \geqslant t_0$ we can pass to the limit in s. □

1.7.13. Definition. *A mapping f of a subset A in a topological vector space E to a topological vector space G is called uniformly continuous on A if for every neighborhood of zero W in G there exists a neighborhood of zero V in E such that if $x_1 - x_2 \in V$, $x_1, x_2 \in A$, then $f(x_1) - f(x_2) \in W$.*

It follows directly from this definition that any uniformly continuous mapping takes Cauchy filters in the space E consisting of subsets of E intersecting A to Cauchy filters in the space G and so this mapping is continuous (takes convergent filters to convergent ones). Certainly, the continuity is even more clear without filters (we leave it to the reader).

Note that every linear mapping of a topological vector space to a topological vector space is uniformly continuous if it is continuous (in particular, if it takes Cauchy filters to Cauchy filters); for multilinear mappings this is not true: the function $f(x, y) = xy$ for $x, y \in \mathbb{R}$ takes Cauchy filters to Cauchy filters (hence is continuous), but is not uniformly continuous. Finally, the continuity of a mapping f of a part of a topological vector space to a topological vector space, in general, does not imply that it takes Cauchy filters to Cauchy filters (an elementary example: $f(x) = \tan x$), but for mappings defined on a complete subset of a topological vector space the continuity is equivalent to the property to take Cauchy filters to Cauchy filters.

1.7.14. Proposition. *Let E be a topological vector space, let H be its everywhere dense subset (not necessarily a vector space), and let f be a mapping from H to a complete Hausdorff topological vector space G. If f is uniformly continuous on H, then it can be uniquely extended to a continuous mapping $F\colon E \to G$. Moreover, F will be automatically uniformly continuous.*

PROOF. The uniqueness of an extension, if it exists, follows from the fact that two continuous mappings f and g that coincide on an everywhere dense set coincide everywhere. Indeed, if $f(a) \neq g(a)$, then, since G is Hausdorff, there exist disjoint open sets $U \ni f(a)$ and $V \ni g(a)$. Taking a neighborhood W of the point a such that $f(W) \subset U$ and $g(W) \subset V$, we obtain that W contains no points where f and g are equal.

We prove the existence of a mapping F. Let us fix a point x. There is a net of points $x_t \in H$ converging to it. By the uniform continuity of the mapping f the net of points $f(x_t)$ is Cauchy in G and hence converges to some element $y \in G$. Set $F(x) := y$. It is readily seen that y does not depend on our choice of a net converging to x. This also yields the uniform continuity of F. □

1.7.15. Remark. Similarly it is proved that if A is a subset of a topological vector space, H is a dense subset of A and f is a uniformly continuous mapping defined on H and taking values in a complete subset G of some topological vector space, then there exists its unique extension to a continuous mapping $F\colon A \to G$; this extension is uniformly continuous.

Note also that one can define the uniform continuity for mappings from metric spaces to metric spaces, from metric spaces to topological vector spaces, and from topological vector spaces to metric spaces. For example, a mapping $f\colon E \to G$

from a metric space (E, ϱ_1) to a metric space (G, ϱ_2) is called uniformly continuous if for every $\varepsilon > 0$ there exists $\delta > 0$ such that the inequality $\varrho_1(x_1, x_2) < \delta$ yields the inequality $\varrho_2\bigl(f(x_1), f(x_2)\bigr) < \varepsilon$. For all these cases, there are analogs of the proven proposition with very similar proofs; one has just to replace in appropriate places the inclusion $x_1 - x_2 \in V$, where V is a suitable neighborhood of zero, with inequalities of the type $\varrho(x_1, x_2) < \varepsilon$.

In fact, all these facts are partial cases of a general result for mappings of uniform spaces to uniform spaces (see Bourbaki [**88**], Kelley [**268**], Engelking [**154**]).

The rest of this section will be devoted to the proof of the theorem on the completion of a topological vector space, according to which every Hausdorff topological vector space can be embedded as an everywhere dense topological vector subspace in a complete Hausdorff topological vector space defined up to an isomorphism.

1.7.16. Definition. *A completion of a topological vector space E is a complete Hausdorff topological vector space $\overline{E}$ with the following property: E is a topological vector subspace of the space $\overline{E}$ and is everywhere dense in $\overline{E}$.*

For a metric space E, such a completion is not the same as a completion in the category of metric spaces, when a dense embedding $E \to \overline{E}$ must preserve the distance, since now also the linearity is required, but the standard construction of a completion of a metric space can be modified.

1.7.17. Theorem. *Every Hausdorff topological vector space E possesses a completion, moreover, if G_1 and G_2 are two its completions, then there exists a unique linear homeomorphism of G_1 onto G_2 leaving all elements of E fixed.*

PROOF. Suppose first that E is metrizable. Let p be a quasi-norm on E defining a translation invariant metric ϱ on E that generates the original topology. According to a standard theorem in mathematical analysis (see, for example, Dieudonné [**130**]), there exists a complete metric space G that is a completion of the metric space (E, ϱ). We show that one can introduce a structure of a vector space in G that agrees with the topology τ defined by its metric such that the original topological vector space E will be a topological vector subspace in the topological vector space (G, τ); this will mean that G is the completion of E.

The operations of the multiplication of elements of G by scalars and the addition of elements of G are defined by the extension by continuity of the operations in the topological vector space E. The existence and uniqueness of such extensions follow by Remark 1.7.15. On the other hand, the metric ϱ_G on the completion is the extension by continuity of the metric ϱ on the space E. Since the quasi-norm p also possesses the extension by continuity to all of G and both extensions are unique, all identities used below in this reasoning that are valid for the functions that we extend will remain valid for their extensions. In particular, the extension p_G of the pseudonorm p will be a pseudonorm on G connected with the metric ϱ_G on G by the equality $\varrho_G(x_1, x_2) = p_G(x_1 - x_2)$. Hence the topology generated by ϱ_G on G agrees with the structure of the additive group of the vector space.

For the proof that it agrees with the vector structure it suffices to verify the continuity of the operation of multiplication. Let $k, q \in \mathbb{K}$ and $x, z \in G$. Then

$(k+q)(x+z) - kx = q \cdot z + q \cdot x + k \cdot z$. Hence for the proof of the continuity of the operation of multiplication by scalars it suffices to verify that for fixed k and x the mappings $z \mapsto k \cdot z$, $G \to G$, $q \mapsto q \cdot x$, $\mathbb{K} \to G$, and also the mapping $(q, z) \mapsto q \cdot z$, $\mathbb{K} \times G \to G$ are continuous at zero of the corresponding space. The continuity (even uniform) of the first of them follows by its definition as the extension by continuity of the uniformly continuous mapping of the multiplication by a fixed scalar. Let us prove the continuity of two remaining mappings.

We start with the mapping $q \mapsto q \cdot x$, $\mathbb{K} \to G$, where x is a fixed element of G. Let a sequence $\{x_n\} \subset E$ converge to x, let V be a closed neighborhood of zero in G, and let W be a circled neighborhood of zero in E such that $W + W \subset V$. The convergent sequence $\{x_n\}$ is a Cauchy sequence, hence there exists a number $n_0 \in \mathbb{N}$ such that $x_s - x_r \in W$ for all $s, r \geqslant r_0$. Since E is a topological vector space, there exists $\varepsilon > 0$ such that if $t \in \mathbb{K}$, $|t| < \varepsilon$, then $tx_{r_0} \in W$. Hence, for all $t \in \mathbb{K}$ such that $|t| < \min(1, \varepsilon)$ if $r \geqslant r_0$, we have

$$tx_r = tx_{r_0} + t(x_r - x_{r_0}) \in W + tW \subset W + W \subset V \cap E.$$

For any fixed t the mapping $x \mapsto tx$ is continuous, as already noted, hence we have $tx = t \lim\limits_{r\to\infty} x_r = \lim\limits_{r\to\infty} (tx_r) \subset \overline{V \cap E} \subset V$ if $|t| < \min(1, \varepsilon)$. The continuity of the mapping $q \mapsto q \cdot x$ is proved.

For the proof of the continuity at zero of the mapping

$$(q, z) \mapsto q \cdot z, \quad \mathbb{K} \times G \to G$$

it suffices to observe that if V is a closed neighborhood of zero in G, then there exist $\varepsilon > 0$ and a neighborhood of zero W in E such that $q \cdot z \in V \cap E$ whenever $|q| < \varepsilon$, $z \in W$. Indeed, then, by the continuity of the mapping $z \mapsto q \cdot z$ (for any fixed q), for each q satisfying the inequality $|q| < \varepsilon$ we have the inclusion $q\overline{W} \subset \overline{V \cap E} \subset V$ and it remains to observe that $\overline{W}$ (the closure of the set W in G) is a neighborhood of zero in G. Thus, under the assumption of metrizability of E the existence of its completion is proved (one can give other justifications using standard explicit constructions of completions of metric spaces).

Let now E be an arbitrary Hausdorff topological vector space and let $\mathcal{P}$ be a set of quasi-norms defining its topology. For every quasi-norm $p \in \mathcal{P}$ let E_p denote the vector space E equipped with the topology defined by the quasi-norm p and let $\widetilde{E}_p$ be the topological vector quotient of the space E_p by its closed subspace $p^{-1}(0)$. Its topology is generated by the quasi-norm p_1 defined as follows: if $z \in \widetilde{E}_p$, then $p_1(z) = \inf\{p(x)\colon\ x \in z\}$. This fact, as well as that p_1 is indeed a quasi-norm, is verified directly. Using that the quotient $\widetilde{E}_p$ is Hausdorff, we see that p_1 vanishes only at the zero element. Hence the topology of the space $\widetilde{E}_p$ is metrizable. Let $\overline{E}_p$ denote its completion; for every element $x \in E$, let x_p denote its canonical image in $\overline{E}_p$.

Set $G = \prod_{p\in\mathcal{P}} \overline{E}_p$. The space G is complete as the product of complete topological vector spaces. The mapping $F\colon\ E \to G$ that to every vector $x \in E$ associates the element $(x_p) \in G$ is a linear homeomorphism of E onto its image in G equipped with the topology induced from G. This follows directly from the definition of the product topology and the topology defined by a family of pseudonorms. We identify E with its image $F(E)$ and assume that E itself is a subspace of the space G. This means that we replace the topological vector space E by the isomorphic topological vector space $E_1 = F(E)$; below we denote it by the same symbol E. Then the topological vector subspace $\overline{E}$ of the space G defined as the closure in G of the subspace E satisfies all conditions needed in order that $\overline{E}$ be a completion of the topological vector space E. Thus, the existence of a completion for an arbitrary Hausdorff topological vector space is proved.

Let us prove its uniqueness. Let E_1 and E_2 be two completions of the topological vector space E and let I be the identical mapping of E into itself. Since E is everywhere dense in E_1 as well as in E_2, the uniformly continuous mappings I and $J = I^{-1}$ extend to continuous mappings $\overline{I}\colon E_1 \to E_2$ and $\overline{J}\colon E_2 \to E_1$. We show that $\overline{I}$ (hence also $\overline{J}$) is an isomorphism of topological vector spaces. In order to see this, it suffices to observe that $\overline{J}\circ\overline{I}$ and $\overline{I}\circ\overline{J}$ are the identity mappings of the spaces E_2 and E_1 onto themselves, respectively, since they are continuous and are identities on everywhere dense sets. □

From the proof one can see the following fact.

1.7.18. Corollary. *Every Hausdorff locally convex space possesses a unique (up to an isomorphism) completion that is a locally convex space.*

In the category of locally convex spaces, one can construct the completion by specifying a certain subspace in $(E')^*$ (see § 3.8). In the category of metrizable topological vector spaces, the completion can be constructed by employing the standard method of completing metric spaces (Exercise 1.12.70).

1.8. Compact and precompact sets

In this section we give a summary of the main results on compact and precompact subsets of topological vector spaces. Some additional results will be presented in § 1.12.

We recall once again that a subset K of a topological vector space E is called precompact (or totally bounded) if it can be covered by the union of finitely many shifts of every neighborhood of zero of this space. As already noted, compact sets are precompact. It is obvious that the precompact sets possess the following properties:

every subset of a precompact set is precompact,

the homothetic images and shifts of precompact sets are precompact,

the union of finitely many precompact sets in the same space is precompact.

It was shown in Lemma 1.5.5 that any *precompact set is bounded.*

Let us establish a bit less obvious properties of precompact sets in topological vector spaces.

1.8.1. Proposition. (i) *The closure of a precompact set is precompact.*

(ii) *The image of a precompact set under a uniformly continuous mapping with values in a topological vector space is precompact.*

(iii) *The vector sum of finitely many precompact sets is precompact.*

(iv) *The circled hull of a precompact set is precompact.*

PROOF. (i) Let S be a precompact set, let $\overline{S}$ be its closure, and let U be a neighborhood of zero. By the continuity of addition there is a neighborhood of zero V such that $v_1+v_2 \in U$ for all $v_1, v_2 \in V$. Let us take points $s_1, \ldots, s_n \in S$ for which $S \subset \bigcup_{i=1}^n (s_i + V)$. We show that $\overline{S} \subset \bigcup_{i=1}^n (s_i + V)$. Let $x \in \overline{S}$. By definition there exists an element $s \in S$ with $x - s \in V$. Taking i such that $s \in s_i+V$, i.e., $s = s_i+v$, $v \in V$, we obtain $x = s+x-s = s_i+v+x-s \in s_i+U$.

Assertion (ii) is easily seen from the definitions, assertions (iii) and (iv) are deduced from it in the following way. If A and B are totally bounded in E, then $A+B$ is the image of the set $A\times B$ in $E\times E$ under the mapping $(u,v) \mapsto u+v$ from $E\times E$ to E. It is obvious that $A\times B$ is precompact and that the indicated mapping is uniformly continuous. Similarly, the circled hull of S is the image of $S\times[-1,1]$ under the mapping $(x,t) \mapsto tx$, which is uniformly continuous on every set of the form $M\times[-1,1]$, where M is bounded in E. □

Let us consider separately convex hulls.

1.8.2. Proposition. *The convex hull, the absolutely convex hull and the closed absolutely convex hull of a precompact set in a locally convex space are precompact.*

PROOF. Let K be a precompact subset of a locally convex space E and let $\operatorname{conv} K$ be its convex hull. We show that for every open absolutely convex neighborhood of zero V in E there exists a collection $\{a_1, \ldots, a_n\} \subset \operatorname{conv} K$ such that $\operatorname{conv} K \subset \bigcup_{i=1}^n (a_i+V)$. By the precompactness of the set K there exist points $c_1, \ldots, c_r \in K$ such that $K \subset \bigcup_{i=1}^r (c_i+V/2)$. Let K_1 be the convex hull of the points $c_1, \ldots, c_r$. The set K_1 is compact in the finite-dimensional subspace of the space E generated by the elements $c_1, \ldots, c_r$, hence also in the whole space E. Hence there exist elements $a_1, \ldots, a_n \in K$ for which $K_1 \subset \bigcup_{i=1}^n (a_i+V/2)$. Then $\operatorname{conv} K \subset \bigcup_{i=1}^n (a_i + V)$.

Indeed, every element in the set $\operatorname{conv} K$ has the form $x = \sum_{j=1}^s t_j b_j$, where $b_j \in K$, $\sum t_j = 1$, $t_j \geqslant 0$. For every $j = 1, 2, \ldots, s$, we choose an index $r(j)$ such that $b_j - c_{r(j)} \in V/2$. Then

$$v := \sum_{j=1}^s t_j c_{r(j)} \in K_1 \quad \text{and} \quad v - x = \sum_{j=1}^s t_j (c_{r(j)} - b_j) \in V/2$$

by the convexity of V. Since the inclusion $v \in K_1$ yields the existence of an element $a_i \in K_1$ such that $v - a_i \in V/2$, we have

$$x - a_i = d - a_i - (v - x) \in V/2 - V/2 = V/2 + V/2 = V,$$

where the last two equalities are valid by the absolute convexity of V. Thus, we have proved that $x \in \bigcup_{i=1}^n (a_i + V)$.

The assertions about the absolutely convex hull and the closed absolutely convex hull follow from what is already proved. □

1.8.3. Example. Let $\mathcal{L}^0$ be the space of all real measurable functions on $[0,1]$ with the topology of convergence in measure (Example 1.3.16). Let $\{\varepsilon_n\}$ be a sequence of real numbers such that $\varepsilon_n \in (0,1)$, $\varepsilon_n \to 0$, $\sum_{n=1}^\infty \varepsilon_n = \infty$.

For every n we define a function $\varphi_n \in \mathcal{L}^0$ as follows: if $S_n = \sum_{r=1}^n \varepsilon_r (\text{mod } 1)$ and $S_n < S_{n+1}$, then $\varphi_n = 10^n \gamma_n$, where γ_n is the indicator of the compact interval $[S_n, S_{n+1}]$. If $S_n > S_{n+1}$, then $\varphi_n = 10^n \chi_n$, where χ_n is the indicator of the union of the intervals $[0, s_{n+1}]$ and $[s_n, 1]$. The set

$$\{0\} \cup \{\varphi_n \colon n \in \mathbb{N}\} \cup \{-\varphi_n \colon n \in \mathbb{N}\}$$

is compact, but its convex closed hull coincides with the whole space $\mathcal{L}^0$, hence is not compact and even is not bounded. Note that the associated Hausdorff space L^0 is metrizable, but not locally convex.

1.8.4. Example. Let c_0 be the Banach space of sequences of real numbers tending to zero equipped with its standard norm $\|(x_n)\| = \max_n |x_n|$ and let E be its linear subspace consisting of all finite sequences (i.e., sequences with finitely many nonzero elements) with the same norm. Let us consider the vectors $e_n = (0, \ldots, 0, 1, 0, \ldots)$, where the unit stands at the nth place and all other coordinates are zero. The set K consisting of the sequence $\{e_n/n\}$, which converges to zero, and zero itself is compact. However, the convex hull $\operatorname{conv} K$ of this set is closed and coincides with the closed convex hull, but is not compact. Regarded as a subset of c_0, the set K, certainly, is also compact, but the set $\operatorname{conv} K$ is not closed in c_0.

1.8.5. Proposition. *A subset of a topological vector space is compact precisely when it is precompact and complete.*

PROOF. It is clear that the compactness yields the precompactness. We show that any compact subset A of a topological vector space is complete. Let Φ be a Cauchy filter in E with $A \in \Phi$ and let Ψ be an ultrafilter majorizing Φ. Then Ψ converges (since every ultrafilter containing a compact set converges to a point of this set, see Exercise 1.12.28). By Proposition 1.7.3 we have convergence of the filter Φ. One can also use nets: if we have an infinite Cauchy net $\{x_t\}$ in A, then it has a limit point in A by compactness; hence it converges to this limit point.

We now show that if a subset A of a topological vector space E is precompact and complete, then it is compact. Let Ψ be an ultrafilter in E with $A \in \Psi$. By the precompactness of A for every neighborhood of zero V in the space E there exist elements $a_1, \ldots, a_n$ in E such that $A \subset \bigcup_{i=1}^n (a_i + V)$. Since Ψ is an ultrafilter, there is i with $a_i + V \in \Psi$ (otherwise $E \backslash (a_i + V) \in \Psi$ for all i, whence $A \notin \Psi$). It follows that Ψ is a Cauchy filter. By the completeness of A it converges to an element of A. □

The proven proposition is an analog of the known result for subsets of metric spaces.

1.8.6. Corollary. *A subset of a topological vector space E is precompact precisely when it is relatively compact in the completion of the space E.*

In terms of filters the following condition is available.

1.8.7. Proposition. *A subset A of a topological vector space E is precompact precisely when every ultrafilter in E containing A is a Cauchy filter.*

PROOF. We have seen above that every ultrafilter containing a precompact set is a Cauchy filter. Let us prove the converse. Suppose that A is not precompact. Then there exists a neighborhood of zero V in E such that no finite family of sets of the form $a+V$, where $a \in E$, covers A. Hence the finite intersections of the sets $A \setminus (a+V)$, where $a \in E$, form a filter basis in E. If Ψ is some ultrafilter majorizing the filter generated by this basis, then by assumption it must be a Cauchy filter. Hence there exists an nonempty set $F \in \Psi$ such that $F - F \subset V$. Hence $F - a \subset V$ for some $a \in F$, i.e., $F \subset a + V$, so that $a + V \in \Psi$. This contradicts the fact that the filter Ψ contains the complements of all sets of the form $a+V$, where $a \in E$. □

1.8.8. Corollary. *A subset A in a topological vector space E is precompact precisely when every filter in E containing A is majorized by a Cauchy filter.*

This is an analog of the known result in the theory of metric spaces according to which a subset of a metric space is precompact (totally bounded) precisely when every sequence of its elements contains a Cauchy subsequence. These assertions can be formulated in terms of nets.

1.8.9. Proposition. *Let E be a topological vector space and $A \subset E$. The set A is totally bounded precisely when every net of its elements contains a Cauchy subnet.*

The set A is compact precisely when every net of its elements contains a subnet converging in A.

Taking into account what we have proved in § 1.7, we obtain the following fact.

1.8.10. Corollary. *Any compact set in a topological vector space is complete. Hence weakly compact sets are complete.*

For compact sets in topological vector spaces there are analogs of assertions proved above for precompact sets.

1.8.11. Proposition. (i) *Any homothetic image, any shift and the circled hull of a compact subset in an arbitrary topological vector space are compact.*

(ii) *The union and the vector sum of finitely many compact subsets of a topological vector space are compact.*

(iii) *The convex hull of the union of finitely many convex compact subsets of a topological vector space is compact.*

(iv) *The vector sum of a closed set and a compact set in a topological vector space is closed.*

PROOF. Assertions (i)–(iii) are proved similarly to the case of precompact sets. We prove only (iv). Let A be compact, let B be closed and let a net $\{w_\alpha\} \subset A + B$ converge to some element x. Since $w_\alpha = a_\alpha + b_\alpha$, where $a_\alpha \in A$ and $b_\alpha \in B$, by compactness of A there exists a subnet $\{a_\beta\}$ converging to an element $a \in A$. Then $\{b_\beta\}$ converges to $x - a$ and $x - a \in B$, since B is closed. Thus, we have $x \in A + B$. □

From Propositions 1.8.2 and 1.8.5 and Corollary 1.8.10 we obtain the following assertion.

1.8.12. Corollary. *In any locally convex space, the closed absolutely convex hull of a precompact set is compact precisely when it is weakly compact.*

1.8.13. Proposition. *The closed convex hull of a precompact subset of a complete or quasi-complete locally convex space is compact.*

PROOF. The closed convex hull of a set K is the closure of its convex hull. The latter is precompact for precompact K. The closure is complete if the whole space is complete or quasi-complete (in the latter case it is also important that any totally bounded set is bounded). However, the sequential completeness is not sufficient (Exercise 5.12.101). □

1.8.14. Proposition. *Every totally bounded set in a metrizable locally convex space lies in the closed convex hull of some sequence converging to zero.*

PROOF. The metrizable topology of the given locally convex space E is given by a sequence of seminorms p_n (see § 1.6). We can always assume that $p_n \leqslant p_{n+1}$ passing to the sums $p_1 + \cdots + p_n$. Let K be totally bounded in E. For every seminorm p_n there is a finite 4^{-n-1}-net $K_n \subset K$, i.e., for every point $k \in K$ there exists $v_k \in K_n$ with $p_n(k - v_k) \leqslant 4^{-n-1}$. The set $\bigcup_{n=1}^{\infty} K_n$ is dense in K. Let $S_1 := 2K_1$. For each $n > 1$ we choose a finite set S_n in E as follows: for every element $v \in K_n$ we find an element $u \in K_{n-1}$ with $p_{n-1}(v - u) \leqslant 4^{-n}$ and take the element $x := 2^n(v - u) \in E$. The cardinality of the set S_n of points x obtained in this way does not exceed the cardinality of K_n. The sequence $\{x_n\}$ obtained by the successive enumeration of the points in $S_1, S_2, \ldots$ converges to zero in every seminorm p_m, since $p_{n-1}\bigl(2^n(v - u)\bigr) \leqslant 2^{-n}$, i.e., $p_{n-1}(x) \leqslant 2^{-n}$ for all $x \in S_n$. It is not difficult to verify that every element $v \in K_n$ has the form $v = 2^{-1}x_{i_1} + \cdots + 2^{-n}x_{i_n}$ and hence belongs to the absolutely convex hull of the sequence $\{x_i\} = \bigcup_{n=1}^{\infty} S_n$. Hence $\bigcup_{n=1}^{\infty} K_n$ belongs to the closed convex hull of $\{x_n\} \cup \{-x_n\}$. Moreover, K lies in the closure of $\bigcup_{n=1}^{\infty} K_n$. See also Example 3.12.35. □

1.8.15. Proposition. *If K is a compact set in a topological vector space E and $\mathcal{V}$ is an open cover of this set, then there exists a neighborhood of zero W in E such that for every $x \in K$ the set $x + W$ is contained in one of the sets of the family $\mathcal{V}$.*

PROOF. For every $x \in K$, let V_x be a neighborhood of zero in E such that $x + V_x \subset V$ for some $V \in \mathcal{V}$. For every $x \in K$ there exists an open neighborhood of zero W_x in E such that $W_x + W_x \subset V_x$. The family of sets $\{x + W_x\colon\, x \in K\}$ forms an open cover of the set K. By the compactness of K this cover contains a finite subcover $\{x_i + W_{x_i}\colon\, i = 1, 2, \ldots, n\}$. Set $W = \bigcap_{i=1}^{n} W_{x_i}$. For every $x \in K$ there exists x_i such that $x \in x_i + W_{x_i}$. Moreover,

$$x + W \subset x + W_{x_i} \subset x_i + W_{x_i} + W_{x_i} \subset x_i + V_{x_i} \subset V$$

for some $V \in \mathcal{V}$. □

1.8.16. Theorem. *If K and F are disjoint sets in a topological vector space E such that K is compact and V is closed, then exists a continuous function $f\colon E \to \mathbb{R}$ with the following properties: if $x \in F$, then $f(x) = 0$; if $x \in K$, then $f(x) = 1$; $0 \leqslant f(x) \leqslant 1$ for all $x \in E$.*

PROOF. Since by Theorem 1.6.5 every topological vector space is completely regular, for every $x \in K$ there exists a continuous function $f_x\colon E \to \mathbb{R}$ equal to 0 on F and equal to 1 at the point x such that $0 \leqslant f_x(z) \leqslant 1$ for all $z \in E$. For every point $x \in E$ the set $V_x = \{z \in E\colon f_x(z) > 1/2\}$ is open and $K \subset \bigcup_{x\in K} V_x$. By the compactness of K, there exists a finite family $\{x_1, \ldots, x_n\} \subset K$ such that $\bigcup_{i=1}^n V_{x_i} \subset K$. The function f on E defined by the equality

$$f(z) = 2\min(2^{-1}, \sum_{i=1}^{n} f_{x_i}(z)),$$

is the required one. □

1.8.17. Corollary. *Under the assumptions of the previous theorem, there exist disjoint open sets V_F and V_K such that $F \subset V_F$, $K \subset V_K$.*

Actually, since in the proof of this theorem we never used the circumstance that E is a topological vector space, the established property holds for every completely regular topological space.

1.8.18. Proposition. *Every continuous mapping from a compact subset of a topological vector space to a topological vector space is uniformly continuous.*

PROOF. Let K be a compact set in a topological vector space E, let f be a continuous mapping from K to a topological vector space G, let W be a neighborhood of zero in G, and let V be a balanced neighborhood of zero in G such that $V + V \subset W$. By the continuity of f, for every point $x \in K$ there exists an open set $V_x \subset E$ such that $f(V_x) \subset f(x) + V$. The sets V_x, where $x \in K$, form a cover of the set K. By Proposition 1.8.15 there exists a neighborhood of zero H in E such that for every $z \in K$ set $z + H$ is contained in some of the sets of this cover. If $x_1, x_2 \in K$, $x_1 - x_2 \in H$, then $x_1 \in x_2 + H \subset V_x$ for some $x \in K$. Hence $f(x_1) \in f(V_x)$, $f(x_2) \in f(V_x)$ (the last inclusion follows from the inclusion $x_2 \in x_2 + H \subset V_x$), whence we obtain $f(x_1) - f(x_2) \in f(V_x) - f(V_x) \subset V - V \subset W$. □

Let us introduce some other useful concepts related to compactness. Let T be a topological space (not necessarily vector) and $K \subset T$.

1.8.19. Definition. *The set K is called countably compact if every infinite subset of K possesses a limit point in K.*

The set K is called relatively countably compact if every infinite subset of K possesses a limit point in T (it can happen that no limit point belongs to K).

The set K is called sequentially compact if every infinite sequence of its elements contains a subsequence converging to an element of K.

The set K is called relatively sequentially compact if every infinite sequence of its elements contains a subsequence converging in T.

It is known that for subsets of metric spaces the compactness, countable compactness and sequential compactness are equivalent and the same is true for the corresponding relative notions. In the general case (as one can easily verify) the countable compactness follows from the sequential compactness and from the compactness, but there are no other possible implications between these properties (see § 3.4).

1.9. Linear operators

Here we present some basic facts related to linear operators on topological vector spaces. We start with a criterion of continuity of linear and multilinear mappings between topological vector spaces (over an arbitrary topological field $\mathbb{K}$).

Let $k \in \mathbb{N}$ and let $E_1,\ldots,E_k$ and G be vector spaces over the same field; a mapping

$$b\colon E_1\times\cdots\times E_k \to G$$

is called *multilinear* (*k-linear*) if for every $j \in \{1,\ldots,k\}$ for all fixed $x_i \in E_i$, where $i \in \{1,\ldots,k\}\setminus\{j\}$, the mapping

$$E_j \to G, \quad x \mapsto b(x_1,\ldots,x_{j-1},x,x_{j+1},\ldots,x_k)$$

is linear. Note that $k = 1$ is not excluded, so that a linear mapping is a partial case of a multilinear mapping.

1.9.1. Proposition. *A multilinear mapping from the product of topological vector spaces $E_1\times\cdots\times E_k$ to a topological vector space is continuous precisely when it is continuous at the point* $(0,\ldots,0)$.

PROOF. For notational simplicity we give a proof for a bilinear mapping. Let $(x_1,x_2) \in E_1\times E_2$; we show that b is continuous at the point (x_1,x_2) once it is continuous at the point $(0,0)$. Let W_0 be a neighborhood of zero in G and let W be a neighborhood of zero in G such that $W+W+W \subset W_0$. There are neighborhoods of zero V_1 and V_2 in E_1 and E_2 such that if $h_1 \in V_1$, $h_2 \in V_2$, then $b(h_1,h_2) \in W$. We now use the equality

$$b(z_1,z_2) - b(x_1,x_2) = b(z_1-x_1,x_2) + b(x_1,z_2-x_2) + b(z_1-x_1,z_2-x_2),$$

which is valid by the bilinearity of b. Let $k_1,k_2 \in \mathbb{K}\setminus\{0\}$ be such that $k_1x_1 \in V_1$, $k_2x_2 \in V_2$, $V' = V_1\cap(k_2V_1)$, and $V'' = V_2\cap(k_1V_2)$. If we are given elements $z_1 \in x_1+V'$, $z_2 \in x_2+V''$, then

$$\begin{aligned} &b(z_1,z_2) - b(x_1,x_2)\\ &= b\bigl(k_2^{-1}(z_1-x_1),k_2x_2\bigr) + b\bigl(k_1x_1,k_1^{-1}(z_2-x_2)\bigr) + b(z_1-x_1,z_2-x_2),\end{aligned}$$

which belongs to $W+W+W\subset W_0$, since $k_2^{-1}(z_1-x_1)\in V_1$, $k_1^{-1}(z_2-x_2)\in V_2$. This means that b is continuous at (x_1,x_2). For $k=1$ the justification is even simpler: $b(x+h) = b(x)+b(h)$. □

A mapping f is called *sequentially continuous* if $f(x) = \lim\limits_{n\to\infty} f(x_n)$ whenever $x = \lim\limits_{n\to\infty} x_n$. This property is weaker than continuity (see Exercise 1.12.51).

1.9.2. Proposition. *Every sequentially continuous multilinear mapping from a finite product of topological vector spaces to a topological vector space takes bounded sets to bounded sets. In particular, this is true for sequentially continuous linear mappings (hence for all continuous linear mappings).*

PROOF. Our assertion is clear from Proposition 1.4.6: if a set B is bounded in $E = E_1 \times \cdots \times E_k$ and a multilinear mapping $f\colon E \to G$ is sequentially continuous, then for every sequence of elements $f(b_n)$ in $f(B)$ and any sequence of numbers $t_n \to 0$ we have $t_n f(b_n) \to 0$ by the sequential continuity, since $t_n f(b_n) = f(|t_n|^{1/k} b_n)\mathrm{sign}\, t_n$ and $|t_n|^{1/k} b_n \to 0$ (in the case of $\mathbb{R}$; the case of $\mathbb{C}$ is similar). □

1.9.3. Theorem. *Let E and G be locally convex spaces whose topologies are defined by collections of seminorms $\{p_\alpha\}$ and $\{q_\beta\}$, respectively. A linear mapping $T\colon E \to G$ is continuous precisely when for every β there exist a finite family $p_{\beta,\alpha_1}, \ldots, p_{\beta,\alpha_n}$ in the collection $\{p_\alpha\}$ and a number C_β such that*

$$q_\beta(Tx) \leqslant C_\beta[p_{\beta,\alpha_1}(x) + \cdots + p_{\beta,\alpha_n}(x)] \quad \forall\, x \in X.$$

In particular, the continuity of a linear functional l on E is equivalent to the estimate $|l| \leqslant C(p_{\alpha_1} + \cdots + p_{\alpha_n})$ for some $C \geqslant 0$ and $\alpha_1, \ldots, \alpha_n$.

PROOF. The indicated estimate yields the continuity of T at zero, and then at every other point. If T is continuous, then for each β the set $\{x\colon q_\beta(Tx) < 1\}$ contains a neighborhood of zero of the form $\{x\colon p_{\alpha_1}(x)<\varepsilon, \ldots, p_{\alpha_n}(x)<\varepsilon\}$. Hence the condition $p_{\alpha_1}(x) + \cdots + p_{\alpha_n}(x) < \varepsilon$ yields the inequality $q_\beta(Tx) < 1$. Therefore, for C_β we can take ε^{-1}. □

This theorem hints some ways of introducing locally convex topologies on spaces of linear operators from a locally convex space E to a locally convex space G. Let $\mathcal{L}(E,G)$ be the set of all *continuous linear operators* from E to G, let $\mathcal{B}(E,G)$ be the space of *bounded linear mappings* from E to G (which is broader in the general case), i.e., taking bounded sets to bounded sets, and let $\overline{\mathcal{L}}(E,G)$ be the subspace in $\mathcal{B}(E,G)$ consisting of sequentially continuous linear mappings (the boundedness of a sequentially continuous linear mapping follows from Proposition 1.9.2). The inclusion $\mathcal{L}(E,G) \subset \mathcal{B}(E,G)$ is obvious from the definitions. In general, $\overline{\mathcal{L}}(E,G) \neq \mathcal{L}(E,G)$ even for $G = \mathbb{R}$ (see Exercise 1.12.51). In the case of general topological vector spaces, the same notions and notation are used.

Let $\mathcal{B}$ be some class of bounded sets in the space E and let $\mathcal{Q}$ be some class of continuous seminorms on G. Then on $\mathcal{L}(E,G)$ we have the following seminorms:

$$p_{B,q}(T) = \sup\{q(Tx)\colon x \in B\}, \quad B \in \mathcal{B},\, q \in \mathcal{Q}.$$

If $\mathcal{Q}$ is the class of all continuous seminorms on Y, then we obtain the *topology of uniform convergence on the class $\mathcal{B}$*. For the class $\mathcal{B}$ of all finite sets this gives the *topology of pointwise convergence* in $\mathcal{L}(E,G)$. Two other frequently applied topologies are the topologies of uniform convergence on totally bounded sets and on compact sets. If X and Y are normed spaces, $\mathcal{B}$ consists of the unit ball in X and the only element in $\mathcal{Q}$ is the norm of Y, then we arrive at the operator norm on $\mathcal{L}(X,Y)$. In Chapter 3 we shall consider an important partial case: a topology on E'.

1.9.4. Definition. *Let E and G be topological vector spaces. A set $\mathcal{F}$ in the space $\mathcal{L}(E,G)$ is called equicontinuous if, for every neighborhood of zero V in G, there is a neighborhood of zero U in E such that $f(u) \in V$ for all $f \in \mathcal{F}$, $u \in U$.*

In particular, a set $\mathcal{F} \subset E'$ is equicontinuous if, for every $\varepsilon > 0$, there exists a neighborhood of zero $U \subset E$ such that $|f(u)| < \varepsilon$ for all $u \in U$, $f \in \mathcal{F}$.

A set $\mathcal{F} \subset \mathcal{L}(E,G)$ is called pointwise bounded if, for each $x \in E$, the set $\{f(x)\colon f \in \mathcal{F}\}$ is bounded in G.

1.9.5. Proposition. *Let E and G be locally convex. The closure of any equicontinuous set $\mathcal{F} \subset \mathcal{L}(E,G)$ in the topology of pointwise convergence is equicontinuous, and if G is quasi-complete, then this closure is complete.*

In addition, the restriction to $\mathcal{F}$ of the topology of pointwise convergence coincides with the restriction of the topology of pointwise convergence on a given set with a dense linear span, and also with the restriction of the topology of uniform convergence on totally bounded sets.

If E is separable, then any equicontinuous set in E' with the topology of pointwise convergence is metrizable.

PROOF. The first assertion is obvious from the definitions. Let $\mathcal{F}$ be closed and let G be quasi-complete. If a net $\{F_\alpha\}$ in $\mathcal{F}$ is Cauchy, i.e., for every $x \in E$ the net $\{F_\alpha x\}$ is Cauchy, then $\{F_\alpha\}$ converges by the quasi-completeness of G, since it is bounded as a subset of the set $\{Fx\colon F \in \mathcal{F}\}$, which is bounded by the equicontinuity of $\mathcal{F}$. The continuity of the limiting linear mapping F also follows from the equicontinuity of $\mathcal{F}$. Finally, $F \in \mathcal{F}$, since $\mathcal{F}$ is closed.

We now show that the restriction to $\mathcal{F}$ of the topology of uniform convergence on totally bounded sets is majorized by the restriction of the topology of pointwise convergence on an arbitrary set A with a dense linear span in E. Let S be totally bounded in E and let a net $\{F_\alpha\}$ in $\mathcal{F}$ converge to $F \in \mathcal{F}$ on every $a \in A$. We verify that convergence is uniform on S. We can assume that $F = 0$. Let us fix a convex neighborhood of zero V in G and find a neighborhood of zero U in E for which $F_\alpha(U) \subset V/2$ for all α. We pick points $b_1, \ldots, b_n$ in the linear span of A such that $S \subset \bigcup_{i=1}^n (U + b_i)$, which is possible under our assumptions.

Let us take α_0 such that $F_\alpha b_i \in V/2$ whenever $\alpha \geqslant \alpha_0$, $i = 1, \ldots, n$. Then $F_\alpha(S) \in V$ for any $\alpha \geqslant \alpha_0$, since for any $s \in S$ we have $s \in b_i + U$ for some index $i \leqslant n$, whence we obtain $F_\alpha s \in F_\alpha b_i + F_\alpha(U) \subset V$.

If E contains a countable everywhere dense set $\{a_n\}$, then on E' we can define a metric $d(f,g) = \sum_{n=1}^\infty 2^{-n} \min\bigl(|f(a_n) - g(a_n)|, 1\bigr)$. This metric generates the topology of pointwise convergence on the elements of $\{a_n\}$, hence on equicontinuous sets it generates the topology of pointwise convergence. A somewhat more general fact will be established in Proposition 1.12.16. □

Here is a criterion of continuity of linear functionals.

1.9.6. Proposition. *A linear function f on a topological vector space E is continuous precisely when its kernel is closed.*

PROOF. If the function f is continuous, then the set $f^{-1}(0)$ is closed. Conversely, suppose that the set $G = \operatorname{Ker} f$ (which is a vector subspace in E) is closed. If $G = E$, then $f(x) = 0$ for all $x \in E$. If $G \neq E$ and $a \in E\backslash G$, then there exists a circled neighborhood of zero V such that $(a+V)\cap G = \varnothing$, because G is closed. We show that the functional f on V is bounded; this will mean that f is continuous. Set $f(a) = \alpha$. If the set of values of the functional f on V is not bounded, then there exists an element $z \in V$ such that $|f(z)| > |\alpha|$. Hence $|\alpha/f(z)| < 1$, whence $z_1 = -\alpha z/f(z) \in V$, since V is circled. Thus, $a + z_1 \in a + V$. This inclusion contradicts the equality $f(a + z_1) = f(a) - \alpha f(z)/f(z) = 0$, which means that $a + z_1 \in G$ (recall that $(a + V) \cap G = \varnothing$). □

1.9.7. Remark. We have just given a direct proof. It can be, however, somewhat simplified by using the quotient. Namely, if $G = \operatorname{Ker} f$ is a closed subspace in E, then f is the composition of two mappings: the continuous canonical mapping from E to the Hausdorff (since G is closed) quotient E/G and some linear functional g on E/G that is continuous, since E/G is one-dimensional.

The next proposition strengthens the previous one.

1.9.8. Proposition. *Let E be a topological vector space, let A be its convex circled subset, and let f be a linear functional on E. In order that the restriction of f to A be continuous in the topology induced from the space E it is necessary and sufficient that the set $A \cap \operatorname{Ker} f$ be closed in A in the induced topology.*

PROOF. The necessity is obvious. Let us prove the sufficiency. If $f = 0$ on A, then there is nothing to prove. Suppose that there exists an element $x \in A$ with $f(x) \neq 0$. We first prove that for every $\varepsilon > 0$ there exists a neighborhood of zero V in E such that $|f(x)| < \varepsilon$ for all $x \in A \cap V$. This will mean the continuity of $f|_A$ (the restriction of f to A) at zero; next we show that the continuity of the functional $f|_A$ at zero yields its continuity everywhere. Since A is circled, there exists $a \in A$ such that $0 < f(a) < \varepsilon$. Then $a \notin (2A) \cap \operatorname{Ker} f$, since $(2A) \cap \operatorname{Ker} f = 2(A \cap \operatorname{Ker} f)$ and the latter set is closed in $2A$. Hence there exists a neighborhood of zero V in E such that the set $a + V$ has no intersection with $(2A) \cap \operatorname{Ker} f$. It follows that whenever $x \in V \cap A$ we have $|f(z)| < |f(a)|$. Indeed, otherwise due to the fact that the set $V \cap A$ is circled there exists an element $z \in V \cap A$ such that $f(z) = -f(a)$. Moreover, $f(z+a) = 0$, so that $z+a \in \operatorname{Ker} f$; at the same time, since the set A is convex, we have $z + a \in (2A) \cap (a + V)$, which contradicts the relation $(a+V)\cap\bigl((2A)\cap \operatorname{Ker} f\bigr) = \varnothing$. Thus, the continuity of the functional $f|_A$ at zero is proved.

We now show that the continuity of $f|_A$ at zero yields the continuity on A. Let V be a circled neighborhood of zero in E such that $|f(x)| < \varepsilon$ for all $x \in V \cap A$ and let W be a circled neighborhood of zero in E for which $W + W \subset V$. We claim that if $a \in A$, $x \in (a + W) \cap A$, then $|f(x) - f(a)| < 2\varepsilon$ (which means the continuity of $f|_A$ at the point a). Indeed, if $x \in (a + W) \cap A$, then

$$x - a \in \bigl((a + W) - (a + W)\bigr) \cap (A - A) \subset V \cap (2A) = 2(2^{-1}V \cap A).$$

Hence the desired estimate holds. □

In the proof of the fact that the continuity of $f|_A$ at zero yields its continuity everywhere on A it is possible (by using neighborhoods of zero) to avoid the condition that f takes scalar values; a suitably modified proof applies to a linear mapping f to any topological vector space. Therefore, for the continuity of the restriction of such a mapping to a circled convex subset it suffices to have the continuity of this restriction at zero. Proposition 1.9.6 remains true for mappings to $\mathbb{R}^n$ (Exercise 1.12.50).

Let us note an interesting property of hypersubspaces. We recall that a set is *nowhere dense* if its closure contains no nonempty open set.

1.9.9. Proposition. *A hypersubspace in a topological vector space is either everywhere dense or closed and then nowhere dense.*

PROOF. Let G be a hypersubspace in a topological vector space E such that we have $G \neq \overline{G}$. Since $\overline{G}$ is a vector subspace in E (Exercise 1.12.34) and $G \subset \overline{G}$, we have $\overline{G} = E$ (a vector subspace in a vector space F containing some hypersubspace but not equal to it must coincide with F).

If G is closed, then it is nowhere dense (as any closed proper vector subspace). Indeed, if V is a nonempty open set in E such that $V \subset G$ and $a \in V$, then $V_1 = V - a$ is an open neighborhood of zero in E contained in G. Hence we have $E = \bigcup_{n=1}^{\infty} nV_1 \subset G$, i.e., $E = G$. □

1.9.10. Example. On every infinite-dimensional metrizable topological vector space there is a discontinuous linear functional: taking linearly independent vectors $v_n \to 0$ and setting $l(v_n) = n$, we can define l by linearity (complementing $\{v_n\}$ to a Hamel basis).

1.10. The Hahn–Banach theorem: geometric form

In different areas of mathematics an important role is played by theorems on extension of mappings. In functional analysis, one often uses theorems on extensions of linear functionals originally defined on vector subspaces of vector (or topological vector) spaces and satisfying certain conditions such as continuity or some inequalities. These theorems can be reformulated as theorems on extensions of vector subspaces not intersecting a given convex subset to hypersubspaces. Various versions of such theorems are usually called Hahn–Banach theorems, and theorems of the first kind are called Hahn–Banach theorems in analytic form, while theorems of the second kind are called Hahn–Banach theorems in geometric form. Close to Hahn–Banach theorems in geometric form is the Kakutani theorem on extensions of disjoint convex subsets of vector spaces (a geometric form of the Hahn–Banach theorem can be obtained as its corollary). All these theorems are discussed below. Many of them even do not use any topology.

In most of the texts on functional analysis, usually a version of the analytic form of the Hahn–Banach theorem is proved and then as a corollary one obtains the Hahn–Banach theorem in the geometric form; sometimes the opposite order is preferred. We have chosen to start with the Kakutani theorem as the most "geometric" one; then as a corollary we obtain the geometric version of the Hahn–Banach theorem from which we deduce the Hahn–Banach theorem in the analytic form. We also give independent proofs of the two last theorems, as well as a derivation of the geometric version of the Hahn–Banach theorem from the analytic version (which, in particular, establishes their equivalence). A similar method of proof of the Hahn–Banach theorem is employed in the book Hille, Phillips [**222**].

If A, B, C, D are some sets in a common space, then we shall say that the pair (A, B) is contained in the pair (C, D) and write $(A, B) \subset (C, D)$ if $A \subset C$ and $B \subset D$.

1.10.1. Theorem. (THE KAKUTANI THEOREM) *Let E be a real vector space and let V and W be two its disjoint convex subsets. Then, there exist disjoint convex subsets V_1 and W_1 in the space E such that $V \subset V_1$, $W \subset W_1$ and $V_1 \cup W_1 = E$.*

PROOF. Our reasoning consists of two parts: one is set-theoretic and the other one is geometric. The set-theoretic part is the proof of the fact that the set $\mathcal{P}$ of all pairs of disjoint convex subsets of the space E containing the pair (V, W) and partially ordered by the relation $\subset$ introduced before the theorem possesses a maximal element; the geometric part is the proof that such an element is a pair of sets whose existence is asserted in the theorem.

For the proof of existence of maximal elements it suffices to verify that $\mathcal{P}$ satisfies the hypotheses of Zorn's lemma: if $\mathcal{P}_1 = \{(V_\alpha, W_\alpha)\}_{\alpha\in\Lambda}$ is a linearly ordered part in $\mathcal{P}$, then the pair (V, W), where $V = \bigcup_\alpha V_\alpha$, $W = \bigcup_\alpha W_\alpha$, is an upper bound for $\mathcal{P}_1$. Indeed, it follows directly from the definition of this pair that we have $(V_\alpha, W_\alpha) \subset (V, W)$ for every $\alpha \in \Lambda$. Thus, it remains to verify that the sets V and W are convex and disjoint. If $a, b \in V$, then there exist indices $\alpha, \beta \in \Lambda$ such that $a \in V_\alpha$ and $b \in V_\beta$; since the set $\mathcal{P}_1$ is linearly ordered, we see that either $V_\alpha \subset V_\beta$ or $V_\beta \subset V_\alpha$. In the first case $a, b \in V_\beta$, hence $[a, b] \subset V_\beta \subset V$; in the second case $a, b \in V_\alpha$ and $[a, b] \subset V_\alpha \subset V$, so that the set V is convex. In the same manner we prove the convexity of the set W.

We now show that $V \cap W = \varnothing$. If $a \in V \cap W$, then $a \in V_\alpha \cap W_\beta$ for some indices $\alpha, \beta\in\Lambda$. Since the set $\mathcal{P}_1$ is linearly ordered, it follows that either $(V_\alpha, W_\alpha) \subset (V_\beta, W_\beta)$ or $(V_\beta, W_\beta) \subset (V_\alpha, W_\alpha)$. Then either

$$a \in V_\alpha \cap W_\beta \subset V_\beta \cap W_\beta \quad \text{or} \quad a \in V_\alpha \cap W_\alpha.$$

Each of these inclusions contradicts the fact that $V_\gamma \cap W_\gamma = \varnothing$ for every $\gamma \in \Lambda$ by the definition of the set $\mathcal{P}$. Thus, for $\mathcal{P}$ the hypotheses of Zorn's lemma are fulfilled. Hence the set of maximal elements in the set $\mathcal{P}$ is nonempty; let (V_∞, W_∞) be one of them. By the definition of $\mathcal{P}$ the sets V_∞ and W_∞ are convex and disjoint. Hence in order to verify that the pair (V_∞, W_∞) is a desired one, it suffices to verify the equality $V_\infty \cup W_\infty = E$.

We show that if this equality does not hold, then the pair (V_∞, W_∞) is not a maximal element in $\mathcal{P}$. Thus, let $x \in E$ and $x \notin V_\infty \cup W_\infty$. Let V_∞^c and W_∞^c denote the convex hulls of the sets $\{x\} \cup V_\infty$ and $\{x\} \cup W_\infty$.

Below we show that at least one of the equalities $V_\infty^c \cap W_\infty = \varnothing$, $V_\infty \cap W_\infty^c = \varnothing$ is valid. In the first case the pair (V_∞^c, W_∞) will belong to the set $\mathcal{P}$ and majorize the pair (V_∞, W_∞), not being equal to it; in the second case the same will be true for the pair (V_∞, W_∞^c). Thus, in any case it turns out that the pair (V_∞, W_∞) is not maximal contrary to its choice. We now proceed to the proof of the fact that $V_\infty^c \cap W_\infty = \varnothing$ or $V_\infty \cap W_\infty^c = \varnothing$. Suppose that this is false, i.e., there exist elements $c \in V_\infty^c \cap W_\infty$ and $d \in V_\infty \cap W_\infty^c$. Since by the last equality the set V_∞ is nonempty, we have the equality $V_\infty^c = \bigcup_{z\in V_\infty}[x,z]$, hence there exists an element $a \in V_\infty$ such that $c \in [x,a]$; similarly, $d \in [x,b]$ for some $b \in W_\infty$.

Without loss of generality we can assume that $x = 0$ (in order to achieve this it suffices to replace all considered sets and elements by their images under the shift $z \mapsto z - x$; the image of x will be zero).

Thus, $c \in [0,a]$, $d \in [0,b]$, so that $c = \lambda a$, $d = \nu b$, $\lambda, \nu \in [0,1]$ (we observe, although this is not used below, that actually $\lambda, \nu \in (0,1)$). Let us show that the intervals $[c,b]$ and $[d,a]$ intersect; since $[c,b] \subset W_\infty$ and $[d,a] \subset V_\infty$ (by the convexity of the sets W_∞ and V_∞), this means that also the sets W_∞ and V_∞ intersect contrary to the inclusion $(V_\infty, W_\infty) \in \mathcal{P}$.

Our next goal is to prove that there exists $z \in [c,b] \cap [d,a]$. The inclusion $z \in [c,b]$ is equivalent to the equality

$$z = tc + (1-t)b = t\lambda a + (1-t)b, \quad t \in [0,1],$$

and the inclusion $z \in [d,a]$ is equivalent to the equality

$$z = \tau d + (1-\tau)a = \tau\nu b + (1-\tau)a, \quad \tau \in [0,1].$$

Thus, for the proof of existence of an element z in $[c,b] \cap [d,a]$ it suffices to prove that there are numbers $t, \tau \in [0,1]$ such that

$$t\lambda a + (1-t)b = \tau\nu b + (1-\tau)a.$$

In turn, it suffices to have simultaneously the equalities

$$t\lambda = 1-\tau, \quad \tau\nu = 1-t.$$

This system of equations coincides with the following one:

$$\tau + t\lambda = 1, \quad \tau\nu + t = 1.$$

This latter system is solvable for any $\nu, \lambda \in [0,1]$, moreover, one can find a solution (τ, t) such that $\tau \in [0,1]$, $t \in [0,1]$. Indeed, if $\nu, \lambda \in [0,1]$, then the determinant $1 - \nu\lambda$ of this system vanishes only in the case $\nu = \lambda = 1$, i.e., where both equations coincide (actually, as it follows from what has been said above this case is impossible), and then $\tau = t = 1/2$ is a solution of the system satisfying the condition $\tau \in [0,1]$, $t \in [0,1]$. If $1 - \lambda\nu \neq 0$, then a solution of the system is given by $\tau = \frac{1-\lambda}{1-\lambda\nu}$, $t = \frac{1-\nu}{1-\lambda\nu}$. Since $\lambda, \nu \in [0,1]$, $0 \leqslant 1 - \lambda \leqslant 1 - \lambda\nu$, $0 \leqslant 1 - \nu \leqslant 1 - \lambda\nu$, we have $\tau, t \in [0,1]$. Thus, the intervals $[c,b]$ and $[d,a]$ intersect. The proof of the theorem is complete. □

1.10.2. Theorem. *Let V and W be nonempty convex subsets in a real topological vector space E such that $V \cap W = \varnothing$, $V \cup W = E$ and $G = \overline{V} \cap \overline{W}$. Then either $G = E$ or G is a hyperplane in E.*

PROOF. We show first that G is a linear manifold. Suppose that $a, b \in G$ and prove that for every $t \in \mathbb{R}$ one has $c(t) = ta + (1-t)b \in G$. If this is not true, then for some $t_0 \in \mathbb{R}$ we obtain $c(t_0) = t_0 a + (1-t_0)b \notin G$; we can assume that

$c(t_0) \notin \overline{V}$ (similarly one can consider the case $c(t_0) \notin \overline{W}$). Thus, $c(t_0)$ is an element of the open set $E\setminus\overline{V}$, which is contained (by the equality $V \cup W = E$) in the interior $\breve{W}$ of the set W, i.e., we have $c(t_0) \in \breve{W}$. We observe that $t_0 \notin [0,1]$, since otherwise we would have $c(t_0) \in G$ (by the convexity of G). We consider two cases: $t_0 > 1$ and $t_0 < 0$. In the first case

$$a = t_0^{-1}c(t_0) - t_0^{-1}(1-t_0)b, \quad t_0^{-1} + t_0^{-1}(-(1-t_0)) = 1,$$

so that $a \in [c(t_0), b)$; in the second case

$$b = (1-t_0)^{-1}c(t_0) - (1-t_0)^{-1}t_0 a, \quad (1-t_0)^{-1} + (1-t_0)^{-1}(-t_0) = 1,$$

so that $b \in \bigl[c(t_0), a\bigr)$. Since $c(t_0) \in \breve{W}$ and $b \in \overline{W}$, by Proposition 1.4.1 in the first case we have $a \in \breve{W}$. Hence by the equality $V \cap W = \varnothing$, which is part of the hypotheses of the theorem, we have $a \notin \overline{V}$, which contradicts the inclusion $a \in G = \overline{V} \cap \overline{W}$. Similarly, in the second case the same proposition and the inclusion $c(t_0) \in \breve{W}$ yield that $b \in \breve{W}$, which contradicts the inclusion $b \in G$. Thus, we have proved that G is a linear manifold.

We now show that if $G \neq E$, then G is a hyperplane in E. Let $a \in E \setminus G$ and let $a \notin \overline{W}$, so that $a \in \breve{V}$, since $V \cup W = E$. The latter equality yields that $G = \overline{V} \cap \overline{W} \neq \varnothing$, because any topological vector space is connected and the sets V and W are nonempty. Let $b \in G$. Without loss of generality we can assume that $b = 0$, so that G is a vector subspace. In order to prove that G is a hyperplane, it suffices to verify that for every $z \in E$ there exist elements $c_0 \in G$ and $t \in \mathbb{R}$ such that $z = c_0 + ta$. Since $a \in \breve{V}$, we have $-a \notin \overline{V}$ (i.e., $-a \in \breve{W}$), because the inclusions $-a \in \overline{V}$ and $a \in \breve{V}$ yield (by the same Proposition 1.4.1) that $a/2 + (-a)/2 = 0 \in \breve{V}$, but in fact we have $0 \in G \subset \overline{W} \subset E \setminus \breve{V}$. We can assume that $z \in \overline{V}$ (the case $z \in \overline{W}$ is considered similarly). Then the set $[-a, z] \cap G$ is nonempty. Indeed, $[-a, z] \cap G = ([-a, z] \cap \overline{V}) \cap ([-a, z] \cap \overline{W})$, where the set in the brackets in the right-hand side are closed in $[-a, z]$ and nonempty, their union coincides with $[-a, z]$, and the interval $[-a, z]$ is a connected set. Let $d \in [-a, z] \cap G$. Then $d = \tau z + (1-\tau)(-a)$ for some $\tau \in [0,1]$. Since $-a \notin \overline{V}$, we have $\tau \neq 0$. Hence $z = d/\tau + (1-\tau)a/\tau$, as required. □

Note that the theorem contains no topological restrictions on V and W.

1.10.3. Remark. At the end of the proof of this theorem we have shown that if in its hypotheses $v_1 \in \overline{W}$, $v_2 \in \overline{V}$, then $[v_1, v_2] \cap G \neq \varnothing$.

1.10.4. Remark. If E is a real vector space, then a set $A \subset E$ is called *finitely open* (see Hille, Phillips [**222**, p. 14]) if its intersection with each finite-dimensional linear subspace is open (in the unique Hausdorff topology of this subspace making it a topological vector space, i.e., in the usual Euclidean topology).

The set of all finitely open subsets of a vector space E forms a topology τ_0 in E called *finitely open*; it is the strongest topology among all topologies that induce the Euclidean topology in every finite-dimensional subspace.

Denote by τ_1 the strongest topology among all locally convex topologies in E and by τ_2 the strongest topology among all topologies in E that agree with the vector structure. One can show that if E possesses an at most countable algebraic basis, then all these three topologies coincide; otherwise they are all distinct (see Exercise 1.12.32 and Example 2.3.3). But also in the latter case the collection of open convex sets is the same in all these three topologies.

1.10.5. Definition. *A point $a \in E$ is called c-inner (or algebraically inner) for a set $A \subset E$ if the following condition is fulfilled:*

$$\forall x \in E\ \exists \varepsilon > 0\colon\ (a - \varepsilon x, a + \varepsilon x) \subset A.$$

The algebraic kernel of a set is the set of all its algebraically inner points.

A set is called *c-open* (or *algebraically open*) if all its point are c-inner (i.e., it coincides with its algebraic kernel). In other words, a set is c-open if its intersection with each one-dimensional linear variety in E in open in this variety equipped with the standard Euclidean topology.

Even in the usual plane $\mathbb{R}^2$ there are c-open sets that are not open. For example, the complement to the set $\{(x_1, x_2) \in \mathbb{R}^2\colon\ x_2 = x_1^2, x_1 > 0\}$ has this property. However, every c-open convex subset of an arbitrary vector space is open in each of the just defined topologies τ_0, τ_1, and τ_2. Moreover, every c-inner point of a convex subset of a vector space is at the same time an inner point in the locally convex topology τ_1. If a convex set has topologically inner points, then its topological interior coincides with the algebraic interior (Exercise 1.12.82).

One can show (we leave as an exercise) that Γ is a hyperplane in E precisely when there exist a (nonzero) linear functional f on E and a scalar α for which $\Gamma = \{x \in E\colon\ f(x) = \alpha\}$. If the hyperplane Γ is not a vector subspace, i.e., if $\alpha \neq 0$, then the latter relation uniquely defines the functional f.

If Γ is a hyperplane in a vector space E and f is the functional mentioned above, then the sets $\{x\colon\ f(x) \geqslant \alpha\}$ and $\{x\colon\ f(x) \leqslant \alpha\}$ are called the *closed halfspaces* defined by the hyperplane Γ and the sets $\{x\colon\ f(x) < \alpha\}$ and $\{x\colon\ f(x) > \alpha\}$ are called the *open halfspaces* defined by it (even if there is no topology in E).

The first two halfspaces are indeed closed and the latter two are open in the strongest locally convex topology on E (in this topology all vector subspaces are closed). However, the property that in a locally convex space E all vector subspaces are closed does not yield that there is no stronger locally convex topology in E.

If E is a nonzero topological vector space and a hyperplane Γ is closed in E, then the closed halfspaces defined by it are closed and the open halfspaces are open also in the original topology of E; moreover, these closed halfspaces are the closures of the open ones (Exercise 1.12.41). We recall (see Proposition 1.9.6) that the continuity of a linear function f is equivalent to the property that its kernel is closed, which is the same that the hyperplane $\{x\colon\ f(x) = \alpha\}$ is closed for some (then for all) α.

Let us introduce one more important concept.

1.10.6. Definition. *Let E be a topological vector space, let Γ be a closed hyperplane in E, and let A and B be subsets of E. We shall say that Γ separates* *the sets A and B if A and B are contained in the different closed halfspaces defined by Γ. If A and B are contained in the different open halfspaces defined by Γ, then we shall say that Γ strictly separates A and B.*

1.10.7. Remark. (i) If the sets A and B are open, then any separating hyperplane strictly separates them.

(ii) It follows from the definition that sets A and B in a topological vector space E can be separated by a closed hyperplane precisely when on the space E there exists a continuous linear functional f such that $f(a) \leqslant f(b)$ for all $a \in A$ and $b \in B$; these sets can be strictly separated by a hyperplane precisely when $f(a) < f(b)$ for all $a \in A$, $b \in B$.

(iii) A closed hyperplane H passing through a point of a set A is called *supporting* if A entirely belongs to one of the two halfspaces defined by H.

Here is the main result on separation of convex sets.

1.10.8. Theorem. *Let A and B be convex sets in a real topological vector space E such that the set B and the interior $\breve{A}$ of the set A are nonempty and disjoint. Then, there exists a closed hyperplane Γ separating A and B.*

PROOF. We know that there exist disjoint convex sets $V, W \subset E$ for which $A \subset V$, $B \subset W$, and $V \cup W = E$. Set $G = \overline{V} \cap \overline{W}$. Since $\breve{V} \cap \overline{W} = \varnothing$, hence $\breve{V} \cap G = \varnothing$, by the condition that $\breve{A}$ is nonempty we obtain that $G \neq E$. It follows from the previous theorem that G is a hyperplane in E (it is important here that W and V are nonempty). We can assume that G is a hypersubspace. Let f be a linear functional the kernel of which is G. Since $\breve{V} \cap G = \varnothing$, on the convex set $\breve{V}$ the functional f has a constant sign; we can assume that $f(x) > 0$ for all $x \in \breve{V}$ (so $f(x) > 0$ for all $x \in \breve{A}$). Since $\breve{V} \cap G = \varnothing$ and $\breve{V} \neq \varnothing$, the hypersubspace G is not everywhere dense, therefore, it is closed. Each point $x \in \overline{A}$ is a limit point of the interval $(x, a]$, where $a \in A$, which is contained in $\breve{A}$, because A is convex. Hence $f(x) \geqslant 0$ for all $x \in \overline{A}$. By the way, this reasoning uses only the continuity of the restriction of f to the interval $[x, a]$ and does not use the continuity of f on E (although it takes place, since G is closed); on the one-dimensional space every linear functional is continuous. Thus, in order to show that G separates A and B it remains to prove that $f(x) \leqslant 0$ for all $x \in B$. If $f(b) > 0$ for some vector $b \in B$, then, for every $x \in \breve{A}$, on the interval $[x, b]$ the functional f has no zero (it is linear and its values at the endpoints of the interval are of the same sign). This means that $[x, b] \cap G = \varnothing$, which contradicts Remark 1.10.3. □

Our assumption that the interior of at least one of the sets A and B is nonempty is essential in this theorem.

1.10.9. Example. Let $\mathbb{R}^{(\infty)}$ be the topological direct sum of a countable family of real lines, i.e., the set of all real sequences of the form $(x_1, \ldots, x_n, 0, 0, \ldots)$, equipped with the strongest locally convex topology generated by all seminorms (general direct sums are discussed in the next chapter). Let A be the subset of $\mathbb{R}^{(\infty)}$ consisting of all vectors (x_n) whose last nonzero component is positive. This set is convex, since if $x, y \in A$ have the last nonzero components $x_n > 0$

and $y_m > 0$, where $n \geqslant m$, then $x_n + y_n > 0$. If G is a hyperplane in $\mathbb{R}^{(\infty)}$ (we observe that every hyperplane in $\mathbb{R}^{(\infty)}$ is closed, since every linear function on $\mathbb{R}^{(\infty)}$ is continuous due to our choice of the topology), V and W are the corresponding open halfspaces, then $A \cap V \neq \varnothing$ and $A \cap W \neq \varnothing$. See also Exercise 1.12.73 for an example of closed convex sets in the spaces l^1 and l^2 that cannot be separated.

Let us give a condition for the strict separation.

1.10.10. Theorem. *Let F and K be nonempty convex subsets of a real locally convex space E such that F is closed, K is compact and $K \cap F = \varnothing$. Then, there exists a closed hyperplane in E strictly separating F and K. In particular, if E is Hausdorff, then any two points can be strictly separated.*

PROOF. Since F is closed, for every point $a \in K$ there exists a convex circled neighborhood of zero V such that

$$(a + V) \cap (F + V) = \varnothing. \tag{1.10.1}$$

Indeed, since F is closed, there exists a convex circled neighborhood of zero W such that

$$(a + W) \cap F = \varnothing. \tag{1.10.2}$$

In order to obtain (1.10.1), it suffices to take $V = W/2$. Indeed, in this case, if there is $z \in (V + F) \cap (a + V)$, then there exist three elements $z_1 \in V$, $z_2 \in V$, $x \in F$ such that $a + z_1 = z = z_2 + x$, i.e., $a + z_1 - z_2 = x$. Since

$$a + z_1 - z_2 \in a + W/2 - W/2 \subset a + W,$$

this contradicts (1.10.2).

For each $a \in K$ we find a circled convex open neighborhood of zero W_a such that $(a + W_a) \cap (F + W_a) = \varnothing$. The sets $a + W_a$ form an open cover of the compact set K. Let $\{a_i + W_{a_i}\}_{i\leqslant n}$ be some finite subcover. For the neighborhood $W_f = \bigcap_{i=1}^n W_{a_i}$ we have $(F + W_f) \cap \Bigl(\bigcup_{i=1}^n (a_i + W_{a_i})\Bigr) = \varnothing$. Thus, $(F + W_f) \cap K = \varnothing$.

Let $W' = W_f/2$. Then $(K + W') \cap (F + W') = \varnothing$; the proof of this is completely analogous to the proof of the relation (1.10.1) for $V = W'/2$. Let G be a closed hyperplane separating the open convex sets $K + W'$ and $F + W'$. According to Remark 1.10.7(i) it strictly separates these sets; hence it strictly separates K and F. □

1.10.11. Remark. By Remark 1.10.7(i) the conclusion of this theorem is equivalent to the existence of a continuous linear functional f on E such that $f(x_1) < f(x_2)$ for all $x_1 \in K$, $x_2 \in F$. Actually, under the hypotheses of this theorem we have a stronger conclusion: in this case there exists a continuous linear functional f on E such that

$$\sup_{x\in K} f(x) < \inf_{x\in F} f(x).$$

Indeed, let G and W' be the same as in the proof of the theorem and let f be a continuous linear functional on E the kernel of which is a shift of G and which is positive on the open halfspace containing F. Set $\gamma = \sup_{x\in K} f(x)$. Let a be an element of K such that $f(a) = \gamma$. There is an element $b \in W'$ such that $f(b) \neq 0$, since the closed hyperplane is nowhere dense. We can assume that $f(b) > 0$, otherwise replacing b by $-b$. Since G separates the sets $K + W'$ and $F + W'$, for all $x_1 \in K$, $x_2 \in F$ we have

$$f(x_1) \leqslant \gamma < \gamma + f(b) = f(a) + f(b) = f(a+b) < f(x_2),$$

where $a + b \in K + W'$, so that the desired inequality holds.

1.10.12. Theorem. (THE HAHN–BANACH THEOREM IN THE GEOMETRIC FORM) *Let E be a real topological vector space, let $V \subset V$ be a nonempty open convex set, and let G be a closed linear variety in E disjoint with V. Then, there exists a closed hyperplane in E containing G and also disjoint with V.*

PROOF. By Theorem 1.10.8, there exists a closed hyperplane H_0 separating V and G. This hyperplane does not intersect V, since V is open.

Let g be a continuous linear functional on E such that

$$H_0 = \{x \in H\colon\ g(x) = \alpha\}, \quad \text{where } \alpha \in \mathbb{R},$$

and, for example, $g(x) > \alpha$ for all $x \in V$ and $g(x) \leqslant \alpha$ for all $x \in G$. Then, there exists $\beta \in \mathbb{R}$ such that $g(x) = \beta$ for all $x \in G$. Indeed, if the linear function g is not constant on the linear manifold G, then the set of its values is $\mathbb{R}^1$, which is impossible, since $g \leqslant \alpha$ on G. Clearly, $\beta \leqslant \alpha$. The set $\{x \in E\colon\ g(x) = \beta\}$ is a hyperplane containing G and disjoint with V.

Let us give yet another ("direct") proof of this theorem. Denote by $\mathcal{P}_0$ the partially ordered set of all vector subspaces of the space E containing the subspace G and disjoint with V. It is readily verified (do this) that the set $\mathcal{P}_0$ satisfies the conditions of the Kuratowski–Zorn theorem; hence $\mathcal{P}_0$ has maximal elements; let G_m be one of them. Thus, G_m is a vector subspace of the space E with the following properties:

(a) $G \subset G_m$; (b) $G_m \cap V = \varnothing$; (c) if L is a vector subspace of the space E containing G_m and disjoint with V, then $L = G_m$.

We show that these three properties yield that G_m is a hypersubspace the existence of which is asserted in the theorem. Let E_0 be the topological vector quotient of the space E by its subspace G_m. Then G_m is a hypersubspace precisely when E_0 is one-dimensional. Thus, in order to prove that G_m is a hypersubspace, it suffices to show that $\dim E_0 = 1$. We now do this. It follows from (b) and (c) that G_m is closed. Indeed, the closure $\overline{G_m}$ of the subspace G_m is a vector subspace of the space E containing G_m and disjoint with V (since the latter is open and G_m does not intersect it). Hence $\overline{G_m} = G_m$ by (b). Therefore, the space E_0 is Hausdorff. Let V_0 denote the image of V in E_0 with respect to the canonical mapping $\Phi\colon\ E \to E_0$; then V_0 is an open convex subset of E_0 and $0 \notin V_0$, since we have $V \cap G_m = \varnothing$.

Suppose now that $\dim E_0 > 1$ and show that E_0 contains a nontrivial vector subspace G_0 disjoint with V_0. Its preimage $\Phi^{-1}(G_0)$ with respect to the canonical mapping will be a vector subspace in E containing G_m and disjoint with V (since $G_0 \cap V_0 = \varnothing$, we have $\Phi^{-1}(G_0) \cap \Phi^{-1}(V_0) = \varnothing$, and since $V \subset \Phi^{-1}(V_0)$, we have $\Phi^{-1}(G_0) \cap V = \varnothing$). Hence by (c) we have $\Phi^{-1}(G_0) = G_m$, but this contradicts the relations $G_m = \Phi^{-1}(0)$, $G_0 \neq \{0\}$.

Thus, it remains to show that if $\dim E_0 > 1$, then E_0 contains a vector subspace G_0 of positive dimension disjoint with V_0. Let e_1 and e_2 be linearly independent elements in E_0 and let $\mathcal{L}(e_1, e_2)$ be the subspace in E_0 generated by them. If $\mathcal{L}(e_1, e_2) \cap V_0 = \varnothing$, then we can take just $G_0 = \mathcal{L}(e_1, e_2)$. Suppose now that $\mathcal{L}(e_1, e_2) \cap V_0 = V_1 \neq \varnothing$ and show that $\mathcal{L}(e_1, e_2)$ contains a one-dimensional subspace disjoint with V_1. For every $\varphi \in [0, 2\pi)$, let $k(\varphi)$ denote the following set (the ray): $\{\lambda(e_1 \cos\varphi + e_2 \sin\varphi)\colon\ \lambda > 0\}$. Since V_1 is convex and $0 \notin V_1$, for every φ we have at least one of the equalities $k(\varphi) \cap V_1 = \varnothing$ or $\bigl(-k(\varphi)\bigr) \cap V_1 = \varnothing$.

Indeed, if this is not true and we have $\lambda_1 a(\varphi) \in V_1$, $\lambda_2 a(\varphi) \in V_1$, where $a(\varphi) = e_1 \cos\varphi + e_2 \sin\varphi$, $\lambda_1 > 0$, $\lambda_2 < 0$ (the equality $\lambda_i = 0$ is excluded, because $0 \notin V_1$), then, since the convex combination

$$\frac{|\lambda_2|}{\lambda_1 + |\lambda_2|}\lambda_1 a(\varphi) + \frac{\lambda_1}{\lambda_1 + |\lambda_2|}\lambda_2 a(\varphi)$$

of the elements $\lambda_1 a(\varphi)$ and $\lambda_2 a(\varphi)$ vanishes, it turns out that $0 \in V_1$, but this is false. In particular, we have at least one of the two equalities $k(0) \cap V_1 = \varnothing$ and $\bigl(-k(0)\bigr) \cap V_1 = \varnothing$. Changing the notation if necessary, we can assume that the first equality holds.

Set now

$$\varphi_m = \sup\bigl\{\varphi \in [0, 2\pi)\colon\ k(\varphi_1) \cap V_1 = \varnothing \ \ \forall \varphi_1 \in [0, \varphi]\bigr\}$$

and show that $\bigl(-k(\varphi_m) \cup k(\varphi_m)\bigr) \cap V_1 = \varnothing$. This will mean that V_1 is disjoint with the one-dimensional space generated by the element $a(\varphi_m)$; this subspace is the set $\{0\} \cup k(\varphi_m) \cup \bigl(-k(\varphi_m)\bigr)$. The assumption that $k(\varphi_m) \cap V_1 \neq \varnothing$ contradicts the definition of φ_m, since V_1 is open. Indeed, if $\lambda a(\varphi_m) \in V_1$ and $\lambda > 0$, then, by the continuity of the mapping $\varphi \mapsto a(\varphi)$, $[0, 2\pi) \to \mathcal{L}(e_1, e_2)$, there exists $\varepsilon > 0$ such that $\lambda a(\varphi_m - \varepsilon) \in V_1$.

Further, if $-\lambda a(\varphi_m) \in V_1$, then, since V_1 is open, there exists $\delta > 0$ such that for all $\varepsilon \in [0, \delta)$ we have $-\lambda a(\varphi_m + \varepsilon) \in V_1$. At the same time, according to the definition of φ_m, among such numbers ε there is ε' for which $k(\varphi_m + \varepsilon') \cap V_1 \neq \varnothing$, i.e., $\lambda_1 a(\varphi_m + \varepsilon') \in V_1$ for some λ_1. The last two inclusions, with the aid of a calculation based on the convexity of V_1 and similar to the one given above, yield the false inclusion $0 \in V_1$. Hence it is proved that the set $\mathcal{L}(e_1, e_2)$ contains a subspace disjoint with the set $V_1 = V_0 \cap \mathcal{L}(e_1, e_2)$; it can be taken for G_0. Thus, the second proof of the theorem is complete. □

In the next section we discuss another approach to extensions.

1.11. The Hahn–Banach theorem: the analytic form

The results presented above on separation of convex sets can be restated in the form of certain inequalities; for this reason one can speak of the Hahn–Banach theorem in the analytic form.

1.11.1. Theorem. *Suppose that E is a real vector space and E_0 is its vector subspace. Let p be a sublinear function (see § 1.4) on E with finite values and let f be a linear functional on E_0 such that for every $x \in E_0$ we have*

$$f(x) \leqslant p(x). \tag{1.11.1}$$

Then, there exists a linear functional $\widetilde{f}$ on E whose restriction to E_1 coincides with f and which everywhere on E satisfies the inequality $\widetilde{f}(x) \leqslant p(x)$. In other words, f can be extended to the whole space E with the preservation of (1.11.1).

PROOF. We set

$$G = E\times\mathbb{R},\ G_0 = E_0\times\mathbb{R},\ V = \{(x,t) \in G\colon\ x \in E,\ p(x) < t\}$$

and denote by τ_E and τ_G the strongest locally convex topologies in E and G, respectively. We verify that V is convex and open in the topology τ_G. If

$$(x_1,t_1),\ (x_2,t_2) \in V,\ \lambda \in [0,1],\ (x_3,t_3) = \lambda(x_1,t_1) + (1-\lambda)(x_2,t_2),$$

then $p(x_3) \leqslant \lambda p(x_1) + (1-\lambda)p(x_2)$, $t_3 = \lambda t_1 + (1-\lambda)t_2$. Since, in addition, $p(x_1) < t_1$, $p(x_2) < t_2$, we have $p(x_3) \leqslant \lambda p(x_1) + (1-\lambda)p(x_2) < t_3$, because $(x_3,t_3) \in V$. Hence V is convex.

The fact that V is open in the topology τ_G follows from the continuity in this topology of the function $(x,t) \mapsto t - p(x)$ from $E\times\mathbb{R}$ to $\mathbb{R}$, which follows from the continuity on (E,τ_E) of the sublinear function p (see Exercise 1.12.39). The established properties of the set V will be needed in order to apply Theorem 1.10.12.

Denote by Γ the graph of f on E_0; Γ is a hypersubspace in G_0. We observe that the estimate $f(x) \leqslant p(x)$ for $x \in E_0$ yields that

$$\Gamma\cap V = \Gamma\cap(V\cap G_0) = \varnothing.$$

Indeed, if $(x,t) \in \Gamma\cap V$, then $p(x) < t$, since $(x,t) \in V$. On the other hand, $t = f(x)$, since $(x,t) \in \Gamma$, so that $f(x) > p(x)$ for some $x \in E_0$, which is impossible.

By Theorem 1.10.12 there exists a hypersubspace Γ_m of the space G containing Γ and disjoint with V (the set Γ is closed in the topology τ_G). We show that Γ_m is the graph of a linear functional $\widetilde{f}$ on E that is an extension of f and satisfies the inequality $\widetilde{f}(x) \leqslant p(x)$ for all $x \in E$.

It suffices to verify the following three things:

(a) that Γ_m is the graph of a function $E \to \mathbb{R}$; then, due to the fact that Γ_m is a vector subspace, this function will be automatically linear;

(b) the equality $\Gamma_m\cap G_0 = \Gamma$; then this functional will be an extension of f;

(c) the required inequality for $\widetilde{f}$ for all $x \in E$.

In order to verify (a) we show that if $(x_0,t_1)\in\Gamma_m$, $(x_0,t_2)\in\Gamma_m$, then $t_1 = t_2$. If this is not true, then $p(x_0) < \lambda t_1 + (1-\lambda)t_2$ for some $\lambda \in \mathbb{R}$. For example, in case $t_1 > t_2$ this inequality holds if

$$\lambda = (p(x_0) - 2t_2 + t_1)/(t_1 - t_2).$$

Hence $(x_0, \lambda t_1 + (1-\lambda)t_2) \in V$, but by the relation $V \cap \Gamma_m = \varnothing$ this contradicts the inclusion $(x_0, \lambda t_1 + (1-\lambda)t_2) \in \Gamma_m$ that follows, in turn, from the inclusion $(x_0, t_1) \in \Gamma_m$, $(x_0, t_2) \in \Gamma_m$ and the linearity of Γ_m.

Let us prove (b). First of all, we have $\Gamma \subset \Gamma_m \cap G_0$, since $\Gamma \subset \Gamma_m$, $\Gamma \subset G_0$. Thus, it suffices to show that $\Gamma \supset \Gamma_m \cap G_0$. Let $x \in (\Gamma_m \cap G_0)\backslash\Gamma$. Since Γ is a hypersubspace in G_0, the smallest subspace T of the space G_1 containing Γ and x coincides with G_0 and hence intersects V (it even contains $V \cap G_0$). However, this contradicts the fact that $\Gamma_m \cap V = \varnothing$, $T \subset \Gamma_m$ (this inclusion follows from the inclusion $x \in \Gamma_m$, $\Gamma \subset \Gamma_m$).

Finally, the estimate $\widetilde{f}(x) \leqslant p(x)$ for all vectors $x \in E$ follows from the equality $V \cap \Gamma_m = \varnothing$, since if $\widetilde{f}(x_0) > p(x_0)$, then $(x_0, \widetilde{f}(x_0)) \in V$, which by the inclusion $(x_0, \widetilde{f}(x_0)) \in \Gamma_m$ contradicts the equality just stated. □

Obviously, the hypothesis of the theorem explicitly employs the condition that we consider real spaces. Complex spaces will be discussed below.

If p is a seminorm, then (1.11.1) is equivalent to the estimate $|f| \leqslant p$, but this case will be specially mentioned below in Theorem 1.11.7.

The proven theorem in most of the texts on functional analysis is called the Hahn–Banach theorem (in full accordance with the historic order of the discovery of these results). For this reason it seems wise to give its direct proof (not based on the geometric version of this theorem). Normally, such a proof is given in standard texts.

Let $\mathcal{P}$ be the partially ordered set consisting of all possible pairs (g, F), where F is a linear subspace in E containing E_1, g is a linear functional on F that is an extension of the functional f and satisfies the inequality $g(x) \leqslant p(x)$ for every point $x \in F$.

The order in $\mathcal{P}$ is introduced as follows: $(g_1, F_1) < (g_2, F_2)$ if $F_1 \subset F_2$ and g_2 is an extension of g_1. It is readily verified that $\mathcal{P}$ satisfies the hypothesis of the Kuratowski–Zorn theorem. Hence $\mathcal{P}$ has maximal elements; denote by (g_0, F_0) one of them and show that $F_0 = E$. This will mean that the domain of definition of the functional g_0 is the whole space E, i.e., that g_0 is the desired extension of the functional f.

If $F_0 \neq E$, then there is $x_0 \in E$ with $x_0 \notin F_0$. Denote by F_1 the subspace in E generated x_0 and F_0, and show that contrary to the maximality of (g_0, F_0) there exists an extension g_1 of the functional g_0 to F_1 such that $(g_1, F_1) \in \mathcal{P}$. If $z \in F_1$, then there exist (uniquely defined) $a \in F_0$ and $\alpha \in \mathbb{R}$ such that $z = a + \alpha x_0$.

Next, if g_1 is a functional on F_1 whose existence we want to prove, then

$$g_1(z) = g_0(a) + \alpha g_1(x_0) \leqslant p(a + \alpha x_0), \tag{1.11.2}$$

since $g_0(x) = g_1(x)$ on F_0 and $a \in F_0$. On the other hand, if we can pick $g_1(x_0)$ such that the latter inequality holds for all α and a, then the functional defined by means of (1.11.2) will be the desired one.

Let us show that such a choice of $g_1(x_0)$ is possible indeed. To this end, we consider (1.11.2) separately for $\alpha > 0$ and $\alpha < 0$ (for $\alpha = 0$ this inequality is fulfilled for any choice of $g_1(x_0)$, since $(g_0, F_0) \in \mathcal{P}$). If $\alpha > 0$, then (1.11.2) is equivalent to the inequality

$$g_1(x_0) \leqslant p\Big(\frac{a}{\alpha}+x_0\Big) - g_0\Big(\frac{a}{\alpha}\Big);$$

if $\alpha < 0$, then it can be rewritten as

$$g_0\Big(\frac{a}{-\alpha}\Big) - g_1(x_0) \leqslant p\Big(\frac{a}{-\alpha}-x_0\Big),$$

which is equivalent to the inequality

$$g_1(x_0) \geqslant -p\Big(\frac{a}{-\alpha}-x_0\Big) - g_0\Big(\frac{a}{\alpha}\Big).$$

Thus, the number $g_1(x_0)$ must be picked in such a way that for all $a_1,\, a_2 \in F_0$ and $\alpha_1,\, \alpha_2 \in (0,\infty)$ the following inequalities be true:

$$p\Big(\frac{a_2}{\alpha_2}+x_0\Big) - g_0\Big(\frac{a_2}{\alpha_2}\Big) \geqslant g_1(x_0) \geqslant -p\Big(\frac{a_1}{-\alpha_1}-x_0\Big) - g_0\Big(\frac{a_1}{\alpha_1}\Big).$$

In order this choice be possible (recall that x_0 is a fixed element of E), it is necessary and sufficient that

$$\inf\bigl\{p(b+x_0) - g_0(b)\colon\ b\in F_0\bigr\} \geqslant \sup\bigl\{-p(-b-x_0) - g_0(b)\colon\ b\in F_0\bigr\},$$

or, which is equivalent, for every $b_1,\, b_2 \in F_0$ the inequality

$$p(b_1+x_0) - g_0(b_1) \geqslant -p(-b_2-x_0) - g_0(b_2)$$

holds. The latter inequality is true, since

$$g_0(b_1) - g_0(b_2) \leqslant p(b_1-b_2) = p(b_1+x_0-b_2-x_0) \leqslant p(b_1+x_0) + p(-b_2-x_0),$$

because p is sublinear.

1.11.2. Remark. The condition that the functional p is finite is essential for the validity of the Hahn–Banach theorem in the analytic form. Indeed, let $E = \mathbb{R}^2$ and let the functional $p\colon\ \mathbb{R}^2 \to \overline{\mathbb{R}}^+$ be defined as follows: $p(x_1,x_2) = 0$ if $x_2 > 0$ or if $x_2 = 0$ and $x_1 \geqslant 0$; in all other cases $p(x_1,x_2) = \infty$ (this is precisely the Minkowski functional of the set $\{(x_1,x_2)\in\mathbb{R}^2\colon\ p(x_1,x_2) < 1\}$). Let us set $E_1 = \{(x_1,0)\colon\ x_1\in\mathbb{R}\}$ and $f(x_1,0) = -x_1$. Then

$$f(x) \leqslant p(x) \quad \text{for } x = (x_1,x_2)\in E_1,$$

however, there is no linear functional on E extending f and satisfying the corresponding inequality. This follows from the fact that E contains no straight line ("hyperplane") passing through the point $(-1,0)$ and not intersecting the set $\{(x_1,x_2)\colon\ p(x_1,x_2) < 1\}$.

We have given two proofs of the Hahn–Banach theorem in the analytic form: a direct proof (the last one) and the one based on the geometric form of this theorem; in turn, this latter form has been proved also in two ways: with the aid of the Kakutani theorem and again "directly". Thus, we have actually given even three different proofs of the Hahn–Banach theorem in the analytic form. Each of these proofs consisted of two parts: a set-theoretic part based on the Kuratowski–Zorn theorem and an elementary part concerned in fact with the geometry of two-dimensional spaces, so that at the level of main ideas all these three different proofs are very close.

Let us now derive the geometric form of the Hahn–Banach theorem from the analytic one.

Let V be a nonempty open convex subset of a topological vector space E and let G be a closed vector subspace in E not intersecting V. We have to prove that E contains a closed hypersubspace F also not intersecting V.

Let $a \in V$ $(a \neq 0)$, $V_a = V - a$, $G_a = G - a$, let p be the Minkowski functional of the set V_a, and let E_0 be the vector subspace in E generated by G_a. In E_0 the subset G_a is a hyperplane (not containing the zero element). Let f_0 be a functional on E_0 for which $G_a = f_0^{-1}(1)$. Then $f_0(x) \leqslant p(x)$ for all $x \in E_0$. Indeed, if for some $x_0 \in E_0$ this is false, i.e., $f_0(x_0) > p(x_0)$, then the vector $z = x_0/f(x_0)$ satisfies the relation $f_0(z) = 1 > p(z)$, but the equality here means that $z \in G_a$, while the inequality says that $z \in V_a$; thus, these inclusions contradict the equality $(G - a) \cap (V - a) = \varnothing$. Let f be an extension of f_0 to all of E satisfying the inequality from the formulation of Theorem 1.11.1, $G' = \{x \in E\colon\ f(x) = 1\}$. Then $G_a \subset G'$ and $G' \cap V_a = \varnothing$. The inclusion is obvious; the equality can be shown as follows. Let $x \in G' \cap V_a$. Then $f(x) = 1$ and $p(x) < 1$ by the condition that V_a is open, which contradicts the inequality $f \leqslant p$. Now for completing the proof it suffices to set $F = G' + a$.

We now proceed to discussing the Hahn–Banach theorem for complex spaces. Here some nuances appear connected with the linearity over $\mathbb{R}$ and $\mathbb{C}$.

Let X be a complex topological vector space. Denote by X_R the real topological vector space which consists of the same elements as X, i.e., X_R and X coincide as sets, but in X_R only the multiplication by real numbers is allowed. Hence if $x \in X_R$, $x \neq 0$, then also $ix \in X_R$, but in X_R the element ix is not the product of x by the number i (this product is not defined in X_R); moreover, the elements x and ix in X_R are linearly independent.

The families of neighborhoods of zero, hence, the closed and open sets in X and X_R are the same. Since the definition of convexity involves only real numbers, any set convex in X is convex in X_R (and conversely). Thus, if X is a locally convex space, then X_R is also a locally convex space.

Note also that any set balanced in X will be balanced in X_R, but the converse can be false. Here is an example: the interval $[-x, x]$.

The space X_R is called the real topological vector space associated with the complex topological vector space X (or the real structure of X). In the space X we can consider complex and real linear manifolds, the latter ones are linear manifolds in X_R. In particular, we shall consider complex and real hyperplanes. For example, $\mathbb{C}^1$ is a complex hyperplane in $\mathbb{C}^2$, but not real (its real codimension is two).

A real linear variety is not, in general, a complex linear manifold. For example, the complex straight line passing through the points 0 and $x \neq 0$ contains the point ix, but the real straight line passing through the same points does not contain the point ix. Nevertheless, there is a close connection between complex and real planes, a partial case of which is described by the following lemma.

1.11.3. Lemma. *Every complex hyperplane H in X is the intersection of two real hyperplanes and the latter are closed in case of closed H.*

PROOF. Let f be a linear functional on X such that

$$H = \{x\colon\ f(x) = \alpha + i\beta\}.$$

Set $\varphi(x) = \operatorname{Re} f(x)$ and $\psi(x) = \operatorname{Im} f(x)$, $x \in X$. It is readily verified that φ and ψ are linear real functionals on X_R. It is clear that $H = H_1 \cap H_2$, where $H_1 = \{x\colon\ \varphi(x) = \alpha\}$ and $H_2 = \{x\colon\ \psi(x) = \beta\}$. In order to show that H_1 and H_2 are closed for closed H we have to show that the continuity of f implies the continuity of φ and ψ. This is also clear, since if there exists a neighborhood of zero V in X such that $|f(x)| < \varepsilon$ for all $x \in V$, then we have $|\varphi(x)| < \varepsilon$ and $|\psi(x)| < \varepsilon$ for all such x. □

1.11.4. Lemma. *For every real linear functional φ on X, there exists a unique complex linear functional f such that $\varphi(x) = \operatorname{Re} f(x)$ for all $x \in X$, moreover, if the functional φ is continuous, then f is also continuous.*

PROOF. Set $f(x) = \varphi(x) - i\varphi(ix)$. Then

$$f(x_1 + x_2) = \varphi(x_1) + \varphi(x_2) - i\varphi(ix_1) - i\varphi(ix_2) = f(x_1) + f(x_2).$$

If λ is real, then it is clear that $f(\lambda x) = \lambda f(x)$. Finally,

$$f(ix) = \varphi(ix) - i\varphi(-x) = i[\varphi(x) - i\varphi(ix)] = if(x),$$

$$f(\alpha + i\beta)x = f(\alpha x) + f(i\beta x) = \alpha f(x) + if(\beta x) = (\alpha + i\beta)f(x).$$

Suppose that there exists another complex linear functional f_1 on X such that $f_1(x) = \varphi(x) + i\psi(x)$. Then

$$f_1(ix) = \varphi(ix) + i\psi(ix) = if_1(x) = i[\varphi(x) + i\psi(x)],$$

i.e., $\psi(x) = -\varphi(ix)$, and the uniqueness of f is proved. If the functional φ is continuous, then for every $\varepsilon > 0$ there exists a circled neighborhood of zero V such that $|\varphi(x)| < \varepsilon/2$ for all $x \in V$. Then $ix \in V$ and hence $|i\varphi(ix)| < \varepsilon/2$, whence we obtain $|f(x)| < \varepsilon$. □

1.11.5. Corollary. *If H_0 is a hypersubspace in X, then $H = H_0 \cap iH_0$ is a complex hypersubspace in X, moreover, H is closed if H_0 is closed.*

PROOF. Let $\varphi(x) = 0$ be an equation defining H_0. If $x \in iH_0$, then $x = iy$, $y \in H_0$. Hence we have $\varphi(ix) = \varphi(-y) = 0$. On the other hand, if $\varphi(ix) = 0$, then $ix = z \in H_0$ and hence $x = -iz$. Therefore, the equation defining the subspace iH_0 has the form $\varphi(ix) = 0$. If the functional f is defined by the equality $f(x) = \varphi(x) - i\varphi(ix)$, then the subspace $H = \{x\colon\ f(x) = 0\}$ consists of exactly those points $x \in X$ for which $\varphi(x) = 0$ and $\varphi(ix) = 0$, i.e., $H = H_0 \cap iH_0$. □

We proceed to the proof of the geometric form of the Hahn–Banach theorem for complex topological vector spaces. The analytic form will be considered below.

1.11.6. Theorem. *Let X be a complex topological vector space, let A be a nonempty open convex set in X, and let P be a complex closed linear variety not intersecting A. Then, there exists a closed complex hyperplane H in X containing P and not intersecting A.*

PROOF. By shifting we can assume that P contains zero. In this case $iP = P$. In the real space X_R associated with X the set A is open and convex and P is a closed linear subspace. Therefore, by the Hahn–Banach theorem for the real case there exists a real closed hyperplane H_0 passing through zero and containing P and not intersecting A. By Corollary 1.11.5 the intersection $H = H_0 \cap iH_0$ is a closed hyperplane in X passing through zero. Clearly, $A \cap H = \varnothing$, and since $P = iP \subset iH_0$, we have $P \subset H$. □

The analytic form of the Hahn–Banach theorem has been proved above only for real spaces. For complex spaces there is a somewhat weaker version of the analytic form of the Hahn–Banach theorem, which will be now proved without using the analytic form of this theorem for real spaces (although there is a proof that uses the latter, see Bogachev, Smolyanov [**72**]). More precisely, the theorem which will be now given is true both for real and complex spaces.

1.11.7. Theorem. *Let E be a vector space over the field of real or complex numbers, let p be a seminorm on E, let E_0 be a vector subspace of the space E, and let f be a linear functional on E_0 such that $|f(x)| \leqslant p(x)$ for all $x \in E_0$. Then, there exists a linear functional $\widetilde{f}$ on E with the following properties:*

$$\widetilde{f}(x) = f(x) \quad \textit{for } x \in E_0, \quad |\widetilde{f}(x)| \leqslant p(x) \quad \textit{for } x \in E.$$

PROOF. We equip E with the topology τ_p generated by the seminorm p. We shall assume that $f \neq 0$ (otherwise the theorem is trivial). Hence there exists an element $a \in E_0$ such that $f(a) = 1$. Then

$$E_0 = \{\lambda a\colon\ \lambda \in \mathbb{R}\} + L, \quad \text{where } L = \operatorname{Ker} f.$$

Let us take the set $A = a + V$, where $V = \{x \in E\colon\ p(x) < 1\}$ is a convex circled absorbent set; the set A is open in the topology τ_p. If $x \in A \cap L$, then we have $x = a + y$, $y \in V$ and $f(x) = f(a) + f(y) = 1 + f(y) \neq 0$, since $|f(y)| \leqslant p(y) < 1$. Hence $L \cap A = \varnothing$. By Theorem 1.10.12 and Theorem 1.11.6 there exists a closed hypersubspace $H_0 \supset L$ not intersecting A. Since $a \notin H_0$, we have $H_0 + \{\lambda a\} = E$.

Let $\widetilde{f}(x) = \lambda$ if $x = \lambda a + z$, $z \in H_0$. Then $\widetilde{f}$ is a continuous linear functional on E. Clearly, $\widetilde{f}(x) = f(x)$ on E_0. It remains to prove that $|\widetilde{f}(x)| \leqslant p(x)$ for all vectors $x \in E$. If $x \in E$ and $p(x) < 1$, i.e., $x \in V$, then $|\widetilde{f}(x)| < 1$, since if $|\widetilde{f}(x)| \geqslant 1$ we can take $\widetilde{x} = -x/\widetilde{f}(x)$ and obtain that $\widetilde{x} \in V$ because V is circled. Therefore, $a + \widetilde{x} \in A$. At the same time we have $\widetilde{f}(a + \widetilde{x}) = 1 - \widetilde{f}(x)/\widetilde{f}(x) = 0$, i.e., $A \cap H_0 \neq \varnothing$, which is impossible.

If now $x \in E$ and $|\widetilde{f}(x)| > p(x)$, then there exists a number $\alpha > 0$ such that $p(x) < \alpha < |\widetilde{f}(x)|$, whence $p(x/\alpha) < 1$, but $|\widetilde{f}(x/\alpha)| > 1$, which is impossible according to what has been proved. □

1.11.8. Corollary. *If E is a locally convex space and E_0 is its vector subspace with the induced topology, then every continuous linear functional f on E_0 has an extension to a functional $\widetilde{f} \in E'$. If E is a normed space, then there exists an extension $\widetilde{f} \in E'$ such that $\|\widetilde{f}\| = \|f\|$.*

PROOF. The first assertion follows from the fact that $|f| \leqslant p$ on E_0, where p is some continuous seminorm (see Theorem 1.9.3). Hence the functional $\widetilde{f}$ on E given by the previous theorem is the desired one. In the second assertion we can assume that $\|f\| = 1$ (if $f \neq 0$). Taking $p(x) = \|x\|$, we obtain $|f(x)| \leqslant p(x)$ for all $x \in E_0$. For any extension with $|\widetilde{f}(x)| \leqslant \|x\|$ we have $\|\widetilde{f}\| \leqslant \|f\|$. Since $\|\widetilde{f}\| \geqslant \|f\|$, we obtain $\|\widetilde{f}\| = \|f\|$. □

1.11.9. Remark. This corollary says that the functional f extends to all of E with the same norm. Such an extension of the functional f is not unique in the general case. Note, however, the following three cases where this extension is unique (a justification is left as Exercise 1.12.46):

(i) E is a Euclidean space and E_0 is its arbitrary vector subspace;

(ii) $E = l_\infty$, $E_0 = c_0$ (see § 1.12(v));

(iii) E is the space of all continuous linear operators on a Hilbert space H equipped with the usual operator norm, E_0 is the subspace of compact operators.

1.11.10. Theorem. *A topological vector space E possesses a nonzero continuous linear functional precisely when E has a convex neighborhood of zero that is not the whole space E.*

PROOF. Let f be such functional. Then the set $V = \{x\colon |f(x)| < 1\}$ is a convex neighborhood of zero by the continuity of f. The inequality $V \neq E$ follows from the fact that if $f(x) \neq 0$, then $|f(\lambda x)| \geqslant 1$ for λ with a sufficiently large absolute value, therefore, $\lambda x \notin V$. Conversely, let V be a convex neighborhood of zero, which without loss of generality can be taken open and symmetric (by passing to $V \cap (-V)$), and let $x_0 \notin V$. On the linear span of x_0 the functional $f_0(tx_0) = t$ is estimated by the Minkowski functional p_V of the set V. The Hahn–Banch theorem gives a linear functional f on E with $f(x_0) = 1$ and $f \leqslant p_V$. For all $x \in V$ we have $|f(x)| \leqslant 1$, which means the continuity of f. □

1.11.11. Theorem. *For every point a in a locally convex space E and every seminorm p on E, there exists a linear functional f such that*

$$f(a) = p(a), \quad |f(x)| \leqslant p(x) \quad \forall\, x \in E.$$

If the seminorm p is continuous, then the functional f is also continuous.

PROOF. On the one-dimensional subspace $D_a = \{\lambda a\}$ we define the linear functional $\varphi(\lambda a) = \lambda p(a)$. Any extension of φ from D_a to all of E by the Hahn–Banach theorem yields a functional f with the desired properties. □

We emphasize that this theorem is not obvious even for the two-dimensional plane.

1.11.12. Corollary. *Let E be a Hausdorff locally convex space and $a \in E$. If for every continuous linear functional f on E we have $f(a) = 0$, then $a = 0$.*

We recall (see Definition 1.3.25) that the dual (the topological dual) to a topological vector space E is the vector space of all linear continuous functionals on E; it is denoted by the symbol E'. We emphasize that the space E' is not a priori given any topology; various topologies on E' will be discussed in Chapter 3. We also recall that the algebraic dual to the vector space E is the space of all linear functionals on E; in this book it is denoted by the symbol E^*.

From Corollary 1.11.12 and Example 1.3.26 we obtain the following fact.

1.11.13. Corollary. *If E is a Hausdorff locally convex space, then the canonical bilinear form*

$$(f, x) \mapsto f(x), \quad E' \times E \to \mathbb{K}$$

establishes a duality between the vector spaces E and E'.

Let us remark that for an arbitrary vector space E the same bilinear form establishes a duality between the spaces E and E^*; this purely algebraic fact is not connected with the Hahn–Banach theorem and is an obvious corollary of the fact that every linear functional defined on a vector subspace of a vector space can be extended to a linear functional on the whole space.

Note also that the local convexity of the space is essential, but not necessary for the validity of the conclusion of Corollary 1.11.13. In particular, on the topological vector space $\mathcal{L}^0$ from Example 1.3.16 there are no nonzero continuous linear functionals (Exercise 1.12.29), so that in this case $(\mathcal{L}^0)' = \{0\}$, hence $\mathcal{L}^0$ and $(\mathcal{L}^0)'$ cannot be put in duality by a bilinear form.

The same is true for the topological vector space $\mathcal{L}^p[0,1]$ with $0 < p < 1$ consisting of measurable functions f on $[0,1]$ such that the function $|f|^p$ is integrable on $[0,1]$; the topology in $\mathcal{L}^p[0,1]$ is defined by the quasi-norm

$$\|f\|_p := \int_0^1 |f|^p \, dt.$$

This space is not Hausdorff, but the associated Hausdorff space also has the trivial zero dual: the convex hull of every ball $U_r = \{f\colon \|f\|_p < r\}$ coincides with the whole space $L^p[0,1]$. Indeed, let $f \in \mathcal{L}^p[0,1]$. Then $f = (f_1 + \cdots + f_n)/n$, where $f_j = nfI_{[s_j, s_{j+1}]}$ and points $0 = s_1 < s_2 < \cdots < s_{n+1} = 1$ are such that the integral of $|f|^p$ over $[s_j, s_{j+1}]$ equals $\|f\|_p^p/n$. Hence $\|f_j\|_p^p = n^{p-1}\|f\|_p^p$ and $f_j \in U_r$ for large n. For an example of a non locally convex topological vector space E for which E and E' are in duality by the canonical bilinear form one can take the space l^p, $0 < p < 1$, consisting of infinite real sequences $x = \{x_n\}$ such that $\sum_{n=1}^\infty |x_n|^p < \infty$; the topology in the space l^p is defined by the quasi-norm $\|x\|_p := \sum_{n=1}^\infty |x_n|^p$.

1.11.14. Corollary. *Let B be a closed absolutely convex subset of a real or complex locally convex space E. Then, for every $a \notin B$, there exists a continuous linear functional f on E such that $|f(a)| > \sup_{x \in B} |f(x)|$.*

PROOF. If E is real, then this is a partial case of Remark 1.10.11. If E is complex, then let $\widetilde{f}$ be a real linear functional on E satisfying the required inequality. Then the complex functional corresponding to it by Lemma 1.11.4 is the one we need. □

1.11.15. Theorem. *Let E be a real or complex locally convex space and let $V \subset E$ be a convex set. Then V is closed in the weak topology $\sigma(E, E')$ precisely when it is closed in the original topology of E. In addition, for every $A \subset E$, the boundedness of A is equivalent to the boundedness in the topology $\sigma(E, E')$.*

PROOF. Since the topology $\sigma(E, E')$ is majorized by the original one, the closedness of V in $\sigma(E, E')$ yields the closedness in the original topology. Suppose now that V is closed in the original topology in E and $a \notin V$. Let us take a functional $f \in E'$ such that $|f(a)| > \alpha$, $\alpha = \sup\{|f(x)|\colon x \in V\}$. Then the set $\{x \in E\colon |f(x)| > \alpha\}$ is an open neighborhood of the point a in the topology $\sigma(E, E')$ and does not intersect V. Similarly we prove the boundedness of A bounded in $\sigma(E, E')$, since the closed absolutely convex hull of A is also bounded in $\sigma(E, E')$. □

Thus, the classes of convex closed sets in the space $\bigl(E, \sigma(E, E')\bigr)$ and in the space E with the original topology coincide (but this does not mean the coincidence of the classes of convex open sets!). In addition, we have the coincidence of the classes of sets bounded in the original topology and in the topology $\sigma(E, E')$; the latter are called *weakly bounded*.

The Hahn–Banach theorem enables us to construct topological complements to finite-dimensional subspaces of locally convex spaces. If we are given a nonzero vector v in an infinite-dimensional topological vector space E, then, certainly, there is a linear subspace E_0 not containing v and generating along with v the whole space E (an algebraic complement of the linear span of v) . However, the arising operator $p\colon E \to E_0$ of the algebraic projection onto E_0 can be discontinuous. For example, this will be the case if on E there are no nonzero continuous linear functions (since if $x = x_0 + tv$, where $v_0 \in E_0$, then the functional $l(x) := t$ is continuous as soon as $p\colon x \mapsto x_0$ is continuous).

1.11.16. Theorem. *Let E be a Hausdorff locally convex space and let E_0 be a finite-dimensional linear subspace in E. Then, there exists a closed linear subspace $E_1 \subset E$ such that $E_0 \cap E_1 = \{0\}$, $E = E_0 \oplus E_1$ and the operators of the algebraic projection to E_0 and E_1 are continuous. Moreover, the latter is true for every closed linear subspace E_1 that gives E in the direct algebraic sum with E_0.*

PROOF. Let $e_1, \ldots, e_n$ be a basis in E_0. According to what has been proved above, we can find functionals $f_1, \ldots, f_n \in E'$ such that $f_i(e_j) = \delta_{ij}$, $i, j \leqslant n$. Set $E_1 = \bigcap_{i=1}^n f_i^{-1}(0)$. Then E_1 is closed and $E = E_0 \oplus E_1$. Clearly, the operators $p_0(x) = f_1(x)e_1 + \cdots + f_n(x)e_n$ and $p_1(x) = x - p_1(x)$ are continuous and are the projection operators to the subspaces E_0 and E_1.

Let now E_1 be some closed linear subspace in E algebraically complementing E_0. Again we take a basis $e_1, \ldots, e_n$ in E_0. Since E_1 is closed, for each e_i the subspace H_i that is the direct sum of E_1 and the linear span of the vectors e_j with

$j \neq i$ is also closed. Moreover, $e_i \notin H_i$. Hence there exists a functional $l_i \in E'$ such that $l_i|_{H_i} = 0$ and $l_i(e_i) = 1$. It is readily seen that $E_1 = \bigcap_{i=1}^n l_i^{-1}(0)$, so the functionals l_i play the role of f_i in the first part of the proof. □

In the general case the projection operators to the closed subspaces giving in the direct sum the whole space can be discontinuous. However, there are important classes of spaces for which such projection operators are automatically continuous (see § 3.9).

Let us give yet another result complementing Proposition 1.9.8 (for a different justification, see Kelley, Namioka [**270**, Section 16.8, p. 144]).

1.11.17. Theorem. *Let A be an absolutely convex set in a locally convex space E. The restriction of a linear function $f \in E^*$ to A is continuous precisely when there exists a sequence of functionals $f_n \in E'$ uniformly converging to f on the set A.*

PROOF. Only the necessity of the indicated condition is not obvious. Passing to the linear span of A, we can assume that E coincides with this linear span. Then the Minkowski functional p_A of the set A is a seminorm on E. Let $\varepsilon > 0$. We have to find $f_\varepsilon \in E'$ with $\sup_{x\in A} |f(x) - f_\varepsilon(x)| \leqslant \varepsilon$. It suffices to do this for $\varepsilon = 1$, passing to f/ε. The continuity of f on A gives an absolutely convex neighborhood of zero U with $|f(x)| \leqslant 1$ for all $x \in A \cap U$. Hence

$$|f(x)| \leqslant p_A(x) + p_U(x), \quad x \in E.$$

Furthermore, $p_U(x-y)+f(y)+p_A(y) \geqslant -p_U(x)+p_U(y)+f(y)+p_A(y)$, which is estimated from below by $-p_U(x)$. Let us consider the function

$$p(x) := \inf_{y\in E} [p_U(x - y) + f(y) + p_A(y)] \geqslant -p_U(x).$$

It is clear that $p(0) = 0$. In addition, the function p is sublinear. Indeed, we have $p(\lambda x) = \lambda p(x)$ for all $\lambda > 0$, since in the formula for $p(\lambda x)$ we can replace the element y by λy. We have $p(x_1 + x_2) \leqslant p(x_1) + p(x_2)$ by the estimate

$$\begin{aligned} &p_U(x_1 + x_2 - y_1 - y_2) + f(y_1 + y_2) + p_A(y_1 + y_2) \\ &\qquad \leqslant p_U(x_1 - y_1) + f(y_1) + p_A(y_1) + p_U(x_2 - y_2) + f(y_2) + p_A(y_2) \end{aligned}$$

for all x_1, x_2, y_1, y_2. Since $p_U(0) = p_A(0) = 0$, we have $p(x) \leqslant p_U(x)$ and also $p(x) \leqslant f(x) + p_A(x)$. By the Hahn–Banach theorem there exists a functional $g \in E^*$ with $g \leqslant p$, moreover, $g \in E'$, since $p \leqslant p_U$. Therefore, $g \leqslant f + p_A$. Hence $|f(x) - g(x)| \leqslant 1$ for all $x \in A$. □

1.12. Complements and exercises

1.12(i). Uniform spaces

Topological vector spaces belong to a broader category of *uniform spaces*. A space X is called uniform if it is endowed with a system $\mathcal{X}$ of subsets of the product $X \times X$, called a *uniformity* or a system of *entourages* of the diagonal $\Delta_X = \{(x,x)\colon\ x \in X\}$, satisfying the following conditions:

(i) $\Delta_X \subset U$ for all $U \in \mathcal{X}$,

(ii) if $U, V \in \mathcal{X}$, then $U \cap V \in \mathcal{X}$, and if $U \in \mathcal{X}$ and $U \subset W \subset X \times X$, then one has $W \in \mathcal{X}$,

(iii) if $U \in \mathcal{X}$, then $U^{-1} := \{(y,x)\colon\ (x,y) \in U\} \in \mathcal{X}$,

(iv) for every $U \in \mathcal{X}$, there exists $V \in \mathcal{X}$ such that, whenever $(x,y) \in V$ and $(y,z) \in V$ for some x, y, z, then $(x,z) \in U$.

A topological vector space E is equipped with the following uniformity: the class $\mathcal{X}$ consists of all sets in $E \times E$ containing a set of the form $\{(x,y)\colon\ x-y \in U\}$, where U is a neighborhood of zero in E. Every metric space (not necessarily vector) also possesses a natural uniformity formed by all sets containing subsets of the form $\{(x,y)\colon\ d(x,y) < r\}$, where $r > 0$ and d is the metric of the given space.

Thus, this new category of spaces embraces both topological vector spaces (and also topological groups) and metric spaces without any algebraic structure.

On the other hand, every uniform space X can be equipped with a topology generated by the uniformity in the following way: the base of neighborhoods (not necessarily open) of the point x in this topology consists of the sets of the form $U(x) := \{y\colon\ (x,y) \in U\}$, where $U \in \mathcal{X}$. Thus, a set W is open if every point $w \in W$ is contained in W along with a neighborhood of such a form. However, not every topology can be obtained from some uniformity: it is known that a topology is generated by a uniformity precisely when the space is completely regular (see Engelking [**154**, Chapter 8]).

A subset A of a uniform space $(X, \mathcal{X})$ is equipped with the *induced uniformity* consisting of the intersections $A \times A$ with the sets from $\mathcal{X}$.

In uniform spaces some concepts encountered above become meaningful: Cauchy (fundamental) nets, completeness, precompactness. A net $\{x_t\}$ in a uniform space $(X, \mathcal{X})$ is called Cauchy if, for every $U \in \mathcal{X}$, there exists an index t_0 such that $(x_t, x_s) \in U$ for all $t, s \geqslant t_0$. Convergence of a net is understood as convergence in the generated topology. Hence one can introduce the notion of a complete uniform space similarly to that of a topological vector space (in the category of uniform spaces also arise completions). Similarly we introduce the notion of a precompact or totally bounded set: this is a set $A \subset X$ such that for every $U \in \mathcal{X}$ there exists a finite cover of the set A by sets $A_1, \ldots, A_n$ with the property that $A_i \times A_i \subset U$.

By means of basis neighborhoods $U(x)$ this can be expressed as follows: there exist $a_1, \ldots, a_n \in A$ such that $A \subset \bigcup_{i=1}^n U(a_i)$.

For a topological vector space these notions coincide with earlier defined. In more detail uniformities are discussed in the books Bourbaki [**88**], Edwards [**150**], Engelking [**154**], and Kelley [**268**]. We mention here only few facts useful in relation with the results discussed, especially connected to completeness and compactness.

As in the case of topological vector spaces, compactness in a uniform space is equivalent to precompactness and completeness.

1.12.1. Lemma. *Let $(X, \mathcal{X})$ be a uniform space.*

(i) *Every set $A \subset X$ with the property that each its infinite sequence has a limit point in X is precompact. In particular, this is true if A is countably compact.*

(ii) *A set A in X has compact closure precisely when its closure is complete and every infinite sequence in A possesses a limit point in X.*

PROOF. (i) If A is not precompact, then there exist an entourage $U \in \mathcal{X}$ and an infinite sequence of points $a_n \in A$ for which $a_{n+1} \notin \bigcup_{i=1}^{n} U(a_i)$, where $U(a) := \{x\colon\ (a, x) \in U\}$. Thus, $(a_i, a_{n+1}) \notin U$ whenever $i \leqslant n$. By assumption this sequence has a limit point $p \in X$. There is an entourage $V \in \mathcal{X}$ with the property that if $(x, y) \in V$ and $(y, z) \in V$, then $(x, z) \in U$. Passing to $V \cap V^{-1}$, we can assume that $V = V^{-1}$. There is also some element a_m in the neighborhood $V(p)$, i.e., $(p, a_m) \in V$, hence $(a_m, p) \in V$. Then for all $n > m$ we obtain $a_n \notin V(p)$, since the inclusion $a_n \in V(p)$, which means that $(p, a_n) \in V$, implies the inclusion $(a_m, a_n) \in U$, which is false if $m < n$.

(ii) It is easy to verify that any compact set in the topology generated by a uniformity is complete in the sense of uniform spaces. In addition, any compact set is countably compact. If the set A has a complete closure and each its infinite sequence possesses a limit point, then A is precompact, which implies the precompactness of its closure, which by completeness is compact. $\square$

In particular, this assertion is true for topological vector spaces. We draw the reader's attention to the important circumstance that assertion (ii) *does not mean* the equivalence of the countable compactness and compactness (unlike the case of metric spaces). The point is that the countable compactness does not imply completeness (but for complete sets equivalence holds by Proposition 1.8.5).

Let us consider such examples (recall that the sequential compactness implies the countable compactness, but does not imply compactness and also does not follow from compactness); we shall be interested in sets in locally convex spaces.

1.12.2. Example. (i) We construct a noncompact absolutely convex closed sequentially compact set V in a sequentially complete locally convex space.

In the space $\mathbb{R}^{\mathbb{R}}$ take the subspace E of all functions that are nonzero in at most countably many points. This subspace is sequentially complete, and the required set V consists of all functions $x \in E$ satisfying the condition $\sup_t |x(t)| \leqslant 1$. The sequential compactness follows from the fact that every sequence of functions in E is concentrated on a countable set, and the noncompactness of V is clear from that the closure of V in $\mathbb{R}^{\mathbb{R}}$ consists of all functions x with $\sup_t |x(t)| \leqslant 1$.

(ii) We construct an absolutely convex sequentially compact set V in a locally convex space whose closure is not countably compact.

Let us take for E the linear subspace in $\mathbb{R}^{\mathbb{R}}$ consisting of all functions x such that for some $n \in \mathbb{N}$ the set $\{t \in [n, \infty)\colon\ x(t) \neq 0\}$ is at most countable. Let V be the set of all functions $x \in E$ that are nonzero in at most countably many points and satisfy the inequality $\sup_t |x(t)| \leqslant 1$. This absolutely convex set is sequentially compact, but its closure is not, moreover, the closure is not countably compact since it contains the sequence of functions $x_n = I_{(-\infty,n]}$ (the indicators of $(-\infty, n]$), which has no limit points in E, being converging in the space $\mathbb{R}^{\mathbb{R}}$ to the function $x(t) = 1$ that does not belong to E.

1.12(ii). Convex compact sets

1.12.3. Definition. *A point x of a set K in a real linear space E is called an extreme point of this set if it is not contained in any interval (a, b) entirely contained in K.*

If the set K is convex, then a vector $x \in K$ is an extreme point precisely when the equality $x = ta + (1-t)b$, where $t \in [0, 1]$ and $a, b \in K$, yields that $x = a$ or $x = b$.

More generally, a subset A of a convex set K is called an *extreme subset* of the set K if the equality $ta + (1-t)b \in A$ for some $t \in (0, 1)$ and some $a, b \in K$ yields that $a, b \in A$. Obviously, if an extreme set consists of a single point, then this point is extreme.

In the three-dimensional Euclidean space the vertices of a closed cube are its extreme points and its edges and faces are its extreme sets. Inner points of edges and faces are not extreme points. An open cube has no extreme points.

1.12.4. Lemma. *Let E be a real locally convex space, let $K \subset E$ be a nonempty compact set, and let f be a continuous linear functional on E. Set*

$$\alpha := \sup\{f(x)\colon\ x \in K\}.$$

Then $A = \{x \in K\colon\ f(x) = \alpha\}$ is a nonempty extreme subset of the set K.

PROOF. Since K is compact and the functional f is continuous, there exists $x \in K$ such that $f(x) = \alpha$, i.e., the set A is nonempty. Suppose that for some $a, b \in K$, $t \in (0, 1)$ we have $ta + (1-t)b \in A$. We show that $a, b \in A$. If, for example, $f(a) < f(b)$, then $f(a) < \alpha$, hence

$$f(ta + (1-t)b) = tf(a) + (1-t)f(b) < t\alpha + (1-t)\alpha = \alpha,$$

which contradicts the inclusion $ta + (1-t)b \in A$. □

The next result is classical in convex analysis.

1.12.5. Theorem. (THE KREIN–MILMAN THEOREM) *Every nonempty convex compact set in a real Hausdorff locally convex space is the closed convex hull of the set of its extreme points.*

PROOF. We show first that every nonempty compact set K in a Hausdorff locally convex space E possesses extreme points. We prove that K contains at least one extreme point. Denote by $\mathcal{M}$ the set of all nonempty closed extreme subsets of K ordered in the following way: if $A, B \in \mathcal{M}$, then $A \leqslant B \iff A \supset B$. Since $K \in \mathcal{M}$, we have $\mathcal{M} \neq \varnothing$.

Furthermore, if $\mathcal{M}_0 \subset \mathcal{M}$ is a linearly ordered part, then the set $\bigcap_{A \in \mathcal{M}_0} A$ is nonempty as the intersection of a family of compact subsets every finite subfamily of which has a nonempty intersection. Thus, $\mathcal{M}$ satisfies the hypotheses of the Kuratowski–Zorn theorem (it follows directly from the definition of an extreme set that the intersection of any family of extreme sets is an extreme set). Hence $\mathcal{M}$ has maximal elements.

Let B be one of them. Thus, B is a nonempty compact extreme subset in K such that no proper closed subset of B is extreme in K. We now prove that B contains precisely one point; this point will be extreme. Suppose that $a, b \in B$, $a \neq b$. The Hahn–Banach theorem yields that there exists a continuous linear functional f on E such that $f(a) < f(b)$. By the lemma proven above the set $B_1 = \{x \in B\colon f(x) = \max_{z\in B} f(z)\}$ is an extreme subset of the extreme subset B of K, hence is an extreme subset of K itself. At the same time $a \notin B_1$, so that B_1 is a proper subset of B. This contradicts the maximality of B.

Let now K be a convex compact set in a locally convex space E and let C be the set of all extreme points of K. We have to prove that $\overline{\operatorname{conv}} C = K$. To this end it suffices to show that $K \subset \overline{\operatorname{conv}} C$ (the inverse inclusion follows immediately by the convexity and closedness of K and the inclusion $C \subset K$). Let $a \in K$, $a \notin \overline{\operatorname{conv}} C$. By the Hahn–Banach theorem, there exists a continuous linear functional g on E such that $g(a) > \max\{g(x)\colon x \in \overline{\operatorname{conv}} C\}$. Then the set

$$\{x \in K\colon g(x) = \max_{x\in K} g(x)\}$$

is extreme in K. As shown above, it contains some extreme point, which at the same time is an extreme point of K, being an extreme point of an extreme subset of K. Let b be one of such points, then $b \notin \overline{\operatorname{conv}} C$, since

$$g(b) \geqslant g(a) > \max\{g(x)\colon x \in \overline{\operatorname{conv}} C\}.$$

Hence $b \notin C$. Thus, we have obtained a contradiction. This theorem is not obvious even in $\mathbb{R}^n$. □

The following result of D.P. Milman is also of interest.

1.12.6. Theorem. *Let K be a compact set in a locally convex space E such that its closed convex hull C is compact (which holds automatically if E is quasi-complete). Then every extreme point of the set C belongs to K.*

PROOF. Let x be an extreme point of C and let U be a convex neighborhood of zero. There exists points $x_1, \ldots, x_n$ in K with $K \subset \bigcup_{i=1}^n (x_i + U)$. Let V_i be the closed convex hull of $K \cap (x_i + U)$. The sets V_i are contained in C and hence are compact. Therefore, the convex hull of their union is also compact and hence equals C. Thus, $x = \sum_{i=1}^n \lambda_i v_i$, where $v_i \in V_i$, $\lambda_i \geqslant 0$ and $\sum_{i=1}^n \lambda_i = 1$. Since x is an extreme point of C, we have $x = v_i$ for some i, whence we obtain $x \in x_i + U \subset K + U$. Since U was arbitrary and K is closed, we obtain $x \in K$. Finally, let us recall that by Proposition 1.8.13 in a quasi-complete space the closed convex hull of any compact set is compact. □

1.12.7. Example. (i) In the space c_0 the closed ball of a positive radius has no extreme points. It follows that c_0 is not the dual to any Banach space. More precisely: there is no Banach space the dual of which can be linearly and isometrically mapped onto c_0. Indeed, if c_0 could be identified as a Banach space with the dual to some Banach space E (equipped with the dual norm), then the unit ball in c_0 would be compact in the topology $\sigma(c_0, E)$ by Theorem 3.1.4, hence the Krein–Milman theorem could be applied to it. Moreover, the space c_0 cannot be linearly homeomorphic to the dual of a Banach space, but this is proved by a different method (see Exercise 1.12.87).

(ii) More generally, an infinite-dimensional Banach space B is not linearly isometric to the dual of a Banach space if its closed unit ball centered at zero has only finitely many extreme points: in this case their closed convex hull is finite-dimensional. Since the unit ball of the space $C[a,b]$ of continuous functions on $[a,b]$ with the norm $\max_t |x(t)|$ has precisely two extreme points ($\varphi(t) = 1$ and $\psi(t) = -1$), then $C[a,b]$ is not a dual Banach space. Moreover, $C[a,b]$ is not even linearly homeomorphic to a dual space (see Exercise 1.12.87).

On convex compact sets, see also § 5.6.

1.12(iii). Fixed point theorems

For mappings of convex compact sets there are important fixed point theorems (see also Exercise 1.12.86), which generalize the classical Bohl–Brouwer theorem for $\mathbb{R}^n$ (see Dunford, Schwartz [**145**, p. 468]).

1.12.8. Theorem. (SCHAUDER–TYCHONOFF THEOREM) *If E is a Hausdorff locally convex space, $K \subset E$ is a convex compact set and $f\colon K \to K$ is a continuous mapping, then there exists an element $a \in K$ such that $f(a) = a$ (this element is called a fixed point of the mapping f).*

PROOF. Let $\mathcal{P}$ be the family of all continuous seminorms on E. We show that for every seminorm $p \in \mathcal{P}$ the set $F_p := \bigl\{z \in K\colon p\bigl(f(z) - z\bigr) = 0\bigr\}$ is nonempty. If this is done, then it remains to observe that the sets F_p are closed (by the continuity of f and p), hence they are compact, moreover, for every finite collection $p_1, \ldots, p_n$ the intersection $F_{p_1} \cap \cdots \cap F_{p_n}$ is nonempty, since $p_1 + \cdots + p_n \in \mathcal{P}$. Hence the intersection of all F_p is nonempty as well, and any its element is a fixed point.

Thus, we fix $p \in \mathcal{P}$. It suffices to verify that for every $\varepsilon > 0$ there exists a point z_ε with $p\bigl(f(z_\varepsilon) - z_\varepsilon\bigr) < \varepsilon$, since then the sequence $\{z_{1/n}\}$ has a limit point $z \in K$, for which by the continuity of f and p we obtain $p\bigl(f(z) - z\bigr) = 0$. By the compactness of K there exist points $a_1, \ldots, a_n \in K$ such that the sets $U_i := \{x\colon p(x - a_i) < \varepsilon/2\}$ cover K. For every $i = 1, \ldots, n$ we define a function ψ_i on E as follows: $\psi_i(x) = \varepsilon - p(x - a_i)$ if $p(x - a_i) \leqslant \varepsilon$, $\psi_i(x) = 0$ if $p(x - a_i) > \varepsilon$. It is readily seen that the functions ψ_i are continuous, moreover, at every point of K at least one of them does not vanish. Therefore, the functions $\alpha_i(x) := \psi_i(x)/\Psi(x)$, $\Psi(x) = \sum_{j=1}^n \psi_j(x)$ are continuous on the set K. We observe that $0 \leqslant \alpha_i(x) \leqslant 1$ and $\sum_{i=1}^n \alpha_i(x) = 1$ on K. The mapping

$$g(x) := \sum_{i=1}^n \alpha_i\bigl(f(x)\bigr) a_i$$

is also continuous and takes K to the finite-dimensional convex set V generated by the points $a_1, \ldots, a_n$. By the aforementioned Bohl–Brouwer theorem for $\mathbb{R}^n$ there exists a point $y \in V \subset K$ with $g(y) = y$. We have

$$f(y) - y = f(y) - g(y) = \sum_{i=1}^{n} \alpha_i(f(y))[f(y) - a_i],$$

since $\sum_{i=1}^{n} \alpha_i\bigl(f(y)\bigr) = 1$. Therefore,

$$p\bigl(f(y) - y\bigr) \leqslant \sum_{i=1}^{n} \alpha_i\bigl(f(y)\bigr)p\bigl(f(y) - a_i\bigr).$$

In the latter sum we have $\alpha_i\bigl(f(y)\bigr) = 0$ whenever $p\bigl(f(y) - a_i\bigr) > \varepsilon$. Therefore, $p\bigl(f(y) - y\bigr) \leqslant \varepsilon$, as required. □

1.12.9. Corollary. *Let V be a closed convex set in a complete (or quasi-complete) locally convex space and let $f\colon V \to V$ be a continuous mapping such that $f(V)$ has compact closure. Then, there is an element $a \in V$ such that $f(a) = a$.*

PROOF. The closed convex hull of $f(V)$ is compact by the completeness of E (quasi-completeness is also enough, see Proposition 1.8.13), moreover, it is mapped into itself by f. By the previous theorem there is a fixed point in it. □

The following Markov–Kakutani theorem deals with affine mappings.

1.12.10. Theorem. *Let K be a convex compact set in a Hausdorff topological vector space E and let G be a family of pairwise commuting continuous mappings from K to K such that*

$$g\bigl(\lambda x + (1-\lambda)y\bigr) = \lambda g(x) + (1-\lambda)g(y) \quad \textit{for all } x, y \in K,\ g \in G.$$

Then, there exists $x_0 \in K$ with $g(x_0) = x_0$ for all $g \in G$.

PROOF. For every $g \in G$ and $n \in \mathbb{N}$ we define a mapping $g_n\colon K \to K$ by the formula $g_n = n^{-1}(I+g+g^2+\cdots+g^{n-1})$. This is a continuous affine mapping. Let us consider the set G_0 of transformations of K that are compositions of finitely many mappings of the form g_n for all possible elements $g \in G$ and $n \in \mathbb{N}$. It is clear that the class G_0 satisfies the same conditions as G. We show that the intersection of the compact sets $f(K)$, where $f \in G_0$, is nonempty. To this end, it suffices to observe that the intersection of every finite collection of such sets is nonempty, since if $f_1, \ldots, f_m \in G_0$, then $f = f_1 \circ \cdots \circ f_m \in G_0$ and $f(K) \subset f_i(K)$ for all $i \leqslant m$ by the commutativity of f_i. Let us take any element x_0 in the indicated intersection and show that it is the required one. Fix $g \in G$. For every n, by construction there exists an element $y \in K$ such that

$$x_0 = g_n(y) = n^{-1}\bigl(y + g(y) + \cdots + g^{n-1}(y)\bigr).$$

Hence $g(x_0) = n^{-1}\bigl(g(y) + g^2(y) + \cdots + g^n(y)\bigr)$, whence we obtain

$$g(x_0) - x_0 = n^{-1}\bigl(g^n(y) - y\bigr) \in n^{-1}(K - K).$$

Since this inclusion is true for all n, we have $g(x_0) = x_0$ by the condition that E is Hausdorff, because for every balanced neighborhood of zero V and every compact set Q there exists n such that $n^{-1}Q \subset V$ by the boundedness of Q. □

Kakutani showed that the commutativity of the mappings can be omitted if they are equicontinuous. This holds if E is a normed space and $G \subset \mathcal{L}(E)$ is bounded. For a proof, see Dunford, Schwartz [**145**, p. 457].

1.12.11. Theorem. *Let K be a convex compact set in a Hausdorff locally convex space E and let G be a group of linear mappings from K to K equicontinuous on K. Then, there exists $x_0 \in K$ with $g(x_0) = x_0$ for all $g \in G$.*

The following theorem due to Kakutani–Ky Fan generalizes the Schauder–Tychonoff theorem to multivalued mappings (the included proof gives also another, though close, justification of the latter).

1.12.12. Theorem. *Let K be a convex compacts set in a Hausdorff locally convex space E and let Φ be a multivalued mapping associating to every $x \in K$ a nonempty convex compact set $\Phi(x) \subset K$. Suppose that Φ is upper semicontinuous in the sense that if $x \in K$ and an open set U contains $\Phi(x)$, then there is a neighborhood V of the point x such that $\Phi(v) \subset U$ for all $v \in V \cap K$. Then, there exists a point $x_0 \in K$ such that $x_0 \in \Phi(x_0)$.*

PROOF. We slightly modify our proof of the Schauder–Tychonoff theorem. For any fixed seminorm $p \in \mathcal{P}$ and $r > 0$ let $K_{p,r}$ denote the set of points $z \in K$ such that $z \in \Phi(z) + r\overline{V}$, $V = \{p < 1\}$. This set is closed, which is easily deduced from the upper semicontinuity of Φ. If all sets $K_{p,r}$ are nonempty, then, as in the proof of the Schauder–Tychonoff theorem, their intersection will be nonempty. Using the closedness of $\Phi(x)$ we then obtain that any element in this intersection is the required one. We now verify that $K_{p,r}$ is nonempty. By the upper semicontinuity there is a neighborhood of zero $W = \{q < 1\}$, where $q \in \mathcal{P}$, such that

$$\Phi(x_0 + w) \subset \Phi(x_0) + rV/2 \quad \text{for all } w \in W.$$

Set $U = W \cap V$. Let $\varepsilon > 0$. Let us take $a_1, \ldots, a_n$ such that

$$K \subset \bigcup_{i=1}^{n} (a_i + \varepsilon 2^{-1} U).$$

Take any $y_i \in \Phi(a_i)$ and continuous bump functions α_i corresponding to the neighborhoods $a_i + \varepsilon 2^{-1} U$ as in the proof of the Schauder–Tychonoff theorem. The mapping $g_\varepsilon(x) = \sum_{i=1}^n \alpha_i(x) y_i$ is continuous and has a fixed point x_ε. Let x_0 be a limit point of $\{x_{1/m}\}$. Let us show that $x_0 \in K_{p,r}$. We shall deal only with indices m for which $x_{1/m} \in x_0 + U$. Among them there exists $m > 4/r$ with $x_{1/m} \in x_0 + U/4$. If $\alpha_i(x_{1/m}) > 0$, then $x_{1/m} \in a_i + m^{-1}U$, whence $x_0 - a_i \in W$ and $y_i \in \Phi(x_0) + rV/2$. Hence $x_{1/m}$ is a convex combination of the vectors $\alpha_i(x_{1/m}) y_i$, where $y_i \in \Phi(x_0) + rV/2$. Hence $x_0 \in \Phi(x_0) + rV$. □

1.12.13. Corollary. *Let S be one more convex compact set in E and let a continuous function f on $K \times S$ be such that the functions $x \mapsto f(x, y)$ are convex and the functions $y \mapsto f(x, y)$ are concave, i.e., the functions $y \mapsto -f(x, y)$ are convex. Then*

$$\min_{x \in K} \max_{y \in S} f(x, y) = \max_{y \in S} \min_{x \in K} f(x, y).$$

PROOF. Let us apply the previous theorem to the map $\Phi\colon (x,y)\mapsto (A_y, B_x)$, where

$$A_y = \{u \in K\colon\ f(u,y) = \min_{k\in K} f(k,y)\},$$
$$B_x = \{z \in S\colon\ f(x,z) = \max_{s\in S} f(x,s)\}.$$

It follows from our assumptions that A_y and B_x are nonempty, convex and closed. We verify that Φ is upper semicontinuous. Let $x \in K$ and let B_x belong to an open set U. The function $g(u) = \max_{s\in S} f(u,s)$ is continuous on K. If there is no neighborhood $V \ni x$ such that $B_v \subset U$ for all $v \in V$, then there exists a net $x_\alpha \to x$ for which there are $y_\alpha \in S\backslash U$ with $f(x_\alpha, y_\alpha) = g(x_\alpha)$. By the compactness of $S\backslash U$ we can assume that $y_\alpha \to y \in S\backslash U$, passing to a subnet. Then $f(x,y) = g(x)$ contrary to the fact that $y \notin U$. The reasoning with A_y is similar. Thus, the previous theorem gives a point (x_0, y_0) for which

$$f(x_0,y_0) = \min_{k\in K} f(k,y_0) = \max_{s\in S} f(x_0,s).$$

This is the required point, since the left side of the equality to be proven is always not smaller than the right side by the inequality $f(x,y) \geqslant \min_{s\in S} f(x,s)$ for all elements $x \in K$, $y \in S$, and the previous relation gives the desired estimates $f(x_0,y_0) \leqslant \max_{s\in S}\min_{k\in K} f(k,s)$, $f(x_0,y_0) \geqslant \min_{k\in K}\max_{s\in S} f(k,s)$. □

1.12(iv). Sequence spaces

In addition to the already encountered spaces of all sequences $\mathbb{R}^\infty$ and rapidly decreasing sequences Σ (Example 1.3.19) and Banach spaces l^p and c_0, in the theory and applications some other locally convex spaces of sequences are used.

1.12.14. Example. The important class of Köthe spaces of sequences is constructed by the following method. A set P of real sequences $\alpha = (\alpha_n)$ is called a Köthe set if $\alpha_n \geqslant 0$ and for every $n \in \mathbb{N}$ there exists $\alpha \in \mathrm{P}$ with $\alpha_n > 0$, and for every two elements $\alpha, \beta \in \mathrm{P}$ there exists $\gamma \in \mathrm{P}$ with $\max(\alpha_n, \beta_n) \leqslant \gamma_n$ for all n. This Köthe set is used to form the Köthe space of sequences (real or complex)

$$\Lambda(\mathrm{P}) := \bigl\{x = (x_n)\colon\ (\alpha_n x_n) \in l^1\ \ \forall \alpha \in \mathrm{P}\bigr\}.$$

A natural collection of seminorms on the Köthe space $\Lambda(\mathrm{P})$ consists of the functions $p_\alpha(x) = \sum_{n=1}^\infty |\alpha_n x_n|$.

For example, if P consists of the single sequence of units, then $\Lambda(\mathrm{P}) = l^1$, and if we take all sequences of the form $\alpha_n = n^k$, $k \in \mathbb{N}$, then we obtain the space Σ from Example 1.3.19.

Various classes of abstract spaces can be often described by means of Köthe spaces and their subspaces (see Jarchow [**237**], Meise, Vogt [**339**], Pietsch [**388**]). Let us give one important and typical example.

Let $C^\infty[0,1]$ be the space of all infinitely differentiable functions on $[0,1]$, let $C_0^\infty[0,1]$ be its subspace consisting of functions vanishing with all derivatives at the endpoints, and let $C_{2\pi}^\infty$ be the space of all infinitely differentiable functions f on $[0,2\pi]$ for which $f^{(k)}(0) = f^{(k)}(2\pi)$ for all $k \geqslant 0$. In all these spaces we consider the topology of uniform convergence of all derivatives.

1.12.15. Theorem. *The spaces $\mathcal{S}(\mathbb{R}^1)$, $C^\infty_{2\pi}$, $C^\infty[0,1]$ and $C^\infty_0[0,1]$ are isomorphic to the space Σ. The same is true for similarly defined spaces of functions of n variables.*

PROOF. For simplicity we consider the one-dimensional case. An isomorphism between Σ and $C^\infty_{2\pi}$ is defined by means of the Fourier expansion:

$$f(t) = a_0 + \sum_{n=1}^{\infty} [a_n \cos nt + b_n \sin nt], \quad j(f) = (a_0, a_1, b_1, a_2, b_2, \ldots).$$

Isomorphisms with other spaces are constructed in a more complicated way, but also by means of suitable bases (see details in Jarchow [**237**], Pietsch [**388**], and Vladimirov [**549**]). For example, in the space $\mathcal{S}(\mathbb{R}^1)$ one can use the basis of Hermite functions. □

1.12(v). Duals to Banach spaces

For most of Banach spaces used in applications sufficiently explicit descriptions of the dual spaces are known. Here we mention a number of such results; their proofs can be found in many texts (including [**65**] and [**72**]).

The following spaces are Banach:

l^∞ is the space of bounded sequences (real or complex) with finite norm $x = (x_n) \mapsto \sup_n |x_n|$;

c_0 is the closed subspace in l^∞ consisting of all sequences with $\lim\limits_{n\to\infty} x_n = 0$;

l^p, where $1 \leqslant p < \infty$, is the space of sequences $x = (x_n)$ (real or complex) for which

$$\|x\|_p := \Bigl(\sum_{n=1}^{\infty} |x_n|^p\Bigr)^{1/p} < \infty;$$

the function $x \mapsto \|x\|_p$ is a norm on l^p;

the space l^2 is even Hilbert with the scalar product

$$(x, y) = \sum_{n=1}^{\infty} x_n \overline{y_n};$$

the space $C_b(T)$ of bounded continuous functions on a topological space T with the norm $\|x\| = \sup_{t\in T} |(x(t)|$;

the space $L^p(\mu)$, where $1 \leqslant p < \infty$ and μ is a nonnegative measure on a space $(T, \mathcal{A})$, consists of the equivalence classes ($f \sim g$ if $f = g$ almost everywhere) of measurable functions f for which the function $|f|^p$ is integrable with respect to μ; the norm on $L^p(\mu)$ is given by the formula

$$\|f\|_p = \Bigl(\int_T |f(t)|^p \, \mu(dt)\Bigr)^{1/p};$$

this expression does not depend on our choice of a representative in a equivalence class; in a special way one defines the space $L^\infty(\mu)$ consisting of equivalence classes of measurable functions having a bounded modification; we set

$$\|f\|_\infty = \inf_{\widetilde{f}\sim f} \sup_{t\in T} |\widetilde{f}(t)|.$$

There are canonical isomorphisms

$$c_0' = l^1,\ (l^1)' = l^\infty,\ (L^p)' = L^q,\ p^{-1}+q^{-1}=1,\ p\in[1,\infty),$$

and the general form of a continuous functional is this:

on c_0: $l(x) = \sum_{n=1}^\infty x_n y_n,\ (y_n)\in l^1$,
on l^1: $l(x) = \sum_{n=1}^\infty x_n y_n,\ (y_n)\in c_0$,
on l^p, $p\in[1,\infty)$: $l(x) = \sum_{n=1}^\infty x_n y_n,\ (y_n)\in l^q,\ p^{-1}+q^{-1}=1$,
on $L^p(\mu)$, $p\in(1,\infty)$: $l(x) = \int_T x(t)y(t)\,\mu(dt),\ y\in L^q(\mu),\ p^{-1}+q^{-1}=1$,
on $L^1(\mu)$ for a finite or σ-finite measure:

$$l(x) = \int_T x(t)y(t)\,\mu(dt), \quad y\in L^\infty(\mu),$$

on $C(K) = C_b(K)$ for compact K: $l(x) = \int_T x(t)\,\mu(dt)$, where μ is a bounded Borel measure on K.

The duals to l^∞ and $L^\infty(\mu)$, where μ is Lebesgue measure, are not isomorphic to l^1 and $L^1(\mu)$, respectively; they can be described by means of finitely additive measures.

1.12(vi). Separability properties

Here we prove two useful results for separable spaces.

1.12.16. Proposition. *Let E and F be topological vector spaces such that F is metrizable and let H be an equicontinuous set in $\mathcal{L}(E,F)$. If E is separable, then the uniform structure of pointwise convergence in H is metrizable. If F is also separable, then H is separable in the topology of pointwise convergence.*

PROOF. Let $\{a_n\}$ be a countable everywhere dense set in E and let d be a translation invariant metric on F defining its topology. Set

$$d_0(f,g) = \sum_{n=1}^\infty 2^{-n}\min\Bigl(1, d\bigl(f(a_n),g(a_n)\bigr)\Bigr), \quad f,g\in H.$$

For a neighborhood of zero $U=\{f\colon\ d\bigl(f(x_i),0\bigr)<\varepsilon,\ i=1,\ldots,n\}$ in $\mathcal{L}(E,F)$, where $x_i\in E$ and $\varepsilon>0$, we can find an open ball W in the metric d_0 centered at zero such that if $f,g\in H$ and $f-g\in W$, then $f-g\in U$. Indeed, the equicontinuity of H gives a neighborhood of zero $V\subset E$ such that $d\bigl(f(v),0\bigr)<\varepsilon/4$ for all $v\in V$ and $f\in H$. Let us take k such that the points $x_1,\ldots,x_n$ are covered by the sets $a_1+V,\ldots,a_k+V$. Finally, let $r=2^{-k}\varepsilon/4$. If now $f,g\in H$ and $d_0(f,g)<r$, then we obtain $d\bigl(f(x_i),g(x_i)\bigr)<\varepsilon$ for all $i\leqslant n$. Indeed, for every $i\leqslant n$ there exists a point a_j with some $j\leqslant k$ such that $x_i\in a_j+V$, whence on account of the equality $d(u,v)=d(u-v,0)$ we find that

$$\begin{aligned} d\bigl(f(x_i),g(x_i)\bigr) &\leqslant d\bigl(f(x_i),f(a_j)\bigr)+d\bigl(f(a_j),g(a_j)\bigr)+d\bigl(g(a_j),g(x_i)\bigr)\\ &\leqslant \varepsilon/4+2^k r+\varepsilon/4<\varepsilon. \end{aligned}$$

It is also clear that every ball W in the metric d_0 centered at zero contains a neighborhood of zero U of the indicated form.

The second assertion follows from the fact that the countable power of a separable metric space is separable. □

1.12.17. Proposition. *Let E be a separable locally convex space E and let p be a continuous seminorm on E. Set $U = \{x\colon\ p(x) \leqslant 1\}$. Then, there exists a sequence $\{f_n\} \subset U^\circ := \{f \in E'\colon\ \sup_{u\in U} |f(u)| \leqslant 1\}$ such that*

$$p(x) = \sup_n |f_n(x)|, \quad x \in E.$$

PROOF. The space E contains an everywhere dense sequence $\{x_n\}$. For every n by the Hahn–Banach theorem we have $p(x_n) = \sup\{|f(x_n)|\colon\ f \in U^\circ\}$. Hence U° contains a sequence f_{nm} such that $p(x_n) = \sup_m |f_{nm}(x_n)|$. The constructed countable set can be enumerated as a single sequence $\{f_n\}$. It is clear that we have

$$q(x) := \sup_n |f_n(x)| \leqslant p(x) \quad \text{for all } x \in E$$

and that $q(x_n) = p(x_n)$ for all n. On the other hand, for every $x \in E$ for any fixed $\varepsilon > 0$ there exists an element x_m with $p(x - x_m) < \varepsilon$, hence

$$\begin{aligned} p(x) &< p(x_m) + \varepsilon = q(x_m) + \varepsilon \leqslant q(x_m - x) + q(x) + \varepsilon \\ &\leqslant p(x_m - x) + q(x) + \varepsilon \leqslant q(x) + 2\varepsilon, \end{aligned}$$

which gives the estimate $p(x) \leqslant q(x)$, since ε was arbitrary.

A shorter justification follows from the results in § 3.2: the set U° is metrizable compact in the topology $\sigma(E', E)$, hence taking a countable everywhere dense subset $\{f_n\}$, we obtain the desired representation. Indeed, $p \leqslant \sup_n |f_n|$. On the other hand, if $x \in U$, then there is $f \in E'$ with $f(x) = p(x)$ and $|f| \leqslant p$, i.e., $f \in U^\circ$, which gives a subsequence $\{n_i\}$ with $p(x) = f(x) = \lim\limits_{i\to\infty} f_{n_i}(x)$. □

1.12.18. Proposition. *Let E be a normed space, let F be a norm separable linear subspace in E', and let ξ be a norm continuous linear function on F. Then, there is a sequence $\{x_n\} \subset E$ such that*

$$\xi(f) = \lim_{n\to\infty} f(x_n) \quad \forall\, f \in F.$$

PROOF. We can assume that $\|\xi\| \leqslant 1$. Let us take a sequence of elements $\{f_k\} \subset F$ of unit norm the linear span of which is dense in F. Next, by induction we construct a sequence $\{x_n\}$ in the unit ball U in E such that $\xi(f_k) = \lim\limits_{n\to\infty} f_k(x_n)$ for each k. Set $x_1 = 0$. If $x_1, \ldots, x_n$ are already found, we take $x_{k+1} \in U$ such that

$$|\xi(f_i) - f_i(x_{k+1})| \leqslant k^{-1} \quad \text{for } i = 1, \ldots, k+1.$$

This is possible, since by the Hahn–Banach theorem there is an element $\eta \in E''$ for which $\|\eta\| \leqslant 1$ and $\xi = \eta|_F$. Moreover, η belongs to the closure of the image of U under the canonical embedding $E \subset E''$ in the topology $\sigma(E'', E')$. Otherwise we could find an element $g \in E'$ such that $|g(u)| \leqslant 1$ for all $u \in U$ and $\eta(g) > 1$; then we would have $\|g\| \leqslant 1$, hence $\|\eta\| > 1$. Therefore, the neighborhood of

η defined by the inequalities $|\zeta(f_i) - \eta(f_i)| < k^{-1}$, $i = 1, \ldots, k+1$, contains a vector from the image of U, so that the corresponding vector from U can be taken for x_{k+1}. Since the linear span of $\{f_k\}$ is dense in F, we obtain that $\xi(f) = \eta(f) = \lim_{n\to\infty} f(x_n)$ for all functionals $f \in F$. □

1.12(vii). Continuous selections and extensions

Here we present a number of useful results connected with constructing inverse mappings to non injective mappings and extensions of continuous mappings. Suppose we are given a surjective mapping $f\colon X\to Y$. A mapping $g\colon Y \to X$ is called a right inverse for the mapping f if $f\bigl(g(y)\bigr) = y$ for all $y \in Y$. It is often of interest to have continuous right inverse mappings for continuous mappings, which is closely related to constructing single-valued branches of multi-valued mappings (such single-valued branches are called selections). A classical result in this area is the following *Michael selection theorem* (see Michael [**342**] or Repovš, Semenov [**415**, p. 190]).

1.12.19. Theorem. *Let M be a metrizable space, let P be a complete metrizable closed subset of a locally convex space E, and let $\Phi\colon M \to 2^P$ be a lower semicontinuous mapping with values in the set of nonempty convex closed subsets of P, i.e., for every open set $U \subset P$ the set $\Phi^{-1}(U) := \{x \in M\colon \Phi(x)\cap U \neq \varnothing\}$ is open. Then, there is a continuous mapping $f\colon M \to P$ such that $f(x) \in \Phi(x)$ for all x.*

It was shown by Filippov [**169**], [**170**] that the closedness of P cannot be omitted even if P is a G_δ-set in l^2 (hence is complete with some metric defining the same topology).

1.12.20. Corollary. *Let $T\colon P \to M$ be a continuous affine mapping from a complete metrizable closed convex set P in a locally convex space to a metrizable set M in a locally convex space such that T is open, i.e., takes open sets to open sets. Then T possesses a continuous right inverse mapping.*

PROOF. We verify that the mapping from the space M to 2^P taking x to the set $\Phi(x) := T^{-1}(x)$ satisfies the hypotheses of Michael's theorem. The sets $\Phi(x)$ are closed in P by the continuity of T and are convex, since T is affine. Let us verify that if $U \subset P$ is open, then the set $\Phi^{-1}(U)$ indicated in the theorem is open in M. For every $x \in \Phi^{-1}(U)$, there is $u \in \Phi(x) \cap U$, i.e., $Tu = x$, but $T(U)$ is open in M by assumption, hence the point x is contained in $T(U)$ along with some neighborhood V, whence $V \subset \Phi^{-1}(U)$, which is easily verified. □

The following corollary is also useful.

1.12.21. Corollary. *Let X_0 be a closed linear subspace of a Fréchet space X and let $\pi\colon X \to X/X_0$ be the canonical projection onto the quotient . Then π has a continuous right inverse mapping (possibly, nonlinear).*

1.12.22. Example. In the situation of the previous corollary, for every compact set $K \subset X/X_0$ there exists a compact set $S \subset X$ such that $\pi(S) = K$. Indeed, taking a continuous right inverse mapping φ to π, we can set $S = \varphi(K)$.

In Exercise 1.12.84 the reader is suggested to obtain the last assertion directly, moreover, even without the condition that the space is locally convex. In Corollary 3.9.14 and Example 3.9.15 the result will be extended to general continuous linear surjections of Fréchet spaces.

The next important result is the Dugundji extension theorem [**142**].

1.12.23. Theorem. *Let Z be a closed subset of a metric space M, let E be a locally convex space, and let $f\colon Z \to E$ be a continuous mapping. Then, there exists a continuous mapping $\widetilde{f}$ from the space M to the convex hull of $f(Z)$ such that $\widetilde{f}$ coincides with f on Z.*

A detailed proof can be found in Borsuk [**85**, Chapter III, § 7].

If E is a Fréchet space and extensions may take values in the closure of the convex hull of $f(Z)$ (no closure is taken in the theorem), then this assertion follows at once from Michael's selection theorem: it suffices to set $\Phi(x) = f(x)$ for all $x \in Z$ and for any $x \notin Z$ take for $\Phi(x)$ the closed convex hull of $f(Z)$. It is easy to verify that Φ is lower semicontinuous.

The following corollary is useful in applications.

1.12.24. Corollary. *For every closed convex set V in a metrizable locally convex space E, there exists a continuous mapping $r\colon E \to V$ that is identical on V (such a mapping is called a retraction and V is called a retract).*

Borges [**84**] extended the Dugundji theorem to the broader class of stratifiable spaces (see § 3.12(v)), whence it follows that every closed convex set in a stratifiable locally convex space is its retract. However, not every convex compact set in an arbitrary locally convex space is its retract (see Sipacheva [**467**]).

Exercises

1.12.25°. Show that the topology of a locally convex space can be defined by a family of seminorms of the minimal possible cardinality.

HINT: using Zorn's lemma in a given family of seminorms generating the topology pick a maximal subfamily $\mathcal{M}$ with the property that no seminorm in $\mathcal{M}$ is dominated by a constant multiplied by a sum of finitely many other seminorms in $\mathcal{M}$. Check that any two such maximal families are of the same cardinality (use the fact that the set of finite subsets of an infinite set has the same cardinality as the given set).

1.12.26°. Justify Example 1.3.6.

1.12.27°. Give an example of a convergent net in a topological space such that this net consists of countably many elements, but does not contain a convergent subsequence.

HINT: in the space of all functions on $[0, 1]$ with the topology of pointwise convergence take the countable set of functions $\sin(nt)$ and observe that the zero function is its limit point, which gives a convergent net of these functions. Show that no subsequence in $\{\sin(nt)\}$ converges pointwise.

1.12.28°. Prove that every ultrafilter containing a compact set converges to a point of this set.

1.12.29°. Prove that on the topological vector space $\mathcal{L}^0$ from Example 1.3.16 there are no nonzero continuous linear functionals.

1.12.30.° Show that on the space $\mathcal{K}(\mathbb{R}^n)$ introduced in Example 1.3.17 there is no metric with respect to which convergence of sequences coincides with convergence of sequences in the topology $\tau_{\mathcal{K}}$.

HINT: take a nonzero element φ and for each $k \in \mathbb{N}$ take the sequence of functions $\varphi_{k,j}(x) = j^{-1}\varphi(x/k)$ converging to zero in $\mathcal{K}(\mathbb{R}^n)$; observe that one cannot choose $j_k \to \infty$ such that $\varphi_{k,j_k} \to 0$.

1.12.31.° Show that the space $\mathcal{D}(\mathbb{R}^n)$ is not metrizable by considering the functions $\varphi_{j,k}(x) = k^{-1}\varphi(x/j)$, where $\varphi \in \mathcal{D}(\mathbb{R}^n)$, $\varphi \not\equiv 0$, and observing that $\varphi_{j,k} \to 0$ as $k \to \infty$, but $\{\varphi_{j,k_j}\}$ does not converge whenever $k_j \to \infty$.

1.12.32. Let E be a vector space, let $\mathcal{P}_0$ be the set of all seminorms on E, and let $\mathcal{P}_1$ be the set of all quasi-norms on E. Prove that if the algebraic dimension of E is finite or countable, then the topologies in E defined by the family of seminorms $\mathcal{P}_0$ and the family of quasi-norms $\mathcal{P}_1$ coincide.

1.12.33.° Prove that the closure of a set A in a topological vector space E coincides with the intersection $\bigcap_{U\in\mathcal{U}}(A+U)$, where $\mathcal{U}$ is a basis of neighborhoods of zero in E

1.12.34.° Prove that the closure of a linear subspace in a topological vector space is a linear subspace.

1.12.35. Let E_0 be a linear subspace in a real linear space E, let p be a seminorm on E, and let p_0 be a seminorm on E_0 such that $p_0(x) \leqslant p(x)$ for all $x \in E_0$. Prove that there exists a seminorm p_1 on E such that $p_1|_{E_0} = p_0$ and $p_1 \leqslant p$.

HINT: consider the Minkowski functional of the absolutely convex hull of the union of the sets $\{p_0 \leqslant 1\}$ and $\{p_1 \leqslant 1\}$.

1.12.36. (i) Let M be a bounded set in a topological vector space E the topology of which is defined by a translation invariant metric. Prove that M is bounded with respect to this metric. (ii) (A.V. Shaposhnikov) Construct a metric d on l^2 defining the same topology as the usual norm on l^2, but having the property that the unit ball with respect to the usual metric is unbounded with respect to the metric d.

HINT: (ii) take continuous functions $f_n\colon l^2 \to [0,n]$ with supports in the balls of radius 1 centered at $3e_n$ such that $f_n(3e_n) = n$, where $\{e_n\}$ is the standard basis; set $f = \sum_{n=1}^{\infty} f_n$ and consider the metric $d(x,y) = \|F(x) - F(y)\|$, where $F\colon l^2 \to l^2$, $F(x) = \bigl(f(x), x_1, x_2, \ldots\bigr)$, $x = (x_n)$.

1.12.37.° Prove that in any topological vector space any Cauchy sequence is bounded.

1.12.38.° Prove that a set in a locally convex space is bounded precisely when every continuous seminorm is bounded on it.

1.12.39.° Show that any sublinear functional on a vector space with finite values is continuous in the strongest locally convex topology.

1.12.40.° Prove that a convex set in $\mathbb{R}^n$ is closed precisely when its intersections with all closed intervals are closed.

1.12.41.° Let E be a nonzero topological vector space and let Γ be a closed hyperplane in E. Prove that the corresponding closed halfspaces are closed and the corresponding open halfspaces are open, moreover, the closed halfspaces are the closures of the open ones.

1.12.42.° Let V be a convex everywhere dense set in a real topological vector space Prove that for every closed hyperplane H the set $H \cap V$ is dense in H.

1.12.43. (i) Prove that in every locally convex space E on which there is a discontinuous linear functional (for example, in every infinite-dimensional metrizable locally convex space) there exist two disjoint everywhere dense convex sets A and B such that $A \cup B = E$.

(ii) Show that the space $\mathbb{R}_0^\infty$ of all finite sequences (the union of all $\mathbb{R}^n$) with the strongest locally convex topology can be also represented as the union of two disjoint everywhere dense convex sets, although all functionals on this spaces are continuous.

(iii) Show that every Hausdorff separable locally convex space admits the above representation.

HINT: (i) take $A = \{f < 0\}$ and $B = \{f \geqslant 0\}$, where f is a discontinuous linear function. (ii) In $\mathbb{R}_0^\infty$ consider the set A of all vectors with a positive last nonzero coordinate and the set B of all vectors with a non-positive last coordinate. (iii) Modify the construction in (ii) as follows (see Bourbaki [**87**, Chapter II, § 1, Exercise 19]): take a linearly independent countable dense set $\{v_n\}$ in E (one can assume that all functionals on E are continuous, so we can equip E with the strongest locally convex topology), consider the set A_1 of finite sums $\sum c_n v_n$ with $c_N > 0$, where N is the maximal number of nonzero coefficients, take B_1 correspondingly. Finally, apply Theorem 1.10.1.

1.12.44. Prove that in any topological vector space the boundary of any convex set with a nonempty interior is nowhere dense.

1.12.45. Let V be a connected closed set in a Hausdorff topological vector space such that every point $x \in V$ has a closed neighborhood W such that $W \cap V$ is convex. Prove that V is convex.

1.12.46. Justify Remark 1.11.9.

1.12.47.° Prove that the set of inner points of any balanced set in a topological vector space is balanced.

1.12.48.° Let X and Y be topological vector spaces and let $T\colon X \to Y$ be a linear mapping taking some neighborhood of zero to a bounded set. Prove that T is continuous.

1.12.49.° Let a seminorm p on a locally convex space be bounded on some nonempty open set. Prove that p is continuous.

1.12.50.° Prove that a linear mapping A from a topological vector space to $\mathbb{R}^n$ is continuous if its kernel $\operatorname{Ker} A$ is closed.

1.12.51. (i) Let $E = C_0^1(\mathbb{R})$ be the space of all continuously differentiable functions with compact support equipped with the family of seminorms $\sup_t |f'(t)|$, $\sup_t |g(t)f(t)|$, where $g \in C(\mathbb{R})$. Show that E is complete and $l(f) = \sum_{n=1}^\infty f'(n)$ is a discontinuous linear function that is weakly sequentially continuous, i.e., takes sequences weakly converging to zero to sequences weakly converging to zero. (ii) Show that this is impossible in a Banach space with the weak topology.

HINT: (i) if $\{f_\alpha\}$ is a Cauchy net, then it converges uniformly along with derivatives to a function $f \in C^1(\mathbb{R})$. Moreover, f has compact support, since otherwise one can find $g \in C(\mathbb{R})$ with $g(t_n)f(t_n) = 1$ for some sequence $t_n > n$. Hence $f_\alpha \to f$ in E. For a similar reason, any convergent sequence in E has uniformly bounded supports, which yields the sequential continuity of l. However, l is not continuous, since it cannot be estimated via finitely many seminorms of the indicated form. One can also use the space $\mathcal{D}(\mathbb{R}^1)$ with the topology τ_2 from Exercise 2.10.49(ii). (ii) A sequentially continuous functional on a Banach space with the weak topology is norm continuous, hence is weakly continuous.

1.12.52. Prove that a locally convex space is sequentially complete precisely when in this space all sequences fundamental with respect to each seminorm from some family generating its topology converge. This is also equivalent to convergence of all sequences fundamental with respect to each continuous seminorm.

1.12.53. Let B be a nonempty closed convex set in a Hausdorff topological vector space E, let $K \subset E$ be a nonempty compact set, and let $A \subset E$ be a set such that $A + K \subset B + K$. Prove that $A \subset B$. In particular, if A is also nonempty, closed and convex and $A + K = B + K$, then $A = B$.

HINT: it suffices to consider the case $A = \{0\}$; then one can take $k_1 \in K$ and find inductively points $k_n \in K$ and $b_n \in B$ such that $k_n = b_n + k_{n+1}$; observe that we have $n^{-1}(b_1 + \cdots + b_n) = n^{-1}(k_1 - k_{n+1}) \to 0$, hence $0 \in B$, since B is convex and closed.

1.12.54. Let A be a convex compact set and let B be a closed bounded convex set in a Hausdorff locally convex space. Prove that the convex hull of $A \cup B$ is closed.

HINT: consider a limit point z of this convex hull, assume that $z = 0$, observe that there are a convex neighborhood of zero V and $r \in (0, 1)$ for which the inclusion $tx + (1 - t)y \in V$ with $x \in K$, $y \in B$, $t \in [0, 1]$ yields that $t \leqslant r$ (otherwise $0 \in K$); finally, for each convex neighborhood of zero W consider the set of triples (t, x, y) such that $tx + (1 - t)y \in W \cap V$, $t \in [0, r]$, $x \in K$, $y \in B$, and pick a net of points of this form converging to zero.

1.12.55. Let S be a compact set of diameter d in a normed space E. Show that the distance between any two supporting hyperplanes to S is not greater than d, and there exist points $a, b \in S$ such that $\|a - b\| = d$ and there are supporting hyperplanes passing through them and having distance d between them.

HINT: the first assertion is obvious. For the proof of the second take $a, b \in S$ with $\|a - b\| = d$ and find $f \in E'$ with $\|f\| = 1$ and $f(b - a) = d$. Let $\alpha = \min_{x \in S} f(x)$, $\beta = \max_{x \in S} f(x)$. Then $L_1 = f^{-1}(\alpha)$ and $L_2 = f^{-1}(\beta)$ are supporting hyperplanes such that the distance between them equals $\beta - \alpha \geqslant f(b) - f(a) = d$. Hence this distance is d and $a \in L_1$, $b \in L_2$.

1.12.56. Show that on every infinite-dimensional normed space there exist norms defining strictly stronger and strictly weaker topologies, respectively.

1.12.57. Prove that the product of any collection of quasi-complete spaces is quasi-complete.

1.12.58. Let E_0 be a linear subspace in a topological vector space E such that every point of E belongs to the closure of some bounded set in E_0. Prove that every continuous linear mapping from E_0 to a quasi-complete Hausdorff topological vector space G extends uniquely to a continuous linear mapping from E to G.

1.12.59.° Suppose that the topology of a locally convex space E is defined by a sequence of seminorms p_n such that $p_n \leqslant p_{n+1}$. Show that this topology can be defined by a single seminorm precisely when there exists k such that for some numbers $C_n > 0$ one has $p_n \leqslant C_n p_k$ for all n.

1.12.60. Let E be a metrizable locally convex space and let S be a balanced set in E absorbing all sequences converging to zero. Prove that S contains a neighborhood of zero

HINT: if $a_n \to 0$, $a_n \notin S$, then there exist $C_n \to +\infty$ with $C_n a_n \to 0$, which leads to a contradiction, since S absorbs $\{C_n a_n\}$.

1.12.61. Show that in a locally convex space a sequence $\{x_n\}$ is Cauchy precisely when for every its subsequence $\{x_{n_k}\}$ the sequence $\{x_{n_{k+1}} - x_{n_k}\}$ converges to zero.

1.12.62. Let V be a convex set with a nonempty interior in a locally convex space E. Prove that V is open precisely when for every nonzero continuous linear functional f on E the set $f(V)$ is open.

HINT: if V is open, then $f(V)$ is open for nonzero f; if $x_0 \in V$ is not an inner point, then take $f \in E'$ with $f(u) \leqslant f(x_0)$ for all u from the interior of V; show that $f(x_0)$ is an inner point of $f(V)$.

1.12.63. (Fichtenholz [**168**]) Let p and q be seminorms on a linear space E and let E_p' and E_q' be the subspaces in E^* corresponding to the functionals continuous with respect to p and q, respectively. Show that $E_p' \subset E_q'$ precisely when $p \leqslant cq$ for some $c > 0$.

1.12.64. Let E and F be locally convex spaces, let two sets $A \subset E$ and $B \subset F$ be absolutely convex, and let f be a real bilinear function on $E \times F$ such that its restriction to $A \times B$ is continuous at zero. (i) Prove that the restriction of f to $A \times B$ is continuous. (ii) Prove that if A is precompact and B is compact, then the restriction of f to $A \times B$ is uniformly continuous.

1.12.65. Prove that a compact space K is metrizable precisely when there is a countable family of continuous functions f_n on K separating points in the following sense: if $x \neq y$, then for some n we have $f_n(x) \neq f_n(y)$. In this case, one can use the metric

$$d(x,y) = \textstyle\sum_{n=1}^{\infty} 2^{-n} \min\bigl(|f_n(x) - f_n(y)|, 1\bigr).$$

1.12.66.° Prove that a Hausdorff compact space K is metrizable precisely when the space $C(K)$ with its usual sup-metric is separable.

1.12.67. Let X be a metrizable compact space and let $f\colon X \to Y$ be a continuous mapping to a Hausdorff space Y. Prove that the compact set $f(X)$ is also metrizable.

HINT: the space $C(X)$ is separable and $C\bigl(f(X)\bigr)$ can be embedded into it isometrically by means of the mapping $\varphi \mapsto \varphi \circ f$. Hence $C\bigl(f(X)\bigr)$ is also separable, which gives the metrizability of $f(X)$.

1.12.68. Prove that a set A in a Hausdorff topological vector space E is precompact precisely when every infinite sequence in A has a limit point in the completion of E.

HINT: see Lemma 1.12.1.

1.12.69. Let A be a locally compact closed convex set in a topological vector space. Prove that if A is bounded, then it is compact.

HINT: show that A is totally bounded arguing as follows: otherwise one can take a balanced neighborhood of zero V such that $a_n - a_k \notin V$ for some sequence $\{a_n\} \subset A$, then the convexity of A can be used.

1.12.70. Let E be a metrizable topological vector space, let d be a translation invariant metric on it defining its topology, and let $\widetilde{E}$ be the space of all equivalence classes of infinite Cauchy sequences $x = (x_n)$ in E, where $x \sim y$ if the sequence $(x_1, y_1, x_2, y_2, \ldots)$ is Cauchy. Let $d(x,y) := \lim\limits_{n\to\infty} d(x_n, y_n)$. Show that $\widetilde{E}$ is a complete metric topological vector space that serves as a completion of E.

1.12.71.° Prove that every countable everywhere dense set in an infinite-dimensional metrizable topological vector space contains a dense linearly independent subset.

1.12.72. Prove that there is no continuous norm on $\mathbb{R}^{\infty}$.

1.12.73. In the Banach space l^1 consider the straight line L defined by the conditions $x_n = 0$, $n \geqslant 2$, and also the set $A := \{x = (x_n)\colon |n^3 x_n - n| \leqslant x_1 \ \forall n \geqslant 2\}$. Show that A is closed and convex and $A \cap L = \varnothing$, but A and L cannot be separated by a closed hyperplane (observe that the set $A - L$ cannot be separated from zero by a closed hyperplane, since this set is everywhere dense). An analogous fact is true for l^2.

1.12.74. Consider the Banach space $l^1 = (c_0)'$ equipped with the topology $\sigma(l^1, c_0)$, in which the closed unit ball U is compact (see § 3.1). Show that no point u in the unit sphere with infinitely many nonzero coordinates has a closed supporting hyperplane to U.

1.12.75. Show that the closed unit ball of the Banach space c_0 of sequences tending to zero (with its natural sup-norm) has no supporting hyperplanes parallel to the closed hyperplane $f^{-1}(0)$, where f is defined by the formula $f(x) = \sum_{n=1}^{\infty} 2^{-n} x_n$.

1.12.76. Let us take the compact set $K := \bigl\{(x_n)\colon \sum_{n=1}^{\infty} \sqrt{n} x_n^2 \leqslant 1\bigr\}$ in l^2 and the point $a \in K$, where $a_n = cn^{-1}$, $c > 0$, $\sum_{n=1}^{\infty} c^2 n^{-3/2} = 1$. Prove that a is an extreme point of K, but there is no supporting hyperplane to K passing through a.

HINT: if this hyperplane has the form $(x, y) = 1$, then $y_n = c\sqrt{n} a_n$.

1.12.77. The space $M = C[0,1]'$ of all Borel measures on $[0,1]$ is equipped with the topology $\sigma\bigl(M, C[0,1]\bigr)$. Let λ be Lebesgue measure, let D be the set of Dirac measures δ_a, $a \in [0,1]$, and let E be the linear subspace in M generated by D and λ and equipped with the induced topology. Show that D is compact in E and λ is an extreme point of the closed convex hull of the set D in E, but does not belong to the convex hull of D.

1.12.78. For each $n \in \mathbb{N}$, let e_n be the sequence of numbers with 1 on the nth place and 0 on all other places. Show that the set K that is the closed convex hull of the sequence $\{(n+1)^{-1} e_n\}$ in the Banach space l^∞ is compact, but does not coincide with the convex hull of its extreme points.

1.12.79. Let K be a convex compact set in a real locally convex space. Prove that every continuous linear functional assumes its minimal and maximal values on K at some extreme points of K.

1.12.80. (i) Let H be a closed hyperplane in a Hausdorff locally convex space E and let A be an absolutely convex set in E. Prove that $\overline{A \cap H} = \overline{A} \cap H$. Give an example of a convex, but not absolutely convex set A in $\mathbb{R}^2$ for which this is false.

(ii) Deduce from (i) that an absolutely convex set in E is closed precisely when its intersections with all closed hyperplanes in E are closed.

HINT: if $h \notin H$ and a net of vectors $h_\alpha + c_\alpha h \in A$, where $h_\alpha \in H$ and $c_\alpha \neq 0$, converges to $b \in H$, then the vectors $a_\alpha = -t_\alpha(h_{\alpha_1} + c_{\alpha_1} h) + (1 - t_\alpha)(h_\alpha + c_\alpha h) \in A$ also converge to b if $t_\alpha \to 0$; observe that $a_\alpha \in H$ if $c_{\alpha_1} t_\alpha = c_\alpha(1 - t_\alpha)$.

1.12.81. (i) Let A be an absolutely convex subset of a locally convex space E equipped with the strongest locally convex topology. Show that $\overline{A \cap L} = \overline{A} \cap L$ for every linear subspace $L \subset E$. (ii) Let A be an absolutely convex subset of $E \times F$, where E and F are locally convex spaces and F is equipped with the strongest locally convex topology. Show that $\overline{A \cap (E \times \{0\})} = \overline{A} \cap (E \times \{0\})$.

HINT: see Pérez Carreras, Bonet [**385**, p. 110].

1.12.82. Let V be a convex set in a topological vector space. Prove that if V has at least one inner point, then its topological interior coincides with the algebraic kernel.

1.12.83. Let E be a normed space and let F be a separable and norm closed linear subspace in E'. Show that there exists a separable closed linear subspace $E_0 \subset E$ for which there exists a linear isometry j of the subspace F to a closed linear subspace in the Banach space E_0'.

HINT: take in F a norm dense sequence $\{f_n\}$ and vectors $a_n \in E$ such that $\|a_n\| \leqslant 1$, $|f_n(a_n)| = (1-n^{-1})\|f_n\|$; let E_0 be the closed linear subspace in E generated by $\{a_n\}$; observe that $\|f\| = \sup_n |f(a_n)|$ for all $f \in F$; for $f \in F$ define $j(f)$ as $f|_{E_0}$, i.e., $j(f)(x) = f(x)$ whenever $x \in E_0$, $\|j(f)\| = \sup\{|f(x)|\colon\, x \in E_0, \|x\| \leqslant 1\} = \|f\|$.

1.12.84. Let X be a Fréchet space, let X_0 be its closed linear subspace, and let $\pi\colon\, X \to X/X_0$ be the canonical projection to the quotient . Prove that for every compact set $K \subset X/X_0$, there exists a compact set $S \subset X$ such that $\pi(S) = K$. Exercise 3.12.148 gives a generalization.

HINT: see Edwards [**150**, Lemma 9.6.9, p. 671].

1.12.85. Let Φ be a multivalued mapping of a topological space T to the set of nonempty subsets of a compact space K having a closed graph. Prove that Φ is upper semicontinuous, i.e., if $t_0 \in T$ and $\Phi(t_0)$ belongs to an open set W, then there exists a neighborhood $U \ni t_0$ such that $\Phi(t) \subset W$ for all $t \in U$.

1.12.86. (Millionschikov [**347**]) Let V be a closed convex set in a complete Hausdorff locally convex space E, let $f_1\colon\, V \to E$ be a contraction, i.e., for each seminorm p in a collection defining its topology there is $\lambda < 1$ such that $p\bigl(f_1(x) - f_1(y)\bigr) \leqslant \lambda p(x-y)$, and let $f_2\colon\, V \to E$ be continuous and such that $f(V)$ is contained in a compact set. Suppose that $f = f_1 + f_2$ maps V to V. Then, there exists $x_0 \in V$ with $f(x_0) = x_0$.

1.12.87. Prove that there is no Banach space whose dual is linearly homeomorphic to the space c_0 of sequences tending to zero (cf. Example 1.12.7). Prove the same for $C[a,b]$.

HINT: find a sequence of vectors $v_n \in c_0$ (and also in $C[a,b]$) such that for each continuous linear functional l there is a finite limit $\lim\limits_{n\to\infty} l(v_n)$, but there is no vector $v \in c_0$ such that this limit equals $l(v)$ for all l.

1.12.88. Prove that on the space of on the spaces of continuous functions on $[0,1]$ convergence almost everywhere cannot be generated by a topology.

HINT: use that there is a sequence of polynomials that converges at no point of $[0,1]$, but converges in measure and also that any sequence converging in measure has a subsequence converging almost everywhere.

1.12.89. Show that there is a sequence of functions f_n on $\mathbb{R}$ for which there are no numbers $\varepsilon_n > 0$ such that $\lim\limits_{n\to\infty} \varepsilon_n f_n(x) = 0$ for every x.

HINT: identify $\mathbb{R}$ with the space of all sequences $\eta = (\eta_n)$ of natural numbers, consider $f_n(\eta) = \eta_n$, and observe that for each sequence of numbers $\varepsilon_n > 0$ there are $\eta_n \in \mathbb{N}$ with $\varepsilon_n \eta_n > 1$.

CHAPTER 2

Methods of constructing topological vector spaces

In this chapter we consider projective limits (in particular, products) of families of topological vector spaces, inductive limits (in particular, topological direct sums) of families of locally convex spaces, including strict inductive limits and inductive limits with compact embeddings, tensor products of locally convex spaces, and nuclear spaces.

Throughout this chapter the symbol $\mathbb{K}$ denotes (if it is not explicitly stated otherwise) the field of complex or the field of real numbers; it is assumed that all considered vector and topological vector spaces are spaces over $\mathbb{K}$.

2.1. Projective topologies

We define projective limits for topological vector spaces, but the construction applies to general topological spaces.

2.1.1. Definition. *Let E be a vector space and let $\mathfrak{A}$ be some index set. Suppose that for every $\alpha \in \mathfrak{A}$ we are given a topological vector space E_α and a linear mapping $g_\alpha \colon E \to E_\alpha$. The projective topology of the family of spaces $\{E_\alpha\}$ with respect to the family of mappings $\{g_\alpha\}$ is the weakest topology in E with respect to which all mappings $\{g_\alpha\}$ are continuous. The projective limit of the family $\{E_\alpha\}$ with respect to the family of mappings $\{g_\alpha\}$ is the vector space E equipped with this topology.*

We verify that the topology in this definition is well-defined, i.e., we prove that the projective topology of the family of spaces $\{E_\alpha\}$ with respect to the family $\{g_\alpha\}$ of mappings exists. The proof is an explicit description of this topology. For every index $\alpha \in \mathfrak{A}$ let $\mathcal{W}_\alpha$ be the class of all sets of the form ${g_\alpha}^{-1}(V)$, where V is an open subset in E_α. Set $\mathcal{W} = \bigcup_\alpha \mathcal{W}_\alpha$. Then the collection $\mathcal{P}$ of the intersections of all possible finite families of sets from $\mathcal{W}$ forms a base of a topology τ in E possessing the required properties, i.e., the weakest topology t in E for which all mappings $g_\alpha \colon (E,t) \to E_\alpha$ are continuous.

In order to see this, we have to verify the following assertions:

(1) $\mathcal{P}$ is a base of some topology τ in E; (2) all mappings $g_\alpha \colon (E,\tau) \to E_\alpha$ are continuous; (3) the topology τ is majorized by every topology t in E for which all mappings $g_\alpha \colon (E,t) \to E_\alpha$ are continuous.

The validity of (1) follows from the fact that the intersection of every finite collection of subsets from $\mathcal{P}$ belongs to $\mathcal{P}$, so that the collection of all subsets

V.I. Bogachev and O.G. Smolyanov, *Topological Vector Spaces and Their Applications*, Springer Monographs in Mathematics,
DOI 10.1007/978-3-319-57117-1_2

of the space E each of which is the union of some family of sets in $\mathcal{P}$ forms a topology. It will be denoted by τ. It is clear that $\mathcal{P}$ is a base of this topology (the family $\mathcal{W}$ is a prebase).

If now $\alpha \in \mathfrak{A}$ and V is an open subset in E_α, then $g_\alpha^{-1}(V) \in \mathcal{W} \subset \mathcal{P} \subset \tau$ by the definition of $\mathcal{W}$, $\mathcal{P}$ and τ, so that the mappings $g_\alpha\colon (E,\tau) \to E_\alpha$ are continuous, i.e., (2) holds.

Let us verify (3). Let t be a topology in E such that for every $\alpha \in \mathfrak{A}$ the mapping $g_\alpha\colon (E,t) \to E_\alpha$ is continuous. Then every set in $\mathcal{W}$ is open in t, i.e., we have $\mathcal{W} \subset t$. Hence $\tau \subset t$ by the definition of τ.

2.1.2. Remark. So far we have never used that for every $\alpha \in \mathfrak{A}$ the topology of the space E_α agrees with the structure of a vector space and that all mappings g_α are linear, as well as that E and E_α are vector spaces. Thus, the definition of the projective limit is meaningful and remains correct without these assumptions (i.e., in the case where E is an arbitrary set, $\{E_\alpha\}$ is an arbitrary family of topological spaces and $\{g_\alpha\}$ is an arbitrary family of mappings from E to E_α).

2.1.3. Proposition. *Under the assumptions of Definition* 2.1.1 *the projective topology τ agrees with the vector structure of the space E.*

PROOF. This follows from Corollary 1.2.9. Indeed, all E_α are topological vector spaces, all mappings g_α are linear. Hence every family of sets $\mathcal{W}_\alpha$ is invariant with respect to translations; then the families $\mathcal{W}$ and $\mathcal{P}$ are also invariant with respect to translations, hence the topology τ is invariant as well.

Further, by the very definition of the topology τ the set $\mathcal{O}$ of all subsets of the space E each of which is the intersection of some finite family of sets of the form $g_\alpha^{-1}(V)$, where V is an open circled neighborhood of zero in E_α, is a base of neighborhoods of zero in the topology τ, moreover, this base possesses properties (1) and (2) from Proposition 1.2.2. Hence by Corollary 1.2.9 the topology τ agrees with the structure of a vector space. □

The term the "projective limit" is often used for a more special construction (which will be discussed below); on the other hand, projective topologies in our sense are also called "initial".

2.1.4. Proposition. *If in Proposition* 2.1.1 *all topological vector spaces E_α are locally convex, then the topology τ is locally convex as well.*

PROOF. The justification is similar to the proof of Proposition 2.1.3; we have only replace the words "an open circled neighborhood" by the words "an open convex circled neighborhood". □

2.1.5. Proposition. *Let E be a topological vector space that is the projective limit of a family $\{E_\alpha\colon\ \alpha \in \mathfrak{A}\}$ of topological vector spaces with respect to linear mappings $\{g_\alpha\colon\ \alpha \in \mathfrak{A}\}$. A mapping f of an arbitrary topological space G to the topological vector space E is continuous at a point $x \in G$ precisely when for every $\alpha \in \mathfrak{A}$ the mapping $g_\alpha \circ f\colon\ G \to E_\alpha$ is continuous at this point.*

PROOF. Since the composition of continuous (at the corresponding points) mappings is continuous, it suffices to prove that the continuity of the mappings

$g_\alpha \circ f$ $(\alpha \in \mathfrak{A})$ implies the continuity of f. Let V be an open neighborhood of the point $f(x)$. We have to show that there exists a neighborhood W of the point x such that $f(W) \subset V$. By the definition of the projective topology there exist indices $\alpha_1, \ldots, \alpha_n \in \mathfrak{A}$ and open subsets $V_1, \ldots, V_n$ of the spaces $E_{\alpha_1}, \ldots, E_{\alpha_n}$ for which $f(x) \in \bigcap_{i=1}^n g_{\alpha_i}^{-1}(V_i) \subset V$. Since every mapping $g_{\alpha_i} \circ f$ is continuous, there exists an open neighborhood W_i of the point x such that $(g_{\alpha_i} \circ f)(W_i) \subset V_i$ for every index $i \in \{1, 2, \ldots, n\}$. This means that $f(W_i) \subset g_{\alpha_i}^{-1}(V_i)$ for every such i. Hence $f(\cap_i W_i) \subset g_{\alpha_i}^{-1}(V_i)$ for all i. Therefore, if $W = \bigcap_{i=1}^n W_i$, then we have $f(W) \subset \bigcap_{i=1}^n g_{\alpha_i}^{-1}(V_i)$. □

If E is a locally convex space and p is a continuous seminorm on E, then the symbol E_p or (E_p, p) denotes the normed space defined as follows: the vector space E_p is the vector factor-space of the vector space E by its subspace $p^{-1}(0)$; in addition, p is the norm on E_p defined as follows: if $x \in E_p$ and x_1 is a representative of the class x, then $p(x) = p(x_1)$.

Note that the canonical mapping $g_p \colon E \to E_p$ is continuous as the composition of two continuous mappings: the identity mapping of the space E equipped with the original topology to E equipped with the topology defined by the seminorm p (the latter is denoted by the symbol (E, p)) and the canonical mapping of (E, p) to the quotient E_p (actually this is the same mapping g_p, but considered as a mapping from (E, p) to the space E_p).

A general projective limit is a very universal object.

2.1.6. Proposition. *Every locally convex space E is the projective limit of the family $\{E_p \colon p \in \mathcal{P}\}$ of normed spaces with respect to the canonical mappings $\{g_p \colon p \in \mathcal{P}\}$, where $\mathcal{P}$ is the set of all continuous seminorms on E.*

PROOF. This follows from the fact that every locally convex topology is defined by the set of all continuous seminorms on this space. □

2.1.7. Remark. A similar proposition is valid for arbitrary topological vector spaces (i.e., spaces that are not locally convex). It suffices to replace the word "seminorm" by the word "quasi-norm" and the word "normed" by the word "metrizable" in the formulation of the proposition above.

2.1.8. Remark. For every seminorm p on E we denote by $\mathcal{E}_p$ the Banach space serving as the completion of the normed space E_p. It follows from the previous proposition that every locally convex space is the projective limit of the family of Banach spaces $\{\mathcal{E}_p\}$ with respect to the corresponding canonical mappings. A similar proposition is valid for arbitrary topological vector spaces, i.e., not necessarily locally convex (in this case the role of Banach spaces is played by complete metrizable topological vector spaces).

2.1.9. Remark. If τ is the projective topology in a vector space E with respect to a certain family $\{E_\alpha \colon \alpha \in \mathfrak{A}\}$ of topological vector spaces and linear mappings $\{g_\alpha \in \mathcal{L}(E, E_\alpha) \colon \alpha \in \mathfrak{A}\}$, then in order that τ be separated, it is necessary and sufficient that for every nonzero element $x \in E$ one could find $\alpha \in \mathfrak{A}$ and a neighborhood of zero V_α in E_α such that $g_\alpha(x) \notin V_\alpha$. In particular, if all topological vector spaces E_α are separated, then in order that E be also separated, it is necessary and sufficient that for every nonzero element $x \in E$ one could find an index $\alpha \in \mathfrak{A}$ such that $g_\alpha(x) \neq 0$. Both assertions follow directly from the definition.

2.2. Examples of projective limits

Let us consider some examples of projective limits.

2.2.1. Example. (*The least upper bound of a family of topologies in a vector space.*) Let E be a vector space and let $\mathfrak{A}$ be an index set such that for every $\alpha \in \mathfrak{A}$ we are given a topology τ_α in E compatible with the structure of a vector space. Then there exists a topology τ in E that is the least upper bound of the set $\{\tau_\alpha\colon\ \alpha \in \mathfrak{A}\}$ in the set of all topologies in E, i.e., the weakest among topologies in E each of which majorizes every topology τ_α. The topology τ agrees with the structure of a vector space and is locally convex if so are all τ_α.

Indeed, the required property holds for the topology of the projective limit of the family of topological vector spaces $\{E_\alpha\colon\ \alpha \in \mathfrak{A}\}$ with respect to the family of mappings $\{g_\alpha\colon\ \alpha \in \mathfrak{A}\}$, where for every $\alpha \in \mathfrak{A}$ we take $E_\alpha = (E, \tau_\alpha)$, and g_α is the identity mapping of the space E. The locally convex case follows from Proposition 2.1.4.

2.2.2. Example. (*Subspaces*, see Example 1.3.10). Let E be a topological vector space, let E_1 be its topological vector subspace (i.e., a vector subspace equipped with the induced topology), and let $g\colon E_1 \to E$ be the canonical embedding. Then E_1 is the projective limit of the one-element family of topological vector spaces $\{E\}$ with respect to the one-element family of mappings $\{g\}$.

2.2.3. Example. (*The product of topological vector spaces.*) Let $\mathfrak{A}$ be a nonempty set and let (E_α, τ_α) be a topological vector space, $\alpha \in \mathfrak{A}$. Let E be the vector space that is the product of the family of vector spaces $\{E_\alpha\colon\ \alpha \in \mathfrak{A}\}$. Thus, the set of elements of E is the set of all functions f on the set $\mathfrak{A}$ with values in the set $\bigcup_{\alpha\in\mathfrak{A}} E_\alpha$ such that $f(\alpha) \in E_\alpha$ for every $\alpha \in \mathfrak{A}$; the structure of a vector space in E is introduced by the relations

$$(\lambda_1 f_1 + \lambda_2 f_2)(\alpha) = \lambda_1 f_1(\alpha) + \lambda_2 f_2(\alpha), \quad \lambda_1, \lambda_2 \in \mathbb{K},\ f_1, f_2 \in E.$$

For every $\alpha \in \mathfrak{A}$ let pr_α denote the projection of E onto E_α defined as follows: if $f \in E$, then $\mathrm{pr}_\alpha(f) = f(\alpha)$. The topology (in E) of the projective limit of the family $\{E_\alpha\colon\ \alpha \in \mathfrak{A}\}$ of topological vector spaces with respect to the family of mappings $\{\mathrm{pr}_\alpha\colon\ \alpha \in \mathfrak{A}\}$ coincides with the topology of Tychonoff's product; this follows from their definitions. Throughout the product of a family of topological vector spaces will be understood as their product of vector spaces equipped with the Tychonoff product topology; if $\{G_\alpha\}$ is the corresponding family of topological vector spaces, then the symbol $\prod_{\alpha\in\mathfrak{A}} G_\alpha$ will denote their product.

2.2.4. Proposition. *Suppose that E is the projective limit of a certain family $\{E_\alpha\colon\ \alpha \in \mathfrak{A}\}$ of topological vector spaces with respect to a certain family of mappings $\{g_\alpha \in \mathcal{L}(E, E_\alpha)\colon\ \alpha \in \mathfrak{A}\}$; assume that its topology is separated. Then*

E is isomorphic — as a topological vector space — to some topological vector subspace in the product G of a family topological vector spaces $\{E_\alpha\colon \alpha \in \mathfrak{A}\}$.

PROOF. The mapping $\Psi\colon E \to G$ defined by: $\Psi(x)(\alpha) = g_\alpha(x)$ is linear by the linearity of all g_α. This mapping is injective, since, due to the assumption that E is separated, for every $x \in E$ there exists $\alpha \in \mathfrak{A}$ such that $g_\alpha(x) \neq 0$. Further, for every $\alpha \in \mathfrak{A}$ the composition $\mathrm{pr}_\alpha \circ \Psi$ coincides with the mapping g_α, hence is continuous. By Proposition 2.1.5 the mapping Ψ is continuous. For completing the proof it remains to verify that the mapping $\Psi^{-1}\colon \Psi(E) \to E$ is also continuous (assuming that the vector subspace $\Psi(E)$ of the space G is equipped with the topology induced by the topology of the space G). For every index $\alpha \in \mathfrak{A}$ the composition of the mapping Ψ^{-1} taking the element $g(x)$ (i.e., the function $\alpha \mapsto g_\alpha(x)$) of the space $\Psi(E)$ to the element $x \in E$ and the mapping g_α coincides with the restriction to $\Psi(E)$ of the projection mapping $\mathrm{pr}_\alpha\colon g(x) \mapsto g_\alpha(x)$ and hence is continuous by the definition of the product topology. Hence — again by Proposition 2.1.5 — the mapping Ψ^{-1} is also continuous. Thus, the mapping Ψ is a linear homeomorphism of E onto the topological vector subspace $\Psi(E)$ of the space $\prod_\alpha E_\alpha$. □

2.2.5. Corollary. *Every Hausdorff locally convex space E is isomorphic to a topological vector subspace of the product $\prod_{p\in\mathcal{P}} E_p$, where $\mathcal{P}$ is the set of all continuous seminorms on E.*

This fact follows from Propositions 2.1.6 and 2.2.4.

2.2.6. Example. (*Weak topologies*, Example 1.3.23). Let E be a vector space, let G be a vector subspace in E^*, and, for every $g \in G$, let E_g be a copy of the field $\mathbb{K}$ considered as a one-dimensional topological vector space (over the field $\mathbb{K}$). Then the topology of the projective limit of the family of topological vector spaces $\{E_g\colon g \in G\}$ with respect to the family of mappings $\{g\colon g \in G\}$ is the weak topology in E defined by the elements of the set G.

2.2.7. Example. (*Limits of inverse spectra of topological vector spaces.*) Let $\mathfrak{A}$ be a directed set. A family $\{E_\alpha\colon \alpha \in \mathfrak{A}\}$ of topological vector spaces is called the *inverse spectre of topological vector spaces* E_α if, for every pair of indices $\alpha, \beta \in \mathfrak{A}$ with $\alpha \leqslant \beta$, a continuous linear mapping $\psi_{\alpha\beta}\colon E_\beta \to E_\alpha$ is given. The limit of such inverse spectre is the topological vector subspace in the product $\prod_\alpha E_\alpha$ denoted by the symbol $\varprojlim E_\alpha$ and consisting of elements $g \in \prod_\alpha E_\alpha$ such that $g(\alpha) = \psi_{\alpha\beta} g(\beta)$ whenever $\alpha, \beta \in \mathfrak{A}$, $\alpha \leqslant \beta$. For example, if $E_\beta \subset E_\alpha$ whenever $\alpha < \beta$ and the natural embedding $E_\beta \to E_\alpha$ is continuous, then $\bigcap_\alpha E_\alpha$ is the limit of the inverse spectre of the spaces E_α.

Every topological vector subspace of the product of an arbitrary family of topological vector spaces is the projective limit of this family with respect to the family of mappings of the regarded space to the factors that are the restrictions of the projection mappings of the product to these factors. Therefore, in particular, the space $\varprojlim E_\alpha$ is the projective limit of the family of topological vector spaces

$\{E_\alpha\colon \alpha \in \mathfrak{A}\}$ with respect to the family of mappings that are the restrictions of the corresponding projections.

Note also that by the continuity of the mappings $\psi_{\alpha\beta}$ the topological vector space $\varprojlim E_\alpha$ is a closed subspace in the product $\prod_\alpha E_\alpha$ (verify this). Since, in addition, any closed subset in a complete topological vector space is complete, and the product of an arbitrary family of complete topological vector spaces is a complete topological vector space, the limit of an inverse spectre of complete topological vector spaces is a complete topological vector space. An analogous assertion is valid also for quasi-complete topological vector spaces.

Let us introduce one more interesting class of spaces.

2.2.8. Example. (*Countably normed spaces.*) This term is used for locally convex spaces that are limits of inverse spectra of Banach spaces possessing the following properties: (a) the index set is the set of natural numbers with its usual order; (b) all mappings ψ_{nj} (defined for $j \geqslant n$) are injective.

The notion of a countably normed space was introduced by Gelfand and Shilov [**193**] by means of another definition: in their book a topological vector space is called countably normed if it is locally convex, metrizable, complete, and its topology can be defined by a countable family of compatible norms. Here two norms on a vector space E are called *compatible* if every sequence of elements in this space that is Cauchy with respect to both norms and converges to zero with respect to one of them must converge to zero with respect to the other.

As an example of incompatible norms on an infinite-dimensional Banach space X we can take the original norm $\|\cdot\|$ on X and the norm $x \mapsto \|x\| + |l(x)|$, where l is any discontinuous linear functional on X. Then one can always find vectors $x_n \in X$ such that $\|x_n\| \to 0$ and $l(x_n) = 1$.

We shall call our definition of countably normed spaces Definition I, and the definition from the book [**193**] will be called Definition II.

We now show that these definitions are equivalent. First we show that the requirements of Definition I imply the requirements of Definition II. Thus, let us consider the inverse spectre $\{E_n\colon n \in \mathbb{N}\}$ of Banach spaces satisfying the conditions in Definition I.

For every $j \in \mathbb{N}$, the norm of E_j will be denoted by the symbol $\|\cdot\|_j$. Since all mappings $\psi_{nj}\colon E_j \to E_n$ are injective, every space E_n can be identified as a vector space with a subspace of each space E_i $(i < n)$ with smaller indices. Thus, replacing the words "is a vector subspace" by the symbol $\subset\subset$, we obtain the following chain of relations:

$$\cdots \subset\subset E_n \subset\subset E_{n-1} \subset\subset \cdots \subset\subset E_1.$$

Here the mappings ψ_{jr} will coincide with the corresponding (identical) embeddings. We emphasize that in general the spaces E_r are not topological vector subspaces of the spaces E_j with smaller indices j, they are merely vector subspaces.

Therefore, the space $\varprojlim E_n$ can be identified as a vector space with the intersection $\bigcap_n E_n$ of the spaces E_n. Indeed, $g \in \varprojlim E_n$ precisely when for all $j \geqslant n$ we have the equality $\psi_{nj}(g(j)) = g(n)$. Since ψ_{nj} is an embedding, these equalities actually mean that $g(1) = g(2) = \cdots = g(n) = \cdots$, so that a natural identification of $\varprojlim E_n$ and $\bigcap_n E_n$ can be defined as follows: $g \in \varprojlim E_n \Longleftrightarrow g(1) \in \bigcap_n E_n$.

Thus, we have shown that $\varprojlim E_n$ coincides with $\bigcap_n E_n$ as a vector space. As we have observed before Example 2.2.8, the topology of the space $\varprojlim E_n$ is the topology of the projective limit of the family of topological vector spaces $\{E_j\colon j \in \mathbb{N}\}$ with respect to the family of mappings that are the restrictions to $\varprojlim E_n$ of the mappings $\mathrm{pr}_j\colon \prod_n E_n \to E_j$. Identifying $\varprojlim E_n$ and $\bigcap_n E_n$, we can assume that the mappings pr_j are defined on $\bigcap_n E_n$; each of these mappings is an embedding into the respective space; say, pr_j coincides on $\bigcap_n E_n$ with the identical embedding $\mathrm{in}_j\colon \bigcap_n E_n \to E_j$. Hence if we equip $\bigcap_n E_n$ with the topology τ of the projective limit of the family of Banach spaces $\{E_n\colon n \in \mathbb{N}\}$ with respect to the mappings in_j, then the identification of the spaces $\varprojlim E_n$ and $\bigcap_n E_n$ described above will be their identification as topological vector spaces.

It follows from Definition I that the topology of the projective limit in $\bigcap_n E_n$ defined above is given by the family of norms $\|\cdot\|_n$ (more precisely, by the restrictions of these norms to $\bigcap_n E_n$). We show that these norms are compatible. Suppose that a sequence $\{x_k\} \subset \bigcap_n E_n$ is Cauchy in both norms $\|\cdot\|_j$, $\|\cdot\|_n$ and $\|x_k\|_n \to 0$. If $n > j$, then the relation $\|x_k\|_j \to 0$ follows from the continuity of the embedding $\psi_{jn}\colon E_n \to E_j$. If $n < j$, then $\{x_k\}$ converges to some x in E_j by the completeness of E_j, which gives the equality $x = 0$ by the continuity and the injectivity of ψ_{nj}. Thus, it is shown that the space $(\bigcap_n E_n, \tau)$, hence also the space $\varprojlim E_n$ that coincides with it, is a countably normed space in the sense of Definition II.

Suppose now that the space E satisfies the conditions of Definition II and let $\{p_j\}$ be compatible norms defining the topology of this space. Replacing, if necessary, the norms p_j by the norms $p_j' = \sum_{n=1}^{j} p_n$ (which define the same topology) and keeping the previous notation we can assume that for all $x \in E$ and n we have $p_n(x) \leqslant p_{n+1}(x)$; the consistence of norms is preserved.

Further, let E_n be the completion of E with respect to the norm p_n, $n \in \mathbb{N}$. For each pair $n, j \in \mathbb{N}$ we define a continuous linear mapping $\psi_{n,n+j}\colon E_{n+j} \to E_n$ as the extension by the continuity of the identity mapping (such an extension exists according to Proposition 1.7.14).

The consistency of our norms yields the injectivity of these mappings, i.e., the equality $\psi_{n,n+j}(x) = 0$, where $x \in E_{n+j}$, yields that $x = 0$. Indeed, there exists a sequence $\{x_i\} \subset E$ converging in E_{n+j} to x. This sequence is fundamental in the norm p_{n+j}, moreover, $p_n(x_i) \to 0$, since $\psi_{n,n+j}(x) = 0$. By the consistency of these norms $p_{n+j}(x_i) \to 0$, whence $x = 0$.

Thus, the family $\{E_n\colon n \in \mathbb{N}\}$ of Banach spaces forms a spectre with respect to the injective mappings $\{\psi_{ns}\colon n \leqslant s\}$ (ψ_{nn} is the identity mapping of E_n to E_n, which, as above, is assumed to be an embedding of vector spaces).

As we have shown above, the space $\varprojlim E_n$ can be identified with the vector space $\bigcap_n E_n$ equipped with the topology defined by the family of norms p_n. Thus, for the proof of the fact that E is a countably normed space in the sense of Definition I, it remains to verify that the set $\bigcap_n E_n$ coincides as a vector space with E.

Since the inclusion $E \subset \bigcap_n E_n$ is true by definition (all spaces E_n are completions of E), we have to prove the opposite inclusion. Let $x \in \bigcap_n E_n$. This means that for every $n \in \mathbb{N}$ there is a sequence $\{x_j^n\} \subset E_n$ converging to x in E_n, i.e., in the norm p_n. Then we can choose a "quasi-diagonal" sequence $\{x_{j(n)}^n\}$ converging in every space E_n, hence fundamental in every norm p_n. Since they define the topology in E, this means that the sequence $\{x_{j(n)}^n\}$ is Cauchy in E and by the completeness of E it converges to some element $z \in E$. This sequence converges to z also in the space E_1 (being converging in E, it converges in every norm p_n). However, $\{x_{j(n)}^n\}$ is a subsequence of the sequence $\{x_j^1\}$ converging to x in E_1 and hence also converges in this space to x. Thus, $x = z$, i.e., $x \in E$. This completes the proof of the equivalence of both definitions.

2.2.9. Remark. Let us note the following fact established in Example 2.2.8. Suppose we are given a sequence of embedded Banach spaces $E_{n+1} \subset\subset E_n$ (as above, the symbol $\subset\subset$ means that the left space is a vector subspace in the right space), where all embeddings are continuous. Such a family can be regarded as the inverse spectre of these spaces the index set of which is the set of natural numbers and the role of the mappings ψ_{nj} is played by the embeddings.

Then the limit of such inverse spectre is (as a vector space) the intersection of all spaces E_n and its topology is defined by means of the restrictions to $\bigcap_n E_n$ of the norms p_n of the Banach spaces E_n.

2.2.10. Example. Certainly, not every Fréchet space is countably normed. For example, the countable product of the real lines $\mathbb{R}^\infty$ is not, since on it there are no continuous norms at all (every neighborhood of zero in this space contains some infinite-dimensional vector subspace).

2.2.11. Example. More interesting is the fact that even a Fréchet space whose topology is defined by a countable collection of norms need not be countably normed (observe by the way that in order the topology of a locally convex space, not necessarily metrizable, could be defined by a family of norms it is sufficient that this space had at least one continuous norm).

A Fréchet space in question can be defined as follows. Let E be the space of all continuously differentiable real functions f on the real line with the following property: $|f(t)| + |f'(t)| \to 0$ as $|t| \to \infty$. For every natural number n denote by p_n the norm on E defined by the equality

$$p_n(f) = \max_{t\in\mathbb{R}^1} |f(t)| + \max_{t\in[-n,n]} |f'(t)| \\ + \max\Bigl\{\bigl|f'\bigl(r + 1/(2k)\bigr)\bigr|:\ r \in \mathbb{Z};\ |r| > n;\ k = 1, 2, \ldots, n\Bigr\}.$$

Let τ be the topology in the space E defined by the countable family of norms $\mathcal{P} = \{p_n\colon\ n \in \mathbb{N}\}$; then (E, τ) is a Fréchet space.

We show that on this space there exists no countable family of compatible norms defining its topology. First of all, no two norms in the family $\mathcal{P}$ are compatible (verify this!).

Suppose now that E possesses a family $\mathcal{P}_1$ of compatible norms defining the topology of E. We can assume that these norms are increasing, so that the set of the corresponding balls $\{x \in E\colon\ q_j(x) < \varepsilon\}$ is a base (not only a prebase) of neighborhoods of zero. Let $q \in \mathcal{P}_1$. Then there exist two distinct norms $p_{j_1}, p_{j_2} \in \mathcal{P}$ such that $q(x) \leqslant Cp_{j_1}(x) \leqslant Cp_{j_2}(x)$ for every $x \in E$, where $C > 0$ is some number. We can assume that $C = 1$. Certainly, the norms p_{j_1} and p_{j_2} are not compatible . We can also find a norm $q' \in \mathcal{P}_1$ and a norm $p_{j_3} \in \mathcal{P}$ for which $q' \leqslant C'p_{j_3}$, so that without loss of generality we can assume that

$$q(x) \leqslant p_{j_1}(x) \leqslant p_{j_2}(x) \leqslant q'(x) \leqslant p_{j_3}(x) \quad \text{for all } x \in E.$$

Certainly, the norms p_{j_1} and p_{j_3} are not compatible as well.

We now show that the norms q and q' are not compatible. Let $\{a_n\}$ be a sequence of elements in E fundamental in the norm p_{j_3}, hence also in the norms p_{j_1} and p_{j_2}, and converging to zero in the norm p_{j_1}, but not converging to zero in the norm p_{j_2} (hence in the norm p_{j_3}). The fact that the topology τ cannot be defined by a family of pairwise compatible norms is implied by the property of $\mathcal{P}$ that, for every three norms p_{j_1}, p_{j_2}, p_{j_3} with $p_{j_1} \leqslant p_{j_2} \leqslant p_{j_3}$, there is a sequence converging to zero in the first norm, fundamental in all the three norms, but not converging to zero in the last two norms. Since $p_{j_1}(a_n) \geqslant q(a_n)$, we have $q(a_n) \to 0$. By the inequality $q' \leqslant p_{j_3}$ the sequence $\{a_n\}$ is fundamental in the norm q'. Finally, the inequality $p_{j_2} \leqslant q'$ yields that it cannot converge to zero in the norm q', since otherwise it would converge to zero in the norm p_{j_2}.

Note also that the relation $|f(t)| + |f'(t)| \to 0$ as $|t| \to \infty$, which is valid for every function $f \in E$, yields that the space (E, τ) is separable.

2.3. Inductive topologies

The concept of an inductive topology is dual to that of a projective topology, but the corresponding results related to one or the other topology are not completely symmetric; the nature of this asymmetry will be clear below.

2.3.1. Definition. *Let E be a vector space and let $\mathfrak{A}$ be a nonempty index set such that for every $\alpha \in \mathfrak{A}$ we are given a locally convex space E_α and a linear mapping $g_\alpha\colon\ E_\alpha \to E$. The inductive topology of the family of spaces $\{E_\alpha\}$ with respect to the family of mappings $\{g_\alpha\}$ (more precisely, the inductive topology in the category of locally convex topologies) is the strongest locally convex topology in E with respect to which all mappings g_α are continuous. The inductive limit of the family $\{E_\alpha\}$ with respect to the mappings $\{g_\alpha\}$ is the vector space E equipped with this topology. Notation*: $E = \mathrm{ind}_\alpha E_\alpha$.

2.3.2. Remark. (i) Let $\mathcal{V}$ be the set of all convex circled absorbent sets V in the space E such that $g_\alpha^{-1}(V)$ is a neighborhood of zero in E_α for each $\alpha \in \mathfrak{A}$.

Then $\mathcal{V}$ is a base of neighborhoods of zero in the inductive topology of the family $\{E_\alpha\}$ with respect to the family of mappings $\{g_\alpha\}$. Indeed, according to Proposition 1.2.11, the set $\mathcal{V}$ is a base of neighborhoods of zero of some locally convex topology τ in E. It follows from the definition of $\mathcal{V}$ that all mappings $g_\alpha\colon E_\alpha \to (E,\tau)$ are continuous. If τ_1 is an arbitrary locally convex topology in E for which all mappings $g_\alpha\colon E_\alpha \to (E,\tau_1)$ are continuous and $\mathcal{V}_1$ is a base of its neighborhoods of zero consisting of circled convex sets (certainly, they are all absorbing), then for every $\alpha \in \mathfrak{A}$ and every $V \in \mathcal{V}_1$ the set $g_\alpha^{-1}(V)$ is a neighborhood of zero in E_α, which by the definition of $\mathcal{V}$ yields that $\mathcal{V}_1 \subset \mathcal{V}$, hence we have $\tau_1 \subset \tau$.

Thus, if E_α is a locally convex space with a basis of absolutely convex neighborhoods of zero $\mathcal{V}_\alpha$, then the set $\mathcal{V}$ of absolutely convex envelopes of all possible sets of the form $\bigcup_\alpha g_\alpha(V_\alpha)$, where $V_\alpha \in \mathcal{V}_\alpha$, is a basis of neighborhoods of zero in the space E.

(ii) If E_α are general topological vector spaces, then the inductive topology on the space E in the category of topological vector spaces is introduced as the strongest vector topology for which all mappings g_α are continuous. This topology also exists. According to Example 2.2.1, the set T of all vector topologies in E for which all maps g_α are continuous is nonempty, moreover, it has the weakest topology. One can verify that T has the required strongest topology, but one can also define explicitly the corresponding base of neighborhoods of zero (see Jarchow [**237**, § 4.1]). For simplicity we assume that E is the union of $g_\alpha(E_\alpha)$. In every space E_α we take a basis $\mathcal{U}_\alpha$ of circled neighborhoods of zero and introduce in E a base of zero $\mathcal{U}$ consisting of the sets of the form

$$U = \bigcup_{n=1}^{\infty} \sum_{k=1}^{n} \bigcup_{\alpha\in\mathfrak{A}} g_\alpha(U_{\alpha,k}), \quad U_{\alpha,k} \in \mathcal{U}_\alpha,$$

where $\sum$ denotes a vector sum. For a countable collection of E_n we can take the sets $U = \bigcup_{K\in\mathcal{K}} \sum_{k\in K} g_k(U_k)$, where $U_k \in \mathcal{U}_k$ and $\mathcal{K}$ is the set of all finite subsets of $\mathbb{N}$. It is clear that U is circled and absorbent and that there exists $V \in \mathcal{U}$ with $V+V \subset U$ (one can take $V_{\alpha,k} \subset U_{\alpha,2k} \cap U_{\alpha,2k-1}$ in $\mathcal{U}_\alpha$). By Corollary 1.2.9 the class $\mathcal{U}$ is a base of neighborhoods of zero of a vector topology τ_2 in E. The continuity of $g_\alpha\colon E_\alpha \to (E,\tau)$ is obvious. The maximality of τ_2 is seen from the fact that if all g_α are continuous when E is equipped with a vector topology τ_0 and W_0 is a balanced neighborhood of zero in it, then we can find $U \in \mathcal{U}$ with $U \subset W$ by taking $W_k \in \tau_0$, $W_k + W_k \subset W_{k-1}$, $U_{\alpha,k} \subset g_\alpha^{-1}(W_k)$. If E_n are locally convex, then so is the topology τ_2, but for uncountable $\mathfrak{A}$ this is false (see an example below).

In addition to inductive topologies in the categories of locally convex spaces and topological vector spaces, the space E can be equipped with the strongest topology (not necessarily vector) with the property that all mappings g_α are continuous. It turns out that even if all E_α are locally convex, these three topologies can be distinct.

2.3.3. Example. Let E be a real vector space with an algebraic basis of cardinality of the continuum, let $\mathcal{F}$ be the set of all its finite-dimensional vector subspaces each of which is equipped with the standard topology, and let g_F be the canonical embedding of F into E for every $F \in \mathcal{F}$ (so here the index set is $\mathcal{F}$ itself). Let τ_1 be the strongest locally convex topology in E for which all g_F are continuous (i.e., the locally convex inductive topology), let τ_2 be the strongest vector topology in which all g_F are continuous, finally, let τ_3 be the strongest topology in E (not necessarily vector) for which the same mappings are continuous.

Then, obviously, $\tau_3 \supset \tau_2 \supset \tau_1$. We show that $\tau_3 \neq \tau_2 \neq \tau_1$.

For the proof of the inequality $\tau_2 \neq \tau_1$ we observe that the topology τ_1 can be defined by the set of all seminorms E and the topology τ_2 can be defined by the set of all quasi-norms on E. Let $\mathcal{P}$ be a Hamel basis in E and let q be the quasi-norm on E defined by the equality

$$q\Bigl(\sum_{e\in\mathcal{P}} c_e \cdot e\Bigr) = \Bigl(\sum_{e\in\mathcal{P}} \sqrt{|c_e|}\Bigr)^2.$$

Certainly, the set of nonzero coefficients in the sum $\sum_{e\in\mathcal{P}} c_e \cdot e$ is finite.

We show that on E there is no seminorm p such that

$$\{x \in E\colon\ p(x) \leqslant 1\} \subset \{x \in E\colon\ q(x) \leqslant 1\}.$$

Indeed, if p is such a seminorm on E, then for every $a \in E$ with $p(a) \leqslant 1$ we have the inequality $q(a) \leqslant 1$. On the other hand, for some $C > 0$ the set $\mathcal{P}_C = \{e \in \mathcal{P}\colon\ p(e) \leqslant C\}$ is infinite, since if such $C > 0$ does not exist, then the set $\mathcal{P}$ is at most countable, which is false by assumption. Thus, for every n we can find n different elements $e_1, \dots, e_n \in \mathcal{P}_C$. Then for $\eta_j = C^{-1}e_j$ we have the estimate $p\bigl(n^{-1}\sum_{j=1}^n \eta_j\bigr) \leqslant 1$, while $q\bigl(n^{-1}\sum_{j=1}^n \eta_j\bigr) = C^{-1}n$. Hence for a sufficiently large n we obtain the inequality $q\bigl(n^{-1}\sum_{j=1}^n \eta_j\bigr) > 1$ contradicting the aforementioned estimate. Therefore, $\tau_2 \neq \tau_1$. A simpler assertion is proven.

Let us proceed to the proof of the inequality $\tau_3 \neq \tau_2$. Let $\|\cdot\|$ be the norm on E defined by the equality $\|\sum_{e\in\mathcal{P}} c_e \cdot e\| = \max_{e\in\mathcal{P}} |c_e|$, where we use the previous notation. For every positive function φ on the set $\mathcal{P}\times\mathcal{P}$ we choose a neighborhood of zero V_φ in the topology τ_3 with the following property: if $\|\lambda(e_1+e_2)\| \geqslant \varphi(e_1, e_2)$, then $\lambda(e_1+e_2)$ is not contained in V_φ. Such a neighborhood of zero in τ_3 exists, since a set is open in the topology τ_3 precisely when its intersection with each finite-dimensional subspace in E is open in the standard topology of this finite-dimensional subspace; for the neighborhood V_φ we can take $E \setminus \bigcup_{e,b\in\mathcal{P}}\{\lambda(e+b)\colon\ \|\lambda(e+b)\| \geqslant \varphi(e,b)\}$.

In order to prove that the topology τ_3 is stronger than τ_2, it suffices to show that there exists a function $\varphi\colon\ \mathcal{P}\times\mathcal{P} \to \mathbb{R}^+$ for which there is no quasi-norm q on E such that if $W = \{x \in E\colon\ q(x) \leqslant 1\}$, then $W + W \subset V_\varphi$. We prove a formally stronger assertion: for some function $\varphi\colon\ \mathcal{P}\times\mathcal{P} \to \mathbb{R}^+$ there is no neighborhood of zero W in the topology τ_3 such that $W + W \subset V_\varphi$. For every neighborhood of zero W in the topology τ_3 we choose some positive function ψ_W on $\mathcal{P}$ with the following property: for every $e \in \mathcal{P}$ one has $\{\lambda e\colon\ \|\lambda e\| < \psi_W(e)\} \subset W$. Since the inclusion $W+W\subset V_\varphi$ yields that $\min\bigl(\psi_W(e_1),\psi_W(e_2)\bigr)\leqslant\varphi(e_1,e_2)$ for all $e_1,e_2\in\mathcal{P}$ (by our choice of V_φ), it suffices to find a function $\varphi\colon\,\mathcal{P}\times\mathcal{P}\to\mathbb{R}^+$ for which the inequality

$$\min\bigl(\psi(e),\psi(b)\bigr)\leqslant\varphi(e,b)$$

cannot hold for all $e,b\in\mathcal{P}$ at once, whatever a function $\psi\colon\,\mathcal{P}\to\mathbb{R}^+$ be.

A function φ with this property can be constructed in the following way. Let $\{e_j\}$ be some countable subset of the set $\mathcal{P}$, let S be the set of all sequences of positive real numbers, and let $\mathcal{P}_0$ be some subset of $\mathcal{P}$ of cardinality of the continuum. Let g be a one-to-one mapping of $\mathcal{P}_0$ onto S. For $e\in\mathcal{P}_0$ we set $\varphi(e,e_j)=g(e)_j$, where $\bigl(g(e)\bigr)_{j=1}^\infty$ is the sequence in S that is the image of the element e under the mapping g. For other pairs $(e,b)\in\mathcal{P}\times\mathcal{P}$ we define $\varphi(e,b)$ in an arbitrary way. Then, given an arbitrary a positive function ψ on $\mathcal{P}$, we obtain that $\varphi(e,b)\leqslant\min\bigl(\psi(e),\psi(b)\bigr)$ for some e,b. This follows from the fact that the set $g(\mathcal{P}_0)$ contains all sequences of positive numbers, in particular, the sequence of the numbers $\alpha_j=\psi(e_j)/j$. It is clear that if $g(e)_j=\alpha_j$, then we have $\varphi(e,e_j)\leqslant\min\bigl(\psi(e),\psi(e_j)\bigr)$ for all sufficiently large j (for arbitrary $\psi(e)$).

Hille and Phillips [**222**] introduced the "finitely open" topology in a vector space in which a set is open precisely when its intersection with every finite-dimensional subspace is open in the standard topology of this subspace (see Remark 1.10.4); thus, this topology coincides with our topology τ_3 (as already noted above). It was mentioned in [**222**] that the authors were unaware whether this topology agrees with the vector structure; a negative answer was given later in Kakutani, Klee [**254**].

The proof of the noncoincidence of the topologies τ_3 and τ_2 given above is actually a proof of the fact that if the algebraic dimension of a vector space is at least of cardinality of the continuum, then the finitely open topology in it is not compatible with the vector structure.

2.3.4. Remark. The term "an inductive limit" is often used for a more special construction (which will be described below). On the other hand, if E is an arbitrary set, $\{E_\alpha\colon\,\alpha\in\mathfrak{A}\}$ is a family of topological spaces such that for every $\alpha\in\mathfrak{A}$ a mapping $g_\alpha\colon\,E_\alpha\to E$ is given, then in E there exists the strongest topology for which all these mapping are continuous. This topology is called the inductive topology (in the class of all topologies) of the family of topological spaces $\{E_\alpha\}$ with respect to the family of mappings $\{g_\alpha\}$; sometimes this topology is also called final. Throughout, however, the terms an inductive limit and an inductive topology are used only in the sense of our definition.

The following simple result is frequently used.

2.3.5. Proposition. *Let a locally convex space E be the inductive limit of some family $\{E_\alpha\colon\,\alpha\in\mathfrak{A}\}$ of locally convex vector spaces with respect to linear mappings $g_\alpha\colon\,E_\alpha\to E$. A linear mapping f from the space E to a locally convex space G is continuous precisely when for every $\alpha\in\mathfrak{A}$ the mapping $f\circ g_\alpha$ is continuous. Therefore, sequentially continuous linear mappings of inductive limits of metrizable spaces are continuous.*

PROOF. As in Proposition 2.1.5, it suffices to show that the continuity of the mappings $f \circ g_\alpha$ implies the continuity of the mapping f. To this end we have to show that if V is a convex circled neighborhood of zero in G, then $f^{-1}(V)$ is a neighborhood of zero in E. The set $f^{-1}(V)$ is convex, circled and absorbent by the linearity of f. Finally, the continuity of all mappings g_α implies that all sets $g_\alpha^{-1}(f^{-1}(V))$ are neighborhoods of zero in the respective spaces. Now it remains to use Remark 2.3.2. □

This proposition is not completely symmetric to Proposition 2.1.5, dual to which it is: here we assume (unlike Proposition 2.1.5) that f is linear and that G is locally convex. Neither of these assumptions can be dropped (the circumstance that now we consider the continuity everywhere and not at a point is not essential, since for linear mappings this is the same). Note also that this proposition does not extend to closed linear subspaces of inductive limits even for linear functionals (see Exercise 2.10.63).

2.4. Examples of inductive limits

We consider some more examples of inductive topologies.

2.4.1. Example. (*The least lower bound of a set of locally convex topologies in a vector space.*) Let E be a vector space such that for every α from some index set $\mathfrak{A}$ we have a locally convex topology τ_α in E. Then among all locally convex topologies in E majorized by the topologies of the family $\{\tau_\alpha\}$ there exists the strongest one – the inductive topology τ of the family of locally convex spaces (E, τ_α) with respect to the family of the "canonical" mappings $(E, \tau_\alpha) \to E$ (each of which is the identical mapping of the vector space E into itself). If $\mathcal{V}_\alpha$ is the class of all neighborhoods of zero in the topology τ_α, then $\bigcap_\alpha \mathcal{V}_\alpha$ is the class of all neighborhoods of zero in the topology τ (Remark 2.3.2).

2.4.2. Example. (*Quotients.*) Let E be a locally convex space and let E_1 be its vector subspace. Then the topology in the quotient E/E_1 making it a topological quotient is the inductive topology of the (one-element) family of locally convex spaces $\{E\}$ with respect to the (one-element) family of mappings from E to E/E_1 whose unique element is the canonical mapping of E onto E/E_1 (see Example 1.3.13).

2.4.3. Example. (*Topological direct sums of locally convex spaces.*) Let (E, τ) be a locally convex space such that the vector space E is a direct sum of a family $\{E_\alpha\}$ of its vector subspaces and every E_α is equipped with the locally convex topology τ_α induced by the topology τ. The space (E, τ) is called the *topological direct sum* of the family of its topological vector subspaces (E_α, τ_α) if the topology τ is the inductive limit of the topological vector spaces (E_α, τ_α) with respect to the canonical embeddings $E_\alpha \to E$.

In the just described situation we assume in advance that the topological vector space (E, τ) is given; so we consider the topological direct sum of a family of topological vector subspaces.

Suppose now that we are given a family $\{(E_\alpha, \tau_\alpha)\}$ of locally convex spaces about which we do not assume in advance that they are subspaces of some vector space.

In this case the topological direct sum of the family $\{(E_\alpha, \tau_\alpha)\}$ of locally convex spaces is the locally convex space (E, τ) defined as follows. The vector space E is the vector subspace in the product $\prod_\alpha E_\alpha$ of the family $\{E_\alpha\}$ of vector spaces consisting of all functions $f \in \prod_\alpha E_\alpha$ (defined on the index set $\mathfrak{A}$) each of which does not vanish in at most finitely many points; τ is the inductive topology of the family of locally convex spaces $\{(E_\alpha, \tau_\alpha)\}$ with respect to the "canonical" embeddings $g_\alpha \colon (E_\alpha, \tau_\alpha) \to E$, where $g_\alpha(x) = f \in E$ for all $\alpha \in \mathfrak{A}$ and $x \in E_\alpha$, and f is defined by $f(\alpha) = x$, $f(\beta) = 0$, whenever $\beta \neq \alpha$.

The topological direct sum of the family of locally convex spaces (E_α, τ_α) is denoted by $\bigl(\bigoplus_\alpha E_\alpha, \bigoplus_\alpha \tau_\alpha\bigr)$. Thus, $\bigl(\bigoplus_\alpha E_\alpha, \bigoplus_\alpha \tau_\alpha\bigr)$ is the topological direct sum of its topological vector subspaces $g_\alpha(E_\alpha)$.

Throughout we identify every locally convex space E_α with its image $g_\alpha(E_\alpha)$ in the sum $\bigoplus_\alpha E_\alpha$, assuming that the spaces E_α are topological vector subspaces in their topological direct sum.

It follows that the topological direct sum of a finite family of locally convex spaces and their product are the same object. For infinite families of topological vector spaces that are not locally convex, topological direct sums are considered very seldom (however, see Jarchow [**237**, § 4.3]).

2.4.4. Definition. *The topological vector space E equal the direct sum of its topological vector subspaces $E_1, \ldots, E_n$ is called the topological direct sum of the family $\{E_1, \ldots, E_n\}$ of its subspaces if E is canonically isomorphic to the product $\prod_{j=1}^n E_j$ of the topological vector spaces E_j.*

This means that the topology of E is the strongest one among all vector topologies in E for which all canonical embeddings $E_j \to E$ are continuous.

2.4.5. Proposition. *Let E_γ, where $\gamma \in \Gamma$, be a collection of locally convex spaces. Then the dual to their direct sum is the product of the dual spaces and the dual to their product is the direct sum of the dual spaces, i.e.,*

$$\Bigl(\bigoplus_\gamma E_\gamma\Bigr)' = \prod_\gamma E'_\gamma, \quad \Bigl(\prod_\gamma E_\gamma\Bigr)' = \bigoplus_\gamma E'_\gamma.$$

Moreover, a set is bounded in $\bigoplus_\gamma E_\gamma$ precisely when it is contained and is bounded in the sum of finitely many spaces E_γ; a set is bounded in $\prod_\gamma E_\gamma$ precisely when it is contained in the product of bounded sets in the factors.

In particular, $(\mathbb{R}^T)'$ is the direct sum of T copies of the real line.

PROOF. The first equality for the dual spaces is obvious. For verifying the second one we observe that if a linear function f on the product of the spaces E_γ is bounded on a basis neighborhood of zero of the form $U \times \prod_{\gamma \notin \Gamma_0}$, where Γ_0 is finite, then $f \in \bigoplus_{\gamma \in \Gamma_0} E'_\gamma$.

Let $A \subset \bigoplus_\gamma E_\gamma$ be bounded. If there are infinitely many indices γ_n for which the projections of A to E_γ contain nonzero vectors v_n, then, taking $f_n \in E'_{\gamma_n}$ such that $f_n(v_n) = n$, we obtain a continuous linear functional unbounded on A. The last assertion of the proposition is obvious. □

2.4.6. Proposition. *Let E be the inductive limit of a family $\{E_\alpha\colon \alpha \in \mathfrak{A}\}$ of locally convex spaces with respect to a family $\{g_\alpha \in \mathcal{L}(E_\alpha, E)\colon \alpha \in \mathfrak{A}\}$ of linear mappings such that the linear span of the set $\bigcup_\alpha g_\alpha(E_\alpha)$ coincides with E. Then E is isomorphic as a topological vector space to some topological vector quotient of the topological direct sum $\bigl(\bigoplus_\alpha E_\alpha, \bigoplus_\alpha \tau_\alpha\bigr)$.*

PROOF. Let $G = \bigl(\bigoplus_\alpha E_\alpha, \bigoplus_\alpha \tau_\alpha\bigr)$ and let $\Phi\colon\ G \to E$ be the linear mapping defined by the formula $\Phi(g) = \sum_\alpha g(\alpha)$; in the latter sum the number of nonzero terms is finite. The surjectivity of Φ follows from the coincidence of the linear span of $\bigcup_\alpha g_\alpha(E_\alpha)$ with E. The mapping Φ is continuous according to Proposition 2.3.5, since $g_\alpha = \Phi\circ \mathrm{im}_\alpha$, where im_α is the canonical embedding of E_α to $\bigoplus_\alpha E_\alpha$ defined in Example 2.4.3, where it is denoted by g_α. Let Ψ_1 denote the canonical mapping of the space G onto $G/\mathrm{Ker}\,\Phi$ and let Ψ_2 be the linear one-to-one mapping of the space $G/\mathrm{Ker}\,\Phi$ onto E defined by the equality $\Psi_2\circ\Psi_1 = \Phi$. We show that Ψ_2 and Ψ_2^{-1} are continuous; this will prove that E is isomorphic (as a topological vector space) to the space $G/\mathrm{Ker}\,\Phi$.

The continuity of Ψ_2 follows from Proposition 2.3.5 and the continuity of Ψ_1 and Φ; here the role of the inductive limit mentioned in Proposition 2.3.5 is played by the quotient $G/\mathrm{Ker}\,\Phi$ (it is the inductive limit of the family of spaces $\{E_\alpha\}$ with respect to the family of mappings $\{\Psi_1\circ g_\alpha\}$). The continuity of Ψ_2^{-1} follows, again according to Proposition 2.3.5, by the fact that $\Psi_2^{-1}\circ g_\alpha = \Psi_1\circ \mathrm{im}_\alpha$, where all mappings $\Psi_1\circ \mathrm{im}_\alpha$ are continuous (now the role of the inductive limit mentioned in Proposition 2.3.5 is played by E itself). □

In Chapter 3 we shall consider the duality between inductive and projective limits under various topologies on the dual space.

2.4.7. Example. (*Limits of direct spectra of locally convex spaces.*) Let $\mathfrak{A}$ be a directed set. A *direct spectre of locally convex spaces* with the index set $\mathfrak{A}$ is a family $\{(E_\alpha, \tau_\alpha)\colon\ \alpha \in \mathfrak{A}\}$ of locally convex spaces (with this index set) provided that to every pair $\alpha, \beta \in \mathfrak{A}$ of indices with $\alpha \leqslant \beta$ a continuous mapping $A_{\beta\alpha}\colon\ E_\alpha \to E_\beta$ is associated.

The limit of such direct spectre is the topological vector quotient of the topological direct sum $\bigl(\bigoplus_\alpha E_\alpha, \bigoplus_\alpha \tau_\alpha\bigr)$ by its vector subspace generated by the set G defined as follows: $f \in G$ if there exist $\alpha, \beta \in \mathfrak{A}$ such that $\alpha \leqslant \beta$, $f(\beta) = A_{\beta\alpha} f(\alpha)$ and $f(\gamma) = 0$ whenever $\gamma \notin \{\alpha, \beta\}$.

Every topological vector quotient of the topological direct sum of an arbitrary family of locally convex spaces is the inductive limit of this family with respect to the family of mappings that are the compositions of the canonical embeddings of the spaces in the family into their sum and the canonical mapping of the latter onto its quotient (this fact was used also in the proof of Proposition 2.4.6). Hence the limit of a direct spectre of locally convex spaces — we shall denote it by the symbol $\varinjlim E_\alpha$ — is the inductive limit of the family $\{(E_\alpha, \tau_\alpha)\colon\ \alpha \in \mathfrak{A}\}$ of locally convex spaces with respect to the just described family of mappings. Note also that here, unlike the case of limits of inverse spectra, in general one cannot assert that the space $\varinjlim E_\alpha$ is complete if all spaces (E_α, τ_α) are complete (see Exercise 2.10.23).

In Schaefer's book [**436**], the object called above "the limit of a direct spectre" is called "an inductive limit"; thus, the meaning of the term an "inductive limit" is broader here than in [**436**] (where it corresponds to an "inductive topology").

Let us now consider one special – but the most important for applications – class of direct spectra.

2.4.8. Example. Suppose that the index set $\mathfrak{A}$ is the set of natural numbers $\mathbb{N}$ with its usual order and that all mappings $A_{ij}\colon E_j \to E_i$ (defined for $j \leqslant i$) are injective. Due to the latter assumption we can assume that every locally convex space E_i is a vector subspace in E_{i+n} (notation: $E_i \subset\subset E_{i+n}$). By the continuity of the mappings A_{ij} the topology induced on E_i by the topology τ_{i+n} of the space E_{i+n} is majorized by the original topology τ_i of the space E_i.

Thus, we assume that $E_1 \subset\subset E_2 \subset\subset \ldots \subset\subset E_n \subset\subset \ldots$ and $E = \bigcup_{j=1}^{\infty} E_j$; the set E has a naturally defined structure of a vector space: if $x_1, x_2 \in E$, then $x_1, x_2 \in E_n$ for some $n \in \mathbb{N}$, and $\lambda_1 x_1 + \lambda_2 x_2$ in E is defined as the corresponding linear combination in E_n; since $E_n \subset\subset E_{n+j}$, this definition does not depend on our choice of E_n.

Under the stated conditions the vector space E equipped with the inductive topology of the family $\{E_n\colon n \in \mathbb{N}\}$ with respect to the embeddings $E_n \to E$ is "canonically isomorphic" to the space $\varinjlim E_n$.

This isomorphism $\Psi\colon E \to \varinjlim E_n$ can be described as follows. Let $x \in E$, i.e., $x \in E_n$ for some $n \in \mathbb{N}$. Then $\Psi(x)$ is an element of the space $\varinjlim E_n$ (that is a quotient of the space $\bigoplus_n E_n$) which is the image under the canonical mapping $\Phi\colon \bigoplus_n E_n \to \varinjlim E_n$ of the element $g_x^n \in \bigoplus_n E_n$ defined as follows: $g_x^n(j) = 0$ if $j \neq n$; $g_x^n(n) = x$. This definition is independent of our choice of n, since if $j \geqslant n$, then $A_{jn}(x) = x$ and hence $\Phi(g_x^n) = \Phi(g_x^j)$. The mapping Ψ is surjective: if $a \in \varinjlim R_n$, $g \in \Phi^{-1}(a)$, then $\Psi\bigl(\sum_n g(n)\bigr) = a$ (the set $\{n\colon g(n) \neq 0\}$ is finite).

Finally, the mappings Ψ and Ψ^{-1} are continuous; the proof of this is similar to the proof of Proposition 2.4.6.

Thus, the space E, equipped with the inductive topology defined above is a realization of a direct spectre. This realization will be denoted by the symbols $\operatorname{ind} E_n$ or $\operatorname{ind}_n E_n$; the locally convex space $\operatorname{ind} E_n$ will be called the *inductive limit of the increasing sequence of locally convex spaces* E_n.

Note that it often happens that it is not E that is constructed by means of a priori defined spaces E_n, but the spaces E_n are defined as suitable vector subspaces (with locally convex topologies) of an a priori given space E (without topology) that is their set-theoretic union. the term the "inductive limit of an increasing sequence of locally convex spaces E_n" means precisely the same as the term the "inductive limit of a family $\{E_n\}$ of locally convex spaces with respect to embeddings $E_n \to E$". Though, once the space $\bigcup_{n=1}^{\infty} E_n$ is introduced, the difference between the latter and the former situations disappears. One should have in mind that even the inductive limit of a sequence of Banach spaces can be non-separated (Exercise 2.10.24). The inductive limit of a sequence of Fréchet spaces is called an *LF-space*.

2.4.9. Example. Let $\mathcal{D}$ be the vector space of all infinitely differentiable (real) functions on $\mathbb{R}^1$ with compact support. For every $n \in \mathbb{N}$ let $\mathcal{D}_n := \mathcal{D}[-n,n]$ denote the subspace consisting of all functions vanishing outside of $[-n,n]$ equipped with the topology defined by the norms

$$p_i(\varphi) = \max_{j\in\{0,1,\dots,i\}} \max_t |\varphi^{(j)}(t)|.$$

Then $\mathcal{D}[-1,1] \subset\subset \mathcal{D}[-2,2] \subset\subset \cdots \subset\subset \mathcal{D}[-n,n] \subset\subset \cdots$ and $\bigcup_n \mathcal{D}[-n,n] = \mathcal{D}$, so that we are in the situation described in the second part of the previous example.

The inductive topology in $\mathcal{D}$ of the family of locally convex spaces $\mathcal{D}[-n,n]$, where $n \in \mathbb{N}$, with respect to the family of embeddings $\mathcal{D}[-n,n] \to \mathcal{D}$ coincides with the topology in $\mathcal{D}$ introduced in Example 1.3.21. Thus, $\mathcal{D} = \varinjlim \mathcal{D}[-n,n]$. This topology in $\mathcal{D}$ is regarded as the standard one; if it is not stated otherwise, it is assumed (in this book and also in other books) that $\mathcal{D}$ is equipped with this topology.

The space $\mathcal{D}[-n,n]$ can be naturally identified with the space of all infinitely differentiable functions on the interval $[-n,n]$ vanishing with all derivatives at the points $\pm n$; to every function in the first space we associate its restriction to the interval $[-n,n]$, and every function in the second space is associated with its extension by zero from $[-n,n]$ to $\mathbb{R}^1$. For the second space we use the same notation $\mathcal{D}[-n,n]$ and equip it with the topology defined by the same norms $p'_j(\varphi) = \max_{i\leqslant j} \max_t |\varphi^{(i)}(t)|$. Although now $\mathcal{D}[-n,n]$ is not formally a subspace in $\mathcal{D}[-n-j,n+j]$, we obtain the same inductive limit.

2.4.10. Remark. (i) For every $j \in \{0,1,2,\dots\}$ and every natural n let $\mathcal{K}^j_{[-n,n]}$ denote the vector space of all j-fold continuously differentiable real functions on $\mathbb{R}^1$ (for $j=0$ we obtain just continuous functions) vanishing outside the interval $[-n,n]$ and equipped with the topology generated by the norms q_j defined by the equality

$$q_j(\varphi) = \max\{\max_t |\varphi^{(i)}(t)|\colon\ i = 0,1,\dots,j\}$$

Then, for every $n \in \mathbb{N}$, the family $\{\mathcal{K}^j_{[-n,n]}\colon\ j = 0,1,\dots\}$ is the inverse spectre with respect to the embeddings $\mathcal{K}^{j+1}_{[-n,n]} \to \mathcal{K}^j_{[-n,n]}$, moreover, the spaces $\varprojlim_j \mathcal{K}^j_{[-n,n]}$ and $\mathcal{D}[-n,n]$ coincide as locally convex spaces (Example 2.4.9). At the same time, for every $j \geqslant 0$ the family $\{\mathcal{K}^j_{[-n,n]}\colon\ n \in \mathbb{N}\}$ forms a direct spectre with respect to the embeddings $\mathcal{K}^j_{[-n,n]} \to \mathcal{K}^j_{[-n-1,n+1]}$. For every $j \geqslant 0$ the set $\mathcal{K}^j = \operatorname*{ind}_n \mathcal{K}^j_{[-n,n]}$ (in place of the symbol $\mathcal{K}^0$ we shall usually use the symbol $\mathcal{K}$). The locally convex spaces $\{\mathcal{K}^j : j \geqslant 0\}$ form an inverse spectre with respect to the embeddings $\mathcal{K}^{j+1} \to \mathcal{K}^j$, and the locally convex space $\underleftarrow{\lim}\, \mathcal{K}^j$ can be identified as a vector space with the space $\bigcap_{j=1}^{\infty} \mathcal{K}^j$, which coincides as a vector space with $\mathcal{D}$. However, as locally convex spaces the space $\mathcal{D}$ with the standard topology and the space $\underleftarrow{\lim}\, \mathcal{K}^j$ are not isomorphic. When we identify them as vector spaces the topology of the space $\mathcal{D}$ turns out to be strictly stronger than the topology of the space $\underleftarrow{\lim}\, \mathcal{K}^j$. Note that no linear mapping of one of these spaces onto the other can be continuous in both directions (i.e., is not an isomorphism of topological vector spaces) at least for the reason that the locally convex space $\mathcal{D}$ with the standard topology is barrelled, but the locally convex space $\underleftarrow{\lim}\, \mathcal{K}^j$ is not (the definition of a barrelled space is given in § 3.5, the proof of the fact that $\underleftarrow{\lim}\, \mathcal{K}^j$ is not barrelled is delegated to Exercise 3.12.55).

Taking into account the definition of the standard topology in $\mathcal{D}$, the observations made above can be summarized as the inequality

$$\underleftarrow{\lim}_j \big(\operatorname*{ind}_n \mathcal{K}^j_{[-n,n]}\big) \neq \operatorname*{ind}_n \big(\underleftarrow{\lim}_j \mathcal{K}^j_{[-n,n]}\big).$$

Thus, the operations of forming direct and inverse spectra do not commute.

(ii) So far we have considered spaces of functions with compact support on $\mathbb{R}^1$ (for simplicity we had in mind real functions, but nothing changes if we consider complex spaces of complex functions). Completely analogous constructions apply to spaces of functions on $\mathbb{R}^n$ and even on domains in $\mathbb{R}^n$.

Namely, let Ω be an open set in $\mathbb{R}^n$. A function on Ω has compact support if it vanishes outside a compact set in Ω (we consider real or complex functions). Let $K_1 \subset K_2 \subset K_3 \subset \cdots$ be an increasing sequence of compact subsets of Ω such that $\bigcup_{m=1}^{\infty} K_m = \Omega$. If in the previous discussion of Example 2.4.9, the symbols $[-n, n]$ and $\mathbb{R}^1$ are replaced with K_m and Ω, the words "compact interval" by the word "compact", and the symbols $\mathcal{K}^j$ and $\mathcal{D}$ with the symbols $\mathcal{K}^j(\Omega)$ and $\mathcal{D}(\Omega)$ denoting the spaces of all j-fold differentiable and all infinitely differentiable functions with compact support in Ω, respectively, and assume that the symbols $|\varphi^{(j)}(x)|$ denote (certain) norms on the spaces to which belong the elements $\varphi^{(j)}(x)$, say, the sums of all mixed partial derivatives of order j, then all our definitions remain meaningful and all our assertions will be valid. For example, $\mathcal{D}(\mathbb{R}^n)$ is the inductive limit of the spaces $\mathcal{D}_m(\mathbb{R}^n)$ of smooth functions with support in the ball $K_m = \{x : |x| \leqslant m\}$.

2.4.11. Example. (*The strongest locally convex topology in a vector space.*) Let E be a vector space. Then the inductive topology in E of the empty family of locally convex spaces with respect to the empty family of mappings is the strongest locally convex topology in E; a base of neighborhoods of zero is, for example, the set of all convex circled absorbent subsets of the space E. This topology can be defined by the set of all seminorms on E. Note that the strongest locally convex topology in a vector space has already been considered in Examples 1.3.18 and 2.4.1. An interesting situation arises if we take for E the space $\mathbb{R}_0^{\infty}$ of all finite sequences, i.e., the union of $\mathbb{R}^n$. On this space the strongest locally convex topology coincides with the topology of the inductive limit of $\mathbb{R}^n$ (this is clear from Proposition 2.3.5).

2.4.12. Example. (*Spaces of germs of continuous, infinitely differentiable and analytic functions at a fixed point.*) We consider only the case of analytic functions, other cases are similar.

Let $z \in \mathbb{C}$ and let E_1 be the set of all complex functions each of which is defined in some neighborhood of the point z and is analytic in this neighborhood and let $\sim$ be the equivalence relation in E_1 defined as follows: $f \sim g$ if and only if there exists a neighborhood V of the point z such that f and g are defined in V and $f(x) = g(x)$ for all $x \in V$.

Finally, let E be the quotient of E_1 with respect to this equivalence relation. Then E is naturally equipped with a vector structure; its elements are called germs of analytic functions at the point z.

For every $n \in \mathbb{N}$ we take the space E^n of all complex functions continuous on the disc $S_n = \{x \in \mathbb{C}\colon |x - z| \leqslant 1/n\}$ and analytic inside S_n and equip it with the topology defined by the norm

$$p(\varphi) = \max\{|\varphi(x)|\colon |x - z| \leqslant 1/n\}.$$

Let g_n be the embedding of E^n into E associating to every function $\varphi \in E^n$ the germ containing it (note that in the remaining three cases mentioned above, similar mappings are not embeddings). We define similarly an embedding $E^n \to E^{n-1}$: to every function in E^n we associate its restriction to S_{n+1}.

Thus, $E^1 \subset\subset E^2 \subset\subset \cdots \subset\subset E^n \subset\subset \cdots$, $E^n \subset\subset E$, and $E = \bigcup_{n=1}^{\infty} E^n$. Let us equip E with the inductive topology of the family of locally convex spaces E^n with respect to the embeddings g_n. This makes E the inductive limit of the increasing sequence of locally convex spaces E_n.

Finally, note that spaces like $\mathcal{D}$ can be defined on smooth manifolds, and spaces of germs of holomorphic functions can be considered on complex manifolds.

In § 2.6 and § 2.7 we describe two classes of inductive limits of increasing sequences of locally convex spaces most often encountered in applications: strict inductive limits and inductive limits of increasing sequences of locally convex spaces with compact embeddings.

2.5. Grothendieck's construction

In this section we consider a method of constructing normed spaces associated with absolutely convex sets in topological vector spaces. This method, which became popular after Grothendieck's works and found numerous applications in the theory of locally convex spaces, the theory of Banach space, the operator theory, and measure theory, consists in the following procedure.

Let E be a Hausdorff topological vector space and let B be its bounded absolutely convex subset. Denote by E_B the vector subspace $\bigcup_{n=1}^{\infty} nB$ (the fact that this is indeed a vector space follows from the absolute convexity of the set B) equipped with the norm p_B that is the Minkowski functional of the set B. The space E_B is also equipped with the topology generated by this norm. The fact that the Minkowski functional is a seminorm on E_B follows from the condition that B is an absolutely convex absorbent subset of the vector space E_B; the condition that B is a bounded subset of a Hausdorff topological vector space yields that the regarded Minkowski functional is a norm.

The set E_B is a vector subspace of the vector space E. However, in the general case, it is not a topological vector subspace in E, i.e., the topology induced on E_B by the topology of the space E does not coincide with the just defined topology of the normed space E_B. The boundedness of the set B implies that the canonical embedding of E_B into E is continuous; this is equivalent to saying that the topology induced in E_B from E is weaker than the norm topology in E_B. See also § 2.10(iii).

2.5.1. Proposition. *If B is an absolutely convex bounded sequentially complete subset of a Hausdorff topological vector space E, then the normed space E_B is complete, i.e., is a Banach space.*

PROOF. Let a sequence $\{a_n\} \subset E_B$ be Cauchy in the norm p_B. This means that for every $\varepsilon > 0$ there exists $n_0 \in \mathbb{N}$ such that $p_B(a_k - a_j) < \varepsilon$ for all $k, j > n_0$. Since the topology induced in E_B by the topology of the space E is majorized by the topology defined by the norm p_B, the sequence $\{a_n\}$ is Cauchy also in E. Hence for every fixed $k \in \mathbb{N}$ the sequence $\{a_k - a_j\}$ is Cauchy in the topology of the space E. Since the inequality $p_B(a_k - a_j) < \varepsilon$ yields that $a_k - a_j \in \varepsilon B$, it follows that $a_k - a_j \in \varepsilon B$ whenever $k, j > n_0$. The set εB is sequentially complete in E, hence the Cauchy sequence $\{a_k - a_j\}_{j=1}^{\infty}$ of its elements converges in the topology of the space E to some element $b_k \in B$. Hence the sequence $\{a_n\}$ converges in the topology of E, moreover, if a is its limit, then $a_k - a = b_k \in \varepsilon B \subset E_B$ for all $k > n_0$.

Since $a_k \in E_B$, we have $a \in E_B$. In addition, the inclusion $a_k - a \in \varepsilon B$ (fulfilled for all $k > n_0$) yields that for such k we have $p_B(a_k - a) \leqslant \varepsilon$. This means that $a_k \to a$ in E_B, because ε was arbitrary. □

Similarly one can prove that if E is a vector space and τ_1 and τ_2 are two Hausdorff topologies in E compatible with the vector structure and τ_1 is stronger than τ_2 and possesses a base of neighborhoods of zero sequentially complete (or complete) in the topology τ_2, then the topological vector space (E, τ_1) is sequentially complete (respectively, complete).

2.5.2. Definition. *A Banach disc in a topological vector space E is an absolutely convex bounded subset B of the space E for which the space E_B is complete.*

The proposition above yields the following fact:
in order that an absolutely convex bounded subset B of a topological vector space be a Banach disc it is sufficient that it be sequentially complete (which holds, in particular, if B is compact); moreover, it is sufficient that B be sequentially complete in a topology in E that agrees with the duality between E and E'; in particular, it is sufficient that this set be compact in some of such topologies.

Neither of the aforementioned conditions is necessary.

2.5.3. Example. (i) The subset

$$\{x=(x_n)\in c_0\colon\ (nx_n)\in c_0,\ |x_n|\leqslant 1/n\ \ \forall n\}$$

of the space c_0 is a Banach disc, although it is not compact in any topology that agrees with the duality between c_0 and $(c_0)'=l^1$.

(ii) the closed unit ball U of $C[0,1]$ is a Banach disc in $L^2[0,1]$ which is not closed and not precompact. The closure of U in $L^2[0,1]$ is also a Banach disc for which the associated Banach space is $L^\infty[0,1]$, where $C[0,1]$ has infinite codimension.

(iii) The space $E=l^1$ is naturally embedded into l^2; let A be its closed unit ball. Let us take the functional $f\in l^\infty$ defined by the sequence $(1,1,\ldots)$. Set $B=A\cap f^{-1}(0)$ and obtain a Banach disc. Certainly, it is easy to describe B explicitly: this is the set of sequences $x=(x_n)$ such that $\sum_{n=1}^\infty |x_n|\leqslant 1$ and $\sum_{n=1}^\infty x_n=0$. The set B is the closed unit ball in the Banach space E_B, that is a closed hyperplane in l^1. But in l^2 the set B is not closed. For example, the vector $h=(2^{-1},0,0,\ldots)\notin B$ is the limit in l^2 of the sequence of vectors $h_k\in B$ of the form $\bigl(2^{-1},-(2k)^{-1},\ldots,-(2k)^{-1},0,0,\ldots\bigr)$, where the number of components $-(2k)^{-1}$ equals k; here $\|h-h_k\|_{l^2}^2=(4k)^{-1}$.

A similar property holds for the non-closed precompact set

$$\{x=(x_n)\in l^2\colon\ |x_n|\leqslant 2^{-n},\ \lim_{n\to\infty} 2^n x_n=0\}.$$

In fact, such examples are quite common.

2.5.4. Example. In every infinite-dimensional Fréchet space F there is a non-closed Banach disc B that serves as the closed unit ball of the Banach space E_B. In addition, we can take B totally bounded in F.

PROOF. The assertion reduces to the case $F=l^2$, since one can find an infinite-dimensional separable Hilbert space continuously embedded into F (then the previous example can be used). To this end it suffices to take in F a bounded sequence $\{v_n\}$ of linearly independent vectors and set $Tx=\sum_{n=1}^\infty 2^{-n}x_n v_n$, $x=(x_n)\in l^2$. The orthogonal complement to the kernel of T is injectively mapped to F and the image of the obtained mapping contains $\{v_n\}$. We observe that the constructed embedding is compact. □

Clearly, a similar example exists in every locally convex space to which one can continuously embed an infinite-dimensional Hilbert space.

It is rather surprising that the closure of a Banach disc is not always a Banach disc (see Pérez Carreras, Bonet [**385**, Remark 8.3.21]).

Let us give a simple application of the notion of a Banach disc.

2.5.5. Example. Let B be a Banach disc in a topological vector space E and let $S\subset E$ be an absolutely convex closed set such that its linear span contains B. Then there exists $t>0$ such that $tB\subset S$.

Indeed, the set $S\cap E_B$ is closed in the Banach space E_B with norm p_B and the latter is the union of the closed sets $n(S\cap E_B)$. By the classical Baire category theorem (see Bogachev, Smolyanov [**72**, § 1.5] or Rudin [**425**, p. 43]) for some n the set $n(S \cap E_B)$ has inner points (in the topology of the norm p_B). Hence this is true also for the set $S \cap E_B$, which by its absolute convexity proves our assertion.

Let us introduce one more important concept.

2.5.6. Definition. *A barrel in a locally convex space E is a closed absolutely convex absorbent set.*

The next simple but useful fact following from the previous example is called the Banach–Mackey theorem.

2.5.7. Theorem. *In every locally convex space every barrel absorbs every Banach disc.*

2.5.8. Proposition. *Let S be a convex subset of a locally convex space E such that every infinite sequence of its elements has a limit point in E. Then S is contained in some Banach disc. In particular, S is absorbed by every barrel.*

PROOF. In analogy with the usual l^1 we introduce the Banach space $l^1(S)$ consisting of real functions ξ on S non-vanishing in at most countably many points and having a finite norm $\|\xi\| = \sum_s |\xi(s)|$.

Let $\widetilde{E}$ denote the completion of E. Consider the mapping $T\colon\, l^1(S) \to \widetilde{E}$, $T\xi = \sum_{s\in S} \xi(s)s$; for every $x \in l^1(S)$ there are only countably many points s_n for which $\xi(s_n) \neq 0$, hence the series $\sum_{n=1}^{\infty} \xi(s_n)s_n$ converges in $\widetilde{E}$ by the completeness of $\widetilde{E}$ and boundedness of S following from our assumptions. It is also seen from this that T is continuous. The desired disc will be the image of the closed unit ball U in $l^1(S)$ if we verify that $S \subset T(U)$ and $T(U) \subset E$. The former is obvious, since $s = T(e_s)$, where e_s is the indicator function of the point s. For proving the latter it suffices to show that $T\xi \in E$ for all $\xi \in U$ such that $\xi(s_n) > 0$ for all points in the support of ξ. In this case $v_N = M_N^{-1} \sum_{n=1}^{N} \xi(s_n)s_n \in S$ by convexity, where $M_N := \sum_{n=1}^{N} \xi(s_n)$. By our assumption, the sequence $\{v_N\}$ has a limit point $v \in E$, but $M_N \to \|\xi\|$ and $M_N v_N \to T\xi$ in $\widetilde{E}$, whence it follows that $\|\xi\| v = T\xi$. Hence $T\xi \in E$. □

2.5.9. Theorem. *Let F be a metrizable locally convex space and let $A \subset F$ be a bounded set. Then there exists a bounded closed absolutely convex set $B \subset F$ such that $A \subset B$ and the normed space E_B induces on A the same topology as F.*

If A is precompact or compact in F, then it will retain this property in E_B.

In addition, for every sequence of bounded sets $A_n \subset F$ there is a bounded closed absolutely convex set D such that all sets A_n are bounded in E_D.

PROOF. Passing to the absolutely convex hull of A, we can assume that A is absolutely convex. In this case it suffices to find a bounded closed absolutely convex set $B \subset F$ containing A for which E_B and F induce the same neighborhoods of zero in A, which reduces to showing that, given a base $\{V_n\}$ of absolutely convex closed neighborhoods of zero in F, for every $\lambda > 0$ we can find n with $A \cap V_n \subset \lambda B$. If this is done, then for every point $a \in A$ and every $\lambda > 0$ we can take V_m such that $A \cap (a + V_m) \subset a + \lambda B$. Indeed, let

V_m be such that $A \cap V_m \subset 2^{-1}\lambda B$. If $a_1 = a + v \in A$, where $v \in V_m$, then $v/2 = (a_1 - a)/2 \in A \cap V_m \subset 2^{-1}\lambda B$ and hence $a_1 - a \in \lambda B$, i.e., $a_1 \in a + \lambda B$.

By the boundedness of A there are numbers $\lambda_n > 0$ for which $A \subset \bigcap_{n=1}^{\infty} \lambda_n V_n$. Clearly, we can assume that $\lambda_n \to +\infty$. Let us set $B := \bigcap_{n=1}^{\infty} \lambda_n^2 V_n$. Then B is absolutely convex, closed and bounded (since we have $\lambda_n^{-2} B \subset V_n$), $A \subset B$.

If $\lambda > 0$, then there exists n_0 such that $\lambda\lambda_n \geqslant 1$ for all $n \geqslant n_0$. Then $A \subset \lambda\lambda_n^2 V_n$ for all $n \geqslant n_0$. Since there exists m such that $V_m \subset \bigcap_{k \leqslant n_0} \lambda\lambda_k^2 V_k$, we have $A \cap V_m \subset \lambda\lambda_n^2 V_n$ for all n, whence we obtain $A \cap V_m \subset \lambda B$, as required.

Let us prove the last assertion. Let $\{V_n\}$ be a basis of closed neighborhoods of zero consisting of absolutely convex sets such that $V_{n+1} \subset V_n$. For every n there exists $c_n > 0$ such that $c_n A_n \subset V_n$. For D we can take the closed absolutely convex hull of the union of $c_n A_n$. □

Note that the spaces E_B and F induce on B also the same uniform structure (see § 1.12(i)).

Even if V is a convex balanced compact set in a Hilbert space, the Banach space (E_V, p_V) is not always separable. For example, this happens with the convex compact set $V = \{(x_n)\colon \ \sup_n |x_n| \leqslant 1\}$ in the weighted Hilbert space

$$E = \Bigl\{x = (x_n)\colon \ \|x\|^2 := \sum_{n=1}^{\infty} n^{-2} x_n^2 < \infty\Bigr\}.$$

In this case E_V coincides with l^∞. However, the following assertion is valid.

2.5.10. Theorem. *Let E be a complete metrizable locally convex space and let K be a compact set in E. Then there exists a convex balanced compact set V containing K such that the Banach space (E_V, p_V) is separable and K is compact with respect to the norm p_V.*

PROOF. Due to the completeness of E and two previous theorems, there exists a balanced convex compact set W_0 containing K such that K is compact as a subset of the Banach space E_{W_0}. Then the linear span of K in E_{W_0} is separable with respect to the norm p_{W_0}. The closure E_0 of this linear span with respect to the norm p_{W_0} in E_{W_0} gives the desired separable Banach space E_W, whose unit ball is the set $W := W_0 \cap E_0$. □

One can go further and obtain the following useful assertion — the Davies–Figiel–Johnson–Pełczyński theorem, a proof of which and of its corollary can be found in Diestel [**120**, Chapter V, § 4], Bogachev, Smolyanov [**72**, Theorem 8.6.24]. On reflexivity, see § 3.7.

2.5.11. Theorem. *Let X be a Banach space and let K be a convex balanced weakly compact set in X. Then there exists a bounded closed balanced convex set W containing K such that the Banach space (E_W, p_W) is reflexive. If K is compact, then W can be chosen compact and E_W can be made separable.*

2.5.12. Corollary. *In the situation of Theorem 2.5.10 the Banach space (E_V, p_V) can be chosen separable reflexive.*

As shown in Fonf, Johnson, Pisier, Preiss [**173**], it is not always possible to take for the space E_W a space with a Schauder basis (see the definition in Chapter 3).

In the case of compact V the topology of the space E_V is much stronger than the original topology of E. Nevertheless, as the following result shows, on the Banach space E_V there are sufficiently many linear functionals continuous with respect to this original topology.

2.5.13. Proposition. *Let V be a compact, convex and balanced set in a locally convex space E and let B^* be the unit ball in the dual to the Banach space E_V. Then the set of all functionals in B^* continuous with respect to the topology induced from E is dense in B^* in the topology of uniform convergence on compact sets in E_V.*

For a proof, see Bogachev, Smolyanov [**72**, Proposition 8.6.26].

2.5.14. Proposition. *Let $\{x_n\}$ be a sequence in a locally convex space converging to zero. Its closed absolutely convex hull K is compact precisely when the space E_K is Banach. In this case the compact K is metrizable.*

PROOF. If K is compact, then E_K is Banach. Conversely, let E_K be complete. Consider the linear mapping $T\colon l^1 \to E_K$, $T(y_n) = \sum_{n=1}^{\infty} y_n x_n$, which is defined and continuous, since the vectors x_n belong to the unit ball of E_K. Let U be the closed unit ball in l^1. Since $l^1 = c_0^*$, the set U is a metrizable compact in the topology $\sigma(l^1, c_0)$, see Theorem 3.1.4.

We show that T is continuous on U with this topology (on all of l^1 this can be false). If vectors $v^i = (v_n^i) \in U$ converge to $v = (v_n) \in U$ coordinate-wise, then for every continuous seminorm p on E and every $\varepsilon > 0$ we find a number m with the property that $p(x_n) < \varepsilon$ for all $n \geqslant m$, and then take a number M such that $|v_n^i - v_n| < \varepsilon\bigl(p(x_1) + \cdots + p(x_m) + 1\bigr)^{-1}$ for all $n = 1, \ldots, m$ and $i \geqslant M$. Then whenever $i \geqslant M$ we obtain the estimate

$$p(Tv^i - Tv) \leqslant \sum_{n=1}^{m} |v_n^i - v_n| p(x_n) + \varepsilon \sum_{n=1}^{\infty} \bigl(|v_n^i| + |v_n|\bigr) \leqslant 3\varepsilon,$$

which proves the continuity of T on U in the topology $\sigma(l^1, c_0)$. Therefore, $T(U)$ is a metrizable compact set in E. This yields at once also the compactness and metrizability of K, since $\{x_n\} \subset T(U)$, but actually we have $K = T(U)$, which is easily verified by observing that the absolutely convex hull of $\{x_n\}$ is dense in $T(U)$. □

2.5.15. Theorem. *In every locally convex space whose algebraic basis is at most of cardinality of the continuum there exists a hyperplane not containing Banach discs with an infinite-dimensional linear span.*

For a proof, see Pérez Carreras, Bonet [**385**, Theorem 6.3.11].

A compact set K in a Banach space X is called s-compact if it is contained in the closed absolutely convex hull of some sequence $\{x_n\}$ in X with the following property: for every $p \in \mathbb{N}$ the sequence $\{n^p x_n\}$ converges to zero.

2.5.16. Theorem. *For every s-compact set K in a Banach space X there exists an s-compact Banach disc C such that E_C is a separable Hilbert space, the embedding $E_C \to X$ is a nuclear operator (see the definition in § 2.9), and K remains s-compact also in E_C.*

For a proof, see [**385**, Theorem 6.5.4].

2.6. Strict inductive limits

Let $\{E_n\}$ be an increasing sequence of locally convex spaces:

$$E_1 \subset\subset E_2 \subset\subset \ldots \subset\subset E_n \subset\subset \ldots$$

such that for each index n the embedding $E_n \subset E_{n+1}$ is continuous, where $\subset\subset$ means "a linear subspace", as above. The union $E = \bigcup_{n=1}^{\infty} E_n$ is equipped with the strongest locally convex topology in which the embeddings $E_n \to E$ are continuous; for a base of neighborhoods of zero in it we can take convex sets $V \subset E$ such that every intersection $V \cap E_n$ is a neighborhood of zero in E_n. The space E is called the inductive limit of the sequence $\{E_n\}$ and denoted by the symbol $\mathrm{ind}_n E_n$ (see § 2.4). We shall see in § 2.10(i) that such inductive limit cannot be an inductive limit in the category of topological spaces, i.e., $\mathrm{ind}_n E_n$ can contain non-closed sets (even convex) whose intersections with all E_n are closed.

Two increasing sequences of locally convex spaces E_n and F_n are called equivalent if for every m there exist p and q such that $E_m \subset F_p$, $F_m \subset E_q$, and these embeddings are continuous.

It is readily seen that $\mathrm{ind}_n E_n = \mathrm{ind}_n F_n$ for two such sequences.

2.6.1. Definition. *If for all n the topology in E_n coincides with the topology induced from E_{n+1} and $E_n \neq E_{n+1}$, then the inductive limit of $\{E_n\}$ is called strict.*

The canonical example is the space $\mathcal{D}$ (see Example 2.4.9) represented as the union of the closed subspaces $\mathcal{D}_n$.

Let us show that the topology in E_n is induced by the topology in E.

2.6.2. Lemma. *If $E = \mathrm{ind}_n E_n$ is a strict inductive limit, then the topology in E_n is induced by the topology in E.*

If all spaces E_n are separated, then E is Hausdorff as well.

PROOF. It is clear that the topology in E_n is not weaker than the induced topology. Conversely, let U be an absolutely convex neighborhood of zero in E_n. Lemma 1.3.12 gives an absolutely convex neighborhood of zero U_{n+1} in E_{n-1} such that $U_{n+1} \cap E_n = U$. By induction we obtain increasing absolutely convex neighborhoods of zero $U_k \subset E_k$ with $k > n$ such that $U_{k+1} \cap E_k = U_k$. Then $V = \bigcup_{k=n+1}^{\infty} U_k$ is a neighborhood of zero in E and $V \cap E_n = U$. It is also seen from this that if all spaces E_n are separated, then so is E. □

2.6.3. Proposition. *Let E be the strict inductive limit of separated locally convex spaces E_n such that every E_n is closed in E_{n+1}.*

(i) *A set A is bounded in E if and only if A is contained in some space E_n and is bounded in it.*

(ii) *Every compact set in E is a compact subset of some E_n, and similarly for totally bounded sets. In particular, every countable sequence converging in E is contained and converges in some of the spaces E_n.*

(iii) *All subspaces E_n are closed in E.*

PROOF. (i) It is clear that if A is bounded in E_n, then it is bounded in E. Conversely, let $A \subset E$ be bounded. If A is contained in no E_n, then there exist increasing numbers k_n and points $a_n \in (E_{k_n} \cap A) \backslash E_{k_{n-1}}$. By using Lemma 1.3.12 and induction we find absolutely convex neighborhoods of zero $V_n \subset E_{k_n}$ such that $n^{-1}a_n \notin V_{n+1}$ and $V_{n+1} \cap E_{k_n} = V_n$ for all n. The set $V = \bigcup_{n=1}^{\infty} V_n$ is a neighborhood of zero in E, but $n^{-1}a_n \notin V$, which contradicts the boundedness of A. Thus, A must belong to one of the sets E_n, and then it is bounded in it due to the coincidence of the topology in E_n with the induced one. Assertion (ii) follows at once by (i). Finally, (iii) follows from the previous lemma, since if a net of elements $x_\alpha \in E_n$ converges to a point $x \in E$, then $x \in E_m$ for some $m \geqslant n$, but E_n is closed in E_m by the closedness of E_n in E_{n+1}. □

Note that for general inductive limits these assertions are false (see Exercise 2.10.25). In addition, one should bear in mind that not every subspace $L \subset E$ is the inductive limit of $L \cap E_n$ (see Proposition 2.10.7 and Exercise 2.10.63).

2.6.4. Corollary. *In the situation of the previous proposition the space E is not metrizable.*

PROOF. If E is metrizable, then, taking a basis of decreasing neighborhoods of zero U_n in E and choosing a vector $a_n \in U_n \backslash E_n$ for each n, we obtain a sequence converging to zero, which contradicts the proven proposition. □

2.6.5. Proposition. *The strict inductive limit E of complete separated locally convex spaces E_n is complete.*

PROOF. Suppose that there is a point $z \notin E$ in the completion $\widetilde{E}$ of the space E. We observe that E_n is closed in $\widetilde{E}$ by the completeness of E_n. Hence for each n there exists an absolutely convex neighborhood of zero W_n such that $(z + W_n) \cap E_n = \varnothing$. These neighborhoods can be chosen in such a way that $W_{n+1} \subset W_n$ for all n.

The convex hull of the set $\bigcup_{n=1}^{\infty}(2^{-1}W_n \cap E_n)$ will be denoted by U. Then U is a neighborhood of zero in E, therefore, its closure $\overline{U}$ in $\widetilde{E}$ is a closed neighborhood of zero in $\widetilde{E}$. Since E is dense in $\widetilde{E}$, the set $(z + \overline{U}) \cap E$ is nonempty; hence for some n there exists $v \in (z + \overline{U}) \cap E_n$, i.e., $v = z + u \in E_n$, where $u \in \overline{U}$. We show that $\overline{U} \subset W_n + E_n$, whence it will follow that $u = w_n + y$, where $w_n \in W_n$, $y \in E_n$, but this will give the inclusion $z + w_n = v - y \in E_n$ contradicting our choice of W_n.

Thus, let $x \in \overline{U}$. Then $x \in U + 2^{-1}W_n$, so $x = u + w_n/2$, where $u \in U$, $w_n \in W_n$. Hence $x = \lambda_1 x_1 + \cdots + \lambda_k x_k + w_n/2$, where $x_j \in 2^{-1}W_j \cap E_j$, $|\lambda_1| + \cdots + |\lambda_k| = 1$. We can assume that $k \geqslant n$. We have

$$\lambda_1 x_1 + \cdots + \lambda_n x_n \in E_n, \quad \lambda_{n+1} x_{n+1} + \cdots + \lambda_k x_k \in 2^{-1}W_n,$$

since $W_j \subset W_n$ for $j > n$ and the neighborhood W_n is absolutely convex. Thus $\lambda_1 x_1 + \cdots + \lambda_k x_k \in E_n + 2^{-1}W_n$, whence we obtain that $x \in E_n + W_n$, as required. □

For example, $\mathcal{D}(\mathbb{R}^n)$ is complete. Some additional interesting information can be found in Pérez Carreras, Bonet [**385**, Chapter 8].

2.7. Inductive limits with compact embeddings

We now turn to another important case where the topology of E_n is a priori strictly stronger than the induced topology of E_{n+1}.

We shall say that an increasing sequence of Hausdorff locally convex spaces E_n is *regular* if, for each n, there is a closed neighborhood of zero in E_n with compact closure in E_{n+1}. If one can find a closed neighborhood of zero in E_n compact in E_{n+1}, then we shall say that we have a *strongly regular* sequence.

It is easy to see that a regular sequence is not always strongly regular.

Recall that the equivalence of sequences of embedded spaces are defined in § 2.6.

2.7.1. Lemma. *Every regular sequence $\{E_n\}$ is equivalent to some strongly regular sequence of separable reflexive Banach spaces.*

PROOF. For every n we find in E_n a closed absolutely convex neighborhood of zero V_n whose closure K_n is compact in E_{n+1} and set $X_n := E_{K_n}$ (see § 2.5). Then X_n is a Banach space, $E_n \subset X_n \subset E_{n+1} \subset X_{n+1}$, and these embeddings are continuous. Hence the closed unit ball of X_n (i.e., K_n) is compact in X_{n-1}. Thus, the sequences $\{E_n\}$ and $\{X_n\}$ are equivalent and $\{X_n\}$ is strongly regular.

Now with the aid of Corollary 2.5.12 we can obtain an equivalent sequence of separable reflexive Banach spaces. □

A sequence of absolutely convex compacts K_n in a locally convex space E will be called regular if $E = \bigcup_{n=1}^{\infty} K_n$, every K_n is contained in the algebraic kernel of K_{n+1} (see § 1.10) and is compact in the Banach space $E_{K_{n+1}}$.

2.7.2. Lemma. *Let $\{K_n\}$ be a regular sequence of absolutely convex compact sets. Then for each m and each scalar λ there exists a number n such that λK_m is contained in the algebraic kernel of K_n.*

PROOF. For $n > m$ let G_n be the algebraic kernel of $K_n \cap E_{K_{m+1}}$. Zero belongs to the topological interior of the set $K_n \cap E_{K_{m+1}}$ in the Banach space $E_{K_{m+1}}$. Hence the algebraic kernel of this set coincides with the topological one in the space $E_{K_{m+1}}$ (Exercise 1.12.82). Thus, the sets G_n are open in $E_{K_{m+1}}$. They cover λK_m, since the algebraic kernels of K_n cover E (since K_n is contained in the algebraic kernel of K_{n+1}), moreover, K_m is compact in this space. Hence λK_m is contained in some G_n and so belongs to the algebraic kernel of K_n. □

2.7.3. Lemma. *Let $\{K_n\}$ be a regular sequence of absolutely convex compact sets. Then each absolutely convex compact set $K \subset E$ is contained in some K_n.*

PROOF. By the continuity of the embedding $E_K \to E$ we obtain that the absolutely convex set set $K_n \cap E_K$ is closed in the Banach space E_K. Hence some intersection $K_m \cap E_K$ contains a neighborhood of zero in E_K. Therefore, there exists $\lambda > 0$ such that $K \subset \lambda K_m$. By the previous lemma K belongs to some K_n. □

2.7.4. Lemma. *In every regular inductive limit of a sequence of locally convex spaces there is a regular sequence of absolutely convex compact sets.*

PROOF. We represent the given space in the form of a regular limit of some sequence of Banach spaces X_n in which increasing closed unit balls U_n are compact in X_{n+1}. Then we can take $K_n = nU_n$. □

2.7.5. Lemma. *Let $E = \mathrm{ind}_n E_n$ be a regular inductive limit of some sequence of locally convex spaces E_n, let $\{K_n\}$ be an arbitrary regular sequence of absolutely convex compacts in E. Then E is the inductive limit of the sequence of Banach spaces E_{K_n}.*

PROOF. Let us take the same X_n and U_n as in the previous proof. Then the sequences $\{X_n\}$ and $\{E_{K_n}\}$ are equivalent by Lemma 2.7.3. □

We shall say that a topological space X is a *free union* of a sequence of its subspaces X_n if $X = \bigcup_{n=1}^{\infty} X_n$ and the closed sets in X are exactly those sets whose intersection which each X_n is closed in X_n in the induced topology. Thus, the space X is the inductive limit of the subspaces X_n in the category of general topological spaces. It is important to emphasize that if X_n are increasing locally convex spaces continuously embedded into embracing spaces, then the topology of the inductive limit in the category of general topological spaces can be strictly stronger than the topology of the inductive limit in the category of locally convex spaces. This happens in the case of the space $\mathcal{D}(\mathbb{R}^1)$: as we shall see below, it *is not* a free union of its closed subspaces $\mathcal{D}[-n, n]$. There is a principal difference between the next theorem and the case of strict inductive limits.

2.7.6. Theorem. *The inductive limit $E = \mathrm{ind}_n E_n$ of a regular sequence of locally convex spaces is a free union of any regular sequence of its absolutely convex compact subsets.*

PROOF. Let $\{K_n\}$ be a regular sequence of absolutely convex compacts in E and let A be a set such that all intersections $A \cap K_n$ are closed. We show that A is closed. Let $x_0 \notin A$. By assumption there exists a number p for which x_0 is contained in K_{p-1}, hence also in the algebraic kernel of K_p. By induction we find an increasing sequence of sets V_n, $n \geqslant p$, with the following properties: 1) V_n is a closed absolutely convex neighborhood of zero in E_{K_n} and is compact in $E_{K_{n+1}}$, 2) $x_0 + V_n \subset K_n$, 3) $(x_0 + V_n) \cap A = \varnothing$.

For the set V_p we take a closed ball of a sufficiently small radius in E_{K_p}; this is possible, since $A \cap (A \cap E_{K_p})$ is closed in E_{K_p}, $x_0 \notin A \cap (A \cap E_{K_p})$ and x_0 is contained in the algebraic kernel of K_p. If increasing absolutely convex compact sets $V_p, \ldots, V_n$ with the desired properties are already picked in the respective spaces up to E_{K_n}, then V_{n+1} is constructed as follows.

The set $A \cap K_{n+1}$ is closed in $E_{K_{n+1}}$ by the closedness in E and continuity of the embedding $E_{K_{n+1}} \to E$. The set $x_0 + V_n$ is disjoint with $A \cap K_{n+1}$ and with the unit sphere S_{n+1} in $E_{K_{n+1}}$ (since along with K_n it is contained in the algebraic kernel of K_{n+1}). Since $x_0 + V_n$ is compact in $E_{K_{n+1}}$ and the set $(A \cap K_{n+1}) \cup S_{n+1}$ is closed in the Banach space $E_{K_{n+1}}$, the distance d between them in the norm of $E_{K_{n+1}}$ is positive.

Therefore, setting $V_{n+1} = V_n + 2^{-1}dK_{n+1}$, we obtain an absolutely convex set compact in $E_{K_{n+2}}$ due to the compactness of V_n and K_{n+1} in $E_{K_{n+2}}$. It is clear that $V_n \subset V_{n+1}$ for all n and V_{n+1} is a closed neighborhood of zero in the space $E_{K_{n+1}}$. Finally, $x_0 + V_{n+1} \subset K_{n+1}$ and $(x_0 + V_{n+1}) \cap A = \varnothing$ due to our choice of d.

Thus, the sets V_n are constructed. Then the set $V = \bigcup_{n=p}^{\infty} V_n$ is a closed absolutely convex neighborhood of zero in E and $(x_0 + V) \cap A = \varnothing$. Therefore, $E \backslash A$ is open in E, i.e., A is closed. □

2.7.7. Corollary. *A set A in a regular inductive limit* $\mathrm{ind}_n E_n$ *is closed precisely when every intersection $A \cap E_n$ is closed in E_n.*

PROOF. Let $A \cap E_n$ be closed in E_n for all n. Taking a regular sequence of absolutely convex compact sets K_n, we obtain that $A \cap K_n$ is closed, which gives the closedness of A. The converse is obvious. □

2.7.8. Corollary. *A set A in a regular inductive limit* $\mathrm{ind}_n E_n$ *is closed precisely when it is sequentially closed, i.e., contains the limits of all its convergent countable sequences.*

PROOF. Let A be sequentially closed in E. We represent E as a regular inductive limit of a sequence of Banach spaces X_n. Then $A \cap X_n$ is sequentially closed in each Banach space X_n, which implies that A is closed in E. The converse is obvious. □

We observe that the previous two corollaries are not true for strict inductive limits (considered in § 2.6), but the next corollary is true for them.

2.7.9. Corollary. *In the situation of the theorem above, every compact set in the space E is contained in some subspace E_n. In addition, every convergent sequence in E is contained and converges in some E_n.*

Moreover, every bounded set in E is contained in some E_n and has compact closure there.

PROOF. It is clear that we can choose a regular sequence of absolutely convex compact sets K_n such that $2K_n \subset K_{n+1}$ for all n. Suppose that for every n there is a point a_n of a given bounded set A not belonging to K_{2n}. Then $2^{-n}a_n \notin K_n$, since $K_n \subset 2^{-n}K_{2n}$. The countable set $\{2^{-n}a_n\}$ has a finite intersection with each K_n and according to the results above turns out to be closed in E. However, this is impossible, since by the boundedness of the set A the sequence $\{2^{-n}a_n\}$ converges to zero that does not belong to this sequence. Thus, the set A is contained in some K_m, but we already know that the set K_m is contained and compact in some E_n. □

If the spaces E_n are distinct, then we see on account of the obtained results that the corresponding regular inductive limit is not metrizable. The proof of the following fact is delegated to Exercise 2.10.28.

2.7.10. Corollary. *Any regular inductive limit is complete.*

2.7.11. Example. Let E be the inductive limit of an increasing sequence of reflexive Banach spaces X_n with compact embeddings. Then E is a regular inductive limit. In particular, this is true if all spaces X_n are Hilbert. If all X_n are finite-dimensional, then this is true also in the case of continuous embeddings.

Indeed, any closed ball in X_n is compact in X_{n+1} by the weak compactness (which follows from the reflexivity) and totally bounded (which follows from the compactness of the embedding).

Here is yet another property of regular inductive limits that is not possessed by strict inductive limits (see Exercise 2.10.63).

2.7.12. Proposition. *Any closed linear subspace of a regular inductive limit* $\mathrm{ind}_n E_n$ *equipped with the induced topology is again a regular inductive limit.*

PROOF. Suppose that F is a closed linear subspace in a regular inductive limit $E = \mathrm{ind}_n E_n$. We can assume that E_n are reflexive Banach spaces such that the closed unit ball K_n of E_n is contained and is compact in the unit ball K_{n+1} of E_{n+1}. We show that $Q_n = K_n \cap F$ is a regular sequence of compact sets in F. It is clear that Q_n is an absolutely convex compact set in F, closed in $E_{Q_{n+1}}$. In addition, the set Q_n is totally bounded in $E_{Q_{n+1}}$. Indeed, it is compact in $E_{K_{n+1}}$, so for every $\varepsilon > 0$ there exist points $x_1, \ldots, x_m \in Q_n$ for which $Q_n \subset \bigcup_{i=1}^m (x_i + \varepsilon K_{n+1})$, whence $Q_n \subset \bigcup_{i=1}^m (x_i + \varepsilon Q_{n+1})$. Therefore, $\{Q_n\}$ is a regular sequence of compact sets in the subspace F. We now verify that F with the induced topology coincides with the regular inductive limit $\mathrm{ind}_n E_{Q_n}$. Let $A \subset F$ be such that all $A \cap Q_n$ are closed. Then $A \cap K_n$ is closed as well, whence it follows that A is closed in E, hence also in F with the induced topology. The converse is also true: if A is closed in F with the induced topology, then all sets $A \cap Q_n$ are closed. Thus, F with the induced topology is a free union of Q_n and hence coincides with $\mathrm{ind}_n E_{Q_n}$. □

2.8. Tensor products

Let E_1 and E_2 be two locally convex spaces. The algebraic tensor product $E_1 \otimes E_2$ is usually defined as the natural quotient of the linear space formally generated by expressions of the form $x \otimes y$, where $x \in E_1$, $y \in E_2$, but here it will be convenient to define $E_1 \otimes E_2$ from the very beginning as the linear subspace in the algebraic dual to the space $B(E_1, E_2)$ of bilinear functions on $E_1 \times E_2$ generated by the elements

$$x \otimes y \colon\ b \mapsto b(x, y), \quad b \in B(E_1, E_2).$$

When using this embedding the necessary factorization is done automatically.

The space $E_1 \otimes E_2$ can be equipped with different locally convex topologies. The *projective topology* τ_π is defined as the strongest locally convex topology in which the canonical bilinear mapping $E_1 \times E_2 \to E_1 \otimes E_2$ is continuous.

It is readily verified (see Schaefer [**436**, p. 93]) that the dual to the space $(E_1 \otimes E_2, \tau_\pi)$ can be identified with the space of all continuous bilinear functions on $E_1 \times E_2$ in such a way that equicontinuous subsets in $(E_1 \otimes E_2, \tau_\pi)'$ will correspond to equicontinuous sets of bilinear functions on $E_1 \times E_2$. In order to define τ_π by means of seminorms the following construction of the tensor product of seminorms p and q on E_1 and E_2 is used. For $w \in E_1 \otimes E_2$ set

$$p \otimes q(w) = \inf\Bigl\{\sum_i p(x_i)q(y_i) \colon\ w = \sum_i x_i \otimes y_i\Bigr\},$$

where inf is taken over all representations of w in the indicated form. It is readily verified that $p \otimes q(x \otimes y) = p(x)q(y)$.

Any locally convex topology can be always defined by a directed family of seminorms $\mathcal{P}$, i.e., in such a way that for any seminorms $p_1, p_2 \in \mathcal{P}$ there exists $p_3 \in \mathcal{P}$ with $p_1 \leqslant p_3$, $p_2 \leqslant p_3$ (for example, one can take all possible finite sums of the original seminorms and their products by positive numbers). Let us take directed families of seminorms $\mathcal{P}_1$ and $\mathcal{P}_2$ on E_1 and E_2 defining the topologies. Then the collection of all seminorms $p_1 \otimes p_2$, where $p_1 \in \mathcal{P}_1$, $p_2 \in \mathcal{P}_2$, defines the *projective topology* τ_π. The completion of the space $(E_1 \otimes E_2, \tau_\pi)$ will be denoted by $E_1 \widetilde{\otimes}_\pi E_2$. The following result was obtained by Grothendieck.

2.8.1. Theorem. *If E_1 and E_2 are metrizable locally convex spaces, then every element $w \in E_1 \widetilde{\otimes}_\pi E_2$ can be represented in the form*

$$w = \sum_{i=1}^{\infty} \lambda_i x_i \otimes y_i, \quad \textit{where} \ \sum_{i=1}^{\infty} |\lambda_i| < \infty,$$

where $x_i \to 0$ and $y_i \to 0$ in E_1 and E_2, respectively.

PROOF. Let us take in E_1 and E_2 increasing sequences of seminorms p_n and q_n generating the topologies. Set $r_n = p_n \otimes q_n$ and extend r_n to $E_1 \widetilde{\otimes}_\pi E_2$. There exist $w_n \in E_1 \otimes E_2$ such that $r_n(w - w_n) \leqslant n^{-2}2^{-n-1}$ and $w_1 = 0$. Let $w_n = \sum_{i=1}^{i_n} \lambda_i x_i \otimes y_i$ and $v_n = w_{n+1} - w_n$. Then

$$\begin{aligned} r_n(v_n) &\leqslant r_n(w - w_n) + r_n(w - w_{n+1}) \\ &\leqslant r_n(w - w_n) + r_{n+1}(w - w_{n+1}) \leqslant n^{-2}2^{-n}. \end{aligned}$$

It follows from the definition of r_n that there exist increasing numbers i_n, $n \in \mathbb{N}$ and representations $v_n = \sum_{i=i_n+1}^{i_{n+1}} \lambda_i \lambda_i x_i \otimes y_i$ such that $\sum_{i=i_n+1}^{i_{n+1}} |\lambda_i| \leqslant 2^{-n}$, $p_n(x_i) \leqslant n^{-1}$, $q_n(y_i) \leqslant n^{-1}$ for all $i_n < i \leqslant i_{n+1}$. Indeed, take a representation of the form $v_n = \sum_{i=i_n+1}^{i_{n+1}} u_i \otimes z_i$ with $\sum_{i=i_n+1}^{i_{n+1}} p_n(u_i) q_n(z_i) \leqslant n^{-2}2^{-n}$, where $p_n(u_i) > 0$ and $q_n(z_i) > 0$. Now we can take $x_i = u_i \bigl(n p_n(u_i)\bigr)^{-1}$, $y_i = z_i \bigl(n q_n(z_i)\bigr)^{-1}$, $\lambda_i = n^2 p_n(u_i) q_n(z_i)$. Thus, the required representation is given by $w = \sum_{n=1}^{\infty} v_n = \sum_{i=1}^{\infty} \lambda_i x_i \otimes y_i$. $\square$

Taking absolutely convex neighborhoods of zero $U\subset E_1$ and $V\subset E_2$, we can define the following seminorm on $E_1\otimes E_2$:

$$p_{U,V}(w)=\sup\Bigl\{\sum_i f(x_i)g(y_i)\colon\ w=\sum_i x_i\otimes y_i,\, f\in U^\circ, g\in V^\circ\Bigr\},$$

where sup is taken over all representations of w in the indicated form. It is readily verified that $p_{U,V}(w)\leqslant p_U\otimes p_V$. Hence the topology τ_ε generated by such seminorms is weaker than τ_π. This weaker topology is called the *topology of equicontinuous convergence.*

The completion of $(E_1\otimes E_2,\tau_\varepsilon)$ will be denoted by the symbol $E_1\widetilde{\otimes}_\varepsilon E_2$.

We shall see in the next section that for a given locally convex space E the equality $E\widetilde{\otimes}_\varepsilon F=E\widetilde{\otimes}_\pi F$ holds for all locally convex spaces F precisely when E is nuclear (which is also defined in the next section). Actually, it suffices to have this equality for all Banach spaces. Since infinite-dimensional Banach spaces are not nuclear, it is easy to give examples where the equality fails. Below we describe these two topologies explicitly in case of Hilbert spaces.

Yet another projective topology on the tensor product $E_1\otimes E_2$ is called *inductive.* It is denoted by the symbol τ_i and is defined as the strongest locally convex topology on $E_1\otimes E_2$ such that the canonical bilinear mapping $E_1\times E_2\to E_1\otimes E_2$ is separately continuous. Hence the projective and inductive topologies coincide if all separately continuous bilinear mappings on $E_1\times E_2$ are continuous. For example, the projective tensor topology on $\mathcal{D}(\mathrm{I\!R})\otimes\mathcal{D}(\mathrm{I\!R})$ is strictly weaker than the inductive topology, since there is a discontinuous separately continuous bilinear form on $\mathcal{D}(\mathrm{I\!R})$, see Exercise 2.10.49.

It is instructive to examine the case of Hilbert spaces E_1 and E_2. The projective tensor topology is generated by the so-called nuclear norm

$$\|w\|_{\mathcal{N}}=\inf\Bigl\{\sum_i \|x_i\|\,\|y_i\|\colon\ w=\sum_i x_i\otimes y_i\Bigr\},$$

the completion with respect to which in case $E_1=E_2=H$ gives the space $\mathcal{N}(H)$ of nuclear (or trace class) operators on H, i.e., $H\widetilde{\otimes}_\pi H=\mathcal{N}(H)$. The inductive topology gives the same. The topology τ_ε is generated by the norm

$$\|w\|_\infty=\sup\Bigl\{\sum_i f(x_i)g(y_i)\colon\ w=\sum_i x_i\otimes y_i,\ \|f\|_{E_1'}\leqslant 1,\ \|g\|_{E_2'}\leqslant 1\Bigr\},$$

which is the operator norm, and in case $E_1=E_2=H$ the completion of $H\widetilde{\otimes}_\varepsilon H$ is the space $\mathcal{K}(H)$ of compact operators on H. Actually, both constructions apply to any Banach spaces (where the elements $x\otimes y$ are regarded as operators from E_1' to E_2 acting by the formula $l\mapsto l(x)y$). However, in the Hilbert case the space $E_1\otimes E_2$ can be equipped with the Hilbert–Schmidt norm (when we identify elements of $E_1\otimes E_2$ with operators from E_1 to E_2), which in case $E_1=E_2=H$ after completing leads to the space $\mathcal{H}(H)$ of Hilbert–Schmidt operators (p. 144). This latter tensor product is called the Hilbert tensor product and is denoted by $E_1\otimes_2 E_2$. Note that in these examples the regarded norms on elements $x\otimes y$ equal $\|x\|\,\|y\|$; such norms on $E_1\otimes_2 E_2$ are called cross-norms. The operator norm and nuclear norm are extreme cross-norms. Using other norms on $E_1 \otimes_2 E_2$ we obtain different tensor products as completions.

Tensor products can be introduced in the language of categories, and they are connected with the so-called representable functors. We make brief remarks in this direction, although categories are never used in this book. In the theory of categories (see Mac Lane [**326**]), it is customary to consider along with usual sets certain "very large sets", called "classes" to distinguish them from sets, in order to use expressions like "the class of all locally convex spaces" (we recall that the concept "the set of all sets" is contradictory). A category is a class $\mathfrak{C}$ whose elements are called objects of the category, and it is assumed that for each pair A, B of objects we are given a set $\hom(A, B)$ whose elements are called morphisms from A to B, provided that they satisfy certain natural axioms.

For example, in the category of sets morphisms are arbitrary mappings. Another relevant example is the category $\mathcal{LCS}$ of real locally convex spaces; morphisms are continuous linear mappings. Certainly, one can also consider the category of complex locally convex spaces. In the category of topological spaces morphisms are continuous mappings.

We say that there is a covariant functor from a category $\mathfrak{C}_1$ to a category $\mathfrak{C}_2$ if to every object A of $\mathfrak{C}_1$ an object $F(A)$ of $\mathfrak{C}_2$ is defined and for every morphism $\varphi \in \hom(A, B)$, where $A, B \in \mathfrak{C}_1$, a morphism $F(\varphi) \in \hom\bigl(F(A), F(B)\bigr)$ is defined such that the identity mappings are taken to identity mappings and F respects compositions (provided they are defined). Contravariant functors are defined similarly, but $F(\varphi) \in \hom\bigl(F(B), F(A)\bigr)$.

For example, let $\mathfrak{S}$ be the category of sets and let $\mathfrak{C}$ be some category. The mapping $h^A \colon C \to \hom(A, C)$, where $A, C \in \mathfrak{C}$, along with the class of naturally defined mappings of sets $\hom(A_1, C)$ to sets $\hom(A_2, C)$ is a covariant functor from $\mathfrak{C}$ to $\mathfrak{S}$.

Isomorphisms of functors are defined in a natural way. A covariant functor F from a category $\mathfrak{C}$ to the category of sets $\mathfrak{S}$ is called representable if, for some object $A \in \mathfrak{C}$, there is an isomorphism of functors $f \colon h^A \to F$; then we say that the object A represents the functor F.

Let us return to the category $\mathcal{LCS}$ of real locally convex spaces. Suppose that we are given $E, G \in \mathcal{LCS}$. For each space $K \in \mathcal{LCS}$ let $B(E \times G, K)$ be the set of all continuous bilinear mappings from $E \times G$ to K. The category $\mathcal{T}$ will be defined as follows: its objects are pairs (K, B), where $K \in \mathcal{LCS}$, $B \in B(E \times G, K)$, and morphisms between objects (K_1, B_1) and (K_2, B_2) are linear operators $\psi \colon K_1 \to K_2$ such that $\psi \circ B_1 = B_2$. Another category is obtained by using the set $B_s(E \times G, K)$ of all separately continuous bilinear mappings.

One can verify that in both cases we obtain a representable covariant functor from the category $\mathcal{LCS}$ to the category of sets $\mathfrak{S}$. In the first case (continuous bilinear mappings) the representing object is the completion of the tensor product $E_1 \otimes E_2$ with the projective topology. The functor isomorphism $f \colon h^{E \otimes G} \to F$ takes the set of continuous linear operators from $E \otimes G$ to K to the set of their compositions with the canonical bilinear mapping from $E \times G$ to $E \otimes G$, which is continuous in this case.

In the second case (separately continuous bilinear mappings) the representing object is the completion of the tensor product $E_1 \otimes E_2$ with the inductive topology. The functor isomorphism is described similarly

2.9. Nuclear spaces

Let X and Y be two normed spaces. An operator $T \in \mathcal{L}(X,Y)$ is called *nuclear* if it can be represented in the form

$$Tx = \sum_{i=1}^{\infty} u_i(x)v_i, \quad \text{where } u_i \in X',\ v_i \in Y,\ \sum_{i=1}^{\infty} \|u_i\|\,\|v_i\| < \infty.$$

The infimum of the sums $\sum_{i=1}^{\infty} \|u_i\|\,\|v_i\|$ over all possible representations of T is called the *nuclear norm* of T and is denoted by the symbol $\|T\|_{\mathcal{N}}$. It is obvious that any nuclear operator between Banach spaces is compact (see the definition in § 3.10).

The concept of nuclear operator is naturally extended to locally convex spaces.

If V is an absolutely convex neighborhood of zero in a locally convex space E, then the Minkowski functional p_V of the set V is a seminorm and generates a norm on the quotient $E/p_V^{-1}(0)$. The completion of this normed space will be denoted by $\widetilde{E}_V$. The norm in $\widetilde{E}_V$ will be also denoted by the symbol p_V. If $p_V^{-1}(0) = 0$, then $\widetilde{E}_V$ is the completion of the usual space (E_V, p_V). We recall that the latter is complete if V is bounded and sequentially complete. However, in typical cases the set V is not bounded and $p_V^{-1}(0) \neq 0$. The natural mapping $j_V : E \to \widetilde{E}_V$ is continuous. Note also that if V contains an absolutely convex neighborhood of zero W, then the operator $\pi_{V,W} : \widetilde{E}_W \to \widetilde{E}_V$ generated by the natural inclusion $p_W^{-1}(0) \subset p_V^{-1}(0)$ and the surjection $E/p_W^{-1}(0) \to E/p_V^{-1}(0)$, is continuous.

Let E and F be two locally convex spaces and let $T\colon E \to F$ be a linear mapping such that for some absolutely convex neighborhood of zero $V \subset E$ the set $T(V)$ is contained in some Banach disc $B \subset F$. Let $\psi_B\colon F_B \to F$ be the natural embedding. Since $Tx = 0$ if $p_V(x) = 0$ (by the boundedness of $T(V)$), the operator T can be written in the form $T = \psi_B \circ \widetilde{T} \circ j_V$, where $\widetilde{T}\colon \widetilde{E}_V \to F_B$ is a continuous linear operator. If we can choose V and B in such a way that the operator $\widetilde{T}$ between the indicated Banach spaces becomes nuclear, then T is called *nuclear*. This yields at once that any nuclear operator is compact also in the case of locally convex spaces (see § 3.10). For normed spaces this gives the previous concept.

One can verify that nuclear mappings are characterized in the following way (see Schaefer [**436**, Theorem 7.1]).

2.9.1. Theorem. *An operator $T \in \mathcal{L}(E,F)$ is nuclear precisely when it has the form*

$$Tx = \sum_{n=1}^{\infty} \lambda_n l_n(x) y_n,$$

where $\sum_{n=1}^{\infty} |\lambda_n| < \infty$, $\{l_n\}$ is an equicontinuous sequence in E' and $\{y_n\} \subset F$ is a sequence contained in some Banach disc.

2.9.2. Definition. *A locally convex space E is called nuclear if it possesses a basis $\mathcal{V}$ of absolutely convex neighborhoods of zero with the property that for every neighborhood $V \in \mathcal{V}$ the canonical mapping $j_V: E \to \widetilde{E}_V$ is nuclear.*

The previous theorem implies that the canonical mappings j_V must have the form

$$j_V x = \sum_{n=1}^{\infty} \lambda_n l_n(x) y_n,$$

where $\sum_{n=1}^{\infty} |\lambda_n| < \infty$, $\{l_n\}$ is an equicontinuous sequence in E' and $\{y_n\} \subset V$.

It is readily verified that a locally convex space is nuclear precisely when its completion is nuclear.

As an example of an infinite-dimensional nuclear space one can take an infinite power $\mathbb{R}^T$ of $\mathbb{R}$; for a basis of neighborhoods of zero we can take the sets $V = \{x\colon |x(t_i)| < \varepsilon, i = 1, \ldots, n\}$, so $p_V^{-1}(0) = \{x\colon x(t_i) = 0, i = 1, \ldots, n\}$, hence the quotient $\mathbb{R}^T / p_V^{-1}(0)$ is finite-dimensional.

2.9.3. Theorem. *The following properties of a locally convex space E are equivalent:*

(i) *E is nuclear;*

(ii) *for every Banach space X, every operator $T \in \mathcal{L}(E, X)$ is nuclear;*

(iii) *every absolutely convex neighborhood of zero V in the space E contains an absolutely convex neighborhood of zero U such that the associated operator $\pi_{V,U}\colon \widetilde{E}_U \to \widetilde{E}_V$ is nuclear.*

PROOF. If E is nuclear and X is a Banach space with the open unit ball U, then $V = T^{-1}(U)$ is a neighborhood of zero in E, so the mapping $j_V: E \to \widetilde{E}_V$ is nuclear. Hence it has the form $\sum_{n=1}^{\infty} \lambda_n l_n(x) y_n$, where $\{l_n\}$ is an equicontinuous sequence in E', $\{\lambda_n\} \in l^1$, $\{y_n\}$ is contained in a Banach disc in $\widetilde{E}_V$. We observe that $Tx = \sum_{n=1}^{\infty} \lambda_n l_n(x) T y_n$, where $\{Ty_n\}$ is contained in a Banach disc in X.

It is clear that (iii) implies (i). Let (ii) hold and let V be an absolutely convex neighborhood of zero in E. Then the operator $j_V: E \to \widetilde{E}_V$ has the form indicated above. By the equicontinuity of $\{l_n\}$ there exists an absolutely convex neighborhood of zero W for which $|l_n(w)| \leqslant 1$ for all $w \in W$, $n \geqslant 1$. Let us take $U = W \cap V$. It is straightforward to verify that the operator $\pi_{V,U}$ is nuclear, since it has the form $\pi_{V,U} z = \sum_{n=1}^{\infty} \widetilde{l}_n(z) y_n$, where the functional $\widetilde{l}_n \in \widetilde{E}_U'$ generated by the functional l_n (the estimate $|l_n(w)| \leqslant 1$ on W means that the norm of $\widetilde{l}_n$ does not exceed 1) and the sequence $\{y_n\}$ is bounded in $\widetilde{E}_V$. □

Property (iii) and the compactness of nuclear operators yield the following important fact.

2.9.4. Corollary. *Every bounded set in any nuclear space is precompact.*

It is obvious from Property (ii) that a normed space is nuclear only when it is finite-dimensional. Nevertheless, as we shall now see, nuclear spaces are closely connected with Hilbert spaces.

2.9.5. Theorem. *Let U be a neighborhood of zero in a nuclear space E. Then there exist an absolutely convex neighborhood of zero $V \subset U$, $W \subset V$ such that the Banach spaces $\widetilde{E}_V$ and $\widetilde{E}_W$ are linearly isometric to l^2 or $\mathbb{R}^n$ and the operator $\pi_{V,W}\colon \widetilde{E}_W \to \widetilde{E}_V$ is nuclear. Hence the topology of E is generated by a family of seminorms defined by nonnegative definite Hermite forms.*

PROOF. We shall show that there exists an operator $A \in \mathcal{L}(E, l^2)$ such that $V = A^{-1}(B) \subset U$, where B is the open unit ball in l^2. We can assume that U is absolutely convex. Then $j_U\colon E \to \widetilde{E}_U$ has the form $j_U x = \sum_{n=1}^{\infty} \lambda_n l_n(x) y_n$, where $\lambda_n > 0$, $\sum_{n=1}^{\infty} \lambda_n = 1$, $\|y_n\| = 1$ in $\widetilde{E}_U$ and $\{l_n\}$ is equicontinuous in E'. Set $Ax = \bigl\{\sqrt{\lambda_n} l_n(x)\bigr\}$. The equicontinuity of $\{l_n\}$ yields that $A(U)$ is bounded in l^2. We have

$$p_U(j_U x) \leqslant \sum_{n=1}^{\infty} \lambda_n |l_n(x)| \leqslant \Bigl(\sum_{n=1}^{\infty} \lambda_n |l_n(x)|^2\Bigr)^{1/2} = \|Ax\|_{l^2}.$$

Thus, $V := A^{-1}(B) \subset U$. Finally, the quotient $E/p_V^{-1}(0)$ with norm p_V is linearly isometric to the Euclidean space $A(E)$, which gives a linear isometry between $\widetilde{E}_V$ and the closure of $A(V)$ in l^2. It remains to apply the previous theorem. □

Note that in place of the Hilbert space l^2 in this theorem one can take any space l^p, $1 \leqslant p \leqslant +\infty$. A justification is similar.

Let Λ be a set that has the cardinality equal to the minimal possible cardinality of a base of neighborhoods of zero in the space E.

2.9.6. Corollary. *If E is nuclear and $\{H_\lambda\}_{\lambda \in \Lambda}$ is a family of infinite-dimensional Hilbert spaces, then there exist linear mappings $T_\lambda\colon E \to H_\lambda$ such that the topology of E is the weakest one in which all T_λ are continuous.*

2.9.7. Corollary. *If a nuclear space E is complete, then it is isomorphic to a projective limit of a family of cardinality Λ of Hilbert spaces.*

A Fréchet space is nuclear precisely when it can be represented in the form of the projective limit $E = \varprojlim H_n$ of a sequence of separable Hilbert spaces H_n with nuclear mappings $\psi_{mn}\colon H_m \to H_n$ for $m < n$. In this case it is separable.

PROOF. The first assertion follows from what we have proved above. Let now E be a nuclear Fréchet space. We already know that E has a base of absolutely convex neighborhoods of zero V_n for which the spaces $H_n = \widetilde{E}_{V_n}$ are Hilbert. We can assume that $V_{n+1} \subset V_n$ and that all canonical mappings $\psi_{n,n+1}\colon \widetilde{E}_{V_{n+1}} \to \widetilde{E}_{V_n}$ are nuclear. This gives the required representation. Conversely, let E have the indicated form. Then E is the subspace in $\prod_{n=1}^{\infty} H_n$ determined by the conditions $\psi_{mn} x_m = x_n$, $m < n$, and its base of neighborhoods of zero is formed by the products $V = \prod_{i=1}^{m} B_i \times \prod_{j=m+1}^{\infty} H_j$, where B_i is a ball centered at the origin in H_i. Hence the canonical mapping $E \to \widetilde{E}_V$ can be identified with the projection of E to $\prod_{i=1}^{m} H_i$. This projection has the form $p = (p_1, \ldots, p_m)$, where p_i is the projection of E to H_i. Since $p_i = \psi_{in} \circ p_n$ for $n > m$, the mapping p is nuclear. □

2.9.8. Theorem. (i) *Any vector subspace and any Hausdorff quotient space of a nuclear space are nuclear.*

(ii) *The product of any family of nuclear spaces and a locally convex direct sum of a countable collection of nuclear spaces are nuclear.*

(iii) *The projective limits of arbitrary families of nuclear spaces and the inductive limits of countable collections of nuclear spaces are nuclear.*

(iv) *The projective tensor product of two nuclear spaces is nuclear.*

For a proof, see Schaefer [**436**, Chapter III, Theorem 7.4 and Theorem 7.5, p. 103, 105]. In terms of projective topologies the following characterization is known (see [**436**, p. 172, 184]).

2.9.9. Theorem. *A locally convex space E is nuclear precisely when we have $E\widetilde{\otimes_\pi} F = E\widetilde{\otimes_\varepsilon} F$ for every locally convex space F. Moreover, it suffices to have this equality for all Banach spaces F.*

Let us consider some examples of nuclear spaces.

2.9.10. Example. The following spaces are nuclear.

(i) Any vector subspace in any product $\mathbb{R}^T$ with the induced topology, in particular, any locally convex space with the weak topology.

(ii) The space Σ of rapidly decreasing sequences (Example 1.3.19).

(iii) The space $C_0^\infty(U)$ of smooth functions vanishing outside of a ball U with the topology of uniform convergence of all derivatives.

(iv) The spaces $\mathcal{D}(\mathbb{R}^n)$ and $\mathcal{S}(\mathbb{R}^n)$.

(v) The space $C^\infty(\mathbb{R}^n)$ of smooth functions on $\mathbb{R}^n$ with the topology of uniform convergence of all derivatives on compact sets. The space $C^\infty(U)$ is also nuclear for any open set U.

(vi) The space $H(U)$ of functions holomorphic in an open set $U \subset \mathbb{C}$ with the topology of uniform convergence on compact sets in U.

PROOF. Assertion (i) follows from what has been said above. The nuclearity of Σ is seen from the fact that the same topology is generated by Euclidean norms q_k (see Example 1.3.19) for which the corresponding operators between Hilbert spaces are nuclear. In (iii) we consider for simplicity the case $n = 1$. For norms on $C_0^\infty[0,1]$ we take the L^2-norms of the even order derivatives. It is straightforward to verify that the corresponding operators are nuclear. It would be also possible to employ an isomorphism with Σ (Theorem 1.12.15). The nuclearity of $\mathcal{D}(\mathbb{R}^n)$ follows from the previous theorem and (iii). The space $\mathcal{S}(\mathbb{R}^n)$ is nuclear, since is isomorphic to the space Σ (Theorem 1.12.15). Finally, $\mathcal{H}(U)$ is a closed subspace in the complex space $C^\infty(U)$. □

Let us mentioned some other classical results on nuclear spaces. Their proofs can be found, e.g., in Jarchow [**237**], where additional information is given.

The following important fact was established by B.S. Mityagin [**350**].

2.9.11. Theorem. *The spaces $C^\infty(\mathbb{R}^n)$ and Σ^∞ are isomorphic.*

The indicated spaces are universal in the following sense.

2.9.12. Theorem. (*Kōmura, Kōmura* [**287**]) *A locally convex space is nuclear precisely when it is isomorphic to a vector subspace with the induced topology in the space Σ^A for some set A.*

A Fréchet space is nuclear precisely when it is isomorphic to a subspace of the space Σ^∞ (and also to a subspace of the space $C^\infty(\mathbb{R}^1)$).

B.S. Mityagin [**350**] obtained a criterion of nuclearity of a Fréchet space in terms of the ε-entropy of compact sets in it.

2.9.13. Theorem. *A Fréchet space E is nuclear precisely when for every compact set and every neighborhood of zero in it one has the equality*

$$\limsup_{\varepsilon\to 0} \log\log N(K,\varepsilon U)/\log\frac{1}{\varepsilon} = 0,$$

where $N(K,\varepsilon U) = \inf\Bigl\{N\colon\ K\subset \bigcup_{i=1}^N (x_i+\varepsilon U),\ x_i\in E\Bigr\}$.

There is also an interesting criterion of nuclearity of a Fréchet space in terms of *unconditionally convergent series*, i.e., series converging for all permutations of its elements, and absolutely convergent series, i.e., series with a general term x_n such that $\sum_{n=1}^\infty p(x_n) < \infty$ for every continuous seminorm p.

2.9.14. Theorem. *A Fréchet space is nuclear precisely when every unconditionally series in it is absolutely converging.*

The next result is known as the Dvoretzky–Rogers theorem.

2.9.15. Corollary. *A Banach space in which every unconditionally series converges absolutely is finite-dimensional.*

The next so-called *Schwartz kernel theorem* gives a characterization of nuclear spaces in terms of bilinear forms.

2.9.16. Theorem. *A locally convex space E is nuclear precisely when for every locally convex space F every continuous bilinear function B on the space $E\times F$ is nuclear, i.e., admits a representation*

$$B(x,y) = \sum_{n=1}^\infty l_n(x) f_n(y),$$

where $l_n\in E'$, $f_n\in F'$ and there exist absolutely convex neighborhoods of zero $U\subset E$ and $V\subset F$ such that $\sum_{n=1}^\infty p_{U^\circ}(l_n)p_{V^\circ}(f_n) < \infty$.

There is another "kernel theorem" also due to Laurent Schwartz for the concrete space $\mathcal{D}(\mathbb{R}^n)$; see Hörmander [**227**, Theorem 5.2.1] for a proof.

2.9.17. Theorem. *A bilinear form B on $\mathcal{D}(\mathbb{R}^n)$ is separately continuous if and only if it admits the representation*

$$B(\varphi,\psi) = F(\varphi\otimes\psi)$$

with $F\in\mathcal{D}'(\mathbb{R}^{2n})$, where $(\varphi\otimes\psi)(x,y) = \varphi(x)\psi(y)$.

In § 3.7 we present some other facts related to the nuclearity of dual spaces. For connections with measure theory, see § 5.10.

2.10. Complements and exercises

(i) Properties of the spaces $\mathcal{D}$ and $\mathcal{D}'$ (139). (ii) Absolutely summing operators (143). (iii) Local completeness (145). Exercises (147).

2.10(i). Properties of the spaces $\mathcal{D}$ and $\mathcal{D}'$

Here we discuss a number of rather exotic properties of the spaces $\mathcal{D}$ and $\mathcal{D}'$. At the beginning of the 1950s it was even unknown whether there exist spaces with such properties at all, then an impressive ingenuity was recruited for constructing some artificial examples, and, finally, it was realized that the classical spaces $\mathcal{D}$ and $\mathcal{D}'$ serve as examples. Though, a verification of this turns out highly non-trivial.

One of the first questions arising at a closer look at the topology of the space $\mathcal{D}$ is this: why do not we introduce in $\mathcal{D}$ the topology $\tau_{\rm top}$ in which closed sets are exactly the sets having closed intersections with all $\mathcal{D}_n$? This topology (of the inductive topological limit) is not weaker than the one we introduced. However, it is different: the indicated stronger topology makes continuous all those functions whose restrictions to all $\mathcal{D}_n$ are continuous. The standard topology of $\mathcal{D}$ has no this property: the quadratic form

$$F(\varphi) = \sum_{n=1}^{\infty} \varphi(n)\varphi^{(n)}(0)$$

is discontinuous in this topology (Exercise 2.10.49), but is obviously continuous on all subspaces $\mathcal{D}_n$. Though, both topologies have equal supplies of continuous linear functions. However, a decisive drawback of the topology $\tau_{\rm top}$ is that it does not make $\mathcal{D}$ a topological vector space: the addition operation is discontinuous. Otherwise it would follow by Remark 2.3.2(ii) that $\tau_{\rm top}$ coincides with the standard topology (since the base of zero mentioned in that remark for countable collections of spaces with bases of absolutely convex sets consists of convex sets). The proofs of the results stated below can be found in Smolyanov [**480**], [**484**], [**486**], [**492**], [**493**], [**494**].

A set is called *sequentially closed* if it contains the limits of all its convergent sequences. A *Fréchet–Uryson space* is a space in which for every set A every point in its closure is the limit of a sequence of points in A (not necessarily distinct). Such spaces belong to the broader class of *sequential spaces* in which the sequential closedness is equivalent to the usual closedness (see Engelking [**154**]). Let $[A]$ denote the closure of A (denoted in other places by $\overline{A}$, but here another notation will be more convenient), let $[A]_s$ denote the sequential closure (the smallest sequentially closed set containing A), and let $[A]_{ss}$ denote the set of all limits of countable sequences of elements in A. For example, every metric space is a Fréchet–Uryson space, but the product of a continuum of real lines is not. As an example of a nonmetrizable Fréchet–Uryson space take any weakly compact set in a Banach space that is not metrizable in the weak topology, e.g., the closed ball in a non-separable Hilbert space (Theorem 3.4.11 and Exercise 3.12.50). Any regular inductive limit (see § 2.7) is a nonmetrizable sequential locally convex space, but this limit is not a Fréchet–Uryson space. A subset S of a locally convex space E will be called a *standard countable sequentially closed non-closed set* if

(i) $0 \notin S, 0 \in [S]$;

(ii) 0 is a unique non-isolated limit point of S (so that, in particular, all points of S are islolated);

(iii) every convergent sequence of elements of the set S is stationary (this yields the sequential closedness of S).

2.10.1. Proposition. *Let*

$$S_1 = \{k^{-1}\delta^{(n)}(\cdot) + \delta^{(k)}(\cdot - n)\colon\ k, n \in \mathbb{N}\} \subset \mathcal{D}'.$$

Then S_1 is a standard countable sequentially closed non-closed subset of the space $\mathcal{D}'$ with the topology of uniform convergence on bounded sets in $\mathcal{D}$, i.e., $\beta(\mathcal{D}', \mathcal{D})$.

PROOF. It is clear that $0 \notin S_1$. Let V be a neighborhood of zero in $\mathcal{D}'$. This means that there exist a bounded subset B in $\mathcal{D}$ and $\varepsilon > 0$ such that

$$V \supset V_1 = \{g \in \mathcal{D}'\colon\ |\langle g, \varphi\rangle| < \varepsilon\ \forall \varphi \in B\}.$$

Since B is bounded, there exists $\alpha > 0$ such that the interval $[-\alpha, \alpha]$ contains the supports of all functions in B. Let $n \in \mathbb{N}$, $n > \alpha$ and let $k \in \mathbb{N}$ be such that $k^{-1}\delta^{(n)} \in V_1$ (such k exists because V_1 is a neighborhood of zero). Then for every $\varphi \in B$ we have

$$(k^{-1}\delta^{(n)}, \varphi) + (\delta^{(k)}(\cdot + n), \varphi) = (k^{-1}\delta^{(n)}, \varphi) < \varepsilon,$$

so that $k^{-1}\delta^{(n)}(\cdot) + \delta^{(k)}(\cdot - n) \in V_1 \subset V$, i.e., $0 \in [S]$. We now verify that 0 is a unique limit point of the set S_1. Let $g \in [S_1]$, $g \notin S_1$, $g \neq 0$. Then

$$g = \sum_{j=1}^{n} \alpha_j \delta^{(j)} + \sum_{n=1}^{\infty} \sum_{r=1}^{r(n)} \beta_{rn} \delta^{(r)}(\cdot - n),$$

since the support of the limit point of the set S_1 must belong to the union of supports of the elements of this set. If $\beta_{rn} \neq 0$ for some r, n, then by the inclusion $g \in [S_1]$ we have $\beta_{rn} = 1$ and $\beta_{ij} = 0$ for every pair (i, j) that differs from the pair (r, n); it follows (again by the inclusion $g \in [S_1]$) that $\alpha_n = 1/r$ and $\alpha_j = 0$ if $j \neq n$. Thus, $g \in S_1$, which contradictions our assumption. For completing the proof of property (ii) it remains to observe that all points of the set S_1 are isolated. Let us now prove property (iii). Suppose that for some j we have the equality

$$a_j = k_j^{-1}\delta^{(n_j)} + \delta^{(k_j)}(\cdot - n_j),$$

and suppose that the sequence $\{a_j\}$ converges in $\mathcal{D}'$. Then we have $\sup_j n_j < \infty$, therefore, $\sup_j k_j < \infty$. Hence the set of different elements in the sequence $\{a_n\}$ is finite. Since this sequence converges, there exists $n \in \mathbb{N}$ such that $a_{j_1} = a_{j_2}$ whenever $j_1, j_2 > n$, i.e., the sequence $\{a_j\}$ is stationary. □

Similarly one can show that also the set

$$S_{1'} = \{k^{-1}\delta^{(n)}(\cdot) + k\delta(\cdot - n)\colon\ k, n \in \mathbb{N}\}$$

possesses the properties analogous to those of S_1. Moreover, both the convex hull of the set S_1 and the convex hull of $S_{1'}$ are sequentially closed, but not closed.

Note also (this will be used below) that the set S_1 (hence its convex span) is contained in a subspace of the space $\mathcal{D}'$ isomorphic to the product of $\mathbb{R}^\infty$ (the countable power of $\mathbb{R}^1$) and the topological direct sum $\mathbb{R}^{(\infty)}$ of a countable family of real lines. Since the space $\mathbb{R}^\infty$ is metrizable and $\mathbb{R}^{(\infty)}$ is sequential, we see that the product of two locally convex spaces, one of which is a Fréchet space and the other one is sequential, need not be a sequential space.

2.10.2. Proposition. *Let $a_{kn}(t) = g(t)(k+1)^{-n}\sin(kt)$, where $k,n \in \mathbb{N}$, $g \in \mathcal{D}_{[-1,1]}$, $g \neq 0$, and $b^1_{nk}(t) = k^{-1}p(t-n)$, where $p \in \mathcal{D}$ and $\operatorname{supp} p \subset [1,\infty)$. Then the set $S_2 = \{a_{kn} + b^1_{nk}\colon\ n,k \in \mathbb{N}\}$ is a standard sequentially closed closed subset of the space $\mathcal{D}$.*

2.10.3. Proposition. *Let G be the closed vector subspace of the space $\mathcal{D}'$ consisting of all generalized functions concentrated at zero, let $E = \mathcal{D}_{[-1,1]}$, and let $k,n \in \mathbb{N}$, $b^2_{nk} = k^{-1}\delta^{(n)} \in G$, $a_{kn} \in E$ be the same as in the previous proposition. Then the set $S_3 = \{a_{kn} + b^2_{nk}\colon\ n,k \in \mathbb{N}\}$ is a standard sequentially closed non-closed subset in the topological direct sum $E \oplus G$.*

2.10.4. Remark. The topological direct sum of a countable collection of real lines is not a space Fréchet–Uryson space, i.e., it contains sets some limit points of which are not limits of convergent sequences of its elements. Among such sets there exist even countable sets.

Indeed, let us realize such topological direct sum as the space G from the previous proposition. Set

$$S_4 = \{n^{-1}\delta + k^{-1}\delta^{(n)}\colon\ n,k \in \mathbb{N}\}.$$

Then $0 \notin S_4, 0 \in [S_4]$, but there is no sequence of elements in the set S_4 converging to zero. Certainly, S_4 is not sequentially closed, since $[S_4] \neq S_4$. Hence G is not sequential. Note also that $[S_4]_{ss} \neq [S_4]_s = [S_4]$. Similarly one can show that the infinite-dimensional Hilbert space with the weak topology is not sequential. Indeed, let $\{e_n\}$ be a countable orthonormal family in this space and let

$$A = \{k^{-1}e_n + ke_{n+1}\colon\ k,n \in \mathbb{N}\}.$$

Then $0 \in [A]$, $0 \notin A$ (verify this!), but no sequence of elements of the set A converges to zero. This example goes back to von Neumann.

Below for a subset A of a locally convex space the symbol $M(A)$ denotes the affine variety generated by A and the symbol $L(A)$ denotes the vector subspace generated by A.

2.10.5. Proposition. *For $k,n,m \in \mathbb{N}$ set*

$$b^3_{nk} = (k+1)^{-1}\delta^{(n)}(\cdot - k^{-1}),\ a^0_{kn} = \delta^{(k)}(\cdot - n),$$
$$C_m = \{a^0_{kn} + b^3_{nk}\colon\ k \in \mathbb{N}, n = 1,2,\ldots,m\} \subset \mathcal{D}',$$
$$M = \bigcup_{m=1}^{\infty} M(C_m).$$

Then $0 \notin M, 0 \in [M], [M] = M\cup\{0\}$ and $[M]_s = M$, so that M is a sequentially closed non-closed affine subvariety in the space $\mathcal{D}'$. Moreover, the vector subspace $L(M)$ of the space $\mathcal{D}'$ is closed and the linear functional $f\colon L(M) \to \mathbb{R}$ defined by the equality $f(M) = 1$ is sequentially continuous, but not continuous in the induced topology.

2.10.6. Proposition. *For $k, n \in \mathbb{N}$ set*

$$b_{nk}^4 = (k+1)^{-1}\delta^{(n)}(\cdot - k^{-1}) \in \mathcal{D}'_{[-1,1]}$$

and take elements a_{kn}^1 of the space $\mathcal{D}_{[-1,1]}$ defined by

$$a_{kn}^1(t) = (k+1)^{-1}\sin(l_{kn}t)f(t),$$

where $f \in \mathcal{D}_{[-1,1]}$, $f \neq 0$, $l_{kn} \in \mathbb{N}$, and $l_{k_1n_1} = l_{k_2n_2}$ precisely when $k_1 = k_2$ and $n_1 = n_2$. For every $m \in \mathbb{N}$ let

$$C_m = \{a_{kn}^1 + b_{nk}^4\colon\ k \in \mathbb{N},\ n = 1, 2, \ldots, m\}$$

and $M_1 = \bigcup_{m=1}^\infty [M(C_m)]$. Then $0 \notin M_1$, $0 \in [M_1]$, $[M_1] = M_1 \cup \{0\}$, $[M_1]_s = M_1$. Therefore, M_1 is a sequentially closed non-closed affine subvariety in the space $\mathcal{D}_{[-1,1]} \times \mathcal{D}'_{[-1,1]}$ and the linear functional f on $L(M_1)$ defined by the equality $f(M_1) = 1$ is discontinuous, but sequentially continuous.

2.10.7. Proposition. *Let a_{kn}^1 be the elements of $\mathcal{D}$ defined by the equality from the previous proposition and let $b_{nk}^5 \in \mathcal{D}$ be defined by $b_{nk}^5(t) = 2^{(-2^k)}g\bigl(2^k(t - n)\bigr)$, where $g \in \mathcal{D}$, $\operatorname{supp} g \in [1, \infty), g \neq 0$. For every $m \in \mathbb{N}$ let*

$$C_m^1 = \{a_{kn}^1 + b_{nk}^5\colon\ k \in \mathbb{N}, n = 1, 2, \ldots, m\}.$$

Then $M_2 = \bigcup_{m=1}^\infty [M(C_m^1)]$ is a sequentially closed non-closed affine variety in $\mathcal{D}$ and the linear functional f on $L(M_2)$ such that $f(M_2) = 1$ is discontinuous, but sequentially continuous.

It follows from this proposition that the subspace $L(M_2)$ is not the inductive limit of the intersections $L(M_2) \cap \mathcal{D}_n$.

2.10.8. Proposition. *Let $\{r_j\}$ be the sequence of all rational numbers, let $(p, j, s) \mapsto n(p, j, s)$ be a one-to-one mapping of $\mathbb{N}^3$ onto $(2, 3, \ldots)$, and let $F_{\mathcal{D}'}$ be the subvariety in the space $\mathcal{D}'$ generated by the set*

$$\Bigl\{r_p\delta(\cdot - r_j) + k^{-1}\delta^{(n(p,j,s))}(\cdot - k^{-1}) + \delta^{(k+1)}\bigl(\cdot - n(p, j, s)\bigr)\colon\ p, k, j, s \in \mathbb{N}\Bigr\}.$$

Then the subspace $F_{\mathcal{D}'}$ is sequentially closed and dense in $\mathcal{D}'$, but is not closed (one has $0 \in [F_{\mathcal{D}'}], 0 \notin F_{\mathcal{D}'}$), so that $F_{\mathcal{D}'} - a$, where $a \in F_{\mathcal{D}'}$, is a sequentially closed non-closed everywhere dense vector subspace in $\mathcal{D}'$.

2.10.9. Proposition. *Let $(k, n) \mapsto l(k, n)$ and $(j, s) \mapsto n(j, s)$ be two bijections of $\mathbb{N}^2$ onto $\mathbb{N}$ and let $\{f_j\}$ be a sequence of elements of $\mathcal{D}$ such that*

(1) if $\min_s n(j, s) = a(j)$ and $\min_{k,s} l\bigl(k, n(j, s)\bigr) = b(j)$, then $f_j^{(a(j))}(x) = 0$ for $x \in \bigl(0, 2\pi/b(j)\bigr)$;

(2) $\operatorname{supp} f_j \subset \bigl(-\infty, a(j)\bigr)$; (3) $[\{f_j\}] = \mathcal{D}$.

The existence of such a sequence follows from the relations $a(j) \to \infty, b(j) \to \infty$.

For each triple $k, j, s \in \mathbb{N}$ *consider the function*

$$\varphi_{k,j,s}(t) = f_j(t) + f(t)(k+1)^{n(j,s)} \sin\Bigl(l\bigl(k, n(j,s)\bigr)t\Bigr) + 2^{-2^k} g\Bigl(2^k\bigl(t - n(j,s)\bigr)\Bigr),$$

where $g \in \mathcal{D}$, $\operatorname{supp} g \subset [1, \infty)$, $f \in \mathcal{D}$, $\operatorname{supp} f \subset (-\infty, 1]$, $f(t) = 1$ *for all points* $t \in (0, 1/2)$. *Then the sequentially closed affine subvariety* $F_{\mathcal{D}}$ *of the space* $\mathcal{D}$ *generated by the set* $\{\varphi_{k,j,s}\colon\ k, j, s \in \mathbb{N}\}$ *is not closed and is everywhere dense, so that* $F_{\mathcal{D}} - a$, *where* $a \in F_{\mathcal{D}}$, *is a non-closed everywhere dense sequentially closed vector subspace in* $\mathcal{D}$.

2.10.10. Remark. A similar method can be applied for constructing a sequentially closed non-closed everywhere dense vector subspace in $\mathcal{D}_{[-1,1]} \oplus \mathcal{D}'_{[-1,1]}$.

2.10(ii). Absolutely summing operators

A series $\sum_{n=1}^{\infty} x_n$ in a Hausdorff locally convex space X is called *unconditionally convergent* if it converges for all permutations of indices. For a scalar series this is equivalent to absolute convergence; thus, the sum does not depend on a permutation. Hence if s is the sum of this series, then for every continuous seminorm q on X and every $\varepsilon > 0$ there exists N such that for every finite set of indices M containing $\{1, \ldots, N\}$ we have $q\bigl(s - \sum_{i \in M} x_i\bigr) < \varepsilon$.

More generally, a family of vectors $x_\gamma \in X$ indexed by some set Γ is called *summable to a vector* $s \in X$ if, for every continuous seminorm q on X and every $\varepsilon > 0$, there exists a finite subfamily $\Gamma_0 \subset \Gamma$ such that $q\bigl(s - \sum_{\gamma \in \Gamma_1} x_\gamma\bigr) < \varepsilon$ for every finite family Γ_1 containing Γ_0.

If $\sum_{\gamma \in \Gamma}^{\infty} q(x_\gamma) < \infty$ for every continuous seminorm q, then the family $\{x_\gamma\}_{\gamma \in \Gamma}$ is called *absolutely summable.* For $\Gamma = \mathbb{N}$ we say of an absolutely convergent series.

In the finite-dimensional space unconditional convergence of a series is equivalent to its absolute convergence. For example, the sign alternating series with a general term $(-1)^n n^{-1}$ converges, but not unconditionally. In the Hilbert space l^2 the series of vectors $x_n = n^{-1} e_n$, where $\{e_n\}$ is the standard basis, does not converges absolutely, but converges unconditionally, since whenever $k_1, \ldots, k_m > n$ we have $\|x_{k_1} + \cdots + x_{k_m}\|^2 < \sum_{k=n}^{\infty} k^{-2}$.

Let $p \in [1, +\infty)$. A family of vectors $\{x_\gamma\}_{\gamma \in \Gamma}$ is called *weakly p-summable* if $\sum_{\gamma \in \Gamma} |l(x_\gamma)|^p < \infty$ for all $l \in X'$.

If $\sum_{\gamma \in \Gamma} q(x_\gamma)^p < \infty$ for every continuous seminorm q on X, then $\{x_\gamma\}_{\gamma \in \Gamma}$ is called *absolutely p-summable.*

2.10.11. Lemma. *Suppose that a sequence* $\{x_n\}$ *in a normed space* X *is weakly p-summable. Then there exists* $C > 0$ *such that*

$$\sup_{l \in X',\, \|l\| \leqslant 1} \sum_{n=1}^{\infty} |l(x_n)|^p \leqslant C.$$

PROOF. Let us consider the linear mapping $S\colon X' \to l^p$, $Sl = \bigl(l(x_n)\bigr)_{n=1}^{\infty}$. We have to show its boundedness. Since X' and l^p are Banach spaces, it suffices to verify that the graph of S is closed (see § 3.9). Let $l_j \to l$ in X' and $Sl_j \to v$ in l^p, where $v = (v_n)$. Then $l_j(x_n) \to l(x_n)$ for every fixed n. On the other hand, we have $l_j(x_n) \to v_n$, whence $l(x_n) = v_n$, i.e., $v = Sl$, as required. □

2.10.12. Definition. *Let X and Y be two locally convex spaces. An operator $T \in \mathcal{L}(X,Y)$ is called absolutely p-summing if it takes weakly p-summing families to absolutely p-summing families.*

An operator T is called absolutely summing if it takes summing families to absolutely summing families.

It is readily seen that in these definitions it suffices to consider countable collections of vectors.

If X and Y are normed spaces, then $T \in \mathcal{L}(X,Y)$ is absolutely p-summing if the condition $\sum_{n=1}^{\infty} |l(x_n)|^p < \infty$ for all $l \in X'$ implies that $\sum_{n=1}^{\infty} \|T(x_n)\|_Y^p < \infty$. In this case there is a number $C > 0$ such that

$$\sum_{n=1}^{\infty} \|Tx_n\|_Y^p \leqslant C \sup_{\|l\|\leqslant 1} \sum_{n=1}^{\infty} |l(x_n)|^p \tag{2.10.1}$$

for every sequence $\{x_n\} \subset X$. Indeed, otherwise for every m we could find a finite set $x_{m,1}, \ldots, x_{m,k}$ for which

$$\sup_{\|l\|\leqslant 1} \sum_{i=1}^{k} |l(x_{m,i})|^p \leqslant 2^{-m} \quad \text{and} \quad \sum_{i=1}^{k} \|Tx_{m,i}\|_Y^p \geqslant 1,$$

which leads to a contradiction. The smallest possible C is denoted by $\pi_p(T)$.

These classes are stable under left and right compositions with bounded operators. By the Dvoretzky–Rogers theorem (Corollary 2.9.15) each infinite-dimensional Banach space X contains an unconditionally, but not absolutely converging series. If X has no subspaces isomorphic to c_0 (and only in this case), then unconditional convergence of the series of x_n is equivalent to the weak 1-summability of $\{x_n\}$ (see Kadec, Kadec [**244**, Chapters 3, 4]). If X and Y are two Hilbert spaces, then the class of absolutely summing operators coincides with the class of absolutely 2-summing operators and with the class of *Hilbert–Schmidt operators*, i.e., operators $T \in \mathcal{L}(X,Y)$ such that $\sum_{\alpha} \|Te_{\alpha}\|_Y^2 < \infty$ for some (and then every) orthonormal basis $\{e_{\alpha}\}$ in X, see Bogachev, Smolyanov [**72**, Proposition 7.10.26] or Pietsch [**388**, § 2.5].

Let us describe one important special absolutely 2-summing embedding (see [**72**, Theorem 7.10.27] or Pietsch [**389**, § 17.3]).

2.10.13. Theorem. *Let μ be a probability measure. Then the identical embedding $L^{\infty}(\mu) \to L^2(\mu)$ is absolutely 2-summing. In particular, if μ is a Radon measure on a topological space, then the embedding $C_b(\Omega) \to L^2(\mu)$ is absolutely 2-summing.*

It is worth noting (see Pietsch [**388**, Lemma 3.3.4]) that every absolutely summing mapping $T\colon X \to Y$ between Banach spaces can be written as a composition $T = T_1 \circ j \circ T_2$, where $j\colon C(\Omega) \to L^2(\mu)$ is the identical embedding for some Radon measure μ on a compact space Ω, $T_1 \in \mathcal{L}(X, C(\Omega))$, $T_2 \in \mathcal{L}(L^2(\mu), Y)$.

As the following theorem due to Pietsch shows, in the case of general Banach spaces absolutely 2-summing operators are also connected with the space L^2 (about measures, see Chapter 5).

2.10.14. Theorem. *Let X and Y be two Banach spaces, let B' be the closed unit ball in the space X' equipped with the topology $\sigma(X',X)$ (making it compact), and let $\sigma\bigl(C(B')\bigr)$ be the σ-algebra generated by continuous functions on the compact space B'. An operator $T \in \mathcal{L}(X,Y)$ is absolutely 2-summing precisely when there exists a bounded nonnegative measure μ on $\sigma\bigl(C(B')\bigr)$ such that*

$$\|Tx\|_Y^2 \leqslant \int_{B'} |\xi(x)|^2\,\mu(d\xi).$$

Any nuclear operator between normed spaces is absolutely summing. Indeed, let $\{x_i\}$ be a weakly absolutely summing sequence and let $Tx = \sum_{n=1}^\infty \lambda_n l_n(x) y_n$ be a nuclear operator, where $\lambda_n \geqslant 0$, $\sum_{n=1}^\infty \lambda_n < \infty$, $\|l_n\| \leqslant 1$, $\|y_n\| \leqslant 1$. As we have shown in the lemma above, there exists $C > 0$ such that

$$\sup_n \sum_{i=1}^\infty |l_n(x_i)| \leqslant C.$$

Therefore,

$$\sum_{i=1}^\infty \|Tx_i\| \leqslant \sum_{i=1}^\infty \sum_{n=1}^\infty \lambda_n |l_n(x_i)|\, \|y_n\| \leqslant C \sum_{n=1}^\infty \lambda_n.$$

In Pietsch [**388**, Chapter 3] one can find the proof of the following theorem due to Grothendieck.

2.10.15. Theorem. *The composition of two absolutely summing operators between normed spaces is nuclear.*

With the aid of absolutely summing operators one can characterize nuclear spaces. For example, according to Theorem 2.9.14, a Fréchet space is nuclear precisely when all unconditionally converging series in it converge absolutely, i.e., the identity operator is absolutely summing.

2.10(iii). Local completeness

In relation to Grothdieck's construction discussed in § 2.5 the following useful property of completeness emerged. For simplicity of terminology, we shall use here the term a *disc* to denote a bounded absolutely convex set in a topological vector space.

2.10.16. Definition. *A Hausdorff locally convex space is called locally complete if every closed disc in it is a Banach disc.*

It follows from the results in § 2.5 that the sequential completeness yields the local completeness, but the converse is false. For example, the space c_0 of all sequences converging to zero equipped with the weak topology $\sigma(c_0, l^1)$ is not sequentially complete, but it is locally complete, since it is Banach with respect to its standard norm and weakly bounded sets are norm bounded.

2.10.17. Definition. *We shall say that a sequence $\{x_n\}$ in a locally convex space E converges to x in the sense of Mackey, or is Mackey convergent, if there is a disc $B \subset E$ such that $p_B(x - x_n) \to 0$. Similarly, a sequence is called Mackey fundamental if for some disc B it is fundamental in E_B.*

2.10.18. Lemma. *A sequence $\{x_n\}$ in a locally convex space E Mackey converges to zero precisely when there exists an increasing sequence of positive numbers $\lambda_n \to +\infty$ for which $\lambda_n x_n \to 0$ in E.*

PROOF. If $\{x_n\}$ is Mackey converging to zero, then such numbers obviously exist: whenever $p_B(x_n) > 0$ we can take $\lambda_n = p_B(x_n)^{-1/2}$. Conversely, if such a sequence $\{\lambda_n\}$ exists, then the closed absolutely convex hull B of the sequence $\{\lambda_n x_n\}$ is a closed disc and $p_B(x_n) = \lambda_n^{-1} \to 0$. □

It follows from this lemma that in any metrizable locally convex space convergence of a sequence is equivalent to its Mackey convergence.

2.10.19. Proposition. *The following conditions for a Hausdorff locally convex space E are equivalent:*

(i) *the space E is locally complete;*

(ii) *every Mackey fundamental sequence in E is Mackey convergent;*

(iii) *every bounded set in E is contained in some Banach disc.*

PROOF. The implications (i)⇒(ii)⇒(iii) are trivial. In order to prove (iii)⇒(i), given a closed disc D we find a Banach disc B containing it. Then the closure of D in E_B will be a Banach disc. We observe that D coincides with this closure, since every limit point for D in E_B will be a limit point in E, but D is closed. □

2.10.20. Corollary. *Any locally complete space E will remain locally complete in all locally convex topologies on E for which the dual space remains E'.*

The description of such topologies that agree with the given duality will be obtained in § 3.2.

2.10.21. Theorem. *The following conditions for a Hausdorff locally convex space E are equivalent:*

(i) *the space E is locally complete;*

(ii) *every sequence in E that is Mackey converging to zero has a compact closed absolutely convex span;*

(iii) *every sequence converging to zero in $\sigma(E, E')$ has a $\sigma(E, E')$-compact closed absolutely convex span;*

(iv) *every sequence converging to zero in E has a compact closed absolutely convex span.*

PROOF. We verify the implication (i)⇒(iii). Let $x_n \to 0$ in $\sigma(E, E')$ and let B be the closed absolutely convex hull of $\{x_n\}$ in the topology $\sigma(E, E')$. Then B is bounded and closed in the original topology, hence is a Banach disc. Proposition 2.5.14 gives the compactness of B in the topology $\sigma(E, E')$.

The implication (iii)⇒(iv) is clear from the fact that the closed absolutely convex hull of any sequence $\{x_n\}$ converging to zero in E is weakly compact by condition (iii), and since it is precompact, by Corollary 1.8.12 we obtain its compactness in the original topology.

The implication (iv)⇒(ii) is trivial. Let us deduce (i) from (ii). Let B be a closed disc in E and let a sequence $\{x_n\}$ be Cauchy in E_B. We show that it has a limit in E_B. Passing to a subsequence, we can assume that $p_B(x_{n+1}-x_n) \leqslant 2^{-2n}$, i.e., $x_{n+1} - x_n \in 2^{-2n}B$. Set $y_n = 2^n(x_{n+1} - x_n)$. Then $p_B(y_n) \leqslant 2^{-n}$. By condition (ii) the absolutely convex hull K of the sequence $\{y_n\}$ is compact in E. The sequence $x_{k+1}-x_1 = \sum_{n=1}^{k} 2^{-n}y_n$ is contained in K, hence has a limit point in K. Hence $\{x_n\}$ has a limit point $x \in K$. We observe that then $p_B(x_n-x) \to 0$. Indeed, let $\varepsilon > 0$. For all $n, k \geqslant m$, where $2^{-m} < \varepsilon$, we have $x_n - x_k \in \varepsilon B$. Since B is closed and x is a limit point of $\{x_k\}$, we obtain $x_n - x \in \varepsilon B$ for all indices $n \geqslant m$, i.e., $p_B(x_n - x) \leqslant \varepsilon$, as required. □

2.10.22. Example. For metrizable locally convex spaces the local completeness is equivalent to completeness. Indeed, if a sequence $\{x_n\}$ is Cauchy in a locally complete metrizable space E, then it converges in the completion of E. Hence in the completion it is Mackey convergent, hence is Mackey fundamental, which yields that it is Mackey fundamental in E, then it is Mackey converging in E.

The Mackey convergence and its desription by Lemma 2.10.18 will be used in § 4.3. See Pérez Carreras, Bonet [**385**, Chapter 5] for additional information

Exercises

2.10.23. Construct an example of an incomplete direct spectre $\underrightarrow{\lim}\, E_\alpha$ of complete spaces (E_α, τ_α).

2.10.24. (i) There exists an incomplete separated inductive limit of an increasing sequence of separable Banach spaces B_n with continuous embeddings $B_n \subset B_{n+1}$.

(ii) There exists a separated inductive limit of an increasing sequence of separable Banach spaces B_n with continuous embeddings $B_n \subset B_{n+1}$ in which there is a bounded set contained entirely in no B_n.

(iii) There exists a separated inductive limit of an increasing sequence of separable Banach spaces B_n with continuous embeddings $B_n \subset B_{n+1}$ in which there is a bounded set contained in B_1 which is unbounded in each B_n.

(iv) There exists an inductive limit of strictly increasing sequence of separable Banach spaces B_n with continuous embeddings $B_n \subset B_{n+1}$ in which there is no neighborhood of zero distinct from the whole space.

HINT: in (i), (ii) consider the following example (see Makarov [**331**], [**332**]). Let B_n be the space of double sequences $x = (x_{i,j})$ with a finite limit $\lim_{j\to\infty} x_{i,j}$ for all $i > n$ such that one has $\lim_{j\to\infty} x_{i,j}/(1+j) = 0$ for $i \leqslant n$, set $\|x\|_n = \sup_{i,j} |x_{i,j}|/c_{i,j,n}$, $c_{i,j,n} = 1+j$ if $i \leqslant n$, $c_{i,j,n} = 1$ if $i > n$; consider the countable set A of elements $a^{k,n}$, where $a^{k,n} = (a^{k,n}_{i,j})$, $a^{k,n}_{i,j} = -1$ if $j = 2m$, $m \leqslant k$, $i \leqslant n$, $a^{k,n}_{i,j} = 1$ in other cases; verify that A is bounded in B_1, but the closure of A contains $\{y^n\}$, where $y^n_{i,j} = -1$ if $j = 2m$, $i \leqslant n$, $y^n_{i,j} = 1$ in other cases, and $y^n \in B_n \backslash B_{n-1}$; verify that the sequence of partial sums of the series of $n^{-1}y^n$ is fundamental, but does not converge. For examples in (iii), (iv), see [**332**].

2.10.25. Generalize the example (ii) as follows. Suppose that for every $k \in \mathbb{N}$ we have a sequence of numbers $a_n^k > 0$ with $a_n^k \leqslant a_n^{k+1}$. Denote by E_k the Banach space of sequences $x = (x_n)$ for which $x_n/a_n^k \to 0$ as $n \to \infty$, equipped with the norm $\|x\|_k = \sup_n |x_n/a_n^k|$. Let $E = \mathrm{ind}_k E_k$. In place of an index n we now use a pair (m, n) and define $a_{(m,n)}^k$ as follows: $a_{(m,n)}^k = n$ if $m < k$, $a_{(m,n)}^k = 1$ in other cases. Let vectors v^k be such that $v_{(m,n)}^1 = 0$ for all m, n, $v_{(m,n)}^k = k^{-1}$ if $m = k-1$, $v_{(m,n)}^k = 0$ in other cases. Show that $v^k \in E_k$, $v^{k+1} \notin E_k$ and $v^k \to 0$ in E. Thus, we obtain a sequence converging in $\mathrm{ind}_k E_k$, but contained in no E_k.

2.10.26. On the product $X = \prod_{t \in T} X_t$ of nonempty topological spaces the *box topology* is generated by a base consisting of the products of the form $\prod_{t \in T} U_t$, where U_t is a nonempty open subset of T (not necessarily equal to X_t for all t except for a finite number as in Tychonoff's topology). (i) Investigate whether an infinite power of a compact interval is compact in the box topology. (ii) Let E_t, where $t \in T$, be an infinite family of separated nonzero topological vector spaces over $\mathbb{R}$ or $\mathbb{C}$. Let us equip their product with the box topology. Show that this topology does not agree with the vector structure. Does it agree with the structure of an additive group?

HINT: (i) observe that the box topology is stronger than Tychonoff's topology; (ii) if $x(t_n) > 0$ for infinitely many $t_n \in T$, then take a box neighborhood of zero W such that $\lambda x \notin W$ for all $\lambda \in (0, 1)$.

2.10.27. Let E be a locally convex space in which there is a closed linear subspace E_0 such that E_0 and E/E_0 are metrizable. Prove that E is metrizable as well. Show that if E_0 and E/E_0 are Fréchet spaces, then E is a Fréchet space as well.

HINT: see Pérez Carreras, Bonet [**385**, p. 51].

2.10.28. Prove Corollary 2.7.10.

2.10.29. Let A be a bounded complete subset of a separated locally convex space E. Show that there exists a closed absolutely convex set D such that A is complete in E_D.

HINT: let D be the closed absolutely convex hull of A and let B be the closure of D in the completion of E; observe that A is closed in the Banach space E_B and E_D is a linear subspace in E_B with the same norm.

2.10.30. Let V be a complete bounded convex set in a Hausdorff locally convex space E. Show that its absolutely convex hull is a Banach disc.

HINT: see Pérez Carreras, Bonet [**385**, p. 88].

2.10.31. Let $m_0(\Omega)$ be the space of all functions with finitely many values on a nonempty set Ω equipped with the norm $\|x\| = \sup_\omega |x(\omega)|$. Show that every Banach disc in $m_0(\Omega)$ is finite-dimensional. Prove that in any infinite-dimensional separable Banach space there exists a hypersubspace F not containing infinite-dimensional Banach discs.

HINT: see Pérez Carreras, Bonet [**385**, p. 90].

2.10.32. Let D be a closed disc in a locally convex space E and let $A \subset D$ be an absolutely convex and precompact set in E_D. Prove that the closures of A in E and E_D coincide.

HINT: see Pérez Carreras, Bonet [**385**, p. 169].

2.10.33. A set A in a locally convex space E will be called hyperprecompact if there exists a closed disc D such that A is precompact in E_D. (i) Prove that the closed absolutely convex hull of any hyperprecompact set is hyperprecompact. (ii) Prove that if a sequence $\{x_n\}$ converges to zero in E_D for some disc D, then its closed absolutely convex hull in E is hyperprecompact. (iii) Prove that every hyperprecompact set is contained in the closed absolutely convex hull of some sequence which, for some disc D, tends to zero in E_D. (iv) Prove that for every hyperprecompact set A there is a closed absolutely convex hyperprecompact set C such that A is precompact in E_C.

HINT: see Pérez Carreras, Bonet [**385**, Proposition 6.1.13].

2.10.34. Let a normed space B be continuously embedded into a sequentially complete locally convex space X such that the closed unit ball of B is closed in X. Prove that B is complete.

HINT: apply Theorem 2.5.1.

2.10.35. Construct an example of a Banach space X and an incomplete normed space E continuously embedded into X by means of an injective operator T for which the extension of T by continuity to the completion of E is not injective.

2.10.36. Let E be the inductive limit of an increasing sequence of Hausdorff locally convex spaces E_n such that in each E_n there is an absolutely convex neighborhood of zero with weakly compact closure in E_{n+1}. Show that if a set A is such that $A \cap E_n$ is weakly closed in E_n for all n, then A is closed in E (if A is convex, then it suffices to have the closedness of $A \cap E_n$ in E_n in the original topology of E_n). Hence if E_n are reflexive Banach spaces and Z is a closed subspace in E, then all sequentially continuous linear functions on Z are continuous.

HINT: see Pérez Carreras, Bonet [**385**, Proposition 8.5.28].

2.10.37. Let E be the inductive limit of an increasing sequence of Hausdorff locally convex spaces E_n such that every bounded set in E is contained and bounded in some E_n. Prove that for every continuous linear operator T from a metrizable locally convex space F to E there exists n such that $T(F) \subset E_n$ and the mapping $T\colon\ F \to E_n$ is continuous.

HINT: see Pérez Carreras, Bonet [**385**, Proposition 8.5.38].

2.10.38. Let E be a Hausdorff inductive limit of an increasing sequence of Hausdorff locally convex spaces E_n such that there exist absolutely convex neighborhoods of zero $U_n \subset E_n$ with $U_n \subset U_{n+1}$. Prove that if a set A is bounded in E, then there exists n such that $A \subset n\overline{U_n}$, where the closure is taken in E.

HINT: see Pérez Carreras, Bonet [**385**, Proposition 8.5.20].

2.10.39. (i) Let E be a topological vector space and let Q be a sequentially closed convex set in E which for some $\beta > 0$ contains all sets αQ with $|\alpha| < \beta$. Prove that if Q absorbs every point of some absolutely convex sequentially complete set A, then Q absorbs A.

(ii) Deduce from (i) that if E is sequentially complete and Q is an absolutely convex sequentially closed absorbent set, then it absorbs every bounded absolutely convex set.

HINT: see Edwards [**150**, Proposition 7.4.1, Corollary 7.4.2].

2.10.40. (i) Suppose that E is a nuclear locally convex space and $V \subset E$ is an absolutely convex neighborhood of zero. Show that the space $\widetilde{E}_V$ is separable. (ii) Show that if E' is metrizable in the strong topology, then E is separable.

2.10.41. Let E and F be two vector spaces in duality, let E be equipped with the topology $\sigma(E,F)$, let V be a barrel in E, and let $\widetilde{E}_V$ be the completion of the normed space (E_V, p_V), where p_V is the Minkowski functional of the set V. Prove that the canonical mapping $j_V\colon\ E \to \widetilde{E}_V$ has a closed graph and $V = j_V^{-1}\bigl(j_V(V)\bigr)$.

HINT: if $p_V(x_\alpha) \to 0$ and $x_\alpha \to x$, where $x \notin V/n$, then x and V/n are separated by a hyperplane.

2.10.42. Let F be a Fréchet space. Prove that F is linearly homeomorphic to a closed subspace in a countable product of Banach spaces.

HINT: let the topology of F be defined by seminorms p_n and let X_n be the completion of $X/p_n^{-1}(0)$ with respect to the norm generated by p_n; the embedding of F into $\prod_{n=1}^{\infty} X_n$ is defined by the formula $x \mapsto (x, x, \ldots)$.

2.10.43. Let X be an infinite-dimensional Banach space. Prove that for every separable Banach space E there exists an injective continuous linear operator $A\colon E \to X$. Prove also that such an operator exists from l^∞ to X.

HINT: construct a continuous linear operator $T\colon l^2 \to X$ with an infinite-dimensional range and restrict it to the orthogonal complement of the kernel; construct explicitly an injective operator from l^∞ to l^2; verify that every separable Banach space can be injectively embedded into l^∞.

2.10.44. Show that the vector sum of two Banach discs is again a Banach disc, the intersection of any family of Banach discs is a Banach disc, and the image of a Banach disc under a continuous linear operator to a Hausdorff locally convex space is a Banach disc.

2.10.45.° Prove that the topology of the spaces $\mathcal{S}(\mathbb{R}^n)$, $H(U)$ and $C^\infty(U)$ from Example 2.9.10 (see also § 1.3) cannot be defined by a norm.

2.10.46. Justify the examples from the propositions in § 2.10(i).

2.10.47.° Let E be a sequentially complete locally convex space and let a sequence $\{x_n\}$ converge to zero in the topology $\sigma(E, E')$. Show that its closed absolutely convex hull coincides with the set $\bigl\{\sum_{n=1}^{\infty} \lambda_n x_n\colon\ \sum_{n=1}^{\infty} |\lambda_n| \leqslant 1\bigr\}$.

2.10.48. Let $F_j \in \mathcal{D}'(\mathbb{R}^1)$, $j \in \mathbb{N}$ and $F_j(\varphi) \to 0$ for every $\varphi \in \mathcal{D}(\mathbb{R}^1)$. Prove that there exist a seminorm p of the form indicated in Example 1.3.21 and numbers $\varepsilon_j \to 0$ for which $|F_j(\varphi)| \leqslant \varepsilon_j p(\varphi)$ for all j.

HINT: show first that this is true for each space $\mathcal{D}_k$ in place of $\mathcal{D}(\mathbb{R})$, since by the Banach–Steinhaus theorem there exist $r \in \mathbb{N}$ and $C > 0$ such that $|F_j(\varphi)|$ does not exceed $C \max_{t\in[-k,k]} |\varphi^{(r)}(t)|$ for all j. By the compactness of the natural embedding $C_0^{r+1}[-k,k] \to C_0^r[-k,k]$ one has $\sup_j |F_j(\varphi)| \to 0$ uniformly on the set of functions $\varphi \in \mathcal{D}_k$ with $\max_{t\in[-k,k]} |\varphi^{(r+1)}(t)| \leqslant 1$. Find functions $\zeta_k \in C_0^\infty(k-2/3, k+2/3)$, $k \in \mathbb{Z}$, with $0 \leqslant \zeta_k \leqslant 1$ and $\sum_{k=-\infty}^{\infty} \zeta_k = 1$. Use numbers $\varepsilon_{k,j} \to 0$ and $r_k \in \mathbb{N}$ for which $|F_j(\varphi)| \leqslant \varepsilon_{k,j} \max_{t\in[-k,k]} |\varphi^{(r_k)}(t)|$ for all $\varphi \in \mathcal{D}_k$ and all j to construct the desired numbers ε_j. To this end, estimate $|F(\varphi)|$ by the series of $|F(\zeta_k\varphi)|$.

2.10.49.° (i) Justify the assertions in § 2.10(i) by proving that the topology τ on the space $\mathcal{D}(\mathbb{R}^1)$ introduced there is strictly weaker than the topology $\tau_{\rm top}$ on $\mathcal{D}(\mathbb{R}^1)$ in which open sets are by definition those sets that give open intersections with all $\mathcal{D}_n$. To this end show that the quadratic function $F(\varphi) = \sum_{n=1}^{\infty} \varphi(n)\varphi^{(n)}(0)$ is discontinuous in the topology τ, but is continuous in the topology $\tau_{\rm top}$.

(ii) Prove that the topology τ is strictly stronger than the topology τ_2 on $\mathcal{D}(\mathbb{R}^1)$ generated by the norms $p_\psi(\varphi) = \sup |\psi(x)\varphi^{(m)}(x)|$, where we take arbitrary nonnegative integer numbers m and positive locally bounded functions ψ. To this end, verify that the linear function $F(\varphi) = \sum_{n=1}^{\infty} \varphi^{(n)}(n)$ is continuous in the topology τ, but is discontinuous in the topology τ_2.

Therefore, the topologies on $\mathcal{D}$ used in Kirillov, Gvishiani [**271**] and Kolmogorov, Fomin [**284**] differ from the standard one and are distinct.

2.10.50. Let L be a linear subspace in $\mathcal{D}(\mathbb{R}^1)$ of finite codimension such that all intersections $L \cap \mathcal{D}_m$ are closed. Prove that L is closed.

HINT: assuming for simplicity that L is of codimension 1, find k such that $L \cap \mathcal{D}_k$ is of codimension 1, i.e., has the form $\mathcal{D}_k \cap f_k^{-1}(0)$ for some $f_k \in \mathcal{D}'$. Next, there is $f_{k+1} \in \mathcal{D}'$ with $L \cap \mathcal{D}_{k+1} = DD_{k+1} \cap f_{k+1}^{-1}(0)$. Hence f_{k+1} and f_k are proportional on $\mathcal{D}_k$ and f_{k+1} can be taken such that $f_{k+1} = f_k$ on $\mathcal{D}_k$. Continuing inductively we find functionals $f_n \in \mathcal{D}'$, $n > k$, such that $f_{n+1} = f_n$ on $\mathcal{D}_n$ and $L \cap \mathcal{D}_n = \mathcal{D}_n \cap f_n^{-1}(0)$. This yields a functional $f \in \mathcal{D}'$ such that L is its kernel.

2.10.51. Show that every complete Hausdorff locally convex space is linearly homeomorphic to a closed linear subspace in some product of Banach spaces.

2.10.52. (i) Let E be a Hausdorff locally convex space and let $C \subset E$ be a closed convex balanced set. Show that if the restriction of a linear function f to C is continuous in the original topology, then it is continuous in the topology $\sigma(E, E')$.

(ii) Construct an example showing that even for a Banach space E the restriction of f to the linear space generated by C need not be continuous.

HINT: (i) use Theorem 1.11.17; (ii) consider the space $E = l^1$ with the standard norm and $C = \{(x_n) \in l^1 \colon |x_n| \leqslant 4^{-n}\}$. Let E_C be the linear span of C and define f by $f(x) = \sum_{n=1}^{\infty} 2^n x_n$ if $x \in E_C$. Extend f by linearity to all of E. Then f is not continuous on E_C in the topology of E, since $f(2^{-n}e_n) = 1$, where $\{e_n\}$ is the standard basis in l^1, but $2^{-n}e_n \to 0$ in l^1. However, f is continuous on C. Indeed, let $x \in C$. Then, whenever $y \in C$ and $\|x - y\| < 4^{-k}$, one has $|f(y) - f(x)| < 3 \cdot 2^{-k}$.

2.10.53°. Let E and F be Hausdorff locally convex spaces and let $S\colon E \to F$ be a sequentially continuous linear mapping. Prove that S takes Cauchy sequences to Cauchy sequences.

2.10.54. Give an example of an absolutely convex precompact set V in l^2 with the closure $\overline{V}$ such that V is not dense in $\overline{V}$ with respect to the norm $p_{\overline{V}}$.

HINT: take the set V of vectors $x = (x_n)$ with finitely many nonzero coordinates such that $|x_n| \leqslant n^{-1}$ for all n; observe that $p_{\overline{V}}(x) = \sup_n |nx_n|$ and $v = (n^{-1}) \in \overline{V}$ does not belong to the closure of V in the norm $p_{\overline{V}}$.

2.10.55. Show that in any locally convex space any sequentially closed convex set with a nonempty interior is closed. Show that this is false without the assumption about interior points.

HINT: assuming that 0 is in the interior of the given set V, verify that $V = \{p_V \leqslant 1\}$; consider the topological dual to l^2 in the algebraic dual.

2.10.56°. Show that a linear functional on a topological vector space bounded on precompact sets is bounded. Verify also that if in two vector topologies precompact sets are the same, then also bounded sets are the same.

2.10.57°. A sequence in a locally convex space converges weakly to zero precisely when zero is contained in the closed convex hull of every subsequence in this sequence.

2.10.58. (Gorin, Mityagin [**198**]) Let F be a Fréchet space whose topology is defined by norms p_n with $p_n \leqslant p_{n+1}$ compatible in the following sense: any sequence Cauchy in p_n and converging to zero in p_{n+1} also converges to zero in p_n. Suppose that bounded sets in F are precompact. Then one can find two collections of norm $\{q_n\}$ and $\{r_n\}$ defining the original topology and possessing the following property: for every $f \in F'$ we have $\lim\limits_{n\to\infty} \|f\|_{q_n^*} = 0$, $\lim\limits_{n\to\infty} \|f\|_{r_n^*} > 0$ if $f \neq 0$, where $\|f\|_{q_n^*}$ and $\|f\|_{r_n^*}$ denote the norms of f on (F, q_n) and (F, r_n).

2.10.59. (Banaszczyk [**41**], [**43**]) Let F be a Fréchet space. Suppose that the series of vectors $\varphi_n \in F$ converges to $s_0 \in F$, G is the subspace in F' consisting of all functionals f with $\sum_{n=1}^{\infty} |f(\varphi_n)| < \infty$, and $\mathfrak{S}(\{\varphi_n\})$ is the set of all sums of convergent rearrangements of the series of the vectors φ_n. Then the nuclearity of F is equivalent to the equality $\mathfrak{S}(\{\varphi_n\}) = s_0 + G_F^{\circ}$ for all series of φ_n converging in F.

2.10.60. (i) (Drewnowski [**136**]) Let E be a Hausdorff locally convex space and let A be its subset that is separable and metrizable in the induced topology. Then on the linear span of A there is a metrizable locally convex topology that is majorized by the original topology and the restriction of which to A coincides with the original topology.

(ii) (Larman, Rogers [**310**]) If, in addition, A is locally bounded (for every $a \in A$ there is a neighborhood of zero U such that $(a + U) \cap A$ is bounded), then on the linear span of A there is a norm generating on A the original topology (but the topology generated by this norm is not always majorized by the original topology on the linear span of A).

2.10.61. Let C be a convex closed set in a normed space with a closed unit ball U and let $qU \subset C + U$ for some $q > 1$. Show that C has a nonempty interior.

HINT: let $x_1 \in U$. By condition there exist vectors $y_1 \in C, x_2 \in U$ for which $qx_1 = y_1 + x_2$,i.e., $x_1 = (y_1 + x_2) = q$. By induction construct $x_n \in U$ and $y_n \in C$ with $qx_n = y_n + x_{n+1}$. Then $x_1 = y_1/q + y_2/q^2 + \cdots + y_n/q^n + x_{n+1}/q^{n+1}$. Set $z_n = y_1/q + y_2/q^2 + \cdots + y_n/q^n + y_1/(q^n(q-1))$. Then $z_n \in C/(q-1)$ since C is convex and the sum of the coefficients at the elements $y_i \in C$ is $1/(q-1)$. The sequence $\{z_n\}$ converges to x_1 by the estimate

$$Z - n - x_1\| = \left\| \frac{y_1}{q^n(q-1)} - \frac{x_{n+1}}{q^n} \right\| \leqslant \frac{\|y_1\|}{q^n(q-1)} + \frac{1}{q^n} \to 0.$$

Since C is closed, $x_1 \in C/(q-1)$. It follows that $U \subset C/(q-1)$, hence C has inner points.

2.10.62. A locally convex space E is locally complete precisely when, for every sequence x_n converging to zero in E and every absolutely convergent series of numbers λ_n, the series $\sum_{n=1}^{\infty} \lambda_n x_n$ converges in E. This is also equivalent to convergence of such series for all bounded sequences $\{x_n\}$. One more equivalent description: convergence of the series $\sum_{n=1}^{\infty} \lambda_n x_n$ for all $\{\lambda_n\} \in l^2$ and all $\{x_n\} \in E$ such that $\{p(x_n)\} \in l^2$ for all continuous seminorms p.

HINT: see Qiu [**401**], Saxon, Sánchez Ruiz [**433**].

2.10.63. (Słowikowski [**470**]) The space $\mathcal{D}(\mathbb{R}^1)$ possesses a closed linear subspace Z on which there is a discontinuous sequentially continuous linear function (hence it cannot be extended to a sequentially continuous linear function on all of $\mathcal{D}(\mathbb{R}^1)$). Therefore, Z is not the inductive limit of $Z \cap \mathcal{D}_n$. Cf. Exercise 2.10.36.

2.10.64. Let $F(\varphi) = \min\Bigl(\sum_{n=1}^{\infty} \sqrt{|\varphi(n)\varphi^{(n)}(0)}, \max_t |\varphi(t)|\Bigr)$, $\varphi \in \mathcal{D}(\mathbb{R})$. Then $0 \leqslant F(\varphi) \leqslant \max_t |\varphi(t)|$, $F(t\varphi) = |t|F(\varphi)$ and F is sequentially continuous. Show that F is not continuous.

HINT: take $\varphi_0 \in \mathcal{D}(\mathbb{R})$ with $|\varphi_0| \leqslant 1$, $\varphi_0(0) = 1$, $\varphi_0(n) = \varphi_0^{(n)}(0) = 0$ for all $n \in \mathbb{N}$. Observe that every neighborhood of φ_0 in $\mathcal{D}(\mathbb{R})$ contains a function φ such that $F(\varphi) > 1/2$.

CHAPTER 3

Duality

A powerful method of proving a great number of results in the theory of locally convex spaces employs passage from some subsets in such spaces to their polars, which are subsets of the dual spaces. Moreover, in place of properties of the original sets certain properties of their polars are studied and then one returns back, more precisely, to the polars of polars (the so-called bipolars), which are absolutely convex closed hulls of the original sets.

About such a method of proof it is customary said that it employs duality in locally convex spaces; sometimes it is also said about the duality theory, although in fact there is no "duality theory" and everything reduces to the procedure described in the previous phrase. Nevertheless a systematic use of this procedure is a typical feature of the theory of locally convex spaces, so that it can be called indeed one of the principal methods of this theory (although, when speaking of "principal methods", one should not forget that one key cannot open many doors).

In this chapter we have collected the fundamental results of the theory of real (or complex) locally convex spaces based on duality.

3.1. Polars

Let E be a vector space and let G be some vector space of linear functionals on E *separating points* in E, i.e., such that for every nonzero element x of the space E there exists a functional $g \in G$ for which $g(x) \neq 0$.

The *polar* in G of a set $A \subset E$ is the subset of the space G denoted by the symbol A°_G and defined by the equality

$$A^\circ_G = \bigl\{g \in G\colon\ |\langle g, x\rangle| \leqslant 1\ \forall x \in A\bigr\},\ \varnothing^\circ := G;$$

usually in place of the symbol A°_G we shall use the symbol A°.

The *polar* in E of a set $B \subset G$ is the subset of the space E denoted by the symbol B°_E (or by the symbol B° without the lower index) defined by

$$B^\circ_E = \bigl\{x \in E\colon\ |\langle g, x\rangle| \leqslant 1\ \forall g \in B\bigr\}.$$

Usually so defined polars are called absolute polars, and polars (not absolute) of the same sets are defined, respectively, as the sets

$$\bigl\{g \in G\colon\ \mathrm{Re}\langle g, x\rangle \leqslant 1\ \forall x \in A\bigr\},\quad \bigl\{x \in E\colon\ \mathrm{Re}\langle g, x\rangle \leqslant 1\ \forall g \in B\bigr\}.$$

V.I. Bogachev and O.G. Smolyanov, *Topological Vector Spaces and Their Applications*, Springer Monographs in Mathematics,
DOI 10.1007/978-3-319-57117-1_3

An advantage of absolute polars is that they do not explicitly involve the complex structure. In addition, in the definition of the absolute polar there is no that element of arbitrariness which is seen in the definition of non-absolute polars: such a polar for a set $A \subset E$ could be also defined as the set $\bigl\{g \in G\colon \operatorname{Re}\langle g, x\rangle \geqslant -1\ \forall x \in A\bigr\}$. Certainly, this second definition will not be equivalent to the previous one. Many of the results given below can be easily transferred to non-absolute polars.

Let us state some simple properties of polars directly following from the definition. We consider here only polars (in G) of subsets of the space E; for polars (in E) of subsets of the space G the picture is completely similar.

1. If $A \subset B$, then $B^\circ \subset A^\circ$; in particular, $\varnothing^\circ = \{0\}^\circ = G$, $E^\circ = \{0\}$.

2. If $t \neq 0$, then $(tA)^\circ = t^{-1}A^\circ$;

3. the polar of any set $A \subset E$ is a convex circled $\sigma(G, E)$-closed subset of G. Indeed, the absolute convexity of A° is obvious; its $\sigma(G, E)$-closedness follows from the fact that it is the intersection of the family of $\sigma(G, E)$-closed sets $\{g \in G\colon |\langle g, x\rangle| \leqslant 1\}$ $(x \in A)$. Note also that for obtaining an analog of this properties for non-absolute polars one should omit the word "circled" in the previous sentence.

4. The polar of the union of any family of sets $\{A_\alpha\}$ is the intersection of their polars:

$$\Bigl(\bigcup_\alpha A_\alpha\Bigr)^\circ = \bigcap_\alpha A_\alpha^\circ.$$

5. If a set A absorbs a set B (i.e., if for all scalars t sufficiently large in absolute value we have the inclusion $tA \supset B$), then B° absorbs the set A°.

6. The polar $A^{\circ\circ} = (A^\circ_G)^\circ_E$ of the polar A° of a set A is called its *bipolar*; similarly we define the notion of a bipolar for subsets of the space G. The following important fact is called the *bipolar theorem*.

3.1.1. Theorem. *The bipolar of every nonempty set A is its $\sigma(E, G)$-closed absolutely convex hull, i.e., the smallest absolutely convex $\sigma(E, G)$-closed set in E containing A. Hence $A = A^{\circ\circ}$ for any absolutely convex $\sigma(E, G)$-closed set A.*

PROOF. By property 3 of polars the set $A^{\circ\circ}$ is absolutely convex and closed. Let $\overline{\operatorname{abs\,conv}}\, A$ be the $\sigma(E, G)$-closed absolutely convex hull of the set A. Since $\overline{\operatorname{abs\,conv}}\, A$ is the intersection of all absolutely convex $\sigma(E, G)$-closed sets containing A, and $A^{\circ\circ}$ is one of them, the inclusion $\overline{\operatorname{abs\,conv}}\, A \subset A^{\circ\circ}$ is obvious.

If $\overline{\operatorname{abs\,conv}}\, A \neq A^{\circ\circ}$, then there exists $a \in A^{\circ\circ} \setminus \overline{\operatorname{abs\,conv}}\, A$, and since $A \neq \varnothing$, one has $a \neq 0$. Hence there exists a linear functional f on E continuous in the topology $\sigma(E, G)$ (i.e., an element of the space G) such that $|f(a)| > 1$ and $|f(x)| \leqslant 1$ for every $x \in \overline{\operatorname{abs\,conv}}\, A$. Since this yields $|f(x)| \leqslant 1$ for every $x \in A$, we obtain $f \in A^\circ$, hence the inequality $|f(a)| > 1$ means that $a \notin A^{\circ\circ}$ contrary to our assumption. □

If we consider non-absolute polars, then the bipolar of a set will be the $\sigma(E, G)$-closed convex hull of the union of this set and zero.

7. It follows from the previous property and property 3 that $A^{\circ\circ\circ} = A^\circ$ (even if $A = \varnothing$).

8. The polar of the intersection of an arbitrary family $\{A_\alpha\}$ of absolutely convex $\sigma(E,G)$-closed subsets of the space E is the $\sigma(E,G)$-closed absolutely convex hull of the union of their polars.

Indeed, by property 6 for every set A_α we have the equality $A_\alpha^{\circ\circ} = A_\alpha$. Hence by property 4 (more precisely, by its analogs for subsets of G) we obtain that $(\bigcup_\alpha A_\alpha^\circ)^\circ = \bigcap_\alpha A_\alpha^{\circ\circ} = \bigcap_\alpha A_\alpha$; now, again applying property 6, we obtain

$$\Bigl(\bigcap_\alpha A_\alpha\Bigr)^\circ = \Bigl(\bigcup_\alpha A_\alpha^\circ\Bigr)^{\circ\circ} = \overline{\mathrm{abs\,conv}}\Bigl(\bigcup_\alpha A_\alpha^\circ\Bigr).$$

9. If A is a linear subspace of the space E, then

$$A^\circ = \bigl\{f \in G\colon\ f(x) = 0\ \forall x \in A\bigr\}.$$

3.1.2. Theorem. *The collection of all barrels (see Definition 2.5.6) in a locally convex space E coincides with the class of the polars of weakly bounded sets in E'.*

PROOF. It is obvious that the polar of a weakly bounded set A in E' is a closed absolutely convex absorbent set in E, i.e., a barrel. Conversely, let B be a barrel in E. Then $B = (B^\circ)^\circ$, where the set B° is weakly bounded in E', because if $a \in \lambda B$, then $|f(a)| \leqslant \lambda$ for all $f \in B^\circ$. □

3.1.3. Remark. It is obvious from Definition 1.9.4 that the equicontinuous sets in the dual E' to a locally convex space E are precisely the subsets of the polars of neighborhoods of zero in E.

The following important property of polars is called the Banach–Alaoglu–Bourbaki theorem.

3.1.4. Theorem. *Let E be a locally convex space. Then the polar of any neighborhood of zero in E is compact in the topology $\sigma(E',E)$. Hence any equicontinuous set in E' is relatively compact in the topology $\sigma(E',E)$. If E is separable, then they are metrizable in this topology.*

PROOF. This assertion is a corollary of Tychonoff's theorem on the compactness of any product of compact spaces. Let V be a neighborhood of zero in E and let p_V be its Minkowski functional. For every $x \in E$ let S_x denote the closed disc in the complex plane centered at the origin and having the radius $p_V(x)$; it is equipped with the standard topology. Let T be the product of the family of compacts $\{S_x\colon x \in E\}$, i.e., the collection of all complex functions f on E such that $f(x) \in S_x$ for every $x \in E$.

The polar V° of the set V also consists of (some) functions on E and if $f \in V^\circ$, then for every $x \in E$ we have $|f(x)| \leqslant p_V(x)$, since this is true on V. Thus, we can assume that $V^\circ \subset T$. In this case the topology induced in V° by the topology of the space T and the topology induced in V° by the topology $\sigma(E',E)$ coincide (verify this fact!).

Due to the fact that by Tychonoff's theorem T is compact, for the proof of compactness of V° it suffices to show that V° is closed in T. Assuming that E is complex (in order to include also the complex case), we observe that

$V^\circ = \{f \in T\colon\ f(t_1x_1 + t_2x_2) = t_1f(x_1) + t_2f(x_2)\ \forall x_1,\, x_2 \in E,\, t_1,\, t_2 \in \mathbb{C}\}.$

Thus, V° is the intersection of all subsets of T of the form

$$\{f \in T\colon\ f(t_1x_1 + t_2x_2) - t_1f(x_1) - t_2f(x_2) = 0\},$$

where $t_1, t_2 \in \mathbb{C}, x_1, x_2 \in E$. Each of these sets is closed in T by the continuity on T of the functions

$$f \mapsto f(t_1x_1 + t_2x_2) - t_1f(x_1) - t_2f(x_2),$$

which follows from the definition of the product topology. Finally, if $\mathcal{F} \subset E'$ is equicontinuous, then $\mathcal{F} \subset U^\circ$, where $U \subset E$ is a neighborhood of zero.

The metrizability of the polars of neighborhoods of zero in the topology $\sigma(E', E)$ in the case of separable E follows from Proposition 1.9.5. □

3.1.5. Corollary. *If E is metrizable and separable, then the space E' is separable in the topology $\sigma(E', E)$.*

PROOF. We have $E' = \bigcup_{n=1}^\infty U_n^\circ$, where $\{U_n\}$ is a base of neighborhoods of zero in E, and each U_n° is a metrizable compact. □

3.1.6. Corollary. *The polars of neighborhoods of zero are compact also in the topology τ_P of uniform convergence on precompact sets. Therefore, this topology coincides with the topology $\sigma(E', E)$ on the polars of neighborhoods of zero, hence on all equicontinuous sets.*

PROOF. Let V be a neighborhood of zero in a locally convex space E. Since the polar V° is compact in E' with the topology $\sigma(E', E)$, it suffices to verify that if a net $\{f_\alpha\}$ converges in V° to some functional f in the topology $\sigma(E', E)$, then convergence holds also in the topology τ_P. Let S be a precompact set in E and let $\varepsilon > 0$. Take points $s_1, \ldots, s_n \in S$ such that $S \subset \bigcup_{i=1}^n (s_i + \varepsilon V)$. Next, we take α_0 such that $|f(s_i) - f_\alpha(s_i)| < \varepsilon$ for all $i = 1, \ldots, n$ and $\alpha \geqslant \alpha_0$. Then we obtain $\sup_{s\in S} |f(s) - f_\alpha(s)| < 3\varepsilon$ whenever $\alpha \geqslant \alpha_0$, since for every $s \in S$ there exists a number i such that $s - s_i \in \varepsilon V$, whence we obtain $|f(s - s_i)| \leqslant \varepsilon$ and $|f_\alpha(s - s_i)| \leqslant \varepsilon$ by the inclusions $f, f_\alpha \in V^\circ$. □

Let us now discuss a property of the weak topology connected with polars. If E and G are two finite-dimensional vector spaces in duality (as usual, depending on the situation we either regard E as a space of linear functionals on G or regard G as a space of linear functionals on E; the bilinear form establishing a duality between these spaces will be denoted by the symbol $\langle\cdot,\cdot\rangle$), then any vector subspace E_1 of the space E is connected by the natural duality with the quotient G/E_1° of the space G; in this case

$$E_1^\circ = \{g \in G\colon\ \langle g, x\rangle = 0\ \forall x \in E_1\}.$$

In other words, the space of all linear functions on E_1 is naturally isomorphic to the space G/E_1° and the space of all linear functions on the quotient G/E_1° is isomorphic to the space E_1 (in the finite-dimensional case the existence of the latter isomorphism follows from the existence of the former one, but we want to describe both isomorphisms explicitly).

If E_1 is a vector subspace in E and g is a linear functional on E_1, then the first of the aforementioned isomorphisms, denoted by F, associates with it the element $F(g)$ of the space G/E_1° having the form $g_1 + E_1^\circ$, where $g_1 \in G$ is a linear functional on E extending g. The fact that the mapping $F\colon E_1' \to G/E_1^\circ$ is indeed an isomorphism is verified directly; in particular, its surjectivity follows from the property that if $g^1 \in G/E_1^\circ$, then $g^1 = F(g)$, where g is the linear functional on E obtained as the restriction to E_1 of an arbitrary representative of the class of functionals on E that represents the element g^1. The second of the considered isomorphisms, denoted by Φ, is defined as follows. If $x^1 \in (G/E_1^\circ)'$, then $\Phi(x^1)$ is the linear functional on G (i.e., in accordance with our agreement, an element of E) which is the composition of x^1 and the canonical mapping of the space G onto G/E_1° (we shall denote this mapping by Ψ); we have $\Phi(x^1) \in E_1$, since $\Phi(x^1)(g) = 0$ for $g \in E_1^\circ$, hence $\Phi(x^1) \in E_1^{\circ\circ}$, and $E_1^{\circ\circ} = E_1$, because E_1 is finite-dimensional. The property that E and G are finite-dimensional has already been used implicitly twice: it yields that G is the set of all linear functionals on E and that E is the set of all linear functionals on G.

It turns out that if we equip the spaces E and G with the topologies $\sigma(E,G)$ and $\sigma(G,E)$, respectively, and consider not all linear functionals, but only continuous (and, in addition, assume that the subspace E_1 is closed), then also in the infinite-dimensional case similar assertions will be valid.

3.1.7. Theorem. *Let E and G be two spaces in duality defined by a bilinear form $\langle\cdot,\cdot\rangle$. The space E is equipped with the topology $\sigma(E,G)$ and G is equipped with the topology $\sigma(G,E)$. Let E_1 be a closed subspace in E and let E_1° be its polar in G equipped with the topology induced from G. Then the space $\bigl(E_1', \sigma(E_1', E_1)\bigr)$ is canonically isomorphic to the topological vector quotient G/E_1° and the space $\bigl(E_1, \sigma(E_1, E_1')\bigr)$ is canonically isomorphic to the dual of the quotient G/E_1° equipped with the weak topology (these isomorphisms will be defined in the proof).*

PROOF. An isomorphism between $\bigl(E_1', \sigma(E_1', E_1)\bigr)$ and G/E_1°, denoted by F, is defined as follows: if $g \in E_1'$, then $F(g) = g_1 + E_1^\circ \in G/E_1^\circ$, where g_1 is a continuous linear functional on E that is an extension of the functional g (an extension exists by the Hahn–Banach theorem). Every element $g_1 + E_1^\circ$ of the space G/E_1° is the image of some element of E_1'. Indeed, if g is the restriction to E_1 of a functional from the set $g_1 + E_1^\circ$, then $F(g) = g_1 + E_1^\circ$. The property that F is injective and linear is also easily verified, which is omitted.

It remains to prove that F is a homeomorphism. To this end we observe that if Ψ is the canonical mapping of G onto G/E_1°, then for all $n \in \mathbb{N}$ and $x_1, \ldots, x_n \in E_1$ we have the equality

$$F\bigl(\{g \in E_1'\colon\ |\langle g, x_j\rangle| < 1, j \leqslant n\}\bigr) = \Psi\bigl(\{g \in G\colon\ |\langle g, x_j\rangle| < 1, j \leqslant n\}\bigr).$$

The set in the brackets in the left side is an arbitrary set from a base of neighborhoods of zero in the topology $\sigma(E_1', E_1)$ and the set in the right side is an arbitrary set from a base of neighborhoods of zero of the space G/E_1°. Thus, we have proved that $(E_1')_\sigma$ and G/E_1° are isomorphic, moreover, an explicit isomorphism (called canonical) between them is described. We now define an isomorphism between the spaces $\bigl((G/E_1^\circ)', \sigma\bigl((G/E_1^\circ)', G/E_1^\circ\bigr)\bigr)$ and E_1, denoted by Φ. This definition actually does not differ from the one used above in the finite-dimensional case; the only difference concerns the verification of the fact that the definition is correct (certainly, the proof of the continuity of the constructed mapping and its inverse is also different).

If $x \in (G/E_1^\circ)'$, then, as above, we define $\Phi(x)$ as the composition of x and Ψ. The mapping $x\circ\Psi$ is continuous in the topology $\sigma(G, E)$ as the composition of two continuous mappings, one is the mapping $\Psi\colon G \to G/E_1^\circ$ and the other one is the functional x. Hence $x \circ \Psi \in E$. In addition, since $\langle g, (x \circ \Psi)\rangle = 0$ for every $g \in E_1^\circ$, we have the inclusion $x \circ \Psi \in E_1^{\circ\circ}$. Since $E_1 = E_1^{\circ\circ}$ by the closedness of E_1, this yields that $\Phi(x) \in E_1$.

The injectivity and linearity of Φ are obvious. In order to show that Φ is surjective, we observe that if $x_1 \in E_1$, then x_1 is a linear functional on G vanishing on the polar E_1°. Hence there exists a unique linear functional x on G/E_1° such that $x_1 = x \circ \Psi$. It follows from the definition of the quotient topology and the continuity of x_1 that the functional x is continuous; the definition of the mapping Φ yields that $\Phi(x) = x_1$.

Let us prove that Φ is a homeomorphism. Since the sets of the form

$$U_{f_1,\ldots,f_n} := \{x \in E_1\colon\ |\langle f_j, x\rangle| < 1, j \leqslant n\},$$

where $f_1, \ldots, f_n \in G$, $n \in \mathbb{N}$, form a base of neighborhoods of zero in E_1, the sets of the form $W_{f_1,\ldots,f_n} := \{x \in (G/E_1^\circ)'\colon\ |\langle f_j, x \circ \Psi\rangle| < 1,\ j \leqslant n\}$, where $f_1, \ldots, f_n \in G$, form a base of neighborhoods of zero in the space $(G/E_1^\circ)'_\sigma$. Now in order to show that Φ is an homeomorphism , it suffices to observe that $\Phi(W_{f_1,\ldots,f_n}) = U_{f_1,\ldots,f_n}$ for all functionals $f_1, \ldots, f_n \in G$. □

3.2. Topologies compatible with duality

Let E and G be vector spaces in the duality defined by a bilinear form $\langle\cdot,\cdot\rangle$ (see Example 1.3.26). We shall identify E with a subspace in G^* and G is a subspace in E^*; thus, the elements of E become functionals on G. We recall (see p. 60) that every family of $\sigma(G, E)$-bounded sets $\mathcal{B}$ in E defines in G the topology $\tau_{\mathcal{B}}$ of uniform convergence on the class $\mathcal{B}$ by means of the seminorms $p_B(g) = \sup_{x\in B} |g(x)|$, $B \in \mathcal{B}$. Similar topologies are defined in E.

3.2.1. Definition. *We shall say that a topology τ in the space E is compatible (or agrees) with the duality between E and G if it is locally convex and the space of all continuous linear functionals on (E, τ) coincides with G.*

We emphasize again that in this definition it is meant that G is identified with some vector space of linear functionals on E by means of the mapping $G \to E^*$ that to an element g associates the functional $x \mapsto \langle g, x\rangle$. Thus, a locally convex topology τ agrees with the duality between E and G precisely when every functional $x \mapsto \langle g, x\rangle$, where $g \in G$, is continuous in this topology and every continuous linear functional on (E, τ) has a such form. Similarly we define topolcgies in G that agree with the duality between G and E.

3.2.2. Theorem. *Every topology τ in E that agrees with the duality between E and G is the topology of convergence on polars in G of all possible neighborhoods of zero in E.*

PROOF. For every neighborhood of zero V in the topology τ we take the seminorm p_V on E defined by the formula

$$p_V(x) := \sup\{|g(x)|\colon\ g \in V^\circ\}.$$

We verify that the obtained system of seminorms defines the topology τ. On the one hand, V is contained in the set $\{x\colon\ p_V(x) \leqslant 1\}$. On the other hand, every neighborhood of zero W contains a set of the form $\{x\colon\ p_V(x) \leqslant 1\}$. To this end it suffices to take an absolutely convex neighborhood of zero $V \subset W$ such that its closure $\overline{V}_\tau$ in the topology τ also belongs to W. By Theorem 1.11.15 the set $\overline{V}_\tau$ is closed in the topology $\sigma(E, G)$. Moreover, as noted in § 3.1, the bipolar of the set V is the absolutely convex $\sigma(E, G)$-closed hull of V. Hence this bipolar, which equals $\{x\colon\ p_V(x) \leqslant 1\}$, coincides with $\overline{V}_\tau$ and is contained in W. □

We recall that by Theorem 3.1.4 the aforementioned pclars are absolutely convex and $\sigma(G, E)$-compact in the space G.

It is shown in Example 1.3.23 that among all topologies in the space E for which the set of linear continuous functionals on E coincides with G there exists the weakest one that is automatically locally convex (see also § 2.1); it follows that among all topology compatible with the duality between E and G there exists the weakest one. It turns out that among them there is also the strongest topology.

3.2.3. Theorem. (THE MACKEY–ARENS THEOREM) *Among all topologies in E that are compatible with the duality between E and G there exists the strongest one. This topology $\tau(E, G)$ is called the Mackey tcpology and is the topology of convergence on the class of all absolutely convex $\sigma(G, E)$-compact subsets of the space G. A similar assertion is true if we interchange the roles of the spaces E and G.*

PROOF. As shown above, every topology τ in E compatible with the duality between E and G is a topology of convergence on polars of some neighborhoods of zero in the space E, moreover, these polars are absolutely convex $\sigma(G, E)$-compact subsets of G. It follows from this that the Mackey tcpology majorizes any such topology τ.

We show that the Mackey topology agrees with the duality between E and G, i.e., $\bigl(E, \tau(E, G)\bigr)' = G$. The consistency of the Mackey topology with the vector structure follows from the fact that every compact set in the topology $\sigma(G, E)$ is bounded in this topology. It is obvious that the Mackey topology is locally convex.

We now prove that if $g \in \bigl(E, \tau(E, G)\bigr)'$, then $g \in G$. We have to show that if a functional g from the space E^* of all linear functions on E is continuous on E in the topology $\tau(E, G)$, then $g \in G$. There is a neighborhood of zero V in the topology $\tau(E,G)$ such that $|g(x)| \leqslant 1$ for all $x \in V$. It follows from the definition of the topology $\tau(E,G)$ that there exists a $\sigma(G,E)$-compact absolutely convex subset K of the space G such that $K^\circ \subset V$. Hence $|g(x)| \leqslant 1$ for all vectors $x \in K^\circ$. The latter means that $g \in (K^\circ)^\circ_{E^*}$.

We emphasize that the polar $(K^\circ)^\circ_{E^*}$ of the set K° is its polar in the algebraic dual space E^*, not in G (because a priori we are not given that $g \in G$; on the contrary, this is what we prove).

We now show that $(K^\circ)^\circ_{E^*} \subset G$; this will prove the inclusion $g \in G$. To this end we observe that the topology on K induced by the topology $\sigma(G,E)$ of the space G coincides with the topology induced by the topology $\sigma(E^*,E)$ of the space E^* (certainly $K \subset E^*$, since $G \subset E^*$). Hence the set K is compact also as a subset of the space $\bigl(E^*, \sigma(E^*,E)\bigr)$. By the fact that the topology $\sigma(E^*,E)$ is Hausdorff it follows that K is closed in this space. Since, in addition, the set K is absolutely convex, by property 6 of polars from § 3.1 we obtain the desired inclusion $(K^\circ)^\circ_{E^*} = K \subset G$. □

3.2.4. Remark. (i) A locally convex space E is called a *Mackey space* if its topology is the Mackey topology $\tau(E,E')$. Examples of topologies that are (and are not) Mackey topologies will be given below. For example, we shall see in § 3.6 that all metrizable locally convex spaces are Mackey spaces. The Mackey topology is also denoted by the symbol $\mu(E,E')$.

(ii) The Mackey topology could be also defined as the topology of convergence on the class of all convex $\sigma(G,E)$-compact subsets of the space G.

Indeed, every such set is contained in an absolutely convex $\sigma(G,E)$-compact set: its absolutely convex hull, since the absolutely convex hull of a convex set K is the convex hull of the union of the sets K and $(-K)$ and the convex hull of finitely many convex compacts is compact, see Proposition 1.8.11.

(iii) If $\dim E > 0$, then in the space E there exists a topology stronger than $\tau(E,G)$ that does not agree with the vector structure, but nevertheless possesses the property that the set of all linear functionals on E continuous in this topology coincides with G. If $E = G = \mathbb{R}^1$, then the discrete topology (in which all sets are simultaneously open and closed) possesses this property.

We observe that Theorem 1.11.15 applied above can be formulated as follows.

3.2.5. Proposition. *In all topologies on E compatible with the duality between E and G closed convex sets are the same.*

A similar fact for bounded (but not necessarily closed) sets is presented in the next theorem; along with the aforementioned Theorem 1.11.15 it belongs to the most frequently used results.

3.2.6. Theorem. *A set $B \subset E$ is bounded in the Mackey topology $\tau(E,G)$ precisely when it is bounded in the weak topology $\sigma(E,G)$. More generally* (without the separation condition), *if a set in a locally convex space E is bounded in the topology $\sigma(E,E')$, then it is bounded.*

PROOF. We first give a proof based on duality. If B is bounded in the topology $\tau(E,G)$, then it is bounded in $\sigma(E,G)$. Let us prove that if B is bounded in $\sigma(E,G)$, then any neighborhood of zero V in $\tau(E,G)$ absorbs B. Let W be a closed absolutely convex neighborhood of zero in $(E,\tau(E,G))$ such that $W \subset V$. By Theorem 1.11.15 the set W is closed in $\sigma(E,G)$. Now it suffices to show that W absorbs B. To this end we show that B° absorbs W°, i.e., $tB^\circ \supset W^\circ$ for all t sufficiently large in absolute value. Then by the properties of polars $t^{-1}B^{\circ\circ} \subset W^{\circ\circ}$. Since $W^{\circ\circ} = W$, we have $B^{\circ\circ} \subset tW$ for all t sufficiently large in absolute value. Since $B \subset B^{\circ\circ}$, we see that $B \subset tW$. It remains to show that B° absorbs W°. By Theorem 3.1.4 the absolutely convex set W° is compact in the topology $\sigma(G,E)$ and hence is complete. By Proposition 2.5.1 it is a Banach disc (see Definition 2.5.2). At the same time B° is a closed absolutely convex set in the space $(G,\sigma(G,E))$ absorbing W°. Hence so is the set $B^\circ \cap E_{W^\circ}$ in the space E_{W°. Therefore, B° absorbs W° according to Example 2.5.5.

Let us now give a direct justification. We show that if q is a seminorm and $\sup_{x\in B}|f(x)| < \infty$ for every functional f with $|f| \leqslant q$, then $\sup_{x\in B}|q(x)| < \infty$. Suppose that this is false. Let U be the set of all functionals f with $|f| \leqslant q$. By the Hahn–Banach theorem $\sup_{f\in U, x\in B}|f(x)| = \infty$. Then $\sup_{f\in rU+g, x\in B}|f(x)| = \infty$ for all $r > 0$ and $g \in U$. We find $f_1 \in U/2$ and $b_1 \in B$ with $f_1(b_1) > 2$. If $f \in U$ and $|f_1 - f| \leqslant q(b_1)^{-1}q$, then $f(b_1) \geqslant 1$, i.e., for $f \in U_1 = f_1 + q(b_1)^{-1}U$ we have $f(b_1) \geqslant 1$. Let is find $f_2 \in U_1$ and $b_2 \in B$ with $f_2(b_2) > 4$ and set $U_2 = f_2 + q(b_2)^{-1}q$. By induction we obtain $f_{n+1} \in U_n = f_n + \varepsilon_n U$ and $b_{n+1} \in B$ such that $\varepsilon_n < 2^{-n}$, $f_{n+1}(b_{n+1}) > 2^{n+1}$. The limit $f(x) = \lim\limits_{n\to\infty} f_n(x)$ exists and defines a functional from U. By construction $f(b_n) \geqslant n$ for all n, which is impossible by condition. □

Since $\tau(E,G)$ and $\sigma(E,G)$ are the strongest and the weakest topologies compatible with the duality between E and G, this theorem can be reformulated as follows.

3.2.7. Corollary. *In all topologies on E compatible with the duality between E and G bounded sets are the same.*

3.2.8. Remark. Bounded sets can be the same also in comparable locally convex topologies with different dual spaces. Let E be the Banach space l^1 with its usual norm and $E' = l^\infty$ and let $G = c_0$. Then bounded sets in the topologies $\sigma(E,E')$ and $\sigma(E,G)$ in E are the same (i.e., norm bounded sets, which follows from the classical Banach–Steinhaus theorem, see Theorem 3.5.4), although the dual space $(E,\sigma(E,G))' = G = c_0$ does not coincide with $(E,\sigma(E,E'))' = l^\infty$.

3.2.9. Theorem. *A linear mapping $T\colon X \to Y$ between locally convex spaces is continuous with respect to the Mackey topologies $\tau(X,X')$ and $\tau(Y,Y')$ precisely when it is continuous with respect to the weak topologies $\sigma(X,X')$ and $\sigma(Y,Y')$ or with respect to the topologies $\tau(X,X')$ and $\sigma(Y,Y')$.*

Proof. If T is continuous with respect to the Mackey topologies, then for every $f \in Y'$ the functional $f \circ T$ is continuous in the topology $\tau(X,X')$, hence also in $\sigma(X',X)$, which gives the weak continuity of T. Conversely, if T is weakly continuous, then for every absolutely convex $\sigma(Y',Y)$-compact set Q

in Y' the set $K := \{l \circ T\colon\ l \in Q\} \subset X'$ is absolutely convex. It is $\sigma(X', X)$-compact, since every net $\{l_\alpha \circ T\}$ in it contains a pointwise convergent subnet, because we can first select such subnet in $\{l_\alpha\}$. Since for all $x \in X$ we have $\sup_{l\in Q} |l(Tx)| = \sup_{f\in K} |f(x)|$, we obtain the continuity of the operator T in the Mackey topologies. The same reasoning applies in the case of the continuity of T from the topology $\tau(X', X)$ to the topology $\sigma(Y', Y)$ in Y, since $K \subset X'$ also in this case. □

One should bear in mind that this theorem does not extend to intermediate topologies (see Exercise 3.12.82).

On the Mackey topologies of products, sums, projective and inductive limits, see Exercises 3.12.137, 3.12.138, and 3.12.139.

3.3. Adjoint operators

An important concept is that of a dual or adjoint operator.

3.3.1. Definition. *Let E and G be two locally convex spaces with the duals E' and G' and let $T\colon\ E \to G$ be a linear mapping. A mapping $T^*\colon\ G' \to E'$ is called dual or adjoint to the mapping T if $g(Bx) = (B^*g)(x)$ for all $x \in E$ and $g \in G$, i.e., $\langle g, Bx\rangle = \langle B^*g, x\rangle$.*

As above, the value of a functional g on a vector x will be occasionally denoted by the symbols $\langle g, x\rangle$ or $\langle x, g\rangle$ (resembling the notation for the inner product).

3.3.2. Proposition. *Let $T\colon\ E \to G$ be a linear mapping between locally convex spaces. A dual mapping $T^*\colon\ G' \to E'$ exists precisely when the mapping T is continuous provided that the spaces E and G are equipped with the topologies $\sigma(E, E')$ and $\sigma(G, G')$, respectively (so that it is sufficient to have continuity of T for the original topologies of E and G). In that case T^* is continuous as a mapping from $\bigl(G', \sigma(G', G)\bigr)$ to $\bigl(E', \sigma(E', E)\bigr)$.*

PROOF. Let $T\colon\ \bigl(G', \sigma(G', G)\bigr) \to \bigl(E', \sigma(E', E)\bigr)$ be continuous. Then the mapping T^* is defined by

$$\langle T^*g, x\rangle = \langle g, Tx\rangle.$$

This definition is meaningful, since the right-hand side of the equality is linear in x and continuous in the topology $\bigl(E', \sigma(E', E)\bigr)$. The linearity of the mapping $T^*\colon\ G' \to E'$ is verified directly.

If a dual mapping $T^*\colon\ G' \to E'$ exists, then it is linear and continuous when the spaces G' and E' are equipped with the topologies $\sigma(G', G)$ and $\sigma(E', E)$, respectively. Indeed, the first of these topologies is defined by the seminorms of the form $q_z(g) = |g(z)|$, where $g \in G'$, $z \in G$, and the second topology is defined by the seminorms of the form $p_x(l) = |l(x)|$, where $l \in E'$, $x \in E$, whence we obtain

$$p_x(T^*g) = |T^*g(x)| = |g(Tx)| = q_{Tx}(g).$$

In addition, the original mapping T is continuous when E and G are equipped with the topologies $\sigma(E, E')$ and $\sigma(G, G')$, respectively. Indeed, the former topology is defined by the system of all seminorms $p_l(x) = |l(x)|$, where $l \in E'$, and the latter by the system of seminorms $q_g(x) = |g(x)|$, where $g \in G'$. The equality $|g(Tx)| = |T^*g(x)|$ for $g \in G'$, $x \in E$ means that the seminorm $q_g(Tx)$ is estimated by the seminorm p_{T^*g} (even coincides with it). □

The property that the mapping $T\colon E \to G$ is continuous when we equip the spaces E and G with the topologies $\sigma(E, E')$ and $\sigma(G, G')$ does not yield that it is continuous also with respect to the original topologies of the spaces E and G: for example, if E is a locally convex space such that its weak topology is strictly weaker than the original one, denoted by the symbol τ (for example, E is an infinite-dimensional Hilbert space), then the identity mapping $E \to E$ is not continuous as a mapping from the space $(E, \sigma(E, E'))$ to the space (E, τ), although it is certainly continuous as a mapping from $(E, \sigma(E, E'))$ to $(E, \sigma(E, E'))$ and for this reason possesses a dual mapping. Nevertheless, the following proposition is valid, for which some additional terminology is needed (see § 1.9). Let X and Z be two locally convex spaces, let $\mathcal{A}$ be some family of bounded subsets of the space X, and let $\mathcal{L}(X, Z)$ be the space of all continuous linear mappings from X to Z. The topology in $\mathcal{L}(X, Z)$ of convergence on the family of sets $\mathcal{A}$ is defined by the family $\mathcal{P}$ of seminorms of the form

$$p(f) = \sup_{x \in A} q\bigl(f(x)\bigr),$$

where q is a continuous seminorm on Z and $A \in \mathcal{A}$.

3.3.3. Proposition. *Let E and G be locally convex spaces and let $T\colon E \to G$ be a linear mapping continuous when we equip E and G with the topologies $\sigma(E, E')$ and $\sigma(G, G')$. If $\mathcal{B}$ is some family of bounded sets in $(G', \sigma(G', G))$, $T^*(\mathcal{B})$ is the family of their images in $(E', \sigma(E', E))$ with respect to the mapping T^*, then T is continuous when we equip the spaces E and G with the topologies of convergence on the families of sets $T^*(\mathcal{B})$ and $\mathcal{B}$, respectively.*

The proof is obvious.

Note that for an arbitrary set $L \subset G'$ we have the equality

$$\begin{aligned}\bigl(T^*(L)\bigr)^{\circ}_{E} &= \bigl\{x \in E\colon\ |\langle f, x\rangle| \leqslant 1\ \forall f \in T^*(L)\bigr\}\\ &= \bigl\{x \in E\colon\ |\langle l, Tx\rangle| \leqslant 1\ \forall l \in L\bigr\}\\ &= T^{-1}\bigl(\{g \in G\colon\ |\langle l, g\rangle| \leqslant 1\ \forall l \in L\}\bigr) = T^{-1}(L^{\circ}_{G}).\end{aligned} \tag{3.3.1}$$

Similarly one easily verifies that for every set $K \subset E$ there holds the equality

$$\bigl(T(K)\bigr)^{\circ}_{G'} = (T^*)^{-1}(K^{\circ}_{E'}). \tag{3.3.2}$$

These relations yield that

$$(\mathrm{Ran}\, T)^{\circ} = \mathrm{Ker}\, T^*, \quad (\mathrm{Ran}\, T^*)^{\circ} = \mathrm{Ker}\, T.$$

Since $\mathrm{Ran}\, T$ and $\mathrm{Ran}\, T^*$ are vector spaces, for their closures (taken in arbitrary topologies compatible with the duality between E and E' and between E' and E) we have $(\mathrm{Ran}\, T)^{\circ\circ} = \overline{\mathrm{Ran}\, T}$, $(\mathrm{Ran}\, T^*)^{\circ\circ} = \overline{\mathrm{Ran}\, T^*}$. Therefore,

$$\overline{\mathrm{Ran}\, T} = (\mathrm{Ker}\, T^*)^{\circ}, \quad \overline{\mathrm{Ran}\, T^*} = (\mathrm{Ker}\, T)^{\circ}. \tag{3.3.3}$$

These equalities are infinite-dimensional analogs of the property that the range of an operator on $\mathbb{R}^n$ is orthogonal to the kernel of its adjoint.

3.4. Weak compactness

Here we obtain criteria of compactness in the weak topology (called also weakened). These criteria will employ the terms "the countable compactness" and "the sequential compactness" introduced in Definition 1.8.19. Although the principal results of this section establish the coincidence of these types of compactness in a number of important cases, we start with examples of their non-coincidence in the general case.

3.4.1. Example. If T is the product of the continuum of copies of the real line, realized as the space of all real functions on the interval $[0,1]$ with the topology of pointwise convergence, then its subset $K = \{g \in T\colon \sup_t |g(t)| \leqslant 1\}$ is compact, hence countably compact, but is not sequentially compact, since the sequence $\{g_n\}$, where $g_n(t) = \sin(nt)$, contains no convergent subsequences. Further, if K_0 is the set formed by all functions in K that do not vanish on an at most countable set of points (depending on the function), then K_0 is sequentially compact, but not compact.

If a subset of a topological space is sequentially or countably compact, then this set equipped with the induced topology is a sequentially (respectively, countably) compact topological space. In particular, the two sets in the example above, one of which is sequentially compact, but not compact, and the other is compact, but not sequentially compact, give examples of topological spaces with the respective properties. The product of these topological spaces is a countably compact space that is neither compact, nor sequentially compact.

Below we prove the theorems of Eberlein and Šmulian showing that under rather broad assumptions for subsets of locally convex spaces equipped with the weak topology the relative versions of compactness, sequential compactness, and countable compactness are equivalent; these theorems find numerous applications (and also considerably ease to verify the conditions for semireflexivity).

The next fact is much more elementary (although rather useful).

3.4.2. Proposition. *Every bounded subset of a locally convex space E is precompact in the topology $\sigma(E, E')$.*

PROOF. It suffices to verify that any bounded set $B \subset E$ can be covered by finitely many shifts of every neighborhood of zero in E of the standard form $U = \{x \in E\colon |f_i(x)| < 1, i = 1, \ldots, n\}$, where $f_i \in E'$ are linearly independent. Since $|f_i(x)| \leqslant M$ for all $x \in B$ and $i \leqslant n$ for some $M > 0$, the assertion follows from the fact that the cube $[-M, M]^n$ can be covered by finitely many shifts of the cube $(-1,1)^n$ by vectors $a_1, \ldots, a_k \in \mathbb{R}^n$: then $B \subset \bigcup_{i=1}^k (b_j + U)$, where vectors $b_j \in E$ are such that $\bigl(f_1(b_j), \ldots, f_n(b_j)\bigr) = a_j$. □

However, even closed in the topology $\sigma(E, E')$ bounded sets are still far from compact sets in this topology, since a bounded $\sigma(E, E')$-closed set is not always $\sigma(E, E')$-complete. For example, the closed unit ball in a non-reflexive Banach space E (say, in l^1) is not complete in the topology $\sigma(E, E')$. The next proposition plays a key role in the proof of the Eberlein theorem. About the topology of pointwise convergence see p. 17, 60.

3.4.3. Proposition. *Let K be compact, let $\mathcal{F}(K)$ be the space of all real functions on K equipped with the topology of pointwise convergence, and let $C_p(K)$ be the topological vector subspace of the space $\mathcal{F}(K)$ consisting of continuous functions (where $C_p(K)$ is equipped with the topology induced from $\mathcal{F}(K)$, i.e., also with the topology of pointwise convergence). If a set $G \subset C_p(K)$ is relatively countably compact in $C_p(K)$, then its closure in $\mathcal{F}(K)$ is contained in $C_p(K)$ and is compact.*

PROOF. The relative countable compactness of a set G in $C = C_p(K)$ yields the relative countable compactness of this set in $\mathcal{F} = \mathcal{F}(K)$, hence its boundedness in $\mathcal{F}$; the boundedness of G in $\mathcal{F}$ implies the relative compactness of G in $\mathcal{F}$, hence it remains to prove that the closure $\overline{G}$ of the set G in $\mathcal{F}$ is contained in C.

Suppose that exists a discontinuous function $g \in \overline{G}$ (i.e., $g \notin C$), and let a be its discontinuity point. Thus, there exists $\varepsilon > 0$ such that every neighborhood V of the point a contains a point $x = x_V$ for which we have $|g(x) - g(a)| > \varepsilon$. Using this we prove below the existence of a sequence of functions $g_n \in G$ and a sequence of points $x_n \in K$ such that $|g_n(x_k) - g(a)| > \varepsilon$ if $k \leqslant n$ and $|g_n(x_k) - g(a)| < \varepsilon/2$ for all $k > n$.

Suppose that this is done and complete the proof of the proposition. Since the set of elements of the sequence $\{x_n\}$ is infinite, by the compactness of K it possesses a limit point b. Since every function g_n is continuous, its value at the point b is a limit point of the sequence $\{g_n(x_k)\}$ (we emphasize that this is not the same as being a limit point of the set of values of the elements of the latter since this set can be finite). Since $|g_n(x_k) - g(a)| < \varepsilon/2$ for all $k > n$, we have $|g_n(b) - g(a)| \leqslant \varepsilon/2$.

Thus, if φ is a function in C that is a limit element of the set $\{g_n\}$ (the existence of φ is ensured by the supposed relative countable compactness of G in C), then $|\varphi(b) - g(a)| \leqslant \varepsilon/2$. On the other hand, since for all $n \geqslant k$ we have $|g_n(x_k) - g(a)| > \varepsilon$, for all k the limit function φ satisfies the inequality $|\varphi(x_k) - g(a)| \geqslant \varepsilon$. By the continuity of the function φ its value at the point b (that is a limit point of the set $\{x_k\}$), must be a limit point of the sequence $\{\varphi(x_k)\}$; the above inequalities show that this is false.

For completing the proof of the proposition it remains to establish the existence of the aforementioned sequences $\{x_k\}$ and $\{g_n\}$. Let V_1 be an arbitrary neighborhood of the point a. Then there exists a point $x_1 \in V_1$ such that $|g(x_1) - g(a)| > \varepsilon$. Since $g \in \overline{G}$, there exists a function $g_1 \in G$ such that

$$|g_1(a) - g(a)| < \varepsilon/2, \ |g_1(x_1) - g(a)| > \varepsilon,$$

and since the function g_1 is continuous, there exists a neighborhood V_2 of the point a such that for all $x \in V_2$ we have

$$|g_1(x) - g(a)| < \varepsilon/2.$$

Further, in the same neighborhood there is a point x_2 for which $|g(x_2)-g(a)| > \varepsilon$. By the inclusion $g \in \overline{G}$ there exists a function $g_2 \in G$ for which

$$|g_2(a) - g(a)| < \varepsilon/2, \ |g_2(x_1) - g(a)| > \varepsilon, \ |g_2(x_2) - g(a)| > \varepsilon.$$

Suppose that for some $n \geqslant 1$ we have already found functions $g_1, \ldots, g_n \in G$ and points $x_1, x_2, \ldots, x_n \in K$ with the following properties:

$$|g_k(a) - g(a)| < \varepsilon/2 \quad \text{for all } k, \quad |g_n(x_k) - g(a)| > \varepsilon \quad \text{for } k \leqslant n,$$

$$|g_n(x_k) - g(a)| < \varepsilon/2 \quad \text{for all } k > n.$$

Then x_{n+1} and g_{n+1} are picked in the following way. Let V_{n+1} be a neighborhood of the point a such that $|g_k(a) - g(a)| < \varepsilon/2$ for all $k \leqslant n$ and $x \in V_{n+1}$. Let x_{n+1} be any point in V_{n+1} for which $|g(x_{n+1}) - g(a)| > \varepsilon$, and let g_{n+1} be any function in G for which $|g_{n+1}(a) - g(a)| < \varepsilon/2$ and $|g_{n+1}(x_k) - g(a)| > \varepsilon$ for all $k > n+1$. The existence of V_{n+1} follows from the continuity of the functions g_j and the existence of g_{n+1} is ensured by the inclusion $g \in \overline{G}$. Thus, we have proved the existence of the desired sequences $\{x_k\}$ and $\{g_n\}$, hence the proposition is proven. □

3.4.4. Theorem. (EBERLEIN'S THEOREM) *Let E be a locally convex space and let $A \subset E$ be relatively countably compact in the topology $\sigma(E, E')$. If the closed convex hull of A is complete in the Mackey $\tau(E, E')$ topology, then the set A is relatively compact in the topology $\sigma(E, E')$.*

PROOF. Let B be the closed convex hull of the set A. We recall that in all topologies that agree with the same duality closed convex sets are the same, so it is makes no difference here with which topology we equip the space E — the original one or some of the topologies $\sigma(E, E')$ and $\tau(E, E')$. Let $\widetilde{E}$ be the locally convex space that is the completion of E in the Mackey topology. Certainly, the spaces E' and $\widetilde{E}'$ coincide. Note that since the set B is supposed to be complete in the space $\bigl(E, \tau(E, E')\bigr)$, it is closed also in $\bigl(\widetilde{E}, \tau(\widetilde{E}, E')\bigr)$.

In addition, since by the countable compactness of $\bigl(E, \sigma(E, E')\bigr)$ the set A is bounded in it, it is bounded also in the space $\bigl(E'^*, \sigma(E'^*, E')\bigr)$, hence is precompact in it by Proposition 3.4.2. Since the space $\bigl(E'^*, \sigma(E'^*, E')\bigr)$ is obviously complete, this means that the closure $\overline{A}$ of the set A in this space is compact.

Now for completing the proof of the theorem it suffices to show that $\overline{A} \subset B$, since $B \subset E$. To this end, in turn, it suffices to verify that $\overline{A} \subset \widetilde{E}$, since B is closed in $\bigl(\widetilde{E}, \tau(\widetilde{E}, E')\bigr)$, hence also in $\bigl(\widetilde{E}, \sigma(\widetilde{E}, E')\bigr)$, and the topology $\sigma(\widetilde{E}, E')$ on $\widetilde{E}$ is induced by the topology $\sigma(E'^*, E')$. Thus, let $a \in \overline{A}$. Then for every neighborhood of zero V in $\bigl(\widetilde{E}, \sigma(\widetilde{E}, E')\bigr)$ the restriction $a|_{V^\circ}$ of the functional a to the compact set $V^\circ = V^\circ_{E'}$ in $\bigl(E', \sigma(E', \widetilde{E})\bigr)$ (its compactness follows from Theorem 3.1.4) belongs to the closure (in the space $\mathcal{F}(V^\circ)$) of the set $A|_{V^\circ}$, consisting of the restrictions to V° of linear functionals belonging to A. All these functionals are continuous in the topology induced on V° by the topology $\sigma(E', \widetilde{E})$. Since in this topology V° is compact, it follows from Proposition 3.4.3 that the function a

is continuous. This proposition is applicable, since the relative countable compactness in $(E, \sigma(E, E'))$ of the set A yields the relative countable compactness in $C_f(V^\circ)$ of the set $A|_{V^\circ}$. Thus, we have proved that for any neighborhood of zero V in $(\widetilde{E}, \tau(\widetilde{E}, E'))$ the restriction of a to V° is continuous in the topology induced on V° by the topology $\sigma(E', \widetilde{E})$. Since the space $(\widetilde{E}, \tau(\widetilde{E}, E'))$ is complete, this implies that $a \in \widetilde{E}$. □

3.4.5. Theorem. (THE ABSTRACT ŠMULIAN THEOREM) *Let E be a vector space and let G be some vector space of linear functions on E such that E contains a countable set everywhere dense in the topology $\sigma(E, G)$, which is not assumed to be separated. Then every relatively countably compact subset of the space $(G, \sigma(E, G))$ is relatively sequentially compact.*

PROOF. Let $\{x_n\}$ be a countable subset of the space E everywhere dense in the topology $\sigma(E, G)$, let A be a relatively countably compact subset of the space $(G, \sigma(G, E))$, and let $\{g_n\} \subset A$. The relative countable compactness of the set A yields that it is bounded in $(G, \sigma(G, E))$. Hence for every $x \in E$ the set of numbers $\{g_n(x)\}$ is bounded. It follows that there exists a subsequence $\{g_{n_j}\}$ in $\{g_n\}$ for which the sequence of numbers $\{g_{n_j}(x_1)\}$ converges. Further, there exists a subsequence $\{g_{n_{j_k}}\}$ in $\{g_{n_j}\}$ for which the sequence $\{g_{n_{j_k}}(x_2)\}$ converges, and so on. Thus, for every $j \in \mathbb{N}$ there exists a subsequence $\{g_n^j\}$ of the sequence $\{g_n\}$ with the following properties: if $j_1 < j_2$, then $\{g_n^{j_2}\}$ is a subsequence in $\{g_n^{j_1}\}$ and for every j the sequence $\{g_n{}^j(x_j)\}$ converges. Hence the "diagonal" sequence $\{g_n^n\}$ possesses the following property: the sequence $\{g_n^n(x_j)\}$ converges for each j. It follows that the set of limit points of this sequence in the space $(G, \sigma(G, E))$ contains precisely one element. Indeed, let f_1 and f_2 be two its limit points. Then we have $f_1(x_j) = \lim\limits_{n\to\infty} g_n^n(x_j) = f_2(x_j)$ for every j, hence the values of the functionals f_1 and f_2, which are continuous in the topology $\sigma(E, G)$, coincide on the elements of the set $\{x_n\}$ that is everywhere dense in this topology. This means that $f_1 = f_2$.

Let f be a (unique) limit point of the sequence $\{g_n^n\}$. We show that $g_n^n \to f$ in $(G, \sigma(G, E))$; this will complete the proof of the theorem. Assume that the sequence $\{g_n^n\}$ does not converge in the space $(G, \sigma(G, E))$ to f. Then there exist an open neighborhood V of the element f and a subsequence $\{g_{n_i}^{n_i}\}$ no element of which is contained in V. Hence the limit points of this subsequence (existing by the relative countable compactness of the set A) do not belong to the neighborhood V, but these limit points are at the same time limit points of the original sequence, hence must coincide with f. □

We observe that this theorem applies to weakly compact sets, which is one of the most important cases for applications.

3.4.6. Corollary. *Let E be a separable quasi-complete locally convex space. A set $S \subset E$ is relatively weakly compact precisely when it is relatively sequentially compact in the topology $\sigma(E, E')$, which is also equivalent to its relative countable compactness in $\sigma(E, E')$.*

PROOF. If S is relatively countably compact in the weak topology, then by Eberlein's theorem 3.4.4 it is weakly relatively compact, because S is bounded, hence its closed convex hull is complete by our assumption that E is quasi-complete. Moreover, by the previous theorem (which requires that E be separable), the set S is relatively sequentially compact in the weak topology. On the other hand, if S is relatively weakly compact, then it is always relatively countably compact in the weak topology. □

3.4.7. Theorem. (ŠMULIAN'S THEOREM) *Let E be a locally convex space such that on E there exists a metrizable locally convex topology weaker than the original one. Then every relatively countably compact subset of the space $\big(E,\sigma(E,E')\big)$ is relatively sequentially compact in it.*

PROOF. It suffices to prove this fact for countable sets. So let A be a countable subset of the space E relatively countably compact in $\sigma(E,E')$. Let G be the closed vector subspace in the space $\big(E,\sigma(E,E')\big)$ generated by the set A (or, equivalently, in the space E equipped with the original topology). Let τ_1 be the topology in G induced by the original topology of the space E and let τ_2 be the topology in G induced by the metrizable locally convex topology in E mentioned in the theorem.

Since the topology $\sigma(G,G')$ in G coincides with the topology induced in G by the topology $\sigma(E,E')$ (this follows from the Hahn–Banach theorem, according to which every element in the space G' is the restriction to G of some functional from E'), for the proof of the theorem it suffices to show that the set A is relatively sequentially compact in the space $\big(G,\sigma(G,G')\big)$. We observe that (again by the coincidence of the topology $\sigma(G,G')$ and the topology induced by the weak topology $\sigma(E,E')$) the set A is relatively countably compact in $\big(G,\sigma(G,G')\big)$.

In addition, as we prove below, the space $\big(G',\sigma(G',G)\big)$ is separable. Thus, the spaces G and G' and the set $A \in G$ satisfy the conditions of the previous theorem (here the role of the space E from the theorem is played by the space G'). It follows that A is relatively sequentially compact in $\big(G,\sigma(G,G')\big)$. Thus, it remains to prove that $\big(G',\sigma(G',G)\big)$ is separable. Let $\{V_n\}$ be a base of neighborhoods of zero in the topology τ_2 consisting of absolutely convex closed sets (in the topology τ_2, hence also in τ_1). We can assume, in addition, that $V_n \subset V_j$ for all $j > n$. According to property 8 of polars, established in § 3.1, we have

$$\Big(\bigcap_{n=1}^{\infty} V_n\Big)^{\circ} = \overline{\bigcup_{n=1}^{\infty} V_n^{\circ}},$$

where the polars are taken in G' and the bar means the closure in the space $\big(G',\sigma(G',G)\big)$. It suffices to prove that each of the sets V_n° contains a countable set everywhere dense in the topology $\sigma(G',G)$. To this end, we first observe that each of these sets is compact in the topology $\sigma(G',G)$ (Theorem 3.1.4). Further, if G_A is a vector subspace in the space G generated by the set A, then the topology $\sigma(G',G_A)$ in G' is metrizable (since the dimension of G_A is at most countable), hence possesses a countable base of neighborhoods of zero. Since G_A is everywhere dense in $\big(G,\sigma(G,G')\big)$, the topology $\sigma(G',G_A)$ is separated.

Since it is majorized by the topology $\sigma(G', G)$, in which the set V_n° is compact, the topology induced by the topology $\sigma(G', G_A)$ on the set V_n° coincides with the topology induced by the topology $\sigma(G', G)$. Thus, V_n° as a topological subspace of the space $(G', \sigma(G', G))$ is metrizable compact, hence is separable. □

3.4.8. Example. The hypotheses of the proven theorem are fulfilled for strict inductive limits of metrizable locally convex spaces (Exercise 3.12.135).

Applying it to the weak topologies of Banach spaces we obtain the following assertion (frequently called the *Eberlein–Šmulian* theorem).

3.4.9. Theorem. *Let A be a set in a Banach space X. Then the following conditions are equivalent*:

(i) *the set A has compact closure in the weak topology*;

(ii) *every sequence in A has a subsequence weakly converging in X*;

(iii) *every infinite sequence in A has a limit point in X in the weak topology (a point each neighborhood of which contains infinitely many elements of this sequence)*.

In particular, for sets in a Banach space with the weak topology, compactness is equivalent to the sequential compactness and also equivalent to the countable compactness.

In [**72**, § 6.10(ii)] there is a direct proof using specific features of this special case.

3.4.10. Lemma. *Let T be compact. If $f_0 \in C(T)$ belongs to the closure of a set $F \subset C(T)$ in the topology of pointwise convergence, then f_0 is contained in the closure of some countable part of F in this topology.*

PROOF. For fixed $n, m \in \mathbb{N}$ and $f \in F$ the set

$$W(m,n,f) = \bigl\{(t_1,\ldots,t_m) \in T^m\colon \sup_{i\leqslant m} |f(t_i) - f_0(t_i)| < n^{-1}\bigr\}$$

is open. Their union in $f \in F$ is T^m, since f_0 belongs to the closure of F in the topology of pointwise convergence. Hence there exists a finite collection $F_{n,m} \subset F$ for which the corresponding sets cover T^m. Then f_0 is contained in the closure of the union $F_{n,m}$ over all n and m in the topology of pointwise convergence. □

3.4.11. Theorem. *Let E be a metrizable locally convex space and let $A \subset E$.*

(i) *If a point x belongs to the closure of A in the weak topology, then x belongs to the weak closure of some countable subset in A.*

(ii) *If A is a subset of a weakly compact set in E and a point x belongs to the closure of A in the weak topology, then x is a limit of some sequence $\{a_n\} \subset A$ in the weak topology.*

PROOF. (i) Let $\{U_n\}$ be a base of neighborhoods of zero in E. Then E' is the union of $\sigma(E', E)$-compact sets U_n°. For each n we apply the previous lemma to A and x, considered as contained in $C(U_n^\circ)$ by means of duality. This gives a countable part $A_n \subset A$ the closure of which in the topology of pointwise convergence on U_n° contains x. It remains to observe that x is contained in the $\sigma(E, E')$-closure of the set $\bigcup_{n=1}^\infty A_n$.

(ii) According to what is already proved we can pass to a countable set A. Then, replacing E by the linear span of A, we arrive at the case of separable E. By Corollary 3.1.5, E' contains a countable $\sigma(E', E)$-dense set $\{f_k\}$. By the diagonal procedure we can extract a sequence $\{a_n\}$ in A for which $f_k(a_n) \to f_k(x)$ for every k. By the Šmulian theorem we can extract a further weakly convergent subsequence, but due to the density of $\{f_k\}$ its limit will be precisely x. □

An important characterization of weak compactness, obtained by James, is given in § 3.12(iv).

3.5. Barrelled spaces

In this section we consider an interesting class of the so-called barrelled spaces related to bounded sets. We recall that it was shown in Proposition 1.9.2 that sequentially continuous linear mappings take bounded sets to bounded sets. *Throughout this section we assume that all considered locally convex spaces are Hausdorff.*

As usual, given two topological vector spaces E and G, the symbol $\mathcal{L}(E, G)$ denotes the space of all linear continuous mappings from E to G. We recall (see Definition 1.9.4) that a set $\mathcal{F}$ is called equicontinuous if for every neighborhood of zero V in G one can find a neighborhood of zero U in the space E such that $\{f(u) \colon f \in \mathcal{F}, u \in U\} \subset V$.

A set $\mathcal{F} \subset \mathcal{L}(E, G)$ is called pointwise bounded if, for every $x \in E$, the set $\{f(x) \colon f \in F\}$ is bounded in G.

3.5.1. Definition. *A locally convex space E is called barrelled if, for every locally convex space G, every pointwise bounded set of continuous linear mappings from E to G is equicontinuous.*

3.5.2. Corollary. *The limit of a pointwise convergent sequence of continuous linear operators from a barrelled space to a locally convex space is continuous.*

This definition is meaningful also in the category of topological vector spaces (not necessarily separated), but in our consideration of barrelled and other close spaces we shall deal with Hausdorff locally convex spaces. Nevertheless, although the requirement that the spaces considered in this section are Hausdorff is always meant, sometimes we repeat it explicitly. The meaning of a somewhat strange term "barrelled space" (espace tonnelé introduced by Bourbaki) will be soon clarified.

3.5.3. Proposition. *If a locally convex space E is barrelled, then for every locally convex space G every pointwise bounded subset $\mathcal{L}(E, G)$ is uniformly bounded on every bounded subset of the space E.*

PROOF. Let $B \subset E$ be bounded, let W be a neighborhood of zero in G, and let V be a neighborhood of zero in E such that $F(V) \subset W$ (the latter exists according to the definition of barrelledness). Since B is bounded, we have $tV \supset B$ for some number $t > 0$. Then we have $tW \supset tF(V) = F(tV) \supset F(B)$. Thus, the boundedness of the set $F(B)$ is proved. □

It will be shown below that every Fréchet space (hence every Banach space) is barrelled. Hence in Proposition 3.5.3 one can take for E an arbitrary Banach space; in this case this proposition becomes the following classical theorem.

3.5.4. Theorem. (THE BANACH–STEINHAUS THEOREM) *Any pointwise bounded family of continuous linear mappings from a Banach space X to a normed space Y is norm bounded, i.e., is uniformly bounded on the unit ball in X.*

Thus, we can say that barrelled spaces are those spaces for which there holds a suitably generalized Banach–Steinhaus theorem. Such a suitable generalization is the content of the foregoing definition of barrelled spaces.

We shall now establish a criterion of barrelledness of a space. We recall (see Definition 2.5.6) that a barrel in a locally convex space is any closed absolutely convex absorbent set. Proposition 1.2.11 shows that in every locally convex space there exists a base of neighborhoods of zero consisting of barrels (but not every barrel must be a neighborhood of zero).

3.5.5. Theorem. *A Hausdorff locally convex space is barrelled precisely when every barrel in it is a neighborhood of zero.*

PROOF. Suppose that every barrel in a Hausdorff locally convex space E is a neighborhood of zero and show that E is barrelled. Let G be one more locally convex space and let $\mathcal{F}$ be a pointwise bounded subset in $\mathcal{L}(E,G)$. We have to show that $\mathcal{F}$ is equicontinuous. Let W be a neighborhood of zero in G and let $W_0 \subset W$ be an absolutely convex closed neighborhood of zero. Set $V = \bigcap_{f\in\mathcal{F}} f^{-1}(W_0)$ and show that V is a barrel in E.

Since all mappings $f \in \mathcal{F}$ are continuous and W_0 is closed, the sets $f^{-1}(W_0)$ are closed, hence their intersection V is closed too. The linearity of all $f \in \mathcal{F}$ and the absolute convexity of the set W_0 yield that V is absolutely convex.

We now show that the set V is absorbing in E. Indeed, if $x \in E$, then the set $\mathcal{F}(x) := \{f(x)\colon\ f \in \mathcal{F}\}$ is bounded in G (this is exactly the pointwise boundedness). Hence there exists a number $\lambda > 0$ such that if $|t| > \lambda$, then $tW_0 \supset \mathcal{F}(x)$, i.e., $W_0 \supset \mathcal{F}(x/t)$. This means that for every mapping $f \in \mathcal{F}$ we have $f(x/t) \in W_0$, i.e., $x/t \in f^{-1}(W_0)$, whence $x/t \in V$. Thus, the set V is absorbent and so is a barrel. The definition of V yields that $f(V) \subset W_0 \subset W$ whenever $f \in \mathcal{F}$; this means the equicontinuity of $\mathcal{F}$.

Suppose now that the space E barrelled and prove that every barrel in it is a neighborhood of zero. To this end we apply the definition in the situation where G is the field of scalars, i.e., $\mathcal{L}(E,G) = E'$. Let V be a barrel in E. Show that its polar is a bounded set in $\bigl(E', \sigma(E',E)\bigr)$, i.e., a pointwise bounded set in E'. We recall that the topology $\sigma(E',E)$ is defined by the system of all seminorms of the form $g \mapsto |g(x)|$, where $x \in E$. Thus, it suffices to show that for every $x \in E$ the function $g \mapsto |g(x)|$ is bounded on V°. However, this follows at once from the fact that V is an absorbent set, since $tx \in V$ for all numbers t sufficiently small in absolute value, whence $|g(tx)| \leqslant 1$ for all $g \in V^\circ$ and hence $|g(x)| \leqslant 1/|t|$. Thus, we have proved that the set V° is pointwise bounded in E'.

The pointwise boundedness of V° and barrelledness of E yield the equicontinuity of V°. Hence there exists an absolutely convex and closed neighborhood of zero W in E such that if $x \in W$ and $g \in V^\circ$, then $|g(x)| \leqslant 1$. In other words, $W^\circ \supset V^\circ$. Then $W = W^{\circ\circ} \subset V^{\circ\circ} = V$, i.e., V is a neighborhood of zero. □

We recall that by Theorem 3.1.2 barrels in E are precisely the polars of bounded sets in $(E', \sigma(E', E))$.

The criterion of barrelledness from the proven theorem is often taken for the definition; then our definition becomes a theorem (which is called the Banach–Steinhaus theorem).

3.5.6. Definition. *A topological vector space is called a Baire space if it cannot be represented as a countable union of nowhere dense sets.*

It is known from the university course that every complete metric space is a Baire space (Baire's theorem, see Bogachev, Smolyanov [**72**]); in particular, all Fréchet spaces, hence all Banach spaces, are Baire spaces.

In § 3.12(i), one can find some additional information about Baire spaces in the case of general topological spaces.

3.5.7. Proposition. *Every Baire Hausdorff locally convex space is barrelled. Hence all Fréchet spaces (so all Banach spaces) are barrelled.*

PROOF. Let V be a barrel in a Baire (separated) locally convex space E. Then we have $E = \bigcup_{n=1}^{\infty} nV$, and since E is a Baire space, there exists $n \in \mathbb{N}$ such that nV is not nowhere dense. Since V is closed, this means that nV contains some open set W, which, in turn, contains a set of the form $a + W_0$, where W_0 is an absolutely convex neighborhood of zero in E. Therefore, by the absolute convexity of nV we have $\frac{1}{2}(a + W_0 - a - W_0) = W_0 \subset nV$. Hence V is a neighborhood of zero. □

The space $\mathcal{D}(\mathbb{R}^1)$ is barrelled, but is not Baire; Theorem 3.12.4 gives a normed barrelled space that is not Baire.

In practice, the criterion of barrelledness from Theorem 3.5.5 is seldom used to verify barrelledness (an exception, by the way, is delivered by the proposition below); usually barrelledness of concrete spaces is discovered by using the fact that the class of barrelled spaces is closed with respect to certain operations.

3.5.8. Proposition. *Let E be a barrelled locally convex space. Then every linear subspace in it of finite or countable codimension is barrelled.*

PROOF. We first consider in E a linear subspace E_0 of codimension 1. Let B be a barrel in E_0. Let us take a vector $v \in E\backslash E_0$. If the closure $\overline{B}$ of the set B in E contains a nonzero vector of the form tv, then $\overline{B}$ turns out to be a barrel in E and hence contains a neighborhood of zero U, which gives the neighborhood of zero $U \cap E_0$ in E_0. Moreover, $U \cap E_0 \subset B$ by the closedness of B in E_0. If there is no such nonzero vector, then $\overline{B} = B$. Then the sum of B and the compact set $\{tv\colon\ |t| \leqslant 1\}$ turns out to be a barrel in E and this sum contains a neighborhood of zero U. As above, $U \cap E_0 \subset B$, since every element in U has the form $b + tv$, where $b \in B$, which gives $t = 0$ in the case where $b + tv \in E_0$. By induction we obtain our assertion in the case of any finite codimension. The case of countable codimension is delegated to Exercise 3.12.58. □

3.5.9. Example. In an infinite-dimensional Banach space, the kernel of any discontinuous linear functional is a barrelled subspace (this subspace is an incomplete barrelled normed space).

3.5.10. Theorem. (i) *Inductive limits and direct sums of barrelled spaces are barrelled.* (ii) *Any Hausdorff locally convex space that is the image of a barrelled space under an open* (*taking open sets to open sets*) *continuous linear mapping is barrelled.* (iii) *Any Hausdorff quotient of a barrelled space is barrelled.*

PROOF. (i) If B is a barrel in the inductive limit E of barrelled spaces E_α with mappings $g_\alpha\colon E_\alpha \to E$, then $g_\alpha^{-1}(B)$ is a barrel in E_α, hence contains a neighborhood of zero in E_α, whence it follows that B contains a neighborhood of zero in E. Thus, in the case of direct sums it suffices to consider finite sums, which reduces everything to the sum of two barrelled spaces E_1 and E_2. Then the intersections of the barrel B with the summands are barrels, hence contain neighborhoods of the origins in the summands, which gives a neighborhood of zero in B.

(ii) The preimage of any barrel under a linear mapping is a barrel, hence contains a neighborhood of zero, and the image of the latter is open under an open mapping. This yields also (iii), since the projection onto the quotient is open. □

3.5.11. Theorem. *Let E be a Hausdorff locally convex space. The following assertions are equivalent:*

(i) *the space E is barrelled;*

(ii) *all $\sigma(E',E)$-bounded sets in E' are equicontinuous* (*i.e., the topology of E coincides with $\beta(E,E')$*);

(iii) *all $\sigma(E',E)$-bounded sets in E' are relatively weakly compact and the topology in E coincides with $\tau(E,E')$.*

Thus, barrelled spaces are Mackey spaces.

PROOF. The implication (i)⇒(ii) is trivial. Let (ii) be fulfilled. Then every weakly bounded set B in E' is relatively weakly compact, since by Proposition 3.4.2 it is precompact in the topology $\sigma(E',E)$, In addition, the equicontinuity ensured by condition (ii) gives completeness, hence compactness of the closure of B according to Proposition 1.9.5. Moreover, the original topology is the Mackey topology, since every absolutely convex $\sigma(E',E)$-compact set Q is weakly bounded, so under condition (ii) it is equicontinuous, which gives a neighborhood of zero U in the original topology with $Q \subset U^\circ$, whence $U \subset Q^\circ$, i.e., the original topology majorizes the Mackey topology. Finally, if (iii) holds, then the polar of any barrel B is weakly compact, which by the properties of polars (see § 3.1) means that B is the polar of an absolutely convex $\sigma(E',E)$-compact set, i.e., a neighborhood of zero in the Mackey topology. By the way, it follows from the theorem that any Fréchet space is a Mackey space, but this is true even without completeness (see Corollary 3.6.12). □

3.5.12. Corollary. *If E is barrelled* (*say, is a Fréchet space*), *then E' is quasi-complete in the weak topology $\sigma(E',E)$.*

PROOF. Closed and bounded in the topology $\sigma(E', E)$ sets are weakly compact by (iii), so they are weakly complete. See also Exercise 3.12.154. □

3.5.13. Corollary. *The product of any family of barrelled spaces is barrelled.*

PROOF. The dual to the product is the direct sum of the duals (Proposition 2.4.5), any bounded set in it is contained in the sum of finitely many summands. □

3.5.14. Proposition. *Suppose that a barrelled space E is the union of an increasing sequence of linear subspaces E_n. Then E is the inductive limit of the spaces E_n with the induced topologies.*

PROOF. The original topology of the space E is majorized by the topology τ_1 of the inductive limit of the spaces E_n. Let us verify the converse. By barrelledness E is a Mackey space, so it suffices to verify that $(E, \tau_1)' = E'$. Let $f \in (E, \tau_1)'$. Then $f|_{E_n} \in E_n'$. By the Hahn–Banach theorem there exists functionals $f_n \in E'$ for which $f|_{E_n} = f_n|_{E_n}$. Since $E = \bigcup_{n=1}^\infty E_n$, the functionals f_n converge pointwise. By the barrelledness of E the sequence $\{f_n\}$ is equicontinuous, which yields the continuity of f in the original topology. □

There is also the following "negative" characterization.

3.5.15. Proposition. *A Hausdorff locally convex space E is barrelled precisely when there is no nowhere dense absolutely convex set B in E such that one has $E = \bigcup_{n=1}^\infty nB$.*

PROOF. If $E = \bigcup_{n=1}^\infty nB$ for some absolutely convex set B, then $\overline{B}$ is a barrel, hence in the case of barrelled E it contains a neighborhood of zero. If E is not barrelled, then there is a barrel B that is not a neighborhood of zero, hence it has no interior points due to its absolute convexity. Hence B is nowhere dense. In addition, we have $E = \bigcup_{n=1}^\infty nB$. □

A locally convex space is called *quasi-barrelled* (or *infrabarrelled*) if every barrel in this space absorbing all bounded sets is a neighborhood of zero.

A topological vector space with a topology t (not necessarily locally convex) is called *ultrabarrelled* if every vector topology in E having a base of neighborhoods of zero consisting of t-closed sets is weaker than t.

One can show (see Edwards [**150**, Theorem 6.2.1]) that a locally convex space (E, t) is barrelled precisely when every locally convex topology in E having a base of neighborhoods of zero consisting of t-closed sets is weaker than t. Hence all ultrabarrelled locally convex spaces are barrelled (but there are examples of barrelled spaces which are not ultrabarrelled). A closed subspace of a barrelled space need not be barrelled (Exercises 3.12.61, 3.12.62, and 3.12.63).

3.5.16. Theorem. *Let E be a metrizable barrelled space, let F be a locally convex space, let T be a metric space, and let M be a set of mappings from $E\times T$ to F satisfying the following conditions*:

(i) *for every $t \in T$, the set of mappings $x \mapsto f(x,t)$, where $f \in M$, is equicontinuous in $L(E, F)$*;

(ii) *for every $x \in E$, the set of mappings $t \mapsto f(x,t)$, where $f \in M$, is equicontinuous.*

Then M is equicontinuous.

For a proof, see Bourbaki [**87**, Chapter III, § 3, Theorem 3].

There is a characterization of barrelledness by means of the local completeness, introduced in § 2.10(iii) (see Pérez Carreras, Bonet [**385**, § 5.1]).

3.5.17. Proposition. *A locally convex space E is barrelled precisely when it is quasi-barrelled and the space $\bigl(E', \sigma(E', E)\bigr)$ is locally complete.*

3.6. Bornological spaces

Here we consider one more interesting (and more narrow than the previous one in the case of quasi-completeness) class of *separated* spaces.

3.6.1. Definition. *A linear mapping between topological vector spaces is called bounded if it maps every bounded set to a bounded set.*

It has already been noted in § 1.9 that continuous (or sequentially continuous) linear mappings are bounded.

3.6.2. Definition. *Let E be a Hausdorff locally convex space. The space E is called bornological if, for every locally convex space G, every bounded linear mapping from E to G is continuous.*

The space E is called ultrabornological if, for every locally convex space G, for the continuity of linear mappings from E to G it is sufficient to have their boundedness on every Banach disc.

The definition of Banach discs is given in § 2.5.

3.6.3. Remark. It is obvious that every ultrabornological space is bornological, so that the introduced terminology is consistent. In addition, any sequentially complete bornological space is ultrabornological, since the closed absolutely convex hull of any bounded set in this space is a Banach disc (see § 2.5). Thus, every quasi-complete bornological space is ultrabornological.

3.6.4. Definition. *A locally convex space is said to satisfy the Mackey convergence condition if, for every sequence $\{a_n\}$ of its elements converging to zero, one can find a real sequence $\{t_n\}$ such that $t_n \to \infty$, but $t_n a_n \to 0$.*

The Mackey convergence condition is equivalent to the property that, for every sequence $\{x_n\}$ in E converging to zero, there exists a disc D such that $\{x_n\}$ converges to zero in E_D. Indeed, if such a disc D exists, then we can take $t_n = p_D(x_n)^{-1/2}$. In the opposite direction, if we have numbers $t_n \to \infty$ for which $t_n a_n \to 0$, then for D we can take the closed absolutely convex hull of $\{t_n a_n\}$; then $p_D(t_n a_n) \leqslant 1$, whence $p_D(a_n) \leqslant |t_n|^{-1}$.

3.6.5. Example. Any metrizable locally convex space satisfies the Mackey convergence condition.

Indeed, let $\{a_n\}$ be a sequence in a metrizable locally convex space E converging to zero and let $\{p_k\}$ be a sequence of seminorms defining the topology of E. We can assume that $p_k \leqslant p_{k+1}$. Let us find increasing numbers n_k such that $p_k(a_n) < 4^{-k}$ whenever $n \geqslant n_k$. Set $c_n = 1$ if $n \leqslant n_1$, $c_n = 2^{-k}$ if $n_k < n \leqslant n_{k+1}$. Then $c_n \to 0$ and $p_k(a_n/c_n) \to 0$ for each fixed k.

An example of a space for which this condition is not fulfilled is the infinite-dimensional Hilbert space H with the weak topology, i.e., the topology $\sigma(H, H)$ defined by the natural duality of H with itself given by the inner product. In this space any orthonormal sequence $\{e_n\}$ (i.e., $\|e_n\| = 1$ and $(e_n, e_m) = 0$ if $m \neq n$) converges to zero, but after multiplication by a sequence tending to infinity it does not converge.

3.6.6. Proposition. *Every sequentially continuous linear mapping from a bornological locally convex space to an arbitrary locally convex space is continuous.*

If a locally convex space E satisfies the Mackey convergence condition and all its sequentially continuous linear mappings to locally convex spaces are continuous, then E is a bornological space.

PROOF. We show that every sequentially continuous linear mapping f from an arbitrary topological vector space E to an arbitrary topological vector space G is bounded. This will prove the first assertion. Let B be a bounded set in E. We have to prove that the set $f(B)$ is bounded in G. For this it suffices to show that if $\{a_n\} \subset f(B)$ and a real sequence $\{t_n\}$ converges to zero, then $t_n a_n \to 0$. We have $a_n = f(b_n)$, where $b_n \in B$. Since B is bounded, we have $t_n b_n \to 0$. Hence $t_n a_n = t_n f(b_n) = f(t_n b_n) \to 0$ by the sequential continuity of f.

Let now E be a locally convex space satisfying the Mackey convergence condition and such that all sequentially continuous linear mappings from E to locally convex spaces are continuous. We have to prove that for every locally convex space G every bounded linear mapping $f\colon E \to G$ is continuous. By our condition it suffices to show that f is sequentially continuous.

Let $a_n \to 0$ in E. Since E satisfies the Mackey convergence condition, there exists a sequence of numbers $t_n > 0$ such that $t_n \to \infty$ and $t_n a_n \to 0$. Since the sequence $\{t_n a_n\}$ is bounded in E, the sequence $\{f(t_n a_n)\}$ is bounded in G. Hence $f(a_n) = t_n^{-1} f(t_n a_n) \to 0$. □

3.6.7. Corollary. *Every metrizable locally convex space is bornological. Every Fréchet space is ultrabornological.*

PROOF. Every sequentially continuous mapping of a metrizable space to a topological space is continuous. □

In particular, the space $\mathbb{R}^\infty$ is ultrabornological, and its subspace $\mathbb{R}^\infty_0$ consisting of finite sequences is bornological, but not ultrabornological, since every Banach disc in it is contained in a finite-dimensional subspace (because an infinite-dimensional Banach space cannot be of countable algebraic dimension).

3.6.8. Theorem. *A locally convex space is bornological precisely when every absolutely convex subset of this space absorbing all bounded sets is a neighborhood of zero. A locally convex space is ultrabornological precisely when every* *absolutely convex subset of this space absorbing every Banach disc is a neighborhood of zero.*

PROOF. We prove only the first part, since the proof of the second is completely similar. Let E be a locally convex space in which every absolutely convex subset absorbing all bounded sets is a neighborhood of zero. Let G be one more locally convex space and let $f\colon E \to G$ be a bounded linear mapping. We prove that f is continuous. Let V be a neighborhood of zero in G; we have to prove that $f^{-1}(V)$ is a neighborhood of zero in E. Let V_0 be an absolutely convex neighborhood of zero in G contained in V. It suffices to prove that the set $f^{-1}(V_0)$ is a neighborhood of zero in E. This set is absolutely convex. The boundedness of f yields that $f^{-1}(V_0)$ absorbs every bounded subset of E. Indeed, if $B \subset E$ is bounded, then $f(B)$ is bounded in G. Hence for numbers t sufficiently large in absolute value we have the inclusion $f(B) \subset tV_0$. Then for such t we have also $B \subset f^{-1}\bigl(f(B)\bigr) \subset tf^{-1}(V_0)$. Thus, by our condition on E, $f^{-1}(V_0)$ is a neighborhood of zero in E.

We prove that if E is a bornological locally convex space, then every absolutely convex set in it absorbing every bounded set is a neighborhood of zero. Let $\mathcal{B}$ be the class of all absolutely convex subsets of E each of which absorbs all bounded sets. According to Proposition 1.2.11, the class $\mathcal{B}$ is a base of neighborhoods of zero of some locally convex topology τ in E (which majorizes the original topology, since every absolutely convex neighborhood of zero in the latter is an element of the class $\mathcal{B}$). The identity mapping of the space E with the original topology to the space (E, τ) is bounded and hence continuous by our assumption that E is bornological. This means that every element of the class $\mathcal{B}$ is a neighborhood of zero in the original topology of E. □

Let us give yet another simple criterion of being a bornological space.

3.6.9. Corollary. *A locally convex space is bornological precisely when all seminorms on it bounded on bounded sets are continuous.*

PROOF. If the given space E is bornological and a seminorm p is bounded on bounded sets, then the set $\{p < 1\}$ absorbs bounded sets and by the previous theorem is a neighborhood of zero, which implies the continuity of p.

Suppose now that the boundedness of a seminorm on bounded sets in E implies its continuity. Then, for every linear mapping T from E to a locally convex space F that is bounded on bounded sets, we obtain the boundedness of the seminorm $p\circ T$ on bounded sets for each seminorm p from a family defining the topology in F. This yields the continuity of $p\circ T$, which means the continuity of the mapping T. □

3.6.10. Remark. (i) The criteria of being bornological and ultrabornological for a locally convex space, contained in the proven theorem, are usually taken as the definitions of the corresponding classes of spaces. The previous corollary is also often used to define bornological spaces.

(ii) The proven theorem is similar to Theorem 3.5.5, but one should pay attention to the following difference in their proofs: in the proof of the fact that in any barrelled space every barrel is a neighborhood of zero, it was sufficient to apply the definition of barrelledness to the family of linear functionals on the considered space. Here the definition of bornological spaces is applied to linear mappings from the considered bornological space to locally convex spaces different from the field of scalars. This difference is essential indeed. While for the barrelledness of a Hausdorff locally convex space E it is sufficient that every pointwise bounded set of linear continuous functionals on it be equicontinuous, the property to be bornological for E does not follow from the fact every bounded linear functional on it is continuous (but it follows from the fact that, for every Banach space G, all bounded linear mappings from E to G are continuous). For example, any infinite-dimensional normed space E equipped with the topology $\sigma(E, E')$ is not bornological, although every bounded functional on it is continuous. However, as observed in Shirota [**451**], the situation changes for the Mackey topology.

3.6.11. Theorem. *A locally convex space E is bornological precisely when its topology is the Mackey topology $\tau(E, E')$ and E' coincides with the space of all linear functions on E bounded on bounded sets.*

PROOF. If E is bornological, then it is a Mackey space, since its identity embedding into E with the Mackey topology is bounded and hence is continuous by the definition of bornological spaces. Conversely, suppose that on E with the Mackey topology all bounded functionals are continuous. Let F be a Banach space and let a linear mapping $T\colon\ E \to F$ be bounded. For every $f \in F'$, the functional $f \circ T$ is bounded, hence continuous. Theorem 3.2.9 gives the continuity of T. □

There are Mackey spaces that are not bornological (see Example 3.6.16).

3.6.12. Corollary. *All metrizable locally convex spaces are Mackey spaces.*

3.6.13. Theorem. *Inductive limits of bornological spaces are bornological spaces.*

The quotient of a bornological space by any vector subspace is a bornological space.

An easy proof of this assertion can be found in Edwards [**150**, Theorem 7.3.3].

3.6.14. Theorem. *All bornological spaces are quasi-barrelled and all ultra-bornological spaces are barrelled. Hence all sequentially complete bornological spaces are barrelled.*

PROOF. The first assertion is obvious, since if E is a bornological space and V is a barrel in it absorbing every bounded set, then by the absolute convexity V is a neighborhood of zero, which means that E is quasi-barrelled. The second assertion follows from Theorem 2.5.7, and the last assertion follows from Remark 3.6.3. □

3.6.15. Example. Let E be an infinite-dimensional normed space having a countable Hamel basis such that the coefficients of the expansion with respect to it are continuous (say, let us take the space of finite sequences in l^2). Then E is a bornological locally convex space that is not barrelled. Indeed, let $\{e_n\}$ be a Hamel basis with the indicated property consisting of unit vectors. Then the set

$$\Bigl\{\sum_{j=1}^{n} t_j e_j\colon\ n\in\mathbb{N},\ |t_j|\leqslant j^{-1}\ \forall j\in\mathbb{N}\Bigr\}$$

is a barrel in E that is not a neighborhood of zero. The described bornological space is not ultrabornological, since it is not barrelled.

3.6.16. Example. Let $F=\mathbb{R}^{[0,1]}$ be the space of all real functions on the interval $[0,1]$ equipped with the topology of pointwise convergence and let F_0 be its subspace consisting of those functions each of which assumes nonzero values only at countably many points. Let also E be the subspace in F generated as a vector space by the space F_0 and an element $x\in F\backslash F_0$. The space E is barrelled, but not bornological (this example is due to M. Valdivia). The barrelledness of E can be proved, for example, as follows: first we verify directly that F_0 is barrelled, then we use the fact that the codimension of F_0 in E equals one. The fact that E is not bornological follows from that the linear functional on E the kernel of which is the subspace F_0 is sequentially continuous, but not continuous.

Note also that the space F_0 not only is barrelled, but is also ultrabornological (Exercise 3.12.66).

3.6.17. Example. It is possible to give an example of a space that is simultaneously barrelled and bornological, but not ultrabornological. Namely, it is possible to prove (Exercise 2.10.31) that in any infinite-dimensional separable Banach space there exists a hypersubspace F not containing infinite-dimensional (i.e., with an infinite-dimensional linear span) Banach disc. This hypersubspace in E with the induced topology is barrelled and bornological, but is not ultrabornological, since Banach discs in F are finite-dimensional, which gives boundedness on them for all linear mapping, including discontinuous linear functionals (obviously, existing on F). The product of the spaces from this example and Example 3.6.16 delivers an example of a quasi-barrelled space that is neither barrelled, no bornological.

It is known (see Pérez Carreras, Bonet [**385**, p. 183–185]), that any closed subspace of finite or countable codimension in an ultrabornological space is ultrabornological, but if an ultrabornological space is not equipped with the the strongest locally convex topology, then it contains a dense hyperplane which is bornological and barrelled, but not ultrabornological.

We recall that a cardinality κ is called two-valued measurable if on the set of all subsets of a space of cardinality κ there is a nonzero measure with values 0 and 1 vanishing on all singletons. The answer to the question about existence of such cardinalities is connected with using additional set-theoretic axioms (the usual axioms do not contradict the assertion that there is no such cardinality). It is known that if a cardinality κ is not two-valued measurable, then so is the cardinality 2^κ (see Bogachev [**65**, Proposition 1.12.43]). Hence the cardinality of the continuum is not two-valued measurable.

A proof of the following fact can be found in Pérez Carreras, Bonet [**385**, Theorem 6.2.23].

3.6.18. Theorem. *The space* $\mathbb{R}^\kappa$ *is bornological precisely when* κ *is not two-valued measurable.*

Note that the classes of bornological and ultrabornological spaces are closed with respect to finite products and arbitrary direct sums and also with respect to formation of separated inductive limits (see [**385**, § 6.2]). In terms of inductive limits these classes can be described as follows (see Schaefer [**436**, Theorem 8.4] or [**385**, Corollary 6.2.3]).

3.6.19. Theorem. *Bornological spaces are precisely the inductive limits of normed spaces, and ultrabornological spaces are precisely the inductive limits of Banach spaces. In particular, every sequentially complete bornological space is an inductive limit of Banach spaces.*

PROOF. We only explain a representation of a bornological space E as an inductive limit of normed spaces (more detailed proofs can be found in the cited books). The family $\mathcal{B}$ of all absolutely convex closed bounded sets in E can be partially ordered by inclusion. Each such set B generates the corresponding normed space (E_B, p_B). The inductive limit of this family coincides with E. In the case of a sequentially complete E the spaces (E_B, p_B) are Banach. □

One can show that normed spaces in these inductive limits can be always taken separable (see Jarchow [**237**, § 13.2]).

Additional information about spaces like bornological can be found in Pérez Carreras, Bonet [**385**, Chapter 6].

3.7. The strong topology and reflexivity

If E is a normed vector space and E' is its topological dual, then the equality $\|f\| = \sup\{|f(x)|\colon\ x \in E, \|x\| \leqslant 1\}$, where $f \in E'$, defines a norm on E'. With this norm E' becomes a Banach space called the *Banach dual* to the space E; the *second Banach dual* to the normed space E is the Banach dual to its Banach dual.

If E is a Hausdorff locally convex space and E' is its dual space, then the canonical bilinear form on $E' \times E$, $(f, x) \mapsto f(x) \equiv \langle f, x\rangle$ puts the spaces E and E' in duality (see Chapter 1). Therefore, we can identify elements of the space E with linear functionals on E' (an element $x \in E$ is identified with the functional $f \mapsto \langle f, x\rangle$).

If E is a normed space and E' and E'' are its first and second Banach duals, then we have an inclusion $E \Subset E''$. If this inclusion is the equality $E = E''$, then E is called *reflexive*; in the case of a strict inclusion the space E is called *nonreflexive*.

3.7.1. Remark. Nonreflexivity means that E'' contains elements not associated by the described procedure with any vectors from E. However, it can happen in this case that E and E'' are isomorphic as Banach spaces (the first example of such situation was constructed by James, see Exercise 3.12.79).

In the case where E is a general Hausdorff locally convex space, the question arises which topology in E' should be regarded as a natural analog of the topology of the Banach dual to a normed space. Since the latter is the topology of convergence on the system of all bounded subsets of the original space, it is natural, at least in the class of topologies of uniform convergence on systems of bounded sets, to take it as a reasonable analog of the topology of the Banach dual. It is this definition that is used in the theorems concerned with conditions for reflexivity of arbitrary locally convex spaces, and the fact that these theorems and their proofs are very elegant and at the same time considerably clarify the situation also in the Banach case shows that this choice of the path from Banach spaces to locally convex spaces has been right.

Thus, let E be a locally convex space and let E' be its dual. It has already been noted that every class $\mathcal{B}$ of bounded sets in E generates the topology $\tau_{\mathcal{B}}$ on E' of convergence on the class $\mathcal{B}$, defined by the seminorms $l \mapsto \sup_{x\in B}|l(x)|$, where $B \in \mathcal{B}$. The largest possible class $\mathcal{B}$ consists of all bounded sets. According to Theorem 3.2.6, the system of $\sigma(E,E')$-bounded sets coincides with the system of sets bounded in every other topology in E compatible with the duality between the spaces E and E'. Certainly, the topology $\tau_{\mathcal{B}}$ does not always agree with the duality between E' and E; examples will be given below.

3.7.2. Definition. *The strong topology in the dual space E' is the topology of convergence on the system of all $\sigma(E,E')$-bounded sets in E. It is denoted by the symbol $\beta(E',E)$ and defined by the seminorms of the form $q(l) = \sup_{x\in B}|l(x)|$, where B is a bounded set in E.*

Below we consider only Hausdorff spaces. Since the situation is completely symmetric with respect to the spaces E and E', we can define the strong topology $\beta(E,E')$ also in the space E. Thus, $\beta(E,E')$ is the topology in the space E of uniform convergence on the class of all bounded subsets of the space $\bigl(E',\sigma(E',E)\bigr)$.

Since the original topology of E is the topology of convergence on the system of all equicontinuous subsets of the space E' and every equicontinuous set in E' is obviously bounded in the topology $\sigma(E',E)$, the topology $\beta(E,E')$ majorizes the original topology in E (this is why it is called strong), and it can happen that $\beta(E,E')$ does not agree with the duality between E and E', i.e., it can happen that $\bigl(E,\beta(E,E')\bigr)' \neq E'$.

Thus, in addition to the original topology in E, one can introduce three more natural topologies: the just defined topology $\beta(E,E')$; the weak topology $\sigma(E,E')$, which, unlike the topology $\beta(E,E')$, always agrees with the duality between E and E'; and also the Mackey topology $\tau(E,E')$.

If we denote the original topology of the space E by the symbol τ_E, then we obtain the inclusions

$$\sigma(E,E') \subset \tau_E \subset \tau(E,E') \subset \beta(E,E');$$

certainly, these inclusions need not be strict.

The space E' can be also equipped with three similar topologies (we recall that no "original" topology is defined in E'): the strong topology $\beta(E',E)$, the Mackey topology $\tau(E',E)$, and the weak topology $\sigma(E',E)$. Again we have the inclusions

$$\sigma(E',E) \subset \tau(E',E) \subset \beta(E',E),$$

the last of which explains why the topology $\beta(E', E)$ is called strong. The topology $\beta(E', E)$ does not always agree with the duality between E' and E, but the other two, certainly, agree. For example, if E is a Banach space c_0, then $E' = l^1$, the topology $\beta(E', E)$ is just the norm topology in l^1, and the dual to l^1 with this topology is l^∞, but not c_0.

Below we use the following notation. If E is a locally convex space, then the space $\bigl(E, \sigma(E, E')\bigr)$ will be denoted by the symbol E_σ, the space $\bigl(E, \tau(E, E')\bigr)$ by the symbol E_τ, and the space $\bigl(E, \beta(E, E')\bigr)$ by E_β. Similarly we set

$$\bigl(E', \sigma(E', E)\bigr) = E'_\sigma, \ \bigl(E', \tau(E', E)\bigr) = E'_\tau, \ \bigl(E', \beta(E', E)\bigr) = E'_\beta.$$

Thus, E' is the dual to the space $E \equiv (E, \tau_E)$, so far not equipped with any topology; the dual to the space E_β will be denoted by the symbol $(E_\beta)'$; again, the latter is not assumed to be equipped with a topology (unlike E'_β, which is a locally convex space; though, as vector spaces they also need not coincide).

The sets in E' bounded in the strong topology $\beta(E', E)$ are called *strongly bounded*. If E is barrelled, then $\sigma(E', E)$-bounded sets are strongly bounded (Proposition 3.5.3).

Although always $(E_\sigma)' = (E_\tau)' = E'$ and $(E_\beta)' \supset E'$, in the general case $(E_\beta)' \neq E'$. Certainly, the space $(E_\beta)'$ can be also equipped with various topologies, but we do not do this here. However, we shall consider topologies in $(E'_\beta)'$; here we again apply our previous conventions, just the role of E is played by E'_β.

For every locally convex space E, the space E'_σ is called the *weak dual*, the space E'_β is called the *strong dual*, and the space E'_τ is called the *Mackey dual*.

The space $(E'_\beta)'_\beta$ is called the *second strong dual*; the similarly defined spaces $(E'_\sigma)'_\sigma$ and $(E'_\tau)'_\tau$ are not called "the second weak dual" and "the second Mackey dual", since these spaces coincide, respectively, with E_σ and E_τ.

In the space E', along with the topology $\sigma(E', E)$ (with it this space is denoted by the symbol E'_σ), it is also useful to consider the topology $\sigma\bigl(E', (E'_\beta)'\bigr)$. This is the "weakened topology in the strong dual"; equipped with it, this space should be denoted, according to our notation, by the symbol $(E'_\beta)_\sigma$.

3.7.3. Definition. *A locally convex space E is called semireflexive if the spaces $(E'_\beta)'$ and E coincide as vector spaces without topologies.*

A locally convex space E is called reflexive if $(E'_\beta)'_\beta = (E, \tau_E)$, i.e., if it is semireflexive and the topology $\beta(E, E')$ coincides with the original topology of E.

In terms of the Mackey topology, the semireflexivity of E is the equality $\tau(E', E) = \beta(E', E)$. If E is semireflexive, then the weak topology $\sigma(E, E')$ in E coincides with the weak topology which can be introduced in the space $(E'_\beta)'$, and the same is true for E'_β. The weak topology $\sigma(E', E)$ in E' is often called the weak-$*$ topology (this is especially popular in the case of normed spaces).

Certainly, to avoid confusion it is preferable to use the detailed notation like $\sigma(E', E)$ and $\sigma(E, E')$.

3.7.4. Theorem. *A Hausdorff locally convex space E is semireflexive precisely when every bounded subset in E is relatively compact in the weak topology $\sigma(E, E')$ (here we mean boundedness in the original topology, but since the* *collection of bounded sets in all topologies compatible with duality between E and E′ is the same, we can assume that boundedness is meant in any of them, say, in the weak topology $\sigma(E, E')$).*

PROOF. This is a corollary of the Mackey–Arens theorem (Theorem 3.2.3), since to say that a locally convex space E is semireflexive is equivalent to saying that the topology $\beta(E', E)$ in the space E' agrees with the duality between E' and E. Since the Mackey topology $\tau(E', E)$ is the strongest among all topologies in E' compatible with the duality between E' and E and the topology $\beta(E', E)$ majorizes it, the topology $\beta(E', E)$ agrees with the duality between E' and E precisely when it coincides with the Mackey topology. Thus, it remains to show that the condition of this theorem is equivalent to the equality $\beta(E', E) = \tau(E', E)$, which, in turn, is equivalent to the inclusion $\beta(E', E) \subset \tau(E', E)$ by the fact that $\beta(E', E)$ majorizes $\tau(E', E)$.

Thus, suppose that every bounded subset of the space E is relatively compact in the topology $\sigma(E, E')$; we prove that $\beta(E', E) \subset \tau(E', E)$. Let V be a neighborhood of zero in the topology $\beta(E', E)$. Then there exists a bounded subset B in E such that $B^\circ \subset V$. However, $B^\circ = (B^{\circ\circ})^\circ$ (see § 3.1), moreover, the set $B^{\circ\circ}$ is absolutely convex and closed (in the topology $\sigma(E, E')$, hence also in every other topology compatible with the duality between E and E') hull of the bounded set B, hence is also bounded, hence is relatively compact in the weak topology $\sigma(E, E')$. Its closedness means that it is compact in this topology. Hence the set $(B^{\circ\circ})^\circ = B^\circ$ is a neighborhood of zero in $\tau(E', E)$. Thus, we have proved that $\beta(E', E) \subset \tau(E', E)$.

We now show that the last inclusion yields that every bounded subset of the space E (or, equivalently, of the space E_σ) is relatively compact in E_σ.

Suppose that this inclusion holds. The polar B° of any bounded set B in E is a neighborhood of zero in the topology $\beta(E', E)$, hence there exists a neighborhood of zero W in the topology $\tau(E', E)$ such that $W \subset B^\circ$. Then the definition of the topology $\tau(E', E)$ yields that there exists an absolutely convex compact subset K of the space E_σ such that $K^\circ \subset W$. This means that $K = K^{\circ\circ} \supset B^{\circ\circ} \supset B$, so that the set B is relatively compact in E_σ. □

Certainly, for the semireflexivity it is sufficient to have relative compactness of bounded sets in the topology $\tau(E, E')$.

3.7.5. Corollary. *Every semireflexive locally convex space is quasi-complete and hence is sequentially complete.*

Moreover, any semireflexive locally convex space E is quasi-complete and sequentially complete in the topology $\sigma(E, E')$.

PROOF. By the proven theorem, every closed bounded subset in any semireflexive space E is contained in a set that is compact in the topology $\sigma(E, E')$, hence is complete in this topology. Clearly, the latter set is complete in the original topology. Hence the original set is also complete in it as a closed subset of a complete set. This is also true for the topology $\sigma(E, E')$ (since it can be regarded as the original one). □

Since for metrizable locally convex spaces the sequential completeness is equivalent to completeness, it follows that semireflexive metrizable locally convex spaces are Fréchet spaces (i.e., are complete).

3.7.6. Theorem. *The topology induced in a space E by the topology of its second strong dual space coincides with the original topology of this space precisely when E is quasi-barrelled (it is meant here that E is considered as a subspace of the space of linear functionals on E' and that $E \subset (E'_\beta)'$).*

PROOF. The quasi-barrelledness of E means that every barrel in E absorbing all bounded sets is a neighborhood of zero. The topology induced in E by the topology of the space $(E'_\beta)'_\beta$, i.e., the topology $\beta\bigl((E'_\beta)', E'\bigr)$, possesses a base of neighborhoods of zero consisting of the polars in E of subsets of the space E' bounded in the topology $\beta(E', E)$ (i.e., the so-called strongly bounded subsets). A set $A \subset E'$ is strongly bounded precisely when it is absorbed by the polar B° of every bounded set $B \subset E$ (since these polars form a base of neighborhoods of zero of the topology $\beta(E', E)$); but if the set A is absorbed by the polar B°, then A° absorbs the set $B^{\circ\circ} \supset B$ and hence absorbs B. Since A° is a barrel in the original topology of E, being the polar of a set bounded in the strong topology $\beta(E', E)$, hence also in the weaker topology $\sigma(E', E)$, the set A° turns out to be a neighborhood of zero in the original topology of the space E.

Thus, every neighborhood of zero from a base of neighborhoods of zero of the topology induced in E by the topology $\beta\bigl((E'_\beta)', E'\bigr)$ is a neighborhood of zero in the original topology of this space, hence the latter is not weaker than the induced topology. However, it is also not stronger, since the polar V° of any neighborhood of zero V of the space E is bounded in E'_β. Indeed, V absorbs every bounded subset $B \subset E$. Hence the polar B° of every such set (these polars form a base of neighborhoods of zero in the topology $\beta(E', E)$) absorbs the set V°, which means that it is bounded. Hence every absolutely convex closed neighborhood of zero in E, being the polar of its polar, is a neighborhood of zero in the topology induced in E by the topology $\beta\bigl((E'_\beta)', E'\bigr)$.

Assume now that the topology induced in E by the topology $\beta\bigl((E'_\beta)', E'\bigr)$ coincides with the original one and prove that E is quasi-barrelled. Let V be a barrel in E absorbing every bounded subset. We show that V must be a neighborhood of zero in E. The polar V° of V is absorbed by the polar of every bounded set, hence is a bounded subset of E'_β. Hence the bipolar of $V^{\circ\circ}$ (which coincides with V) is a neighborhood of zero in the topology induced in E by the topology $\beta\bigl((E'_\beta)', E'\bigr)$, hence is a neighborhood also in the original topology of E. □

There is an efficient criterion of reflexivity.

3.7.7. Theorem. *A locally convex space E is reflexive precisely when E is quasi-barrelled and every bounded set in it is relatively compact in the weak topology $\sigma(E, E')$. Every reflexive locally convex space is barrelled.*

PROOF. The first assertion follows from the two previous theorems. It remains to prove that if E is reflexive, then it is barrelled. Since E is semireflexive, the weakly and strongly bounded subsets of the space E' are the same (the weak boundedness is boundedness in the space E'_σ, the strong boundedness is boundedness in E'_β). Therefore, if V is a barrel in E, then its polar V°, being a bounded subset in E'_σ, is bounded in E'_β. Hence $V^{\circ\circ} = V$ is a neighborhood of zero in the topology $\beta(E, E')$, then also in the original topology. □

3.7.8. Corollary. *If a space E is semireflexive, then its topological dual space E' is barrelled in the strong topology (which coincides with $\tau(E', E)$). Conversely, if $\bigl(E', \tau(E', E)\bigr)$ is barrelled, then E is semireflexive. If the space E is reflexive, then E' with the strong topology is also reflexive.*

PROOF. This follows from the previous results taking into account that the polar of any bounded set in E is a barrel in E'. □

The obtained results can be represented in a more symmetric form.

3.7.9. Theorem. *For a locally convex space E with the dual space E' the following conditions are equivalent:*
(i) *the space $\bigl(E, \tau(E, E')\bigr)$ is reflexive;*
(ii) *the space $\bigl(E', \tau(E', E)\bigr)$ is reflexive;*
(iii) *both spaces $\bigl(E, \sigma(E, E')\bigr)$ and $\bigl(E', \sigma(E', E)\bigr)$ are semireflexive;*
(iv) *both spaces $\bigl(E, \tau(E, E')\bigr)$ and $\bigl(E', \tau(E', E)\bigr)$ are barrelled.*

3.7.10. Remark. (i) Since the class of bounded subsets in a locally convex space E is not determined directly by its topology, but rather by its dual space E' as a vector space, the property to be (or not to be) semireflexive and also the topology induced in E from its second strong dual depend only on the pair (E, E') of vector spaces. Thus, for all topologies E compatible with the duality between E and E', the space E is (or is not) simultaneously semireflexive, and for all these topologies, the topologies induced in E from the second strong dual to E coincide.

(ii) Since any normed space is bornological and hence quasi-barrelled, the topology induced in it by the topology of its second strong dual coincides with the original one, so that for normed spaces the reflexivity and semireflexivity are the same property. However, for the weak topology this is not true: l^2 with the weak topology is semireflexive, but not reflexive. By the way, note that the topology of the first strong dual to a normed space is the topology of its Banach dual; the same is true for the second strong dual. The reflexivity of a Banach space is equivalent to the weak compactness of its closed unit ball.

3.7.11. Example. (i) The spaces L^p with $1 < p < \infty$ are reflexive, but the spaces c_0, l^1, $C[0,1]$, and $L^1[0,1]$ are not reflexive.

(ii) The space that is the strict inductive limit of an increasing sequence of its closed reflexive subspaces is itself reflexive. The inductive limits of increasing sequences of spaces with compact embeddings are reflexive (see § 2.6, § 2.7).

(iii) The spaces $\mathcal{D}_n$, $\mathcal{D}$, $\mathcal{D}'$, $\mathcal{S}$, $\mathcal{S}'$, $\mathcal{E}$, $\mathcal{E}'$, and $\mathbb{R}^T$ are reflexive.

The requirement of the relative compactness of bounded sets in the weak topology, contained in Theorems 3.7.4 and 3.7.7, is rather subtle, since every bounded set in an arbitrary locally convex space is weakly precompact (Proposition 3.4.2). In order to verify weak compactness, the Eberlein and Šmulian theorems discussed in § 3.4 are useful.

There are incomplete reflexive locally convex spaces, even with relatively compact bounded sets (Knowles and Cook [**277**] corrected an example from Amemiya, Kōmura [**16**]; see also Kōmura [**288**]). Some information about the dual space E' with the strong topology $\beta = \beta(E', E)$ is given in Edwards [**150**, § 8.4]; there the notion of a *distinguished* space is used, i.e., a Hausdorff locally convex space X such that every $\sigma(X'', X)$-bounded set in X'' is contained in the closure of some $\sigma(X'', X)$-bounded set in X. For example, (E', β) is barrelled precisely when it is distinguished; this is equivalent to saying that E is distinguished. If E is a Fréchet space with a decreasing base of closed absolutely convex neighborhoods of zero V_n, then E is distinguished precisely when the topology β on E' coincides with the topology of the inductive limit of the increasing Banach spaces $E_{V_n^\circ}$.

The next result follows from Theorem 2.9.12, but is easily proved directly.

3.7.12. Theorem. *Every nuclear Fréchet space is reflexive.*

The strong dual to a nuclear space need not be nuclear, but the following facts are true.

3.7.13. Theorem. *A metrizable locally convex space is nuclear precisely when its strong dual is nuclear. Similarly, a locally convex space with a metrizable strong dual is nuclear precisely when its strong dual is nuclear.*

3.7.14. Theorem. *If a locally convex space E has a nuclear strong dual and a locally convex space F is nuclear, then the space $\mathcal{L}_b(E, F)$ of continuous linear operators from E to F with the topology of uniform convergence on bounded sets is nuclear.*

For proofs, see Schaefer [**436**, Chapter 3, § 7, Chapter 4, § 9].

3.8. Criteria for completeness

In this section, we prove a number of results containing criteria for completeness of locally convex spaces. These results were first obtained by Banach for normed spaces and later extended by Grothendieck to arbitrary locally convex spaced.

Suppose that in a space E dual to a locally convex space G, we are given a family γ of bounded (in the topology $\sigma(E, G)$) subsets such that the family of polars of the elements of γ is a prebase of neighborhoods of zero in G. We shall called γ a *saturated family* if it covers E and the inclusions $A \in \gamma$ and $B \subset A$ imply that $B \in \gamma$. For any space F dual to E (for example, for G), the topology of uniform convergence on the elements of γ (generally speaking, not compatible with the vector structure in F, but compatible with the structure of a vector group) is denoted by the symbol τ_γ.

3.8.1. Theorem. *Let E be a locally convex space, let E' be its topological dual equipped with the topology τ_γ of convergence on some saturated family γ*

of bounded sets in E. The space (E', τ_γ) is complete precisely when a linear functional on E is continuous provided that its restriction to every set in γ is continuous in the topology induced from E.

PROOF. The sufficiency is almost obvious. Indeed, let $\{f_\alpha\}$ be a Cauchy net in E'. Since every singleton belongs to γ, for every $x \in E$, there exists $\lim\limits_\alpha f_\alpha(x) = f(x)$. The function f defined by this equality is linear on E. Since the net $\{f_\alpha\}$ converges to f uniformly on the sets from the family γ, the restriction of f to any set from γ is continuous. By the hypothesis of the theorem this means that f is continuous on E. Hence $f \in E'$ and $f_\alpha \to f$ in E'.

We prove the necessity. Suppose that E' is complete and f is a linear functional on E the restriction of which to every set from γ is continuous. We verify that $f \in E'$. To this end, we show that every neighborhood of the function f, considered as an element of the space E^* with the topology of convergence on the family γ, contains some element of the space E'. Since E', being complete, is closed in E^* (the topology induced in E' by the topology that we introduced in E^* coincides with the one introduced in E'), this will mean that $f \in E'$.

Thus, let V be some neighborhood of f in the space E^*. Then, there exists a set $B \in \gamma$ such that $f + B^\circ_{E^*} \subset V$ (since the sets of the type $B^\circ_{E^*}$ form a base of neighborhoods of zero in the space E^* in the topology τ_γ). We can assume that B is closed and absolutely convex due to the equality of the polars of B and $\overline{\operatorname{abs\,conv}}\, B$. By the continuity of the function $f|_B$ there is a closed and absolutely convex neighborhood of zero W in the space E such that $|f(x)| < 1$ whenever $x \in B \cap W$, i.e., $f \in (B \cup W)^\circ_{E^*}$. We have

$$(B \cup W)^\circ_{E^*} = \overline{\operatorname{conv}}\,(B^\circ_{E^*} \cup W^\circ_{E^*}) \subset B^\circ_{E^*} + W^\circ_{E^*},$$

so that $f \in B^\circ_{E^*} + W^\circ_{E^*}$, i.e., there exists an element $a \in W^\circ_{E^*}$ such that f is contained in $B^\circ_{E^*} + a$, or, which is the same, $a \in f + B^\circ_{E^*}$. Since $W^\circ_{E^*} \subset E'$, the point $a \in E'$ is the required element in the neighborhood of f. □

Note that the property of the space (E', τ_γ) to be complete depends not directly on the topology of the space E, but only on the pair (E, E') and, certainly, on the system of sets γ. The proof gives the following description of the completion of the space (E', τ_γ).

3.8.2. Theorem. *Let E be a locally convex space, let E' be its dual equipped with the topology τ_γ of convergence on some family γ of bounded subsets of the space E. Then, the completion of the space E' can be realized as the vector space Γ of all linear functionals on E continuous on every set in the system γ with respect to the topology in E, where Γ is equipped with the topology of convergence on the family of sets γ.*

PROOF. It has been shown in the proof of the second part of the previous theorem that if f is a linear functional on E continuous on an absolutely convex closed bounded set $B \subset E$ (in the topology induced from E), then there exists an element $a \in E'$ such that $a \in f + B^\circ_{E^*}$. This means that such a functional is bounded on B. Indeed, since $a - f \in B^\circ_{E^*}$, we see that the absolute value of the functional $a - f \in E^*$ is bounded by one on B. In addition, the continuous functional a is bounded on B.

Thus, every functional in Γ is bounded on every subset $B \in \gamma$. Hence the topology of convergence on the system of sets γ agrees with the vector structure in Γ. Now the completeness of Γ is proved precisely in the same way as the first part of the previous theorem (with the difference that now there is no need to use the fact that the continuity of a functional $f \in E^*$ on every set in γ implies its continuity on all of E).

Next, the fact used above that for every functional $f \in E^*$ continuous on an absolutely convex closed bounded set $B \subset E$, there exists a functional $a \in E'$ for which $a - f \in B^\circ_{E^*}$ means that the set E' is everywhere dense in (Γ, τ_γ). By the completeness of the space (Γ, τ_γ) we conclude that (Γ, τ_γ) is a completion of the space (E', τ_γ). □

3.8.3. Remark. In a similar way one can prove a completeness criterion and describe the completion for the space $\mathcal{L}_\gamma(E, G)$ of all linear continuous mappings from an arbitrary locally convex space E to a complete locally convex space G with respect to the topology of convergence on a system γ of bounded subsets in E. Namely, this space is complete precisely when the continuity of a linear mapping $f\colon E \to G$ on every set in the system γ implies the continuity on all of E. Furthermore, the completion of $\mathcal{L}_\gamma(E, G)$ can be realized as the vector space of all linear mappings from E to G continuous on every set in the system γ, where $\mathcal{L}_\gamma(E, G)$ is equipped with the topology of convergence on the system γ (prove this).

The next result is important in applications. It deals with completeness, not merely sequential completeness, and is not trivial even for Fréchet spaces.

3.8.4. Theorem. *Let E be a Hausdorff bornological space. Then E' is complete with the strong topology $\beta(E', E)$. If E is sequentially complete, then the dual space E' is complete also in the Mackey topology $\tau(E', E)$. For example, this is true if E a Fréchet space.*

PROOF. The first assertion follows from Theorem 3.6.11. Let E be sequentially complete and let a linear function $f \in E^*$ be the limit of a net of elements $f_\alpha \in E'$ converging uniformly on absolutely convex weakly compact sets in E. We show that $f \in E'$. Since E is bornological, it suffices to verify the boundedness of f. Assuming that f is not bounded, we obtain a bounded sequence $\{a_n\}$ with $f(a_n) \to +\infty$. Passing to a subsequence, we can assume that $f(a_n) \geqslant 4^n$, i.e., $f(2^{-n}a_n) \geqslant 2^n$. This leads to a contradiction, since $\{2^{-n}a_n\}$ is contained in the absolutely convex compact set that is the image of the closed unit ball in l^2 under the operator $A\colon l^2 \to E$, $(x_n) \mapsto \sum_{n=1}^\infty 2^{-n}x_n a_n$; convergence of the series follows from the boundedness of $\{a_n\}$ and the sequential completeness of E. □

Regarding the original locally convex space as the vector space of linear functionals on E' equipped with the topology of convergence on the system β_e of all equicontinuous subsets of the space E', we obtain the following criterion of completeness of a locally convex space.

3.8.5. Theorem. *A locally convex space E is complete precisely when every linear functional on E' continuous in the topology $\sigma(E',E)$ on every equicontinuous set in E' is an element of E.*

The completion of the locally convex space E can be realized as the space of all linear functionals on E' having continuous restrictions to all equicontinuous subsets, where E is equipped with the topology of convergence on all equicontinuous sets in E'.

Since for separable E the topology $\sigma(E',E)$ is metrizable on equicontinuous subsets in E' (Proposition 1.9.5), the following useful fact holds.

3.8.6. Corollary. *If a locally convex space E is separable and complete, then for the continuity of a linear function Λ on E' in the topology $\sigma(E',E)$ it is sufficient that, for every sequence $\{f_n\}\subset E'$ pointwise converging to zero, we have $\Lambda(f_n)\to 0$.*

3.8.7. Proposition. *Let E be a locally convex space. For any $A\subset E$, let $(A^\circ)^\circ_{E'^*}$ denote the polar in the space E'^* (the algebraic dual of E') of the polar of A in E'. If $\widetilde{E}$ is the completion of the space E, then we have the following equality of vector spaces:*

$$\widetilde{E}=\bigcap_{V\in\tau_0}\bigl(E+(V^\circ)^\circ_{E'^*}\bigr),$$

where τ_0 is the set of all neighborhoods of zero in E. Similarly, the completion of the space E' equipped with the topology of convergence on some class $\mathcal{A}$ of bounded sets in E the union of which covers E is the vector subspace

$$\widetilde{E'_{\mathcal{A}}}=\bigcap_{A\in\mathcal{A}}(E'+A^\circ)$$

in the algebraic dual E^* equipped with the topology of convergence on $\mathcal{A}$, where polars are taken in E^*.

PROOF. This follows from Theorem 3.8.1 and Theorem 3.8.5. Indeed, if $f\in\widetilde{E}$, then f is a functional on E' the restriction of which to every set V°, where $V\in\tau_0$, is continuous in the topology induced on V° by the weak topology $\sigma(E',E)$. Therefore, according to the reasoning given at the very end of the proof of Theorem 3.8.1, $f\in a+(V^\circ)^\circ_{E'^*}$, where $a\in E$, i.e., $f\in E+(V^\circ)^\circ_{E'^*}$. Since this is true for every $V\in\tau_0$, we have $f\in\bigcap_{V\in\tau_0}\bigl(E+(V^\circ)^\circ_{E'^*}\bigr)$.

Assume now that the latter inclusion holds. This means that for every neighborhood of zero in E'^* of the form $(V^\circ)^\circ_{E'^*}$, where $V\in\tau_0$ (such neighborhoods form a base of zero in the space E'^* in the topology of convergence on the family of all subsets of E' of the type V°, where $V\in\tau_0$), there exists an element $a\in E$ such that $a\in f+(V^\circ)^\circ_{E'^*}$. Thus, every neighborhood of the functional f contains elements from E. In other words, for any set $V\in\tau_0$, the restriction of f to V° can be uniformly approximated by the restrictions to V° of functionals that are elements of E (verify that this follows from the previous sentence). Hence by Theorem 3.8.5 we obtain $f\in\widetilde{E}$. The second assertion of the theorem obviously follows from the first one. □

3.8.8. Remark. Theorem 3.8.5 yields the fact, established in Example 1.7.11, that no infinite-dimensional Banach space B (hence no normed space) is complete in the weak topology $\sigma(B, B')$, because there exist discontinuous functionals on the Banach dual B', which are discontinuous in the topology $\sigma(B', B'')$, hence in the topology $\sigma(B', B)$. Indeed, the polar of any neighborhood of zero in the space $(B, \sigma(B, B'))$, taken in the space B', is contained in a finite-dimensional subspace of the latter, so the restriction to it of every linear functional on E' is continuous in the topology induced by an arbitrary Hausdorff locally convex topology on the space B'.

Similarly, the fact that on every infinite-dimensional metrizable locally convex space there exist discontinuous linear functionals yields that the dual to such a space equipped with the weak topology cannot be complete.

In relation to these facts we mention that in the old literature (for example, in the book by Dunford and Schwarz [**145**]), the term a "weakly complete Banach space" means a "Banach space sequentially complete in the weak topology".

All reflexive Banach spaces are sequentially complete in the weak topology (why?); nonreflexive Banach spaces can be sequentially complete or incomplete in this topology. An example of the first type is the space l^1, an example of the second is the space c_0.

3.8.9. Theorem. *A locally convex space is complete precisely when every hypersubspace of its dual possessing weakly closed intersections with the polars of all neighborhoods of zero in the original space is itself closed in the weak topology.*

For the proof it suffices to apply Theorem 3.8.5 and Proposition 1.9.6.

In the formulation of this theorem the term "hypersubspace" can be replaced by the term "hyperplane".

3.8.10. Definition. *Let E be a locally convex space and let E' be its dual. A set $A \subset E'$ is called almost closed if, for every neighborhood of zero V in E, the set $A \cap V^\circ$ is closed in the topology $\sigma(E', E)$.*

Using this definition, the previous theorem can be reformulated as follows.

3.8.11. Theorem. *A locally convex space E is complete precisely when every almost closed hypersubspace of its dual E' is closed in the topology $\sigma(E', E)$ (or every almost closed hyperplane in its dual is closed in the same topology).*

3.8.12. Definition. *A locally convex space E is called fully complete if every almost closed vector subspace of its dual space is closed in the topology $\sigma(E', E)$. The classes of B_r-complete spaces, hypercomplete spaces, and the Krein–Šmulian spaces are defined by replacing the term "almost closed subspace" by the terms "almost closed everywhere dense in the topology $\sigma(E', E)$ vector subspace", "almost closed absolutely convex subset", and "almost closed convex subset", respectively.*

It follows from Theorem 3.8.9 that every B_r-complete space is complete. In addition, the definition given above yields that the classes of spaces introduced in it have the following relations:

B_r-complete locally convex spaces $\supset$ fully complete locally convex spaces $\supset$ hypercomplete locally convex spaces $\supset$ Krein–Šmulian locally convex spaces.

Any infinite-dimensional Banach space with the strongest locally convex topology is complete, but not B_r-complete (Exercise 3.12.142); the spaces $\mathcal{D}$ and $\mathcal{D}'$ are also complete, but not B_r-complete, since by Proposition 2.10.8 and Proposition 2.10.9 they contain non-closed dense sequentially closed linear subspaces (sequentially closed sets in these spaces are almost closed, since weakly compact sets in them are metrizable). The product of the countable power of the real line and the countable direct sum of the real lines is hypercomplete, but is not a Krein–Šmulian space (Exercise 3.12.164). A long standing problem of distinguishing the classes of B_r-complete and fully complete spaces was solved by Valdivia [**544**], who constructed an example of a B_r-complete space that is not fully complete (this example is also described in the book Pérez Carreras, Bonet [**385**, § 7.4]). It is still not known whether the classes of fully complete and hypercomplete spaces are different.

Fully complete locally convex spaces are also called *B-complete* or *Ptak spaces*. Significance of B_r-complete and B-complete spaces is explained by their role in the closed graph theorem and the open mapping theorem; this is discussed below in § 3.9.

The next result is classical in the theory of locally convex spaces.

3.8.13. Theorem. (THE KREIN–ŠMULIAN THEOREM) *Every Fréchet space is a Krein–Šmulian space.*

PROOF. Let E be a Fréchet space and let $\mathfrak{T}$ be the topology in E' defined as follows: a set $A \subset E'$ is closed precisely when it is almost closed (verify that this definition is correct!). To prove the theorem, it suffices to verify that this topology is locally convex and agrees with the duality between E' and E. If this is done and $A \subset E'$ is an almost closed convex set, then A is closed in the topology $\mathfrak{T}$, hence, by its convexity, in every other topology in E' compatible with the duality between E' and E, in particular, in the topology $\sigma(E', E)$.

In order to prove that the topology $\mathfrak{T}$ is locally convex, it suffices to show that, for every $\mathfrak{T}$-open set $W \subset E'$ and every point $a \in W$, there exists a $\mathfrak{T}$-open convex set W_a such that $a \in W_a \subset W$.

The set $F = E' \setminus W$ is closed in the topology $\mathfrak{T}$, i.e., is almost closed. Our goal is to prove the existence of a $\mathfrak{T}$-open convex set B such that $a \in B$ and $B \cap F = \varnothing$.

Let $V_1 \supset V_2 \supset V_3 \supset \cdots$ be a base of absolutely convex neighborhoods of zero in E; then $\bigcup_{j=1}^{\infty} V_j^\circ = E'$. Thus, $a \in V_n^\circ$ for some n; clearly, we can assume that $n = 1$. Let $B_1 \ni a$ be a $\sigma(E', E)$-open convex set in E' the closure $\overline{B_1}$ of which in the topology $\sigma(E', E)$ is disjoint with $V_1^\circ \cap F$. Such a set B_1 exists, since $V_1^\circ \cap F$ is closed in the topology $\sigma(E', E)$. We have

$$\overline{B_1} \cap (V_1^\circ \cap F) = (\overline{B_1} \cap V_1^\circ) \cap F,$$

moreover, $\overline{B_1} \cap V_1^\circ = \overline{B_1 \cap V_1^\circ}$ by the compactness of the polar V_1° in the topology $\sigma(E', E)$. Thus, $W_1 = B_1 \cap V_1^\circ$ is a neighborhood of a in the space V_1° with the topology induced by the topology $\sigma(E',E)$, and the closure of this neighborhood in V_1° is disjoint with F.

Suppose that we have constructed a finite number $W_1 \subset W_2 \subset \cdots \subset W_n$ of convex sets in E' with the following properties: for every $j \leqslant n$, the set W_{j+1} is an open neighborhood of the set W_j in the space V_{j+1}° equipped with the topology induced by the topology $\sigma(E',E)$, and the closure of the set W_j in V_j° is disjoint with F. Then W_{n+1} is the intersection of the set V_{n+1}° with some open convex neighborhood of the set $\overline{W_n}$ in the space $\bigl(E',\sigma(E',E)\bigr)$ the closure of which in the topology $\sigma(E',E)$ is disjoint with F. Hence the set $\bigcup_{n=1}^\infty W_n$ is convex and open in the topology $\mathfrak{T}$ (since its complement in E' is almost closed).

It remains to prove that the topology $\mathfrak{T}$ agrees with the duality between E' and E. This topology majorizes $\sigma(E',E)$ (every set closed in the topology $\sigma(E',E)$ is automatically almost closed), so that every element of the space E is continuous as a functional on $(E',\mathfrak{T})$. On the other hand, if a linear function f on E' is continuous in the topology $\mathfrak{T}$, then its kernel is closed in this topology, i.e., is almost closed. Hence f is continuous on every set V_j° with the topology $\sigma(E',E)$ (Proposition 1.9.8). By the completeness of E we have $f \in \bigl(E',\sigma(E',E)\bigr)' = E$ by Theorem 3.8.5. □

In a Banach space, it suffices to check the polar of a single ball, which gives the following fact.

3.8.14. Corollary. *Let X be a Banach space and let $V \subset X'$ be convex. If the intersection of V with every closed ball of radius n centered at zero is closed in the topology $\sigma(X',X)$, then V is also closed in $\sigma(X',X)$.*

If X is separable, then in order the set V be closed in the topology $\sigma(X',X)$, it is sufficient that V contain the limits of all its weak-$*$ convergent countable sequences.

Note that the definition of the topology $\mathfrak{T}$ and the proof of its local convexity do not use the completeness of the metrizable locally convex space E. The question arises whether metrizability is essential. There is an example due to Collins (see Exercise 3.12.146) showing that for general locally convex spaces the topology $\mathfrak{T}$ may fail to be locally convex. However, even without the completeness assumption of the metrizable space E the topology $\mathfrak{T}$ admits a very simple description, which is given by the following Banach–Dieudonné theorem.

Let $\mathfrak{P}$ be the class of all precompact sets in a locally convex space E and let $\mathfrak{S}$ be the class of all sequences in E converging to zero. It is clear that $\mathfrak{S} \subset \mathfrak{P}$.

3.8.15. Theorem. *Let E be a metrizable locally convex space. Then, on the space E', the topologies of uniform convergence on the classes $\mathfrak{P}$ and $\mathfrak{S}$ coincide with the strongest topology $\mathfrak{T}$ on E' that coincides with $\sigma(E',E)$ on all equicontinuous sets.*

If E is complete, then the indicated identical topologies are compatible with the duality between E' and E.

PROOF. The topologies $\tau_{\mathfrak{P}}$ and $\tau_{\mathfrak{S}}$ of convergence on the classes $\mathfrak{P}$ and $\mathfrak{S}$ coincide by Proposition 1.8.14. On the polars of neighborhoods of zero all the three considered topologies coincide (see Corollary 3.1.6). Hence $\tau_{\mathfrak{P}}$ is majorized by the topology $\mathfrak{T}$. We show that the inverse relation is also true (for complete E this is clear from the previous theorem). We already know that $\mathfrak{T}$ is locally convex. Let W be a $\mathfrak{T}$-neighborhood of zero in E'. We construct a sequence S in E converging to zero such that $S^\circ \subset W$. By induction we find nonempty finite sets $F_n \subset V_n$ such that $V_{n+1}^\circ \cap H_n^\circ \subset W$, where $H_n = F_1 \cup \cdots \cup F_n$. Then for S we shall take the elements of the sets F_n numbered consecutively; since $S^\circ \cap V_{n+1}^\circ \subset W$ for all n and $E' = \bigcup_{n=1}^\infty V_n^\circ$, we shall obtain $S^\circ \subset W$. A set F_1 exists, since $W \cap V_1^\circ$ is open in the topology $\sigma(E', E)$ and contains the intersection of V_1° with the polar of some finite set. Suppose that $F_1, \ldots, F_n$ are already picked. We need a nonempty finite set $F_{n+1} \subset V_{n+1}$ such that $V_{n+2}^\circ \cap (H_1 \cup F_{n+1})^\circ \subset W$. Suppose that for every nonempty finite set $F \subset V_{n+1}$, the set $V_{n+2}^\circ \cap (H_n \cup F)^\circ \backslash W$ is nonempty. This set is $\sigma(E', E)$-closed in V_{n+2}° by the coincidence of the considered topologies on V_{n+2}°, hence it is $\sigma(E', E)$-compact by the compactness of V_{n+2}°. The collection of the obtained nonempty compact sets possesses the property that the intersection of every its finite subcollection is nonempty. Therefore, the intersection of all these compact sets is nonempty. Since $(H_n \cup F)^\circ = H_n^\circ \cap F^\circ$ and the intersection of the sets F° over all finite $F \subset V_{n+1}$ is V_{n+1}°, the set $H_n^\circ \cap V_{n+1}^\circ \backslash W$ is nonempty. This contradicts the assumption that $H_n^\circ \cap V_{n+1}^\circ \subset W$. In the case of complete E we can take only absolutely convex compact sets, since in that case every totally bounded set is contained in an absolutely convex compact set. □

Note that by Theorem 3.8.5 the completeness of the space E is also necessary in order that the indicated topologies agree with the duality between E' and E.

3.9. The closed graph theorem

Here we briefly discuss the closed graph theorem and two close results: the inverse mapping theorem and the open mapping theorem. These classical results go back to the works of Banach and Schauder and were originally obtained for complete metrizable topological vector spaces. The formulations given in university courses of functional analysis are these.

We recall that the graph of a mapping $T\colon X \to Y$ is the set

$$\Gamma_T = \{(x, Tx)\colon x \in X\} \subset X \times Y.$$

3.9.1. Theorem. *Let X and Y be complete metrizable topological vector spaces and let $T\colon X \to Y$ be a linear mapping.*

(i) *If T is continuous and surjective, then, for every open set $U \subset X$, the set $T(U)$ is open in Y.*

(ii) *If T is continuous and one-to-one, then T^{-1} is also continuous.*

(iii) *The continuity of T is equivalent to the closedness of its graph.*

3.9.2. Corollary. (i) *If a complete metrizable topological vector space E is equipped with a stronger vector topology with respect to which E is also complete and metrizable, then both topologies coincide.*

(ii) *If a complete metrizable topological vector space is a direct algebraic sum of two its closed linear subspaces, then the algebraic projections onto them are continuous, hence the sum is topological.*

We do not repeat the proofs given in many books (see, e.g., [**72**]), but prove instead more general results for locally convex spaces. We just note some implications between assertions (i)–(iii). It is clear that (ii) follows at once from (i). On the other hand, assertion (ii) yields (i) by passing to the quotient $X/\mathrm{Ker}\,T$, which in the case of continuous T is mapped one-to-one onto Y. Certainly, assertion (iii) implies (ii) due to the fact that the graph of a one-to-one operator T is obtained from the graph of T^{-1} under the linear homeomorphism $X\times Y \to Y\times X$, $(x, y) \mapsto (y, x)$. Finally, in the case of complete metrizable spaces X and Y assertion (iii) follows from (ii), since $X\times Y$ is the space of the same class, hence the same is true for its closed subspace Γ_T, which gives the continuity of the operator inverse to a one-to-one projection $P\colon \Gamma_T \to X$, hence also the continuity of T (observe that T is the composition of P^{-1} and the continuous projection onto Y). However, in this last assertion we have used the closedness of the given class of spaces with respect to finite products and passage to closed subspaces, and for the implication (ii)⇒(i) it has been important to pass to quotients. This circumstance becomes essential in generalizations of all the three assertions. Such generalizations are very numerous (see also Exercise 3.12.87–3.12.92) and have been obtained in the following directions.

1) One is interested in two possibly broader classes of locally convex spaces $\mathcal{X}$ and $\mathcal{Y}$ for elements X and Y of which some of the stated assertions are true. 2) One fixes some class $\mathcal{Y}$ of spaces Y and is interested in a possibly broader class $\mathcal{X}$ of spaces X for mappings of which to all spaces $Y \in \mathcal{Y}$ such assertions are true. 3) One fixes some class $\mathcal{X}$ of spaces X and is interested in a possibly broader class $\mathcal{Y}$ of spaces Y for mappings to which from all spaces $X \in \mathcal{X}$ some of the above assertions are true.

Certainly, one can expect in advance that for assertions (ii) (or (i)) and (iii) different results will be obtained. Indeed, in case (ii) we shall conclude that the spaces X and Y are isomorphic, while the closed graph theorem does not expect this at all. For this reason, we shall mainly concentrate on this theorem.

We shall further consider Hausdorff spaces.

3.9.3. Theorem. (i) *Let X be a barrelled space and let Y be B_r-complete. Then any linear operators from X to Y with a closed graph is continuous. In particular, every linear mapping with a closed graph from a barrelled space to a Fréchet space is continuous.*

(ii) *If a locally convex space X is such that for every Banach space Y every linear mapping with a closed graph from X to Y is necessarily continuous, then X is barrelled.*

For a proof, see Schaefer [**436**, § IV.8], where one can find related results about open mappings.

Thus, if a locally convex space E is not barrelled, then there exist a Banach space F and a discontinuous linear mapping $A\colon E \to F$ with a closed graph.

This theorem exhibits asymmetry between properties of X and Y. Say, it is impossible to interchange the properties of the spaces: the theorem fails for mappings of Banach spaces to barrelled spaces. For example, if X is an infinite-dimensional Banach space and Y is the same space with the strongest locally convex topology, then Y is a complete barrelled space (Example 1.7.12 and Exercise 3.12.53), the identity mapping from X to Y has a closed graph (since the inverse is continuous), but is discontinuous.

If we wish to apply this theorem to an operator $T\colon X \to X$ in a space X that is not a Fréchet space, then so far we can only do this under the assumption that it is barrelled and B_r-complete. As we have already noted, the classical spaces $\mathcal{D}$ and $\mathcal{D}'$ are not B_r-complete. However, it turns out that the closed graph theorem is applicable to them.

Let $\mathcal{UF}$ denote the class of all Hausdorff locally convex spaces Y representable in the form $Y = \bigcup_{n=1}^{\infty} Y_n$, where Y_n is a linear subspace in Y possessing a topology of a Fréchet space majorizing the topology induced from Y. Thus, Y becomes a countable union of continuously embedded Fréchet spaces. Note that we do not require that Y be the inductive limit of the spaces Y_n (although these spaces can be made increasing). For example, in the inductive limit of an increasing sequence of Fréchet spaces we can take a weaker locally convex topology.

Grothendieck [**206**], [**207**, p. 147–149] proved the following important result (a proof can be read also in Raĭkov [**409**], an appendix to the Russian edition of Robertson, Robertson [**420**]). We recall that ultrabornological spaces are precisely the inductive limits of Banach spaces, hence the continuity of linear operators on them follows from the continuity of the restrictions to the respective Banach spaces.

3.9.4. Theorem. *The closed graph theorem holds for mappings from ultrabornological spaces to spaces of class $\mathcal{UF}$.*

In addition, if $Y = \bigcup_{n=1}^{\infty} Y_n$, where each Y_n is a Fréchet space continuously embedded into Y, then any Fréchet space continuously embedded into Y is contained in some of the subspaces Y_n.

This theorem (in which the hypotheses about the domain of definition and range are also asymmetric) already applies to operators from $\mathcal{D}$ to $\mathcal{D}$, but is still unapplicable to $\mathcal{D}'$. Raikov [**408**] substantially enlarged the class of spaces for mappings into which of Banach spaces the closed graph theorem is true, so that $\mathcal{D}'$ also belongs to this class. Let us give a partial case of his result, a proof of which can be read in [**409**].

3.9.5. Theorem. *Suppose that a Hausdorff space Y is the image under an open continuous linear mapping of a closed linear subspace in a countable product of spaces of class $\mathcal{UF}$. Then the closed graph theorem holds for mappings from all ultrabornological spaces to the space Y.*

In § 3.12(ii) we consider a generalization of the closed graph theorem in another direction, where Y is a Souslin space (Definition 5.2.9). For separable spaces, the classes considered above belong to the class of Souslin spaces (separability is also necessary for that), since the latter is closed with respect to countable products and passage to closed subspaces and continuous images. Let us give here a particular case of Corollary 3.12.9.

3.9.6. Theorem. *A linear mapping with a closed graph from a Fréchet space to a Souslin locally convex space is continuous.*

Analogs of this result for spaces built by means of more general schemes similar to Souslin schemes are obtained in De Wilde [**110**], [**112**], [**111**], Ferrando, Kąkol, López Pellicer [**161**], Ferrando, Kąkol, López Pellicer, Śliwa [**164**], Kąkol, Kubiś, López-Pellicer [**245**], and Smirnov [**471**].

A Hausdorff locally convex space E is called a *webbed space* if it is equipped with a family of absolutely convex sets $W_{n_1 n_2 \ldots n_k}$ indexed by all possible finite collections of natural numbers $(n_1, \ldots, n_k)$ such that the following conditions are fulfilled:

(i) the set $\bigcup_{n=1}^{\infty} W_n$ is absorbing,

(ii) for every multiindex $(n_1, \ldots, n_k)$, all sets $W_{n_1 n_2 \ldots n_k j}$ with $j \geqslant 1$ are contained in the set $W_{n_1 n_2 \ldots n_k}/2$ and the set $\bigcup_{j=1}^{\infty} W_{n_1 n_2 \ldots n_k j}$ absorbs $W_{n_1 n_2 \ldots n_k j}$,

(iii) whenever we take increasing collections n_1, $(n_1, n_2), \ldots$ and arbitrary vectors $x_k \in W_{n_1, \ldots, n_k}$, the series $\sum_{k=1}^{\infty} x_k$ converges in E.

A family of discs with the same number of indices is called a slice (so that there is the first slice $\{W_n\}$, the second slice $\{W_{nk}\}$ and so on), and a thread is a sequence of sets of the form W_{n_1}, $W_{n_1 n_2}$, $W_{n_1 n_2 n_3}$ and so on. Thus, in (iii) we are concerned with convergence of the series of elements of a thread.

In a Hausdorff topological vector space E, a web is defined similarly, just in place of the absolute convexity of the sets $W_{n_1 n_2 \ldots n_k}$ we require that they must be circled, and condition (ii) is written in the following form: for all indices j one has $W_{n_1 n_2 \ldots n_k j} + W_{n_1 n_2 \ldots n_k j} \subset W_{n_1 n_2 \ldots n_k}$.

3.9.7. Example. (i) Any Banach space is a webbed space with the web in which all discs of the first slice with number k are the same disc $2^{-k}U$, where U is the closed unit ball. Similarly, an arbitrary Fréchet space F possesses a web in which $W_{n_1 \ldots n_k} = U_k$, $U_k = \{p_1 + \cdots + p_k \leqslant 2^{-k}\}$, where $\{p_k\}$ is a sequence of seminorms defining the topology such that $p_k \leqslant p_{k+1}$. In the category of topological vector spaces, all complete metrizable spaces possess the corresponding webs.

(ii) The strong dual to a metrizable locally convex space E has a web of the form $W_n = U_n^\circ$, $W_{nk} = 2^{-1}U_n^\circ$, $W_{n_1 \ldots n_k} = 2^{-k}U_n^\circ$, where $\{U_n\}$ is a decreasing base of absolutely convex neighborhoods of zero in E.

A web exists also in the strong dual of any strict inductive limit of a sequence of metrizable locally convex spaces (Exercise 3.12.165). Other properties of webbed spaces are mentioned in Exercise 3.12.166. Baire webbed spaces are precisely complete metrizable spaces (Exercise 3.12.167).

3.9.8. Theorem. *Any linear mapping T with a sequentially closed graph from a Fréchet space X to a webbed locally convex space Y is continuous.*

PROOF. Let us take a base of absolutely convex neighborhoods of zero U_n in X such that $2U_{n+1} \subset U_n$. Let $\{W_{n_1 \ldots n_k}\}$ be a web in Y. It follows from the definition of a web that $\bigcup_{n=1}^{\infty} T^{-1}(W_n)$ is an absorbent set in X. By Baire's theorem there exists n_1 for which the closure of $T^{-1}(W_{n_1})$ has inner points. Next, $\bigcup_{k=1}^{\infty} T^{-1}(W_{n_1 k})$ absorbs $T^{-1}(W_{n_1})$. Hence there is n_2 for which $T^{-1}(W_{n_1 n_2})$ has inner points. Continuing by induction, we obtain a thread w_{n_1}, $W_{n_1 n_2}, \ldots$, for which the closures of the sets $T^{-1}(W_{n_1 \ldots n_k})$ have inner points, but owing to its absolute convexity the closure of every such set contains some neighborhood of zero U_{m_k}.

We observe that, for every thread $\{W_{n_1 \ldots n_k}\}$, every closed neighborhood of zero $V \subset Y$ contains entirely some $W_{n_1 \ldots n_k}$. Indeed, if there are vectors $y_1 \in V \backslash W_{n_1}$, $y_2 \in V \backslash W_{n_1 n_2}$, $y_3 \in V \backslash W_{n_1 n_2 n_3}$ and so on, then the series of y_k must converge, which is impossible, since $y_k \not\to 0$. Thus, the closure of $T^{-1}(V)$ has a nonempty interior. We show that actually $T^{-1}(V)$ itself has a nonempty interior, whence the continuity of T follows at once. To this end, we take $W_{n_1 \ldots n_k} \subset V$ and verify that $U_{m_{k+1}} \subset T^{-1}(V)$. It is readily verified that for all l the closure of $T^{-1}(W_{n_1 \ldots n_l})$ is contained in $T^{-1}(W_{n_1 \ldots n_l}) + U_{m_{l+1}}$. Hence we have $U_{m_l} \subset T^{-1}(W_{n_1 \ldots n_l}) + U_{m_{l+1}}$.

Let us fix $z \in U_{k_{m+1}}$ and show that $z \in T^{-1}(V)$, i.e., $Tz \in V$. There are vectors $x_1 \in T^{-1}(W_{n_1 \ldots n_{k+1}})$ and $z_1 \in U_{m_{k+2}}$ with $z = x_1 + z_1$.

Next we find $x_2 \in T^{-1}(W_{n_1 \ldots n_{k+2}})$ and $z_2 \in U_{m_{k+3}}$ with $z_1 = x_2 + z_2$, i.e., $z = x_1 + x_2 + z_2$. By induction we find $z_j \in T^{-1}(W_{n_1 \ldots n_{k+j}})$ and $x_j \in U_{m_{k+1+j}}$ with $z = z_{j+1} + x_1 + \cdots + x_j$. Since $z_j \to 0$, the series of x_j converges to z. In addition, the series of Tx_j converges to some $y \in Y$ (property (ii) of webs). Since $W_{n_1 \ldots n_{l+1}} \subset W_{n_1 \ldots n_l}/2$, the partial sums of the series of Tx_j belong to V, so $y \in V$. Finally, for completing the proof it remains to observe that $y = Tz$. It is here that we need the sequential closedness of the graph. □

It is seen from the proof that in place of completeness of X it is sufficient that X be Baire and metrizable. The theorem easily yields the next fact. Moreover, the metrizability can be dropped if we require the usual closedness of the graph (see below).

3.9.9. Corollary. *Any linear mapping with a sequentially closed graph from an ultrabornological space (i.e., an inductive limit of Banach spaces) to a webbed locally convex space is continuous.*

In the category of topological vector spaces the situation is completely similar, moreover, also a justification is similar (see Jarchow [**237**, § 5.4]).

3.9.10. Theorem. *If a topological vector space X is the inductive limit of a family of Baire topological vector spaces X_α and Y is a webbed topological vector space, then any linear mapping from X to Y with a closed graph is continuous.*

If all X_α are metrizable, then it suffices to that the graph be just sequentially closed.

It is seen from the presented results that the requirement of the validity of the closed graph theorem for all operators on a space X with a too large class of range spaces (say, all Banach spaces) turns out to be rather restrictive. This becomes especially obvious in those cases where we take for X the dual to a nonnormable space. In applications one usually deals with a rather narrow class of range spaces. Let us mention some results useful in this situation (about local completeness, see § 2.10(iii)).

3.9.11. Theorem. *Let X be locally convex. The space $\bigl(X',\sigma(X',X)\bigr)$ is locally complete precisely when every linear mapping $T\colon X\to l^2$ with a closed graph is continuous.*

The proof is given in Valdivia [**539**].

3.9.12. Theorem. *Let X be a locally complete locally convex space (for example, sequentially complete), let Y be a locally convex space such that Y' is metrizable in the Mackey topology $\tau(Y',Y)$ (which is fulfilled if Y is a reflexive Banach space), and let $T\colon X'\to Y$ be a linear mapping. Then T is continuous for X' equipped with the Mackey topology $\tau(X',X)$ if the graph of T is closed in $\bigl(X',\tau(X',X)\bigr)\times Y$. For example, the latter is true if Y possesses a weaker Hausdorff topology τ_0 such that T is continuous from $\bigl(X',\tau(X',X)\bigr)$ to (Y,τ_0). We emphasize that τ_0 need not be a vector topology.*

PROOF. It is clear that the continuity of T from $\bigl(X',\tau(X',X)\bigr)$ to (Y,τ_0) ensures the closedness of the graph in $\bigl(X',\tau(X',X)\bigr)\times Y$. The space X' with the topology $\tau(X',X)$ will be denoted by G. By Theorem 3.2.9, it suffices to verify the continuity of the mapping $T\colon G\to\bigl(Y,\sigma(Y,Y')\bigr)$. Let L be the set of all functionals $l\in Y'$ for which the functional $l\circ T$ is continuous in the topology $\sigma(G,X)$, i.e., is given by an element of X (we recall that by the Mackey theorem $G'=X$). We have to show that $L=Y'$.

The set L is a linear subspace. It is everywhere dense in Y' with the metrizable (by assumption) topology $\tau(Y',Y)$, since it is dense in the topology $\sigma(Y',Y)$. Indeed, otherwise we could find a nonzero element $y_0\in Y$ with the property that $l(y_0)=0$ for all $l\in L$. The element $(0,y_0)$ does not belong to the graph of T, which is a closed linear subspace in $G\times Y$. By the Hahn–Banach theorem there is a continuous linear functional on $G\times Y$ vanishing on the graph of T and equal to 1 at $(0,y_0)$. This functional has the form $(g,y)\mapsto g(a)+f(y)$, where $a\in X=G'$ and $f\in Y'$, and we have $f(y_0)=1$ and $g(a)+f(Tg)=0$ for all $g\in G$. Therefore, $f\circ T\in L$, but then $f(y_0)=0$ contrary to the equality $f(y_0)=1$.

Thus, on the everywhere dense linear subspace $L\subset Y'$ we obtain a mapping $A\colon l\mapsto v_l$ with values in X, where $v_l\in X$ is a vector generating the functional $l\circ T$. This mapping is linear. We verify its continuity in the topology induced by $\tau(Y',Y)$. By the assumed metrizability of the latter, it suffices to show that convergence of a sequence $\{l_n\}$ to zero yields convergence of $\{Al_n\}$ to zero in X, and for this it suffices to verify that A takes every bounded set V in L to a bounded set in X. This amounts to showing the boundedness of the set of numbers $\{g(Al)\colon l\in V\}$ for every fixed $g\in G$. The latter holds indeed, since $g(Al)=l(Tg)$ and the set V is bounded also in the topology $\sigma(G,Y)$.

It is now time to use the local completeness of X, which enables one to extend A to a continuous linear operator $\widetilde{A}\colon Y'\to X$ (Exercise 3.12.136). Finally, we observe that for every $l\in Y'$ we have $l(Tg)=g(\widetilde{A}l)$ for all $g\in G$, since for any fixed g both sides are continuous in l on Y' and are equal on the dense set L. Thus, we have $L = Y'$. □

3.9.13. Example. Let X be a locally complete locally convex space and let a linear mapping $T\colon X' \to L^p(\mu)$, where $p \in (1, +\infty)$ and μ is a finite measure, be continuous on X' with the topology $\tau(X', X)$ provided that $L^p(\mu)$ is equipped with the topology of convergence in measure. Then T is continuous also when $L^p(\mu)$ is equipped with the standard norm.

Note an interesting corollary of the open mapping theorem and Michael's selection theorem 1.12.19 (a partial case has already been given in Corollary 1.12.21).

3.9.14. Corollary. *Let $T\colon X \to Y$ be a continuous linear surjection of Fréchet spaces. Then T has a continuous right inverse (possibly, nonlinear).*

3.9.15. Example. In the situation of this corollary, for every compact set $K \subset Y$, there exists a compact set $S \subset X$ such that $T(S) = K$. Indeed, taking a continuous right inverse R for T, we can set $S = R(K)$.

3.10. Compact operators

Here we consider an important class of linear operators: the so-called compact operators. Their theory in case of Banach spaces goes back to Riesz and Schauder, who, in turn, developed the classical theory of integral operators due to Fredholm, Hilbert, Carleman, and others.

3.10.1. Definition. *A linear mapping A from a topological vector space E to a topological vector space G is called compact if it maps some neighborhood of zero V in the space E to a relatively compact subset of the space G.*

Since relatively compact sets are bounded, for every neighborhood of zero W in G, there exists a number $\alpha > 0$ such that $A(\alpha V) = \alpha A(V) \subset W$. Hence any compact linear mapping is continuous at the origin of the space E, therefore, also everywhere. Certainly, not every continuous linear mapping is compact (an example is the identity mapping of any Hausdorff infinite-dimensional topological vector space). Compact mappings are also called completely continuous (note that some authors use these two terms — compact and completely continuous — in different meanings; see, for example, Pietsch [**389**]). The class of compact operators is denoted by $\mathcal{K}(E, G)$.

Throughout this section E is a Hausdorff topological vector space over $\mathbb{R}$ or $\mathbb{C}$ (in some results it will be locally convex), I is the *identity operator* from E to E, A is a compact linear operator from E to E, $\lambda \in \mathbb{C}$ or $\lambda \in \mathbb{R}$ in the real case,

$$T_\lambda = A - \lambda I, \quad T := A - I.$$

The mapping T possesses the properties close to the properties of linear mappings in finite-dimensional spaces. Our next goal is the proof of the following theorem.

3.10.2. Theorem. *If* $\operatorname{Ker} T = \{0\}$, *then T is a linear homeomorphism from E onto E.*

For general operators in infinite-dimensional spaces the injectivity does not imply the surjectivity. In the next section we reinforce the stated assertion by proving that for the operators of the considered form the surjectivity implies the injectivity.

The proof of this theorem is rather involved and is based on the following lemmas.

3.10.3. Lemma. *The subspace $T(E)$ is closed in E and T is an open mapping from E onto $T(E)$ (the images of open sets are open in $T(E)$). If T is injective, then T is a linear homeomorphism from E onto $T(E)$.*

PROOF. Everything reduces to the case of injective T, since the kernel of T is finite-dimensional (on the kernel of T, the compact operator A is the identity operator), hence there is a closed linear subspace E_1 for which $E = E_1 \oplus \operatorname{Ker} T$ and the projections onto the summands are continuous. Thus, let $\operatorname{Ker} T = 0$. Suppose that a net $\{Tx_\alpha\}$ converges to some y. By assumption, there is a quasi-norm d in some family defining the topology such that the corresponding balls centered at the origin are balanced and $A(U)$, where $U = \{d \leqslant 1\}$, has compact closure. Suppose first that some subnet in $\{x_\alpha\}$ is contained in kU for some $k > 0$. We can assume that this is the original net. Then $\{Ax_\alpha\}$ is contained in a compact set, hence we can pass to a convergent subnet, which gives the subnet $x_\beta = Ax_\beta - Tx_\beta$ converging to some $x \in E$. Hence $y = Tx$.

It remains to consider the case where $p_U(x_\alpha) \to +\infty$. We show that this is impossible. Indeed, in this case for the net $z_\alpha = x_\alpha / p_U(x_\alpha)$ we have $Tz_\alpha \to 0$ and $p_U(z_\alpha) = 1$. As shown at the previous step, this yields the existence of a subnet $\{z_\beta\}$ converging to some $z \in E$ such that $Tz = 0$. By the injectivity of T we obtain $z = 0$, which contradicts the equality $p(z) = 1$. Thus, we have proved that the set $T(E)$ is closed. Moreover, it follows from what has been said that T^{-1} is continuous on $T(E)$. Indeed, we have seen that for every convergent net of vectors $y_\alpha = Tx_\alpha$ in $T(E)$ the net $x_\alpha = T^{-1}y_\alpha$ has a convergent subnet. However, by the injectivity of T all such subnets must have a common limit, which means convergence of the whole net $\{x_\alpha\}$. The continuity of T^{-1} yields that T is open. □

3.10.4. Lemma. *Let G be a vector subspace of a topological vector space F and let W be a neighborhood of zero in F with $G + W \neq F$. Then, there exists an element $a \in 2W$ such that $a \neq G + W$, i.e., $2W \not\subset G + W$.*

PROOF. If $2W \subset G + W$, then for every $n \in \mathbb{N}$ we have

$$2^n W = 2^{n-1} 2W \subset 2^{n-1}(G + W) \subset G + 2^{n-1} W,$$

which is contained in $G + G + 2^{n-2}W$. Continuing, we obtain $2W \subset G + W$. Since $\bigcup_{n=1}^{\infty} 2^n W = F$, it follows that $F = G + W$, which contradicts our assumption. □

3.10.5. Lemma. *Let F be a topological vector space, G its proper closed vector subspace, $B\colon\ F \to F$ a compact linear operator such that $(B - I)(F) \subset G$,* *and let W be a neighborhood of zero in F such that the set $B(W)$ is relatively compact. Then $G+W \neq F$.*

PROOF. If $G+W=F$, then for every $\varepsilon>0$ we have

$$F=\varepsilon F=\varepsilon G+\varepsilon W=G+\varepsilon W=G-\varepsilon(B-I)(W)+\varepsilon B(W)$$
$$\subset G-\varepsilon G+\varepsilon B(W)=G+\varepsilon B(W).$$

Since the set $B(W)$ is relatively compact and hence bounded, for every neighborhood of zero U in the space F we have $\varepsilon B(W)\subset U$ for sufficiently small numbers $\varepsilon>0$. Hence for such ε we have the inclusion $F\subset G+U$. Since G is closed, this means that $F=G$. □

PROOF OF THEOREM 3.10.2. By Lemma 3.10.3 it suffices to prove the equality $T(E)=E$. Suppose that $T(E)\neq E$. By the injectivity of the operator T, in the chain of subspaces $E\supset T(E)\supset T^2(E)\supset\cdots$ all inclusions are strict.

Let V be a neighborhood of zero in E such that the set $\overline{A(V)}$ is compact. We show that for every natural number n there exists an element $a_n\in(2V)\cap T^{n-1}(E)$ not contained in $T^n(E)+V$, where we set $T^0(E)=E$. To this end, we verify that, for every fixed n, the subspace $F=T^{n-1}(E)$ of the space E, the subspace $G=T^n(E)$ of the space F and the neighborhood of zero $W=V\cap F$ in the space F satisfy the conditions in Lemma 3.10.4.

Clearly, it suffices to verify that $G+W\neq F$. The latter inequality follows from Lemma 3.10.5. Indeed, by Lemma 3.10.3 the set $T(E)$ is closed and T is a linear homeomorphism. Hence $G=T^n(E)$ is a closed vector subspace in $F=T^{n-1}(E)$. In order to apply Lemma 3.10.5, we have to take $B=A|_{T^{n-1}(E)}$, i.e., B is the restriction of A to $T^{n-1}(E)$. As noted above, the assumption that $T(E)\neq E$ made at the very beginning yields that $F\neq G$. Thus, all conditions in Lemma 3.10.5 are fulfilled. Thus, there exists a sequence $\{a_n\}$ mentioned at the beginning of the proof.

The vectors $A(a_n)$ with $n>1$ are contained in the relatively compact set $A(V)$. In addition, for all $n,r\in\mathbb{N}$ with $n>r$ we have

$$A(a_n)-A(a_r)=a_n+Ta_n-a_r-Ta_r\in -a_r+T^rE,$$

i.e., $a_r\in A(a_r)-A(a_n)+T^rE$. This yields that $A(a_r)-A(a_n)\notin V$, since $a_r\notin T^rE+V$. We obtain a contradiction with the fact that the set $\{A(a_n)\}$ is precompact. The theorem is proven.

3.10.6. Definition. *The spectre of a continuous linear operator A in a complex topological vector space E is the set $\sigma(A)$ consisting of $\lambda\in\mathbb{C}$ such that $A-\lambda I$ is not a topological isomorphism of E.*

If E is a Fréchet space, then $\sigma(A)$ consists of the *eigenvalues*, i.e., numbers λ for which $\operatorname{Ker}(A-\lambda I)\neq 0$, and numbers λ such that $(A-\lambda I)(E)\neq E$ (see Theorem 3.9.1). A nonzero vector v is called an *eigenvalue* if $Av=\lambda v$.

3.10.7. Corollary. *The spectre of a compact linear operator in an infinite-dimensional topological vector space consists of the nonzero eigenvalues (which may be absent) and the zero point.*

PROOF. The main theorem is also true for the operator $T_\lambda = A - \lambda I$ with any $\lambda \neq 0$, since it differs from the operator $\lambda^{-1}T_\lambda$ by a constant factor. This means that if $\lambda \neq 0$ is not an eigenvalue, then this operator has a continuous inverse. Finally, it is clear that a compact operator in an infinite-dimensional space cannot be a homeomorphism, hence its spectre contains the zero point. □

A more complete description of the spectre of a compact operator is contained in the following theorem.

3.10.8. Theorem. *The set of eigenvalues of a compact operator in a Hausdorff topological vector space is either finite or they form a sequence converging to zero; the multiplicity of every nonzero eigenvalue is finite.*

PROOF. We first observe that the eigenvectors corresponding to different eigenvalues are linearly independent (Exercise 3.12.168). Let us proceed to the proof of the theorem. Suppose that it is false. Then either the multiplicity of some nonzero eigenvalue is infinite or there exists $\varepsilon > 0$ such that the set of eigenvalues with absolute values greater than ε is infinite. The first possibility is ruled out simply: the nonzero operator λI cannot be compact in an infinite-dimensional space.

We now show that the existence of infinitely many (different) eigenvalues with absolute values greater than ε is also impossible. Indeed, let $\{\lambda_n\}$ be a sequence of pairwise different numbers with absolute values greater than ε and let $\{x_n\}$ be a sequence of nonzero vectors in E with $Ax_n = \lambda_n x_n$. Let H_n denote the subspace generated by $x_1, \ldots, x_n$. By the linear independence of vectors corresponding to different eigenvalues we have $H_n \subset H_{n+1}$, $H_n \neq H_{n+1}$ for every n, and all spaces H_n are finite-dimensional, hence closed.

Let V be a circled neighborhood of zero in E such that the set $A(V)$ is relatively compact. By using a procedure similar to the one employed in the proof of the main theorem of this section we show that, for every $n \in \mathbb{N}$, there exists an element $a_n \in (2V \cap H_{n+1})\backslash(H_n + V)$.

Set $G = H_n$, $F = H_{n+1}$, $W = F \cap V$ and show that all conditions of Lemma 3.10.4 are fulfilled. Let us use Lemma 3.10.5. Let $B = (\lambda_{n+1}^{-1}A)\big|_{H_{n+1}}$; one has $T(H_{n+1}) \subset H_n$. Then G, being finite-dimensional, is closed in F, and the set $B(W)$ is relatively compact in H_{n+1}, hence $G+W \neq F$ by Lemma 3.10.5, so that Lemma 3.10.4 is applicable. Thus, the existence of elements a_n is proved.

Now, for all $n, r \in \mathbb{N}$ with $n > r$ we have

$$\begin{aligned} A(a_n) - A(a_r) &= \lambda_{n+1}B(a_n) - \lambda_{n+1}B(a_r) \\ &= \lambda_{n+1}a_n - \lambda_{n+1}T(a_n) - \lambda_{n+1}a_r + \lambda_{n+1}T(a_r) \in \lambda_{n+1}a_n + H_n, \end{aligned}$$

i.e., $\lambda_{n+1}a_n \in \big(A(a_n) - A(a_r)\big) + H_n$. This means that $A(a_n) - A(a_r) \notin \varepsilon V$, since $\lambda_{n+1}a_n \notin H_n + \varepsilon V$ (this relation follows from the relation $a_n \notin H_n + V$ and the condition that V is circled). The obtained relation contradicts the condition that $A(V)$ is precompact. □

We now turn to locally convex spaces, where one can use adjoint operators.

3.10.9. Theorem. *If A is a compact linear mapping from a locally convex space E to a locally convex space G, then its adjoint operator is compact provided that we equip the spaces E' and G' with the topologies of uniform convergence on the classes of all precompact subsets in the spaces E and G. This is also true for the topologies of uniform convergence on the classes of all compact subsets.*

PROOF. Let τ_E and τ_G be the topologies in E' and G' of one of the two types mentioned in the formulation. Let V be a neighborhood of zero in E such that $A(V)$ has compact closure $\overline{A(V)}$ in G. Then $W = \bigl(A(V)\bigr)^\circ$ is a neighborhood of zero in the topology τ_G of the space G'. By (3.3.2) we have the equality $(A^*)^{-1}(V^\circ) = \bigl(A(V)\bigr)^\circ = W$, so that $A^*(W) \subset V^\circ$, but the last set is compact in the topology τ_E by Corollary 3.1.6. □

It is worth noting that if we equip the dual spaces with other topologies, then the adjoint to a compact linear operator can fail to be compact.

For example, let $E = H \oplus H_1$, where H is an infinite-dimensional Hilbert space with the topology generated by its norm, let H_1 be the same vector space equipped with the weak topology, and let $A\colon E \to E$, $A(x, y) = (0, x)$. Then A is compact by the weak compactness of the unit ball, but A^* is not compact if we equip E' with the strong topology (which in the considered case coincides with the Mackey topology). Indeed, $A^*(f, g) = (g, 0)$. Note that A^* is continuous, but for the compact identity embedding $H \to H_1$ the adjoint operator is not even continuous from the weak to the norm topology.

It follows from this example that the classical Schauder theorem, according to which a continuous linear operator between Banach spaces is compact precisely when its adjoint is compact (acting between the Banach duals to the original spaces), does not extend to arbitrary locally convex spaces. Nevertheless, some analog of the Schauder theorem for arbitrary locally convex spaces, containing this theorem as a partial case, still exists. This analog is as follows.

3.10.10. Proposition. *Let E and G be locally convex spaces, let $T\colon E \to G$ be a continuous linear operator, and let $\mathcal{A}$ and $\mathcal{B}'$ be the families of all bounded sets in the spaces $\bigl(E, \sigma(E, E')\bigr)$ and $\bigl(G', \sigma(G', G)\bigr)$, respectively. Then all sets $T(A)$, where $A \in \mathcal{A}$, are precompact in the topology of convergence on the family $\mathcal{B}'$ precisely when all sets $T^*(B)$, where $B \in \mathcal{B}'$, are precompact in the topology of convergence on the family $\mathcal{A}$.*

PROOF. If the set $T(A)$, where $A \in \mathcal{A}$, is precompact in the topology of $\mathcal{B}'$-convergence in G, then for every $B \in \mathcal{B}'$ there exist elements $a_1, \ldots, a_n$ in E such that

$$T(A) \subset \bigcup_{j=1}^{n} \bigl(g(a_j) + B^\circ\bigr).$$

It follows from this inclusion (show this) that

$$A \subset \bigcup_{j=1}^{n} \Bigl(a_j + \bigl(g^*(B)\bigr)^\circ\Bigr).$$

This means that every set $A \in \mathcal{A}$ is precompact in the topology of convergence on the family $\{T^*(B)\colon\ B \in \mathcal{B}'\}$ of subsets of the space E'. According to Corollary 3.10.12 proved below, every set in the latter family is precompact in the topology of convergence on the family $\mathcal{A}$. Thus, one implication in the proposition is proved; the reverse implication is proved similarly. □

3.10.11. Lemma. *Let E be a locally convex space and let sets $A \subset E$ and $B \subset E'$ be bounded in $\bigl(E, \sigma(E,E')\bigr)$ and $\bigl(E', \sigma(E',E)\bigr)$, respectively, and possess the following property: for every $\varepsilon > 0$ there exists a finite set $K_\varepsilon \subset E$ such that $A \subset K_\varepsilon + \varepsilon B^\circ$. Then, for every $\varepsilon > 0$, there exists a finite set $L_\varepsilon \subset B$ such that $B \subset L_\varepsilon + \varepsilon A^\circ$.*

PROOF. According to Proposition 3.4.2, the boundedness of B in the topology $\sigma(E',E)$ implies that for the finite set $K_{\varepsilon/4} \subset E$ there exists a finite set $L_\varepsilon \subset E'$ such that $B \subset L_\varepsilon + 2^{-1}\varepsilon K^\circ_{\varepsilon/4}$; it is clear that here we can take $L_\varepsilon \subset B$. Let us verify that $B \subset L_\varepsilon + \varepsilon A^\circ$. Let $b \in B$. Choose $l \in L_\varepsilon$ such that $b - l \in 2^{-1}\varepsilon K^\circ_{\varepsilon/4}$. By assumption, for every $a \in A$ there is a vector $k \in K_{\varepsilon/4}$ such that we have $v := a - k \in 4^{-1}\varepsilon B^\circ$. Therefore,

$$|\langle b-l, a\rangle| = |\langle b-l, v+k\rangle| \leqslant |\langle b, v\rangle| + |\langle l, v\rangle| + |\langle b-l, k\rangle| < \frac{\varepsilon}{4} + \frac{\varepsilon}{4} + \frac{\varepsilon}{2},$$

since $b, l \in B$, $v \in 4^{-1}\varepsilon B^\circ$, $b - l \in 2^{-1}\varepsilon K^\circ_{\varepsilon/4}$, $k \in K_{\varepsilon/4}$. Thus, $b - l \in A^\circ$, as required. □

We now justify the assertion used in Proposition 3.10.10.

3.10.12. Corollary. *Let E be a locally convex space and let $\mathcal{A}$ and $\mathcal{B}$ be some families of bounded subsets of the spaces E and E', respectively. Let α be the topology of convergence on the system of sets $\mathcal{A}$ in E' and let β be the topology of convergence on the system of sets $\mathcal{B}$ in E. Then every set in the system $\mathcal{A}$ is precompact in the topology β precisely when every set in the system $\mathcal{B}$ is precompact in the topology α. Moreover, on the sets in $\mathcal{A}$ the topology β coincides with $\sigma(E,E')$ and similarly for $\mathcal{B}$.*

PROOF. This follows directly from the lemma and the definition of the topology of convergence on a system of sets. □

We now show that the Schauder theorem (formulated above) is indeed a corollary of what has been just proved. Thus, let E and G be two Banach spaces and let $T\colon E \to G$ be a compact linear operator. Then T is continuous, hence the adjoint operator $T^*\colon G' \to E'$ is defined, where E' and G' are Banach duals to the spaces E and G and are equipped with the corresponding norms and the topologies generated by these norms.

It follows from the compactness of the operator T that it takes every bounded subset of the space E to a relatively compact subset of the space G. According to what has been proved above, the mapping T^* takes every norm bounded subset of the space G' to a precompact subset of the space E'; we use that the topology of the space E is the topology of uniform convergence on the family of all balls in the space E', and the topology in G is the topology of convergence on the set of all balls in G' (certainly, in place of the family of all balls one can deal with the single unit ball). Since E' is complete, every its precompact subset is contained in some compact set (for example, its closure). Thus, one of the two parts of the Schauder theorem is proved; the second one is proved similarly. Certainly, there is also an easy direct proof of the Schauder theorem for Banach spaces.

3.11. The Fredholm alternative

Let us apply the results presented above to finding conditions for solvability of linear equations containing compact linear operators in locally convex spaces. In particular, we prove here that for the operators of the form $T = I - A$ with compact A the so-called *Fredholm alternative* is valid, i.e., precisely one of the following two assertions is true: either the homogeneous equation $Ax - x = 0$ has a nonzero solution in E or the non-homogeneous equation $Ax - x = z$ with respect to x is solvable in E for every right-hand side in E (and then a solution is unique). In other words, for T the injectivity is equivalent to surjectivity. For arbitrary continuous linear operators these assertions are not mutually excluding (for example, T can be injective, but not surjective, and can be surjective not being injective). We have seen in the previous section that if the operator $(A - I)^{-1}$ exists and is everywhere defined, then it must be continuous. Throughout this section E is a Hausdorff locally convex space.

3.11.1. Theorem. *For a given vector $z \in E$, the equation*

$$Ax - \lambda x = z$$

is solvable with respect to x precisely when for every $g \in \mathrm{Ker}\,(A^* - \lambda I)$ we have the equality $\langle g, z\rangle = 0$.

PROOF. Our assertion is a direct corollary of the equality

$$\mathrm{Ran}\,(A - \lambda I) = \bigl(\mathrm{Ker}\,(A^* - \lambda I)\bigr)^\circ,$$

which is valid by the closedness of the subspace $\mathrm{Ran}\,(A - \lambda I)$, which follows from the compactness of the operator A. □

3.11.2. Theorem. *For a given functional $f \in E'$, the equation*

$$A^*g - \lambda g = f$$

is solvable with respect to g precisely when for every element $x \in \mathrm{Ker}\,(A - \lambda I)$ we have the equality $\langle f, x\rangle = 0$.

PROOF. The equality

$$\overline{\mathrm{Ran}\,(A - \lambda I)^*} = \bigl(\mathrm{Ker}\,(A - \lambda I)\bigr)^\circ$$

yields the equality

$$\mathrm{Ran}\,(A - \lambda I)^* = \bigl(\mathrm{Ker}\,(A - \lambda I)\bigr)^\circ.$$

Indeed, the operator A^* is compact if we equip E' with the topology of uniform convergence on compact sets (Theorem 3.10.9). Hence the range of $\lambda I - A^*$ is closed in the indicated topology in E', whence by linearity we obtain its closedness in the topology $\sigma(E', E)$. □

3.11.3. Corollary. *If the equation $Ax - \lambda x = 0$ has only zero solution, then the non-homogeneous equation $A^*g - \lambda g = f$ is solvable for every right-hand side $f \in E'$; if the equation $A^*g - \lambda g = 0$ has only zero solution, then the equation $Ax - \lambda x = z$ is solvable for every right-hand side $z \in E$.*

3.11.4. Theorem. *If $\lambda \neq 0$, then the linear subspaces*

$$N_\lambda = \mathrm{Ker}\,(A - \lambda I) \quad \textit{and} \quad N'_\lambda = \mathrm{Ker}\,(A^* - \lambda I)$$

have the same finite dimension.

PROOF. Since the operator A is compact, we have $\dim N_\lambda < \infty$. By Theorem 3.10.9, if we equip E' with the topology of uniform convergence on precompact sets, the adjoint operator A^* becomes compact. Hence $\dim N'_\lambda < \infty$. For the proof of the equality $\dim N_\lambda = \dim N'_\lambda$ it suffices to prove the inequality $\dim N_\lambda \geqslant \dim N'_\lambda$. Indeed, suppose we have proved that $\dim N_\lambda \geqslant \dim N'_\lambda$. Then for the operator $A^{**} - \lambda I$ we obtain $\dim \mathrm{Ker}\,(A^{**} - \lambda I) \leqslant \dim N'_\lambda$. Since A is the restriction of the operator A^{**} to E (Exercise 3.12.77), we have $N_\lambda \subset \mathrm{Ker}\,(A^{**} - \lambda I)$, hence $\dim N_\lambda \leqslant \dim N'_\lambda$.

Let $\dim N_\lambda = n < \dim N'_\lambda$. Choose in N_λ a basis $e_1, \ldots, e_n$. There is a closed linear subspace $E_0 \subset E$ such that E is the direct sum of E_0 and N_λ and the natural projections onto these subspaces are continuous (Theorem 1.11.16). Let us define a finite-dimensional operator $K\colon\, E \to E$ as follows: if $x = y + z$, $y \in E_0$, $z \in N_\lambda$, then $Kx = z$, i.e., $K = 0$ on E_0 and $K = I$ on N_λ. It is clear that the operator K is compact. Then the operator $A_1 = A + K$ is compact as well. Obviously, the kernel of $A_1 - \lambda I$ is trivial. Hence $(A_1 - \lambda I)(E) = E$. On the other hand,

$$(A_1 - \lambda I)(E) = (A - \lambda I)(E) \oplus K(E) = (A - \lambda I)(E) \oplus N_\lambda.$$

Thus, the range of $A - \lambda I$ has codimension n, which leads to a contradiction, since this range is the annihilator of the kernel of $A^* - \lambda I$ and hence has codimension $\dim N'_\lambda$. □

3.11.5. Example. For $\lambda = 0$ the equality $\dim N'_\lambda = \dim N_\lambda$ is false. Indeed, let $E = l^2$,

$$A\colon\, (x_1, x_2, \ldots) \mapsto (x_2, x_3/3, x_4/4, \ldots).$$

Then the kernel of A is one-dimensional, but the kernel of A^* is trivial.

It has been actually shown in the proof of Lemma 3.10.3 above that the chain $T_\lambda E \supset T_\lambda^2 E \supset \cdots$ of closed subspaces of the space E (with $\lambda \neq 0$) must stabilize, i.e., there exists a number $k \in \mathbb{N}$ such that $T_\lambda^k E = T_\lambda^{k+j} E$ for all numbers $j \in \mathbb{N}$. Similarly one can show that also the following chain of closed subspaces stabilizes: $T_\lambda^{-1}(0) \subset T_\lambda^{-2}(0) \subset \cdots$ $(\lambda \neq 0)$.

Let now k be the minimal natural number for which $T_\lambda^k E = T_\lambda^{k+1} E$ (then also $T_\lambda^k E = T_\lambda^{k+j} E$ for all $j \in \mathbb{N}$) and let n be the minimal natural number for which $T_\lambda^{-n}(0) = T_\lambda^{-n-1}(0)$ (then also $T_\lambda^{-n}(0) = T_\lambda^{-n-j}(0)$ for all $j \in \mathbb{N}$). Then $n = k$ and E is the algebraic direct sum of its subspaces $T_\lambda^{-n}(0)$ and $T_\lambda^n E$.

The last fact is purely algebraic (its proof will be given below). It follows from Lemma 3.10.3 that all subspaces $T_\lambda^k E$ are closed: this is seen from the fact that by the equality

$$T_\lambda^k = A^k - k\lambda A^{k-1} + \cdots + (-\lambda)^k I$$

the mapping T_λ^k, as well as the mapping T_λ, is the sum of two mappings (compact and a multiple of the identity). Hence the algebraic direct sum $T_\lambda^{-n}(0) \oplus T_\lambda^n E$ of two subspaces, one of which is closed and finite-dimensional, is also the topological direct sum.

Using again that the operators T_λ^j with $\lambda \neq 0$ are sums of compact operators and multiples of the identity, we obtain that for every natural j one has

$$\dim T_\lambda^{-j}(0) = \dim (T_\lambda^*)^{-j}(0).$$

We now prove that $E = T_\lambda^{-n}(0) \oplus T_\lambda^n(E)$. First we observe that the equalities

$$T^s(E) = T^{s+r}(E), \tag{3.11.1}$$

$$T^r(E) \oplus T^{-s}(0) = E \tag{3.11.2}$$

are equivalent. Indeed, if the first one is fulfilled and $x \in E$, then there exists $z \in E$ such that $T^s x = T^{s+r} z$, but this means that $T^s(x - T^r z) = 0$, i.e., we have $x - T^r z \in T^{-s}(0)$. Thus, $x = T^r z + (x - T^r z)$, so that (3.11.2) holds.

We verify that (3.11.2) implies (3.11.1). If (3.11.2) is valid and $x \in E$, then there exist elements $z, z_1 \in E$ such that $x = T^r z + z_1$, where $T^s z_1 = 0$, so that we have $T^s x = T^{s+r} z$; this means that (3.11.1) holds.

Let us prove the equivalence of the equalities

$$T^{-r}(0) = T^{-s-r}(0), \tag{3.11.3}$$

$$T^r(E) \cap T^{-s}(0) = \{0\}. \tag{3.11.4}$$

Let (3.11.3) be valid and $x \in T^r(E) \cap T^{-s}(0)$. Then $T^s x = 0$. On the other hand, there exists $z \in E$ such that $x = T^r z$, so we have $T^{r+s} z = 0$. Hence we obtain $z \in T^{-s-r}(0)$. Therefore, by equality (3.11.3) we obtain $z \in T^{-r}(0)$, which means that $x = T^r z = 0$. Thus, (3.11.4) is proved.

Suppose now that (3.11.4) holds and prove that (3.11.3) is true. It suffices to show that if $z \in T^{-s-r}(0)$, then $z \in T^{-r}(0)$. The first inclusion means that we have $T^{r+s} z = 0$, i.e., $T^r z \in T^{-s}(0)$. It follows from equality (3.11.4) that $T^r z = 0$, so we have $z \in T^{-r}(0)$.

We proceed to the proof of the equalities $k = n$ and $T^n(E) \oplus T^{-n}(0) = E$. If the first one is proved, then the second will be a direct corollary of the equalities $T^n(E) + T^{-n}(0) = E$, $T^n(E) \cap T^{-n}(0) = \{0\}$ (which hold in this case; they are partial cases of equalities (3.11.2) and (3.11.4)).

The equality $k = n$ is proved in the following way. Since by the definition of n we have $T^{-n}(0) = T^{-n-1}(0)$, we obtain $T^n(E) \cap T^{-1}(0) = \{0\}$.

If $k \leqslant n$, then again by the definition (this time of the number k) we have the equality $T^k(E) = T^n(E)$. Hence the previous equality can be replaced by the following one: $T^k(E) \cap T^{-1}(0) = \{0\}$. According to (3.11.3) and (3.11.4), this means that $T^{-k}(0) = T^{-k-1}(0)$; the last equality means, in turn, that $k \geqslant n$. Thus we obtain that $k = n$.

Let $n \leqslant k$. By the definition of k we have $T^k(E) = T^{k+1}(E)$. Since (3.11.1) implies (3.11.2), we have $T(E) + T^{-k}(0) = E$. Since now $n \leqslant k$, by definition we have $T^{-n}(0) = T^{-k}(0)$, hence $T(E) + T^{-n}(0) = E$. Since (3.11.2) implies (3.11.1), we have $T^n(E) = T^{n+1}(E)$, so that $n \geqslant k$. Hence again we have $n = k$.

We recall that a nonzero vector $a \in E$ is called an *adjoint vector* of a linear operator $A\colon E \to E$ associated to a number λ if $(A-\lambda I)^n a = 0$ for some $n \in \mathbb{N}$; the smallest number n with this property is called the order of the adjoint vector. In particular, all eigenvectors are adjoint of the first order. Usually, eigenvectors are not included into adjoint vectors.

It is known that the Jordan form of the matrix of a linear operator in a finite-dimensional space is constructed by using a basis consisting of the eigenvectors and adjoint vectors of this operator. To every Jordan block there corresponds a collection of the basis vectors consisting of one eigenvector and several adjoint vectors associated with the same eigenvalue; the maximal order of these adjoint vectors coincides with the number of the diagonal elements in this Jordan block.

As one can show, it follows from what has been said above about the subspaces $T_\lambda^{-n}(0)$ and $T_\lambda^n(E)$ that they are invariant subspaces of the operator A, moreover, the subspace $T_\lambda^{-n}(0)$ is finite-dimensional and contains a basis consisting of eigenvectors and adjoint vectors of the operator A of order not greater than n, associated with the eigenvalue λ. However, this does not imply that in an infinite-dimensional space there is a basis (at least algebraic) consisting of eigenvectors and adjoint vectors of the operator A. Even in a separable Hilbert space not every compact operator possesses a Hamel basis consisting of eigenvectors and adjoint vectors; the simplest example is the Volterra operator

$$Af(t) = \int_0^t f(s)\,ds$$

in $L^2[0,1]$, which has no eigenvectors and adjoint vectors.

3.11.6. Remark. Almost all results on compact operators presented above remain valid for any operator a nonzero polynomial of which is compact. An exception is the structure of its spectre: if the operator $\varphi(A)$, where φ is a nonzero polynomial, is compact, then the spectre of $\varphi(A)$ is still finite or countable, but the set of its limit points must belong to the set of zeros of the polynomial φ (say, zero need not be a limit point of the spectre of such an operator). The proof of this (which does not differ much from the reasoning given above) is left as an exercise.

3.12. Complements and exercises

3.12(i). Baire spaces

Here we present some additional information about Baire spaces. In general topology, Baire spaces are defined as follows.

3.12.1. Definition. *A topological space X is called Baire if no its nonempty open subset can be represented as a countable union of nowhere dense sets.*

Countable unions of nowhere dense sets are called *first category sets*, all other sets are called *second category sets*. Thus, in a Baire space all nonempty open sets (hence the space itself) are second category sets. A somewhat weaker condition is that the whole space is a second category set. Though, for topological vector spaces both properties are equivalent, since if a neighborhood of a point is a first category set, then the neighborhood of zero obtained by shifting this point to zero will be also a first category set, then the whole space will be a first category set as well. Certainly, in general spaces there is no such equivalence: for example, the union of $[0, 1]$ with the set of rational numbers in $[2, 3]$ is not a Baire space in the sense of the above definition, although it is a second category set.

The space is Baire precisely when the intersection of every sequence of everywhere dense open sets is everywhere dense. Indeed, if open sets G_n are everywhere dense in a Baire space X, then $\bigcap_{n=1}^{\infty}(G_n \cap U) \neq \varnothing$ for every nonempty open set U, since otherwise the closed sets $X \backslash G_n$ are nowhere dense and cover U. In the opposite direction, if an open set U is the union of nowhere dense sets F_n, then the sets $G_n = X \backslash \overline{F_n}$ are open and everywhere dense (the closures of F_n are nowhere dense too), but their intersection does not meet U, hence this intersection is not everywhere dense.

More detailed information about Baire spaces can be found in Haworth, Mccoy [**216**] (see also Bourbaki [**88**], Pérez Carreras, Bonet [**385**], and Valdivia [**543**]). We only mention some facts related to products.

Let X and Y be two Baire spaces. Then $X \times Y$ is Baire in either of the following cases:

(i) Y has a countable topology base;

(ii) X and Y are metrizable and Y is complete.

In addition, the following fact holds.

3.12.2. Theorem. *The product of any collection of metrizable separable Baire spaces is Baire. The product of any collection of complete metric spaces is Baire. For example, $\mathbb{R}^T$ is Baire.*

Oxtoby [**377**] constructed an example of a completely regular Baire space such that its square is not Baire (he used the continuum hypothesis, but without it an example was constructed in Cohen [**103**]). Later examples of metric spaces with such a property were found (see Krom [**300**]), and Arias de Reyna [**23**] constructed an example of two Euclidean Baire spaces X and Y for which $X \times Y$ is not Baire. Let us give a result from Saxon [**430**].

3.12.3. Theorem. *A topological vector space E is Baire precisely when every absorbing closed balanced set in E has an inner point.*

PROOF. If V be a set with the indicated properties in a Baire space E, then the closed sets nV cover E, hence some of them has an inner point.

Suppose now that E is not Baire. Then E can be represented as the union of a sequence of increasing closed sets F_n without inner points. We can assume that E

is real (the complex case is slightly more difficult, see Pérez Carreras, Bonet [**385**, p. 14]). We can also assume that $F_n = -F_n$, since the sets $F_n \cup (-F_n)$ are also closed and nowhere dense. Let us take a nonzero vector v_0 and a balanced closed neighborhood of zero W not containing v_0 (which always exists in a Hausdorff nonzero space, but in our special case we actually do not need the assumption that E is Hausdorff, since otherwise $F_n = E$). We can find a balanced closed neighborhood of zero U with $U + U \subset W$.

Let us show that the set $A = \bigcup_{n=1}^{\infty} n^{-1}(F_n \cap U)$ is nowhere dense and is absorbing. Then the set B that equals the closure of the set of all centrally symmetric closed intervals contained in A will be closed, nowhere dense, balanced and absorbent (for every $x \in E$, there is a closed interval of the form $[-\varepsilon x, \varepsilon x]$ contained in A).

If A is not nowhere dense, then the closure of A is dense in some open set G. For every $k \geqslant 1$, the closure of A is the union of the sets $n^{-1}(F_n \cap U)$ over all $n \leqslant k-1$ and the closure of $A_k := \bigcup_{n=k}^{\infty} n^{-1}(F_n \cap U)$. Then the open set $G \backslash \overline{A_k}$ is contained in the union of the sets $n^{-1}(F_n \cap U)$ with $n \leqslant k-1$, which is nowhere dense. Hence this open set is empty, i.e., $G \subset \overline{A_k} \subset k^{-1}U$ for all k.

Taking any element $g \in G$, we obtain that $g + tv_0, g - tv_0 \in G$ with some number $t > 0$, whence $2tv_0 \in k^{-1}U + k^{-1}U \subset k^{-1}W$ for all k. Taking k such that $2tk \geqslant 1$, we obtain from the inclusion $2tkv_0 \in W$ and the condition that W is balanced that $v_0 \in W$, which is a contradiction. Thus, A is nowhere dense.

Let us show that A is absorbent. Let $e \in E$, $e \neq 0$. The closed interval $U \cap \mathbb{R}^1 e$ is covered by the closed sets $F_n \cap U$, hence there exist $p \in \mathbb{N}$, $z \in \mathbb{R}^1$ and $\varepsilon > 0$ such that $te \in F_p \cap U$ for all $t \in [z - \varepsilon, z + \varepsilon]$. If $z = 0$, then we obtain that $se \in A$ whenever $|s| < \delta$, where a positive number $\delta < \varepsilon/p$ is such that $(-\delta e, \delta e) \in p^{-1}U$.

Let $z \neq 0$. We can assume that $\varepsilon \geqslant (p+1)^{-1}z$ (increasing p) and that $z > 0$. Let $0 < s \leqslant p^{-1}\varepsilon$. Choose $n \geqslant p$ such that $(n+1)^{-1}z \leqslant s \leqslant n^{-1}z$. Then $z - \varepsilon \leqslant ns \leqslant z + \varepsilon$, since $z - \varepsilon \leqslant n(n+1)^{-1}z$. Hence $nse \in F_p \subset F_n$, whence we obtain $sz \in n^{-1}(F_n \cap U) \subset A$. Since A is centrally symmetric, we have $se \in A$ also for $s \in [-p^{-1}\varepsilon, 0)$. Finally, $0 \in A$. □

For the proof of the next theorem, see Pérez Carreras, Bonet [**385**, p. 17].

3.12.4. Theorem. *Let E be a separable infinite-dimensional Baire topological vector space. Under the assumption of the continuum hypothesis (or Martin's axiom, according to which the union of less than the continuum of sets of measure zero on the real line has also measure zero), there exists an everywhere dense hyperplane in E that is not a Baire space.*

We shall need one more concept related to Baire's name: we shall say that a set A in a topological space X has the Baire property if there exists an open set U such that $U \triangle A$ is a first category set.

It is clear that all open sets have the Baire property and that the class $\mathfrak{B}(X)$ of all sets in X with this property is closed with respect to taking complements and countable unions. Therefore, the class $\mathfrak{B}(X)$ is a σ-algebra containing the Borel σ-algebra, which is the smallest σ-algebra containing all open sets (see § 5.1).

Actually, this is the smallest σ-algebra containing all Borel sets and all first category sets.

3.12.5. Lemma. *Let A be a second category set with the Baire property in a topological vector space. Then $A - A$ contains a neighborhood of zero.*

PROOF. By assumption, there is an open set U for which $U \triangle A$ is a first category set. In addition, $U - U$ is a neighborhood of zero.

We show that $U - U \subset A - A$. Let $x \in U - U$. Then $x = u - v$, where $u, v \in U$, whence we obtain $v \in (x + U) \cap U$. Since the set $U \backslash A$ is a first category set, its shift $(U + x)\backslash(A + x)$ is also a first category set. It readily follows from this (Exercise 3.12.71) that $A \cap (A + x) \neq \varnothing$, which gives the inclusion $x \in A - A$. □

By the way, it follows from this result that a second category set with the Baire property on the real line must have the cardinality of the continuum.

3.12.6. Example. Let us consider an example of a closed subspace in a Baire space that is not Baire.

In the space $X = \prod_{n=1}^{\infty} X_n$, where $X_n = \mathbb{R}^{\infty}$, take some linear subspace D algebraically complementing the sum of X_n; the latter sum consists of all sequences $(x_1, \ldots, x_n, 0, 0, \ldots)$, $x_n \in X_n$, where we regard X_n embedded into X as the set of vectors with zero components on the positions with numbers $k \neq n$. Let E_n be the direct sum of D and $X_1, \ldots, X_n$. Since the union of E_n is X, we obtain by Exercise 3.12.68 that for some n the subspace E_n is dense and Baire. Let us take in X_{n+1} a subspace L of countable dimension and set $E = E_n + L$. Then E is Baire (since E_n is dense in E and is Baire), $L = E \cap X_{n+1}$ is closed in E by the closedness of X_{n+1} in X, but is not Baire.

3.12(ii). The Borel graph theorem

For the definition of a Borel mapping, see § 5.2. Here we discuss an interesting generalization of the closed graph theorem.

3.12.7. Theorem. *Let E be a Baire topological vector space, let F be a locally convex space, and let $T\colon E \to F$ be a Borel linear mapping. Then T is continuous. In particular, every Borel linear function on E is continuous.*

PROOF. Let $V \subset F$ be a closed absolutely convex neighborhood of zero. Then $B = T^{-1}(V)$ is an absorbent absolutely convex Borel set in E. Being Borel, the set B possesses the Baire property. By Lemma 3.12.5 the set $2B = B - B$ contains a neighborhood of zero. Hence B also does, which gives the continuity of T. □

3.12.8. Theorem. *Let E and F be Souslin locally convex spaces, where E is Baire, and let $T\colon E \to F$ be a linear mapping with a Borel graph. Then T is continuous.*

PROOF. In order to apply the previous theorem we verify that T is Borel. First observe that $E \times F$ is also a Souslin space, hence the graph Γ of the mapping T is a Souslin set (see Bogachev [**65**, Corollary 6.6.7]).

Let $B \subset F$ be a Borel set. Let π_1 and π_2 denote the projection operators from $E \times F$ onto E and F. These projection operators are continuous and hence are Borel measurable. Hence $\pi_2^{-1}(B) \cap \Gamma$ is a Borel set, then also a Souslin set in Γ. Hence $T^{-1}(B) = \pi_1\bigl(\pi_2^{-1}(B) \cap \Gamma\bigr)$ is a Souslin set in E. Since this is true also for $E \backslash T^{-1}(B) = T^{-1}(F \backslash B)$, we see that $T^{-1}(B)$ is a Borel set by a classical result (see [**65**, Corollary 6.6.10]). □

This theorem is applicable if, say, E is a separable Fréchet space. However, in this case separability is not needed.

3.12.9. Corollary. *Let E be a Fréchet space, let F be a Souslin locally convex space, and let $T\colon\ E \to F$ be a linear mapping with a Borel graph. Then T is continuous.*

PROOF. It suffices to verify the continuity for restrictions to separable closed subspaces $E_0 \subset E$, and the graph of this restriction is a Borel set, being the intersection of the original graph with $E_0 \times F$. □

3.12.10. Corollary. *Let E and F be Souslin locally convex spaces such that E is Baire and let $\Lambda\colon\ F \to E$ be a continuous one-to-one linear mapping. Then $\Lambda^{-1}\colon\ E \to F$ is also continuous.*

PROOF. The graph of the continuous mapping Λ is Borel in $F \times E$, hence the graph of its inverse (the same set, but in $E \times F$) is Borel. □

3.12.11. Corollary. *Let E be a Fréchet space and let E_0 be a Souslin linear subspace of finite codimension. Then E_0 is closed. The same is true if E_0 is of countable codimension.*

PROOF. Let E_1 be a finite-dimensional subspace in E complementing E_0. We observe that $E = E_0 \bigoplus E_1$ is separable. The space $E_0 \times E_1$ is Souslin, the mapping $(u, v) \mapsto u + v$ from $E_0 \times E_1$ to E is one-to-one and continuous. By the previous corollary the inverse is also continuous, which implies that E_0 is closed.

If E_0 has countable codimension and $\{a_n\}$ is a countable basis in the algebraic complement, then, according to what we have proved, all subspaces E_n generated by E_0 and a_i with $i \geqslant n$ are closed . Their intersection is E_0. □

It is clear from the proof that it is sufficient that E be Baire (it will be also Souslin, since $E = E_0 \bigoplus E_1$).

It follows from the last corollary that any Souslin hyperplane in a Fréchet space is closed.

3.12(iii). Bounding sets

3.12.12. Definition. *A subset B of a topological space T is called bounding if every continuous function on T is bounded on B.*

The space T itself is bounding if it is *pseudocompact*, i.e., all continuous functions on T are bounded.

It is clear that if every infinite sequence in B has a limit point in T, then B is bounding. Every part of a compact set is a bounding set. The space of all countable ordinal numbers with the order topology is not compact, but is countably compact and hence pseudocompact. The Tychonoff plane $([1,\omega_0]\times[1,\omega_1])\backslash(\omega_0,\omega_1)$, where ω_0 is the smallest countable ordinal and ω_1 is the smallest uncountable ordinal, where the intervals of ordinals are equipped with the order topology, gives an example of a pseudocompact space that is not countably compact.

It is of interest to find conditions for relative compactness of bounding sets in locally convex spaces. We include some results on this topic, their proofs and related references can be found in Floret [**171**]. Note that the notion of bounding sets involves nonlinear functions.

For a general completely regular space T, the Shirota–Nachbin theorem asserts that every bounding set in T is relatively compact precisely when the space $C(T)$ of continuous functions on T with the topology of uniform convergence on compact sets is barrelled (see, e.g., Schmets [**438**]). M. Valdivia proved the following result.

3.12.13. Theorem. *Let E be a locally convex space such that E' is separable in the topology $\sigma(E',E)$. Then every $\sigma(E,E')$-bounding set in E is relatively compact in $\sigma(E,E')$.*

3.12.14. Corollary. *In a quasi-complete locally convex space E, a set is bounding in the topology $\sigma(E,E')$ precisely when it is relatively compact in this topology.*

This yields the result of Pták on the relative weak compactness of weakly pseudocompact sets in quasi-complete locally convex spaces.

3.12.15. Corollary. *A quasi-complete locally convex space E is semireflexive precisely when every $\sigma(E,E')$-continuous function is bounded on all bounded sets.*

3.12.16. Corollary. *If a bounding set A in a locally convex space E has a complete in the Mackey topology $\tau(E,E')$ closed convex envelope or if E' is separable in the topology $\sigma(E',E)$, then A is relatively compact.*

3.12(iv). The James theorem

Let us give without proof the important James theorem [**234**] (extended to locally convex spaces by James himself and also by Pryce [**396**]), a proof of which can be also found in Floret [**171**, p. 59].

3.12.17. Theorem. *A weakly closed set A in a quasi-complete locally convex space is weakly compact precisely when every continuous linear functional has a maximum on A.*

3.12.18. Corollary. *A Banach space X is reflexive precisely when every continuous linear functional on X has a maximum on the closed unit ball.*

3.12.19. Corollary. *Let E be a quasi-complete locally convex space, let $\mathcal{A}$ be some σ-algebra of subsets of a set Ω, and let $\nu\colon \mathcal{A}\to E$ be a vector measure on $\mathcal{A}$ i.e., $\nu(A)=\sum_{n=1}^{\infty}\nu(A_n)$ if $A_n\in\mathcal{A}$ are pairwise disjoint and $A=\bigcup_{n=1}^{\infty}A_n$. Then the set of values of ν is relatively weakly compact. In particular, this is true for Banach spaces E.*

PROOF. Since the set of values of every countably additive real measure on the σ-algebra $\mathcal{A}$ is bounded, the set $M = \{\nu(A)\colon A \in \mathcal{A}\}$ is weakly bounded. Let C be its weak closure. In order to verify weak compactness of C we apply the James theorem and show that every functional $f \in E'$ has a maximum on C. This reduces to showing that the measure $f(\nu)$ on $\mathcal{A}$ assumes its maximal value; but this is a known fact in measure theory (see Bogachev [**65**, § 3.1]). □

A derivation of the next assertions from the James theorem can be found in Floret [**171**, p. 87, 88].

3.12.20. Theorem. *A quasi-complete locally convex space is semireflexive precisely when every two disjoint closed bounded convex sets can be strictly separated by a closed hyperplane.*

3.12.21. Theorem. *An absolutely convex bounded closed set in a quasi-complete locally convex space is weakly compact precisely when it can be strictly separated by a closed hyperplane from every closed bounded convex set disjoint with it.*

Hence a Banach space is reflexive precisely when its closed unit ball can be strictly separated by a closed hyperplane from every closed bounded convex set disjoint with it. For example, in $C[0,1]$ there is a bounded closed convex set which is disjoint with the closed unit ball, but cannot be strictly separated from it by a hyperplane.

The James theorem gives the following nonlinear characterization of weak compactness (see Floret [**171**, p. 91]).

3.12.22. Proposition. *Let V be a bounded convex set in a locally convex space that is complete in the Mackey topology. Weak compactness of V is equivalent to that every mapping $\Phi\colon V \to V$ continuous in the weak topology has a fixed point.*

A close result of D.P. Milman and V.D. Milman asserts that weak compactness of V is equivalent to the property that for every convex closed subset $W \subset V$ all affine continuous mappings from W to W have fixed points.

3.12(v). Topological properties of locally convex spaces

General topology considers various properties of topological spaces generalizing the properties of metric spaces. In this way, normal, perfectly normal, paracompact, and other spaces arise. Here we give some basic information about locally convex spaces in these classes. It becomes often necessary to apply a result in general topology to a concrete locally convex space, which requires membership in a particular class of general topological spaces. Usually, this concrete space is obtained from some simpler spaces (say, metrizable) by operations like inductive or projective limits in the category of locally convex space. Since such topologies are weaker than those in the category of general topological spaces (recall that $\mathcal{D}$ is not the topological inductive limit of $\mathcal{D}_n$), such questions turn out to be rather subtle. Another example: spaces (even Banach) with weak topologies.

We already know that Hausdorff topological vector spaces are Tychonoff. The next by power property is normality. A Hausdorff space T is called *normal* if for every pair of disjoint closed sets A and B in T there exists a pair of disjoint open sets U and V such that $A \subset U$, $B \subset V$. Such spaces are completely regular. A more narrow class consists of *perfectly normal spaces*: Hausdorff spaces in which every closed set has the form $f^{-1}(0)$ for some continuous function f. Any uncountable power of the real line is not a normal space (see Arkhangel'skiĭ, Ponomarev [**22**, Chapter II, § 5, Problem 392] or Engelking [**154**, Problem 2.7.16]). A topological space is called *Lindelöf* if every its cover by open sets contains an at most countable subcover. For metric spaces, this is equivalent to separability. Any Lindelöf Hausdorff topological vector space is normal (Exercise 3.12.189). A Hausdorff space is called *paracompact* if in every its open cover one can inscribe a locally finite open subcover. It is known (see Engelking [**154**, § 5.1]) that the class of such spaces contains all compact spaces, metrizable spaces, Lindelöf spaces and is contained in the class of normal spaces. In addition, it is contained in a more narrow class of *collectively normal spaces*, i.e., Hausdorff spaces in which, for every discrete collection of closed sets F_s, there is a discrete collection of open sets $U_s \supset F_s$ (a collection of sets is discrete if every point has a neighborhood meeting at most one of them).

The results of D.P. Baturov and E.A. Reznicheno (see Arkhangelskii [**21**], Reznichenko [**418**]) yield the following remarkable fact.

3.12.23. Theorem. *For a metrizable locally convex space with the weak topology, the normality and the Lindelöf property are equivalent.*

In addition, it is shown in [**418**] that if in a locally convex space E there is a stronger metrizable locally convex topology, then for the space $(E, \sigma(E, E'))$ normality and collective normality are equivalent (then also equivalent to paracompactness). These matters are closely connected with the study of the space $C_p(T)$ of continuous functions on a topological space T equipped with the topology of pointwise convergence.

For example, a Banach space X with the weak topology is linearly homeomorphic to a closed linear subspace in $C_p(T)$ with compact T equal the unit ball in X' equipped with the topology $\sigma(X', X)$. Similarly, a metrizable locally convex space E with the weak topology can be embedded as a closed linear subspace of $C_p(T)$ with the σ-compact space T equal the union of the polars of the elements of a countable base of neighborhoods of zero in E.

The following sufficient condition for the Lindelöf property in the weak topology is due to Talagrand [**510**]. Certainly, the simplest sufficient condition is separability with respect to the norm.

3.12.24. Theorem. *If a Banach space X contains a weakly compact set with a dense linear span, then X with the weak topology is Lindelöf.*

The hypothesis of this theorem is fulfilled for every Hilbert space (and every reflexive Banach space).

In relation to the problem of extending mappings (see Theorem 1.12.23), let us mention the following interesting class of *stratifiable* spaces: these are Hausdorff spaces such that for every closed set Z one can choose a sequence of open sets $U_n(Z) \supset Z$ with

$$Z = \bigcap_{n=1}^{\infty} U_n(Z) = \bigcap_{n=1}^{\infty} \overline{U_n(Z)},$$

and if $Z \subset Z'$, where Z' is closed, then $U_n(Z) \subset U_n(Z')$.

For example, in a metric space for $U_n(Z)$ one can take the open $1/n$-neighborhood of Z.

A functional characterization of stratifiable spaces (from which their perfect normality follows at once) is this: to every open set U one can associate a continuous function f_U with values in $[0,1]$ such that $U = \{f_U \neq 0\}$ and if $V \subset U$ is also open, then $f_V \leqslant f_U$.

A Hausdorff space is called *monotonically normal* if, to every pair of sets $F \subset U$, where F is closed and U is open, one can associate an open set $V(F,U)$ such that $F \subset V(F,U) \subset \overline{V(F,U)} \subset U$, and for every pair of sets $F' \subset U'$ such that F' is closed and U' is open and $F' \subset F$, $U' \subset U$ we have $V(F',U') \subset V(F,U)$.

It is known (see Gruenhage [**208**]) that a space is stratifiable precisely when its product with any convergent sequence is monotonically normal. As observed by S.A. Shkarin [**461**], this yields that for locally convex spaces both properties are equivalent. His paper [**458**] contains the following important result giving broad classes of nonmetrizable stratifiable locally convex spaces.

3.12.25. Theorem. (i) *A direct sum of stratifiable locally convex spaces is stratifiable.* (ii) *A strict inductive limit of a sequence of metrizable locally convex space is stratifiable.* (iii) *If in the dual F' of a Fréchet space F all bounded sets are metrizable, then F' with the Mackey topology is stratifiable.*

Thus, the class of stratifiable (hence also perfectly normal) spaces contains $\mathcal{D}$, $\mathcal{D}'$, $\mathcal{S}$, and $\mathcal{S}'$.

However, the situation with weak topologies is more complicated: a Banach space with the weak topology can be stratifiable only if it is finite-dimensional; the space $C_p(T)$ of continuous functions on a topological space T with the topology of pointwise convergence is stratifiable only for countable T. However, if T is a complete separable metric space, then the space $C_k(T)$ of continuous functions on T is stratifiable in the topology of uniform convergence on compact sets. It remained open for quite a long time whether the space $C_k(\mathbb{Q})$ is stratifiable, but Reznichenko [**419**] proved that a separable metrizable space X is Polish if and only if the function space $C_k(X)$ is stratifiable. For a survey of results on this topic and further references, see Gruenhage [**208**].

Let us also mention the following characterization of Fréchet–Urysohn spaces (in which all limit points of a set are the limits of convergent sequences of its elements) obtained in Kąkol, López Pellicer, Todd [**249**] (see also Cascales, Kąkol, Saxon [**96**] and Kąkol, Kubiś, López-Pellicer [**245**]).

3.12.26. Theorem. *For topological vector space E, the following conditions are equivalent*: (i) *E is a Fréchet–Urysohn space*; (ii) *if a point belongs to the closure of a set $A \subset E$, then it belongs to the closure of some bounded set $B \subset A$;*

(iii) *if we are given a sequence of sets* $A_n \subset E$ *the closures of which contain zero, then there exists a sequence of sets* $B_n \subset A_n$ *for which* $\bigcup_{n=1}^{\infty} B_n$ *is bounded and zero belongs to the closure of the union* $\bigcup_{n=m}^{\infty} B_n$ *for all* m.

The existence of nonmetrizable separable locally convex Fréchet–Urysohn spaces depends on additional set-theoretic assumptions. Such spaces exist if we assume Martin's axiom; for broad classes of locally convex spaces metrizability follows from the Fréchet–Urysohn property (see [**245**] and [**249**]). It is known that all sequentially complete Fréchet–Urysohn spaces are Baire; it follows from this, for example, that the inductive limit of the increasing sequence of $\mathbb{R}^n$ is not a Fréchet–Urysohn space (although is a sequential space).

The topological classification of Fréchet spaces is completely established. An outstanding achievement here is the following result of Kadec [**242**] and Anderson [**17**] (the former proved that all infinite-dimensional separable Banach spaces are homeomorphic, and the latter constructed a homeomorphism between l^2 and $\mathbb{R}^\infty$). A proof is also given in Bessaga, Pełczyński [**52**], and Toruńczyk [**528**].

3.12.27. Theorem. *Every infinite-dimensional separable Fréchet space is homeomorphic to* $\mathbb{R}^\infty$.

The proof employs the next fact that is of independent interest.

3.12.28. Theorem. *Every infinite-dimensional separable Banach space is homeomorphic to its closed unit ball and also to the unit sphere.*

In any infinite-dimensional Fréchet space F, every closed convex set with a nonempty interior is homeomorphic to F. Certainly, it is important to have a nonempty interior (a compact set is not homeomorphic to F). As shown in Banakh, Cauty [**39**], every nonseparable convex set in F that is metrizable by a complete metric is homeomorphic to some Hilbert space. For convex compact sets, there is another fact establish by Keller [**267**]: every convex compact set in a Fréchet space with an infinite-dimensional linear span is homeomorphic to the Hilbert cube $[0,1]^\infty$ (actually, in the cited paper this is proved for compact sets in l^2, to which the general case easily reduces, see Exercise 3.12.191). The same work contains the following remarkable result: for every two points x, y in the Hilbert cube $[0,1]^\infty$, there exists a homeomorphism $h\colon Q \to Q$ such that $Q(x) = y$. In the case of arbitrary Fréchet spaces Toruńczyk obtained the following nice result (see Toruńczyk [**528**]).

3.12.29. Theorem. *Each Fréchet space is homeomorphic to a Hilbert space.*

Thus, two Fréchet spaces are homeomorphic precisely when they have equal minimal cardinalities of families of everywhere dense sets. An example of a closed hyperplane in an incomplete infinite-dimensional Euclidean space that is not homeomorphic to the whole space is constructed in van Mill [**344**].

The following classical theorem is due to A.A. Milyutin (see [**343**]; a proof can be also found in Pełczyński [**382**]).

3.12.30. Theorem. *Let* K *be an uncountable metric compact space. Then the spaces* $C(K)$ *and* $C[0,1]$ *are linearly homeomorphic.*

3.12(vi). The Eberlein–Šmulian properties

Here we give some additional results related to the Eberlein–Šmulian theorem.

3.12.31. Definition. *A Hausdorff topological space T is called angelic if every set $S \subset T$ with the property that every infinite sequence of its elements has a limit point in T, possesses also the following properties: S is relatively compact and each point in the closure of S is the limit of some sequence in S.*

In angelic spaces, the properties of compactness, countable compactness and sequential compactness coincide. In addition, the closure of a relatively compact set is exhausted by the limits of sequences of points in this set. Certainly, not every compact is an angelic space, but it is interesting that there is an example of two angelic compact spaces the product of which is not angelic. It is also known that if a regular space X can be continuously injected into an angelic space Y, then X is also angelic. The proofs of these assertions and the following theorem can be found in Floret [**171**, § 3].

3.12.32. Theorem. *Let a topological space X be the closure of the countable union of sets K_n with the property that every infinite sequence in K_n has a limit point in X. Then, for every metric space Z, the space $C(X,Z)$ with the topology of pointwise convergence is angelic.*

If all K_n are compact, then, for every set $A \subset C(X,Z)$, every point in the closure of A in the indicated topology belongs to the closure of some at most countable part of A.

When applied to metrizable locally convex spaces, this gives the assertion already known from Theorem 3.4.11: if A is a set in a metrizable locally convex space, then every point in the weak closure of A is contained in the weak closure of some at most countable part of A.

By using the results presented above, one can also prove the following facts.

3.12.33. Theorem. (i) *If a locally convex space E is metrizable or admits a weaker metrizable locally convex topology, then it is angelic for the weak topology $\sigma(E,E')$. In particular, this is true for all subspaces of $\mathbb{R}^\infty$ with locally convex topologies majorizing the topology of coordinate-wise convergence.*

(ii) *Strict inductive limits and inductive limits with weakly compact embeddings of sequences of Fréchet spaces are angelic for the weak topology.*

The following theorem is due to Grothendieck.

3.12.34. Theorem. *Let K be compact. A set $F \subset C(K)$ is weakly compact precisely when it is norm bounded and compact in the topology of pointwise convergence.*

PROOF. One implication is obvious. Let F be bounded and compact in the topology of pointwise convergence. Since $C(K)$ is angelic in both considered topologies, it suffices to verify that every uniformly bounded pointwise converging sequence in $C(K)$ is weakly converging. This is obvious from the Lebesgue dominated convergence theorem and the fact that $C(K)^*$ is identified with the space of Radon measures on K. □

Recall that by Proposition 1.8.14 any compact set in a metrizable space is contained in the closed absolutely convex envelope of a sequence converging to zero. This fact does not extend to metrizable compact sets in nonmetrizable spaced.

3.12.35. Example. The unit ball in the space l^2 with the weak topology is compact and metrizable, but cannot belong to the closed absolutely convex envelope of a sequence $\{a_n\}$ weakly converging to zero. Indeed, otherwise by Exercise 2.10.47 the operator $A\colon (\lambda_n) \mapsto \sum_{n=1}^{\infty} \lambda_n a_n$ from l^1 to l^2 would be surjective, hence the range of A^* would be closed (Exercise 3.12.181), but we have $A^*(l^2) \subset c_0$, since $(A^*x)_n = (x, a_n) \to 0$, which is impossible, since c_0 contains no infinite-dimensional closed Hilbert subspaces (see Lindenstrauss, Tzafriri [**318**, Proposition 2.a.2]).

3.12(vii). Schauder bases

It is known from the university course that the most important attribute of Hilbert spaces are orthogonal bases. For more general spaces the concept of a topological basis is introduced. We shall consider here only Hausdorff locally convex spaces.

3.12.36. Definition. *A sequence $\{h_n\}$ in a locally convex space X is called a topological basis if, for every $x \in X$, there is a unique sequence of numbers $\{c_n(x)\}$ such that $x = \sum_{n=1}^{\infty} c_n(x)h_n$, where the series converges in X. If all functionals $x \mapsto c_n(x)$ are continuous, then $\{h_n\}$ is called a Schauder basis.*

Topological bases and Schauder bases in the weak topology are called respectively weak topological bases and weak Schauder bases.

It is clear that any topological basis is a linearly independent set. In an infinite-dimensional Banach space a topological basis cannot be an algebraic basis (a Hamel basis), since the latter is always uncountable. In the Banach space $C[0,1]$ a topological basis was constructed by Faber already in 1910. After appearance of general Banach spaces such bases became an object of intensive investigations (Banach, Mazur, Schauder and other eminent mathematician were involved in these investigations). In some spaces the existence of bases remained unknown for a long time. For example, in 1974 S.V. Bochkarev solved Banach's problem of constructing a Schauder basis in the space of functions analytic in the open disc and continuous on the closed disc equipped with the sup-norm. For several decades it was an open problem whether a Schauder basis exists in every separable Banach space. This was one of the most famous problems in the theory of Banach spaces. Finally, in 1973 P. Enflo published his celebrated counterexample. Simultaneously he solved another old difficult problem of existence of a separable Banach spaces without the approximation property. A Banach space X is said to possess the *approximation property* if, for every compact set $K \subset X$ and every $\varepsilon > 0$, there exists a continuous finite-dimensional operator T such that $\|x - Tx\| < \varepsilon$ for all $x \in K$. This is equivalent to the following property: for every Banach space Z, the set of finite-dimensional operators is norm dense in the space $\mathcal{K}(Z, X)$ of compact operators. If X has a Schauder basis $\{h_n\}$, then the projection operators $x \mapsto \sum_{i=1}^{n} x_i h_i$ converge to the identity operator uniformly on compact sets, hence X has the approximation property. However, there is a space with the approximation property, but without Schauder bases. It is known that spaces without the approximation property (hence without Schauder bases) exist even among closed subspaces of c_0 and l^p with $p > 2$ (certainly, they exist also among closed subspaces of the universal space $C[0,1]$). Some brief information about Schauder bases can be found in Bogachev, Smolyanov [**72**, § 6.10(iv)], where the corresponding references are given. Here we only mention several key results for more general spaces. In this circle of problems, one of the most important positive results is the following fact (going back to Banach for Banach spaces). For its proof, see Edwards [**150**, § 6.8], Jarchow [**237**, Chapter 14].

3.12.37. Theorem. *In any Fréchet space and in any inductive limit of a sequence of Fréchet spaces, every weak topological basis is a Schauder basis in the original topology.*

It is easy to construct an example of an incomplete normed space with a topological basis that is not a Schauder basis: one can take a space with a countable Hamel basis $\{h_n\}$ for which the first coefficient c_1 (or even each coefficients c_n) is discontinuous. Say, take a discontinuous linear functional l, a vector h_1 with $l(h_1) = 1$, and a basis $h_2, h_2, \ldots$ in $l^{-1}(0)$.

Dieudonné [**127**] proved that any weak Schauder basis in a barrelled space is a Schauder basis (his proof can be also found in Jarchow [**237**, Chapter 14]).

A Schauder basis $\{h_n\}$ in X is called absolute if, for every continuous seminorm p on X, there is a continuous seminorm q such that

$$\sum_{n=1}^{\infty} |c_n(x)| p(h_n) \leqslant q(x) \quad \text{for all } x.$$

An infinite-dimensional Banach space with an absolute basis is isomorphic to l^1. This is false for Fréchet spaces. Dynin and Mityagin [**146**] (see also Mityagin [**350**]) proved that in any nuclear Fréchet space every Schauder basis is absolute.

3.12.38. Theorem. *Let F be a Fréchet space with a Schauder basis. The following conditions are equivalent:* (i) *F is nuclear,* (ii) *F and F'_β possess absolute bases,* (iii) *if $\{h_n\}$ is a Schauder basis in F with the respective coordinate functionals $\{c_n\}$, then $\{h_n\}$ is an absolute basis in F and $\{c_n\}$ is an absolute basis in the space F'_β.*

For a proof, see [**237**, § 21.10]. It is known (Wojtyński [**572**]) that if in a Fréchet space with a Schauder basis every such basis is absolute, then this space is nuclear. It was unknown for a long time whether there exist nuclear Fréchet spaces without Schauder bases. The first example of such a space was constructed by Mityagin and Zobin [**582**]. Later their construction was simplified (see [**237**, § 21.10]). About bases in Fréchet spaces, see Dragilev [**134**], Dragilev, Chalov [**135**], Korobeinik [**290**], Dubinsky [**138**], [**139**], Taskinen [**519**], Valdivia [**546**], and Vogt [**555**], [**556**].

3.12(viii). Minimal spaces and powers of the real line

A Hausdorff topological vector space E is called *minimal* if it has no strictly weaker Hausdorff vector topology. Similarly in the category of locally convex spaces one defines minimal locally convex spaces. Below we consider only Hausdorff spaces.

3.12.39. Example. Any power of the real line $\mathbb{R}^T$ is minimal.

PROOF. Suppose that a Hausdorff locally convex topology τ on $\mathbb{R}^\infty$ is majorized by the product topology. In order to show the coincidence of these topologies, it suffices to establish the continuity in the topology τ of all coordinate functionals $l_n(x) = x_n$. In other words, we have to prove the following fact of the infinite-dimensional linear algebra: if the linear space $\mathbb{R}_0^\infty$ generated by all l_n contains a linear subspace L separating points in $\mathbb{R}^\infty$, then $L = \mathbb{R}_0^\infty$. Suppose that this is false. Then, there exists a nonzero functional l on $\mathbb{R}_0^\infty$ vanishing on L. As any functional on $\mathbb{R}_0^\infty$, l is given by a sequence $(y_n) \in \mathbb{R}^\infty$. Hence $y_n = 0$, since L separates points. The obtained contradiction means that $L = \mathbb{R}_0^\infty$. The general case follows from the proven fact. Indeed, if T is uncountable, then it also suffices to verify the continuity of the coordinate functionals $l_t\colon x \mapsto x(t)$ in every Hausdorff locally convex topology τ on $\mathbb{R}^T$. If the set T_0 of all $t \in T$ such that the functionals l_t are discontinuous in τ is nonempty, then it is infinite (otherwise already on $\mathbb{R}^n$ we obtain a separated locally convex topology different from the standard one), hence we can pass to a countable part of T_0 and apply the proven fact. □

As we shall see below, there are no other minimal locally convex spaces.

3.12.40. Lemma. *A locally convex space is minimal precisely when all its one-to-one continuous linear mappings onto locally convex space are homeomorphisms. A similar assertion is true in the class of topological vector spaces.*

PROOF. Let E be minimal and let $T\colon E \to F$ be a continuous one-to-one linear mapping onto a locally convex space F. If T^{-1} is not continuous, then the collection of seminorms of the form $q\circ T$, where $q \in \mathcal{Q}$ and $\mathcal{Q}$ is some collection of seminorms defining the topology of F, generates in E a Hausdorff locally convex topology that is majorized by the original topology. Hence this topology coincides with the original one, which gives the continuity of the operator T^{-1}.

Conversely, let E be not minimal, i.e., on E there is a strictly weaker Hausdorff locally convex topology τ. Then the identity mapping from E to (E, τ) cannot be a homeomorphism. A similar reasoning applies in the category of topological vector space. □

3.12.41. Proposition. *Let E be a Hausdorff topological vector space.*

(i) *If E is minimal, then E is complete.*

(ii) *If E is minimal, then every closed vector subspace in E is minimal.*

(iii) *If F is a closed linear subspace in E and G is a minimal vector subspace in E such that $F \cap G = 0$, then $F + G$ is closed and its topology coincides with the topology of the direct sum of F and G.*

PROOF. (i) Let $\widetilde{E}$ be the completion of E. Suppose that there is $v \in \widetilde{E}\backslash E$. Let $L = \mathbb{R}^1 v$. We obtain a continuous linear operator of the canonical projection $P\colon\ \widetilde{E} \to \widetilde{E}/L$, which is injective on E. By the minimality of E this operator gives a topological isomorphism between E and $P(E)$. However, taking a net $\{v_\alpha\}$ in E converging to v, we obtain the net $\{Pv_\alpha\}$ in $P(E)$ converging to zero, which means convergence to zero of the net $\{v_\alpha\}$, hence we obtain a contradiction.

(ii) If a closed subspace F in E is not minimal, then on F there is a strictly weaker vector topology s, by using which one can equip E with the following vector topology τ: a base of neighborhoods of zero in this topology is formed by the sets $U + V$, where U is a neighborhood of zero in E in the original topology and V is a neighborhood of zero in F in the topology s. It is readily verified that we have obtained a separated vector topology the restriction of which to F is s; thus, τ is strictly weaker than the original topology, contrary to its minimality.

(iii) Let $P\colon\ E \to E/F$ be the canonical projection. Then $P(G)$ is a minimal subspace in E/F, hence by assertion (i) it is complete, hence is closed. Hence $G + F = P^{-1}\bigl(P(G)\bigr)$ is closed in E. Let Q be the canonical projection of $F + G$ onto $(F + G)/F$ and let q be the restriction of Q to G. Then q is continuous, injective, surjective and linear. By the minimality of G the operator q is an isomorphism. □

3.12.42. Theorem. (i) *The minimal real Hausdorff locally convex space are precisely the powers of* $\mathbb{R}^1$.

(ii) *Every closed linear subspace in* $\mathbb{R}^T$ *is isomorphic to some* $\mathbb{R}^S$.

(iii) *If a linear subspace of a Hausdorff locally convex space is minimal, then it is complemented.*

PROOF. (i) We know that any power of $\mathbb{R}^1$ is a minimal space. By the previous theorem a minimal real Hausdorff locally convex space E with the weak topology w is complete. The algebraic dual space to E' is isomorphic to $\mathbb{R}^T$, where T is the index set for a Hamel basis $\{f_t\}$ in E'. The natural embedding $(E, w) \to (E')^*$ is linear and continuous, hence by the minimality of E it must be a topological isomorphism between (E, w) and E with the topology from $\sigma\bigl((E')^*, E'\bigr)$. Since (E, w) is complete (as noted above) and E is dense in $(E')^*$ with the indicated topology, we obtain that $E = (E')^*$, i.e., E is isomorphic to $\mathbb{R}^T$. Assertion (ii) follows from (i) and the previous theorem.

(iii) Suppose that a linear subspace F in a Hausdorff locally convex space E is minimal. We know that F is isomorphic to $\mathbb{R}^T$ for some nonempty set T. For every $t \in T$, the functional $x \mapsto x(t)$ on F obtained from this isomorphism extends to a continuous linear functional f_t on E by the Hahn–Banach theorem. The mapping $P\colon\ E \to F = \mathbb{R}^T$, $x \mapsto \bigl(f_t(x)\bigr)_{t\in T}$ is linear and continuous, on F it coincides with the identity operator. Hence $G = P^{-1}(0)$ is a closed subspace in E algebraically complementing F. Applying assertion (iii) of the previous theorem, we complete the proof. □

3.12.43. Corollary. *If a locally convex space E is complete in the weak topology $\sigma(E, E')$, then it is minimal and isomorphic to a power of the real line.*

PROOF. Let $\Lambda = E'$ and let $T\colon E \to \mathbb{R}^\Lambda$ be the embedding defined by the formula $Tx(\lambda) = \lambda(x)$. As it follows from the definition of the weak topology $\sigma(E, E')$, this embedding is a topological isomorphism. By the completeness of E the subspace $T(E)$ is closed in $\mathbb{R}^\Lambda$. Indeed, if a net $\{Tx_\alpha\}$ converges to $y \in \mathbb{R}^\Lambda$, then the net $\{x_\alpha\}$ is Cauchy in the topology $\sigma(E, E')$ and hence converges weakly to some element $x \in E$, which gives $y = Tx$. It remains to apply Proposition 3.12.41(ii). □

3.12.44. Theorem. *For any infinite-dimensional Fréchet space F the following conditions are equivalent:*

(i) *there are no continuous norms on F;*

(ii) *there is a closed linear subspace in F isomorphic to the space $\mathbb{R}^\infty$;*

(iii) *there is a complemented closed linear subspace in F isomorphic to $\mathbb{R}^\infty$.*

PROOF. It is clear that (ii) implies (i), since there are no continuous norms on $\mathbb{R}^\infty$ (Exercise 1.12.72).

Suppose that (i) is true. Let us take an increasing sequence of seminorms p_n defining the topology of F. We can assume that $p_n^{-1}(0) \neq F$ and the set $p_{n+1}^{-1}(0)$ is strictly contained in $p_n^{-1}(0)$. Choose vectors $h_n \in p_n^{-1}(0)\backslash p_{n+1}^{-1}(0)$. We observe that for every real sequence $\{x_n\}$ the series $\sum_{n=1}^\infty x_n h_n$ converges in F, since $p_k(h_n) = 0$ whenever $n \geqslant k$. Thus, we have a linear mapping $T\colon \mathbb{R}^\infty \to F$, $(x_n) \mapsto \sum_{n=1}^\infty x_n h_n$. This mapping is continuous by the continuity of all finite sums. In addition, it is obviously injective. It follows from Theorem 3.12.42 that T is a homeomorphism between $\mathbb{R}^\infty$ and $T(\mathbb{R}^\infty)$. Finally, $T(\mathbb{R}^\infty)$ is closed, since if a sequence of vectors $y_n = Tv_n$ converges in $\mathbb{R}^\infty$ to y, then it is fundamental, hence by the continuity of T^{-1} the sequence $\{v_n\}$ is Cauchy as well. Therefore, $v_n \to v$ in $\mathbb{R}^\infty$, whence $y = Tv$. Thus, we have proved (ii), but actually also (iii) is proved, since identifying $\mathbb{R}^\infty$ with a subspace in F we can extend the coordinate functionals to elements $u_n \in F'$, which gives a mapping $P\colon x \mapsto \sum_{n=1}^\infty u_n(x)h_n$. This mapping is continuous and is a projection on $\mathbb{R}^\infty$. □

3.12.45. Theorem. (Eidelheit [152]) *Let F be a nonnormable Fréchet space. Then there exists a continuous linear surjection of the space F onto $\mathbb{R}^\infty$.*

PROOF. Let us take a sequence of seminorms p_n on F defining the topology such that $p_n \leqslant p_{n+1}$, but p_{n+1} is not estimated by Cp_n (the latter is possible, because the topology of F is not defined by a single norm). Let us take a nonzero functional $f_1 \in F'$ for which $f_1 \leqslant p_1$. We observe that the space E_n' of all linear functionals on F continuous with respect to the seminorm p_n (i.e., admitting estimates $|f| \leqslant \text{const}\, p_n$) is strictly contained in E_{n+1}'. Hence we can construct a sequence of functionals $f_{n+1} \in E_{n+1}'\backslash E_n'$. It is clear that $f_{n+1} \in F'$ and the function f_{n+1} is unbounded on the set $\{p_n \leqslant 1\}$.

We show that the mapping $T\colon x \mapsto \bigl(f_n(x)\bigr)_{n\geqslant 1}$, which is linear and continuous, maps F onto the whole $\mathbb{R}^\infty$. Let us fix $\{y_n\} \in \mathbb{R}^\infty$. By induction we construct vectors $x_n \in F$ with the following properties:

1) $f_1(x_1) = 1$, $f_n(x_n) = z_n - \sum_{i=1}^{n-1} f_n(x_i)$, $n > 1$;

2) $x_n \in \bigcap_{i=1}^{n-1} f_i^{-1}(0) \cap \{x\colon p_{n-1}(x) \leqslant 2^{-n}\}$, $n > 1$.

Let us take a vector $x_1 \in F$ such that $f_1(x_1) = 1$. Since the function f_2 is not bounded on the set $\{x\colon p_1(x) \leqslant 2^{-1}\}$ and its kernel does not coincide with the kernel of f_1 (otherwise f_2 would have the form cf_1), there exists a vector x_2 in $f_1^{-1}(0) \cap \{x\colon p_1(x) \leqslant 2^{-1}\}$ for which $f_1(x_2) = z_2 - f_2(x_1)$. Suppose that $x_1, \ldots, x_n$ with the indicated properties are already chosen. The function f_{n+1} is not bounded on the set $\{x\colon p_n(x) \leqslant 2^{-n-1}\}$ and does not vanish on the intersection of the kernels $f_1, \ldots, f_n$ (otherwise f_{n+1} would be their linear combination). It is readily seen that it is not bounded on the absolutely convex set $\bigcap_{i=1}^{n} f_i^{-1}(0) \cap \{x\colon p_n(x) \leqslant 2^{-n-1}\}$ and assumes on it all real values. Hence this set contains a vector x_{n+1} for which $f_{n+1}(x_{n+1}) = z_{n+1} - \sum_{i=1}^{n} f_{n+1}(x_i)$. We now observe that the series $\sum_{n=1}^{\infty} x_n$ converges in F, since for any fixed k we have $p_k(x_m) \leqslant p_m(x_m) \leqslant 2^{-m-1}$ for all $m \geqslant k$. For the sum x of this series we obtain $f_k(x) = z_k$ for all k, since $f_k(x_n) = 0$ if $n \geqslant k+1$, $f_k(x_1 + \cdots + x_k) = z_k$ by our construction of the vectors x_n. □

Certainly, this result does not assert that any nonnormable Fréchet space is isomorphic to $\mathbb{R}^\infty$ (which is obviously false even for separable spaces). See also Moscatelli [**353**].

Exercises

3.12.46° Let E be a locally convex space. The Arens topology $\kappa(E', E)$ is the topology in E' a base of neighborhoods of zero in which are the polars of absolutely convex compacts in E. Show that this topology is locally convex and $\sigma(E', E) \leqslant \kappa(E', E) \leqslant \tau(E', E)$, but it can be strictly weaker than the Mackey topology $\tau(E', E)$.

3.12.47. Prove that the completion of any Hausdorff barrelled space is barrelled.

3.12.48° Give an example of a pointwise convergent sequence in $C[0, 1]$ that is not uniformly bounded (so does not converge weakly).

3.12.49. Show that on the unit sphere in a Hilbert space the weak topology coincides with the norm topology, although this is not true for the ball.

3.12.50° Give an example of a weakly compact set in a Banach space that is not metrizable in the weak topology.

3.12.51. Prove that a Banach space is reflexive if it is the union of a countable collection of weakly compact sets.

HINT: apply Baire's theorem.

3.12.52. Let E be an infinite-dimensional separable Hausdorff locally convex space. Prove that there exist sequences $\{x_n\} \subset E$ and $\{f_n\} \subset E'$ such that $f_n(x_k) = \delta_{nk}$ and the linear span of $\{x_n\}$ is dense in the space E.

HINT: see Pérez Carreras, Bonet [**385**, p. 44].

3.12.53. Let E be a vector space and let E^* be the space of all linear functions on E. Prove that the Mackey topology $\tau(E, E^*)$ coincides with the strongest locally convex topology on E and that E is barrelled in this topology. Show that if E is infinite-dimensional, then $\tau(E, E^*) \neq \sigma(E, E^*)$.

HINT: see Edwards [**150**, subsection 8.3.7].

3.12.54. Let E be a Hausdorff quasi-complete bornological locally convex space. Then for the coincidence of the topologies $\sigma(E',E)$ and $\tau(E',E)$ in E', it is necessary and sufficient that the topology of E be its strongest locally convex topology.

HINT: show that every bounded subset of E is fnite-dimensional.

3.12.55. Show that the space $\varprojlim \mathcal{K}^j$ described in Remark 2.4.10 is not barrelled.

3.12.56.° Show that the space l^1 with the norm from l^2 is not barrelled, although is bornological.

3.12.57. Show that a Hausdorff locally convex space with a countable algebraic basis is barrelled precisely when its topology coincides with the strongest locally convex topology.

HINT: it follows by Theorem 1.11.16 that the given space E has a Hamel basis such $\{e_n\}$ that the coefficients of the expansion $x = \sum_n x_n e_n$ are continuous linear functionals. Let E be equipped with its strongest locally convex topology. We show that it is barrelled by verifying that each pointwise bounded set $F \subset E'$ is equicontinuous. For each n we have $c_n = \sup_{f\in F} |f(e_n)| < \infty$. The seminorm $p(x) = \sum_n c_n|x_n|$ is continuous in the strongest locally convex topology and $|f(x)| \leqslant p(x)$ for all $f \in F$, which yields the equicontinuity. Conversely, let E be barrelled. In order to verify that the original topology is the strongest locally convex one, it suffices to show that each seminorm $q(x) = \sum_n \alpha_n|x_n|$ is continuous. Let F be the set of all functionals of the form $f(x) = \varepsilon_1\alpha_1x_1+\cdots+\varepsilon_n\alpha_1x_n$, where $\varepsilon_i \in \{-1,1\}$. This set is pointwise bounded, hence there is a continuous seminorm q_0 such that $\sup_{f\in F} |f(x)| \leqslant q_0$. Then $q(x) \leqslant q_0(x)$, hence q is continuous.

3.12.58. Show that a vector subspace of countable codimension in a barrelled space is itself barrelled.

HINT: see Pérez Carreras, Bonet [**385**, Theorem 4.3.6].

3.12.59. Show that in any barrelled space the closed linear subspace of countable codimension is complemented.

HINT: see Wilansky [**567**, p. 222].

3.12.60. (i) Let us equip the space $\mathbb{R}^{[0,1]}$ with the box topology τ_0, a base of which are all possible products of open sets in the factors (this topology is not vector, see Exercise 2.10.26). Show that the subspace $X \subset \mathbb{R}^{[0,1]}$ of all functions that are nonzero only on finite sets equipped with the topology τ_0 is a complete locally convex space that is not barrelled. To this end, observe that the topology in X is given by the seminorms of the form $p_\psi(x) = \sup_t |\psi(t)x(t)|$, where ψ is a positive function; consider the following set: $A = \bigl\{x \in X\colon\ \sum_t |x(t)| \leqslant 1\bigr\}$.

(ii) Show that in X with the topology τ_0 the family of bounded sets is the same as in the strongest locally convex topology on X and that (X,τ_0) is not infrabarrelled.

(iii) Show that the strong dual to X is not complete.

3.12.61. Construct an example of a closed linear subspace in a barrelled space that is not barrelled.

HINT: use the space X from Exercise 2.10.51 and the fact that every locally convex space is isomorphic to a subspace in the product of Banach spaces, which follows from Corollary 2.2.5, bearing in mind that any product of Banach spaces is barrelled.

3.12.62. Show that every Hausdorff locally convex space E is isomorphic to a closed linear subspace in some barrelled space.

HINT: take a basis $\mathcal{V}$ of absolutely convex closed neighborhoods of zero in E, for every $V \in \mathcal{V}$ take the Banach space $E_{(V)}$ obtained by factorizing and completing the space

E with the seminorm p_V (the Minkowski functional of V); embed E into the barrelled space $F = \prod_{V \in \mathcal{V}} E_{(V)}$, take a Hamel basis $\{e_\alpha\}$ in the algebraic complement of E in F and for every α consider the hyperplane H_α in F obtained as the linear span of E and all e_β with $\beta \neq \alpha$. Finally, observe that the space E is isomorphic to a closed subspace in the space $\prod_\alpha H_\alpha$, which is barrelled by the barrelledness of all H_α.

3.12.63. Let m_0 be the linear span in l^∞ of all sequences of zeros and ones. Then m_0 with the norm from l^∞ is barrelled, but its closed subspace $m_0 \cap c_0$ is not barrelled.

HINT: for proving the barrelledness of m_0 one has to verify boundedness of every $\sigma(\Lambda, m_0)$-bounded set B in $\Lambda = (l^\infty)'$; identifying Λ with the space of additive measures of bounded variation on $\mathbb{N}$, this can be easily deduced from the Phillips lemma, according to which for any sequence of bounded additive set functions m_n on the σ-algebra of all subsets of $\mathbb{N}$ such that $\lim\limits_{n\to\infty} m_n(A) = 0$ for each $A \subset \mathbb{N}$, one has $\lim\limits_{n\to\infty} \|m_n\| = 0$, where $\|\mu_n\|$ is the total variation (see Bogachev [**65**, Lemma 4.7.41]); in order to see that the normed space $m_0 \cap c_0$ consisting of finite sequences is not barrelled, it suffices to verify that the set $A = \{(x_n)\colon\ |x_n| \leqslant 1/n\}$ is a barrel.

3.12.64. Give an example of a bounded absolutely convex set A in a Banach space for which the normed space E_A is barrelled, but is not complete.

HINT: take a barrelled incomplete normed space, for example, a non-closed hyperplane in a Banach space.

3.12.65. (Antosik, Burzyk [**18**]) Let E be a locally convex space such that all sequentially continuous seminorms on E are continuous and every sequence converging to zero contains a subsequence of vectors the series of which converges. Prove that E is barrelled and bornological. Cf. Exercise 3.12.70.

3.12.66. Show that the space of all real functions on the interval $[0, 1]$ each of which vanishes outside of some countable set, equipped with the topology of pointwise convergence, is ultrabornological.

3.12.67. Suppose that a topological vector space E has a dense linear subspace that is Baire. Show that E itself is Baire.

HINT: suppose that E_0 is a dense Baire linear subspace and $E = \cup_{n=1}^\infty A_n$, where the sets A_n are nowhere dense. One can take them closed. Since E_0 is Baire, there is n such that $A_n \cap E_0$ contains $U \cap E_0$ for some open set U in E. Then A_n is dense in U. Indeed, let $u \in U$ and let V be a neighborhood of zero. Since the subspace E_0 is dense, there is a vector $y \in E_0 \cap (x + V) \cap U$. Then $y \in A_n$.

3.12.68. Suppose that a Baire topological vector space is covered by a sequence of vector subspaces. Prove that at least one of them is Baire and everywhere dense.

3.12.69. Deduce from the previous exercise that in every infinite-dimensional Baire topological vector space there is a proper Baire everywhere dense vector subspace.

3.12.70. Let E be a metrizable topological vector space such that, for every sequence $\{x_n\}$ converging to zero in E, there exists a subsequence $\{x_{n_k}\}$ for which the series of x_{n_k} converges. Prove that E is Baire.

HINT: see Pérez Carreras, Bonet [**385**, p. 20].

3.12.71. Let E be a Baire topological vector space, $A, B \subset E$, and let $U, V \subset E$ be open sets such that $U \cap V \neq \varnothing$, $U\backslash A$ and $V\backslash B$ are first category sets. Prove that $A \cap B \neq \varnothing$.

3.12.72. Show that any sequentially closed hyperplane in a bornological space is closed.

HINT: see Pérez Carreras, Bonet [**385**, Proposition 6.2.15].

3.12.73. Show that any subspace of finite codimension in a bornological space is bornological.

HINT: see [**385**, Corollary 6.3.4]; for countable codimension this is false (Dierolf, Lurje [**119**]).

3.12.74. Show that a Hausdorff locally convex space E is bornological precisely when for every locally convex space F all linear mappings from E to F bounded on compact sets are continuous.

3.12.75. Let E be a sequentially complete locally convex space. (i) Show that every barrel in E absorbs every absolutely convex bounded set. (ii) Show that if E infrabarrelled, then it is barrelled.

HINT: apply Theorem 2.5.7.

3.12.76. Show that on l^∞ with the topology $\tau(l^\infty, l^1)$ all sequentially continuous linear functionals are continuous, although this space is not bornological.

HINT: l^∞ with the topology $\tau(l^\infty, l^1)$ is not bornological since its dual is l^1, but the Banach dual of l^∞ contains a functional L not belonging to l^1, and this functional is bounded on $\tau(l^\infty, l^1)$-bounded sets (which are norm bounded sets by the Banach–Steinhaus theorem). Any sequentially continuous functional F in the topology $\tau(l^\infty, l^1)$ is continuous in this topology, i.e., is generated by an element of l^1. This follows from the fact that for every $x = (x_n) \in l^\infty$ the vectors $x^n = (x_1, \ldots, x_n, 0, 0, \ldots)$ converge to x in $\tau(l^\infty, l^1)$. The latter follows from the property of l^1 that weakly compact sets are norm compact.

3.12.77. Let $A\colon E \to G$ be a continuous linear mapping of separated locally convex spaces and let $A^{**}\colon E'' \to G''$ be the operator adjoint to $A^*\colon G' \to E'$, where G' and E' are equipped with the strong topologies $\beta(G', G)$ and $\beta(E', E)$. Show that the restriction of A^{**} to E coincides with A.

3.12.78. Suppose that A and B are bounded operators in a Hilbert space H such that $|(Ax, Bx)| \leqslant k\|Ax\|\,\|Bx\|$ for all x some $k < 1$. Prove that the operator ranges satisfy the equality $(A^* + B^*)(H) = A^*(H) + B^*(H)$.

HINT: see Edwards [**150**, Exercise 8.37, p. 613].

3.12.79. (*The James space* [**231**]) Let J be the linear subspace in c_0 consisting of all elements with finite norm

$$\|x\|_J := \sup\bigl((x_{j_2} - x_{j_1})^2 + \cdots + (x_{j_{2m}} - x_{j_{2m-1}})^2 + (x_{j_{2m+1}})^2\bigr)^{1/2},$$

where sup is taken over all finite sets of indices $1 \leqslant j_1 < j_2 < \cdots < j_{2m+1}$. Prove that the space J under the canonical embedding into J'' is of codimension 1 and hence is nonreflexive, but is linearly isometric to J''. Deduce from this that J cannot be isomorphic to $X \oplus X$ for a Banach space X. In particular, J is not isomorphic to $J \oplus J$.

3.12.80. (The Goldstine theorem) Let X be a normed space, let U_X and $U_{X''}$ be the closed unit balls in X and X'', respectively, and let $J\colon X \to X''$ be the canonical embedding. Prove that the set $J(U_X)$ is everywhere dense in $U_{X''}$ in the topology $\sigma(X'', X')$. Hence $J(X)$ is everywhere dense in X'' in the topology $\sigma(X'', X')$.

HINT: see Bogachev, Smolyanov [**72**, Theorem 6.7.6] or Day [**109**, Chapter II, § 5, Theorem 4].

3.12.81. Let E be a Banach space. Show by example that the unit ball $U_{E''}$ in E'' can contain elements that are not limits of sequences from the unit ball U_E in E in the weak topology $\sigma(E'', E')$, although by the Goldstine theorem the ball U_E is always dense in $U_{E''}$ in this topology.

HINT: consider the space $E = l^1$ with $E' = l^\infty$ and use the fact that weakly fundamental sequences in l^1 are norm convergent.

3.12.82. (i) Let X be a normed space and let τ be some locally convex topology on X intermediate between the weak topology and the norm topology. Show that if a linear mapping $A\colon X \to X$ is continuous as a mapping from (X, τ) to (X, τ), then it is a bounded operator.

(ii) Show that the assertion converse to (i) is not always true, by constructing a unitary operator A in a Hilbert space H that is discontinuous as a mapping from (H, τ) to (H, τ) for some locally convex topology on H intermediate between the weak topology and the norm topology.

HINT: (i) follows from the proof of Theorem 3.2.9. In (ii) we can take $H = L^2(\mathbb{R}^1)$ and the Fourier transform for A, defining the topology τ by adding to the seminorms of the weak topology $x \mapsto |(x, y)|$ one more seminorm $p(x) := \|I_{[0,1]}x\|$. On the subspace H_0 of functions with support in $[0, 1]$, for no $C > 0$ and $y_1, \ldots, y_n \in L^2(\mathbb{R}^1)$, the estimate $p(Ax) \leqslant C[p(x) + |(x, y_1)| + \cdots + |(x, y_n)|]$ can be valid, since $p(x) = 0$ on H_0, $p(Ax) > 0$ for nonzero $x \in H_0$, because Ax is an analytic function, and for x one can take a function orthogonal to $y_1, \ldots, y_n$.

3.12.83. (Bogachev, Smolyanov [**71**]) Let f be a measurable function on $\mathbb{R}^1$ such that the linear subspace $L := \{\varphi \in \mathcal{D}(\mathbb{R}^1)\colon f\varphi \in L^1(\mathbb{R}^1)\}$ is of finite codimension. Prove that there exists a generalized function $F \in \mathcal{D}'(\mathbb{R}^1)$ acting on $\varphi \in L$ as the integral of $f\varphi$.

HINT: considering separately f^+ and f^-, we can assume that $f \geqslant 0$; let us set $f_n = \min(f, n)$ and let E be the set of all $\varphi \in \mathcal{D}(\mathbb{R}^1)$ for which there exists a finite limit $l(\varphi)$ of the integrals of $f_n\varphi$ as $n \to \infty$; then E is a linear subspace of finite codimension, since $L \subset E$. Observe that the intersections $E \cap D_m$ are Borel sets in $\mathcal{D}_m$; apply to them Corollary 3.12.11, which gives their closedness and the continuity of l.

3.12.84. Let X be an infinite-dimensional normed space. Let us consider the topology bw^* on X' (related to Corollary 3.8.14) defined as follows: a set is bw^*-closed in X' if its intersection with each closed ball is closed in the weak-$*$ topology.

(i) Prove that the topology bw^* is strictly stronger than the topology $\sigma(X', X)$ and strictly weaker than the norm topology, and if X is Banach, then it is not defined by a metric.

(ii) Give an example of a locally convex topology τ on X' such that it does not majorize $\sigma(X', X)$, but on all bounded sets induces the same topology as $\sigma(X^*, X)$ (hence the same as bw^*); for this purpose consider c_0' with the topology $\sigma(c_0', L)$, where L is the space of finite sequences.

(iii) Let X, Y be Banach spaces and $T \in \mathcal{L}(X, Y)$. Prove that the operator T is compact precisely when the operator T^* is continuous from the topology bw^* on Y' to the norm topology on X'.

HINT: see Megginson [**338**, § 2.7, § 3.4].

3.12.85. Suppose that a Banach space E is continuously embedded into the space of all real functions on $[0, 1]$ with the topology of pointwise convergence and contains $C^\infty[0, 1]$. Prove that E contains $C^k[0, 1]$ for some k.

3.12.86. Let X and Y be Hausdorff topological spaces and let a mapping $f\colon X \to Y$ have a closed graph. Prove that, for every compact set $K \subset Y$, the set $f^{-1}(K)$ is closed in X and that, for every open set $U \subset Y$, the set $f^{-1}(K) \cap f^{-1}(U)$ is open in $f^{-1}(K)$.

3.12.87. Let E and F be locally convex spaces such that E is Souslin and F is Baire, and let $T\colon E \to F$ be a continuous linear surjection. Show that T is an open mapping, i.e., takes open sets to open sets.

HINT: reduce to the case of an injective mapping and consider the graph of the inverse mapping (see also Pérez Carreras, Bonet [**385**, p. 27]).

3.12.88. Let E be a complete metrizable topological vector space and let F be a Baire topological vector space. Prove that every continuous linear surjection $T\colon E \to F$ is open.

3.12.89. (i) (The Grothendieck theorem) Let E be a Hausdorff topological vector space over $\mathbb{R}$. Suppose that for every $n \in \mathbb{N}$ we are given a continuous linear mapping T_n from a complete metrizable topological vector space E_n to E, and that $E = \bigcup_{n=1}^{\infty} T_n(E_n)$. Then every continuous linear surjection of the space E onto a Fréchet space is open.

(ii) Every continuous linear surjection T of a Souslin locally convex space E onto a Fréchet space F is open. In particular, this yields (i) for separable Fréchet spaces E_n.

HINT: (ii) observe that the quotient space $E_0 = E/\mathrm{Ker}\, T$ is Souslin (as the projection of E) and apply Theorem 3.9.6 to the mapping from F to E_0 generated by T.

3.12.90. (The Raikov–Pták theorem) Suppose that E and F are Hausdorff locally convex spaces such that E is a quasi-complete space in which the closedness of a set follows from the closedness of its intersections with all absolutely convex compact sets and F is the union of an increasing sequence of absolutely convex compact sets. Then every linear mapping $T\colon E \to F$ with a closed graph is continuous.

HINT: see Pérez Carreras, Bonet [**385**, Theorem 8.3.59].

3.12.91. Let E be a Hausdorff locally convex space such that all linear mappings from E to Banach spaces with sequentially closed graphs are continuous. Prove that E is barrelled.

3.12.92. (i) (McIntosh [**337**]) Let E and F be Mackey spaces such that E is sequentially complete, E'_β is complete, and F is semireflexive. Then, for the continuity of a linear operator $T\colon E \to F$ with a closed graph, it is sufficient that either F be Souslin or F be an inductive limit of a sequence of fully complete spaces.

(ii) (Kalton [**256**]) The closed graph theorem is also true if E is a Mackey space with sequentially complete $\bigl(E', \sigma(E', E)\bigr)$ and F is separable and B_r-complete (see also Wilansky [**567**, p. 203]).

3.12.93. Suppose that E and F are reflexive Fréchet spaces, $L\colon F' \to E'$ is a linear mapping the graph of which is sequentially closed in $F' \times E'$ with respect to the product of the strong topologies. Show that L is continuous with respect to the strong topologies.

HINT: apply Exercise 3.12.123; see Edwards [**150**, Subsection 8.4.19].

3.12.94. Let X be a nonreflexive Banach space such that X' is separable. Show that the identity mapping of X' from the topology $\tau(X', X)$ to the norm topology has a closed graph, but is discontinuous.

3.12.95. Let $(X, \|\cdot\|)$ be an infinite-dimensional Banach space. Prove that there is a norm q on X such that $q(x) \leqslant \|x\|$, but X is not complete with this norm.

HINT: consider first the case of separable X and use an injective operator from X to l^2 with a dense range not equal to l^2.

3.12.96. Let E be a vector space over $\mathbb{R}$ or $\mathbb{C}$ and let E^* be its algebraic dual. Equip E with the topology $\sigma(E, E^*)$.

(i) Prove that bounded sets in this topology are finite-dimensional and that all linear subspaces are closed.

(ii) Prove that if E is infinite-dimensional, then the completion of $\big(E, \sigma(E, E^*)\big)$ contains a compact set not contained in the closure of any bounded set in E.

(iii) Prove that $\tau(E, E^*)$ is the strongest locally convex topology in E and that $\tau(E^*, E)$ coincides with $\sigma(E^*, E)$.

HINT: see Bourbaki [**87**, Chapitre IV, § 1, Exercice 11, p. 60, § 2, Exercice 6, p. 80] (in the English translation TVS IV.48).

3.12.97. Let $F = \mathbb{R}_0^\infty$ be the space of finite sequences with the topology $\sigma(F, l^1)$. Show that in this topology all convex compact sets are finite-dimensional, but there exist infinite-dimensional nonconvex compact sets. Deduce from this that $\sigma(l^1, F) = \tau(l^1, F)$ and $\tau(l^1, F)$ does not coincide with the topology of convergence on weakly compact sets in the space F.

3.12.98. Let E be an infinite-dimensional metrizable locally convex space over $\mathbb{R}$. Prove that the topologies $\sigma(E, E')$ and $\tau(E, E')$ in E coincide i and only if E is isomorphic to an everywhere dense subspace in $\mathbb{R}^\infty$.

HINT: the latter is sufficient for the equality $\sigma(E, E') = \tau(E, E')$ since the weak topology of $\mathbb{R}^\infty$ is metrizable, hence is the Mackey topology. Conversely, if $\sigma(E, E')$ coincides with $\tau(E, E')$, then these topologies coincide with the metrizable original topology. Therefore, there is a countable family of functionals $f_n \in E'$ for which the sets $\{x \colon |f_i(x)| < 1/k,\, i = 1, \ldots, n\}$ form a topology base of at zero. It follows that E' equals the linear span of $\{f_n\}$, hence it has a countable Hamel basis $\{g_i\}$. The mapping $x \mapsto \big(g_i(x)\big)_{i=1}^\infty$ to $\mathbb{R}^\infty$ has a dense range and establishes a linear homeomorphism between E and its image.

3.12.99. Let F be a separable Fréchet space and let $W \subset F'$ be a convex set. Prove that W is closed in the topology $\sigma(F', F)$ precisely when W contains the limits of all sequences in W converging in the topology $\sigma(F', F)$.

HINT: use the Krein–Šmulian theorem and that the polar of any neighborhood of zero in F is metrizable in the topology $\sigma(F', F)$, since F is separable.

3.12.100. Let E be a locally convex space such that every convex set absorbing all bounded sets contains a neighborhood of zero. Prove that the strong dual to E is complete. In particular, this is true if E is metrizable.

HINT: given a net of functionals $f_\alpha \in E'$ Cauchy in the topology $\beta(E', E)$, verify that its pointwise limit f has the property that the set $f^{-1}(-1, 1)$ absorbes bounded sets.

3.12.101. Prove that a quasi-complete Hausdorff locally convex space is semireflexive precisely when every closed separable subspace in it is semireflexive.

HINT: use Theorem 3.7.4 and Corollary 3.4.6.

3.12.102. Prove that if a quasi-complete Hausdorff locally convex space is not semireflexive, then it contains two convex closed bounded sets the convex envelope of the union of which is not closed.

HINT: see Bourbaki [**87**, Chapitre IV, § 3, Exercice 9, p. 93].

3.12.103. Let E be a metrizable locally convex space with the strong dual E'_β.

(i) Prove that $\big(E'', \beta(E'', E')\big)$ is a complete metrizable locally convex space.

(ii) Let E'_β be separable. Prove that E is separable and E' is barrelled and bornological in the strong topology.

HINT: (i) see Schaefer [**436**, Chapter IV, § 6.5, Corollary 2, § 6, Theorem 6.6]; (ii) see [**436**, Chapter IV, § 6, Theorem 6.6, Corollary 2].

3.12.104. Let X be a Banach space and let U be its closed unit ball.

(i) Let F_1 and F_2 be two linear subspaces in X' everywhere dense in the weak topology $\sigma(X', X)$. Prove that the topologies $\sigma(X, F_1)$ and $\sigma(X, F_2)$ induce on U the same topology precisely when F_1 and F_2 have equal norm closures. In particular, the restriction of $\sigma(X, F_1)$ to U coincides with the restriction of $\sigma(X, X')$ to U precisely when F_1 is norm dense in X'.

(ii) Show that every point in U has a countable fundamental system of neighborhoods in the topology $\sigma(X, X')$ on U only in the case where X' is norm separable.

(iii) Suppose that a linear subspace $F \subset X'$ is everywhere dense in the topology $\sigma(X', X)$ and let G be the linear subspace in X' generated by the closure of the set $F \cap \{x' \in X' \colon \|x'\| \leqslant 1\}$ in the topology $\sigma(X', X)$. Prove that $G = X'$ precisely when the indicated closure contains a ball of positive radius with respect to the norm in X'.

HINT: see Bourbaki [**87**, Chapitre IV, § 5, p. 121] (or TVS IV.55 in the English translation).

3.12.105. Suppose that E is a locally convex space such that E' is separable in the topology $\sigma(E', E)$. Prove that if $z \in (E')^* \backslash E$ and F is the linear space generated by E and z, then E' is separable also in the topology $\sigma(E', F)$.

HINT: see Floret [**171**, p. 22].

3.12.106. In the subspace E of the space $\mathbb{R}^{\mathbb{R}}$ formed by all functions with finite or countable supports consider the subset A formed by the indicator functions of at most countable sets. Show that A is sequentially compact, but the closure of A is noncompact.

3.12.107. Prove that in a uniform space every bounding set is precompact.

3.12.108. Prove that on a pseudocompact space every continuous function attains its maximum.

3.12.109. Show that in any quasi-complete locally convex space all pseudocompact sets are relatively compact.

3.12.110. Prove that every pseudocompact normal space is countably compact.

3.12.111. Show that there is a locally convex space having a subset that is pseudocompact and closed in the weak topology, but not weakly countably compact. Deduce from this that there are locally convex spaces that are not normal in the weak topology.

3.12.112. Show that a set in a locally convex space is precompact precisely when it is bounded and the induced uniform structure on it coincides with the uniform structure induced by the weak topology.

HINT: see Edwards [**150**, Proposition 9.1.2].

3.12.113. Let E be a Hausdorff locally convex space such that E' is separable in the topology $\sigma(E', E)$ and let $A \subset E$ be a set such that every infinite sequence of its elements has a weak limit point in E. Show that the weak closure of A is metrizable in the weak topology $\sigma(E, E')$.

HINT: passing to a completion, one can assume that E is complete; by Eberlein's theorem the closure of A in the topology $\sigma(E, E')$ is compact; use a countable set $\{f_n\}$ dense in $\bigl(E', \sigma(E', E)\bigr)$ to define a metric on this closure generating the topology $\sigma(E, E')$.

3.12.114. Let E be a separable locally convex space and let $\mathcal{F} \subset E'$ be a set such that every infinite sequence of its elements has a limit point in E' in the strong topology $\beta(E', E)$. Prove that every sequence in $\mathcal{F}$ contains a subsequence converging in $\beta(E', E)$.

3.12.115. Let μ be a measure such that the space $L^1(\mu)$ is separable. Prove that the space $L^\infty(\mu)$ with the topology $\sigma\bigl(L^\infty(\mu), L^1(\mu)\bigr)$ is angelic (see definition on p. 218).

3.12.116. Let $A\colon H \to H$ be a nuclear operator on a separable Hilbert space H. Prove that for every infinite-dimensional separable Banach space X there exist continuous linear operators $T\colon H \to X$ and $S\colon X \to H$ such that $A = ST$.

HINT: use the following fact (see Lindenstrauss, Tzafriri [**318**, p. 44]): there is a biorthogonal sequence $\{x_n\} \subset X$, $\{f_n\} \subset X'$ (i.e., $f_i(x_j) = \delta_{ij}$) for which the linear span of $\{x_n\}$ is dense in X, the linear span of $\{f_n\}$ is dense in X' with the topology $\sigma(X', X)$, $\|x_n\| = 1$, $\sup_n \|f_n\| < \infty$; $Ah = \sum_{n=1}^\infty \alpha_n(h, e_n)\varphi_n$, where $\{e_n\}$ and $\{\varphi_n\}$ are orthonormal bases in H, $\lambda_n \geqslant 0$, $\sum_{n=1}^\infty \lambda_n < \infty$. Finally, set $Th = \sum_{n=1}^\infty \lambda_n^{1/2}(h, e_n)x_n$ and $Sx = \sum_{n=1}^\infty \lambda_n^{1/2} f_n(x)\varphi_n$.

3.12.117. (Schreier [**439**]) Prove that there is a sequence of continuous functions f_n on $[0, 1]$ with the following properties: $0 \leqslant f_n \leqslant 1$, $\lim\limits_{n\to\infty} f_n(x) = 0$ for each $x \in [0, 1]$, but there is no increasing sequence of indices n_k such that $\bigl(f_{n_1}(x) + \cdots + f_{n_k}(x)\bigr)/n_k \to 0$ for all x.

3.12.118. (i) Let U be the closed unit ball in c_0 and let L be the closed subspace in c_0 consisting of all sequences $x = (x_n)$ for which $\sum_{n=1}^\infty 2^{-n}x_n = 0$. Prove that the set $U + (U \cap L)$ is not closed.

(ii) Let U be the closed unit ball in $C[0, 1]$. Prove that in $C[0, 1]$ there exist closed linear subspaces L_1 and L_2 such that the set $(U \cap L_1) + (U \cap L_2)$ is not closed.

3.12.119. A vector space E is equipped with two topologies each of which makes it a Fréchet space such that there is a collection of linear functions on E separating points and continuous in both topologies. Prove that these topologies coincide by applying the closed graph theorem.

3.12.120. Let E be a locally convex space and let $B \subset E$ be a bounded set such that its polar B° is separable in the strong topology in E'. Prove that B is separable in the original topology.

3.12.121. Let E and F be locally convex spaces, let $T\colon E \to F$ be a linear mapping, and let T^* be its algebraic adjoint. Consider the following conditions: (1) T is continuous; (2) T is bounded; (3) $T^*(F') \subset E'$.

Prove that (1) implies (2) and (3) and that (3) implies (2). Deduce from this that for a bornological space E all the three conditions are equivalent.

3.12.122. Let E be a Fréchet space that is not semireflexive. Prove that E in E'' is a countable union of nowhere dense sets.

HINT: otherwise the embedding $E \to E''$ is surjective by the open mapping theorem.

3.12.123. Let $\{U_\alpha\}$ be a base of absolutely convex neighborhoods of zero in a locally convex space E, $W_\alpha = U_\alpha^\circ$, and let X_α be the Banach space obtained by equipping the linear span of W_α by the norm p_{W_α} (the gauge function of W_α). Introduce in E' the topology λ of the inductive limit of Banach spaces X_α with respect to the natural embeddings $X_\alpha \to E'$.

(i) Show that the strong topology $\beta(E', E)$ is majorized by the topology λ.

(ii) Show that if E is infrabarrelled, then $\beta(E', E)$ and λ have the same bounded sets.

(iii) Show that if E is infrabarrelled, then $\lambda = \beta(E', E)$ precisely when E' is bornological with the topology $\beta(E', E)$.

(iv) Show that if E' with the topology $\beta(E', E)$ is barrelled and has a base of $\beta(E', E)$-closed neighborhoods of zero in the topology λ, then $\lambda = \beta(E', E)$.

(v) Show that E' with the topology $\beta(E', E)$ is barrelled precisely when E is distinguished, i.e., every $\sigma(E'', E')$-bounded set in E'' is contained in the $\sigma(E'', E')$-closure of some bounded set in E.

(vi) Let F be a Fréchet space. Show that F is distinguished precisely when we have $\lambda = \beta(F', F)$. In this case $\bigl(F', \beta(F', F)\bigr)$ is an inductive limit of an increasing sequence of Banach spaces. For example, this is true if F is reflexive.

HINT: see Edwards [**150**, § 8.4, p. 511, 512].

3.12.124. The topological dual F' to a Fréchet space F is fully complete in every locally convex topology between the Arens topology $\kappa(F', F)$ and the Mackey topology $\tau(F', F)$.

HINT: see Edwards [**150**, Proposition 8.10.7, p. 540].

3.12.125. Let F be a reflexive Fréchet space. Show that F' is fully complete in the strong topology and that every strongly closed linear subspace in F' is B_r-complete in the induced topology.

3.12.126. Let E be a separable Banach space. Prove the equivalence of the following two assertions:

(i) every weakly compact set in E is compact;

(ii) if $x_n \to 0$ in the topology $\sigma(E, E')$ and $f_n \to 0$ in the topology $\sigma(E', E)$, then one has $f_n(x_n) \to 0$.

An example of a space in which these assertions are true is l^1, but in l^2 they are false. See also Exercises 3.12.151 and 4.10.53.

3.12.127. Let E be a Banach space such that E' is norm separable. Prove that there exist two sequences $\{x_n\} \subset E$ and $\{f_n\} \subset E'$ such that $x_n \to 0$ in the topology $\sigma(E, E')$, $f_n \to 0$ in the topology $\sigma(E', E)$, but $f_n(x_n) = 1$. The separability of E' is important, since for $E = l^1$ there are no such sequences.

HINT: let $\{g_n\}$ be a norm dense sequence in E'. For each n there is a vector $x_n \in E$ with $\|x_n\| = 1$ and $g_1(x_n) = \cdots = g_n(x_n) = 0$. Then $x_n \to 0$ in $\sigma(E, E')$. Indeed, let $f \in E'$ and $\varepsilon > 0$. There is k such that $\|f - g_k\| < \varepsilon$. For all $n \geqslant k$ we have $|f(x_n)| = |f(x_n) - g_k(x_n)| < \varepsilon$. Next, by the separability of E' the space E is also separable and contains a dense sequence $\{y_i\}$. For each n take $i_n > n$ such that x_{i_n} does not belong to the linear span E_n of $y_1, \ldots, y_n$. By the Hahn–Banach theorem there is $f_n \in E'$ such that $\|f_n\| = 1$, $f_n|_{E_n} = 0$, $f_n(x_{i_n}) = 1$. Then $\{f_n\}$ and $\{x_{i_n}\}$ are required sequences. For l^1 such sequences do not exist since weak convergence to zero implies norm convergence.

3.12.128. Let X be a Banach space. Prove that every reflexive closed subspace in X' is closed in the topology $\sigma(X', X)$.

HINT: verify the closedness in the indicated topology of all closed balls in this subspace.

3.12.129. Let E be a complete metrizable topological vector space and let C be a closed cone (i.e., C is convex and $\lambda C \subset C$ for all numbers $\lambda \geqslant 0$) such that $E = C - C$. Prove that if a linear function f on E is nonnegative on C, then f is continuous.

HINT: take a quasi-norm p defining the topology such that (E,p) is complete and $p(tx) \leqslant p(x)$ if $|t| \leqslant 1$. Set

$$q(x) = \inf\{p(x_1) + p(x_2)\colon x = x_1 - x_2,\ x_1, x_2 \in C\}.$$

Then q is a quasi-norm and $p \leqslant q$. Moreover, $p = q$ on C. It is not difficult to verify that (E,q) is complete, whence it follows that q defines the same topology. Finally, f is continuous on (E,q). Indeed, otherwise there are x_n such that $q(x_n) \to 0$ and $f(x_n) \geqslant 1$. Hence one can find $z_n \in C$ with $\sum_{n=1}^{\infty} p(z_n) \leqslant 1$ and $f(z_n) \geqslant 1$. Since (E,p) is complete, the series of z_n converges to some $z \in E$. Since C is a convex closed cone, $z \in C$ and $\sum_{n=N+1}^{\infty} z_n \in C$ for all N. By assumption $f \geqslant 0$ on C, hence we have the bound $f(z) \geqslant f(z_1 + \cdots + z_N) \geqslant N$, which is impossible.

3.12.130. Prove that in every infinite-dimensional Banach space there exist two closed linear subspaces X_1 and X_2 such that $X_1 \cap X_2 = 0$ and $X_1 + X_2$ is not closed.

HINT: see Pérez Carreras, Bonet [**385**, p. 48].

3.12.131. Prove that if a locally convex space E is sequentially complete, then all $\sigma(E', E)$-bounded sets are strongly bounded.

HINT: apply Proposition 2.5.1 and Theorem 3.5.4.

3.12.132. Let E and F be two locally convex spaces such that E is sequentially complete. Show that if a family $\mathcal{T} \subset \mathcal{L}(E,F)$ is pointwise bounded, then, for every bounded set $B \subset E$, the set $\{Tx\colon T \in \mathcal{T}, x \in B\}$ is bounded in F. In particular, every weakly bounded set in E' is strongly bounded (Exercise 3.12.131).

HINT: see Exercise 3.12.131 or Edwards [**150**, Theorem 7.4.4].

3.12.133. Let X and Y be Banach spaces and let $A\colon X \to Y$ be a continuous linear operator such that $A(X)$ be dense in Y. Then $A(X) = Y$ precisely when $A^{**}(X'') = Y''$.

HINT: see Edwards [**150**, Corollary 8.7.4].

3.12.134. (A. Grothendieck) Let μ be a probability measure, let $V \subset L^\infty(\mu)$ be a convex set, and let $f \in L^\infty(\mu)$ belong to the closure of V in the topology $\sigma(L^\infty, L^1)$. Prove that there is a sequence $\{f_n\} \subset V$ such that $f_n \to f$ in $L^p(\mu)$ for all $p \in [1, +\infty)$.

Prove that if V is bounded, then the converse is also true. Observe also that one can take $\{f_n\}$ converging to f almost everywhere.

HINT: see Edwards [**150**, Example 8.3.6, p. 506, Exercises 8.5, 8.6, 8.7, p. 608], Grothendieck [**207**, p. 67].

3.12.135. Let E be a strict inductive limit of a sequence of metrizable locally convex spaces. Prove that there exists a weaker metrizable locally convex topology on E, hence for E the conclusion of Theorem 3.4.7 is true.

3.12.136. Let E be a metrizable locally convex space, let G be a locally complete locally convex space, let $L \subset E$ be an everywhere dense linear subspace, and let $A\colon L \to G$ be a continuous linear mapping. Prove that A extends to a continuous linear mapping of E to G.

HINT: extend A to a continuous linear operator $\widetilde{A}$ from E to the completion $\widetilde{G}$ of the space G; then $\widetilde{A}x \in G$ for all $x \in E$. Indeed, we can take $x_n \in L$ such that $x_n \to x$, then take numbers $C_n \to +\infty$ for which $C_n(x_n - x) \to 0$ in E (which is possible by the metrizability of E). Then $C_n\widetilde{A}(x_n - x) \to 0$ in $\widetilde{G}$, hence $p_B\bigl(\widetilde{A}(x_n - x)\bigr) \to 0$, where B is the closed absolutely convex hull of $\{\widetilde{A}(x_n - x)\}$ in $\widetilde{G}$. Therefore, the sequence $\{Ax_n\}$ is Cauchy with respect to the norm p_C, where $C = B \cap G$ is a closed disc in G. By the local completeness of G there is a limit $y \in G$ of this sequence with respect to the norm p_C; then $Ax_n \to y$ in G, whence $y = Ax \in G$.

3.12.137. Let E_t be locally convex spaces and $E = \prod_t E_t$, $E_0 = \bigoplus_t E_t$. Then for the Mackey topologies the following equalities hold:

$$\tau(E, E') = \prod_t \tau(E_t, E'_t), \ \tau(E', E) = \bigoplus_t \tau(E'_t, E_t),$$

$$\tau(E_0, E'_0) = \bigoplus_t \tau(E_t, E'_t), \ \tau(E'_0, E_0) = \prod_t \tau(E'_t, E_t).$$

HINT: see Schaefer [**436**, § IV.4].

3.12.138. (i) Let E be the inductive limit of locally convex spaces E_α with mappings $h_\alpha\colon E_\alpha \to E$; assume that E is separated. Then E' as a vector space can be identified with the projective limit of the spaces E'_α with mappings $h^*_\alpha\colon E' \to E'_\alpha$.

(ii) If in (i) every space E_α is equipped with the topology of convergence on some class $\mathcal{A}_\alpha$ of bounded sets covering E_α, then the topology of the projective limit of the spaces E'_α is the topology of convergence in E' on the class $\mathcal{A}$ of all sets of the form $h_{\alpha_1}(A_1) \cup \cdots \cup h_{\alpha_n}(A_n)$, where $A_i \in \mathcal{A}_{\alpha_i}$. Hence if the projective limit of the spaces E'_α is a Mackey space, then it coincides with the projective limit of $\bigl(E'_\alpha, \tau(E'_\alpha, E_\alpha)\bigr)$.

For example, if E is the strict inductive limit of increasing Mackey spaces E_n such that E_n is closed in E_{n+1} (or if E is the inductive limit of such spaces with compact embeddings), then the Mackey topology $\tau(E', E)$ is the projective limit of the spaces $\bigl(E'_n, \tau(E'_n, E_n)\bigr)$, since every absolutely convex weakly compact set in E is contained and weakly compact in some of E_n (and conversely).

(iii) The space E' with the strong topology $\beta(E', E)$ need not be the projective limit of $\bigl(E'_\alpha, \beta(E'_\alpha, E_\alpha)\bigr)$.

HINT: (i), (ii) see Robertson, Robertson [**420**, Chapter V, § 4, Proposition 15, p. 85], Schaefer [**436**, § IV.4]; (iii) see Kelley, Namioka [**270**, p. 196].

3.12.139. (i) Let E be the projective limit of locally convex spaces E_α with mappings $h_\alpha\colon E \to E_\alpha$. Assume that E is separated. Then E' as a vector space can be identified with the inductive limit of the spaces E'_α with mappings $h^*_\alpha\colon E'_\alpha \to E'$.

(ii) The inductive limit of $\bigl(E'_\alpha, \tau(E'_\alpha, E_\alpha)\bigr)$ is $\bigl(E', \tau(E', E)\bigr)$.

(iii) The space E' with the strong topology $\beta(E', E)$ need not be the inductive of limit of $\bigl(E'_\alpha, \beta(E'_\alpha, E_\alpha)\bigr)$.

HINT: (i), (ii) see Robertson, Robertson [**420**, p. 85], Schaefer [**436**, § IV.4]; (iii) see Kelley, Namioka [**270**, p. 221].

3.12.140. (i) Let E be a locally convex space and let E' be equipped with the topology of convergence on some class $\mathcal{A}$ of bounded sets covering E. Let $\widetilde{E'_{\mathcal{A}}}$ denote the completion of E' in this topology. Show that the topologies $\sigma(E, E')$ and $\sigma(E, \widetilde{E'_{\mathcal{A}}})$ coincide on all sets in $\mathcal{A}$.

(ii) Deduce from this that if we equip E' with the Mackey topology $\tau(E', E)$, then the topology of the completion will be $\tau(\widetilde{E'_\tau}, E)$.

(iii) Show that if E is a Hausdorff bornological space and every compact set is contained in some set from the class $\mathcal{A}$, then E' with the topology of convergence on $\mathcal{A}$ is complete.

3.12.141. Use Proposition 3.8.7 to show that if E is a locally convex space and E' is equipped with the strong topology $\beta(E', E)$, then the topology of its completion $\widetilde{E'_\beta}$ is exactly $\beta(\widetilde{E'_\beta}, E)$.

3.12.142. Show that any infinite-dimensional Banach space with its strongest locally convex topology is complete, but not B_r-complete.

3.12.143. Let E be a locally convex space such that its dual space E' is the union of increasing convex sets A_n. Prove that every bounded linear functional f on E is contained in some set nA_n^*, where A_n^* is the closure of A_n in the algebraic dual space E^* with the topology $\sigma(E^*, E)$.

HINT: increasing convex sets $A_n \cap (-A_n)$ also cover E'; so we can assume that $A_n = -A_n$; if $f \notin nA_n^*$, then there exists $x_n \in E$ such that $f(x_n) = 1$, $|g(x_n)| < 1$ for all $g \in nA_n^*$, whence $f(nx_n) = n$, which contradicts the boundedness of f by the boundedness of the sequence $\{nx_n\}$. The latter follows from the fact that for every $l \in E'$ we have $l \in B_n$ for all $n \geqslant n_l$, then $|l(nx_n)| = |nl(x_n)| < 1$, since $nl \in nA_n^*$ if $n \geqslant n_l$.

3.12.144. Let $\widetilde{E}$ be the completion of a locally convex space E. Suppose that every sequence in $\bigl(E', \sigma(E', E)\bigr)$ Mackey converging to zero is equicontinuous. Prove that every element in $\widetilde{E}$ is a bounded linear functional on $\bigl(E', \sigma(E', E)\bigr)$.

3.12.145. (i) Suppose that a barrelled space E is the union of an increasing sequence of its absolutely convex closed subsets A_n and that the completion of E is a Baire space (say, E is metrizable). Prove that there exists A_n with a nonempty interior.

(ii) Suppose that a locally convex space E is the union of an increasing sequence of its convex subsets A_n. Suppose that every sequence in $\bigl(E', \sigma(E', E)\bigr)$ Mackey converging to zero is equicontinuous. Prove that the completion of E is the union of the completions of A_n.

(iii) Suppose that a Mackey space E is the union of an increasing sequence of its convex complete subsets A_n. Suppose that E' is locally complete. Prove that E is complete.

HINT: see Valdivia [**543**, Chapter 1, § 3].

3.12.146. (Collins [**104**]) Let us consider the strongest locally convex topology τ on the space $E = l^2$. Let us equip the space $E^* = (E, \tau)'$ with the topology $\mathfrak{T}$ from Theorem 3.8.13. Show that it is not locally convex.

HINT: E' is $\mathfrak{T}$-closed in E^*; if the topology $\mathfrak{T}$ is locally convex, then it agrees with the duality between E and E^*, which follows from the completeness of (E, τ); then E' is closed in the topology $\sigma(E^*, E)$, which is impossible, since E' is dense in E^* in this topology.

3.12.147. Let E be a metrizable locally convex space and let E'_β be its dual with the strong topology.

(i) Show that if E'_β is metrizable, then E is normable. (ii) Give an example of a Fréchet space E for which E'_β is not barrelled (take $C^\infty(\mathbb{R}) \cap L^1(\mathbb{R})$). (iii) Prove that if a sequence of convex neighborhoods of zero V_n in E'_β has the property that $V = \bigcap_{n=1}^\infty V_n$ absorbs strongly bounded sets, then V is a neighborhood of zero. (iv) Show that every countable bounded set in the second strong dual E'' is equicontinuous. (v) Prove that the second strong dual E'' is a Fréchet space and is sequentially complete in the topology $\sigma(E'', E''')$. (vi) Show that E'_β is barrelled precisely when it is quasi-barrelled and that this is also equivalent to the property that it is bornological. (vii) Show that if E is a reflexive Fréchet space, then E'_β is bornological. In addition, if E'_β is separable, then it is bornological.

HINT: see Schaefer [**436**, § IV.6, 6.5, 6.6], Bierstedt, Bonet [**54**].

3.12.148. Let $T\colon X \to Y$ be a continuous linear surjection of complete metrizable topological vector spaces. Prove that for every compact set $K \subset Y$ there exists a compact set $S \subset X$ such that $T(S) = K$.

HINT: apply Exercise 1.12.84 and the fact that T generates an isomorphism between Y and $X/\operatorname{Ker} T$.

3.12.149. Following Grothendieck, a locally convex space E is called a $(\mathcal{DF})$-space if it possesses a fundamental sequence of bounded sets B_n (so that every bounded set is contained in some B_n) and every strongly bounded countable union of equicontinuous sets in E' is equicontinuous.

(i) Show that this class of spaces contains the strong duals to metrizable spaces, normed spaces, and quasi-barrelled spaces with fundamental sequences of bounded sets.

(ii) Show that a convex set V in a $(\mathcal{DF})$-space is a neighborhood of zero precisely when for every absolutely convex bounded set B the intersection $B \cap V$ is a neighborhood of zero in B.

HINT: see Schaefer [**436**, § IV.6, 6.7].

3.12.150. (Grothendieck [**205**, Theorem 9]) Let E be the inductive limit of a sequence of normed spaces E_i with respect to mappings $g_i\colon E_i \to E$. Then every bounded set in E is contained in the closed convex hull of finitely many images of balls in E_i.

More generally, let E_i be a $(\mathcal{DF})$-space. Then E is also a $(\mathcal{DF})$-space, the strong topology in E' is the topology of uniform convergence on sets of the form $g_i(B_i)$, where B_i is bounded in E_i, and every bounded set in E is contained in the closed convex hull of finitely many such sets.

3.12.151. (I.M. Gelfand [**189**]) Let X be a separable Banach space. Prove that a set A in X is totally bounded precisely when every sequence of functionals $f_n \in X'$ pointwise converging to zero converges uniformly on A.

HINT: show first that A is bounded; next, assuming that A is contained in the unit ball, suppose that there is a sequence of elements $a_i \in A$ with $\|a_i - a_j\| \geqslant r > 0$ for $i \neq j$, take a dense countable set $\{x_n\}$ in the unit ball and pick functionals $f_n \in X'$ and numbers i_n for which $f_n(x_i) = 0$ for $i = 1, \dots, n$, $f_n(a_{i_n}) = 1$, $\|f_n\| \leqslant 2/r$; to this end observe that for every finite-dimensional subspace L there exists an index i_n with $\operatorname{dist}(a_{i_n}, L) \geqslant r/2$. See also Exercise 4.10.53.

3.12.152. Suppose that a sequence of continuous functions f_n on a Baire space converges pointwise. Prove that it is uniformly bounded on some nonempty open set.

3.12.153. Let E be a complete locally convex space and let $\{a_{i\,n}\}$, $\{a_n\}$, $\{b_i\}$ be its countable subsets such that $\lim\limits_{n\to\infty} a_{i,n} = b_i$ in the weak topology for all i, $\lim\limits_{i\to\infty} a_{i,n} = a_n$ in the original topology uniformly in n. Prove that there exists an element $a \in E$ such that $\lim\limits_{n\to\infty} a_n = a$ in the weak topology and $\lim\limits_{i\to\infty} b_i = a$ in the original topology.

HINT: see Edwards [**150**, § 6.8, Lemma 1, p. 454].

3.12.154. (i) Let E be a barrelled space. Prove that E' with the strong topology $\beta(E', E)$ is quasi-complete.

(ii) Let E be a barrelled space, let F be a Hausdorff quasi-complete space, and let $\mathcal{B}$ be some class of bounded sets covering E. Prove that the space $\mathcal{L}_{\mathcal{B}}(E, F)$ of continuous linear mappings from E to F with the topology of uniform convergence on $\mathcal{B}$ is Hausdorff and quasi-complete. In particular, $\bigl(E', \sigma(E', E)\bigr)$ is quasi-complete; for example, this is true if E is a Fréchet space.

HINT: see Definition 3.5.1 or Bourbaki [**87**, Chapter III, § 3.8, Chapter III, §.3.2] and Schaefer [**436**, Chapter IV, § 6].

3.12.155. Let E be a locally convex space and let τ_0 be a locally convex topology on E' such that $(E', \tau_0)' \subset E$. Prove that τ_0 is majorized by the Mackey topology $\tau(E', E)$.

3.12.156. Let E and F be two metrizable topological vector spaces and let E be barrelled. Prove that every bilinear mapping $\Psi\colon E\times F\to G$ to a locally convex space G that is continuous in every variable separately is continuous.

HINT: see Schaefer [**436**, Chapter III, § 5.1, p. 88].

3.12.157. Let E and F be two Fréchet spaces and let E be nuclear. Prove that every separately continuous bilinear function on $E'_\beta\times F'_\beta$ is continuous.

HINT: see Schaefer [**436**, Chapter IV, Section 9.9].

3.12.158. Deduced from Theorem 3.12.17 the following Dieudonné–Šmulian theorem: a weakly closed set A in a quasi-complete locally convex space is weakly compact precisely when for every sequence of closed convex sets C_n such that all sets $A\cap C_1\cap\cdots\cap C_n$ are nonempty, the intersection of A with $\bigcap_{n=1}^\infty C_n$ is also nonempty.

HINT: see Bourbaki [**87**, Chapitre IV, § 2, Exercice 16].

3.12.159. A *Montel* space is a Hausdorff barrelled space E such that every bounded set in E is contained in a compact set.

(i) Prove that $\mathbb{R}^\infty$, $\mathcal{D}(\mathbb{R}^n)$, $\mathcal{S}(\mathbb{R}^n)$, and the space $H(U)$ of holomorphic functions on an open set $U\subset\mathbb{C}^1$ with the topology of uniform convergence on compact sets are Montel spaces.

(ii) Prove that every Montel space E is reflexive and E' with the strong topology $\beta(E',E)$ is also Montel.

(iii) (Dieudonné [**128**]) Every Montel Fréchet space is separable.

HINT: (ii) see Schaefer [**436**, Chapter IV, § 5.6, p. 145, § 5.9, p. 147]; (iii) see Köthe [**292**, p. 370].

3.12.160. A Hausdorff locally convex space E is called a *Schwartz space* if in this space for every absolutely convex neighborhood of zero U there exists an absolutely convex neighborhood of zero V that is a totally bounded set with respect to the seminorm p_U. (i) Show that in any Schwartz space every bounded set is totally bounded. (ii) Show that a complete quasi-barrelled Schwartz space is Montel. (iii) There exists a Montel Fréchet space that is not a Schwartz space.

HINT: see Meise, Vogt [**339**, p. 286, 338].

3.12.161. (Bonet, Lindström, Valdivia [**77**]) Let F be a Fréchet space. (i) F is Montel if and only if every sequence in F' converging to zero in the topology $\sigma(F',F)$ converges in the topology $\beta(F',F)$ (for the case of separable F, see Köthe [**292**, p. 370]). (ii) F is a Schwartz space precisely when every sequence in F' converging to zero in the topology $\sigma(F',F)$ converges uniformly on some neighborhood of zero in F. (iii) The space F does not contain a copy of l^1 precisely when convergence of a sequence of functionals in the Mackey topology $\tau(F',F)$ implies convergence in the strong topology $\beta(F',F)$.

3.12.162. (Grothendieck [**205**, p. 84]) Let E be a metrizable locally convex space. Then every bounded separable set A in its completion $\widetilde{E}$ is contained in the closure in $\widetilde{E}$ of some bounded set from E. This assertion is false without the assumption about separability of A (Dieudonné [**129**], Amemiya [**15**], Köthe [**292**, v. 1, p. 404]).

3.12.163. (Dieudonné [**129**]) Under the assumption of the continuum hypothesis there exists a nonseparable metrizable locally convex space in which all bounded sets are separable.

3.12.164. Show that the product of the countable power of the real line and the countable sum of the real lines is hypercomplete, but is not a Krein–Šmulian space.

3.12.165. Prove that in the strong dual to the strict inductive limit of a sequence of metrizable locally convex spaces there is a web (see Definition on p. 196).

3.12.166. Show that the class of webbed spaces (in the category of locally convex spaces as well as in the category of topological vector spaces) is closed with respect to the following operations: (i) passage to sequentially closed subspaces, (ii) passage to separated images under sequentially continuous linear mappings, (iii) countable products and countable direct sums, (iv) projective limits of countable sequences, (v) separated inductive limits of countable sequences.

HINT: see Jarchow [**237**, § 5.4].

3.12.167. Prove that a topological vector space is complete metrizable precisely when it is Baire and has a web.

HINT: see Jarchow [**237**, § 5.4, Theorem 4].

3.12.168°. Show that eigenvectors corresponding to different eigenvalues of a linear mapping in a linear space are linearly independent.

3.12.169. Let E be a Baire topological vector space and let F be a topological vector space. Show that every pointwise bounded family of continuous linear operators from E to F is equicontinuous.

3.12.170. (Drewnowski [**137**]) Let K be a compact set in a topological vector space X such that its closed absolutely convex hull $\overline{\operatorname{abs\,conv} K}$ is compact and contained in the linear span of K. If a linear mapping T from X to a topological vector space Y has a continuous restriction to K, then its restriction to $\overline{\operatorname{abs\,conv} K}$ is continuous as well.

3.12.171. (Drewnowski [**137**]) A resolution of a set A is its cover by subsets A_α indexed by elements $\alpha \in \mathbb{N}^\infty$ and satisfying the following condition: if $\alpha \leqslant \beta$ coordinatewise, then $A_\alpha \subset A_\beta$. Let X be a complete metrizable topological vector space with some resolution $\{X_\alpha\}$. If a linear mapping T from X to a topological vector space Y has a continuous restriction to every set X_α, then T is continuous.

3.12.172. Let E be a locally convex space and let a set A in E' be bounded in the topology $\sigma(E', E)$. Give an example in which A is not bounded in $\beta(E', E)$.

HINT: consider the space l^2_0 of finite sequences with the norm from l^2 and the functionals $l_n(x) = nx_n$.

3.12.173. Let V be a closed convex set without inner points in a Hausdorff locally convex space E such that the convex hull of A and the zero element has inner points. Show that V is contained in a closed hyperplane.

HINT: Valdivia [**543**, p. 18].

3.12.174. Show that every infinite-dimensional Baire topological vector space contains an everywhere dense linear subspace that is not Baire.

HINT: take a Hamel basis Γ, pick its countable part $\{v_n\}$, consider the linear span of $\Gamma\backslash\{v_n\}$ and $v_1, \ldots, v_n$.

3.12.175. Let $E = l^\infty_0$ be the subspace of finite sequences with the norm from l^∞, $E' = l^1$, $S = \{nl_n\} \cup \{0\}$, $l_n(x) = x_n$. Show that S is compact in the topology $\sigma(E', E)$, but S° is not a neighborhood of zero in the topology $\tau(E, E')$, since $S^{\circ\circ}$ is not compact in $\sigma(E', E)$.

3.12.176. (i) (Grothendieck [**205**]) There exists a reflexive (even Montel) Fréchet space possessing a quotient isomorphic to the space l^1.

(ii) (Valdivia [**545**]) Let F be a Fréchet space. All its quotients are reflexive precisely when F is isomorphic to a closed subspace in a countable product of reflexive Banach spaces.

3.12.177. (Kadec, Kadec [**243**]) Let F be a Fréchet space and let p be a continuous norm on F such that every continuous linear functional bounded on $U = \{p \leqslant 1\}$ attains its maximum on U. Then F is a Banach space.

3.12.178. (i) (Smolyanov [**478**]) There is a locally convex space of countable algebraic dimension that is not nuclear, but has the property that all bounded sets in it are finite-dimensional. (ii) (Gutnik [**213**]) There is a complete locally convex space not containing the direct sum of countably many one-dimensional spaces such that all its bounded subsets are finite-dimensional. In addition, there exists a locally convex Mackey space in which all bounded sets are finite-dimensional, but its completion has no such a property.

3.12.179. (Raikov [**403**]) Let F be a nuclear countably normed Fréchet space (see Example 2.2.8) the topology of which is defined by an increasing sequence of compatible norms p_n, and let $f_n \in F'$ be such that $\sum_{n=1}^{\infty} |f_n(x)| < \infty$ for all x. Then there exists m such that $\sum_{n=1}^{\infty} \|f_n\|_m < \infty$, where $\|f\|_m = \sup\{|f(x)|\colon\ p_n(x) \leqslant 1\}$. This property of series was considered in Gelfand, Kostyuchenko [**192**] and Gelfand [**190**] (in [**190**] this property of series is given as the definition of nuclear spaces attributed to Grothendieck, although he used another definition).

3.12.180. (Retakh [**416**]) Let X and Y be locally convex spaces and let $A \in \mathcal{L}(X,Y)$ be a surjective operator. Suppose that the space $K = A^{-1}(0)$ is metrizable in the induced topology and that for every neighborhood of zero $U \subset K$ there exists a neighborhood of zero $V \subset K$ with the property that for every neighborhood of zero W and every $\varepsilon > 0$ there is $\alpha > 0$ for which $V \subset \varepsilon U + \alpha W$. Then $A^*(Y') = X'$. In particular, this is true if X is a metrizable Schwartz space, which reinforces the result of Grothendieck obtained in [**205**].

3.12.181. Let X and Y be Fréchet spaces and $T \in \mathcal{L}(X,Y)$. Then the closedness of $T(X)$ is equivalent to the closedness of $T^*(Y')$ in the topology $\sigma(X',X)$, and if X and Y are Banach spaces, then in the norm of X'.

HINT: see Schaefer [**436**, Chapter IV, Theorem 7.7, p. 160], Bogachev, Smolyanov [**72**, p. 291].

3.12.182. (Shkarin [**454**]) Let F be a Fréchet space, $\{f_n\} \subset F'$, and let L be the linear span of $\{f_n\}$. The following conditions are equivalent:

(i) for arbitrary real numbers c_n there is a vector $x \in F$ with $f_n(x) = c_n$ for all n;

(ii) for every sequence of real numbers $c_n > 0$ there exists a vector $x \in F$ with $f_n(x) \neq 0$ and $|f_{n+1}(x)/f_n(x)| \geqslant c_n$ for all n;

(iii) the sequence $\{f_n\}$ is linearly independent and there are no linearly independent sequence $\{g_n\}$ in the subspace L and a sequence of real numbers $m_n > 0$ such that $\sup_n |g_n(x)|/a_n < \infty$ for all x;

(iv) the topology in F can be defined by an increasing sequence of seminorms p_n such that every functional f_n is continuous with respect to the seminorm p_{n+1}, i.e., satisfies the estimate $|f_n| \leqslant C_n p_{n+1}$, but is discontinuous with respect to p_n;

(v) the sequence $\{f_n\}$ is linearly independent and every functional $f \in F'$ vanishing on $\bigcap_{n=1}^{\infty} f_n^{-1}(0)$ belongs to L.

3.12.183. (Valdivia [**542**]) Suppose that a Hausdorff locally convex space E is the strict inductive limit of an increasing sequence of its closed linear subspaces E_n. Then the following is true.

(i) If a linear subspace $L \subset E$ has the property that all intersections $L \cap E_n$ are weakly complete, then L is the inductive limit of $L \cap E_n$ and is complementable, i.e., there is a closed linear subspace K such that E is a direct topological sum of L and K.

(ii) There is a complementable closed subspace in E isomorphic to the direct countable topological sum of one-dimensional subspaces.

(iii) If an absolutely convex set A has the property that all intersections $A \cap E_n$ are weakly complete, then A is closed. As a corollary we obtain that the product of $\mathbb{R}^\infty$ and the direct countable sum of the real lines is hypercomplete. See also Exercise 3.12.164.

3.12.184. Show that an infinite-dimensional Banach space cannot be covered by a sequence of balls with radii tending to zero.

3.12.185. Prove that every weakly fundamental sequence in l^1 converges in the norm. Deduce from this the norm compactness of weakly compact sets in l^1. Does the weak topology in l^1 coincide with the norm topology on the unit ball?

3.12.186. Show that every continuous linear operator $T\colon\ l^2 \to l^1$ is compact.

HINT: use that the closed unit ball U in l^2 is weakly compact, hence $T(U)$ is weakly compact in l^1, apply the previous exercise.

3.12.187. Prove that every continuous linear operator $T\colon\ c_0 \to l^2$ is compact, hence there is no continuous linear surjection from c_0 onto l^2.

HINT: apply the previous exercise to the adjoint operator from l^2 to l^1.

3.12.188. Prove that there exists a continuous linear surjection $T\colon\ C[0,1] \to L^2[0,1]$.

HINT: $H = L^2[0,1]$ is isomorphic to a closed linear subspace in $L^1[0,1]$ (take the closure of the linear span of a sequence of independent standard Gaussian random variables); hence H is isomorphic to a closed linear subspace in $C[0,1]'$; this subspace is closed in the topology $\sigma\bigl(C[0,1]', C[0,1]\bigr)$ by Exercise 3.12.128; take the operator dual to this embedding.

3.12.189. Prove that every Lindelöf regular topological space is normal.

HINT: see Engelking [**154**, Lemma 1.5.14 and Theorem 3.8.2].

3.12.190. Show that there exists a Banach space that is not Lindelöf in the weak topology. Consider l^∞.

3.12.191. Let K be a convex compact set in a topological vector space E such that there is a sequence $\{f_n\} \subset E'$ separating points in K. Prove that K is affine homeomorphic to a convex compact set in l^2.

HINT: take $x \mapsto \bigl(f_n(x)\bigr)_{n=1}^\infty$, where $f_n \in E'$, $\sup_K |f_n(x)| \leqslant n^{-1}$.

3.12.192. (Grothendieck's theorem, see Floret [**171**, p. 15]) Let V be a convex and complete set in a locally convex space E. A set $A \subset V$ has $\sigma(E,E')$-compact closure precisely when it is bounded and for every equicontinuous set $S \subset E'$ the equality

$$\lim_{m\to\infty}\lim_{n\to\infty} f_m(x_n) = \lim_{n\to\infty}\lim_{m\to\infty} f_m(x_n)$$

holds for all sequences $\{x_n\} \subset A$ and $\{f_m\} \subset S$ such that both limits exist.

3.12.193. Let E be an infinite-dimensional locally convex space and let $W = \{f > 0\}$, where $f \in E'$, $f \neq 0$. Prove that there exists a homeomorphism $h\colon\ W \cup \{0\} \to E$ such that $h(x) = x$ whenever $f(x) \geqslant 1$.

HINT: see Bessaga, Pełczyński [**52**, p. 114].

3.12.194. (Smolyanov [**476**]) Suppose that a topological vector space E is the union of a strictly increasing sequence of its linear subspaces E_n and that every sequence converging in E is contained in some of E_n. Then E contains a set A such that not all its limit points are limits of sequences of points in A.

3.12.195. (Mankiewicz [**333**]) Every LF-space (the inductive limit of an increasing sequence of Fréchet spaces) is homeomorphic to one of the following spaces:
(i) $\mathbb{R}^{(\infty)}$ (the inductive limit of $\mathbb{R}^n$),
(ii) $l^2(\kappa)$ for some cardinality κ,
(iii) $l^2(\kappa)\times\mathbb{R}^{(\infty)}$ for some infinite cardinality κ,
(iv) $\bigoplus_{i=1}^{\infty} l^2(\kappa_i)$ for some strictly increasing sequence of infinite cardinality κ_i.
See also Banakh, Repovš [**40**], Sakai [**427**].

3.12.196. (Corson [**108**], Haydon [**217**]) Suppose that K is a convex compact set in a Hausdorff locally convex space such that the set of its extreme points is the image of a separable metric space under a continuous mapping. Then K is metrizable.

3.12.197. (Schlüchtermann, Wheeler [**437**]) An infinite-dimensional Banach space with the weak topology is not a k-space.

3.12.198. (Gabriyelyan, Kąkol, Kubzdela, López-Pellicer [**182**]) Let E be a metrizable locally convex space. Every bounded subset of E is a Fréchet-Urysohn space in the weak topology if and only if every bounded sequence in E has a Cauchy subsequence in the weak topology.

3.12.199. (Ruess [**426**]) Let E be a complete locally convex space in which every bounded set is metrizable. Then E has no closed subspaces linearly homeomorphic to l^1 if and only if every bounded sequence in E has a Cauchy subsequence in the weak topology.

3.12.200. (Gabriyelyan, Kąkol, Plebanek [**183**]) A Banach space X has no closed subspaces linearly homeomorphic to l^1 if and only if every real sequentially continuous function on the unit ball with the weak topology is continuous in the weak topology.

3.12.201. (Banakh [**38**]) (i) The space $\mathcal{D}'(\mathbb{R}^1)$ is homeomorphic to the countable power of the inductive limit of the spaces $\mathbb{R}^n$.

(ii) The strong dual to a nuclear LF-space is homeomorphic to one of the spaces $\mathbb{R}^\infty$, $\mathbb{R}^{(\infty)}$, $\mathbb{R}^\infty\times\mathbb{R}^{(\infty)}$, $(\mathbb{R}^{(\infty)})^\infty$, and $[0,1]^\infty\times\mathbb{R}^{(\infty)}$.

CHAPTER 4

Differential calculus

The concept of differentiable mapping from a topological vector space to a topological vector space was worked out relatively recently. In the mid 60s of the XX century the number of existing definitions of differentiability of mappings of topological vector spaces was very large and was comparable (if not greater) with the number of papers devoted to the study of such mappings.

Nevertheless no differential calculus for mappings of such spaces did actually exist that time, and it seemed natural to expect that no unified differential calculus for them could exist and that to every definition of differentiability its own differential calculus must correspond. However, it was soon discovered that the collection of non-equivalent definitions of (one-fold) differentiability in topological vector spaces is quite manageable and can be arranged in a simple scheme. Moreover, it turned out that there actually exist precisely two types of infinitely differentiable mappings of locally convex spaces (the same is true also for $n > 1$ times differentiable mappings if we ignore a possible decrease of the order of differentiability by one).

It was this discovery that enabled one to construct a unified and in essence very simple differential calculus for mappings of topological vector spaces. The basics of this calculus are presented in this chapter. A key role is played by differentiability with respect to a system of sets. In this approach two series of definitions of differentiability arise that cover in total all principal known definitions; each of these series contains precisely one definition of infinite differentiability. A description of differentiable mappings corresponding to one of these series involves filters (or nets) and the language of the theory of pseudo-topological spaces (but not this theory itself). A detailed presentation is given in Smolyanov [**489**], here we only give some minimal information in § 4.10(vi). In the main part of the chapter we consider only the results related to the other series of definitions of differentiability whose description does not require any use of pseudo-topologies. Nevertheless we observe that the language using concepts of the theory of pseudo-topological spaces, especially the notion of a filter, is rather flexible and expressive and is not more difficult than the standard language of the theory of topological vector spaces (employing the concepts of bounded and open sets and also convergent sequences). At the same time, all proofs using the language of pseudo-topologies are completely similar to the proofs in the case of the series of definitions of differentiability in the main text which do not use that language, and,

V.I. Bogachev and O.G. Smolyanov, *Topological Vector Spaces and Their Applications*, Springer Monographs in Mathematics,
DOI 10.1007/978-3-319-57117-1_4

with some training, can be obtained by a simple translation from one language to another (certainly, the meaning of proofs will change in this procedure). Probably, it is even possible to prove a "meta-theorem" formalizing this.

Throughout this chapter, we assume, if it is not explicitly stated otherwise, that the field of scalars is the field of real numbers.

If P and Q are two sets, then by $\mathcal{F}(P,Q)$ or by Q^P we denote the set of all mappings of the set P to the set Q; if Q is a vector space, then we assume that the set Q^P is equipped with its natural vector structure. If we are given a topology $\tau(P,Q)$ in $\mathcal{F}(P,Q)$, then $\mathcal{F}_\tau(P,Q)$ will denote the topological space obtained by equipping $\mathcal{F}(P,Q)$ with the topology $\tau(P,Q)$. If S is a part of $\mathcal{F}(P,Q)$, then S_τ will denoted the set S with the topology $\tau(P,Q)$ restricted to it. Mainly $\tau(P,Q)$ will be the topology of uniform convergence on sets from some class of sets $\mathcal{A}$ in a topological vector space P; then we shall use the notation $\mathcal{F}_{\mathcal{A}}(P,Q)$.

If P and Q are vector spaces, then $Lin(P,Q)$ will denote the vector space of all linear mappings from P to Q. If P and Q are topological vector spaces, then the symbol $\mathcal{L}(P,Q)$ will denote the vector space of all continuous linear mappings from P to Q.

The symbol $\overline{\mathcal{L}}(P,Q)$ denotes the vector space of all sequentially continuous linear mappings from P to Q and the symbol $\mathcal{B}(P,Q)$ will denote the space of all bounded linear mappings from P to Q. Finally, if $\mathcal{A}$ is some class of bounded sets in P, then $\mathcal{L}_{\mathcal{A}}(P,Q)$ will denote the class of all linear mappings from E to F bounded on the sets from $\mathcal{A}$. We shall equip $\mathcal{L}_{\mathcal{A}}(P,Q)$ with the topology of uniform convergence on the class $\mathcal{A}$. Thus, $\mathcal{B}(P,Q) \subset \mathcal{L}_{\mathcal{A}}(P,Q)$.

Note that $\mathcal{B}(P,Q) \supset \overline{\mathcal{L}}(P,Q) \supset \mathcal{L}(P,Q)$ for any topological vector spaces P and Q; $\mathcal{B}(P,Q) = \overline{\mathcal{L}}(P,Q) = \mathcal{L}(P,Q)$ if P is a bornological space.

For the space of scalars $E = \mathbb{R}$ we shall usually identify the space $\mathcal{L}(E,G)$ $(= \overline{L}(E,G) = \mathcal{B}(E,G))$ with the target space G; in this case every linear mapping $f \in \mathcal{L}(E,G)$ is identified with the element $f(1)$ of the space G such that we have $f(a) = af(1)$ for all $a \in \mathbb{R}$.

4.1. Differentiability with respect to systems of sets

In this section we introduce the series of definitions of differentiability of mappings of topological vector spaces that does not employ the language of pseudo-topologies (the approach based on pseudo-topologies will be briefly outlined in § 4.10(vi)).

Let E and G be topological vector spaces and let $\mathcal{H}(E,G)$ and $\mathcal{R}(E,G)$ be some vector subspaces in the space $\mathcal{F}(E,G)$ of all mappings from E to G satisfying the following conditions:

(C) if U is a neighborhood of zero in E, $L \in \mathcal{H}(E,G)$, $r \in \mathcal{R}(E,G)$ and $L(x) = r(x)$ for all $x \in U$, then $L = 0$.

Concrete examples of such pairs will be given below. As a trivial example, note that (C) is fulfilled if $\mathcal{R}(E,G)$ consists of the zero element and $\mathcal{H}(E,G)$ is the class of all bounded linear mappings.

4.1.1. Definition. *A mapping f of an open set V in a topological vector space E to a topological vector space G is called $\mathcal{HR}$-differentiable at a point $x_0 \in V$ if there exist an element of the space $\mathcal{H}(E,G)$, called an $\mathcal{HR}$-derivative of the mapping f at the point x_0 and denoted by the symbol $f'(x_0)$, a neighborhood of zero $V_0 \subset E$ and a mapping $r \in \mathcal{R}(E,G)$, called $\mathcal{R}$-small, such that $x_0 + V_0 \subset V$ and for all $h \in V_0$ we have*

$$f(x_0 + h) - f(x_0) = f'(x_0)h + r(h). \tag{4.1.1}$$

Thus, the symbols $\mathcal{H}$ and $\mathcal{R}$ are not shown in the notation for the corresponding derivatives.

The requirements imposed on the pair $\bigl(\mathcal{H}(E,G), \mathcal{R}(E,G)\bigr)$ yield that for, every point $a \in V$, any linear combination of mappings that are $\mathcal{HR}$-differentiable at the point a (and defined on the intersection of the their domains of definition, so also defined in a neighborhood of a) is also $\mathcal{HR}$-differentiable at a and its derivative is the corresponding linear combination of the derivatives of the given mappings. Furthermore, a mapping can have at most one derivative at a given point and every mapping $r \in \mathcal{R}(E,G)$ is $\mathcal{HR}$-differentiable at the origin with $r'(0) = 0$. Certainly, the converse is also true: if a mapping f from E to G is $\mathcal{HR}$-differentiable at zero and $f'(0) = 0$, then $f \in \mathcal{R}(E,G)$.

Thus, the problem of differentiation is a kind of the approximation problem: the derivative of a mapping at a given point is a mapping from a specific class approximating the increment of the regarded mapping up to a mapping from some other class. For defining a differentiation (on a set of mappings between two topological vector spaces) it suffices to define two spaces of mappings of the first space to the second: a space of mappings (it is denoted above by the symbol $\mathcal{H}(E,G)$) that will serve as derivatives, and also a space of mappings (denoted by the symbol $\mathcal{R}(E,G)$) that will be considered as "small". This scheme applies also to global Sobolev type differentiabilities not discussed here (see Bogachev [**66**]).

Below, if it is not explicitly stated otherwise, we assume that, given two topological vector spaces E and G, the space $\mathcal{H}(E,G)$ coincides with one of the operator spaces $\mathcal{L}(E,G)$, $\overline{\mathcal{L}}(E,G)$ and $\mathcal{B}(E,G)$; sometimes we shall also use the space $\mathcal{L}_{\mathcal{A}}(E,G)$. For specifying spaces $\mathcal{R}(E,G)$ in this section we use topologies of uniform convergence on various families of bounded sets in E; these topologies are introduced in the space $\mathcal{F}(E,G)$.

Let σ be some class of bounded subsets of the space E containing all finite subsets.

4.1.2. Definition. *Let $n \in \mathbb{N}$. A mapping r from a topological vector space E to a topological vector space G is called σ-small of order n if for every set $B \in \sigma$ we have $r(th)/t^n \to 0$ in G uniformly in $h \in B$ as $t \to 0$. A mapping that is σ-small of the first order will be called σ-small.*

The set of all σ-small of order n mappings from E to G will be denoted by $\mathcal{R}^{n,\sigma}(E,G)$. Thus, $r \in \mathcal{R}^{n,\sigma}(E,G)$ precisely when for every neighborhood of zero $V \subset G$ and for every set $B \in \sigma$ there exists $\varepsilon > 0$ such that if $t \in (-\varepsilon, \varepsilon)\backslash\{0\}$ and $h \in B$, then $t^{-n}r(th) \in V$. The symbol $\mathcal{R}^{\sigma}(E,G)$ will denote the same object as the symbol $\mathcal{R}^{1,\sigma}(E,G)$.

The symbol $\mathcal{F}_\sigma(E, G)$ will denote the space $\mathcal{F}(E, G)$ equipped with the topology of uniform convergence on the sets of the system σ (if G is Hausdorff and is of positive dimension, then $\mathcal{F}_\sigma(E, G)$ is a topological vector space precisely when the family σ contains no infinite sets).

The next proposition is a reformulation of the definition that justifies what has been said above about defining spaces of small mappings by means of topologies.

Given two vector spaces E_1 and E_2, a mapping $r \in \mathcal{F}(E_1, E_2)$, and a number $t \in \mathbb{R}\backslash\{0\}$, the symbol r_t will denote the mapping of E_1 to E_2 defined by the equality $r_t(x) = t^{-1}r(tx)$ $(x \in E_1)$, and the symbol r_0 will denote the zero mapping of E_1 to E_2, i.e., $r_0(x) = 0$ for all $x \in E_1$.

4.1.3. Proposition. *A mapping r from a topological vector space E to a topological vector space G is σ-small if and only if the mapping $t \mapsto r_t$ from $\mathbb{R}$ to $\mathcal{F}_\sigma(E, G)$ is continuous at zero.*

The next simple proposition will be frequently used below.

4.1.4. Proposition. *In order that a mapping r from a topological vector space E to a topological vector space G be σ-small it is necessary and sufficient that it satisfy the following condition: whatever be a set $A \in \sigma$, if $\{a_n\}$ is a sequence of its elements and $\{t_n\}$ is a sequence of nonzero numbers converging to zero, we have $t_n^{-1}r(t_n a_n) \to 0$ in the space G as $n \to \infty$.*

This proposition follows at once from the previous one and the following assertion.

4.1.5. Proposition. *If Q is a topological commutative group, P is a set, and σ is a system of subsets of the set P, then a sequence of elements g_n of the space $\mathcal{F}_\sigma(P, Q)$ converges in this space to $g \in \mathcal{F}(P, Q)$ precisely when for every set $A \in \sigma$ and every sequence $\{h_n\} \subset A$ we have $g_n(h_n) - g(h_n) \to 0$ in Q.*

PROOF. It is clear that convergence of a sequence $\{g_n\}$ in $\mathcal{F}_\sigma(P, Q)$ implies the property stated above. If a sequence $\{g_n\}$ in $\mathcal{F}_\sigma(P, Q)$ does not converge in this space, then for every $g \in \mathcal{F}_\sigma(P, Q)$ there exist a neighborhood of zero V in Q and a set $A \in \sigma$ such that for every $n \in \mathbb{N}$ there is an element $h_n \in A$ with $g_n(h_n) - g(h_n) \notin V$. Thus, $g_n(h_n) - g(h_n) \not\to 0$ in the group Q. □

Until the end of this section the symbols E and G denote Hausdorff topological vector spaces and the symbol σ is used to denote a family of bounded subsets of the space E containing all finite sets.

If $\mathcal{R}(E, G) = \mathcal{R}^\sigma(E, G)$, $\mathcal{H}(E, G)$ is one of the three spaces indicated above, then all requirements on the pair $\big(\mathcal{H}(E, G), \mathcal{R}(E, G)\big)$ are fulfilled (this is verified directly), so that Definition 4.1.1 applies.

In place of the term the $\mathcal{HR}^\sigma$-derivative (or differentiability) we shall use the term the $\mathcal{H}_\sigma$-derivative and same with differentiability. Additional abbreviations will be used for $\mathcal{L}_\sigma$- and $\overline{\mathcal{L}}_\sigma$-derivatives: the first will be called the σ-derivative and the second will be called the $\bar{\sigma}$-derivative; similarly the terms σ- and $\bar{\sigma}$-differentiability will be used.

Let us give a straightforward definition of σ- and $\bar{\sigma}$-differentiability and also $\mathcal{A}$-differentiability generated by a class $\mathcal{A}$ of bounded sets taken for σ when for the class of operators $\mathcal{H}(E,G)$ we take the class $\mathcal{L}_{\mathcal{A}}(E,G)$ of all linear operators bounded on the sets in $\mathcal{A}$.

4.1.6. Definition. *A mapping f from a nonempty set $V \subset E$ in G is called σ-differentiable at a point $x_0 \in V$ if there exist a continuous linear mapping from E to G, called a σ-derivative of the mapping f at the point x_0 and denoted by the symbol $f'(x_0)$, a neighborhood of zero $V_0 \subset E$ and a σ-small mapping $r\colon E \to G$ such that $x_0 + V_0 \subset V$ and for all $h \in V_0$ equality* (4.1.1) *holds.*

Similarly we introduce the $\bar{\sigma}$-differentiability and $\bar{\sigma}$-derivatives, but in place of continuity of $f'(x_0)$ only its sequential continuity is required. If $\sigma = \mathcal{A}$ is some class of bounded sets and in place of continuity of the derivative we require only that $f'(x_0) \in \mathcal{L}_{\mathcal{A}}(E,G)$, then we obtain $\widetilde{\mathcal{A}}$-differentiability.

Thus, in order that a mapping f of a subset of E to G be σ-differentiable at a point $x_0 \in V$ in the sense of this definition it is necessary that this point be interior; if this is the case, then the mapping f is σ-differentiable at the point x_0 precisely when every extension of f to a mapping on all of E is σ-differentiable at this point; if $\bar{f}$ is such an extension, then $\bar{f}'(x_0) = f'(x_0)$. Hence it is natural to introduce the following less restrictive definition of σ-differentiability.

4.1.7. Definition. *Let $V \subset E$. A mapping $f\colon V \to G$ is called σ-differentiable (respectively, $\bar{\sigma}$-differentiable) at a point $x_0 \in V$ if at this point every extension of this mapping to a mapping of the whole space E is σ-differentiable (respectively, $\bar{\sigma}$-differentiable) in the sense of Definition* 4.1.6. *In this case the σ-derivative (the $\bar{\sigma}$-derivative) of the mapping f at x_0 is defined as the σ-derivative (respectively, the $\bar{\sigma}$-derivative) at this point of an arbitrary extension of f to all of E.*

4.1.8. Remark. If f is differentiable at a point $x_0 \in V$ in the sense of Definition 4.1.7, then its derivative at this point is defined uniquely (prove this). Note also that for every inner point x_0 in the domain of definition of the function f Definitions 4.1.6 and 4.1.7 are equivalent; however, a mapping can be differentiable in the sense of Definition 4.1.7 also at a point that is not inner. At the same time, if E is a locally convex space, τ is its original topology and τ_1 is a locally convex topology in E (different from τ) such that $(E,\tau)' = (E,\tau_1)'$, then a real function g defined on a subset V of E is σ-differentiable at a point $x_0 \in V$ in the sense of Definition 4.1.7 as a mapping on a subset of the space (E,τ) precisely when it is σ-differentiable at this point in the sense of the same definition as a mapping of the space (E,τ_1) (why?). If E is an infinite-dimensional normed space, $\tau = \sigma(E,E')$, τ_1 is the topology in E defined by its norm, V is the unit ball centered at the origin in E, and f is the restriction to V of a continuous linear functional on E, then the function f considered as a function on the set V in (E,τ) turns out to be σ-differentiable at the origin in the sense of Definition 4.1.7, but not in the sense of Definition 4.1.6 (the ball has no interior in the weak topology). At the same time f is differentiable both in the sense of Definition 4.1.6 and in the sense of Definition 4.1.7 as a function on the subset V of the space (E,τ_1).

Below, if it is not explicitly stated otherwise, σ- or $\bar{\sigma}$-differentiability will mean σ- or $\bar{\sigma}$-differentiability in the sense of Definition 4.1.7. When speaking of differentiability of functions defined on proper subsets of the space E, where it becomes necessary in accordance with Definition 4.1.7 to consider extensions of these functions to all of E, we shall use without additional comments the same notation for such extensions as for the original functions.

In the next three definitions we consider differentiability on sets (not only at points), higher derivatives, and the differentiability along subspaces.

4.1.9. Definition. *A mapping f from a subset of E to G is called σ-differentiable (respectively, $\bar{\sigma}$-differentiable) on a set V if it is σ-differentiable (respectively, $\bar{\sigma}$-differentiable) at each point $x \in V$.*

A mapping $f'\colon x \mapsto f'(x)$, $V \to \mathcal{L}_\sigma(E,G)$ (respectively, $V \to \overline{\mathcal{L}}_\sigma(E,G)$) is called a σ-derivative (respectively, a $\bar{\sigma}$-derivative) of the mapping f on the set V.

Define inductively the spaces $\mathcal{L}_\sigma^k(E,G) = \mathcal{L}_\sigma\big(E, \mathcal{L}_\sigma^{k-1}(E,G)\big)$ by setting $\mathcal{L}_\sigma^1(E,G) = \mathcal{L}\sigma(E,G)$, $\mathcal{L}_\sigma^2(E,G) = \mathcal{L}_\sigma\big(E, \mathcal{L}_\sigma(E,G)\big)$.

4.1.10. Definition. *Let $n > 1$ and $V \subset E$. A mapping $f\colon V \to G$ is called n times σ-differentiable at a point $x_0 \in V$ if it is $n-1$ times σ-differentiable on some set V_0 containing x_0 and its $(n-1)$th σ-derivative $f^{(n-1)}\colon V_0 \to \mathcal{L}_\sigma^{n-1}(E,G)$ is σ-differentiable at the point x_0. In this case the nth σ-derivative of the mapping f at the point x_0 is defined as the first order σ-derivative of the mapping $f^{(n-1)}$ at this point: $f^{(n)}(x_0) = (f^{(n-1)})'(x_0)$. The mapping f is called infinitely σ-differentiable at the point x_0 if f is differentiable at this point n times for every n.*

Similarly one defines $\bar{\sigma}$- and $\widetilde{\mathcal{A}}$- *derivatives of order* n and the respective n*-fold* $\bar{\sigma}$- and $\widetilde{\mathcal{A}}$-differentiability.

Everywhere the difference between $\bar{\sigma}$ and σ is concerned with the sequential continuity of operators in place of the usual continuity.

The symbol $\widetilde{A}$ in the notation for differentiability and derivative indicates that we deal with a certain class $\mathcal{A}$ of bounded sets and that linear mappings arising as derivatives must be bounded on sets in $\mathcal{A}$. If $\mathcal{A}$ is the class of all bounded sets, then these are bounded linear mappings (sequentially continuous operators are bounded, but not conversely).

We shall always assume that $f^{(0)} = f$. Suppose now that a topological vector space E is a vector subspace in some vector space E_0 and W is a subset of E_0.

4.1.11. Definition. *A mapping $f\colon W \to G$ is called n-fold σ-differentiable at a point x_0 along the subspace E, or σE-differentiable, if the mapping f_{x_0} from the set $(W - x_0) \cap E$ to G defined by the equality $f_{x_0}(x) = f(x_0 + x)$ is n-fold σ-differentiable at zero. In this case the nth σ-derivative of f at x_0 along the subspace E is defined as the nth σ-derivative of the mapping f_{x_0} at zero; the nth σ-derivative of the mapping f at the point x_0 along the subspace E will be denoted by $f_E^{(n)}(x_0)$.*

A mapping $f\colon W \to G$ is called *n-fold σ-differentiable on a set $W_1 \subset W$ along the subspace E* (or *σE-differentiable*) if it is n-fold σ-differentiable at every point in W_1; in this case the mapping

$$f_E^{(n)}\colon x \mapsto f_E^{(n)}(x), \quad W \to \mathcal{L}_\sigma^n(E, G)$$

is called the *nth σ-derivative of the mapping f on the set W_1 along the subspace E*. Similarly one defines the n-fold *$\bar{\sigma}$-differentiability*.

Thus, by our definition, $f_E^{(n)}(x_0) = (f_{x_0})^{(n)}(0)$ for each n. In particular, if $E = E_0$, then the nth σ-derivative of the mapping f along the subspace E coincides with its nth σ-derivative (at the same point) in the sense of Definition 4.1.10; below in this case we shall use a shorter term and the corresponding notation (i.e., $f^{(n)}$ in place of $f_E^{(n)}$).

Let σ_{fin}^E be the family of all finite subsets of the space E, let σ_b^E be the family of all bounded sets, let σ_c be the family of all compact sets, and let σ_s^E be the family of all sets consisting of elements of convergent sequences. If this does not lead to ambiguity, in place of the symbols σ_b^E, σ_c^E, and σ_s^E we shall use the symbols σ_b, σ_c, and σ_s, or even b, c, and s. In place of the term σ-differentiability it is customary to use the term *differentiability with respect to the system of sets σ*. For example, b-differentiability is differentiability with respect to the system of all bounded sets and $\bar{b}$-differentiability is its version with the sequential continuity of derivatives.

The terms σ_{fin}-differentiability, b-differentiability, c-differentiability, and s-differentiability will be often replaced by the terms *Gâteaux differentiability*, *Fréchet differentiability*, *differentiability with respect to the system of compact sets* and *sequential differentiability* (we emphasize that the latter is not the same as differentiability with respect to the system of all compact sets if the space is not metrizable); $\bar{b}$-differentiability will be also called *bounded differentiability*. Similar conventions are in force for the terms containing the word "derivative".

In metrizable spaces, c-differentiability coincides with s-differentiability. In the case of normed (or metrizable) spaces differentiability with respect to the system of compact sets is called *Hadamard differentiability*.

It follows from the definition that σ_b-differentiability yields σ_{fin}-differentiability and the corresponding derivatives coincide. Below the symbol $E \Subset E_0$ means that E is a vector subspace of the space E_0.

In the case of differentiability with respect of a system $\mathcal{A}$ of bounded sets in E it is convenient to introduce inductively defined spaces

$$L_k := \mathcal{L}_{\mathcal{A}}(E, L_{k-1}), \quad \text{where } L_0 = G,$$

with the topology of uniform convergence on $\mathcal{A}$. Let $\mathcal{L}_{\mathcal{A}}^k(E,G)$ be the linear space of k-linear mappings from E to G with the topology of uniform convergence on the sets of the form $A_1 \times \cdots \times A_k$, where $A_i \in \mathcal{A}$. The space L_k can be identified with $\mathcal{L}_{\mathcal{A}}^k(E,G)$. For example, an element $\Psi \in L_2$ defines an element $\Psi_2 \in \mathcal{L}_{\mathcal{A}}^2(E,G)$ by the formula $\Psi_2(u,v) = \Psi u(v)$, where the right-hand side is bounded in $u \in A$, $v \in B$ for all $A, B \in \mathcal{A}$, which follows from the definition of the topology in $\mathcal{L}_2$. Conversely, every element $\Psi_2 \in \mathcal{L}_{\mathcal{A}}^2(E,G)$ defines $\Psi \in L_2$ by the formula $\Psi u(v) = \Psi_2(u,v)$.

The $\mathcal{A}$-derivative of order k of a mapping f at a point x will be denoted by the symbol $D^k_{\mathcal{A}}f(x)$; it can be regarded as a k-linear mapping from X^k to Y. Then $D^k_{\mathcal{A}}f(x) \in \mathcal{L}^k_{\mathcal{A}}(X, Y)$.

We recall that a mapping F between normed spaces is *Lipschitzian* if

$$\|F(x) - F(y)\| \leqslant L\|x - y\|,$$

where L is a constant. For locally Lipschitzian (i.e., Lipschitzian in a neighborhood of every point) mappings of normed spaces, Gâteaux and Hadamard differentiabilities coincide.

4.1.12. Theorem. *Let X and Y be normed spaces and let $F\colon X \to Y$ be locally Lipschitzian. If F is Gâteaux differentiable at a point x, then at this point F is Hadamard differentiable and the corresponding derivatives coincide.*

PROOF. Let K be compact in X and let $\varepsilon > 0$. Let F satisfy the Lipschitz condition with constant L on the ball $B(x, r)$ with $r > 0$, let K be contained in the ball $B(0, R)$, and let $M := \max(L, R, \|DF(x)\|)$. We find a finite ε-net $h_1, \ldots, h_m$ in K. There is a number $\delta \in (0, r/R)$ such that, whenever $|t| < \delta$, for every $i = 1, \ldots, m$ we have

$$\|F(x + th_i) - F(x) - tDF(x)(h_i)\| \leqslant \varepsilon|t|.$$

Then, whenever $|t| < \delta$, for every $h \in K$ we obtain

$$\|F(x + th) - F(x) - tDF(x)(h)\| \leqslant \varepsilon|t| + 2M\varepsilon|t|,$$

since there exists h_i with $\|h - h_i\| \leqslant \varepsilon$, whence

$$\|F(x + th) - F(x + th_i)\| \leqslant M\|th - th_i\| \leqslant M\varepsilon|t|$$

and $\|tDF(x)h - tDF(x)h_i\| \leqslant M\varepsilon|t|$. Thus, F is Hadamard differentiable at x. Clearly, the Hadamard derivative serves as the Gâteaux derivative, since the latter is unique. □

In infinite-dimensional Banach spaces (and many other spaces), Fréchet differentiability is strictly stronger than Hadamard differentiability.

4.1.13. Proposition. *Let E be a separable locally convex space in which there is a bounded sequence $\{h_n\}$ that is not precompact. Then, for every sequence of nonzero numbers $t_n \to 0$, there exists a continuous $\bar{c}$-differentiable function f on E (even with a derivative f' that is continuous in the topology of convergence on compact sets in E') such that $f(0) = 0$, $f'(0) = 0$, but there is no limit of the sequence $f(t_n h_n)/t_n$.*

PROOF. Applying Exercise 4.10.53 and passing to a subsequence in $\{h_n\}$, we find an equicontinuous sequence $\{f_n\} \subset E'$ pointwise converging to zero such that $f_n(h_n) = 1$, $f_n(h_i)$ when $i < n$. This sequence converges to zero uniformly on compact sets. Choosing a subsequence in $\{t_n\}$, we can also assume that

$$|t_n| \sum_{i=1} |t_i|^{-1} < 4^{-n} \sup\{|f_j(h_i)|\colon\ i, j \in \mathbb{N}\}.$$

Let us take $\varphi_1, \varphi_2 \in C^\infty(\mathbb{R})$ such that $|\varphi_1| \leqslant 1$, $\varphi_1(t) = 1$ if $|t| \leqslant 1/4$, $\varphi_1(t) = 0$ if $|t| \geqslant 1/2$, $\varphi_2(t) = 0$ if $|t| \leqslant 1/2$, $\varphi_2(1) = 1$. Set

$$f(x) = \sum_{n=1}^{\infty} (-1)^n t_n \varphi_1\bigl(f_1(x)/t_1\bigr) \cdots \varphi_1\bigl(f_{n-1}(x)/t_{n-1}\bigr)\varphi_2\bigl(f_n(x)/t_n\bigr).$$

By the disjointness of supports of φ_1 and φ_2, for every x only one term can be nonzero; the same is true for the series of the derivatives along any fixed vector h. For the same reason these series converge uniformly in x. Moreover, for every compact set $K \subset E$, the series of the derivative along h converges uniformly in $x \in E$, $h \in K$, which is easily deduced from the indicated conditions on f_n, t_n. Thus, we obtain the continuity of f and the continuity of $f'\colon E \to E'$ when E' is equipped with the topology of convergence on compact sets. Finally, we have $f(0) = 0$, $f'(0) = 0$, $f(t_n h_n) = (-1)^n t_n$. □

4.1.14. Corollary. *On every infinite-dimensional separable normed space there is an Hadamard differentiable function that has no Fréchet derivative at the origin.*

It should be noted, however, that if a real function f on a Hilbert space X is everywhere Hadamard differentiable, then its derivative is bounded on some ball (this follows from Exercise 3.12.152), hence f is Lipschitzian on this ball, therefore, by a famous result of Preiss, has points of Fréchet differentiability (see Preiss [**392**] or Benyamini, Lindenstrauss [**50**]).

4.2. Examples

Here we consider a collection of illuminating examples of functions possessing (or failing to possess) various differentiability properties. By E and G we denote topological vector spaces (locally convex when this is appropriate).

4.2.1. Example. If $E \Subset E_0$ is one-dimensional, then σ_b-differentiability and $\sigma_{\rm fin}$-differentiability along the subspace E of a mapping from E_0 to an arbitrary topological vector space are equivalent. Hence in the case $\dim E = 1$ we shall call σ-differentiable (along the subspace E) mappings just differentiable.

4.2.2. Example. Let E_0 be a vector space, let $h \in E_0$, and let E be the one-dimensional subspace generated by h (as usual, equipped with the standard topology). A mapping $f\colon E_0 \to G$ is called *differentiable along* the vector h if it is differentiable along the subspace E; the derivative of f along h is the element $\partial_h f(x_0) := f'(x_0)h$ of the space G; it is clear that it is given by $\partial_h f(x_0) = \lim\limits_{t\to 0}[f(x_0+th) - f(x_0)]/t$. If $E = E_0 = \mathbb{R}$, then in accordance with the above convention the space $\mathcal{L}(E, G)$ is identified with G; the derivative $f'(x_0)$ is identified with the element $f'(x_0)1 \in G$. Since

$$f'(x_0)1 = \lim_{\tau\to 0}[f(x_0+\tau) - f(x_0)]/\tau,$$

this means that for $E = E_0 = G = \mathbb{R}$ the derivative $f'(x_0)$ is identified with the derivative (in the sense of elementary calculus) of a real function of a real argument. Thus, the usual derivative of a real function of a real argument is the derivative of a mapping from $\mathbb{R}$ to $\mathbb{R}$ along 1.

4.2.3. Example. If $E = \mathbb{R}^n$ and $G = \mathbb{R}$, then compact differentiability of a mapping $f\colon \mathbb{R}^n \to \mathbb{R}$, its Fréchet differentiability and differentiability in the sense of elementary calculus are the same.

4.2.4. Remark. In elementary calculus, along with a derivative, one defines also the so-called differential; its standard definition in old textbooks is this: "a differential is the main linear part of the increment of the function". Thus, the differential in the sense of old textbooks of calculus for a given value of the increment of the argument $h = (\Delta x_1, \dots, \Delta x_n) \in \mathbb{R}^n$ (we now consider the case of a finite-dimensional space E) is the value of the Fréchet derivative at the element h, so that it seems to be preferable, when using the term "differential" in this sense, to apply it in combinations like "differential at a given increment of the argument".

The term "differential" is employed also in modern mathematical texts, but mostly as a synonym for a derivative (usually, for mappings of normed spaces); yet another synonym is a *tangent mapping*.

Though, both terms are usually used not in the theory of smooth mappings of infinite-dimensional spaces, but in the theory of smooth manifolds. However, the term a "mapping tangent to zero" is used in the theory of smooth mappings (between normed spaces) as a synonym of a "σ_b-small mapping".

4.2.5. Example. If φ is a constant mapping of E to G (i.e., a mapping assuming the same value at all points), then φ is everywhere σ_b-differentiable and its derivative equals zero (i.e., the mapping taking every element of E to the zero element of G).

4.2.6. Example. If $\varphi \in \mathcal{L}(E, G)$, then φ is everywhere σ_b-differentiable and $\varphi'(x) = \varphi$ for all $x \in E$. If $\varphi \in \overline{\mathcal{L}}(E, G)$, then the operator φ is everywhere $\bar{\sigma}_b$-differentiable and $\varphi'(x) = \varphi$.

4.2.7. Example. Let φ be a linear mapping from E to G that is not continuous. Then φ is σ_b-differentiable at every point along every finite-dimensional subspace of E, but at no point is $\sigma_{\rm fin}$-differentiable (along the whole space E).

4.2.8. Example. Let g be a multilinear continuous real function on the product $E \times \cdots \times E$ of several copies of E. Then the function $g_1\colon x \mapsto g(x, \dots, x)$ is σ_b-differentiable at every point and

$$g_1'(x)h = g(h, x, \dots, x) + g(x, h, \dots, x) + \cdots + g(x, x, \dots, h).$$

4.2.9. Example. Let r, φ be polar coordinates on the Euclidean plane. The function $(r, \varphi) \mapsto r \sin^3 \varphi$, defined everywhere on $\mathbb{R}^2$, is differentiable at every point along every vector, but at the origin is not even Gâteaux differentiable.

4.2.10. Example. Define a function $g\colon \mathbb{R}^2 \to \mathbb{R}$ as follows. For any vector $x = (x_1, x_2) \in \mathbb{R}^2$ with $x \neq 0$ and $x_2 = x_1^2$ let $g(x) = 1$; else $g(x) = 0$. Then at the origin the function g is Gâteaux differentiable (and $g'(0) = 0$), but is not Fréchet differentiable.

4.2.11. Example. Let r, φ be polar coordinates on the Euclidean plane and let g be the real function on $\mathbb{R}^2$ defined by

$$g(r,\varphi) = \frac{r^2}{|\sin\varphi|}\exp\Bigl(-\frac{r}{|\sin\varphi|}\Bigr) \quad \text{if } \varphi \neq \pi k,$$

$$g(r,\varphi) = 0 \text{if } \varphi = \pi k \quad (k \in \mathbb{Z}).$$

This function is Gâteaux differentiable at every point, but is not Fréchet differentiable at zero.

4.2.12. Example. Let H be a real Hilbert space. The real function f on H defined by the equality $f(x) = \|x\|$ is Fréchet differentiable at every point excepting zero and $f'(x) = x/\|x\|$, i.e., $f'(x)h = (x,h)/\|x\|$.

4.2.13. Example. Let f be the function on a separable Hilbert space H defined as follows: if $x = te_n$, where $\{e_n\}$ is an orthonormal basis in H, then $f(x) = t^{1+1/n}$; $f(x) = 0$ at all other points $x \in H$. The function f is compactly differentiable at the point $x = 0$, but is not Fréchet differentiable at this point.

4.2.14. Example. The function

$$f\colon\ L^1[0,1] \to \mathbb{R}^1, \quad f(x) = \int_0^1 \sin x(s)\,ds \tag{4.2.1}$$

is everywhere Hadamard differentiable, but nowhere Fréchet differentiable. The same is true for the mapping

$$F\colon\ L^2[0,1] \to L^2[0,1], \quad F(x)(s) = \sin x(s). \tag{4.2.2}$$

PROOF. When proving differentiability it is often useful to find a candidate for the derivative, which is done by evaluating partial derivatives. For the function f we have the following equality:

$$f(x+th) = \int_0^1 \sin[x(s) + th(s)]\,ds.$$

It can be differentiated in t with the aid of the Lebesgue dominated convergence theorem, which gives

$$\partial_h f(x) = \int_0^1 h(s)\cos x(s)\,ds.$$

Clearly, the Gâteaux derivative exists and is given by the functional

$$Df(x)h = \int_0^1 h(s)\cos x(s)\,ds.$$

Since $\|Df(x)\| \leqslant 1$, by the aid of the mean value theorem for functions on the real line we conclude that the function f is Lipschitzian (certainly, this can be verified directly). By Theorem 4.1.12 we obtain its Hadamard differentiability. For the mapping F the reasoning is similar. Here we have the operators $DF(x)$ in $L^2[0,1]$ such that

$$\bigl(DF(x)h\bigr)(s) = \bigl(\cos x(s)\bigr)h(s).$$

Let us investigate whether f and F are Fréchet differentiable. Let $x = 0$. Then $f(x) = 0$. For Fréchet differentiability, we have to check if the relation $f(h) - Df(0)h = o(\|h\|)$ is true. The left-hand side equals

$$\int_0^1 [\sin h(s) - h(s)]\, ds.$$

Since the Taylor expansion of $\sin h(s) - h(s)$ begins with h^3 and our space is L^1, we may suspect that there is no Fréchet differentiability here. To justify this, we take for h elements of the unit ball on which $f(th) - tDf(0)h$ is not uniformly $o(t)$. Namely, let $h_k(s) = k$ if $0 \leqslant s \leqslant 1/k$ and $h_k(s) = 0$ if $s > 1/k$. Then

$$f(th_k) - tDf(0)h_k = k^{-1} \sin kt - t.$$

This quantity is not $o(t)$ uniformly in k: it suffices to set $t = k^{-1}$, which gives the value $t(\sin 1 - 1)$. For an arbitrary point x our reasoning is similar. We fix a version of x and consider

$$\int_0^1 \Bigl(\sin\bigl(x(s) + th(s)\bigr) - \sin x(s) - th(s) \cos x(s)\Bigr)\, ds.$$

The functions $\cos x(s)$ and $\sin x(s)$ have a common Lebesgue point $s_0 \in (0, 1)$. For every ε this point is a Lebesgue point also for $\sin\bigl(x(s) + \varepsilon\bigr) - \sin x(s) - \varepsilon \cos x(s)$. Choose a number $\varepsilon \in (0, 1)$ such that $\sin\bigl(x(s_0) + \varepsilon\bigr) - \sin x(s_0) - \varepsilon \cos x(s_0) \neq 0$. Now let us set $h_k = kI_{E_k}$, where $E_k = (s_0 - k^{-1}, s_0 + k^{-1})$. For $t = \varepsilon k^{-1}$ we obtain the quantity

$$\int_{E_k} \Bigl(\sin\bigl(x(s) + \varepsilon\bigr) - \sin x(s) - \varepsilon \cos x(s)\Bigr)\, ds$$

of order of smallness $Lk^{-1} = L^{-1}t$, where $L \neq 0$ is some number, since the limit of this quantity multiplied by $k/2$ equals

$$\sin\bigl(x(s_0) + \varepsilon\bigr) - \sin x(s_0) - \varepsilon \cos x(s_0) \neq 0 \quad \text{as } k \to \infty.$$

Similar estimates work in the case of F. □

It is interesting to note that if the function f is considered not on L^1, but on L^2, then it becomes Fréchet differentiable.

4.2.15. Example. The function f defined by formula (4.2.1) on the space $L^2[0, 1]$ is everywhere Fréchet differentiable. The mapping F defined by formula (4.2.2) on $C[0, 1]$ is everywhere Fréchet differentiable.

PROOF. A nuance making difference between the properties of f on the spaces L^1 and on L^2 is that, due to the bound $|\sin(x + h) - \sin x - h\cos x| \leqslant h^2$, the quantity $|f(x + h) - f(x) - Df(x)h|$ is estimated by the integral of h^2, which equals the square of the L^2-norm (infinite for some h in L^1). A similar reasoning applies to the mapping F on the space $C[0, 1]$. Here the quantity $\|F(x + h) - F(x) - DF(x)h\|$ is estimated by $\|h\|^2$ in the case of the sup-norm, but not in the case of the L^2-norm, where the indicated estimate leads to the integral of h^4. □

Let us consider yet another instructive infinite-dimensional example. It employs a function that is frequently used in applications: the distance to a set.

4.2.16. Example. Let X be an infinite-dimensional normed space and let K be a compact set. Set

$$f(x) = \operatorname{dist}(x, K) = \inf\{\|x - y\|: \ y \in K\}.$$

Then the function f satisfies the Lipschitz condition, but is not Fréchet differentiable at the points of K.

If K has the property that $\alpha K \subset K$ whenever $|\alpha| \leqslant 1$ and the set $\bigcup_{n=1}^{\infty} nK$ is everywhere dense in X, then f has a zero Gâteaux derivative at the point $0 \in K$. For example, one can take for K the ellipsoid

$$K = \Bigl\{(x_n) \in l^2: \ \sum_{n=1}^{\infty} n^2 x_n^2 \leqslant 1\Bigr\}$$

in the Hilbert space l^2.

PROOF. Let $x \in K$. Then $f(x) = 0$. Suppose that at x there exists a Fréchet derivative $f'(x)$. This derivative can be only zero, since, for any nonzero vector h, the function $t \mapsto f(x + th)$ has minimum at $t = 0$. We arrive at a contradiction if we show that $f(x+h) - f(x) - f'(x)h = f(x+h)$ is not $o(\|h\|)$. For every $n \in \mathbb{N}$ we find a vector h_n such that $\|h_n\| \leqslant 1/n$ and the ball of radius $\|h_n\|/4$ centered at $x + h_n$ does not intersect K. This will give the estimate $f(x + h_n) \geqslant \|h_n\|/4$. The compact set K can be covered by finitely many balls of radius $(4n)^{-1}$ with centers at some points $a_1, \ldots, a_k$. Let L be the finite-dimensional linear space generated by these centers. There is a vector h_n with $\|h_n\| = 1/n$ and $\operatorname{dist}(h_n, L) = 1/n$. This vector is what we look for. Indeed, if there is a vector $y \in K \cap B(x + h_n, \|h_n\|/4)$, then we obtain the following decomposition: $x = u + l_1$, $y = v + l_2$, where $l_1, l_2 \in L$, $\|u\| \leqslant (4n)^{-1}$, $\|v\| \leqslant (4n)^{-1}$ and $\|x + h_n - y\| \leqslant \|h_n\|/4$. Therefore,

$$\|h_n - (l_2 - l_1) + u - v\| \leqslant (4n)^{-1}$$

and hence $\|h_n - (l_2 - l_1)\| \leqslant 3(4n)^{-1}$, contrary to our choice of h_n, since we have $l_2 - l_1 \in L$.

Suppose now that K satisfies the indicated additional conditions. We show that at the point $0 \in K$ the Gâteaux derivative exists and equals zero. To this end we have to verify that for any fixed $h \in X$ we have the equality $\lim_{t\to 0} t^{-1} f(th) = 0$. Let $\varepsilon > 0$. By assumption there exists a vector $v \in nK$ such that $\|h - v\| \leqslant \varepsilon$. Since $tv \in K$ whenever $|t| \leqslant n^{-1}$ by our condition, we have $f(tv) = 0$ for such t. Hence $|t^{-1} f(th)| \leqslant \varepsilon$ by the estimate $|f(th) - f(tv)| \leqslant \|th - tv\| \leqslant |t|\varepsilon$, that holds by the Lipschitzness of f. □

It is suggested in Exercise 4.10.43 to verify that if the set K is convex, then f has zero Gâteaux derivative at all points of $\bigcup_{0\leqslant t<1} tK$.

4.2.17. Example. Let E be the product of the continuum copies of $\mathbb{R}$ realized as the space of all real functions on $[0, 1]$ with the topology of pointwise convergence and let $K = \{x_n\}$ be the countable subset of E consisting of the indicators of sets that are unions of finitely many intervals in $[0,1]$ with rational endpoints and having Lebesgue measure greater than $1/2$. The set K is relatively compact in E; at the same time, there is a sequence of elements of the set K not containing Cauchy subsequences. If $x = tx_n$, then we set $f(x) = t^{1+1/n}$; at all other points $x \in E$ we set $f(x) = 0$. The function f is sequentially differentiable (i.e., s-differentiable) at the point $x = 0$, but is not differentiable at this point with respect to the system of compact sets.

4.2.18. Example. Let $\mathcal{D}$ be the space of all infinitely differentiable functions with compact support on the real line, let τ_1 be its standard topology, and let τ_2 be the topology $\sigma(\mathcal{D}, \mathcal{D}')$. Then the quadratic mapping $f\colon (\mathcal{D}, \tau_2) \to (\mathcal{D}, \tau_1)$ taking the function φ from $\mathcal{D}$ to the function $\varphi^2 = [t \mapsto \varphi^2(t)]$ is Fréchet differentiable at the point $0 \in \mathcal{D}$, but is continuous at no point.

4.2.19. Example. Let $\mathcal{D}$ and τ_1 be the same as in the previous example and let $\varphi \in \mathcal{D}$, $\varphi \neq 0$. For all numbers $m, k \in \mathbb{N}$ and $t \in \mathbb{R}$ we set $x_{mk}(t) = m^{-1}\varphi(t) + k^{-1}\varphi(t-m)$. Then the indicator function of the set $S = \{x_{mk}\}$ is a real function on the space $(\mathcal{D}, \tau_1)$ that is Fréchet differentiable at the point 0, but is not continuous at this point.

4.2.20. Example. Let $\mathcal{D}$ and τ_1 be the same as in the previous two examples and let $\mathcal{D}'$ be the dual to $(\mathcal{D}, \tau_1)$ equipped with the strong topology; in place of the symbol $(\mathcal{D}, \tau_1)$ we shall now use the symbol $\mathcal{D}$. The bilinear function $\Phi\colon \mathcal{D}' \times \mathcal{D} \to \mathbb{R}$, $(g, x) \mapsto g(x)$ is discontinuous at every point, but is everywhere infinitely Fréchet differentiable. Moreover, we have the easily verified equalities $\Phi'(g, x) = [(h, k) \mapsto h(x) + g(k)]$, $\Phi''(g, x) = \Phi'$, $\Phi^{(n)} = 0$ if $n > 2$, so that all derivatives are continuous as mappings from $\mathcal{D}' \times \mathcal{D}$ to $\mathcal{L}_b^n(\mathcal{D}' \times \mathcal{D}, \mathbb{R})$.

4.2.21. Example. Let E be a topological vector space that is a vector subspace of the vector space E_0, E_1 a topological vector subspace in E, f a mapping of a set W in E_0 to a topological vector space G, σ and σ_1 some systems of bounded subsets of E containing all finite subsets, and $\sigma' = E_1 \cap \sigma$. Then the σE-differentiability of the mapping f at a point $x \in W$ yields its $\sigma_1 E$-differentiability at the same point if σ_1 is one of the following systems of sets:

1) a subsystem of σ;
2) the system of unions of some finite collections of sets in σ;
3) the system of images of the sets in σ under all homotheties of the space E;
4) the system of circled envelopes of sets in σ.

It is worth noting that it is not always possible to replace σ by the system of closures of all sets in σ or by the system of their convex envelopes.

If $\sigma_2 \subset \sigma'$, then the σE-differentiability (at a point $x \in W$) of the mapping f yields its $\sigma_2 E_1$-differentiability at this point.

4.2.22. Example. Let E be an infinite-dimensional barrelled locally convex space, let E_σ be the same vector space equipped with the weak topology $\sigma(E, E')$, and let E' be the dual to E equipped with the topology $\sigma(E', E)$. The mapping $E' \times E \to \mathbb{R}$, $(g, x) \mapsto g(x)$ is everywhere discontinuous, everywhere Fréchet differentiable, everywhere infinitely $\bar{b}$-differentiable, but is twice Fréchet differentiable at no point.

4.3. Differentiability and continuity

In the previous section some examples have been given demonstrating that even everywhere infinitely Fréchet differentiable real functions can be everywhere discontinuous. Here, among other things, we describe the class of all spaces for mappings of which to an arbitrary topological vector space Fréchet differentiability at a given point yields the continuity at this point. This class contains all metrizable spaces, but not only such spaces.

4.3.1. Proposition. *Let* $\dim G > 0$. *Every mapping from* E *to* G *that is* $\bar{s}$*-differentiable at a given point is sequentially continuous at this point precisely when* E *satisfies the following condition*:

(A_1) *for every sequence* $\{a_n\}$ *of elements of the space* E *converging to zero, there exist its subsequence* $\{a_{n_k}\}$ *and a sequence* $\{\lambda_k\}$ *of real numbers such that* $\lambda_k \to \infty$ *and* $\lambda_k a_{n_k} \to 0$.

PROOF. Let $f\colon E \to G$ be $\bar{c}$-differentiable at a point $x_0 \in E$. We show that if condition (A_1) is fulfilled, then f is sequentially continuous at x_0. We can assume that $f \in \mathcal{R}^s(E,G)$ (i.e., that $x_0 = 0$, $f(0) = 0$, $f'(0) = 0$). Let $\{h_n\}$ be a sequence of elements of E converging to zero. If $f(h_n) \not\to 0$, then there exists a strictly increasing sequence $\{n_k\}$ of natural numbers such that no subsequence in the sequence $\{f(h_{n_k})\}$ converges to zero. However, by condition (A_1) there exist a subsequence $\{n_{k(i)}\}$ in $\{n_k\}$ and a sequence $\{\lambda_i\}$ of real numbers such that $\lambda_i \to \infty$ and $\lambda_i h_{n_{k(i)}} \to 0$. These relations and the inclusion $f \in \mathcal{R}^s(e,G)$ yield that

$$f(h_{n_{k(i)}}) = \lambda_i^{-1}\lambda_i f\bigl(\lambda_i^{-1}(\lambda_i h_{n_{k(i)}})\bigr) \to 0 \quad \text{as } i \to \infty,$$

which contradicts our choice of the sequence $\{n_k\}$.

Suppose now that condition (A_1) is not fulfilled and that $\{h_n\}$ is a sequence of nonzero elements of E converging to zero such that if a sequence of real numbers $\{\lambda_i\}$ tends to infinity and $\{n_i\}$ is a strictly increasing sequence of natural numbers, then $\lambda_i h_{n_i} \not\to 0$. Let S be the set of all elements of the sequence $\{h_n\}$, I_S its indicator function, $a \in G$, $a \neq 0$, and $g(x) = I_S(x)a$, $x \in E$. Then the mapping $g\colon E \to G$ is $\bar{s}$-differentiable (and even $\bar{b}$-differentiable) at the origin, but is not sequentially continuous at this point. □

4.3.2. Remark. As the proof of the proposition above shows, it remains valid if in the formulation the term "$\bar{s}$-differentiability" is replaced by the term "$\bar{b}$-differentiability".

Note that a topological vector space E satisfies condition (A_1) if it satisfies the "Mackey convergence condition" from Definition 3.6.4 (and Lemma 2.10.18):

for every sequence $\{x_n\}$ *of elements of* E *converging to zero, there exists a sequence* $\{r_n\}$ *of real numbers such that* $r_n \to \infty$, *but* $r_n x_n \to 0$.

We do not know whether these two conditions are equivalent.

We recall (see § 2.10(i)) that a topological space is called a *Fréchet–Urysohn space* if, for every part A of this space and every point a in the closure of A, there exists a sequence of elements of the set A converging to a.

The sequential continuity of a mapping from a Fréchet–Urysohn space implies its continuity, since the preimage of every closed set is closed.

4.3.3. Lemma. *Let E be a Fréchet–Urysohn topological vector space. Suppose that for every $k \in \mathbb{N}$ we are given a sequence $\{a_n^k\}$ of elements in E converging to zero. Then, there exist strictly increasing sequences of natural numbers $\{k(r)\}$ and $\{n(r)\}$ for which $a_{n(r)}^{k(r)} \to 0$ as $r \to \infty$.*

PROOF. We can assume that $\dim E \neq 0$. Let $x_0 \in E$, $x_0 \neq 0$. For any $n, k \in \mathbb{N}$ we set $x_{nk} = k^{-1}x_0 + a_{n+k}k$ if $k^{-1}x_0 \neq -a_{n+k}^k$ and $x_{nk} = k^{-1}x_0$ otherwise. Let also $A = \{x_{nk}\}$. Then $0 \in \overline{A}\backslash A$. It is obvious that $0 \notin A$.

Let now V be a neighborhood of zero in E. Then, there exists a neighborhood of zero V_0 in E such that $V_0 + V_0 \subset V$. Let k and n be natural numbers for which $kx_0 \in V_0$ and $a_{n+k}^k \in V_0$ (the existence of n for which the latter inclusion is valid follows from the fact that $a_{n+k}^k \to 0$ as $n \to \infty$). Then $x_{nk} = k^{-1}x_0 + a_{n+k}^k \in V$.

Since E is a Fréchet–Urysohn space, there are sequences $\{k_1(r)\}$ and $\{n_1(r)\}$ of natural numbers such that $x_{n_1(r)k_1(r)} \to 0$. We have $k_1(r) \to \infty$. Indeed, otherwise, passing to a subsequence in $\{k_1(r)\}$, we could assume that $k_1(r) = k_0$ is constant. Then we would get $n_1(r) \to \infty$, hence $a_{n_1(r)}^{k_0} \to -k_0x_0 \neq 0$, which contradicts the fact that by condition we have $a_{n+k_0}^{k_0} \to 0$. Passing to a subsequence, we can assume that $k_1(r) < k_1(r+1)$ and $n_1(r) + k_1(r) < n_1(r+1) + k_1(r+1)$ for all r. Set now $k(r) = k_1(r)$ and $n(r) = k_1(r) + n_1(r)$. Then $a_{n(r)}^{k(r)} \to 0$ as $r \to \infty$, and the sequences of indices strictly increase. □

A sequence $\{a_{n(r)}^{k(r)}\}$ for which $n(1) < n(2) < \cdots$ and $k(1) < k(2) < \cdots$ is naturally called *quasi-diagonal*. Thus, the lemma asserts that from any sequence of sequences converging to zero one can extract a quasi-diagonal sequence also converging to zero. Not every space has this property.

4.3.4. Corollary. *Every Fréchet–Urysohn topological vector space satisfies condition* (A_1) *from Proposition* 4.3.1.

PROOF. Let $\{x_n\}$ be a sequence of elements of a topological vector Fréchet–Urysohn space converging to zero. In order to prove that there exist its subsequence $\{x_{n_i}\}$ and a sequence $\{\lambda_i\}$ of real numbers for which $\lambda_i \to \infty$ and $\lambda_i x_{n_i} \to 0$, it suffices to apply the lemma to the sequence of sequences $\{kx_n\}$ converging to zero (for each fixed k). □

4.3.5. Theorem. *Let E and G be topological vector spaces,* $\dim G > 0$*, and let β be some system of bounded subsets E such that $\beta \supset \sigma_s^E$. In order that every mapping from E to G that is β-differentiable at some point be continuous at this point it is necessary and sufficient that E be a Fréchet–Urysohn space.*

PROOF. Let E be a Fréchet–Urysohn space and let a mapping f from E to G be β-differentiable at a point $x_0 \in E$. Since $\beta \supset \sigma_s^E$, this mapping is

$\bar{s}$-differentiable at the point x_0. By the above corollary, the space E satisfies condition (A$_1$) from Proposition 4.3.1, therefore, the mapping f is sequentially continuous at the point x_0. Since E is a Fréchet–Urysohn space, it follows that f is continuous at this point.

Suppose that E is not a Fréchet–Urysohn space. Let $A \subset E$ have a limit point x_0 that is not a limit of a convergent sequence of elements of A. Let I_A be the indicator of A, $b \in G$, $b \neq 0$. Then the function

$$g\colon x \mapsto I_A(x)b, \quad E \to G,$$

is obviously discontinuous at the point x_0 and is β-differentiable at this point. Since $\sigma_b^E \supset \beta$, it is sufficient to prove that f is b-differentiable at x_0. Let $\{t_n\}$ be a sequence of nonzero real numbers converging to zero and let $\{h_n\}$ be a bounded sequence of elements of E. Since the set A can contain only finitely many points of the sequence $\{x_0+t_nh_n\}$, we have $t_n^{-1}\bigl(g(x_0+t_nh_n)-g(x_0)\bigr) \to 0$ as $n \to \infty$. This means that the mapping g is b-differentiable at the point x_0 (and has the derivative at this point). □

Yet another condition close to condition (A_1) and the Mackey condition will be used in § 4.10(i).

It has been shown above that only for mappings defined on Fréchet–Urysohn topological vector spaces s-differentiability at a point implies the continuity at this point (hence this is also true for differentiability with respect to the system of bounded sets). However, it can happen that a topological vector space E has the following property: c-differentiability of an arbitrary mapping from E to an arbitrary topological vector space G at a single point does not imply its continuity, but s-differentiability at every point of an arbitrary open set already implies the continuity on this set. One class of topological vector spaces possessing such a property is described below.

4.3.6. Theorem. *Let E be the inductive limit of an increasing sequence of Banach spaces E_n such that the embeddings $E_n \to E_{n+1}$ are compact. Then every mapping f from the space E to a topological vector space G that is s-differentiable in some neighborhood of a point $x \in E$ is continuous at this point.*

PROOF. Suppose that a mapping $f\colon E \to G$ is s-differentiable at all points of some neighborhood of the point x_0. Let us take in this neighborhood a closed neighborhood V of x_0. We prove that the restriction of f to V is continuous (this will give the continuity of the original mapping in every inner point of V).

Suppose the contrary. Then G contains a closed set B such that the set $A = f^{-1}(B) \cap V$ is not closed in V; then it is not closed in E. Under the assumptions of the theorem, a subset F of the space E is closed precisely when for every n the intersection $F \cap E_n$ is closed in E_n (Corollary 2.7.7). Hence for some n the set $A \cap E_n$ is not closed in E_n. This means that the set $A \cap (V \cap E_n)$ is not closed in $V \cap E_n$, therefore, the restriction f_n of the mapping f to the subspace E_n is not a continuous function on the set $V \cap E_n$. However, the mapping f_n is obviously s-differentiable at every point of the set $V \cap E_n$ in the Banach space E_n.

By the previous theorem it must be continuous on the set $V \cap E_n$. The obtained contradiction completes the proof. □

The space E in this theorem is a Fréchet–Urysohn space precisely when it is finite-dimensional. Hence inductive limits of infinite-dimensional strictly increasing Banach spaces with compact embeddings form a class of spaces mentioned before the theorem. Indeed, if E is infinite-dimensional, then $E_n \neq E_{n+1}$ for infinitely many indices n (otherwise not all canonical embeddings $E_n \to E_{n+1}$ are compact). We can assume that $E_n \neq E_{n+1}$ for all n. Let $x_n \in E_{n+1} \backslash E_n$ for every natural number n. If E is a Fréchet–Urysohn space, then by Lemma 4.3.3 applied to the sequence of sequences $\{k^{-1}x_n\}_{k=1}^{\infty}$ converging to zero, there exist strictly increasing sequences $\{n(i)\}$ and $\{k(i)\}$ of natural numbers such that $k(i)^{-1}x_{n(i)} \to 0$ in E. Then there exists m such that $k(i)^{-1}x_{n(i)} \in E_m$ for all i (Corollary 2.7.9), which contradicts the fact that $n(i) \to \infty$.

4.3.7. Remark. We have discussed above a possibility to obtain the usual continuity from rather weak (in nonnormable spaces) forms of differentiability; it does not come as a surprise that this is not always possible. However, the situation changes if we weaken the concept of continuity. Suppose we are given a class $\mathcal{A}$ of bounded sets in a topological vector space X. A sequence $\{x_n\}$ will be called $\mathcal{A}$-converging to x if there is a sequence of numbers $t_n \to 0$ such that $(x_n - x)/t_n \in A$ for some $A \in \mathcal{A}$.

Notation: $x_n \xrightarrow{\mathcal{A}} x$.

For example, the Mackey convergence from Definition 2.10.17 corresponds to $\mathcal{A}$ equal to the class of all Banach discs (see Lemma 2.10.18). In a metrizable space, the usual convergence is equivalent to $\mathcal{A}$-convergence for every system $\mathcal{A}$ containing all compact sets (this is true for the broader class of spaces satisfying the Mackey convergence condition mentioned on page 257). In the general case, $\mathcal{A}$-convergence obviously yields convergence in the topology of X.

A mapping f from X to a locally convex space Y will be called $\mathcal{A}$-continuous at x_0 if $f(x_n) \to f(x_0)$ as $x_n \xrightarrow{\mathcal{A}} x_0$. For spaces satisfying the Mackey convergence condition, $\mathcal{A}$-continuity is equivalent to the sequential continuity if $s \subset \mathcal{A}$. On an infinite-dimensional Hilbert space equipped with the weak topology, the function $f(x) = (x, x)$ is discontinuous, but is $\mathcal{A}$-continuous for the class $\mathcal{A}$ of all bounded sets.

Note that $\widetilde{\mathcal{A}}$-differentiability of f at x_0 implies $\mathcal{A}$-continuity at this point. Indeed, assuming that $x_0 = 0$ and $f(x_0) = 0$, whenever $h_n \xrightarrow{\mathcal{A}} 0$, we obtain $f(h_n) = f(t_n h_n/t_n) = f'(0)(t_n h_n/t_n) + r(t_n h_n/t_n)$ for the corresponding numbers $t_n \to 0$ such that $h_n/t_n \in A$, where $A \in \mathcal{A}$. Since $f'(0) \in \mathcal{L}_{\mathcal{A}}(X, Y)$, we obtain that $f(h_n) \to 0$.

For an open set $U \subset E$, by induction we introduce the classes $C^n_{\mathcal{A}}(U, G)$ of mappings n-fold $\widetilde{\mathcal{A}}$-differentiable in U and having $\mathcal{A}$-continuous derivatives. For example, $C^1_{\mathcal{A}}(U, G)$ consists of $\widetilde{\mathcal{A}}$-differentiable mappings f such that the mapping $x \mapsto f'(x)$ from U to $\mathcal{L}_{\mathcal{A}}(E, G)$ is $\mathcal{A}$-continuous. Then $f \in C^n_{\mathcal{A}}(U, G)$ provided that $f \in C^1_{\mathcal{A}}(U, G)$ and $f' \in C^{n-1}_{\mathcal{A}}(U, L_{n-1})$.

For normed spaces and the class $\mathcal{A} = b$, this leads to the usual class of mappings $C^n(U, G)$ possessing n continuous Fréchet derivatives.

4.4. Differentiability and continuity along a subspace

Let E_0 be a vector space and let E and G be topological vector spaces such that E is a vector subspace in E_0.

4.4.1. Definition. *A mapping $f\colon E_0 \to G$ is called continuous at a point $x \in E_0$ along the subspace E (or E-continuous) if the mapping $h \mapsto f(x+h)$ from E to G is continuous at zero. A similar terminology is introduced for the sequential continuity.*

4.4.2. Definition. *A mapping $f\colon E_0 \to G$ is called σ-continuous at a point $x \in E_0$ along the subspace E (or σE-continuous at x) if, for every set $B \in \sigma$, we have $f(x+th) \to f(x)$ as $t \to 0$ uniformly in $h \in B$.*

If $E = E_0$, then in place of the term "σE-continuity" we use the term "σ-continuity".

Note that σE-continuity at a point $x \in E_0$ is equivalent to the continuity at the point $t = 0$ of the mapping $t \mapsto [h \mapsto f(x+th)]$, $\mathbb{R} \to \mathcal{F}_\sigma(E, G)$. In the language of the theory of pseudotopological spaces one can similarly reformulate Definition 4.4.1; this will be noted below.

Definitions 4.4.1 and 4.4.2 are actually local and can be easily formulated for mappings defined on the set $V+x$, where $V \subset E$ is a neighborhood of zero. Note also that the mappings σ-continuous at a given point form a vector space and that the same is true for the mappings σE-continuous at a given point.

The next assertions are analogs of the results obtained above for differentiable mappings.

4.4.3. Proposition. *In order a mapping $f\colon E_0 \to G$ be E-continuous at a point $x \in E_0$, the following conditions are necessary and sufficient: whatever be a set $B \in \sigma$, a sequence $\{h_n\}$ of its elements and a sequence of real numbers $\{t_n\}$ converging to zero, one has $f(x+t_nh_n) \to f(x)$ in G as $n \to \infty$.*

The proof is similar to the proof of Proposition 4.3.1.

4.4.4. Proposition. (i) *Every sequentially continuous mapping from E to G is σ-continuous.* (ii) *Every E-continuous mapping from E_0 to G is σE-continuous.* (iii) *Every bounded linear mapping from E to G is σ-continuous.*

All these assertions follow immediately from the definitions (for the proof of (ii) one should also use that the continuity implies the sequential continuity).

The next fact is also a straightforward corollary of the definition.

4.4.5. Proposition. *Any mapping $f\colon E_0 \to G$ that is σE-differentiable at some point is σE-continuous at this point.*

4.4.6. Proposition. *Let* $\dim G > 0$. *Then all mappings from E_0 to G that are sE-continuous at a given point are E-sequentially continuous at this point precisely when E satisfies Condition (A_1) from Proposition* 4.3.1.

The proof of this proposition is similar to the proof of Proposition 4.3.1 above.

4.4.7. Theorem. *Let* $\dim G > 0$ *and* $\sigma \supset \sigma_s^E$*. Then every mapping from* E_0 *to* G *that is* σE*-continuous at a given point is* E*-continuous at this point precisely when* E *is a Fréchet–Urysohn space.*

The proof is similar to the proof of Theorem 4.3.5.

In connection with continuity, let us prove a result on differentiability of the inverse mapping. Some additional remarks will be made in the last section. For arbitrary topological vector spaces, the theorem on existence of a (local) mapping inverse for a differentiable mapping and the theorem on differentiability of the inverse mapping are false (for the regarded definitions of differentiability); the corresponding examples will be given later. Here we prove a simple assertion the main advantage of which is that it is valid for all topological vector spaces.

Let E be a topological vector space, σ some system of its bounded subsets containing all singletons. A mapping $g\colon E \to E$ is called *strongly* σ*-continuous* at a point $x \in E$ if, for every set $B \in \sigma$, every sequence of real numbers $\{t_n\}$ converging to zero and every sequence $\{h_n\} \subset B$, there exists a set $P \in \sigma$ such that we have the inclusion $g(x + t_n h_n) \in g(x) + t_n P$.

If $\sigma \supset \sigma_s^E$, then every mapping that is σ-differentiable at a point $x \in E$ is strongly σ-continuous at this point; in addition, the strong σ-continuity implies σ-continuity. However, even for the space $E = \mathbb{R}$ the strong σ-continuity does not follow from continuity.

4.4.8. Theorem. *Let* g *be a one-to-one mapping from an open subset* V *of a topological vector space* E *onto an open subset* W *of the same space. Suppose that the mapping* g *is* σ*-differentiable at a point* $x \in V$*, that* $g'(x)$ *is a linear homeomorphism of the space* E*, and that the mapping* g^{-1} *inverse to* g *is strongly* σ*-continuous at the point* $z = g(x)$*. Then the mapping* g^{-1} *is* σ*-differentiable at the point* z *and* $(g^{-1})'(z) = \bigl(g'(x)\bigr)^{-1}$.

PROOF. Without loss of generality we can assume that $x = z = 0$ and that $g'(x)$ is the identity mapping. Thus, $g(h) = h + r(h)$ and $r \in \mathcal{R}^\sigma(E, E)$. Set $g^{-1}(p) = p + s(p)$. We have to show that $s \in \mathcal{R}^\sigma(E, E)$. Clearly, we have $s(p) = -r\bigl(g(p)\bigr)$. Let $Q \in \sigma$, $p_n \in Q$, $t_n \in \mathbb{R}$, $t_n \neq 0$, $t_n \to 0$. Since the mapping g^{-1} is strongly σ-continuous at zero, there exists $B \in \sigma$ such that $t_n^{-1} g^{-1}(t_n p_n) \in B$ for all n. Therefore,

$$t_n^{-1} s(t_n p_n) = -t_n^{-1} r\bigl(t_n \bigl(t_n^{-1} g(t_n p_n)\bigr)\bigr) \to 0 \quad \text{as } n \to 0.$$

This proves that $s \in \mathcal{R}^\sigma(E, E)$. □

4.4.9. Remark. If $\sigma = \sigma_b^E$ and E is a normed space, then the assumption of the strong σ-continuity of the mapping g^{-1} in the theorem can be replaced by the assumption of its continuity. In the general case, as we shall see in § 4.10(i), this is impossible.

Note also that the conditions of the theorem do not imply the continuity of the mapping g.

4.4.10. Example. As in Example 4.2.19, let

$$x_{mk}(t) = m^{-1}\varphi(t) + k^{-1}\varphi(t-m),\ \varphi \in \mathcal{D},\ \varphi \neq 0,\ m, k \in \mathbb{N},\ t \in \mathbb{R},$$

$$z_{mk}(t) = k^{-1}\varphi(t-m).$$

Let $\Phi\colon\ \mathcal{D} \to \mathcal{D}$ be defined as follows: $\Phi(x_{mk}) = z_{mk}$, $\Phi(z_{mk}) = x_{mk}$, $\Phi(x) = x$ at other points. Then Φ is a one-to-one mapping of $\mathcal{D}$ onto $\mathcal{D}$, but Φ and Φ^{-1} are σ_b-differentiable at zero and discontinuous at zero.

4.5. The derivative of a composition

In this section, given a topological vector space E, the symbol σ_E will be used to denote a family of its bounded subsets containing all finite subsets (not necessarily all bounded subsets, which is the largest possible family) and the symbol $\mathcal{E}$ will be used to denote a fixed class of topological vector spaces. We shall say that in the class $\mathcal{E}$ the chain rule holds for σ-derivatives of order n if the following condition is fulfilled: whatever are $E, G, F \in \mathcal{E}$, a mapping $f\colon\ E \to G$ that is n-fold σ-differentiable at a point $x \in E$ and a mapping $g\colon\ G \to F$ that is n-fold σ_G-differentiable at the point $f(x)$, the composition $g\circ f\colon\ E \to F$ is n-fold σ_E-differentiable at the point x. Below in place of the symbol σ_E we usually employ the shortened symbol σ. Our discussion of compositions will be continued in § 4.9, where we consider the converse problem of characterization of differentiability of mappings through differentiability of its compositions (left or right).

(i) First order derivatives

We start with a result related to differentiability with respect to the system of compact sets (with an additional requirement of the sequential continuity of derivatives) that is readily verified directly (see Bogachev, Smolyanov [**72**, Theorem 12.2.3]).

4.5.1. Theorem. *Let X, Y and Z be three Hausdorff locally convex spaces, $\Psi = G\circ F\colon\ X \to Z$, where $F\colon\ X \to Y$ and $G\colon\ Y \to Z$, $x_0 \in X$ and $y_0 = F(x_0)$. Suppose that F and G are $\bar{c}$-differentiable at points x_0 and y_0, respectively, and the operator $F'(x_0)$ takes compact sets to compact sets. Then Ψ is also $\bar{c}$-differentiable at x_0, and $\Psi'(x_0) = G'(y_0)F'(x_0)$.*

A similar assertion is true if both mappings are b-differentiable and the operator $F'(x_0)$ takes bounded sets to bounded sets. Finally, if F is Gâteaux differentiable at x_0 and G is $\bar{c}$-differentiable at y_0, then Ψ is Gâteaux differentiable at x_0 and the same formula holds for $\Psi'(x_0)$.

A similar assertion is true in the case of differentiability with respect to the system of sequentially compact sets, since $F'(x_0)$ is sequentially continuous (but in this case we do not require that the operator $F'(x_0)$ must take compact sets to compact sets).

The composition of two Gâteaux differentiable mappings need not be Gâteaux differentiable (Exercise 4.10.36).

4.5.2. Proposition. *Let $\mathcal{E}$ be some class of topological vector spaces. The following conditions are sufficient in order that the chain rule be valid in the class $\mathcal{E}$ for the first order σ-derivatives:*

(a) *if $E,\ G,\ F \in \mathcal{E}$, $r \in \mathcal{R}^\sigma(E,G)$, $l \in \mathcal{L}(G,F)$, then $l\circ r \in \mathcal{R}^\sigma(E,F)$;*

(b) *if $E,\ G,\ F \in \mathcal{E}$, $r \in \mathcal{R}^\sigma(G,F)$ and a mapping $\theta\colon\ E \to G$ is σ-differentiable at the point 0 with $\theta(0)=0$, then $r\circ\theta \in \mathcal{R}^\sigma(E,F)$.*

If a mapping $f\colon\ E \to G$ is σ-differentiable at the point $x_0 \in E$, a mapping $g\colon\ G \to F$ is σ-differentiable at the point $f(x_0)$, where $E,G,F \in \mathcal{E}$, then

$$(g\circ f)'(x_0) = g'\bigl(f(x_0)\bigr)\circ f'(x_0).$$

PROOF. We verify that if $f\colon\ E \to G$, $g\colon\ G \to F$, where both mappings are σ-differentiable at zero (in the corresponding spaces), $f(0)=0$, $g(0)=0$, then the composition $g\circ f$ is σ-differentiable at the point $0 \in E$. Under the stated assumptions, if $h \in E$, then

$$\begin{aligned} g\bigl(f(h)\bigr) &= g'(0)f(h) + r_g\bigl(f(h)\bigr) \\ &= g'(0)f'(0)h + g'(0)r_f(h) + r_g\bigl(f'(0)h + r_f(h)\bigr), \end{aligned}$$

where $r_g \in \mathcal{R}^\sigma(G,F)$, $r_f \in \mathcal{R}^\sigma(E,G)$, therefore, by Conditions (a) and (b)

$$\bigl[h \mapsto g'(0)r_f(h) + r_g\bigl(f'(0)h + r_f(h)\bigr)\bigr] \in \mathcal{R}^\sigma(E,G).$$

The last assertion is obvious from the proof. □

4.5.3. Remark. Similar assertions are valid for $\bar{\sigma}$-derivatives (in which in place of continuous linear operators we consider sequentially continuous ones) and $\mathcal{A}$-derivatives.

4.5.4. Corollary. *For first order b-, c-, s-, $\bar{\text{b}}$- and $\bar{\text{s}}$-derivatives the chain rule is valid in the class of all topological vector spaces.*

PROOF. Let us verify, for example, Conditions (a) and (b) for b-differentiability. Let $F,\ E,\ G \in \mathcal{E}$, $r \in \mathcal{R}^b(E,G)$, $l \in \mathcal{L}(G,F)$, $Q \in \sigma_b^E$, $\{h_n\} \subset Q$, and let $\{t_n\}$ be a sequence of nonzero numbers converging to zero. Then

$$t_n^{-1}\bigl[l\bigl(r(t_nh_n)\bigr)\bigr] = l\bigl(t_n^{-1}r(t_nh_n)\bigr) \to 0$$

as $n \to 0$, hence $l\circ r \in \mathcal{R}^b(E,F)$. If θ is a b-differentiable mapping from E to G and $\theta(0)=0$, $r \in \mathcal{R}^b(G,F)$, $\{h_n\} \subset Q \in \sigma_b^E$, then $\theta = A + r_\theta$, $A \in \mathcal{L}(E,G)$, $r_\theta \in \mathcal{R}^b(E,G)$ and

$$t_n^{-1}r\bigl(\theta(t_nh_n)\bigr) = t_n^{-1}r\bigl[t_n\bigl(Ah_n + t_n^{-1}r_\theta(t_nh_n)\bigr)\bigr] \to 0,$$

since $\{Ah_n\} \in \sigma_b^E$ and $t_n^{-1}r_\theta(t_nh_n) \to 0$. Thus, $r\circ\theta \in \mathcal{R}^b(E,G)$. □

Note that this list does not include $\bar{\text{c}}$-derivative, because the image of a compact set under a sequentially continuous operator need not be compact (certainly, in nonmetrizable spaces). For example, the identity operator from $\mathcal{D}(\mathbb{R})$ with the weak topology to $\mathcal{D}(\mathbb{R})$ with the standard topology is not continuous, but is sequentially continuous.

(ii) Higher order derivatives. The case of arbitrary topological vector spaces

4.5.5. Lemma. *Let E, F, and G be topological vector spaces, η_{EG} and η_{GF} some families of bounded subsets of the spaces $\overline{\mathcal{L}}_\sigma(E,G)$, and $\overline{\mathcal{L}}_\sigma(G,F)$, respectively, and η the family of all subsets of the space $\overline{\mathcal{L}}_\sigma(E,G)\times\overline{\mathcal{L}}_\sigma(G,F)$ each of which is contained in some set (depending on the considered subset) of the form $C_1\times C_2$, where $C_1\in\eta_{EG}$, $C_2\in\eta_{GF}$. Suppose also that, for any sets $C_{EG}\in\eta_{EG}$, $C\in\sigma_E$ and any mapping $g\in\overline{\mathcal{L}}_\sigma(E,G)$, we have $C_{EG}(C)\in\sigma_G$, $g(C)\in\sigma_G$. Then the mapping*

$$\psi\colon (f_1,f_2)\mapsto f_2\circ f_1,\quad \overline{\mathcal{L}}_\sigma(E,G)\times\overline{\mathcal{L}}_\sigma(G,F)\to\overline{\mathcal{L}}_\sigma(E,F)$$

is σ-differentiable at every point and for all pairs (f_1,f_2), (g_1,g_2) in the space $\overline{\mathcal{L}}_\sigma(E,G)\times\overline{\mathcal{L}}_\sigma(G,F)$ we have

$$\psi'(f_1,f_2)(g_1,g_2)=\psi(f_1,g_2)+\psi(g_2,f_1)=g_2\circ f_1+f_2\circ g_1.$$

PROOF. Since

$$\psi(f_1+g_1,f_2+g_2)=\psi(f_1,f_2)+\psi(f_1,g_2)+\psi(g_1,f_2)+\psi(g_1,g_2),$$

it suffices to show that, for every fixed f_1 and f_2, the linear mapping

$$\Phi_1\colon (g_1,g_2)\mapsto(g_2\circ f_1+f_2\circ g_1),\quad \overline{\mathcal{L}}_\sigma(E,G)\times\overline{\mathcal{L}}_\sigma(G,F)\to\overline{\mathcal{L}}_\sigma(E,F)$$

is sequentially continuous and the mapping

$$\psi\colon (g_1,g_2)\mapsto g_2\circ g_1,\ \overline{\mathcal{L}}_\sigma(E,G)\times\overline{\mathcal{L}}_\sigma(G,F)\to\overline{\mathcal{L}}_\sigma(E,F)$$

is η-small. In order to prove the sequential continuity of Φ_1, we show the sequential continuity of the mappings

$$\Phi_2\colon\ g_2\mapsto g_2\circ f_1,\ \overline{\mathcal{L}}_\sigma(G,F)\to\overline{\mathcal{L}}_\sigma(E,F),$$

$$\Phi_3\colon\ g_1\mapsto f_2\circ g_1,\ \overline{\mathcal{L}}_\sigma(E,G)\to\overline{\mathcal{L}}_\sigma(E,F).$$

Let $\varphi_n\to 0$ in $\overline{\mathcal{L}}_\sigma(G,F)$. In order to justify convergence $\Phi_2(\varphi_n)\to 0$ in $\overline{\mathcal{L}}_\sigma(E,F)$ it suffices to verify that if $Q\in\sigma_E$ and $\{h_n\}\subset Q$, then we have $\Phi_2(\varphi_n)(h_n)\to 0$ in F (see Proposition 4.1.5). Since under these conditions $\{f_1(h_n)\}\in\sigma_G$ and $\varphi_n\to 0$ in $\overline{\mathcal{L}}_\sigma(G,F)$, we obtain convergence $\Phi_2(\varphi_n)(h_n)=\varphi_n\bigl(f_1(h_n)\bigr)\to 0$ in F (see Proposition 4.1.5). Thus, the sequential continuity of Φ_2 is proved.

Next, if $\gamma_n\to 0$ in $\overline{\mathcal{L}}_\sigma(E,G)$ and $\{h_n\}\subset B\in\sigma_E$, then, again by Proposition 4.1.5, we have convergence $\gamma_n(h_n)\to 0$ in the space G, whence it follows that $\Phi_3(\gamma_n)(h_n)=f_2\bigl(\gamma_n(h_n)\bigr)\to 0$ in F by the sequential continuity of f_2. Hence $\Phi_3(\gamma_n)\to 0$ in $\overline{\mathcal{L}}_\sigma(E,F)$ (we use again the same proposition). The sequential continuity of Φ_1 is proved.

For the proof of the inclusion

$$\psi\in\mathcal{R}^\eta\bigl(\overline{\mathcal{L}}_\sigma(E,G)\times\overline{\mathcal{L}}_\sigma(G,F),\overline{\mathcal{L}}_\sigma(E,F)\bigr)$$

it suffices to show that if $\{\gamma_n,\varphi_n\}$ is a sequence of elements of some set in the system η and $\{t_n\}$ is a sequence of nonzero real numbers, then $t_n\varphi_n\circ\gamma_n\to 0$ in $\overline{\mathcal{L}}_\sigma(E,F)$, i.e., that $t_n\varphi_n\bigl(\gamma_n(h_n)\bigr)\to 0$ in F for every sequence $\{h_n\}$ in $Q\in\sigma_E$.

The inclusion indicated in the hypotheses of the lemma and the fact that the elements γ_n belong to some of the sets η_{EG} (since $\{(\gamma_n, \varphi_n)\}$ is contained in some of the sets of the system η) yield that the elements $\gamma_n(h_n)$ are contained in some of the sets of the system σ_G. Since the elements φ_n belong to some of the sets of the system η_{GF} and for this reason the sequence $\{\varphi_n\}$ is bounded in $\overline{\mathcal{L}}_\sigma(G,F)$, the sequence $\{\varphi_n(\gamma_n(h_n))\}$ is bounded in F, hence $t_n\varphi_n(\gamma_n(h_n)) \to 0$ in F. □

4.5.6. Remark. The proposition obtained by replacing in the formulation of the lemma of all symbols $\overline{\mathcal{L}}_\sigma(\,\cdot\,)$ by the symbols $\mathcal{L}_\sigma(\,\cdot\,)$ is also true. We leave the proof of this assertion as an exercise.

4.5.7. Proposition. *For $\mathfrak{b}$-, $\mathfrak{c}$-, $\mathfrak{s}$-, $\bar{\mathfrak{b}}$- and $\bar{\mathfrak{s}}$-derivatives of the second order the chain rule is valid in the class of all topological vector spaces.*

PROOF. Let $f\colon E \to G$ and $g\colon G \to F$ be two mappings such that f is twice $\mathfrak{b}$-differentiable at a point $x \in E$ and g is twice $\mathfrak{b}$-differentiable at $f(x)$. Hence the mapping $x \mapsto f'(g(x))$, $E \to \mathcal{L}_\sigma(G,F)$ is one-fold $\mathfrak{b}$-differentiable at x (as a composition of two differentiable mappings: the mapping $x \mapsto g(x)$, $E \to G$, and the mapping $z \mapsto f'(z)$, $G \to \mathcal{L}_\sigma(G,F)$). Since the mapping $x \mapsto g'(x)$, $E \to \mathcal{L}_\sigma(E,G)$ is also differentiable at x (because g is twice differentiable at x), the mapping

$$x \mapsto \Big(f'(g(x)), g'(x)\Big), \quad E \to \mathcal{L}_\sigma(G,F)\times\mathcal{L}_\sigma(E,G) \tag{4.5.1}$$

is differentiable as well (this follows from Exercise 4.10.34, but can be also verified directly). The mapping $x \mapsto f'(g(x))\circ g'(x)$, $E \to \mathcal{L}_\sigma(E,F)$, i.e., the derivative of the mapping $f\circ g$, is the composition of the mapping (4.5.1) and the mapping the differentiability of which was proved in Lemma 4.5.5. Hence by Proposition 4.5.2 this mapping is also differentiable at x and its derivative is given by the equality

$$(f\circ g)''(x) = \Big[h_1 \mapsto \big[h_2 \mapsto f''(g(x))g'(x)h_1 g'(x)h_2 + f'(g(x))g''(x)h_1h_2\big]\Big].$$

For other types of differentiability mentioned in the formulation proofs are completely similarly. □

4.5.8. Lemma. *Let E, G, and F be topological vector spaces such that the space G satisfies Condition (A_1) in Proposition* 4.3.1. *Then the mapping ψ from Lemma* 4.5.5 *is infinitely $\bar{\mathfrak{b}}$-differentiable.*

PROOF. By Lemma 4.5.5 it suffices to show that the linear mapping

$$\psi'\colon \overline{\mathcal{L}}_b(E,G)\times\overline{\mathcal{L}}_b(G,F) \to \overline{\mathcal{L}}_b\big(\overline{\mathcal{L}}_b(E,G)\times\overline{\mathcal{L}}_b(G,F), \overline{\mathcal{L}}_b(E,F)\big)$$

defined by the equality

$$(f_1, f_2) \mapsto \big[(g_1,g_2) \mapsto g_2\circ f_1 + f_2\circ g_1\big]$$

is sequentially continuous. Let $\{\varphi_n\}$ and $\{\gamma_n\}$ be sequences of elements of the spaces $\overline{\mathcal{L}}_b(E,G)$ and $\overline{\mathcal{L}}_b(G,F)$ converging to zero in these spaces. We have to prove that the sequence of mappings $\big[(g_1,g_2) \mapsto g_2\circ\varphi_1 + \gamma_n\circ g_1\big]$ converges to zero in the space $\overline{\mathcal{L}}_b\big(\overline{\mathcal{L}}_b(E,G)\times\overline{\mathcal{L}}_b(G,F), \overline{\mathcal{L}}_b(E,F)\big)$. We use again Proposition 4.1.5.

Let $\{\alpha_n\}$ and $\{\chi_n\}$ be bounded sequences in the spaces $\overline{\mathcal{L}}_b(E,G)$ and $\overline{\mathcal{L}}_b(G,F)$. It suffices to show that the sequence of sums $\chi_n\circ\varphi_n+\gamma_n\circ\alpha_n$ of elements of the space $\overline{\mathcal{L}}_b(E,F)$ converges to zero in this space. In turn, it suffices to prove that if $\{h_n\}$ is a bounded sequence in E, then

$$\chi_n\bigl(\varphi_n(h_n)\bigr)+\gamma_n\bigl(\alpha_n(h_n)\bigr)\to 0 \quad \text{in } F.$$

The boundedness of $\{\alpha_n\}$ and $\{h_n\}$ yields that the sequence $\{\alpha_n(h_n)\}$ is bounded in G. Hence by convergence $\gamma_n\to 0$ in $\overline{\mathcal{L}}_b(G,F)$ we have $\gamma_n\bigl(\alpha_n(h_n)\bigr)\to 0$ in F.

Thus, it remains to show that $\chi_n\bigl(\varphi_n(h_n)\bigr)\to 0$ in F. If this is not true, then there exists a subsequence in the latter sequence containing no convergent subsequence. We shall see that this leads to a contradiction. By condition (A_1) in Proposition 4.3.1 and convergence $\varphi_n(h_n)\to 0$ in G, for every sequence $\{n_j\}$ of natural numbers there are its subsequence $\{n_{j(i)}\}$ and a sequence $\{t_i\}$ of nonzero real numbers such that $t_i\to\infty$, but $t_i\varphi_{n_{j(i)}}(h_{n_{j(i)}})\to 0$. Then the sequence of elements $\chi_{n_{j(i)}}\bigl(t_i\varphi_{n_{j(i)}}(h_{n_{j(i)}})\bigr)$ is bounded in F. Therefore,

$$\chi_{n_{j(i)}}\bigl(\varphi_{n_{j(i)}}(h_{n_{j(i)}})\bigr)=t_i^{-1}\Bigl(\chi_{n_{j(i)}}\bigl(t_i\varphi_{n_{j(i)}}(h_{n_{j(i)}})\bigr)\Bigr)\to 0,$$

which is a contradiction. □

4.5.9. Lemma. *If E, G, and F are topological vector spaces, then the mapping*

$$\psi\colon\ \overline{\mathcal{L}}_s(E,G)\times\overline{\mathcal{L}}_s(G,F)\to\overline{\mathcal{L}}_s(E,F),\quad (f_1,f_2)\mapsto f_2\circ f_1$$

is infinitely $\overline{s}$-differentiable.

The proof is completely similar to the proof of the previous lemma.

4.5.10. Remark. It is interesting that the natural analogs of the last two lemmas for b- and s-differentiable mappings are false.

4.5.11. Theorem. *In the class of all topological vector spaces the chain rule is valid for $\overline{c}$-derivatives of any order. In the class of topological vector spaces satisfying Condition (A_1) in Proposition* 4.3.1 *the chain rule is valid for $\overline{b}$-derivatives of any order.*

PROOF. These assertions are deduced by a standard reasoning from Corollary 4.5.4 and Lemmas 4.5.8 and 4.5.9. For example, let us prove the second assertion.

Let E, G, and F be three topological vector spaces such that G satisfies Condition (A_1), let $f\colon E\to G$ be n-fold $\overline{b}$-differentiable at a point x_0, and let $g\colon G\to F$ be n-fold $\overline{b}$-differentiable at the point $f(x_0)$; we show that the composition $g\circ f$ is n-fold $\overline{b}$-differentiable at the point x_0.

By Corollary 4.5.4 the composition $g\circ f$ is one-fold $\overline{b}$-differentiable at the point x_0. Hence we can use induction on n. Let $n>1$ and let the mapping $x\mapsto g'\bigl(f(x)\bigr)$ from E to $\overline{\mathcal{L}}_b(G,F)$ be $(n-1)$-fold $\overline{b}$-differentiable at this point. Since this is certainly true also for the mapping $x\mapsto f'(x)$ from E to $\overline{\mathcal{L}}_b(E,G)$, the mapping

$$x\mapsto\bigl\langle f'(x),g'\bigl(f(x)\bigr)\bigr\rangle,\quad E\to\overline{\mathcal{L}}_b(E,G)\times\overline{\mathcal{L}}_b(G,F)$$

is $(n-1)$-fold $\overline{b}$-differentiable at the point x_0 Since by Lemma 4.5.8 the mapping ψ of taking the composition is everywhere infinitely differentiable, by the inductive assumption it follows that the mapping $x \mapsto g'(f(x)) \circ f'(x)$, $E \to \overline{\mathcal{L}}_b(E,F)$, is $(n-1)$-fold $\overline{b}$-differentiable at the point x_0, but this mapping is (where it is defined) the first $\overline{b}$-derivative of the mapping $x \mapsto g\bigl(f(x)\bigr)$, $E \to F$. This means that the latter mapping is n-fold $\overline{b}$-differentiable at x_0. □

Our reasoning yields the following fact.

4.5.12. Corollary. *Let E, G, and F be topological vector spaces such that G satisfies Condition (A_1) and E satisfies the following condition:*

(A_2) *all sequentially continuous mappings from E to all topological vector spaces are continuous.*

If a mapping $f\colon\ E \to G$ is n-fold b-differentiable at a point x_0 and a mapping $g\colon G \to F$ is n-fold b-differentiable at the point $f(x_0)$, then the composition $g \circ f$ is n-fold b-differentiable at x_0. A similar assertion is valid for c-derivatives.

We do not know whether the chain rule is valid for b-, $\overline{b}$- and s-derivatives in the class of all topological vector spaces; in the class of all locally convex spaces, it holds in all these cases, as we show in the next subsection.

(iii) Higher order derivatives. Spaces possessing sufficient supplies of continuous linear functionals

Let E, F and G be topological vector spaces and $l \in \overline{\mathcal{L}}(G,F)$. For every $n \in \mathbb{N}$, let l_*^n denote the linear mapping from the space $\overline{\mathcal{L}}_\sigma^n(E,G)$ to $\overline{\mathcal{L}}_\sigma^n(E,F)$ defined as follows: if $f \in \overline{\mathcal{L}}_\sigma^n(E,G)$ and $x_1,\ldots,x_n \in E$, then

$$\bigl(l_*^n(f)\bigr)(x_1,\ldots,x_n) = l\bigl(f(x_1,\ldots,x_n)\bigr).$$

In addition, we set $l_*^0 = l$.

Next, if E and G are topological vector spaces and $n \in \mathbb{N}$, then $\mathcal{B}_\sigma^n(E,G)$ will denote the subspace of the space $\overline{\mathcal{L}}_\sigma^n(E,G)$ consisting of mappings which correspond to symmetric mappings under the canonical embedding of this space to the space of all n-linear mappings from $E \times \ldots \times E$ to G. This subspace is equipped with the induced topology.

Thus, $\Phi \in \mathcal{B}_\sigma^n(E,G)$ precisely when $\Phi \in \overline{\mathcal{L}}_\sigma^n(E,G)$ and

$$\Phi(x_1,\ldots,x_n) = \Phi(x_{s(1)},\ldots,x_{s(n)})$$

for each permutation s of $\{1,2,\ldots,n\}$. In addition, let $\mathcal{B}_\sigma^0(E,G) = \overline{\mathcal{L}}_\sigma^0(E,G) = G$.

4.5.13. Lemma. *If $n \in \mathbb{N}$ and $g\colon\ E \to G$ is n-fold $\overline{\sigma}$-differentiable at a point $x_0 \in E$, then, for every $l \in \overline{\mathcal{L}}(G,F)$, the composition $l \circ g$ is also n-fold $\overline{\sigma}$-differentiable at x_0 and*

$$(l \circ g)^{(n)}(x_0) = l_*^n\bigl(g^{(n)}(x_0)\bigr).$$

The proof of this lemma is straightforward. A similar assertion is valid for σ-differentiable mappings.

4.5.14. Proposition. *Let E be a topological vector space and let G be a Hausdorff locally convex space. Assume that a given system σ_E includes the class of bounded subsets of E contained in finite-dimensional subspaces of E. If $g\colon E \to G$ is n-fold $\bar{\sigma}$-differentiable at a point $x_0 \in E$, then*

$$g^{(n)}(x_0) \in \mathcal{B}_\sigma^n(E, G).$$

PROOF. We can assume that $x_0 = 0$. Let $h_1, \ldots, h_n \in E$, let $\widetilde{g}$ be the restriction of g to the subspace $E_n \subset E$ generated by $h_1, \ldots, h_n$, and let $l \in \overline{\mathcal{L}}(G, \mathbb{R})$. By the previous lemma the composition $l \circ \widetilde{g}$ is a real function defined on a finite-dimensional space and n-fold $\bar{\sigma}$-differentiable, i.e., differentiable in the classical sense. Hence by the corresponding classical theorem

$$(l \circ \widetilde{g})^{(n)}(0)(h_1, \ldots, h_n) = (l \circ \widetilde{g})^{(n)}(0)(h_{s(1)}, \ldots, h_{s(n)}) \qquad (4.5.2)$$

for every permutation s of the numbers $1, 2, \ldots n$. Indeed, if $\psi\colon \mathbb{R}^n \to \mathbb{R}$ is defined by the equality $\psi(t_1, \ldots, t_n) = l\Bigl(g\bigl(\sum_{i=1}^n t_i h_i\bigr)\Bigr)$, then, as it follows from the definition of $\bar{\sigma}$-differentiability, this mapping is n-fold differentiable in the classical sense and

$$\frac{\partial^n \psi(0)}{\partial t_1 \ldots \partial t_n} = \frac{\partial^n \psi(0)}{\partial t_{s(1)} \ldots \partial t_{s(n)}}.$$

The left-hand side of this equality coincides with the left-hand side of (4.5.2), and its right-hand side coincides with the right-hand side of equality (4.5.2) (this follows directly from the definition of $\bar{\sigma}$-differentiability). It follows from (4.5.2) and the previous lemma that

$$l\bigl(g^{(n)}(0)(h_1, \ldots, h_n)\bigr) = l\bigl(g^{(n)}(0)(h_{s(1)}, \ldots, h_{s(n)})\bigr).$$

Since l was an arbitrary element of $\overline{\mathcal{L}}(G, \mathbb{R})$, we have

$$g^{(n)}(0)(h_1, \ldots, h_n) = g^{(n)}(0)(h_{s(1)}, \ldots, h_{s(n)})$$

by the Hahn–Banach theorem. □

4.5.15. Remark. An analog of the proven proposition remains valid with the same proof also for σ-differentiable mappings. The proposition will be also true if the assumption that the space G is locally convex is replaced by the weaker condition of the existence of a set of continuous linear functionals on G separating points.

We need two rather technical lemmas.

4.5.16. Lemma. *Let E, G, and F be topological vector spaces such that the space F is Hausdorff locally convex, let $k \in \mathbb{N}$, and let $n_0, n_1, \ldots, n_k$ be nonnegative integer numbers such that $n_1 + \cdots + n_k = n_0$. Suppose that the following condition is fulfilled*:

(A_3) *if $A^i \in \sigma\bigl(\overline{\mathcal{L}}_\sigma^{n_i}(E, G)\bigr)$, $A_E^1 \in \sigma_E, \ldots, A_E^{n_i} \in \sigma_E$ for all $i \in \{1, 2, \ldots, k\}$, $b \in \overline{\mathcal{L}}_\sigma^n(E, G)$, then $A^i(A_E^1, \ldots, A_E^{n_i}) \in \sigma_G$, $b(A_E^1, \ldots, A_E^{n_i}) \in \sigma_G$.*

Finally, let

$$\beta = \Big\{B \,|\, \forall i \in \{0,1,\ldots,k\}\ \exists A^i \in \sigma\big(\overline{\mathcal{L}}_\sigma^{n_i}(E,G)\big),$$
$$A_0 \in \sigma\big(\mathcal{B}_\sigma^k(G,F)\big)\colon\ B \subset A_0 \times \prod_{i=0}^{k} A^i\Big\}.$$

If $n_i = 0$, *then we set* $A^i(A_E^1,\ldots,A_E^{n_i}) := A^i$, $b^{n_i}(\,\cdot\,) := b^{n_i}$; *we also use other similar relations. Then the mapping*

$$s\colon\ \mathcal{B}_\sigma^k(G,F)\times\prod_{i=0}^{k}\overline{\mathcal{L}}_\sigma^{n_i}(E,G) \to \mathcal{B}_\sigma^{n_0}(E,F),$$

$$s\colon\ (\tilde b, b^{n_1},\ldots,b^{n_k}) \mapsto \big[(x_1,\ldots,x_n)\mapsto \tilde b\big(b^{n_1}(x_1,\ldots,x_{n_1}),$$
$$b^{n_2}(x_{n_1+1},\ldots,x_{n_1+n_2}),\ldots,b^{n_k}(x_{n_1+\cdots+n_{k-1}+1,\ldots,x_{n_k}})\big)\big]$$

is β*-differentiable at every point and the value of its derivative at* $(\tilde b_1, b_1^{n_1},\ldots,b_1^{n_k})$ *is the mapping*

$$s'(\tilde b_1, b_1^{n_1},\ldots,b_1^{n_k})\colon\ (\tilde b_2, b_2^{n_1},\ldots,b_2^{n_k}) \mapsto s(\tilde b_2, b_1^{n_1},\ldots,b_1^{n_k})$$
$$+\, s(\tilde b_1, b_2^{n_1},\ldots,b_1^{n_k}) + \cdots + s(\tilde b_1, b_1^{n_1},\ldots,b_2^{n_k}).$$

PROOF. We show first that, for every number $i \in \{1,2,\ldots,k\}$ for any fixed elements $\tilde b_1, b_1^{n_1},\ldots,b_1^{n_i-1}, b_1^{n_i+1},\ldots,b_1^{n_k}$, the mapping

$$s_i'\colon\ b_2^{n_i} \mapsto s(\tilde b_1, b_1^{n_1},\ldots,b_1^{n_i-1}, b_2^{n_i}, b_1^{n_i+1}\ldots,b_1^{n_k}),$$
$$\overline{\mathcal{L}}_\sigma^{n_i}(E,G) \to \overline{\mathcal{L}}_\sigma^{n}(E,F).$$

is sequentially continuous. Suppose that a sequence $\{\varphi_r\}$ in $\overline{\mathcal{L}}_\sigma^{n_i}(E,G)$ converges to zero. We show that the sequence $s_i'(\tilde b_1, b_1^{n_1},\ldots,\varphi_r,\ldots,b_1^{n_k})$ in the space $\overline{\mathcal{L}}_\sigma^{n}(E,F)$ converges to zero, i.e., that

$$s_i'(\tilde b_1, b_1^{n_1},\ldots,\varphi_r,\ldots,b_1^{n_k})(\alpha_r^1,\alpha_r^2,\ldots,\alpha_r^n) \to 0 \tag{4.5.3}$$

in F for every all $j \in \{1,2,\ldots,n\}$, $\{\alpha_r^j\} \subset Q \in \sigma_E$. Using the symmetry of b_1, we can rewrite the expression in (4.5.3) in the following form:

$$\tilde b_1\big(\varphi_r(\alpha_r^{n_1+\cdots+n_{i-1}+1},\ldots,\alpha_r^{n_1+\cdots+n_i}), b_1^{n_1}(\alpha_r^1,\ldots,\alpha_r^{n_1}),\ldots$$
$$\ldots, b_1^{n_k}(\alpha_r^{n_1+\cdots+n_k+1},\ldots,\alpha_r^{n}\big).$$

Since $\varphi_r(\alpha_r^{n_1+\cdots+n_{i-1}+1},\ldots,\alpha_r^{n_1+\cdots+n_i}) \to 0$ in G and for every i the sequence $b_i^{n_i}(\,\cdot\,)$ is an element of some set in σ_G, relation (4.5.3) holds indeed.

We show that for every fixed collection $b_1^{n_1}, b_1^{n_2},\ldots,b_1^{n_k}$ the mapping

$$s_0'\colon\ \tilde b_2 \mapsto s(\tilde b_2, b_1^{n_1},\ldots,b_1^{n_k}),\quad \mathcal{B}_\sigma^k(G,F) \to \mathcal{B}_\sigma^k(E,F)$$

is sequentially continuous. If a sequence $\{g_r\}$ in the space $\mathcal{B}^k_\sigma(G,F)$ converges to zero and for every $i=1,2,\ldots,n$ we are given a sequence $\{\alpha^i_r\}\subset E$ contained in some set from σ_E, then

$$g_r\bigl(b_1^{n_1}(\alpha_r^1,\ldots,\alpha_r^{n_1}),\ldots,b_1^{n_k}(\alpha_r^{n_1+\cdots+n_k+1},\ldots,\alpha_r^n)\bigr)\to 0$$

in F, since by the assumption of the lemma for every i the sequence $b_1^{n_i}(\,\cdot\,)$ is contained in one of sets of the system σ_G. This means that the mapping s_0' is sequentially continuous.

The sequential continuity of the mappings s_i' with $i\leqslant k$ yields the continuity of the mapping $\sum_{i=0}^k s_i'$. Thus, for completing the proof of the lemma it remains to show that, for fixed $\tilde b_1, b_1^{n_1}, b_1^{n_2},\ldots,b_1^{n_k}$, the mapping

$$\mathcal{B}^k_\sigma(G,F)\times\prod_{i=1}^k\overline{\mathcal{L}}_\sigma^{n_i}(E,G)\to\mathcal{B}^n_\sigma(E,F)$$

sending the element $(\tilde b_2, b_2^{n_1}, b_2^{n_2},\ldots,b_2^{n_k})$ to

$$\begin{aligned}s(\tilde b_1+\tilde b_2, b_1^{n_1}+b_2^{n_1},\ldots,b_1^{n_k}+b_2^{n_k})-s(\tilde b_1,b_1^{n_1},\ldots,b_1^{n_k})\\ -s(\tilde b_1,b_2^{n_1},b_1^{n_2},\ldots,b_1^{n_k})-s(\tilde b_1,b_1^{n_1},b_2^{n_2},\ldots,b_1^{n_k})\\ -\cdots-s(\tilde b_1,b_1^{n_1},b_1^{n_2},\ldots b_2^{n_k})-s(\tilde b_2,b_1^{n_1},\ldots,b_1^{n_k}),\end{aligned}$$

is β-small. This mapping is the sum of finitely many mappings of the form

$$(\tilde b_2, b_2^{n_1},\ldots b_2^{n_k})\mapsto s(\tilde b_1,p_1,\ldots p_k) \tag{4.5.4}$$

or

$$(\tilde b_2, b_2^{n_1},\ldots b_2^{n_k})\mapsto s(\tilde b_2,p_1,\ldots p_k), \tag{4.5.5}$$

where $p_i=b_2^{n_i}$ or $p_i=b_1^{n_i}$, in case of (4.5.4) the last equality is valid for at least two values of the index i, and in case of (4.5.5) for at least one. Thus, it remains to establish that the mappings (4.5.4) and (4.5.5) are β-small. The proof of this is completely similar to the just given justification of the sequential continuity of the mappings $s_0', s_1',\ldots,s_k'$. □

Let us proceed to the second technical assertion about differentiability of compositions of derivatves.

4.5.17. Lemma. *Let E, G, F, k, and n_i be the same as in the previous lemma. Suppose that for $i=0,1,\ldots k$ the mapping $f_i\colon E\to G$ is (n_i+1)-fold $\overline{\mathrm{b}}$-differentiable at a point $x_0\in E$, the mapping $g\colon G\to F$ is $(k+1)$-fold $\overline{\mathrm{b}}$-differentiable at the point $f_0(x_0)$. Then the mapping*

$$\begin{aligned}\Psi\colon x\mapsto\bigl[(h_1,\ldots,h_n)\mapsto g^{(k)}\bigl(f_0(x)\bigr)\bigl(f_1^{(n_1)}(x)(h_1,\ldots,h_{n_1}),\ldots\\ \ldots,f_k^{(n_k)}(x)(h_{n_1+n_2+\cdots+n_{k-1}+1},\ldots,h_{n_k})\bigr)\bigr]\end{aligned}$$

from E *to* $\mathcal{B}_\sigma^n(E,F)$ *is* $\overline{b}$*-differentiable at the point* x_0 *and if* $h\in E$*, then*

$$\begin{aligned}\Psi'(x_0)h = \bigl[(h_1,\dots,h_n)\mapsto g^{(k+1)}\bigl(f_0(x_0)\bigr)\bigl(f_0'(x_0)h, f_1^{(n_1)}(x_0)(h_1,\dots,h_{n_1}),\\ \dots, f_k^{(n_k)}(x_0)(h_{n_1+n_2+\dots+n_{k-1}+1},\dots,h_{n_k})\bigr)\\ + g^{(k)}\bigl(f_0(x_0)\bigr)\bigl(f_1^{(n_1+1)}(x_0)(h,h_1,\dots,h_{n_1}),\\ \dots, f_k^{(n_k)}(x_0)(h_{n_1+n_2+\dots+n_{k-1}+1},\dots,h_{n_k})\bigr)\\ \dots + g^{(k)}\bigl(f_0(x_0)\bigr)\bigl(f_1^{(n_1)}(x_0)(h_1,\dots,h_{n_1}),\\ \dots, f_k^{(n_k+1)}(x_0)(h,h_{n_1+n_2+\dots+n_{k-1}+1},\dots,h_{n_k})\bigr)\bigr].\end{aligned}$$

This follows from Lemma 4.5.16, Proposition 4.5.14 and Corollary 4.5.4.

4.5.18. Theorem. *Let* E*,* G*, and* F *be topological vector spaces such that* F *is Hausdorff locally convex. Let* $f\colon E\to G$ *be* n*-fold* $\overline{b}$*-differentiable at a point* $x_0\in E$ *and let* $g\colon G\to F$ *be* n*-fold* $\overline{b}$*-differentiable at the point* $f(x_0)$*. Then the composition* $g\circ f$ *is* n*-fold* $\overline{b}$*-differentiable at the point* x_0 *and*

$$\begin{aligned}(g\circ f)^{(n)}(x_0)(h_1,\dots,h_n) = \sum_{k=1}^n\sum g^{(k)}\bigl(f(x_0)\bigr)\bigl(f^{(r_1)}(x_0)(h_{i_1^1},\dots,h_{i_{r_1}^1}),\\ \dots, f^{(r_k)}(x_0)(h_{i_1^k},\dots,h_{i_{r_k}^k})\bigr), \quad (4.5.6)\end{aligned}$$

where the inner sum is taken over all partitions of the set $\{h_1,\dots,h_n\}$ *into* k *disjoint sets* $\{h_{i_1^1},\dots,h_{i_{r_1}^1}\}$*, ...,* $\{h_{i_1^k},\dots,h_{i_{r_k}^k}\}$*, and* $r_1+\dots+r_k=n$*,* $r_j>0$*,* $i_1^j<i_2^j<\dots<i_{r_j}^j$*,* $j=1,2,\dots,k$*,* $i_1^1<i_1^2<\dots<i_1^k$*.*

PROOF. For $n=1$ this is true by Corollary 4.5.4. We can pass from n to $n+1$ for $\overline{b}$- and $\overline{s}$-derivatives by Lemma 4.5.17. The validity of this theorem for $\overline{b}$- and $\overline{s}$-derivatives yields that it is valid also for b- and s-derivatives. Similar assertions are true in case of b-, $\overline{s}$- and s-differentiabilities. □

4.5.19. Remark. In all previous propositions the requirement of local convexity of the space F can be replaced by the requirement of existence of a set of continuous linear functionals on F separating points.

Equality (4.5.6) for $E=F=G=\mathbb{R}^1$ becomes the so-called Faà di Bruno formula.

4.5.20. Proposition. *Let* φ *be a continuous bilinear mapping from the product* $E_1\times E_2$ *of two topological vector spaces to a topological vector space* F*. Suppose that mappings* $g_i\colon G\to E_i$*,* $i=1,2$*, are* n*-fold* σ*-differentiable at a point* $x_0\in G$*. Then the mapping* $f\colon x\mapsto\varphi\bigl(g_1(x),g_2(x)\bigr)$*,* $G\to F$ *is* n*-fold* σ*-differentiable at the point* $x_0\in G$ *and*

$$f^{(n)}(x_0)(h_1,\dots,h_n)=\sum\varphi\bigl(g_1^{(k_1)}(x_0)(h_{i_1},\dots,h_{i_{k_1}}), g_2^{(k_2)}(x_0)(h_{j_1},\dots,h_{j_{k_2}})\bigr),$$

where the sum is taken over all partitions of $\{h_1,\dots,h_n\}$ *in disjoint pairs of sets* $\{h_{i_1},\dots,h_{i_{k_1}}\}$ *and* $\{h_{j_1},\dots,h_{j_{k_2}}\}$*,* $k_1+k_2=n$*,* $i_1<\dots<i_{k_1}$*,* $j_1<\dots<j_{k_2}$*.*

This proposition is proved by induction on the order of differentiability n.

4.6. The mean value theorem

A mapping f of a set V on the real line to a topological vector space G is called *right differentiable* at a point $x_0 \in V$ if the point x_0 belongs to the closure of the set $V \cap (x_0, \infty)$ and there exists a limit

$$\lim_{t\downarrow 0,\, x_0+t\in V} t^{-1}\big(f(x_0+t) - f(x_0)\big);$$

this limit denoted by $f'_+(x_0)$ is called *the right derivative* of the mapping f at the point x_0. Similarly one defines *left differentiability* and *left derivatives*.

4.6.1. Theorem. *Let S be an at most countable subset of $[a,b]$, G a Hausdorff locally convex space, B is a convex closed subset, $f\colon [a,b] \to G$ a continuous mapping, and g a continuous nonincreasing real function on $[a,b]$ such that at every point $t \in [a,b]\backslash S$ the functions f and g are right differentiable and*

$$f'_+(t) \in g'_+(t)B. \tag{4.6.1}$$

Then

$$f(a) - f(b) \in \big(g(b) - g(a)\big)B. \tag{4.6.2}$$

This theorem follows (as we shall see below) from the next lemma.

4.6.2. Lemma. *Let a, b and S be the same as in the theorem above and let φ and ψ be continuous real functions on $[a,b]$ right differentiable ar every point in the set $[a,b]\backslash S$. Suppose that ψ is increasing and at each point $t \in [a,b]\backslash S$ the inequality $\varphi'_+(t) \leqslant \psi'_+(t)$ holds. Then $\varphi(b) - \varphi(a) \leqslant \psi(b) - \psi(a)$.*

PROOF. Let $S = \{t_n\}$. For every point t in $[a,b]$ let N_t denote the set of all natural numbers n for which $t_n < t$. It suffices to show that for every $\varepsilon > 0$ we have

$$\varphi(b) - \varphi(a) \leqslant \psi(b) - \psi(a) + \varepsilon \sum_{n\in N_t} 2^{-n} + \varepsilon(t-a) + \varepsilon. \tag{4.6.3}$$

Suppose that for some $\varepsilon > 0$ inequality (4.6.3) is false. By continuity, for any $t \in [a,b]$ sufficiently close to b we have

$$\varphi(t) - \varphi(a) > \psi(t) - \psi(a) + \varepsilon \sum_{n\in N_t} 2^{-n} + \varepsilon(t-a) + \varepsilon. \tag{4.6.4}$$

Let c be the infimum of the set T of points t for which inequality (4.6.4) is true. Then $c \notin T$, since, by the continuity of both functions, along with every point t the set T contains all sufficiently close points smaller than t. If now $c \in S$, then again by continuity there exists $\delta > 0$ such that for every point t in the interval $[c, c+\delta]$ we have

$$\varphi(t) - \varphi(a) \leqslant \psi(t) - \psi(a) + \varepsilon \sum_{n\in N_t} 2^{-n} + \varepsilon(t-a) + \varepsilon.$$

which contradicts the equality $c = \inf\{t\colon\ t \in T\}$.

If $c \notin S$, then at c both functions are right differentiable and $\varphi'_+(c) \leqslant \psi'_+(c)$ by assumption. Therefore, there exists $\eta > 0$ such that if $t \in (c, c+\eta]$, then

$$\varphi'_+(c) = \frac{\varphi(t)-\varphi(c)}{t-c} - \frac{\varepsilon}{2}, \quad \psi'_+(c) \leqslant \frac{\psi(t)-\psi(c)}{t-c} - \frac{\varepsilon}{2}.$$

The last three inequalities yield that

$$\varphi(t)-\varphi(c) \leqslant \psi(t)-\psi(c)+\varepsilon(t-c).$$

Since $c \notin T$, we have

$$\varphi(c)-\varphi(a) \leqslant \psi(c)-\psi(a)+\varepsilon\sum_{n\in N_t} 2^{-n}+\varepsilon(t-a)+\varepsilon.$$

The last two inequalities imply that

$$\varphi(t)-\varphi(a) \leqslant \psi(t)-\psi(a)+\varepsilon\sum_{n\in N_t} 2^{-n}+\varepsilon(t-a)+\varepsilon.$$

Thus, if $t \in (c, c+\eta]$, then $t \notin T$ contrary to the definition of c. Hence inequality (4.6.3) is true for every $\varepsilon > 0$. □

PROOF OF THEOREM 4.6.1. If the inclusion (4.6.2) is false, then by the Hahn–Banach theorem there exists a continuous linear functional l on G such that

$$l\bigl(f(b)-f(a)\bigr) > \sup_{z\in B} l\Bigl(\bigl(g(b)-g(a)\bigr)z\Bigr). \tag{4.6.5}$$

Set $d = \sup_{z\in B} l(z)$, $\psi(t) = dg(t)$, $\varphi(t) = l\bigl(f(t)\bigr)$. By (4.6.1) the functions φ and ψ satisfy the conditions of the lemma proved above, whence

$$l\bigl(f(b)-f(a)\bigr) > \bigl(g(b)-g(a)\bigr)\sup_{z\in B} l(z).$$

Since $g(b)-g(a) \geqslant 0$, the right-hand side of the last inequality equals the right-hand side of inequality (4.6.5). The obtained contradiction proves the theorem.

4.6.3. Remark. Certainly, a similar assertion is valid for the left differentiability. To show this it suffices to observe that the change of variable $t \mapsto -t$ takes left differentiable functions to right differentiable functions.

4.6.4. Corollary. (THE MEAN VALUE THEOREM) *Let S be an at most countable subset of the interval $[a,b]$, let G be a Hausdorff locally convex space, and let $f\colon [a,b] \to G$ be a continuous mapping differentiable at each point of $[a,b]\backslash S$. Then*

$$f(b)-f(a) \in \overline{\mathrm{conv}}\,\{f'(\theta)(b-a)\colon\ \theta\in(a,b)\backslash S\}.$$

PROOF. It suffices to apply the theorem, taking the mapping $g(t) = t(b-a)$ and the set $B = \overline{\mathrm{conv}}\,\{f'(\theta)(b-a)\colon\ \theta\in(a,b)\backslash S\}$. □

4.6.5. Corollary. *Let V be an open connected set in a topological vector space E and let f be a mapping from V to a Hausdorff locally convex space G that is Gâteaux differentiable at every point of the set V. If $f'(x) = 0$ for every point $x \in V$, then $f(x) \equiv c \in G$.*

The mean value theorems proved in this section belong to the most important results in the differentiation theory, in spite of their trivial justifications.

4.7. Taylor's formula

We first consider Taylor's formula with the remainder in the Lagrange form.

4.7.1. Theorem. *Let S be an at most countable subset of the interval $[a,b]$, $n \in \{0\} \cup \mathbb{N}$, G a Hausdorff locally convex space, and $f\colon [a,b] \to G$ a mapping which is n-fold continuously differentiable everywhere on $[a,b]$ and $(n+1)$-fold differentiable at every point of the set $[a,b]\backslash S$. Then the following relation (called Taylor's formula with the remainder in the Lagrange form) is valid:*

$$f(b) - f(a) - \sum_{i=1}^{n} \frac{1}{i!} f^{(i)}(a)(b-a)^i$$

$$\in \overline{\mathrm{conv}} \Bigl\{ \frac{1}{(n+1)!} f^{(n+1)}(\theta)(b-a)^{n+1}\colon \theta \in (a,b)\backslash S \Bigr\}. \quad (4.7.1)$$

PROOF. For $n = 0$ our assertion coincides with the mean value theorem from the previous section. Suppose that it is already proven for some integer $n = m \geqslant 0$ and prove that it remains valid for $n = m+1$. By the mean value theorem for the function

$$t \mapsto f(t) - \sum_{i=1}^{m+1} \frac{1}{i!} f^{(i)}(a)(t-a)^i$$

we have the inclusion

$$f(b) - f(a) - \sum_{i=1}^{m+1} \frac{1}{i!} f^{(i)}(a)(b-a)^i \in \overline{\mathrm{conv}} \biggl\{ \Bigl(f'\bigl(a + \theta(b-a)\bigr) - f'(a)$$

$$- \sum_{i=1}^{m} \frac{1}{i!} f^{(i+1)}(a) \bigl(\theta(b-a)\bigr)^i \Bigr)(b-a)\colon 0<\theta<1 \biggr\}.$$

By the inductive assumption applied to the map $f'\colon [a, a+\theta(b-a)] \to G$, which satisfies the corresponding conditions, we obtain

$$f'\bigl(a+\theta(b-a)\bigr) - f'(a) - \sum_{i=1}^{m} \frac{1}{i!} f^{(i+1)}(a) \bigl(\theta(b-a)\bigr)^i$$

$$\in \overline{\mathrm{conv}} \bigl\{ f^{(m+2)}\bigl(a+\alpha\theta(b-a)\bigr) \bigl(\theta(b-a)\bigr)^{m+1}\colon a+\alpha\theta(b-a) \notin S,\, 0<\alpha<1 \bigr\}.$$

The last two inclusions yield

$$f(b) - f(a) - \sum_{i=1}^{m+1} \frac{1}{i!} f^{(i)}(a)(b-a)^i$$

$$\in \overline{\mathrm{conv}} \bigl\{ f^{(m+2)}\bigl(a+\alpha\theta(b-a)\bigr) \cdot \bigl(\theta^{m+1}(b-a)^{m+2}\bigr)\colon$$

$$a + \alpha\theta(b-a) \notin S,\, 0<\alpha<1, 0<\theta<1 \bigr\},$$

as required. □

4.7.2. Corollary. *Suppose that E is a topological vector space, G is a Hausdorff locally convex space, $x_1, x_2 \in E$, $[x_1, x_2]$ is the compact interval joining x_1 and x_2, $n \in \mathbb{N}$, and $f\colon E \to G$ is a mapping continuous at every point of $[x_1, x_2]$ and $(n+1)$-fold weakly differentiable at every point of the interval $(x_1, x_2) = [x_1, x_2]\backslash\{x_1, x_2\}$ such that the mapping $x \mapsto f^{(n)}(x_1 - x_2)$, $[x_1, x_2] \to G$ is continuous. Then*

$$f(x_2) - f(x_1) - \sum_{i=1}^{n} \frac{1}{i!} f^{(i)}(x_1)(x_2 - x_1)\dots(x_2 - x_1)$$
$$\in \overline{\text{conv}}\Bigl\{\frac{1}{(n+1)!} f^{(n+1)}(x)(x_2 - x_1)\dots(x_2 - x_1)\colon\, x \in (x_1, x_2)\Bigr\}.$$

PROOF. We apply the theorem to the mapping $t \mapsto f\bigl(x_1 + t(x_2 - x_1)\bigr)$ from the interval $[0,1]$ to G. □

We now turn to Taylor's formula with the remainder in the Peano form. Taylor's formula proved above with the remainder in the Lagrange form is applicable to $(n+1)$-fold σ-differentiable mappings of arbitrary topological vector spaces to arbitrary Hausdorff locally convex spaces; the properties of the remainder do not depend on σ. In contrast, the properties of the remainder in the Peano form explicitly depend on the system of sets σ; this remainder is a σ-small mapping of the corresponding order.

4.7.3. Theorem. *Suppose that a mapping g from a topological vector space E to a Hausdorff locally convex space G is n-fold σ-differentiable at a point $x_0 \in E$. Then there exists a σ-small mapping $r\colon E \to G$ of order n such that for all $h \in E$ one has*

$$g(x_0 + h) - g(x_0) = \sum_{i=1}^{n} \frac{1}{i!} g^{(i)}(x_0)\underbrace{(h, \dots, h)}_{i} + r(h). \tag{4.7.2}$$

PROOF. We observe first of all that if $n > 1$ and s is a mapping from E to $\mathcal{L}_\sigma^{n-1}(E, G)$ equal to the nth derivative (at some point) of some mapping from E to G, then $\varphi_s\colon x \mapsto s(x, \dots, x)$ is a σ-small mapping. This follows from the fact that all spaces $\mathcal{L}_\sigma^n(E, G)$ (to which the values of derivatives belong) are topological vector, hence in each of them every neighborhood of zero is an absorbent set; for this reason the mapping

$$\mathbb{R}^1\backslash\{0\} \ni t \mapsto [x \mapsto t^{-1}\varphi_s(tx)] = [x \mapsto t^{n-1}\varphi_s(x)],\ 0 \mapsto 0$$

from $\mathbb{R}^1$ to $\mathcal{F}_\sigma(E, G)$ is continuous at zero.

We show that if $s = g^{(n)}(x_0)$, then the mapping

$$\varphi_s\colon x \mapsto s(x, \dots, x), \quad E \to G$$

is infinitely σ-differentiable at every point and for every natural number $i \leqslant n$ one has the equality (recall that in accordance with our agreement $\mathcal{L}_\sigma^0(E, G) = G$, $\varphi_s^{(0)}\colon x \mapsto s(x, \dots, x) \in G$ and so on)

$$(\varphi_s)^{(i)}(x) = \bigl[(h_1, \dots, h_i) \mapsto n(n-1)\cdots(n-i+1)s(x, \dots, x, h_1, \dots, h_i)\bigr],$$

and for every natural number $i > n$ we have $\varphi_s^{(i)} = 0$. First of all, the mapping φ_s is one-fold σ-differentiable and for $i = 1$ we have equality (4.7.2). Indeed, the mapping

$$\psi\colon\, h \mapsto \varphi_s(x+h) - \varphi_s(x) - ns(\underbrace{x,\ldots,x}_{n-1},h)$$

is (by the symmetry of the mapping s proved in Proposition 4.5.14) a linear combination with positive integer coefficients of the mappings

$$s_x^k\colon\, h \mapsto s(\underbrace{x,\ldots,x}_{n-k},\underbrace{h,\ldots,h}_{k}),$$

where $k \in \{2,3,\ldots,n\}$. The fact that each of these mappings is σ-small follows from what has been said at the beginning of the proof, because

$$s_x^k(h) = \bigl(g^{(n-k)}\bigr)^{(k)}(x_0)(\underbrace{h,\ldots,h}_{k})(\underbrace{x,\ldots,x}_{n-k})$$

and the mapping

$$h \mapsto \bigl(g^{(n-k)}\bigr)^{(k)}(x_0)(h,\ldots,h), \quad E \to \mathcal{L}_\sigma^{n-k}(E,G)$$

is σ-small, since it is the derivative of the mapping $g^{(n-k)}$. Hence it is σ-small as a mapping from E to $\mathcal{L}_f^{n-k}(E,G)$ (the topology of convergence on the sets of the system σ_E is stronger than the topology of pointwise convergence in the space of mappings from E to an arbitrary topological vector space). Further, suppose that for some $i \leqslant n-2$ ($n > 2$) the assertions about the i-fold σ-differentiability of the mapping φ_s and the value of its ith σ-derivative are already proven. Then the differentiability of the mapping $\varphi_s^{(i)}$ follows from the above reasoning taking into account the equality

$$\bigl[(h_1,\ldots,h_i) \mapsto s(\underbrace{x,\ldots,x}_{n-i},h_1,\ldots,h_i)\bigr] = \bigl(g^{(i)}\bigr)^{(n-i)}(x_0)(\underbrace{x,\ldots,x}_{n-i})$$

and the σ-smallness of the mapping that is the σ-derivative of order $j > 1$ of some other mapping at the given point. Finally, if our assertion about differential properties of φ_s is already established for $i = n-1$, then its validity for all larger natural i follows from (4.7.2) and the linearity and continuity of the mapping

$$h \mapsto \bigl(g^{(n-1)}\bigr)'(x_0)h, \quad E \to \mathcal{L}_\sigma^{n-1}(E,G).$$

Thus, for every number $i \in \{1,2,\ldots,n\}$ the mapping $x \mapsto g^{(i)}(x_0)(x,\ldots,x)$ is infinitely σ-differentiable and all its derivatives, excepting the ith derivative, vanish at the point 0 and its ith derivative equals $i!g^{(i)}(x_0)$. Hence for the mapping $r\colon E \to G$ defined by equality (4.7.2) we have $r(0)=0, r'(0)=0,\ldots,r^{(n)}(0)=0$ (the symbols of differentiation denote σ-derivatives of the corresponding orders), therefore, as we shall now show, $r \in \mathcal{R}^{n\sigma}(E,G)$. Since for $n = 1$ this is true by definition, we can use induction on n. Suppose that $n \geqslant 2$ and the previous assertion is true for $n-1$. The n-fold σ-differentiability of the mapping r at the point 0 yields that for all $h \in E$ we have

$$t^{-n}r(th) \in \overline{\operatorname{conv}}\,\{t^{-(n-1)}r'(\theta th)h\colon\ 0<\theta<1\} \tag{4.7.3}$$

for all nonzero $t \in \mathrm{IR}^1$ with sufficiently large absolute values. By the inductive assumption, if $\{t_j\}$ is a sequence of nonzero real numbers converging to zero and $\{h_j\}$ is a sequence of elements of the set $Q \in \sigma$, then $t_j^{-(n-1)}r'(t_jh_j) \to 0$ in the space $\mathcal{L}_\sigma(E,G)$. But this and inclusion (4.7.3) yield by the local convexity of the space G that $t_j^{-1}r(t_jh_j) \to 0$ in G, i.e., that $r \in \mathcal{R}^{n\sigma}(E,G)$. □

4.8. Partial derivatives

The next theorem is an analog of the classical assertion that for differentiability of a real function of n real variables at a given point (in the classical sense, i.e., in the sense of Fréchet) it is sufficient that all its partial derivatives exist in some neighborhood of the given point and be continuous at this point.

4.8.1. Theorem. *Let $E = E_1\times\ldots\times E_n$, where $E_1,\ldots,E_n$ are topological vector spaces, and let β be some family of bounded subsets of E such that $\beta \supset \sigma^E_{\mathrm{fin}}$. For every $i \in \{1,2,\ldots,n\}$ let β_i denote the family of all subsets E_i that are projections onto E_i of the sets in β. Let G be a vector space and let τ_1 and τ_2 be two Hausdorff locally convex topologies on G possessing the same supplies of closed convex sets.*

If $g\colon E \to G$ is β-differentiable along the subspace E_1 at the point x_0 as a mapping from E to (G,τ_1) and for $i=1,2,\ldots,n$ is Gâteaux differentiable along the subspace E_i in some neighborhood of x_0 as a mapping from E to (G,τ_2), moreover, $g'(x) \in \overline{\mathcal{L}}_{\beta_i}(E,(G,\tau_1))$ for every point $x \in E$ and the mappings $x \mapsto g'_{E_i}(x)$ from E to $\overline{\mathcal{L}}_{\beta_i}(E,(G,\tau_1))$ are sequentially continuous at the point x_0, then g is β-differentiable at x_0 and

$$g'(x_0)h = \sum_{i=1}^{n} g'_{E_i}(x_0)h_i.$$

PROOF. We consider the case $n=2$; the general case can be obtained by induction on n. If $h=(h_1,h_2)\in E$, then

$$\begin{aligned} r(h) &= g(x_0+h) - g(x_0) - g'_{E_1}(x_0)h_1 - g'_{E_2}(x_0)h_2 \\ &= \bigl[g\bigl(x_0+(h_1,h_2)\bigr) - g(x_0+(h_1,0)) - g'_{E_2}(x_0)h_2\bigr] \\ &\qquad + \bigl[g\bigl(x_0+(h_1,0)\bigr) - g(x_0) - g'_{E_1}(x_0)h_1\bigr]. \end{aligned}$$

Let $r_2(h)$ be the expression in the square brackets. It follows from the hypotheses of the theorem that $r_2 \in \mathcal{R}^\beta(E,G)$. Hence it remains to prove that if $r_1(h)$ is the expression in the first square brackets, then $r_1 \in \mathcal{R}^\beta(E,G)$. By the mean value theorem, as $t\in\mathrm{IR}^1$, $t\neq 0$, we have

$$t^{-1}r(th) \in \overline{\operatorname{conv}}\,\{g'_{E_2}(x_0+t(h_1,0)+\theta t(0,h_2))h_2 - g'_{E_2}(x_0)h_2\colon\ 0<\theta<1\}.$$

Taking into account the sequential continuity of the mapping $x\mapsto g'_{E_2}(x)$ from E to $\overline{\mathcal{L}}_\beta(E_2,G)$, we obtain that $r_1 \in \mathcal{R}^\beta(E,G)$. □

It is clear from the proof that an analog of this theorem is also true for β-differentiable mappings.

4.9. Inversion of Taylor's formula and the chain rule

It is well-known (see also Exercise 4.10.57) that the continuity of the mapping $f\colon X \to Y$ between normed spaces is equivalent to the continuity of all compositions $f\circ\varphi$ for continuous mappings $\varphi\colon \mathbb{R} \to X$ and is also equivalent to the continuity of the compositions $\psi\circ f$ for all continuous functions $\psi\colon Y \to \mathbb{R}$. Here we discuss analogs of such descriptions of differentiability. In particular, it will be shown that for a number of important types of differentiability a mapping f on an infinite-dimensional space X belongs to the class C^n if the composition $f\circ\varphi$ belongs to the class C^n for every C^∞-mapping φ from $\mathbb{R}^2$ to X. However, we first give some examples showing that the situation with differentiability is more subtle.

4.9.1. Example. Let $f\colon \mathbb{R}^2 \to \mathbb{R}^1$, $f(x,y) = y^3/(x^2+y^2)$, $f(0,0)=0$. Then the function f is not Gâteax differentiable at the origin, although for every mapping $\varphi \in C^1(\mathbb{R}^1,\mathbb{R}^2)$ we have $f\circ\varphi \in C^1(\mathbb{R}^1)$.

PROOF. We have $\partial_{e_1} f(0,0) = 0$, $\partial_{e_2} f(0,0) = 1$, $\partial_{e_1+e_2} f(0,0) = 1/2$ for the standard basis vectors e_1 and e_2, hence f has no Gâteax derivative at the origin. Let $\varphi \in C^1(\mathbb{R}^1,\mathbb{R}^2)$, $\varphi = (\varphi_1,\varphi_2)$. Let us verify the continuous differentiability of $f\circ\varphi$ in a neighborhood of the origin. We can assume that $\varphi(0) = (0,0)$, otherwise there is no problem. If $\varphi_2'(0) \neq 0$, then by a straightforward evaluation we obtain the equality

$$(f\circ\varphi)'(0) = \bigl(\varphi_2'(0)\bigr)^3\bigl(|\varphi_1'(0)|^2 + |\varphi_2'(0)|^2\bigr)^{-1}.$$

In this case $\varphi_2(t) \neq 0$ if $t \neq 0$ in a neighborhood of zero, whence we can easily obtain the claim. Suppose now that $\varphi'(0) = (0,0)$. Then $(f\circ\varphi)'(0) = (0,0)$, since $|f(\varphi(t))/t| \leqslant |\varphi_2(t)/t| \to 0$. Thus, we have established differentiability of composition in all cases. If in a neighborhood of zero we have $\varphi(t) \neq (0,0)$ whenever $t \neq 0$, then

$$(f\circ\varphi)' = \partial_x f(\varphi)\varphi_1' + \partial_y f(\varphi)\varphi_2',$$

which tends to zero as $t \to 0$, since a simple verification shows the boundedness of $\partial_x f$ and $\partial_y f$. Hence the composition has a continuous derivative. If the origin is not an isolated zero of φ, then, as above, the derivative of the composition at zero is zero. Hence we have to verify that $(f\circ\varphi)'(t_n) \to 0$ as $t_n \to 0$. Here it suffices to consider only those points at which $(f\circ\varphi)'(t_n) \neq 0$. It follows that $\varphi_2'(t_n) \neq 0$. If $\varphi(t_n) \neq (0,0)$, then

$$(f\circ\varphi)'(t_n) = \partial_x f\bigl(\varphi(t_n)\bigr)\varphi_1(t_n)' + \partial_y f\bigl(\varphi(t_n)\bigr)\varphi_2'(t_n) \to 0,$$

since $\varphi_1'(t_n) \to 0$ and $\varphi_2'(t_n) \to 0$ by the equality $\varphi_1'(0) = \varphi_2'(0) = 0$ (the origin is not an isolated point). In the remaining case $\varphi(t_n) = (0,0)$ we obtain as above

$$(f\circ\varphi)'(t_n) = \bigl(\varphi_2'(t_n)\bigr)^3\bigl(|\varphi_1'(t_n)|^2 + |\varphi_2'(t_n)|^2\bigr)^{-1} \to 0,$$

which completes the proof. □

4.9.2. Example. Let $X = L^1[0,1]$ and let

$$f(x) = \int_0^1 \sin x(t)\,dt.$$

Then f has no Fréchet derivative at any point (Example 4.2.14), but for all $n \in \mathbb{N}$ for every $\varphi \in C^1(\mathbb{R}^n, X)$ the inclusion $f\circ\varphi \in C^1(\mathbb{R}^1)$ holds, which is verified directly.

We now consider three examples with outer compositions (i.e., left compositions).

4.9.3. Example. Let $f\colon \mathbb{R}^1 \to c_0$, $f(t) = \bigl(n^{-1}\sin(nt)\bigr)_{n=1}^{\infty}$. Then f is not differentiable at any point even in the weak topology, although $\psi\circ f \in C^1(\mathbb{R}^1)$ for all $\psi \in C^1(c_0)$, in particular, $l(f) \in C^1(\mathbb{R}^1)$ for all $l \in c_0' = l^1$.

PROOF. If f is differentiable at least in the weak topology, then we have the equality $f'(0) = \bigl(\cos(nt)\bigr)_{n=1}^{\infty}$, which is not an element of c_0 for any t. For $\lambda = (\lambda_n) \in l^1 = c_0'$ we have $l(f)(t) = \sum_{n=1}^{\infty} n^{-1}\lambda_n \sin(nt)$, which is a continuously differentiable function on the real line. It follows from Corollary 4.5.4 that $\psi\circ f \in C^1(\mathbb{R}^1)$ for all functions $\psi \in C^1(c_0, \mathbb{R}^1)$. □

4.9.4. Example. Let us define a mapping $f\colon \mathbb{R}^1 \to l^2$ by $f(t) = \bigl(t\eta(2^n t)\bigr)_{n=1}^{\infty}$, where $\eta \in C^\infty(\mathbb{R}^1)$, $\eta(0) = 0$ if $t \notin [3/4, 5/4]$, $\eta(1) = 1$. Then f is not differentiable at zero, is continuously differentiable in the weak topology and $\psi\circ f \in C^1(\mathbb{R}^1)$ for all $\psi \in C^1(l^2, \mathbb{R}^1)$.

If we set $f(t) = \bigl(t^m\eta(2^n t)\bigr)_{n=1}^{\infty}$, where $m \in \mathbb{N}$, $m \geqslant 2$, then $\psi\circ f \in C^m(\mathbb{R}^1)$ for all $\psi \in C^m(l^2, \mathbb{R}^1)$, moreover, $f \in C^{m-1}(\mathbb{R}^1, l^2)$, but $f \notin C^m(\mathbb{R}^1, l^2)$.

PROOF. For $\lambda = (\lambda_n) \in l^2$ we obtain a real function

$$l(f)(t) = \sum_{n=1}^{\infty} \lambda_n t\eta(2^n t).$$

This convergent series can be differentiated term-by-term, which produces the series $\sum_{n=1}^{\infty} \lambda_n[\eta(2^n t) + t2^n\eta'(2^n t)]$, since the latter series (as the original one) converges uniformly, because the functions $\eta(2^n t)$ have disjoint supports for different numbers n and $\lambda_n \to 0$. As in the previous example, this gives the continuous differentiability of $\psi\circ f \in C^1(\mathbb{R}^1)$ for all $\psi \in C^1(c_0, \mathbb{R}^1)$. However, there is no derivative at zero in the norm topology, because for $t_n = 2^{-n}$ we have $\|f(t_n) - f(0)\|/t_n = 1$, although the derivative at zero in the weak topology exists and is zero: the ratio $f(t)/t$ tends to zero coordinate-wise and is norm bounded. The case $m > 1$ is similar. □

In the last example both spaces are infinite-dimensional Hilbert.

4.9.5. Example. Let $X = L^2[0,1]$ and let $F\colon X \to X$ be defined by the formula $F(x)(t) = \sin x(t)$. Then F is Fréchet differentiable at no point (Example 4.2.15), but $\psi\circ F \in C^1(X)$ for all $\psi \in C^1(X)$ and $F\circ\varphi \in C^1(\mathbb{R}^n, X)$ for all $\varphi \in C^1(\mathbb{R}^n, X)$ for all n, which is verified directly.

Let us proceed to positive results. They admit a unified description in terms of differentiability with respect to a system $\mathcal{A}$ of subsets of a Hausdorff locally convex space X satisfying the following conditions: $\mathcal{A}$ contains all finite sets (hence $\mathcal{A}$-differentiability implies Gâteax differentiability; if $\mathcal{A}$-continuity yields the continuity of linear operators, then also $\widetilde{\mathcal{A}}$-differentiability implies Gâteax differentiability), is closed with respect to multiplication by scalars, and the union of any two sets in $\mathcal{A}$ is contained in a set from $\mathcal{A}$. These conditions are fulfilled for the systems of finite sets, bounded sets, compact sets, sequentially compact sets, and also Banach discs. The class of convergent sequences does not satisfy the latter condition, but the results presented below can be extended also to this class by using sequential compactness of all sets in this class. For metrizable spaces, one can have in mind differentiability with respect to the system of bounded sets or the system of compact sets.

We recall (see § 4.1) that the symbol $\mathcal{L}_{\mathcal{A}}(X,Y)$ denotes the linear space of all linear $\mathcal{A}$-continuous operators from X to Y equipped with the topology of uniform convergence on the class $\mathcal{A}$, and that the definition of $\mathcal{A}$-differentiability of a mapping $f\colon X\to Y$ at a point x_0 is introduced by replacing the requirement that the operator $f'(x_0)$ belongs to the class $\mathcal{L}(X,Y)$ of continuous operators or to the class $\overline{\mathcal{L}}(X,Y)$ of sequentially continuous operators by the inclusion $f(x_0) = f'(x_0) \in \mathcal{L}_{\mathcal{A}}(X,Y)$.

For an open set $U \subset X$, the class $C^n_{\mathcal{A}}(U,Y)$ consists of n-fold $\mathcal{A}$-continuously differentiable mapping. Let $D^n_{\mathcal{A}}f$ denote $\widetilde{\mathcal{A}}$-derivative of order n. In the case of normed spaces and the class $\mathcal{A}$ of all bounded sets, $\widetilde{\mathcal{A}}$-differentiability is Fréchet differentiability and the class $C^n_{\mathcal{A}}(U,Y)$ is the usual class of n-fold Fréchet differentiable mappings with continuous derivatives up to order n. We now obtain the inversion of Taylor's formula.

For a k-linear mapping Ψ we set $\Psi h^k := \Psi(h,\dots,h)$.

4.9.6. Theorem. *Suppose that f is an $\mathcal{A}$-continuous mapping from U to a Hausdorff locally convex space Y and, for some $n\in\mathbb{N}$, for each $x\in U$ and each $k=1,\dots,n$ there is a symmetric k-linear mapping $\Psi_k(x)\colon X^k\to Y$ such that the mapping r_n defined by the formula*

$$f(x+h)-f(x)=\sum_{k=1}^{n}\frac{1}{k!}\Psi_k(x)h^k+r_n(x,h),$$

has the following properties: $r_n(x,0)=0$ and for every $A,B\in\mathcal{A}$ uniformly in $v\in A$ and $h\in B$ as $t\to 0$ we have $r_n(x+tv,th)/t^n\to 0$. Then we have $f\in C^n_{\mathcal{A}}(U,Y)$ and $D^k_{\mathcal{A}}f(x)=\Psi_k(x)$.

PROOF. Taking n different numbers $\alpha_j\in(0,1)$, we can linearly express $\Psi_k(x)h^k$ through $f(x+\alpha_j h)-f(x)-r_n(x,\alpha_j h)$ from the equalities

$$f(x+\alpha_j h)-f(x)-r_n(x,\alpha_j h)=\sum_{k=1}^{n}\alpha_j^k\Psi_k(x)h^k/k!.$$

The condition implies $\mathcal{A}$-continuity of $f(x+\alpha_j h)$ and $r_n(x,\alpha_j h)$ in h at zero. This gives $\mathcal{A}$-continuity of $\Psi_k(x)h^k$ in h, whence by the symmetry we obtain

$\mathcal{A}$-continuity of the mapping $\Psi_k(x)$. Assuming that the theorem is true for $n-1$, we show that it holds for n. Writing expansions for $f(x+h+w)$ in two ways for $x+(h+w)$ and for $(x+h)+w$, we obtain

$$\sum_{k=1}^{n} \frac{1}{k!}\Psi_k(x+h)w^k + r_n(x+h,w) = \sum_{k=1}^{n} \frac{1}{k!}\Psi_k(x)(h+w)^k + r_n(x,h+w),$$

whence we find

$$\begin{aligned}\sigma(x,h,w) :&= r_n(x+h,w) - r_n(x,h+w)\\ &= g_1(x,h)w + \cdots + g_n(x,h)w^n, \text{ where } g_k(x,h)\in\mathcal{L}^k_{\mathcal{A}}(X,Y).\end{aligned}$$

This is verified by using that Ψ_k is k-linear. For example,

$$g_1(x,h)w = \Psi_1(x+h)w - \sum_{k=1}^{n} \frac{1}{k!}\Psi_k(x)(h^{k-1},w). \tag{4.9.1}$$

Now we have to verify that the inductive assumption is applicable to the mapping $\Psi_1\colon U \to \mathcal{L}_{\mathcal{A}}(X,Y)$. To this end, we have to show that this mapping is $\mathcal{A}$-continuous and that for all $A,B\in\mathcal{A}$ uniformly in $v\in A$ and $h\in B$ we have $g_1(x+tv,th)/t^{n-1}\to 0$ in $\mathcal{L}_{\mathcal{A}}(X,Y)$ as $t\to 0$. To do this, we observe that, as $t\to 0$ we have

$$\frac{\sigma(x+tv,th,tw)}{t^n} = \frac{r_n(x+tv+th,tw) - r_n(x+tv,th+tw)}{t^n} \to 0$$

uniformly in $v\in A$, $h\in B$, $w\in C$ for any fixed $A,B,C\in\mathcal{A}$. As above, the elements $g_k(x,h)w^k$ can be linearly express through $\sigma(x,h,\alpha_j w)$ from the system $\sum_{k=1}^{n}\alpha_j^k g_k(x,h)w^k = \sigma(x,h,\alpha_j w)$. Due to the noted property of σ this yields uniform convergence of

$$g_1(x+tv,th)w/t^{n-1} = g_1(x+tv,th)(tw)/t^n$$

to zero as $t\to 0$. Along with (4.9.1) and the continuity of the k-linear mappings $\Psi_k(x)$ this yields the continuity of the mapping $x\mapsto\Psi_1(x)$. Finally, it is easy to verify by definition that $f'=\Psi_1$. □

Let X and Y be Hausdorff locally convex spaces, $n\in\mathbb{N}$, $C^n_b(X,Y)$ the class of mappings $g\colon X\to Y$ that are n-fold differentiable with respect to the system of bounded sets and have the property that for each $k\leqslant n$ the mapping $x\mapsto g^{(k)}(x)$ is sequentially continuous if we equip the space of k-linear sequentially continuous mappings from X^k to Y with the topology of convergence on bounded sets. Let $\mathcal{A}$ be a class of bounded sets in X satisfying the conditions indicated above.

4.9.7. Theorem. *Let* $n\in\mathbb{N}$. (i) *If* $f\colon U\to Y$ *is such that for every infinitely Fréchet differentiable mapping* $\varphi\colon \mathbb{R}^2\to U$ *the composition* $f\circ\varphi$ *is* $(n+1)$*-fold continuously differentiable, then* $f\in C^n_b(U,Y)$.

(ii) *If for all* $A,B\in\mathcal{A}$ *the set* $\bigcup_{a\in A, b\in B, t\in[0,1]}(a+tb)$ *is contained in* $\mathcal{A}$ *and every infinite sequence in* A *contains an* $\mathcal{A}$*-converging subsequence, then the condition that* $f\circ\varphi\in C^n(\mathbb{R}^2,Y)$ *for all functions* $\varphi\in C^\infty(\mathbb{R}^2,U)$ *implies that we have* $f\in C^n_{\mathcal{A}}(U,Y)$.

PROOF. In order to apply the previous theorem we construct the required k-linear mappings Ψ_k. Let D_1 and D_2 be the operators of differentiation with respect to the first and second variables for functions on $\mathbb{R}^2$ and let D be the differentiation operator for functions on the real line. Fix $h_1, \ldots, h_{n+1} \in X$. Set

$$\Psi_1(x)h_1 = D_1(f\circ\varphi)(0,0), \quad \text{where } \varphi(t_1,t_2) = x + t_1 h_1,$$

$$\Psi_2(x)(h_1,h_2) = D_2 D_1(f\circ\varphi)(0,0), \quad \text{where } \varphi(t_1,t_2) = x + t_1 h_1 + t_2 h_2.$$

We observe that $\Psi_2(h_1,h_2) = D\Psi_1(x+th_2)h_1(0)$. The mapping $\Psi_1(x)$ is linear for every x, because letting $\tau_1 = \lambda_1 t_1$, $\tau_2 = \lambda_2 t_2$, $\alpha(t_1,t_2) = x + t_1(\lambda_1 h_1 + \lambda_2 h_2)$, $\beta(t_1,t_2) = x + \tau_1 h_1 + \tau_2 h_2$ we have

$$\begin{aligned}\Psi_1(x)(\lambda_1 h_1 + \lambda_2 h_2) &= D_1(f\circ\alpha)(0,0)\\ &= D_1(f\circ\beta)(0,0)D_1\tau_1(0,0) + D_2(f\circ\beta)(0,0)D_1\tau_2(0,0)\\ &= \lambda_1\Psi_1(x)h_1 + \lambda_2\Psi_1(x)h_2.\end{aligned}$$

The mapping $h_1 \mapsto \Psi_2(x)(h_1,h_2)$ is linear as the derivative of a mapping linear in h_1, hence it is linear also in h_2 by the symmetry of the mixed derivative.

Suppose that for some number $k > 1$ we have multilinear mappings $\Psi_j(x)$, where $j = 1, \ldots, k$, such that

$$\Psi_k(x)(h_1,\ldots,h_k) = D\Psi_{k-1}(x+th_k)(h_1,\ldots,h_{k-1})(0),$$

$$\Psi_j(x)(h_1,\ldots,h_k) = \sum_{i=1}^{n_k}\lambda_i\Psi_k(x)v_i^k, \tag{4.9.2}$$

where v_i is some linear combination of $h_1, \ldots, h_k$ with constant coefficients (depending along with coefficients λ_i on j), and, moreover, similar relations hold for all smaller k. Then, as one can readily verify, we have the equality $\Psi_k(x)v_i^k = D_1^k(f\circ\alpha)(0,0)$, where $\alpha(t_1,t_2) = x + t_1 v_i + t_2 h_{k+1}$. Hence the mapping $t \mapsto \Psi_k(x+th_{k+1})v_i^k$ is differentiable and

$$D\Psi_k(x+th_{k+1})v_i^k(0) = D_2 D_1^k(f\circ\alpha)(0,0). \tag{4.9.3}$$

Set

$$\Psi_{k+1}(x)(h_1,\ldots,h_{k+1}) = D\Psi_k(x+th_{k+1})(h_1,\ldots,h_k)(0).$$

We observe that the operator $D_2 D_1^k$ can be written as a linear combination of operators of differentiation of the form $\partial_{u_i}^m$, where $u_i \in \mathbb{R}^2$. For this purpose we can use the identity

$$\prod_{i=1}^m z_i = \frac{1}{m!}\sum_{r=1}^m (-1)^{m-r}\sum_{i_1<i_2<\cdots<i_r}(z_{i_1}+\cdots+z_{i_r})^m,$$

which holds for commuting elements of an arbitrary ring. This identity applies to the differential operators $z_1 = D_2$, $z_i = D_1$, $i > 1$, which gives in the right-hand side some powers of the differentiations along vectors of the form $k_1 e_1 + (r-k_1)e_2$. From this on account of (4.9.2) and (4.9.3) we obtain the representation

$$\Psi_{k+1}(x)(h_1,\ldots,h_{k+1}) = \sum_{i=1}^{n_{k+1}}\mu_i\Psi_{k+1}(x)w_i^{k+1},$$

where w_i is some linear combination of $h_1, \ldots, h_{k+1}$ with constant coefficients. Thus, (4.9.2) is fulfilled for $k+1$. The linearity of $\Psi_k(x)$ in $h_1, \ldots, h_k$ yields the linearity of $\Psi_{k+1}(x)$ with respect to these variables, and the linearity in h_{k+1} follows from (4.9.3) and the linearity of $D_2(f \circ \alpha)(0)$ in h_{k+1}. Thus, the inductive step is justified. Hence for all $k \leqslant n+1$ we can determine the element $\Psi_k(x)$, moreover,

$$\Psi_k(x + th_k)(h_1, \ldots, h_k) = D\Psi_{k-1}(x + th_k)(h_1, \ldots, h_{k-1}),$$

whence $\Psi_k(x + th)h^k = D^k f(x + th)$.

Let us prove the $\mathcal{A}$-continuity of the mapping $(x, h) \mapsto \Psi_k(x)h^k$ (here $\mathcal{A}$ is the class of all bounded sets in case (i)). Let $x_i \overset{\mathcal{A}}{\to} x_0$ and $h_i \overset{\mathcal{A}}{\to} h_0$. Suppose that the elements $\Psi_k(x_i)h_i^k$ do not converge to $\Psi_k(x_0)h_0^k$. Passing to a subsequence, we can assume that they all do not belong to some neighborhood of $\Psi_k(x_0)h_0^k$. Using Exercise 4.10.44, we obtain infinitely differentiable mappings $x, h\colon \mathbb{R}^1 \to X$ and numbers $t_i \to 0$ such that $x(t_i) = x_i$ and $h(t_i) = h_i$ for infinitely many i. Let $\alpha(\tau, t) = x(t) + \tau h(t)$. Then $\Psi_k\bigl(x(t)\bigr)h^k(t) = D_1^k(f \circ \alpha)(0, t)$. By assumption, $f \circ \alpha \in C^{n+1}(\mathbb{R}^2, Y)$, therefore, the mapping $t \mapsto \Psi_k\bigl(x(t)\bigr)h^k(t)$ is continuous and $\Psi_k\bigl(x(t_i)\bigr)h^k(t_i)$ converges to $\Psi_k\bigl(x(t_0)\bigr)h^k(t_0)$, which contradicts our assumption. Thus, we have proved the $\mathcal{A}$-continuity of $(x, h) \mapsto \Psi_k(x)h^k$.

Let us now apply the inversion of Taylor's formula. To this end, setting

$$r_n(x, h) = f(x + h) - f(x) - \sum_{k=1}^{n} \frac{1}{k!}\Psi_k(x)h^k,$$

we have to verify that, as $t \to 0$, we have $r_n(x + tw, th)/t^n \to 0$ uniformly in w and h in any fixed sets in $\mathcal{A}$. The function $t \mapsto r_n(x, th)$ belongs to $C^n(\mathbb{R}^1, Y)$. Taylor's formula for functions of a real variable gives the inclusion

$$r_n(x, th) \in \frac{t^n}{n!}\overline{\mathrm{conv}}\,\bigl\{\Psi_n(x + \theta th)h^n - \Psi_n(x)h^n\colon\ \theta \in (0, 1)\bigr\},$$

whence in case (i), where the compositions belong to $C^{n+1}(\mathbb{R}^2, U)$, we obtain the inclusion

$$r_n(x + tw, th)/t^n \in \frac{t^n}{n!}\overline{\mathrm{conv}}\,\bigl\{\Psi_{n+1}(x + tw + \theta_1 th)(\theta th, h^n)\colon\ \theta, \theta_1 \in (0, 1)\bigr\}.$$

Taking into account the $\mathcal{A}$-continuity of Ψ_{n+1}, we obtain the desired uniform relation. In case (ii) our conditions on $\mathcal{A}$ (including the existence of an $\mathcal{A}$-converging subsequence in every sequence in a fixed set in $\mathcal{A}$) show that, as $t \to 0$,

$$\Psi_n(x + tw + \theta th)h^n - \Psi_n(x + tw)h^n \to 0$$

uniformly in w and h in any fixed sets in $\mathcal{A}$ and uniformly in $\theta \in (0, 1)$. Thus, the conditions of the previous theorem are fulfilled, hence we conclude that f belongs to $C^n_{\mathcal{A}}(U, Y)$. □

Our conditions on $\mathcal{A}$ are fulfilled if X is metrizable and $\mathcal{A}$ is the class of all compact sets. It is seen from Examples 4.9.2 and 4.9.5 that in case (i) we must admit that the order of differentiability can be lower by one, even if we admit all possible φ of class C^1, not only C^∞ (hence the major role in (i) is played by the condition of $\mathcal{A}$-sequential compactness of the sets in $\mathcal{A}$). Example 4.9.2 shows that we cannot replace $\mathbb{R}^2$ by $\mathbb{R}^1$ (however, see Exercise 4.10.37). Exercise 4.10.47 shows that it is not sufficient to consider the restrictions to all possible closed hyperplanes.

The next result follows immediately from the theorem.

4.9.8. Corollary. (i) *If the space X is metrizable, then $f \in C^n_c(X,Y)$ precisely when we have $f\circ\varphi\in C^n(\mathbb{R}^2,Y)$ for every infinitely Fréchet differentiable mapping $\varphi\colon \mathbb{R}^2\to X$.*

(ii) *If X is normable, then $f \in C^n(U,Y)$ precisely when $f\circ\varphi \in C^{n+1}(\mathbb{R}^2,Y)$ for every infinitely Fréchet differentiable mapping $\varphi\colon \mathbb{R}^2 \to X$. Hence the inclusion $f \in C^\infty(U,Y)$ is equivalent to the inclusion $f\circ\varphi \in C^\infty(\mathbb{R}^2,Y)$ for every infinitely Fréchet differentiable mapping $\varphi\colon \mathbb{R}^2 \to X$.*

Note that the obtained result is not obvious at all even for mappings of finite-dimensional Euclidean spaces.

Let us turn to outer compositions $\psi\circ f$. We first consider an important partial case of continuous linear functionals.

4.9.9. Theorem. *Let Y be sequentially complete for the Mackey topology and let $n \in \mathbb{N}$. If $l(f) \in C^n_{\mathcal{A}}(U)$ for all $l \in Y'$, then $f \in C^{n-1}_{\mathcal{A}}(U,Y)$, and if Y is sequentially complete for the topology $\sigma(Y,Y')$, then $f \in C^n_{\mathcal{A}}(U,Y_\sigma)$. If bounded sets are relatively compact in Y, then $f \in C^n_{\mathcal{A}}(U,Y)$.*

PROOF. Let $x \in U$, $A \in \mathcal{A}$, $l \in Y'$. By the mean value theorem for $l(f)$ we have

$$\Bigl\langle l, \frac{f(x+th)-f(x)}{t} - \frac{f(x+\tau h)-f(x)}{\tau}\Bigr\rangle = (l\circ f)'(x+\alpha th)h - (l\circ f)'(x+\beta\tau h)h \to 0 \quad (4.9.4)$$

as $t,\tau \to 0$ uniformly in $h \in A$ (here $\alpha,\beta \in (0,1)$). Hence for any sequence of nonzero numbers $t_n \to 0$ the sequence $y_n = \bigl(f(x+t_nh)-f(x)\bigr)/t_n$ is weakly fundamental uniformly in $h_n \in A$. Thus, the set of its elements is bounded in the Mackey topology. Therefore, $f(x+th) \to f(x)$ as $t \to 0$ uniformly in $h \in A$, which shows the $\mathcal{A}$-continuity of f.

If Y is weakly sequentially complete, then $\{y_n\}$ has a weak limit $y \in Y$ uniformly in $h \in A$, moreover, this limit does not depend on $\{t_n\}$, because $l(y) = (l\circ f)'(x)h$ for all $l \in Y'$. Set $f'(x)h := y$. The mapping $h \mapsto f'(x)h$ is linear and weakly bounded on all sets in $\mathcal{A}$, i.e., $f'(x) \in \mathcal{L}_{\mathcal{A}}(X,Y_\sigma)$. The equality $\langle l, f'(x)h\rangle = (l\circ f)'(x)h$ yields the inclusion $f \in C^1_{\mathcal{A}}(U,Y_\sigma)$.

Let $n \geqslant 2$. Applying in (4.9.4) the mean value theorem once again, for $t \neq \tau$ we find that

$$\frac{1}{t-\tau}\Bigl\langle l, \frac{f(x+th)-f(x)}{t} - \frac{f(x+\tau h)-f(x)}{\tau}\Bigr\rangle = \frac{\alpha t-\beta\tau}{t-\tau}(l\circ f)''\bigl(x+\beta\tau h+\theta(\alpha t-\beta\tau)h\bigr)h^2,$$

which is bounded uniformly in $h \in A$ as $t, \tau \to 0$. Hence for every sequence of nonzero numbers $t_n \to 0$ the sequence of vectors

$$y_n = \big(f(x + t_n h) - f(x)\big)/t_n$$

is Cauchy in the Mackey topology uniformly in $h \in A$. If Y is sequentially complete in the Mackey topology, then $\{y_n\}$ has a limit $y \in Y$ uniformly in $h \in A$ and this limit does not depend on $\{t_n\}$. This gives $\mathcal{A}$-differentiability of f and the equality $y = f'(x)h$. Taking into account the inclusion $l(f) \in C^2_{\mathcal{A}}(U)$ for all $l \in Y'$ and the boundedness of weakly bounded sets, we obtain that, for every fixed $A, B \in \mathcal{A}$, as $t \to 0$, the ratio $\big(f'(x+tw)h - f'(x)h\big)/t$ is bounded uniformly in $h \in A$, $w \in B$. This gives the $\mathcal{A}$-continuity of f'. Thus, $f \in C^1_{\mathcal{A}}(U, Y)$.

In order to extend this to all n we shall apply again Theorem 4.9.6. To this end, we construct mappings $\Psi_k(x) \in \mathcal{L}^k_{\mathcal{A}}(X, Y)$ for $k \leqslant n - 1$, and, if Y_σ is sequentially complete, also for $k = n$, by setting $\Psi_1(x)h_1 = \frac{d}{dt} f(x + th_1)\big|_{t=0}$,

$$\Psi_k(x)(h_1, \ldots, h_k) = \frac{d}{dt}\Psi_{k-1}(x + th_k)(h_1, \ldots, h_{k-1})\big|_{t=0}, \quad k > 1.$$

This construction is well-defined, since for any $l \in Y'$ we have

$$l(g_{k-1})(x) = (l \circ f)^{(k-1)}(x)(h_1, \ldots, h_{k-1}),$$

where g_{k-1} is defined by $g_{k-1}(x) = \Psi_{k-1}(x)(h_1, \ldots, h_{k-1})$. By construction we have the equality $\langle l, \Psi_k(x)h^k\rangle = (l \circ f)^{(k)}(x)h^k$. Hence if we write

$$f(x + h) - f(x) = \sum_{k=1}^{n-1} \Psi_k(x)h^k/k! + r_{n-1}(x, h),$$

then for any $l \in Y'$ the usual Taylor formula with the remainder in the Lagrange form yields that $\langle l, r_{n-1}(x, h)\rangle = (l \circ f)^{(n)}(x)h^n/n!$. Therefore, as $t \to 0$, the elements $t^{-n}\langle l, r_{n-1}(x + tw, th)\rangle = (l \circ f)^{(n)}(x + tw + \theta th)h^n/n!$ are bounded uniformly in w and h in any fixed sets of class $\mathcal{A}$. Theorem 4.9.6 yields that we have $f \in C^{n-1}_{\mathcal{A}}(U, Y)$.

If Y_σ is sequentially complete, we obtain also the inclusion $f \in C^n_{\mathcal{A}}(U, Y_\sigma)$ similarly to the case $n = 1$.

Finally, if all bounded sets in the space Y are relatively compact, then we have $f \in C^n_{\mathcal{A}}(U, Y)$. Indeed, in this case Y is weakly sequentially complete, hence we have $f \in C^n_{\mathcal{A}}(U, Y)$. By Taylor's formula for the weak topology we obtain

$$f(x + h) - f(x) = \sum_{k=1}^{n} f^{(k)}(x)h^k + r_n(x, h),$$

where $r_n(x, h)$ belongs to the closed convex hull of the vectors

$$[f^{(n)}(x + \theta h)h^n - f^{(n)}(x)h^n]/n!, \quad \theta \in (0, 1).$$

This implies that $r_n(x + tw, th)/t^n \to 0$ uniformly in w and h in any fixed sets in $\mathcal{A}$, since this holds for the weak topology, but in the considered case every weakly bounded set in Y is contained in a compact set in the original topology and on compact sets the weak topology coincides with the original one. Similarly we verify the $\mathcal{A}$-continuity of $f^{(n)}$. □

4.9.10. Corollary. *If X the satisfies the Mackey convergence condition, Y is sequentially complete, and for each $l \in Y'$ the function $l(f)$ is $(n+1)$-fold $\overline{c}$-differentiable in U, then f has a $\overline{b}$-derivative of order n in U.*

PROOF. The theorem gives the inclusion $f \in C^n_{\overline{c}}(U, Y_\sigma)$. Let us verify that the $\overline{c}$-derivative of order n at the point $x_0 \in U$ serves as the $\overline{b}$-derivative of order n. We can assume that for the derivatives of lower orders this is already known. We have to show that

$$f^{(n-1)}(x_0+th)(h_1,\dots,h_{n-1}) - f^{(n-1)}(x_0)(h_1,\dots,h_{n-1}) \\ = tf^{(n)}(x_0)(h_1,\dots,h_{n-1},h) + r(th),$$

where $r(th)/t \to 0$ uniformly in $h, h_1, \dots, h_{n-1}$ in every fixed bounded set. By the mean value theorem the vector $r(th)/t$ is contained in the closed convex hull of the vectors

$$[f^{(n)}(x_0+\theta th) - f^{(n)}(x_0)](h_1,\dots,h_{n-1},h), \quad \text{where } \theta \in (0,1).$$

Suppose that there are a sequence of nonzero numbers $t_j \to 0$, numbers $\theta_j \in (0,1)$, and vectors $h^j, h^j_1, \dots, h^j_{n-1}$ in a bounded set A for which

$$[f^{(n)}(x_0+\theta_j t_j h^j) - f^{(n)}(x_0)](h^j_1,\dots,h^j_{n-1},h^j) \not\to 0.$$

We can assume that $t_j > 0$. The condition yields the weak boundedness of the sequence

$$[f^{(n)}(x_0+\theta_j t_j^{1/2} t_j^{1/2} h^j) - f^{(n)}(x_0)](\varepsilon_j h^j_1,\dots,\varepsilon_j h^j_{n-1},\varepsilon_j h^j)/(\theta_j t_j^{1/2}),$$

where $\varepsilon_j^n = t_j^{1/4}$, since $\varepsilon_j h^j_i \to 0$. Since $\theta_j t_j^{1/4} \to 0$, we arrive at a contradiction. Finally, by the Mackey convergence condition for X and Exercise 4.10.54 we obtain the inclusion $f^{(n)}(x_0) \in \mathcal{L}^n_b(X,Y)$. □

This corollary can be proved directly without using the previous theorem. In addition, for the existence of the last $\overline{b}$-derivative at the point x_0 it suffices to have the existence of the $\overline{c}$-derivative of order $n+1$ at this point for the functions $l(f)$ (Exercise 4.10.55).

Note also that in these results in place of open sets U it is sufficient that U be the complement of a sequentially closed set.

4.9.11. Corollary. *Let X be bornological and satisfy the Mackey convergence condition. In order that f be n-fold b-differentiable in U it is sufficient that f be $(n+1)$-fold $\overline{c}$-differentiable as a mapping to Y_σ.*

4.9.12. Corollary. *For mappings between Fréchet spaces, the infinite Fréchet differentiability is equivalent to the infinite Hadamard differentiability.*

In the case of the one-fold Hadamard differentiability there is no loss of differentiability as compared to compositions.

4.9.13. Theorem. *Let Y be separable and sequentially complete and let $\mathcal{A}$ be the class of all compact sets in X. If $\psi\circ f \in C^1_{\mathcal{A}}(U)$ for all functions $\psi \in C^1_c(Y)$, then $f \in C^1_{\mathcal{A}}(U,Y)$.*

PROOF. Let $x \in U$. We can assume that $x = 0$ and $f(x) = 0$. The condition implies that $l(f) \in C_c^1(U)$ for all $l \in Y'$. Therefore, for every sequence of nonzero numbers $t_n \to 0$ and every fixed $h \in X$, the sequence $y_n = f(t_n h)/t_n$ is weakly fundamental and hence is bounded. If it is precompact, then it converges in the original topology of Y by the sequential completeness of Y. This shows the differentiability of f along h. If the precompactness fails, we arrive at a contradiction, since Proposition 4.1.13 gives a function $\psi \in C_c^1(Y)$ for which $\psi(0) = 0$, $\psi'(0) = 0$, but $\psi(t_n y_n)/t_n$ has no limit, which contradicts the existence of a limit of $\psi(f(t_n h))/t_n$. We observe that in this reasoning in place of $t_n h$ one can take $t_n h_n$ for an arbitrary sequence of vectors h_n in any fixed compact set. This proves the $\mathcal{A}$-continuity of f. The inclusion $l(f) \in C_c^1(U)$ yields the linearity of $\partial_h f(x)$ in h and also the inclusion $f \in C_c^1(U, Y_\sigma)$.

We show that for every compact set $A \subset X$ and every sequence $\{x_n\} \subset U$ that $\mathcal{A}$-converges to $x_0 \in U$, the set $S = \{f'(x_n)h\colon\ h \in A, n \in \mathbb{N}\}$ is precompact in Y. This set is weakly bounded, hence bounded in the Mackey topology. If it is not precompact, then by using Exercise 4.10.53 we obtain an equicontinuous sequence $\{f_n\} \subset Y'$ converging to zero uniformly on precompact sets, and also vectors $y_m = f'(x_{n_m})h_m$, $h_m \in A$ such that $f_m(y_m) = 1$, $f_m(y_k) = 0$ whenever $k < m$. Below we assume that $n_m = m$ and $x_0 = 0$, $f(x_0) = 0$. Let $z_n = f(x_n)$. Here we can pass to the case $\tau_n = f_n(z_n) \neq 0$, $f_n(y_n) > 1/2$. Indeed, if $f_n(z_n) = 0$ for some number, then the equality $f(x_n + th_n) = f(x_n) + tf'(x_n)h_n + r(x_n, th_n)$ and the properties of f_n enable us to choose a sequence $t_n \to 0$ such that for the vectors $\widehat{z}_n = f(x_n + t_n h_n)$ and $\widehat{y}_n = f'(x_n + t_n h_n)h_n$ we obtain $f_n(\widehat{z}_n) \neq 0$ and $f_n(\widehat{y}_n) = f_n(y_n) + [f_n(\widehat{y}_n) - f_n(y_n)] > 1/2$. The condition that $x_n \stackrel{\mathcal{A}}{\to} 0$ is preserved, since $t_n \to 0$ and $h_n \in A$. Below we assume that we have passed to this case. We observe that the sequence $Z = \{z_n = f(x_n)\}$ is precompact, since it converges to $f(0) = 0$ by the established $\mathcal{A}$-continuity of f. Hence $\tau_n \to 0$. We now use a reasoning similar to the proof of Proposition 4.1.13. Picking a subsequence in $\{t_n\}$, we can assume that

$$|t_n| \sum_{i=1} |t_i|^{-1} < 4^{-n} \sup\{|f_j(z)|\colon\ j \in \mathbb{N}, z \in Z\}.$$

For the same functions φ_1 and φ_2 as in Proposition 4.1.13, but with the additional condition $\varphi_2'(1) = 1$, let

$$\psi(y) = \sum_{n=1}^{\infty} (-1)^n \tau_n \varphi_1\bigl(f_1(y)/\tau_1\bigr) \cdots \varphi_1\bigl(f_{n-1}(y)/\tau_{n-1}\bigr)\varphi_2\bigl(f_n(y)/\tau_n\bigr).$$

As in the cited proposition, we have $\psi \in C_c^1(Y)$. Differentiating the series termwise, it is easy to verify that $\psi'(y) \in Y'$, $\psi'(0) = 0$ and $\psi'(z_n)y_n > 1/2$. Since by construction

$$\psi'(z_n)y_n = \psi'\bigl(f(x_n)\bigr)f'(x_n)h_n = (\psi\circ f)'(x_n)h_n$$

and $(\psi\circ f)'(0) = \psi'(0)f'(0) = 0$, we obtain a contradiction with the condition that $\psi \circ f \in C_c^1(U)$. Thus, precompactness of the set S is established. If we now admit that the sequence of vectors $f'(x_n)h - f'(x_0)h$ does not converge to zero uniformly in $h \in A$, then there exists a sequence $\{h_n\} \subset A$ such that the sequence of vectors $z_n = f'(x_n)h_n f'(x_0)h_n$ does not converge to zero in Y, but is precompact and has only zero limit point (since $z_n \to 0$ weakly), which is impossible. Thus, the mapping $f'\colon U \to \mathcal{L}_{\mathcal{A}}(X,Y)$ is $\mathcal{A}$-continuous. □

Lobanov [**321**] showed that there is no gain in Theorem 4.9.9 if in place of linear functionals we use nonlinear outer functions. More precisely, the following is true.

4.9.14. Theorem. *If in Theorem 4.9.9 we have $l(f) \in C^n_{\mathcal{A}}(U)$ for every bounded linear functional l on Y, then $\psi\circ f \in C^n_{\mathcal{A}}(U)$ for all $\psi \in C^n_b(Y)$, where b is the class of all bounded sets.*

If $n > 1$, the sets in $\mathcal{A}$ are $\mathcal{A}$-sequentially compact and $l(f) \in C^n_{\mathcal{A}}(U)$ for all functionals $l \in Y'$, then $\psi\circ f \in C^n_{\mathcal{A}}(U)$ for all $\psi \in C^n_c(Y)$, where c is the class of all compact sets.

In the same paper the inversion of Taylor's formula was employed to prove a result on differentiability of compositions, giving an important addition to the results in § 4.5. Let Z be one more Hausdorff locally convex space, V an open set in the space Y, $f\colon U \to V$, $g\colon V \to Z$.

4.9.15. Theorem. *Let $n \geqslant 1$, $f \in C^n_{\mathcal{A}}(U,Y)$, $g \in C^n_c(V,Z)$. If $n > 1$ and all sets in $\mathcal{A}$ are $\mathcal{A}$-sequentially compact, then $g\circ f \in C^n_{\mathcal{A}}(U,Z)$. In case $n = 1$, the same is true under the additional condition that either Y' coincides with the space of bounded linear functionals or Y satisfies the Mackey convergence condition.*

It should be noted that our conditions on f and g are not symmetric: say, if $n = 1$, $X = Z = \mathbb{R}^1$, $Y = \mathbb{R}^2$ and $\mathcal{A}$ is the class of all bounded sets, then the condition $g \in C^1_{\mathcal{A}}(V,Z)$ is not sufficient, which is seen from Example 4.2.11.

4.10. Complements and exercises

4.10(i). The inverse function theorem

It is well-known that the classical inverse function theorem remains true for Fréchet differentiable mappings of Banach spaces (see [**72**, Chapter 12]). A typical formulation is this: if a continuously differentiable mapping on a domain in a Banach space has an invertible derivative at some point, then in a neighborhood of this point the mapping is invertible (moreover, the inverse mapping is differentiable). A weaker version of this theorem (which can be called the theorem on differentiability of the inverse function) asserts that if an a priori invertible mapping f of a Banach space is differentiable at a point a and the inverse mapping f^{-1} is continuous at $f(a)$ and the operator $f'(a)$ is invertible, then f^{-1} is differentiable at $f(a)$. Though, this version is not a corollary of the former theorem, but in applications it is often important to have a local invertibility of a nonlinear mapping with an invertible derivative. Naturally, the question arises about analogs of these results for more general space. Certainly, here one can also consider different formulations. For example, one can seek for conditions for invertibility of nonlinear mappings with invertible derivatives or for differentiability of inverse mappings under the assumption of their existence. It turns out that for nonnormable spaces the situation becomes more complicated in all settings of the problem. First of all, the invertibility of the derivative does not always ensure a local invertibility of the mapping itself.

4.10.1. Example. Let φ be a smooth function on the real line with support in $[1/4, 3/4]$ such that $\varphi(1/2) = 1$. We define a mapping $f\colon \mathbb{R}^\infty \to \mathbb{R}^\infty$ by the formula $\bigl(f(x)\bigr)_n = x_n - \varphi(x_n)$. It is everywhere differentiable with respect to the system of bounded sets (which coincides here with the system of precompact sets), $f'(0) = I$, and the mapping $x \mapsto f'(x)$ is continuous if we equip the space of operators with the topology of convergence on bounded sets; it suffices to observe that if for each n we have $x_{n,k} \to x_n$ and $h_{n,k} \to h_n$ as $k \to \infty$, then $\varphi'(x_{n,k})h_{n,k}$ converges to $\varphi'(x_n)h_n$. However, in no neighborhood of zero can f be injective, since in any such neighborhood there is a point x with $x_n = 1/2$ and $x_k = 0$ for all $k \neq n$.

The invertibility of a derivative does not imply even a local surjectivity of f (Exercise 4.10.42).

Secondly, even for differentiable homeomorphisms with invertible derivatives the inverse mappings can be nondifferentiable. Let us mention an impressive result due to Yu.I. Prostov [**395**].

4.10.2. Theorem. *For every nonnormable Fréchet space X there exists a homeomorphism $f\colon X \to X$ such that $f(0) = 0$, at the origin f has an invertible derivative with respect to the system of bounded sets, but the mapping f^{-1} has no derivative at the origin even with respect to the system of compact sets. For $X = \mathbb{R}^\infty$ such a homeomorphism can be made even real-analytic.*

In Smolyanov [**487**], there is a description of the class of all spaces in which the theorem on differentiability of the inverse function is true.

4.10.3. Theorem. *The following conditions are equivalent for any Hausdorff locally convex space X:*

(i) *for every sequence of vectors $a_n \neq 0$ converging to zero in X there exist numbers λ_n such that the sequence $\{\lambda_n a_n\}$ is bounded, but does not converge to zero;*

(ii) *if a one-to-one mapping $f\colon X \to X$ is differentiable at a point x with respect to the system of bounded sets, the mapping f^{-1} is continuous at the point $f(x)$, and the operator $f'(x)$ is a linear homeomorphism, then f^{-1} is differentiable at $f(x)$ with respect to the system of bounded sets.*

The class of spaces satisfying condition (i) contains all normed spaces and is closed with respect to formation of strict inductive limits of increasing sequences of spaces. On the contrary, nonnormable metrizable spaces do not satisfy this condition (Exercise 4.10.35).

It is shown in Shavgulidze [**450**] that every infinite-dimensional real locally convex space is C^∞-diffeomorphic to the complement of some its closed subspace, and if the dual of the given space is weakly separable, then for such a subspace one can take any finite-dimensional subspace. For nonnormable Fréchet spaces a number of deep theorems on inverse functions is known oriented towards applications to concrete spaces (see Nirenberg [**372**] and Ekeland [**153**]).

4.10(ii). Polynomials

Let E and F be linear spaces over the field $\mathbb{R}$. In this subsection we consider only *separated* locally convex spaces.

4.10.4. Definition. *A mapping $f\colon E \to F$ is called homogeneous polynomial (or a homogeneous polynomial) of degree (or order) $k \in \mathbb{N}$ if*

$$f(x) = W_k(x, \dots, x),$$

where the mapping $W_k\colon E^k \to F$ is linear in every variable separately. A homogeneous polynomial of degree $k = 0$ is a constant mapping.

A polynomial (or a polynomial mapping) of degree n on E is a mapping of the form $f = f_n + \cdots + f_1 + f_0$, where f_k is a homogeneous polynomial of degree k for all $k \leqslant n$. The smallest possible n in such a representation is the exact degree of the polynomial.

It is clear that a k-linear mapping W_k generating a homogeneous polynomial of degree k can be chosen symmetric, passing to the mapping

$$V_k(x_1, \dots, x_k) = \frac{1}{k!} \sum W_k(x_{i_1}, \dots, x_{i_k}),$$

where summation is taken over all permutations of $1, \dots, k$.

4.10.5. Proposition. *Let f_k be a homogeneous polynomial of degree k. Then the symmetric k-linear mapping generating this polynomial is uniquely determined by the formula*

$$V_k(x_1, \dots, x_k) = \frac{1}{k!} \sum_{\varepsilon_1, \dots, \varepsilon_k \in \{0,1\}} (-1)^{k - \varepsilon_1 - \cdots - \varepsilon_k} f_k(x_0 + \varepsilon_1 x_1 + \cdots + \varepsilon_k x_k), \tag{4.10.1}$$

where $x_0 \in E$ is an arbitrary point.

The proof is delegated to Exercise 4.10.45. The reader is offered to prove also the following elementary, but not quite trivial fact.

4.10.6. Lemma. *If a function f on $\mathbb{R}^n$ is polynomial in every variable for all fixed values of other variables, then it is a polynomial. In particular, the degrees of polynomials obtained by fixing $n-1$ variables are uniformly bounded.*

4.10.7. Lemma. *Suppose that a mapping $f\colon E \to F$ is polynomial on every affine line. Then there exists a sequence of polynomial mappings $f_k\colon E \to F$, $k = 0, 1, \dots$, homogeneous of order k, such that $f(x) = \sum_{k=0}^\infty f_k(x)$, where for each x only finitely many terms are nonzero.*

PROOF. For every vector x the mapping $t \mapsto f(tx)$ is polynomial. Hence we have $f(tx) = \sum_{k=0}^{\infty} t^k f_k(x)$, where $f_k(x) \in F$ and only finitely many terms are nonzero. Since

$$f(stx) = \sum_{k=0}^{\infty} s^k t^k f_k(x) = \sum_{k=0}^{\infty} s^k f_k(tx),$$

we have $f_k(tx) = t^k f_k(x)$. We show that f_k is a homogeneous polynomial of order k. Let us define V_k via f_k according to formula (4.10.1). Then we have $f_k(x) = V_k(x, \dots, x)$, hence it suffices to verify that V_k is k-linear. This follows from Lemma 4.10.6, since the proof of the linearity of the mapping V_k in every variable reduces everything to some finite-dimensional subspace, but on this subspace f is a polynomial. □

We need the following interesting algebraic fact.

4.10.8. Lemma. *Let $f\colon E \to F$ be a homogeneous polynomial mapping, let $U \subset E$ be a balanced set, and let $V \subset F$ be an absolutely convex set such that $f(U + a) \subset V$ for some $a \in E$. Then $f\bigl((2e)^{-1}U\bigr) \subset V$.*

PROOF. Let f be homogeneous of order k. Then (4.10.1) yields the equality

$$f(x) = \frac{1}{k!}\sum_{m=0}^{k} (-1)^{k-m} C_k^m f(a + mx).$$

For $x \in U/k$ and $m \leqslant k$ we have $(-1)^{k-m} f(a + mx) \in V$. Hence $f(x) \in \lambda V$ if $x \in U/k$, $\lambda = \sum_{m=0}^{k} C_k^m/k! = 2^k/k!$. Hence for $x \in U$ we have the inclusion $f(x) \in (2k)^k V/k! \subset (2e)^k V$. In place of $(2e)^{-1}$ one can take $1/2$, for the proof one has to apply an estimate due to Chebyshev (see [**145**, Exercise 78, p. 369]). □

We shall now see that the continuity of a polynomial is equivalent to its continuity at one point.

4.10.9. Proposition. *Let E and F be real locally convex spaces and let $f\colon E \to F$ be a homogeneous polynomial mapping of order k. Then the following assertions are equivalent*:

(i) *the mapping f is continuous,*

(ii) *the mapping f is continuous at some point,*

(iii) *for every continuous seminorm q on F, the function $q \circ f$ is bounded on some nonempty open set.*

If E is metrizable, then the continuity of f is equivalent to the continuity of the compositions $l \circ f$ for all $l \in F'$.

PROOF. With the aid of the previous lemma it is easy to verify that the boundedness of $q \circ f$ on some neighborhood gives boundedness on a neighborhood of zero, whence it is easy to obtain the continuity at zero. The remaining implications are trivial.

If E is metrizable, then the absence of a neighborhood of zero on which the function $q \circ f$ is bounded gives a sequence $a_n \to 0$ for which $\{f(a_n)\}$ is not bounded, but then there exists a functional $l \in F'$ for which the sequence of numbers $l\bigl(f(a_n)\bigr)$ is not bounded. □

4.10.10. Corollary. *A real polynomial f on a locally convex space E is continuous precisely when it is bounded on some neighborhood of zero. This is also equivalent to the continuity of its homogeneous components (and of their generating symmetric k-linear functions). The latter is also true for polynomial mappings to locally convex spaces F.*

PROOF. Suppose that a polynomial f of degree k is bounded on an absolutely convex neighborhood of zero U. Then the family of polynomials $p_x(t) = f(tx)$ of degree k, where $x \in U$, is uniformly bounded on $[-1, 1]$. Hence their derivatives at zero are uniformly bounded, which gives the boundedness on U of the homogeneous part f of order k, whence its continuity follows. The continuity of the generating k-linear functions is clear from (4.10.1). If f takes values in F and q is a continuous seminorm on F, then, taking a neighborhood of zero U on which the function $q(f)$ is bounded, we obtain the boundedness on U of the polynomials $l(f)$, where $l \in F'$ and $|l| \leqslant q$. For the homogeneous components V_k of the mapping f this gives the boundedness on U of the functions $l(V_k)$, hence also of $q(V_k)$. □

4.10.11. Corollary. *Any continuous real polynomial on a dense linear subspace in a locally convex space extends to a continuous polynomial on the whole space. This is also true for polynomial mappings with values in complete locally convex spaces.*

PROOF. The previous corollary reduces this to extensions of k-linear functions generating the homogeneous components, hence to extensions of continuous linear functionals or operators. □

4.10.12. Theorem. *Let E and F be two real locally convex spaces such that E is Baire and let a mapping $f\colon E \to F$ be continuous and polynomial on every affine line. Then f is a polynomial mapping. In addition, the homogeneous components of f are continuous.*

PROOF. By Lemma 4.10.7 there are homogeneous of order k polynomial mappings f_k for which $f = \sum_{k=0}^{\infty} f_k$ pointwise. We observe that each f_k is continuous. This is verified by induction. Indeed, if $f_1, \ldots, f_{k-1}$ are continuous, then we have $f_k(x) = \lim\limits_{n\to\infty} g_n(x)$,

$$g_n(x) = n^k\bigl(f(n^{-1}x) - f(0) - f_1(n^{-1}x) - \cdots - f_{k-1}(n^{-1}x)\bigr).$$

The mappings g_n are continuous. Hence, by the Baire property of E, for every continuous seminorm q on F the sequence of functions $q \circ g_n$ is uniformly bounded on some neighborhood (Exercise 3.12.152). The previous proposition gives the continuity of f_n. The sets $M_n = \{x \in E\colon\ f_k(x) = 0 \ \forall\, k > n\}$ are closed and cover E. Using the Baire property of E once again, we find M_n with a nonempty interior. It is clear from Lemma 4.10.8 that $f_k \equiv 0$ if $k > n$. □

Certainly, the Baire property of E is important. For example, on the space of finite sequences the function $f(x) = \sum_{k=0}^{\infty} x_k^k$ is not polynomial, although it is polynomial on every affine line.

4.10(iii). Ordinary differential equations in locally convex spaces

Here we give a very short survey of the principal results related to ordinary differential equations in Hausdorff locally convex spaces. A much more detailed account and extensive bibliography can be found in Bogachev [**62**] and Lobanov, Smolyanov [**322**].

Let E be a locally convex space and let $f\colon E \to E$ be continuous. Consider the ordinary differential equation

$$x'(t) = f\bigl(x(t)\bigr), \quad x(0) = x_0 \in E. \tag{4.10.2}$$

Its solution is a continuous mapping $f\colon [0,\delta) \to E$ on some semiopen interval $[0,\delta)$ that is differentiable in $(0,\delta)$ and satisfies (4.10.2). Certainly, one can consider solutions on any interval $[t_0, t_0+\delta)$ in place of $[0,\delta)$. The classical result (Peano's theorem) asserts that in the case $E = \mathbb{R}^n$ this equation is solvable on some $[0,\delta)$. The situation is completely different in infinite-dimensional spaces.

4.10.13. Example. (i) (Dieudonné's example) Let $E = c_0$ and

$$f(x) = \bigl(f_n(x_n)\bigr), \ f_n(z) = |z|^{1/2}.$$

Then f is continuous in E, but (4.10.2) has no solutions for $x_0 = (a_n)$ with infinitely many positive coordinates. Indeed, if $x(t) = (x_n(t))$ is a solution on the interval $[0,\delta)$, then we have $t = 2|x_n(t)|^{1/2} - 2a_n^{1/2} \to 0$ as $n \to \infty$, where we take only n with $a_n > 0$.

(ii) Let $E = \mathbb{R}^\infty$ and $f(x) = \bigl(f_n(x_n)\bigr)$, $f_n(z) = z^2 + n^2$. Then the interval of existence of a solution to the one-dimensional ordinary differential equation $y'(t) = y(t)^2 + n^2$, $y(0) = 0$ tends to zero, hence there is no common interval of existence of all these solutions.

(iii) Let $E = C_0^\infty[-1,1]$ be the space of all infinitely differentiable functions φ with support in the closed interval $[-1,1]$, equipped with the system of norms $p_n(\varphi) = \sup_t |\varphi^{(n)}(t)|$. Then E is a nuclear Fréchet space, the differential operator $f\colon \varphi \mapsto \varphi''$ is linear and continuous, but the equation (4.10.2) is solvable only for $x_0 = 0$. Indeed, if $\varphi(t)$ is a solution with nonzero $\varphi(0)$, then $g(t,x) = \varphi(t)(x)$ satisfies the equation

$$\frac{\partial g}{\partial t} = \partial_x^2 g, \ g(0,x) = \varphi(0).$$

This follows from the fact that convergence in E implies pointwise convergence. However, all solutions of the latter equation are analytic and cannot have bounded support, with the exception of the case $g \equiv 0$.

It turns out that example (i) is typical for Banach spaces. A.N. Godunov [**196**] established a remarkable fact: on every infinite-dimensional Banach space there is a continuous mapping f for which equation (4.10.2) has no solutions. Later S.A. Shkarin [**455**] observed that this theorem extends to all Fréchet spaces, moreover, the case of nonnormable Fréchet spaces turned out even much simpler. By Theorem 3.12.44, every such space E has a complemented subspace E_0 isomorphic to $\mathbb{R}^\infty$; let $P\colon E \to E_0$ be the corresponding projection. By Corollary 3.9.14 of Michael's selection theorem there exists a continuous mapping $S\colon E_0 \to E$

such that $PS(x) = x$ for all $x \in E_0$. This gives a mapping $F\colon E \to E$, $F(x) = S\bigl(f(Px)\bigr)$, where f is the mapping on $E_0 = \mathbb{R}^\infty$ constructed above, for which (4.10.2) has no solutions. Then the equation $x'(t) = F\bigl(x(t)\bigr)$ with zero initial condition has no solutions in E, since otherwise the projection of a solution to E_0 would be a solution in E_0 due to our choice of S. Moreover, Shkarin [**456**] modified this construction in such a way that the following final result was obtained.

4.10.14. Theorem. *On every infinite-dimensional Fréchet space there exists a continuous mapping f such that the equation $x'(t) = f\bigl(x(t)\bigr)$, $x(t_0) = x_0$ has no solutions at all for any t_0 and x_0.*

Let us mention a number of positive results. The first general existence theorems were obtained by A.N. Tychonoff for continuous mappings of the space $\mathbb{R}^T$ with bounded image and by R. Phillips for continuous mappings of Fréchet spaces with precompact image. The existence and uniqueness theorem for solutions to equations with Lipschitzian f extends obviously to Banach spaces. The following theorem from Astala [**28**] has the interesting feature that it covers all continuous mappings of a given space.

4.10.15. Theorem. *If a sequentially complete locally convex space E contains a compact or sequentially compact barrel, then Peano's theorem holds in E.*

S.V. Shkarin conjectured that the validity of Peano's theorem in a sequentially complete locally convex space E is equivalent to the property that E possesses a countably compact barrel, but this conjecture remains open. An important general result is the following theorem due to V.M. Millionstchikov [**347**], which combines Lipschitzness and compactness (we formulate it for autonomous equations).

4.10.16. Theorem. *Let E be a sequentially complete locally convex space and let $f\colon E \to E$.*

(i) *Suppose that for every seminorm p in some collection $\mathcal{P}$ defining the topology, we have $p\bigl(f(x) - f(y)\bigr) \leqslant Kp(x - y)$ in a neighborhood U of a point x_0. Then equation* (4.10.2) *has a solution on some interval.*

(ii) *The same is true if f is continuous and has the form $f = f_1 + f_2$, where f_1 satisfies the conditions in* (i), *f_2 is continuous in U and $f_2(U)$ is contained in a compact set.*

In particular, Peano's theorem is true for continuous mappings of sequentially complete spaces with relatively compact images.

It is interesting that in the space $\mathbb{R}^\infty$, unlike $C_0^\infty[-1,1]$, every linear equation (4.10.2) is solvable. This nontrivial result with a misleadingly elementary formulation was proved by Shkarin [**452**].

4.10.17. Theorem. *Let A be a continuous linear operator in $\mathbb{R}^\infty$ and let a mapping $f\colon [0,1] \to \mathbb{R}^\infty$ be continuous. Then the equation $x'(t) = Ax(t) + f(t)$ is solvable for every initial condition.*

It should be noted that there might be no uniqueness of solution in this case.

4.10.18. Example. Let $(Ax)_n = x_{n+1}$. Then the linear equation $x'(t) = Ax$, $x(0) = 0$ has nonzero solutions in $\mathbb{R}^\infty$. Indeed, for every infinitely differentiable function φ on the real line the formula $x_n(t) = \varphi^{(n)}(t)$ defines a solution to the equation $x'(t) = Ax(t)$ with the initial condition $(\varphi'(0), \ldots, \varphi^{(n)}(0), \ldots)$. It remains to take a nonzero function φ such that $\varphi^{(n)}(0) = 0$ for all n.

The following useful fact was first observed by A.N. Godunov.

4.10.19. Proposition. *Let A be a continuous linear operator on a locally convex space E and let $z \in E$. Suppose that the Cauchy problem $x'(t) = Ax(t)$, $x(0) = z$ has a solution in the weak topology $\sigma(E, E')$. Then this solution is a solution in the Mackey topology.*

PROOF. Let $x\colon [0, c) \to E$ be a solution in the weak topology. We have to verify the differentiability of the mapping x in the Mackey topology. Let $t \in (0, c)$. The set $x([t - r, t + r])$, where $[t - r, t + r] \subset (0, c)$, is weakly compact. If $0 < |h| < r$, then by the mean value theorem

$$\frac{x(t+h) - x(t)}{h} - Ax(t) \in \overline{\operatorname{abs\,conv}}\,\{Ax(t+s) - Ax(t)\colon\ s \in [t - r, t + r]\},$$

and, for every neighborhood of zero V in the Mackey topology, the set in the right-hand side is contained in V for sufficiently small r. Therefore, $(x(t+h)-x(t))/h$ converges to $Ax(t)$ in the Mackey topology as $h \to 0$. □

We shall say that a locally convex space E possesses property (P) if convergence $x_n \to 0$ in E and convergence $y_n \to 0$ in E' in the topology $\sigma(E', E)$ yield convergence $\langle y_n, x_n \rangle \to 0$.

Property (P) is shared by all barrelled spaces (more generally, by Mackey spaces for which E' is sequentially complete in the topology $\sigma(E', E)$, see Exercise 4.10.52).

4.10.20. Proposition. *If E has property (P) and $u \in E'$ is such that the Cauchy problem $y'(t) = A^*y(t)$, $y(0) = u$ in E' is solvable, then $\langle u, x(t) \rangle = 0$ for every solution x to the Cauchy problem $x'(t) = Ax(t)$, $x(0) = 0$. Hence if the Cauchy problem for A^* has a solution for all $u \in E'$, then the Cauchy problem for A has at most one solution.*

We shall say that an operator $A \in \mathcal{L}(E, E)$ has exponent if for every $x \in E$ for all $t \in \mathbb{R}$ the series $\exp(tA)x := \sum_{n=1}^\infty t^n A^n x/n!$ converges in the space E. If E is barrelled, then the operator $\exp(tA)$ is continuous. For convergence of this series in the Mackey topology for all t it suffices to have weak convergence (Exercise 4.10.40). It is verified directly that $\exp(tA)z$ is a solution to the Cauchy problem $x'(t) = Ax(t)$, $x(0) = z$. However, the unique solvability of these Cauchy problems does not imply the existence of exponent.

4.10.21. Example. Let $E = \mathcal{D}(\mathbb{R}^1)$, $Af = f'$. Then, for every $z \in \mathcal{D}(\mathbb{R})$, the Cauchy problem $x'(t) = Ax(t)$, $x(0) = z$ has a unique solution defined by the formula $x(t)(s) = z(t + s)$, but $\exp(tA)z$ exists only for $z = 0$. Indeed, it is readily verified that a similar formula gives a solution to the Cauchy problem for A^* in $\mathcal{D}'$, which by the previous proposition ensures the unique solvability of the Cauchy problem for A. The existence of $\exp(tA)z$ yields the analyticity of $z(t+s)$ in t, which is only possible for $z = 0$ due to compactness of support.

4.10.22. Proposition. *If E has property (P) (say, is barrelled) or is locally complete (say, is sequentially complete), then the existence of $\exp(tA)z$ for all t and z implies the unique solvability of the Cauchy problem for A for all initial conditions.*

PROOF. If E is barrelled, then it suffices to observe that $R(t)u = \exp(tA)^*u$ gives a solution to the Cauchy problem $y'(t) = A^*y(t)$, $y(t) = u$. In the general case we cannot use the existence of $R(t)$, since the operator $\exp(tA)$ need not be continuous. Let $x'(t) = Ax(t)$, $x(0) = 0$. We show that $x(t) = 0$. We observe that it suffices to verify that for every $\tau > 0$ and every $y \in E'$ the function

$$t \mapsto \sum_{n=1}^{\infty} t^n \langle (A^*)^n y, x(\tau - t)\rangle / n!$$

has the zero derivative (this function equals 0 at $t = \tau$ and $\langle y, x(\tau)\rangle$ at $t = 0$). By the equality $x(\tau - t)' = -Ax(\tau - t)$ the formal differentiation gives the desired result, but we have to justify this. To this end, it suffices to verify uniform convergence of the series of the derivatives on compact intervals. The derivative of the nth partial sum of the series equals

$$-t^n \langle (A^*)^{n+1} y, x(\tau - t)\rangle / n! = -t^n \langle (A^*)^n y, Ax(\tau - t)\rangle / n!.$$

If $t \in [-c/4, c/4]$, where $c > 0$, then the values $x(\tau - t)$ belong to a compact set in E, which in the case of locally complete E is contained in a Banach disc B. The existence of $\exp(tA)$ yields that for every $v \in E$ there is a number $k(v) > 0$ such that

$$|\langle (A^*)^n y, v\rangle| \leqslant k(v) c^{-n} n!.$$

The functionals $(A^*)^n y$ are continuous on E_B, hence the Banach–Steinhaus theorem shows that the numbers $k(v)$ can be chosen uniformly bounded for all $v \in B$, hence for all $v = x(\tau - t)$ with $t \in [-c/4, c/4]$. This ensures uniform convergence of the derivatives of the terms of our series. If in place of the local completeness we have property (P), then the functionals $t^n (A^*)^n y/n!$ converge pointwise to zero by the existence of $\exp(tA)$. Hence for every convergent sequence of numbers $t_n \to t$ one has $t_n^n \langle (A^*)^n y, Ax(\tau - t_n)\rangle / n! \to 0$, whence the desired uniform convergence follows. □

For additional results, see Leonov, Shkarin [**313**], Shkarin [**460**], [**462**], [**463**]. Smolyanov [**490**] developed an analog of Holmgren's method of proving uniqueness of solutions to nonlinear differential equations in locally convex spaces.

4.10(iv). Passage to the limit in derivatives

Here we discuss convergence of differentiable mappings.

4.10.23. Theorem. *Let E be a topological vector space, let G be a locally convex space, and let β be some system of bounded subsets of E with the following properties*:

(1) *if* $\{h_n\}$ *is a sequence converging in* E, *then the set of its elements is an element of the system* β;

(2) *if* $B \in \beta$, $\{\lambda_n\}$ *is a convergent real sequence and* $\{h_n\} \subset B$, *then the set of all elements of the sequence* $\{\lambda_n h_n\}$ *is an element of the system* β;

(3) *if* $x \in E$ *and* $B \in \beta$, *then* $x + B \in \beta$.

Let $f_n \colon E \to G$, *where* $n \in \mathbb{N}$, *be* $\bar{\beta}$*-differentiable mappings and let* f_n' *be the* $\bar{\beta}$*-derivative of* f_n. *Then the following assertions are true*:

(I) *if the sequence* $\{f_n\}$ *converges in* $\mathcal{F}_\beta(E,G)$ *to a mapping* f *and the sequence* $\{f_n'\}$ *converges in* $\mathcal{F}_\beta\bigl(E, \overline{\mathcal{L}}_\beta(E,G)\bigr)$ *to some mapping* g, *then the mapping* f *is* $\bar{\beta}$*-differentiable and* $f' = g$;

(II) *if the sequence* $\{f_n'\}$ *converges in* $\mathcal{F}_\beta\bigl(E, \overline{\mathcal{L}}_\beta(E,G)\bigr)$ *to a mapping* g *and the sequence* $\{f_n(x_0)\}$ *converges in* G *for some* $x_0 \in E$, *then, in case of sequentially complete* G, *there exists a* $\bar{\beta}$*-differentiable mapping* $f \colon E \to G$ *such that* $f' = g$ *and one has convergence* $f_n \to f$ *in* $\mathcal{F}_\beta(E,F)$.

PROOF. We first prove (II). Let $\{h_k\}$ be a sequence of elements of some set $B \in \beta$, $\{t_k\}$ a bounded sequence of nonzero real numbers. For every collection (m, n, k) of natural numbers we set

$$b_{m,n;k} = t_k^{-1}\bigl(f_m(x_0 + t_k h_k) - f_m(x_0)\bigr) - t_k^{-1}\bigl(f_n(x_0 + t_k h_k) - f_n(x_0)\bigr).$$

Then

$$b_{m,n;k} \in \overline{\mathrm{conv}}\,\bigl\{f_m'(x_0 + \theta t_k h_k)h_k - f_n'(x_0 + \theta t_k h_k)h_k \colon\ 0 < \theta < 1\bigr\}$$

for all admissible m, n, k. Since by condition (2) on the system β the set of all elements of the sequence $\{\lambda_n h_n\}$ is an element of the system β for every convergent real sequence $\{\lambda_n\}$ and since $f_m' - f_n' \to 0$ in $\mathcal{F}_\beta\bigl(E, \overline{\mathcal{L}}_\beta(E,G)\bigr)$ as $n, m \to \infty$, we have

$$f_m'(x_0 + \lambda_k h_k)h_k - f_n'(x_0 + \lambda_k h_k)h_k \xrightarrow[m,n,k\to\infty]{} 0 \qquad (4.10.3)$$

in G, provided that the sequence $\{\lambda_k\}$ converges in $\mathbb{R}^1$. Hence for every neighborhood of zero V in the space G, there exist natural numbers n_0, m_0, k_0 such that if $n > n_0$, $k > k_0$ and $m > m_0$, then

$$c_{m,n;k} \equiv \{f_m'(x_0 + \theta t_k h_k)h_k - f_n'(x_0 + \theta t_k h_k)h_k \colon\ 0 < \theta < 1\} \subset V, \quad (4.10.4)$$

since otherwise for some neighborhood of zero V_0 in G we could find strictly increasing sequences $\{n_i\}$, $\{m_i\}$, $\{k_i\}$ of natural numbers and a sequence $\{\theta_i\}$ of real numbers such that the real sequence $\{t_{k_i}\theta_i\}$ converges and

$$f_{m_i}'(x_0 + \theta_i t_{k_i} h_{k_i})h_{k_i} - f_{n_i}'(x_0 + \theta_i t_{k_i} h_{k_i})h_{k_i} \notin V_0$$

for all i, but this contradicts (4.10.3).

If the set V is closed and convex, then (4.10.4) yields that $\overline{\mathrm{conv}}\, c_{m,n;k} \subset V$ for such m, n, k; since the space G is locally convex, it follows that in G we have

$$b_{m,n;k} \xrightarrow[m,n,k\to\infty]{} 0. \qquad (4.10.5)$$

Since by our hypothesis there exists a point $z_0 \in G$ such that $f_n(x_0) \to z_0$, we have $a_{mn} \equiv f_m(x_0) - f_n(x_0) \to 0$ in G as $m, n \to \infty$. It follows from this convergence along with (4.10.5) (if we set $t_k = 1$ for all k) that

$$f_m(x_0 + h_k) - f_n(x_0 + h_k) \xrightarrow[m,n,k\to\infty]{} 0. \tag{4.10.6}$$

Therefore, for every $x \in E$ we obtain

$$f_m(x) - f_n(x) \xrightarrow[m,n\to\infty]{} 0, \tag{4.10.7}$$

since we can assume that $h_k = x - x_0$ for every k.

It follows from (4.10.7) by the sequential completeness of G that there exists a mapping $f\colon E \to G$ such that $f_m(x) - f(x) \to 0$ for every vector $x \in E$. In particular, for every k we have

$$f_m(x_0 + h_k) - f(x_0 + h_k) \xrightarrow[m\to\infty]{} 0. \tag{4.10.8}$$

Relations (4.10.6) and (4.10.8) and Corollary 4.5.4 yield that

$$f_m(x_0 + h_k) - f(x_0 + h_k) \to 0 \quad \text{as } m, k \to \infty.$$

This means that one has $f_m \to f$ in $\mathcal{F}_\beta(E, G)$.

We now prove that f is β-differentiable at every point $x \in E$ and $f' = g$. Since it is already proved that the sequence $\{f_m(x)\}$ converges for all $x \in E$, we can assume without loss of generality that $x = x_0$.

For every pair i, j of natural numbers set

$$c_{ij} = t_j^{-1}\bigl(f_i(x_0 + t_j h_j) - f_i(x_0)\bigr), \quad c_j = t_j^{-1}\bigl(f(x_0 + t_j h_j) - f(x_0)\bigr).$$

According to (4.10.5) we have $c_{mk} - c_{nk} \to 0$ as $m, n, k \to \infty$ and according to (4.10.8) we have $c_{mk} - c_k \to 0$ as $m \to \infty$ for each k. Hence

$$c_{mk} - c_k \xrightarrow[m,k\to\infty]{} 0. \tag{4.10.9}$$

This fact is verified directly.

Now, for every pair i, j of natural numbers we set

$$d_{ij} = c_{ij} - f_i'(x_0)h_j, \quad d_j = c_j - g(x_0)h_j.$$

Since by convergence of the sequence $\{f_i'\}$ in the space $\mathcal{F}\bigl(E, \overline{\mathcal{L}}_\beta(E, G)\bigr)$ we have the relation

$$f_i'(x_0)h_j - g(x_0)h_j \xrightarrow[i,j\to\infty]{} 0,$$

it follows from (4.10.9) that

$$d_{ij} - d_j \xrightarrow[i,j\to\infty]{} 0. \tag{4.10.10}$$

In addition, the differentiability of every function f_i implies that

$$d_{ij} \xrightarrow[j\to\infty]{} 0. \tag{4.10.11}$$

It follows from (4.10.10) and (4.10.11) that $d_j \to 0$. This means that the mapping $r\colon h \mapsto f(x_0 + h) - f(x_0) - g(x_0)h$ is β-small. Thus, in order to establish β-differentiability of the mapping f at the point x_0 and the equality $f'(x_0) = g(x_0)$, it suffices to show that $g(x_0) \in \overline{\mathcal{L}}(E, G)$. This inclusion is a corollary of the fact that the limit in $\mathcal{F}_\beta(E,G)$ of any sequence of sequentially continuous mappings is a sequentially continuous mapping. Let us prove the latter assertion. Let $s_n \to s$ in $\mathcal{F}_\beta(E,G)$, let $\{h_k\}$ be a convergent sequence of elements of E, and let $x_0 \in E$. Since the set of all elements of the sequence $\{h_k\}$ belongs to the family β, we have

$$f_m(x_0+h_k)-f(x_0+h_k)\xrightarrow[m,k\to\infty]{}0,$$

$$f_m(x_0+h_k)-f(x_0)\xrightarrow[k\to\infty]{}0\ \forall m,\quad f_m(x_0)\xrightarrow[m\to\infty]{}f(x_0).$$

By assertion (II) proved above these relations yield that $f(x_0+h_k)\to f(x_0)$. This means that f is sequentially continuous. Thus, assertion (II) of the theorem is proven. The part of our reasoning concerned with establishing $\overline{\beta}$-differentiability of f proves at the same time assertion (I). □

4.10.24. Remark. The conditions imposed in Theorem 4.10.23 on the system β are satisfied for the systems σ_b^E and σ_c^E.

4.10(v). Completeness of spaces of smooth mappings

If E and G are topological vector spaces and β is some class of bounded subsets of E, then the symbol $\widetilde{C}^0_\beta(E,G)$ will denote the set of all sequentially continuous mappings from E to G equipped with the topology induced from $\mathcal{F}_\beta(E,G)$; it is easy to see that $\widetilde{C}^0_\beta(E,G)$ is a topological vector group. Let $\widetilde{C}^k_\beta(E,G)$, where $k\in\mathbb{N}$, denote the topological vector group defined as follows:

$\widetilde{C}^k_\beta(E,G)$ consists of mappings $f\in\widetilde{C}^0_\beta(E,G)$ such that f is k-fold β-differentiable at every point x and the mapping $x\mapsto f^{(j)}(x)$, $E\to\overline{\mathcal{L}}^j_\beta(E,G)$ is sequentially continuous for every $j\in\{1,2,\dots,k\}$. The topology in $\widetilde{C}^k_\beta(E,G)$ is the projective topology with respect to the mappings

$$\widetilde{C}^k_\beta(E,G)\to\mathcal{F}(E,\overline{\mathcal{L}}^j_\beta(E,G)),\quad f\mapsto f^{(j)}.$$

Let $C^k_\beta(E,G)$, where $k\in\{0,1,2,\dots\}$, denote the subspace of the topological vector space $\widetilde{C}^k_\beta(E,G)$ consisting of all mappings f such that for every j in $\{0,1,2,\dots,k\}$ the mapping $x\mapsto f^{(j)}(x)$, $E\to\overline{\mathcal{L}}^j_\beta(E,G)$ is bounded in the sense that it takes bounded sets to bounded and is equipped with the induced topology. It is obvious that all $C^k_\beta(E,G)$ are topological vector spaces.

Let $C^\infty_\beta(E,G)$ denote the projective limit of the sequence of spaces $C^k_\beta(E,G)$ with respect to the canonical mappings.

4.10.25. Theorem. *Let E and G be topological vector spaces such that G is sequentially complete. Then the topological vector space $C^k_\beta(E,G)$ is sequentially complete. If β is a class of bounded subsets of E satisfying the hypotheses of Theorem* 4.10.23, *then the topological vector group $\widetilde{C}^k_\beta(E,G)$ is sequentially complete, where $k=0,1,2,\dots$.*

PROOF. The first assertion is easily verified directly and the second one follows from Theorem 4.10.23. □

Let us study the differentiability of the composition mapping. We need the following lemma.

4.10.26. Lemma. *Let E_0 and G_0 be topological vector spaces, where G_0 satisfies the Mackey convergence condition from Remark 4.3.2. and let T_0 be a Hausdorff locally convex space. Suppose that L is a topological vector subspace in $C_b^0(E_0, G_0)$, a mapping $F \in C_b^1(G_0, T_0)$ is everywhere twice $\bar{b}$-differentiable and the mapping F_0'' is bounded in the sense above. Let $F_*\colon L \to \widetilde{C}_b^0(G_0, T_0)$ be defined as follows: if $\varphi \in L$, then $F_*(\varphi)(x) = F\bigl(\varphi(x)\bigr)$. Then F_* is everywhere sequentially continuous and $\bar{b}$-differentiable and for any φ, $h \in L$ and $x \in E$ have*

$$(F_*)'(\varphi)(h)(x) = F'\bigl(\varphi(x)\bigr)\bigl(h(x)\bigr). \tag{4.10.12}$$

PROOF. By the Mackey convergence condition imposed on G_0, the boundedness and sequential continuity of the mapping $z \mapsto F'(z)$, $G_0 \to \overline{\mathcal{L}}_b^1(G_0, T_0)$ yield that, for every mapping $\varphi \in C_b^0(E_0, G_0)$, the mapping

$$h \mapsto \bigl[x \mapsto F'\bigl(\varphi(x)\bigr)\bigl(h(x)\bigr)\bigr], \quad L \to \widetilde{C}_b^0(E_0, T_0)$$

is well-defined and sequentially continuous. Suppose now that $\varphi \in L$, $\{x_n\}$ is a bounded sequence in E, $\{h_n\}$ is a bounded sequence in L and $\{t_n\}$ is a sequence of nonzero numbers converging to zero. Then the elements

$$t_n^{-1}\Bigl(F\bigl(\varphi(x_n) + t_n h_n(x_n)\bigr) - F\bigl(\varphi(x_n)\bigr)\Bigr) - F'\bigl(\varphi(x_n)\bigr)\bigl(h_n(x_n)\bigr)$$

belong to the set

$$\begin{aligned}&\overline{\text{conv}}\,\bigl\{F'\bigl(\varphi'(x_n) + \theta t_n h_n(x_n)\bigr)h_n(x_n) - F'\bigl(\varphi(x_n)h_n(x_n)\bigr) \colon 0 < \theta < 1\bigr\}\\ &\quad\subset \overline{\text{conv}}\,\bigl\{F''\bigl(\varphi(x_n) + \lambda\theta t_n h_n(x_n)\bigr) \cdot t_n h_n(x_n) \cdot h_n(x_n) \colon\ 0 < \lambda, \theta < 1\bigr\}.\end{aligned}$$

The boundedness of the sequences $\{h_n\}$ and $\{x_n\}$ yield that, for any choice of sequences $\{\lambda_n\}$ and $\{\theta_n\}$ in $[0,1]$, the sequences $\{\varphi(x_n) + \lambda_n\theta_n t_n h_n(x_n)\}$ and $\{h_n(x_n)\}$ are bounded in G. Since $t_n \to 0$, it follows that for such sequences $\{\lambda_n\}$ and $\{\theta_n\}$ we have

$$F''\bigl(\varphi(x_n) + \lambda_n\theta_n t_n h_n(x_n)\bigr) \cdot t_n h_n(x_n) \cdot h_n(x_n) \to 0$$

in T_0 as $n \to \infty$. This means that we have

$$t_n^{-1}\bigl(F_*(\varphi + t_n h_n) - F_*(\varphi)\bigr) - F'\bigl(\varphi(\cdot)\bigr)h_n(\cdot) \to 0 \text{ as } n \to \infty$$

in $C_b^0(E_0, T_0)$, i.e., the mapping F_* is $\bar{b}$-differentiable at the point φ and equality (4.10.12) holds. □

Let E, G, and T be topological vector spaces, where T is locally convex, and let Com be the composition mapping:

$$\text{Com}\colon\ (f,g) \mapsto f \circ g, \quad \mathcal{F}(E,G) \times \mathcal{F}(G,T) \to \mathcal{F}(E,T).$$

Suppose that p is a natural number, q is a nonnegative integer number, c_1 is the restriction of the mapping Com to $C_c^q(E,G) \times C_s^{q+p+1}(G,T)$ and c_2 is the restriction of the mapping Com to $C_b^q(E,G) \times C_b^{q+p+1}(G,T)$.

4.10.27. Theorem. *The mapping c_1 is sequentially continuous and p-fold $\bar{\mathfrak{s}}$-differentiable as a mapping from its domain of definition to the space $C^q_{\mathfrak{s}}(E,T)$. The mapping c_2 is sequentially continuous and p-fold Bb-differentiable as a mapping from its domain of definition to the space $C^q_{\mathfrak{b}}(E,T)$, and if the space G satisfies the Mackey convergence condition from Remark* 4.3.2, *then the mapping c_2 is p-fold $\bar{\mathfrak{b}}$-differentiable. If in this situation $r \in \{1,2,\ldots,p\}$ and*

$$(g,f),(k_1,h_1),\ldots,(k_r,h_r) \in C^q_{\mathfrak{b}}(E,G)\times C^{q+p+1}_{\mathfrak{b}}(G,T),$$

then with the natural notation

$$c_2^{(r)}(g,f)\bigl((k_1,h_1),\ldots,(k_r,h_r)\bigr) \tag{4.10.13}$$

$$= \sum_{i=1}^{r} h_i^{(r-1)}(g)(k_r,\ldots,k_{i+1},k_{i-1},\ldots,k_1) + f^{(r)}(g)(k_r,\ldots,k_1);$$

if $(g,f),(k_1,h_1),\ldots,(k_r,h_r) \in C^q_{\mathfrak{s}}(E,G)\times C^{q+p+1}_{\mathfrak{s}}(G,T)$, then there holds the equality obtained from equality (4.10.13) *by replacing the symbol c_2 by c_1.*

PROOF. The fact that the images of the mappings c_2 and c_1 belong to the corresponding spaces follows from the results in § 4.5. The sequential continuity of the mapping

$$c_2\colon\ C^q_{\mathfrak{b}}(E,G)\times C^{q+p+1}_{\mathfrak{b}}(G,T) \to C^q_{\mathfrak{b}}(E,T)$$

follows from Theorem 4.5.18 (verify this).

It follows almost directly from the lemma above that the mapping c_2 is one-fold $\bar{\mathfrak{b}}$-differentiable for $p=1$, $q=0$. Indeed, this lemma asserts that the mapping c_2 in this case is $\bar{\mathfrak{b}}$-differentiable along the subspace $\widetilde{C}^0_{\mathfrak{b}}(E,G)$. In addition, the linearity and continuity of the mapping

$$\widetilde{C}^2_{\mathfrak{b}}(G,T) \ni \varphi \mapsto \varphi\circ f \in \widetilde{C}^0_{\mathfrak{b}}(E,T)$$

mean that c_2 is $\bar{\mathfrak{b}}$-differentiable along the subspace $\widetilde{C}^2_{\mathfrak{b}}\bigl(C^q_{\mathfrak{b}}(E,G)\times C^{q+p+1}_{\mathfrak{b}}(G,T)\bigr)$. Hence the sequential continuity of the mapping

$$\widetilde{C}^0_{\mathfrak{b}}(E,G)\times\widetilde{C}^2_{\mathfrak{b}}(G,T) \to \overline{\mathcal{L}}^1_{\mathfrak{b}}\bigl(\widetilde{C}^0_{\mathfrak{b}}(E,G)\times\widetilde{C}^2_{\mathfrak{b}}(G,T),\widetilde{C}^0_{\mathfrak{b}}(E,T)\bigr),$$

$$(f_1,f_2)\mapsto \bigl[(\varphi_1,\varphi_2)\mapsto \varphi_2\circ f_1\bigr]$$

(which is a corollary of the fact that the space G satisfies the Mackey convergence condition) and Theorem 4.8.1 imply that the mapping c_2 is $\bar{\mathfrak{b}}$-differentiable and for the corresponding variables we have

$$(c_2)'(g,f)(k,h) = h\bigl(g(\cdot)\bigr) + f\bigl(g(\cdot)\bigr)\bigl(k(\cdot)\bigr). \tag{4.10.14}$$

Let now $s \geqslant 1$, $p=1$. Using Lemma 4.5.17, one can show that the linear mapping

$$\widetilde{C}^q_{\mathfrak{b}}(E,G)\times\widetilde{C}^{q+2}_{\mathfrak{b}}(G,T) \to \widetilde{C}^q_{\mathfrak{b}}(E,T),\quad (k,h)\mapsto h\circ g + f'\bigl(g(\cdot)\bigr)k(\cdot)$$

is sequentially continuous for any fixed g and f in the corresponding spaces. Hence for the proof of $\bar{\mathfrak{b}}$-differentiability of the mapping c_2 and equality (4.10.13) it suffices to show that if $k_n, g \in \widetilde{C}^q_{\mathfrak{b}}(E,G)$, $h_n, f \in \widetilde{C}^{q+2}_{\mathfrak{b}}(G,T)$, $n\in\mathbb{N}$ and the sequences $\{k_n\}$ and $\{h_n\}$ are bounded in the corresponding spaces, then with the natural notation we have

$$t_n^{-1}\big((f+t_nh_n)\circ(g+t_nk_n)-f\circ g\big)-h_n\circ g-f'(g(\cdot))\circ k_n(\cdot)\to 0$$

in $\widetilde{C}_b^q(E,T)$, provided that $t_n\to 0$, $t_n\in\mathbb{R}^1\backslash\{0\}$. For this it suffices to show (by (4.10.14)) that for every $j\in\{1,2,\ldots,q\}$ the sequence of $\overline{b}$-derivatives of order j of the mappings

$$x\mapsto \frac{(f+t_nh_n)\big((g+t_nk_n)(x)\big)-f(g(x))}{t_n}-h_n(g(x))-f'(g(x))k_n(x)$$

converges to some mapping from E to $\overline{\mathcal{L}}_b^j(E,T)$ uniformly on bounded sets in E, since by Theorem 4.10.23 and relation (4.10.14) this mapping will be identically zero. The latter assertion will be proved if we show that (for each indicated j) the mapping

$$\Phi\colon\ (g,f)\mapsto (f\circ g)^{(j)},\quad \widetilde{C}_b^q(E,G)\times\widetilde{C}_b^{q+2}(G,T)\to\widetilde{C}_b^0\big(E,\overline{\mathcal{L}}_b^j(E,T)\big)$$

is $\overline{b}$-differentiable. In this case (again by Theorem 4.10.23) the $\overline{b}$-derivative of this mapping at the point (g,f) at the increment (k,h) will automatically equal the jth derivative of the mapping in the right-hand side of (4.10.14).

By Theorem 4.5.18 the mapping Φ is the sum of finitely many mappings; each of these mappings is determined by natural numbers $m\in\{1,2,\ldots,j\}$ and $n_1,\ldots,n_m$, $n_1+\cdots+n_m=j$; the mapping determined by these numbers is defined as follows:

$$(g\circ f)\mapsto f^{(m)}(g(\cdot))\big(g^{(n_1)}(\cdot),\ldots,g^{(n_m)}(\cdot)\big)\in\widetilde{C}_b^0\big(E,\overline{\mathcal{L}}_b^j(E,T)\big).$$

Let us denote the latter mapping by Ψ; thus, everything reduces to proving its $\overline{b}$-differentiability.

Let $V=G\times\overline{\mathcal{L}}_b^{n_1}(E,G)\times\cdots\times\overline{\mathcal{L}}_b^{n_m}(E,G)$, $f\in\widetilde{C}^{q+2}(G,T)$, and let the mapping F_{Vf} from V to $\overline{\mathcal{L}}_b^j(E,T)$ be defined as follows:

$$F_{Vf}(z,l_1,\ldots,l_m)=f^{(m)}(z)\big(l_1(\cdot),\ldots,l_m(\cdot)\big).$$

By Lemma 4.5.16 we obtain that the mapping F_{Vf} is infinitely $\overline{b}$-differentiable along the subspace $\overline{\mathcal{L}}_b^{n_1}(E,G)\times\cdots\times\overline{\mathcal{L}}_b^{n_m}(E,G)$. The two-fold $\overline{b}$-differentiability of the mapping $f^{(m)}\colon\ G\to\overline{\mathcal{L}}_b^m(G,T)$ and the sequential continuity and boundedness (on bounded sets) of its first two derivatives yield that the derivative of the mapping F_{Vf} along the subspace $\overline{\mathcal{L}}_b^{n_1}(E,G)\times\cdots\times\overline{\mathcal{L}}_b^{n_m}(E,G)$ is $\overline{b}$-differentiable along the subspace G and that the mapping F_{Vf} itself is twice $\overline{b}$-differentiable along the subspace G. Moreover, again by Lemma 4.5.16, the $\overline{b}$-derivative of the mapping F_{Vf} along the subspace G turns out to be $\overline{b}$-differentiable along the subspace $\overline{\mathcal{L}}_b^{n_1}(E,G)\times\cdots\times\overline{\mathcal{L}}_b^{n_m}(E,G)$.

All the listed derivatives are sequentially continuous as mappings to the respective spaces. The indicated properties of the mapping F_{Vf} and Theorem 4.8.1 yield that this mapping is twice $\overline{b}$-differentiable. Finally, the fact that the values of the first two derivative of the mapping $f^{(m)}$ are bounded on bounded sets implies that the same is true for the first two derivative of F_{Vf} (to observe this one has to use the explicit expressions for these derivatives established by using Lemma 4.5.16).

It follows from what has been said in the last sentence that the mapping Ψ is $\overline{b}$-differentiable along the subspace $\widetilde{C}^q(E,G)$. Indeed, if $f \in \widetilde{C}^{q+2}(G,T)$, then the mapping $g \mapsto \Psi(g,f)$ is the composition of two mappings:

$$\widetilde{C}^q_b(E,G) \xrightarrow{\Lambda} \widetilde{C}^{q-m}_b(E,V) \xrightarrow{(F_{Vf})_*} C^0_b(E,\overline{\mathcal{L}}^j_b(E,T));$$

here $\Lambda = \big[g \mapsto (g(\cdot), g^{(n_1)}(\cdot), \ldots, g^{(n_m)}(\cdot)\big]$ and the symbol $(\cdot)_*$ is defined in the lemma above.

The mapping Λ is linear and continuous, hence it is infinitely $\overline{b}$-differentiable. By the cited lemma the mapping $(F_{Vf})_*$ is also $\overline{b}$-differentiable. Hence by Theorem 4.5.11 the mapping $g \mapsto \Psi(g,f)$ is $\overline{b}$-differentiable as well. Since the mapping Ψ is linear and continuous in the second argument, it is $\overline{b}$-differentiable along the subspace $\widetilde{C}^{q+2}(G,T)$. The Mackey convergence condition, imposed on the space G, implies, as one can easily see, that the mapping

$$(g,f) \mapsto \Psi'_2(g,f),$$

$$\widetilde{C}^q_b(E,G)\times\widetilde{C}^{q+2}_b(G,T)\to\overline{\mathcal{L}}^1_b\Big(\widetilde{C}^q_b(E,G)\times\widetilde{C}^{q+2}_b(G,T),\widetilde{C}^0_b\big(E,\overline{\mathcal{L}}^j_b(E,T)\big)\Big),$$

$$\Psi'_2(\cdot,\cdot) = \Psi'_{\widetilde{C}^{q+2}_b(G,T)}(\cdot,\cdot)$$

is sequentially continuous. By Theorem 4.8.1 the mapping Φ is $\overline{b}$-differentiable. Thus, it is proved that the mapping c_2 is one-fold $\overline{b}$-differentiable for every s and also that for $r = 1$ equality (4.10.13) holds.

Suppose now that $p \geqslant 2$, $j \in \{1,2,\ldots,p-1\}$ and that it is already proved that the mapping

$$c_3\colon\ \widetilde{C}^q_b(E,G)\times\widetilde{C}^{q+p+1}_b(G,T) \to \widetilde{C}^q_b(E,T)$$

is j-fold $\overline{b}$-differentiable and for $r = j$ and the corresponding values of variables we have equality (4.10.13). Let us prove that the same will be true for $j+1$.

It suffices to show that the mapping

$$G_1\colon\ \widetilde{C}^q_b(X,Y)\times\widetilde{C}^{q+p+1}_b(Y,Z) \to \overline{L}^j\big(\widetilde{C}^q_b(X,Y)\times\widetilde{C}^{q+p+1}_b(Y,Z), \widetilde{C}^q_b(X,Z)\big),$$

$$G_1(g,f)\colon\ (k_1,\ldots,k_j) \mapsto f^{(j)}(g)(k_j,\ldots,k_1)$$

and the mappings

$$Q_i\colon\ \widetilde{C}^q_b(X,Y) \to \overline{L}^j\big(\widetilde{C}^q_b(X,Y)\times\widetilde{C}^{q+p+1}_b(Y,Z), \widetilde{C}^q_b(X,Z)\big),$$

$$Q_i(g)\colon\ (k_1,\ldots,k_{i-1},k_{i+1},\ldots,k_j,h_i) \mapsto h_i^{(j-1)}(g)(k_j,\ldots,k_1)$$

with $i \leqslant j$ are one-fold $\overline{b}$-differentiable and

$$G'_1(g,f)(k,h)\colon\ (k_1,\ldots,k_j) \mapsto h^{(j)}(g)(k_j,\ldots,k_1) + f^{(j+1)}(g)(k,k_j,\ldots,k_1),$$

$$Q'_1(g)(k)\colon\ (k_1,\ldots,k_{i-1},k_{i+1},\ldots,k_j,h_i) \mapsto h_i^{(j)}(g)(k,k_j,\ldots,k_1).$$

For the proof of $\overline{b}$-differentiability of the mappings Q_i and G_1 we use that the assertion of the theorem is already proven for $p = 1$. Namely, the mapping G_1

is the composition of the following mappings: $(g, f) \mapsto (g, f^{(j)}) \mapsto f^{(j)}(g)$; $\bar{b}$-differentiability of the first of these mappings is obvious and $\bar{b}$-differentiability of the second one follows from the already proven part of the theorem. Suppose now that $i \leqslant j$ and h_i is a fixed element of the space $\widetilde{C}_b^{q+p+1}(Y, Z)$. Then the mapping

$$g \mapsto [(k_1, \ldots, k_{i-1}, k_{i+1}, \ldots, k_j) \mapsto h_i^{(j-1)}(g)(k_j, \ldots, k_1)]$$

from $\widetilde{C}_b^q(X, Y)$ to $\overline{L}_b^j(\widetilde{C}_b^q(X, Y), \widetilde{C}_b^q(X, Z))$ is $\bar{b}$-differentiable by the already proven part of the theorem. This means that Q_1 is also $\bar{b}$-differentiable as a mapping to the space that coincides with $\overline{L}_b^j(\widetilde{C}_b^q(X, Y) \times \widetilde{C}_b^{q+p+1}(Y, Z), \widetilde{C}_b^q(X, Z))$ as a set, but is equipped with the topology of pointwise convergence. It follows from Theorem 4.8.1 that Q_i is also $\bar{b}$-differentiable (the verification of one of the conditions of this theorem, namely, the continuity of the corresponding derivative, employs the Mackey convergence condition). The assertion of the theorem on $\bar{b}$-differentiability is proven.

The assertion on $\bar{s}$-differentiability is proved similarly; here the assertions analogous to those which proofs employed the Mackey convergence condition turn out to be true without this condition.

Finally, the proof of the remaining part of the theorem on Bb-differentiability is also close to the given proof of its last part; some difference is that in place of the Mackey convergence condition one has to verify directly membership of the arising mappings in the required spaces. In addition, the linear mappings that are derivatives (of the corresponding mappings) are merely bounded, but not necessarily sequentially continuous. □

Let us consider the so-called exponential law for spaces of smooth mappings. Let X and Y be two topological vector spaces and let $\beta(X)$ be some class of bounded subsets of X.

We say that β-differentiable mappings of the spaces X and Y satisfy the exponential law if, for every locally convex space Z, the mapping from $\widetilde{C}_\beta^\infty(X \times Y, Z)$ to $\widetilde{C}_\beta^\infty(X, \widetilde{C}_\beta^\infty(Y, Z))$ taking a mapping f to the mapping $x \mapsto F_f$ defined by $F_f(x)(y) = f(x, y)$, is an isomorphism of these spaces.

4.10.28. Theorem. *The exponential law is fulfilled for $\bar{b}$-differentiable mappings of arbitrary topological vector spaces satisfying the Mackey convergence condition.*

The proof can be read in Smolyanov [**489**, p. 57].

Diverse results connected with constructions and extensions of smooth functions on locally convex spaces can be found in Atkin [**29**], Colombeau [**105**], Shkarin [**457**], [**459**].

4.10(vi). Differentiability via pseudotopologies

Here we briefly explain the approach to differentiability based on pseudotopologies mentioned at the beginning of the chapter. The set of all filters on a nonempty set P will be denoted by $\Phi(P)$; $2^{\Phi(P)}$ is the set of all subsets of $\Phi(P)$.

4.10.29. Definition. *A pseudotopology on P is a mapping $\tau\colon P \to 2^{\Phi(P)}$ such that* (i) *the filter Φ_x of all subsets containing x belongs to $\tau(x)$ for all x,* (ii) *for every point x we have $\varphi \cap \psi \in \tau(x)$ for all $\varphi, \psi \in \tau(x)$ and if a filter $\eta \in \Phi(P)$ contains an element of $\tau(x)$, then $\eta \in \tau(x)$.*

A space equipped with a pseudotopology is called *pseudotopological*. Every topological space becomes pseudotopological if $\tau(x)$ is defined as the collection of all filters converging to x.

A filter φ in a pseudotopological space P converges to a point $p \in P$ if $\varphi \in \tau(p)$. A mapping $f\colon P \to Q$ of pseudotopological spaces is called continuous at a point p if it takes filters converging to p to filters converging to $f(p)$.

The product of two pseudotopological spaces (P, τ_P) and (Q, τ_Q) is equipped with the pseudotopology $\tau_{P\times Q} = \tau_P \times \tau_Q$ that is the weakest pseudotopology among all pseudotopologies on $P\times Q$ for which the projections on the factors are continuous. Convergence of a filter η in this pseudotopology to (p, q) means that η contains $\varphi\times\psi$, where a filter φ converges to p in τ_P and a filter ψ converges to q in τ_Q.

The notion of a *pseudotopological linear* space (X, τ) is now introduced as a linear space for which the linear structure agrees with the pseudotopology in the sense that the operations $(x, y) \mapsto x-y$ and $(\lambda, x) \mapsto \lambda x$ are continuous on $X\times X$ and $\mathbb{R}^1\times X$, respectively. Let $\tau_{\rm tv}$ denote the strongest vector topology on X for which the generated pseudotopology is majorized by the pseudotopology τ and let $X_{\rm tv} = (X, \tau_{\rm tv})$.

Suppose we are given two pseudotopological linear spaces X and Y. Suppose that the space $\mathcal{F}(X, Y)$ of mappings from X to Y is also equipped with some pseudotopology τ. We shall say that a mapping $r\colon X \to Y$ is τ-small of the first order if the following mapping $t \mapsto r_t$, $\mathbb{R}^1 \to \mathcal{F}(X, Y)$ is continuous at zero: $r_t(x) = t^{-1}r(tx)$ if $t \neq 0$, $r_0(x) = 0$. Certainly, we can assume that we consider only mappings in a subset of $\mathcal{F}(X, Y)$ equipped with a pseudotopology.

Suppose also that we are given linear subspaces $\mathcal{H}(X, Y)$ and $\mathcal{R}(X, Y)$ in $\mathcal{F}(X, Y)$ such that $\mathcal{R}(X, Y)$ consists of mapping that are τ-small of the first order and if $L \in \mathcal{H}(X, Y)$, $r \in \mathcal{R}(X, Y)$ and $L(x) = r(x)$ for all x in some neighborhood of zero in $\tau_{\rm tv}$, then $L = 0$. The role of $\mathcal{H}(X, Y)$ is usually played by some space of linear mappings. Let $x_0 \in V \subset X$. Assume that there is a neighborhood of zero $V_0 \in \tau_{\rm tv}$ such that $x_0 + V_0 \subset V$.

4.10.30. Definition. *A mapping $f\colon V \to Y$ is called $\mathcal{H}_\tau$-differentiable at x_0 if there is a mapping $f'(x_0) \in \mathcal{H}(X, Y)$ with $f(x_0 + h) - f(x_0) = f'(x_0)(h) + r(h)$, where $r \in \mathcal{R}(X, Y)$.*

Various kinds of differentiability can be obtained by varying parts of $\mathcal{F}(X, Y)$, on which different pseudotopologies τ are defined, and choosing different classes $\mathcal{H}(X, Y)$ and $\mathcal{R}(X, Y)$. In detail these questions are discussed in the book Smolyanov [**489**]. It is shown in Smolyanov [**488**] that for a broad class of locally convex spaces a large part of the known definitions of infinite differentiability splits into two series, each consisting of equivalent definitions. Note that such differentiabilities do not always imply continuity.

4.10(vii). Smooth functions on Banach spaces

Let us mention some interesting facts connected with smooth functions on Banach spaces. This case is so special that one could hardly find appropriate to discuss it in a book on general topological vector spaces, but still some information is worth including. We touch three questions: approximations by smooth functions, existence of smooth functions with bounded support and Sard's theorem. In much more detail this direction is discussed in the books Benyamini, Lindenstrauss [**50**], Deville, Godefroy, Zizler [**118**], Fabian et al. [**155**], [**156**], Llavona [**319**], and the papers Nemirovskii, Semenov [**368**], Tsar'kov [**531**], and Bogachev [**61**], where one can find also the proofs of the presented facts and further references.

We start with some remarks on approximations by differentiable mappings in infinite-dimensional spaces. Let X be a separable Banach space with the closed unit ball U.

4.10.31. Theorem. *Every uniformly continuous real function f on U is uniformly approximated by Lipschitzian functions that are Hadamard differentiable.*

However, on the space $C[0,1]$ even the norm is not approximated uniformly on U by Fréchet differentiable functions.

4.10.32. Theorem. *On a Hilbert space, uniformly continuous functions are uniformly approximated by functions with bounded and continuous second Fréchet derivatives, but there is a Lipschitzian function on l^2 which is not uniformly approximated on U by functions with uniformly continuous second derivatives.*

Thus, even in the case of a Hilbert space the border between positive and negative results runs between the continuity and the uniform continuity of bounded second derivatives of approximating functions.

The situation with approximation of infinite-dimensional mappings is even more complicated. There are uniformly continuous mappings from separable Banach spaces to l^2 which cannot be uniformly approximated by Lipschitzian mappings. However, uniformly continuous mappings between Hilbert spaces possess uniform Lipschitzian approximations with bounded Fréchet derivatives. Some problems in constructing approximations are connected with the existence of smooth functions with bounded support. Let us mention some interesting facts.

4.10.33. Theorem. (i) *There are no nonzero Fréchet differentiable functions on $C[0,1]$ with bounded support.*

(ii) *If a Banach space X and its dual possess nonzero functions with bounded support and locally Lipschitzian derivatives, then X is isomorphic to a Hilbert space.*

(iii) *The existence of nonzero functions with bounded support and Lipschitzian derivatives is equivalent to the existence of an equivalent norm with a Lipschitzian derivative on the unit sphere.*

(iv) *On c_0 there is a nonzero C^∞-function with bounded support* (*on c_0 there is even an equivalent norm that is real-analytic outside of the origin*).

(v) *If X possesses a nonzero C^k-function with bounded support, then X contains an isomorphic copy either of c_0 or of l^k.*

On Rolle's theorem in infinite-dimensional spaces (including some counterexamples), see Shkarin [**453**]. We end this excursion by mentioning Sard's theorem, according to which a smooth mapping of finite-dimensional spaces takes the set of its critical points (i.e., points where the derivative is not a surjective operator) to a measure zero set. This theorem has a number of infinite-dimensional analogs, but these analogs require various additional conditions. For example, Exercise 4.10.48 exhibits a polynomial of the third degree on l^2 for which the image of the set of critical points fills in the whole interval, and also a polynomial mapping of the third degree in $C[0,1]$ for which the image of the set of critical points has a nonempty interior. See also Bates [**45**] and Exercise 4.10.51.

Exercises

4.10.34° Prove the differentiability of the mapping (4.5.1).

4.10.35. Let X be a nonnormable metrizable locally convex space. Show that X contains a sequence of nonzero vectors a_n converging to zero such that, for any sequence of numbers λ_n, the sequence of vectors $\lambda_n a_n$ either converges to zero or is unbounded.

4.10.36° Give an example of two Gâteaux differentiable mappings of the plane the composition of which is not Gâteaux differentiable.

4.10.37. Prove that the mapping $F\colon X \to Y$ of normed spaces is Hadamard differentiable at $x_0 \in X$ precisely when there exists an operator $L \in \mathcal{L}(X,Y)$, such that, for every differentiable mapping $\varphi\colon \mathbb{R}^1 \to X$ with $\varphi(0) = x_0$ differentiable at zero, the composition $F\circ\varphi\colon \mathbb{R}^1 \to Y$ is differentiable at the point 0 and $(F\circ\varphi)'(0) = L\varphi'(0)$.

HINT: see Bogachev, Smolyanov [**72**, Theorem 12.2.5].

4.10.38° (i) Prove that the intersection of affine subspaces of a vector space is an affine subspace.

(ii) Let X and Y be two linear spaces and M let be an affine subspace in $X\times Y$. Prove that the set $M_y = \{x \in X\colon (x,y) \in M\}$ is an affine subspace in X for every $y \in Y$.

4.10.39. (S.A. Shkarin) Let E be a sequentially complete locally convex space and let $F\colon E\times[0,1] \to E$ be such that for every $x \in E$ the mapping $t \mapsto F(x,t)$ is Lipschitzian (in every seminorm on E from some family defining the topology) and for every $t \in [0,1]$ the mapping $x \mapsto F(x,t)$ is linear and sequentially continuous. Prove that F is sequentially continuous in both variables.

4.10.40° Let E be a locally convex space and let $\{v_n\} \subset E$ be such that for every point $t \in (-1,1)$ the series $\sum_{n=1}^{\infty} t^n v_n$ converges in the topology $\sigma(E,E')$. Prove that it converges in the Mackey topology.

4.10.41. (Astala [**28**]) Prove that for a locally convex space E the following conditions are equivalent: (i) the space E is sequentially complete and contains a compact barrel; (ii) it has the form (X',τ), where X' is the dual to the barrelled normed space X equipped with the locally convex topology τ intermediate between $\sigma(X',X)$ and the topology of convergence on precompact sets in X.

4.10.42° (i) Let $X = C(\mathbb{R}^1)$ be equipped with the topology of uniform convergence on compact sets and $f(x)(t) = \exp x(t)$. Show that f a has derivative f' with respect to the system of compact sets and this derivative is continuous with respect to the topology of convergence on compact sets, $f'(0) = I$, but no neighborhood of $f(0)$ belongs completely to the image of f.

(ii) Construct a similar example of a mapping $f\colon \mathbb{R}^\infty \to \mathbb{R}^\infty$.

4.10.43.° Suppose that in the situation of Example 4.2.16 the set K is compact, convex and balanced and that $\bigcup_{n=1}^\infty nK$ is dense in X. Show that f has zero Gâteaux derivative at all points of $\bigcup_{0\leqslant t<1} tK$.

4.10.44. Suppose that $\{x_n\}$ is a sequence in a locally convex space E such that there exist numbers $\lambda_n \to +\infty$ for which $\lambda_n x_n \to 0$. Show that there exist a function $f \in C^\infty(\mathbb{R}^1, E)$ and numbers $t_n \to 0$ such that $f(t_n) = x_n$ for infinitely many n.

HINT: note that passing to a subsequence one can assume that $2^{n^2} x_n \to 0$; consider the mapping $f(t) = \sum_{n=1}^\infty \varphi_1(2^{n-1}t)\varphi_2(2^n t)x_n$, where $\varphi_1, \varphi_2 \in C^\infty(\mathbb{R}^1)$, $\varphi_1(t) = 1$ if $t \leqslant 1/2$, $\varphi_1(t) = 0$ if $t \geqslant 3/4$, $\varphi_2(t) = 0$ if $t \leqslant 3/4$, $\varphi_2(t) = 1$ if $t \geqslant 1$; then we have $f(2^{-n}) = x_n$.

4.10.45. Prove Proposition 4.10.5 and Lemma 4.10.6.

4.10.46. (Lobanov [**321**]) Suppose that a separable locally convex space E contains a bounded set that is not precompact. Then, for every $m \in \mathbb{N}$ and every precompact set $K \subset E$, there exists a function $f \in C_c^m(E)$ for which the derivative of order m with respect to the system of bounded sets exists at no point of K.

HINT: modify the proof of Proposition 4.1.13 as follows: take

$$\tau_n = \sup\{|f_i(x)|\colon\ i \geqslant n, x \in K\},$$

replace 4^{-n} by 8^{-n} in the estimate for t_n, pass to the case $\tau_n < t_n/8$, in the formula for $f(x)$ replace t_n by t_n^m, and take the function φ_2 with the additional condition that $\varphi_2^{(m-1)}(t) = 1$ if $|t-1| \leqslant 1/8$.

4.10.47. (Lobanov [**320**]) Let $\varphi\colon \mathbb{R} \to l^2$, $t \mapsto \{t^n/n!\}_{n=1}^\infty$ and let U_k be the ball in l^2 of radius 8^{-k^2} centered at the point $a_k = \varphi(2^{-k})$. Then the function f on l^2 that vanishes outside of all U_k and on each U_k is infinitely Fréchet differentiable with support strictly inside U_k such that $f(a_k) = 1$, is discontinuous at zero, but its restriction to every closed hyperplane is infinitely Fréchet differentiable.

4.10.48. (Bonic [**79**]) Let $\varphi(t) = 3t^2 - 2t^3$, $\varphi_n(t) = 2^{1-n}\varphi(2^{n/4}t)$. (i) Show that the function f on l^2 defined by the formula $f(x) = \sum_{n=1}^\infty \varphi_n(x_n)$ is a continuous polynomial of the third degree and the image of the set of its critical points contains $[0,2]$.

(ii) Verify that the mapping $f\colon x \mapsto x^3$ of the space $C[0,1]$ has for the set of critical points the set S of functions x having zeros, S possesses inner points and $f(S) = S$.

4.10.49. (Bonic [**79**]) Let $F\colon l^2 \to l^2$, $F(x) = (x_1^2, x_n^2, \ldots)$. Show that F is a continuous second order polynomial mapping, but it is not true that it takes weakly convergent sequences to norm convergent, although all derivatives $F'(x)$ are compact operators.

4.10.50. Let F be a Fréchet space, let $u_0 \in F$ and suppose that for every $n \in \mathbb{N}$ we are given a continuous n-linear symmetric mapping $u_n\colon \mathbb{R}^n \to F$. Prove that there exists an infinitely differentiable mapping $u\colon \mathbb{R}^1 \to F$ such that $u^{(n)}(0) = u_n$ for all $n \geqslant 0$. For more general results, see Colombeau [**105**] and Shkarin [**454**].

4.10.51. (Bates [**46**]) Let X be an infinite-dimensional Banach space. Prove that there exists an infinitely Fréchet differentiable surjective mapping $f\colon X \to \mathbb{R}^2$ such that the derivative $f'(x)$ is surjective at no point x.

4.10.52. Show that if E is a Mackey space such that E' is sequentially complete in the topology $\sigma(E',E)$ (which is true for barrelled E), then convergence $x_n \to 0$ in E and convergence $y_n \to 0$ in the topology $\sigma(E',E)$ imply convergence $\langle y_n, x_n\rangle \to 0$.

4.10.53. Suppose that a set A is bounded, but not precompact in a separable locally convex space E. Prove that there exist a sequence $\{a_n\} \subset A$ and an equicontinuous sequence $\{f_k\} \subset E'$ pointwise converging to zero for which $f_n(a_n) = 1$ and $f_m(a_n) = 0$ whenever $m > n$. Cf. Exercise 3.12.151.

HINT: there exist an absolutely convex neighborhood of zero W and a sequence $\{a_n\} \subset A$ for which $a_i - a_j \notin W$ for all $i \neq j$; let $\{b_i\}$ be an everywhere dense sequence in E; passing to a subsequence in $\{a_n\}$, we can assume that $p_W(a_n - a_i) \geqslant 1$ and $p_W(a_n - b_i) \geqslant 1$ if $n > i$; on the linear span of $a_1,\ldots,a_n$ and $b_1,\ldots,b_{n-1}$ there is a linear functional $f_n \leqslant p_W$ such that $f_n(a_n) = 1$, $f_n(a_i) = f_n(b_i) = 0$ if $i < n$; one can extend f_n to E and observe that $\{f_n\} \subset W^\circ$ and $f_n(b_i) \to 0$ for every fixed i.

4.10.54. Suppose that a locally convex space X satisfies the Mackey convergence condition. Show that for every locally convex space Y every n-linear mapping $L\colon X \to Y$ that is bounded on sequences converging to zero is also bounded on bounded sets.

HINT: if sequences $\{h_1^j\},\ldots,\{h_n^j\}$ are bounded in X, but $\{L(h_1^j,\ldots,h_n^j)\}$ is unbounded in Y, then there exist numbers $c_j \to 0$ for which the sequence $\{L(c_j h_1^j,\ldots,h_n^j)\}$ is unbounded; taking numbers $T_j \to \infty$ such that $T_j c_j h_1^j \to 0$ in X, observe that $T_j^{-1/n} h_k^j \to 0$, whence one obtains that $L(c_j T_j T_j^{-1/n} h_1^j, T_j^{-1/n} h_2^j,\ldots,T_j^{-1/n} h_n^j) \to 0$.

4.10.55. Prove the following analog of Corollary 4.9.10: if X satisfies the Mackey convergence condition and Y is sequentially complete, then for the existence of $\overline{b}$-derivative of $f\colon X \to Y$ of order n at the point x_0 it suffices that all functions $l(f)$, where $l \in Y'$, have $\overline{c}$-derivatives of order $n+1$ at this point.

4.10.56. Let us equip the space E of finite sequences with the topology of the strict inductive limit of the spaces $\mathbb{R}^n$. Let $e_n \in E$, $e_n = (0,\ldots,0,1,0,\ldots)$ with 1 in the nth place. Let a functional f_n be defined by the formula $f_n(x) = x_n$, $x = (x_n)$. Show that the operator $Ax = \sum_{n=1}^\infty f_n(x)e_{n+1}$ is continuous in E, but the equation $x'(t) = Ax(t)$, $x(0) = z$ is solvable only for $z = 0$.

HINT: use the fact that every compact set in E is contained in some $\mathbb{R}^n$.

4.10.57.° Let X and Y be normed spaces. Show that the continuity of a mapping $f\colon X \to Y$ is equivalent to the continuity of all compositions $f\circ\varphi$ for continuous mappings $\varphi\colon \mathbb{R} \to X$ and is also equivalent to the continuity of the compositions $\psi\circ f$ for all continuous functions $\psi\colon Y \to \mathbb{R}$.

4.10.58. (Tsar'kov, Shavgulidze [**532**]) (i) Let S be the unit sphere in l^2. Then there exists an infinitely Fréchet differentiable isometry $F\colon l^2 \to S$ with bounded derivatives of all orders.

(ii) For every separable Banach space X, there exists an infinitely Fréchet differentiable isometry from X to the unit sphere in $C[0,1]$ with bounded derivatives of all orders.

4.10.59. (The Peetre theorem) A linear mapping $L\colon C^\infty(\mathbb{R}^n) \to C^\infty(\mathbb{R}^n)$ such that $\operatorname{supp} Lf \subset \operatorname{supp} f$ is a differential operator of a locally finite order with smooth coefficients.

HINT: see Peetre [**381**], Narasimhan [**363**, Theorem 3.3.3].

4.10.60. (Ekeland [**153**]) Let $f\colon X \to Y$ be a continuous Gâteaux differentiable mapping of Banach spaces such that $f(0) = 0$ and, whenever $\|x\| \leqslant 1$, the operators $f'(x)$ are invertible and $\|(f'(x))^{-1}\| \leqslant M$. Then, for every y with $\|y\| < 1/M$, there exists x with $\|x\| < 1$ and $f(x) = y$.

CHAPTER 5

Measures on linear spaces

In this chapter we give a brief account of measure theory on linear spaces. We assume some acquaintance with basics of the Lebesgue theory of measure and integral (see, for example, Chapters 2 and 3 in [**72**]). We present the fundamental facts of the theory of Gaussian measures, discuss weak convergence of measures and the Fourier transform of measures.

5.1. Cylindrical sets

Let E and G be two real linear spaces in duality; the bilinear form defining the duality will be denoted by $\langle\,,\,\rangle$; i.e., the action of a functional $g \in G$ on an element $x \in E$ is $\langle g, x\rangle$, and the same formula makes x a linear functional on G.

Sets of the form

$$C_{g_1,\ldots,g_n,B} := \bigl\{x \in E\colon\ \bigl(g_1(x),\ldots,g_n(x)\bigr) \in B\bigr\},$$

where $g_1,\ldots,g_n \in G$ and B is contained in the Borel σ-algebra $\mathcal{B}(\mathbb{R}^n)$ (see § 5.2), are called *G-cylindrical sets* or *G-cylinders* in the space E generated by the duality with G. If E is a locally convex space and $G = E'$, we use the terms a *cylindrical set* or a *cylinder* in E without indicating the space of functionals.

Certainly, sets of such a form can be considered for an arbitrary collection G of linear functions on E, not necessarily forming a linear space.

The class of all G-cylinders is denoted by the symbol $\mathfrak{A}_G$. It is readily seen that $\mathfrak{A}_G$ is an algebra. Indeed, the complement of a cylinder $C_{g_1,\ldots,g_n,B}$ is the cylinder $C_{g_1,\ldots,g_n,\mathbb{R}^n\backslash B}$ and the union of any two cylinders $C_{g_1,\ldots,g_n,B_1}$ and $C_{f_1,\ldots,f_k,B_2}$, where $B_1 \in \mathcal{B}(\mathbb{R}^n)$ and $B_2 \in \mathcal{B}(\mathbb{R}^k)$, can be written in the following form: we first observe that

$$C_{g_1,\ldots,g_n,B_1} = C_{g_1,\ldots,g_n,f_1,\ldots,f_k,B_1\times\mathbb{R}^k},$$
$$C_{f_1,\ldots,f_k,B_2} = C_{g_1,\ldots,g_n,f_1,\ldots,f_k,\mathbb{R}^n\times B_2},$$

where the sets $B_3 := B_1 \times \mathbb{R}^k$ and $B_4 := \mathbb{R}^n \times B_2$ are Borel in $\mathbb{R}^{n+k}$; then it is clear that

$$C_{g_1,\ldots,g_n,B_1} \cup C_{f_1,\ldots,f_k,B_2} = C_{g_1,\ldots,g_n,f_1,\ldots,f_k,B_3\cup B_4}.$$

The σ-algebra generated by all G-cylindrical sets is denoted by the symbol $\sigma(G)$.

Given a linear mapping $(g_1,\ldots,g_n)\colon\ E \to \mathbb{R}^n$, denoted by $P_{g_1,\ldots,g_n}$, we can take a finite-dimensional linear space $H \subset E$ which gives E in the direct

V.I. Bogachev and O.G. Smolyanov, *Topological Vector Spaces and Their Applications*, Springer Monographs in Mathematics,
DOI 10.1007/978-3-319-57117-1_5

sum with the closed subspace $L := \bigcap_{i=1}^{n} g_i^{-1}(0) = \mathrm{Ker}\, P_{g_1,\ldots,g_n}$. The cylinder $C_{g_1,\ldots,g_n,B_1} = P_{g_1,\ldots,g_n}^{-1}(B)$ can be written in the following geometrically transparent form:

$$C_{g_1,\ldots,g_n,B_1} = P_{g_1,\ldots,g_n}^{-1}(B) = C + L,$$

where C is a Borel set in the finite-dimensional space H (equipped with its unique locally convex topology). Indeed, let d be the dimension of $P_{g_1,\ldots,g_n}(E)$; then the dimension of H also equals d and $P_{g_1,\ldots,g_n}|_H$ is a linear isomorphism between H and $P_{g_1,\ldots,g_n}(E)$. Now for C we take $P_{g_1,\ldots,g_n}|_H^{-1}(B)$.

Conversely, every set of the form $C + L$, where $C \in \mathcal{B}(H)$, is a cylinder and has the form $C_{g_1,\ldots,g_n,B}$, where $B = P_{g_1,\ldots,g_n}(C)$.

Thus, a "geometric" description of G-cylindrical sets is this: these are sets of the form

$$C + L,$$

where L is the intersection of kernels of finitely many elements of G and C is a Borel set in a finite-dimensional space algebraically complementing L. If E is a locally convex space and $G = E'$, then the subspaces L of the indicated form are all possible closed linear subspaces of finite codimension.

The set C is called a base of the considered cylinder.

It is clear that the sets L and C are not uniquely determined by the cylinder $C + L$, but if we take for L a space of the minimal possible codimension, then L is defined uniquely. Certainly, a choice of the set C still remains non-unique, since L can be complemented by different subspaces.

5.1.1. Example. Let T be a nonempty set, $E = \mathbb{R}^T$, $G = E'$. Then $E' = \mathbb{R}_0^T$ and the cylinders in E are all sets of the form

$$\Bigl\{x \in \mathbb{R}^T\colon\ \bigl(x(t_1), \ldots, x(t_n)\bigr) \in B\Bigr\},$$

where $B \in \mathcal{B}(\mathbb{R}^n)$, $t_1, \ldots, t_n \in T$.

5.1.2. Example. Let E be a Hilbert space and let $G = E'$. Then the cylinders in E are all sets of the form

$$\{x \in E\colon\ Px \in B\},$$

where P is an orthogonal projection on some finite-dimensional linear subspace in E and B is a Borel set in this subspace.

5.1.3. Remark. Note that if a family G of linear functionals on E is not a linear space, then the generated algebra of cylinders coincides with the algebra of cylinders generated by the linear span $\mathrm{Lin}\, G$ of the set G. Indeed, every set $C_{f_1,\ldots,f_n,B}$ in $\mathfrak{A}_{\mathrm{Lin}\, G}$ can be written as $C_{g_1,\ldots,g_n,C}$, where $g_i \in G$ and $C \in \mathcal{B}(\mathbb{R}^n)$, since there exist elements $g_1, \ldots, g_n \in G$ such that the functionals f_i are their linear combinations, i.e., $f_i = \alpha_{i,1} g_1 + \cdots + \alpha_{i,n} g_n$, which enables us to take for C the Borel set $A^{-1}(B)$, where A is the linear operator in $\mathbb{R}^n$ defined by the matrix $(\alpha_{i,j})$.

5.1.4. Proposition. *Let $F \subset G$ be such that every functional in G is a limit of some pointwise convergent sequence of functionals from F. Then $\sigma(G) = \sigma(F)$.*

PROOF. It is known that the measurability with respect to a σ-algebra is preserved by pointwise limits of sequences, hence for every $g \in G$ the set $\{x\colon\ g(x) < c\}$ belongs to $\sigma(F)$. Hence the σ-algebra generated by such functionals is contained in $\sigma(F)$. □

We shall see below that the σ-algebra generated by cylinders plays an important role in measure theory.

5.2. Measures on topological spaces

We recall that a σ-algebra $\mathcal{A}$ in a space X is a class of sets containing X and closed with respect to taking complements and countable unions and intersections. Below the term a "measure on a measurable space $(X, \mathcal{A})$" will mean a real (finite) countably additive measure on a σ-algebra $\mathcal{A}$ in X (see [**72**, Chapter 2]). When speaking of cylindrical measures that need not be countably additive we shall always use the extra term "cylindrical".

A real-valued function f is called measurable with respect to a σ-algebra $\mathcal{A}$ if for each $c \in \mathbb{R}^1$ we have $\{x\colon\ f(x) < c\} \in \mathcal{A}$.

For every measure μ on a σ-algebra $\mathcal{A}$ in X there exists a decomposition

$$\mu = \mu^+ - \mu^-,$$

called the Hahn–Jordan decomposition, in which the measures μ^+ and μ^- are nonnegative and concentrated on disjoint sets $X^+, X^- \in \mathcal{A}$ with $X = X^+ \cup X^-$. It is clear that the latter condition uniquely determines the measure μ^+ and μ^-. Let $|\mu|$ denote the *variation* of μ, i.e., the sum

$$|\mu| := \mu^+ + \mu^-.$$

If $\mu \geqslant 0$ and $\mu(X) = 1$, then μ is called a *probability* measure. If $\mu \geqslant 0$, then

$$\mu^*(S) := \inf\{\mu(A)\colon\ S \subset A, A \in \mathcal{A}\}$$

is called the *outer measure* generated by μ.

Set $\|\mu\| := |\mu|(X)$. The space of all bounded measures on $\mathcal{A}$ is Banach with the norm $\mu \mapsto \|\mu\|$.

For a nonnegative measure μ let $\mathcal{A}_\mu$ denote the Lebesgue completion of $\mathcal{A}$ with respect to μ, i.e., the class of all sets of the form $A \cup C$, where $A \in \mathcal{A}$ and C has outer measure zero. Sets in $\mathcal{A}_\mu$ are called μ-measurable. For a signed measure μ the class $\mathcal{A}_\mu$ is defined as $\mathcal{A}_{|\mu|}$. A set E is called a *set of full μ-measure* if $|\mu|(X\backslash E) = 0$. A measure $\mu \geqslant 0$ is called *atomless* if every set of positive measure has a subset of a strictly smaller nonzero measure.

Usually $\mathcal{A}_\mu$ is larger than $\mathcal{A}$, so there are more $\mathcal{A}_\mu$-measurable functions than $\mathcal{A}$-measurable functions. It is useful to introduce a bit broader concept of μ-measurable function: this is a function f on X that is defined and finite on a set $E \subset X$ of full μ-measure and is $\mathcal{A}_\mu$-measurable on E, i.e., $\{x \in E\colon\ f(x) < c\} \in \mathcal{A}_\mu$ for all $c \in \mathbb{R}^1$; outside of E the function f need not be defined at all or can assume infinite values. The terms "almost everywhere", "a.e." or "μ-a.e." mean "outside of a measure zero set".

The class of all μ-measurable functions is denoted by $\mathcal{L}^0(\mu)$, and the subclass of functions integrable to power $p \in (0, +\infty)$ is denoted by $\mathcal{L}^p(\mu)$. The class $\mathcal{L}^\infty(\mu)$ consists of bounded *everywhere defined* μ-measurable functions. If $f = g$ almost everywhere, then f and g are called equivalent and also *versions* or *modifications* of each other; the notation $f \sim g$ is used. The space of all equivalence classes of functions from $\mathcal{L}^p(\mu)$ is denoted by $L^p(\mu)$, $p \in (0, +\infty]$. For a signed measure μ, the integrability of f with respect to μ is understood as the $|\mu|$-integrability.

We set by definition

$$\int_X f(x)\,\mu(dx) := \int_{X^+} f(x)\,\mu^+(dx) - \int_{X^-} f(x)\,\mu^-(dx).$$

For $p \in [1, +\infty)$ the space $L^p(\mu)$ is equipped with the norm

$$\|f\|_{L^p(\mu)} := \|f\|_p := \left(\int_X |f(x)|^p\,|\mu|(dx)\right)^{1/p}.$$

The standard norm on $L^\infty(\mu)$ is introduced by $\|f\|_\infty := \inf_{g\sim f} \sup_{x\in X} |g(x)|$, where inf is taken over all functions g equivalent to f.

For a signed measure μ the symbol $(f, g)_{L^2(\mu)}$ is understood as $(f, g)_{L^2(|\mu|)}$.

If X is a Banach space, then $\mathcal{L}^p(\mu, X)$ denotes the space of μ-measurable mappings f with values in X such that f takes values in a separable subspace (dependent of f) and $\|f(\cdot)\| \in L^p(\mu)$, where $p \in [1, +\infty)$. The space $L^p(\mu, X)$ of equivalence classes in $\mathcal{L}^p(\mu, X)$ is equipped with the norm

$$\|f\|_p = \left(\int \|f(x)\|^p\,|\mu|(dx)\right)^{1/p},$$

with respect to which it is Banach.

Suppose we are given two measures μ and ν on a σ-algebra $\mathcal{A}$ in a space X. The measure ν is called *absolutely continuous* with respect to μ, which is denoted by $\nu \ll \mu$, if $\nu(A) = 0$ for all sets $A \in \mathcal{A}$ with $|\mu|(A) = 0$. According to the Radon–Nikodym theorem, this is equivalent to the property that ν has the form $\nu = \varrho \cdot \mu$, $\varrho \in \mathcal{L}^1(\mu)$, i.e., one has

$$\nu(A) = \int_A \varrho(x)\,\mu(dx), \quad A \in \mathcal{A}.$$

The function ϱ is called a *density* of ν with respect to μ (or the *Radon–Nikodym density*) and is denoted by $d\nu/d\mu$. If $\nu \ll \mu$ and $\mu \ll \nu$, then the measures μ and ν are called *equivalent*; notation: $\mu \sim \nu$. This is equivalent to the property that $\nu \ll \mu$ and $|\mu|$-a.e. $d\nu/d\mu \neq 0$.

If there is a set $A_0 \in \mathcal{A}$ such that $|\mu|(A_0) = |\mu|(X)$, $|\nu|(X\backslash A_0) = |\nu|(X)$, then μ and ν are called *mutually singular*, which is denoted by $\mu \perp \nu$. In the general case there is a decomposition $\nu = \nu_1 + \nu_2$, where $\nu_1 \ll \mu$ and $\nu_2 \perp \mu$.

Let $\lambda \geqslant 0$ be a measure on $\mathcal{A}$ such that $\mu = f \cdot \lambda$ and $\nu = g \cdot \lambda$ (for example, we can take $\lambda = |\mu| + |\nu|$). Then

$$\|\mu - \nu\| = \int_X |f - g|\,d\lambda.$$

The *Borel* σ-algebra $\mathcal{B}(X)$ of a topological space X is the smallest σ-algebra containing all open sets. Functions measurable with respect to $\mathcal{B}(X)$ are called *Borel* or *Borel measurable*. Sometimes the *Baire* σ-algebra $\mathcal{B}a(X)$ of the topological space X is used, i.e., the smallest σ-algebra with respect to which all continuous functions on X are measurable.

If X is a locally convex space, it is customary to use also the σ-algebra $\sigma(X')$ generated by all sets of the form $\{x\colon\ l(x) \leqslant c\}$, where $l \in X'$ and $c \in \mathbb{R}^1$, i.e., the smallest σ-algebra with respect to which all functionals $l \in X'$ are measurable. This is the σ-algebra generated by all cylinders. Below we shall deal with such σ-algebras.

We shall also deal with σ-algebras generated by families of sets: for every family of sets $\mathcal{M}$ in a given space there is the smallest σ-algebra containing $\mathcal{M}$, denoted by the symbol $\sigma(\mathcal{M})$ and called the σ-algebra generated by $\mathcal{M}$ (one should not try to imagine elements of $\sigma(\mathcal{M})$ somehow constructively expressed via sets in $\mathcal{M}$). Every set $A \in \sigma(\mathcal{M})$ actually belongs to the more narrow σ-algebra $\sigma(\{M_n\})$ generated by some countable family $\{M_n\} \subset \mathcal{M}$; this follows from the observation that the union of all such countably generated σ-algebras is a σ-algebra. The product of two σ-algebras $\mathcal{A}$ and $\mathcal{B}$ in two spaces X and Y is the σ-algebra $\mathcal{A}\otimes\mathcal{B}$ in $X\times Y$ generated by the sets $A\times B$, where $A \in \mathcal{A}$, $B \in \mathcal{B}$.

For every class $\mathcal{F}$ of functions on a given space, there exists the smallest σ-algebra $\sigma(\mathcal{F})$ with respect to which all functions from $\mathcal{F}$ are measurable: this is the σ-algebra generated by the sets $\{x\colon\ f(x) < c\}$ (it suffices to take only rational c). It follows from what has been said above that every function g measurable with respect to $\sigma(\mathcal{F})$ is measurable with respect to $\sigma(\{f_n\})$ for some countable collection $\{f_n\} \subset \mathcal{F}$ and it has the form $g = \psi(f_1, f_2, \ldots)$, where ψ is a Borel function on $\mathbb{R}^\infty$ (Exercise 5.12.77).

For example, every set from $\sigma(X')$ belongs to $\sigma(\{l_n\})$ for some countable collection $\{l_n\} \subset X'$; it has the form $\bigl\{x \in X\colon\ \bigl(l_1(x), l_2(x), \ldots\bigr) \in B\bigr\}$, where B is a Borel set in $\mathbb{R}^\infty$. Another example: the Baire σ-algebra $\mathcal{B}a(X)$ is generated by the class $C_b(X)$ of bounded continuous functions on X, and every function f measurable with respect to this σ-algebra has the form $f(x) = \psi\bigl(f_1(x), f_2(x), \ldots\bigr)$, where ψ is a Borel function on $\mathbb{R}^\infty$ and $\{f_n\} \subset C_b(X)$.

If X is a metric space, then the Borel and Baire σ-algebras coincide, since every closed set can be represented as the set of zeros of a continuous function (for example, the distance to this set). In the general case the Borel σ-algebra is broader than the Baire one even for compact spaces. For example, if $X = [0,1]^c$ is the product of the continuum of copies of the unit interval, then every continuous function on X is a function of countably many coordinates, i.e., it has the form $f(x) = f_0\bigl(\pi(x)\bigr)$, where π is the projection from X to some countable product of intervals and f_0 is a continuous function on this countable product (this is clear from the Stone–Weierstrass theorem, since the functions of the indicated type form a closed algebra and separate points). Therefore, such a simple Borel set as a singleton is not a Baire set here. One can also verify that on the product of the continuum of real lines the Borel σ-algebra is broader than the Baire one for a similar reason.

If X is a locally convex space, then $\sigma(X')$ is contained in the Baire σ-algebra, but the inclusion can be strict. For example, if X is a nonseparable Hilbert space, its singletons and closed balls do not belong to $\sigma(X')$. Indeed, if $B \in \sigma(X')$, then $B \in \sigma(\{l_n\})$ for some countable collection of functionals l_n. The subspace L equal to the intersection of the kernels of l_n is infinite-dimensional, since the functionals l_n have the form $l_n(x) = (x, h_n)$, where $h_n \in X$, so L is the orthogonal complement to the linear span of $\{h_n\}$, which is infinite-dimensional, since otherwise X would be separable. Now observe that $B = B + L$ (if B is empty, then $B + L$ is also). This follows from the fact that the indicator of B is a function of $\{l_n\}$ (as noted above). Thus, $\sigma(X')$ has no nonempty bounded sets.

5.2.1. Proposition. *Let E be a separable metrizable locally convex space. Then $\sigma(E') = \mathcal{B}(E)$, and there is a countable set $\Gamma \subset E'$ for which $\sigma(\Gamma) = \mathcal{B}(E)$. If E is complete, then this is true for any collection $\Gamma \subset E'$ separating points in the space E.*

PROOF. Passing to the completion we can assume that E is complete. By the separability of E there is a countable part of E' separating points. Moreover, for every such countable part Γ we have $\sigma(\Gamma) = \mathcal{B}(E)$, which follows from a more general result contained in [**65**, Theorem 6.8.9] (see also Theorem 5.12.74 below). □

5.2.2. Definition. *A Borel measure on a topological space is a measure on the Borel σ-algebra. A Baire measure on a topological space is a measure on the Baire σ-algebra.*

5.2.3. Definition. *A Borel measure μ on a Hausdorff space is called a Radon measure if, for every Borel set B and every $\varepsilon > 0$, there is a compact set K in B such that $|\mu|(B \backslash K) \leqslant \varepsilon$.*

Every Radon measure has a *topological support*, i.e., the smallest closed set of full measure. This follows from the fact that for any nonnegative Radon measure the union of any family of open sets of measure zero is also a measure zero set (since every compact in this union is covered by finitely many of these open sets of measure zero).

The set of all Radon measures on a Hausdorff space X is denoted by $\mathcal{M}_r(X)$, and its subset consisting of probability measures is denoted by $\mathcal{P}_r(X)$.

Any Radon measure μ is *regular*: for every $B \in \mathcal{B}(X)$ and every $\varepsilon > 0$, there exist a closed set $Z \subset B$ and an open set $U \supset B$ with $|\mu|(U \backslash Z) < \varepsilon$. However, the regularity is not equivalent to the Radon property.

5.2.4. Proposition. *Every Borel measure μ on an arbitrary metric space (X, d) is regular.*

PROOF. We can assume that $\mu \geqslant 0$. Let $\mathcal{E}$ be the class of all sets $B \in \mathcal{B}(X)$ for which for every $\varepsilon > 0$ there exist a closed set Z and an open set U with $Z \subset B \subset U$ and $\mu(U \backslash Z) < \varepsilon$. All closed set belong to $\mathcal{E}$, because for a closed B we can take $Z = B$ and for an open set we can take $U_n := \{x\colon\ \mathrm{dist}(x, B) < 1/n\}$ for some n, since the sets U_n decrease to B. The class $\mathcal{E}$ is obviously closed with respect to complements. In addition, $B = \bigcup_{n=1}^{\infty} B_n \in \mathcal{E}$ if $B_n \in \mathcal{E}$. Indeed, for any $\varepsilon > 0$ we can find open sets $U_n \supset B_n$ with $\mu(U_n \backslash B_n) < \varepsilon 4^{-n}$ and take $U = \bigcup_{n=1}^{\infty} U_n$. Next, we find N such that $\mu\bigl(B \backslash \bigcup_{n=1}^{N} B_n\bigr) < \varepsilon/4$ and take the closed set $Z_n \subset B_n$ with $\mu(B_n \backslash Z_n) < \varepsilon 4^{-n}$. The set $Z = \bigcup_{n=1}^{N} Z_n \subset B$ is closed and we have $\mu(U \backslash Z) < \varepsilon$. Hence $\mathcal{E}$ is a σ-algebra containing all closed sets, i.e., $\mathcal{E} = \mathcal{B}(X)$. □

5.2.5. Remark. It is seen from the proof that the class $\mathcal{E}$ is a σ-algebra also in the case of a topological space. The metrizability of X is needed in order to conclude that $\mathcal{E}$ contains all closed sets.

Let us take a set $X \subset [0,1]$ that is nonmeasurable with respect to Lebesgue measure λ and $\lambda^*(X) = 1$. Then the formula $\mu(B \cap X) := \lambda(B)$ defines a probability Borel measure on X. Indeed, it is easy to verify that any Borel subset of X has the form $B \cap X$, where B is a Borel set in $[0,1]$. In addition, we have $\lambda(B_1) = \lambda(B_2)$ if $B_1 \cap X = B_2 \cap X$, since $\lambda^*(X) = 1$. As shown above, the measure μ is regular, but it is not Radon, since X is nonmeasurable in $[0,1]$, but compact subsets of X are compact also in $[0,1]$.

Any Radon measure μ is *tight* in the sense of the following definition.

5.2.6. Definition. *A nonnegative set function ν on some domain of definition $\mathfrak{A}$ in a topological space X is called tight if for every $\varepsilon > 0$ there exists a compact set $K_\varepsilon \subset X$ such that $\nu(A) < \varepsilon$ for every set $A \in \mathfrak{A}$ disjoint with K_ε. A signed measure μ on a σ-algebra is called tight if its total variation is tight.*

However, again this is not equivalent to the Radon property: there are examples of nonregular Borel measures on nonmetrizable compact spaces. Only the combination of the regularity and tightness is equivalent to being Radon. On most of the spaces encountered in applications all Borel measures are Radon (although there are exceptions, for example, the product of the continuum of intervals). In particular, the following Ulam theorem holds.

5.2.7. Theorem. *On every complete separable metric space all Borel measures are Radon.*

PROOF. Let μ be a nonnegative Borel measure on a complete separable metric space X. We already know that this measure is regular. By using the separability and completeness of X we establish its tightness. Let $\varepsilon > 0$ and $\{x_n\}$ be a countable everywhere dense set in X. For every $k \in \mathbb{N}$ the union of the open balls $B(x_n, 1/k)$ of radius $1/k$ is the whole space X. Hence there exists N_k such that $\mu\bigl(X \backslash \bigcup_{n=1}^{N_k} B(x_n, 1/k)\bigr) < \varepsilon 2^{-k}$. Let us now set $B := \bigcap_{k=1}^{\infty} \bigcup_{n=1}^{N_k} B(x_n, 1/k)$. Then $\mu(X \backslash B) \leqslant \varepsilon$. The set B is totally bounded, since for every k it is covered by finitely many balls of radius $1/k$. Hence, by completeness of X, B has compact closure with the desired property. □

5.2.8. Theorem. (LUSIN'S THEOREM) *If μ is a Radon measure on a Tychonoff space X and f is a μ-measurable function, then, for every $\varepsilon > 0$, there is a continuous function f_ε such that $|\mu|\bigl(x\colon f(x) \neq f_\varepsilon(x)\bigr) < \varepsilon$.*

Among the most important results in measure theory on topological spaces one should mention the following theorem on the universal measurability of Souslin sets, going back to Lusin.

5.2.9. Definition. *Souslin sets in a Hausdorff topological space X are defined to be the images of complete separable metric spaces under continuous mappings with values in X.*

Borel sets in complete separable metric spaces are Souslin, but in every uncountable complete separable metric space there is a Souslin set that is not Borel (for the proofs of all these facts, see Bogachev [**65**, Chapters 6, 7]). Souslin sets in complete separable metric spaces can be also described by means of the Souslin operation over closed (or open) sets.

Let $\mathcal{E}$ be some class of subsets of a space X. If to every finite sequence of natural numbers $(n_1,\ldots,n_k)$ we associate a set $A_{n_1,\ldots,n_k}\in\mathcal{E}$, then we say that a Souslin scheme (or table) of sets $\{A_{n_1,\ldots,n_k}\}$ with values in $\mathcal{E}$ is defined. The Souslin operation (or A-operation) over the class $\mathcal{E}$ is the mapping which to every Souslin table $\{A_{n_1,\ldots,n_k}\}$ with values in $\mathcal{E}$ associates the set

$$A=\bigcup_{(n_i)\in\mathbb{N}^\infty}\bigcap_{k=1}^\infty A_{n_1,\ldots,n_k}.$$

Sets of this form along with the empty set are called $\mathcal{E}$-Souslin or $\mathcal{E}$-analytic. Their collection is denoted by $\mathcal{S}(\mathcal{E})$. It turns out that if we apply the Souslin operation to the class of closed (or open) sets in a complete separable metric space, then we obtain exactly the class of Souslin sets. In the case of the real line it suffices to apply the Souslin operation to the class of intervals. Yet another equivalent description of $\mathcal{E}$-Souslin sets characterizes them as the projections to X of the sets in $X\times[0,1]$ representable as countable intersections of countable unions of products of the form $E\times[a,b]$, where $E\in\mathcal{E}$ and $[a,b]\subset[0,1]$. The same class coincides with the projections of the sets in $\mathcal{E}\otimes\mathcal{B}([0,1])$. For example, a set $A\subset\mathbb{R}^1$ is Souslin precisely when it is the projection of a Borel set on the plane. It is important to bear in mind that such projections need not belong to the class of Borel sets.

5.2.10. Theorem. (i) *Let μ be a measure on a measurable space $(X,\mathcal{A})$. Then, applying the Souslin operation to sets from $\mathcal{A}_\mu$ we obtain sets from $\mathcal{A}_\mu$, i.e., the Souslin operation preserves measurability.*

(ii) *In any Hausdorff space X all Souslin sets are measurable with respect to every Radon measure on X. If X is Souslin, then every Borel measure on X is Radon.*

(iii) *All Borel mappings between Souslin spaces take Souslin sets to Souslin sets.*

Assertion (i) is obviously nontrivial: in the formation of a set by means of the Souslin operation we use an uncountable union over all infinite sequences (n_i) of natural numbers! Let us give one more useful result in this area (see Bogachev [**65**, Corollary 6.10.10]).

5.2.11. Theorem. *Let $(X, \mathcal{E})$ be a measurable space and let Y be a Souslin space. Then the projection onto X of every set from $\mathcal{S}(\mathcal{E}\otimes\mathcal{B}(Y))$ belongs to $\mathcal{S}(\mathcal{E})$ and hence is measurable with respect to each measure on $(X, \mathcal{E})$.*

For example, if X is a Hausdorff topological space and $\mathcal{E} = \mathcal{B}(X)$, then for any Souslin space Y we have the equality (see [**65**, Lemma 6.4.2])

$$\mathcal{B}(X\times Y) = \mathcal{B}(X)\otimes\mathcal{B}(Y). \tag{5.2.1}$$

Hence for every set B from $\mathcal{B}(X\times Y)$, its projection onto X is measurable with respect to all Borel measures on X, although even in the case $X = Y = [0,1]$ this projection need not be Borel.

It is known that if μ and ν are two Radon measures on a Tychonoff space X such that

$$\int_X f(x)\,\mu(dx) = \int_X f(x)\,\nu(dx)$$

for all functions f from the class $C_b(X)$ of bounded continuous functions on X, then $\mu = \nu$. A bit stronger assertion is valid.

5.2.12. Lemma. *If the foregoing equality, where μ and ν are Radon measures, is true for all functions from some class $\mathcal{F} \subset C_b(X)$ with the property that for every pair of distinct points $x, y \in X$ there is a function $f \in \mathcal{F}$ with $f(x) \neq f(y)$, then $\mu = \nu$.*

PROOF. The weakest topology on X in which all functions from $\mathcal{F}$ are continuous is completely regular and the measures μ and ν are Radon in it, since compact sets in the original topology are compact also in the new topology. Hence the values of these measures coincide on all compact sets in the new topology, in particular, on the compact sets in the original topology, whence it follows that our measures coincide on all Borel sets in the original topology. □

In conclusion we briefly discuss an important question about the existence of Radon extensions of measures originally defined on more narrow σ-algebras or even algebras. The next theorem and its corollary are very useful in applications. The proof, which can be found in Bogachev [**65**, § 7.3], employs the inner measure μ_* constructed from a nonnegative additive set function μ on an algebra $\mathcal{A}$ by the formula

$$\mu_*(E) = \sup\{\mu(A)\colon\ A \in \mathcal{A}, A \subset E\}.$$

5.2.13. Theorem. *Suppose that $\mathcal{A}$ is some algebra of subsets of a Hausdorff space X containing a base of the topology and μ is a regular additive set function of bounded variation on $\mathcal{A}$. If the measure μ is tight, then it admits a unique extension to a Radon measure on X.*

In the case where $\mu \geqslant 0$, the extension for all sets $B \in \mathcal{B}(X)$ is given by the formula

$$\widehat{\mu}(B) = \inf\Bigl\{\mu_*(U)\colon\ U \text{ open in } X \text{ and } B \subset U\Bigr\}. \tag{5.2.2}$$

It is important to bear in mind that this theorem principally differs from the known theorem in the university course on the extension of a countably additive measure from an algebra to the generated σ-algebra: the latter theorem can be insufficient for obtaining an extension to the whole Borel σ-algebra. Let us consider the following simple example of a tight Baire measure for which the Radon extension to the Borel σ-algebra exists, but cannot be obtained by means of the Lebesgue completion of $\mathcal{B}a(X)$.

5.2.14. Example. Let $X = \mathbb{R}^T$, where T is an uncountable set (say, an interval of the real line) and let x_0 be any element of X (for example, the identically zero function). Let ν be the measure on the σ-algebra $\mathcal{B}a(X)$ defined by the formula $\nu(B) = 1$ if $x_0 \in B$ and $\nu(B) = 0$ otherwise (i.e., ν is Dirac's measure δ_{x_0} at x_0). Clearly, this measure is tight and by the same formula can be extended to $\mathcal{B}(X)$. However, the singleton set $\{x_0\}$ is not measurable with respect to the Lebesgue completion of the measure ν on $\mathcal{B}a(X)$. Indeed, otherwise this set would be the union of some set from $\mathcal{B}a(X)$ and some set of outer measure zero with respect to ν on $\mathcal{B}a(X)$, which is impossible, since no singleton is Baire in our space, while the point x_0 has outer measure 1.

Here is a typical application of Theorem 5.2.13.

5.2.15. Example. Every tight Baire measure on a Tychonoff space X has a unique Radon extension. This follows from the fact that Baire sets of the type $\{x\colon\ f(x) > 0\}$, where f is a continuous function on X, form a topology base, since X is completely regular.

In particular, this example applies to Hausdorff locally convex spaces, but here it would be nice to have Radon extensions of all tight cylindrical measures (such measures are discussed in § 5.4). A straightforward application of Theorem 5.2.13 does not imply this, since the algebra of cylinders seldom contains a topology base (if this is not the weak topology). However, one can derive from this theorem the following useful fact (see Bogachev [**65**, § 7.3]).

5.2.16. Corollary. *Let X be a Tychonoff space and let Γ be some family of continuous functions on X separating the points in X. Then, every tight measure μ on the σ-algebra $\sigma(\Gamma)$ generated by Γ admits a unique extension to a Radon measure on X. Moreover, the same is true if μ is a regular and tight additive set function of bounded variation on the algebra $\mathfrak{A}(\Gamma)$ generated by Γ.*

This result obviously applies to the algebra of cylinders, which will be done in Theorem 5.4.10.

It should be noted that a tight Baire measure on a Tychonoff space can have Borel extensions that are not Radon (this happens even for compact spaces, on which all Baire measures are tight); if a Baire measure is not tight, then it can fail to have Borel extensions (see [**65**, Example 7.3.9]). Although the Lebesgue completion of the algebra of cylinders need not contain all Borel sets, the following important fact is true.

5.2.17. Proposition. *Let μ be a Radon measure on a Hausdorff locally convex space E and let $B \in \mathcal{B}(E)$. Then, for every $\varepsilon > 0$, there exists an open cylindrical set C_ε such that $|\mu|(B \bigtriangleup C_\varepsilon) < \varepsilon$.*

If B is compact, then C_ε can be taken to contain B and $\mu(B) = \inf \mu(B+H)$, where inf *is taken over all closed subspaces H of finite codimension.*

PROOF. Clearly, it suffices to verify the second assertion, where B is compact. We can find a compact set $K_\varepsilon \supset B$ such that $|\mu|(E\backslash K_\varepsilon) < \varepsilon/2$. In addition, we can find an open set $U_\varepsilon \supset B$ for which $|\mu|(U\backslash B) < \varepsilon$. On the compact set K_ε, the original topology coincides with $\sigma(E, E')$. Hence there exists a $\sigma(E, E')$-open set V_ε such that $V_\varepsilon \cap K_\varepsilon = U_\varepsilon \cap K_\varepsilon$. The set V_ε is the union of some collection of open cylindrical sets, hence by the compactness of B one can take a finite union of these cylinders contained in V_ε and covering B. This union C_ε is the desired set, since

$$|\mu|(C_\varepsilon\backslash B) \leqslant |\mu|(V_\varepsilon\backslash B) \leqslant |\mu|(U_\varepsilon\backslash B) + \varepsilon/2 < \varepsilon.$$

Finally, if a cylinder C contains the compact set B, we take a finite-dimensional subspace L containing the base of C and a continuous projection P onto L. Let $H = P^{-1}(0)$, $K = P(B)$. Then the cylinder $K+H = P(K)+H = P^{-1}\bigl(P(K)\bigr)$ with a compact base contains B and is contained in C. □

It would be interesting to know whether one can take for C_ε a convex set if B is convex (if B is convex and compact, then this is true). Note that the justification of this offered in Bogachev [**63**, Lemma 2.6.7] contains a gap.

5.2.18. Corollary. *For a Radon measure μ on a locally convex space E, the set of bounded cylindrical functions is everywhere dense in $L^p(\mu)$ for all $p \in [1, +\infty)$. Moreover, the set of functions of the form $\varphi(l_1, \ldots, l_n)$, where $\varphi \in C_0^\infty(\mathbb{R}^n)$ and $l_i \in E'$, is everywhere dense.*

5.2.19. Remark. A useful abstract generalization of the concept of a Radon measure is a measure with a compact approximating class. A class $\mathcal{K}$ of subsets of a set X is called *compact* if, for every sequence of sets $K_n \in \mathcal{K}$ with the property that all finite intersections $\bigcap_{n=1}^N K_n$ are nonempty, the set $\bigcap_{n=1}^\infty K_n$ is also nonempty. Any class consisting of compact sets in a topological space is compact, but this concept does not employ topologies.

A compact class $\mathcal{K}$ in X is called *approximating* for an additive function $\mu \geqslant 0$ on an algebra of sets $\mathcal{A}_0$ in X if, for every $A \in \mathcal{A}_0$ and $\varepsilon > 0$, there exist $K_\varepsilon \in \mathcal{K}$ and $A_\varepsilon \in \mathcal{A}_0$ such that $A_\varepsilon \subset K_\varepsilon \subset A$ and $\mu(A\backslash A_\varepsilon) < \varepsilon$. It is easy to verify that this yields the countable additivity of μ on $\mathcal{A}_0$. Although there exist countably additive measures on σ-algebras without approximating compact classes (see [**65**, Example 7.5.3 and Theorem 7.5.6]), for the measures encountered in real applications such classes usually exist.

5.3. Transformations and convergence of measures

Let $(X, \mathcal{A})$ and $(Y, \mathcal{B})$ be two measurable spaces, i.e., spaces equipped with σ-algebras. A mapping $f\colon X \to Y$ is measurable with respect to the pair of

σ-algebras $(\mathcal{A}, \mathcal{B})$ (or $(\mathcal{A}, \mathcal{B})$-measurable) if we have

$$f^{-1}(B) \in \mathcal{A} \quad \text{for all} \quad B \in \mathcal{B}.$$

If $(Y, \mathcal{B})$ is the real line $\mathbb{R}^1$ with its Borel σ-algebra $\mathcal{B} = \mathcal{B}(\mathbb{R}^1)$, then $(\mathcal{A}, \mathcal{B})$-measurable functions are precisely $\mathcal{A}$-measurable functions: $\mathcal{A}$-measurability of a real function f is equivalent to the inclusion $f^{-1}(B) \in \mathcal{A}$ for all Borel sets $B \subset \mathbb{R}^1$. If X and Y are topological spaces, then $(\mathcal{B}(X), \mathcal{B}(Y))$-measurable mappings are called *Borel* mappings.

The image of a measure μ on $(X, \mathcal{A})$ under a μ-measurable mapping f from X to $(Y, \mathcal{B})$, i.e., $(\mathcal{A}_\mu, \mathcal{B})$-measurable (for example $(\mathcal{A}, \mathcal{B})$-measurable), is denoted by $\mu \circ f^{-1}$ and defined by the equality

$$\mu \circ f^{-1}(B) = \mu\bigl(f^{-1}(B)\bigr).$$

This definition yields the following change of variables formula:

$$\int_X \varphi\bigl(f(x)\bigr)\, \mu(dx) = \int_{\mathbb{R}^1} \varphi(t)\, \mu \circ f^{-1}(dt), \tag{5.3.1}$$

valid for every bounded $\mathcal{B}$-measurable function φ; more generally, this formula is true for every $\mathcal{B}$-measurable function φ integrable with respect to the nonnegative measure $|\mu| \circ f^{-1}$.

The product of two measure spaces $(X, \mathcal{A}, \mu)$ and $(Y, \mathcal{B}, \nu)$ is usually denoted by $(X \times Y, \mathcal{A} \otimes \mathcal{B}, \mu \otimes \nu)$, see Bogachev [**65**, Chapter 3].

Transformations of measures are related to the concept of conditional measures, which we discuss for the case of Souslin spaces (proofs and additional details can be found in [**65**, Chapter 10]).

Let X and Y be two Hausdorff spaces such that X is Souslin (for example, complete separable metrizable), let μ be a nonnegative Radon measure on $X \times Y$, and let μ_Y be the projection of μ to Y. Then, for every $y \in Y$, there exists a Radon probability measure μ^y on $X \times \{y\}$ such that for every set $A \in \mathcal{B}(X \times Y)$ the function $y \mapsto \mu^y\bigl(A \cap (X \times \{y\})\bigr)$ is measurable with respect to μ_Y and

$$\mu(A) = \int_Y \mu^y\bigl(A \cap (X \times \{y\})\bigr)\, \mu_Y(dy). \tag{5.3.2}$$

A similar equality is true also for signed measures with the difference that in place of the measure μ_Y we take the projection $|\mu|_Y$ of the measure $|\mu|$ on Y, and measures μ^y can be signed; they satisfy the condition $\|\mu^y\| = 1$. To this end, it suffices to write the measure μ as $\mu = \varrho \cdot |\mu|$, where $|\varrho| = 1$, and take the measures $\mu^y = \varrho \cdot |\mu|^y$. Measures μ^y are called *conditional* or *regular conditional* measures. The conditional measures μ^y on $X \times \{y\}$ are defined μ_Y-uniquely: two families of such measures coincide μ_Y-a.e. If the space Y is also Souslin (which is the case in most of applications), then the conditional measures μ^y can be chosen in such a way that the functions $y \mapsto \mu^y\bigl(A \cap (X \times \{y\})\bigr)$ will be Borel measurable for all Borel sets $A \in \mathcal{B}(X \times Y)$.

In place of measures μ^y on the slices $X \times \{y\}$ it is possible to define conditional measures μ_y on X, see (5.3.6) below.

For any sets $A \in \mathcal{B}(X)\otimes\mathcal{B}(Y)$ and $B \in \mathcal{B}(Y)$, (5.3.2) yields the equality

$$\mu(A\cap(X\times B)) = \int_B \mu^y(A\cap(X\times\{y\}))\,\mu_Y(dy). \tag{5.3.3}$$

It also follows from (5.3.2) that for every $|\mu|$-integrable Borel function f one has the equality

$$\int_{X\times Y} f\,d\mu = \int_Y \int_{X\times\{y\}} f(x,y)\,\mu^y(dx)\,|\mu|_Y(dy), \tag{5.3.4}$$

where the repeated integral exists in the following sense: for $|\mu|_Y$-a.e. $y\in Y$ the function $x\mapsto f(x,y)$ is integrable with respect to the conditional measure μ^y and the obtained integral is a $|\mu|_Y$-integrable function.

Let us note a useful corollary of the last formula. Let ν be a Radon measure on $X\times Y$ such that $\nu\ll\mu$. Then ν can be represented in the form

$$\nu(A) = \int_Y \int_{X\times\{y\}} \sigma^y(A\cap(X\times\{y\}))\,|\mu|_Y(dy), \tag{5.3.5}$$

where σ^y is a Radon measure on the space $X\times\{y\}$ for each $y\in Y$ and the functions $y\mapsto\|\sigma^y\|$ and $y\mapsto\sigma^y(A)$ for any $A\in\mathcal{B}(X)$ are integrable with respect to $|\mu|_Y$. Indeed, we can assume that $\mu\geqslant 0$. By the Radon–Nikodym theorem $\nu=\varrho\cdot\mu$, where $\varrho\in\mathcal{L}^1(\mu)$ is a Borel function. By formula (5.3.4) we find that

$$\begin{aligned}\nu(A) &= \int_{X\times Y} I_A\varrho\,d\mu = \int_Y\int_{X\times\{y\}} I_A(x,y)\varrho(x,y)\,\mu^y(dx)\,\mu_Y(dy)\\ &= \int_Y \sigma^y(A\cap(X\times\{y\}))\,\mu_Y(dy),\end{aligned}$$

where $\sigma^y := \varrho(\,\cdot\,,y)\cdot\mu^y$. The measures σ^y with the indicated properties are also defined uniquely up to a redefinition of such measures for points y from some set of $|\mu|_Y$-measure zero.

Sometimes it becomes useful to represent a measure μ on $X\times Y$ in the form

$$\mu(B) = \int_Y \mu^y(B_y)\,\sigma(dy), \quad B_y := \{x\in X\colon\ (x,y)\in B\}, \tag{5.3.6}$$

where μ^y are Borel measures on X, σ is some nonnegative Radon measure on Y for which $\mu_Y\ll\sigma$, and the functions $y\mapsto\mu^y(B_y)$ and $y\mapsto\|\mu^y\|$ are integrable with respect to the measure σ. Measures μ^y are defined uniquely up to a redefinition for points y from a set of σ-measure zero. If we are given two measures μ and ν on the product $X\times Y$ and the measure $\zeta := |\mu|+|\nu|$ possesses conditional measures ζ^y on X, then we can take $\sigma=\zeta_Y$ and obtain representations of the form (5.3.6) for μ and ν with a common measure σ. A similar representation is possible for a countable collection of measures.

5.3.1. Proposition. (i) *The relation $\mu\ll\nu$ is equivalent to the property that $\mu^y\ll\nu^y$ for σ-a.e. y.* (ii) *The relation $\mu\perp\nu$ is equivalent to the property that $\mu^y\perp\nu^y$ for σ-a.e. y.*

PROOF. Let μ, ν, σ be probability measures. If $\mu^y \ll \nu^y$ for σ-a.e. y, then we have $\mu \ll \nu$. Let $\mu \ll \nu$ and $f = d\mu/d\nu$. Then the measure μ is represented both by means of the measures $f(\,\cdot\,, y)\cdot \nu^y$ and by means of the measures μ^y, whence $\mu^y = f(\,\cdot\,, y)\cdot\nu^y$ for σ-a.e. y by the essential uniqueness of representing measures. If $\mu \perp \nu$, then there exists a set B with $\mu(B) = 0$ and $\nu(B) = 1$. Then $\mu^y(B_y) = 0$ and $\nu^y(B_y) = 1$ σ-a.e., i.e., $\mu^y \perp \nu^y$. Conversely, let $\mu^y \perp \nu^y$ σ-a.e. If it is not true that $\mu \perp \nu$, then $\mu = \mu_1 + \mu_2$, where $\mu_1 \ll \nu$, $\mu_2 \perp \nu$. Then $\mu^y = \mu_1^y + \mu_2^y$ by uniqueness, moreover, $\mu_1^y \ll \nu^y$ and $\mu_2^y \perp \nu^y$ as shown above. Since $\mu^y \perp \nu^y$, we have $\mu_1^y = 0$ σ-a.e., i.e., $\mu_1 = 0$. The case of signed measures reduces to the considered one by using the Hahn decomposition. □

An important partial case arises when a locally convex space X with a Radon measure μ is represented as a direct sum $X = X_0 \oplus Y$, where Y is a finite-dimensional linear subspace and X_0 is a closed linear subspace complementing Y; such X_0, as we known, always exists for any finite-dimensional Y.

The results about conditional measures can be represented in a more general form. For simplicity suppose that $\pi\colon X \to Y$ is a Borel mapping of Souslin spaces. Let μ be a Borel probability measure on X and let $\nu = \mu\circ\pi^{-1}$. Then, on the sets $\pi^{-1}(y)$, where $y \in Y$, there exist probability measures μ^y such that for every set $B \in \mathcal{B}(X)$ the function $y \mapsto \mu^y\bigl(B\cap\pi^{-1}(y)\bigr)$ is measurable with respect to ν and

$$\mu(B) = \int_Y \mu^y\bigl(B\cap\pi^{-1}(y)\bigr)\,\nu(dy). \tag{5.3.7}$$

For any μ-integrable Borel function f we have the equality

$$\int_X f\,d\mu = \int_Y\int_X f(x)\,\mu^y(dx)\,\nu(dy).$$

One can make all functions $y \mapsto \mu^y(B)$ Borel measurable if the measures μ^y are allowed to concentrate on $\pi^{-1}(y)$ not for all y, but just for ν-a.e. y. In the general case, it is impossible to combine the requirement of the Borel measurability of all such functions with the equality $\mu^y\bigl(\pi^{-1}(y)\bigr) = 1$ for all y even for Borel functions π on $[0,1]$. If X and Y are Borel sets in Polish spaces, then both requirements can be combined only when $\pi(X)$ is a Borel set and π possesses a Borel selection, i.e., there exists a Borel mapping $g\colon \pi(X) \to X$ with $\pi\bigl(g(y)\bigr) = y$ (see the results of Blackwell and Ryll-Nardzewski described in [**65**, Exercise 10.10.55]). The existence of the required conditional measures in the presence of the indicated mapping g is proved simply. Let us take conditional measures μ^y such that the function $y \mapsto \mu^y(B)$, $B \in \mathcal{B}(X)$, are Borel and $\mu^y\bigl(\pi^{-1}(y)\bigr) = 1$ for all $y \in Y_0$, where $Y_0 \subset \pi(X)$ is a Borel set of full ν-measure. For every $y \in \pi(X)\backslash Y_0$ let us set $\mu^y := \delta_{g(y)}$. It is readily seen that the functions $y \mapsto \delta_{g(y)}(B)$, $B \in \mathcal{B}(X)$, are Borel. A Borel selection does not always exist (see [**65**, § 6.9]). If X and Y are Polish space, then for its existence it is sufficient that the sets $\pi^{-1}(y)$ be countable unions of compact sets.

The Hahn–Jordan decomposition gives (5.3.7) for a signed measure μ with $\nu = |\mu|\circ\pi^{-1}$, and the measures μ^y on $\pi^{-1}(y)$ are defined uniquely ν-a.e. (for uniqueness a.e. it is sufficient that $\mathcal{B}(X)$ be countably generated).

The case of products considered above is a partial case in which π is a projection from the product to a factor. In turn, it is easy to derive the general case from this partial one. To this end we pass to the space $X\times Y$ equipped with the measure μ_0 on the graph of π obtained as the image of the measure μ under the mapping $x\mapsto (x,\pi(x))$ from X to $X\times Y$. The projection of this measure to Y equals $\nu=\mu\circ\pi^{-1}$. According to (5.3.2), there are conditional measures μ_0^y on the slices $X\times\{y\}$. For ν-a.e. y, these measures are concentrated on the intersections of the slices with the graph of π, i.e., on the sets $\pi^{-1}(y)\times\{y\}$, which enables one to define the required measures μ^y on $\pi^{-1}(y)$ for ν-a.e. y. In the case of arbitrary spaces these formulations are not equivalent, moreover, conditional measures need not exist.

To the most important concepts of measure theory on topological spaces belongs also weak convergence of measures. Let us give some basic facts (for proofs, see Bogachev [**65**, Chapter 8], [**69**]). Let $C_b(X)$ be the space of bounded continuous functions on a topological space X. A sequence of Radon measures μ_n on X is called *weakly converging* to a Radon measure μ if for every function $f\in C_b(X)$ we have

$$\lim_{n\to\infty}\int_X f(x)\,\mu_n(dx)=\int_X f(x)\,\mu(dx).$$

Similarly one defines weak convergence of a net of measures μ_α to μ. These concepts are meaningful for completely regular topological spaces and can be defined also for Baire measures. Weak convergence is convergence in the topology $\sigma(\mathcal{M},C_b(X))$ on the space of measures $\mathcal{M}$ (Baire, Borel or Radon), which is called *weak*, where each $f\in C_b(X)$ defines a functional on the space $\mathcal{M}$ just as the integral of f against measures from $\mathcal{M}$.

For probability measures, there is the following criterion of weak convergence, obtained by A.D. Alexandroff.

5.3.2. Theorem. *Suppose we are given a Tychonoff topological space X, a sequence of probability Radon measures $\{\mu_n\}$ and a probability Radon measure μ on X. Then the following conditions are equivalent:*

(i) *the sequence $\{\mu_n\}$ converges weakly to μ;*

(ii) *for every closed set F one has the inequality* $\limsup\limits_{n\to\infty}\mu_n(F)\leqslant\mu(F)$;

(iii) *for every open set U one has the inequality* $\liminf\limits_{n\to\infty}\mu_n(U)\geqslant\mu(U)$.

The same is true for nets of measures.

Note that in the sufficiency part in conditions (ii) and (iii), in place of arbitrary closed and open sets one can take functionally closed and functionally open sets, i.e., sets of the form $\{x\colon\varphi(x)=0\}$ and $\{x\colon\varphi(x)\neq 0\}$, respectively , where φ belongs to $C_b(X)$. Actually, this theorem is true also for Baire measures if in its formulation we consider functionally open and closed sets. It is easy to modify this theorem for nonnegative measures not necessarily of the same total mass.

Another important result of A.D. Alexandroff asserts the sequential completeness of the space of Baire measures $\mathcal{M}_\sigma$ in the weak topology $\sigma(\mathcal{M}_\sigma,C_b(X))$.

5.3.3. Theorem. *If a sequence of Baire measures μ_n on X is weakly fundamental (i.e., the integrals of every function $f \in C_b(X)$ against the measures μ_n have a finite limit), then $\{\mu_n\}$ converges weakly to some Baire measure μ.*

The space of Radon (or tight) measures is not always weakly sequentially complete: the weak limit of a sequence of Radon measures need not be a tight measure (Exercise 5.12.79). Certainly, such problems do not arise if all Baire measures on X have Radon extensions.

Since $C_b(X)$ is a Banach space with respect to the sup-norm, the question arises about the connection of its Banach dual with the space of measures (of course, its dual for the topology $\sigma(C_b(X), \mathcal{M})$ will be always $\mathcal{M}$). The principal fact here is the following Riesz theorem.

5.3.4. Theorem. *For any compact space X, the dual to the space $C_b(X)$ with the sup-norm can be identified with the space of Radon measures on X, i.e., every continuous linear functional on $C_b(X)$ is given as the integral with respect to some Radon measure.*

The compactness of the space X is important.

5.3.5. Example. There is a continuous linear functional on the space $C_b(\mathbb{R}^1)$ that cannot be represented as the integral with respect to a Borel measure on $\mathbb{R}^1$. For example, one can take an extension with the preservation of norm of the functional $l(f) = \lim\limits_{t\to+\infty} f(t)$ on the subspace of functions having a limit at infinitity.

In the general case, the following fact is true.

5.3.6. Theorem. (i) *Let X be a topological space. A continuous linear functional l on $C_b(X)$ is represented as the integral with respect to a Baire measure precisely when $l(f_n) \to 0$ for every sequence $\{f_n\} \subset C_b(X)$ pointwise decreasing to zero.*

(ii) *Let X be a Tychonoff space. A continuous linear functional l on $C_b(X)$ is represented as the integral with resect to a Radon measure precisely when for every $\varepsilon > 0$ there exists a compact set K_ε such that $|l(f)| \leqslant \varepsilon \sup_x |f(x)|$ for every function $f \in C_b(X)$ vanishing on K_ε.*

Assertion (ii) yields that for every functional $l \in C_b(X)'$, where X is a Tychonoff space, there is a Radon measure ν on the Stone–Čech compactification βX of the space X (see Engelking [**154**, § 3.6]) for which $l(f)$ is the integral against the measure ν of the extension of f to βX.

For applications it is very important to have efficient criteria for a given sequence of measures to contain a weakly convergent subsequence. The main general result here was obtained by Yu.V. Prohorov.

5.3.7. Theorem. *Let X be a complete metric space and let $\{\mu_n\}$ be a sequence of Radon measures on X. Every subsequence in $\{\mu_n\}$ contains a further subsequence weakly converging to a Radon measure if and only if the sequence $\{\mu_n\}$ is uniformly bounded in variation and uniformly tight, i.e., for every $\varepsilon > 0$, there exists a compact set K_ε such that $|\mu_n|(X\backslash K_\varepsilon) < \varepsilon$ for all n.*

5.3.8. Theorem. *Let X be a Fréchet space and let $\mathcal{M}$ be a uniformly tight family of Radon measures on X. Then, there exists a separable reflexive Banach space E compactly embedded into X such that $|\mu|(X\backslash E) = 0$ for all $\mu \in \mathcal{M}$ and the restrictions of the measures from $\mathcal{M}$ to E form a uniformly tight set of measures with respect to the norm of the space E. Hence for any weakly convergent sequence of measures μ_n on X, the space E with the stated properties can be chosen in such a way that the measures $\mu_n|_E$ will be weakly converging on the Banach space E.*

For Tychonoff spaces, we have the following fact.

5.3.9. Theorem. *Any uniformly bounded and uniformly tight family of Radon measures on a Tychonoff space is contained in a set compact in the weak topology.*

5.4. Cylindrical measures

Let E be a real vector space, let G be a subset of its algebraic dual, and let $\mathfrak{A}_G$ be the algebra of G-cylindrical sets in E.

5.4.1. Definition. *A cylindrical measure on $(E, \mathfrak{A}_G)$, or a G-cylindrical measure, on E is a function ν: $\mathfrak{A}_G \to \mathbb{R}^1$ the restriction of which to every σ-algebra $\mathfrak{A}_F$, where $F \subset G$ is finite, is countably additive.*

If $\nu \geqslant 0$ and $\nu(E) = 1$, then ν is called a probability G-cylindrical measure.

Similarly one defines complex cylindrical measures and also cylindrical measures with values in a locally convex space T.

We shall be mostly concerned with bounded (i.e., having bounded sets of values) cylindrical measures.

Note that any cylindrical measure on $(E, \mathfrak{A}_G)$ is additive. This is obvious from the fact that any two sets in $(E, \mathfrak{A}_G)$ belong to one of the algebras $\mathfrak{A}_F$. However, it is very important that this measure need not be countably additive on the whole algebra $\mathfrak{A}_G$ (the countable additivity holds only for cylinders with bases in a common finite-dimensional space).

We recall a general fact (see Dunford, Schwartz [**145**, Chapter IV]) valid for any bounded additive real function ν on an algebra $\mathfrak{A}$: the Jordan decomposition theorem according to which $\nu = \nu^+ - \nu^-$, where for every $A \in \mathfrak{A}$ we have

$$\nu^+(A) = \sup\{\nu(B)\colon\ B \subset A,\ B \in \mathfrak{A}\},\ \nu^-(A) = (-\nu)^+(A), \tag{5.4.1}$$

and the functions ν^+ and ν^- on $\mathfrak{A}$ are additive.

Every finite collection $g_1, \ldots, g_n \in G$ generates the finite-dimensional operator $Px = \bigl(g_1(x), \ldots, g_n(x)\bigr)$ and the projection μ_P of the cylindrical measure μ on $\mathbb{R}^n$ by the formula

$$\mu_P(B) = \mu\bigl(P^{-1}(B)\bigr), \quad B \in \mathcal{B}(\mathbb{R}^n).$$

This projection is countably additive by definition.

5.4.2. Lemma. *Let ν be a bounded real G-cylindrical measure on E and let $\nu = \nu^+ - \nu^-$ be its the Jordan decomposition . Then ν^+ and ν^- are bounded G-cylindrical measures on E.*

PROOF. The cylindrical measure ν is bounded, hence the Jordan decomposition holds for it, but the question arises about the countable additivity of its finite-dimensional projections. For every set $A \in \mathfrak{A}_G$, we have equality (5.4.1) with $\mathfrak{A} = \mathfrak{A}_G$. We have to show that for every mapping $P\colon E \to \mathbb{R}^k$, where P has the form $P = (g_1, \ldots, g_k)$, $g_i \in G$, and every sequence of pairwise disjoint sets $B_n \in \mathcal{B}(\mathbb{R}^k)$ with the union B we have the inequality

$$\nu^+\bigl(P^{-1}(B)\bigr) \leqslant \sum_{n=1}^{\infty} \nu^+\bigl(P^{-1}(B_n)\bigr).$$

Let us fix $\varepsilon > 0$. By definition, there exists a G-cylinder $C \subset P^{-1}(B)$ such that $\nu^+\bigl(P^{-1}(B)\bigr) \leqslant \nu(C)+\varepsilon$. Increasing k, we arrive at the case where $C = P^{-1}(D)$, $D \in \mathcal{B}(\mathbb{R}^k)$. Then by the countable additivity of $\nu\circ P^{-1}$ we obtain

$$\nu(C) = \sum_{n=1}^{\infty} \nu\bigl(P^{-1}(D \cap B_n)\bigr) \leqslant \sum_{n=1}^{\infty} \nu^+\bigl(P^{-1}(B_n)\bigr),$$

which, since ε was arbitrary, gives the desired inequality proving the countable additivity of $\nu^+\circ P^{-1}$. The countable additivity of the projections of ν^- follows from the equality $\nu^- = (-\nu)^+$. □

The function $\nu \mapsto \|\nu\| := \nu^+(E)+\nu^-(E)$ is a norm on the space of bounded cylindrical measures, with which this space is complete.

5.4.3. Definition. *A mapping f from E to a locally convex space Y is called G-cylindrical if there exist a number $n \in \mathbb{N}$, functionals $g_1, \ldots, g_n \in G$ and a map $\varphi\colon \mathbb{R}^n \to Y$ such that $f(x) = \varphi\bigl(g_1(x), \ldots, g_n(x)\bigr)$ for all $x \in E$.*

Real or complex cylindrical mappings are called *cylindrical functions*.

If the mapping φ from this definition is Borel, i.e., measurable as a mapping from the measurable space $\bigl(\mathbb{R}^n, \mathcal{B}(\mathbb{R}^n)\bigr)$ to the measurable space $\bigl(Y, \mathcal{B}(Y)\bigr)$, then the cylindrical mapping is measurable as a mapping from the measurable space $(E, \mathfrak{A}_{\{g_1,\ldots,g_n\}})$ to the measurable space $\bigl(Y, \mathcal{B}(Y)\bigr)$; such cylindrical mappings will be called Borel cylindrical.

Since the restriction of any G-cylindrical measure to the σ-algebra $\mathfrak{A}_F$ with finite $F \subset G$ is countably additive, for real (or complex) Borel G-cylindrical functions the notion of the Lebesgue integral with respect to a real or complex G-cylindrical measure is meaningful. For every bounded cylindrical function f one has the equality

$$\int_E f(x)\, \mu(dx) = \int_{\mathbb{R}^n} \varphi(y)\, \mu_P(dy), \tag{5.4.2}$$

where $P = (g_1, \ldots, g_n)$, μ_P is the corresponding projection of μ and φ is the function used in the above representation of f. This equality can be taken as the definition of the integral in the left side.

In the case $E = \mathbb{R}^T$, to define a cylindrical measure μ means, in the language of the theory of random processes, to define "a consistent system of finite-dimensional distributions", i.e., to define, for every finite collection of different points $t_1, \ldots, t_n$ in T, a finite countably additive measure $\mu_{t_1,\ldots,t_n}$ on $\mathbb{R}^n$ satisfying the following *consistency conditions*:

(i) the projection of the measure $\mu_{t_1,\ldots,t_n}$ to $\mathbb{R}^{n-1}$ equals $\mu_{t_1,\ldots,t_{n-1}}$,

(ii) if a collection of points $s_1, \ldots, s_n \in T$ is a permutation of the collection $t_1, \ldots, t_n$, then the measure $\mu_{s_1,\ldots,s_n}$ is the image of $\mu_{t_1,\ldots,t_n}$ under the mapping $\mathbb{R}^n$ generated by the corresponding permutation of coordinates.

For example, the measure μ_{t_2,t_1} is the image of the measure μ_{t_1,t_2} under the mapping $(x_1, x_2) \mapsto (x_2, x_1)$.

The same notions are meaningful if we consider the product of arbitrary nonempty measurable spaces $(X_t, \mathcal{A}_t)$. Certainly, in this case the product (consisting of all collections of the form (x_t), where $x_t \in X_t$) need not be a linear space or topological space, but we still have the σ-algebra $\mathcal{A} = \bigotimes_t \mathcal{A}_t$ generated by the products $\prod_t A_t$, where only finitely many sets A_t can differ from X_t and $A_t \in \mathcal{A}_t$ for all t.

If a measure μ is defined on the σ-algebra $\mathcal{A}$, then its projections $\mu_{t_1,\ldots,t_n}$ are defined on $X_{t_1,\ldots,t_n} = X_{t_1} \times \cdots \times X_{t_n}$ in place of $\mathbb{R}^n$. These projections satisfy the consistency conditions (i) and (ii), which are introduced similarly; certainly in condition (ii), permutations of indices $\sigma\colon \{t_1, \ldots, t_n\} \to \{t_1, \ldots, t_n\}$ yields mappings of the spaces $X_{t_1,\ldots,t_n} \to X_{\sigma(t_1),\ldots,\sigma(t_n)}$.

In this area an important role is played by the following celebrated Kolmogorov theorem on consistent distributions. The definition of a compact approximating class in its second part (obtained by E. Marczewski) is given in Remark 5.2.19.

5.4.4. Theorem. *Suppose that T is a nonempty set and we are given a consistent family of probability distributions $\mu_{t_1,\ldots,t_n}$ on the spaces $\mathbb{R}^n$. Then, there exists a probability measure μ on the σ-algebra generated by the cylindrical sets in $\mathbb{R}^T$ such that, for every collection of different points $t_1, \ldots, t_n$ in T, the measure $\mu_{t_1,\ldots,t_n}$ coincides with the image of the measure μ under the corresponding projection $x \mapsto (x(t_1), \ldots, x(t_n))$.*

A similar assertion is true for measures on products of abstract measurable spaces $(X_t, \mathcal{A}_t)$ under the additional condition that the measures μ_t on X_t possess compact approximating classes.

A relatively simple proof of this classical theorem can be found in many texts on the theory of random processes (see, for example, Gihman, Skorohod [**194**], Neveu [**371**, Section III.3], Wentzel [**564**]), so we do not include it.

Some comments are in order. The Kolmogorov theorem deals with different points t_i, but this is not important. In place of collections of different points one can consider finite subsets $S \subset T$ and the corresponding measures μ_S on the finite-dimensional spaces $\mathbb{R}^S$. Then only one condition remains: if $S_1 \subset S_2$, then the measure μ_{S_1} must coincide with the projection of μ_{S_2} under the natural projection of $\mathbb{R}^{S_2}$ onto $\mathbb{R}^{S_1}$, which associates to a function on S_2 its restriction to S_1. The second condition in the given formulation tells exactly this: one can define the measure μ_S by numbering the points of S in some order $s_1, \ldots, s_n$ and the measure μ_S will not depend on this numeration.

Certainly, in the described alternative approach in terms of finite subsets the points are automatically different. However, having the measures $\mu_{t_1,\ldots,t_n}$ for distinct points t_i, one can easily define the measures $\mu_{s_1,\ldots,s_m}$ for all collections with possibly repeating points: if, for example, the points $s_1,\ldots,s_n$ are distinct and $s_n = s_{n+1} = \cdots = s_m$, then for $\mu_{s_1,\ldots,s_m}$ one should take the image of the measure $\mu_{s_1,\ldots,s_n}$ under the mapping $\mathbb{R}^n \to \mathbb{R}^m$ that takes $(x_1,\ldots,x_n)$ to $(x_1,\ldots,x_n,x_n,\ldots,x_n)$. Thus, we arrive at the situation considered above, when we could take mappings $(g_1,\ldots,g_n)$ with possibly repeating $g_i \in G$; in the present case for the collection of functionals G on $\mathbb{R}^T$ we take the evaluation functionals $x \mapsto x(t)$ for all possible points $t \in T$.

Moreover, since every continuous linear functional on $E = \mathbb{R}^T$ (with the product topology) has the form

$$x \mapsto c_1x(t_1) + \cdots + c_nx(t_n)$$

with some $c_i \in \mathbb{R}$ and $t_i \in T$, the measures $\mu_{t_1,\ldots,t_n}$ enable one to uniquely reconstruct the measures $\mu_{l_1,\ldots,l_k}$, $l_i \in E'$. Indeed, if $l_i = c_{i,1}t_1 + \cdots + c_{i,n}t_n$, where $t_1,\ldots,t_n \in T$ are all points used for defining $l_1,\ldots,l_k$, then for $\mu_{l_1,\ldots,l_k}$ we take the image of $\mu_{t_1,\ldots,t_n}$ under the linear mapping from $\mathbb{R}^n$ to $\mathbb{R}^k$ defined by the formula $P(x_1,\ldots,x_n) = (y_1,\ldots,y_k)$, $y_i = c_{i,1}x_1 + \cdots + c_{i,n}x_n$. Thus, we obtain a E'-cylindrical measure on $E = \mathbb{R}^T$.

We emphasize that the aforementioned facts are also valid for signed measures. Thus, the definition by means of families of consistent measures $\mu_{t_1,\ldots,t_n}$ turns out to be equivalent to the definition of a cylindrical measure on $\mathbb{R}^T$. The boundedness of this cylindrical measure is equivalent to the uniform boundedness of the variations of the measures $\mu_{t_1,\ldots,t_n}$; this is seen from the fact that the variation of the measure $\mu_{l_1,\ldots,l_k}$ considered above does not exceed the variation of the measure $\mu_{t_1,\ldots,t_n}$.

5.4.5. Corollary. *The Kolmogorov theorem remains valid for any consistent system of countably additive measures $\mu_{t_1,\ldots,t_n}$ on the spaces $\mathbb{R}^n$ having uniformly bounded variations.*

PROOF. As explained above, the system of measures $\mu_{t_1,\ldots,t_n}$ generates a cylindrical measure μ on $\mathbb{R}^T$ with projections $\mu_{t_1,\ldots,t_n}$. Moreover, μ is bounded by the uniform boundedness of variations of the given measures, as we have noted. Hence we obtain nonnegative cylindrical measures μ^+ and μ^-, which are countably additive by the Kolmogorov theorem. Then their difference μ is countably additive as well. □

The Kolmogorov theorem deals with the situation in which each bounded cylindrical measure on a given space is countably additive. This is due to the fact that the space $\mathbb{R}^T$ is very large. Let us give an example of a probability cylindrical measure that is not countably additive on the algebra of cylinders.

5.4.6. Example. (i) Let us take for E the Hilbert space l^2 and for G take the family of coordinate functionals $x \mapsto x_n$. A probability G-cylindrical measure λ on E will be defined as follows: if $C = P^{-1}(B)$, where $B \in \mathcal{B}(\mathbb{R}^n)$ and P is the orthogonal projection onto $\mathbb{R}^n$, then $\lambda(C) := \lambda_n(B \cap [0,1]^n)$, where λ_n is the usual Lebesgue measure on $\mathbb{R}^n$. Actually, the G-cylindrical measure λ on l^2 is generated by a true probability measure on $\mathbb{R}^\infty$ that is the countable power of Lebesgue measure on $[0,1]$.

The measure λ is not countably additive on the algebra of G-cylinders. Indeed, otherwise it could be extended to a probability Borel measure on l^2, but the outer measure with respect to λ of every ball B_R of radius R centered at the origin in l^2 is zero, since for any fixed R we have

$$\lambda_{2n}(B_R) = R^{2n}\lambda_{2n}(B_1) = \pi^n R^{2n}/n! \to 0,$$

hence the extension would be the identically zero measure. In place of λ one can take any countable product of probability measures on the real line that vanishes on l^2 (it is countably additive on $\mathbb{R}^\infty$).

(ii) Let us take again $E = l^2$ and identify $G = E'$ with E in the usual way. A cylindrical measure ν on E will be defined as follows: if $C = P^{-1}(B)$, where P is the orthogonal projection onto an n-dimensional subspace $E_n \subset E$ and $B \in \mathcal{B}(E_n)$, then we set $\nu(C) := \gamma_n(B)$, where γ_n is the standard Gaussian measure on E_n, i.e., the measure with density $(2\pi)^{-n/2}\exp\bigl(-(x,x)^2/2\bigr)$ with respect to the standard Lebesgue measure on E_n corresponding to the given inner product. It is straightforward to verify that the cylindrical measure ν is well-defined (actually this follows from the fact that the projection of γ_{n+1} to $\mathbb{R}^n$ coincides with γ_n). However, as above, there is no countable additivity: again the outer measure of every ball in E is zero (Exercise 5.12.80). The defined cylindrical measure is called "the canonical cylindrical Gaussian measure on the Hilbert space". Below in § 5.7 we consider in more detail Gaussian cylindrical measures.

5.4.7. Remark. A general G-cylindrical measure ν on E can be also described as a system of consistent (in the same sense as in the Kolmogorov theorem) measures $\nu_{g_1,\ldots,g_n}$ on the spaces $\mathbb{R}^n$. Indeed, we simply set

$$\nu(C_{g_1,\ldots,g_n,B}) := \nu_{g_1,\ldots,g_n}(B).$$

The specific feature of the situation in the Kolmogorov theorem is that the space E there turns out to be sufficiently large in order to support the corresponding countably additive measure, but in the general case this is not always true.

Suppose we are given vector spaces E_1 and E_2 with certain spaces of linear functions G_1 and G_2 defined on them such that there exists a linear mapping $T\colon E_1 \to E_2$ for which $g\circ T \in G_1$ for every $g \in G_2$. Then, for every $A \in \mathfrak{A}_{G_2}$, we have $T^{-1}(A) \in \mathfrak{A}_{G_1}$. Thus, every cylindrical measure ν_1 on E_1 generates the cylindrical measure $\nu_2 := \nu_1\circ T^{-1}$ on $\mathfrak{A}_{G_1}$ called the *image of the cylindrical measure ν_1 under the mapping T*.

Cylindrical measures can fail to be countably additive on cylindrical algebras. It turns out, however, that every vector space with a cylindrical algebra can be enlarged in such a way that all cylindrical measures on the extension generated by bounded cylindrical measures on the original space will be countably additive on this extension. A precise formulation is this.

5.4.8. Theorem. *Let E be a vector space and let G be some linear space of linear functions on E separating points. Let j be the natural injective embedding of E in $\mathbb{R}^G$ defined by the formula $j(x)(g) = \langle g, x\rangle$. Then, for every bounded cylindrical measure ν on $\mathfrak{A}_G$, the measure $\nu\circ j^{-1}$ on the cylindrical algebra in $\mathbb{R}^G$ is countably additive.*

PROOF. Since the cylindrical measure $\nu\circ j^{-1}$ on $\mathbb{R}^G$ is also bounded, it is countably additive by the Kolmogorov theorem. □

5.4.9. Remark. This theorem leads to the following universal procedure of constructing bounded cylindrical measures. Suppose we are given two linear spaces E and G in duality and a measurable space $(\Omega, \mathcal{B})$ with a bounded (countably additive) real measure μ on $\mathcal{B}$. Suppose that we are also given a mapping ξ from G to the space $\mathcal{L}_0(\mu) = \mathcal{L}_0(|\mu|)$ of μ-measurable real functions on Ω such that ξ is linear in the following sense: for every $g_1, \ldots, g_n \in G$ and $\alpha_1, \ldots, \alpha_n \in \mathbb{R}^1$, almost everywhere

$$\xi(\alpha_1 g_1 + \cdots + \alpha_n g_n) = \alpha_1\xi(g_1) + \cdots + \alpha_n\xi(g_n).$$

Actually this means that we have defined a linear mapping from G to $L^0(\mu)$, the space of equivalence classes. Then we can set

$$\nu_{g_1,\ldots,g_n} := \mu\circ F_n^{-1}, \quad F_n(\omega) = \bigl(\xi(g_1)(\omega), \ldots, \xi(g_n)(\omega)\bigr).$$

It is obvious that the measures $\nu_{g_1,\ldots,g_n}$ on $\mathbb{R}^n$ are consistent, hence generate a G-cylindrical measure ν on E. If μ is a probability measure, then ξ is called "a random linear function on E" (in this case it is customary to say that we have "a weak distribution on E").

Universality of this construction is seen from the fact that every bounded G-cylindrical measure ν on E can be obtained in this way from some countable additive measure. To this end, it suffices to take for μ the countably additive measure $\nu\circ j^{-1}$ on $\mathbb{R}^G$ from the previous theorem (extended to a countably additive measure on the generated σ-algebra) and set $\xi(g)(\omega) := \omega(g)$.

Let us now give a general result about Radon extensions of cylindrical measures that follows directly from Corollary 5.2.16.

5.4.10. Theorem. *Let E be a topological vector space and G let be a family of continuous linear functions on E separating points in E. Then, every tight bounded measure on the algebra of G-cylinders admits a unique extension to a Radon measure on E.*

PROOF. The above corollary applies, since every G-cylindrical measure is regular due to regularity of all Borel measures on $\mathbb{R}^n$. □

5.4.11. Example. Let X be a normed space and let μ be a measure on the σ-algebra $\mathcal{E}$ of the space X' generated by the elements of X. Then μ possesses a unique extension to a Radon measure on X' with the weak-$*$ topology. Indeed, by the Banach–Alaoglu theorem all closed balls in X' are compact in the weak-$*$ topology. Hence the measure μ is tight.

In conclusion let us mention yet another very useful theorem due to Kolmogorov that enables one to restrict measures originally defined on the space of all paths to the space of continuous paths. First we observe that the set of all continuous functions $C[0,1]$ does not belong to the σ-algebra $\sigma(\mathbb{R}^{[0,1]})$ generated by cylinders in the space $\mathbb{R}^{[0,1]}$ of all functions on $[0,1]$, since every set in the latter σ-algebra is determined by the values of functions on some countable set. Moreover, even if we take some Borel probability measure P on $C[0,1]$ and construct the corresponding measure on $\sigma(\mathbb{R}^{[0,1]})$ by the formula $P(B) = P\big(B \cap C[0,1]\big)$, $B \in \sigma(\mathbb{R}^{[0,1]})$, then $C[0,1]$ will not belong even to the Lebesgue completion of $\sigma(\mathbb{R}^{[0,1]})$ with respect to P on $\sigma(\mathbb{R}^{[0,1]})$. Indeed, the set $C[0,1]$ contains no nonempty set from $\sigma(\mathbb{R}^{[0,1]})$, hence if it belongs to the Lebesgue completion of $\sigma(\mathbb{R}^{[0,1]})$ with respect to P, it must have the outer P-measure zero, which is impossible in our situation.

Thus, even extending P to $\sigma(\mathbb{R}^{[0,1]})$, one cannot restrict the extension back to $C[0,1]$ in the usual way. However, there is a bit more tricky procedure of restricting a measure to a set of full outer measure, not necessarily measurable. Namely, if $P^*(\Omega^*) = 1$ for some set Ω^* in a measurable space $(\Omega, \mathcal{F})$ with a probability measure P, then we can make Ω^* a measurable space with a probability measure Q, taking the σ-algebra of all sets of the form $\Omega^* \cap F$, where $F \in \mathcal{F}$, and letting $Q(\Omega^* \cap F) := P(F)$. This definition is correct, since if $\Omega^* \cap F = \Omega^* \cap G$, where $F, G \in \mathcal{F}$, then $P(F) = P(G)$, because $F \triangle G \subset \Omega\backslash\Omega^*$, whence we obtain $P(F \triangle G) = 0$ by the equality $P^*(\Omega^*) = 1$ (this equality means that the complement of Ω^* contains no sets from $\mathcal{F}$ of positive measure).

5.4.12. Theorem. *Let $T = [a,b]$ and let P be a probability measure on $\mathbb{R}^T$ such that, for some $M > 0$, $\alpha > 0$ and $\beta > 1$, for all $t, s \in T$ we have*

$$\int |x(t) - x(s)|^\alpha \, P(dx) \leqslant M|t-s|^\beta.$$

Then $P^*\big(C[a,b]\big) = 1$. If X is a Banach space, then a similar assertion is true for measures on the space X^T if we replace $C[a,b]$ by the space $C([a,b],X)$ of continuous functions with values in X.

For a proof, see Neveu [**371**, Section III.5], where the first case is considered, but the case of $C([a,b],X)$ is completely similar. In place of the interval $[a,b]$ one can take any set on the real line. There are analogous results for metric spaces.

5.5. The Fourier transform

Here we consider the Fourier transform of measures on infinite-dimensional spaces introduced by Kolmogorov [**280**] (he used the term "the Laplace transform", following the French mathematical tradition of the epoch).

The *Fourier transform* (or the *characteristic functional*) of a bounded Borel measure μ on $\mathbb{R}^n$ is the complex function

$$\widetilde{\mu}(y) = \int_{\mathbb{R}^n} \exp\big(i(x,y)\big) \, \mu(dx).$$

If the measure μ is defined by a density ϱ with respect to Lebesgue measure, then $\widetilde{\mu}$ differs from the inverse Fourier transform of the function ϱ just by the missing factor $(2\pi)^{-n/2}$.

Below we consider again pairs of spaces E and G in duality.

5.5.1. Definition. *The Fourier transform of a real or complex G-cylindrical measure ν on E is the complex function $\widetilde{\nu}$ on G defined by*

$$\widetilde{\nu}(g) = \int_E \exp(i\langle g, x\rangle)\, \nu(dx).$$

If E is a Hilbert space, $G = E'$, and ν is a cylindrical measure on E, then, identifying E' with E by means of the Riesz theorem, its Fourier transform $\widetilde{\nu}$ can be defined on E itself by the formula

$$\widetilde{\nu}(x) = \int_E e^{i(x,z)}\, \nu(dz),$$

where $(\,\cdot\,,\,\cdot\,)$ is the inner product in E. This Fourier transform can be called "the Hilbert Fourier transform".

Let μ_P be the measure on $\mathbb{R}^n$ that is the image of μ with respect to the operator $P = (g_1, \ldots, g_n)\colon E \to \mathbb{R}^n$, where $g_i \in G$. Formula (5.4.2) for the vector $\alpha = (\alpha_1, \ldots, \alpha_n)$ yields

$$\begin{aligned}\widetilde{\mu}(\alpha_1 g_1 + \cdots + \alpha_n g_n) &= \int_E \exp\bigl(i\alpha_1 g_1 + \cdots + i\alpha_n g_n\bigr)\, d\mu \\ &= \int_{\mathbb{R}^n} \exp\bigl(i\alpha_1 y_1 + \cdots + i\alpha_n y_n\bigr)\, \mu_P(dy) = \widetilde{\mu_P}(\alpha). \quad (5.5.1)\end{aligned}$$

In particular, if ν_g is the measure on the real line that is the image of the measure ν with respect to a functional $g \in G$, then

$$\widetilde{\nu}(g) = \int_{\mathbb{R}^1} e^{it}\, \nu_g(dt).$$

Below we usually consider the following two situations. In one of them the role of E is played by some locally convex space and the role of G is played by its topological dual E'; in the other situation the role of E is played by the topological dual to some locally convex space G.

5.5.2. Proposition. *If μ and ν are cylindrical measures on $(E, \mathfrak{A}_G)$ and $\widetilde{\mu}(g) = \widetilde{\nu}(g)$ for all $g \in G$, then $\mu = \nu$, i.e., any cylindrical measure is uniquely determined by its Fourier transform.*

PROOF. Suppose first that $E = \mathbb{R}^n$; here the algebra $\mathfrak{A}_E$ coincides with the σ-algebra of all Borel sets in $\mathbb{R}^n$ and the measures μ and ν are countably additive. In order to prove that $\mu = \nu$, it suffices to show that both measures assign equal integrals to every continuous function f with compact support.

Let $\varepsilon > 0$. We can assume that f has support in the cube $K = [-\pi k, \pi k]^n$, where k is so large that $\mathbb{R}^n \backslash K$ has measure less than ε with respect to the measure $|\mu| + |\nu|$. In addition, we can assume that $|f| \leqslant 1$. There is a $2\pi k$-periodic (in every variable) function φ of the form $\varphi(x) = c_1 \exp\bigl(i(y_1, x)\bigr) + \cdots + c_m \exp\bigl(i(y_m, x)\bigr)$

such that $|f(x) - \varphi(x)| \leqslant \varepsilon$ for all $x \in K$. Then $|\varphi(x)| \leqslant 1 + \varepsilon$ for all x, whence $|f(x) - \varphi(x)| \leqslant 1 + \varepsilon$ for all x. Therefore, taking into account the equality of the integrals of φ against the measures μ and ν, we obtain

$$\left|\int_{\mathbb{R}^n} f\,d\mu - \int_E f\,d\nu\right| \leqslant \int_{\mathbb{R}^n} |f-\varphi|\,d(|\mu|+|\nu|)$$
$$\leqslant \int_K |f-\varphi|\,d(|\mu|+|\nu|) + (1+\varepsilon)(|\mu|+|\nu|)(\mathbb{R}^n\backslash K)$$
$$\leqslant \varepsilon(\|\mu\|+\|\nu\|) + \varepsilon(1+\varepsilon)(\|\mu\|+\|\nu\|).$$

Letting $\varepsilon \to 0$, we obtain the desired equality. Another possible justification can be given with the aid of the Parseval equality (see below).

The general case reduces to the considered one: for any $g_1, \ldots, g_n \in G$, the operator $P = (g_1, \ldots, g_n)\colon E \to \mathbb{R}^n$ takes the measures μ and ν to their projections μ_P and ν_P on $\mathbb{R}^n$, moreover, $\widetilde{\mu_P} = \widetilde{\nu_P}$, since by formula (5.5.1) we have the equalities

$$\widetilde{\mu_P}(\alpha) = \widetilde{\mu}(\alpha_1 g_1 + \cdots + \alpha_n g_n), \quad \widetilde{\nu_P}(\alpha) = \widetilde{\nu}(\alpha_1 g_1 + \cdots + \alpha_n g_n),$$

where $\alpha = (\alpha_1, \ldots, \alpha_n)$. Hence $\mu_P = \nu_P$, whence $\mu = \nu$, because P was arbitrary. □

5.5.3. Proposition. *Let E and G be two vector spaces in duality and let μ and ν be countably additive measures on the σ-algebras σ_G in E and σ_E in G, respectively, such that the function $(g,x) \mapsto \langle g,x\rangle$ is measurable with respect to the product of the σ-algebras σ_G and σ_E. Then the following Parseval equality is valid*:

$$\int_E \widetilde{\nu}(x)\,\mu(dx) = \int_G \widetilde{\mu}(g)\,\nu(dg). \tag{5.5.2}$$

PROOF. By assumption, there exists the integral

$$\int_{E\times G} \exp\bigl(i\langle g,x\rangle\bigr)\,\mu\otimes\nu(dx\,dg).$$

Evaluating this integral by Fubini's theorem in two ways we obtain the desired equality. □

The Parseval equality also yields the coincidence of measures with equal Fourier transforms, since if $\widetilde{\mu_1} = \widetilde{\mu_2}$, then μ_1 and μ_2 assign equal integrals to the Fourier transform of every measure ν, which implies the equality of the integrals against these measures for every function $f \in C_0^\infty(\mathbb{R}^n)$, since every such function f is the Fourier transform of the measure $g\,dx$, where g differs from the direct Fourier transform of the function f by a constant factor (this follows from the known inversion formula for the Fourier transform).

The Parseval equality gives a method to recover a measure by using its Fourier transform. It is natural to ask about a possibility to decide directly in terms of a given function χ on $\mathbb{R}^n$ whether it is the Fourier transform of a bounded measure. Unfortunately, there is no constructive answer to this question (as well as there is no explicit description of the class of the Fourier transforms of integrable functions), but there are constructive necessary and sufficient conditions in the very important partial case of nonnegative measures on $\mathbb{R}^n$.

5.5.4. Definition. *A complex function φ on a linear space E is called positive definite if, for all vectors $x_1, \ldots, x_n \in E$ and complex numbers $c_1, \ldots, c_n$, one has*

$$\sum_{j,k=1}^{n} c_j \overline{c}_k \varphi(x_j - x_k) \geqslant 0.$$

The following classical Bochner–Khintchin theorem is valid (its proof can be found, for example, in [**65**, § 7.13]).

5.5.5. Theorem. *A complex function on $\mathbb{R}^n$ is continuous and positive definite precisely when it coincides with the Fourier transform of a nonnegative bounded Borel measure on $\mathbb{R}^n$.*

Certainly, one can say that the Fourier transforms of bounded measures are the differences of continuous positive definite functions, but there is no much use in such "characterizations". The continuity condition can be formally weakened, by requiring it only at the origin, but it cannot be completely removed (see below).

5.5.6. Corollary. *A complex function φ on the space G is the Fourier transform of some E-cylindrical nonnegative measure on G precisely when this function is positive definite and its restriction to every finite-dimensional subspace in G is continuous (with the standard topology).*

PROOF. The necessity follows from the fact that, for any $g_1, \ldots, g_n \in G$, the function $(t_1, \ldots, t_n) \mapsto \widetilde{\nu}(t_1 g_1 + \cdots + t_n g_n)$ on $\mathbb{R}^n$ is the Fourier transform of the image of the measure ν under the map $P_{g_1,\ldots,g_n} = (g_1, \ldots, g_n) \colon E \to \mathbb{R}^n$. The sufficiency follows from the previous theorem. Indeed, for any $g_1, \ldots, g_n \in G$, the function indicated above is continuous and positive definite, hence it coincides with the Fourier transform of some bounded nonnegative Borel measure $\nu_{g_1,\ldots,g_n}$. The obtained family of measures turns out to be consistent, since the image of the measure $\nu_{g_1,\ldots,g_n}$ with respect to the natural projection from $\mathbb{R}^n$ onto $\mathbb{R}^{n-1}$ has the Fourier transform evaluated by the formula

$$(y_1, \ldots, y_{n-1}) \mapsto \int_{\mathbb{R}^n} \exp\Bigl(i \sum_{j=1}^{n-1} y_j x_j\Bigr) \, \nu_{g_1,\ldots,g_n}(dx)$$
$$= \widetilde{\nu_{g_1,\ldots,g_n}}(y_1, \ldots, y_{n-1}, 0) = \varphi(y_1 g_1 + \cdots + y_{n-1} g_{n-1}),$$

which coincides with the Fourier transform of the measure $\nu_{g_1,\ldots,g_{n-1}}$. Hence this image is $\nu_{g_1,\ldots,g_{n-1}}$. □

It is worth noting that there exist discontinuous positive definite functions on the real line, but the continuity at zero of a positive definite function implies its continuity everywhere by the following inequality (which can be easily derived from the definition of nonnegative definite functions):

$$|\varphi(x) - \varphi(y)| \leqslant 2\varphi(0)\bigl(1 - \operatorname{Re}\varphi(x - y)\bigr). \tag{5.5.3}$$

5.5.7. Theorem. *A sequence of probability measures μ_j on $\mathbb{R}^T$ equipped with the σ-algebra $\sigma\bigl((\mathbb{R}^T)'\bigr)$ converges weakly to a probability measure μ precisely when $\widetilde{\mu}_j(l) \to \widetilde{\mu}(l)$ for all $l \in (\mathbb{R}^T)'$.*

For $\mathbb{R}^n$ this is the Lévy theorem (see [**65**, p. 210]), which yields the case of $\mathbb{R}^\infty$ by [**65**, Example 8.2.16], and Theorem 5.12.71 implies the general case. In § 5.11, we discuss conditions for the countable additivity of cylindrical measures in terms of their Fourier transforms. Here we give only one simple general result ensuring the continuity of Fourier transforms in some topology.

5.5.8. Proposition. *Let μ be a Radon measure on a locally convex space E. Suppose that E' is equipped with the topology of uniform convergence on a system $\mathcal{S}$ of Borel sets in E such that $\|\mu\| = \sup\{|\mu|(S)\colon\ S \in \mathcal{S}\}$. Then the function $\widetilde{\mu}$ is continuous in this topology.*

PROOF. With the aid of the Jordan decomposition the assertion reduces to nonnegative measures. Hence we can assume that μ is a probability measure. By inequality (5.5.3) it suffices to verify the continuity of $\mathrm{Re}\,\widetilde{\mu}$ at the origin. Let us fix $\varepsilon \in (0, 1/2)$. Let us take a set $S \in \mathcal{S}$ such that $\mu(E\backslash S) < \varepsilon/4$. The set $V = \{f \in E'\colon\ \sup_{x\in S} |f(x)| < \varepsilon\}$ is a neighborhood of zero in the considered topology on E'. For all $f \in V$, by the estimate $1 - \cos t \leqslant t^2$ we have

$$\int_E [1 - \cos f(x)]\, \mu(dx) \leqslant 2\mu(E\backslash S) + \int_S |f(x)|^2\, \mu(dx) \leqslant \frac{1}{2}\varepsilon + \varepsilon^2 \leqslant \varepsilon,$$

which gives the desired continuity. □

5.5.9. Example. If E is quasi-complete, then the function $\widetilde{\mu}$ is continuous in the Mackey topology $\tau(E', E)$. Moreover, it is sufficient that every compact set in E be contained in a convex compact set.

It should be noted that Kwapien and Tarieladze [**309**] constructed an example of a Radon probability measure on a metrizable locally convex space X such that its Fourier transform is not continuous in the Mackey topology $\tau(X', X)$. Thus, in the previous example, certain additional conditions on the space (or on the measure) are necessary.

5.6. Covariance operators and means of measures

Throughout this section X is a Hausdorff locally convex space and the considered measures are nonnegative.

5.6.1. Definition. (i) *A measure μ on $\sigma(X')$ is called a measure with weak moment of order $r > 0$ (or a measure of weak order r) if X' is contained in $L^r(\mu)$.*

(ii) *A Borel (or Baire) measure μ on X is called a measure with strong moment of order $r > 0$ (or of strong order r) if $\psi \in L^r(\mu)$ for every continuous seminorm ψ on X.*

The measure μ on the Hilbert space l^2 with values n^{-2} at the points ne_n, where $\{e_n\}$ is the standard basis, has weak first moment, since $\sum_{n=1}^\infty n^{-1}|y_n| < \infty$ whenever $(y_n) \in l^2$, but has no strong first moment, since $\sum_{n=1}^\infty n^{-1} = \infty$.

5.6.2. Definition. *We say that a measure μ on X of weak order* 1 *has a mean (or barycenter) $m_\mu \in X$ if for every $l \in X'$ we have*

$$l(m_\mu) = \int_X l(x)\,\mu(dx).$$

Barycenters of signed measures are defined similarly.

In the general case the existence of weak moments does not guarantee the existence of barycenters. For example, the atomic measure μ defined on the Banach space c_0 by $\mu(2^n e_n) = 2^{-n}$, where $\{e_n\}$ is the standard basis in c_0, has weak first moment, but has no mean, since otherwise all coordinates of its mean m would be equal to 1 by the equality $l_k(m) = \sum_{n=1}^{\infty} l_k(e_n) = 1$ for every coordinate functional l_k. It is worth noting that such examples are impossible in spaces not containing c_0.

5.6.3. Proposition. *If a complete metrizable locally convex space X has no subspace linearly homeomorphic to c_0, then every Radon measure μ on X of weak order* 1 *has a mean m_μ.*

A proof can be found in Vakhania, Tarieladze [**537**].

5.6.4. Lemma. *If a measure μ on $\sigma(X')$ has a mean m_μ and a continuous seminorm p is integrable with respect to μ, then*

$$p(m_\mu) \leqslant \int_X p\,d\mu. \tag{5.6.1}$$

PROOF. By the Hahn–Banach theorem there is a functional $f \in X'$ such that $f(m_\mu) = p(m_\mu)$ and $f(x) \leqslant p(x)$ for all $x \in X$. Then

$$p(m_\mu) = f(m_\mu) = \int_X f(x)\,\mu(dx) \leqslant \int_X p(x)\,\mu(dx),$$

as required. □

Let us give a sufficient condition for the existence of a mean ensuring also a possibility to express it as a vector integral

$$\int_X x\,\mu(dx), \tag{5.6.2}$$

understood in the following sense: there is a net of Borel mappings $F_\alpha\colon X \to X$ each of which assumes finitely many values $z_{\alpha,1}, \ldots, z_{\alpha,k}$ on sets $A_{\alpha,1}, \ldots, A_{\alpha,k}$, where $k = k(\alpha)$ and $\lim\limits_{\alpha} \max_{i\leqslant k} \mu(A_{\alpha,i}) = 0$, such that

$$\lim_{\alpha} \int_X p\bigl(x - F_\alpha(x)\bigr)\,\mu(dx) = 0$$

for every continuous seminorm p and, moreover, the integrals of F_α defined as sums $\sum_{i=1}^{k} \mu(A_{\alpha,i}) z_{\alpha,i}$ converge in X to m_μ.

In case of Banach spaces this gives the *Bochner integral.*

Note that the mapping $x - F_\alpha(x)$ is Borel measurable, since on every Borel set $A_{\alpha,i}$ it differs from the identity by a constant shift. Therefore, the function $p\bigl(x - F_\alpha(x)\bigr)$ is also Borel measurable.

If the space X is complete, then the integrals of F_α form a Cauchy net, hence converge automatically; the same is true if the indicated integrals belong to a compact set. Indeed, for every continuous seminorm p, the integrals m_α of F_α satisfy the estimate

$$p(m_\alpha - m_\beta) \leqslant \int_X p\bigl(F_\alpha(x) - F_\beta(x)\bigr)\,\mu(dx),$$

moreover, $p\bigl(F_\alpha(x) - F_\beta(x)\bigr) \leqslant p\bigl(x - F_\alpha(x)\bigr) + p\bigl(x - F_\beta(x)\bigr)$. In the general case, convergence of the integrals must be required additionally.

Suppose we are given a set S in a locally convex space X and a directed index set $\mathfrak{A}$ and that for every index $\alpha \in \mathfrak{A}$ we are given a finite partition $\pi_\alpha = \{S_{\alpha,i}\}$ of the set S into disjoint sets $S_{\alpha,i}$, where $i = 1, \ldots, k(\alpha)$. We shall say that the net of partitions π_α is refining if, for every continuous seminorm p on X, we have

$$\limsup_\alpha \sup_i \sup_{x,y\in S_{\alpha,i}} p(x-y) = 0.$$

5.6.5. Lemma. *In case of a totally bounded set S, a refining net of finite partitions always exists, and in case of a metrizable compact set there is a countable refining sequence of finite partitions. For Borel sets, the elements of partitions can be chosen also Borel.*

PROOF. For an index set $\mathcal{P}$ we take the family of all continuous seminorms on X equipped with the natural partial ordering defined by the pointwise comparison. This set is directed, since the sum of continuous seminorms is a continuous seminorm. For any $p \in \mathcal{P}$, by assumption there exist points $s_1, \ldots, s_k \in S$ such that S is covered by the sets $s_i + V$, where $V = \{x\colon p(x) \leqslant 1\}$. Hence we can partition S into disjoint parts $S_{p,1}, \ldots, S_{p,k}$ such that $p(x-y) \leqslant 1$ if $x, y \in S_{p,i}$. If S is Borel, then $S_{p,i}$ can be chosen Borel as well.

We now show that we have obtained a refining net of partitions. Let $p \in \mathcal{P}$ and $\varepsilon > 0$. Then $q_\varepsilon = \varepsilon^{-1}p \in \mathcal{P}$. In addition, for all $q \geqslant q_\varepsilon$ and all $x, y \in S_{q,i}$, where $S_{q,i}$ are constructed by q as indicated above, we have

$$p(x-y) = \varepsilon q_\varepsilon(x-y) \leqslant \varepsilon q(x-y) \leqslant \varepsilon,$$

which proves our claim.

Suppose now that S is compact and metrizable by a metric d. For every n we partition S into finitely many parts $S_1, \ldots, S_k$ of diameter less than n^{-1}, where $k = k(n)$. We show that for every continuous seminorm p and every $\varepsilon > 0$ there exists n_0 such that, for every $n \geqslant n_0$ and all $i \leqslant k(n)$ we have $p(x-y) \leqslant \varepsilon$ whenever $x, y \in S_i$. To this end, it suffices to prove that there exists $\delta > 0$ such that $p(x-y) \leqslant \varepsilon$ whenever $d(x,y) \leqslant \delta$. The existence of such $\delta > 0$ follows from the fact that the function $(x,y) \mapsto p(x-y)$ is continuous on the compact set $S \times S$ with the metric $d_1\bigl((x_1,y_1),(x_2,y_2)\bigr) = d(x_1,x_2) + d(y_1,y_2)$, hence is uniformly continuous, but at the diagonal it vanishes. □

5.6.6. Proposition. *Suppose that a Radon probability measure μ on X has compact support contained in a convex compact set K. Then μ has a mean $m_\mu \in K$, moreover, for every refining net of partitions of K into disjoint Borel* *parts $A_{\alpha,i}$ and every choice of points $x_{\alpha,i} \in A_{\alpha,i}$, the means of the measures $\mu_\alpha := \sum_i \mu(A_{\alpha,i})\delta_{x_{\alpha,i}}$ converge to m_μ. In addition, the vector integral* (5.6.2) *exists.*

PROOF. Let us set $m_\alpha := \sum_i \mu(A_{\alpha,i})x_{\alpha,i}$. By the convexity of K we have $m_\alpha \in K$. Mappings F_α are defined as follows: $F_\alpha(x) = x_{\alpha,i}$ if $x \in A_{\alpha,i}$. Let p be a continuous seminorm on X and $\varepsilon > 0$. Choose an index α_0 such that

$$\sup_i \sup_{x,y\in A_{\alpha,i}} p(x-y) \leqslant \varepsilon$$

whenever $\alpha \geqslant \alpha_0$. Then for any $x \in A_{\alpha,i}$ we have $p(x - x_{\alpha,i}) \leqslant \varepsilon$ for all $i \leqslant k$. Hence the integral of $p(x-F_\alpha(x))$ over the set $A_{\alpha,i}$ with respect to the measure μ does not exceed $\varepsilon\mu(A_{\alpha,i})$. Thus, the integral of $p(x - F_\alpha(x))$ over the whole set K does not exceed ε. This yields convergence of $\{m_\alpha\}$ in K to some vector m. It is clear that the integral of f over K equals $f(m)$. □

5.6.7. Corollary. *If a Radon probability measure on a weakly sequentially complete and quasi-complete locally convex space has weak first moment, then it has a mean in this space.*

PROOF. Let μ be a Radon probability measure on a space X with the stated properties. Quasi-completeness of the space gives convex compact sets K_n for which $\mu(X\backslash K_n) \to 0$. The measures $I_{K_n} \cdot \mu$ have means m_n. If μ has weak first moment, then, for every $f \in X'$, the sequence of numbers $f(m_n)$, which are equal to the integrals of f over K_n, converges to the integral of f. By assumption, there exists a vector $m \in X$ for which the limit of this sequence is $f(m)$. Hence m is the mean of μ. □

We shall see below that the same is true even without the assumption of the weak sequential completeness for measures of weak order $p > 1$.

5.6.8. Corollary. *Suppose that a Radon probability measure μ on a sequentially complete space X is concentrated on a countable union of convex compact sets (which holds if X is quasi-complete) and possesses strong first moment. Then μ has a mean $m_\mu \in X$.*

PROOF. Let μ be concentrated on the union of convex compact sets K_n. We already know that the measures $\mu_n = I_{K_n} \cdot \mu$ have means $m_n \in K_n$. For every continuous seminorm p and all $n \geqslant l$ we have

$$p(m_n - m_l) \leqslant \int_{K_n\backslash K_l} p\, d\mu,$$

whence by the integrability of p we obtain that the sequence $\{m_n\}$ is Cauchy. Since X is sequentially complete, this sequence converges to some vector $m \in X$. It is obvious that m is the mean of the measure μ. □

For metrizable compact sets, the condition of completeness of X can be slightly weakened. The proof of the next assertion can be extracted from the above reasoning.

5.6.9. Proposition. *If X is sequentially complete and $K \subset X$ is compact metrizable, then every Borel measure on K has a mean $I(\mu)$ in X. In addition, $I(\mu)$ belongs to the closed convex envelope of K, provided that μ is a probability measure.*

Note that the Jordan decomposition $\mu = \mu^+ - \mu^-$ of a signed measure μ gives the mean $I(\mu)$ contained in the closed absolutely convex envelope of K, provided that $\|\mu\| \leqslant 1$.

5.6.10. Theorem. *The closed absolutely convex envelope of a metrizable compact set in a locally convex space X is metrizable, and if X is sequentially complete, then it is compact.*

PROOF. The first assertion follows from the second, since X has a completion (which is, certainly, sequentially complete). Let X be sequentially complete. By the Riesz theorem $C(K)^*$ is the space of signed Borel measures on the given compact space K. The closed unit ball U in $C(K)^*$ is compact in the weak-$*$ topology. Since the space $C(K)$ is separable (Exercise 1.12.66), the ball U is metrizable and compact in the weak topology (Theorem 3.1.4). Let us consider the mapping $I\colon U \to X$ from the previous proposition (taking into account our remark about signed measures). It is readily seen that this mapping is continuous, provided that U is equipped with the weak-$*$ topology and X is equipped with the weak topology. Hence the absolutely convex set $I(U)$ is weakly compact in X. By the metrizability of U this set is also metrizable (Exercise 1.12.67). It is clear that $I(U)$ contains the closed convex envelope A of the set K, since $K \subset I(U)$ by the equality $k = I(\delta_k)$, where δ_k is the Dirac probability measure at the point k. Hence A is metrizable and compact, being a closed subset of a metrizable compact space (actually, as is easily verified, $I(U)$ coincides with the set A). □

It is often useful in applications to have similar facts for weakly compact sets (but without the assumption of weak completeness of X).

5.6.11. Theorem. *Let X be a separable complete locally convex space and let K be a weakly compact set in X. Then its closed convex envelope is weakly compact and coincides with the set of barycenters of Radon probability measures on K (with the weak topology).*

PROOF. Let $\mathcal{P}(K)$ be the set of Radon probability measures on K (with the weak topology) also equipped with the weak topology. For each $\mu \in \mathcal{P}(K)$, consider the functional in $(X')^*$ taking $l \in X'$ to the integral of l with respect to the measure μ. By Corollary 3.8.6 this functional is given by some vector $b(\mu) \in X$, since if a sequence $\{l_n\} \subset X'$ converges to zero pointwise, it is uniformly bounded on K (by weak compactness), hence we have convergence of the integrals. Since S is convex and closed, one has $b(\mu) \in S$. The mapping $m \mapsto b(\mu)$ is continuous in the weak topology. Thus, the image of the convex compact set $\mathcal{P}(K)$ is a convex weakly compact set, dense in S. This image contains the convex envelope of K, since for every $k \in K$ we have $k = b(\delta_k)$. Therefore, S coincides with this image. □

5.6.12. Corollary. (KREIN'S THEOREM) *Let K be a compact set in a Hausdorff locally convex space E. Its closed convex envelope S is compact precisely when it is complete in the Mackey topology. In particular, if the set K is weakly compact and its closed convex envelope is complete in the Mackey topology, then it is weakly compact.*

PROOF. If S is compact, then it is complete in the Mackey topology. Let S be complete in the Mackey topology. Since S is precompact, it suffices to show that it is complete in the original topology, and the weak completeness is sufficient for this. Hence we can assume from the very beginning that the original topology is the weak topology. Moreover, we can assume that E is complete, since S is complete also in the completion of E. In addition, by convexity, S is weakly closed. Thus, we have to show relative weak compactness of S, which by the Eberlein theorem reduces to showing that every sequence in S has a weak limit point in E. Therefore, everything reduces to showing that S is relatively weakly compact in a separable complete space E, so that it remains to apply the previous theorem. □

These questions are related to the important Choquet representations of points of the closed convex envelopes of compact sets in the form of barycenters of probability measures on these sets. Let X be a separated locally convex space, let K be a compact set in X, and let μ be a Radon probability measure on K. The results obtained above yield that if μ has a mean b_μ, then $b_\mu \in \overline{\mathrm{conv}}\, K$. However, there might be no mean unless we impose additional conditions on X (such as quasi-completeness or compactness of the closed convex envelopes of all compact sets). Nevertheless, the following fact is true.

5.6.13. Theorem. *The closed convex envelope of any compact set K coincides with the set of barycenters of Radon probability measures on K (having barycenters).*

PROOF. If X is complete, then every Radon measure μ on K has a barycenter $b_\mu \in \overline{\mathrm{conv}}\, K$. Therefore, considering X in its completion and using the compactness of K in the completion, it suffices to prove the theorem for complete X. A justification is only needed for a representation of every element in $\overline{\mathrm{conv}}\, K$ in the form of a barycenter of a Radon probability measure on K. The points of K itself are the barycenters of the Dirac measures at these points. Hence all points in $\mathrm{conv}\, K$ admit the required representation. We now observe that the mapping $I\colon \mu \mapsto b_\mu$ from the set $P(K)$ of Radon probability measures on K to X is continuous if we equip the space of measures with the weak topology and the space X with the weak topology $\sigma(X, X')$. By compactness of K the set $P(K)$ is also compact, hence its image is compact as well. Since we already know that the image of $P(K)$ is dense in $\overline{\mathrm{conv}}\, K$, we obtain the equality $I\bigl(P(K)\bigr) = \overline{\mathrm{conv}}\, K$. Certainly, here one can use a more direct reasoning: for any $b \in \overline{\mathrm{conv}}\, K$, we can take the net of the barycenters $b_\alpha \in \mathrm{conv}\, K$ of discrete probability measures μ_α on K converging to b, by choosing in $\{\mu_\alpha\}$ a subnet converging in the topology $\sigma\bigl(C(K), C(K)'\bigr)$ with a limit μ (belonging to $P(K)$), which obviously gives the equality $b = b_\mu$. □

A measure μ is called a representing measure for b if $b = b_\mu$. For a convex compact set, it is useful to have representing measures concentrated on the set of its extreme points. The existence of such measures is ensured by the following Choquet–Bishop–de Leeuw theorem. Choquet proved this theorem for metrizable K. In this case the set $\mathrm{ext}K$ of extreme points of the compact K is a G_δ-set, in particular, is a Borel set. In the general case this is not true, which leads to a modified formulation.

5.6.14. Theorem. *If K is a convex compact set in X, then, for any point $k \in K$, there exists a Radon probability measure μ on K representing k and vanishing on all Baire sets in $K\backslash\mathrm{ext}K$. If K is metrizable, then $\mu(\mathrm{ext}K) = 1$.*

On the Choquet representations, see Alfsen [**13**], Edwards [**150**], Meyer [**340**], Phelps [**387**], Simon [**464**] (where one can additional information related to convexity), von Weizsäcker [**561**], von Weizsäcker, Winkler [**563**], where additional references can be found.

For a measure μ of weak order p we have the natural embedding operator

$$T_\mu\colon\ X' \to L^p(\mu).$$

5.6.15. Lemma. *Suppose that a measure μ on a locally convex space X has weak moment of order p. Then the operator $T_\mu\colon\ X' \to L^p(\mu)$ has closed graph provided that X' is equipped with any locally convex topology τ stronger than $\sigma(X',X)$. Hence this operator is continuous if for (X',τ) the closed graph theorem holds (for example, if (X',τ) is a Fréchet space or is a strict inductive limit of a sequence of such spaces).*

PROOF. If $f_n, f \in X'$ and $f_n(x) \to f(x)$ pointwise and $f_n \to g$ in $L^p(\mu)$, then the sequence $\{|f_n|^p\}$ is uniformly integrable, hence $f_n \to f$ in $L^p(\mu)$ and $f = g$ a.e. □

For example, if X is a normed space, then X' is Banach with its natural norm, so the closed graph theorem holds for X'. It is also sufficient that X' be barrelled in the Mackey topology $\tau(X',X)$, but this yields the semireflexivity of X (see § 3.7).

Let us give a modification of Lemma 5.6.7.

5.6.16. Lemma. *If a Radon probability measure μ on a quasi-complete locally convex space X satisfies the condition $X' \subset L^p(\mu)$, where $p > 1$, then the operator $T_\mu\colon\ X' \to L^p(\mu)$ is continuous when X' is equipped with the Mackey topology $\tau(X',X)$ and μ has a mean.*

PROOF. Since quasi-completeness implies local completeness (see § 2.10(iii)), it follows from Example 3.9.13 that it suffices to verify that T_μ is continuous with respect to the topology $\tau(X',X)$ and the topology of convergence in measure on $L^p(\mu)$. Using quasi-completeness once again, for every $\varepsilon > 0$ we can find a convex compact set $K_\varepsilon \subset X$ such that $\mu(X\backslash K_\varepsilon) < \varepsilon$. If now $f \in X'$ and $\sup_{x\in K} |f(x)| \leqslant \varepsilon$, then obviously $\mu(x\colon\ |f(x)| > \varepsilon) < \varepsilon$. This is the required continuity.

Since X coincides with the dual to X' with the topology $\tau(X',X)$, there is the adjoint operator $T^*\colon\ L^q(\mu)\to X$, where $p^{-1}+q^{-1}=1$, which gives a mean by the formula $m=T^*1\in X$. Indeed,

$$f(T^*1)=\langle 1,Tf\rangle=\int_X f(x)\,\mu(dx),\quad f\in X',$$

which follows from the definition of T^*. □

5.6.17. Definition. *Let μ be a measure of weak order 2 on a locally convex space X. Its covariance is the function $C_\mu\colon\ X'\times X'\to\mathbb{R}$ defined by the formula*

$$C_\mu(l_1,l_2)=\int_X l_1(x)l_2(x)\,\mu(dx)-\int_X l_1(x)\,\mu(dx)\int_X l_2(x)\,\mu(dx).$$

The covariance operator R_μ from X' to the algebraic dual to X' is defined by the equality

$$R_\mu\colon\ X'\to(X')^*,\quad R_\mu(f)(g)=C_\mu(f,g).$$

Let us also consider the operator

$$K_\mu\colon\ X'\to(X')^*,\quad K_\mu(f)(g)=(f,g)_{L^2(\mu)}.$$

If the measure μ of weak second order has a mean m, then its covariance operator coincides with the covariance operator of the shifted measure $\mu_m(B)=\mu(B+m)$ having zero mean, i.e., $R_\mu=K_{\mu_m}$.

It is clear that any covariance operator R has the following properties:

1) linearity,

2) nonnegativity, i.e., $\langle f,R(f)\rangle\geqslant 0$ for all $f\in X'$,

3) symmetry, i.e., $\langle R(f),g\rangle=\langle R(g),f\rangle$, $f,g\in X'$.

Under broad assumptions about the measure or about the space the covariance operator takes values in more narrow subspaces of the algebraic dual to X' (for example, in X'' or X) and is continuous in reasonable topologies. This question is studied in detail in Vakhania, Tarieladze [**537**]. Let us give some typical results.

5.6.18. Theorem. *Let μ be a Radon probability measure on a complete (or quasi-complete) locally convex space X with second weak moment. Then we have $R_\mu(X')\subset X$.*

PROOF. By Lemma 5.6.16 the measure μ has a mean, which reduces our assertion to measures with zero mean. It remains to apply the cited lemma and observe that $R_\mu f=T_\mu^*f$, $f\in X'$, where $T^*\colon\ L^2(\mu)\to X$ is the dual operator for T_μ from the lemma. □

If X is a Hilbert space with the inner product $(\,\cdot\,,\,\cdot\,)$, then it is convenient to identify X' with X and consider the covariance operator of the measure μ as the operator in X defined by the formula

$$(R_\mu u,v)=\int_X (u,x)(v,x)\,\mu(dx)-\int_X (u,x)\,\mu(dx)\int_X (v,x)\,\mu(dx).$$

For measures with zero mean we obtain the formula

$$(R_\mu u, v) = \int_X (u, x)(v, x)\,\mu(dx).$$

5.6.19. Theorem. *The class of covariance operators of measures of second weak order on a separable Fréchet space X coincides with the class of all symmetric nonnegative operators from X' to X.*

For a proof, see Vakhania, Tarieladze [**537**]. Usually the class of covariance operators of measures of strong second order is smaller.

5.6.20. Proposition. *Let H be a separable Hilbert space and let μ be a measure of weak second order. Then μ has strong second moment if and only if its covariance operator R_μ is nuclear.*

PROOF. Let $\{e_n\}$ be an orthonormal basis in H. If μ has strong second moment, then, as proved above, it has a mean m and the measure μ_m also has strong second moment. Hence we can assume that μ has zero mean. Then

$$\int_H (x, x)\,\mu(dx) = \sum_{n=1}^{\infty} \int_H (x, e_n)^2\,\mu(dx) = \sum_{n=1}^{\infty} (R_\mu e_n, e_n).$$

Thus, R_μ is a nuclear operator. Conversely, suppose that R_μ is a nuclear operator. If the measure μ has zero mean, then the previous equality gives at once the integrability of (x, x) by the monotone convergence theorem. In the general case we observe that μ has mean m (by Corollary 5.6.7). Hence the measure μ_m has strong second moment and the same is true for μ. □

Let us mention one more result from Vakhania, Tarieladze [**537**].

5.6.21. Theorem. *If X is a Banach space, then the following conditions are equivalent:*

(i) *X is linearly homeomorphic to a Hilbert space,*

(ii) *for every two Radon probability measures μ and ν on X with $R_\mu = R_\nu$, the existence of strong second moment of the measure μ implies the existence of strong second moment of the measure ν.*

5.7. Gaussian measures

A thorough discussion of Gaussian measures and an extensive bibliography can be found in the book Bogachev [**63**]. Here we only mention some basic concepts and facts necessary for our further exposition.

5.7.1. Definition. *A Gaussian measure on $\mathbb{R}^1$ is a probability Borel measure that is either concentrated at some point a, i.e., is Dirac's measure δ_a at a, or is defined by density $(2\pi\sigma)^{-1/2}\exp\bigl(-(2\sigma)^{-1}(x-a)^2\bigr)$ with respect to Lebesgue measure, where $a \in \mathbb{R}^1$ is called the mean and $\sigma > 0$ is the called the variance of this measure.*

The measure for which $a = 0$ and $\sigma = 1$ is called *standard Gaussian*. A direct calculation shows that the Fourier transform of the Gaussian measure with mean a and variance σ has the form

$$\exp\Bigl(iax - \frac{\sigma}{2}x^2\Bigr).$$

Suppose now that we are given two linear spaces E and G in duality.

5.7.2. Definition. *A probability measure γ on the σ-algebra $\sigma(G)$ is called Gaussian if for every $g \in G$ the induced measure $\gamma\circ g^{-1}$ is Gaussian on the real line.*

A Radon probability measure γ on a locally convex space X is called Gaussian if so is its restriction to $\sigma(X')$.

From the change of variables formula and the formula for the Fourier transform of a one-dimensional Gaussian measure we obtain the following expression for the Fourier transform of a Gaussian measure on $\sigma(G)$:

$$\widetilde{\gamma}(g) = \exp\Bigl(ia_g - \frac{\sigma_g}{2}\Bigr), \quad g \in G,$$

$$a_g = \int_E g(x)\,\gamma(dx), \quad \sigma_g = \int_E \bigl(g(x) - a_g\bigr)^2\,\gamma(dx).$$

It is clear that the function $g \mapsto a_g$ is linear on G and the function $g \mapsto \sigma_g$ is a nonnegative quadratic form on G. This quadratic form is generated by the bilinear function

$$Q(f, g) = \int_E \bigl(f(x) - a_f\bigr)\bigl(g(x) - a_g\bigr)\,\gamma(dx),$$

called the covariance of the measure γ.

Conversely, if a measure γ on $\sigma(G)$ is such that there exist a linear function a and a quadratic function Q on G for which

$$\widetilde{\gamma}(g) = \exp\Bigl(ia(g) - \frac{1}{2}Q(g)\Bigr), \tag{5.7.1}$$

then the measure γ is Gaussian. Indeed, for every $g \in G$ the function $t \mapsto \widetilde{\gamma}(tg)$ has the form $\exp\bigl(ita_g - t^2 Q(g)/2\bigr)$, i.e., is the Fourier transform of the Gaussian measure with mean a_g and variance σ_g. But this function is the Fourier transform of the measure $\gamma\circ g^{-1}$.

It is important here that we are given in advance that the right-hand side of (5.7.1) is the Fourier transform of some measure. Not for every nonnegative quadratic form Q on G the function $\exp(-Q/2)$ is the Fourier transform of a countably additive measure. We return to this question below in § 5.11.

It is known that any Radon Gaussian measure γ has mean $m \in X$, i.e., there is a vector m such that

$$f(m) = \int_X f(x)\,\gamma(dx) \quad \forall f \in X^*.$$

If $m = 0$, i.e., the measure $\gamma\circ f^{-1}$ for each $f \in X^*$ have mean 0, then γ is called *centered*. Any Radon Gaussian measure γ is the shift of the centered Gaussian measure γ_m defined by the formula $\gamma_m(B) := \gamma(B + m)$.

The most important example of a Gaussian measure is the countable power of the standard Gaussian measure on the real line. This measure is defined on the space $\mathbb{R}^\infty$. Its Fourier transform has the form

$$(y_1, \dots, y_n, 0, 0, \dots) \mapsto \exp\Bigl(-\frac{1}{2}(y_1^2 + \dots + y_n^2)\Bigr).$$

Another key example is the Wiener measure P_W on the space $\mathbb{R}^{[0,T]}$ of all functions on $[0,T]$ or on the space of continuous functions $C[0,T]$. On the space $\mathbb{R}^{[0,T]}$ the Wiener measure is defined by its finite-dimensional projections $P_{t_1,\dots,t_n}$, which for $0 < t_1 < \dots < t_n \leqslant T$ are given by densities $p_{t_1,\dots,t_n}(x_1,\dots,x_n)$ with respect to Lebesgue measure on $\mathbb{R}^n$ of the following special form:

$$\frac{1}{\sqrt{2\pi t_1}} \exp\Bigl(-\frac{x_1^2}{2t_1}\Bigr) \times \frac{1}{\sqrt{2\pi(t_2-t_1)}} \exp\Bigl(-\frac{(x_2-x_1)^2}{2(t_2-t_1)}\Bigr) \times \dots \times \\ \times \frac{1}{\sqrt{2\pi(t_n-t_{n-1})}} \exp\Bigl(-\frac{(x_n-x_{n-1})^2}{2(t_n-t_{n-1})}\Bigr).$$

In addition, it is required that $P_0 = \delta_0$. With the aid of Kolmogorov's theorem on consistent finite-dimensional distributions one can easily verify the existence of the Wiener measure on $\mathbb{R}^{[0,T]}$. Next, by using another Kolmogorov's theorem 5.4.12, one can verify that the set $C[0,T]$ of continuous functions has outer measure 1 and hence the measure P_W can be defined on $C[0,T]$, which is called the *classical Wiener measure*. To apply the cited theorem we have to observe that

$$\int |x(t) - x(s)|^4 \, P_W(dx) = 3|t-s|^3.$$

There are ways of defining the Wiener measure directly on $C[0,T]$ (as did Wiener himself). One can also take an orthonormal basis $\{\varphi_n\}$ in $L^2[0,1]$, a sequence $\{\xi_n\}$ of independent standard Gaussian random variables and consider the random series $\sum_{n=1}^\infty \xi_n(\omega) e_n(t)$, where e_n is the indefinite integral of φ_n on $[0,1]$. It turns out that such a series converges uniformly on $[0,1]$ for almost all ω and the generated measure is the Wiener measure.

The Fourier transform of the classical Wiener measure is defined on the dual space to $C[0,T]$, i.e., on the space of bounded Borel measures on $[0,T]$, and is given by the formula

$$\nu \mapsto \exp\Bigl(-\frac{1}{2}\int_{[0,T]}\int_{[0,T]} \min(t,s)\, \nu(dt)\, \nu(ds)\Bigr).$$

This equality can be verified as follows. First we check it directly for measures of the form $\nu = \sum_{i=1}^n c_i \delta_{t_i}$, where $t_i \in [0,T]$. To do this, we use the indicated finite-dimensional distributions to find an explicit expression for the image of the Wiener measure with respect to the linear functional $x \mapsto \sum_{i=1}^n c_i x(t_i)$, which coincides with the image of the measure with density $p_{t_1,\dots,t_n}$ on $\mathbb{R}^n$ under the linear function $x \mapsto \sum_{i=1}^n c_i x_i$. Next, by a limit procedure, we extend the result to all measures (any measure ν is approximated by a weakly convergent sequence of discrete measures).

For a centered Radon Gaussian measure γ let X_γ^* denote the closure of X' in $L^2(\gamma)$. Elements of X_γ^* are called *γ-measurable linear functionals*. There is an operator $R_\gamma\colon X_\gamma^* \to X$, called the covariance operator of the measure γ, such that

$$f(R_\gamma g) = \int_X f(x)g(x)\,\gamma(dx), \quad f \in X',\, g \in X_\gamma^*.$$

Set $\widehat{h} := g$ if $h = R_\gamma g$. Then $\widehat{h}$ is called the γ-measurable linear functional generated by h. We have also the vector equality (if X is a Banach space, then in the sense of Bochner's integral)

$$R_\gamma g = \int_X g(x)x\,\gamma(dx), \quad g \in X_\gamma^*.$$

Thus, this operator extends the covariance operator defined on X'.

For example, if γ is a centered Gaussian measure on a separable Hilbert space X, then there exists a nonnegative nuclear operator K on X for which $Ky = R_\gamma y$ for all $y \in X$, provided that we identify X' with X. Then we obtain

$$(Ky, z) = \int_X (y,x)(z,x)\,\gamma(dx), \quad \widetilde{\gamma}(y) = \exp\Bigl(-\frac{1}{2}(Ky,y)\Bigr).$$

Let us take the orthonormal eigenbasis $\{e_n\}$ of the operator K with eigenvalues $\{k_n\}$. Then γ is the image of the countable power γ_0 of the standard Gaussian measure on $\mathbb{R}^1$ under the mapping $\mathbb{R}^\infty \to X$, $(x_n) \mapsto \sum_{n=1}^\infty \sqrt{k_n}x_n e_n$. This series converges γ_0-a.e. in X by convergence of the series $\sum_{n=1}^\infty k_n x_n^2$, which follows from convergence of the series of k_n and the fact that the integral of x_n^2 against the measure γ_0 equals 1. In this way X_γ^* is identified with the completion of X with respect to the norm $x \mapsto |\sqrt{K}x|_X$, i.e., the embedding $X = X' \to X_\gamma^*$ is a Hilbert–Schmidt operator.

Returning to the general case, note the equality

$$\widetilde{\gamma}(l) = \exp\Bigl(-\frac{1}{2}|R_\gamma(l)|_H^2\Bigr), \quad l \in X',$$

which can be now extended to $l \in X_\gamma^*$.

The space $H(\gamma) = R_\gamma(X_\gamma^*)$ is called the *Cameron–Martin space* of the measure γ. It is Hilbert with respect to the inner product

$$(h,k)_H := \int_X \widehat{h}(x)\widehat{k}(x)\,\gamma(dx).$$

The corresponding norm is given by the formula

$$|h|_H := \|\widehat{h}\|_{L^2(\gamma)}.$$

Moreover, it is known that $H(\gamma)$ with the indicated norm is separable. If X is metrizable, then this is a simple fact following from the separability of $L^2(\mu)$ for every Radon measure μ on a metric space; in the general case the proof is highly nontrivial. Note that the same norm on $H(\gamma)$ is given by the formula

$$|h|_H = \sup\{f(h)\colon\ f \in X',\, \|f\|_{L^2(\gamma)} \leqslant 1\}. \tag{5.7.2}$$

One should have in mind that if $\dim H(\gamma) = \infty$, then $\gamma\bigl(H(\gamma)\bigr) = 0$.

In the above example of a Gaussian measure γ on a Hilbert space we have $H(\gamma) = \sqrt{K}(X)$. Indeed, here X_γ^* is the weighted Hilbert space of sequences (y_n) such that $\sum_{n=1}^\infty k_n y_n^2 < \infty$, and the operator R_γ takes (y_n) to $(k_n y_n)$ and has the range $\sqrt{K}(X)$.

5.7.3. Theorem. *The set $H(\gamma)$ is the collection of all $h \in X$ such that $\gamma_h \sim \gamma$, where $\gamma_h(B) := \gamma(B+h)$, moreover, the Radon–Nikodym density of the measure γ_h with respect to γ is given by the following Cameron–Martin formula:*

$$\frac{d\gamma_h}{d\gamma} = \exp\bigl(-\widehat{h} - |h|_H^2/2\bigr).$$

For every $h \notin H(\gamma)$ we have $\gamma \perp \gamma_h$.

Any centered Radon Gaussian measure is uniquely determined by its Cameron–Martin space (with the indicated norm!): if μ and ν are two centered Radon Gaussian measures such that $H(\mu) = H(\nu)$ and $|h|_{H(\mu)} = |h|_{H(\nu)}$ for all vectors $h \in H(\mu) = H(\nu)$, then $\mu = \nu$. The Cameron–Martin space is also called the reproducing Hilbert space.

If γ is a measure on $X = \mathbb{R}^\infty$ that is the countable product of the standard Gaussian measures on the real line, then X^* can be identified with the space of all sequences of the form $f = (f_1, \ldots, f_n, 0, 0, \ldots)$. In this case

$$(f,g)_{L^2(\gamma)} = \sum_{i=1}^\infty f_i g_i.$$

Hence X_γ^* can be identified with l^2; any element $l = (c_n) \in l^2$ defines an element of $L^2(\gamma)$ by the formula $l(x) := \sum_{n=1}^\infty c_n x_n$, where the series converges in $L^2(\gamma)$. Therefore, the Cameron–Martin space $H(\gamma)$ coincides with the space l^2 with its natural inner product. An element l is a continuous linear functional precisely when only finitely many numbers c_n are nonzero. The restriction of l to l^2 is obviously continuous on l^2 with its Hilbert norm, since it has the form of the inner product with the vector (c_n).

For the Wiener measure on $C[0,1]$, the Cameron–Martin space coincides with the Sobolev class $W_0^{2,1}[0,1]$ of all absolutely continuous functions h on $[0,1]$ such that $h(0) = 0$ and $h' \in L^2[0,1]$; the inner product is defined by the formula

$$(h_1, h_2)_H := \int_0^1 h_1'(t) h_2'(t)\, dt.$$

The general form of a measurable linear functional for the Wiener measure is given by the stochastic integral (see [**564**])

$$l(x) = \int_0^1 h'(t)\, dx(t).$$

Such functional is continuous on $C[0,1]$ precisely when h' has bounded variation.

The mapping $h \mapsto \widehat{h}$ establishes an isomorphism between $H(\gamma)$ and X_γ^* preserving the inner product. In addition, $R_\gamma \widehat{h} = h$. If $\{e_n\}$ is an orthonormal basis in $H(\gamma)$, then $\{\widehat{e_n}\}$ is an orthonormal basis in X_γ^* and $\widehat{e_n}$ are independent random variables. One can take in X^*_γ an orthonormal basis consisting of elements $\xi_n \in X^*$. Any element $l \in X^*_\gamma$ has the form $l = \sum_{n=1}^\infty c_n \xi_n$, where the series converges in $L^2(\gamma)$. Since ξ_n are independent Gaussian random variables, the series converges γ-a.e. The domain of its convergence L is a Borel linear subspace of full measure (one even has $L \in \sigma(X)$). One can take a version of l linear on all of X in the usual sense; it is called a *proper linear version*. For example, let l on L be the sum of the indicated series; on all of X we define l by linearity (taking a linear subspace L_1 algebraically complementing L and setting $l(x+y) = l(x)$ for $x \in L$, $y \in L_1$). Such a version is not unique in the infinite-dimensional case, but any two properly linear versions coincide on the subspace $H(\gamma)$ (although it has measure zero!). This is seen from the fact that if $l_1(x) = l_2(x)$ a.e. and $h \in H(\gamma)$, then $l_1(x-h) = l_2(x-h)$ a.e. by the equivalence of the measures γ and γ_h.

Thus, every γ-measurable linear functional f has a version linear on the whole space. This version is automatically continuous on $H(\gamma)$ with its norm $|\cdot|_H$.

Conversely, every continuous linear functional l on the Hilbert space $H(\gamma)$ admits a unique extension to a γ-measurable properly linear functional $\widehat{l}$ such that $\widehat{l}$ coincides with l on $H(\gamma)$. For every $h \in H(\gamma)$, such an extension of the functional $x \mapsto (x,h)_H$ is precisely $\widehat{h}$. If $h = \sum_{n=1}^\infty c_n e_n$, then $\widehat{h} = \sum_{n=1}^\infty c_n \widehat{e_n}$. Two γ-measurable linear functionals are equal almost everywhere precisely when their proper linear versions are equal on $H(\gamma)$. It is known that every proper linear γ-measurable function belongs to X^*_γ.

5.7.4. Definition. *A Radon Gaussian measure γ on a locally convex space X is called nondegenerate if, for every nonzero functional $f \in X^*$, the measure $\gamma\circ f^{-1}$ is not concentrated at a point.*

5.7.5. Theorem. *Nondegeneracy of a Radon Gaussian measure γ is equivalent to the property that $\gamma(U) > 0$ for every nonempty open set $U \subset X$. This is also equivalent to the property that the Cameron–Martin space $H(\gamma)$ is dense in the space X.*

For any degenerate Radon Gaussian measure γ there exists the smallest closed linear subspace $L \subset X$ for which $\gamma(L+m) = 1$, where m is the mean of γ. If $m = 0$, then on L the measure γ is nondegenerate.

It follows from this theorem that the aforementioned smallest closed affine (linear for a centered measure) subspace of full measure coincides with the topological support of the measure γ, i.e., the smallest closed set of full measure. In addition, we see that in the case of a measure with zero mean every neighborhood of zero has positive measure.

The role of the countable product of the standard Gaussian measures is clear from the following important and difficult theorem due to Tsirelson.

5.7.6. Theorem. *Let γ be a centered Radon Gaussian measure on a locally convex space X, let $\{e_n\}$ be an orthonormal basis in $H(\gamma)$, and let $\{\xi_n\}$ be independent standard Gaussian random variables (for example, the sequence of the coordinate functions on $\mathbb{R}^\infty$ with the countable product of the standard Gaussian measures on the real line). Then the series $\sum_{n=1}^\infty \xi_n(\omega) e_n$ converges in X for* *a.e. ω and the distribution of its sum is γ. In particular, this is true if $\xi_n = \widehat{e_n}$. In addition, there exists a Souslin linear subspace $S \subset X$ such that $\gamma(S) = 1$.*

This theorem shows that the countable power of the standard Gaussian measure on the real line is the main (and essentially unique) example of a centered Radon Gaussian measure, since every centered Radon Gaussian measure γ is the image of this countable product under a measurable linear mapping T (however, T need not be continuous); the mapping T is given by the series indicated in the formulation and its restriction to l^2 is an isometry between l^2 and $H(\gamma)$.

In analogy with functionals, a mapping T from X to a locally convex space Y is called a *γ-measurable linear operator* if it is measurable with respect to the pair of σ-algebras $\mathcal{B}(X)_\gamma$ and $\mathcal{B}(Y)$ and has a linear (in the usual sense) version (called *proper linear*).

5.7.7. Theorem. *Let γ be a centered Radon Gaussian measure on a locally convex space X with the Cameron–Martin space H. Then, for every operator $T \in \mathcal{L}(H)$, there exists a γ-measurable proper linear mapping $\widehat{T}\colon X \to X$ with the following properties:*

(i) *on H the mapping $\widehat{T}$ coincides with T,*

(ii) *the image of the measure γ under the mapping $\widehat{T}$ is a centered Radon Gaussian measure μ with the Cameron–Martin space $H(\mu) = T(H)$.*

Any two such mappings are equal γ-a.e. If the measure γ is the distribution of the series $\sum_{n=1}^{\infty} \xi_n(\omega)e_n$ from Theorem 5.7.6, then μ is the distribution of the series $\sum_{n=1}^{\infty} \xi_n(\omega)Te_n$, which converges a.e.

Note that the Fourier transform of the measure μ has the form

$$\widetilde{\mu}(l) = \exp\bigl(-|TR_\gamma(l)|_H^2/2\bigr).$$

Using this theorem, one can obtain a somewhat more general result (passing to the space $X \times Y$).

5.7.8. Corollary. *Let γ and μ be centered Radon Gaussian measures on locally convex spaces X and Y, respectively . Let $A\colon H(\gamma) \to H(\mu)$ be a continuous linear operator. Then A extends to a γ-measurable linear mapping $\widehat{A}\colon X \to Y$ such that the image of γ under this mapping is the centered Radon Gaussian measure with the Cameron–Martin space $A\bigl(H(\gamma)\bigr)$.*

Conversely, if $A\colon X \to Y$ is a γ-measurable linear mapping such that the measure $\mu = \gamma\circ A^{-1}$ is Radon, then for its proper linear version one has

$$Ax = \sum_{n=1}^{\infty} \widehat{e_n}(x)Ae_n \quad \gamma\text{-a.e.},$$

where the series converges in Y, moreover, $A\bigl(H(\gamma)\bigr) = H(\mu)$, $A = \widehat{A|_{H(\gamma)}}$.

In particular, for the identity operator A we obtain

$$x = \sum_{n=1}^{\infty} \widehat{e_n}(x)e_n \quad \gamma\text{-a.e.},$$

which is also not trivial.

For two Hilbert spaces H_1 and H_2, let $\mathcal{H}(H_1, H_2)$ denote the space of Hilbert–Schmidt operators from H_1 to H_2 equipped with its natural structure of a Hilbert space (see [**72**, Chapter 7]) defined by the inner product

$$(A, B)_{\mathcal{H}} = \sum_{\alpha} (Ae_\alpha, Be_\alpha)_{H_2},$$

where $\{e_\alpha\}$ is an orthonormal basis in H_1.

5.7.9. Corollary. *Let X and Y be locally convex spaces and let γ be a centered Radon Gaussian measure on X. A continuous linear operator $A\colon H(\gamma)\to Y$ is the restriction to $H(\gamma)$ of a measurable linear operator $\widehat{A}\colon X \to Y$ such that the measure $\gamma\circ\widehat{A}^{-1}$ is Radon precisely when there exists a Radon Gaussian measure ν on Y such that $A\bigl(H(\gamma)\bigr) \subset H(\nu)$.*

If Y is a Hilbert space, then this condition is equivalent to the inclusion $A \in \mathcal{H}\bigl(H(\gamma), Y\bigr)$, hence to the existence of a Hilbert–Schmidt operator T on the space Y with $A\bigl(H(\gamma)\bigr) \subset T(Y)$.

An important property of Gaussian measures is the so-called 0–1 law, according to which certain sets can have measure either 0 or 1. Actually, there are several results of this type.

5.7.10. Theorem. *Let γ be a Radon Gaussian measure on a locally convex space X.* (i) *For every γ-measurable affine subspace $L \subset X$ we have either $\gamma(L) = 0$ or $\gamma(L) = 1$.*

(ii) *Let $\{e_n\}$ be an orthonormal basis in $H(\gamma)$. Let E be a γ-measurable set such that, for every n and every rational number r, the sets E and $E+re_n$ coincide up to a set of measure zero. Then either $\gamma(E) = 0$ or $\gamma(E) = 1$. In particular, this is true if the γ-measurable set E is invariant with respect to translations by the vectors re_n.*

Another classical alternative in the theory of Gaussian measures is the Hajek–Feldman theorem on equivalence and singularity.

5.7.11. Theorem. *If μ and ν are Radon Gaussian measures on the same space, then either $\mu \sim \nu$ or $\mu \perp \nu$.*

One more important fact is the following Fernique theorem.

5.7.12. Theorem. *If γ is a centered Radon Gaussian measure and a γ-measurable function q is a seminorm on a γ-measurable linear subspace of full measure, then* $\exp(\varepsilon q^2) \in L^1(\gamma)$ *for some* $\varepsilon > 0$.

In the theory of Gaussian measures an important role is played by Hermite (or Chebyshev–Hermite) polynomials H_n defined by

$$H_0 = 1, \quad H_n(t) = \frac{(-1)^n}{\sqrt{n!}} e^{t^2/2} \frac{d^n}{dt^n}\bigl(e^{-t^2/2}\bigr), \quad n > 1.$$

They have the following properties:

$$H_n'(t) = \sqrt{n}H_{n-1}(t) = tH_n(t) - \sqrt{n+1}H_{n+1}(t).$$

In addition, the system of functions $\{H_n\}$ is an orthonormal basis in $L^2(\gamma)$, where γ is the standard Gaussian measure on the real line.

For the *standard Gaussian* measure γ_n on $\mathbb{R}^n$ (the product of n standard Gaussian measures on the real line) an orthonormal basis is $L^2(\gamma_n)$ is formed by the polynomials of the form

$$H_{k_1,\ldots,k_n}(x_1,\ldots,x_n) = H_{k_1}(x_1)\cdots H_{k_n}(x_n), \quad k_i \geqslant 0.$$

If γ is a centered Radon Gaussian measure on a locally convex space X and $\{l_n\}$ is an orthonormal basis in X_γ^*, then the polynomials

$$H_{k_1,\ldots,k_n}(x) = H_{k_1}\big(l_1(x)\big)\cdots H_{k_n}\big(l_n(x)\big), \quad k_i \geqslant 0, n \in \mathbb{N}$$

form a basis in $L^2(\gamma)$. For example, for the countable power of the standard Gaussian measure on the real line these polynomials are $H_{k_1,\ldots,k_n}(x_1,\ldots,x_n)$.

Let us group the polynomials $H_{k_1,\ldots,k_n}$ according to the degrees $k_1+\cdots+k_n$. For $k = 0,1,\ldots$, let $\mathcal{X}_k$ denote the closed linear subspace generated by the functions $H_{k_1,\ldots,k_n}$ with $k_1+\cdots+k_n = k$. The space $\mathcal{X}_0$ is one-dimensional and consists of constants and $\mathcal{X}_1 = X_\gamma^*$. One can show that every element $f \in \mathcal{X}_2$ is represented in the form $f = \sum_{n=1}^\infty \alpha_n(l_n^2 - 1)$, where $\{l_n\}$ is an orthonormal basis in X_γ^* and $\sum_{n=1}^\infty \alpha_n^2 < \infty$ (i.e., the series for f converges in $L^2(\gamma)$).

We now clarify the role of the space H in the situation where the standard cylindrical Gaussian measure on H "extends" to a true Gaussian measure on an enlargement of H. The formulation employs a useful general concept.

5.7.13. Definition. *We say that (E', H, E) is a rigged Hilbert space (Gelfand's triple) if E is a Hausdorff locally convex space and H is a separable Hilbert space continuously embedded into E as a dense linear subspace, which generates also a dense embedding $j_H\colon E' \subset H$ defined by the formula $\big(j_H(l), h\big)_H = \langle l, h\rangle$, where $l \in E'$ and $h \in H$.*

5.7.14. Lemma. *In the situation of the previous definition we have*

$$H = \Big\{x \in E\colon \ \sup\{l(x)\colon l \in E',\ |j_H(l)|_H \leqslant 1\} < \infty\Big\}. \tag{5.7.3}$$

PROOF. Indeed, it is clear that if $h \in H$, then

$$l(h) = \big(j_H(l), h\big)_H \leqslant |j_H(l)|_H |h|_H,$$

hence h belongs to the right-hand side of the equality to be proved.

Let $x \in E\backslash H$. Hence the element x belongs to no set nU, where U is the closed unit ball in H. The sets nU are closed in E by their weak compactness in H. Hence for each n there exists a functional $f_n \in E'$ such that $f_n(x) > 1$ and $f_n(u) \leqslant 1$ for all $u \in nU$. Hence $\big(j_H(f_n), u\big)_H \leqslant n^{-1}$ for $u \in U$, whence we obtain $|j_H(f_n)| \leqslant n^{-1}$. Thus, for $l_n = nf_n$ we have $l_n(x) > n$, but $|j_H(l_n)| \leqslant 1$. This means that x does not belong to the right-hand side of (5.7.3). □

5.7.15. Proposition. *If (E', H, E) is a rigged Hilbert space such that the image of the standard cylindrical Gaussian measure ν_H on H under the embedding of H into E extends to a Radon measure γ on E, then H coincides with the Cameron–Martin space of this extension.*

PROOF. For every $l \in E'$, by the definition of an extension of a cylindrical measure we have

$$\widetilde{\gamma}(l) = \widetilde{\nu_H}\bigl(j_H(l)\bigr) = \exp\Bigl(-\frac{1}{2}|j_H(l)|_H^2\Bigr),$$

where for ν_H we use its Hilbert Fourier transform (defined on H identified with H'). This means that $|j_H(l)|_H = |R_\gamma(l)|_{H(\gamma)}$, whence it follows that $H = H(\gamma)$, since $H(\gamma)$ is the set of all vectors with finite norm (5.7.2) and H is characterized by equality (5.7.3). □

There is extensive literature on covariance operators of Gaussian measures (see references in Bogachev [**63**], Vakhania, Tarieladze [**537**], and Vakhania, Tarieladze, Chobanyan [**538**]).

Various aspects of the theory of Gaussian measures are discussed also in Bogachev [**66**], [**67**], Fernique [**159**], Kuo [**305**], and Lifshits [**316**].

5.8. Quasi-measures

The concept of cylindrical measure introduced above does not cover constructions connected with the so-called Feynman measure. In order to cover also this case in the considered theory, in place of the algebra of all cylindrical sets we shall consider its subalgebra that we now define.

Let E be a vector space and let G be some vector space of linear functionals on E. For every $n \in \mathbb{N}$ and every collection $g_1, \ldots, g_n$ of elements of G, the symbol $P_{g_1,\ldots,g_n}$ will denote the mapping $E \to \mathbb{R}^n$ defined by the formula

$$P_{g_1,\ldots,g_n}(x) = \bigl(g_1(x), \ldots, g_n(x)\bigr).$$

Let $\mathcal{B}_n^b$ be the algebra of subsets of $\mathbb{R}^n$ defined as follows: $A \in \mathcal{B}_n^b$ if either A or $\mathbb{R}^n \backslash A$ is a bounded Borel set in $\mathbb{R}^n$.

Let $\mathfrak{A}_{g_1,\ldots,g_n}^b$ denote the algebra of all sets $P_{g_1,\ldots,g_n}^{-1}(\mathcal{B}_n^b)$ and let

$$\mathfrak{A}_G^b = \bigcup_{n,g_1,\ldots,g_n} \mathfrak{A}_{g_1,\ldots,g_n}^b.$$

The class $\mathfrak{A}_G^b$ is an algebra of subsets of E; any element $A \in \mathfrak{A}_G^b$ with a nonempty bounded base B in a representation of the form $A = P_{g_1,\ldots,g_n}^{-1}(B)$, will be called a *scalarly bounded* G-cylindrical subset of the space E.

5.8.1. Definition. *An additive complex function ν on $\mathfrak{A}_G^b$ is called a G-cylindrical quasi-measure on $\mathfrak{A}_G^b$ if it has the following property: for any $n \in \mathbb{N}$, $g_1, \ldots, g_n \in G$ and any bounded Borel set $B \in \mathbb{R}^n$, the restriction of ν to the σ-algebra of subsets of $P_{g_1,\ldots,g_n}^{-1}(B)$ consisting of all sets of the form $P_{g_1,\ldots,g_n}^{-1}(V)$, where V is a Borel subset of B, is countably additive.*

In all examples below we consider quasi-measures whose value on E is 1, hence it suffices to define their values only on scalarly bounded elements of $\mathfrak{A}_G^b$: for every cylindrical set $A \in \mathfrak{A}_G^b$ whose complement $E \backslash A$ is scalarly bounded we set $\nu(A) = 1 - \nu(\mathbb{R}^n \backslash A)$. We observe that for a signed measure this does not imply the boundedness of its range.

5.8.2. Example. Let $E = \mathbb{R}^n$ be canonically identified with the dual space. A quasi-measure ν on $\mathfrak{A}^b_{\mathbb{R}^n}$ is called a *nondegenerate classical Feynman quasi-measure* if there exist a symmetric positive operator Q in $\mathbb{R}^n$, called the *correlation operator*, and an element $a \in \mathbb{R}^n$, called the *expectation or mean* of the quasi-measure ν such that for every bounded Borel set $A \in \mathbb{R}^n$ one has

$$\nu(A) = (2\pi)^{-n/2}(\det Q)^{-1/2} \int_A \exp\Bigl[-i\frac{1}{2}(Q^{-1}(x-a), x-a))\Bigr]\, dx.$$

For other scalarly bounded cylindrical sets of the form $A = P^{-1}A_1$, where the operator $P\colon\, E \to P(E)$ is an orthogonal projection in E and A_1 is a bounded set in $P(E)$, we set

$$\begin{aligned}\nu(A) =& (2\pi)^{-\dim P(E)/2}(\det PQ)^{-1/2}\\ &\times \int_{A_1} \exp\Bigl[-i\frac{1}{2}\Bigl((PQ)^{-1}(x-P(a)), x-P(a)\Bigr)\Bigr]\, dx.\end{aligned}$$

It is clear that PQ is a nondegenerate positive operator from $P(E)$ to $P(E)$.

5.8.3. Example. Suppose again that $E = \mathbb{R}^n = G$ and $n = 2k$ is even. The *simplectic Feynman quasi-measure* on $\mathbb{R}^n$ is the quasi-measure ν on $\mathbb{R}^n$ whose value on every bounded Borel set $A \subset \mathbb{R}^n$ is defined by

$$\nu(A) = \int_A \exp\Bigl(i\sum_{j=1}^k x_j x_{k+j}\Bigr)\, dx \equiv \iint_A \exp\bigl(i(y,z)_{\mathbb{R}^k}\bigr)\, dy\, dz,$$

where $x = (x_1, \ldots x_{2k})$, $y = (x_1, \ldots x_k)$, $z = (x_{k+1}, \ldots x_{2k})$; for other scalarly bounded cylindrical sets that are preimages of bounded Borel sets under linear factorizations the value of this measure is defined as the value of the image of ν on the image of the original set with respect to this factorization.

5.8.4. Example. Let H be a real infinite-dimensional Hilbert space with the inner product $(\,\cdot\,,\,\cdot\,)$ identified with its dual. The *standard Feynman quasi-measure* on H is the H-cylindrical quasi-measure ν_Φ on H defined as follows: if P is an orthogonal projection of H onto a finite-dimensional subspace H_P of the space H and A is a bounded Borel set in H_P, then

$$\nu_\Phi(P^{-1}A) = (2\pi)^{-\dim H_P/2} \int_A \exp\Bigl[-i\frac{1}{2}(x,x)\Bigr]\, dx.$$

We now describe cylindrical functions that can be integrated over G-cylindrical quasi-measures. A G-cylindrical function f will be called *finite* if

$$f(x) = \varphi\bigl(g_1(x), \ldots, g_n(x)\bigr),$$

where $g_1, \ldots, g_n \in G$ and φ is a Borel function on $\mathbb{R}^n$ vanishing outside of some ball.

The integral of a cylindrical finite Borel function f with respect to a cylindrical quasi-measure ν is defined as the integral of the function φ with respect to the measure $\nu_{g_1,\ldots,g_n}$ (if the latter exists, which holds if, say, the function φ is bounded on all balls).

Let $\mathcal{D}_G$ denote the set of all finite G-cylindrical functions f on E of the analogous form $f(x) = \varphi\bigl(g_1(x),\dots,g_n(x)\bigr)$, where $n \in \mathbb{N}$, $g_1,\dots,g_n \in G$ and $\varphi \in \mathcal{D}(\mathbb{R}^n)$.

It is straightforward to verify that each cylindrical function of this form is the Fourier transform of a measure concentrated on the subspace L of the space G generated by the functionals $g_1,\dots,g_n$. In the case where these functionals are linearly independent, this measure possesses a density with respect to Lebesgue measure on L (arising if we take the elements $g_1,\dots,g_n$ as an orthonormal basis); this density up to a constant factor is the (finite-dimensional) direct Fourier transform of the function φ.

This circumstance enables us to define the Fourier transform for G-cylindrical quasi-measures on E.

5.8.5. Definition. *The Fourier transform of a G-cylindrical quasi-measure ν is a function $\widetilde{\nu}$ on G with the following property: for any $n \in \mathbb{N}$, linearly independent elements $g_1,\dots,g_n \in G$, vectors $e_1,\dots,e_n \in E$ with $\langle g_j, e_k\rangle = \delta_{jk}$, and any function $\varphi \in \mathcal{D}(\mathbb{R}^n)$, one has*

$$\int_{\mathbb{R}^n} \overline{\widetilde{\nu}\Bigl(\sum_{k=1}^n z_k g_k\Bigr)}\, \widetilde{\varphi}\Bigl(\sum_{k=1}^n z_k g_k\Bigr)\, dz = \int_E \overline{\varphi\bigl(g_1(x),\dots,g_n(x)\bigr)}\, \nu(dx),$$

where for $z = (z_1,\dots,z_n) \in \mathbb{R}^n$ we set

$$\widetilde{\varphi}\Bigl(\sum_{k=1}^n z_k g_k\Bigr) := \int_{\mathbb{R}^n} \exp\Bigl(i\sum_{k=1}^n z_k g_k(x_k e_k)\Bigr)\varphi\Bigl(\sum_{k=1}^n x_k e_k\Bigr)\, dx_1\cdots dx_n.$$

In other words, the Fourier transform is defined by means of the Parseval equality. It should be emphasized that not every quasi-measure possesses the Fourier transform in this sense. The Fourier transform of an arbitrary quasi-measure on $\mathbb{R}^n$ can be naturally defined as a functional on the space $\widetilde{\mathcal{D}(\mathbb{R}^n)}$ of the Fourier transforms of functions in $\mathcal{D}(\mathbb{R}^n)$.

The standard Feynman quasi-measure on H has the Fourier transform defined by

$$\widetilde{\nu}(g) = \exp\Bigl(-\frac{i}{2}(g,g)\Bigr).$$

For cylindrical quasi-measures, some constructions defined above for cylindrical measures remain meaningful. In particular, their images under linear mappings are defined, which is connected with the Fourier transform by the same formula as in Exercise 5.12.83 for cylindrical measures.

In applications, it is convenient to define the Feynman quasi-measure by means of its Fourier transform. In particular, if E is a vector space, G is a vector subspace in its algebraic dual and b is a bilinear function on $G\times G$, the G-cylindrical Feynman quasi-measure on E with the correlation function b (and zero expectation) is by definition the G-cylindrical quasi-measure w the Fourier transform of which is defined by the equality

$$\widetilde{w}(g) = \exp\bigl(-ib(g,g)/2\bigr).$$

Similarly we define Feynman quasi-measures with nonzero expectations. If E is a Hilbert space and $G = E' = E$ with the usual Riesz identification, then in place of the correlation functional it is more convenient to deal with the correlation operator Q defined by the equality $(Qx, x) = b(x, x)$.

5.9. Sufficient topologies

In this section E is a real vector space and G is some vector space of linear functionals on E.

A topology τ in a E (not assumed to be compatible with the vector structure) is called *positively sufficient* if the continuity in this topology of the Fourier transform of a nonnegative E-cylindrical measure on G implies that this measure is countably additive. If the continuity of the Fourier transform is sufficient for the countable additivity of signed bounded E-cylindrical measures on G, then the topology is called *sufficient*.

It is obvious that every sufficient topology is also positively sufficient; in this section we show that in the class of topologies invariant with respect to translations the converse is also true. In the next section we describe two sufficient topologies: the Sazonov topology and the Gross–Sazonov topology.

A topology τ in a vector space E is called *necessary* if the Fourier transform of every countably additivity bounded real E-cylindrical measure on G is continuous in the topology τ.

Similarly one could define the positively necessary topology, however, the Jordan theorem, according to which every countably additive real measure of bounded variation defined on an algebra of subsets of an arbitrary space is the difference of two nonnegative countably additive measures on the same algebra, yields that every positively necessary topology is necessary; since the inverse implication is also obviously true, the class of positively necessary topologies (would we introduce it) will coincide with the class of necessary topologies.

The main result of this section is the following theorem due to Tarieladze [**516**] (our proof employs the method suggested in Smolyanov, Shavgulidze [**496**]).

5.9.1. Theorem. *Every translation invariant positively sufficient topology in a vector space E is sufficient.*

PROOF. Let τ be a positively sufficient topology in E invariant with respect to translations and let ν be a bounded cylindrical measure on $(G, \mathfrak{A}_E)$ such that the function $\widetilde{\nu}$ is continuous on (E, τ). Let $\nu = \nu^+ - \nu^-$ be the Jordan decomposition of the measure ν. According to Lemma 5.4.2, ν^+ and ν^- are also bounded nonnegative cylindrical measures.

For the proof of the countable additivity of ν it suffices to prove the countable additivity of ν^+ and ν^-; to this end, in turn, it suffices to prove that the functions $\widetilde{\nu^+}\colon (E,\tau) \to \mathbb{C}$ and $\widetilde{\nu^-}\colon (E,\tau) \to \mathbb{C}$ are continuous. The latter will be proved if we show that $\widetilde{\nu^+}$ and $\widetilde{\nu^-}$ can be uniformly approximated by linear combinations of translations of the Fourier transform of the original measure ν; this will be done in the next proposition in the general case. The continuity of the functions

$x \mapsto \widetilde{\nu}(x+x_r)$ is seen from the continuity of $\widetilde{\nu}$ and the invariance of the topology τ with respect to translations. □

We now proceed to the aforementioned proposition.

5.9.2. Proposition. *Let ν be a bounded cylindrical measure on $(G, \mathfrak{A}_E)$ and let $\nu = \nu^+ - \nu^-$ be its Jordan decomposition. Then the functions $\widetilde{\nu^+}$ and $\widetilde{\nu^-}$ can be uniformly approximated by functions of the form*

$$x \mapsto \sum_{j=1}^{n} c_j \widetilde{\nu}(x + x_j), \quad \textit{where } c_j \in \mathbb{C} \textit{ and } x_j \in E.$$

PROOF. We give a proof for $\widetilde{\nu^+}$; the assertion for $\widetilde{\nu^-}$ is obtained by passing to the measure $-\nu$. Let $\varepsilon > 0$ and let $A \in \mathfrak{A}_E$ be such that $\nu(A) > \nu^+(G) - \varepsilon$. Since $\nu(A) = \nu^+(A) - \nu^-(A)$, we have

$$\nu^+(G\backslash A) < \varepsilon + \nu(A) - \nu^+(A) = \varepsilon\nu^-(A) \leqslant \varepsilon,$$

whence by the inequalities $\nu^+ \geqslant 0$, $\nu^- \geqslant 0$ we obtain

$$\nu^+(G\backslash A) < \varepsilon, \quad \nu^-(A) < \varepsilon.$$

Our set A has the form $A = P^{-1}(B)$, where $B \in \mathcal{B}(\mathbb{R}^n)$,

$$P = (\psi_1, \ldots, \psi_n)\colon G \to \mathbb{R}^n, \quad \psi_i \in E.$$

The measure $|\nu| = \nu^+ + \nu^-$ is also bounded cylindrical. By the countable additivity of the measure $|\nu|\circ P^{-1}$, there exist numbers $c_1, \ldots, c_k \in \mathbb{C}$ and linear functions $l_1, \ldots, l_k$ on $\mathbb{R}^n$ such that the complex function

$$\Phi(u) = \sum_{j=1}^{k} c_j \exp\bigl(il_j(u)\bigr)$$

approximates I_B in $L^1(|\nu|\circ P^{-1})$ up to ε, i.e.,

$$\int_{\mathbb{R}^n} |I_B - \Phi|\, d(|\nu|\circ P^{-1}) < \varepsilon.$$

Let $F(g) := \sum_{j=1}^{k} c_j \exp\bigl(il_j(Pg)\bigr)$. By the change of variables formula we obtain

$$\int_G |I_A(g) - F(g)|\, |\nu|(dg) < \varepsilon.$$

This integral is defined, since the function under the integral sign is bounded and Borel E-cylindrical.

Let us estimate the function $|\widetilde{\nu^+} - \widetilde{F\cdot\nu}|$, where $F\cdot\nu$ is the E-cylindrical measure on $\mathfrak{A}_E$ defined by the equality

$$(F\cdot\nu)(C) = \int_C F(g)\,\nu(dg);$$

similarly we define the measure $I_A\cdot\nu$. Using the equality

$$\nu^+ - I_A\cdot\nu = I_{G\backslash A}\cdot\nu^+ - I_A\cdot\nu^-,$$

we obtain the chain of inequalities

$$\begin{aligned}|\widetilde{\nu^+}(x) - \widetilde{F\cdot\nu}(x)| &\leqslant |\widetilde{\nu^+}(x) - \widetilde{I_A\cdot\nu}(x)| + |\widetilde{I_A\cdot\nu}(x) - \widetilde{P\cdot\nu}(x)| \\ &\leqslant \|\nu^+ - I_A\cdot\nu\| + \int_G |I_A(g) - F(g)|\,|\nu|(dg) \\ &\leqslant \nu^+(G\backslash A) + \nu^-(A) + \varepsilon \leqslant 3\varepsilon.\end{aligned}$$

In addition, we have

$$\begin{aligned}\widetilde{F\cdot\nu}(x) &= \int_G \sum_{j=1}^k c_j e^{il_j(Pg)} e^{ig(x)}\,\nu(dg) \\ &= \int_G \sum_{j=1}^k c_j e^{ig(x_j)} e^{ig(x)}\,\nu(dg) = \sum_{j=1}^k c_j \widetilde{\nu}(x+x_j),\end{aligned}$$

where the vectors $x_j \in E$ are constructed as follows: writing the functionals l_j in the form $l_j(x) = \lambda_{j,1}u_1 + \cdots + \lambda_{j,n}u_n$, we set $x_j := \lambda_{j,1}\psi_1 + \cdots + \lambda_{j,n}\psi_n$, $j = 1,\ldots,k$, where $Pg = (\psi_1(g),\ldots,\psi_n(g))$, $\psi_1,\ldots,\psi_n \in E$. With this choice we have $l_j(Pg) = g(x_j)$ whenever $j = 1,\ldots,k$. Thus, the existence of the required uniform approximations is established. □

We emphasize that no countable additivity of ν was assumed. However, the boundedness of variation of the measure ν was essential. Note that the boundedness of variation of ν does not follow from the boundedness of $\widetilde{\nu}$ even under the additional condition of continuity of $\widetilde{\nu}$ (see Exercise 5.12.96).

5.10. The Sazonov and Gross–Sazonov topologies

In this section we introduce the Sazonov and Gross–Sazonov topologies and prove that they are necessary. We also show that although the Sazonov and Gross–Sazonov topologies do not coincide, the Fourier transforms of bounded cylindrical measures that are continuous in one of them are continuous in the other.

Throughout this section the symbol H denotes a separable real Hilbert space with the inner product $(\cdot,\cdot)_H$ and norm $|\cdot|_H$; for every $r > 0$, let S_r denote the open ball of radius r centered at zero in H and let w_r denote the Gaussian cylindrical measure in H the Fourier transform of which has the form $x \mapsto \exp\bigl(-\frac{1}{2}r(x,x)\bigr)$; in particular, w_1 is the canonical Gaussian measure in H (see Example 5.4.6(ii)).

For every closed vector subspace F of the space H let P_F denote the orthogonal projection onto F. The set of all finite-dimensional vector subspaces in H is denoted by the symbol $\mathrm{Fin}(H)$; let $J(H)$ denote the set of all nonnegative self-adjoint nuclear operators in H. As above, $\mathcal{H}(H)$ is the set of all Hilbert–Schmidt operators in H.

For every operator $T \in J(H)$ let $p(T)$ denote the seminorm on H defined by $p_T(x) = \sqrt{(Tx,x)_H}$.

5.10.1. Definition. *The Sazonov topology in the space H is the locally convex topology defined by the family of seminorms $S_H = \{p_T\colon\ T \in J(H)\}$.*

The seminorms of the indicated form on H are precisely the seminorms of the form $p_S(x) = |Sx|_H$, where S is a Hilbert–Schmidt operator in H. Indeed, if $T \in J(H)$, then we can take $S = \sqrt{T}$; conversely, for every Hilbert–Schmidt operator S, the operator $T = S^*S$ belongs to the class $J(H)$ and we have the equality $(Tx, x)_H = (Sx, Sx)_H$.

Let now E be an arbitrary real Hausdorff locally convex space.

5.10.2. Definition. *The Sazonov topology τ^S in E associated with the original topology of the locally convex space E is generated by the following family of seminorms S^E on E: $p \in S^E$ precisely when there exist a continuous linear operator $\Lambda\colon E \to H$ (where H is equipped with the topology generated by its norm) and a seminorm $q \in S_H$ for which $p(x) = q(\Lambda x)$.*

It is readily seen that if E is a separable Hilbert space, then τ^S coincides with the Sazonov topology defined above for this special case. Indeed, if $p \in S_E$, then obviously $p \in S^E$, since one can take the identity mapping Λ. Conversely, if $p \in S^E$, then we can assume that $E = H$, and then the inclusion $p \in S_H$ follows from the fact that the composition of a continuous linear operator and a Hilbert–Schmidt operator is a Hilbert–Schmidt operator.

It turns out that for defining a base of the Sazonov topology there is no need to take the intersections of balls with respect to different seminorms in S^E.

5.10.3. Proposition. *The family of subsets of E of the form $\{x \in E\colon p(x) < 1\}$, where $p \in S^E$, is a base of neighborhoods of zero in the Sazonov topology.*

PROOF. We have to prove that for every $p, q \in S^E$ there exists a seminorm $p_0 \in E$ such that

$$\{x \in E\colon p_0(x) < 1\} \subset \{x\colon p(x) < 1\} \cap \{x\colon q(x) < 1\}.$$

By definition, there are operators $T_1, T_2 \in J(H)$ and $\Lambda_1, \Lambda_2 \in \mathcal{L}(E, H)$ such that $p(x) = (T_1\Lambda_1 x, \Lambda_1 x)_H^{1/2}$, $q(x) = (T_2\Lambda_2 x, \Lambda_2 x)_H^{1/2}$ for all $x \in E$. Let us set $H_0 = H \times H$ and define $\Lambda \in \mathcal{L}(E, H_0)$ and $T \in J(H_0)$ by the equalities $\Lambda(x) := \bigl(\Lambda_1(x), \Lambda_2(x)\bigr)$, $T(h_1, h_2) := (T_1 h_1, T_2 h_2)$. If we now take the seminorm $p_0(x) := \bigl(Tg(x), g(x)\bigr)^{1/2}$, then we obtain the desired inclusion. □

Note that it follows from what has been said in § 2.9 (see Theorem 2.9.5) that a separated locally convex space is nuclear if and only if its topology coincides with the associated Sazonov topology. We recall that every locally convex space E is nuclear in its weak topology $\sigma(E, E')$ (in particular, every finite-dimensional locally convex space is nuclear). Thus, such spaces are equipped with their Sazonov topologies.

Let us proceed to the definition of the Gross–Sazonov topology. This topology is also first defined for a separable Hilbert space H. Recall that w_1 is the standard cylindrical Gaussian measure on H.

5.10.4. Definition. *A seminorm q on H is called measurable if the following condition is fulfilled:*

$$\forall \varepsilon > 0\ \exists C > 0\colon\ w_1\bigl(x \in H\colon\ q(P_F x) > C\bigr) < \varepsilon \quad \forall F \in \mathrm{Fin}(H).$$

Note that if $\dim H < \infty$, then every seminorm on H is measurable in this sense. Observe also that the definition of measurability of seminorms does not use any algebra (or σ-algebra) of subsets.

Let us give an example of a measurable seminorm.

5.10.5. Example. If T is a nonnegative selfadjoint nuclear operator in H, then the seminorm $x \mapsto (Tx, x)^{1/2}$ on H is measurable.

Indeed, let $F \in \mathrm{Fin}(H)$, let P_F be an orthogonal projection from H onto F, and let w^F be the canonical Gaussian measure in F. Then by the Chebyshev inequality

$$w_1\bigl(h\in H\colon\ |\sqrt{T}P_F h|_H > C\bigr) = w^F\bigl(z\in F\colon\ (P_F T z, z)_H > C^2\bigr)$$
$$\leqslant C^{-2}\int_F (P_F T h, h)\, w^F(dh) = C^{-2} = C^{-2}\mathrm{tr}\,(P_F T) \leqslant C^{-2}\mathrm{tr}\, T,$$

where tr denotes the trace, whence our claim follows.

5.10.6. Remark. If $\{e_n\}$ is an orthonormal basis in H, P_n is the projection onto the linear span of $e_1, \ldots, e_n$ and q is a continuous seminorm on H, then for its measurability it suffices to have the required estimate just for the countable set of projections P_n, i.e., it suffices, for every $\varepsilon > 0$, to have $C > 0$ such that

$$w_1\bigl(x \in H\colon\ q(P_n x) > C\bigr) < \varepsilon \quad \forall n.$$

Indeed, in this case, for every finite-dimensional projection P, by using the continuity of q, one can find a projection P_F onto a subspace F contained in $P_n(H)$ for a sufficiently large n such that

$$w_1\bigl(x \in H\colon\ q(Px) > C\bigr) < w_1\bigl(x \in H\colon\ q(P_F x) > C\bigr) + \varepsilon.$$

Since $w_1\bigl(x \in H\colon\ q(P_F x) > C\bigr) \leqslant w_1\bigl(x \in H\colon\ q(P_n x) > C\bigr)$, the left-hand side is less than 2ε.

5.10.7. Definition. *The Gross–Sazonov topology τ_{GS} in H is generated by the family Γ_H of all measurable seminorms.*

The Gross–Sazonov topology τ_{GS} in an arbitrary locally convex space E associated with the original topology of this space is generated by the family of seminorms Γ^E defined as follows: $p \in \Gamma^E$ precisely when there exist a continuous linear mapping $\Lambda\colon E \to H$ and a seminorm $q \in \Gamma_H$ such that $p(x) = q(\Lambda x)$.

Let us note two obvious corollaries of the definition: if a seminorm p is measurable and a seminorm q satisfies the estimate $q \leqslant kp$ with some k, then q is measurable as well; in addition, the sum of finitely many measurable seminorms is measurable. For example, if two seminorms p_1 and p_2 are measurable, then for every $C > 0$ we have

$$\{x\colon\ p_1(P_F x) + p_2(P_F x) > C\} \subset \{x\colon\ p_1(P_F x) > C/2\} \cup \{x\colon\ p_2(P_F x) > C/2\},$$

hence the measure of the left-hand side does not exceed ε if the measure of every term in the right-hand side does not exceed $\varepsilon/2$.

As for the Sazonov topology, in the case where E is a separable Hilbert space the two arising definitions of the Gross–Sazonov topology are equivalent, i.e., we have $\Gamma_H = \Gamma^H$. The inclusion $\Gamma_H \subset \Gamma^H$ is true for the same reasons as a similar inclusion for the Sazonov topology. The inverse inclusion follows immediately from the next lemma (which will be also used later for another purpose).

5.10.8. Lemma. *For every measurable seminorm q on H and every continuous linear operator $\Lambda\colon H \to H$ the seminorm $q\circ\Lambda$ is also measurable.*

PROOF. For every $F \in \mathrm{Fin}(H)$ let w^F denote the canonical Gaussian measure on F. Set $V := \{x \in H\colon q(x) \leqslant 1\}$. The measurability of q is equivalent to the following property of the set V:

$$\forall \varepsilon > 0\ \exists C > 0\ \forall F \in \mathrm{Fin}(H)\colon \quad w^F(C\cdot V \cap F) \geqslant 1 - \varepsilon. \tag{5.10.1}$$

Let $W := \{x \in H\colon q(\Lambda x) \leqslant 1\} = \Lambda^{-1}(V)$. In order to prove that the seminorm $q \circ \Lambda$ is measurable, it suffices to verify that the set W possesses the property analogous to property (5.10.1) of the set V. We observe that it suffices to prove our assertion for invertible operators Λ, since the operator $\Lambda_\lambda := \Lambda - \lambda I$ is invertible if $|\lambda| > \|\Lambda\|$ and $p(\Lambda x) \leqslant p(\Lambda_\lambda x) + |\lambda| p(x)$, but the sum of measurable seminorms is measurable. Furthermore, as it follows directly from the definition of this property, we can assume that $\|\Lambda\| \leqslant 1$. Since V possesses property (5.10.1) by the measurability of q, for proving that W possesses the same property, it suffices to verify that for every $F \in \mathrm{Fin}(H)$ there exists $Z \in \mathrm{Fin}(H)$ such that $w^F(W \cap F) \geqslant w^Z(V \cap Z)$. We show that this is true if $Z = \Lambda(F)$.

Let A be a linear isometry from Z onto F. It is obvious from the definitions of the measures w^F and w^Z that $w^Z(V \cap Z) = w^F\bigl(A(V \cap Z)\bigr)$, i.e., it suffices to show that $w^F\bigl(A(V\cap Z)\bigr) \leqslant w^F(W \cap F)$. Since $A(V \cap Z) = A\Lambda(W \cap F)$, where $\|A\Lambda\| \leqslant 1$, the required inequality follows from the following general fact, a proof of which can be found in [**63**, § 1.8] or obtained as Exercise 5.12.81: if U is an absolutely convex set in $\mathbb{R}^n$ and γ is the standard Gaussian measure on $\mathbb{R}^n$, then for every linear operator S on $\mathbb{R}^n$ with $\|S\| \leqslant 1$ one has $\gamma\bigl(S(U)\bigr) \leqslant \gamma(U)$. □

The next fact is not obvious from the definition.

5.10.9. Lemma. *Every measurable seminorm on H is continuous.*

PROOF. Let q be a measurable seminorm that is not continuous at zero. Then there exist vectors $h_n \in H$ such that $|h_n|_H \to 0$ and $q(h_n) \to +\infty$. By condition there exists $C > 0$ such that $w_1\bigl(x \in H\colon q(P_F x) > C\bigr) < 1/4$ for all subspaces $F \in \mathrm{Fin}(H)$. This means that the standard Gaussian measure of the intersection of F with the set $V := \{x\colon q(x) \leqslant C\}$ is not less than $3/4$. In particular, this is true for the one-dimensional subspaces. For them the indicated intersection is either the whole line or a symmetric compact interval. However, whenever $q(h_n) > C$, the point h_n does not belong to V, hence $V \cap \mathbb{R}^1 h_n$ is an interval of length less than $2|h_n|_H$. It is clear that the one-dimensional Gaussian measure of such interval tends to zero as $n \to \infty$, which leads to a contradiction. □

5.10.10. Lemma. *Let E be a locally convex space. Then the family*

$$\bigl\{(q\circ\Lambda)^{-1}([0,1))\colon\ q \in \Gamma_H, \Lambda \in \mathcal{L}(E,H)\bigr\}$$

is a base of neighborhoods of zero in the topology τ^{SG}.

PROOF. Directly from the definition of the topology τ^{SG} we only see that this family of sets is a prebase. Let $\Lambda_k \in \mathcal{L}(E,H)$, $q_k \in \Gamma_H$, $k=1,2$. Set $H_1 = H\times H$ and consider the operator $\Lambda \in \mathcal{L}(E,H_1)$ defined by the equality $\Lambda x = \Lambda_1 x + \Lambda_2 x$; finally, set $q(x)=q_1(x)+q_2(x)$. Then the seminorm q on H_1 is measurable and $q_k(\Lambda_k x) \leqslant q(\Lambda x)$, $k=1,2$. □

The role of the introduced concept of measurability of seminorms is related to the following circumstance.

5.10.11. Theorem. *Suppose that q is a continuous norm on H such that the image of the canonical cylindrical Gaussian measure on H under the natural embedding of H into the Banach space X obtained by completing H with respect to the norm q is a countably additive measure. Then the norm q is measurable.*

PROOF. Suppose the contrary. Then there exists $\varepsilon>0$ such that for every $n\in\mathbb{N}$ there is a finite-dimensional subspace F_n in H with the following property:

$$w_1\bigl(x\in H\colon\ q(P_{F_n}x)\leqslant n\bigr)\leqslant 1-\varepsilon.$$

Pick $C>0$ such that $\gamma(x\in X\colon\ q(x)\leqslant C)>1-\varepsilon$, where γ is the Borel extension of the countably additive image of the measure w_1 under the natural embedding of H into X. According to Proposition 5.7.15, the set H is the Cameron–Martin space of the measure γ. We further consider only numbers $n\geqslant C$; for them we have

$$w_1\bigl(x\in H\colon\ q(P_{F_n}x)\leqslant C\bigr)\leqslant 1-\varepsilon.$$

Let us consider the increasing subspaces H_n that are the linear spans of $F_1,\ldots,F_n$. We observe that $w_1\bigl(x\in H\colon\ q(P_{H_n}x)\leqslant C\bigr)\leqslant 1-\varepsilon$ for all $n\geqslant C$. This follows from the fact that the standard Gaussian measure of the convex balanced set Q in $\mathbb{R}^n$ does not exceed the measure of $Q\cap\mathbb{R}^k$ with respect to the standard Gaussian measure on $\mathbb{R}^k$ whenever $k\leqslant n$ (Exercise 5.12.81). Moreover, increasing subspaces H_n, we can assume that their linear span is dense in H. Let S_n denote the measurable linear operator on X corresponding to the finite-dimensional projection P_{H_n} (see § 5.7). The Tsirelson theorem yields that $S_nx\to x$ almost everywhere with respect to γ. Hence

$$\gamma\bigl(x\colon\ q(S_nx)\geqslant C\bigr)\to\gamma\bigl(x\colon\ q(x)\geqslant C\bigr)>1-\varepsilon.$$

Since $\gamma\bigl(x\colon\ q(S_nx)\geqslant C\bigr)=w_1\bigl(x\in H\colon\ q(P_{H_n}x)\leqslant C\bigr)$, this is a contradiction, which completes the proof. □

We shall see below that the converse assertion is not true: the measurability of a norm does not imply the countable additivity of the image of the standard Gaussian measure under the embedding into the completion with respect to this norm.

5.10.12. Theorem. *The Gross–Sazonov topology on an infinite-dimensional Hilbert space is strictly stronger than the Sazonov topology. In addition, there exists a measurable norm on H which cannot be majorized by a norm of the form $h\mapsto (Th,h)_H^{1/2}$, where T is a nonnegative definite nuclear operator in H.*

PROOF. Let $H = W_0^{2,1}([0,1])$ and let q be the classical norm

$$q(h) = \|h\|_C \equiv \max_{t\in[0,1]} |h(t)|, \quad h \in H.$$

The norm q is measurable, since the completion of H with respect to the norm q is the space $C_0[0,1]$ and the image of the canonical cylindrical Gaussian measure w_1 on H with respect to the canonical embedding $H \to C[0,1]$ is the Wiener measure ν on the algebra of cylindrical sets.

We now prove that there is no nonnegative symmetric nuclear operator T in H such that for all $h \in H$ one has $\|h\|_C \leqslant (Th,h)_H^{1/2}$. Indeed, otherwise for every orthonormal system $\{e_n\}$ in H the series $\sum_{n=1}^\infty \|e_n\|_C^2$ must converge. Hence for the proof of nonexistence of a nuclear operator with the stated property it suffices to construct an orthonormal system $\{e_n\}$ in H such that $\sum \|e_n\|_C^2 = \infty$.

Let g be the function on $[0,\infty)$ defined as follows: $g(t) = t$ if $t \in [0,\frac{1}{2}]$; $g(t) = 1-t$ if $t \in [\frac{1}{2},1]$; $g(t) = 0$ if $t > 1$. For every natural n and every $m = 1,2,\ldots,2^{n-1}$, we define the function $e_{n,m}$ on $[0,1]$ by

$$e_{n,m} = 2^{-\frac{n-1}{2}} g\Bigl(2^{n-1}\bigl(t - 2^{-n+1}(m-1)\bigr)\Bigr).$$

It is readily seen that $\{e_{n,m}\}$ is an orthonormal system in H and that we have $\|e_{n,m}\|_C^2 = 2^{-n-1}$ for every $n \in \mathbb{N}$ for all admissible m. Therefore, we obtain that $\sum_{n,m} \|e_{n,m}\|_C^2 = \sum_{n=1}^\infty \sum_{m=1}^{2^{n-1}} 2^{-n-1} = \infty$. □

5.10.13. Remark. In the book Kuo **[305]**, a seminorm q on a Hilbert space is called measurable (and we shall call it *Gross measurable*) if the following condition is fulfilled: for every $\varepsilon > 0$ there exists $F \in \mathrm{Fin}(H)$ such that for every subspace $G \in \mathrm{Fin}(H)$ orthogonal to F one has

$$w_1\bigl(z \in H\colon\ q(P_G z) > \varepsilon\bigr) < \varepsilon.$$

5.10.14. Proposition. *Every Gross measurable seminorm on a Hilbert space is measurable.*

PROOF. Let $\varepsilon \in (0,1)$. By definition there exists a subspace $F_0 \in \mathrm{Fin}(H)$ such that for every subspace $F_1 \in \mathrm{Fin}(H)$ orthogonal to F_0 one has

$$w_1\bigl(z \in H\colon\ q(P_{F_1} z) > \varepsilon/2\bigr) < \varepsilon/2.$$

Let us take $C > 0$ for which $w_1\bigl(z \in H\colon\ q(P_{F_0} z) > C\bigr) < \varepsilon/2$; this is possible, because the restriction of the function w_1 to the σ-algebra $\mathfrak{A}_{F_0}$ is countably additive. The inequality $q(h_1+h_2) \leqslant q(h_1) + q(h_2)$ yields that

$$\{z\in H\colon\ q(z) \leqslant C+1\} \supset \{h\in F_0\colon\ q(h) \leqslant C\} + \{h\in F_0^\perp\colon\ q(h) < \varepsilon/2\}.$$

Now, for arbitrary $F \in \mathrm{Fin}(H)$, we take the subspace F_2 generated by F and F_0 denote by F_1 the orthogonal complement to F_0 in F_2. Then

$$\begin{aligned}\{z \in H\colon\ q(P_F z) > C+1\} &\subset \{z \in H\colon\ q(P_{F_2} z) > C+1\}\\ &\subset \{z \in H\colon\ q(P_{F_0} z) > C\} \cup \{z \in H\colon\ q(P_{F_1} z) > \varepsilon/2\},\end{aligned}$$

so that $w_1\bigl(z \in H\colon\ q(P_F z) > C+1\bigr) < \varepsilon$. Since $\varepsilon > 0$ was arbitrary, this means that q is measurable. □

We shall now see that the converse is not true: not every measurable seminorm on H is Gross measurable. In the same book Kuo [**305**], the "Gross topology" was introduced in a Hilbert space: this is the topology in H generated by the set of all Gross measurable seminorms on H. It follows from our previous discussion that this topology is strictly weaker than the Gross–Sazonov topology in H introduced above (the Gross topology was not considered in [**305**] for arbitrary locally convex spaces).

5.10.15. Example. The norm $q(x) = \sup_n |x_n|\bigl(\ln(n+1)\bigr)^{-1/2}$ on $H = l^2$ is measurable, but not Gross measurable. Indeed, for the sequence of independent standard Gaussian random variables ξ_n almost surely we have

$$\limsup_{n\to\infty} |\xi_n\bigl(\ln(n+1)\bigr)^{-1/2}| = 1.$$

To see this, we verify that $P\bigl(\sup_{n\geqslant k} |\xi_n| \leqslant \lambda\bigl(\ln(n+1)\bigr)^{1/2}\bigr) = 0$ for each k and each $\lambda < 1$ and that for each $\lambda > 1$ the probability that for infinitely many n one has $|\xi_n| \geqslant \lambda\bigl(\ln(n+1)\bigr)^{1/2}$ is zero (here we use that $P(|\xi_1| > R)$ is asymptotically $CR^{-1}\exp(-R^2/2)$ for large R). Hence Remark 5.10.6 yields the measurability of q. The aforementioned fact shows also that there is no number m such that $w_1\bigl(z \in H\colon\, q(P_G z) > 1/4\bigr) < 1/4$ for all projections P_G onto subspaces orthogonal to $P_m(H)$. One can derive from this the absence of any projections P_F required by the Gross definition. An alternative justification is possible: if the norm q were Gross measurable, then, as shown below, the image of the standard cylindrical Gaussian measure on H under the embedding to the completion of H with respect to the norm q would be countably additive. Since this completion is separable and any centered Gaussian measure on a separable Banach space is positive on all balls centered at zero (see § 5.7), we would obtain that $\mathrm{P}(\sup_n |\xi_n| < r) > 0$ for all $r > 0$, which is false.

5.10.16. Proposition. *If a norm q on H is Gross measurable and B is the completion of H with respect to q, then the image of the standard Gaussian cylindrical measure under the natural embedding of H into B is countably additive. Conversely, for the Gross measurability the analog of Theorem* 5.10.11 *holds.*

PROOF. It follows from the definition that there exists a sequence of projections P_n onto increasing finite-dimensional subspaces $H_n \subset H$ such that the union of H_n is dense in H and $w_1(x\colon\, q(Px) > 2^{-n}) < 2^{-n}$ whenever $P \perp P_n$. Let $\{e_i\}$ be an orthonormal basis in H such that $e_1, \ldots, e_{k_n}$ is a basis in H_n. Let us take a sequence of independent standard Gaussian random variables ξ_i. The sequence of random elements $\eta_n = \sum_{i=1}^{k_n} \xi_i e_i$ in B is fundamental in probability, since

$$\begin{aligned}\mathrm{P}\bigl(q(\eta_n - \eta_k) > 2^{-n}\bigr) &= w_1\bigl(x\colon\, q(P_n x - P_k x) > 2^{1-n}\bigr)\\ &\leqslant w_1\bigl(x\colon\, q(P_n x) > 2^{-n}\bigr) + w_1\bigl(x\colon\, q(P_k x) > 2^{-n}\bigr) < 2^{1-n} \quad \forall\, k \geqslant n.\end{aligned}$$

Hence some subsequence $\{\eta_{n_k}\}$ converges in B almost everywhere. It is clear that the distribution of the limit coincides with the image of the standard Gaussian cylindrical measure on H. The converse assertion is proved similarly to Theorem 5.10.11. □

Our discussion of measurable norms is continued in § 5.12(xii).

5.11. Conditions for countable additivity

Here we give sufficient conditions for countable additivity of measures in terms of their Fourier transforms. In particular, the continuity of the Fourier transform in the Gross–Sazonov topology is sufficient. This condition is not necessary.

5.11.1. Remark. (i) The countable additivity of a nonnegative E-cylindrical measure is not enough for the continuity of its Fourier transform in the original topology of the space E. For example, let E be an infinite-dimensional separable Hilbert space with its weak topology. Let us take a centered Gaussian measure on E with the Fourier transform $\exp\bigl(-(Qx,x)\bigr)$, where Q is a nonnegative nuclear operator in E with an infinite-dimensional range. This function is discontinuous in the weak topology, because the norm $(Qx,x)^{1/2}$ cannot be estimated by finitely many seminorms of the form $|(x,v)|$.

At the same time the Fourier transform of an arbitrary E-cylindrical countably additive measure is always sequentially continuous; this follows from the Lebesgue dominated convergence theorem.

(ii) Since the Gross–Sazonov and Sazonov topologies associated with the weak topology of an arbitrary locally convex space coincide with this weak topology, it follows from (i) that the countable additivity of a E-cylindrical measure on E' does not imply the continuity of its Fourier transform in the Sazonov or Gross–Sazonov topology (though, as we shall see below, for the Fourier transforms of cylindrical measures the continuity in one of these topologies is equivalent to the continuity in the other).

(iii) The following question arises: if the space E is equipped with the Mackey topology $\tau(E,E')$, will the countable additivity of a E-cylindrical measure on E' imply the continuity of its Fourier transform in the Sazonov topology associated with $\tau(E,E')$ (or, which is equivalent, in the Gross–Sazonov topology associated with $\tau(E,E')$)? The answer remains negative. For the corresponding counterexample we can take the space $C[0,1]$ with the classical Wiener measure. Indeed, let $E = \bigl(C', \tau(C', C[0,1])\bigr)$, where C' is the dual to the Banach space $C[0,1]$. Then, owing to Theorem 5.10.12, the Fourier transform of the Wiener measure is not continuous in the Sazonov topology (hence also in the Gross–Sazonov topology) associated with the original topology of the space E.

In this example the space $C[0,1]$ is separable Banach, hence the Fourier transform of the Wiener measure is continuous in the Mackey topology $\tau(E,E')$ on $E = C'$. However, as already noted, Kwapien and Tarieladze [**309**] constructed a very subtle example of a Radon probability measure on a metrizable locally convex space X whose Fourier transform is not continuous in the Mackey topology $\tau(X',X)$.

5.11.2. Remark. Let E be a locally convex space, let ν be a nonnegative E-cylindrical measure on E' with the Fourier transform $\widetilde{\nu}$, and let τ be an arbitrary topology on E compatible with the vector structure (but not necessarily connected with the original topology). If the function $\widetilde{\nu}$ is continuous in the topology τ at zero, then it is continuous in this topology everywhere. Indeed, by (5.5.3) we have

$$|\widetilde{\nu}(x_1)-\widetilde{\nu}(x_2)|^2 \leqslant 2\widetilde{\nu}(0)\bigl(1-\operatorname{Re}\widetilde{\nu}(x_1-x_2)\bigr).$$

Let us proceed to positive results, which give even more than the countable additivity of single measures and ensure the uniform tightness of families of cylindrical measures.

5.11.3. Theorem. *Let E be a locally convex space, let $\mathcal{P}$ be some set of probability E-cylindrical measures on E'. In order that $\mathcal{P}$ be uniformly tight when E' is equipped with the topology $\sigma(E',E)$, it is sufficient that the family of the Fourier transforms of the measures from $\mathcal{P}$ be equicontinuous at zero in the Gross–Sazonov topology associated with the original topology of E.*

This is a corollary of the following main theorem of this section.

A set S in a real vector space X is called an *ellipsoid* if it has the form $S=\{x\in X\colon Q(x)\leqslant 1\}$, where Q is a nonnegative quadratic form on X, i.e., $Q(x)=B(x,x)$, where B is a symmetric bilinear function and $Q\geqslant 0$.

5.11.4. Theorem. *Let E be a locally convex space, let $\mathcal{O}(E)$ be the set of all neighborhoods of zero in the Gross–Sazonov topology τ_{GS}^E associated with the original topology of E, and let $\mathcal{K}$ be the set of all compact ellipsoids in the space $\bigl(E',\sigma(E',E)\bigr)$. Then, for every $\varepsilon>0$ and $V\in\mathcal{O}(E)$, there is an ellipsoid $K\in\mathcal{K}$ such that whenever a probability E-cylindrical measure ν on E' satisfies the estimate $|\widetilde{\nu}(x)-1|<\varepsilon$ for all $x\in V$, one has $\nu(A)<6\varepsilon$ for every $A\in\mathfrak{A}(E',E)$ with $A\cap K=\varnothing$.*

PROOF. Suppose first that E is a Hilbert space. Let $V\in\mathcal{O}(E)$, $C>0$, $F\in\operatorname{Fin}(E)$, $\nu\in\mathcal{P}(E)$. Let $\|\cdot\|$ denote the original norm in E and let U_C denote the closed ball in E of radius C with respect to this norm centered at zero. Then the following chain of equalities and inequalities holds (below we justify those which are not completely obvious):

$$\begin{aligned}
\nu\bigl(x\in F\colon \|P_F x\|>C\bigr) &= \nu\circ P_F^{-1}(F\backslash U_C)\\
&\leqslant \sqrt{e}(\sqrt{e}-1)^{-1}\int_F\Bigl(1-e^{-\frac{(x,x)}{2C^2}}\Bigr)\,\nu\circ P_F^{-1}(dx)\\
&= \sqrt{e}(\sqrt{e}-1)^{-1}\int_F(1-\widetilde{\nu}(z))\,(w_{1/C^2}\circ P_F^{-1})(dz)\\
&\leqslant \sqrt{e}(\sqrt{e}-1)^{-1}\int_{V\cap F}|1-\widetilde{\nu}(z)|\,(w_{1/C^2}\circ P_F^{-1})(dz)\\
&\qquad + \sqrt{e}(\sqrt{e}-1)^{-1}\int_{F\backslash V}|1-\widetilde{\nu}(z)|\,(w_{1/C^2}\circ P_F^{-1})(dz).
\end{aligned}$$

The first inequality in this chain follows from the estimate

$$1\leqslant \frac{\sqrt{e}}{\sqrt{e}-1}\Bigl(1-e^{-\frac{(x,x)}{2C^2}}\Bigr)\quad \text{if } x\in F\backslash U_C,\ \text{i.e., if } \|x\|>C,$$

and the subsequent replacement of the domain of integration $F\backslash U_C$ by the whole space F, which is possible since the function under the integral sign is nonnegative.

The equality following this inequality is Parseval's identity: $\widetilde{\nu}$ is the Fourier transform of the measure $\nu\circ P_F^{-1}$, and the measure $w_{1/C^2}\circ P_F^{-1}$ on F, i.e., the Gaussian measure on F with the correlation operator $C^{-2}I$, is the measure whose Fourier transform is the function $x\mapsto e^{-\frac{(x,x)}{2C^2}}$ on F; in addition, we use that the measures ν and w_{1/C^2} are probability measures.

By condition we have $|1-\widetilde{\nu}(x)|<\varepsilon$ for $x\in V$, so that

$$\int_{V\cap F} |1-\widetilde{\nu}(z)|\,(w_{1/C^2}\circ P_F^{-1})(dz)<\varepsilon.$$

In addition, for any given $\varepsilon>0$ there exists $C>0$ such that

$$\int_{F\backslash V} |1-\widetilde{\nu}(z)|\,(w_{1/C^2}\circ P_F^{-1})(dz)<\varepsilon.$$

Indeed, using the change $y=Cz$, we find that

$$\int_{F\backslash V} |1-\widetilde{\nu}(z)|(w_{1/C^2}P_F{}^{-1})(dz)=\int_{F\backslash CV} |1-\widetilde{\nu}(C^{-1}y)|\,(w_1\circ P_F^{-1})(dy).$$

Since V is a neighborhood of zero in the topology τ_{GS}^E, it contains the unit ball $V_0=\{x\in E\colon\ q(x)\leqslant 1\}$ with respect to some measurable seminorm q. According to the definition of measurable seminorms, for all $C\geqslant C(\varepsilon,V_0)$ we have

$$(w_1\circ P_F^{-1})(F\backslash CV_0)<\varepsilon/4.$$

Since $V_0\subset V$, we have $(w_1\circ P_F^{-1})(F\backslash CV)<\varepsilon/4$, whence on account of the estimate $|1-\widetilde{\nu}|\leqslant 2$ for such C we obtain the inequality

$$\int_{F\backslash CV} |1-\widetilde{\nu}(C^{-1}y)|\,(w_1\circ P_F^{-1})(dy)<\varepsilon/2.$$

We finally obtain $\nu(x\in F\colon\ \|P_F x\|>C)<2\varepsilon\sqrt{e}(\sqrt{e}-1)^{-1}$. It is easy to verify that the right-hand side does not exceed 6ε. Thus, for Hilbert spaces the theorem is proven: our reasoning shows that its conclusion holds if we take $K=U_{C(\varepsilon,V_0)}$.

Suppose now E is a general locally convex space, $\varepsilon>0$, a continuous linear operator $\Lambda\colon\ E\to H$, a neighborhood $V\in\mathcal{O}(H)$ and a probability E-cylindrical measure ν on E' are such that $|1-\widetilde{\nu}(x)|<\varepsilon$ if $\Lambda x\in V$. We can assume that $V=\{q<1\}$, $q\in S_H$. We show that the set $K=\Lambda^*(U_{C(\varepsilon,V)})$ is suitable, where Λ^* is regarded as a mapping from H to E' (the space H' is identified with H in the usual manner). It suffices to show that for every finite-dimensional subspace $F\subset E$ the estimate $(\nu\circ\psi^{-1})(E_F\backslash G_F)<6\varepsilon$ holds, where ψ is the canonical mapping from E' onto quotient $E_F:=E'/F^\circ$ and $G_F:=\psi\big(\Lambda^*(U_{C(\varepsilon,V)})\big)$.

Let E_1 denote the vector subspace in E_F generated by the set G_F and let E_2 be its algebraic complement in E_F. Let $(\cdot\,,\cdot)_1$ be the inner product in E_1 with respect to which $G_F^C\equiv\big(C(\varepsilon,V)\big)^{-1}G_F$ is the unit ball, and let $(\cdot\,,\cdot)_2$ be an arbitrary nondegenerate inner product in E_2. For every $\delta>0$, let q_δ denote the norm in E_F defined by the equality

$$q_\delta^2(x_1+x_2)=(x_1,x_1)_1+\delta(x_2,x_2)_2,\quad \text{where } x_j\in E_j.$$

Further, let B_δ be the unit ball in E_F with respect to this norm and let q_δ^1 be the norm in F with respect to which the unit ball is the polar B_δ° in F of the set B_δ.

Since $G_F = \bigcap_{\delta>0} C(\varepsilon, V)B_\delta$, for the proof of the desired inequality it suffices to show that for all $\delta > 0$ we have

$$\nu\circ\psi^{-1}\bigl(E_F\backslash C(\varepsilon, V)B_\delta\bigr) < 6\varepsilon.$$

Let Λ_0 denote the restriction of Λ to the polar E_2° of the subspace $E_2 \subset E_F$ taken in F (E_F and F are in the natural duality). Then Λ_0 is an isometric embedding to H of the space $(E_2^\circ, q_\delta^1|_{E_2^\circ})$, where $q_\delta^1|_{E_2^\circ}$ is the restriction of q_δ^1 to E_2°. Let Λ_1 be an extension of Λ_0 to an isometric embedding into H of the space F equipped with the norm q_δ^1 and let $H_1 = \Lambda_1(F)$. We denote by $T\colon\ E_F \to H_1$ the operator dual to $\Lambda_1^{-1}\colon\ H_1 \to F$. Let ν_1 be the H-cylindrical measure in H defined as follows:

$$\nu_1 = \nu\circ\psi^{-1}\circ T^{-1}.$$

Its Fourier transform satisfies the equality

$$\widetilde{\nu}_1(y) = \widetilde{\nu}(\Lambda_1^{-1}y), \quad y \in H_1,$$

so that $|1 - \widetilde{\nu}_1(y)| < \varepsilon$ if $y \in V \cap H_1$. Therefore,

$$\nu\circ\psi^{-1}\bigl(E_F\backslash C(\varepsilon, V)B_\delta\bigr) = \nu_1\bigl(H_1\backslash U_{C(\varepsilon,V)}\bigr) < 6\varepsilon,$$

as required. □

A direct corollary of this theorem is the next assertion.

5.11.5. Theorem. *Let E be a locally convex space and let ν be a probability E-cylindrical measure on E'. For countable additivity of ν it is sufficient that its Fourier transform be continuous at zero in the Gross–Sazonov topology associated with the original topology of E.*

We shall see below that although the Gross–Sazonov topology is strictly stronger than the Sazonov topology, the continuity of the Fourier transform of a E-cylindrical measure in the Gross–Sazonov topology implies its continuity also in the Sazonov topology.

The next theorem, concretizing Proposition 5.5.8, is a partial converse to Theorem 5.11.4.

5.11.6. Theorem. *Let E be a locally convex space and let ν be a countably additive real measure on the σ-algebra of Borel subsets of the space $\bigl(E', \sigma(E', E)\bigr)$ possessing the following property: for every $\varepsilon > 0$ there exists a compact ellipsoid K in the space $\bigl(E', \sigma(E', E)\bigr)$ such that $\|\nu\|(E'\backslash K) < \varepsilon$. Then the Fourier transform $\widetilde{\nu}$ of this measure is continuous in the Sazonov topology τ_S associated with the Mackey topology.*

PROOF. Passing from ν to $\|\nu\|$ we can assume that the original measure ν is nonnegative; in addition, we can assume that it is probability, so that $\widetilde{\nu}(0) = 1$. By inequality (5.5.3) it suffices to establish the continuity of the function $\mathrm{Re}\,\widetilde{\nu}$ on (E, τ_S) at zero. In turn, to this end it suffices to show that for every $\varepsilon > 0$ there are a Hilbert seminorm p_ε on the space $\bigl(E,\tau(E',E)\bigr)$ and a nonnegative Hilbert–Schmidt operator $A_\varepsilon\colon \overline{E}_{p_\varepsilon}\to\overline{E}_{p_\varepsilon}$, where $\overline{E}_{p_\varepsilon}$ is the completion of the quotient E_{p_ε} of the pre-Hilbert space (E,p_ε) with respect to its subspace ${p_\varepsilon}^{-1}\{0\}$, having the following properties. If the symbol $\overline{p}_\varepsilon$ denotes the extension to $\overline{E}_{p_\varepsilon}$ of the norm on E_{p_ε} generated by the norm p_ε, then the inequality $\overline{p}_\varepsilon(A_\varepsilon Q_{p_\varepsilon}x)<1$, where Q_{p_ε} is the canonical operator $E\to\overline{E}_{p_\varepsilon}$, yields the inequality $|\mathrm{Re}\,\widetilde{\nu}(x)-1|<\varepsilon$. We recall that the set $\bigl\{x\in E\colon\, p_\varepsilon(A_\varepsilon Q_{p_\varepsilon}x)<1\bigr\}$ is a neighborhood of zero in the topology τ_S.

Thus, let $\varepsilon>0$ be fixed and let K be a compact ellipsoid in $\bigl(E',\sigma(E',E)\bigr)$ for which $\nu(E'\backslash K)<\varepsilon/4$. As we have verified in the proof of Proposition 5.5.8, under these conditions $1-\mathrm{Re}\,\widetilde{\nu}(x)\leqslant\varepsilon$ for every x such that $|x(g)|\leqslant\varepsilon$ for all $g\in K$. Set $p(x):=\sup\{|g(x)|\colon\, g\in K\}$, $x\in E$.

Since K is a compact ellipsoid in $\bigl(E',\sigma(E',E)\bigr)$, the function p is a Hilbert seminorm on E; the absolute convexity and compactness of the set K yield that p is continuous in the Mackey topology $\tau(E,E')$. The Hilbert space $(E')_K$, i.e., the linear subspace in E' generated by K with norm p_K, for which the closed unit ball is K, can be naturally identified with the Hilbert space dual to $\overline{E}_p$. For this, for every $\widetilde{x}\in E_p$ and every $g\in(E')_K$, it suffices to set $g(\widetilde{x})=g(Q_P^{-1}\widetilde{x})$, where Q is the canonical mapping from E to E_p (this definition is correct, since for any $g\in(E')_K$ the function $x\mapsto g(x)$ is constant on every set $Q_p^{-1}x$); next, for every $g\in(E')_K$ we can extend the function $\widetilde{x}\mapsto g(\widetilde{x})$ from the space E_p to the space $\overline{E}_p$ as a continuous function with respect to the norm $\overline{p}$.

Using the described identification, we define a nonnegative quadratic function ψ on $\overline{E}_p$ by the equality

$$\psi(\widetilde{x})=\int_K |g(x)|^2\,\nu(dg).$$

There is a nonnegative operator A in $\overline{E}_p$ such that

$$p(A\widetilde{x})^2=\psi(\widetilde{x})$$

for all $\widetilde{x}\in E_p$. Finally, set

$$\widetilde{A}:=\varepsilon^{-1/2}A.$$

If $\overline{p}(\widetilde{A}\widetilde{x})<1$, then $\psi(\widetilde{x})<\varepsilon$; hence (by the definition of ψ), for any $x\in E$ satisfying the inequality $\overline{p}(\widetilde{A}Q_px)<1$ we have

$$1-\mathrm{Re}\,\widetilde{\nu}(x)\leqslant\varepsilon.$$

We now show that A (hence also $\widetilde{A}$) is a Hilbert–Schmidt operator. Let $\{e_\alpha\}$ be an orthonormal basis in $\overline{E}_p$. Then

$$\begin{aligned}\sum\overline{p}^2(Ae_\alpha)&=\sum\psi(e_\alpha)=\sum_K\int_K|g(e_\alpha)|^2\,\nu(dg)\\&\leqslant\int_K\sum|g(e_\alpha)|^2\,\nu(dg)\leqslant\int_K p_K(g)^2\,\nu(dg)\leqslant 1,\end{aligned}$$

Since by the definition of the norm p_K for any $g\in K$ we have $p_K(g)\leqslant 1$ (in the above chain of relations we pass from equalities to an inequality, because $\{e_\alpha\}$

can be uncountable). Thus, the norm $p_\varepsilon = p$ and the operator $A_\varepsilon = \widetilde{A}$ possess all required properties. □

Similarly one proves the following theorem.

5.11.7. Theorem. *Let E be a barrelled locally convex space and let $\mathcal{M}$ be a set of real Borel measures on the space $\bigl(E', \sigma(E', E)\bigr)$ with the following two properties*:

(i) $\sup\{\|\nu\|(E')\colon\ \nu \in \mathcal{M}\} < \infty$,

(ii) *for every $\varepsilon > 0$, there is a compact set K in the space $\bigl(E', \sigma(E', E)\bigr)$ such that for every measure $\nu \in \mathcal{M}$ one has $\|\nu\|(E'\backslash K) < \varepsilon$.*

Then the set $\widetilde{\mathcal{M}}$ of the Fourier transforms of measures in $\mathcal{M}$ is equicontinuous in the topology of the space E.

PROOF. Without loss of generality we can assume that all measures in $\mathcal{M}$ are probability measures. Let $\varepsilon > 0$ and let K be the corresponding compact set from the hypotheses of the theorem. Then, as in the proof of Proposition 5.5.8, for every measure $\nu \in \mathcal{M}$ we have

$$1 - \operatorname{Re} \widetilde{\nu}(x) \leqslant \varepsilon$$

if $|x(g)| \leqslant \varepsilon$ for all $g \in K$. Let K° be the polar in E of the set K. Since E is barrelled, the set K° is a neighborhood of zero in E. Let $V = \varepsilon^{1/2} K^\circ$. If $x \in V$, then by the previous inequality we obtain the equicontinuity of functions in $\widetilde{\mathcal{M}}$ at zero in the space E. The inequality

$$|\widetilde{\nu}(x_2) - \widetilde{\nu}(x_1)|^2 \leqslant 2\bigl(1 - \operatorname{Re} \widetilde{\nu}(x_2 - x_1)\bigr)$$

yields that for every $\varepsilon > 0$ there exists a neighborhood of zero V in E such that $|\widetilde{\nu}(x_2) - \widetilde{\nu}(x_1)| < \varepsilon$ if $\widetilde{\nu} \in \widetilde{\mathcal{M}}$ and $x_1 - x_2 \in V$. Thus, the set of the Fourier transforms of the measures in $\mathcal{M}$ is equicontinuous on the space E. □

From Theorem 5.11.6 we immediately obtain the following fact (note that compactness of K is required in the weak-$*$ topology, not the strong one).

5.11.8. Theorem. *Let E be a locally convex space, let ν be a real E-cylindrical measure on E' such that for every $\varepsilon > 0$ there exists a compact ellipsoid K in the space $\bigl(E', \sigma(E', E)\bigr)$ such that $|\nu(C)| < \varepsilon$ for every E-cylindrical set C in E' disjoint with K. Then the Fourier transform of the measure ν is continuous in the Sazonov topology.*

Taking into account Theorem 5.11.4, we obtain the following assertion.

5.11.9. Corollary. *If on the dual E' to a locally convex space E we are given a nonnegative E-cylindrical measure ν whose Fourier transform is continuous at zero in the Gross–Sazonov topology on E, then it is continuous everywhere in the Sazonov topology.*

Note that the converse assertion is, certainly, also true, since the Gross–Sazonov topology is stronger than the Sazonov topology.

5.11.10. Theorem. *Let E be a barrelled locally convex space the topology of which can be defined by a family of Hilbert seminorms. A function φ on E is the Fourier transform of some nonnegative Radon measure on the space $\bigl(E', \sigma(E', E)\bigr)$ precisely when this function is positive definite and continuous in the Gross–Sazonov topology or in the Sazonov topology associated with the original topology in E.*

PROOF. If φ is the Fourier transform of a nonnegative Radon measure ν, then the function φ is positive definite. Let us prove its continuity. For every $\varepsilon > 0$ there exists a compact set K in the space $\bigl(E', \sigma(E', E)\bigr)$ such that $\nu(E'\backslash K) < \varepsilon$. Let B_K be the closed convex circled hull of this set; then $\nu(E'\backslash B_K) < \varepsilon$. Since the topology of a barrelled space is its Mackey topology, the polar B_K° in the space E of the set B_K is a neighborhood of zero in E. Since the topology of this space by assumption is generated by a family of Hilbert seminorms, there exists a continuous Hilbert seminorm p on E such that the set $S = \{x \in E\colon\, p(x) < 1\}$ is contained in B_K°; then the polar S° in E' of the set S is a compact ellipsoid in $\bigl(E', \sigma(E', E)\bigr)$, moreover, $S^\circ \supset B$. Hence $\nu(E'\backslash S^\circ) < \varepsilon$. Thus, ν satisfies the condition of Theorem 5.11.8, which gives the continuity of φ in the Sazonov topology. The converse follows from the results above. □

As a corollary we obtain the classical Minlos theorem.

5.11.11. Corollary. *If E is a nuclear locally convex space, then every positive definite continuous function on E is the Fourier transform of a nonnegative Radon measure on $\bigl(E', \sigma(E', E)\bigr)$ (which is Radon also in the strong topology $\beta(E', E)$). If E is barrelled, then the converse is true.*

Applying the theorem to Hilbert spaces we obtain another classical result: the Sazonov theorem.

5.11.12. Corollary. *A bounded cylindrical measure on a separable Hilbert space is countably additive precisely when its Hilbert Fourier transform is continuous in the Sazonov topology (or in the Gross–Sazonov topology).*

For example, this assertion implies the already known fact: in order that a Gaussian cylindrical measure on a separable Hilbert space be countably additive, it is necessary and sufficient that its correlation operator be nuclear.

5.12. Complements and exercises

5.12(i). Convolution

Throughout this section we consider Hausdorff spaces. If μ and ν are two measures defined on the σ-algebra $\sigma(X')$ in a locally convex space X, then their product $\mu\otimes\nu$ is a measure on $\sigma\bigl((X\times X)'\bigr)$. In case of Radon measures μ and ν, the product $\mu\otimes\nu$ is a tight measure, hence admits a unique extension to a Radon measure on $X\times X$. The same is true if μ and ν are Radon measures on a Hausdorff topological vector space. By a product of Radon measures we shall always mean this extension.

5.12.1. Definition. *Let μ and ν be Radon measures on a Hausdorff locally convex (or Hausdorff topological vector) space X. Their convolution $\mu*\nu$ is defined as the image of the Radon extension of the measure $\mu\otimes\nu$ on $X\times X$ under the mapping $X\times X\to X$, $(x,y)\mapsto x+y$. The convolution of measures on $\sigma(X')$ is defined similarly on $\sigma(X')$.*

Below we consider Hausdorff spaces.

5.12.2. Theorem. *Let μ and ν be Radon measures on a locally convex space X. Then, for every Borel set $B\subset X$, the function $x\mapsto\mu(B-x)$ is Borel measurable and*

$$\mu*\nu(B)=\int_X\mu(B-x)\,\nu(dx).$$

In addition, $\mu*\nu=\nu*\mu$ and $\widetilde{\mu*\nu}=\widetilde{\mu}\widetilde{\nu}$.

PROOF. The set $A=\{(x,y)\in X\times X\colon\ x+y\in B\}$ is Borel. According to a general result (see Bogachev [**65**, Theorem 7.6.5]), the measure of A with respect to the Radon extension of the product of μ and ν equals the integral with respect to ν of the function $x_2\mapsto\mu(A_{x_2})$, where

$$A_{x_2}=\{x_1\colon\ (x_1,x_2)\in A\}=\{x\colon\ x_1+x_2\in B\}=B-x_2,$$

moreover, the indicated function is Borel. The same is true for these measures in the opposite order. □

It is clear that similarly one can defined the convolution of two bounded cylindrical measures.

5.12.3. Proposition. *Let μ and λ be two Radon probability measures on a locally convex space X. Suppose that there exists a positive definite function $\varphi\colon\ X^*\to\mathbb{C}$ such that $\widetilde{\lambda}=\varphi\widetilde{\mu}$. Then there exists a Radon probability measure ν on X with $\widetilde{\nu}=\varphi$. In addition, $\lambda=\nu*\mu$.*

PROOF. It follows from our condition that the restrictions of the function φ to finite-dimensional subspaces are continuous at zero, hence at every point. Therefore, φ is a characteristic functional of a nonnegative quasi-measure ν on the algebra of cylindrical sets. It remains to show that the set function ν is tight, since in that case the equality $\widetilde{\lambda}=\widetilde{\nu}\widetilde{\mu}$ will imply the equality $\lambda=\nu*\mu$. Let $\varepsilon>0$ and let S be a compact set such that $\mu(X\backslash S)+\lambda(X\backslash S)<\varepsilon/2$. We can assume that $0\in S$. The set $K:=S-S$ is compact and $S\subset K$. Let C be a cylindrical set with $C\cap K=\varnothing$. The set C has the form $C=P^{-1}(B)$, where $B\in\mathcal{B}(\mathbb{R}^n)$ and $P\colon\ X\to\mathbb{R}^n$ is a continuous linear mapping. We observe that $B\cap P(K)=\varnothing$, since if $x\in C$, then $x+h\in C$ for all $h\in\operatorname{Ker}P$. In particular, $B\cap P(S)=\varnothing$, i.e., $C \cap P^{-1}(P(S)) = \varnothing$. For the cylindrical set $C_0 := P^{-1}(P(S))$ we have $S \subset C_0$ and

$$1 - \varepsilon/2 \leqslant \lambda(S) \leqslant \lambda(C_0) = \int_X \nu(C_0 - x)\,\mu(dx) \leqslant \int_S \nu(C_0 - x)\,\mu(dx) + \varepsilon/2,$$

whence we obtain the existence of $x_0 \in S$ such that $\nu(C_0 - x_0) \geqslant 1 - \varepsilon$. Moreover, $(C_0 - x_0) \cap C = \varnothing$, since $P(C_0 - x_0) \subset P(S - S)$, because $x_0 \in S$. Thus, we have $\nu(C) \leqslant \varepsilon$, i.e., the measure ν is tight. □

The proof of the following proposition can be read in Vakhania, Tarieladze, Chobanyan [**538**, § VI.3].

5.12.4. Proposition. *Let μ_1 and μ_2 be two nonnegative cylindrical measures on the algebra of cylindrical sets in a locally convex space X such that the measure μ_1 is symmetric, i.e., $\mu_1(A) = \mu_1(-A)$. If $\mu = \mu_1 * \mu_2$ admits a Radon extension, then both measures μ_1 and μ_2 admit Radon extensions.*

The assumption that μ_1 is symmetric cannot be omitted. Indeed, let l be a discontinuous linear functional on X^* (which exists, for example, if X is an infinite-dimensional Banach space). Then the functionals $\exp(il)$ and $\exp(-il)$ are the Fourier transforms of two cylindrical measures on $Cyl(X, X^*)$ without Radon extensions, but their convolution is Dirac's measure δ. This example is typical: according to Rosiński [**423**], if μ and ν are nonnegative cylindrical measures on $Cyl(X, X^*)$ such that $\mu * \nu$ is tight, then there exists an element l in the algebraic dual X^* with the property that the cylindrical measures $\mu * \delta_l$ and $\nu * \delta_{-l}$ (where δ_l and δ_{-l} are the cylindrical measures with the Fourier transforms $\exp(il)$ and $\exp(-il)$, respectively) are tight on X (and hence have Radon extensions). These results can be extended to families of measures in the following way (see Vakhania, Tarieladze, Chobanyan [**538**, Proposition I.4.8]).

5.12.5. Proposition. *Let $\{\mu_\lambda\}$ and $\{\nu_\lambda\}$ be two families of Radon probability measures on a topological vector space X. Suppose that the family $\{\mu_\lambda * \nu_\lambda\}$ is uniformly tight, i.e., for every $\varepsilon > 0$ there is a compact set K_ε such that $\mu_\lambda * \nu_\lambda(X \backslash K_\varepsilon) < \varepsilon$ for all λ. Then there exists a family $\{x_\lambda\}$ of points in X such that $\{\mu_\lambda * \delta_{x_\lambda}\}$ is a uniformly tight family. If, in addition, the measures μ_λ are symmetric, then both families $\{\mu_\lambda\}$ and $\{\nu_\lambda\}$ are uniformly tight.*

For a Radon measure μ let S_μ denote its topological support (the smallest closed set of full measure, see § 5.2).

5.12.6. Proposition. *Let μ and ν be two Radon probability measures on a locally convex space X. Then $S_{\mu * \nu}$ is the closure of $S_\mu + S_\nu$.*

PROOF. If $a \in S_\mu$, $b \in S_\nu$, then for every absolutely convex neighborhood of zero U we have $\mu(a + U) > 0$, $\nu(b + U) > 0$, whence $\mu * \nu(a + b + 2U) > 0$, because $\mu(a + b + 2U - x) > 0$ for $x \in b + U$ by the inclusion $U \subset 2U - x$ for all $x \in U$. Hence $S_\mu + S_\nu \subset S_{\mu * \nu}$, which implies the inclusion also for the closure by the closedness of $S_{\mu * \nu}$. Let now $c \in S_{\mu * \nu}$. If c does not belong to the closure of $S_\mu + S_\nu$, then there exists an absolutely convex neighborhood of zero U

such that $c \notin S_\mu + S_\nu + U$. Then $(c+U-x)\cap S_\mu = \varnothing$ for all $x \in S_\nu$, whence $\mu(c+U-x) = 0$. Thus, $\mu * \nu(c+U) = 0$, which is a contradiction. □

5.12.7. Definition. *A random vector in a locally convex space X is a measurable mapping ξ from $(\Omega, \mathcal{F}, \mathrm{P})$ to $\bigl(X, \sigma(X')\bigr)$, where $(\Omega, \mathcal{F}, \mathrm{P})$ is a probability space. The distribution of the random vector ξ is the probability measure μ on the σ-algebra $\sigma(X')$ that is the image of* P, *i.e.,* $\mu(B) = \mathrm{P}\bigl(\xi^{-1}(B)\bigr)$. *If μ has zero mean, then ξ is called centered.*

Two random vectors ξ and η with values in X are called *independent* if

$$\mathrm{P}(\xi \in A, \eta \in B) = \mathrm{P}(\xi \in A)\mathrm{P}(\eta \in B) \quad \forall\, A, B \in \sigma(X').$$

This is equivalent to the condition that the distribution of the random vector (ξ, η) in $X \times X$ is the product of the distributions of the vectors ξ and η. Hence, for independent random vectors ξ and η with distributions P_ξ and P_η, the distribution of the random vector $\xi + \eta$ is the convolution of the distributions of ξ and η, i.e., we have $P_{\xi+\eta} = P_\xi * P_\eta$.

An important area of measure theory on topological vector spaces is related to the study of convergence of series of random vectors. Let us mention a typical result. For a sequence of random vectors η_n in a locally convex space X we can consider its convergence to a random vector η almost everywhere, in probability (i.e., convergence in probability to zero of the sequence $q(\eta - \eta_n)$ for every continuous seminorm q on X) and in distribution, i.e., weak convergence of the distributions of η_n to the distribution of η.

5.12.8. Theorem. *Let $\{\xi_n\}$ be a sequence of independent random vectors in a separable Fréchet space F. Then, for the series of ξ_n, convergence almost everywhere, convergence in probability and convergence in distribution are equivalent.*

The proof of this fact, easily reducing to the case of Banach spaces due to the existence of a common Banach support of the distributions of ξ_n, can be found in Buldygin [**90**, p. 90], Vakhania, Tarieladze, Chobanyan [**538**, Chapter V, §2, Theorem 2.3], Kruglov [**301**, p. 121]. In the case of symmetric distributions a somewhat stronger assertion holds.

5.12.9. Theorem. *Let $\{\xi_n\}$ be a sequence of independent random vectors with symmetric distributions in a separable Fréchet space F. Then, convergence of the series of ξ_n almost everywhere is equivalent to the uniform tightness of the distributions of the partial sums of this series.*

In addition, it is equivalent to the existence of a random element ζ with values in F such that, for every functional g from some set $G \subset F'$ separating points in F, the series of $g(\xi_n)$ converges almost everywhere to $g(\zeta)$.

In this circle of problems the following Lévy inequality is often used (for a proof, see [**538**, p. 264]).

5.12.10. Theorem. *Suppose that a sequence of random vectors ξ_n in a locally convex space X is sign invariant, i.e., for every collection of numbers $\theta_n = \pm 1$ the* *sequence* $\{\theta_n\xi_n\}$ *has the same distribution in* X^∞ *as* $\{\xi_n\}$. *Let* $S_n = \xi_1+\cdots+\xi_n$. *Then, for every set* $B \in \sigma(X')$, *the following inequalities hold*:

$$\mathrm{P}\Bigl(\bigcup_{k=1}^{n}\{S_k \notin (B+B)/2\}\Bigr) \leqslant 2\mathrm{P}(S_n \notin B),$$

$$\mathrm{P}\Bigl(\bigcup_{k=1}^{n}\{\xi_k \notin (B+B)/2\}\Bigr) \leqslant 2\mathrm{P}(S_n \notin B).$$

5.12(ii). 0–1 **laws**

We have already encountered the so-called 0–1 laws for Gaussian measures (Theorem 5.7.10). The assertion for affine spaces is a particular case of a more general result for stable measures discussed below, but can be also deduced from the second assertion about invariant sets, which, in turn, can be obtained from the following Kolmogorov theorem concerning product-measures, i.e., products of measures. Suppose we are given probability spaces $(X_i, \mathcal{A}_i, \mu_i)$. Their product $X = \prod_{i=1}^\infty X_i$ is equipped with the σ-algebra $\mathcal{A} = \bigotimes_{i=1}^\infty \mathcal{A}_i$ and the measure $\mu = \bigotimes_{i=1}^\infty \mu_i$. Let $\mathcal{X}_n := \bigotimes_{i=n+1}^\infty \mathcal{A}_i$ and $\mathcal{X} := \bigcap_{n=1}^\infty \mathcal{X}_n$, where sets in $\mathcal{X}_n$ are naturally identified with subsets of X (by writing them as $X_1\times\cdots\times X_n\times A$).

5.12.11. Theorem. *For every* $E \in \mathcal{X}$ *we have either* $\mu(E) = 1$, *or* $\mu(E) = 0$. *In particular, every* $\mathcal{X}$*-measurable function is equal* μ*-a.e. to some constant.*

In the case of identical measures the following result due to Hewitt and Savage holds (the proofs of both theorems can be found in Bogachev [**65**, § 10.10(iv)]).

5.12.12. Theorem. *Let* $(X_n, \mathcal{A}_n, \mu_n) = (X_1, \mathcal{A}_1, \mu_1)$ *for all* n, *let* $\mathcal{S}_n$ *be the* σ*-algebra generated by all* $\mathcal{A}$*-measurable functions invariant with respect to all permutations of the variables* $x_1, \ldots, x_n$, *and let* $\mathcal{S} := \bigcap_{n=1}^\infty \mathcal{S}_n$. *Then for all* $E \in \mathcal{S}$ *we have either* $\mu(E) = 1$, *or* $\mu(E) = 0$. *In particular, every* $\mathcal{S}$*-measurable function is equal* μ*-a.e. to some constant.*

5.12.13. Corollary. *Suppose that in the previous theorem* f *is a* μ*-measurable function such that for every permutation* σ *of* $\mathbb{N}$ *changing only finitely many elements we have* $f(x) = f\bigl(\sigma(x)\bigr)$ μ*-a.e., where* $\sigma(x) = (x_{\sigma(1)}, x_{\sigma(2)}, \ldots)$. *Then this function is equal to some constant* μ*-a.e.*

PROOF. The set Σ of all permutations of the indicated form is countable. For every $\sigma \in \Sigma$ there is a set $\Omega_\sigma \in \mathcal{A}$ of full measure on which $f(x) = f\bigl(\sigma(x)\bigr)$ and the restriction of the function f to which is $\mathcal{A}$-measurable. It is readily verified that the class $\mathcal{A}$ is invariant with respect to the transformations of X generated by permutations in Σ and that the measure μ is also invariant with respect to such transformations. Hence the intersection of countably many sets obtained from all Ω_σ by all possible transformations by means of different permutations in Σ belongs to $\mathcal{A}$, has full measure and is invariant with respect to permutations in Σ. On this intersection Ω the function f is $\mathcal{A}$-measurable and invariant with respect to permutations in Σ. Outside of Ω we set $f = 0$. The obtained function coincides with a constant μ-a.e., whence our claim follows. □

These assertions will be applied in the situation where there is a vector structure. The following result was obtained in Smolyanov [**481**]. Suppose we are given a family of linear spaces E_α, $\alpha \in \mathfrak{A}$, equipped with σ-algebras $\mathcal{A}_\alpha$ and probability measures μ_α on $\mathcal{A}_\alpha$. Note that the linear structure in E_α need not be compatible with the measurable structure. Set $E = \prod_\alpha E_\alpha$, $\mu = \bigotimes_\alpha \mu_\alpha$, $\mathcal{A} = \bigotimes_\alpha \mathcal{A}_\alpha$. We recall that an affine subspace of a linear space E is a set of the form $L + v$, where L is a linear subspace in E, $v \in E$.

5.12.14. Theorem. *Suppose that every affine subspace in* $(\mathcal{A}_\alpha)_{\mu_\alpha}$ *has measure either* 0 *or* 1. *Then the same is true for affine subspaces in* $\mathcal{A}_\mu$.

PROOF. Let $M \in \mathcal{A}_\mu$ be an affine subspace with $\mu(M) > 0$. We first consider the case $\mathfrak{A} = \mathbb{N}$. We show that for every n there exists an affine subspace M_n in E such that $M_n = \prod_{i=1}^\infty L_i$, where $L_i \in (\mathcal{A}_i)_{\mu_i}$ and $\mu_i(L_i) = 1$, and the restriction of the function I_M to M_n does not depend on the first n variables. For $n = 1$ we write E as $E_1 \times Y$ and μ as $\mu_1 \otimes \nu$. By Fubini's theorem, there exists $y \in Y$ for which $\mu_1(M_y) > 0$, $M_y = \{x \in E_1\colon (x, y) \in M\}$. Since M_y is an affine subspace (a simple proof is delegated to Exercise 4.10.38), by assumption we have $\mu_1(M_y) = 1$. Set $M_1 = M_y \times Y$. If $(x, z) \in M_1 \cap M$ and $(x', z) \in M_1$, then $(x, y) \in M$, $(x', y) \in M$, hence $(x' - x, 0)$ belongs to a linear space L representing M in the form $M = L + v$, $v \in E$. Hence $(x', z) = (x' - x, 0) + (x, z) \in M$. Thus, the restriction of I_M to M_1 does not depend on x_1. We find such an affine subspace $M_{(i)}$ for every $i \leqslant n$ and set $M_n = M_{(1)} \cap \cdots \cap M_{(n)}$, $M_{(1)} := M_1$. We have obtained an affine subspace of the required form.

Independence of the restriction of I_{M_n} to M on $x_1, \ldots, x_n$ is verified as follows. Suppose that $x \in M_n \cap M$, $y \in M_n$, $x_i = y_i$ if $i > n$. Let us consider the point $x' = (y_1, x_2, \ldots, x_n, \ldots)$. Since $x, y \in M_{(1)}$, where $M_{(1)}$ is the product of an affine subspace in E_1 with the remaining spaces E_i, we have $x' \in M_{(1)}$, whence $x' \in M$, since $x' = x - \widetilde{x_1} + \widetilde{y_1}$, $\widetilde{x_1} = (x_1, 0, 0, \ldots)$, $\widetilde{y_1} = (y_1, 0, 0, \ldots) \in M_{(1)}$. Continuing by induction, we obtain that $y \in M$, hence $I_{M_n}(y) = I_{M_n}(x)$.

The set $M_\infty = \bigcap_{n=1}^\infty M_n$ is an affine subspace of full measure. Considering the measure μ on M_∞ and applying Kolmogorov's 0–1 law, we obtain that $\mu(M) = 1$, since $\mu(M) > 0$.

Let us turn to uncountable $\mathfrak{A}$. Then there exist a countable part $\mathfrak{S} \subset \mathfrak{M}$ and a set $B \in \bigotimes_{\beta \in \mathfrak{S}} \mathcal{A}_\beta$ for which $M \subset B \times Y$, $Y = \prod_{\alpha \notin \mathfrak{S}} E_\alpha$, $\mu(M) = \mu(B \times Y)$. Set $X = \prod_{\beta \in \mathfrak{S}} E_\beta$ and denote by P the natural projection of E onto X. Consider the set $M' = P^{-1}\bigl(P(M)\bigr)$. It is clear that $M \subset M' \subset B \times Y$, hence M' is measurable and $\mu(M') = \mu(M)$. In addition, $M' = P(M) \times Y$, hence $P(M)$ coincides with B up to a set of measure zero with respect to the projection of the measure μ on X. Thus, $P(M)$ has a positive measure with respect to this projection. It is readily seen that $P(M)$ is an affine subspace in X; according to what has been proved above it has full measure, hence the same is true for M. □

The hypothesis of this theorem is satisfied by any Borel probability measures on finite-dimensional spaces equivalent to Lebesgue measures, by Gaussian measures, by convex measures considered below, and by symmetric stable measures. The situation becomes especially simple when all spaces E_α are one-dimensional and equipped with Borel probability measures μ_α. Then a necessary and sufficient condition for the validity of the 0–1 law for measurable affine subspace in E is that every measure μ_α either has no points of positive measures or is Dirac's measure at a point.

We also observe that if for a measure μ on a linear space E with a σ-algebra $\mathcal{E}$ the 0–1 law holds for measurable affine subspaces and T is a measurable affine mapping from E to a linear space F with a σ-algebra $\mathcal{F}$, then this law also holds for the measure $\mu\circ T^{-1}$. In this way one can considerably enlarge the number of examples, using the aforementioned ones.

Let us give one more result related to linear supports and following from Proposition 7.14.58 in Bogachev [**65**].

5.12.15. Proposition. *If a probability measure μ on the σ-algebra $\sigma(X')$ in a locally convex space X is not concentrated on a closed hyperplane, then, for every set $M \subset X'$ that is convex and compact in the topology $\sigma(X', X)$, this topology is metrizable on M and coincides with the topology of convergence in measure μ (e.g., this is true for the polars of neighborhoods of zero in the Mackey topology on X).*

Since in the case of a Radon measure the intersection of all closed linear subspaces of full measure also has full measure, we can pass to the situation described in this proposition.

5.12.16. Corollary. *If X is a metrizable locally convex space and μ is a Radon measure with respect to the weak topology on X, then it is tight with respect to the original metric on X.*

PROOF. As noted above, we can pass to the case of a measure not concentrated on a hyperplane. Taking a basis of convex neighborhoods of zero U_n in X, we obtain from the previous proposition that their polars U_n° are metrizable compact sets in the topology $\sigma(X', X)$, which gives separability of X' in this topology. Hence there is a countable collection $\{f_n\} \subset X'$ separating points in X. Hence weakly compact sets in X are metrizable. The measure μ is concentrated on some countable union of such sets K_n. Let us take in each K_n a countable set dense in the weak topology. The obtained countable set S is dense in X with respect to the original metric, since otherwise we could find a closed hyperplane containing S, hence also all K_n, which is impossible due to our assumption about the measure. Therefore, the Borel σ-algebra of X with respect to the metric coincides with $\sigma(X')$ according to Proposition 5.2.1. The Borel extension of μ to the completion of X (as a metrizable locally convex space) is a Radon measure by separability of X, moreover, the sets K_n are closed in the completion by their weak compactness in X. Hence the measure μ is Radon with respect to the metric. □

5.12(iii). Convex measures

Here we briefly discuss an important class of measures called convex or logarithmically concave that includes Gaussian measures. Convexity of a probability measure on $\mathbb{R}^n$ is equivalent to the property that on some affine subspace it is given by a density of the form $\exp(-V)$, where V is a convex function with values in $(-\infty, +\infty]$ (see Borell [**80**]).

5.12.17. Definition. *A probability measure μ defined on the σ-algebra $\sigma(X')$ in a locally convex space X is called convex (or logarithmically concave) if for all nonempty sets $A, B \in \sigma(X')$ and all $\alpha \in [0,1]$ one has*

$$\mu_*(\alpha A + (1-\alpha)B) \geqslant \mu(A)^\alpha \mu(B)^{1-\alpha}.$$

A Radon probability measure μ is convex if its restriction to $\sigma(X')$ is convex.

5.12.18. Lemma. *A Radon probability measure μ is convex if (and only if) all its finite-dimensional projections are convex.*

PROOF. Taking for A and B cylindrical sets, we obtain convexity of all finite-dimensional projections. Conversely, let all such projections be convex and let $A, B \in \mathcal{B}(X)$. Since μ is Radon, it suffices to consider the case where A and B are compact. Then A and B are compact in the weak topology, and since μ is Radon, for any $\varepsilon > 0$ one can find an open cylindrical set C such that $\alpha A + (1-\alpha)B \subset C$ and $\mu(C) < \mu(\alpha A + (1-\alpha)B) + \varepsilon$. Using compactness of A and B once again, we can find a convex cylindrical neighborhood of zero V such that $\alpha(A+V) + (1-\alpha)(B+V) \subset C$. The sets $A+V$ and $B+V$ are cylinders. By convexity of finite-dimensional projections

$$\mu(C) \geqslant \mu(\alpha(A+V) + (1-\alpha)(B+V)) \geqslant \mu(A+V)^\alpha \mu(B+V)^{1-\alpha} \\ \geqslant \mu(A)^\alpha \mu(B)^{1-\alpha},$$

which gives the desired estimate, since ε was arbitrary. □

5.12.19. Corollary. (i) *If μ is a convex Radon probability measure on a locally convex space X and $T\colon X \to Y$ is a continuous linear mapping to a locally convex space Y, then the measure measure $\mu\circ T^{-1}$ is convex.*

(ii) *If μ is a convex Radon probability measure on a locally convex space X and ν is a convex Radon probability measure on a locally convex space Y, then $\mu\otimes\nu$ is a convex measure on $X\times Y$.* (iii) *If $X = Y$, then $\mu * \nu$ is a convex measure.*

Let us also mention the following results from Borell [**80**], [**82**] (their proofs can be also found in Bogachev [**66**, § 4.3]).

5.12.20. Theorem. *Let μ be a convex Radon measure on a locally convex space X and let p be a seminorm on X measurable with respect to μ. Then there is $c > 0$ such that $\exp(cp) \in L^1(\mu)$. In particular, $p \in L^r(\mu)$ for all $r \in (0, \infty)$.*

5.12.21. Theorem. *Let μ be a convex Radon measure on a locally convex space X, let $h \in X$ be a nonzero vector, and let Y be a closed hyperplane for which $X = Y \oplus \mathbb{R}^1 h$. Then one can choose convex probability measures μ^y on the straight lines $y + \mathbb{R}^1 h$, $y \in Y$, such that*

$$\mu(B) = \int_Y \mu^y(B)\, \nu(dy), \quad B \in \mathcal{B}(X),$$

where ν is the image of μ under the natural projection $X \to Y$.

Borell [**80**] (see also [**83**]) established the following 0–1 law for convex measures.

5.12.22. Theorem. *Let μ be a convex Radon measure on a locally convex space X.*

(i) *Let G be an additive subgroup in X. Then either $\mu_*(G)=0$ or $\mu_*(G)=1$. If $G\in\mathcal{B}(X)_\mu$, then either $\mu(G)=0$ or $\mu(G)=1$.*

(ii) *Suppose that for every $f\in X'$ the support of the measure $\mu\circ f^{-1}$ is either a point or the whole real line. Then the topological support of μ is a closed affine subspace.*

PROOF. (i) Let $0<\mu_*(G)<1$. Let us take a compact set $K_0\subset G$ with $\mu(K_0)>0$. Set $K=K_0\cup(-K_0)$ and $H=\bigcup_{n=1}^\infty K_n$, $K_n=K_{n-1}+K$, $K_1=K$, i.e., H is the smallest additive subgroup containing K. Then $H\subset G$ and $\mu(H)\leqslant\mu_*(G)<1$. Let $\varepsilon=2^{-1}\min\bigl(\mu(K),1-\mu(H)\bigr)$. Let us take compact sets $Q\subset X\backslash H$ and $S\subset X$ such that

$$\mu(Q)>1-\mu(H)-\varepsilon,\quad \mu(X\backslash S)<2^{-1}\bigl(1-\mu(H)-\varepsilon\bigr). \tag{5.12.1}$$

It is clear that $0\notin K+Q$, since otherwise $Q\cap K\neq\varnothing$. Hence there exists $n\in\mathbb{N}$ such that

$$H_n:=(n-1)K+nQ\subset X\backslash S. \tag{5.12.2}$$

Indeed, if there are $k_n\in K$ and $q_n\in Q$ such that $(n-1)k_n+nq_n\in S$, then by the boundedness of S and K we obtain $k_n+q_n\to0$, whence $0\in K+Q$ by the closedness of $K+Q$. We now use the inclusion

$$n^{-1}\bigl(X\backslash(H\cup H_n)\bigr)+(1-n^{-1})K\subset X\backslash(H\cup Q).$$

It is verified as follows. If $x\in H\cup Q$ has the form $n^{-1}y+(1-n^{-1})k$, where $y\in X\backslash(H\cup H_n)$, then in case $x\in H$ we obtain $y+(n-1)k\in H$, whence $y\in H$, which is impossible, and in case $x\in Q$ we have $y\in nQ-(n-1)K=H_n$, which is also impossible. The establish inclusion and convexity of μ yield the estimate

$$\mu\bigl(X\backslash(H\cup Q)\bigr)\geqslant\min\Bigl(\mu\bigl(X\backslash(H\cup H_n)\bigr),\mu(K)\Bigr).$$

Since the left side is $1-\mu(H)-\mu(Q)<\varepsilon<\mu(K)$ by (5.12.1), we obtain the inequality

$$\varepsilon>\mu\bigl(X\backslash(H\cup Q)\bigr)\geqslant\mu\bigl(X\backslash(H\cup H_n)\bigr)\geqslant1-\mu(H)-\mu(H_n),$$

giving the estimate $\mu(H_n)\geqslant1-\mu(H)-\varepsilon>0$. However, (5.12.1) and (5.12.2) imply the inequality $\mu(H_n)\leqslant\mu(X\backslash S)<2^{-1}\bigl(1-\mu(H)-\varepsilon\bigr)$. The obtained contradiction means that $\mu(H)=1$. Hence $\mu_*(G)=1$.

(ii) Being a Radon measure, μ has topological support S. Convexity of μ yields convexity of S: if $a,b\in S$, then for every convex neighborhood of zero V the sets $V+a$ and $V+b$ have positive measures, whence

$$\mu\bigl(V+(a+b)/2\bigr)\geqslant\mu(V+a)^{1/2}\mu(V+b)^{1/2}>0,$$

i.e., $(a+b)/2\in S$. Therefore, the whole interval $[a,b]$ is contained in S. If $S\neq X$, then for every point $z\notin S$ there exists a functional $f\in X'$ with $f|_S\leqslant1$ and $f(z)>1$. Our assumption implies that f coincides almost everywhere with some constant c_f. Hence S is contained in the closed affine subspace $H_f = f^{-1}(c_f)$. It follows from this that S coincides with the intersection of all closed affine subspaces of full measure, since this intersection H has full measure by the Radon property of μ, and the restriction of μ to H cannot have a smaller support. □

5.12(iv). The central limit theorem

Several interesting classes of measures on infinite-dimensional spaces are introduced by means of independent random vectors or convolutions. In this and the next subsections we consider some examples; here we discuss the central limit theorem (abbreviation: CLT).

Let X be a locally convex space and let $\{\xi_n\}$ be a sequence of X-valued independent centered random vectors with a common Radon distribution μ. Set

$$S_n = \frac{\xi_1 + \ldots + \xi_n}{\sqrt{n}}.$$

Note that the distribution of ξ_n coincides with the measure μ^{*n} defined by the equality

$$\mu^{*n}(A) = (\mu * \ldots * \mu)(n^{-1/2}A),$$

where the convolution is n-fold. The first problem the central limit theorem deals with concerns convergence of the sequence of random vectors S_n (in a suitable sense).

5.12.23. Definition. (i) *A probability measure μ with mean m on $\sigma(X')$ is called pre-Gaussian if it has weak second moment and there is a Gaussian measure γ with mean m on X such that*

$$\int_X f\, g\, d\mu = \int_X f\, g\, d\gamma \quad \forall\, f, g \in X'.$$

(ii) *A Radon probability measure μ with zero mean on X satisfies the central limit theorem* (CLT) *if the sequence $\{\mu^{*n}\}$ is uniformly tight. A Radon probability measure μ with mean m is said to satisfy the* CLT *if the measure μ_{-m} with zero mean satisfies the* CLT.

(iii) *A space X is called a space with the* CLT *property if every Radon probability measure μ on X with zero mean and strong second moment satisfies the* CLT; *X is called a space with the strict* CLT *property if the* CLT *holds for every Radon probability measure μ on X with zero mean and weak second moment.*

Locally convex spaces with the strict CLT property were introduced in Bogachev [**60**]. Note that even for measures on the real line the CLT requires finite second moments (see Bogachev [**69**, Chapter 1]), so in relation to the CLT we consider only measures with weak second moment.

5.12.24. Lemma. *Let μ be a Radon probability measure with zero mean on X and weak second moment. If the sequence $\{\mu^{*n}\}$ is uniformly tight, then it converges weakly to some centered Radon Gaussian measure γ. In addition, μ is a pre-Gaussian measure.*

PROOF. The uniform tightness of $\{\mu^{*n}\}$ implies the existence of a Radon probability measure γ that is a limit point for the set $\{\mu^{*n}\}$ in the weak topology. We verify that γ is a Gaussian measure. It suffices to verify that all measures $\mu\circ f^{-1}$, where $f \in X'$, are Gaussian on the real line. Thus, the assertion reduces to the one-dimensional case, in which it is well known (see [**69**]). □

On $\mathbb{R}^n$ every probability measure with weak second moment satisfies the CLT. Certainly, such a measure has strong second moment. The situation is completely different in the infinite-dimensional case. For example, the space $C[0,1]$ does not have the CLT property. Moreover, there exists a pre-Gaussian measure with compact support in $C[0,1]$ not satisfying the CLT. On the other hand, there exists a measure with compact support in $C[0,1]$ which is not pre-Gaussian. Finally, there exists a measure on $C[0,1]$ with weak second moment which satisfies the CLT, but has no strong second moment (concerning such examples, see Paulauskas, Rachkauskas [**380**]). It is known that Hilbert spaces have the CLT property. Since in a Hilbert space the covariance operator of a probability measure μ is nuclear precisely when μ has strong second moment, here the class of pre-Gaussian measures coincides with the class of measures satisfying the CLT (and also with the class of probability measures having strong second moment). As the example of $C[0,1]$ shows, these three classes of measures can be different in general Banach spaces. The coincidence of these classes characterizes Hilbert spaces. In other words, a Banach space is linearly homeomorphic to a Hilbert space if and only if the existence of strong second moment of a probability measure is equivalent to the property that it satisfies the CLT. It is known that every probability measure with strong second moment on a Banach space X satisfies the CLT precisely when X is a space of type 2 (for the definitions of type and cotype, see Vakhania, Tarieladze, Chobanyan [**538**]). Thus, on non-Hilbert spaces of type 2 there are measures having weak second moment and satisfying the CLT, but not having strong second moment. If every measure on X satisfying the CLT and having weak second moment has strong second moment, then X is a space of cotype 2, moreover, this property completely characterizes spaces of cotype 2. The existence of cotype 2 is equivalent also to the property that every pre-Gaussian measure on X satisfies the CLT. The proofs of these assertions and the corresponding references can be found in [**380**, Chapter 3] and Ledoux, Talagrand [**311**, Chapter 10].

Let us mention two results from Bogachev [**60**]; their proofs can be found in Bogachev [**69**, § 4.8(iii)] (see also Exercise 5.12.93).

5.12.25. Theorem. (i) *A Banach space has the strict* CLT *property precisely when it is finite-dimensional.*

(ii) *The strict* CLT *property is inherited by closed subspaces and preserved under formation of strict inductive limits of increasing sequences of closed subspaces, countable products, arbitrary direct sums, and countable projective limits.*

5.12.26. Example. Let X be the dual to a complete nuclear barrelled locally convex space Y. Then X with the strong topology possesses the strict CLT property. For example, this is true if X is the dual to a nuclear Fréchet space.

The following spaces have the strict CLT property: $C_0^\infty[a,b]$, $\mathcal{S}(\mathbb{R}^k)$, $\mathcal{S}'(\mathbb{R}^k)$, and $\mathbb{R}^\infty$.

In the next subsection we consider another normalization of sums of independent random vectors.

5.12(v). Infinitely divisible and stable measures

5.12.27. Definition. *A random vector ξ in a locally convex space X with distribution μ is called stable of order $\alpha \in (0,2]$ if, for every n, there exists a vector $a_n \in X$ such that, for independent random vectors $\xi_1,\ldots,\xi_n$ with the same distribution μ, the random vector $n^{-1/\alpha}(\xi_1+\cdots+\xi_n)-a_n$ also has distribution μ. The distribution μ is called stable of order α. If one can take $a_n = 0$, then μ is called strictly stable.*

A probability measure μ on $\sigma(X')$ is called infinitely divisible if, for each n, there exists a probability measure ν_n on $\sigma(X')$ such that μ coincides with the n-fold convolution of ν_n.

Every stable of order α measure μ is infinitely divisible, since it equals the n-fold convolution of the distribution of the random vector $n^{-1/\alpha}(\xi - n^{1/\alpha-1}a_n)$. Stable of order 2 random vectors are precisely Gaussian random vectors.

In terms of the Fourier transform of μ the property to be infinitely divisible means that $\widetilde{\mu} = (\widetilde{\nu_n})^n$ for every n, and the property to be stable is equivalent to the relation $\widetilde{\mu}(l) = e^{-il(a_n)}\widetilde{\mu}(n^{-1/\alpha}l)^n$.

Let us consider a measure of the form

$$\mu = e(\nu) := e^{-\nu(X)} \sum_{n=0}^{\infty} \frac{\nu^{*n}}{n!},$$

where ν is a finite nonnegative Radon measure on X, ν^{*n} is the n-fold convolution of ν, $\nu^{*0} := \delta_0$. Clearly, for the Fourier transform of this measure we obtain the equality $\widetilde{\mu} = e^{-\nu(X)}e^{\widetilde{\nu}}$. Thus, $\mu = \nu_k^{*k}$ for all k, where $\nu_k = e(\nu/k)$, hence μ is infinitely divisible.

General infinitely divisible measures are described by means of the so-called Poisson measures of the following type. Let B be an absolutely convex compact set in X and let G be a nonnegative Borel measure on X with values in $[0,+\infty]$ concentrated on the linear span of B, finite on all sets $X\backslash rB$, $r>0$, and such that $f^2/(1+f^2) \in L^1(G)$ for all $f \in X'$. Set

$$K(f,x) = e^{if(x)} - 1 - if(Tx),\ Tx = x \ \text{ if } x\in B,\ Tx = x/p_B(x) \ \text{ if } x\notin B.$$

Suppose that the function

$$\exp\biggl(\int_X K(f,x)\,G(dx)\biggr)$$

is the Fourier transform of a Radon probability measure Π on X. Then the measure Π is called the *Poisson* measure with spectral measure G (or with the *Lévy measure* G) and is denoted by the symbol $\Pi(G)$. Under broad conditions on the space X, infinitely divisible measures on X are precisely measures of the form $\mu = \gamma * \Pi(G)$, where γ is a Gaussian measure and G is some Lévy measure. For example, this is true for Fréchet spaces (the case of Banach spaces is considered in Kruglov [**301**], and the case of Fréchet spaces reduces to that of Banach spaces due to the existence of a Banach support, see § 5.12(vi)).

Tortrat [**526**] obtained the following representation of the Fourier transform of a stable measure μ.

5.12.28. Theorem. *Let μ be a stable of order α Radon measure on a quasi-complete locally convex space X. Then there exist a vector $a \in X$ and a nonnegative measure Γ with compact support in X for which*

$$\widetilde{\mu}(l) = \exp\Bigl(il(a) - \int |l(x)|^\alpha\,\Gamma(dx) + iQ(\alpha,\Gamma,l)\Bigr), \quad l \in X', \tag{5.12.3}$$

$$Q(\alpha,\Gamma,l) = \tan\frac{\pi\alpha}{2}\int l(x)|l(x)|^{\alpha-1}\,\Gamma(dx), \quad \alpha \neq 1,$$

$$Q(1,\Gamma,l) = -\frac{2}{\pi}\int l(x)\ln|l(x)|\,\Gamma(dx).$$

The measure Γ is called spectral. It is clear that it is not uniquely defined. For example, for $\alpha \neq 1$ it can be replaced by the measure $\lambda^{-\alpha}\Gamma_\lambda$, where Γ_λ is the image of Γ under the homothety with coefficient λ. Thus, one can take for Γ a probability measure. In Bogachev [**59**] a modified representation

$$\widetilde{\mu}(l) = \exp\Bigl(il(a) - \int |l(x)|^\alpha\,\Gamma(dx) + i\beta Q(\alpha,\Gamma,l)\Bigr), \quad \beta \in [0,1], \tag{5.12.4}$$

of the Fourier transform of the measure μ was considered, which better reflects symmetry properties of the measure. For example, if the measure μ is symmetric, i.e., $\mu(B) = \mu(-B)$, then $Q(\alpha,\Gamma,l) = 0$, hence one can find representation (5.12.4) with $\beta = 0$.

The *asymmetry index* $\beta(\mu)$ of a stable of order α measure μ is defined as the infimum of numbers $\beta \in [0,1]$ for which representation (5.12.4) exists with some $a \in X$ and some measure Γ with bounded support. The next result is proved in [**59**].

5.12.29. Theorem. *The asymmetry index of a stable measure μ equals the supremum of the asymmetry indices of finite-dimensional projections of μ. Moreover, there exist $a \in X$ and a measure Γ with compact support such that* (5.12.4) *holds with $\beta = \beta(\mu)$.*

If $\beta(\mu) = 0$, then the measure μ is symmetric and coincides with its symmetrization $\mu_s(B) = \mu * \overline{\mu}(2^{1/\alpha}B)$, where $\overline{\mu}(B) = \mu(-B)$. It is clear that $\widetilde{\mu_s} = |\mu|$. If $\beta(\mu) = 1$, then μ is called completely asymmetric. Finally, if $0 < \beta(\mu) < 1$, then there exists a completely asymmetric stable of order α measure ν such that $\mu = \mu_1 * \nu$, where $\mu_1(B) = \mu_s(t^{-1/\alpha}B)$, $t = 1 - \beta(\mu)$.

For stable measures, the following version of the 0–1 law was obtained in Dudley, Kanter [**141**].

5.12.30. Theorem. *Let μ be a stable Radon measure on a locally convex space X. Then, for every μ-measurable affine subspace $L \subset X$, either $\mu(L) = 0$ or $\mu(L) = 1$.*

PROOF. Passing to a shift of the measure, we can assume that L is a linear subspace. Suppose first that μ is symmetric. Let $\mu(L) = a > 0$. We can assume that L is Borel, since L contains a compact set of positive measure, hence we can deal with its linear span. Let μ^t denote the image of μ under the homothety $x \mapsto tx$, $t > 0$. We observe that $\widetilde{\mu^t} = \widetilde{\mu}^{t^\alpha}$. Hence for all $t \in (0,1)$ we have the equality $\mu = \mu^t * \mu^s$, where $s^\alpha = 1 - t^\alpha$. Taking $t = n^{-1}$, we obtain

$$\begin{aligned}\mu(L - n^{-1}x) &= \int_X \mu(nL - x - ns^{-1}y)\,\mu(dy)\\ &\geqslant \int_L \mu(nL - x - ns^{-1}y)\,\mu(dy) = \mu(L - x)\mu(L)\end{aligned}$$

for all x. If $x \notin L$, then the sets $L - n^{-1}x$ are disjoint, whence it follows that $\mu(L - x) = 0$ for $x \notin L$. Then by the equality $\mu * \mu = \mu^\theta$, where $\theta = 2^{1/\alpha}$, we obtain

$$a = \mu(L) = \mu(\theta L) = \mu * \mu(L) = \int_X \mu(L - x)\,\mu(dx) = \int_L \mu(L - x)\,\mu(dx) = a^2,$$

whence $a = 1$. Let us now drop the assumption that the measure is symmetric. Let $\nu = \mu * \overline{\mu}$, $\overline{\mu}(B) = \mu(-B)$. Since $\nu(L) \geqslant \mu(L)^2$, by the proven assertion we have $\nu(L) = 1$. Hence $\mu(L + x) = 1$ for μ-a.e. x. In particular, such x exists in the set L of positive measure, which gives $\mu(L) = 1$. □

A similar assertion is true for stable measures on $\sigma(X')$.

5.12.31. Lemma. *Let μ be a Radon probability measure on a locally convex space X, let m be a Radon probability measure on a topological space $(T, \mathcal{B})$, and suppose that for every $t \in T$ we are given a cylindrical probability measure ν_t on X with the following property: for every $l \in X'$ the function $t \mapsto \widetilde{\nu_t}(l)$ is measurable with respect to m and*

$$\widetilde{\mu}(l) = \int_T \widetilde{\nu_t}(l)\,m(dt).$$

Suppose also that for any $\varepsilon > 0$ there is a compact set $S_\varepsilon \subset T$ with $m(S_\varepsilon) > 1 - \varepsilon$ such that for every cylinder C with a compact base the function $t \mapsto \nu_t(C)$ on S_ε is upper semicontinuous (this holds automatically if there is a countable set in X' separating points in X). Then, for m-a.e. t, the measure ν_t has a Radon extension such that for every $B \in \mathcal{B}(X)$ the function $t \mapsto \nu_t(B)$ is measurable with respect to m and

$$\mu(B) = \int_T \nu_t(B)\,m(dt). \tag{5.12.5}$$

PROOF. It follows from our condition that for any function f which is a linear combination of $\exp(il)$, where $l \in X'$, the integral of f against the measure ν_t is a measurable function of t and

$$\int_X f\,d\mu = \int_T \int_X f(x)\,\nu_t(dx)\,m(dt).$$

It is easy to extend this assertion to bounded functions f of the form $g\circ P$, where P is a finite-dimensional projection and g is a Borel function on $P(X)$. Hence equality (5.12.5) is true for cylinders B.

Let $n \in \mathbb{N}$. Let us fix a compact set $K \subset X$. Let us consider the set $\mathcal{H}$ of all closed subspaces in X of finite codimension. It is directed by the natural partial order relation: $H_1 \leqslant H_2$ if $H_2 \subset H_1$. The net of functions $f_H(t) = \nu_t(K+H)$ decreases to $f(t) = \inf_H f_H(t)$. On every compact set S_ε from the condition of the theorem the functions f_H are upper semicontinuous. Hence so is f. By [**65**, Lemma 7.2.6] we have

$$\int_{S_\varepsilon} f\, dm = \lim_H \int_{S_\varepsilon} f_H\, dm.$$

Since ε was arbitrary, we obtain the measurability of f and the equality

$$\int_T f\, dm = \lim_H \int_T f_H\, dm = \lim_H \mu(K+H) = \mu(K),$$

where the last relation follows from Proposition 5.2.17. Taking compact sets K_n with $\mu(K_n) \geqslant 1-n^{-2}$, for $T_n := \{t\colon\ \inf_H \nu_t(K_n+H) \geqslant 1-n^{-1}\}$ we obtain the estimate

$$m(T_n) + (1-n^{-1})\bigl(1-m(T_n)\bigr) \geqslant \int_T f\, dm \geqslant 1-n^{-2},$$

whence we see that $m(T_n) \geqslant 1-n^{-1}$. The set $T_0 = \bigcup_{n=1}^\infty T_n$ has full m-measure and $\inf_H \nu_t(K_n+H) \geqslant 1-n^{-1}$ for $t \in T_0$. This means that the measures ν_t are tight for $t \in T_0$: for every Borel cylinder C disjoint with K_n, we have $\nu_t(C) \leqslant n^{-1}$, since there exists $H \in \mathcal{H}$ with $(K_n+H)\cap C = \varnothing$. The Radon extensions of the measures ν_t will be denoted by the same symbols. According to what has been proved above, formula (5.12.5) is true for compact sets B, since $\nu_t(B) = \inf_H \nu_t(B+H)$ for $t \in T_0$. Using that μ is Radon, it is easy to extend this to closed sets B. It is clear that the class $\mathcal{B}$ of all sets $B \in \mathcal{B}(X)$ for which (5.12.5) is true admits countable unions of disjoint sets and that if $B_1, B_2 \in \mathcal{B}$ and $B_1 \subset B_2$, then $B_2\backslash B_1 \in \mathcal{B}$. Since the class of closed sets belongs to $\mathcal{B}$ and admits finite intersections, by the known theorem on σ-additive classes (see [**65**, Theorem 1.9.3]) we have $\mathcal{B} = \mathcal{B}(X)$.

Finally, if there is a countable collection $\{l_n\} \subset X'$ separating points in X, then the indicated additional condition of upper semicontinuity is automatically fulfilled. Indeed, for every n choose a countable set $\mathcal{F}_n$ of smooth functions on $\mathbb{R}^n$ with compact support by which one can uniformly approximate all continuous functions with compact support. It will be convenient to assume below that $\mathcal{F}_n$ is closed with respect to the operations $\min(f,1)$, $\max(f,0)$, and $\min(f_1,\ldots,f_n)$, where $f, f_i \in \mathcal{F}_n$. The set $\mathcal{F}$ of functions of the form $f(l_{i_1},\ldots,l_{i_n})$, $f \in \mathcal{F}_n$, is countable. Hence for every $\varepsilon > 0$ there exists a compact set $S_\varepsilon \subset T$ such that for all $f \in \mathcal{F}$ the functions

$$t \mapsto \int f\, d\nu_t$$

are continuous on S_ε. Let C be a cylinder with a compact base in a finite-dimensional subspace L and let $H \in \mathcal{H}$. We verify the upper semicontinuity of the function $g\colon t \mapsto \nu_t(C+H)$ on S_ε. By the continuity on S_ε of all functions in $\mathcal{F}$, it suffices to find a sequence of functions $f_n \in \mathcal{F}$ pointwise decreasing to g. This reduces to the following: given a compact set $K \subset \mathbb{R}^k$, we have to find functions $f_n \in \mathcal{F}_n$ pointwise decreasing to I_K. Let K_n be the open n^{-1}-neighborhood of K. Let us take a continuous function ψ_n with $0 \leqslant \psi_n \leqslant 1$, equal to 1 on K and 0 outside of K_n. Find $\varphi_n \in \mathcal{F}_n$ with $\sup_x |\psi_n(x) - \varphi_n(x)| \leqslant n^{-1}$; we can assume that $0 \leqslant \varphi_n \leqslant 1$, passing to the functions $\min(\varphi_n, 1)$ and then to $\max(\varphi_n, 0)$. Finally, take $\min(\varphi_1, \ldots, \varphi_n)$. □

The following result from Sztencel [**508**] shows that the distributions of symmetric stable vectors are mixtures of Gaussian measures.

5.12.32. Theorem. *Let μ be a Radon probability measure on a locally convex space X that is symmetric and stable of order α. Then, there exist a Radon probability measure m on the space $T = X^\infty \times \mathbb{R}^\infty$ and symmetric Radon Gaussian measures γ_t on X, where $t \in R$, such that for every $B \in \mathcal{B}(X)$ the function $t \mapsto \gamma_t(B)$ is measurable with respect to m and*

$$\mu(B) = \int_T \gamma_t(B)\, m(dt), \quad B \in \mathcal{B}(X). \tag{5.12.6}$$

PROOF. We shall assume that $\alpha < 2$, since for $\alpha = 2$ the measure μ is Gaussian. In addition, we first suppose that X is quasi-complete. By the Tortrat theorem we have (5.12.3) with some probability measure Γ concentrated on a compact set K and $a = 0$, $Q = 0$. Let $\{\xi_i\}$ be a sequence of independent centered Gaussian random variables with $\mathbb{E}|\xi_i|^\alpha = 1$, $\{\eta_i\}$ a sequence of independent random variables with exponential distribution, $Z, Z_1, \ldots$ a sequence of independent random vectors in X with distribution Γ such that all these sequences are jointly independent. We can assume that they are defined on the space $\Omega = \mathbb{R}^\infty \times \mathbb{R}^\infty \times X^\infty$ with the product of the corresponding one-dimensional distributions. Let us set $s_k = \eta_1 + \cdots + \eta_k$. Then $\mathbb{E}s_k = k$. For a measure m we shall take $m_1 \otimes m_2$, where m_1 is the distribution of $\{Z_i\}$ and m_2 is the distribution of $\{s_k\}$. Thus, our probability space is $T = X^\infty \times \mathbb{R}^\infty$, where we identify $\mathbb{R}^\infty \times \mathbb{R}^\infty$ and $\mathbb{R}^\infty$. Let

$$C_\alpha = \int_0^\infty s^{-\alpha} \sin s\, ds.$$

The following fact is known (see LePage, Woodroffe, Zinn [**314**]): given a sequence of equidistributed symmetric random variables ζ_k jointly independent with the sequence $\{\eta_k\}$ and having finite moment of order α, the series $\sum_{k=1}^\infty s_k^{-1/\alpha}\zeta_k$ converges almost surely and its sum has a stable of order α distribution with the Fourier transform given by $y \mapsto \exp\bigl(-C_\alpha^{-1}|y|^\alpha \mathbb{E}|\zeta_1|^\alpha\bigr)$. We now consider a formal series

$$C_\alpha^{1/\alpha} \sum_{k=1}^\infty s_k^{-1/\alpha} \xi_k Z_k.$$

This formal series "converges" in the following sense: for every $l \in X'$ the series

$$C_\alpha^{1/\alpha} \sum_{k=1}^{\infty} s_k^{-1/\alpha} \xi_k l(Z_k)$$

of real random variables converges almost surely and for the Fourier transform of its sum Λ_l the aforementioned fact yields the equality

$$\mathbb{E}\exp(i\Lambda_l) = \exp\Bigl(-\mathbb{E}|l(Z_1)|^\alpha\Bigr) = \exp\Bigl(-\int |l(x)|^\alpha\, \Gamma(dx)\Bigr) = \widetilde{\mu}(l). \quad (5.12.7)$$

By the law of large numbers there is a set $S \subset \mathbb{R}^\infty$ of full m_2-measure such that $\lim\limits_{k\to\infty} s_k/k \to 1$ for all $(s_k) \in S$. For every element $t \in X^\infty \times \mathbb{R}^\infty$ of the form $t = ((x_k), (s_k))$, where $x_k \in K$, we would like to define a centered Gaussian measure γ_t as the distribution of the random vector $C_\alpha^{1/\alpha} \sum_{k=1}^\infty s_k^{-1/\alpha} \xi_k x_k$.

If $\alpha < 1$, then it is easy to show that this series converges in X almost surely and its distribution is a centered Radon Gaussian measure. Indeed, we shall assume that the distribution of $\{\xi_k\}$ is the measure on $\mathbb{R}^\infty$ equal to the countable power of the one-dimensional distribution of ξ_1. We observe that the linear subspace $L \subset \mathbb{R}^\infty$ of all sequences $\xi = (\xi_k)$ with $\limsup_{k\to\infty} |\xi_k||\ln(1+k)|^{-1/2} < \infty$ has full measure. For all elements $t \in K^\infty \times S$ and $(\xi_k) \in L$, the series $\sum_{k=1}^\infty s_k^{-1/\alpha} \xi_k x_k$ converges in X, since for every continuous seminorm p on X the series $\sum_{k=1}^\infty s_k^{-1/\alpha} |\xi_k| p(x_k)$ converges, because

$$s_k^{-1/\alpha} \leqslant Ck^{-1/\alpha}, \quad |\xi_k| \leqslant C|\ln(k+1)|^{1/2}, \quad \sup_k p(x_k) < \infty.$$

In the general case verification is technically more involved, for this reason we shall define γ_t as a cylindrical measure and then verify that it is tight. The former is very simple, since for every functional $l \in X'$ the series $C_\alpha^{1/\alpha} \sum_{k=1}^\infty s_k^{-1/\alpha} \xi_k l(x_k)$ of independent Gaussian random variables converges almost surely by convergence of the series of $s_k^{-2/\alpha} |l(x_k)|^2$, which follows from the boundedness of l on K and the estimate $s_k^{-2/\alpha} \leqslant Ck^{-2/\alpha}$, where $2/\alpha > 1$.

We now verify that m-almost all cylindrical measures γ_t are tight and satisfy equality (5.12.6). We first verify (5.12.6) for cylindrical sets. To this end, it suffices to show that the integral of $\widetilde{\gamma_t}(l)$ against the measure m equals $\widetilde{\mu}(l)$ for all $l \in X'$. A straightforward calculation shows that

$$\widetilde{\gamma_t}(l) = \exp(-Q_t(l)/2), \quad Q_t(l) = C_\alpha^{2/\alpha} \sum_{k=1}^{\infty} s_k^{-2/\alpha} |l(x_k)|^2 \mathbb{E}\xi_1^2.$$

It is obvious that in this formula the function $t \mapsto Q_t(l)$ is Borel. The integral of $\widetilde{\gamma_t}(l)$ against the measure m coincides with the quantity

$$\mathbb{E}\exp\Bigl(iC_\alpha^{1/\alpha} \sum_{k=1}^{\infty} s_k^{-1/\alpha} \xi_k l(Z_k)\Bigr)$$

equal to $\widetilde{\mu}(l)$, which follows from formula (5.12.7) and Fubini's theorem applied on account of independence of the considered random variables.

Thus, equality (5.12.6) is true for cylindrical sets. To complete the proof by applying the previous lemma we have to verify the required additional condition of semicontinuity (fulfilled automatically if there is a countable collection in X' separating points). For this we observe that the set $S \subset \mathbb{R}^\infty$ of full m_2-measure introduced above contains the subsets $S_M = \{(s_k)\colon\, s_k \geqslant M^{-1}k\}$ and $m_2(S_M) \to 1$ as $M \to \infty$. This gives the estimate $s_k^{-2/\alpha} \leqslant M^{2/\alpha}k^{-2/\alpha}$ on S_M, which yields the continuity of the restrictions of the functions $t \mapsto Q_t(l)$ to $S_M \times K^\infty$ for all $l \in X'$. It is easy to see that this gives the continuity of the restrictions to $S_M \times K^\infty$ of the integrals against the measures ν_t for all cylindrical functions of the form $f(l_1, \dots, l_n)$, $f \in C_b(\mathbb{R}^n)$, $l_i \in X'$. As one can see from the proof of the last assertion of the lemma, we obtain the required upper semicontinuity. □

Let us give a condition for nondegeneracy of the Gaussian measures γ_t.

5.12.33. Corollary. *Suppose that X' contains a countable collection of functionals separating points in X (for example, X is a separable Fréchet space). If the measure μ from the theorem is not concentrated on a proper closed linear subspace, then this is also true for m-almost all measures γ_t.*

PROOF. The Radon measure μ is concentrated on the linear subspace X_0 that is the union of an increasing sequence of compact sets K_n, which are metrizable due to the existence of a sequence $\{l_j\} \subset X'$ separating points, moreover, for the corresponding metric we can take $d(x, y) = \sum_{j=1}^\infty 2^{-j} \min\bigl(|l_j(x-y)|, 1\bigr)$. Then almost all measures γ_t are concentrated on X_0 as well, hence we can deal with the space $X = X_0$.

We show that for m-a.e. t the measures $\gamma_t \circ l$, $l \in X'$, $l \neq 0$, have positive variances. It is seen from the proof of the theorem that we have to show that for m_1-a.e. $(x_k) \in K^\infty$ we have $\sup_k |l(x_k)| > 0$ for all nonzero $l \in X'$. To each element $z = (z_k) \in K^\infty$ we associate the closed linear span $E(z)$ of the sequence of vectors z_k. It suffices to establish that $E(z) = X$ for m_1-a.e. z.

In each K_n we choose a countable set $\{x_{n,i}\}$ everywhere dense with respect to the metric d. For every fixed pair of numbers n, i consider the function

$$\psi_{n,i}(z) = \inf\bigl\{d(x_{n,i}, x)\colon\, x \in E(z) \cap K_n\bigr\}, \ z \in K^\infty.$$

This function is Borel, since it is the limit of functions

$$\psi_{n,i,k}(z) = \inf\bigl\{d(x_{n,i}, x)\colon\, x \in \langle z_1, \dots, z_k\rangle\bigr\} \cap K_n,$$

where $\langle z_1, \dots, z_k\rangle$ is the linear span of $z_1, \dots, z_k$, and the functions $\psi_{n,i,k}$ are continuous. Since $E(z)$ does not change under permutations of components of z, the function $\psi_{n,i,k}$ also has this property, hence by Theorem 5.12.12 the function $\psi_{n,i,k}$ is equal to a constant m_1-a.e. Hence there is a set Ω_0 of full m_1-measure on which all functions $\psi_{n,i}$ are constant. This means that on Ω_0 the space $E(z)$ is constant. Indeed, if $E(z) \neq E(z')$, then $K_n \cap E(z) \neq K_n \cap E(z')$ for some n. We can assume that there is a point x in $\bigl(K_n \cap E(z)\bigr)\backslash E(z')$. Hence there exists a ball of radius $\varepsilon > 0$ centered at x in which there are no points of $K_n \cap E(z')$. Next, there exists a point $x_{n,i}$ such that $d(x, x_{n,i}) < \varepsilon/4$. Then $\psi_{n,i}(z') > \varepsilon/2$, although $\psi_{n,i}(z) < \varepsilon/4$ due to the inclusion $x \in K_n \cap E(z)$.

Finally, $E(z) = X$ for all $z \in \Omega_0$. Indeed, otherwise there exists a nonzero functional $l \in X'$ the kernel of which contains all $E(z)$ for $z \in \Omega_0$, i.e., $l(z_k) = 0$ for all k if $z = (z_k) \in \Omega_0$. Therefore, we obtain the equality $\Gamma(l^{-1}(0)) = 0$, since $m_1 = \Gamma^\infty$. It is clear that this yields the equality $\mu(l^{-1}(0)) = 0$, which contradicts the condition. □

In place of a countable separating collection of functionals one could require a more general condition: the metrizability of compact sets in X; however, this reduces to the considered case by the same reasoning as in the proof of the previous corollary.

In Acosta [**1**], Louie, Rajput, Tortrat [**323**], Rajput [**410**], [**411**], [**412**], Rajput, Rama-Murthy, Żak [**413**], and Tortrat [**524**], [**525**], [**527**], where some additional references can be found, topological supports of infinitely divisible and stable measures are studied; in particular, it is shown that the topological support S_μ of a symmetric stable measure μ is a linear subspace and the support of an infinitely divisible measure is a convex cone (see [**412**]).

5.12.34. Theorem. *Let μ be a symmetric stable of order α Radon measure. Then its topological support is a linear space. If $\alpha \geqslant 1$, then the topological support of a Radon α-stable measure is an affine subspace. If $\alpha < 1$, then the topological support is a convex cone.*

PROOF. We verify only the first assertion (for the proof of the second, see the paper [**412**]), following from the previous corollary. Suppose that the space X is separable metrizable. Passing to the smallest closed linear subspace of full measure, we obtain that almost all Gaussian measures γ_t, the mixture of which is μ, are positive on nonempty open sets. Hence the same is true for μ.

Let us consider the general case. Suppose that there is a point a not belonging to S_μ. Then there exists a closed absolutely convex neighborhood of zero V with $\mu(a+V) = 0$. Consider the normed space $E_{(V)}$ obtained by the factorization of X with respect to the kernel of the Minkowski functional of the set V. Let ν denote the image of the measure μ under the natural projection $\pi\colon X \to E_{(V)}$, under which V is mapped onto the closed unit ball in $E_{(V)}$. The Radon measure ν is symmetric, stable and is not concentrated on a proper closed subspace (otherwise μ would be concentrated on the preimage of this subspace), hence on account of the proven case we obtain that its support is the whole space $E_{(V)}$. However, the ball of radius 1 centered at $\pi(a)$ must have zero ν-measure, since its preimage under the projection π is $a + V$, because the preimage of the unit ball is V. The obtained contradiction shows that $S_\mu = X$. □

It is not known whether a Radon stable measure on a general locally convex space has always a metrizable compact set of positive measure (as is the case for Gaussian measures). Additional information about infinite-dimensional stable and infinitely divisible measures can be found in the books Hazod, Siebert [**218**], Kruglov [**301**], and Linde [**317**] and the papers Acosta [**1**], Acosta, Samur [**2**], Chung, Rajput, Tortrat [**102**], Dettweiler [**117**], Dudley, Kanter [**141**], Fernique [**158**], and Kanter [**258**].

5.12(vi). Banach supports of measures

In many applications it is useful to have results asserting that certain measures on infinite-dimensional spaces are actually concentrated on some more special or simpler spaces. For example, having a measure on a general locally convex space, it can be useful to know that it is actually concentrated on a smaller Banach space, and once this is achieved, one might think of a Hilbert support. Certainly, this is not always possible, but some positive results are available.

The following result was obtained by Buldygin [**91**] for Banach spaces and later extended in Bogachev [**60**] to Fréchet spaces (earlier some less general results were obtained in Kuelbs [**302**], Sato [**428**], and Ostrovskii [**376**]).

5.12.35. Theorem. *Let μ be a Radon measure on a Fréchet space X. Then there exists a reflexive separable Banach space $E \subset X$ such that $|\mu|(X\backslash E) = 0$ and the closed unit ball in E is compact in X.*

PROOF. It suffices to consider probability measures. The topology of X is generated by a metric ϱ. For each n take a compact set K_n with $\mu(X\backslash K_n)<1/n$. Then $\mu\bigl(\bigcup_{n=1}^\infty K_n\bigr) = 1$. Pick numbers $c_n > 0$ such that c_nK_n belongs to the ball of radius $1/n$ centered at zero. It is readily seen that the closure S of the set $\bigcup_{n=1}^\infty c_nK_n$ is compact (every sequence in S has a limit point). The closed absolutely convex hull K_0 of the set S is compact, but this set is not always suitable, since E_{K_0} need not be even separable. However, by Theorem 2.5.11 one can take a larger absolutely convex compact set W such that E_W will be a separable reflexive space and K_0 will be compact as a subset of E_W. The measure μ can be restricted to $E = E_W$, since all Borel sets in E_0 are Borel in X due to the injectivity of the embedding (see Bogachev [**65**, Theorem 6.8.6]). □

In the general case, one cannot find E with a Schauder basis (see Fonf, Johnson, Pisier, Preiss [**173**]; for a sufficient condition, see Theorem 5.12.63). In particular, it cannot be always made Hilbert, however, for measures on $\mathbb{R}^\infty$ this can be easily done (Exercise 5.12.82). The following interesting fact is known (see Mushtari [**359**], Sato [**429**]).

5.12.36. Theorem. *If every Radon measure on a Banach space X is concentrated on a continuously embedded separable Hilbert space, then X is isomorphic to a Hilbert space.*

It is clear that it is sufficient that the condition be fulfilled for measures with bounded support. As the space $\mathbb{R}^\infty$ shows, this theorem does not extend to arbitrary Fréchet spaces.

Recall that a measure μ on a Banach space has *strong moment of order r* if we have $\|\cdot\|^r \in L^1(\mu)$. More generally, a measure μ on a locally convex space has strong moment of order r if every continuous seminorm belongs to $L^r(\mu)$.

For such measures Theorem 5.12.35 can be sharpened (Bogachev [**64**]).

5.12.37. Theorem. *Let μ be a probability Borel measure on a separable Banach space X with strong moment of some order $r > 0$. Then there exists a linear subspace $E \subset X$ with the following properties:*

(i) *E with some norm $\|\cdot\|_E$ is a separable reflexive Banach space the closed balls of which are compact in X;*

(ii) *$\mu(E) = 1$ and*

$$\int_E \|z\|_E^r \,\mu(dz) < \infty.$$

If μ on X has all strong moments, then E can be chosen with this property. Finally, these assertions are true for separable Fréchet spaces.

PROOF. We modify the reasoning from Theorem 5.12.35. Let $\varphi \geqslant 0$ be a decreasing function on $[0,\infty)$ and $\sum_{n=1}^\infty \varphi(n) < \infty$. There are numbers $\alpha_n \downarrow 0$ for which $\alpha_n n \to \infty$ and $\sum_{n=1}^\infty \varphi(\alpha_n n) < \infty$. Let $\varphi(R) = \mu(x\colon \|x\| > R^{1/r})$. Then $\sum_{n=1}^\infty \varphi(n) < \infty$. Let us take α_n as above.

For every n, there is a compact set K_n in the ball U_n of radius $n^{1/r}$ centered at zero such that

$$\mu(\alpha_n^{1/r} K_n) \geqslant \mu(\alpha_n^{1/r} U_n) - 2^{-n}.$$

The set $K = \bigcup_{n=1}^\infty c_n K_n$, where $c_n := \alpha_n^{1/r} n^{-1/r}$, has compact closure. The closed convex hull V of the set K is also compact. Let p_V and E_V be the Minkowski functional of V and the associated Banach space. Since $\{p_V \leqslant c\} = cV$ for all $c \geqslant 0$, the function p_V is measurable. Furthermore,

$$\alpha_n^{1/r} K_n \subset n^{1/r} K \subset \{p_V \leqslant n^{1/r}\} = n^{1/r} V.$$

Due to our choice of the sets K_n we obtain $p_V^r \in L^1(\mu)$, since

$$\mu(x\colon\ p_V^r(x) > n) = 1 - \mu(x\colon\ p_V(x) \leqslant n^{1/r}) \leqslant 1 - \mu(\alpha_n^{1/r} K_n)$$
$$\leqslant 1 + 2^{-n} - \mu(\alpha_n^{1/r} U_n) = 2^{-n} + \mu(x\colon\ \|x\| > \alpha_n^{1/r} n^{1/r}) = 2^{-n} + \varphi(\alpha_n n).$$

It is clear that $\mu(E_V) = 1$ and $\mu(\alpha_n^{1/r} K_n) \to 1$, since the balls $\alpha_n^{1/r} U_n$ have radii $(\alpha_n n)^{1/r} \to \infty$ by the condition that $\alpha_n n \to \infty$. Next we argue as in Theorem 5.12.35. The case of a Fréchet space reduces to the one considered above by passing to the subspace $X_0 := \{q < \infty\}$, where $q^r := \sum_{n=1}^\infty c_n q_n^r$ and q_n are seminorms defining the topology, $c_n = 2^{-n}(\|q_n^r\|_{L^1(\mu)} + 1)^{-1}$. □

Note that this theorem does not extend to weak moments (see Exercise 5.12.84 and Exercise 5.12.85).

5.12.38. Corollary. *Let μ be a probability measure with strong moment of order r on a separable Banach space X with the approximation property, i.e., for every compact set K in X and every $\delta > 0$ there exists a finite-dimensional operator $S \in \mathcal{L}(X)$ with $\sup_{x\in K} \|x - Sx\| \leqslant \delta$. Then, for every $\varepsilon > 0$, there exists a finite-dimensional operator $T \in \mathcal{L}(X)$ such that*

$$\int_X \|x - Tx\|^r \,\mu(dx) < \varepsilon.$$

PROOF. Let E be the space from the previous theorem and let K be its unit ball. We find $\varepsilon_0 > 0$ such that the integral of the function $\|z\|_E^r$ over E is less than $\varepsilon/\varepsilon_0$. Let us take a finite-dimensional operator T with $\sup\limits_K \|z - Tz\| \leqslant \varepsilon_0$.

Then we have $\|z - Tz\| \leqslant \varepsilon_0 \|z\|_E$ if $z \in E$. Thus, the integral of $\|z - Tz\|^r$ does not exceed the integral of $\|z\|_E^r$ over E multiplied by ε_0, which is less than ε. □

This corollary does not extend to arbitrary Banach spaces (see Fonf, Johnson, Pisier, Preiss [**173**]). However, it would be interesting to study the existence of such approximations for measures from various special classes.

5.12(vii). Infinite-dimensional Wiener processes

Let X be a Hausdorff locally convex space, let H be a separable Hilbert space continuously and densely embedded into X, and let $j_H \colon X^* \to H$ be the associated embedding defined in the following way. For every $k \in X^*$, the functional $h \mapsto \langle k, h\rangle$ is continuous on H. Hence there is a vector $j_H(k) \in H$ such that for all $h \in H$ we have

$$\bigl(j_H(k), h\bigr)_H = \langle k, h\rangle. \tag{5.12.8}$$

Let $(\Omega, \mathcal{F}, P)$ be a probability space. For the definition of a real Wiener process, see Bogachev [**63**] or Wentzell [**564**].

5.12.39. Definition. *A continuous random process $(W_t)_{t\geqslant 0}$ on $(\Omega, \mathcal{F}, P)$ with values in X is called a Wiener process associated with H if, for every $k \in X^*$ with $|j_H(k)|_H = 1$, the one-dimensional process $\langle k, W_t\rangle$ is Wiener.*

Let $\mathcal{F}_t \subset \mathcal{F}$, $t \geqslant 0$, be an increasing family of σ-algebras. A Wiener process $(W_t)_{t\geqslant 0}$ is called an $\mathcal{F}_t$-Wiener process if, for all $\tau \geqslant 0$, $t, s \geqslant \tau$, the random vector $W_t - W_s$ is independent of $\mathcal{F}_\tau$ and the random vector W_t is $\mathcal{F}_t$-measurable.

Note that the only relation of the process to the topology of X concerns the continuity of paths (and, of course, the continuity of the embedding $H \to X$). In particular, if X is a Banach space, then replacing its norm by an equivalent one, we do not influence Wiener processes associated with H (but changing the norm of H, we alter the class of the associated processes).

Simple examples of infinite-dimensional Wiener processes are: any sequence $(w_{t,n})_{n=1}^\infty$ of independent real Wiener processes, which defines a Wiener process in $\mathbb{R}^\infty$ with $H = l^2$, and the sequence $(\alpha_n w_{t,n})_{n=1}^\infty$ with $\alpha_n > 0$ and $(\alpha_n) \in l^2$, which defines a Wiener process in l^2 for which H is the weighted Hilbert space of sequences $h = (h_n)$ with $\sum_{n=1}^\infty \alpha_n^2 h_n^2 < \infty$.

A rather obvious necessary and sufficient condition for the existence of a Wiener process in X is the existence of a separable Hilbert space continuously and densely embedded into X (see Bogachev [**63**, § 7.2]). Certainly, not every locally convex space satisfies this condition. For example, it is not fulfilled for strict inductive limits of increasing closed subspaces. For separable Fréchet spaces it is fulfilled. A much more delicate question is the existence of a Wiener process with a prescribed space H. Let us give a sufficient condition (see [**63**, § 7.2]).

5.12.40. Proposition. *Let X be a sequentially complete locally convex space on which there exists a centered Gaussian measure Radon γ with the Cameron–Martin space $H = H(\gamma)$ dense in X. Then there exists a Wiener process $(W_t)_{t\geqslant 0}$ associated with H for which the distribution of W_1 coincides with γ.*

A process $(W_t)_{t\geqslant 0}$ can be constructed in the form

$$W_t = \sum_{n=1}^{\infty} w_{t,n} e_n,$$

where $(w_{t,n})_{n=1}^{\infty}$ is a sequence of independent real Wiener processes and $\{e_n\}$ is an orthonormal basis in H. The hypotheses of the proposition ensure convergence of this series in X and the continuity of its sum.

As noted in [**63**, § 7.2], the condition of sequential completeness can be replaced by the existence of a convex compact set of positive γ-measure. However, it remains unknown whether the only assumption of the existence of the aforementioned measure γ is sufficient.

5.12(viii). Prohorov locally convex spaces

Here we consider an important class of spaces connected with weak convergence of measures and Theorems 5.3.7 and 5.3.9. The symbol $\mathcal{M}_r(X)$ will denote the set of all Radon measures on a topological space X. Let $\mathcal{M}_t^+(X)$ denote the set of all nonnegative tight Baire measures on a topological space X. These spaces will be equipped with the weak topology.

5.12.41. Definition. (i) *A Tychonoff topological space X is called Prohorov if every set in the space of measures $\mathcal{M}_t^+(X)$ that is compact in the weak topology is uniformly tight.*

(ii) *A Tychonoff topological space X is called sequentially Prohorov if every sequence of nonnegative tight Baire measures weakly converging to a tight measure is uniformly tight.*

In this definition we could speak of Radon measures, i.e., replace $\mathcal{M}_t^+(X)$ by $\mathcal{M}_r^+(X)$, since every tight Baire measure on X has a unique Radon extension (see § 5.2).

If in the definition we admit signed measures, then we arrive at the classes of *strongly Prohorov* and *strongly sequentially Prohorov* spaces.

Every complete separable metric space is a Prohorov space; an arbitrary metric space is sequentially Prohorov (see [**65**, § 8.10(ii)]). Clearly, any Prohorov space is sequentially Prohorov. However, it is known that the space $\mathbb{Q}$ of rational numbers is sequentially Prohorov, but not Prohorov. We observe that the sequential Prohorov property is weaker than the requirement that all weakly convergent sequences of tight measures be uniformly tight (since their limits need not be tight measures). But for spaces on which all Baire measures are tight, this is the same.

The proof of the following result from Hoffmann-Jørgensen [**223**] can be found also in Bogachev [**69**, Chapter 4].

5.12.42. Theorem. *The class of Prohorov spaces is preserved by taking* (i) *countable products,* (ii) *countable intersections,* (iii) *closed subspaces and open subspaces, hence by passing to G_δ-subsets.*

In addition, a space is Prohorov provided that every point has a neighborhood which is a Prohorov space (for example, if it admits a locally finite cover by closed Prohorov subspaces).

We recall that a space X is called hemicompact if it possesses a fundamental sequence of compact sets K_n (i.e., every compact set in X is contained in some of K_n). If the continuity of a function on X follows from its continuity on all compact sets, then X is called a k_R-space. The latter property holds for any k-space, i.e., a space in which a set is closed precisely when it has closed intersections with all compact sets. A nontrivial example: an inductive limit with compact embeddings (see § 2.7).

5.12.43. Corollary. *The class of Prohorov spaces contains all locally compact spaces and all hemicompact k_R-spaces.*

The union of two Prohorov subspaces, even if one of them is a point, need not be Prohorov. In addition, a countable union of closed Prohorov subspaces is not always Prohorov.

Let us mention some results and examples that enable one to construct broader classes of Prohorov and sequentially Prohorov spaces by means of the operations listed in Theorem 5.12.42.

5.12.44. Proposition. *Let X be a Tychonoff space possessing a countable family of closed subspaces X_n with the following property: a function on X is continuous if and only if its restriction to every X_n is continuous.*

(i) *If every X_n is Prohorov, then X is Prohorov as well.*

(ii) *If all X_n are either Polish spaces or compact spaces, then every weakly fundamental sequence in $\mathcal{M}_r(X)$ is uniformly tight. In particular, X is strongly sequentially Prohorov.*

A proof can be found in Bogachev [**65**, § 8.10(ii)] or [**69**, Chapter 4].

5.12.45. Example. In the following cases every weakly fundamental sequence of tight measures on X is uniformly tight:

(i) X is a hemicompact k_R-space;

(ii) X is a locally convex space that is the inductive limit of an increasing sequence of separable Banach spaces E_n such that the embeddings of E_n to E_{n+1} are compact operators (by definition X is equipped with the strongest locally convex topology for which all embeddings $E_n \to X$ are continuous).

PROOF. Case (i) is clear from Proposition 5.12.44. (ii) The results proved in § 2.7 yield that X is a k-space with a fundamental sequence of compact sets. □

5.12.46. Lemma. *Let $X = \bigcup_{n=1}^\infty X_n$ be a separated locally convex space that is the strict inductive limit of an increasing sequence of closed subspaces X_n. If a sequence $\{\mu_i\}$ of nonnegative Radon measures on X converges weakly to a Radon measure μ, then, for every $\varepsilon > 0$, there exists $n \in \mathbb{N}$ such that $\mu_i(X\backslash X_n) < \varepsilon$ for all $i \in \mathbb{N}$.*

Moreover, for every family $\{\mu_\alpha\}$ of nonnegative Radon measures on X which has compact closure in the weak topology in the space $\mathcal{M}_r(X)$ and every $\varepsilon > 0$, there exists $n \in \mathbb{N}$ such that $\mu_\alpha(X\backslash X_n) < \varepsilon$ for all α.

PROOF. Without loss of generality we can assume that μ_i and μ are probability measures (if $\mu_i(X) \to 0$, then our assertion is trivial). If the assertion is false, then for every $n \in \mathbb{N}$ there exists $i(n) \in \mathbb{N}$ with $\mu_{i(n)}(X_n) < 1 - \varepsilon$. Passing to a new sequence of measures, we can assume that $i(n) = n$. Choose $m \in \mathbb{N}$ such that $\mu(X_m) > 1-\varepsilon/2$. Set $k_1 := m$. Find $k_2 > m$ with $\mu_m(X_{k_2}) > 1-\varepsilon/2$. Next we find a convex symmetric open set U_1 in the space X_{k_2} such that $X_m \subset U_1$ and $\mu_m(U_1) < 1 - \varepsilon$. Such a set U_1 exists. Indeed, by the Hahn–Banach theorem the subspace X_m is the intersection of all closed hyperplanes containing it. By the Radon property of μ_m there exists a finite collection of closed hyperplanes $L_1, \ldots, L_p$ in X_{k_2} such that $X_m \subset \bigcap_{i=1}^p L_i$ and $\mu_m\bigl(\bigcap_{i=1}^p L_i\bigr) < 1 - \varepsilon$. Then $L_i = l_i^{-1}(0)$ for some $l_i \in X_{k_2}^*$, and the set $\bigcap_{i=1}^p l_i^{-1}(-\delta, \delta)$ can be taken for U_1 for sufficiently small $\delta > 0$. Next we take $k_3 \geqslant k_2$ with $\mu_{k_2}(X_{k_3}) > 1 - \varepsilon/2$. There exists a convex symmetric neighborhood of zero $W \subset X_{k_3}$ such that $W \cap X_{k_2} = U_1$ (see Lemma 1.3.12). As above, there exists a convex symmetric open set V in the space X_{k_3} such that $X_{k_2} \subset V$ and $\mu_{k_2}(V) < 1 - \varepsilon$. Set $U_2 := W \cap V$. Continuing the described process by induction, we obtain an increasing sequence of indices $k_n \geqslant n$ such that every space $X_{k_{n+1}}$ contains a convex symmetric open set U_n for which

$$1)\ U_n \cap X_{k_n} = U_{n-1}, \quad 2)\ \mu_{k_n}(U_n) < 1 - \varepsilon,\ \mu_{k_n}(X_{k_{n+1}}) > 1 - \varepsilon/2.$$

By the definition of a strict inductive limit, the set $U = \bigcup_{n=1}^\infty U_n$ is a neighborhood of zero in X. By construction, for every n we have

$$\mu_{k_n}(U) < \mu_{k_n}(U \cap X_{k_{n+1}}) + \varepsilon/2 = \mu_{k_n}(U_n) + \varepsilon/2 < 1 - \varepsilon/2,$$

which contradicts weak convergence (see Theorem 5.3.2), since $\mu(U) > 1 - \varepsilon/2$. In case of a relatively weakly compact family $\{\mu_\alpha\}$ the reasoning is similar. We choose a sequence $\{\mu_{\alpha(n)}\}$ as above and denote by μ some of its weak limit points. The previous choice of U leads again to a contradiction with Theorem 5.3.2, since there exists a subnet $\{\mu_\beta\}$ in $\{\mu_{\alpha(n)}\}$ weakly converging to μ. □

It is not clear whether this lemma is true for signed measures.

5.12.47. Proposition. *Suppose that a locally convex space X is the strict inductive limit of an increasing sequence of its closed subspaces X_n. Then X is a Prohorov space if all spaces X_n are Prohorov. In particular, if each X_n is a separable Fréchet space, then every weakly fundamental sequence of nonnegative Baire measures on X is uniformly tight.*

PROOF. By Lemma 5.12.46 for every $\varepsilon > 0$ the measures from every weakly compact family M of nonnegative Radon measures on X are ε-concentrated on some subspace X_n. According to Bogachev [**65**, Corollary 8.10.2], the restrictions of the measures from M to X_n form a set with compact closure.

In order to obtain the last assertion, it suffices to recall that the union of a sequence of separable Fréchet spaces is Souslin and that Baire measures on it are Radon (see § 5.2). □

Obviously, one can multiple the number of such examples by constructing countable products and passing to closed subsets. We observe that many classical spaces of functional analysis, such as $\mathcal{D}(\mathbb{R}^d)$, $\mathcal{D}'(\mathbb{R}^d)$, $\mathcal{S}(\mathbb{R}^d)$, and $\mathcal{S}'(\mathbb{R}^d)$, are Prohorov spaces, since they can be obtained by means of the indicated operations.

5.12.48. Remark. The space $\mathcal{D}(\mathbb{R}^1)$ is a Prohorov space, but is neither a k_R-space (Exercise 5.12.86), nor a hemicompact space (in addition, it is not σ-compact). The absence of a countable family of compact sets which is either fundamental or exhausting follows from Baire's theorem applied to the subspaces $\mathcal{D}_n(\mathbb{R}^1)$ and the fact that every compact set in $\mathcal{D}(\mathbb{R}^1)$ is contained in some of the subspaces $\mathcal{D}_n(\mathbb{R}^1)$.

Let us discuss a property close to Prohorov's property. It is related to the following important result due to A.D. Alexandroff [**12**, §18], a proof of which can be found in [**65**, Proposition 8.1.10]. We recall that a *functionally closed* set is the set of zeros of a continuous function.

5.12.49. Proposition. *If a sequence of Baire measures μ_n on a topological space converges weakly to a measure μ, then this sequence has no eluding load:*

$$\lim_{n\to\infty} \sup_k |\mu_k|(Z_n) = 0 \tag{5.12.9}$$

for every sequence of pairwise disjoint functionally closed sets Z_n with the property that the union of every subfamily in $\{Z_n\}$ is functionally closed.

If all measures μ_n are Radon and the space is completely regular, then the following fact is true.

5.12.50. Proposition. *Suppose that a sequence of Radon measures μ_n on a completely regular space X converges weakly to a measure μ. Then*

$$\lim_{n\to\infty} \sup_k |\mu_k|(U_n) = 0 \tag{5.12.10}$$

for every locally finite sequence of disjoint open sets U_n.

PROOF. Here one can apply the same reasoning which proves Alexandroff's result (this case is even simpler). Assuming the opposite and passing to a subsequence, we obtain compact sets $K_n \subset U_n$ such that $|\mu_n|(K_n) \geqslant c > 0$. There exist continuous functions $\psi_n\colon X \to [-1,1]$ such that $\psi_n = 0$ outside of U_n and the integral of ψ_n against the measure μ_n is not less than $c/2$. For every bounded real sequence $\{c_k\}$ the function $\varphi_{\{c_k\}} := \sum_{k=1}^\infty c_k\psi_k$ is bounded on X and continuous due to the local finiteness of $\{U_n\}$. By condition, the sequence of the integrals of such function converges. This means that the sequence of vectors $v^n = (v_1^n, v_2^n, \ldots)$, where

$$v_k^n = \int_X \psi_k \, d\mu_n,$$

is weakly fundamental in the Banach space l^1. Indeed, $v^n \in l^1$ by the disjointness of supports of the functions ψ_k. Applying to v^n the functional on l^1 defined by the bounded sequence $\{c_k\}$ we obtain

$$\sum_{k=1}^\infty c_k \int_X \psi_k \, d\mu_n = \int_X \varphi_{\{c_k\}} \, d\mu_n.$$

Since any weakly fundamental sequence in the space l^1 must converge in norm (Exercise 3.12.185), there exists a number N such that $|v_n^n| < c/4$ for all $n > N$, which contradicts our assumption. □

Note that the limit measure μ is not assumed to be Radon (and can fail indeed to be Radon).

Using the terminology of A.D. Alexandroff, one can call property (5.12.10) the *absence of eluding load on open sets*.

5.12.51. Remark. For every sequence of Radon measures μ_n on a normal space property (5.12.10) is equivalent to the property that (5.12.9) holds for every sequence of pairwise disjoint closed sets Z_n with the property that the union of every subfamily in $\{Z_n\}$ is closed.

Indeed, if $\{U_n\}$ is a locally finite sequence of disjoint open sets, then for any compact sets $K_n \subset U_n$ we obtain a sequence of sets all unions of which are closed. Conversely, suppose that for every sequence of locally finite disjoint open sets U_n we have (5.12.10).

It suffices to verify (5.12.9) for compact sets. Suppose we are given a sequence of disjoint compact sets Z_n all unions of which are closed. Suppose that $\mu_n(Z_n) \geqslant c > 0$. By induction we construct pairwise disjoint open sets $U_n \supset Z_n$. For example, U_1 is constructed as follows: the compact set Z_1 and the closed (by assumption) set $\bigcup_{n=2}^{\infty} Z_n$ can be separated by open sets U_1 and V_1; next, in V_1 we find a neighborhood U_2 of the compact set Z_2 that is disjoint with some neighborhood of the closed set $\bigcup_{n=3}^{\infty} Z_n$, and so on. Since the space is normal, there is a continuous function $f\colon X \to [0,1]$ that equals 1 on the closed union of Z_n and equals 0 outside of the open union of U_n. Finally, let $\psi_n = f$ on U_n and $\psi_n = 0$ outside of U_n. The continuity of the function ψ_n and the function $\varphi_{\{c_n\}}$ (see the proof of Proposition 8.1.10 in Bogachev [**65**]) is verified directly. Thus, for fully normal spaces (i.e., spaces in which all closed sets are functionally closed) and Radon measures we have equivalence of property (5.12.9) for general closed sets and property (5.12.10).

One can consider classes of spaces for which (5.12.9) or (5.12.10) imply uniform tightness. Let $\mathcal{AL}$ be the class of Tychonoff spaces X for which every sequence of tight Baire measures μ_n not having eluding load is uniformly tight; such spaces can be called *Alexandroff spaces*. Let $\mathcal{AL_U}$ be the class of Tychonoff spaces X for which every sequence of Radon measures μ_n not having eluding load is uniformly tight; such spaces can be called *$\mathcal{U}$-Alexandroff spaces*.

By the previous propositions these classes are contained in the class of strongly sequentially Prohorov spaces. In addition, the spaces of Radon measures on these spaces are weakly sequentially complete. For fully normal spaces both introduced properties coincide. It is not difficult to verify that complete metric spaces belong to $\mathcal{AL}$ and $\mathcal{AL_U}$: actually, this fact is usually established in the proof of the Prohorov theorem for complete metric spaces (see the proof of Theorem 8.6.2 in the book [**65**]).

However, an incomplete separable metric space need not be Alexandroff (although it is always sequentially Prohorov). Say, we can take a Lebesgue nonmeasurable set X in $[0,1]$ of outer measure 1 and inner measure 0, on which there is a sequence of measures with finite supports weakly converging to a non-Radon measure on E induced by Lebesgue measure (see Exercise 5.12.79 or Example 8.4.6

in [**65**]); this sequence is not uniformly tight, but has no eluding load. It is easy to verify the following properties of the classes $\mathcal{AL}$ and $\mathcal{AL}_{\mathcal{U}}$ (Exercise 5.12.87).

5.12.52. Proposition. (i) *Closed subsets of fully normal spaces of class* $\mathcal{AL}$ *belong to this class.*

(ii) *The classes* $\mathcal{AL}$ *and* $\mathcal{AL}_{\mathcal{U}}$ *admit countable products.*

5.12.53. Proposition. *Every k-space with a countable fundamental system of compact sets belongs to* $\mathcal{AL}$ *and* $\mathcal{AL}_{\mathcal{U}}$.

PROOF. We can assume that the given space X is the union of a strictly increasing sequence of compact sets K_n and that every compact set in X is contained in some K_n. If X does not belong to $\mathcal{AL}_{\mathcal{U}}$, then one can construct a sequence of Radon measures μ_k and a sequence of disjoint compact sets Q_k with the following properties: $\mu_k(Q_k) \geqslant c > 0$ and each K_n has a nonempty intersection only with finitely many sets Q_k. Hence every union of Q_k is closed. Since X is normal (Exercise 5.12.88), it remains to apply Remark 5.12.51. □

5.12.54. Corollary. *Inductive limits of sequences of locally convex spaces with compact embeddings belong to* $\mathcal{AL}$ *and* $\mathcal{AL}_{\mathcal{U}}$.

No explicit descriptions of the class of Prohorov spaces or the class $\mathcal{AL}$ are known.

5.12(ix). Measurable linear and multilinear functions

Let μ be a Radon probability measure on a Hausdorff locally convex space X with the topological dual X'. A function $l\colon X \to \mathbb{R}^1$ is called proper linear μ-measurable if it is linear on all of X in the usual sense and is μ-measurable. The set of all such functions will be denoted by $\Lambda(\mu)$. Let $\widetilde{\Lambda}(\mu)$ be the class of all functions possessing modifications from $\Lambda(\mu)$.

There is also another natural way of defining measurable linear functions. Let $\Lambda_0(\mu)$ denote the closure of X' in $L^0(\mu)$, i.e., $l \in \Lambda_0(\mu)$ if there exists a sequence of functions $l_n \in X'$ converging to l in measure. Since $\{l_n\}$ contains an almost everywhere convergent subsequence, we can assume that $l_n \to l$ a.e.

5.12.55. Lemma. *The inclusion* $\Lambda_0(\mu) \subset \widetilde{\Lambda}(\mu)$ *is true.*

PROOF. Let $l_n \to l$ in measure, where $l_n \in X'$. As already noted, we can assume that $l_n \to l$ a.e., then the linear subspace X_0 of all x for which there exists a finite limit $l_0(x) = \lim\limits_{n\to\infty} l_n(x)$ has full measure. It is clear that on X_0 the function l_0 is linear and measurable. It remains to take its arbitrary linear extension to X. □

Several examples are known in which $\widetilde{\Lambda}(\mu) \neq \Lambda_0(\mu)$ even for a symmetric measure μ, see Kanter [**258**], [**259**], Skorohod [**469**], Smoleński [**473**], and Urbanik [**534**]. One of such examples is the distribution of any process with independent increments that is stable of order $\alpha < 2$.

Similarly one can introduce two classes of measurable linear mappings from X to a locally convex space Y. The class $\widetilde{\Lambda}(\mu, Y)$ consists of μ-measurable mappings to Y, equipped with the σ-algebra $\sigma(Y')$, possessing a linear version. The class $\Lambda_0(\mu, Y)$ consists of mappings that are μ-a.e. limits of sequences of continuous linear operators. Certainly, one can even take the smaller class $\Lambda_1(\mu, Y)$ of mappings which μ-a.e. have the form $Ax = \lim\limits_{n\to\infty} A_n x$, where A_n is a continuous linear operator with a finite-dimensional range.

Let us consider the following property of a mapping A:

(RB) *there exists a separable reflexive Banach space* $(E, \|\cdot\|_E)$ *of full measure such that it is compactly embedded into* X *and* A *has a version that is a continuous linear operator from the space* $(E, \|\cdot\|_E)$ *to the space* Y.

5.12.56. Theorem. (Yurova [**576**]) *Let* X *and* Y *be separable Fréchet spaces, let* μ *be a Borel probability measure on* X*, and let* $A_n\colon X \to Y$ *be a sequence of continuous linear operators such that* $A_n x \to Ax$ *almost everywhere. Then there exist a separable reflexive Banach space* $(E, \|\cdot\|_E)$ *of full measure compactly embedded into* X *and a continuous linear operator* $\widetilde{A}\colon E \to Y$ *that almost everywhere coincides with* A*, i.e., the operator* A *has property* (RB).

PROOF. We know (see Theorem 5.12.35) that there exists a separable reflexive Banach space $(E_0, \|\cdot\|_{E_0})$ of full measure compactly embedded into X. It is obvious that on $(E_0, \|\cdot\|_{E_0})$ the operators A_n are continuous and $A_n x \to Ax$ almost everywhere on E_0. Hence it suffices to establish the existence of a Banach space $E \subset E_0$ with all necessary properties with the exception of reflexivity (since in that space we can find a smaller reflexive space). Suppose first that Y is Banach. We show that there exists a subsequence of indices $\{n_k\}$ such that

$$\sum_{k=1}^{\infty} 2^k \|A_{n_{k+1}}x - A_{n_k}x\|_Y^2 < \infty \tag{5.12.11}$$

almost everywhere. We observe that for this it is sufficient to have almost everywhere

$$\sum_{k=1}^{\infty} 2^k \|A_{n_{k+1}}x - Ax\|_Y^2 < \infty.$$

Let us prove the existence of $\{n_k\}$. Pointwise convergence $A_n \to A$ yields convergence in measure, therefore, for every k there is n_k such that $\mu(\Omega_k) < 2^{-k}$, where

$$\Omega_k = \{x\colon \|A_{n_k}x - Ax\|_Y^2 > 2^{-2k}\}.$$

Set $\Lambda = \bigcap_{N=1}^{\infty} \bigcup_{k=N}^{\infty} \Omega_k$. Since $\mu\bigl(\bigcup_{k=N}^{\infty} \Omega_k\bigr) \leqslant 2^{1-N}$, we have $\mu(\Lambda) = 0$. If $x \in X\backslash\Lambda$, then

$$\sum_{k=1}^{\infty} 2^k \|A_{n_{k+1}}x - A_{n_k}x\|_Y^2 < \infty,$$

as required. Let us consider the space $E \subset E_0$ of all vectors x such that (5.12.11) holds. This linear subspace has full measure; we define the following norm on it:

$$\|x\|_E = \|x\|_{E_0} + \Bigl(\sum_{k=1}^{\infty} 2^k \|A_{n_{k+1}}x - A_{n_k}x\|_Y^2\Bigr)^{1/2}.$$

The space $(E, \|\cdot\|_E)$ is complete. Indeed, let $\{x_n\}$ be a Cauchy sequence in E. Then, as it follows from the expression for the norm on E, it is Cauchy in E_0. Since E_0 is complete, this sequence has some limit x in E_0. Let us show that $\|x_n - x\|_E \to 0$. Let $\varepsilon > 0$. Then, for some N, for all $m, n > N$ we have

$$\sum_{k=1}^{\infty} 2^k \|(A_{n_{k+1}} - A_{n_k})(x_n - x_m)\|_Y^2 < \varepsilon.$$

Passing to the limit as $m \to \infty$ and taking into account the continuity of the operators A_k, we obtain the estimate

$$\sum_{k=1}^{\infty} 2^k \|(A_{n_{k+1}} - A_{n_k})(x_n - x)\|_Y^2 \leqslant \varepsilon.$$

Hence $x \in E_0$ and $\|x_n - x\|_E \to 0$. Thus, $(E, \|\cdot\|_E)$ is a Banach space continuously embedded into $(E_0, \|\cdot\|_{E_0})$ and compactly embedded into X. It follows from (5.12.11) that $\mu(E) = 1$. Let us consider the space $Z = E_0 \bigoplus \bigoplus_{l_2} Y_k$, where $Y_k = (Y, 2^k\|\cdot\|_Y)$ and the l^2-sum of the spaces Y_k is the space of sequences $z = (z_k)_{k=1}^{\infty}$, where $z_k \in Y_k$, with finite norm $\bigl(\sum_{k=1}^{\infty} \|z_k\|_{Y_k}^2\bigr)^{1/2}$. It is readily seen that it separable, since E_0 and Y are separable. The mapping

$$T\colon\ E \to Z,$$

$$Tx = \bigl(x, (A_{n_2} - A_{n_1})x, (A_{n_3} - A_{n_2})x, (A_{n_4} - A_{n_3})x, \ldots\bigr)$$

establishes a linear isometry between the space E and a subspace in the separable space Z, therefore, E is separable as well. For every $x \in E$, the sequence $\{A_{n_k}x\}$ is obviously Cauchy in Y, hence it converges in Y. We further consider the version of A equal to the limit of $\{A_{n_k}\}$ on E. For this version we have the equality $\|Ax\|_Y = \lim\limits_{k\to\infty} \|A_{n_k}x\|_Y$, whence on account of the equality

$$A_{n_k}x = (A_{n_k}x - A_{n_{k-1}}x) + \cdots + (A_{n_2}x - A_{n_1}x) + A_{n_1}x$$

we obtain that

$$\begin{aligned}
\|Ax\|_Y &\leqslant \|A_{n_1}x\|_Y + \sum_{k=1}^{\infty} \|A_{n_{k+1}}x - A_{n_k}x\|_Y \\
&\leqslant \|A_{n_1}x\|_Y + \Bigl(\sum_{k=1}^{\infty} 2^{-k}\Bigr)^{1/2} \Bigl(\sum_{k=1}^{\infty} 2^k \|A_{n_{k+1}}x - Ax\|_Y^2\Bigr)^{1/2} \\
&\leqslant \|A_{n_1}x\|_Y + \|x\|_E.
\end{aligned}$$

Therefore, the operator A is continuous from $(E, \|\cdot\|_E)$ to $(Y, \|\cdot\|_Y)$.

We now turn to a general Fréchet space Y and explain necessary changes in our reasoning. We can assume that the topology in Y is defined by a sequence of seminorms p_n with $p_n \leqslant p_{n+1}$. Now in place of estimate (5.12.11) we can obtain that almost everywhere

$$\sum_{k=1}^{\infty} 2^k p_k^2(A_{n_{k+1}}x - A_{n_k}x) < \infty. \tag{5.12.12}$$

A norm on the space E will be defined by the formula

$$\|x\|_E = \|x\|_{E_0} + \Bigl(\sum_{k=1}^{\infty} 2^k p_k^2(A_{n_{k+1}}x - A_{n_k}x)\Bigr)^{1/2}.$$

As above, E with this norm is complete and separable. The operator $A\colon E \to Y$ is now defined by the equality $Ax = \lim\limits_{k\to\infty} A_{n_k}x$. Since for any fixed m and all $k > m$ we have the estimates

$$\begin{aligned} p_m(A_{n_{k+1}}x) &\leqslant p_m(A_{n_{k+1}}x - A_{n_k}x) + \cdots + p_m(A_{n_{m+1}}x - A_{n_m}x) + p_m(A_{n_m}x) \\ &\leqslant p_k(A_{n_{k+1}}x - A_{n_k}x) + \cdots + p_m(A_{n_{m+1}}x - A_{n_m}x) + p_m(A_{n_m}x) \\ &\leqslant \|x\|_E + p_m(A_{n_m}x), \end{aligned}$$

the operator A is continuous. □

The proven result can be slightly generalized.

5.12.57. Corollary. *In place of the separability of the spaces X and Y it is enough to require that the measure μ be Radon on X.*

PROOF. In this case the measure μ is concentrated on some separable closed subspace X_0 and the closure of the linear span of the separable sets $A_n(X_0)$ is a separable Fréchet space. □

The following fact is obvious from the definition of the norm in E.

5.12.58. Corollary. *If Y is a Banach space, then the space E mentioned in the theorem can be taken such that some subsequence $\{A_{n_k}\}$ will converge on E in the operator norm.*

5.12.59. Theorem. (Yurova **[576]**) *Let $(E, \|\cdot\|)$ be a separable reflexive Banach space compactly embedded into a locally convex space X and let A be a continuous linear operator from $(E, \|\cdot\|)$ to a Banach space $(Y, \|\cdot\|_Y)$ with a Schauder basis. Then there exists a sequence of finite-dimensional continuous linear operators $A_n\colon X \to (Y, \|\cdot\|_Y)$ pointwise converging to A on the space E.*

PROOF. Let us first consider the case $Y = \mathbb{R}$. Let V be the closed unit ball in $(E, \|\cdot\|)$ and let B^* be the closed unit ball in the dual space E^*; then $E = \bigcup_{n=1}^{\infty} nV$. Since E is reflexive and compactly embedded into X, the set V is compact in X. It is obvious that the norm on E coincides with the Minkowski functional p_V of the set V. Applying Proposition 2.5.13, we obtain that the set of all functionals from B^* that are continuous with respect to the topology induced from X is dense in B^* with respect to the topology of pointwise convergence on V. Since E is separable, the topology of pointwise convergence on V is metrizable. Therefore, any element of the closure is the limit of some sequence. In our case, if $f \in E^*$, then $f/\|f\| \in B^*$, hence there exists a sequence $g_n \in B^*$ converging to $f/\|f\|$ pointwise and consisting of functionals continuous on E in the topology induced from X. It is clear that $\|f\|g_n \to f$. The functionals $\|f\|g_n$ can be extended to all of X by continuity. The case $Y = \mathbb{R}^k$ is easily deduced from the proven result, which gives the case where Y is finite-dimensional. Let now Y be a space with a Schauder basis $\{e_n\}$. Let P_k be the projection from Y onto the linear span Y_k of the vectors $e_1, \ldots, e_k$. This is a continuous operator, hence for every finite-dimensional continuous operator $P_k A$ there exists a sequence of continuous operators $A_{k,n}\colon X \to Y$ pointwise converging to $P_k A$ on E. It remains to a choose a subsequence $\{A_{k,n_k}\}$ converging to A pointwise on E. For example, we can use the metrizability of pointwise convergence on V. □

Two previous theorems imply the following assertion.

5.12.60. Corollary. *If X is a Fréchet space and Y is a Banach space with a Schauder basis, then the inclusion to $\Lambda_0(\mu, Y)$ is equivalent to property* (RB). *In this case $\Lambda_0(\mu, Y) = \Lambda_1(\mu, Y)$.*

5.12.61. Remark. (i) The space E in Theorem 5.12.56 can be taken Hilbert if X and Y are Hilbert. For this we first take a Hilbert space E_0.

(ii) It is readily seen that the last theorem and its corollary remain valid for any Fréchet space Y with a basis. It also is clear that in Theorem 5.12.59 it suffices to have a sequence of continuous finite-dimensional operators on Y pointwise converging to the identity operator. On the other hand, as observed by A.N. Plichko, taking a separable reflexive Banach space $E = Y$ without the latter property and a compact injective operator $j\colon E \to l^2$, we see that there are no finite-dimensional operators $A_n\colon l^2 \to Y$ whose restrictions to E embedded into l^2 (more precisely, to $j(E)$) converge pointwise to the identity operator $A\colon E \to E = Y$ (more precisely, $j^{-1}\colon j(E) \to E$). Actually, if on E the identity operator is not a pointwise limit of a sequence of compact operators, then there are no even continuous operators A_n pointwise converging to A, since the compositions $j(A_n)$ are compact.

Herer [**219**] and Okazaki [**373**] considered the so-called stochastic bases in a separable Fréchet space X with a Borel probability measure μ, i.e., systems of vectors $\varphi_n \in X$ such that there exist $f_n \in X'$ with $f_n(\varphi_k) = \delta_{nk}$ such that for $P_n x := \sum_{i=1}^n f_i(x)\varphi_i$ we have $P_n x \to x$ for μ-a.e. x. The next two theorems were obtained in [**373**].

5.12.62. Theorem. *If a Radon probability measure μ on a separable Fréchet space X has a stochastic basis, then there exists a Banach space of full measure continuously embedded into X and possessing a Schauder basis.*

PROOF. By Egorov's theorem there exist increasing compact sets K_n on which convergence $P_j x \to x$ is uniform. Set $P_0 = I$. The topology of the space X can be defined by some increasing sequence of seminorms q_n. By uniform convergence the quantities $C_n = 1 + \sup_{x\in K_n, j\geqslant 0} q_n(P_j x)$ are finite. Let us consider the space X_0 of all vectors $x \in X$ with finite norm

$$\|x\| = \sum_{n=1}^{\infty} 2^{-n} C_n^{-1} \sup_{j\geqslant 0} q_n(P_j x).$$

This is a norm indeed, since the functions $q_n(P_j x)$ are seminorms and if $\|x\| = 0$, then $q_n(x) = 0$ for all n, hence $x = 0$. It is clear that $K_n \subset X_0$ for all n, hence $\mu(X_0) = 1$. The continuity of q_n and P_j implies that the set $\{x\colon \|x\| \leqslant 1\}$ is closed in X, hence X_0 is Banach. The embedding of X_0 into X is continuous, since for each fixed n we have $q_n(x) \leqslant 2^n C_n \|x\|$.

We observe that $\|P_m x\| \leqslant \|x\|$ for all m, since $P_j P_m x = P_m x$ for all $j \geqslant m$, whence we obtain $\sup_{j\geqslant 0} q_n(P_j P_m x) \leqslant \sup_{j\geqslant 0} q_n(P_j x)$. Thus, the operators P_m are finite-dimensional projections in X_0 with unit norm.

The set $E = \{x \in X_0\colon \lim\limits_{n\to\infty} \|P_n x - x\| = 0\}$ is a closed linear subspace in X_0 and hence is also a Banach space. Moreover, $K_m \subset E$ for all m. Indeed, for any given $\varepsilon > 0$ there exists a number $n_1 > m$ with $\sum_{n>n_1} 2^{-n} < \varepsilon/4$, next, there is a number j_1 such that $q_n(P_j x - x) < \varepsilon/2$ for all $j \geqslant j_1$ for every $n = 1, \ldots, n_1$. Then for any $j \geqslant j_1$ we obtain $\|P_j x - x\| \leqslant \varepsilon$, so that $q_n(P_j x) \leqslant C_n$ for all $j \geqslant 0$ and $n \geqslant m$ by inclusion $x \in K_m \subset K_n$. Therefore, E is a Banach space of full measure possessing a Schauder basis. □

It is not clear whether such a space with a Schauder basis can be found reflexive and compactly embedded. In addition, it is not clear whether it can be chosen such that the measure μ on it would have finite strong moment of order r, provided this is the case on the original space.

5.12.63. Theorem. *Let μ be a Radon probability measure on a Fréchet space X such that all continuous seminorms belong to $L^2(\mu)$, all elements of X' possess zero means and there is a sequence of functionals $f_n \in X'$ such that its elements are independent random variables with respect to μ and their linear span is dense in X' with respect to the norm in $L^2(\mu)$. Then there is a stochastic basis in X. Therefore, there is a Banach space of full measure having a Schauder basis and continuously embedded into X.*

PROOF. It is clear from what we have proved above that we can pass to the case of a separable reflexive Banach space X. Let us take a sequence $\{f_n\}$ in X' that is an orthonormal basis in X' with the inner product from $L^2(\mu)$. Let $K_\mu\colon X' \to X$ be the covariance operator of μ. We set $\varphi_k = K_\mu f_k$ and show that this is a stochastic basis. The σ-algebra generated by the functionals $f_1, \ldots, f_n$ will be denoted by $\mathcal{F}_n$.

Let us observe that by our assumptions the element $P_n x := \sum_{i=1}^n f_i(x)\varphi_i$ is the conditional expectation of the identity mapping with respect to $\mathcal{F}_n$. Indeed, for every function ψ of the form $\psi(x) = g\bigl(f_1(x), \ldots, f_n(x)\bigr)$, where the function g is bounded and continuous on $\mathbb{R}^n$, and every $l \in X'$ we have

$$\int_X l(x)\psi(x)\,\mu(dx) = \int_X l(P_n x)\psi(x)\,\mu(dx),$$

since by independence of f_n this equality is true for $l = f_j$, and for any element of X' it remains true by the density of the linear span of $\{f_n\}$ in X' with respect to the norm in $L^2(\mu)$. In addition, the Borel σ-algebra $\mathcal{B}(X)$ is generated by the union of $\mathcal{F}_n$. It is known (see Vakhania, Tarieladze, Chobanyan [**538**, Theorem 4.2b]) that this yields convergence $\|x - P_n x\| \to 0$ almost everywhere. □

The result from Fonf, Johnson, Pisier, Preiss [**173**] mentioned in § 5.12(vi) implies that stochastic bases do not always exist. Such bases exist for Gaussian measures in general spaces.

5.12.64. Example. Let γ be a centered Radon Gaussian measure on a locally convex space X and A a measurable linear operator on X with values in a separable Fréchet space Y. Then (see § 5.7), there is an orthonormal basis $\{e_n\}$ in the Cameron–Martin space $H(\gamma)$ of γ and a sequence $\{\xi_n\} \subset X^*$, orthonormal in $L^2(\gamma)$, such that the vectors $A_n x = \sum_{k=1}^n \xi_k(x) A e_k$ converge to Ax almost everywhere. It is clear that A_n are continuous linear operators. If X is also a Fréchet space, then by Theorem 5.12.56 there exist a separable reflexive Banach space $(E, \|\cdot\|_E)$ of full measure compactly embedded into X and a continuous linear operator $\widetilde{A}\colon E \to Y$ almost everywhere equal to A. In particular, this implies the well-known fact that for any Radon Gaussian measure on a Fréchet space X the space of measurable linear functionals $\Lambda_0(\gamma)$ (which coincides with $\widetilde{\Lambda}(\gamma)$) coincides also with the space of linear functions possessing continuous restrictions to compactly embedded reflexive separable Banach spaces of full measure.

We observe that in the space $\Lambda_0(\mu)$ one can consider the smaller subspaces $\Lambda_p(\mu)$ consisting of the limits of sequences of continuous linear functionals converging in $L^p(\mu)$ (certainly, this can be done for measures μ with $X' \subset L^p(\mu)$). In the case of a Gaussian measure, all spaces $\widetilde{\Lambda}(\mu)$, $\Lambda_0(\mu)$, $\Lambda_p(\mu)$ coincide. It would be interesting to enlarge the class of measures for which such coincidence holds. It is not clear whether the classes $\widetilde{\Lambda}(\mu)$ and $\Lambda_0(\mu)$ can differ for a convex Radon measure μ (but for it we always have $\Lambda_0(\mu) = \Lambda_p(\mu)$).

The situation with measurable polynomials of orders greater than one is more complicated. Here it is natural to introduce the following spaces of polynomials (for another approach, see Smolyanov [**479**] and Exercise 5.12.113):

1) the space $\mathcal{P}_k(\mu)$ consists of the limits in measure of sequences of real polynomials of the form $f(l_1, \ldots, l_n)$, where $l_i \in X'$ and f is a polynomial on $\mathbb{R}^n$ of order k;

2) the space $\mathcal{P}_{k,p}(\mu)$ consists of the limits in $L^p(\mu)$ of sequences of real polynomials of the form $f(l_1, \ldots, l_n)$, where $l_i \in X'$ and f is a polynomial on $\mathbb{R}^n$ of order k (this is meaningful for measures μ with $X' \subset L^{kp}(\mu)$);

3) $\widetilde{\mathcal{P}}_k(\mu)$ consists of μ-measurable functions possessing versions that are polynomials of order k on X in the usual algebraic sense.

In addition, in 1) and 2) in place of finite-dimensional polynomials one could take any continuous polynomials of order k. In the Gaussian case this does not enlarge the classes $\mathcal{P}_k(\mu)$ and $\mathcal{P}_{k,p}(\mu)$, but the general case has not been studied so far. In particular, it would be interesting to study the question about the relation between the spaces $\mathcal{P}_k(\mu)$ and $\widetilde{\mathcal{P}}_k(\mu)$ for convex measures. The problem is open even for $k = 1$. However, the following important result has been obtained in Arutyunyan, Yaroslavtsev [**26**]. It shows that always $\widetilde{\mathcal{P}}_k(\mu) \subset \mathcal{P}_k(\mu)$.

5.12.65. Theorem. *Let X be a linear space.* (i) *Let $n \in \mathbb{N}$ and let $f_k\colon X^n \to \mathbb{R}$ be a sequence of multilinear functions converging on some set $\Omega \subset X^n$ to a function f_0. Then there is a multilinear function $f\colon X^n \to \mathbb{R}$ such that $f|_\Omega = f_0$.*

(ii) *Let $\{f_k\}$ be a sequence of polynomials of degree d on E converging to a function f_0 on some set $\Omega \subset E$. Then there is a polynomial g of degree d such that $g|_\Omega = f_0|_\Omega$.*

Therefore, if μ is a Radon probability measure on a locally convex space X and $\{f_k\}$ is a sequence of μ-measurable polynomials of degree d converging in measure to a function f, then f has a version that is a polynomial of degree d.

PROOF. (i) Let $\{e_\alpha\}_{\alpha\in A}$ be a Hamel basis in X. Let us consider the space $\mathbb{R}^{(A^n)}$ of all real functions on A^n and its linear subspace Y generated by the indicator functions $I_{\alpha_1,\ldots,\alpha_n}$ of points $(\alpha_1,\ldots,\alpha_n)\in A^n$. The family of these indicator functions is a Hamel basis in Y. We now define linear functionals F_k on Y by setting $F_k(I_{\alpha_1,\ldots,\alpha_n}) = f_k(e_{\alpha_1},\ldots,e_{\alpha_n})$. Let us consider the subspace $Y_0\subset Y$ of all points at which the sequence $\{F_k\}$ has a finite limit. It is clear that Y_0 is a linear subspace and the formula $F_0(y) = \lim\limits_{k\to\infty} F_k(y)$ defines a linear functional F_0 on it. Let F be a linear extension of the functional F_0 to all of Y. We define a multilinear function f on the space X^n by means of the equalities $f(e_{\alpha_1},\ldots,e_{\alpha_n}) = F(I_{\alpha_1,\ldots,\alpha_n})$. Let us show that this is the desired extension of f_0. Let $x^1,\ldots,x^n\in\Omega$. The elements x^i can be represented in the form $x^i=\sum_{j=1}^m x^i_j e_j$, where $\{e_1,\ldots,e_m\}\subset\{e_\alpha\}$ is a finite set and we shall assume that $\{1,\ldots,m\}\subset A$. Then, by the linearity of f_k in every variable, we obtain

$$f_k(x^1,\ldots,x^n) = \sum_{j_1,\ldots,j_n\leqslant m} x^1_{j_1}\cdots x^n_{j_n} f_k(e_{j_1},\ldots,e_{j_n}),$$

and hence

$$f_0(x^1,\ldots,x^n) = \lim_{k\to\infty} \sum_{j_1,\ldots,j_n\leqslant m} x^1_{j_1}\cdots x^n_{j_n} f_k(e_{j_1},\ldots,e_{j_n}).$$

On the other hand, by the definition of the functions F_k we have

$$F_k\Bigl(\sum_{j_1,\ldots,j_n\leqslant m} x^1_{j_1}\cdots x^n_{j_n} I_{j_1,\ldots,j_n}\Bigr) = \sum_{j_1,\ldots,j_n\leqslant m} x^1_{j_1}\cdots x^n_{j_n} f_k(e_{j_1},\ldots,e_{j_n}),$$

whence it follows that

$$y := \sum_{j_1,\ldots,j_n\leqslant m} x^1_{j_1}\cdots x^n_{j_n} I_{j_1,\ldots,j_n}\in Y_0,$$

in addition, $F_0(y) = \lim\limits_{k\to\infty} F_k(y) = f_0(x^1,\ldots,x^n)$. Finally, by the definition of the function f and the linearity of F we have

$$f(x^1,\ldots,x^n) = \sum_{j_1,\ldots,j_n\leqslant m} x^1_{j_1}\cdots x^n_{j_n} F(I_{j_1,\ldots,j_n}) = F(y) = F_0(y),$$

therefore, $f(x^1,\ldots,x^n) = f_0(x^1,\ldots,x^n)$.

(ii) There are i-linear forms l^i_k such that $f_k(x) = \sum_{i=0}^d l^i_k(x,\ldots,x)$. For each i consider the space $\mathbb{R}^{(A^i)}$ and its subspace Y^i constructed in the same manner as Y in the proof of (i). Set $Z=\mathbb{R}\times Y^1\times\cdots\times Y^d$. For every k and $i=1,\ldots,d$ consider the linear functional $L^i_k\colon Z\to\mathbb{R}$ such that its restriction to the space Y^i, regarded as a subspace in Z, is defined as follows:

$$L^i_k|_{Y^i}(I_{e_{\alpha_1},\ldots,e_{\alpha_i}}) = l^i_k(e_{\alpha_1},\ldots,e_{\alpha_i}),$$

where $I_{e_{\alpha_1},\dots,e_{\alpha_i}}: A^i \to \mathbb{R}$ is the indicator function of the point $(e_{\alpha_1},\dots,e_{\alpha_i}) \in A^i$ in the space Y^i. This functional is defined by zero on the algebraic complement to the space Y^i and is extended to Z by linearity. Let $L_k^0\colon \mathbb{R} \to \mathbb{R}$, $L_k^0(x) = l_k^0 x$, where l_k^0 is a constant. We assume further that L_k^0 is a linear function on Z equal to zero on the algebraic complement of the subspace $Y_0 := \mathbb{R}$. Finally, set

$$F_k(z) := \sum_{i=0}^{d} L_k^i(z),$$

which gives a linear functional on Z.

Let us consider the subspace $Z_0 \subset Z$ of all points at which the sequence $\{F_k\}$ has a finite limit. It is clear that Z_0 is a linear subspace and the formula $F_0(z) = \lim\limits_{k\to\infty} F_k(z)$ defines a linear functional on Z_0. We extend F_0 to a linear functional F on all of Z. The functional F is decomposed into a sum of linear functionals depending only on the projections onto Y^i:

$$F(z) = \sum_{i=0}^{d} L_i(z) = \sum_{i=0}^{d} L_i|_{Y^i}(z_i),$$

where $z_i \in Y^i$, $z = (z_0,\dots,z_d)$. For $i \leqslant d$, by means of the equalities

$$l_i(\alpha_1,\dots,\alpha_i) = L_i|_{Y_i}(I_{\alpha_1,\dots,\alpha_i}), \qquad l_0 = L_0|_{Y_0}(1)$$

we define the multilinear functions l_i on X^n. Finally, we define a polynomial on the space X as follows: $f(x) = \sum_{i=0}^d l_i(x,\dots,x)$. Let us show that this is the desired extension of f_0. Every element $x \in \Omega$ can be represented in the form $x = \sum_{s=1}^m x_s e_s$, where $e_s \in \{e_\alpha\}$ (as above, we assume that $\{1,\dots,m\} \subset A$). We show that $f(x) = f_0(x)$. By the linearity of l_k^i in every variable we obtain

$$l_k^i(x,\dots,x) = \sum_{j_1,\dots,j_i \leqslant m} x_{j_1}\cdots x_{j_i} l_k^i(e_{j_1},\dots, e_{j_i}),$$

whence it follows that

$$f_0(x) = \lim_{k\to\infty} \sum_{i=0}^{d} \sum_{j_1,\dots,j_i \leqslant m} x_{j_1}\cdots x_{j_i} l_k^i(e_{j_1},\dots,e_{j_i}).$$

On the other hand, by the definition of the functions L_k^i we have

$$\sum_{i=0}^{d} L_k^i\Bigl(\sum_{j_1,\dots,j_i} x_{j_1}\cdots x_{j_i} I_{\alpha_{j_1},\dots,\alpha_{j_i}}\Bigr) = \sum_{i=0}^{d} \sum_{j_1,\dots,j_i \leqslant m} x_{j_1}\cdots x_{j_i} l_k^i(e_{j_1},\dots,e_{j_i}),$$

whence it follows that for $y = (y_0,\dots,y_d)$, where

$$y_i := \sum_{j_1,\dots,j_i} x_{j_1}\cdots x_{j_i} I_{\alpha_{j_1},\dots,\alpha_{j_i}} \in Y^i,$$

we have the equality

$$F_k(y) = \sum_{i=0}^{d} \sum_{j_1,\dots,j_i \leqslant m} x_{j_1}\cdots x_{j_i} l_k^i(e_{j_1},\dots,e_{j_i}) = f_k(x).$$

Hence $f_0(x) = \lim_{k\to\infty} f_k(x) = \lim_{k\to\infty} F_k(y) = F_0(y)$ and $y \in Z_0$. Finally, by the definition of the function F we obtain

$$f(x) = \sum_{i=0}^{d} \sum_{j_1,\ldots,j_i \leqslant m} x_{j_1} \cdots x_{j_i} L_k^i(I_{j_1,\ldots,j_i}) = F(z) = F_0(z),$$

therefore, $f(x) = f_0(x)$. □

Note that this result was a long standing problem even for a Gaussian measure μ on a separable Hilbert space (and even for $d = 2$). In relation to this result, see also Arutyunyan, Kosov, Yaroslavtsev [**25**].

Let us mention the following important result due to Bobkov [**56**].

5.12.66. Theorem. *Set $C := 22/\ln 2$. Let ν be a convex measure on $\mathbb{R}^n$ and let f be a polynomial on $\mathbb{R}^n$ of degree at most k. Then for all $p \in [1,\infty)$ one has*

$$\|f\|_{L^p(\nu)} \leqslant p^{Ck} \|f\|_{L^1(\nu)}.$$

In particular, on the space $\mathcal{P}_k(\mathbb{R}^n)$ of polynomials on $\mathbb{R}^n$ of degree at most k all $L^p(\nu)$-norms are equivalent with constants independent of dimension n, but dependent only on k and p.

Bobkov's estimate yields the following result (obtained in Berezhnoy [**51**]).

5.12.67. Theorem. *Let μ be a convex probability measure on a Hausdorff locally convex space X. Then the following assertions are true.*

(i) *For every $p \in [1,\infty)$ we have $\mathcal{P}_k(\mu) = \mathcal{P}_{k,p}(\mu)$ and*

$$\|f\|_{L^p(\nu)} \leqslant p^{Ck} \|f\|_{L^1(\nu)}, \quad f \in \mathcal{P}_k(\mu). \tag{5.12.13}$$

(ii) *On the space $\mathcal{P}_k(\mu)$, the norms of all $L^p(\mu)$ with $p \in [1,+\infty)$ are equivalent, moreover, it is complete with respect to each of them.*

(iii) *If a sequence $\{f_j\} \subset \mathcal{P}_k(\mu)$ converges in measure, then it converges in all spaces $L^p(\mu)$, $p \in [1,+\infty)$.*

PROOF. Let $\mathcal{P}_k^{fin}$ be the class of finite-dimensional polynomials on X of the form $f = f_0(l_1,\ldots,l_n)$, where $l_i \in X'$ and f_0 is a polynomial on $\mathbb{R}^n$ of degree k. Since every convex measure on $\mathbb{R}^n$ has all moments, we have $\mathcal{P}_k^{fin} \subset L^p(\mu)$ for all $p < \infty$. Suppose that $\varphi_j \in \mathcal{P}_k^{fin}$ converge in measure to φ. The previous theorem gives (5.12.13) for all $\psi \in \mathcal{P}_k^{fin}$. This estimate holds for $p = 2$ and $\psi = \varphi_j$, which yields the uniform boundedness of $\|\varphi_j\|_{L^1(\mu)}$. Indeed, otherwise we can assume that $\|\varphi_j\|_{L^1(\mu)} \to \infty$. Then the functions $\varphi_j/\|\varphi_j\|_{L^1(\mu)}$ converge to zero in measure. Their norms in $L^1(\mu)$ equal 1 and the norms in $L^2(\mu)$ are dominated by 2^{Ck}. Hence they converge to zero in $L^1(\mu)$, which is impossible. The boundedness in $L^p(\mu)$ along with convergence in measure gives convergence in $L^r(\mu)$ for any $r < p$. Since this is true for all $p < \infty$, the sequence $\{\varphi_j\}$ converges to φ in all $L^p(\mu)$. Thus, we have $\varphi \in L^p(\mu)$ and (5.12.13) for all $\psi \in \mathcal{P}_k(\mu)$. Applying this reasoning to the whole class $\mathcal{P}_k(\mu)$, we obtain assertions (i)–(iii). The equivalence of all L^p-norms follows from (5.12.13) and the bound $\|f\|_{L^1(\mu)} \leqslant \|f\|_{L^p(\mu)}$. The completeness of $\mathcal{P}_k(\mu)$ with respect to all these norms is clear from the proven facts. □

The following result from Arutyunyan, Kosov [**24**] has a more involved proof.

5.12.68. Theorem. *Let μ be a convex Radon measure on a locally convex space E and let f be a μ-measurable polynomial of degree d. Then $f \in L^p(\mu)$ for every $p > 0$. Moreover, if f does not coincide with a constant a.e., the following estimate is true*:

$$\|f\|_{L^p(\mu)} \leqslant k(f)c^d\bigl(pd(1+\Gamma(pd))\bigr)^{1/p},$$

where $k(f)$ is the number from the previous corollary. If $pd \geqslant 1$, the following estimate is also true:

$$\|f\|_{L^p(\mu)} \leqslant k(f)(c_1 pd)^d,$$

where $c_1 > 1$ is a universal constant independent of d and p.

Yet another result from [**24**] says that for limits in measure of continuous polynomials the L^p-norms on the whole space are equivalent to the L^p-norms on arbitrary subsets of positive measure.

5.12.69. Theorem. *Let μ be a convex Radon measure on a locally convex space E and let U be a measurable set of a positive measure. Let f be a limit in measure of a sequence of continuous polynomials of degree d. Then*

$$\mu(U)^{d+1}\int_E |f|\,d\mu \leqslant (Cd)^{2d}\int_U |f|\,d\mu,$$

where C is a universal constant. In other words,

$$\|f\|_{L^1(\mu)} \leqslant \frac{(Cd)^{2d}}{\mu(U)^d}\|f\|_{L^1(\mu_U)}, \quad \textit{where} \quad \mu_U(A) = \frac{\mu(A\cap U)}{\mu(U)}.$$

The following $0-1$ law for polynomials is also proved in [**24**].

5.12.70. Theorem. *Let μ be a convex Radon measure on a locally convex space, and let f and f_n, where $n \in \mathbb{N}$, be measurable polynomials of degree d. Then, for every constant C, $\mu(\{x\colon\ f(x) = C\})$ equals either 0 or 1. Moreover, $\mu(\{x\colon\ \exists \lim\limits_{n\to\infty} f_n(x)\})$ also equals either 0 or 1.*

See also the survey Bogachev [**68**] and Yurova [**577**].

5.12(x). Relations between different σ-algebras

Let X be a locally convex space. As we have seen, for the purposes of measure theory it is useful to consider the σ-algebra $\sigma(X')$ generated by cylinders in X, the Borel σ-algebra $\mathcal{B}(X)$, and the Baire σ-algebra $\mathcal{B}a(X)$. If we equip X with other locally convex topologies, the corresponding triples of σ-algebras arise (certainly, for topologies that agree with duality $\sigma(X')$ does not change). If X is a separable Fréchet space, then these three σ-algebras coincide, and in metrizable spaces $\mathcal{B}(X)$ and $\mathcal{B}a(X)$ coincide. In the general case these σ-algebras can differ. Hence it is useful to have nontrivial cases of their coincidence. We mention some of them. The following important result goes back to Bokshtein [**58**] and Mibu [**341**] (see also Arkhangel'skii, Ponomarev [**22**, Chapter II, § 5, Exercise 386]) and describes the structure of Baire sets in product spaces (not necessarily locally convex).

5.12.71. Theorem. *Let $\{X_t\}_{t\in T}$ be a collection of separable spaces, let Y be a separable metric space, and let $F\colon \prod_{t\in T} X_t \to Y$ be a continuous mapping. Then there exist a finite or countable set $S\subset T$ and a continuous map $F_0\colon \prod_{s\in S} X_s \to Y$ such that $F = F_0\circ\pi_S$, where $\pi_S\colon \prod_{t\in T} X_t \to \prod_{s\in S} X_s$ is the natural projection. In particular, $\mathcal{B}a\bigl(\prod_{t\in T} X_t\bigr)$ is generated by the coordinate mappings to the spaces $\bigl(X_t, \mathcal{B}a(X_t)\bigr)$.*

This assertion is based on the following interesting fact (it was this fact that was discovered by M.F. Bokshtein), which we need below (for the proof, see [**22**, Chapter II, § 5, Exercise 385]).

5.12.72. Proposition. *The closure of every open set U in the product of separable spaces X_t, $t\in T$, depends only on countably many coordinates, i.e., for some at most countable set $S\subset T$ it has the form $Z\times\prod_{t\notin S} X_t$, where Z is closed in $\prod_{s\in S} X_s$. An equivalent assertion: for any disjoint open sets U and V in $\prod_{t\in T} X_t$, there exists an at most countable set $S\subset T$ such that $\prod_{s\in S} X_s$ contains disjoint open sets $\widetilde{U}$ and $\widetilde{V}$ with $U\subset \pi_S^{-1}(\widetilde{U})$, $V\subset\pi_S^{-1}(\widetilde{V})$.*

The next result (following from Theorem 5.12.71) is due to Edgar [**148**].

5.12.73. Theorem. *If X is equipped with the topology $\sigma(X,X')$, then the corresponding Baire σ-algebra coincides with $\sigma(X')$.*

PROOF. It is clear that $\sigma(X')$ is contained in the Baire σ-algebra of the space X with the weak topology. To verify the inverse inclusion it suffices to show that, for every function F on X continuous in the topology $\sigma(X,X')$, the set $\{x\in X\colon F(x)>0\}$ belongs to $\sigma(X')$. Let us embed X into $\mathbb{R}^T$ as a dense linear subspace, taking for T a Hamel basis in X'; the density follows from the fact that for every finite collection $t_1,\ldots,t_n\in T$ and any numbers $c_1,\ldots,c_n$ there exists a vector $v\in X$ such that $t_i(v)=c_i$. Moreover, after this embedding, the weak topology in X coincides with the one induced from $\mathbb{R}^T$. For any rational number r let $U_r=\{x\in X\colon F(x)>r\}$, $V_r=\{x\in X\colon F(x)<r\}$. Let us take open sets $\widetilde{U}_r,\widetilde{V}_r$ in $\mathbb{R}^T$ with $\widetilde{U}_r\cap X=U_r$, $\widetilde{V}_r\cap X=V_r$. We have $\widetilde{U}_r\cap\widetilde{V}_r=\varnothing$, since X is dense in $\mathbb{R}^T$. Now we apply Proposition 5.12.72, which gives a countable set S and open sets U_r', V_r' in $\mathbb{R}^S$ such that $U_r'\cap V_r'=\varnothing$, $\widetilde{U}_r\subset\pi_S^{-1}(U_r')$, $\widetilde{V}_r\subset\pi_S^{-1}(V_r')$. The open sets U_r', V_r' in the metrizable space $\mathbb{R}^S$ are Baire, hence $X\cap\pi_S^{-1}(U_r')$ and $X\cap\pi_S^{-1}(V_r')$ belong to $\sigma(X')$. It remains to observe that $\{x\in X\colon F(x)>0\}$ coincides with the union of the sets $X\cap\pi_S^{-1}(U_r')$ over all rational numbers $r>0$, which is verified directly. □

Let us mention the following fact (see Bogachev [**65**, Chapter 6]).

5.12.74. Theorem. *If X is a Souslin locally convex space, then we have the equalities $\sigma(X')=\mathcal{B}(X)=\mathcal{B}a(X)$.*

5.12.75. Proposition. *Suppose that a Banach space X with the closed unit ball U and unit sphere S is equipped with a locally convex topology τ such that the ball U is closed in τ. Then the mapping $(t,u)\mapsto tu$ is a Borel isomorphism of the spaces $\bigl((0,+\infty)\times S, \mathcal{B}\bigl((0,+\infty)\times(S,\tau)\bigr)\bigr)$ and $\bigl(X\backslash\{0\}, \mathcal{B}(X\backslash\{0\},\tau)\bigr)$.*

PROOF. Since the topology τ is locally convex, the indicated mapping is continuous with respect to this topology on X, hence it is Borel measurable. The inverse mapping has the form $x \mapsto (\|x\|, x/\|x\|)$. It follows from our assumption that the norm is Borel measurable on (X,τ), whence it follows that the inverse mapping is Borel measurable. □

Note that we have the equality

$$\mathcal{B}\big((0,+\infty)\times(S,\tau)\big) = \mathcal{B}\big((0,+\infty)\big)\otimes\mathcal{B}(S,\tau),$$

and the same equality is true for any separable metric space in place of $(0,+\infty)$, see, e.g., [**65**, Lemma 6.4.2].

5.12.76. Corollary. *Let τ_1 and τ_2 be two locally convex topologies on a Banach space X such that the closed unit ball is closed in both of them and they coincide on the unit sphere. Then $\mathcal{B}(X,\tau_1) = \mathcal{B}(X,\tau_2)$.*

For example, if X is Hilbert, then for τ_1 and τ_2 one can take the norm topology and the weak topology (Exercise 3.12.49).

Let us make a remark on measurability of operations in a locally convex space E equipped with a σ-algebra $\mathcal{E}$. The following subtlety arises here: the product $\mathcal{E}\otimes\mathcal{E}$ can be smaller than the corresponding σ-algebra in $E\times E$. For example, this can happen for the Borel σ-algebra $\mathcal{B}(E)$, see [**65**, v. 2, p. 15]. As a result, for a nonseparable Banach space X the addition operation $X\times X \to X$ can fail to be measurable with respect to $\mathcal{B}(X)\otimes\mathcal{B}(X)$ and $\mathcal{B}(X)$. Note that it is shown in Talagrand [**513**] that X is a measurable vector space, i.e., the operations $(t,x,y) \mapsto tx+y$, $\mathbb{R}^1\times X\times X \to X$ are measurable with respect to $\mathcal{B}(\mathbb{R}^1)\otimes\mathcal{B}(X)\otimes\mathcal{B}(X)$ and $\mathcal{B}(X)$, precisely when $\mathcal{B}(X)\otimes\mathcal{B}(X) = \mathcal{B}(X\times X)$. In the same paper, the existence of a nonseparable Banach space X satisfying this equality is proved. In addition, it is shown there that the continuum hypothesis implies the measurability (in the indicated sense) of the space l^∞. Certainly, the operations, being continuous, become measurable if $X\times X$ is considered with its Borel σ-algebra. It is also worth noting that, according to Talagrand [**512**], in the space l^∞ the Borel σ-algebras corresponding to the weak topology and the norm topology are different. However, there are no such measurability problems for the σ-algebra $\sigma(E')$ generated by E': for all locally convex spaces X and Y we have

$$\sigma\big((X\times Y)'\big) = \sigma(X')\otimes\sigma(Y'),$$

Since $(X\times Y)' = X'\times Y'$, the σ-algebra $\sigma\big((X\times Y)'\big)$ is generated by sets of the form $\{(x,y)\colon\ f(x)+g(y) < c\}$, where $f\in X'$, $g\in Y'$, but sets of this form can be written as a countable unions of sets $\{x\colon\ f(x)\in C_j\}\times\{y\colon\ g(y)\in D_j\}$, by writing the half-plane $\{(s,t)\colon\ s+t<c\}$ as the union of a countable collection of squares $C_j\times D_j$.

According to Fremlin [**177**], there exists a Banach space X in which the Borel σ-algebra is not generated by balls. However, as shown in Talagrand [**511**], there exists a nonseparable metric space the Borel σ-algebra of which is generated by balls.

5.12(xi). Radonifying operators

Let X and Y be locally convex spaces. An operator $T \in \mathcal{L}(X,Y)$ is called *radonifying* if it transforms every probability cylindrical measure on X to a measure on Y possessing a Radon extension. If this is true for all cylindrical measures with weak moment of some order $p \in (0,\infty)$, then T is called *p-radonifying*; finally, if this property holds for Gaussian cylindrical measures, then the operator T is called *γ-radonifying*. The classes of radonifying, p-radonifying and γ-radonifying operators are denoted by the symbols $\mathfrak{R}(X,Y)$, $\mathfrak{R}_p(X,Y)$, and $\mathfrak{R}_\gamma(X,Y)$, respectively. The class $\mathfrak{R}_{p,q}(X,Y)$, where $0 \leqslant q \leqslant p$, consists of the operators $T \in \mathcal{L}(X,Y)$ transforming every probability cylindrical measure with weak moment of order p to a measure with a Radon extension and strong moment of order q.

For Hilbert spaces X and Y, the class of radonifying operators coincides with the class of γ-radonifying operators and with the class of Hilbert–Schmidt operators. A nontrivial example of a noncompact radonifying operator is the identity embedding $l^1 \to l^2$.

If X and Y are Banach spaces, then $\mathfrak{R}_{p,p}(X,Y) = \mathfrak{R}_p(X,Y)$ for $p > 0$, in addition, every p-radonifying operator is p-summing and every radonifying operator is p-summing for all $p > 0$. In the case $1 < p < \infty$ the classes of p-radonifying and p-summing operators coincide. If $0 < p < 1$, then one has the inclusion $\mathfrak{R}_p(X,Y) \subset \mathfrak{R}(X,Y)$.

There is a vast literature on these classes in Banach spaces, see, for example, Diestel, Jarchow, Tonge [**121**], van Neerven [**367**], Pietsch [**389**], Schwartz [**442**], Vakhania, Tarieladze, Chobanyan [**538**], and Vladimirskii [**550**], [**551**], [**552**]. Many results for Banach spaces apply directly to locally convex spaces by means of factorizations of operators through Banach or Hilbert spaces. For example, every continuous linear operator from a Banach space to a nuclear space is radonifying, since it is a composition of a continuous operator and a Hilbert–Schmidt operator in a Hilbert space.

5.12(xii). Measurable norms

As we have seen in § 5.10, the class of measurable norms introduced there is broader than the class of Gross measurable norms (Remark 5.10.13). The latter definition was extended in Dudley, Feldman, Le Cam [**140**] to general nonnegative cylindrical measures on locally convex spaces. Let $Y = E'$ be the dual to a locally convex space E and let m be a nonnegative E-cylindrical measure on Y. A seminorm q on Y will be called (DFL)-measurable with respect to m if the following condition is fulfilled: for every $\varepsilon > 0$, there exists a finite-dimensional subspace $L \subset Y$ such that for every finite-dimensional subspace K contained in the polar of L taken in E (i.e., in the annihilator of L) we have

$$m\big(y \in Y\colon\ \inf\{q(y-z),\, z \in K^\circ\} > \varepsilon\big) \leqslant \varepsilon.$$

According to [**140**], for a given seminorm q continuous in the Mackey topology $\tau(Y,E)$, its (DFL)-measurability with respect to m is equivalent to the following property: the image of m under the natural mapping from Y to the Banach space generated by q has a Radon extension. In addition, if a continuous linear operator T from Y to a locally convex space Z with a dense range takes m to a measure with a Radon extension, then, for every continuous seminorm p on Z, the seminorm $p\circ T$ is (DFL)-measurable with respect to m. Finally, if $Y = l^2$ and m is the standard cylindrical Gaussian measure on Y, then the (DFL)-measurability coincides with the Gross measurability. However, Maeda [**329**] and Maeda, Harai, Hagihara [**330**] constructed examples of non-Gaussian cylindrical measures m on a Hilbert space for which there are (DFL)-measurable norms that are not Gross measurable (such a norm can be even made measurable with respect to the standard Gaussian cylindrical measure).

Exercises

5.12.77. Let $\mathcal{F}$ be some family of real functions on a set X and let $\sigma(\mathcal{F})$ be the σ-algebra in X generated by this family. Prove that every function g measurable with respect to $\sigma(\mathcal{F})$ has the form $g = \psi(f_1, f_2, \ldots)$, where ψ is a Borel function on $\mathbb{R}^\infty$ and $\{f_n\} \subset \mathcal{F}$.

5.12.78° Prove that the linear span of a compact set in a topological vector space is a Borel set.

5.12.79° Let $E \subset [0,1]$ be a Lebesgue nonmeasurable set of outer measure 1 and let μ be the restriction of Lebesgue measure to the Borel σ-algebra of E; then μ is a non-Radon Borel measure. Show that there exists a sequence of measures with finite supports in E weakly converging to μ.

5.12.80° Let γ_n be the standard Gaussian measure on the space $\mathbb{R}^n$, i.e., the measure with density $(2\pi)^{-n/2}\exp\bigl(-(x,x)^2/2\bigr)$ with respect to Lebesgue measure. Show that $\lim\limits_{n\to\infty} \gamma_n(B_R) = 0$ for the ball B_R of radius R centered at the origin in $\mathbb{R}^n$.

5.12.81. (i) Let ν be a centered Gaussian measure on $\mathbb{R}^n$ and let C be an absolutely convex set. Prove that $\nu(C-h) \leqslant \nu(C)$ for all vectors $h \in \mathbb{R}^n$.

(ii) Let ν_1 and ν_2 be two centered Gaussian measures on $\mathbb{R}^n$ with covariance operators R_1 and R_2 such that $R_1 \leqslant R_2$, i.e., $(R_1 v, v) \leqslant (R_2 v, v)$. Prove that for every absolutely convex set C in $\mathbb{R}^n$ one has $\gamma_1(C) \geqslant \gamma_2(C)$.

(iii) Let C be an absolutely convex set in $\mathbb{R}^n$ and let γ_n be the standard Gaussian measure on $\mathbb{R}^n$. Prove that for every linear operator S on $\mathbb{R}^n$ such that $\|S\| \leqslant 1$ one has $\gamma_n\bigl(S(C)\bigr) \leqslant \gamma_n(C)$. In addition, show that $\gamma_n(C) \leqslant \gamma_k(C \cap \mathbb{R}^k)$ whenever $k < n$.

HINT: (i) pass to the standard Gaussian measure and use polar coordinates; (ii) observe that $\gamma_1 = \gamma_2 * \gamma_3$, where γ_3 is a centered Gaussian measure, apply (i); deduce (iii) from (ii).

5.12.82. Prove that every Borel measure on $\mathbb{R}^\infty$ is concentrated on a compactly embedded Hilbert space.

HINT: observe that such a measure is concentrated on the linear span of a product of compact intervals.

5.12.83° Let ν be a cylindrical measure on a locally convex space X and let T be a continuous linear operator from E to a locally convex space Y. Show that the Fourier transform of the cylindrical measure $\eta = \nu\circ T^{-1}$ is given by $\widetilde{\eta}(f) = \widetilde{\nu}(T^* f)$, $f \in Y'$.

5.12.84. Let μ be the probability measure on l^2 concentrated on the countable set of vectors ne_n such that $\mu(ne_n) = cn^{-2}$. Show that there is no Banach space E of full measure compactly embedded into l^2 such that $E' \subset L^2(\mu)$, although $(l^2)' \subset L^2(\mu)$.

HINT: verify that otherwise one could find a larger Hilbert space E with the same property, which would imply compactness of the covariance operator of the measure μ on l^2, but this operator is $c \cdot I$.

5.12.85. Let μ be a probability measure on c_0 with first weak moment and without mean, for example, $\mu(2^n e_n) = 2^{-n}$, where $\{e_n\}$ is the standard basis in c_0. Show that there is no Banach space E of full measure compactly embedded into c_0 such that $E' \subset L^1(\mu)$.

HINT: verify that otherwise one could find a larger reflexive Banach space E with the same property, which would imply the existence of the mean of μ.

5.12.86. Prove that the space $\mathcal{D}(\mathbb{R}^1)$ with its standard topology is not a k_R-space (for the definition, see § 5.12(viii)).

5.12.87. Prove Proposition 5.12.52.

5.12.88. Prove that every k-space possessing a countable fundamental system of compact sets is normal.

HINT: assuming that X is the union of increasing compact sets K_n with the property that every compact set in X is contained in some of K_n, for every disjoint closed sets A and B, one constructs inductively a continuous function f with $f|_A = 0$, $f|_B = 1$; such a function can be constructed on the compact set K_1; if it is already constructed on K_n, it extends to K_{n+1} in such a way that $f|_{A\cap K_{n+1}} = 0$, $f|_{B\cap K_{n+1}} = 1$; to this end, we set $f = 0$ on $A \cap K_{n+1}$, $f = 1$ on $B \cap K_{n+1}$, then observe that we have obtained a continuous function on the compact set $K_n \cup (A \cap K_{n+1}) \cup (B \cap K_{n+1})$ and extend it to K_{n+1}.

5.12.89°. Suppose that X and Y are locally convex spaces, linear operators $A_n\colon X \to Y$ are continuous and on some set $S \subset X$ these operators uniformly converge to zero. Prove that one has uniform convergence on the closed absolutely convex envelope of S.

5.12.90. Let K be a compact set in a locally convex space E, let $\{f_n\} \subset E'$ be uniformly bounded on K, and let $f_n(x) \to 0$ for all $x \in K$. Prove that $f_n(y) \to 0$ for all vectors y in the closed convex envelope of K.

HINT: apply Theorem 5.6.13 and the Lebesgue dominated convergence theorem.

5.12.91. Let ξ and η be independent random vectors in a locally convex space X such that the distribution of η is symmetric. Prove that for every absolutely convex set $C \in \sigma(X')$ one has $\mathrm{P}(\xi \notin C) \leqslant 2\mathrm{P}(\xi + \eta \notin C)$, and if the distributions of ξ and η are Radon, then this is true for all Borel absolutely convex sets C.

HINT: let $D = X \backslash C$ and observe that $D \subset (D - x) \cup (D + x)$ for every x.

5.12.92. Suppose that a Hausdorff locally convex space E is the strict inductive limit of its increasing closed subspaces E_n and μ is a Radon probability measure on E such that E' belongs to $L^1(\mu)$. Prove that there exists n such that $\mu(E_n) = 1$.

HINT: if there are compact sets $K_n \subset E \backslash E_n$ with $\mu(K_n) > 0$, then we can assume that $K_n \subset E_{n+1}$; there exists a functional $f \in E'$ such that the integral of $|f(x)|$ over K_n is not less than 1; to this end, we construct inductively functionals $f_n \in E'$ with $f_n|_{E_n} = 0$ for which the integral of the function $|f_n(x)|$ over K_n is greater than the integral of the function $|f_1(x) + \cdots + f_{n-1}(x)| + 1$ over K_n; finally, take $f = \sum_{n=1}^{\infty} f_n \notin L^1(\mu)$.

5.12.93. Prove assertion (ii) in Theorem 5.12.25 using the previous exercise.

5.12.94°. Let K be a convex compact set in a locally convex space. Suppose that a sequence of Radon probability measures μ_n on K converges to a measure μ in the weak-$*$ topology on $C(K)'$. Prove that the barycenters of the measures μ_n converge to the barycenter of μ.

HINT: on K the weak topology coincides with the original one.

5.12.95. (Talagrand [**515**, p. 184]) Let E be the Banach space of all bounded functions on $[0,1]$ that do not vanish in at most countably many points and let E be equipped with the norm $\sup_t |f(t)|$. Then, on the weak Borel σ-algebra $\mathcal{B}((E,\sigma(E,E')))$ there exists a probability measure that assumes only the values 0 and 1, but is not Radon.

5.12.96. Construct an example of a cylindrical quasi-measure of unbounded variation on l^2 the characteristic functional of which is bounded and continuous in the Sazonov topology.

HINT: let $p_n(t) = \pi^{-1}2^n(1+2^{2n}t^2)^{-1}$; take the product of the measures $p_n\,dt$ and the function $\varphi(y) = -i(y,h)\widetilde{\mu}(y)$, see Bogachev [**66**, Example 3.6.9].

5.12.97. (Kwapień [**308**]) Let ξ_n be random vraiables on a probability space $(\Omega,\mathcal{F},P)$ such that the series $\sum_{n=1}^\infty \lambda_n\xi_n$ converges in probability for every sequence of numbers λ_n converging to zero. Prove that $\sum_{n=1}^\infty |\xi_n|^2 < \infty$ a.e. Deduce from this that the embedding $l^1 \to l^2$ is a radonifying operator, i.e., takes every nonnegative cylindrical quasi-measure on l^1 with a continuous characteristic functional to a Radon measure on l^2.

5.12.98. (Borell [**81**]) A linear function on a Fréchet space measurable with respect to all Radon Gaussian measures is continuous. Hence the measurability of a linear function on a Banach space with respect to all Radon measures implies its continuity (Schwartz [**441**]).

HINT: see Bogachev [**63**, Proposition 3.9.9]; see also related results in Christensen [**101**] and Kats [**265**].

5.12.99. (Talagrand [**514**]) Under Martin's axiom, in every infinite-dimensional separable Banach space X there exists a non-closed hyperplane X_0 that is measurable with respect to every Borel measure on X. Deduce from the previous exercise that X_0 cannot be a kernel of a universally measurable linear function.

5.12.100. Let T be a topological space and let $\mathcal{M}_\sigma(T)$ be the space of all Baire measures on T with the weak topology $\sigma(\mathcal{M}_\sigma(T), C_b(T))$. Show that the set of finite linear combinations of Dirac measures is everywhere dense in $\mathcal{M}_\sigma(T)$.

HINT: given a measure $\mu \in \mathcal{M}_\sigma(T)$, functions $f_1,\ldots,f_n$ in the unit ball of the space $C_b(X)$ and $\varepsilon > 0$, one can partition T into disjoint parts $T_1,\ldots,T_N$ of the form $\bigcap_{i=1}^n \{c \leqslant f_i \leqslant c+\varepsilon\}$; assuming that $\|\mu\| \leqslant 1$, take arbitrary points $t_i \in T_i$ and consider the measure $\sum_{i=1}^N \mu(T_i)\delta_{t_i}$.

5.12.101. (Ostling, Wilansky [**375**]) In a sequentially complete locally convex space the closed convex envelope of a compact set is not always compact. Show that for such a compact set one can take $K = \{0,1\}^\kappa$, where the cardinality κ is greater than the continuum, so that K is nonseparable, K is embedded into the space $\mathcal{M}$ of Radon measures on K with separable supports equipped with the weak topology $\sigma(\mathcal{M}, C(K))$, and the embedding takes $k \in K$ to the Dirac measure δ_k.

HINT: the image of K in $\mathcal{M}$ is compact (the embedding of K is a homeomorphism); $\mathcal{M}$ is sequentially complete, since a weak limit of any sequence of measures with separable topological supports also has a separable support; in addition, $\mathcal{M}$ is dense in the space $\mathcal{M}_r$ of all Radon measures with the weak topology and the closed convex envelope of K in $\mathcal{M}_r$ is the set of all Radon probability measures on K and contains the measure ν^κ, where $\nu(-1) = \nu(1) = 1/2$; finally, the measure ν does not belong to $\mathcal{M}$, since its support is the whole set K.

5.12.102. (von Weizsäcker [**562**]) (i) Let X be the space of all Borel measures on $[0,1]$ equipped with the weak topology and let K be the convex compact set in X consisting of all probability measures. Let λ be Lebesgue measure and let μ be the image of λ under the continuous (in the indicated topology) mapping $\pi\colon t \mapsto \delta_t$, $[0,1] \to K$, where δ_t is the Dirac measure at the point t. Set

$$C := \bigcap_{n=1}^{\infty} \{m \in K\colon\ \lambda + n^{-1}(\lambda - m) \notin K\}.$$

Prove that C is a convex G_δ-set in K and $\mu(C) = 1$, but $\mu(S) = 0$ for every convex compact set $S \subset C$.

(ii) Let K be a convex compact set in a locally convex space X having an infinite-dimensional linear span. Prove that there exist a convex set $C \subset K$ and a Radon probability measure μ on K such that C is a G_δ-set in some metrizable convex compact set $K_0 \subset K$, $\mu(C) = 1$, but $\mu(S) = 0$ for every convex compact set $S \subset C$.

HINT: (i) It is easy to verify that C is convex and can be represented as the intersection of a sequence of open sets in K with the weak topology, moreover,

$$C = K \backslash \bigcup_{\varepsilon > 0} \{m \in K\colon\ \lambda + \varepsilon(\lambda - m) \in K\}.$$

Let D be the set of all Dirac measures. Then D is compact in K and $\mu(D) = 1$. If $S \subset C$ is a convex compact set with $\mu(C) > 0$, then $\mu(S \cap D) > 0$. Hence the set $A := \pi^{-1}(S \cap D)$ is compact and $\lambda(A) > 0$. Since $\delta_t \in S$ for $t \in A$, by the convexity and closedness of S we obtain that every probability measure ν on A belongs to S. In particular, $\nu := \lambda(A)^{-1}\lambda|_A \in S$. The measure $\lambda + \lambda(A)(\lambda - \nu)$ is probability, i.e., belongs to K. According to the indicated equality for C we obtain that $\nu \notin C$. This contradicts the fact that $\nu \in S \subset C$. Assertion (ii) can be deduced from (i) with the aid of a suitable mapping (see details in [**562**]).

5.12.103. (Fremlin [**178**, § 461E]). Let X be a complete locally convex space and let μ be a τ-additive probability measure on X (i.e., if $\{U_\alpha\}$ is an increasing net of open sets, then $\mu\bigl(\bigcup_\alpha U_\alpha\bigr) = \lim_\alpha \mu(U_\alpha)$) with bounded support. Then μ has a barycenter.

5.12.104. Let μ and ν be τ-additive measures on a locally convex space X with equal Fourier transforms. Prove that $\mu = \nu$.

HINT: let p be a continuous seminorm on X, let X_p be the normed space obtained by factorization of X over $p^{-1}(0)$, and let $\pi_p\colon X \to X_p$ be the natural projection. The sets of the form $\pi_p^{-1}(U)$, where p is a continuous seminorm and U is open in X_p, form a topology base in X. Hence it suffices to show the coincidence of μ and ν on these sets. The measures $\mu \circ \pi_p^{-1}$ and $\nu \circ \pi_p^{-1}$ on X_p have equal Fourier transforms and are τ-additive. Both properties are preserved for the natural extensions of both measures to the completion of X_p. On a Banach space, any τ-additive measure is Radon, which gives the equality of the indicated extensions on the completion of X_p, hence also the equality $\mu \circ \pi_p^{-1} = \nu \circ \pi_p^{-1}$. Therefore, $\mu\bigl(\pi_p^{-1}(U)\bigr) = \nu\bigl(\pi_p^{-1}(U)\bigr)$ for all open sets $U \subset X_p$.

5.12.105. Let X be a Hausdorff topological vector space, let μ and ν be Radon probability measures such that $\mu = \mu * \nu$. Show that ν is the Dirac measure at the origin.

HINT: see Vakhania, Tarieladze, Chobanyan [**538**, Proposition I.4.7].

5.12.106. (Grothendieck [**207**, p. 229]) Suppose that K is a nonempty compact space and a set $M \subset \mathcal{M} = \mathcal{M}_r(K) = C(K)'$ has compact closure in the Mackey topology $\tau\bigl(\mathcal{M}, C(K)\bigr)$. Show that M has compact closure in the topology $\sigma(\mathcal{M}, \mathcal{M}')$.

HINT: according to the Eberlein–Šmulian theorem and Theorems 4.7.24 and 8.10.58 in [**65**], it suffices to show that $\lim\limits_{n\to\infty} \mu_n(U_n) = 0$ for every sequence of measures $\mu_n \in M$ and every sequence of disjoint open sets $U_n \subset K$. If this is false, then there exist functions

$f_n \in C(K)$ such that $|f_n| \leqslant 1$, $f_n = 0$ outside of U_n and the integral of f_n against μ_n is not less than some $\varepsilon > 0$; $\{f_n\}$ converges to zero pointwise, hence in the weak topology in the space $C(K)$. It is readily seen that the closed convex envelope of $\{f_n\}$ is weakly compact. This contradicts compactness of the closure of M in the topology of uniform convergence on convex weakly compact sets.

5.12.107. Construct an example of a probability measure on a locally convex space (X, τ) that is defined on $\sigma(X')$ and is tight in the weak topology $\sigma(X, X')$, but is not tight in the original topology τ.

HINT: let $E = C[0,1]$, let $X := E'$ be equipped with the topology $\sigma(E', E)$, and let μ be the image of Lebesgue measure on $[0,1]$ under the mapping $t \mapsto \delta_t$. Then μ is a tight Baire measure in the topology $\sigma(E', E)$. For τ take the Mackey topology $\tau(E', E)$. If μ were tight in this topology, it would be also tight in the topology $\sigma(E', E'')$ according to Exercise 5.12.106. Then μ would have a Radon extension for the norm topology, hence would have a norm separable support. This leads to a contradiction, since $\|\delta_t - \delta_s\| = 2$ whenever $t \neq s$.

5.12.108. (Bogachev **[63]**) Let P be a probability measure and let γ be a centered Radon Gaussian measure on $L^2(P)$ such that the embedding of its Cameron–Martin space into $L^2(P)$ is not nuclear. Show that $\gamma(E) = 0$ for every separable Hilbert space E continuously embedded into $L^\infty(P)$. This yields the fact proved in Smolyanov, Uglanov **[498]** that the Wiener measure on $C[0,1]$ has no Hilbert support.

HINT: use the fact that the embedding $E \to L^2(P)$ is a Hilbert–Schmidt operator (see Bogachev, Smolyanov **[72**, Corollary 7.10.28]).

5.12.109. (cf. Wiśniewski **[568]**) Let X and Y be separable Fréchet spaces, where Y has the approximation property, let μ be a probability Borel measure on X, and let $A\colon X \to Y$ be a mapping such that there exists a sequence of continuous linear operators $A_n\colon X \to Y$ such that for every $l \in Y'$ one has μ-almost everywhere $l(A_n x) \to l(Ax)$. Then there exists a sequence of continuous finite-dimensional linear operators $Q_n\colon X \to Y$ that converge to A μ-almost everywhere.

HINT: it is easy to see that A is measurable; let $\{p_k\}$ be an increasing sequence of seminorms on Y defining the topology; it suffices to show that, for every index k and every number $\varepsilon > 0$, there exists a continuous finite-dimensional operator $Q\colon X \to Y$ such that

$$\mu\bigl(x\colon\ p_k(Ax - Qx) > \varepsilon\bigr) < \varepsilon;$$

let $\nu = \mu \circ A^{-1}$; take a compact set K with $\nu(Y \backslash K) < \varepsilon/2$ and find $T \in \mathcal{L}(Y)$ of the form $Ty = \sum_{i=1}^m f_i(y) y_i$, where $f_i \in Y'$, $y_i \in Y$, $p_k(y_i) \leqslant 1$, such that $p_k(y - Ty) < \varepsilon/2$ for all $y \in K$; next, find n such that $\mu\bigl(x\colon\ |f_i(Ax) - f_i(A_n x)| > \varepsilon(2m)^{-1}\bigr)$, $i \leqslant m$; let $Qx = \sum_{i=1}^m f_i(A_n x) y_i$. Then $\mu\bigl(x\colon\ p_k(Ax - Qx) > \varepsilon\bigr)$ does not exceed the sum of the quantities $\mu\bigl(x\colon\ p_k(Ax - TAx) > \varepsilon/2\bigr)$ and $\mu\bigl(x\colon\ p_k(TAx - Qx) > \varepsilon/2\bigr)$, where the first number is $\nu\bigl(y\colon\ p_k(y - Ty) > \varepsilon/2\bigr) < \varepsilon/2$ and the second one is estimated by the series $\sum_{i=1}^m \mu\bigl(x\colon\ |f_i(Ax) - f_i(A_n x)| > \varepsilon(2m)^{-1}\bigr) < \varepsilon/2$.

5.12.110. Let B be a Borel set in a separable Fréchet space and let f be a Borel function on B such that $f(z) = \alpha f(x) + \beta f(y)$ if $x, y, z \in B$, $z = \alpha x + \beta y$. Prove that f extends to a Borel linear function on the linear span L of the set B.

HINT: the sets $L \times \mathrm{I\!R}^1$ and $S = \bigcup_{n=1}^\infty (B^n \times \mathrm{I\!R}^n)$ are Souslin, the mapping

$$g\colon\ B^n \times \mathrm{I\!R}^n \to L \times \mathrm{I\!R}^1, \quad (b_1, \ldots, b_n, t_1, \ldots, t_n) \mapsto \Bigl(\sum_{i=1}^n t_i b_i,\ \sum_{i=1}^n t_i f(b_i)\Bigr)$$

is continuous; verify that its image is the graph of a linear function on L; use the fact that a mapping of Souslin spaces with a Souslin graph is Borel (see [**65**, Lemma 6.7.1]).

5.12.111. Let $X = \prod_{n=1}^{\infty} X_n$, $\mu = \bigotimes_{n=1}^{\infty} \mu_n$, $\mu^{(n)} = \bigotimes_{k=n+1}^{\infty} \mu_k$ for probability spaces $(X_n, \mathcal{A}_n, \mu_n)$. Show that, for every function $\varphi \in L^1(\mu)$, the function $\varphi_n(x_1, \ldots, x_n)$ obtained by integrating $\varphi(x_1, \ldots, x_n, y_{n+1}, \ldots)$ in $(y_{n+1}, \ldots)$ against the measure $\mu^{(n)}$ is the conditional expectation of f with respect to the σ-algebra generated by $x_1, \ldots, x_n$. This yields convergence of φ_n to φ almost everywhere (see [**65**, § 10.2]).

5.12.112. (Smolyanov [**481**]) Let μ be the countable power of a symmetric probability Borel atomless measure μ_1 on the real line, let $\lambda_k \in \mathbb{R}$. Show that if the set of convergence of the series $\sum_{k=1}^{\infty} \lambda_k x_k^2$ contains a linear subspace of full μ-measure, then $\sum_{k=1}^{\infty} |\lambda_k|$ is finite. If the measure μ_1 has finite second moment, then the converse is also true.

5.12.113. (Smolyanov [**479**]) Let m be a symmetric probability Borel atomless measure on the real line, let μ be the measure on $\mathbb{R}^\infty$ equal to the countable power of m, and let f be a μ-measurable linear function on $\mathbb{R}^\infty$. (i) Prove that $\sum_{n=1}^{\infty} |f(e_n)|^2 < \infty$, where e_n is the vector with 1 at the nth place and 0 at other places, and also that $f(x) = \sum_{n=1}^{\infty} f(e_n) x_n$ a.e., where the series converges almost everywhere.

(ii) A function F that is $\mu \otimes \mu$-a.e. defined and measurable on $\mathbb{R}^\infty \times \mathbb{R}^\infty$ will be called measurable bilinear if, for each y for which the function $x \mapsto F(x, y)$ is defined μ-a.e., it is linear on some measurable linear subspace of μ-measure 1, and similarly for the other variable. Show that two such functions either differ a.e. or are equal a.e., and the latter holds precisely when they are equal on all vectors with finitely many nonzero components.

HINT: the function $\exp\bigl(itf(x)\bigr)$ for every fixed t is a.e. the limit of the integrals in y of the functions $\exp\bigl(itf(P_n x + S_n y)\bigr)$, where $P_n x = x_1 e_1 + \cdots + x_n e_n$, $S_n x = x - P_n x$; denoting by $\varphi_n(t)$ the integral of $\exp\bigl(itf(S_n y)\bigr)$ in y, we obtain that $\exp\bigl(it \sum_{j=1}^{n} f(e_j) x_j\bigr) \varphi_n(t)$ converges to $\exp\bigl(itf(x)\bigr)$ a.e., whence one can readily obtain convergence of the series of $f(e_j) x_j$ a.e.; since x_j are independent random variables with respect to μ, by the Kolmogorov three series theorem (see Gihman, Skorohod [**194**, Chapter II, § 3, Theorem 5]) and the symmetry of m we obtain that for every $c > 0$ the series of the integrals of the functions η_i against the measure m converges, where $\eta_j(s) = |f(e_j)s|^2$ if $|f(e_j)s| \leqslant c$, $\eta_j(s) = 0$ if $|f(e_j)s| > c$; this yields convergence of the series of $f(e_j)^2$; finally, we observe that the series of $f(e_j) x_j$ converges a.e. to $f(x)$; derive (ii) from (i) and Fubini's theorem.

5.12.114. (Wiśniewski [**569**]) Let X and Y be separable Fréchet spaces and let μ be a symmetric Borel probability measure on X (i.e., $\mu(B) = \mu(-B)$), not concentrated on a proper closed linear subspace. Suppose that X has a stochastic basis $\{\varphi_n\}$ with the corresponding sequence of functionals $f_n \in X'$ that are independent random variables on (X, μ). Then, for every μ-measurable linear operator $A\colon X \to Y$, almost everywhere $Ax = \sum_{n=1}^{\infty} f_n(x) A\varphi_n$.

5.12.115. (Kosov [**291**]) Let μ be a Borel probability measure on a separable Banach space X and let $X' \subset L^1(\mu)$. Then the following conditions are equivalent: (i) there exists a compactly embedded separable Banach space $Y \subset X$ with $\mu(Y) = 1$ and $Y' \subset L^1(\mu)$, (ii) the restriction of μ to every Borel set has a mean.

If $X' \subset L^p(\mu)$, where $p > 1$, then the existence of a compactly embedded separable Banach space $Y \subset X$ with $\mu(Y) = 1$ and $Y' \subset L^p(\mu)$ is equivalent to the following: the natural embedding operator $X' \to L^p(\mu)$ is compact and for every $\psi \in L^q(\mu)$, where $q = p/(p-1)$, the measure $\psi \cdot \mu$ has a mean. If $p = 2$, then this is equivalent to compactness of the covariance operator $K_\mu\colon X' \to X$.

Comments

The history of development of the theory of topological vector spaces is inseparable from the general history of functional analysis (see Dunford, Schwarz [**145**], Birkhoff, Kreyszig [**55**], Dieudonné [**131**], Lützen [**324**], Mauldin [**334**], Monna [**351**], Narici [**364**], Pietsch [**390**], and Taylor [**521**]). Its prehistory was concerned with various infinite-dimensional spaces of sequences and functions such as $\mathbb{R}^\infty$, considered at the beginning of the XX century by Fréchet, Hilbert, Volterra, F. Riesz, Weyl, and other researchers, with problems of convex analysis (Minkowski, Hahn, Helly, Banach), with weak convergence in Hilbert and more general functional spaces (Hilbert, F. Riesz, Weyl, von Neumann), and also with general topology (especially works of Hausdorff and Tychonoff). The birth of the proper theory of topological vector spaces and locally convex spaces is dated by the 30–40s of the XX century, when a number by now classical works of Kolmogorov, von Neumann, Tychonoff, Šmulyan, Krein, Milman, Dieudonné, Eberlein, and Bourbaki appeared. This direction was deeply influenced (and is still being influenced) by the theory of Banach spaces, created in the outstanding monograph of its founder (see Banach [**37**]). Kolmogorov [**279**] turned to topological linear spaces in 1934 (apparently, thinking of measures in such spaces) and obtained his criterion of normability; in 1935 von Neumann published his paper [**370**] devoted to such spaces (completely bounded sets were introduced there); the same year is marked by Tychonoff's paper [**533**] on fixed points of continuous mappings of convex compacta, where the term a "locally convex space" was first used. In the original paper in German, the term "lokal-konvexer Raum" was used. The modern definition of a locally convex space by means of seminorms was given by Dieudonné [**125**].

The first real achievements of the theory of locally convex spaces were connected exactly with weak and weak-$*$ topologies of Banach spaces. In 1929, the weak topology was explicitly introduced in Hilbert spaces by von Neumann [**369**], in 1934 in Köthe, Toepliz [**293**] it was defined on some spaces of sequences, Wehausen [**560**] extended von Neumann's definition to normed spaces, finally, in 1939 Šmulyan [**501**] considered the weak topology generated by a linear space of functionals (i.e., $\sigma(E, G)$ in modern notation; about his works, see Raikov [**407**]), and a year later Dieudonné [**123**] introduced dual pairs (pairs in duality), making the definition symmetric. In 1938 Goldstine [**197**] proved that the elements of the ball of the second dual to a Banach space are "weak" limits of nets of elements of the ball of the original space (see Exercise 3.12.80), and Alaoglu [**7**] and Bourbaki [**86**] (detailed expositions were given in Alaoglu [**8**] and Dieudonné [**125**]) established the very important property of weak-$*$ compactness of balls in the dual space (somewhat later this was also proved by Šmulyan [**501**] and Kakutani [**253**]); certainly, an important role was played here by compactness of products of compact intervals, discovered shortly before that by Tychonoff, and also by the case of separable spaces considered earlier by Banach (many

V.I. Bogachev and O.G. Smolyanov, *Topological Vector Spaces and Their Applications*, Springer Monographs in Mathematics,
DOI 10.1007/978-3-319-57117-1

researchers note with surprise that Banach did not use concepts of general topology already known by that time and rapidly gaining popularity; his results related to weak convergence were formulated in terms of transfinite sequences). To this circle of problems belong also generalizations and geometric versions of the Hahn–Banach theorem on extensions of linear functionals; Mazur [**335**] obtained a result on separation of convex sets in normed spaces, close results were obtained in Eidelheit [**152**], Kakutani [**252**], Krein, Šmulian [**297**], and Dieudonné [**124**]; Suhomlinov [**507**] considered the complex field case (and also the case of quaternions). In relation to the study of weak compactness one should mention the papers Šmulyan [**499**], [**500**], [**502**], [**503**] (see also Gantmacher, Šmulyan [**185**]) and Eberlein [**147**]. Fichtenholz [**168**] considered sequential continuity of functionals not connected with a norm or metric. Note that not every reasonable convergence in a linear space is generated by a topology. For example, convergence almost everywhere of measurable functions on $[0,1]$ (even of polynomials) cannot be generated by a topology, see Exercise 1.12.88.

However, it became possible to consider this theory as an independent area, apparently, only in the 50s of the XX century after works of Laurent Schwartz [**440**] in the theory of distributions (strongly influenced by works of S.L. Sobolev on generalized derivatives) and subsequent works of Grothendieck, Dieudonné, Mackey, Arens, Köthe, Klee, Pták, and other mathematicians from many countries stimulated partly by these investigations and partly by the study of weak topologies in Banach spaces. The main themes of this period were "duality theory" and weak topologies, special classes of nonnormable spaces (such as inductive and projective limits), and tensor products. The most important achievements of these years (until the 60s of the XX century) are well documented in the books Bourbaki [**87**], Edwards [**150**], Floret, Wloka [**172**], Grothendieck [**207**], Kantorovich, Akilov [**261**] (in the second edition [**262**] this material was reduced), Kelley, Namioka [**270**], Köthe [**292**], Robertson, Robertson [**420**], Schaefer [**436**], and Trèves [**530**]. The first detailed expositions of the theory were given already in the 1950s by the author groups Bourbaki [**87**] and a bit later Kelley, Namioka [**270**] (the latter included also W.F. Donoghue, K.R. Lucas, B.J. Pettis, E.T. Poulsen, G.B. Price, W. Robertson, W.R. Scott, and K.T. Smith), and also by Köthe [**292**]; extensive material was presented in the books Grothendieck [**207**] and Gelfand, Shilov [**193**], but they had a more special character. One should also mention a number of papers of the same period that played a very important role in later investigations: Grothendieck [**205**] (a survey on the status of the problems posed by Grothendieck is given in Cakić et al. [**93**]), Arens [**19**], Dieudonné, Schwartz [**132**], Grothendieck [**202**]–[**206**], Mackey [**327**], [**328**], and Sebastião e Silva [**444**]. There were also surveys Dieudonné [**126**], Hyers [**230**], and Nakano [**361**]. Note that Kakutani [**251**] proved the existence of a translation invariant metric in a metrizable Abelian group (in particular, in a metrizable topological vector space, see Theorem 1.6.1), and Klee [**273**] solved Banach's problem about completeness of such an invariant metric in the case of the existence of some complete metric generating the topology.

Substantial contributions to the study of weak compactness were made by the papers Grothendieck [**202**], James [**232**]–[**236**]. Note also the series of papers Klee [**272**], [**274**], [**275**], devoted to convexity. In relation to the condition for the equality of double limits and its applications to weak compactness mentioned in Exercise 3.12.192, we refer to the papers Pták [**400**], [**399**]. In the latter, connections of this condition with some other problems are discussed. In particular, it is proved there that for a bounded separately continuous function f on the product $S\times T$ of Tychonoff spaces the equality $\lim\limits_{m\to\infty}\lim\limits_{n\to\infty} f(s_m,t_n) = \lim\limits_{n\to\infty}\lim\limits_{m\to\infty} f(s_m,t_n)$ is true for all sequences $\{s_m\}\subset S$ and $\{t_n\}\subset T$ for which both limits exist precisely when f extends to a separately continuous function on the product of the Stone–Čech compactifications of S and T, moreover, this is equivalent to the property that the set of functions $s \mapsto f(t,s)$, where $t \in T$, is relatively weakly compact in $C_b(S)$.

In the subsequent years the main progress was concerned with solving special important problems (for example, connected with the properties of the spaces $\mathcal{D}$ and $\mathcal{D}'$ and other concrete spaces of smooth and generalized functions) and also with applications, among which one should mention in the first place mathematical problems in quantum physics (see Bogolubov et al. [**73**]), the theory of distributions with applications to partial differential equations, development of Kolmogorov's ideas on estimating complexity of functional spaces (see Kolmogorov, Tikhomirov [**283**]), optimal control and extremal problems, measure theory on infinite-dimensional spaces, and infinite-dimensional stochastic analysis. Important sources of development were investigations of operator ideals, nuclear spaces, bases and approximations. Finally, it is necessary to note the continuing influence of general nonlinear infinite-dimensional analysis, constantly generating new interesting objects of study and important problems. The achievements of the 60–80s of the XX century are well presented in very informative monographs Jarchow [**237**], Pérez Carreras, Bonet [**385**], Valdivia [**543**], and Wilansky [**567**]. There is a more recent book Narici, Beckenstein [**365**].

There is a huge number of papers on inductive limits of locally convex spaces. In addition to the works already cited in exercises and comments note the papers Akhmedova [**4**], Bierstedt [**53**], Makarov [**332**], Raikov [**402**], [**404**], [**406**], Retakh [**417**], and Smolyanov [**477**].

Efforts of many researchers were directed towards generalizations of the fundamental theorems connected with the name of Banach and concerned with the conditions for the continuity of linear operators (the closed graph theorem, the inverse mapping theorem, the open mapping theorem, and so on). One direction of research followed the line of the properties like completeness and category, see Grothendieck [**205**], Dieudonné, Schwartz [**132**], Kalton [**256**], McIntosh [**337**], Pták [**397**], [**398**]; for this purpose various classes of complete spaces were invented (Pták spaces and other spaces; hypercompleteness was introduced by Kelley [**269**]). These types of completeness were considered in many papers, references to which can be found in the books cited above; see also Raikov [**405**], Smolyanov [**480**], [**482**]–[**486**], [**485**], and Shavgulidze [**446**]–[**449**]. However, it was shown by Smolyanov [**482**] that the most important among nonmetrizable spaces the space of smooth functions $\mathcal{D}$ is not fully complete; in this work a stronger assertion was proved: $\mathcal{D}$ possesses an incomplete factor space isomorphic to a dense subspace in the countable product of the real lines (hence even metrizable). The same method applies for constructing an incomplete factor space in $\mathcal{D}'$, whence it follows that $\mathcal{D}'$ is not fully complete; the latter result had been earlier obtained in Słowikowski [**470**] (see Remark on p. 249 in Raikov's supplement to [**420**]) and Smolyanov [**480**]. As was later shown by Valdivia [**540**], [**541**], the same result yields that $\mathcal{D}$ and $\mathcal{D}'$ are not B_r-complete; the latter was independently proved by Shavgulidze [**447**], who used not the result, but the method of [**480**], [**482**]. These results, unexpected for experts, lead to decrease of interest in such types of completeness. In § 2.10(i) we present the results obtained by the method from [**482**], in particular, we construct a closed linear subspace in the space $\mathcal{D}'$ and a sequentially continuous discontinuous functional on it, whence it follows that the Hahn–Banach theorem does not extend to sequentially continuous functionals (for $\mathcal{D}$ such an example is given in [**470**]), and also that the so-called sequential topology on $\mathcal{D}$ is not compatible with the vector structure.

Another direction of research was connected with the idea of equipping the space with some covers and is related to the methods of the descriptive set theory and topology such as Souslin schemes and analytic spaces, see Schwartz [**441**], Raikov [**408**], De Wilde [**110**]–[**112**], Zabreiko, Smirnov [**578**], Smirnov [**471**], Ferrando, Kąkol, López Pellicer [**161**],

Ferrando, Sánchez Ruiz [**166**], and Robertson [**421**]. It appears that for practical applications the approaches of Schwartz and De Wilde, presented in Chapter 3, have proved the most useful.

Large sections devoted to various aspects of the theory of locally convex spaces are included in many books on functional analysis and partial differential equations, see Bogachev, Smolyanov [**72**], where an extensive bibliography is given, and also Choquet [**100**], Edwards [**150**], Garnir, De Wilde, Schmets [**186**], Lelong [**312**], Meise, Vogt [**339**], and Reed, Simon [**414**]. On bornology, see Hogbe-Nlend [**226**], Palamodov [**378**], Waelbroeck [**559**], and Wilansky [**566**].

Distributions, partial differential operators and related questions in the theory of locally convex spaces are discussed in Duistermaat, Kolk [**143**], Garsoux [**187**], Horvath [**228**], Hörmander [**227**], Kanwal [**264**], Trèves [**529**], and Vladimirov [**549**]. However, we do not discuss the theory of distributions in this book, since the topological aspects at the present stage of the development of this theory play definitely a secondary role, while in the first place are analytic problems. Nevertheless it is necessary to emphasize that a good number of important concrete problems of the theory of distributions and differential equations for distributions was concerned exactly with problems of a topological character. For example, the study of solvability of differential equations in $\mathcal{D}'$ and related equations in convolutions lead to problems of the existence of extensions of sequentially continuous functionals on subspaces to distributions (not always solvable, see Exercise 2.10.63).

Ordered vector spaces are considered in Akilov, Kutateladze [**6**], Kantorovich, Vulikh, Pinsker [**263**], Aliprantis, Burkinshaw [**14**], Luxemburg, Zaanen [**325**], Peressini [**384**], and Schaefer [**436**].

On geometry and topology of Banach spaces, see Albiac, Kalton [**10**], Beauzamy [**49**], Benyamini, Lindenstrauss [**50**], Bourgin [**89**], Carothers [**94**], Day [**109**], Defant, Floret [**113**], Deville, Godefroy, Zizler [**118**], Diestel [**120**], Diestel, Uhl [**122**], van Dulst [**144**], Fabian et al. [**155**], [**156**], Godefroy, Lancien, Zizler [**195**], Guerre-Delabrière [**209**], Guirao, Montesinos, Zizler [**212**], Johnson, Lindenstrauss [**238**], Kadec, Kadec [**244**], Li, Queffélec [**315**], Lindenstrauss, Tzafriri [**318**], Megginson [**338**], Pietsch [**390**], Pisier [**391**], and Wojtaszczyk [**571**]. On operator ideals, see Defant, Floret [**113**], Diestel, Jarchow, Tonge [**121**], and Junek [**239**]. There is an extensive literature on the weak topology of Banach spaces; in addition to the already cited papers, we mention the influential paper Corson [**107**] and also Edgar, Wheeler [**149**],

An interesting model locally convex space, on which many important general concepts and phenomena are considered, is the space $C_p(T)$ of continuous functions on a topological space T equipped with the topology of pointwise convergence; see Arhangel'skii [**21**], McCoy, Ntantu [**336**], and Tkachuk [**522**], [**523**]. Topological spaces with certain structures of infinite-dimensional manifolds are studied in Bessaga, Pełczyński [**52**], Fedorchuk, Chigogidze [**157**], Chapman [**99**], and van Mill [**345**], [**346**]. The theory of topological groups is presented in the fundamental monograph Arhangel'skii, Tkachenko [**20**].

About nonlinear equations and fixed points, see Akhmerov et al. [**5**], Deimling [**115**], Granas, Dugundji [**199**], Namioka [**362**], and Nirenberg [**372**]. Note that only recently Cauty [**98**] solved positively the problem (remaining open for 70 years) of the validity of the Schauder–Tychonoff theorem for every topological vector space (not necessarily locally convex).

Certainly, it is not possible even to briefly mention here all modern directions of the development of the theory of topological vector spaces; some additional information can be found in Akbarov [**3**], Aytuna et al. [**33**], Banaszczyk [**42**], Beattie, Butzmann [**47**], Beatty,

Schaefer [**48**], Bierstedt [**53**], Bierstedt, Bonet [**54**], Bonet, Cascales [**74**], Cascales, Orihuela [**97**], Cooper [**106**], Ferrando, Kąkol, López Pellicer [**160**], Ferrando et al. [**163**]–[**165**], Fetisov, Filippenko, Kozobrod [**167**], Gabriyelyan, Kąkol, Kubzdela, López-Pellicer [**182**], Guirao, Montesinos [**210**], Guirao, Montesinos, Zizler [**211**], Kadelburg, Radenović [**241**], Kąkol, Kubiś, López-Pellicer [**245**], Kąkol, López Pellicer [**246**], Kąkol, López Pellicer, Okunev [**247**], Kąkol, López Pellicer, Sliwa [**248**], Kąkol, Saxon [**250**], Kalenda, Spurný [**255**], Kalton, Peck, Roberts [**257**], Klee, Maluta, Zanco [**276**], Kolesnikov [**278**], Kondakov [**289**], Kraaij [**295**], Kunzinger [**304**], Peralta et al. [**383**], Perez-Garcia, Schikhof [**386**], Rolewicz [**422**], Saxon [**431**], [**432**], Saxon, Tweddle [**434**], Valdivia [**547**], Vogt [**554**], Waelbroeck [**558**], and Zahariuta [**579**]. In particular, on Fréchet spaces, see [**165**], [**257**], [**304**], [**422**]. An efficient way of finding information on modern research is searching in the database MathSciNet by first finding the authors from our bibliography and then finding the authors citing them. Among actively developing directions of the theory of locally convex spaces one could mention the study of general topological questions (in particular, connected with compactness), measures on infinite-dimensional spaces, and various classes of operators and tensor products at the junction with geometry of Banach spaces and concrete functional spaces (such as spaces of smooth functions).

It is worth noting that not so long ago the answers to the well-known questions posed by Grothendieck in the 1950s have been found. Taskinen [**517**], [**518**] constructed an example of a Montel Fréchet space F with the approximation property for which in the tensor product $F\widetilde{\otimes}_\pi F$ there is a bounded set not contained in the closed absolutely convex hull of a set $A\otimes B$ with bounded A and B. Hence such a tensor product is not Montel. On this topic, see also Bonet, Díaz, Taskinen [**75**], Bonet, Galbis [**76**], Bonet, Taskinen [**78**], and Defant, Floret, Taskinen [**114**].

Investigation of differentiability of mappings on infinite-dimensional spaces of curves or functions goes back to early classics in variational calculus (Newton, Euler, Lagrange, and others). Systematic studies began at the end of the XIX century and the beginning of the XX century by Volterra [**557**] and his school (see Arzelà [**27**]) and Hadamard (his book [**214**] has considerable impact) and continued, among many other researchers, by Gâteaux [**188**] and Fréchet [**175**], whose names are connected with important special types of differentiation. There are many works devoted to differentiation theory and differential calculus in infinite-dimensional spaces and giving also some historic information; we mention only surveys and books Averbuch, Smolyanov [**30**], [**31**], Balabanov [**36**], Bastiani [**44**], Cartan [**95**], Dieudonné [**130**], Deimling [**115**], Eells [**151**], Frölicher, Bucher [**180**], Frölicher, Kriegl [**181**], Gaehler [**184**], Graves [**200**], Hildebrandt, Graves [**221**], Keller [**266**], Krasnosel'skii, Zabreiko [**296**], Kriegl, Michor [**298**] (see also Kriegl, Michor, Rainer [**299**]), Milnor [**348**], Mujica [**356**], Neeb [**366**], Omori [**374**], Smolyanov [**489**], Sova [**504**], [**505**], Sukhinin [**506**], Vainberg [**535**], and Yamamuro [**573**]. Extensive historical comments can be found in [**30**], Pietsch [**390**], and Taylor [**520**]. On subdifferentials and convex analysis, see Kusraev, Kutateladze [**307**], Mordukhovich [**352**], Singer [**465**], [**466**], and Zălinescu [**580**]. The results in § 4.9 were obtained by Lobanov [**321**], who found some reinforcements and generalizations of the results for Banach spaces obtained in Albrecht, Diamond [**11**] and Hain [**215**] and also in Kats, Kurato [**240**], Smolyanov [**488**], where compositions with linear functions on locally convex spaces were considered. Our discussion of polynomials on locally convex spaces follows Bochnak, Siciak [**57**].

A topic strongly related to locally convex spaces and differential equations in such spaces that we do not touch in this book concerns operator semigroups, see the book Yosida [**575**], and the following papers where numerous additional references can be found:

Albanese, Jornet [**9**], Babalola [**34**], Dembart [**116**], Frerick, Jordá, Kalmes, Wengenroth [**179**], Komatsu [**285**], Kōmura [**286**], and Kraaij [**294**].

On general questions of measure theory in infinite-dimensional spaces, see Badrikian, Chevet [**35**], Bogachev [**63**], [**65**], [**66**], Buldygin [**90**], Buldygin, Solntsev [**92**], Gelfand, Vilenkin [**191**], Gihman, Skorohod [**194**], Hazod, Siebert [**218**], Heyer [**220**], Hoffmann-Jørgensen [**224**], Kruglov [**301**], Kuo [**305**], Ledoux, Talagrand [**311**], Linde [**317**], Mushtari [**359**], Schwartz [**443**], Skorohod [**468**], Smolyanov, Fomin [**495**], Smolyanov, Shavgulidze [**497**], Vakhania [**536**], Vakhania, Tarieladze, Chobanyan [**538**], Vershik, Sudakov [**548**], and Yamasaki [**574**]. A very important role in the creation and development of measure theory on infinite-dimensional linear spaces was played by the works and ideas of Wiener, Kolmogorov, Fréchet , Prohorov, Gelfand (in particular, one should mention the papers [**565**], [**280**], [**176**], [**393**], [**394**], [**191**]). Close to that period were the papers Fortet, Mourier [**174**], Gross [**201**], Mourier [**355**], and Segal [**445**]. In the book [**65**, § 7.14(vii)] one can find some information on the so-called Radon spaces, in which all Borel measures are Radon. Though, it is worth noting that a practical value of this concept is not high, since in applications very seldom there are measures defined on the whole Borel σ-algebra and not obviously Radon. It is much more often that one has to verify the tightness of a measure defined on some smaller σ-algebra (for example, generated by cylinders), with the purpose of extending it to a Radon measure. On vector integration, see Dinculeanu [**133**], Panchapagesan [**379**], and Roth [**424**]. A vector function ψ on a measurable space $(X, \mathcal{A}, \mu)$ with values in a locally convex space X is called Pettis integrable if for every $l \in X'$ the function $\langle l, \psi \rangle$ is integrable and its integral $l(m)$, where $m \in X$ does not depend on l; then m is called the Pettis integral of ψ. This terminology is often used under the stronger assumption that in the first sense the restrictions of ψ to all sets $A \in \mathcal{A}$ are integrable. The mean of a measure introduced in the text is the Pettis integral of $\psi(x) = x$ in the first weaker sense.

For applications, the most convenient conditions for countable additivity of cylindrical measures in terms of their Fourier transforms are the theorems due to Sazonov [**435**] and Minlos [**349**], which appeared under the strong influence of the ideas of Prohorov [**393**] and Gelfand [**190**]. An important observation on this question was made by Kolmogorov [**281**]. The question about extending these results to signed measures was raised by Smolyanov at the end of the 1960s, but was completely resolved only 20 years later by Tarieladze [**516**] (intermediate results were obtained in Shavgulidze [**448**] and Smolyanov [**491**]). A study of topologies on infinite-dimensional spaces in relation to Fourier transforms of measures was conducted by Mushtari [**357**]–[**360**], [**354**]. As one can see from these and many other works (see Vladimirskii [**553**]), sufficient conditions for countable additivity of a cylindrical measure can be expressed not only in terms of the Fourier transform; for example, various operators associated with the measure can be also used.

On limit theorems in Banach spaces, see also Kuelbs, Zinn [**303**], Ledoux, Talagrand [**311**], and Johnson, Lindenstrauss [**238**, p. 1177–1200]. On 0–1 laws in linear spaces, see Hoffmann-Jørgensen [**225**], Takahashi, Okazaki [**509**], and Zinn [**581**]. Measurable linear functions and operators are also studied in Kanter [**260**], Smoleński [**472**], [**474**], [**475**], and Wiśniewski [**570**]. Differentiability connected with measures on infinite-dimensional spaces is studied in Averbuch, Smolyanov, Fomin [**32**], Bogachev [**66**], and Bogachev, Smolyanov [**70**].

References

[1] Acosta A. de. *Stable measures and seminorms*. Ann. Probab. 1975. V. 3. P. 865–875. [*390*][1]

[2] Acosta A. de, Samur J. *Infinitely divisible probability measures and the converse Kolmogorov inequality in Banach spaces*. Studia Math. 1978. V. 66. P. 143–160. [*390*]

[3] Akbarov S.S. *Pontryagin duality in the theory of topological vector spaces and in topological algebra*. J. Math. Sci. (New York). 2003. V. 113, №2. P. 179–349. [*422*]

[4] Akhmedova P.D. *Sequentially continuous linear functionals in strict inductiv limits of sequences of Fréchet spaces*. Uspekhi Mat. Nauk. 1972. V. 27, №2. P. 163–164 (in Russian). [*421*]

[5] Akhmerov R.R., Kamenskii M.I., Potapov A.S., Rodkina A.E., Sadovskii B.N. Measures of noncompactness and condensing operators. Birkhäuser, Basel, 1992; viii+249 pp. [*422*]

[6] Akilov G.P., Kutateladze S.S. Ordered vector spaces. Nauka, Novosibirsk, 1978; 368 pp. (in Russian). [*422*]

[7] Alaoglu L. *Weak convergence of linear functionals*. Bull. Amer. Math. Soc. 1938. V. 44. P. 196. [*419*]

[8] Alaoglu L. *Weak topologies of normed linear spaces*. Ann. Math. 1940. V. 41. P. 252–267. [*419*]

[9] Albanese A.A., Jornet D. *Dissipative operators and additive perturbations in locally convex spaces*. Math. Nachr. 2016. V. 289, №8-9. P. 920–949. [*424*]

[10] Albiac F., Kalton N.J. Topics in Banach space theory. Springer, New York, 2006; xii+373 pp. [*422*]

[11] Albrecht F., Diamond H. *A converse of Taylor's theorem*. Indiana Univ. Math. J. 1971/1972. V. 21, №4. P. 347–350. [*423*]

[12] Alexandroff (Aleksandrov) A.D. *Additive set functions in abstract spaces*. Mat. Sbornik. 1940. V. 8(50). P. 307–348; ibid. 1941. V. 9(51). P. 563–628; ibid. 1943. V. 13(55). P. 169–238. [*397*]

[13] Alfsen E.M. Compact convex sets and boundary integrals. Springer-Verlag, Berlin – New York, 1971; 210 pp. [*343*]

[14] Aliprantis C.D., Burkinshaw O. Locally solid Riesz spaces with applications to economics. 2nd ed. Amer. Math. Soc., Providence, Rhode Island, 2003; xii+344 pp. [*422*]

[15] Amemiya I. *Some examples of (F) and (DF) spaces*. Proc. Japan Acad. 1957. V. 33. P. 169–171. [*238*]

[16] Amemiya I., Kōmura Y. *Über nicht-vollständige Montelräume*. Math. Ann. 1968. B. 177. S. 273–277. [*186*]

[17] Anderson R.D. *Hilbert space is homeomorphic to the countable infinite product of lines*. Bull. Amer. Math. Soc. 1966. V. 72. P. 515–519. [*217*]

[18] Antosik P., Burzyk J. *Sequential conditions for barrelledness and bornology*. Bull. Pol. Acad. Sci., Math. 1987. V. 35. P. 457–459. [*226*]

[19] Arens R. *Duality in linear spaces*. Duke Math. J. 1947. V. 14. P. 787–794. [*420*]

[20] Arhangel'skii A., Tkachenko M. Topological groups and related structures. Atlantis Press, Paris; World Sci. Publ., Hackensack, New Jersey, 2008; xiv+781 pp. [*422*]

[21] Arkhangel'skiĭ A.V. Topological function spaces. Kluwer, Dordrecht, 1992; x+205 pp. [*215, 422*]

[22] Arkhangel'skiĭ A.V., Ponomarev V.I. Fundamentals of general topology. Problems and exercises. Reidel, Dordrecht, 1984; xvi+415 pp. (Russian ed.: Moscow, 1974). [*6, 215, 409*]

[1] In square brackets we indicate all page numbers of citing the publication.

V.I. Bogachev and O.G. Smolyanov, *Topological Vector Spaces and Their Applications*, Springer Monographs in Mathematics,
DOI 10.1007/978-3-319-57117-1

[23] Arias de Reyna J. *Normed barely Baire spaces*. Israel J. Math. 1982. V. 42. P. 33–36; correction: ibid., 1985. V. 50. P. 264. [*209*]

[24] Arutyunyan L.M., Kosov E.D. *Estimates for integral norms of polynomials on spaces with convex measures*. Matem. Sbornik. 2015. V. 206, №8. P. 3–22 (in Russian); English transl.: Sbornik Math. 2015. V. , №8. P. 1030–1048. [*409*]

[25] Arutyunyan L.M., Kosov E.D., Yaroslavtsev I.S. *On some properties of polynomials measurable with respect to a Gaussian measure*. Dokl. Ross. Akad. Nauk. 2014. V. 457, №2. P. 131–135 (in Russian); English transl.: Dokl. Math. 2014. V. 90, №1. P. 419–423. [*408*]

[26] Arutyunyan L.M., Yaroslavtsev I.S. *On measurable polynomials on infinite-dimensional spaces*. Dokl. Ross. Akad. Nauk. 2013. V. 449, №6. P. 627–630 (in Russian); English transl.: Dokl. Math. 2013. V. 87, №2. P. 214–217. [*405*]

[27] Arzelà C. *Funzioni di linei*. Atti Accad. Lincei Rend. 1889. V. 5, №1. P. 342–348. [*423*]

[28] Astala K. *On Peano's theorem in locally convex spaces*. Studia Math. 1982. V. 73, №3. P. 213–223. [*295*]

[29] Atkin C.J. *Extension of smooth functions in infinite dimensions. I. Unions of convex sets*. Studia Math. 2001. V. 146, №3. P. 201–226. [*305*]

[30] Averbuch V.I., Smolyanov O.G. *Theory of differentiation in linear topological spaces*. Uspekhi Mat. Nauk. 1967. V. 22, №6. P. 201–260 (in Russian); English transl.: Russian Math. Surveys. 1967. V. 22, №6. P. 201–258. [*423*]

[31] Averbuch V.I., Smolyanov O.G. *The various definitions of the derivative in linear topological spaces*. Uspekhi Mat. Nauk. 1968. V. 23, №4. P. 67–116 (in Russian); English transl.: Russian Math. Surveys. 1968. V. 23, №4. P. 67–113; An addendum: ibid., №5. P. 223–224. [*423*]

[32] Averbuh V.I., Smolyanov O.G., Fomin S.V., *Generalized functions and differential equations in linear spaces. I, Differentiable measures*. Trudy Moskovsk. Matem. Ob. 1971. V. 24. P. 133–174 (in Russian); English transl.: Trans. Moscow Math. Soc. 1971. V. 24. P. 140–184. [*424*]

[33] Aytuna A., Djakov P.B., Goncharov A.P., Terzioglu T., Zahariuta V.P. *Some open problems in the theory of locally convex spaces*. Linear Topol. Spaces Complex Anal. 1994. V. 1. P. 147–165. [*422*]

[34] Babalola V.A. *Semigroups of operators on locally convex spaces*. Trans. Amer. Math. Soc. 1974. V. 199. P. 163–179. [*424*]

[35] Badrikian A., Chevet S. Mesures cylindriques, espaces de Wiener et fonctions aléatoires gaussiennes. Lecture Notes in Math. 1974. V. 379. P. 1–383. [*424*]

[36] Balabanov M.V. Differentiation of mappings in infinite-dimensional vector spaces. KomKniga, Moscow, 2006; 176 pp. (in Russian). [*423*]

[37] Banach S. Théorie des opérations linéaires. Warszawa, 1932; vii+254 pp. [*419*]

[38] Banakh T. *Topological classification of strong duals to nuclear* (LF)*-spaces*. Studia Math. 2000. V. 138, №3. P. 201–208. [*242*]

[39] Banakh T., Cauty R. *Topological classification of closed convex sets in Fréchet spaces*. Studia Math. 2011. V. 205, №1. P. 1–11. [*217*]

[40] Banakh T., Repovš D. *A topological characterization of* LF*-spaces*. Topology Appl. 2012. V. 159, №5. P. 1475–1488. [*242*]

[41] Banaszczyk W. *The Steinitz theorem on rearrangement of series for nuclear spaces*. J. Reine Angew. Math. 1990. B. 403. S. 187–200. [*152*]

[42] Banaszczyk W. Additive subgroups of topological vector spaces. Lecture Notes in Math. V. 1466. Springer-Verlag, Berlin, 1991; viii+178 pp. [*422*]

[43] Banaszczyk W. *Rearrangement of series in nonnuclear spaces*. Studia Math. 1993. V. 107. P. 213–222. [*152*]

[44] Bastiani A. *Applications différentiables et variétés différentiables de dimension infinie*. J. Anal. Math. 1964. V. 13. P. 1–114. [*423*]

[45] Bates S.M. *On the image size of singular maps*. Duke Math. J. 1992. V. 68, №3. P. 463–476. [*308*]

[46] Bates S.M. *On smooth rank-*1 *mappings of Banach spaces onto the plane*. J. Differential Geom. 1993. V. 37. P. 729–733. [*309*]

[47] Beattie R., Butzmann H.-P. Convergence structures and applications to functional analysis. Kluwer, Dordrecht, 2002; xiv+264 pp. [*422*]

[48] Beatty T.A., Schaefer H.H. *Semi-bornological spaces*. Math. Z. 1996. B. 221, №1. S. 337–351. [*423*]

[49] Beauzamy B. Introduction to Banach spaces and their geometry. 2nd ed. North-Holland, Amsterdam, 1985; xv+338 pp. [*422*]

[50] Benyamini Y., Lindenstrauss J. Geometric nonlinear functional analysis. Amer. Math. Soc., Providence, Rhode Island, 2000; 488 pp. [*251, 307, 422*]

[51] Berezhnoy V.E. *On the equivalence of integral norms on the space of measurable polynomials with respect to a convex measure*. Theory Stoch. Process. 2008. V. 14, №1. P. 7–10. [*408*]

[52] Bessaga C., Pełczyński A. Selected topics in infinite-dimensional topology. Polish Scientific Publ., Warszawa, 1975; 353 pp. [*217, 241, 422*]

[53] Bierstedt K.D. *An introduction to locally convex inductive limits*. In: Functional Analysis and its Applications (H. Hogbe-Nlend, ed.), World Sci., 1988. P. 35–133. [*421, 423*]

[54] Bierstedt K.D., Bonet J. *Some aspects of the modern theory of Fréchet spaces*. Rev. R. Acad. Cien. Ser. A Mat. 2003. V. 97. P. 159–188. [*423*]

[55] Birkhoff G., Kreyszig E. *The establishment of functional analysis*. Historia Math. 1984. V. 11, №3. P. 258–321. [*419*]

[56] Bobkov S.G. *Remarks on the growth of L^p-norms of polynomials*. Lecture Notes in Math. 2000. V. 1745. P. 27–35. [*408*]

[57] Bochnak J., Siciak J. *Polynomials and multilinear mappings in topological vector spaces*. Studia Math. 1971. V. 39, №1. P. 59–76. [*423*]

[58] Bockshtein M. *Un théorème de séparabilité pour les produits topologiques*. Fund. Math. 1948. V. 35. P. 242–246. [*409*]

[59] Bogachev V.I. *Indices of asymmetry of stable measures*. Matem. Zametki. 1986. V. 40, №1. P. 127–138 (in Russian); English transl.: Math. Notes. 1986. V. 40. P. 569–575. [*384*]

[60] Bogachev V.I. *Locally convex spaces with the CLT property and supports of measures*. Vestnik Moskovsk. Univ. 1986. №6. P. 16–20 (in Russian); English transl.: Moscow Univ. Math. Bull. 1986. V. 41, №6. P. 19–23. [*381, 382, 391*]

[61] Bogachev V.I. *Smooth measures, the Malliavin calculus and approximations in infinite dimensional spaces*. Acta Math. Univ. Carolinae, Math. et Phys. 1990. V. 31, №2. P. 9–23. [*307*]

[62] Bogachev V.I. *Deterministic and stochastic differential equations in infinite dimensional spaces*. Acta Appl. Math. 1995. V. 40. P. 25–93. [*294*]

[63] Bogachev V.I. Gaussian measures, Amer. Math. Soc., Providence, Rhode Island, 1998; 433 pp. [*321, 345, 354, 362, 393, 394, 415, 417*]

[64] Bogachev V.I. *Average approximations and moments of measures*. J. Complexity. 2000. V. 16. P. 390–410. [*391*]

[65] Bogachev V.I. Measure theory. V. 1, 2. Springer, Berlin, 2007; xvii+500 pp., xiii+575 pp. [*179, 211, 214, 226, 316, 318 – 320, 322, 325, 337, 376, 378, 386, 391, 395, 398, 410, 418, 424*]

[66] Bogachev V.I. Differentiable measures and the Malliavin calculus. Amer. Math. Soc., Rhode Island, Providence, 2010; xvi+488 pp. [*354, 379, 415, 424*]

[67] Bogachev V.I. Gaussian measures on infinite-dimensional spaces. In: Real and Stochastic Analysis: Current Trends (M.M. Rao ed.), pp. 1–83. World Sci., Singapore, 2014. [*354*]

[68] Bogachev V.I. *Distributions of polynomials on multidimensional and infinite-dimensional spaces with measures*. Uspekhi Mat. Nauk. 2016. V. 71, №4. P. 107–154 (in Russian); English transl.: Russian Math. Surveys. 2016. V. 71, №4. P. 703–749. [*409*]

[69] Bogachev V.I. Weak convergence of measures. Inst. Computer Sci., Moscow – Izhevsk, 2016; 396 pp. (in Russian); English transl. in preparation. [*381, 382, 394*]

[70] Bogachev V.I., Smolyanov O.G. *Analytic properties of infinite dimensional distributions*. Uspekhi Mat. Nauk. 1990. V. 45, №3. P. 3–83 (in Russian); English transl.: Russian Math. Surveys. 1990. V. 45, №3. P. 1–104. [*424*]

[71] Bogachev V.I., Smolyanov O.G. *Generalized functions obtained by the regularization of non-integrable functions*. Dokl. Akad. Nauk 2008. V. 419, №6. P. 731–734 (in Russian); English transl.: Dokl. Math. 2008. V. 77, №2. P. 302–305. [*228*]

[72] Bogachev V.I., Smolyanov O.G. Real and functional analysis. A university course. 2nd ed. Regular and Chaotic Dynamics, Moscow – Izhevsk, 2011; 728 pp. (in Russian); English transl. in preparation. [*6, 28, 77, 122, 123, 124, 144, 169, 172, 175, 227, 240, 263, 289, 308, 417, 422*]

[73] Bogolubov N.N., Logunov A.A., Oksak A.I., Todorov I.T. General principles of quantum field theory. Kluwer, Dordrecht, 1990; xx+694 pp. [*421*]

[74] Bonet J., Cascales B. *Non-complete Mackey topologies on Banach spaces*. Bull. Aust. Math. Soc. 2010. V. 81, №3. P. 409–413. [*423*]

[75] Bonet J., Díaz J.C., Taskinen J. *Tensor stable Fréchet and (DF)-spaces*. Collect. Math. 1991. V. 42, №3. P. 199–236. [*423*]

[76] Bonet J., Galbis A. *A note on Taskinen's counterexamples on the problem of topologies of Grothendieck*. Proc. Edinburgh Math. Soc. (2). 1989. V. 32, №2. P. 281–283. [*423*]

[77] Bonet J., Lindström M., Valdivia M. *Two theorems of Josefson–Nissenzweig type for Fréchet spaces*. Proc. Amer. Math. Soc. 1993. V. 117, №2. P. 363–364. [*238*]

[78] Bonet J., Taskinen J. *Quojections and the problem of topologies of Grothendieck*. Note Mat. 1991. V. 11. P. 49–59. [*423*]

[79] Bonic R.A. *Four brief examples concerning polynomials on certain Banach spaces*. J. Differential Geom. 1968. V. 2. P. 391–392. [*309*]

[80] Borell C. *Convex measures on locally convex spaces*. Ark. Math. 1974. V. 12, №2. P. 239–252. [*379*]

[81] Borell C., *Gaussian Radon measures on locally convex spaces*. Math. Scand. 1976. V. 38, №2. P. 265–284. [*415*]

[82] Borell C. *A note on conditional probabilities of a convex measure*. Lecture Notes in Math. 1978. V. 644. P. 68–72. [*379*]

[83] Borell C. *Convexity of measures in certain convex cones in vector space σ-algebras*. Math. Scand. 1983. V. 53, №1. P. 125–144. [*380*]

[84] Borges C.J.R. *On stratifiable spaces*. Pacif J. Math. 1966. V. 17, №1. P. 1–16. [*94*]

[85] Borsuk K. Theory of retracts. Panstwowe Wydawnictwo Naukowe, Warsaw, 1967; 251 pp. [*94*]

[86] Bourbaki N. *Sur les espaces de Banach*. C. R. Acad. Sci. Paris. 1938. T. 206. P. 1701–1704. [*419*]

[87] Bourbaki N. Espaces vectoriels topologiques. Hermann & Cie, Paris, 1953; ii+124+iv pp.; 1955; ii+191 pp. 2nd. ed: Espaces vectoriels topologiques. Chapitres 1 à 5. Masson, Paris, 1981; vii+368 pp. (in French); English transl.: Topological vector spaces. Chapters 1–5. Springer-Verlag, Berlin, 1987; viii+364 pp. [*ix, 96, 230, 237, 420*]

[88] Bourbaki N. Topologie générale. Chapitres 1 à 4. Hermann, Paris, 1971; xv+357 pp. (in French); English transl.: General topology. Chapters 1–4. Springer-Verlag, Berlin, 1989; viii+437 pp. [*45, 51, 209*]

[89] Bourgin R.D. Geometric aspects of convex sets with the Radon–Nikodým property. Lecture Notes in Math. V. 993. Springer-Verlag, Berlin, 1983; xii+474 pp. [*422*]

[90] Buldygin V.V. Convergence of random elements in topological spaces. Naukova Dumka, Kiev, 1980; 239 pp. (in Russian). [*375, 424*]

[91] Buldygin V.V. *Supports of probability measures in separable Banach spaces*. Teor. Verojatn. i Primen. 1984. V. 29, №3. P. 528–532 (in Russian); English transl.: Theory Probab. Appl. 1984. V. 29. P. 546–549. [*391*]

[92] Buldygin V.V., Solntsev S.A. Asymptotic behaviour of linearly transformed sums of random variables. Kluwer, Dordrecht, 1997; xvi+500 pp. [*424*]

[93] Cakić N., Kadelburg Z., Rajović M., Radenović S. *On some problems of Grothendieck concerning (F) and (DF) spaces*. Numer. Funct. Anal. Optim. 2009. V. 30, №1-2. P. 37–45. [*420*]

[94] Carothers N.L. A short course on Banach space theory. Cambridge University Press, Cambridge, 2005; xii+184 pp. [*422*]

[95] Cartan H. Calcul différentiel. Formes différentielles. Applications élémentaires au calcul des variations et à la théorie des courbes et des surfaces. Hermann, Paris, 1967; 178 pp., 186 pp. English transl.: Differential calculus. Differential forms. Houghton Mifflin, Boston, Massachusetts, 1970; 167 pp., 160 pp. [*423*]

[96] Cascales B., Kąkol J., Saxon S.A. *Metrizability vs. Fréchet–Urysohn property*. Proc. Amer. Math. Soc. 2003. V. 131, №11. P. 3623–3631 (electronic). [*216*]

[97] Cascales B., Orihuela J. *On compactness in locally convex spaces*. Math. Z. 1987. B. 195. S. 365–381. [*423*]

[98] Cauty R. *Rétractes absolus de voisinage algébriques*. Serdica Math. J. 2005. V. 31, №4. P. 309–354. [*422*]

[99] Chapman T.A. Lectures on Hilbert cube manifolds. Amer. Math. Soc., Providence, Rhode Island, 1976; x+131 pp. [*422*]

[100] Choquet G. Lectures on analysis. V. I: Integration and topological vector spaces. W.A. Benjamin, New York, 1969; xx+360+xxi pp. [*422*]

[101] Christensen J.P.R. Topology and Borel structure. North-Holland, Amsterdam – London, Amer. Elsevier, New York, 1974; 133 pp. [*415*]

[102] Chung D.M., Rajput B.S., Tortrat A. *Semistable laws on topological vector spaces*. Z. Wahrscheinlichkeitstheorie verw. Geb. 1982. B. 60. S. 209–218. [*390*]

[103] Cohen P.E. *Products of Baire spaces*. Proc. Amer. Math. Soc. 1976. V. 55, №1. P. 119–124. [*209*]

[104] Collins H.S. *Completeness and compactness in linear topological spaces*. Trans. Amer. Math. Soc. 1955. V. 79. P. 256–280. [*236*]

[105] Colombeau J.F. *Infinite dimensional C^∞-mappings with a given sequence of derivatives at a given point*. J. Math. Anal. Appl. 1979. V. 71. P. 95–104. [*305, 309*]

[106] Cooper J.B. Saks spaces and applications to functional analysis. North-Holland, Amsterdam – New York, 1978; x+325 pp. (2nd ed.: 1987). [*423*]

[107] Corson H. *The weak topology of a Banach space*. Trans. Amer. Math. Soc. 1961. V. 101. P. 1–15. [*422*]

[108] Corson H.H. *Metrizability of compact convex sets*. Trans. Amer. Math Soc. 1970. V. 151. P 589–596. [*242*]

[109] Day M.M. Normed linear spaces. 3d ed. Springer-Verlag, New York – Heidelberg, 1973; viii+211 pp. [*227, 422*]

[110] De Wilde M. *Théorème du graphe fermé et espaces à réseau absorbant*. Bull. Math. Soc. Sci. Math. R. S. Roum. 1967. V. 11. P. 225–238. [*196, 421*]

[111] De Wilde M. *Réseaux dans les espaces linéaires à semi-normes*. Mém. Soc. Roy. Sci. Liège. 1969. V. 18, №2. P. 1–144. [*196, 421*]

[112] De Wilde M. Closed graph theorems and webbed spaces. Pitman, London, 1978; viii+158 pp. [*196, 421*]

[113] Defant A., Floret K. Tensor norms and operator ideals. North-Holland, Amsterdam, 1993; xii+566 pp. [*422*]

[114] Defant A., Floret K., Taskinen J. *On the injective tensorproduct of (DF)-spaces*. Arch. Math. 1991. B. 57, №2. P. 149–154. [*423*]

[115] Deimling K. Nonlinear functional analysis. Springer-Verlag, Berlin, 1985; xiv+450 pp. [*422, 423*]

[116] Dembart B. *On the theory of semigroups of operators on locally convex spaces*. J. Funct. Anal. 1974. V. 16. P. 123–160. [*424*]

[117] Dettweiler E. *Grenzwertsätze für Wahrscheinlichkeitsmasse auf Badrikianschen Räumen*. Z. Wahrscheinlichkeitstheorie verw. Geb. 1976. B. 34. S. 285–311. [*390*]

[118] Deville R., Godefroy G., Zizler V. Smoothness and renormings in Banach spaces. Longman Scientific & Technical, Harlow; John Wiley & Sons, New York, 1993; xii+376 pp. [*307, 422*]

[119] Dierolf S., Lurje P. *Deux exemples concernant des espaces (ultra) bornologiques*. C. R. Acad. Sci. Paris. Sér. A-B. 1976. T. 282, №23. P. A1347–A1350. [*227*]

[120] Diestel J. Geometry of Banach spaces – selected topics. Lecture Notes in Math. V. 485. Springer-Verlag, Berlin – New York, 1975; xi+282 pp. [*123, 422*]

[121] Diestel J., Jarchow H., Tonge A. Absolutely summing operators. Cambridge University Press, Cambridge, 1995; xvi+474 pp. [*412, 422*]

[122] Diestel J., Uhl J.J. Vector measures. Amer. Math. Soc., Providence, 1977; xiii+322 pp. [*422*]

[123] Dieudonné J. *Topologies faibles dans les espaces vectoriels*. C. R. Acad. Sci. Paris. 1940. T. 211. P. 94–97. [*419*]

[124] Dieudonné J. *Sur le théorème de Hahn–Banach*. Revue Sci. 1941. V. 70. P. 642–643. [*420*]

[125] Dieudonné J. *La dualité dans les espaces vectoriels topologiques*. Ann. Sci. École Norm Super. (3). 1942. V. 59. P. 107–139. [*419*]

[126] Dieudonné J. *Recent developments in the theory of locally convex vector spaces*. Bull. Amer. Math. Soc. 1953. V. 59. P. 495–512. [*420*]

[127] Dieudonné J. *On biorthogonal systems*. Michigan Math. J. 1954. V. 2. P. 7–20. [*220*]

[128] Dieudonné J. *Sur les espaces de Montel métrisables*. C. R. Acad. Sci. Paris. 1954. T. 238. P. 194–195. [*238*]

[129] Dieudonné J. *Bounded sets in (F)-spaces*. Proc. Amer. Math. Soc. 1955. V. 6. P. 729–731. [*238*]

[130] Dieudonné J. Foundations of modern analysis. Academic Press, New York – London, 1960; xiv+361 pp. [*51, 423*]

[131] Dieudonné J. History of functional analysis. North-Holland, Amsterdam – New York, 1981; vi+312 pp. [*419*]

[132] Dieudonné J., Schwartz L. *La dualité dans les espaces* $(\mathcal{F})$ *et* $(\mathcal{LF})$. Ann. Inst. Fourier Grenoble. 1949. V. 1. P. 61–101. [*420, 421*]

[133] Dinculeanu N. Vector integration and stochastic integration in Banach spaces. Wiley–Interscience, New York, 2000; xvi+424 pp. [*424*]

[134] Dragilev M.M. Bases in Köthe spaces. Rostov Gos. Univ., Rostov-on-Don, 2003; 143 pp. (in Russian). [*220*]

[135] Dragilev M.M., Chalov P.A. *On Fréchet spaces with an unconditional basis*. Mat. Zametki. 2006. V. 80, №1. P. 29–32 (in Russian); English transl.: Math. Notes. 2006. V. 80, №1-2. P. 27–30. [*220*]

[136] Drewnowski L. *The metrizable linear extensions of metrizable sets in topological linear spaces*. Proc. Amer. Math. Soc. 1975. V. 51. P. 323–329. [*152*]

[137] Drewnowski L. *Resolutions of topological linear spaces and continuity of linear maps*. J. Math. Anal. Appl. 2007. V. 335, №2. P. 1177–1194. [*239*]

[138] Dubinsky E. The structure of nuclear Fréchet spaces. Lecture Notes in Math. V. 720. Springer, Berlin, 1979; ii+187 pp. [*220*]

[139] Dubinsky E. *Nuclear Fréchet spaces without the bounded approximation property*. Studia Math. 1981/82. V. 71, №1. P. 85–105. [*220*]

[140] Dudley R.M., Feldman J., Le Cam L. *On seminorms and probabilities, and abstract Wiener spaces*. Ann. Math. (2). 1971. V. 93. P. 390–408. Corrections: ibid. 1976. V. 104, №2. P. 391. [*412*]

[141] Dudley R.M., Kanter M. *Zero–one laws for stable measures*. Proc. Amer. Math. Soc. 1974. V. 45, №2. P. 245–252; Correction: ibid. 1983. V. 88, №4. P. 689–690. [*384*]

[142] Dugundji J. *An extension of Tietze's theorem*. Pacif J. Math. 1951. V. 1. P. 353–367. [*94*]

[143] Duistermaat J.J., Kolk J.A.C. Distributions: theory and applications. Springer, New York, 2010; xvi+445 pp. [*422*]

[144] Dulst D. van. Reflexive and superreflexive Banach spaces. Mathematisch Centrum, Amsterdam, 1978; v+273 pp. [*422*]

[145] Dunford N., Schwartz J.T. Linear operators, I. General Theory. Interscience, New York, 1958; xiv+858 pp. [*86, 88, 292, 327, 419*]

[146] Dynin A., Mitiagin B. *Criterion for nuclearity in terms of approximative dimension*. Bull. Acad. Polon. Sci. Sér. Sci. Math. Astronom. Phys. 1960. V. 8. P. 535–540. [*220*]

[147] Eberlein W.F. *Weak compactness in Banach spaces. I*. Proc. Nat. Acad. Sci. U. S. A. 1947. V. 33. P. 51–53. [*420*]

[148] Edgar G.A. *Measurability in a Banach space. I*. Indiana Math. J. 1977. V. 26, №4. P. 663–677; *II*. ibid., 1979. V. 28, №4. P. 559–579. [*410*]

[149] Edgar G.A., Wheeler R.F. *Topological properties of Banach spaces*. Pacific J. Math. 1984. V. 115. P. 317–350. [*422*]

[150] Edwards R.E. Functional analysis. Theory and applications. Holt, Rinehart and Winston, New York – London, 1965; xiii+781 pp. [*ix, 6, 82, 100, 149, 174, 186, 220, 224, 227, 229, 231, 234, 237, 343, 422*]

[151] Eells J. *A setting for global analysis*. Bull. Amer. Math. Soc. 1966. V. 72, №5. P. 751–807. [*423*]

[152] Eidelheit M. *Zur Theorie der konvexen Mengen in linearen normierten Räumen*. Studia Math. 1936. V. 6. P. 104–111. [*223, 420*]

[153] Ekeland I. *An inverse function theorem in Fréchet spaces*. Annales Inst. H. Poincaré (C) Non Linear Anal. 2011. V. 28, №1. P. 91–105. [*291, 310*]

[154] Engelking P. General topology. Polish Sci. Publ., Warszawa, 1977; 626 pp. [*6, 45, 51, 82, 139, 215, 241, 326*]

[155] Fabian M., Habala P., Hájek P., Montesinos Santalucía V., Pelant J., Zizler V. Functional analysis and infinite-dimensional geometry. Springer-Verlag, New York, 2001; x+451 pp. [*307, 422*]

[156] Fabian M., Habala P., Hájek P., Montesinos V., Zizler V. Banach space theory. The basis for linear and nonlinear analysis. Springer, New York, 2011; xiv+820 pp. [*307, 422*]

[157] Fedorchuk V.V., Chigogidze A.Ch. Absolute retracts and infinite-dimensional manifolds. Nauka, Moscow, 1992; 232 pp. (in Russian). [*422*]

[158] Fernique X. *Une démonstration simple du théorème de R.M. Dudley et M. Kanter sur les lois zero–un pour les mesures stables*. Lecture Notes in Math. 1974. V. 381. P. 78–79. [*390*]

[159] Fernique X. Fonctions aléatoires gaussiennes, vecteurs aléatoires gaussiens. Université de Montréal, Centre de Recherches Mathématiques, Montréal, 1997; iv+217 pp. [*354*]

[160] Ferrando J.C., Kąkol J., López Pellicer M. *Necessary and sufficient conditions for precompact sets to be metrisable*. Bull. Austral. Math. Soc. 2006. V. 74, №1. P. 7–13. [*423*]

[161] Ferrando J.C., Kąkol J., López Pellicer M. *A revised closed graph theorem for quasi-Suslin spaces*. Czech. Math. J. 2009. V. 59. P. 1115–1122. [*196, 421*]

[162] Ferrando J.C., Kąkol J., López Pellicer M., Saxon S.A. *Tightness and distinguished Fréchet spaces*. J. Math. Anal. Appl. 2006. V. 324, №2. P. 862–881. [*423*]

[163] Ferrando J.C., Kąkol J., López Pellicer M., Saxon S.A. *Quasi-Suslin weak duals*. J. Math. Anal. Appl. 2008. V. 339. P. 1253–1263. [*423*]

[164] Ferrando J.C., Kąkol J., López Pellicer M., Śliwa W. *Web-compact spaces, Fréchet–Urysohn groups and a Suslin closed graph theorem*. Math. Nachr. 2010. B. 283, №5. S. 704–711. [*196, 423*]

[165] Ferrando J.C., López Pellicer M., Sánchez Ruiz L.M. Metrizable barrelled spaces. Pitman Research Notes in Math. Ser. V. 332. Longman, Harlow; Wiley, New York, 1995; 238 pp. [*423*]

[166] Ferrando J.C., Sánchez Ruiz L.M. *On C-Suslin spaces*. Math. Nachr. 2015. B. 288, №8-9. S. 898–904. [*422*]

[167] Fetisov V.G., Filippenko V.I., Kozobrod V.N. Operators and equations in linear topological spaces. VNC RAN, Vladikavkaz, 2006; 432 pp. (in Russian). [*423*]

[168] Fichtenholz G. *Sur les fonctionnelles linéaires, continues au sens généralisé*. Matem. Sbornik. 1938. V. 4, №1. P. 193–214. [*98, 420*]

[169] Filippov V.V. *On a question of E.A. Michael*. Comm. Math. Univ. Carol. 2004. V. 45, №4. P. 735–738. [*93*]

[170] Filippov V.V. *On a problem of E. Michael*. Mat. Zametki. 2005. V. 78, №4. P. 638–640 (in Russian); English transl.: Math. Notes. 2005. V. 78, №3-4. P. 597–599. [*93*]

[171] Floret K. Weakly compact sets. Lecture Notes in Math. V. 801. Springer-Verlag, Berlin – New York, 1980; vii+123 pp. [*213, 214, 218, 231, 241*]

[172] Floret K., Wloka J. Einführung in die Theorie der lokalkonvexen Raume. Lecture Notes in Math. V. 56. Springer-Verlag, Berlin – New York, 1968; vii+194 S. [*420*]

[173] Fonf V.P., Johnson W.B., Pisier G., Preiss D. *Stochastic approximation properties in Banach spaces*. Studia Math. 2003. V. 159, №1. P. 103–119. [*124, 391, 393, 494*]

[174] Fortet R., Mourier E. *Les fonctions aléatoires comme éléments aléatoires dans les espaces de Banach*. Studia Math. 1955. V. 15. P. 62–73. [*424*]

[175] Fréchet M. *la notion de différentielle dans l'analyse générale*. Ann. Sci. École Norm. Sup. 1925. V. 42. P. 293–323. [*423*]

[176] Fréchet M. *Les éléments aléatoires de nature quelconque dans un espace distancié*. Ann. Inst. H. Poincaré. 1948. V. 10. P. 215–310. [*424*]

[177] Fremlin D.H. *Borel sets in non-separable Banach spaces*. Hokkaido Math. J. 1980. V. 9. P. 179–183. [*411*]

[178] Fremlin D. Measure theory. V. 1–5. University of Essex, Colchester, 2000–2003. [*416*]

[179] Frerick L., Jordá E., Kalmes Th., Wengenroth J. *Strongly continuous semigroups on some Fréchet spaces*. J. Math. Anal. Appl. 2014. V. 412, №1. P. 121–124. [*424*]

[180] Frölicher A., Bucher W. Calculus in vector spaces without norm. Lecture Notes in Math. V. 30. Springer-Verlag, Berlin – New York, 1966; x+146 pp. [*423*]

[181] Frölicher A., Kriegl A. Linear spaces and differentiation theory. John Wiley & Sons, Chichester, 1988; xvi+246 pp. [*423*]

[182] Gabriyelyan S.S., Kąkol J., Kubzdela A., López-Pellicer M. *On topological properties of Fréchet locally convex spaces with the weak topology*. Topology Appl. 2015. V. 192. P. 123–137. [*242, 423*]

[183] Gabriyelyan S., Kąkol J., Plebanek G. *The Ascoli property for function spaces and the weak topology of Banach and Fréchet spaces*. Studia Math. 2016. V. 233, №2. P. 119–139. [*242*]

[184] Gaehler W. Grundstrukturen der Analysis. B. 1, 2. Birkhäuser, Basel – Stuttgart, 1977, 1978; viii+412 S., viii+623 S. [*423*]

[185] Gantmacher V., Šmulian V. *Über schwache Kompaktheit im Banachschen Raum*. Rec. Math. [Mat. Sbornik] N.S. 1940. V. 8 (50), №3. P. 489–492 (in Russian, German summary). [*420*]

[186] Garnir H.G., De Wilde M., Schmets J. Analyse fonctionnelle. T. I–III. Birkhäuser, Basel–Stuttgart, 1968, 1972, 1973; 562 pp., 287 pp., 375 pp. [*422*]

[187] Garsoux J. Espaces vectoriels topologiques et distributions. Dunod, Paris, 1963; xiii+324 pp. [*422*]

[188] Gâteaux R. *Sur les fonctionnelles continues et les fonctionnelles analytiques*. Compt. Rend. Acad. Sci. (Paris). 1913. T. 157. P. 325–327. [*423*]

[189] Gelfand I.M. *Sur un lemme de la théorie des espaces linéaires*. Commun. Inst. Sci. Math. et Mecan., Univ. Kharkoff et Soc. Math. Kharkoff, IV. Ser. 13. P. 35–40. [*237*]

[190] Gel'fand I.M. *On some problems of functional analysis*. Uspekhi Mat. Nauk. 1956. V. 11, №6. P. 3–12 (in Russian). [*240, 424*]

[191] Gel'fand I.M., Vilenkin N.Ya. Generalized functions. V. 4. Applications of harmonic analysis. Translated from the Russian. Academic Press, New York – London, 1964; xiv+384 pp. (Russian ed.: Moscow, 1961). [*424*]

[192] Gel'fand I.M., Kostyučenko A.G. *Expansion in eigenfunctions of differential and other operators*. Dokl. Akad. Nauk SSSR (N.S.). 1955. V. 103, №3. P. 349–352 (in Russian). [*240*]

[193] Gel'fand I.M., Shilov G.E. Generalized functions. V. 2. Spaces of fundamental and generalized functions. Academic Press, New York – London, 1968; x+261 pp. (Russian ed.: Moscow, 1958) [*106, 420*]

[194] Gihman I.I., Skorohod A.V. The theory of stochastic processes. V. 1. Springer-Verlag, New York – Heidelberg, 1974; viii+570 pp. (Russian ed.: Moscow, 1971). [*329, 418, 424*]

[195] Godefroy G., Lancien G., Zizler V. *The non-linear geometry of Banach spaces after Nigel Kalton*. Rocky Mountain J. Math. 2014. V. 44, №5. P. 1529–1583. [*422*]

[196] Godunov A.N. *The Peano theorem in Banach spaces*. Funkc. Anal. i Pril. 1975. V. 9, №1. P. 59–60 (in Russian); English transl.: Functional Anal. Appl. 1975. V. 9, №1. P. 53–55. [*294*]

[197] Goldstine H.H. *Weakly complete Banach spaces*. Duke Math. J. 1938. V. 4. P. 125–131. [*419*]

[198] Gorin E.A., Mitjagin B.S. *Systems of norms in a countably-normed space*. Uspekhi Mat. Nauk. 1958. V. 13, №5. P. 179–184 (in Russian). [*151*]

[199] Granas A., Dugundji J. Fixed point theory. Springer, New York, 2003; xv+690 pp. [*422*]

[200] Graves L.M. *Topics in the functional calculus*. Bull. Amer. Math. Soc. 1935. V. 41, №10. P. 641–662. [*423*]

[201] Gross L. *Harmonic analysis on Hilbert space*. Mem. Amer. Math. Soc. 1963. V. 46. P. 1–62. [*424*]

[202] Grothendieck A. *Critères de compacité dans les espaces fonctionnels généraux*. Amer. J. Math. 1952. V. 74. P. 168–186. [*420*]

[203] Grothendieck A. *Résumé des résultats essentiels dans la théorie des produits tensoriels topologiques et des espaces nucléaires*. Ann. Inst. Fourier Grenoble. 1952. V. 4. P. 73–112. [*420*]

[204] Grothendieck A. *Résumé de la théorie métrique des produits tensoriels topologiques*. Bol. Soc. Mat. Sao Paulo. 1953. V. 8. P. 1–79. [*420*]

[205] Grothendieck A. *Sur les espaces* (F) *et* (DF). Summa Brasil. Math. 1954. V. 3. P. 57–123. [*237, 238, 420, 421*]

[206] Grothendieck A. Produit tensoriels topologiques et espaces nucléaires. Mem. Amer. Math. Soc. 1955. №16. 140 pp.; erratum: Ann. Inst. Fourier, Grenoble. 1955–1956. V. 6. P. 117–120. [*195, 420*]

[207] Grothendieck A. Espaces vectoriels topologiques. Inst. Mat. Pura Apl., Univ. de São Paulo, São Paulo, 1954; 240 pp.; English transl.: Topological vector spaces. Gordon and Breach, New York – London – Paris, 1973; x+245 pp. [*ix, 195, 234, 416, 420*]

[208] Gruenhage G. *Metrizable spaces and generalizations*. Recent Progress in General Topology-II (M. Hušek, J. van Mill, eds.). P. 201–225. Elsevier, Amsterdam, 2002. [*216*]

[209] Guerre-Delabrière S. Classical sequences in Banach spaces. Marcel Dekker, New York, 1992; xvi+207 pp. [*422*]

[210] Guirao A.J., Montesinos V. *Completeness in the Mackey topology*. Funct. Anal. Appl. 2015. V. 49, №2. P. 97–105. [*423*]

[211] Guirao A.J., Montesinos V., Zizler V. *A note on Mackey topologies on Banach spaces*. J. Math. Anal. Appl. 2017. V. 445, №1. P. 944–952. [*423*]

[212] Guirao A.J., Montesinos V., Zizler V. Open problems in the geometry and analysis of Banach spaces. Springer, Cham, 2016; xii+169 pp. [*422*]

[213] Gutnik L.A. *Locally convex spaces that do not contain infinite-dimensional bounded sets*. Uspekhi Mat. Nauk. 1977. V. 32, №6. P. 251–252 (in Russian). [*240*]

[214] Hadamard J. Leçons sur le calcul des variations. Hermann, Paris, 1910; 534 pp. [*423*]

[215] Hain R.M. *A characterization of smooth functions defined on a Banach space*. Proc. Amer. Math. Soc. 1979. V. 77, №1. P. 63–67. [*423*]

[216] Haworth R.C., Mccoy R.A. Baire spaces. Dissert. Math. CXLI. Warszawa, 1977; 73 pp. [*209*]

[217] Haydon R. *An extreme point criterion for separability of a dual Banach space, and a new proof of a theorem of Corson*. Quart. J. Math. Oxford (2). 1976. V. 27, №107. P. 379–385. [*242*]

[218] Hazod W., Siebert E. Stable probability measures on Euclidean spaces and on locally compact groups. Structural properties and limit theorems. Kluwer, Dordrecht, 2001; xviii+612 p. [*390, 424*]

[219] Herer W. *Stochastic bases in Fréchet spaces*. Demonstr. Math. 1981. V. 14. P. 719–724. [*403*]

[220] Heyer H. Structural aspects in the theory of probability. 2nd ed. World Sci., Hackensack, New Jersey, 2010; xii+412 pp. [*424*]

[221] Hildebrandt T.H., Graves L.M. *Implicit functions and their differentials in general analysis*. Trans. Amer. Math. Soc. 1927. V. 29, №1. P. 127–153. [*423*]

[222] Hille E., Phillips R.S. Functional analysis and semi-groups. Amer. Math. Soc., Providence, Rhode Island, 1974; xii+808 pp. [*64, 66, 112*]

[223] Hoffmann-Jørgensen J. *Weak compactness and tightness of subsets of* $M(X)$. Math. Scand. 1972. V. 31, №1. P. 127–150. [*394*]

[224] Hoffmann-Jørgensen J. Probability in Banach spaces. Lecture Notes in Math. 1976. V. 598. P. 1–186. [*424*]

[225] Hoffmann-Jørgensen J. *Integrability of seminorms, the* 0–1 *law and the affine kernel for product measures*. Studia Math. 1977. V. 61, №2. P. 137–159. [*424*]

[226] Hogbe-Nlend H. Théorie des bornologies et applications. Lecture Notes in Math. V. 213. Springer-Verlag, Berlin – Heidelberg – New York, 1971; iv+168 pp. [*422*]

[227] Hörmander L. The analysis of linear partial differential operators. I–IV. Springer-Verlag, Berlin – New York, 1983, 1985; ix+391 pp., viii+391 pp., viii+525 pp., vii+352 pp. [*138, 422*]

[228] Horvath J. Topological vector spaces and distributions. V. I. Addison-Wesley, Reading – London – Don Mills, 1966; xii+449 pp. [*422*]

[229] Hrbacek K., Jech T. Introduction to set theory. 3d ed. Marcel Dekker, New York, 1999; ix+291 pp. [*6*]

[230] Hyers D.H. *Linear topological spaces*. Bull. Amer. Math. Soc. 1945. V. 51, №1. P. 1–21. [*420*]

[231] James R.C. *A non-reflexive Banach space isometric with its second conjugate space*. Proc. Nat. Acad. Sci. U. S. A. 1951. V. 37. P. 174–177. [*227*]

[232] James R.C. *Reflexivity and the supremum of linear functionals*. Ann. Math. (2). 1957. V. 66. P. 159–169. [*420*]

[233] James R.C. *Characterizations of reflexivity*. Studia Math. 1963/1964. V. 23. P. 205–216. [*420*]

[234] James R.C. *Weakly compact sets*. Trans. Amer. Math. Soc. 1964. V. 113. P. 129–140. [*213, 420*]

[235] James R.C. *Weak compactness and reflexivity*. Israel J. Math. 1964. V 2. P. 101–119. [*420*]

[236] James R.C. *Reflexivity and the sup of linear functionals*. Israel J. Math. 1972. V. 13. P. 289–300. [*420*]

[237] Jarchow H. Locally convex spaces. B.G. Teubner, Stuttgart, 1981; 548 pp. [*ix, 89, 110, 114, 137, 180, 197, 220, 239, 421*]

[238] Johnson W.B., Lindenstrauss J. (ed.) Handbook of the geometry of Banach spaces. V. I, II. North-Holland, Amsterdam, 2001, 2003; x+1005 pp., xii+1007–1866 pp. [*422, 424*]

[239] Junek H. Locally convex spaces and operator ideals. Teubner, Leipzig, 1983; 180 pp. [*422*]

[240] Kac M.P., Kurato A. *A sufficient condition for the differentiability of vector-valued functions*. Vestnik Moskov. Univ. Ser. I Mat. Meh. 1976. №4. P. 44–51 (in Russian); English transl.: Moscow Univ. Math. Bull. 1976. V. 31, №3-4. P. 83–89. [*423*]

[241] Kadelburg Z., Radenović S. Subspaces and quotients of topological and ordered vector spaces. University of Novi Sad, Inst. Math., Novi Sad, 1997; iv+122 pp. [*423*]

[242] Kadets (Kadec) M.I. *Proof of the topological equivalence of all separable infinite-dimensional Banach spaces*. Funkc. Anal. i Pril. 1967. V. 1, №1. P. 61–70 (in Russian); English transl.: Funct. Anal. Appl. 1967. V. 1, №1. P. 53–62. [*217*]

[243] Kadets V.M., Kadets M.I. *A condition for normability of Frechet spaces*. Mat. Zametki. 1985. V. 38, №1. P. 142–147 (Russian); English transl.: Math. Notes. 1985. V. 38, №1. P. 592–595. [*240*]

[244] Kadets M.I., Kadets V.M. Series in Banach spaces. Conditional and unconditional convergence. Birkhäuser, Basel, 1997; viii+156 pp. [*144, 422*]

[245] Kąkol J., Kubiś W., López-Pellicer M. Descriptive topology in selected topics of functional analysis. Springer, Berlin, 2011; xii+493 pp. [*196, 216, 423*]

[246] Kąkol J., López Pellicer M. *On realcompact topological vector spaces*. Rev. R. Acad. Cienc. Exactas Fiis. Nat. Ser. A Math. RACSAM. 2011. V. 105, №1. P. 39–70. [*423*]

[247] Kąkol J., López Pellicer M., Okunev O. *Compact covers and function spaces*. J. Math. Anal. Appl. 2014. V. 411, №1. P. 372–380. [*423*]

[248] Kąkol J., López Pellicer M., Sliwa W. *Weakly K-analytic spaces and the three-space property for analyticity*. J. Math. Anal. Appl. 2010. V. 362, №1. P. 90–99. [*423*]

[249] Kąkol J., López Pellicer M., Todd A.R. *A topological vector space is Fréchet–Urysohn if and only if it has bounded tightness*. Bull. Belg. Math. Soc. Simon Stevin. 2009. V. 16, №2. P. 313–317. [*216*]

[250] Kąkol J., Saxon S.A. *Montel (DF)-spaces, sequential (LM)-spaces and the strongest locally convex topology*. J. London Math. Soc. (2). 2002. V. 66, №2. P. 388–406. [*423*]

[251] Kakutani S. *Über die Metrisation der topologischer Gruppen*. Proc. Imper. Acad. Tokyo. 1936. V. 12. P. 82–84. [*420*]

[252] Kakutani S. *Ein Beweis des Satzes von M. Eidelheit über konvexe Mengen*. Proc. Imper. Acad. Tokyo. 1937. V. 13. P. 93–94. [*420*]

[253] Kakutani S. *Weak topology, bicompact set and the principle of duality*. Proc. Imper. Acad. Tokyo. 1940. V. 16. P. 63–67. [*419*]

[254] Kakutani S., Klee V. *The finite topology of a linear space*. Arch. Math. (Basel). 1963. V. 14. P. 55–58. [*112*]

[255] Kalenda O.F.K., Spurný J. *Boundaries of compact convex sets and fragmentability*. J. Funct. Anal. 2009. V. 256, №3. P. 865–880. [*423*]

[256] Kalton N.J. *Some forms of the closed graph theorem*. Proc. Cambridge Philos. Soc. 1971. V. 70. P. 401–408. [*229, 421*]

[257] Kalton N.J., Peck N.T., Roberts J.W. An F-space sampler. Cambridge University Press, Cambridge, 1984; xii+240 pp. [*423*]

[258] Kanter M. *Linear sample spaces and stable processes*. J. Funct. Anal. 1972. V. 9, №4. P. 441–459. [*390, 399*]

[259] Kanter M. *Random linear functionals and why we study them*. Lecture Notes in Math. 1978. V. 645. P. 114–123. [*399*]

[260] Kanter M. *Completion measurable linear functionals on a probability space*. Colloq. Math. 1978. V. 38, №2. P. 277–304. [*424*]

[261] Kantorovich L.V., Akilov G.P. Functional analysis in normed spaces. Macmillan, New York, 1964; xiii+771 pp. [*420*]

[262] Kantorovich L.V., Akilov G.P. Functional analysis. 2nd edition. Pergamon Press, Oxford– Elmsford, New York, 1982; xiv+589 pp. [*420*]

[263] Kantorovič L.V., Vulih B.Z., Pinsker A.G. Functional analysis in partially ordered spaces. GITTL, Moscow, 1950; 548 pp. (in Russian). [*422*]

[264] Kanwal R.P. Generalized functions. Theory and applications. 3d ed. Birkhäuser Boston, Boston, 2004; xviii+476 pp. [*422*]

[265] Kats M.P. *Continuity of universally measurable linear mappings*. Sibirsk. Mat. Zh. 1982. V. 23, №3. P. 83–90; correction: ibid. 1983. V. 24, №3. P. 217 (in Russian); English transl.: Sib. Math. J. 1982. V. 23, №3. P. 358–364. [*415*]

[266] Keller H.H. Differential calculus in locally convex spaces. Lecture Notes in Math. V. 417. Springer-Verlag, Berlin – Heidelberg – New York, 1974; 143 pp. [*423*]

[267] Keller O.-H. *Die Homoiomorphie der kompakten konvexen Mengen im Hilbertschen Raum*. Math. Ann. 1931. B. 105, №1. S. 748–758. [*217*]

[268] Kelley J.L. General topology. Van Nostrand, Toronto –New York – London, 1955; xiv+298 pp. [*6, 45, 51, 82*]

[269] Kelley J.L. *Hypercomplete linear topological spaces*. Michigan Math. J. 1958. V. 5. P. 235–246. [*421*]

[270] Kelley J.L., Namioka I. Linear topological spaces. 2nd ed. Springer-Verlag, New York – Heidelberg, 1976; xv+256 pp. [*ix, 81, 235, 420*]

[271] Kirillov A.A., Gvishiani A.D. Theorems and problems in functional analysis. Springer-Verlag, New York – Berlin, 1982; ix+347 pp. [*150*]

[272] Klee V.L. *Convex sets in linear spaces*. Duke Math. J. 1951. V. 18. P. 443–466. [*420*]

[273] Klee V.L. *Invariant metrics in groups (solution of a problem of Banach)*. Proc. Amer. Math. Soc. 1952. V. 3. P. 484–487. [*420*]

[274] Klee V.L. *Boundedness and continuity of linear functionals*. Duke Math. J. 1955. V. 22. P. 263–269. [*420*]

[275] Klee V.L. *Some topological properties of convex sets*. Trans. Amer. Math. Soc. 1955. V. 78. P. 30–45. [*420*]

[276] Klee V., Maluta E., Zanco C. *Basic properties of evenly convex sets*. J. Convex Anal. 2007. V. 14, №1. P. 137–148. [*423*]

[277] Knowles R.J., Cook T.A. *Incomplete reflexive spaces without Schauder bases*. Proc. Cambridge Philos. Soc. 1973. V. 74. P. 83–86. [*186*]

[278] Kolesnikov A.P. Functional splines in topological vector spaces. URSS, Moscow, 2008; 440 pp. (in Russian). [*423*]

[279] Kolmogoroff A.N. *Zur Normierbarkeit eines allgemeinen topologischen linearen Raumes*. Studia Math. 1934. V. 5. P. 29–33 (English transl in [**282**]). [*40, 419*]

[280] Kolmogoroff A. *La transformation de Laplace dans les espaces linéaires*. C. R. Acad. Sci. Paris. 1935. T. 200. P. 1717–1718 (English transl. in [**282**]). [*333, 424*]

[281] Kolmogorov A.N. *A note on the papers of R.A. Minlos and V. Sazonov*. Teor. Verojant. i Primen. 1959. V. 4, №2. P. 237–239 (in Russian); English transl.: Theory Probab. Appl. 1959. V. 4. P. 221–223. [*424*]

[282] Kolmogorov A.N. Selected works. V. I,II. Kluwer, Dordrecht, 1991, 1992. [*435*]

[283] Kolmogorov A.N., Tihomirov V.M. *ε-entropy and ε-capacity of sets in functional spaces*. Uspekhi Mat. Nauk. 1959. V. 14, №2. P. 3–86 (in Russian); English transl.: Amer. Math. Soc. Transl. (2). 1961. V. 17. P. 277–364. [*421*]

[284] Kolmogorov A.N., Fomin S.V. Introductory real analysis. V. 1. Metric and normed spaces. V. 2. Measure. The Lebesgue integral. Hilbert space. Transl. from the 2nd Russian ed. Corr. repr. Dover, New York, 1975; xii+403 pp. (Russian ed.: Moscow, 1954, 1960). [*x, 5, 150*]

[285] Komatsu H. *Semi-groups of operators in locally convex spaces*. J. Math. Soc. Japan. 1964. V. 16. P. 230–262. [*424*]

[286] Kōmura T. *Semigroups of operators in locally convex spaces*. J. Funct. Anal. 1968. V. 2. P. 258–296. [*424*]

[287] Kōmura T., Kōmura Y. *Über die Einbettung der nuklearen Räume in $(s)^A$*. Math. Ann. 1966. B. 162. S. 284–288. [*138*]

[288] Kōmura Y. *Some examples on linear topological spaces*. Math. Ann. 1964 B. 153. S. 150–162. [*186*]

[289] Kondakov V.P. Three main principles of functional analysis, their generalizations and applications. VNC RAN, Vladikavkaz, 2007; 208 pp. (in Russian). [*423*]

[290] Korobeinik Ju.F. *Representing systems*. Uspekhi Mat. Nauk. 1981. V. 36, №1. P. 73–126 (Russian); English transl.: Russian Math. Surveys. 1981. V. 36, №1. P. 75–137. [*220*]

[291] Kosov E.D. *Supports of measures with weak moments*. Dokl. Akad. Nauk. 2012. V. 447, №3. P. 254–258 (Russian); English transl.: Dokl. Math. 2012. V. 86, №3. P. 776–780. [*418*]

[292] Köthe G. Topological vector spaces. V. I, II. Springer-Verlag, Berlin, 1969, 1979; xv+456 pp., xii+331 pp. (German original: Topologische lineare Räume. I. Springer-Verlag, Berlin, 1960; xii+456 S.) [*ix, 238, 238, 420*]

[293] Köthe G., Toepliz O. *Lineare Räume mit unendlichvielen Koordinaten und Ringe unendlicher Matrizen*. J. Reine Angew. Math. 1934. B. 171. S. 193–226. [*419*]

[294] Kraaij R. *Strongly continuous and locally equi-continuous semigroups on locally convex spaces*. Semigroup Forum. 2016. V. 92, №1. P. 158–185. [*424*]

[295] Kraaij R.C. *A Banach–Dieudonne theorem for the space of bounded continuous functions on a separable metric space with the strict topology*. Topology Appl. 2016. V. 209. P. 181–188. [*423*]

[296] Krasnosel'skii M.A., Zabreiko P.P. Geometrical methods of nonlinear analysis. Springer-Verlag, Berlin, 1984; xix+409 pp. [*423*]

[297] Krein M., Šmulian V. *On regulary convex sets in the space conjugate to a Banach space*. Ann. Math. (2). 1940. V. 41. P. 556–583. [*420*]

[298] Kriegl A., Michor P.W. The convenient setting of global analysis. Amer. Math. Soc., Providence, Rhode Island, 1997; x+618 pp. [*423*]

[299] Kriegl A., Michor P.W., Rainer A. *The exponential law for spaces of test functions and diffeomorphism groups*. Indag. Math. (N.S.). 2016. V. 27, №1. P. 225–265. [*423*]

[300] Krom M.R. Cartesian products of metric Baire spaces. Proc. Amer. Math. Soc. 1974. V. 42. P. 588–594. [*209*]

[301] Kruglov V.M. Supplementary chapters of probability theory. Vysš. Škola, Moscow, 1984; 264 pp. (in Russian). [*375, 384, 390, 424*]

[302] Kuelbs J. *Some results for probability measures on linear topological vector spaces with an application to Strassen's LogLog Law*. J. Funct. Anal. 1973. V. 14, №1. P. 28–43. [*391*]

[303] Kuelbs J., Zinn J. *Another view of the CLT in Banach spaces*. J. Theoret. Probab. 2008. V. 21, №4. P. 982–1029. [*424*]

[304] Kunzinger M. Barrelledness, Baire-like- and (LF)-spaces. Pitman Research Notes in Math. Ser. V. 298 Longman, Harlow; Wiley & Sons, New York, 1993; xiii+160 pp. [*423*]

[305] Kuo H. Gaussian measures in Banach spaces. Lecture Notes in Math. V. 463. Springer-Verlag, Berlin – New York, 1975; vi+224 pp. [*354, 364*]

[306] Kurosh A.G. Lectures in general algebra. Pergamon Press, Oxford– Edinburgh – New York, 1965; x+364 pp. [*1, 5*]

[307] Kusraev A.G., Kutateladze S.S. Subdifferentials: theory and applications. Kluwer, Dordrecht, 1995; x+398 pp. [*423*]

[308] Kwapień S. *Complément au théorème de Sazonov–Minlos*. C. R. Acad. Sci. Paris Sér. A-B. 1968. T. 267. P. A698–A700. [*415*]

[309] Kwapień S., Tarieladze V. *Mackey continuity of characteristic functionals*. Georg. Math. J. 2002. V. 9, №1. P. 83–112. [*337, 366*]

[310] Larman D.G., Rogers C.A. *The normability of metrizable sets*. Bull. London Math. Soc. 1973. V. 5. P. 39–48. [*152*]

[311] Ledoux M., Talagrand M. Probability in Banach spaces. Isoperimetry and processes. Springer-Verlag, Berlin, 1991; xii+480 pp. [*382, 424*]

[312] Lelong P. Introduction à l'analyse fonctionnelle. I: Espaces vectoriels topologiques. Les cours de Sorbonne. Centre de Documentation Universitaire, Paris, 1971; ii+230 pp. [*422*]

[313] Leonov A., Shkarin S. *On weak and strong Peano theorems*. Russ. J. Math. Phys. 2004. V. 11. P. 77–80. [*297*]

[314] LePage R., Woodroofe M., Zinn J. *Convergence to a stable distribution via order statistics*. Ann. Probab. 1981. V. 9, №4. P. 624–632. [*387*]

[315] Li D., Queffélec H. Introduction à l'étude des espaces de Banach. Soc. Math. de France, Paris, 2004; xxiv+627 pp. [*422*]

[316] Lifshits M.A. Gaussian random functions. Translated from the Russian. Kluwer, Dordrecht, 1995; xii+333 pp. [*354*]

[317] Linde W. Probability in Banach spaces – stable and infinitely divisible distributions. Wiley, New York, 1986; 195 pp. [*390, 424*]

[318] Lindenstrauss J., Tzafriri L. Classical Banach spaces. V. I,II. Springer, Berlin – New York, 1996; 424 pp. [*219, 232, 422*]

[319] Llavona J.G. Approximation of continuously differentiable functions. North-Holland, Amsterdam, 1986; xiv+241 pp. [*307*]

[320] Lobanov S.G. *Sufficient conditions for differentiability of mappings of locally convex spaces*. Mat. Zametki. 1986. V. 39, №1. P. 70–82 (in Russian); English transl.: Math. Notes. 1986. V. 39, №1-2. P. 39–46. [*309*]

[321] Lobanov S.G. *The chain rule and its converse for mappings of locally convex spaces*. Mat. Zametki. 1989. V. 45, №1. P. 43–56 (in Russian); English transl.: Math. Notes. V. 45, №1-2. P. 30–38. [*289, 309*]

[322] Lobanov S.G., Smolyanov O.G. *Ordinary differential equations in locally convex spaces*. Uspekhi Mat. Nauk. 1994. V. 49, №3. P. 93–168 (in Russian); English transl.: Russian Math. Surveys. 1994. V. 49, №3. P. 97–175. [*294*]

[323] Louie D., Rajput B.S., Tortrat A. *A zero-one dichotomy theorem for r-semistable laws on infinite-dimensional linear spaces*. Sankhyā, Ser. A. 1980. V. 42. P. 9–18. [*390*]

[324] Lützen J. The prehistory of the theory of distributions. Springer-Verlag, New York – Berlin, 1982; viii+232 pp. [*419*]

[325] Luxemburg W.A.J., Zaanen A.C. Riesz spaces. V. I, II. North-Holland, Amsterdam – London; American Elsevier, New York, 1971, 1983; xi+514 pp., xi+720 pp. [*422*]

[326] Mac Lane S. Categories for the working mathematician. 2nd ed. Springer, New York, 1998; xii+314 pp. [*133*]

[327] Mackey G. *On infinite-dimensional linear spaces*. Trans. Amer. Math. Soc. 1945. V. 57. P. 155–207. [*420*]

[328] Mackey G. *On convex topological linear spaces*. Trans. Amer. Math. Soc. 1946. V. 60. P. 519–537. [*420*]

[329] Maeda M. *Some examples of measurable norms*. J. Math. Anal. Appl. 1984. V. 98, №1. P. 158–165. [*413*]

[330] Maeda M., Harai K., Hagihara R. *Some examples and the connection between cylindrical measures and measurable norms*. J. Math. Anal. Appl. 2003. V. 288, №2. P. 556–564. [*413*]

[331] Makarov B.M. *Inductive limits of normed spaces*. Dokl. Akad. Nauk SSSR. 1958. V. 119. P. 1092–1094 (in Russian). [*147*]

[332] Makarov B.M. *Some pathological properties of inductive limits of B-spaces*. Uspekhi Mat. Nauk. 1963. V. 18, №3. P. 171–178 (in Russian). [*147, 421*]

[333] Mankiewicz P. *On topological, Lipschitz, and uniform classification of LF-spaces*. Studia Math. 1974. V. 52. P. 109–142. [*242*]

[334] Mauldin R.D. (ed.) The Scottish Book. Mathematics from the Scottish Cafe. Birkhäuser, Boston, 1981; xiii+268 pp. [*419*]

[335] Mazur S. *Über konvexe Mengen in linearen normierten Räumen*. Studia Math. 1933. V. 4. P. 70–85. [*420*]

[336] McCoy R.A., Ntantu I. Topological properties of spaces of continuous functions. Lecture Notes in Math. V. 1315. Springer, 1988; iv+124 pp. [*422*]

[337] McIntosh A. *On the closed graph theorem*. Proc. Amer. Math. Soc. 1969. V. 20. P. 397–404. [*229, 421*]

[338] Megginson R.E. An introduction to Banach space theory. Springer-Verlag, New York, 1998; xx+596 pp. [*228, 422*]

[339] Meise R., Vogt D. Introduction to functional analysis. The Clarendon Press, Oxford University Press, New York, 1997; x+437 pp. [*89, 238, 422*]

[340] Meyer P.-A. Probability and potentials. Blaisdell, Waltham – Toronto – London, 1966; xiii+266 pp. [*343*]

[341] Mibu Y. *On Baire functions on infinite product spaces*. Proc. Imp. Acad. Tokyo. 1944. V. 20. P. 661–663. [*409*]

[342] Michael E. *A selection theorem*. Proc. Amer. Math. Soc. 1966. V. 17. P. 1404–1406. [*93*]

[343] Miljutin A.A. *Isomorphism of the spaces of continuous functions over compact sets of the cardinality of the continuum*. Teor. Funkcii Funkcional. Anal. i Pril. Khar'kov. 1966. V. 2. P. 150–156 (in Russian). [*217*]

[344] Mill J. van. *Domain invariance in infinite-dimensional linear spaces*. Proc. Amer. Math. Soc. 1987. V. 101, №1. P. 173–180. [*217*]

[345] Mill J. van. Infinite-dimensional topology. Prerequisites and introduction. North-Holland, Amsterdam, 1989; xii+401 pp. [*422*]

[346] Mill J. van. The infinite-dimensional topology of function spaces. North-Holland, Amsterdam, 2001; xii+630 pp. [*422*]

[347] Millionščikov V.M. *On the theory of differential equations in locally convex spaces*. Mat. Sb. (N.S.). 1962. V. 57, №4. P. 385–406 (in Russian). [*100, 295*]

[348] Milnor J. *Remarks on infinite-dimensional Lie groups*. In: Relativity, groups and topology, II (Les Houches, 1983). P. 1007–1057. North-Holland, Amsterdam, 1984. [*423*]

[349] Minlos R.A. *Generalized random processes and their extension to a measure*. Trudy Moskov. Mat. Obsc. 1959. V. 8. P. 497–518 (in Russian); English transl.: Math. Stat. Probab. 1959. V. 3. P. 291–314. [*424*]

[350] Mitiagin B.S. *Approximate dimension and bases in nuclear spaces*. Uspekhi Mat. Nauk. 1961. V. 16, №4. P. 63–132 (Russian); English transl.: Russian Math. Surveys. 1961. V. 16, №4. P. 59–127. [*137, 220*]

[351] Monna A.F. Functional analysis in historical perspective. John Wiley & Sons, New York – Toronto, 1973; viii+167 pp. [*419*]

[352] Mordukhovich B.S. Variational analysis and generalized differentiation. I. Basic theory. II. Applications. Springer-Verlag, Berlin, 2006; xxii+579 pp., xxii+610 pp. [*423*]

[353] Moscatelli V.B. *Fréchet spaces without continuous norms and without bases*. Bull. London Math. Soc. 1980. V. 12, №1. P. 63–66. [*224*]

[354] Mouchtari D. *La topologie du type de Sazonov pour les Banach et les supports hilbertiens*. Ann. Univ. Clermont. 1976. V. 61. P. 77–87. [*424*]

[355] Mourier E. *Eléments aléatoires dans un espace de Banach*. Ann. Inst. H. Poincaré. 1953. T. 19. P. 161–244. [*424*]

[356] Mujica J. Complex analysis in Banach spaces. Holomorphic functions and domains of holomorphy in finite and infinite dimensions. North-Holland, Amsterdam, 1986; xii+434 pp. [*423*]

[357] Muštari (Mushtari) D.H. *Certain general questions of the theory of probability measures in linear spaces*. Teor. Verojatnost. i Primenen. 1973. V. 18, №1. P. 66–77 (in Russian); English transl.: Theory Probab. Appl. 1973. V. 18, №1. P. 64–75. [*424*]

[358] Mushtari D.Kh. *Topological questions of the theory of probability measures in Fréchet spaces*. Izv. Vyssh. Uchebn. Zaved. Mat. 1983. №12. P. 43–51 (in Russian); English transl.: Soviet Math. (Iz. VUZ). 1983. №12. P. 50–60. [*424*]

[359] Mushtari D.Kh. Probabilities and topologies on linear spaces. Kazan Mathematics Foundation, Kazan', 1996; xiv+233 pp. [*391, 424*]

[360] Mushtari D.Kh., Chuprunov A.N. *Sufficient topologies and norms*. Teor. Verojatn. i Primen. 1983. V. 28, №4. P. 700–714 (in Russian); English transl.: Theory Probab. Appl. 1984. V. 28, №4. P. 736–751. [*424*]

[361] Nakano H. Topology of linear topological spaces. Maruzen, Tokyo, 1951; viii+281 pp. [*420*]

[362] Namioka I. *Kakutani-type fixed point theorems: a survey*. J. Fixed Point Theory Appl. 2011. V. 9, №1. V. 1–23. [*422*]

[363] Narasimhan R. Analysis on real and complex manifolds. North-Holland, Amsterdam, 1985; xiv+246 pp. [*310*]

[364] Narici L. *On the Hahn–Banach theorem*. Advanced courses of mathematical analysis. II, p. 87–122, World Sci. Publ., Hackensack, New Jersey, 2007 (http://at.yorku.ca/p/a/a/o/58.htm). [*419*]

[365] Narici L., Beckenstein E. Topological vector spaces. 2nd ed. CRC Press, Boca Raton, Florida, 2011; xviii+610 pp. [*ix, 421*]

[366] Neeb K.-H. *Towards a Lie theory of locally convex groups*. Japan. J. Math. 2006. V. 1, №2. P. 291–468. [*423*]

[367] Neerven J. van. *γ-radonifying operators: a survey*. Proc. Centre Math. Appl. Austral. Nat. Univ., 44, Canberra, 2010. P. 1–61. [*412*]

[368] Nemirovskiĭ A.S., Semenov S.M. *On polynomial approximation of functions on Hilbert space*. Mat. Sb. (N.S.) 1973. V. 92, №2. P. 257–281 (in Russian); English transl.: Math. USSR-Sbornik. 1973. V. 21, №2. P. 255–277. [*307*]

[369] Neumann J. von. *Zur Theorie der Gesellsschaftsspiele*. Math. Ann. 1929. B. 100. S. 295–320. [*419*]

[370] Neumann J. von. *On complete topological linear spaces*. Trans. Amer. Math. Soc. 1935. V. 37. P. 1–20. [*419*]

[371] Neveu J. Bases mathématiques du calcul des probabilités. Masson et Cie, Paris, 1964; xiii+203 pp. English transl.: Mathematical foundations of the calculus of probability. Holden-Day, San Francisco, 1965; 231 pp. [*329, 333*]

[372] Nirenberg L. Topics in nonlinear functional analysis. New York University, Courant Institute of Mathematical Sciences, New York; Amer. Math. Soc., Providence, Rhode Island, 2001; xii+145 pp. [*291, 422*]

[373] Okazaki Y. *Stochastic basis in Fréchet space*. Math. Ann. 1986. B. 274. S. 379–383. [*403*]

[374] Omori H. Infinite-dimensional Lie groups. Amer. Math. Soc., Providence, Rhode Island, 1997; xii+415 pp. [*423*]

[375] Ostling E.G., Wilansky A. *Locally convex topologies and the convex compactness property*. Proc. Camb. Phil. Soc. 1974. V. 75. P. 45–50. [*415*]

[376] Ostrovskiĭ E.I. *On the supports of probability measures in separable Banach spaces*. Dokl. Akad. Nauk SSSR. 1980. V. 225, №6. P. 1319–1320 (in Russian); English transl.: Soviet Math. Dokl. 1980. V. 22, №3. P. 836–838. [*391*]

[377] Oxtoby J.C. *Cartesian products of Baire spaces*. Fund. Math. 1960/1961. V. 49. P. 157–166. [*209*]

[378] Palamodov V.P. *Homological methods in the theory of locally convex spaces*. Uspekhi Mat. Nauk. 1971. V. 26, №1. P. 3–65 (in Russian); English transl.: Russian Math. Surveys. 1971. V. 26, №1. P. 1–64. [*422*]

[379] Panchapagesan T.V. The Bartle–Dunford–Schwartz integral. Integration with respect to a sigma-additive vector measure. Birkhäuser, Basel, 2008; xvi+301 pp. [*424*]

[380] Paulauskas V., Račkauskas A. Approximation theory in the central limit theorem. Exact results in Banach spaces. Kluwer, Dordrecht, 1987; xviii+156 pp. [*382*]

[381] Peetre J. *Rectifications à l'article "Une caractérisation abstraite des opérateurs différentiels"*. Math. Scand. 1960. V. 8. P. 116–120. [*310*]

[382] Pełczyński A. Linear extensions, linear averagings, and their applications to linear topological classification of spaces of continuous functions. Dissert. Math. (Rozprawy Mat.) 1968. V. 58. 92 pp. [*217*]

[383] Peralta A.M., Villanueva I., Wright J.D.M., Ylinen K. *Weakly compact operators and the strong∗ topology for a Banach space*. Proc. Roy. Soc. Edinburgh Sect. A. 2010. V. 140, №6. P. 1249–1267. [*423*]

[384] Peressini A.L. Ordered topological vector spaces. Harper & Row, New York - London, 1967; x+228 pp. [*422*]

[385] Pérez Carreras P., Bonet J. Barrelled locally convex spaces. North-Holland, Amsterdam, 1987; xvi+512 pp. [*ix, 99, 121, 124, 127, 147, 148, 149, 175, 179, 180, 191, 209, 224, 227, 229, 421*]

[386] Perez-Garcia C., Schikhof W.H. Locally convex spaces over non-Archimedean valued fields. Cambridge University Press, Cambridge, 2010; xiv+472 pp. [*423*]

[387] Phelps R.R. Lectures on Choquet's theorem. Van Nostrand, Princeton, 1966; v+130 pp.; 2nd ed.: Lecture Notes in Math. V. 1757. Springer-Verlag, Berlin – New York, 2001. [*343*]

[388] Pietsch A. Nuclear locally convex spaces. Springer-Verlag, New York – Heidelberg, 1972; ix+193 pp. [*89, 144*]

[389] Pietsch A. Operator ideals. North-Holland, Amsterdam – New York, 1980; 451 pp. [*144, 412*]

[390] Pietsch A. History of Banach spaces and linear operators. Birkhäuser Boston, Boston, 2007; xxiv+855 pp. [*419, 422*]

[391] Pisier G. Introduction to operator space theory. Cambridge University Press, Cambridge, 2003; vii+478 pp. [*422*]

[392] Preiss D. *Differentiability of Lipshitz functions on Banach spaces*. J. Funct. Anal. 1990. V. 91. P. 312–345. [*251*]

[393] Prohorov Yu.V. *Convergence of random processes and limit theorems in probability theory*. Teor. Verojatn. i Primen. 1956. V. 1, №2. P. 177–238 (in Russian); English transl.: Theory Probab. Appl. 1956. V. 1. P. 157–214. [*424*]

[394] Prohorov Yu.V. *The method of characteristic functionals*. Proc. 4th Berkeley Symp. Math. Statist. and Probab. V. 2. P. 403–419. University of California Press, Berkeley, 1960. [*424*]

[395] Prostov Yu.I. *For homeomorphisms of nonnormable Fréchet spaces the theorem on the differentiability of an inverse function is invalid*. Mat. Zametki. 1990. V. 47, №2. P. 78–88 (Russian); English transl.: Math. Notes. 1990. V. 47, №2. P. 170–178. [*290*]

[396] Pryce J.D. *Weak compactness in locally convex spaces*. Proc. Amer. Math. Soc. 1966. V. 17. P. 148–155. [*213*]

[397] Pták V. *On complete topological linear spaces*. Čehoslovack. Mat. Zh. 1953. V. 3 №4. P. 301–364 (in Russian). [*421*]

[398] Pták V. *Completeness and the open mapping theorem*. Bull. Soc. Math. France. 1958. V. 86. P. 41–74. [*421*]

[399] Pták V. *An extension theorem for separately continuous functions and its application to functional analysis*. Czech. Math. J. 1964. V. 14, №4. P. 562–581. [*420*]

[400] Pták V. *A combinatorial lemma on the existence of convex means and its application to weak compactness*. Proc. Sympos. Pure Math., Vol. VII, pp. 437–450. Amer. Math. Soc., Providence, Rhode Island, 1963. [*420*]

[401] Qiu J. *Local completeness and dual local quasi-completeness*. Proc. Amer. Math. Soc. 2000. V. 129, №5. P. 1419–1425. [*152*]

[402] Raĭkov D.A. *On two classes of locally convex spaces which are important in applications*. Voronez. Gos. Univ. Trudy Sem. Funkcional. Anal. 1957. V. 5. P. 22–34 (in Russian). [*421*]

[403] Raĭkov D.A. *On a property of nuclear spaces*. Uspekhi Mat. Nauk. 1957. V. 12, №5. P. 231–236 (in Russian). [*240*]

[404] Raĭkov D.A. *Inductive and projective limits with completely continuous mappings*. Dokl. Akad. Nauk SSSR (N.S.). 1958. V. 113. P. 984–986 (in Russian). [*421*]

[405] Raĭkov D.A. *A criterion of completeness of locally convex spaces*. Uspekhi Mat. Nauk. 1959. V. 14, №1. P. 223–229 (in Russian). [*421*]

[406] Raĭkov D.A. *An exponential law for spaces of continuous linear mappings*. Mat. Sb. (N.S.) 1965. V. 67, №2. P. 279–302 (in Russian). [*421*]

[407] Raĭkov D.A. *The work of V.L. Shmul'yan on topological linear spaces*. Uspekhi Mat. Nauk. 1965. V. 20, №2. P. 135–147 (in Russian); English transl.: Russian Math. Surveys. 1965. V. 20, №2. P. 130–141. [*419*]

[408] Raĭkov D.A. *Double closed-graph theorem for topological linear spaces*. Sibirsk. Mat. Zh. 1966. V. 7, №2. P. 353–372 (in Russian); English transl.: Sib. Math. J. 1966. V. 7. P. 287–300. [*195, 421*]

[409] Raĭkov D.A. *Supplements to the Russian translation of* [**420**]. Moscow, Mir, 1967 (in Russian). [*195, 421*]

[410] Rajput B.S. *On the support of certain symmetric stable probability measures on TVS*. Proc. Amer. Math. Soc. 1977. V. 63. P. 306–312. [*390*]

[411] Rajput B.S. *On the support of symmetric infinitely divisible and stable probability measures on LCTVS*. Proc. Amer. Math. Soc. 1977. V. 66. P. 331–334. [*390*]

[412] Rajput B.S. *Supports of certain infinitely divisible probability measures on locally convex spaces*. Ann. Probab. 1993. V. 21, №2. P. 886–897. [*390*]

[413] Rajput B.S., Rama-Murthy K., Żak T. *Supports of semi-stable probability measures on locally convex spaces*. J. Theor. Probab. 1994. V. 7, №4. P. 931–942. [*390*]

[414] Reed M., Simon B. Methods of modern mathematical physics. I. Functional analysis. 2nd ed. Academic Press, New York, 1980; xv+400 pp. [*422*]

[415] Repovš D., Semenov P.V. Continuous selections and multivalued mappings. Kluwer, Dordrech – Boston – London, 1998; viii+356 pp. [*93*]

[416] Retah V.S. *The adjoint homomorphism of locally convex spaces*. Funkcional. Anal. i Pril. 1969. V. 3, №4. P. 63–68 (in Russian); English transl.: Funct. Anal. Appl. 1969. V. 3. P. 307–311. [*240*]

[417] Retah V.S. *The subspaces of countable inductive limits*. Dokl. Akad. Nauk SSSR. 1970. V. 194. P. 1277–1279 (in Russian); English transl.: Soviet Math. Dokl. 1970. V. 11. P. 1384–1386. [*421*]

[418] Reznichenko E.A. *Normality and collective normality of function spaces*. Vestnik Moskov. Univ. Ser. I Mat. Mekh. 1990. №6. P. 56–58 (in Russian); English transl.: Moscow Univ. Math. Bull. 1990. V. 45, №6. P. 25–26. [*215*]

[419] Reznichenko E. *Stratifiability of $C_k(X)$ for a class of separable metrizable X*. Topology Appl. 2008. V. 155. P. 2060–2062. [*216*]

[420] Robertson A.P., Robertson W. Topological vector spaces. 2nd ed. Cambridge University Press, London – New York, 1973; viii+172 pp. [*ix, 195, 235, 420*]

[421] Robertson W. *On the closed graph theorem and spaces with webs*. Proc. London Math. Soc. (3). 1972. V. 24. P. 692–738. [*422*]

[422] Rolewicz S. Metric linear spaces. 2nd ed. Reidel, Dordrecht; PWN – Polish Sci. Publ., Warsaw, 1985; xii+459 pp. [*423*]

[423] Rosiński J. *On the convolution of cylindrical measures*. Bull. Acad. Polon. Sci. 1982. V. 30, №7–8. P. 379–383. [*374*]

[424] Roth W. Operator-valued measures and integrals for cone-valued functions. Lecture Notes in Math. V. 1964. Springer, Berlin, 2009; x+356 pp. [*424*]

[425] Rudin W. Functional analysis. 2nd ed. McGraw-Hill, New York, 1991; xv+424 pp. [*x, 122*]

[426] Ruess W. *Locally convex spaces not containing ℓ_1*. Funct. Approx. Comment. Math. 2014. V. 50. P. 531–558. [*242*]

[427] Sakai K. *On $\mathbb{R}^\infty$-manifolds and Q^∞-manifolds*. Topology Appl. 1984. V. 18, №1. P. 69–79. [*242*]

[428] Satô H. *Banach support of a probability measure in a locally convex space*. Lecture Notes in Math. 1976. V. 526. P. 221–226. [*391*]

[429] Satô H. *Hilbertian support of a probability measure on a Banach space*. Lecture Notes in Math. 1979. V. 709. P. 195–205. [*391*]

[430] Saxon S.A. *Two characterizations of linear Baire spaces*. Proc. Amer. Math. Soc. 1974. V. 45, №2. P. 204–208. [*209*]

[431] Saxon S.A. *Problems from the Perez Carreras/Bonet book*. In: Progress in functional analysis (Peniscola, 1990), pp. 233–237. North-Holland, Amsterdam, 1992. [*423*]

[432] Saxon S.A. *Non-Baire hyperplanes in nonseparable Baire spaces*. J. Math. Anal. Appl. 1992. V. 168, №2. P. 460–468. [*423*]

[433] Saxon S.A., Sánchez Ruiz L.M. *Dual local completeness*. Proc. Amer. Math. Soc. 1997. V. 125. P. 1063–1070. [*152*]

[434] Saxon S.A., Tweddle I. *The fit and flat components of barrelled spaces*. Bull. Austral. Math. Soc. 1995. V. 51, №3. P. 521–528. [*423*]

[435] Sazonov V.V. *On characteristic functionals*. Teor. Veroyatn. i Primen. 1958. V. 3, №2. P. 201–205 (in Russian); English transl.: Theory Probab. Appl. 1958. V. 3. P. 188–192. [*424*]

[436] Schaefer H.H. Topological vector spaces. Springer-Verlag, Berlin – New York, 1971; xi+294 pp. [*ix, 19, 116, 131, 134, 137, 180, 186, 194, 231, 235–238, 240, 420, 422*]

[437] Schlüchtermann G., Wheeler R.F. *The Mackey dual of a Banach space*. Noti de Matem. 1991. V. 11. P. 273–287. [*242*]

[438] Schmets J. Espaces de fonctions continues. Lecture Notes in Math. V. 519. Springer-Verlag, Berlin – New York, 1976; xii+150 pp. [*213*]

[439] Schreier J. *Ein Gegenbeispiel zur Theorie der schwachen Konvergenz*. Studia Math. 1930. V. 2. P. 58–62. [*232*]

[440] Schwartz L. Théorie des distributions. T. I,II. Hermann & Cie, Paris, 1950, 1951; 148 pp., 169 pp. (2e ed.: Hermann, Paris, 1966; xiii+420 pp.) [*420*]

[441] Schwartz L. *Sur le théorème du graphe fermé*. C. R. Acad. Sci. Paris. 1966. T. 263. P. 602–605. [*415, 421*]

[442] Schwartz L. *Probabilités cylindriques et applications radonifiantes*. J. Fac. Sci. Univ. Tokyo, Sect. I A. 1971. V. 18. P. 139–286. [*412*]

[443] Schwartz L. Radon measures on arbitrary topological spaces and cylindrical measures. Oxford University Press, London, 1973; xii+393 pp. [*424*]

[444] Sebastião e Silva J. *Su certe classi di spazi localmente convessi importanti per le applicazioni*. Rend. Mat. Appl. (5). 1955. V. 14. P. 388–410. [*420*]

[445] Segal I. *Distributions in Hilbert space and canonical system of operators*. Trans. Amer. Math. Soc. 1958. V. 88, №1. P. 12–41. [*424*]

[446] Shavgulidze E.T. *On the hypercompleteness of locally convex spaces*. Mat. Zametki. 1973. V. 13, №2. P. 297–302 (in Russian); English transl.: Math. Notes. 1973. V. 13. P. 180–182. [*421*]

[447] Shavgulidze E.T. *On B_r-completeness*. Funk. Anal. i Pril. 1975. V. 9, №4. P. 95–96 (in Russian); English transl.: Funct. Anal. Appl. V. 9, №4. P. 363–364. [*421, 421*]

[448] Shavgulidze E.T. *The Minlos theorem for measures of indefinite sign*. Uspekhi Mat. Nauk. 1976. V. 31, №3. P. 222 (in Russian). [*424*]

[449] Shavgulidze E.T. *Conditions for certain forms of completeness in the class of projective limits of sequences of inductive limits of sequences of Fréchet spaces*. Funk. Anal. i Pril. 1977. V. 11, №1. P. 91–92 (in Russian); English transl.: Funct. Anal. Appl. 1977. V. 11, №1. P. 82–83. [*421*]

[450] Shavgulidze E.T. *A diffeomorphism of a locally convex space*. Uspekhi Mat. Nauk. 1979. V. 34, №5. P. 231–232 (in Russian). [*291*]

[451] Shirota T. *On locally convex vector spaces of continuous functions*. Proc. Japan Acad. 1954. V. 30. P. 294–299. [*178*]

[452] Shkarin S.A. *Some results on the solvability of ordinary linear differential equations in locally convex spaces*. Mat. Sb. 1990. V. 181, №9. P. 1183–1195 (in Russian); English transl.: Math. USSR-Sb. 1992. V. 71, №1. P. 29–40. [*295*]

[453] Shkarin S.A. *On the Rolle theorem in infinite-dimensional Banach spaces*. Mat. Zametki. 1992. V. 51, №3. P. 128–136 (in Russian); English transl.: Math. Notes. V. 51, №3-4. P. 311–317. [*308*]

[454] Shkarin S.A. *The moment problem in Fréchet spaces*. Mat. Zametki. 1993. V. 54, №1. P. 110–123 (in Russian); English transl.: Math. Notes. 1993. V. 54, №1-2. P. 739–746. [*240, 309*]

[455] Shkarin S.A. *Peano's theorem in infinite-dimensional Fréchet spaces is invalid*. Funkt. Anal. i Pril. 1993. V. 27, №2. P. 90–92 (in Russian); English transl.: Funct. Anal. Appl. 1993. V. 27, №2. P. 149–151. [*294*]

[456] Shkarin S.A. *Peano's theorem is invalid in infinite-dimensional F'-spaces*. Mat. Zametki. 1997. V. 62, №1. P. 128–137 (in Russian); English transl.: Math. Notes. 1997. V. 62, №1-2. P. 108–115. [*294*]

[457] Shkarin S.A. *Extending smooth functions from linear subspaces of locally convex spaces*. J. Math. Sci. (New York). 1998. V. 91. P. 3323–3338. [*305*]

[458] Shkarin S.A. *On stratifiable locally convex spaces*. Russ. J. Math. Phys. 1999. V. 6. P. 435–460. [*216*]

[459] Shkarin S.A. *Whitney's type theorems for infinite dimensional spaces*. Infin. Dimens. Anal. Quantum Probab. Relat. Top. 2000. V. 3. P. 141–160. [*305*]

[460] Shkarin S.A. *Sequentially continuous non-linear fundamental systems of solutions of affine equations in locally convex spaces*. Demonstratio Math. 2003. V. 36. P. 611–626. [*297*]

[461] Shkarin S.A. *Monotonically normal topological vector spaces are stratifiable*. Topology Appl. 2004. V. 136, №1. P. 129–134. [*216*]

[462] Shkarin S.A. *On solvability of linear differential equations in* R^N. Demonstratio Math. 2005. V. 38. P. 85–99. [*297*]

[463] Shkarin S.A. *Compact perturbations of linear differential equations in locally convex spaces*. Studia Math. 2006. V. 172, №3. P. 203–227. [*297*]

[464] Simon B. Convexity. An analytic viewpoint. Cambridge Univ. Press, Cambridge, 2011; x+345 pp. [*343*]

[465] Singer I. Abstract convex analysis. John Wiley & Sons, New York, 1997; xxii+491 pp. [*423*]

[466] Singer I. Duality for nonconvex approximation and optimization. Springer, New York, 2006; xx+355 pp. [*423*]

[467] Sipacheva O.V. *On a class of free locally convex spaces*. Mat. Sb. 2003. V. 194, №3. P. 25–52 (in Russian); English transl.: Sb. Math. V. 194, №3-4. P. 333–360. [*94*]

[468] Skorohod A.V. Integration in Hilbert space. Springer-Verlag, Berlin – New York, 1974; xii+177 pp. [*424*]

[469] Skorohod A.V. *Linear and almost linear functionals on a measurable Hilbert space*. Teor. Verojatnost. i Primenen. 1978. V. 23, №2. P. 397–402 (in Russian); English transl.: Theor. Probability Appl. 1978. V. 23, №2. P. 380–385. [*399*]

[470] Słowikowski W. *Extensions of sequentially continuous linear functionals in inductive sequences of* $(\mathcal{F})$*-spaces*. Studia Math. 1966. V. 26. P. 193–221. [*152, 421*]

[471] Smirnov E. Hausdorff spectra in functional analysis. Springer, London, 2002; viii+209 pp. [*196, 421*]

[472] Smoleński W. *Pre-supports and kernels of probability measures in Fréchet spaces*. Demonstr. Math. 1977. V. 10, №3-4. P. 751–762. [*424*]

[473] Smoleński W. *An abstract form of a counterexample of Marek Kanter*. Lecture Notes in Math. 1984. V. 1080. P. 288–291. [*399*]

[474] Smoleński W. *On the approximation of measurable linear functionals*. Statist. Probab. Lett. 1985. V. 3, №4. P. 205–207. [*424*]

[475] Smoleński W. *On the kernel of probability measures on a Banach space*. Demonstr. Math. 1988. V. 21. P. 569–572. [*424*]

[476] Smolyanov (Smoljanov) O.G. *On linear topological spaces not satisfying the first axiom of countability*. Uspekhi Mat. Nauk. 1964. V. 19, №6. P. 199–200 (in Russian). [*242*]

[477] Smolyanov O.G. *On the topology of inductive limits of infinite sequences of linear topological spaces satisfying the first axiom of countability*. Vestnik Moskov. Univ. Ser. I Mat. Meh. 1965. №1. P. 26–29 (in Russian). [*421*]

[478] Smolyanov O.G. *On locally-convex topologies in the space K and in some other spaces of finite functions*. Vestnik Moskov. Univ. Ser. I Mat. Meh. 1965. №3. P. 5–11 (in Russian). [*240*]

[479] Smolyanov O.G. *Measurable polylinear and power functionals in certain linear spaces with measure*. Dokl. Akad. Nauk SSSR. 1966. V. 170. P. 526–529 (in Russian); English transl.: Soviet Math. Dokl. 1966. V. 7. P. 1242–1246. [*405, 418*]

[480] Smolyanov O.G. *Almost closed linear subspaces of strict inductive limits of sequences of Fréchet spaces*. Matem. Sb. (N.S.) 1969. V. 80, №4. P. 513–520 (in Russian); English transl.: Math. USSR-Sbornik. 1969. V. 9, №4. P. 479–485. [*139, 421*]

[481] Smolyanov O.G. *Measurable linear varieties in products of linear spaces with measure*. Matem. Zametki. 1969. V. 5, №5. P. 623–634 (in Russian). English transl.: Math. Notes. 1969. V. 5. P. 374–379. [*377*]

[482] Smolyanov O.G. *The space D is not hereditarily complete*. Izv. Akad. Nauk SSSR Ser. Mat. 1971. V. 35, №3. P. 682–696 (Russian); English transl.: Math. USSR-Izvestiya. 1971. V. 5, №3. P. 696–710. [*421*]

[483] Smolyanov O.G. *Some results on fully complete and hereditary complete spaces*. Uspekhi Mat. Nauk. 1972. V. 27, №2. P. 181–182 (in Russian). [*421*]

[484] Smolyanov O.G. *Sequentially closed subsets of products of locally convex spaces*. Funkcional. Anal. i Pril. 1973. V. 7, №1. P. 88–89 (in Russian); English transl.: Functional Anal. Appl. 1973. V. 7. P. 80–81. [*139, 421*]

[485] Smolyanov O.G. *The size of the classes of hypercomplete spaces and spaces that satisfy the Krein–Šmul'jan condition*. Uspekhi Mat. Nauk. 1975. V. 30, №1. P. 259–260 (in Russian). [*421*]

[486] Smolyanov O.G. *Almost closed subsets of countable products of locally convex spaces*. Trudy Moskov. Mat. Obsc. 1975. V. 32. P. 61–76 (in Russian). [*139, 421*]

[487] Smolyanov O.G. *The class of spaces in which the theorem on the bounded differentiability of the inverse function is valid*. Mat. Zametki. 1975. V. 17, №5. P. 703–709 (in Russian); English transl.: Math. Notes. 1975. V. 17, №5. P. 418–421. [*290*]

[488] Smolyanov O.G. *Higher derivatives of mappings of locally convex spaces*. Mat. Zametki. 1977. V. 22, №5. P. 729–743 (in Russian); English transl.: Math. Notes. 1977. V. 22, №5-6. P. 899–906. [*306, 423*]

[489] Smolyanov O.G. Analysis on topological linear spaces and its applications. Moskov. Gos. Univ., Moscow, 1979; 86 pp. (in Russian). [*37, 243, 305, 306, 423*]

[490] Smolyanov O.G. *A method for proving uniqueness theorems for evolution differential equations*. Mat. Zametki. 1979. V. 25, №2. P. 259–269 (in Russian); English transl.: Math. Notes. 1979. V. 25, №1-2. P. 135–140. [*297*]

[491] Smolyanov O.G. *The Gross–Sazonov theorem for sign-variable cylindrical measures*. Vestnik Moskovsk. Univ. 1983. №4. P. 4–12 (in Russian); English transl.: Moscow Univ. Math. Bull. 1983. V. 38. P. 1–9. [*424*]

[492] Smolyanov O.G. *Topology of the spaces $\mathcal{D}$ and $\mathcal{D}'$*. (Russian) Vestnik Moskov. Univ. Ser. I Mat. Mekh. 1984. №1. P. 66–68 (in Russian); English transl.: Moscow Univ. Math. Bull. 1984. V. 39, №1. P. 57–60. [*139*]

[493] Smolyanov O.G. *Nonclosed sequentially closed subsets of locally convex spaces and applications*. Note di Matematica. 1992. V. 12. P. 237–244. [*139*]

[494] Smolyanov O.G. *Solutions of D.A. Raikov's problems in the theory of topological vector spaces*. Russian J. Math. Phys. 2008. V. 15, №4. P. 522–529. [*139*]

[495] Smolyanov O.G., Fomin S.V. *Measures on topological linear spaces*. Uspekhi Mat. Nauk. 1976. V. 31, №4. P. 3–56 (in Russian); English transl.: Russian Math. Surveys. 1976. V. 31, №4. P. 1–53. [*424*]

[496] Smolyanov O.G., Shavgulidze E.T. *A simple proof of Tarieladze's theorem on sufficiency of positively sufficient topologies*. Teor. Verojatn. i Primen. 1992. V. 37, №2. P. 421–424 (in Russian); English transl.: Theory Probab. Appl. 1992. V. 37, №2. P. 402–404. [*357*]

[497] Smolyanov O.G., Shavgulidze E.T. Continual integrals. 2nd ed. URSS, Moscow, 2015; 336 pp. (in Russian). [*424*]

[498] Smolyanov O.G., Uglanov A.V. *Every Hilbert subspace of a Wiener space has measure zero*. Matem. Zamet. 1973. V. 14, №3. P. 369–374 (in Russian); English transl.: Math. Notes. 1973. V. 14. P. 772–774. [*417*]

[499] Šmulian V. (Šmul'yan V.L.) *Sur les ensembles régulièrement fermés et faiblement compacts dans l'espace du type (B)*. C. R. (Dokl.) Acad. Sci. URSS. 1938. V. 18, №7. P. 405–407. [*420*]

[500] Šmulian V. *On the principle of inclusion in the space of the type (B)*. Rec. Math. Moscou. 1939. V. 5, №2. P. 317–328 (in Russian). [*420*]

[501] Šmulian V. *Über lineare topologische Räume*. Rec. Math. Moscou. 1940. V. 7. P. 425–444. [*419*]

[502] Šmul'yan V.L. *Sur les espaces linéaires topologiques. II*. Mat. Sb., N. Ser. 1941. V. 9, №3. P. 727–730 (in Russian. French summary). [*420*]

[503] Šmulian V. *Sur les ensembles compacts et faiblement compacts dans l'espace du type (B)*. Mat. Sb., N. Ser. 1943. V. 12, №1. P. 91–98. [*420*]

[504] Sova M. *The general theory of differentiability in linear topological spaces*. Czechoslovak Math. J. 1964. V. 14, №3. P. 485–508 (in Russian). [*423*]

[505] Sova M. *Conditions for differentialbility in linear topological spaces*. Czechoslovak Math. J. 1966. V. 16, №3. P. 339–362 (in Russian). [*423*]

[506] Sukhinin M.F. Selected chapters in nonlinear analysis. Izd-vo RUDN. Moscow, 1992; 302 pp. (in Russian). [*423*]

[507] Sukhomlinov G.A. *On extension of linear functionals in a complex and quaternion linear space*. Mat. Sbornik. 1938. V. 3, №2. P. 353–357 (in Russian). [*420*]

[508] Sztencel R. *On the lower tail of stable seminorm*. Bull. Acad. Polon. Sci. 1984. V. 32, №11–12. P. 715–719. [*387*]

[509] Takahashi Y., Okazaki Y. 0–1 *laws of a probability measure on a locally convex space*. Publ. Res. Inst. Math. Sci. 1986. V. 22, №1. P. 97–102. [*424*]

[510] Talagrand M. *Sur une conjecture de H.H. Corson*. Bull. Sci. Math. (2). 1975. V. 99, №4. P. 211–212. [*215*]

[511] Talagrand M. *Les boules peuvent engendrer la tribu borélienne d'un espace métrizable non séparable?* Séminaire Choquet, 17e anée. 1977–1978. F. 5, №2, 2 pp. Paris, 1978. [*411*]

[512] Talagrand M. *Comparaison des boreliens d'un espace de Banach pour les topologies fortes et faibles*. Indiana Univ. Math. J. 1978. V. 27, №6. P. 1001–1004. [*411*]

[513] Talagrand M. *Est-ce que l^∞ est un espace mesurable?* Bull. Sci. Math. 1979. T. 103, №3. P. 255–258. [*411*]

[514] Talagrand M. *Hyperplans universellement mesurables*. C. R. Acad. Sci. Paris, Sér. A. 1980. T. 291. P. A501–A502. [*415*]

[515] Talagrand M. Pettis integral and measure theory. Mem. Amer. Math. Soc. 1984. V. 51, №307. P. 1–224. [*415*]

[516] Tarieladze V.I. *On a topological description of characteristic functionals*. Dokl. Akad. Nauk SSSR. 1987. V. 295, №6. P. 1320–1323 (in Russian); English transl.: [*357, 424*]

[517] Taskinen J. *Counterexamples to "problème des topologies" of Grothendieck*. Ann. Acad. Sci. Fenn. Ser. A I Math. Dissertationes. 1986. №63. 25 pp. [*423*]

[518] Taskinen J. *The projective tensor product of Fréchet–Montel spaces*. Studia Math. 1988. V. 91, №1. P. 17–30. [*423*]

[519] Taskinen J. *A Fréchet–Schwartz space with basis having a complemented subspace without basis*. Proc. Amer. Math. Soc. 1991. V. 113, №1. P. 151–155. [*220*]

[520] Taylor A.E. *The differential: Nineteenth and twentieth century developments*. Arch. Hist. Exact Sci. 1974. V. 12. P. 355–383. [*423*]

[521] Taylor A.E. *A study of Maurice Fréchet. I–III*. Arch. Hist. Exact Sci. 1982. V. 27, №3. P. 233–295; ibid., 1985. V. 34, №4. P. 279–380; ibid., 1987. V. 37, №1. P. 25–76. [*419*]

[522] Tkachuk V.V. *A selection of recent results and problems in C_p-theory*. Topology Appl. 2007. V. 154, №12. P. 2465–2493. [*422*]

[523] Tkachuk V.V. A C_p-theory problem book. Topological and function spaces. Special features of function spaces. Compactness in function spaces. Functional equivalencies. Springer, New York, 2011; xvi+485 pp.; Springer, Cham, 2014–2016; xiv+583 pp., xvi+524 pp., xiv+727 pp. [*422*]

[524] Tortrat A. *Structure des lois indéfiniment divisibles ($\mu \in \mathfrak{I} = \mathfrak{I}(X)$) dans un espace vectoriel topologique (séparé) X*. Lecture Notes in Math. 1967. V. 31. P. 299–327. [*390*]

[525] Tortrat A. *Sur la structure des lois indéfiniment divisibles (classe $\mathfrak{T}(X)$) dans les espaces vectoriels X (sur le corps réel)*. Z. Wahrscheinlichkeitstheorie verw. Geb. 1969. B. 11. S. 311–326. [*390*]

[526] Tortrat A. *Lois $e(\lambda)$ dans les espaces vectoriels et lois stables*. Z. Wahrscheinlichkeitstheorie verw. Geb. 1976. B. 37, №2. S. 175–182. [*384*]

[527] Tortrat A. *Sur le support des lois indéfiniment divisibles dans les espaces vectoriels localment convexes*. Ann. Inst. H. Poincaré. 1977. V. 13. P. 27–43; Complément: ibid., P. 293–298; Second complément: ibid., P. 349–354. [*390*]

[528] Toruńczyk H. *Characterizing Hilbert space topology*. Fund. Math. V. 1981. V. 111. P. 247–262. [*217*]

[529] Trèves J.F. Linear partial differential equations with constant coefficients: existence, approximation and regularity of solutions. Gordon and Breach, New York, 1966; x+534 pp. [*422*]

[530] Trèves F. Topological vector spaces, distributions and kernels. Academic Press, New York – London, 1967; xvi+624 pp. [*ix, 420*]

[531] Tsar'kov I.G. *Smoothing of Hilbert-valued uniformly continuous mappings*. Izv. Ross. Akad. Nauk Ser. Mat. 2005. V. 69, №4. P. 149–160 (in Russian); English transl.: Izv. Math. 2005. V. 69, №4. P. 791–803. [*307*]

[532] Tsar'kov I.G., Shavgulidze E.T. *Smooth isometric embedding into an infinite-dimensional sphere*. Funkt. Anal. i Pril. 1999. V. 33, №3. P. 93–95 (in Russian); English transl.: Funct. Anal. Appl. 1999. V. 33, №3. P. 239–240. [*310*]

[533] Tychonoff A. *Ein Fixpunktsatz*. Math. Ann. 1935. B. 111. S. 767–776. [*419*]

[534] Urbanik K. *Random linear functionals and random integrals*. Colloq. Math. 1975. V. 38, №2. P. 255–263. [*399*]

[535] Vainberg M.M. Variational method and method of monotone operators in the theory of nonlinear equations. Halsted Press, New York – Toronto, 1973; xi+356 pp. [*423*]

[536] Vakhania N.N. Probability distributions on linear spaces. Tbilisi, 1971 (in Russian); English transl.: North-Holland, Amsterdam, 1981; xv+123 pp. [*424*]

[537] Vakhania N.N., Tarieladze V.I. *Covariance operators of probability measures in locally convex spaces*. Teor. Verojatn. i Primen. 1978. V. 23. P. 3–26 (in Russian); English transl.: Theor. Probab. Appl. 1978. V. 23. P. 1–21. [*338, 344, 345*]

[538] Vakhania N.N., Tarieladze V.I., Chobanyan S.A. Probability distributions in Banach spaces. Translated from the Russian. Kluwer, 1991; xxvi+482 pp. (Russian ed.: Moscow, 1985) [*354, 374, 375, 382, 404, 412, 416, 424*]

[539] Valdivia M. *Mackey convergence and the closed graph theorem*. Arch. Math. 1974. B. 25. S. 649–656. [*198*]

[540] Valdivia M. *The space of distributions $\mathcal{D}'(U)$ is not B_r-complete*. Math. Ann. 1974. B. 211. S. 145–149. [*421*]

[541] Valdivia M. *The space $\mathcal{D}(U)$ is not B_r-complete*. Ann. Inst. Fourier (Grenoble). 1977. V. 27, №4. P. 29–43. [*421*]

[542] Valdivia M. *On weakly complete sets in a locally convex space*. Arch. Math. (Basel). 1977. B. 28, №6. S. 638–643. [*240*]

[543] Valdivia M. Topics in locally convex spaces. North-Holland, Amsterdam, 1982; xiii+510 pp. [*209, 236, 421*]

[544] Valdivia M. *B_r-complete spaces which are not B-complete*. Math. Z. 1984. B. 185, №2. S. 253–259. [*191*]

[545] Valdivia M. *A characterization of totally reflexive Fréchet spaces*. Math. Z. 1989. B. 200, №3. S. 327–346. [*240*]

[546] Valdivia M. *Bases and quasi-reflexivity in Fréchet spaces*. Math. Nachr. 2005. B. 278, №6. S. 712–729. [*220*]

[547] Valdivia M. *On certain (LB)-spaces*. Bull. Belg. Math. Soc. Simon Stevin. 2007. V. 14, №3. P. 565–575. [*423*]

[548] Vershik A.M., Sudakov V.N. *Probability measures in infinite-dimensional spaces*. Zap. Nauchn. Sem. LOMI. 1969. V. 12. P. 7–67 (in Russian); English transl.: Seminars in Math. V.A. Steklov Math. Inst., Leningrad. 1971. V. 12. P. 1–28. [*424*]

[549] Vladimirov V.S. Methods of the theory of generalized functions. Taylor & Francis, London, 2002; xiv+311 pp. [*90, 422*]

[550] Vladimirskiĭ Yu.N. *Cylindrical measures and p-summing operators*. Teor. Veroyatnost. i Primenen. 1981. V. 26, №1. P. 59–72 (in Russian); English transl.: Theory Probab. Appl. 1981. V. 26, №1. P. 56–68. [*412*]

[551] Vladimirskiĭ Yu.N. *On a probabilistic characterization of some classes of locally convex spaces*. Teor. Veroyatnost. i Primenen. 1983. V. 28, №3. P. 521–532 (in Russian); English transl.: Theory Probab. Appl. 1983. V. 28, №3. P. 549–560. [*412*]

[552] Vladimirskiĭ Yu.N. *Compactness of γ-summing operators*. Mat. Zametki. 1985. V. 37, №5. P. 743–750 (in Russian); English transl.: Math. Notes. 1985. V. 37, №5-6, P. 410–414. [*412*]

[553] Vladimirskiĭ Yu.N. *Conditions for the countable additivity of a cylindrical measure in a conjugate locally convex space*. Mat. Zametki. 1994. V. 56, №3. P. 13–19 (in Russian); English transl.: Math. Notes. 1994. V. 56, №3-4. P. 884–888. [*424*]

[554] Vogt D. *Regularity properties of (LF)-spaces*. Progress in functional analysis. P. 57–84. North-Holland, Amsterdam, 1992. [*423*]

[555] Vogt D. *A nuclear Fréchet space of C^∞-functions which has no basis*. Note Mat. 2005/06. V. 25, №2. P. 187–190. [*220*]

[556] Vogt D. *A nuclear Fréchet space consisting of C^∞-functions and failing the bounded approximation property*. Proc. Amer. Math. Soc. 2010. V. 138, №4. P. 1421–1423. [*220*]

[557] Volterra V. *Sopra le funzioni che dipendeno da altre funzioni*. Atti Realle Accad. Lincei Rend. Ser. 4. 1887. V. 3, №2. P. 97–105, 141–146, 153–158. [*423*]

[558] Waelbroeck L. Topological vector spaces and algebras. Lecture Notes in Math. V. 230. Springer-Verlag, Berlin-New York, 1971; vii+158 pp. [*423*]

[559] Waelbroeck L. Bornological quotients. Académie Royale de Belgique. Classe des Sciences, Brussels, 2005; 251 pp. [*422*]

[560] Wehausen J. *Transformations in linear topological spaces*. Duke Math. J. 1938. V. 4. P. 157–169. [*419*]

[561] Weizsäcker H. von. *Der Satz von Choquet–Bishop-de Leeuw für konvexe nicht kompakte Mengen straffer Maße über beliebigen Grundräumen*. Math. Z. 1975. B. 142. S. 161–165. [*343*]

[562] Weizsäcker H. von. *A note on infinite dimensional convex sets*. Math. Scand. 1976. V. 38, №2. P. 321–324. [*415*]

[563] Weizsäcker H. von, Winkler G. *Integral representation in the set of solutions of a generalized moment problem*. Math. Ann. 1979. B. 246. S. 23–32. [*343*]

[564] Wentzell A.D. A course in the theory of stochastic processes. Translated from the Russian. McGraw-Hill International Book, New York, 1981; x+304 pp. (Russian ed.: Moscow, 1975). [*329, 393*]

[565] Wiener N. *Differential space*. J. Math. Phys. 1923. V. 2. P. 131–174. [*424*]

[566] Wilansky A. Functional analysis. Blaisdell – Ginn, New York – Toronto – London, 1964; xvi+293 pp. [*422*]

[567] Wilansky A. Modern methods in topological vector spaces. McGraw-Hill, New York, 1978; xiii+298 pp. [*ix, 225, 229, 421*]

[568] Wiśniewski A. *Measurable linear operators on Banach spaces*. Colloq. Math. 1987. V. 54, №2. P. 261–265. [*417*]

[569] Wiśniewski A. *Measurable linear functionals and operators on Fréchet spaces*. Proc. Amer. Math. Soc. 1992. V. 114, №4. P. 1079–1085. [*418*]

[570] Wiśniewski A. *Measurable linear operators induced by stochastic processes*. Demonstr. Math. 2002. V. 35, №4. P. 873–878. [*424*]

[571] Wojtaszczyk P. Banach spaces for analysts. Cambridge University Press, Cambridge, 1991; xiv+382 pp. [*422*]

[572] Wojtyński W. *On conditional bases in non-nuclear Fréchet spaces*. Studia Math. 1970. V. 35. P. 77–96. [*220*]

[573] Yamamuro S. Differential calculus in topological linear spaces. Lecture Notes in Math. V. 374. Springer-Verlag, Berlin – Heidelberg – New York, 1974; iv+179 pp. [*423*]

[574] Yamasaki Y. Measures on infinite-dimensional spaces. World Sci., Singapore, 1985; x+256 pp. [*424*]

[575] Yosida K. Functional analysis. 6th ed. Springer-Verlag, Berlin, 1980; xii+501 pp. [*423*]

[576] Yurova E.V. *On continuous restrictions of measurable linear operators*. Dokl. Akad. Nauk. 2012. V. 443, №3. P. 300–303 (in Russian); English transl.: Dokl. Math. 2012. V. 85, №2. P. 229–232. [*400, 402*]

[577] Yurova E.V. *On continuous restrictions of measurable multilinear mappings*. Matem. Zametki. 2015. V. 98, №6. P. 930–936 (in Russian); English transl.: Math. Notes. 2015. V. 98, №6. P. 977–981. [*409*]

[578] Zabreiko P.P., Smirnov E.I. *On the closed graph theorem*. Sibirsk. Mat. Z. 1977. V. 18, №2. P. 304–313 (Russian); English transl.: Siberian Math. J. 1977. V. 18, №2. P. 218–224. [*421*]

[579] Zahariuta V. *Linear topologic invariants and their applications to isomorphic classification of generalized power spaces*. Turk. J. Math. 1996. V. 20, №2. P. 237–289. [*423*]

[580] Zălinescu C. Convex analysis in general vector spaces. World Scientific, River Edge, New Jersey, 2002; xx+367 pp. [*423*]

[581] Zinn J. *Zero-one laws for non-Gaussian measures*. Proc. Amer. Math. Soc. 1974. V. 44, №1. P. 179–185. [*424*]

[582] Zobin N.M., Mitjagin B.S. *Examples of nuclear metric linear spaces without a basis*. Funkcional. Anal. i Pril. 1974. V. 8, №4. P. 35–47 (in Russian); English transl.: Functional Anal. Appl. 1974. V. 8, №4. P. 304–313. [*220*]

Author Index

V.I. Bogachev and O.G. Smolyanov, *Topological Vector Spaces and Their Applications*, Springer Monographs in Mathematics,
DOI 10.1007/978-3-319-57117-1

Subject Index

Notation:

V.I. Bogachev and O.G. Smolyanov, *Topological Vector Spaces and Their Applications*, Springer Monographs in Mathematics,
DOI 10.1007/978-3-319-57117-1

STAKEHOLDER PERSPECTIVES ON WORLD HERITAGE AND DEVELOPMENT IN AFRICA

Stakeholder Perspectives on World Heritage and Development in Africa argues that World Heritage Sites (WHS) across the African continent should adopt practical, innovative, creative, and alternative management approaches that bring greater socio-economic benefits to society, whilst protecting their Outstanding Universal Value.

Drawing on empirical evidence gathered in conversation with stakeholders at WHS across Africa, the book explores the challenges involved in implementing conservation and socio-economic development as a stakeholder-driven process. Demonstrating that heritage can no longer be viewed as totally separate from its socio-economic context, Pascall argues that decisions about the management of heritage need to make sense at the local level if they are to be supported by stakeholders. As the book shows, heritage is still viewed and managed through systems, approaches, and strategies inherited from the colonial period, despite the increasing availability of inclusive governance systems. Stakeholders offer alternative, creative, and innovative approaches that capitalize on the potential of World Heritage to contribute to socio-economic development, whilst ensuring that its credibility and integrity are maintained.

Stakeholder Perspectives on World Heritage and Development in Africa offers unique insights into local perspectives on World Heritage and development in Africa. The book will be essential reading for academics, students, development partners, and practitioners around the world who are interested in museums and heritage, conservation, development, and the African continent.

Also, the book will be useful in the preparation of nomination dossiers, management plans, development plans, and in disaster risk management at WHS.

Pascall Taruvinga has a Doctoral degree in Archaeology from the University of Cape Town in South Africa and has over 25 years of experience in heritage management in various leadership and executive positions in the Africa region. Currently, he is Lecturer/Coordinator: Postgraduate Diploma in Heritage Management (Rhodes University, South Africa) and Director and Founder of IRAHA and Consultancy Pvt Ltd. His former positions include Chief Heritage Officer of Robben Island World Heritage Site (South Africa), Senior Consultant (African World Heritage Fund), World Heritage Specialist (National Heritage Council of South Africa), Programme Manager (AFRICOM-Kenya), Director Research and Development (Zimbabwe), and Head of Archaeology Department-Zimbabwe.

STAKEHOLDER PERSPECTIVES ON WORLD HERITAGE AND DEVELOPMENT IN AFRICA

Pascall Taruvinga

Routledge
Taylor & Francis Group
LONDON AND NEW YORK

Cover image: © Pascall Taruvinga

First published 2023
by Routledge
4 Park Square, Milton Park, Abingdon, Oxon, OX14 4RN

and by Routledge
605 Third Avenue, New York, NY 10158

Routledge is an imprint of the Taylor & Francis Group, an informa business

Library of Congress Cataloging-in-Publication Data
A catalog record for this title has been requested.

ISBN: 978-0-367-71113-9 (hbk)
ISBN: 978-0-367-69480-7 (pbk)
ISBN: 978-1-003-14936-1 (ebk)

DOI: 10.4324/9781003149361

Typeset in Bembo
by KnowledgeWorks Global Ltd.

This book is dedicated to my parents, Mathews and Sarah Taruvinga. Unfortunately, my Father, Mathews Taruvinga passed away while I was writing this book on the 1st of February 2021. To both of them, you are my love, inspiration, and fountain of wisdom.

CONTENTS

FIGURES

PLATES

TABLES

ABBREVIATIONS

AMA	Adaptive Management Approach
ASAPA	Association of Southern African Professional Archaeologists
AU	African Union
CAMPFIRE	Communal Areas Management Programme for Indigenous Resources
CDF	Conservation Development Framework
CoAL	Coal of Africa
COHWHS	Cradle of Humankind World Heritage Site
DDI	Destination Development Initiative
DFFE	Department of Forestry, Fisheries and Environmental
DMR	Department of Mineral Resources
DRDLR	Department of Rural Development and Local Restitution
DSAC	Department of Sport, Arts and Culture
DWAS	Department of Water Affairs and Sanitation
EIA	Environmental Impact Assessment
EPWP	Expanded Public Works Programme
GDP	Gross Domestic Product
GMTFCA	Greater Mapungubwe Transfrontier Conservation Area
GZWHS	Great Zimbabwe World Heritage Site
HDI	Human Development Index
HIA	Heritage Impact Assessment
ICCROM	International Centre for the Study of the Preservation and Restoration of Cultural Property
ICT	Information Communication Technology
ICOMOS	International Council on Monuments and Sites
IDCs	Indigenous and Descendent Communities
IUCN	International Union on Conservation of Nature

MAG	Mapungubwe Action Group
MCLWHS	Mapungubwe Cultural Landscape World Heritage Site
METT	Monitoring Effectiveness Tracking Tool
MMSF	Multiple and Multi-layered Stakeholder Framework
MMST	Multiple and Multi-layered Stakeholder Theory
MoA	Memorandum of Agreement
MRPRDA	Minerals and Petroleum Resources Development Act
NCA	Ngorongoro Conservation Area
NCCA	Ngorongoro Conservation Area
NDT	National Department of Tourism
NDP	National Development Plans
NDPs	National Development Plans
NGOs	Non-Governmental Organizations
NHRA	National Heritage Resources Act of 1999
NMMZ	National Museums and Monuments of Zimbabwe
OGWHC	Operational Guidelines on the Implementation of the World Heritage Convention
OUV	Outstanding Universal Value
PA	Protected Areas
PHRA	Provincial Heritage Resources Act
RIWHS	Robben Island World Heritage Site
SADC	Southern African Development Community
SAHRA	South African Heritage Resources Agency
SANParks	South African National Parks
SAWHCC	South African World Heritage Convention Committee
SBMS	State-Based Management Systems
SD	Sustainable Development
SDGs	Sustainable Development Goals
SMC	Save Mapungubwe Coalition
SOC	State of Conservation
SWHCC	South African World Heritage Convention Committee
TFCA	Transfrontier Conservation Area
TMS	Traditional Management Systems
UNESCO	United Nations Educational, Scientific and Cultural Organisation
WHC	World Heritage Convention
WHS	World Heritage Site

ACKNOWLEDGEMENTS

I would like to thank Professors Shadreck Chirikure, Weber Ndoro, Gilbert Pwiti, Munyaradzi Manyanga, George Abungu, and Janette Deacon and Doctors Albino Jopela, Godfrey Mahachi, Foreman Bandama, Robert Nyamushosho, Thabo Manetsi, Kendi Borona, Jesmael Mataga, Godfrey Maringira, Moleboheng Mohapi (my "office motivator"); and John Zulu, Thabo Seshoka, Ayanda Woji (my "office field assistant"), Gershon Manana and Souayibou Varissou, Patience and Philip Kwedza, the Indigenous and Descendant Communities, and stakeholders for supporting the writing of this book, and last but not the least, my wife, Thabani Taruvinga, and our children (Nyasha, Shaun, and Shane) for enduring my 'absence' while I was writing the book. I would like also to acknowledge the support of the following institutions: National Museums and Monuments of Zimbabwe (NMMZ), South African National Parks (SANParks), Robben Island Museum (RIM), African World Heritage Fund (AWHF), UNESCO, University of Cape Town (UCT), and many others that allowed me to access their information. I thank the University of Rhodes (RU) for granting me time to complete this book after joining them in 2022. More importantly, I would like to give glory to God and honour my parents, Mathews and Sarah Taruvinga, and Spiritual Father, all as my inspiration. Last but not the least, I would like to thank the many friends and relatives I cannot mention by their names in this acknowledgement.

PREFACE

A few months after I joined the National Museums and Monuments of Zimbabwe (NMMZ) as a curator of archaeology in 1997, I became intrigued by the frequency at which I encountered stakeholder "minefields" at heritage sites and museums scattered throughout Zimbabwe. Fast forward, the Domboshava site museum, north of Harare (Zimbabwe) is repeatedly burnt down, and eventually the rock art paintings at the site are defaced using enamel paint. Elsewhere, east of Harare, in the Mutoko area, black granite quarrying was becoming rampant and destroying diverse heritage sites, while communities continued to live in poverty. This signalled the overriding power of mines and minerals legislation over heritage legislations. Also, at many heritages sites open to the public in Zimbabwe, competition to control tourism initiatives was becoming rife between NMMZ and local communities, with no clear partnership solutions in sight! Also, while undertaking Reactive Monitoring Missions (on behalf of UNESCO), I witnessed strained, conflicted, and divisive relations among management authorities and stakeholders in promoting both conservation and livelihoods at WHS in Africa. Science appeared to override everything else, including muffling alternative views from stakeholders. At UNESCO meetings, I observed that conservation and development were becoming polarized and politicized issues, and at the same time, systematically peripheralized other interested and affected stakeholders. Similarly, most WHS I visited had visible tensions between conservation and development, all yearning for a home-grown solution. The cases of coal and uranium extraction at Mapungubwe (South Africa) and Selous Game Reserve (Tanzania) respectively triggered my desires to understand the stakeholder perspectives on these matters. Their experiences, characterized by varied and conflicting views, showed that neither conservation nor development plan for each other at WHS. African heritage sites have become widely contested, and the battle for control is not only at the national level but also in the boardrooms of international agencies. It was clear that the scientific value proposition was not considering the social value one advanced by stakeholders at WHS. I became interested in finding the underlying cause and possible solutions, and understanding who are these stakeholders, their views, and future of WHS in Africa.

Organizing a conference on "Living with World Heritage in Africa" in 2013, which was attended by varied stakeholders, among them 18 African ministers, community representatives, academics, private sector, and developers, spurred the idea of research and a book on understanding stakeholder views on conservation and development from an empirical perspectives at WHS in Africa. The Ngorongoro meeting on World Heritage and sustainable development (SD) and the international seminar on the same at Robben Island (South Africa) further pushed the idea and became launching sources into writing this book. It was obvious that the concept of heritage has remained a privy of heritage experts and the role of stakeholders remained unexplored. Heritage value proposition had not been evaluated against the opinions of stakeholders, who are beneficiaries one way or another. I began to view heritage experts as compliance officers rather than as co-creators of knowledge, co-drivers of conservation and development practices at heritage sites. With information gathered through interviews, seminars, workshops, field research, and national and international meetings attended since 2010, I began writing the book in phases from 2017. The book aims at understanding conservation and development as a stakeholder-driven process at WHS in Africa. Writing the book while fully employed, including keeping track of fast-changing discussions on conservation, development, and stakeholder landscape in Africa, was challenging. Writing this book has made me realize that heritage governance is no longer a one-way street.

This book intends to make the reader understand how conservation and livelihoods have been handled from pre-colonial times to the present, who are stakeholders and what is their proposition, what is the future of conservation and development, as well as what can be done to improve heritage governance at WHS. These areas are minefields that require serious and systematic navigation. In the process of listening to stakeholders as illustrated in this book, solutions begin to emerge, thereby turning challenges and contestations into opportunities at WHS. This is critical to understanding the conservation-development nexus as a stakeholder process at WHS, a subject of interest to varied readers in Africa and across the world.

1

INTRODUCTION

Stakeholders, Conservation, and Development

Introduction: The Idea!

This book presents conservation and development as stakeholder-driven processes at World Heritage Sites (WHS) in Africa. It presents the outcomes of a research study conducted using mixed field methodologies between 2015 and 2019 in South Africa and other selected State Parties (SPs) of Africa, which include Zimbabwe, Ghana, Tanzania, and Senegal. This included attending World Heritage Committee sessions. This discussion acknowledges that while a plethora of recommendations on conservation and development have been made in the last 20 years, including inferences to stakeholders, little empirical evidence has been presented on the alternatives they can offer in managing heritage sites (Taruvinga, 2020b). Furthermore, the success of World Heritage on the African continent should not be measured using conservation indicators, number of inscribed sites, and those removed from the list of endangered sites only, but rather and in addition, its successfulness in promoting sustainable socio-economic development alongside conservation at WHS (Albert, 2012). The relationship between conservation, development, and stakeholders has evolved over the last two decades, warranting further empirical research to assist in shaping both theory, pragmatic, and policy changes in heritage management. Subsequently, an alternative management thinking anchored on practical and localized approaches at WHS is put forward in this book, fully understanding that these are evolving dimensions. Overall, the discussion offers pragmatic approaches on how to involve stakeholders in balancing conservation and development. It advances the notion of doing things differently, rather than doing things the same way and expecting alternative results! How WHS and sustainable development (SD) value propositions negotiate with each other shall determine the future of WHS.

DOI: 10.4324/9781003149361-1

Post-colonial conservation approaches in Africa need to bring WHS closer to the aspirations of stakeholders in Africa. For a long time, heritage has been viewed as a privy of experts and State institutions, yet it directly and indirectly belongs to all stakeholders. Stakeholders are defined as those interested and affected by WHS, including but not limited to local communities, academics, politicians, public and private sectors, students, universities, experts, and others. This book strongly advocates for stakeholder engagement and involvement at WHS. This book argues for greater and effective stakeholder role in heritage management in Africa, including facilitating this through adaptive management approach (AMA) at the local levels of WHS. Localization of conservation and development is viewed as an adaptive process of addressing the stand-off between and among the local, national, and international perspectives of the stakeholders at WHS. Managing these stakeholders is like the conductor of an orchestra band who creates space for every player and different instruments towards a shared value and output. An orchestra gives a voice to all for their own mutual benefit and partnership. Pursuant to this goal, the following definitions are provided from the onset:

- Adaptative management: defined as a "learning by doing approach" or continual learning process with its roots in Protected Areas (PA) advancing decision-making controls in ever-changing circumstances of WHS.
- Cultural heritage: refers to the legacy inherited from the past and present societies, which is expressed through the tangible and intangible, moveable and immoveable attributes, processes, and social memory defining both culture and nature sites, including cultural landscapes.
- Governance: refers to how institutions are managed as defined by power, legitimacy, and resource matrix, including defining the role of other players in the process.
- Localization: refers to the engagement between the local and international experiences taking into consideration the dynamics, socio-economic contexts, and experiences at the local levels of WHS.
- Multiple and Multilayered Stakeholder Framework (MMSF): refers to the preferred governance approach for managing the multiple and multilayered stakeholders as it happens at the local, national, regional, and international levels of a WHS.
- Multiple and Multilayered Stakeholders (MMS): refers to the many and overlaid stakeholders affected and interested in WHS management in its broadest sense for their individual and collective benefit based on their defined aspirations.
- Stakeholders: refers to the multiple actors affected and interested in decisions made at WHS, including but not limited to local communities, academics, politicians, public and private sectors, students, universities, experts, and others.
- State-Based Management Systems (SBMS): refers to the management systems used by State Party (national government and related institutions) largely informed by legislation and other administrative protocols, and otherwise known as modern management systems with their roots in colonialism.

- SD: defined as "development that meets the needs of the present without compromising the ability of future generations to meet their own needs" (Brundtland Report, 1987) and currently has four pillars: economic growth, environmental protection, social equality, and peace and security.
- SPs: refers to countries that have ratified/accepted the 1972 World Heritage Convention (WHC) as an international instrument for the protection of World Heritage considered to be of outstanding universal value (OUV).
- Traditional Management Systems (TMS): refers to cumulative customs, beliefs, and practices representing the worldview or religious perspectives of African communities often expressed through Indigenous Knowledge Management Systems (IKS).
- World Heritage site: refers to a site (whether culture, nature, cultural landscape or mixed) inscribed on the World Heritage List (WHL) by the World Heritage Committee in terms of the 1972 WHC.
- World Heritage: refers to sites considered to be of OUV and meeting the conditions of authenticity, and/or integrity, as well as protection and management systems in terms of the 1972 WHC and the associated Operational Guidelines.

Heritage Management Philosophy: Protection for Posterity

The view that heritage is solely protected for posterity is entrenched in colonial, national, and World Heritage notions advancing conservation that stirs the approach of SBMS (Mumma, 2003; Ndoro & Wijesuriya, 2014). Yet, Africa, as a developing continent, is faced with cardinal challenges of poverty, food and water insecurity, population growth, unemployment and inequality, the negative impact of external geo-political environments, poor infrastructure, energy crisis, and civil conflicts that exacerbate their already dire socio-economic conditions (Nkoana-Mashabane, 2013: 19; Meskell, 2013b). SBMS simply refers to management systems enforced through enacted legislation that gives power to state-based institutions. SBMS emerged in the colonial period, thereby marginalizing the already existing TMS in Africa. Examples of SBMS in Africa include but are not limited to South African Heritage Resources Agency (SAHRA) in South Africa, National Museums and Monuments of Zimbabwe (NMMZ), and the National Museums of Kenya (NMK). All are enacted and backed up by respective legislations. On the other hand, the World Heritage concept is enshrined in the "Convention concerning the Protection of the World Cultural and Natural Heritage of 1972" ("the 1972 WHC" or "the Convention"). The Convention is supported by the "Operational Guidelines on the Implementation of the World Heritage Convention" (OGWHC). The Convention inscribes and protects heritage considered to be of "OUV" as WHS. The Convention has a futuristic desire centred on saving heritage from multiple threats and the incapacity of SPs who are unable to fully implement protection and conservation measures.

Both SBMS and the Convention have largely retained a narrow and exclusionary conservation approach, however this is being continuously challenged by different stakeholders as it is not inclusive and neither does it meet their needs. Additionally, heritage is first and foremost a local resource with multiple use values to different stakeholders. This can be traced back to the pre-colonial Africa in which local communities preserved and used heritage simultaneously, including extracting other resources within the locality of sacred sites for their own benefit (Igboin, 2011; Keitumetse, 2007). This was done through TMS approaches binding local communities. The colonial systems, the SBMS and the WHC, have continuously interfered with the functioning of TMS as a communally driven approach in Africa (Ndoro & Wijesuriya, 2014).

While both SBMS and WHC have made attempts in post-colonial Africa to adopt strategies re-positioning conservation and development for the benefit of stakeholders at WHS, this has been a very slow process and remains in its infancy (Chirikure *et al.*, 2016). Neither are the stakeholders formally involved in the process and in heritage governance, including at the WHC level. In this context, WHS are being challenged to offer benefits beyond conservation to stakeholders (Meskell, 2013b; Ndoro & Wijesuriya, 2014). Heritage must meaningfully contribute in a measurable way to socio-economic development aspirations of stakeholders at WHS. Socio-economic development is concerned with meeting the needs of the society using both renewable and non-renewable resources within their localities (Nkoana-Mashabane, 2013). This is what is now commonly referred to as SD. Africa, as a developing continent, is need to deal with multiple challenges affecting the socio-economic conditions of her people (Nkoana-Mashabane, 2013: 19; Meskell, 2016).

Socio-economic needs are localized aspirations manifesting in the same physical locations as WHS themselves. These aspirations are concerned with ensuring that local communities have access to livelihoods irrespective of whether they are sustainable or not sustainable. Such developments include agriculture, dam constructions, infrastructural development, extractive industries, urbanization, industrialization, energy projects, housing projects, tourism initiatives, among many others (Nkoana-Mashabane, 2013; UNDP, 2020). In the current understanding of SBMS and WHS, what seems to be clear but contradicts the above is that the generation of today is expected to bequeath resources to future generations without adequately meeting its own contemporary needs. This generation is unable to fully capitalize on heritage to solve their socio-economic challenges. For these reasons, implementing WHC to achieve socio-economic development, alongside its aspiration of conservation, is now a priority in Africa as a stakeholder-driven process.

The notion that World Heritage must contribute to socio-economic development is now deeply entrenched worldwide, and it is challenging the futuristic perspective of World Heritage. However, this notion lacks empirical evidence on how it is influenced by stakeholders. Equally, understanding how WHC and SBMS have practically adapted from a governance perspective lacks empirical analysis. Stakeholders are simply defined as the interested and affected role players at WHS. These are supposed to be the beneficiaries of both conservation and development. There is need for clarity on how conservation and socio-economic development should influence each other from a broader planning perspective. The coexistence of conservation and socio-economic development at WHS has created tensions that now characterize heritage management in Africa. While SD principles assist in mitigating this tension, practical solutions have remained elusive to enablers, implementers, and the supposed beneficiaries (who are the stakeholders). Also, empirical data on how conservation and development engage each other as a stakeholder-driven process is still lacking. Stakeholder influence in this process remains in the theoretical realm of the enabling policies and numerous "politically correct" recommendations of the SPs on these matters. Therefore, it is important to examine the opinions and views of stakeholders towards finding localized solutions to the "inseparable" (indissoluble) and "inexorable" (unavoidable) relationship between conservation and socio-economic development at WHS. However, this must be contextualized in the evolving heritage management systems in Africa.

For SBMS and WHC to embrace conservation and SD as stakeholder-driven processes at WHS, there is a need to explore the potential of AMA. AMA is defined as a "learning by doing approach" with its roots in Protected Areas (PA). "Learning by doing" capitalizes on strengths and, equally, converts challenges into opportunities in pushing a progressive management approach to achieve predetermined outputs. Through AMA, heritage governance has the potential of progressively and proactively responding to the challenges bedeviling the practical coexistence of conservation and socio-economic developments as a stakeholder-driven process at WHS. This also allows the "local belief influenced conservation practices" to be "implemented alongside international practices" (Chirikure *et al.*, 2016: 167). Whether this adaptive strategy is in sync with the WHC ratified by African SPs and their SBMS is debatable. However, AMA has been slow in embracing heritage governance, which remains an optional approach, yet it has the potential to provide empirical perspectives on pragmatically aligning conservation and development as long-evolved mutual processes.

Africa and the Evolving Heritage Management Systems

The experiences of TMS, as a local approach tested and validated through time, has been ignored in Africa, yet the history of the African continent shows that local communities cared for their heritage before being colonized (Ndoro & Wijesuriya, 2014; Jopela, 2016a). They utilized resources around their heritage places for their own survival, spirituality and well-being as guided by TMS (Igboin, 2011; Jopela, 2016a). Colonialism resulted in the dispossession, social dislocation, and social transition of local communities from being owners to servants of colonial repressive governance systems (Ndoro & Wijesuriya, 2014). Foreign "economic and religious values" were promoted in colonies resulting in the "exploitation" and "vitrification" of traditional values (Igboin, 2011; Okoduwa, 2008: 18). Colonialism did not take into consideration the long-defined experiences, aspirations, and use values of local communities relating to heritage (Pwiti & Ndoro, 2001). Colonial heritage practices were not inclusive but rather exclusionary. Local communities remain marginalized through a complex matrix of colonial and post-colonial legislations and policies (Mumma, 2002, 2003; Ndoro & Wijesuriya, 2014). Inclusivity and adopting local approaches have eluded heritage management in post-colonial Africa (Chirikure *et al.*, 2016).

Heritage management approach in the colonial and post-colonial period has been beset with principles and a mindset of safeguarding the irreplaceable past, which is facing multiple threats (Logan & Reeves, 2009). Colonial environmentalism concluded that "natives did not understand their own heritage" and if left in their custody, it would be destroyed, hence the creation of PA for conservation as "gardens of Eden" (Nelson, 2003: 71). This is the same rationale advanced by the WHC too. This colonial "authoritarian conservation perspective" has persisted until today with minimal changes in Africa (Nelson, 2003: 71). It influenced the concepts of heritage, its ownership, identification, and protection, including the professional ethos and approaches in Africa (Goh, 2008; Ndoro & Wijesuriya, 2014). SBMS now own heritage and alienate local communities, except in cases where TMS are strongly established, such as the Mijikenda Forests (Kenya) and Kasubi Tombs (Uganda) (Abungu, 2016; Jopela, 2016a; Goh, 2008; Michael & Negri, 2000). Heritage legislations in Africa are not adequately providing modalities on how conservation and development can coexist (Ndoro & Wijesuriya, 2014). Current legislative review efforts are slow and frustrating stakeholders. In the colonial period, where heritage was identified and tourism development occurred, this was established as an enclave for foreign tourists under the management of district commissioners as was the case in Zimbabwe. Commissioners were legally empowered to look after heritage sites in their districts, especially the British Pioneer Column forts, rock art sites, and stonewalled sites associated with Zimbabwe culture, but initially attributed to foreign origins. These legal arrangements were and are not meant to benefit local communities in Africa. Rather they enforced an exclusionary approach, which remains incorporated in SBMS and WHC systems until today. Through political independence, democracy and good governance have become prevalent approaches in Africa, but these have not benefitted "heritage management practices" (Ndoro & Chirikure, 2009: 72). These practices are still suffering from colonial hangovers and systematically maintained exogenous concepts that are not sympathetic to how heritage has been managed and used in the pre-colonial period (Taruvinga & Varrissou, 2013). Post-colonial heritage experts in Africa still have 'an uncut umbilical cod' with their colonial authorities on how they connect with local communities (Ndoro & Wijesuriya, 2014; Taruvinga & Varrissou, 2013). When this situation is reviewed against the application of the 1972 World Heritage Convention in Africa, SBMS approaches are reinforced, with their exclusionary tendencies maintained.

World Heritage Concept and Africa

The WHC aspires to inscribe and protect sites with values considered to be illustrating "OUV". However, not all values of a site qualify for this inscription (Meskell, 2014). OUV is "cultural and/or natural significance" which is "exceptional as to transcend national boundaries and to be of common importance for present and future generations of all humanity" (OGWHC, 2015: 49). Such sites should meet the conditions of authenticity and/or integrity, including having conservation, protection and management systems in place. Overall, the Convention raises "the moral obligation of humanity as a whole to respect and safeguard natural and cultural properties which are of OUV" (Von Droste, 2012a: 10, 2012b). However, the management of WHS remains the responsibility of the SPs under the guidance of national laws, OGWHC, and the Convention itself. Using this Convention, Africa has gradually increased the number of sites inscribed on the WHL. Africa, as of 2021, contributes around 10% of WHS inscribed by United Nations Educational, Scientific and Cultural Organization (UNESCO). Despite this, Africa remains underrepresented on the WHL, and also has the second highest representation on the List of Endangered World Heritage sites. Critics of the WHC argue that it reinforces the conservation mantra of the SBMS, however, and recently it is deliberately moving towards embracing SD (Abungu, 2016; Meskell, 2013b; Ndoro & Wijesuriya, 2014).

Furthermore, the WHC has a futuristic desire to save heritage from multiple negative impacts arising from human-made developments, natural, biological, disasters, and other factors. The adoption of the WHC was foreshadowed by a dam construction which affected the Abu Simbel Temples in Egypt (Donnachie, 2010; Abu-Zeid & El-Shibini, 2010; McCartney, 2009; Kashef, 1981). As mitigation, the temple was relocated to higher ground in order to pave way for the Aswan Dam in 1960 (Donnachie, 2010). The relocation, though out of its original physical location but generally within the same geographical area, preserved heritage for the future and allowed development to take place. This is comparable to museum collections completely detached from their original contexts. The relocation represents a solution to a local and temporal issue (shortage of water) that had to be addressed. Such trade-offs (relocations) have not been consistently nurtured and successfully repeated in Africa (Donnacie, 2010). As a result, a tensional mode has characterized the relationship between conservation and development (Jokiletho, 2011; Jokiletho & Cameron, 2008; Labadi, 2013; Meskell, 2013b; Meskell & Brumann, 2015). In the process, WHC has been consistently challenged to offer stakeholders benefits beyond the conservation mantra (Meskell, 2013b; Ndoro & Wijesuriya, 2014).

Traditionally, the WHC is concerned with conservation, but of late (in the last two decades), the use of heritage (SD) in meeting the needs (well-being) of society has taken centre stage in discussions (Meskell, 2012, 2013). What lacks in these discussions is the empirical understanding of the views and opinions of stakeholders in balancing conservation and SD at WHS. These stakeholders are still ill-defined, not properly understood, and remain at the periphery of decision-making at WHS. Both SBMS and WHC assume that the needs of the world community should override the needs of the locals in order to maintain the OUV (Meskell, 2013b; Harrison, 2010b). This global and local contrast in aspirations and needs demonstrates the challenge of integrating World Heritage within the broader territorial development agendas of SPs in Africa. Some of the international strategies and decisions at WHS are not sympathetic to local aspirations and expectations. The social and political construct of cultural heritage reflects and validates the "identity of nations, communities, families and even individuals", including aspirations beyond conservation thinking (Labadi & Logan, 2015: xiii). As such, heritage can no longer be viewed as totally separate from its socio-economic context and the intended beneficiaries (Berkes *et al.*, 2006). Stakeholders living with WHS in Africa are not benefiting in a meaningful and measurable way from their own heritage. There is a growing feeling that the WHC and the attendant OGWHC are unfriendly towards development. They make it difficult for developing nations and continents to exploit renewable and non-renewable resources at WHS (DAC & AWHF, 2014).

In addition, Africa has not sufficiently harnessed its WHS to promote development for the benefit of the society (Ndoro & Wijesuriya, 2014). The obvious benefits of WHS include being part of a global community dedicated to conservation, access to funding for research, conservation and developing management systems, international cooperation and technical support, capacity building, and developing highly controlled tourism initiatives (Donnachie, 2010: 118). However, debates on the relationship between conservation and development in Africa, including benefits thereof, are now characterized by deepening differences between views and opinion of stakeholders and the dilemma as to which development model should be pursued (Meskell & Brumann, 2015; Szirmai, 2005). This has its roots in the long-standing historical problems associated with establishing PA and gazetting of monuments in Africa (Ndoro & Wijesuriya, 2014; Ndoro & Chirikure, 2009a; Chirikure & Pwiti, 2008). This has continued with WHS being carved out of the same appropriated community territories, thereby reinforcing colonial approach in the current (Borona, 2015). While the Convention enforces compliance with conservation principles and protocols to mitigate threats of developments, SPs, fearing political retribution from the electorate, try to balance international commitments on conservation and local development aspirations. While UNESCO appreciates this dilemma, it has been very slow to offer a conciliatory position to ensure WHC is relevant to the local context in which it is applied.

The above may not have an impact if not anchored on the understanding that the protection and use of heritage for the social well-being of society is as old as humanity itself (Taruvinga, 2020a). Societies, in their dynamic state, have been utilizing both renewable and non-renewable resources available to them since time immemorial (Taruvinga, 2020a). This defines the pioneering spirit and approach of local communities (to be referred to as Indigenous and

Descendant Communities, "IDCs") in maintaining and using their heritage and diverse resources around it (Taruvinga, 2020a). Both SBMS and the WHC were thus introduced in the shadows of this ignored pioneering spirit and approach of IDCs in Africa. This constitutes the "localness" of heritage in Africa. While SBMS have appropriated the localness of heritage and its values, the WHC has universalized the same into global processes supported by values perceived outside the local context of the African society but popularized as good practices. Therefore, the contribution, if any, of these local sites (now WHS) to development is rarely a topic of empirical investigation in Africa.

Development Dimensions and World Heritage

When the current global discrepancies in Human Development Index (HDI) indicators are taken into consideration, relationship between conservation and development at WHS becomes politically charged in developing continents such as Africa. The HDI shows that developed nations are always ranked as having Very High Human Development index (VHHDI). As of 2020, the top ten countries with VHHDI were Norway, Ireland, Switzerland, Hong Kong, China (SAR), Iceland, Germany, Sweden, Australia, Netherlands, and Denmark (UNDP, 2020). On the other hand, developing nations are ranked to be either having a High Human Development Index (HDDI), Medium Human Development Index (MHDI) or Low Human Development Index (LHDI) (UNDP, 2020). Africa is in these last categories of the HDI, in particular the LHDI save for a few nations. African nations with an HHDI ranking include Seychelles, Tunisia, Botswana, South Africa, and Gabon, while the rest are in MHDI or LHDI (UNDP, 2020). Developing nations must address this LHDI ranking, and this is where development comes in (UNDP, 2020). Development becomes a necessity in improving the well-being of the society.

HDI indicators include Inequality-adjusted HDI (IHDI), Gender Development Index, Gender Inequality Index and the Multidimensional Poverty Index (UNDP, 2020). The last two are highly pronounced in the categories HHDI, MHDI, and LHDI (UNDP, 2020). Contribution of socio-economic and heritage to HDI at national levels is often linked to the gross domestic product (GDP), life expectancy, levels of literacy and employment, and poverty reduction (UNDP, 2020). The contribution of WHS in this is currently unknown and not measured.

The paradox is that the LHDI of African nations demands accelerated developments, but at the same time, heritage should be protected in tandem with good conservation practices. Therefore, the WHC can no longer operate in a social vacuum given the needs emanating from the local context of the WHS (Meskell, 2014). Neither can this be completely understood without understanding the term "SD". SD is defined as "development that meets the needs of the present without compromising the ability of future generations to meet their own needs" (WCED, 1987: 43; Brundtland Report, 1987). The original and three main pillars of SD are economic growth, environmental protection, and social equality (WCED, 1987: 43; Brundtland Report, 1987). UNESCO attempted to make culture the fourth pillar, but this was not successful; rather peace and security was added (Bokova, 2013; UN, 2014). This denotes how heritage remains at the periphery of development agenda, yet it has always been playing a role in Africa since pre-colonial times. However, World Heritage is expected to contribute to Sustainable Development Goals (SDGs), in particular No. 11.4, which aims to "protect and safeguard the world's cultural and natural heritage" (UN, 2015). Furthermore, the African Union's (AU) 2063 Agenda adopted in 2013 highlights the vision of Africa as "The Future We Want for Africa". In this discussion, SD is treated as a framework ensuring that both conservation and development embrace principles of sustainability. Though SD makes scientific sense, not all developments are sustainable in their nature. Hence, localizing development at WHS in Africa remains a grey area, even though the 2015 World Heritage Policy on Sustainable Development exists (Eyong & Foy, 2006; Meskell, 2014).

What is not clear is how World Heritage interfaces with this local context and what "happens when the reverence for heritage collides with other value orientations and livelihood needs" at sites (Brumann & Berliner, 2016: 1)? Heritage sites retain their own social dynamics, and the error in the World Heritage processes is that these dynamics are subjected to globalization without an opportunity to express themselves (Brumann & Berliner, 2016). Additionally, humanity is given universal rights over a WHS via a globally distributed language of the WHC, while local communities are kept at a distance, yet they have used and coexisted with this heritage for many centuries (Brumann & Berliner, 2016; Jopela, 2016a). As such, heritage discourse remains "constrained to state-level conversations", thereby excluding other multiple and alternative voices that could bring benefits to conservation and development discussions (Joy, 2016: 64; Blacik, 2007).

Positioning conservation into the future without considering social dynamics at the local levels is a narrow way of reading heritage given its multiple social dimensions in the past, present, and future. The impact of the WHC should not be defined by the number of sites inscribed or demarcated boundaries but rather "linchpins of global imaginaries and networks" creating room for "local and the trans-local" interconnections (Brumann & Berliner, 2016: 3). The broader use value of heritage to different stakeholders should be considered as a function of time and context. The World Heritage processes are not systematically, practically, and adequately considering this dimension (Brumann & Berliner, 2016). How World Heritage is placed in the wider geography and social space with stakeholders having varying interests is also not clearly understood (Brumann & Berliner, 2016). Hence, the need to understand the opinions and views of the stakeholders on this issue against the evolving approach of WHC on how to engage development at WHS. These stakeholders are defined as including but not limited to archaeological communities, IDCs, farmers, extractive industries, hoteliers, hunters, academics and academic institutions, politicians, learners or students, lawyers, provincial and national government departments, UNESCO, Advisory Bodies, non-governmental organizations (NGOs), community pressure groups, professional associations, and private sector. All these are broadly classified as both interested and affected stakeholders at WHS, who often have differing or splitting local opinions on the relationship between conservation and development (Brumann & Berliner, 2016: 10).

When the SPs and WHC have consensus on conservation and development, they always prioritize the former, thereby creating tension with stakeholders preferring the latter (Brumann & Berliner, 2016). In the process, SPs become double-edged swords that either favour conservation or development or both, depending on reasons beyond the understanding of the WHC. In a situation where there is domestic consensus on development between SPs and stakeholders, development is authorized to proceed. In this case, the WHC becomes powerless and often resorts to statutory decisions of the World Heritage Committee, compelling the SPs to prioritize conservation (Brumann & Berliner, 2016). Sometimes the SPs support conservation and development, in the process making some trade-offs. In such situations, the SPs decide which aspect to trade-off, and this is influenced by factors outside the WHC ambit. Therefore, not understanding the views and opinions of stakeholders at WHS under any of these circumstances is a disservice to the governance and future of WHC, including failing to understand the dilemma and complexities SPs find themselves in when dealing with these matters. In the process, what is also worrisome is how stakeholders are considered "permanent observers" at World Heritage Committee meetings. Indigenous rights of local communities are neither being formally respected nor binding to both SPs and WHC at WHS (Brumann & Berliner, 2016; Blacik, 2007).

The above is made worse by the fact that the SBMS manages heritage through the colonial lenses, yet there are opportunities of doing this differently, including embracing good elements of TMS (Joy, 2016; Matthias, 2017; Ndoro & Wijesuriya, 2014). Colonial lenses and their bureaucratic systems do not accommodate the multivocality associated with stakeholders. Equally, the opportunity that comes with systemic and multi-directional communication with stakeholders has not been explored by SBMS (Matthias, 2017). Stakeholders, in their diversity and multidimensional perspectives, are a "Heritage Practice Laboratory" that offers alternative approaches beyond the colonially infected lenses of SBMS and WHC. While a plethora of recommendations have been made on the relationship between conservation and development, including peripheral inferences to stakeholder approaches, very little research has been carried out to learn from the "Heritage Practice Laboratory" towards doing things differently at WHS in Africa. The major question is why keep heritage sites if they cannot benefit stakeholders? There is a need to develop a pragmatic synergy between conservation and development as mutually non-exclusive processes driven by different stakeholders living with WHS. This requires gathering empirical experiences and practices of stakeholders at WHS.

The success of the WHC on the African continent should not only be measured using conservation indicators; rather, and additionally, how successful it is in supporting development (the well-being of society) at WHS should be equally considered (Albert, 2012). WHS have a role to play in the fight against poverty in Africa (Albert, 2012: 41). The question is how can WHS improve their stakeholder governance framework to achieve this? What are the challenges bedeviling stakeholders at WHS that need to be solved? For instance, stakeholder perspectives on the evolving land use systems, landownership and the resultant contestations is not adequately understood and mitigated at WHS. Africa and her quest for equitable development are inseparable, and this shall remain so into the future, yet WHC has not been paying attention to this until recently. In this context, the seemingly dysfunctional relationship between conservation and development needs further discussion as a stakeholder-driven process.

Debunking Heritage Governance and Stakeholders in Africa

For a long time, heritage was considered to be within the ambits of few experts, academics or professionals. These few people are viewed as having studied heritage concepts and are or were practising in various positions. By virtue of this, they become authorized experts/players in heritage management and discourses while other stakeholders are seen as passive players. Heritage governance and engagement techniques have been developed from this mindset and applied as such. SBMS and WHC governance value the opinions of authorized experts more than of those considered not competent in this profession. However, some devolution approaches have been tried, and this includes Community Based-Resources Management (CBRM), Community Management of Protected Areas Conservation (COMPACT), Communal Areas Management Programme for Indigenous Resources (CAMPFIRE) and recently People Centred and the Nature-Culture Journey approaches. While all these approaches emphasized participation, involvement, sustainability, beneficiation, and conservation, their empirical journey demonstrates that other stakeholders shall not be equal role players. The setback is that the authorized experts still stir and dominate these approaches meant to empower other stakeholders. The authorized players are always leading and pointing others towards a particular agenda, as if these stakeholders do not have own agendas. Other stakeholders have become victims of this "technical bureaucracy" and science-based approaches, all supported by the SBMS. This has contributed to the progressive failure of heritage around stakeholder management.

Contrary to this strict scientific and rigid approach advanced by authorized players, the social and political construct of cultural heritage is manifesting and reflects that the "identity of nations, communities, families and even individuals" and aspirations are all beyond conservation thinking (Labadi & Logan, 2015: xiii). SBMS, authorized players and WHS have isolated and insulated themselves from this thinking through legal systems and their decision-making powers.

They have further reduced stakeholder engagement to the level of stakeholders being "subjects of compliance process" ('ticking the box approach' for auditing purposes). In the process, SBMS have failed to effectively harness the opportunities of partnering and working together with diverse stakeholders, including converting their challenges into positive, creative, and innovative bridges at WHS in Africa. The excluded and marginalized stakeholders are equally interested and affected by what authorized role players are managing, including access to other resources within the physical spaces in which heritage sites are located. Why would SBMS and WHC alienate themselves from the stakeholders that they are meant to serve? How does it benefit SBMS and WHC to exclude stakeholders who may provide alternative solutions at heritage sites? Why does it appear like the pioneering spirit and approaches of indigenous communities were and are ignored like an abrupt, once-off phenomenon or process, yet it has all the ingredients of innovation, resilience, effectiveness, evolving governance, and in-built adaptive capabilities on conservation and development? This points to a situation where neither conservation nor development plan for each other at WHS, yet both cannot be disassociated from their broader territorial dimensions, environment, and related stakeholders. Stakeholders are an extremely important part of managing any heritage site, and if they are not taken into consideration, then there is "no need" to have and preserve these sites for future generations.

With the above understanding, it is clear that heritage is now perceived as representing a system of diverse entities with varied uses of heritage through time and space (Kalman, 2014). Cultural heritage cannot be separated from its day-to-day use by stakeholders (Siravo, 2014; Berkes & Davidson-Hunt, 2006). Understandably, daily use threatens heritage sites; however, this can be mitigated together with the users (being the stakeholders). These stakeholders have to be considered in the governance matrixes of heritage in Africa. Such consideration should go beyond the legal fixation, past unpleasant engagement experiences, and fears associated with SBMS handing power to other stakeholders. Surely, the contrast in stakeholder aspirations at global and local contexts of WHS requires an inclusive governance approach, which is missing at the moment.

Novelty: Stakeholders, Conservation, and Development

While World Heritage is a universal concept, heritage itself is a function and manifestation of society-driven processes at local levels The same applies to socio-economic aspirations of the society, which are also local needs. The localness of both heritage and its use is in tandem with the experiences of Africa. From the onset, stakeholders and their experiences in localizing both conservation and development aspirations at heritage sites has been practised over many centuries in Africa. Sites such as Great Zimbabwe World Heritage Site (GZWHS), Mapungubwe Cultural Landscape World Heritage Site (MCLWHS) and many others are symbols of development associated with historic urban centres supported by varied socio-economic initiatives such as agriculture, hunting, mining and processing of minerals, quarrying of stones, construction, and long-distance trade while maintaining their cultural practices and significance. The manifestation of these processes is what is now recognized as intrinsic values illustrating OUV. On the other hand, these processes are more pronounced due to population increases in Africa. That alone makes the "bureaucratic forms of protection" associated with heritage governance in Africa insufficient in providing guidance on conservation and development (Meskell & Brumann, 2015; Selman, 2004; Taruvinga, 2019).

Conservation and development are non-exclusive to each other, so is heritage governance and stakeholders since pre-colonial times in Africa. Neither is the past, present, and the future not exclusive to each other. This gives birth to the notion of an AMA, which is lacking in heritage management. AMA is simplified as "learning by doing", and this is relevant in furthering pragmatic approaches which balance conservation and development as stakeholder-driven processes at WHS. If adopted, this approach may assist SBMS and WHC to maintain the localness of heritage and its relevance to society. Currently, the management of WHS in Africa needs practical and alternative adaptive approaches, in which stakeholders can partner with SBMS and WHC to promote both conservation and development at WHS. When this discussion is contextualized in the existing literature on stakeholder perspectives on conservation and development, it brings out add-value arguments to this discourse.

Keitumetse (2011) discusses cultural heritage resources within a SD concept framework and the strategies thereof. Emphasis is placed on sustainability principles relating to community participation but does not consider the role of other stakeholders in its broader perspective. Ndoro, Chirikure and Deacon (2018) present a wide range of issues on the state of heritage management in Africa with an emphasis on archaeology, living traditions, sacred sites, heritage of pain (slavery), international conventions, cultural landscapes, heritage in conflict areas and heritage versus development. However, the stakeholder perspectives are not discussed in detail. Sinamai (2019) highlights how ignoring intangible values and other forms of remembrance and celebrations at Khami World Heritage site (Zimbabwe) has led to management challenges. The discussion highlights the challenges that arise when the local views of what is significant heritage varies among stakeholders (Sinamai, 2019). On the other hand, Lynam *et al.* (2017) illustrate the actions and representations of stakeholders in heritage management in conflict settings, which are prevalent at sites due to diverse and competing interest, while Peterson *et al.* (2015) demonstrate how the lifeways of the past are catalysts for political and cultural entrepreneurs in Africa. Furthermore, Peterson *et al.* (2015) bring out the political mode in which heritage finds itself as a result of failing to address local needs in Africa.

Authors such as Basu and Modest (2015) discuss the relationship between museums, heritage and development, which has received little academic scrutiny. This book advances a critical conversation between these interdisciplinary fields to bring them closer to each other in providing solutions to the challenges bedeviling stakeholders. This is taken further through detailed empirical studies at different WHS. On the other hand, Brumann and Berliner (2016) demonstrate how local communities are restricted by global heritage policies and approaches, yet heritage is supposed to benefit them. These restrictions need to be unlocked through practical and adaptive approaches as proposed in this discussion. While Waterton and Watson (2015) provide a localized vision of heritage resulting from considering a range of perspectives and disciplines related to heritage, and as advanced by different stakeholders, among students, practitioners and scholars, this book adds depth to the diversity and range of stakeholders that need to be involved in the management of WHS. While SD principles have been adopted recently by the World Heritage community, the 2015 World Heritage Policy on World Heritage Convention, a lot of practical work to implement this at sites still needs to be undertaken, including creating a conducive governance environment. Stakeholders have become part of the solutions at WHS.

From the onset, greater engagement and involvement of stakeholders in management processes at WHS does not necessarily undermine commitment to conservation as erroneously perceived by SBMS themselves. This erroneous assertion is further compounded by the little empirical research on the contribution of conservation and development to each other as stakeholder-driven processes at the local levels of WHS. Furthermore, research is weaker on establishing the potential benefits that might flow from greater stakeholder involvement in heritage management. This discussion advocates inclusive heritage governance, which can help both SBMS and WHC to partner with perceived opponents or enemies of conservation for mutual benefit. This is an untapped opportunity for coexisting and co-creating together with stakeholders at WHS. If this does not happen, the cyclical and problematic perpetrator and victims' debate shall become a permanent challenge at WHS. Additionally, it is challenging to define who is the perpetrator or the victim in this case without understanding the dynamics among SMBS, WHC and stakeholders on matters of conservation and development at WHS. This book asserts that coexisting and co-creating while safeguarding heritage sites is possible at heritage sites! For this to happen, there is a need to shift the scales of power from global to local levels of WHS. Heritage has become a widely contested resource, to the extent that the local voices have become subservient to global interests. There is now a battle to control these scales of power for contradictory purposes; SBMS want to maintain status quo while stakeholders want to change the status quo. The question is how can this situation be brokered for mutual benefit and how are the interests of local stakeholders being balanced against those of international ones?

Conservation and development in Africa are no longer linear scientific issues but rather stakeholder-driven processes. Neither is stakeholder management also a linear process. All these dimensions are dynamic, complex, and come with their own challenges. What is missing is the linking framework for their mutual benefit at WHS. While WHS is for conservation, development is about meeting the well-being of the society, and stakeholders the "oil" for both. All these dimensions now require elements of humanity, respect, integrity, unity, inclusivity, and continuously building bridges for their mutual benefit. This book advances the notion of conservation and development as stakeholder-driven processes, without which the future of World Heritage in Africa is at risk. These stakeholder perspectives will be derived from engaging various stakeholders at WHS in Africa. World Heritage governance cannot remain controlled by SPs (Brumann, 2009; Esposito & Gaulis, 2010; Joy, 2016; Matthias, 2017; Meskell, 2014; Rico, 2008).

Empirical Perspectives: Introducing the Main and Supportive Case Studies

With the above understanding, and to discuss in depth the juxtaposed conservation and development as a stakeholder-driven process at WHS, case studies are identified for use in this book. These case studies are not exhaustive but rather laboratory environments that can be abstracted and applied in the African region desiring to balance conservation and development at WHS. The book uses a main case study approach; however, this is supported by other selected WHS in Africa. The selection of the case studies was influenced by time, resource constraints and language barriers. I would have liked to look at sites in the Oceanic Islands, West, North, and Central Africa. This could be the subject of future research to enrich this discussion.

Main Case Study: MCLWHS, located in the Limpopo Province of South Africa, is used as the main case study (Figure 1.1). MCLWHS illustrates an important and traceable interchange of human values and different land uses dating from the pre-colonial period to the present (Pikirayi, 2016; DEA, 2014b; Carruthers, 2006; Hall & Smith, 2000; Huffman, 2000, 2005). MCLWHS presents a complex system to analyze given its "scale, different land use systems and multiple-ownerships" (Selman, 2004: 366; Brown & Mitchell, 2000). The presence of renewable and non-renewable resources is an important dimension of the past, present and future land uses at MCLWHS (Chirikure *et al.*, 2010). The choice of MCLWHS as the primary case study was influenced by its four-tier status as a (i) National Park, (ii) National Heritage site (iii) WHS, and lastly (iv), as Transfrontier Conservation Area (TFCA), forming the Greater Mapungubwe Transfrontier Conservation Area (GMTFCA) covering Botswana, South Africa, and Zimbabwe (SANParks, 2015). The GMTFCA involves a complex network of actors who have interest in the land and its resources (Sinthumule, 2016, 2017). The subsequent stakeholder opinions on conservation and development, including their own roles at MCLWHS, lack empirical studies that can bring out salient features that can improve how these matters can be better handled (Chirikure *et al.*, 2010).

The MCLWHS setup provides an opportunity to seek alternative voices on how conservation and development should be handled at WHS. These alternative voices have no representation, nor are they well understood in heritage

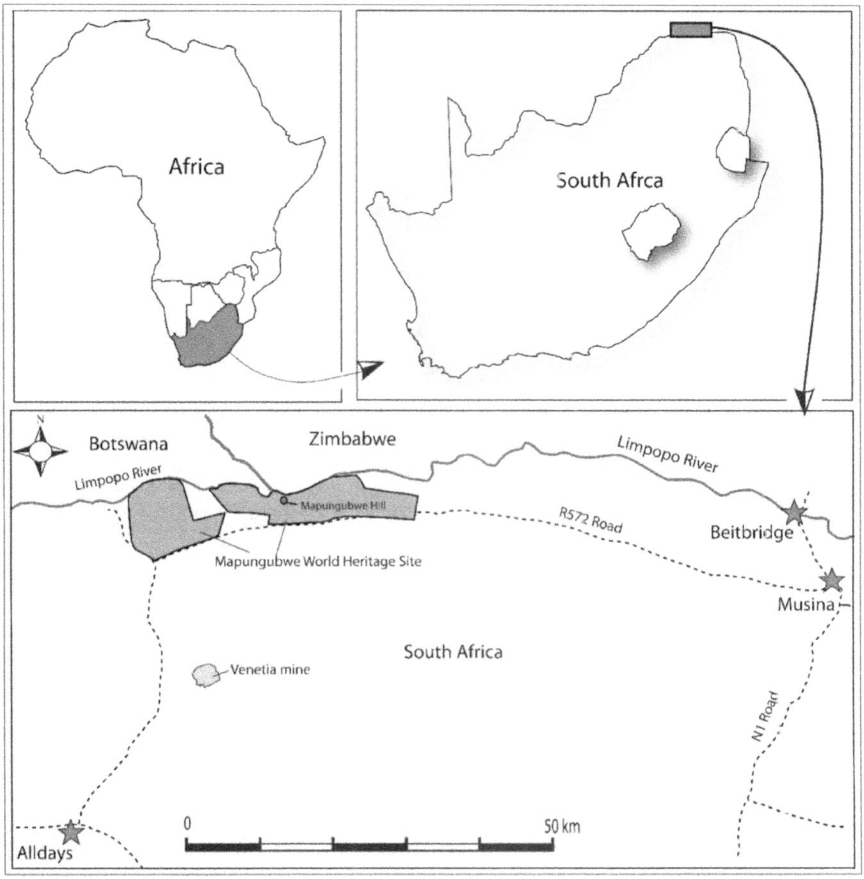

FIGURE 1.1 Map showing location of Mapungubwe Cultural Landscape, Limpopo Province, South Africa

management (Chirikure *et al.*, 2010; Forssman, 2014; Meskell, 2014; Pikirayi, 2016). Additionally, MCLWHS is under threat from multiple socio-economic activities, among them extractive industries, tourism, agriculture, and related water abstraction. Furthermore, MCLWHS, with its history of land alienation in the past, is now facing multiple land claims by IDCs. All these processes have implications on the future use of the site (SANParks, 2015).

Supporting Case Studies: In this discussion, experiences from other WHS in Africa are also used (Figure 1.2). The WHS selected include but are not limited to GZWHS in Zimbabwe; Robben Island World Heritage site (RIWHS), South Africa; Ngorongoro Conservation Area (NCA) in Tanzania; the Forts and Castles, Volta, Greater Accra, Central and Western Regions in Ghana; and Goree Island WHS (Senegal). The details of these case studies are outlined in Figure 1.2 These WHS were selected because they offer varying experiences, different typologies and management contexts, as well as competing conservation

World Heritage site	Location	Summary of Outstanding Universal Values (based on criterion)
Forts and Castles, Volta, Greater Accra, Central and Western Regions World Heritage sites (vi)	Located along the coast of Ghana between Keta and Beyin.	The Castles and Forts of Ghana, with its 28 components, illustrate the events that shaped not only Ghana's history but that of the world over four centuries as the focus of first the gold trade and then the slave trade. They are a significant and emotive symbol of European-African encounters and of the starting point of the African Diaspora (vi).
Goree Island World Heritage site (vi)	Located off the coast, opposite Dakar in Senegal.	Goree Island World Heritage site illustrates an exceptional testimony to one of the greatest tragedies in the history of human societies: the slave trade. The various elements of this "memory island" – fortresses, buildings, streets, squares, etc. – recount, each in its own way, the history of Gorée which, from the 15th to the 19th century, was the largest slave-trading centre of the African coast (vi).
Great Zimbabwe World Heritage site (i, iii, vi)	Located in southern Zimbabwe and is approximately 30 km from Masvingo town in Masvingo	Great Zimbabwe World Heritage site illustrates a unique artistic achievement, this great city has struck the imagination of African and European travellers since the Middle Ages, as evidenced by the persistent legends which attribute to it a Biblical origin (i); bear a unique testimony to the lost civilisation of the Shona between the 11th and 15th centuries (iii) and the Zimbabwe nation has identified with this historically symbolic ensemble and has adopted as its emblem the steatite bird, which may have been a royal totem (vi).
Ngorongoro Conservation Area World Heritage site (iv, viii, ix, x)	Located in Arusha region of northern Tanzania, southeast of Serengeti National Park and is part of the Eastern (Great) Rift Valley of eastern Africa	Ngorongoro Conservation Area, a mixed World Heritage sites, illustrates an exceptionally long sequence of crucial evidence related to human evolution and human-environment dynamics, including physical evidence of the most important benchmarks in human evolutionary development stretching from four million years ago (iv); spectacular concentration of wildlife on the planet (viii); the largest unbroken caldera in the world, several overlapping ecosystems and distinct habitats, with short grass plains, highland catchment forests, savanna woodlands, montane long grass plains and high open moorlands (ix); several overlapping ecosystems and distinct habitats; a population of some 25,000 large animals, mostly ungulates, alongside the highest density of mammalian predators in Africa including the densest known population of lion (x).
Robben Island World Heritage site (iii, vi)	Located in the Table Bay, some 6km west of Bloubergstrand in Cape Town, Western Cape in South Africa	Robben Island, used at various times between the 17th century and the 20th century as a prison, a hospital for socially unacceptable groups, maximum security prison for political prisoners and a military base, bears eloquent witness to its sombre history (iii) the Island and its prison buildings symbolize the triumph of the human spirit, of freedom and of democracy over oppression (vi).

FIGURE 1.2 Supporting World Heritage sites from Africa Region

and development issues. In addition, they have been subject to reactive monitoring missions and technical assessments by UNESCO over the years on the same issues. For instance, NCA has been riddled with threats emanating from tourism pressure, conflicts between people and nature, and land issues, with the Maasai increasingly protesting against them being 'forced out of their land'. GZWHS has been discussed in the context of proposed construction of the Great Zimbabwe University on the edges of the site, while RIWHS, Goree Island and the Forts and Castles have similar tourism development, conservation, and stakeholder pressures. All these WHS continue to experience stakeholder, conservation, and development issues, just like MCLWHS. Furthermore, while research on the significance and importance of these sites has advanced, the discussion of their management in relation to SD is lagging. Both the main and supporting case studies are representatives of similar WHS on the African continent Both offer "laboratory" experiences that can be analyzed and extrapolated for application at other WHS.

Structure of the Book

The chapters of this book are structured as follows: *Chapter 2* reviews conservation and development within the geo-political and socio-economic context of Africa with a view of understanding the evolving heritage management systems, whether there is tension or no tension between conservation and development and the emerging challenges in practically integrating the two at WHS in Africa. *Chapter 3* defines the Multiple and Multi-layered Stakeholder Framework (MMST) as the preferred approach in managing stakeholders at WHS. It defines heritage governance, stakeholder theory and management, and the ensuing local-global nexus central to identifying the multiple and multi-layered stakeholders at WHS. *Chapter 4* discusses the history, values, significance, and management of the main case study: Mapungubwe Cultural Landscape World Heritage. It highlights the local, regional, and international significance of the site, as well as the management complexities arising from its four tiers of protection as a national heritage site, National Park, WHS and TFCA. *Chapter 5* unmasks the multiple and multi-layered stakeholders at WHS using historical context, values, significances, legislative environment, interests and relationships among stakeholders, evolving land uses,, and the emerging socio-economic opportunities. It systematically identifies, categorizes, profiles, and discusses stakeholders at MCLWHS as the main representative case study.

Chapter 6 presents the empirical perspective of stakeholders using both quantitative and qualitative methods to analyze data collected through a mixed research methodology applied at MCLWHS. It highlights empirical perspectives on the state of conservation, stakeholder profiles, aspirations, and perspectives on various issues, spirituality and scaredness, and the emerging relationship between conservation and development at MCLWHS. *Chapter 7* provides a critical analysis of emerging stakeholders, conservation, and development issues at WHS, with reference to other sites selected in Africa. Emphasis is placed on engaging and involving stakeholders as a good governance approach, understanding stakeholders aspirations and perspectives on SD, camaraderie between conservation and development, opportunities and challenges regarding crossing stakeholders, conservation, and development frontiers at WHS in Africa. These frontiers need to engage, negotiate, and integrate each other at WHS. *Chapter 8* makes overarching recommendations on deconstructing stakeholders and their profiles as a governance issue, localization of SD at WHS, implementing AMA to improve the effectiveness of heritage governance, converting challenges into opportunities, creativity, and innovation as a function of multidisciplinary mindsets at WHS. Furthermore, it argues that conservation must abandon its comfort zone by embracing development value propositions at WHS as stakeholder-driven processes.

Conclusion

This chapter introduces the idea of an integrated perspective of stakeholders, conservation, and development at WHS in Africa. While the WHS value proposition is about promoting conservation, there is growing demand for a practical coexistence with SD value proposition geared to provide for the well-being of society living with WHS. The paucity of data on how this can happen at WHS needs fusing theoretical and empirical perspectives on stakeholders, conservation, and SD. The chapter, therefore, motivates the need to understand the empirical perspectives of conservation and SD as stakeholder-driven and adaptive processes at WHS in Africa. The evolving interplay between conservation and SD can be viewed as a function of the pioneering spirit, experiences, wisdom, and practices of different stakeholders living with WHS. All this can be traced from pre-colonial to recent times in Africa. The ideas of stakeholders and adaptive management are advanced as having potential to promote and localize new transversal or multidimensional approaches, including governance, towards making WHS become responsive to the evolving needs of the society. The future of WHS lies in its ability to move beyond focusing only on conservation value proposition to include caring for the social well-being of society too.

2

GEOPOLITICS, SOCIO-ECONOMIC CONTEXT, AND HERITAGE MANAGEMENT IN AFRICA

Introduction: Irony of Conservation and Development in Africa

The African region is endowed with diverse heritage and natural resources that have a role to play in continuously meeting the needs of the society. However, at the same time, Africa has long chronicled positive benefits and negative challenges in implementing both conservation and development at World Heritage Sites (WHS). This defines the irony of implementing both yet they are necessary but equally challenging to each other. It is important, therefore, to analyze the regional and international perspectives of conservation and development at play in Africa. This includes tracing the engagement and involvement of local communities and other role players towards defining the term "stakeholder", which is adopted for this discussion. This provides understanding as to whether the evolving heritage management and governance systems, stakeholder roles, geo-politics, and socio-economic dimensions are or not promoting conservation and development at WHS in Africa. Overall, it is important to discuss whether there is or no tension between conservation and development in Africa. The is traced from the pre-colonial, colonial, and post-colonial periods of Africa, a continent of juxtaposing prosperity and poverty, hope and pain, peace and conflict, as well as laughter and cries! Her people desire better, brighter and fulfilling lifestyles! It is in this contrasting tales of Africa that conservation is expected to play a catalytic role in facilitating the caring for heritage and the well-being of society.

Africa: Geopolitics and Socio-Economic Needs

Africa, as a developing continent, faces cardinal challenges of poverty, food, and water shortages, high levels of unemployment, energy deficits, infrastructure, and inequality due to underdevelopment (Askew, 2010; Keitumetse, 2011;

DOI: 10.4324/9781003149361-2

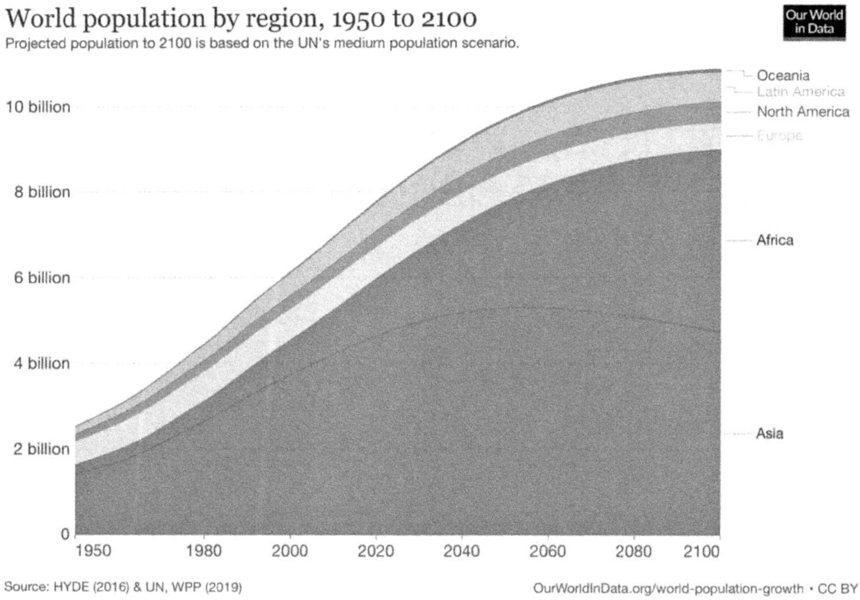

FIGURE 2.1 Actual and projected population growth in Africa (1950–2100)

Nkoana-Mashabane, 2013). The systemic root of this underdevelopment is colonialism, and the post-colonial economic models have retained sub-Saharan Africa as the primary source of raw materials extraction while value-added production processes are undertaken in industrialized and developed nations. Furthermore, Africa is challenged with dealing with pandemics (i.e., COVID-19, HIV and AIDS, and tuberculosis infections), conflicts and political instability, unprecedented corruption, and deteriorating public service delivery. Also, Africa's population has increased from 9% (1950) to 17.2% (2020) and is predicted to reach 39.4% in 2100. Africa is the "second-most populous continent behind Asia since the year 1996", and her population is estimated at 1.34 billion as of July 2020 (UN, 2020: 1). Overall, Africa has been experiencing the fastest population growth since 1967 as illustrated in Figure 2.1 (UN, 2020). Despite this phenomenal population growth, Africa ranks at the bottom in terms of gross domestic product (GDP) per capita. Africa has GDP per capita of US1,890, compared to Oceania (US54,220), North America (US47,750), Europe (US29,450), South America (US8,510), and Asia (US7,090) (IMF, 2018).

The relationship between the population growth and GDP per capita ranking reflects the disproportionate relationship between the abundant renewable and non-renewable resources and levels of poverty in Africa. Africa currently has the fastest growing youthful age category which needs jobs and livelihoods. Available resources should be used to mitigate underdevelopment in Africa, apart from grossly benefiting developed nations as a result of the skewed economic models retaining Africa as a supplier of raw materials using affordable labour. Africa is home to a large number of natural resources, among them gold, diamonds, iron, uranium, platinum, silver, bauxite, crude oil, cobalt, salt, woods, water, as well as other agricultural resources such as sugar, cocoa, beans, tropical fruits, cotton, and maize (Moyo, 2009). In specific terms, and as an example, Africa has contributed mineral production at world scale as follows: bauxite (9%), aluminium (5%), chromite (44%), cobalt 57%, copper 5%, gold 21%, iron ore 4%, steel 2%, lead (Pb) 3%, manganese 39%, zinc 2%, cement 4%, natural diamond 46%, graphite 2%, phosphate rock 31%, coal 5%, mineral fuels (including coal) and petroleum 13%, and uranium 16% (Yager *et al.*, 2010). Also, the African economy is anchored on agriculture, mining, trade, industry, and human resources.

While this makes Africa a rich-resource continent, she remains unable to mitigate her cardinal challenges of underdevelopment. The reasons for this are multiple, and among them is the impact of colonialism, lack of infrastructure, challenges associated with language diversity, economic models characterized by dependency theory, governance challenges, social and economic inequality, lack of local incomes, and reliance on foreign aid with stringent conditions, which further reinforces dependency on developed nations (Kayizzi-Mugerwa, 1999; Moyo, 2009; Rodney, 1982).

While arable land and fresh water (65% and 10%, respectively, at world scale) exists in abundance (Moukala, 2019: 2), technological support is lacking to maximize these resources for the well-being of the society. The energy generation potential of Africa is curtailed by lack of technical development and infrastructure. Though Africa has ports, lack of connecting and support infrastructure affects

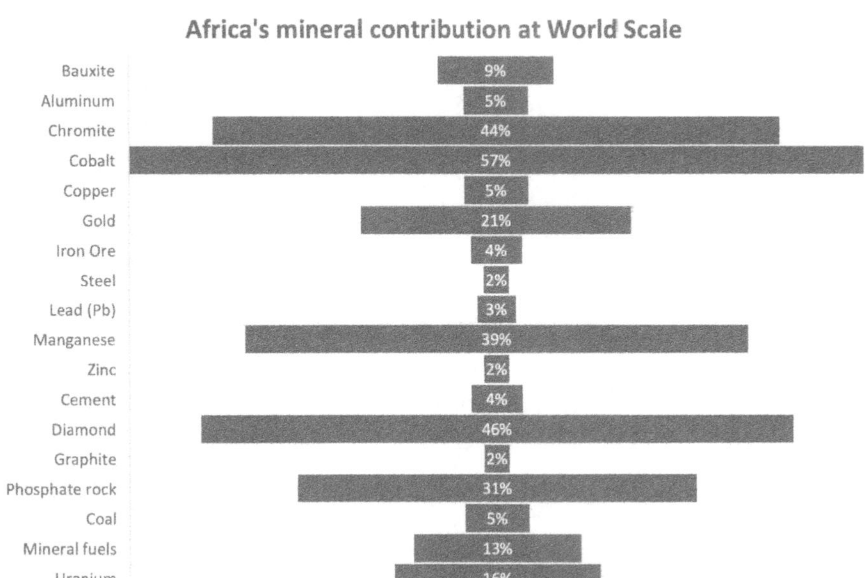

FIGURE 2.2 Africa's mineral contribution at world scale (Source: Yager *et al.*, 2010)

the economy and systems supporting socio-economic activities. As such, most African countries are always lowly ranked on the Human Development Index (HDI) list. There is a combined 1.5 billion people in 102 developing countries who are living in multidimensional poverty (UNDP, 2020). Multidimensional poverty is measured against factors such as access to health, education, and the standard of living (Our World in Data, 2017; UNDP, 2020). The net effect of population growth is that national governments are expected to meet the increasing socio-economic needs of their people (Taruvinga, 2019. This pattern has been consistent from the pre-colonial times to the present as society always has needs (Igboin, 2011). For example, addressing energy security issues is central to the development of Africa (Sparks, *et al.*, 2014). The energy options available (e.g., solar or wind) have their own negative environmental impacts, which still need to be mitigated (Chirikure *et al.*, 2010, 2013; Sparks, *et al.*, 2014; MacEachern, 2010). However, these mitigation measures are not always successful (Brida *et al.*, 2011). An illustrative example is the contamination of underground water by acid seeping from abandoned and disused mines around the Cradle of Humankind, part of the serial Fossil Hominid World Heritage Site in South Africa (Taruvinga, 2019). The contaminated underground water is likely affecting hominid fossils at the site through acid corrosion (Taruvinga, 2019). The full nature and extent of this impact is yet to be quantified and assessed in many similar regions and heritage sites across the world (Brida *et al.*, 2011; Chirikure *et al.*, 2010, 2013; MacEachern, 2010).

In response to these increasing human needs, the United Nations has successively developed the MDGs, Sustainable Development Goals (SDGs), and the RIO 21 recommendations for implementation through State Parties (SPs) (Sachs, 2012). The MDGs (2000–2015) were aimed at eradicating extreme poverty, promote access to education, healthcare, political, and environmental sustainability (Labadi, 2018). The SDGs cover social, economic, and ecological development goals (Figure 2.3). The African Union (AU) 2063 Agenda, "The Future We Want for Africa", embraces the SDGs while addressing human needs in Africa (Taruvinga, 2019). The AU Agenda has seven aspirations meant to address the socio-economic challenges of Africa as reflected in Figure 2.3 (Taruvinga, 2019). The agenda also seeks to create self-sufficiency without over-reliance on developed nations (Taruvinga, 2019). It also emphasizes a people-centred approach in Africa (AU, 2013: 3). This inculcates a locally driven development in order to avert the impact of unfavourable technical support and financial conditions in Africa (AU, 2013). All these goals are intertwined and inseparable parts of a solution to the increasing development challenges of the World (see Figure 2.3).

The Agenda builds on the Lagos Plan of Action (Bujra, 2002), the Abuja Treaty, the Minimum Integration Programme (MIP), the Programme for Infrastructural Development in Africa (PIDA), the Comprehensive Africa Agricultural Development Programme (CAADP), and the New Partnership for Africa's Development (NEPAD). The AU 2063 Agenda and SDGs are finding

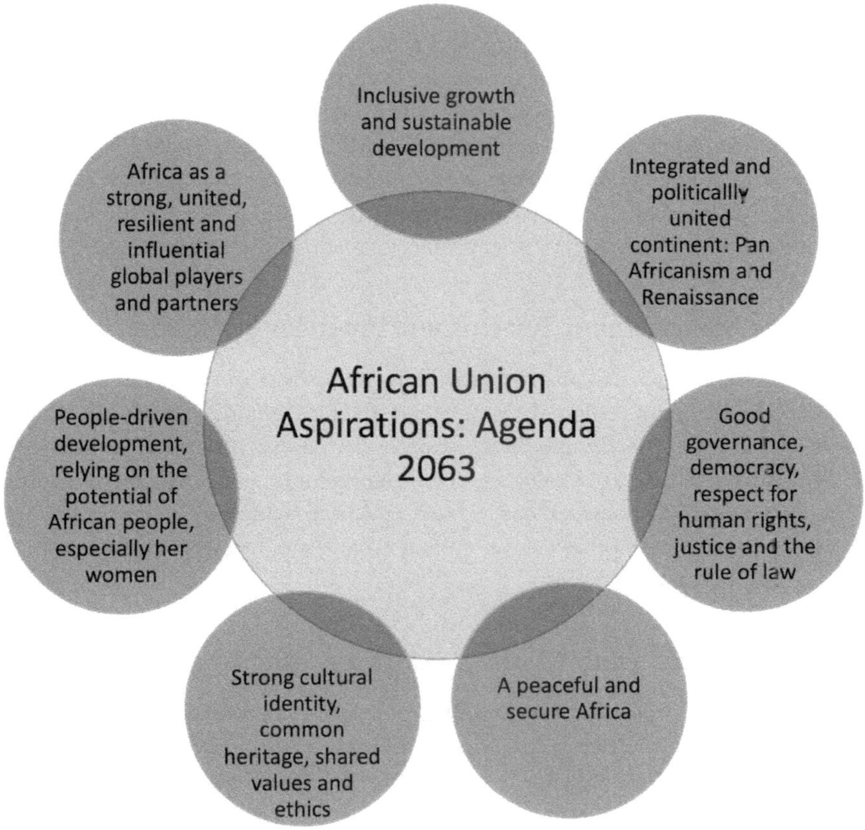

FIGURE 2.3 Sustainable development (SD) indicators and AU 2063 Agenda

expression in sub-regional and National Development Plans (NDPs) of African countries (Taruvinga, 2019). These NDPs are supposed to create jobs and address income inequality, attain energy, food, and water security, as well as expand trade opportunities and urbanization (Nkoana-Mashabane, 2013: 19). Also, NDPs are expected to build inclusive economies that benefit society in its broader diversity (South African National Development Planning Commission, 2016). Overall, SDGs and NDPs assist in achieving local, regional, and international development targets (Sachs, 2012). All these initiatives are thus linked to the United Nations Agenda 2030, which is being implemented using varying socio-economic development approaches (Szirmai, 2005; Taruvinga, 2019). However, studies are sometimes uncertain as to which development approach is best applicable (Szirmai, 2005). This is worse in Africa where there is a dearth of empirical data, let alone a deeper understanding of stakeholders perspectives on these matters (Taruvinga, 2019). Also, development approaches are viewed as a euphemism for Western penetration and domination of the world (Szirmai, 2005). Others view development approaches as a normative concept involving making choices and values influenced by dominant cultures and international power relationships (Szirmai, 2005). But, research that has not been fully explored is on who makes these choices and values in Africa (Taruvinga, 2019). However, development does not mean that all societies ought to develop in the same manner, or that they need to agree on some common standard in their approach (Szirmai, 2005). Development takes multiple forms (multi-dimensional) and should not be only benchmarked on Western values or models (Labadi, 2018). It should also be benchmarked with the experiential dimensions of the target society.

Heritage Management Systems and Practices in Africa

The evolving heritage management systems and practices of Africa cannot be divorced from the pre-colonial, colonialism, the post-colonial periods in Africa, as well as the socio-economic context described above (Ndoro & Wijesuriya, 2014; Taruvinga, 2019). Analyzing these experiences assists in understanding how conservation framework has evolved in Africa, which then sets the framework for understanding how development issues have been handled from pre-colonial period to the present.

Traditional Management Systems (TMS) and Africa

TMS have been in place from time immemorial (pre-colonial times) and continue to persist on the African continent (Eboreime, 2008, 2009; Jopela, 2010). This demonstrates the resilience of local cultures despite the rapid changes Africa has gone through in the ambit of colonialism. TMS are "as old as humans on earth" (Abungu, 2016: 9) and are the unwritten (informal) legal instruments in Africa (Mahachi & Kamuhangeri, 2008; Mumma, 2002). These are "known behaviours and practices that have been experienced, tested and accepted" through time which "govern human practices and ensure responsible utilization of resources and harmonious co-existence" between nature and culture (Abungu, 2016: 9). These cumulative customs, beliefs, and practices define TMS in Africa (Jopela, 2010). Overall, TMS represent worldview or religious perspectives of African communities (Berkes *et al.*, 2000; Jopela, 2010, 2011). Also, TMS can be viewed as a community-based philosophy of conservation, supported by communal (institution) ownership and legitimacy where common values are passed from one generation to another (Berkes *et al.*, 2000; Mumma, 2003, 2005).

Pre-colonial institutions of Africa [e.g., MCLWHS (South Africa), Kingdoms of Buganda and Ankole (East Africa), Great Zimbabwe kingdoms (Zimbabwe), Songhai Empire (West Africa), and Luba kingdom in Central Africa) are powerful examples of TMS (Igboin, 2011; Mahachi & Kamuhangeri, 2008; Murimbika, 2006). TMS protect places or landscapes considered culturally and ecologically significant to Indigenous and Descendant Communities (IDCs) of Africa (Jopela, 2016a; Ndoro *et al.*, 2008; Sheridan & Nyamweru, 2008). Examples of such places include the Mijikenda Kayas sacred forest (Kenya), Royal Palaces of Abomey (Benin), Kasubi Tombs (Uganda), Sukur Cultural landscapes and Osun Oshogbo Sacred Groves (Nigeria), and the many thousands of Rock Art sites in Africa. For instance, the Kasubi Tombs have witnessed effective and constant maintenance of the site through time using TMS, including the restorations post the 2010 fires at the site (Mwanja, 2014). This is testimonial to how local communities have the ability to manage their own heritage, control related socio-economic uses, including partnering with other stakeholders such as the tourism industry and government (Taruvinga, 2019).

Pre-colonial ethnic institutions of Africa shaped regional development through promoting local industries and the livelihoods of their respective communities (Michalopoulos & Papaioannou, 2012). They demonstrated a strong association between "political centralization and regional development" (Michalopoulos & Papaioannou, 2012: 1). Communities in Africa have experiences in resource utilization in order to meet their needs, while maintaining their cultures (Taruvinga, 2019). This birthed intra- and inter-kingdom trade, and the long-distance trade with the Far East and Arab region through the African coastline gateway (Chirikure *et al.*, 2010; Ndoro, 2005). Great Zimbabwe and Mapungubwe, revered as religious and economic centres, supported long-distance trade with the Far East (Ndoro, 2001; Pikirayi *et al.*, 2016). The existence of grain bins and domestic animal bones at the sites point to agriculture and animal husbandry (Ndoro, 2001), while the massive engineering structures at Great Zimbabwe, Khami (Zimbabwe), Toutswe (Botswana), and Thimlich Ohinga (Kenya) archaeological sites represent stone quarrying industries that thrived for many centuries without possibly causing undesirable environmental scars on the landscape (Taruvinga, 2019). The water holes within the greater landscape of Great Zimbabwe site denote traditional water management systems practised by local communities (Pikirayi *et al.*, 2016). With the San communities (spread across sub-Saharan Africa, but concentrated around Kalahari Desert, South Africa, Namibia, Botswana, and Lesotho), hunting and extracting resources such as tubers, trees, and medicinal plants among many others were part of their daily lives controlled by their consciousness of ecosystems as they responded to seasonality, thereby allowing regeneration of resources (Lewis-Williams, 1994; Loubser & Laurens, 1994).

The large quantities of potsherds recovered at Great Zimbabwe, Mapungubwe, and many other archaeological sites in Africa are testimonial to the past pottery industries, and this has thrived in the present as the backbone of the curio industries (Ndoro, 2001; Taruvinga, 2019). The gold and silver objects, including related metal instruments recovered at archaeological sites also demonstrate ancient mining and processing of minerals into secondary products by local communities (Desai & Lee-Thorp, 2000; Swan, 1994, 2007: 12). The same applies to the Okavango Delta in Botswana where local communities have been harvesting fish and extracting minerals for centuries (Taruvinga, 2019). All these are experiential perspectives demonstrating that local communities are capable of caring for their heritage while meeting their needs. It further confirms that regional development is synonymous with complex and socially structured societies of Africa (Igboin, 2011: 23).

While stakeholders behind TMS are mainly the creators and custodians of the cultural values at heritage sites, their role at present is riddled with challenges, including recognition apathy across decision-making structures of state-based management systems (SBMS). These stakeholders have historical, social, spiritual, and socio-economic connections with individual or territorial places connected with their complex history (Pikirayi, 2016). Such local communities are increasingly being called IDCs, reconstructed right from the pre-colonial period to the present, including understanding their relevance and legitimacy in relation to conservation and development at heritage sites. However, IDCs is a contested terrain with colonial overhangs, giving SBMS competitive advantage over them. In such cases, SBMS encourage them to resolve their internal conflicts while they continue making decisions without them at heritage sites. Furthermore, the legitimacy of TMS publics was weakened by colonial practices and legislations, which marginalized them, including providing punitive measures in case of violations. Currently, IDCs are relegated to ceremonial status for political reasons by SBMS, otherwise their governance power on heritage matters remains weaned since colonialism. Their current locations in relation to the original landscapes they are connected to is a major constrain for them. Additionally, migration and socio-economic opportunities have scattered IDCs wide and afar, including being concentrated in urban areas. Urbanized IDCs now have new lifestyles, some which of contradict their beliefs, customs, practices, and affinity to their own cultural heritage. It would appear that TMS is at risk of becoming "essentialized" (stereotyped as a public act covering past suppression and omissions, yet maybe offensive in nature) and "romanticized" as part of the post-colonization reaction. The experiential perspective of IDCs should find space and as part of innovation in conservation and development. As a result, there may be large groups in the wider public who have no voices at all, implying that no one can speak for them.

Overall, TMS resemble contemporary scientific practices for ecosystem-based management and illustrate elements of succession management, landscape patchiness management, resource use, and rotational approaches (Berkes& Davidson-Hunt, 2006). These elements empowered indigenous communities in situations where communal decision-making was required (Berkes & Davidson-Hunt, 2006). TMS can be considered as adaptable social institutions and mechanisms that regulate the use of resources at the local level through time and space (Jopela, 2016a). This is community wisdom in action. If this wisdom can be tapped into, it has the potential to build a more inclusive decision-making process and robust constituency in balancing conservation and development at WHS (Berkes & Davidson-Hunt, 2006).

Emergence of SBMS in Africa

Colonialism imposed a Eurocentric approach to heritage management in Africa, resulting in establishment of SBMS (Hulme & Shepherd, 2003; Jopela, 2016;

Ndoro, 2001; Ndoro & Wijesuriya, 2014). This approach values "western science" at the "expense of local values" often viewed as "superstitious and irrelevant to conservation and research" (Chirikure & Pwiti, 2008: 470). There was no effort to contextualize TMS experiences and discourses within this emerging SBMS framework (Nelson, 2003). SMBS are largely in charge of heritage in Africa, and they emphasize the role of experts and defined scientific approaches with minimum involvement of local communities (Ndoro & Chirikure, 2009; Ndoro & Wijesuriya, 2014; Pwiti & Ndoro, 1999). This also includes use of management plans, value or significance-based, desired state, adaptive, and integrated planning frameworks (Taruvinga, 2019). Management plans outline protection measures to retain values considered to be of significance (Fredheim & Khalaf, 2016; Meskell, 2011; Munjeri, 2004a; Smith, 2006), but are thin on developments supposed to benefit other stakeholders. In the process, experts have more affinity to notions of untouched, pristine conditions and unaltered landscapes defining environmental colonialism in action, which are rarely reconciled with TMS perspectives (Taruvinga, 2019).

According to environmental colonialism, the "natives did not understand their own heritage" and they would destroy heritage if they were left in charge (Nelson, 2003: 65, 71). It was systematically enforced through the establishment of Protected Areas (PA) (i.e., national parks) and the declaration of sites as monuments in colonial Africa, while communities were forcefully relocated to marginal and unproductive lands (Keitumetse, 2007, 2009; Kigongo & Reid, 2007; Nelson, 2003; Ndoro & Wijesuriya, 2014). These parks and monuments became enclaves of colonial administrators and scientists in an effort to "saving Africa from Africans" (Nelson, 2003: 65). National parks such as the Selous Game Reserve and Serengeti (Tanzania), Hwange National Park and Matobo Hills (Zimbabwe), the Okavango Delta (Botswana), Kruger and Mapungubwe National Parks (South Africa), and many others in Africa demonstrate this unfortunate colonial enclaving process (Taruvinga, 2019). In South Africa, separatist politics under apartheid influenced this process, while precluding IDCs from power, authority, and influence in the management and use of their own heritage (Meskell, 2011). Park officials have always conceived their relationship with their "neighbours" as that of mainly policing and maintaining fences to prevent them from encroaching into designated conservation areas (Cock & Fig, 2000; Smith, 2004). This authoritarian conservation perspective persists with minimal changes in Africa (Nelson, 2003). However, entrenchment that separates heritage and people has no future in the evolving social context of Africa, where conservation and well-being of society have always been intersecting as a going concern since pre-colonial times.

The colonial period gave birth and credence to post-colonial heritage management approaches in Africa. The resultant SBMS emphasize formal legislation, expertise, museology, and monumentalism at the continued exclusion of TMS and communities (Taruvinga, 2019). The "idea of materiality and the authenticity of fabric" is emphasized, and this negates the inseparable connection between the tangible and intangible aspects of heritage (Chirikure *et al.*, 2016: 165). This

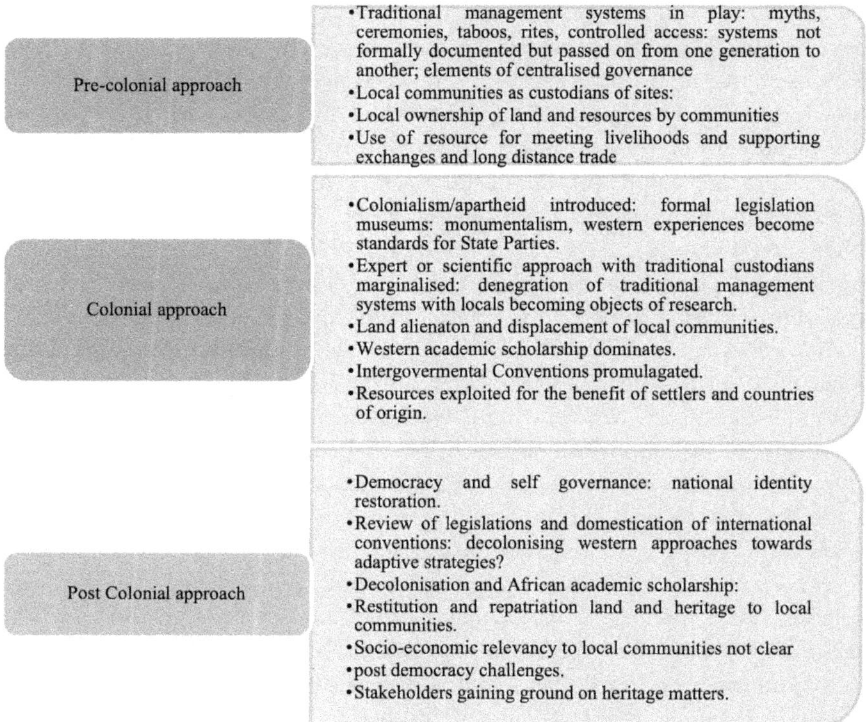

FIGURE 2.4 Evolving heritage management systems in Africa

further includes the link between nature and culture. Also, the definition and dimensions of heritage in the present do not adequately address the appropriation of the localness of heritage through the use of western concepts. Parallel to this was the orchestrated looting and plundering of local heritage in Africa (Chirikure *et al.*, 2016). These evolving SBMS (Figure 2.4) are further reinforced by the Convention and related Operational Guidelines in Africa.

World Heritage Management System and Africa

The Convention is a popularized international instrument worldwide. It is central to the protection of heritage considered to have an Outstanding Universal Value (OUV) (UNESCO, 1972). It is "a vast international bureaucracy that decides what is world heritage" (Donnachie, 2010: 115). The Convention has an influence over the management of WHS as an irreplaceable resource (Labadi, 2018). It is premised on the realization that cultural and natural heritage is under increasing threats from destruction by traditional causes of decay, changing social, environmental and economic conditions (Labadi, 2018). Furthermore, it acknowledges that protection of heritage is incomplete due to insufficient economic, scientific, and technological resources available at national levels (UNESCO, 1972).

The Convention defines OUV as the "cultural and/or natural significance which is so exceptional as to transcend national boundaries and to be of common importance for present and future generations of all humanity. As such, the permanent protection of this heritage is of the highest importance to the international community as a whole" (Operational Guidelines, 2019). OUV is amplified by a criterion covering both nature and culture values, as the test of authenticity and/or integrity, and protection and management systems. Authenticity refers to how credibly and truthfully the attributes conveying the significance of the site are expressed at the site, while integrity refers to the completeness/wholeness, the adequacy of the size to protect all the elements, and the threats the site may be facing (Alberts & Hazen, 2010; Stovel, 2007). Authenticity is relevant to culture while integrity is for nature; however, both apply to mixed sites or cultural landscapes. Protection and management systems refer to the legal protection in place, and the forms of management associated with a WHS, either management plans or TMS (Stovel, 2007). Properties are inscribed either as single sites, mixed sites, or cultural landscapes. Cultural landscapes are either organically evolved or associative landscapes (Jones & Cloke, 2002; Selman, 2004). Organically evolved landscapes are derived from active economic land uses, while associative landscapes are linked to historic events (Selman, 2004). Cultural landscapes illustrate an "interplay between human and non-human realms" and the "social construction and geographic spaces (Brace, 2003; Matless, 1998; Selman, 2004: 366).

The Convention framework further outlines the responsibilities and obligations of SPs involved in its implementation. The identification and protection

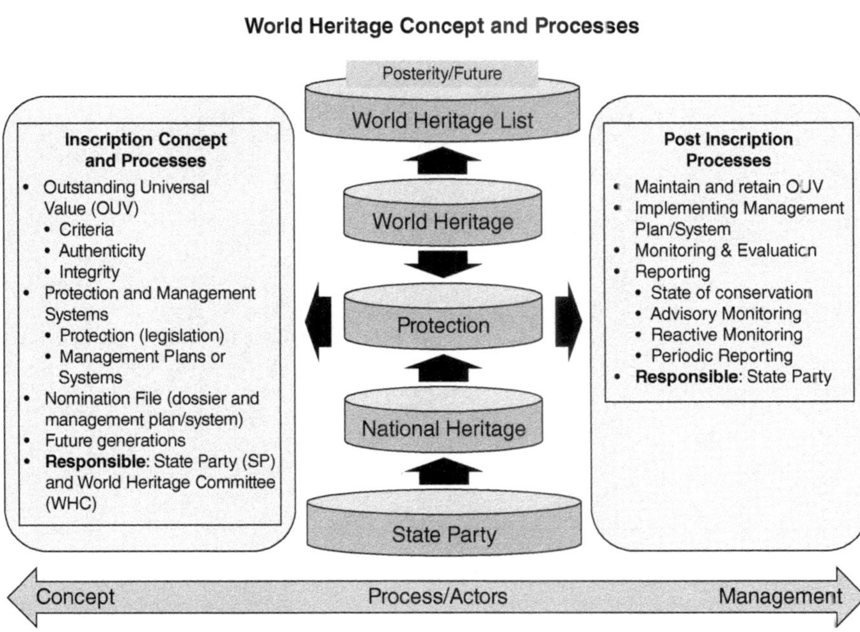

FIGURE 2.5 World Heritage concept and processes

of WHS remains a responsibility of the respective SPs. This is done through a wide range of management actions and policies by SPs while endeavouring to be aligned with precepts of the Convention.

The implementation of the Convention is coordinated by UNESCO supported by the World Heritage Centre, SPs and Advisory Bodies (ABs) (explained in the box below). These stakeholders constitute an Intergovernmental Structure known as the World Heritage Committee. ABs are scientifically and politically independent of governments, but sometimes these are funded by SPs and the Intergovernmental structures of the Convention (Donnachie). The Committee delimits and runs complex protocols and governance procedures for the Convention (Blacik, 2007; Rössler, 2007).

The Committee is responsible for inscribing or not inscribing new sites, placing sites on the List of World Heritage in Danger, disbursing resources, and

	Advisory Body	**Function**	**Implementation Matrix**
1	International Union on Conservation of Nature (IUCN)	IUCN evaluates new nominations, monitors state of conservation, promotes the Convention as an instrument of valuing and conserving nature, sustainable use and supports capacity building programmes on nature conservation.	Partially supported by UNESCO in their functions as Abs. Both are supported by a network of professionals across the world. Both advise on mixed sites and cultural landscapes.
2	International Council on Monuments and Sites (ICOMOS)	ICOMOS plays the same role as IUCN but for culture values and the conservation of tangible and intangible, immoveable and moveable heritage.	
3	International Centre for the Study of the Preservation and Restoration of Cultural Property (ICCROM).	ICCROM advises the World Heritage Committee on the conservation of all forms of heritage and supports capacity building, information dissemination, cooperation, research and advocacy with the support of Member States.	Supported by Member States. Also supported by network of professionals across the world.

FIGURE 2.6 Roles and functions of the Advisory Bodies [International Union on Conservation of Nature (IUCN), International Council on Monuments and Sites (ICOMOS), and International Centre for the Study of the Preservation and Restoration of Cultural Property (ICCROM)]

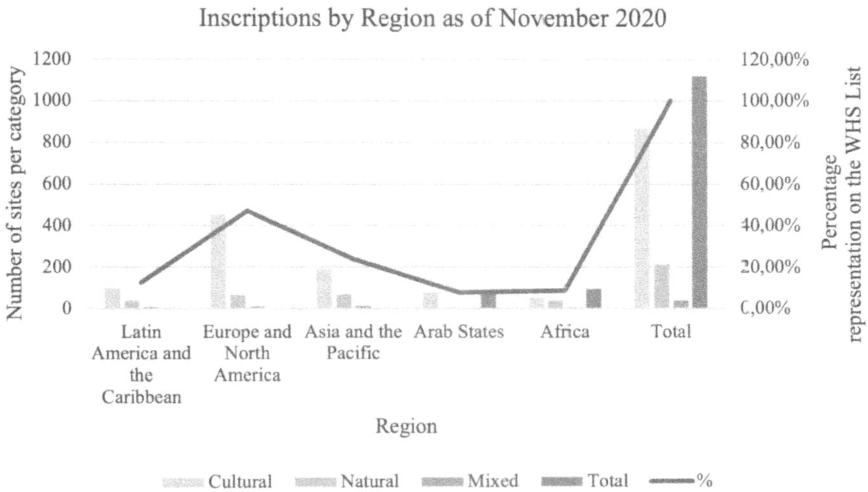

FIGURE 2.7 World Heritage List as of 2020 (Source: UNESCO, 2020)

examining state of conservation reports, including monitoring the implementation of the Convention at large (Rössler, 2007). However, the governance system of the WHC has suffered from politicization by SPs over the years. This has witnessed political decisions interfering with scientific rational of the Committee in its decision-making process (Beracchini et al, 2016).

The Convention has benefits, which include being part of a global community dedicated to conservation, access to funding, international cooperation, professional networks, and capacity-building opportunities (Cassel & Pashkevich, 2013; Donnachie, 2010; Edroma, 2004). Also, the Convention has increased awareness on heritage issues worldwide (Di Giovine, 2008; Labadi, 2018). As of 2020, the World Heritage List had 1121 sites from 165 countries (Africa, 35; Arab States, 16; Asia and the Pacific, 36; Europe and North America, 50; the Latin America and the Caribbean, 28).

Africa (as defined by UNESCO as a region for execution) represents 8.56% of all inscribed sites, with the Arab States contributing 7.67%, Europe and North America (47.19%), Asia and Pacific (23.91%), and Latin America and the Caribbean contributing 12.67%. While the Convention has been successful in inscribing sites on the World Heritage List, another negative pattern is the increasing number of sites on the List of World Heritage in Danger (see Figure 2.8). The threats emanate from both natural and human-induced factors.

From the above chart, Africa accounts for 30% of sites on the List of World Heritage in Danger, and if the Arab region (40%) is considered as part of Africa as per AU geographical and political boundaries, it is a shocking 70%. Regardless, this demonstrates increasing threats to sites in Africa and the need to mitigate them. While the Convention was premised on mitigating such threats, it has taken a very long time to find a sustainable formula for doing this (Cassel &

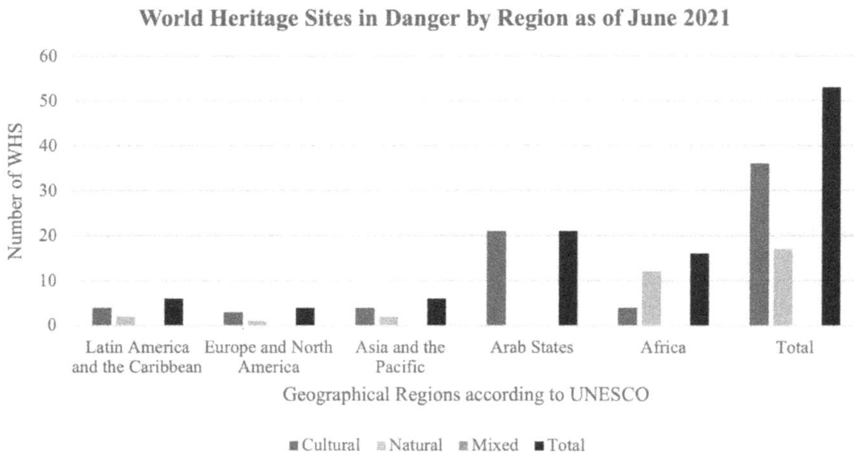

FIGURE 2.8 Sites on the List of World Heritage in Danger as of 2020 (Source: http://www.whc.unesco.org/end/list/stat; UNESCO, 2020)

Pashkevich, 2013; Labadi, 2018). To experts and WHC, the SPs are failing to live up to the original Convention objectives, yet for stakeholders, it is failing to support their well-being aspirations (Taruvinga, 2019). Being included on the endangered list is like being in an "intensive care unit", where the WHC, through statutory decisions, instructs the SPs to undertake corrective measures to retain the OUV.

The perceived insensitivity of the Convention to the plight of the local communities has led to some sites being delisted at the instigation of SPs who are in favour of development, as was the case with Oman in 2007 (Taruvinga, 2019). The WHC disagreed with the decision of the SPs to reduce the size of the PA (Arabian Oryx Sanctuary) by 90% in favour of a hydrocarbon prospection project. The World Heritage Committee views such SPs as failing in their legal obligation to protect WHS in line with the statutes and Operational Guidelines of the Convention. WHS in similar conflicting situations are subjected to Reactive Monitoring Missions by the WHC as was the case at MCLWHS due to threats posed by coal extraction, Selous Game Reserve World Heritage site (Tanzania) threatened by uranium extraction, and Timbuktu (Mali) suffering from the impact of prolonged conflict among many others. The views of stakeholders in the process remain at the periphery of Committee decisions.

On the African continent, it is clear that heritage has not been sufficiently harnessed to support development (Ndoro & Wijesuriya, 2014). Stakeholders living with WHS in Africa argue that they are not benefiting in a meaningful and measurable way (Taruvinga, 2019). Most WHS are carved out of territories once owned by communities now under SBMS control (Ndoro & Chirikure, 2009). Just like SBMS, the Convention overrides the local cultural context with "a language and method of management" focused on universalism (Harrison, 2010a: 191, 2010b). It "operationalizes" western notions in non-western heritage contexts (Harrison, 2010a: 170). The Convention was also slow in recognizing the intangible values of heritage, preferring the long-established monumentalism. It was only through the collective effort of Asia and Africa that the Document on Authenticity (NARA), and the recent additions, NARA+20, that forced the Convention to recognize the inseparable connection between intangible and tangible values (Ndoro *et al.*, 2018; Taruvinga, 2020a). This slowness created tension between universal and local perspectives of heritage (Byrne, 1991; Harrison, 2010a: 170).

While the Convention is not necessarily incompatible with controlled resource development approaches (Blacik, 2007; Nicholson, 2002), it is just that powerful SPs on the global politics arena find it easier to have their ideas accepted than those of the weaker and impoverished communities of developing nations (Harrison, 2010a). This reflects an increasing politically complex field as the Convention is linked to the international community and geo-politics from its inception (Labadi, 2018; Meskell, 2014). In the process, developing nations are vulnerable to global processes and political dimensions of the Convention, in particular its slowness in facilitating development at local levels (Meskell, 2014; Ndoro & Wijesuriya, 2014). Furthermore, SBMS and WHC decision-making processes exclude stakeholders and do not give them any voting powers. Stakeholders are accommodated as observers or are represented as an extension of SPs, but with "muffled" voices (Taruvinga, 2019). Those who are excluded, or exclude themselves from it, are not involved in the decision-making process of WHC (Donnachie, 2010: 117; Meskell, 2014). However, recently, IUCN has supported the establishment of the Indigenous Community Forum, which is yet to be officially recognized by the WHC. It is slowly becoming an official avenue for Indigenous communities to be heard by the international community (Taruvinga, 2019).

Though the Convention recognized development at inception, it has been "unrushed' to embrace SD or other alternative approaches to development. The Convention does not mention the term "sustainable development" in its doctrinal text, further demonstrating how development aspirations of stakeholders are not adequately addressed in such instruments (Labadi, 2018). World Heritage processes only acknowledged SD in 2002, followed by its inclusion in the Operational Guidelines in 2005 (Labadi, 2018). The adoption of the 2015 World Heritage Policy on Sustainable Development by UNESCO further reinforced this inclusion (Labadi, 2018; UNCSD, 2012). This policy still requires empirical testing at WHS as an integration gap between national heritage laws and the Convention exists. However, the latter recognizes the former to prove the existence of protection system at a national level for a WHS site. National heritage laws are considered insufficient warranting the Convention to be used as the overarching tool at local levels, for instance, insistence on the OUV-Based Impact Assessment yet Environmental Impact Assessment (EIA) and Heritage Impact Assessment (HIA) frameworks already exist. Furthermore, not many African countries have domesticated the Convention as a way of bridging this integration gap. One example is South Africa, which has enacted the South African World Heritage Convention Act of 1999. In addition, the use of the broader environment of the site for development has always been interpreted from a conservation perspective rather than from a development perspective (Taruvinga, 2019). Overall, the Convention is criticized for its strong conservation agenda, a pattern that persists in the present despite efforts to accommodate SD (Nelson, 2003; Ndoro *et al.*, 2018; Ndoro & Wijesuriya, 2014).

Emerging Relationship: Conservation and Development

As briefly argued in Chapter 1, conservation and development are as old as humanity itself; therefore, balancing them remains an ongoing process and a matter of concern (Taruvinga, 2020b). While TMS regulated interplay between the two in the past, the Convention has adopted modern tools such as SD. SD is now a bridge and preferred approach for aligning heritage to development, though not all developments are sustainable in nature (Labadi, 2018; Taruvinga, 2020b). SD links "mounting environmental problems, poverty, inequality" and "concerns about a healthy future for humanity" (Hopwood *et al.*, 2005: 4).

SD ushers cross-cutting principles covering futurity (inter-generational equity), social justice (intra-generational equity), transfrontier responsibility (geographical equity), procedural equity (people treated openly and fairly), and inter-species equality (importance of biodiversity) in order to be effective (Haughton, 1999; Hopwood *et al.*, 2005: 9). SD has four principles: economic, social, environmental, and security and peace (Barton, 2000; Du Plessis, 2000; Giddings *et al.*, 2002; Hardi & Zdan, 1997; Labadi, 2018; UN, 1992; UNCSD, 2012). The social equality pillar focuses on the social well-being of people, while the environmental protection pillar advocates for people to invest in green technologies in order to make sure that the environment remains diverse and productive. The economic growth pillar focuses on attempting to attain more sustainable efforts and development, while security and peace advocates for mitigating conflicts and the associated vulnerability of society which is against development. However, these principles are not equally but selectively applied in country policies and practices worldwide, with the environmental remaining dominant among all (Barton, 2000; Brundtland Report, 1987; Daly & Cobb, 1990; Du Plessis, 2000; Giddings *et al.*, 2002; Hardi & Zdan, 1997). Furthermore, the attempt to have culture as a fourth principle of SD by the World Commission on Culture and Sustainable Development Conference (1995) and the Johannesburg Declaration (2002) failed (Labadi, 2018). The MDGs never mentioned culture and heritage, while the 2015 SDGs recognized culture and heritage but did not highly profile it (Labadi, 2018). Therefore, culture plays a catalytic role in SD until such time its recognized as a full pillar (Taruvinga, 2019).

SD is considered a "contested concept, with theories shaped by people and organizations with different worldviews" (Giddings *et al.*, 2002: 187). Also, it is argued to cover a "multitude of sins" committed by developers (Giddings *et al.*,

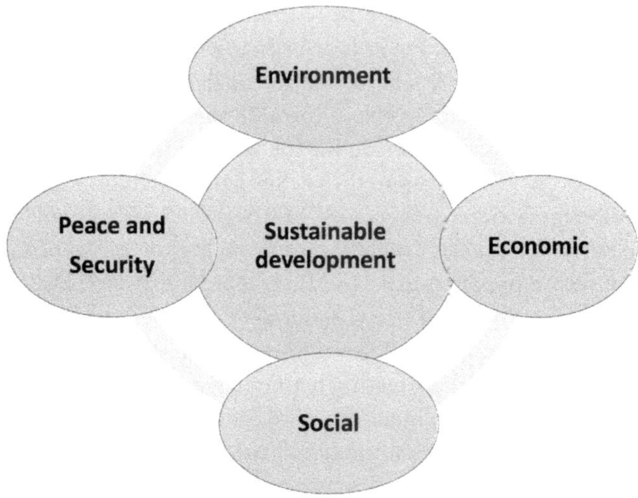

FIGURE 2.9 Principles of sustainable development

2002: 188). Additionally, SD pillars are considered "fractured and multi-layered" as they can be applied at "different spatial levels" in real situations (Giddings *et al.*, 2002: 187). Practically, SD principles are intertwined, integrated, and dependent on each other (Giddings *et al.*, 2002). An integrated approach would help in embracing the "multitude of environments, societies, economies and cultures (Giddings *et al.*, 2002: 188).

The emerging link between World Heritage and SDGs is important in that goal eleven (11) on sustainable cities and communities, in particular, Target 11.4, strengthens "efforts to protect and safeguard the world's cultural and natural heritage to make our cities inclusive, safe, resilient and sustainable" (ICOMOS, 2017). This places cultural heritage as a driver and enabler of SD (ICOMOS, 2017). While there are a few success stories on the impact of World Heritage on local communities, the act of balancing the competing objectives of conservation and development remains a challenge at WHS (Donnachie, 2010: 149). World Heritage has slowly and cautiously negotiated with the development through a plethora of policies and adopted recommendations (Labadi, 2018; Meskell, 2014).

In applying SD principles, there is an erroneous perception that supports thinking globally but acting locally (World Commission on Culture and Sustainabkle Development Conference, 1995). In this process, the experiences of European and North American countries serve as a reference point for developing countries, but this seldom offers neat solutions to policy makers, politicians, entrepreneurs, or development partners in different geo-cultural contexts (Szirmai, 2005; Maddison, 2001; Landes, 1998). This accounts for the varied interpretation and application of SD in real situations (Daly & Cobb, 1990; Hopwood *et al.*, 2005). Borrowing principles and experiences from other contexts and making them fit within the local context does not always give desired or same results. It is suggested that it is better to move away from the term SD to sustainability, or sustainable livelihoods, as SD is seen as prioritizing human needs over everything else (Giddings *et al.*, 2002: 187). SD is viewed as an ambiguous concept and that it is "inconceivable to talk about the sustainable use of non-renewable resources as exploitation of this finite stock will inexorably lead to its exhaustion" (Labadi, 2018: 41). However, it is undisputed that SD has a catalytic role in the "integration of different actions and sectors" dealing with development, including overcoming barriers between disciplines in resolving the needs of society without excluding future opportunities (Giddings *et al.*, 2002: 192; Daly, 1992; Rees *et al.*,, 1996; Turner, 1988). With or without SD principles, development remains a reality for centuries to come in developing continents or global south (Beel *et al.*, 2016). This requires interrogating heritage management systems to find out how conservation and development can co-exist (Taruvinga, 2019).

A number of regional and international meetings, workshops, and conferences were held over the years to discuss the link between World Heritage and SD (Taruvinga, 2019). In 1999, the WHC session held in Marrakesh (Morocco) examined mining in PA, and this was followed by an IUCN workshop in September 2000 (Gland in Switzerland). Both meetings argued for conservation to contribute to the "social and economic development and the quality of life our communities" (UNESCO, 2010: 2). Furthermore, in 2002, the "Johannesburg Declaration" (South Africa) on SD recognized heritage as a tool for promoting development and poverty alleviation (Johannesburg Declaration, 2002). It also recognized the much-needed "long-term perspective and broad-based participation in policy formulation, decision-making and implementation at all levels" (Johannesburg Declaration, 2002: 3). Furthermore, in 2008, IUCN issued a statement that WHS should be off limits for all extractive industries (Marton-Lefevre, 2008). As a result, the International Council on Mining and Metals (ICMM) committed to refrain from undertaking exploration or mining activities at WHS (ICMM, 2008). The Minerals Advisory Council of Canada and Shell made similar commitments as part of moving towards green credentials (Brida *et al.*, 2011). However, commitments of large oil and gas companies on the same do not match those of their mining counterparts and activities at WHS (Affolder, 2009).

The Paraty Action Plan of 2010 recognized the "contribution of World Heritage to sustainable development while noting that securing sustainable development is – almost by definition- an essential condition to guarantee the conservation of heritage" (Decision 34 COM 5D, 2010). The Plan suggested policies and procedures for integrating SD into the processes of the Convention (UNESCO, 2011). As a result, Operational Guidelines of the Convention were amended in 2011, in particular paragraphs 112, 119, 132 and annexes: paragraph 5: 4b and 5e. Furthermore, the 40th Anniversary of the Convention in Africa (held in 2012, South Africa) under the theme "World heritage and Sustainable Development; Living with World Heritage in Africa" bridged the "understanding" gap existing among different stakeholders on these matters. It was attended by 16 African Ministers of Culture, Tourism, Environment and Home Affairs, 32 local community representatives living at WHS (from 16 African countries), academics, heritage institutions, information technology sector, extractive and tourism industries, UNESCO, and ABs.

The 2012 conference recommended the coexistence of conservation and development at WHS in Africa (Cooper *et al.*, 2008; DAC & AWHF, 2014). Stakeholders highlighted that neither conservation nor development should be sacrificed for the benefit or survival of the other (DAC & AWHF, 2014). The stakeholders considered effective inclusion and engagement of all stakeholders in the decision-making process of World Heritage as important, as well as promoting transparency in the process (DAC & AWHF, 2014). This underpins a paradigm shift in how heritage management needs to be re-articulated to reflect its relevancy to the local context in which World Heritage finds itself (Taruvinga, 2019). This is in line with Rio +20 Recommendations (United Nations Conference on Sustainable Development) that recognized stakeholders in their diversity as being at the centre of developments. Furthermore, the 20th anniversary of the NARA document on Authenticity recognized SD and the role of multiple and multi-layered stakeholders as important in the implementation of the Convention (Taruvinga, 2020b). The WHC acknowledges the importance of local investments, sustainable forms of inclusive and equitable economic growth in the process (Labadi, 2018). Failure to take into consideration definitions and manifestations of heritage as a local process, as well as realities of the inescapable socio-economic context, is a major stumbling block in the application of international conventions in the post-colonial Africa (Ndoro & Abungu, 2008).

The challenge and complexity of balancing conservation and development in Africa has been equally daunting and continues to be highlighted through successive position papers developed by the African World Heritage Fund (AWHF). This includes hosting side events on the same during WHC. More importantly, the Ngorongoro Declaration of 2016 demonstrates the political and social urgency of addressing this matter in Africa. In the process, it is evident that World Heritage has been cautiously confined to the role of being an active guardian of the past while aligning itself to SDGs (Taruvinga, 2019). Also, the Charter for African Cultural Renaissance adopted by the AU at its session held in Khartoum (Sudan) in 2006 highlights that "people have the inalienable right to organize their cultural life in full harmony with their political, economic, social, philosophical and spiritual ideas" and promotes the social liberation of people (Charter for African Cultural Renaissance, African Union, 2006: 2). It further acknowledges that "cultural diversity and unity are a factor of equilibrium and strength in African economic development, conflict resolution and reducing inequality" (Charter for African Cultural Renaissance, African Union, 2006: 2). The Charter further recognizes non-institutional actors (such as designers, private developers, associations, local governments, and the private sector), who can play a role in heritage management (Charter for African Cultural Renaissance: Part III: Article 11). Although this Charter empowers AU and Member States in promoting Pan-Africanism, cultural, and national policies renewal, not many African governments have ratified it. This is worrisome in the sense that the embracing of home-grown solutions by the continent itself is slow and does not help in changing heritage management approaches in Africa (Taruvinga, 2019).

Tension or No Tension: Conservation and Development?

The relationship between conservation and socio-economic development cannot be divorced from its pre-colonial, colonial, and post-colonial contexts, national sovereignty rights of SPs in caring for heritage, and meeting the well-being of the society. The pre-colonial use of heritage and the 1959 trade-off relating to the Aswan Dam Project (Egypt) sets an interesting precedence; however, this has not been systematically investigated or replicated to establish a pragmatic approach for handling such matters at WHS (Taruvinga, 2019). Hence, the conflicting decisions and punitive measures by the WHC on matters of conservation and development.

Pre-Colonial Understanding of Sustainability

During the pre-colonial period, protection and use of heritage sites, including exploitation of resources within the immediate environs, was balanced through TMS. There are linkages between local experiences and SD concept in Africa. Perpetuating global needs without integrating them into local ones is likely to be catastrophic and a scientific error. Heritage remains a local phenomenon before it becomes a universal one. Furthermore, the abstract transfer of knowledge on SD in the present lacks the social and empirical context provided by TMS, in which the medium (language and approach) of expression and the opportunity to practice the acquired knowledge was in real time.

An example of how TMS encouraged sustainable approaches is how for centuries, local communities in Africa practised shifting (rotational) cultivation and nomadism. The latter is still prevalent with some communities in Africa, although this is being curtailed by governments through forced sedentary lifestyles. With shifting cultivation, the land would be utilized for subsistence farming for years and then left fallow for about 10–25 years to regenerate for future uses (Grogan *et al.*, 2013). Alternatively, it is described as "infield-outfield system by transporting nutrients from the extensively utilized outfield to the intensively cultivated infield" (Grogan *et al.*, 2013: 78: Stromgard, 1989). Various terms are used to refer to this approach, among them is the *chitemene* (Zambia) and *ntemele* in Tanzania (Itani, 2007; Stromgard, 1989). Understandably, this system would have induced negative impacts such as deforestation, nutrient leaching, and soil erosion, but the principle of shifting demonstrates the ingrown mentality of sustainability in TMS. It could have also led to the "transition of shifting cultivation towards more intensive land use practices" as population increased and colonialism forced communities to create reserves outside areas fertile for agriculture (Grogan *et al.*,

2013: 77). This demonstrates adaptation capabilities embodied in TMS and communities themselves facing changing lifestyles due to colonialism, external factors, and pressure from national governments enforcing development across the nation. External factors may include "growing population, government policies, and increasing commercialization/market integration" (Grogan *et al.*, 2013: 77).

Clearly, sustainable use embedded in TMS was overtaken by sedentary and commercial farming approaches starting in colonial times, which, to a limited extent, applies the fallowing approach as a rotational system. However, artificial nutrients (fertilizers) replaced traditional manures (e.g., cow dung) in order to increase yields per hectare. While traditional manures are friendly to soils, fertilizers eventually affect soil and water qualities due to their chemical nature. These artificial developments are often hailed as improving living standards (Grogan *et al.*, 2013), however they have negative effects on the well-being of society such as diseases resulting from dietary patterns associated with genetically modified food. This further impoverishes the very communities whose well-being should be maintained. The question is, what is the integrative point between the past and present on sustainability? Commercial agriculture caused large-scale invasion of pristine landscapes, destruction of heritage sites, disruption of biodiversity systems, and animal corridors. The point being made is that development is double edged sword with both negative and positive dimensions. The numerous private conservancies and farm-based sanctuaries that equally emerged are benefiting the rich and international trophy hunters at the expense of local communities. Of course, campfire projects meant to benefit local communities were tried but with limited success in Zimbabwe. Such projects remain as peripheral activities on the edges of either public or private PA.

The failure in harnessing and using sustainable practices from TMS as a foundation of SD discourse at the local level remains a missed opportunity. Departing from a known point (an experience) into the unknown (new future or concepts) is an opportunity of having everyone on-board the SD "flight". The known point denotes the knowledge, processes, and experiences of TMS. Local communities are not mere recipients of new concepts, but rather knowledge and experience producers in their own context. Therefore, localization of SD cannot be effective without negotiating and integrating itself with the knowledge and experiences of such producers, especially if they are to embrace contemporary developments. Neither should the contemporary approach be imposed on them. Local languages and traditional approaches can be used to enrich development of new approaches for a new future. Doing this together with them creates a support base and ownership of new concepts. The hidden opportunity in connecting TMS and SD is that sustainable use of resources is not a new phenomenon in Africa. It is erroneous to always consider everything from the past as "backward" without deliberately extracting principles and guidelines associated with this past towards integrating it into emerging concepts. Such an approach facilitates steps of adjustment, cooperation, and social exchanges in implementing SD as a "tradition-science process" in Africa. TMS offer an opportunity to study how society behaves and influences the world around it. According to Grogan *et al.* (2013: 91), TMS provide a "form of resilience in a highly changeable environment", which points to adaptive capabilities. Such capabilities are opportunities for connecting past and present practices in sustainable use of resources in Africa.

The only challenge to the above connecting opportunity is that SBMS and their experts still have more in common with colonial authorities than with the stakeholders at the local levels of the heritage site (Chirikure & Pwiti, 2008; Mahachi & Ndoro, 1997; Munjeri, 2004; Ndoro & Wijesuriya, 2014; Taruvinga, 2008). This includes their allegiance to the Convention (Ndoro & Wijesuriya, 2014). This colonial legacy represents a mosaic of intertwined and sometimes conflicting factors, which have to be disentangled to create a conducive environment for decolonized models that involve and empower other stakeholders (Munjeri, 2002; Muringaniza, 1998; Pwiti & Ndoro, 2006; Taruvinga, 2019; Taruvinga & Ndoro, 2003).

National Sovereignty and Decision-Making

National sovereignty assumes that matters at local levels are domestic issues (Nicholson, 2002; Shackelford, 2008). In terms of the United Nations Charter of 1945, there was no "legal right at international law for other states to intrude into matters of purely domestic concern" (Nicholson, 2002: 182). For example, in interpreting how the law of mining is applied to the concept of "common heritage of mankind", there are vastly different and controversial approaches taken in mining in high seas, objects in space, in the Antarctic, and World Heritage areas (Nicholson, 2002). Ongoing debate over this aspect is twofold. One side argues that cultural property belongs to the sphere of national law often called the "cultural nationalism approach", while the other argues that cultural property is part of the common heritage of mankind, thereby justifying the World Heritage Convention approach (Müller, 1998: 396; Shackelford, 2008). Cultural nationalism is based on legal instruments and interests at national level, while the common heritage of mankind is driven by international redistribution instruments that create universal access in favour of those who were not able to access such sites (Nicholson, 2002). This creates tension between the local and global perspective frameworks in dealing with local challenges. Some scholars argue that World Heritage brings about economic restrictions for developing nations through its prioritization of conservation over other interests, while others argue that international cultural heritage laws are "powerless paper tigers that lack enforcement mechanisms" at national levels (Müller, 1998: 404). Therefore, the idea of common heritage should be related to national legal status and interests (Müller, 1998; Nicholson, 2002).

Both the nation-state and the international community have legitimate stakes in cultural heritage protection, and both could be suitable actors on different issues (Müller, 1998). Any sovereign state has the power to define and enact appropriate regulations to protect her heritage, including how it is used (Müller,

1998). But sometimes governments are more concerned with their relative position on the international level and do not care so much about morals or issues of legitimacy (Müller, 1998). Also, what is still lacking between the opposing sides of this debate are the views and opinions of stakeholders that will guide both politics and jurisprudence (Shackelford, 2008). Without a binding resolution or legal effect on the above matters, the international rule of law could be subverted and replaced by dissident state action, confrontation, and disunity (Nicholson, 2002). This is complicated by the fact that developing nations are seeking a new and more equitable global economic order through mutually beneficial investments (Nicholson, 2002). Understandably, the modern world is reliant on commercial resource development, but parameters within which this can occur are yet to be finalized against competing global, national, and individual priorities (Nicholson, 2002).

Tradeoff Perspectives; "Aswan High Ground"

The 1959 Aswan Dam trade-off incident offers interesting dimensions on protection and developments at heritage. The dam was constructed along the Nile river to support agriculture and hydropower and improve navigation in Egypt (Abu-Zeid & El-Shibini, 2010). The project threatened to flood the Abu Simbel temples which were located along the Nile river and within the flood zone of the dam (Donnachie, 2010). In response, Egypt and Sudan appealed to UNESCO for assistance, and as a result, a UNESCO multi-million-dollar rescue conservation programme to save this heritage was undertaken (AKTC, 2005; Donnachie, 2010). The intensive archaeological rescue excavation recommended "the removal of the monuments that were under threat from the flood line" (Donnachie, 2010: 118). This decision paved the way for the dam construction. The rescued monuments were "re-constructed" on a higher ground but in the general environment of their original location (Donnachie, 2010). This relocation demonstrates that heritage can facilitate development without sites being lost. SBMS should not always freeze development in favour of conservation without exploring possible trade-off bridges like what happened with Aswan Dam.

Furthermore, the Aswan Dam is a testament to how such and related projects are associated with numerous technical, social, and political problems that affect stakeholders despite the benefit accruing from intensified agriculture (Kashef, 1981). The Aswan Dam is argued to have increased transport productivity, summer crops productions, and created many downstream jobs in Egypt (Strzepek *et al.*, 2008). However, the negative impacts of the dam after its construction remain a concern to various authorities (Kashef, 1981). Many technical and non-technical measures have been developed to ameliorate the negative impacts of dams post this Aswan Dam project (McCartney, 2009). The successes of these measures have not been properly assessed due to either technical reasons or as a consequence of a variety of socio-economic constraints (McCartney, 2009). Despite this, Aswan dam symbolizes a local need that became an international crisis requiring the intervention of all affected and interested stakeholders, and this was done without prior positions on the matter. UNESCO was fully behind this progressive solution and was supposed to build on this experience with the Convention right from the beginning. This negates that heritage has always been sensitive to development needs since pre-colonial period. Furthermore, this recorded large-scale trade-off between conservation and development has not been replicated with success at any other sites facing similar challenges as SBMS have become much stricter, if not narrow minded, in their approach on such issues. Neither has an attempt been made to build on the stakeholder consultation processes that resulted in the trade-off at Aswan Dam to deal with similar local challenges. In the process, voices of the state actors and global communities are more pronounced on these matters, while the opinions and views of non-state actors (stakeholders) are hardly given a chance in this process.

Though relocation of heritage outside its original setting to a new place has its own implications, the bottom line is that the re-assembled temple represents the original values and related attributes, but in a different setting within the same broader environmental settings. This is similar to moveable African collections that have found second homes in museums worldwide. Such collections are very far from their owners and original places, yet their significance has diminished as evidenced by increasing repatriation and restitution processes back to Africa.

Similarly, proposed water abstraction (dam) along the rift valley system (Lake Turkana) connecting Ethiopia and Kenya has faced the "resistance" from UNESCO. International community considers the project to be a threat to the Lake Turkana ecosystems (38th Session, UNESCO, 2014). Therefore, the WHS status of Lake Turkana is in danger according to UNESCO. From a societal perspective, the dam is critical to the livelihoods of stakeholders, among them communities and farmers. This demonstrates tensions between local realities and World Heritage in developing nations. Local needs are still superseded by conservation in Africa. This approach is replicating itself and continues to manifest on the African continent through the implementation of the Convention. Such water harnessing developments are now being used to persuade SPs and communities that conservation overrides other local needs (Lane & Ersson, 2016). The Mohenjo-Daro project (Pakistan), Borobudur temple (Indonesia), and the Cultural Triangle of Sri Lanka have faced similar international pressure to prioritize conservation (Donnachie, 2010; Ndoro & Wijesuriya, 2014). In China, the construction of the Three Gorges dams to reduce dependence on dirty coal and to address increasing energy demands has also faced similar international pressure (Ndoro & Wijesuriya, 2014).

Exclusionist Approach: Marginalizing Stakeholders

While regional and international recommendations on conservation and SD have been made over the years, their implementation has been very slow (Taruvinga, 2019). For decades, the World Heritage Committee has been engrossed with debates on conservation, which continues to frustrate stakeholders interested in development initiatives at WHS. The 2015 World Heritage Policy on Sustainable Development now bridges this gap by advocating coexistence between the two aspects. This frustration among stakeholders denotes that there have been no systematic and inclusive engagement processes involving them in developing pragmatic solutions addressing such issues (Taruvinga, 2019). Stakeholders expect benefits beyond the concept of World Heritage (Galla, 2012). This implies that there are many actors and forces, as well as aspirations, existing outside the limits of conservation (Brumann & Berliner, 2016; Chirikure, 2014) Failure to embrace actors, forces, and aspirations makes "local communities become heritage victims" (Meskell 2009: 11). It is clear that national authorities, WHC, local communities, extractive industries, private sector, agriculturalists, financiers, bankers, politicians, and civil society do not have a common approach on conservation and development (Taruvinga, 2020b).

It is clear that WHS are failing to involve stakeholders, either through deliberate exclusion or just benign neglect (Jokiletho & Cameron, 2008). This has caused local communities to sometimes campaign against World Heritage designation because it is a "threat to various aspects of their ways of life" (Jones & Shaw, 2012: 83). Furthermore, stakeholders are not motivated to participate due to lack of confidence in the SBMS. Furthermore, stakeholders are frustrated with the laborious and drawn-out consultative process of SBMS, including the gradual erosion of collective values of society in comparison with an increasingly marked individualism, and varied interests regarding land uses at WHS (Albert *et al.*, 2012; Lane & Ersson, 2016). This has led to conflicting claims and interpretations on the role of World Heritage that occur outside the formal discussions of World Heritage (Cameroun, 2014). Failing to consider the realities and needs of stakeholders is a major stumbling block in the proper application of World Heritage approach in Africa (Chirikure, 2014; Taruvinga & Souayibou, 2013).

Culture Statistics: Towards Measuring Impact of Conservation and Development

The absence of quantitative data on the relationship between conservation and development is a challenge for making informed decisions, in particular how they affect and contribute to each other at WHS in Africa (Taruvinga, 2019). SBMS reports on WHS are mainly qualitative in nature and largely report positives on conservation while highlighting development threats (Taruvinga, 2019, 2020). For instance, the Second Cycle of Periodic Reporting of the Africa region conducted in 2011 stated that 21% of the cultural properties in Africa were affected by extractive processes (WHC 35, COM 7C). The Third Cycle of Periodic Reporting of the Africa region of 2020 maintains this negative qualitative trajectory. Furthermore, the 2013 IUCN study on World Heritage and Extractive Industries observed that seven sites were "spatially coincident or proximal to producing extractive operations", while eight sites were spatially coincident or proximal to active explorations/development operations" (Turner, 2013: 31). However, the IUCN study did not consider "local environmental conditions nor the impact of operating on them" in order to give a total picture of what is actually happening (Turner, 2013: 32). Overall, the quantitative data on World Heritage has a strong bias towards the impact of development on conservation indicators. Statistics are scarce on how conservation contributes to socio-economic development and vice versa. This matter is generalized in tourism statistics projected against set development indicators of an SP. The absence of quantitative statistics to back up generalized qualitative statements creates an environment that pits conservation against development in developing nations (Taruvinga, 2019). Quantifying the reciprocal benefits of conservation and development through "culture statistics" (quantitative indicators) is now a priority at WHS (Turner, 2013: 22).

Conclusion

The discussion demonstrates that conservation and SD are inseparable stakeholder-driven processes, which are also as old as humanity itself as illustrated by pre-colonial Africa. The post-colonial heritage management approaches in Africa have not significantly differed from those of the colonial period, which have equally failed to embrace the experiences and approaches of the pre-colonial era. Pre-colonial period witnessed protection and use of heritage, including other resources around sites through TMS. Today, this process has become the privy of SBMS backed by multiple legal frameworks. This skewed approach, though being slowly addressed, is further reinforced by international instruments such as the Convention. While conservation remains the agenda of SBMS, there is growing interest in socio-economic benefits, and this cannot be ignored anymore. This pattern, rooted in pre-colonial practices, is unlikely to change in the future, and can only get intensified with the growing population of Africa. The application of bridging concepts such as SD has not entirely created a conducive environment for involving stakeholders outside the conservation realm, largely because the top-down governance approach of SBMS and UNESCO persists unabated. Neither has conservation and SD been fully integrated into each other as stakeholder-driven processes in Africa. This, coupled with not embracing stakeholders in their broadest sense, is threatening to alienate conservation from the well-being of the society. The stakeholders, supposed beneficiaries of both conservation and development at WHS, cannot continue to be marginalized anymore. As such, the multiplier effect of WHS on the economy of SPs and its link to the localization of SDGs is not well understood nor is it effectively facilitated in a manner that make stakeholders become partners and beneficiaries at the same time (Taruvinga, 2019). Ignoring the wider context of development, environmental implications, and consequences for the lives and aspirations of stakeholders, WHS can be misconstrued as being against people (Lane & Ersson, 2016). Heritage governance is persuaded not to automatically work against both development and stakeholders at WHS given the increasing calls for heritage to assist in meeting the well-being of the society.

3

MULTIPLE AND MULTI-LAYERED STAKEHOLDER FRAMEWORK AND WORLD HERITAGE GOVERNANCE

Introduction: Towards Stakeholders Approach

Recognizing the existing local and global stakeholder tensions resulting from their varying perspectives on conservation and development at World Heritage Sites (WHS), the "Multiple and Multi-Layered Stakeholder Framework" (MMSF) is introduced in this chapter as the preferred governance approach for managing stakeholders at WHS. It is developed from stakeholder management theory and experiences at WHS. "Multiple" is defined as the engagement and involvement of more than one individual or group of stakeholders, while "Multi-layered" is defined as the hierarchical levels of engagement by multiple stakeholders at WHS. Stakeholder engagement is understood to happen at the local, regional, national, and international levels of WHS. Simplified, MMSF refers to the "many" and "overlaid" stakeholders with multiple interests and operating at different levels of the WHS. With this understanding, stakeholder perceptions, interests, legal mandates, power matrix, roles, responsibility, relations, and ability to influence decision-making processes at WHS are contextualized in this chapter to show its effectiveness as an alternative governance approach. The terms "Multiple" and "Multi-layered" are used in their combinative nature to denote the complexity of stakeholders and their role in balancing conservation and development at WHS. MMSF illustrates that WHS governance is confronted with growing complexities and needs to align with the socio-economic context in which it operates. This lays the foundation of gathering empirical perspectives of stakeholders on conservation and development at WHS in Africa.

DOI: 10.4324/9781003149361-3

Heritage Governance "Scaffolding" in Africa

The interpretation that heritage sites are only relevant to local communities changed with colonialism, and there is a growing list of other players interested and affected by heritage sites in Africa. In discussing role players at heritage sites, different scholars have used different terms such as local communities, traditional custodians, communities, value-based stakeholders, indigenous communities, archaeology communities, and many others (see Jopela, 2016a; Ndoro & Wijesuriya, 2014; Pikirayi, 2016; Sinamai, 2003; Taruvinga, 2007). All these have become stakeholders operating at varying levels and, equally, with varying powers and legitimacies at heritage sites. Stakeholders are defined as "individuals or groups who depend on the organization to fulfil their own goals" (Johnson, 2002: 206). In South Africa, some legislations define stakeholders as a person, an organ of state, or a community contemplated in section 82(1)(a), or an indigenous community contemplated in section 82(1)(b) of the National Environmental Management: Biodiversity Act (Act No. 10 of 2004) (NEM: BA). The term "stakeholder" has its origins in the 1930s with four key parties being identified: "customers, employees, community, and stockholders" (Preston & Sapienza, 1990: 362). In this book, local communities are considered to be a part of stakeholders. Stakeholders are connected to each other through their geographic location, shared interests, values, experiences, politics, or traditions or cultures (Pikirayi, 2016). Overall stakeholders "have interests, rights or ownership in an organization or its activities" and a claim for resources (Bryson, 2004; Hellriegel *et al.*, 1999). Nowadays, geographic location of stakeholders has a digital dimension supported by Information Communication Technologies (ICT). As such, stakeholders are "fluid, continuously replicating and transforming", meaning they are not fossilized in the past (Pikirayi, 2016: 125). Therefore, managing stakeholders requires good governance approaches.

The term Indigenous and Descendant Communities (IDCs) is deliberately used to acknowledge value-based communities who are connected with the multiple and multi-layered values of a heritage site (Pikirayi, 2016). IDCs are those people with "ancestral connections to a particular site or landscape" (Chirikure *et al.*, 2015; Pikirayi, 2016: 116). These are considered "local descendant communities" if they still reside at the site (for example, the custodians of the Mijikenda Forests in Kenya), while "non-local descendant communities" refers to those who are now living in another geographical location due to other reasons, among them impact of colonialism and creation of Protected Areas (Pikirayi, 2016: 116). The latter characterizes much of Africa, given the history of colonialism, land alienation (forced removals), and the failure of post-colonial land redistributions (Huffman, 2014; Ndoro & Wijesuriya, 2014; Pikirayi, 2016). Both local and non-local descendant communities have developed a "corpus of knowledge" through interacting with the landscape over time (Evans, 1995; Pikirayi, 2016). This makes IDCs indispensable to the process of building knowledge and managing their indigenous knowledge systems at heritage sites (Pikirayi, 2016).

Before delving into stakeholder management approaches and practices, it is important to understand how stakeholder management has been perceived from pre-colonial to present times. IDCs in Africa had systematic and community-endorsed decision-making practices, but these was disrupted by colonialism, which legitimized State-Based Management Systems (SBMS), and heritage experts. Colonial governments assumed there was no experience on good governance prior to colonialism in Africa. This is an erroneous interpretation and conclusion, as colonialism weakened community-driven governance systems which already existed in Africa. This African experience should not just be viewed as an African historiography of the pre-colonial times but rather a learning curve on how community-driven decisions were sustained through generations. Of course, one shortcoming of colonialism was the language barrier between the colonialists/imperialists and the colonized, which made it difficult for them to have mutually beneficiary dialogues. This barrier exacerbated the observed historical injustice (Africa as backward), yet this is resulting from the failure of the colonialists to understand Africa in its evolving context.

IDCs remain peripheral players and victims of both colonial and post-colonial governance systems. Post-colonial governance approaches emerged as reforms designed to address bad leadership and poor choices of African leaders (Kauzya & Balogun, 2005). Poor leadership syndrome is often cited as one of reasons why Africa remains poor (Poncian & Mgaya, 2015). These reforms include establishing the African Leadership Council, introduction of the Code of African Leadership with 23 commandments, Mombasa Declaration on better Leadership, and innumerable governance training programmes on the African continent (Rotberg, 2004). Of late, this has been augmented with developments such King Codes (I-IV) on Governance in South Africa, as well as the intensive capacity-building programmes on the same, instilling good governance in Africa. Demonstrating good governance is a condition of multilateral and bilateral partners (such as the International Monetary Fund, World Bank, United Nations, European Union, and recently Brazil, Russia, India, China and South Africa alliance (BRICS) among many others) in disbursing financial and aid support. Good governance has become a development condition from developed nations investing in Africa, yet it is disconnected from the history and valuable experiences of Africa in the pre-colonial era (Poncian & Mgaya, 2015).

Stakeholder management is underpinned with evolving concepts and principles of good governance. However, stakeholder management is an evolving process providing both endless opportunities and challenges. In order to address contextualizing stakeholder management in this discussion, the MMSF is proposed. MMSF combines pre-colonial experiences and evolving stakeholder concepts. The pioneering spirit and wisdom of the pre-colonial experiences, which were sidelined in the wake of colonialism, should have been adapted by SBMS to give birth to MMSF deeply connected to community-driven governance systems and principles of the pre-colonial times. This could have become the "governance scaffold" underpinning heritage management in Africa, meaning building from past to the present experience. Pre-colonial, colonial, and post-colonial experiences are viewed as steps of progressive and evolving heritage governance systems: "scaffolding" being a teaching notion derived from an approach where successive or incremental steps/ladders are used to induce total understanding of a phenomenon. Our understanding of the successive pre-colonial, colonial, and post-colonial governance systems advances MMSF, being an inclusive stakeholder management approach capable of bringing together stakeholders to a point of common understanding, even though varying and opposing decisions may be taken.

Pioneering Spirit and Experiences: Pre-Colonial Governance in Africa

Pre-colonial Africa remains a rich didactic source of important lessons, which if were adapted to the evolving operational environment and external influences, the continent would be way ahead in balancing conservation and development. There are numerous and credible examples of traditional leadership and governance that evolved over time, "developing institutions and styles of leadership suited to the realities of the time" in Africa (Mkapa, 2010: 25). This was supported by "social and cultural norms, political structures, and processes, as well as philosophical underpinnings of governance, that could find relevance and applicability in modern Africa" (Mkapa, 2010: 26). While post-colonial leadership challenges may be complex and multifaceted compared to the pre-colonial period, lessons could have been drawn from the latter to benefit the former (Poncian & Mgaya, 2015). Colonial governance, backed up with draconian legislation and highly structured administrative systems, weakened and impoverished communally driven governance systems of the pre-colonial era, and sometimes, practicing them became a punishable crime in secular and civic courts. The powers of traditional leaders and their support systems were drastically reduced to induce their submission to colonial governments. Furthermore, communally driven conflict resolution approaches gradually lost their niche to secularized options.

Communally driven governance systems of the pre-colonial had evolving principles and pillars. Leaders in pre-colonial Africa exhibited "high level of thinking and planning" for the protection and development of their respective societies (Poncian & Mgaya, 2015: 109). For example, emperors and empire building in West Africa (Ghana, Mali, and Songhai) illustrate "a high level of sophistication of pre-colonial leadership and institutions" (Falola & Aderinto, 2010; Poncian & Mgaya, 2015: 109). Emperors controlled and participated in the Trans Saharan Trade taking advantage of their natural resources such as gold, thereby creating an enabling environment which gave birth to inter and intra-empire, as well as intraregional trade (Poncian & Mgaya, 2015). The intraregional trade birthed the interregional Trans Saharan Trade, which assisted in revenue generation for these empires (Poncian & Mgaya, 2015). This trade attracted "foreign expertise, technology and knowledge" which benefited their societies (Poncian & Mgaya, 2015: 110). The use of knowledge and expertise from outside by pre-colonial leaders allowed for the transfer, exchange, and adaptation of skills for the benefit of pre-colonial kingdoms (Poncian & Mgaya, 2015). For example, Mansa Musa from Mali is credited for bringing Muslim scholars, artists, scientists, and architects that led to the "expansion and administration of the empire" (Poncian & Mgaya, 2015: 111). Learning and adapting skills from outside Africa should benefit the local situation rather than perpetuating stereotypes at the expense of contributing to the well-being of the society in Africa.

Similarly, in Southern Africa, Great Zimbabwe (Zimbabwe), Manyikeni (Mozambique), and Mapungubwe (South Africa), played similar but critical roles for the protection and development of their own societies. Furthermore, pre-colonial leaders diversified economic base beyond trade to include but not limited to crafts, fishing, crop cultivation, metal extraction, livestock keeping, and cotton weaving (Craig *et al.*, 2012). Revenue collection from taxes ensured that pre-colonial empires had resources to continuously improve the well-being of the societies (Chijoriga, 2012; Poncian & Mgaya, 2015). Economic diversification guaranteed prosperity of societies and their pre-colonial kingdoms (Conrad, 2005; Poncian & Mgaya, 2015).

Another important, but unexplored dimension of pre-colonial leadership is how leaders avoided autocratic decision-making styles (Poncian & Mgaya, 2015). Pre-colonial leaders, despite wielding "enormous powers", exhibited democratic elements (Poncian & Mgaya, 2015: 111; Chimakonam *et al.*, 2014). This was achieved through balance and checks by citizens, as well as their involvement in the governance systems though defined structures (Ayittey, 1992). Pre-colonial leaders were bound to "serve their people, observe customs and traditions", failure to which the very same people they are leading could remove them from positions of authority (Poncian & Mgaya, 2015: 111). Traditional councils acted in response to public opinions while ensuring fairness and transparency (Ayittey, 1992). Also, successors to the positions of authority in pre-colonial period were not allowed to lead the process of their own installation, lest they manipulated it against community democracy (Ayittey, 1992). An opportunity for defending oneself before a judgement was passed was equally provided for (Ayittey, 1992). Failure to observe these communally accepted governance principles would result in rebellions as a recourse to failed justice (Ayittey, 1992). One can safely argue that "democratic governance in Africa predates the western based governance models" (Poncian & Mgaya, 2015: 111). Therefore, colonial governance models should have considered the pre-colonial experiences of Africa. Rarely do you find contemporary governance manuals referring to this rich context and wisdom-enshrined community-driven governance system characterized with consensus, inclusive and unanimity principles (Chimakonam *et al.*, 2014).

Traditional Management Systems (TMS) have a unique strength and practicality in engaging and involving young people as they became of age to ensure that they grow wisely into "decision makers of their own future". This demonstrates how intergenerational gap between the elders and young people was reduced through structured and practical oriented dialogues. Because of this latitude, young people would not find it difficult to consult elders on critical issues but still make their final decisions. TMS used expressive idioms and proverbs in developing a wisdom-filled mindset in these young people, yet today, SBMS and their experts suppress this approach, largely because they view young people as inexperienced, competitors, and fast decision-makers. This has created an intergenerational skills gap in Africa, hence the highly polarized governance system, as the older generation is unwilling to give their decision-making powers to younger people. Leaders were appointed to lead people into "prosperity, freedom, equality and development", meaning power resides in people rather than leaders (Poncian & Mgaya, 2015: 112). The collapse of great empires in Africa should not eclipse the lessons and values enshrined in pre-colonial governance systems.

Appropriation and Exclusionary Governance: Perpetuating Colonial Approaches

As argued in the previous chapter, colonialism disrupted and usurped pre-colonial governance systems and ushered marginalized governance approaches supporting SBMS. There was accelerated trans-positioning of western governance models in Africa. This includes colonial privileges and powers that were left intact in former colonies on the eve of democracy (Cock & Fig, 2000; Eboreime, 2009; Hughes & Harrison, 2010). Though democracy was both ballot and negotiated settlements, many of the explicit and implicit compromises on privileges and governance power in Africa have not been fulfilled at all (Taruvinga, 2019). SBMS governance cannot remain technical, standardized, and non-flexible, yet it can benefit from other stakeholders (Chirenje *et al.*, 2013; Waterton, 2010). These stakeholders are not afforded a chance to say what will work and not work for them in their local conditions (Simmons, 1994). Their potential to become part of the solution to local issues is not fully exploited by SMBS (Cole, 2006; Tosun & Timothy, 2003). As such, state actors retain a monopoly over heritage and its use in Africa (Hughes & Harrison, 2010: 247; Ndoro & Chirikure, 2009; Ndoro & Wijesuriya, 2014).

Within the ambit of SBMS governance systems, stakeholders are consulted on a need rather than an ongoing basis. Such consultations do not consider the merit of their input in the decision making processes, except to rubber-stamp what has already been decided by state actors. The "need consultations" are "legally fortified" and used to funnel information for noting to stakeholders. This constitutes the long-established and sustained biased top-down approach, where stakeholders play second fiddle to SBMS. This means stakeholders are not included from conceptualization stage, nor in the implementation phases of heritage management unless there is legislative and political pressure inducing such an approach. As a solution, the bottom-up approach is viewed as facilitating decision-making as it takes into consideration the local needs (Chirenje *et al.*, 2013). Heritage reforms in Africa are slower than the manifesting aspirations of stakeholders and their eagerness to be part of the solution to their own local challenges (Taruvinga, 2020b). Furthermore, they are slower than the pace at which good or corporate governance has developed in other sectors in Africa. However, good heritage laws without the holistic and inclusive heritage reforms are not necessarily a panacea to conservation (Munjeri, 2004b) and development issues at heritage sites (Chirikure & Pwiti, 2008; Ndoro & Chirikure, 2009, 2014; Ndoro *et al.*, 2018). However, in younger democracies such as South Africa, the reforms were faster as the National Heritage Resources Act (NHRA) of 1999 was introduced only 5 years after apartheid was dismantled, whereas in Zimbabwe, despite attaining independence in 1980, reforms only commenced in 2018. However, most reforms are not drastic, opting to offer cosmetic changes, which reinforces the already known governance stereotypes and their exclusionary tendencies (Taruvinga, 2019). The objective and intensity of participation, consultation, decision-making framework, initiation action, and the instruments used, influence the level of stakeholder participation (Blanchard & Trotter, 2001; Charron, 2007; Fowler, 2002; Maradze, 2003; Mazibuko, 2000; Paul, 1987). These processes and instruments of engaging should recognize and stakeholders, as well as their level of operation: local, district, national, continental, and international scales (Fowler, 2002: 82; Samantha, 2012). These levels vary based on the legal status and context of a heritage place.

Stakeholder participation has different meanings to different actors, resulting in many terms being coined to explain what it means (Chirenje *et al.*, 2013). These terms include participation, involvement, integration, consultation, community-based management, co-opted, cooperation, collaborating, co-learning, and collective action (Chirikure *et al.*, 2010; Chirenje *et al.*, 2013).

In the last 20 years, cultural heritage has become significant in different ways to a much broader range of stakeholders beyond local communities (Meskell, 2014). Such stakeholders cut across the socio-economic and political spectrum of the society, including the still ill-defined digital global communities that did not exist prior to 1994 (Prieto *et al.*, 2014). These stakeholders sometimes agree or disagree over any aspect of the heritage processes, especially when management plans limit or even preclude "non-conforming" types of socio-economic development at WHS (Jones & Shaw, 2012; Tunbridge & Ashworth, 1996). Also, working with such different stakeholders is not always straightforward, and can create both ethical and practical dilemmas in the process (Prieto *et al.*, 2014). Ethics of consultation, identifying different legitimate voices, weighing the competing interests, and maintaining connections remain as challenges for SBMS (Prieto *et al.*, 2014). Another challenge is how far back one can go in defining or identifying stakeholders given that they operate at very different levels of social and political complexity in time and space (Jones & Shaw, 2012). In addition, determining the geographic or political boundaries of stakeholders is challenging due to the globalization of socio-economic systems (Harper, *et al.*, 2003). It is evident that physical, social, political, and economic boundaries of WHS are becoming fluid and dynamic in nature, as many stakeholders identify with sites in one way or another (Harper, *et al.,* 2003).

Universalism: World Heritage Governance

While the governance framework of the Convention recognizes stakeholders, it does not give them decision-making powers and same representation as States Parties (SPs) and Advisory Bodies (ABs). The OG identifies local communities, indigenous peoples, SPs, local and regional governments, non-governmental organizations (NGOs) and private organizations, as well as landowners as stakeholders (Operational Guidelines, 2019: para: 12, 40, 64, 119, 123). As alluded in the previous chapter, the World Heritage Committee is the decision-maker. This decision-making system is further supported by some NGOs with an appetite for conservation (Labadi, 2018; Meskell, 2014). However, the Convention bestows the responsibility of managing World Heritage on SPs (Fowler, 2003). All other stakeholders are viewed as observers by the World Heritage Convention (WHC) or are supposedly represented by the SPs (Fowler, 2002). This makes it difficult for stakeholders to be taken aboard on World Heritage governance (Meskell, 2014). This decision-making system somehow contradicts one of the five Cs of the Convention, which refers to "communities" as these are synonymous with stakeholders (Rossler, 2006, 2010). The other four Cs are: credibility, conservation, communication, and capacity. With mounting pressure from "stakeholders" at grassroots levels in Africa, including how it is increasingly manifesting at WHC sessions, the decision not to involve them is becoming increasingly difficult. Stakeholders are widely accepted as "institutional political economies and capillary networks of power" that underlie World Heritage processes (Meskell, 2014: 221).

While WHC is the administrator and convener of the Convention, it is increasingly becoming clear that real power lies with the SPs, who can decide to comply or not comply with its decisions. Bilateral and multilateral political networks are now used to politicize the WHC decision-making process (Meskell, 2014). In the process, the SPs abuse their privilege of representing all stakeholders through their fortified and jealously guarded corridors of power, which excludes the greater society (Taruvinga, 2019). What SPs exhibit at Committee meetings mirrors the approach at national levels. Neither can heritage governance be separated from politics as an inherent dimension of social complexity in Africa. As such, discerning voices at the local levels are usually muffled, while those with resources embark on litigation to have their opinion considered by WHC, SPs, and UNESCO. The question that remains is, if the SBMS cannot effectively represent other stakeholders, how many of them can sustain such representation at their own cost, given the scarcity of resources compared to the annually replenished budgets of SBMS? Observer status and indirect representation come with limited participation and non-existing voting rights in the process.

Where matters from stakeholders are too sensitive and have political repercussions, the WHC often encourages SPs to deal with them, similar to how UNESCO refers back to the SPs all complaints lodged with it by stakeholders. This is against a legal interpretation aligned to an understanding that the SPs are the signatories to the Convention and have a legal obligation to resolve domestic issues at WHS, however, without compromising the spirit of the Convention (Taruvinga, 2020b). The question then becomes, is UNESCO only interested in the scientific thread of outstanding universal value (OUV) and not the politics and needs associated with the greater society on the same matter? Or is it a situation whereby the governance and management of WHS have not developed to a level where they can accommodate the greater society as rights or shareholders to cultural and natural values now being perceived as World Heritage? Or is it a case where UNESCCO has to maintain global networks and funding partners through governance exclusions and universalism, while risking the future of the WHC? However in this debacle, SPs can be double-edged swords depending on the matter under discussion. It is argued that "within the UN family, nation-states are impossible to bypass or dislodge whether one is talking about human rights or heritage rights, and, in cases of doubt, political sovereignty often takes precedence over the protection of people or things" (Meskell & Brumann, 2015: 24). The conduit approach of UNESCO and WHC may not work well in the long term, as stakeholders want an equal voice in the governance of the Convention. While "multivocality muffing" is taking place at both local and international levels, stakeholders are always exploring opportunities to push their agendas (Taruvinga, 2019). This is because the "one-blanket-fits-all approach to World Heritage" stifles alternative and localized voices in the governance of WHS (Chirikure, 2014: 11). This approach imposes decisions rather than bringing together the views and expectations of different role players to a point of congruency on conservation and development at WHS (Taruvinga, 2019).

Stakeholders require capillary action over broader territories that are "representing one among other legitimate societal interests, all deeply connected within the bio-cultural continuum" (Boccardi, 2015: 94). This includes considering people-centred and rights-based approaches as part of this capillary action (Larsen & Wijesuriya, 2015). Capillary action amongst stakeholders promotes cohesion, sharing common and varying issues while dissecting the intersecting challenge(s) for the benefit of all concerned. The social fabric, common human needs, and moral values of the society should be the foundation of this capillary action.

As such, the Convention is a complex global governance system, however, with SPs, who often exercise political sovereignty (Meskell & Brumann, 2015). This is escalated by emerging "new geopolitical alliances and power bases" that "challenge the perceived hegemony of European nations within the organization" (Meskell & Brumann, 2015: 33). Such alliances have short-, medium-, and long-term cooperating objectives, thereby resulting in shifting alliances influenced by political, regional, religious, or economic linkages among and between

SPs (Meskell & Brumann, 2015). The geographical axis of representation within the UNESCO family has dramatically shifted and witnessed Asian and African SPs coming on board and challenging the old order (Meskell & Brumann, 2015). In this matrix, and equally complicating decisions at Committee sessions is that, SPs are first and foremost local and regional before they become 'universal' members of UNESCO. Also, they have a dual responsibility of balancing conservation and development needs at national levels. On the other hand, UNESCO has limited powers in imposing effective punitive measures "upon a recalcitrant nation state, even when government promises made before inscription are not honored afterwards" (Meskell & Brumann, 2015: 27). Historically, and within the punitive measures available to the Committee, only two sites delisted at the instigation of respective SPs due to conflict between conservation and development. Recently, the Committee exercised its power by delisting the Liverpool Waterfront (United Kingdom) for the same reasons. UNESCO has no power except to oblige to such requests or enforce perennial advisory and statutory reactive monitoring missions to the affected site. UNESCO always recommends remedial actions as a conciliatory position to assist the SPs in complying and remaining committed to upholding the spirit of the Convention, but the SPs have the final say, exercising sovereign rights.

SBMS, UNESCO, WHC, and ABs should not remain isolated and insulated against stakeholders. The question is why would one alienate themselves from the stakeholders that they are meant to serve? How does it benefit heritage to exclude them, yet they may provide alternative and creative solutions at heritage sites? Why does it appear that the pioneering spirit and approaches of indigenous communities seem ignored like an abrupt, once-off phenomenon or process, yet they have all the ingredients of resilience, effectiveness, innovation and in-built adaptive capabilities? This demands a different governance system as a sign of deliberate fundamental and inclusive change to the current exclusionary one.

Inclusivity and Transparency: Democracy and Human Development

Democracy is believed to bring about inclusivity and transparency in governance systems of previously oppressed nations, which should be devolved to the lowest decision-making unity of the social structures. Good governance should always promote rather than retard human development and should be "pragmatic and flexible" in order to remain relevant to the "evolving realities" of the context in which it is applied (Qudrat-I Elahi, 2009: 1169). The United Nations Development Programme (UNDP) defines human development as "a process of enlarging people's choices, achieved through expanding human capabilities and functionings" towards attaining political, social, and economic aspirations and targets (Qudrat-I Elahi, 2009: 1169). UNDP further points out that human development is a goal, while governance is the means to attain the goal and the state, private sector, and civil society constitute the political and non-political role players involved in the process (Qudrat-I Elahi, 2009: 1171; UNDP, 1997). As such, "Governance and human development are indivisible" in eradicating poverty, hence its potential role in balancing the caring of heritage and facilitating the well-being of the society (Qudrat-I Elahi, 2009: 1168). Human development, governance, and key role players need to be an interlinked system if good results are to be produced (Qudrat-I Elahi, 2009).

Therefore, poor governance is a reflection of a failed system of integration between and among the State, private sector, and civil society in mobilizing citizens around common policies and activities regarding human development (Qudrat-I Elahi, 2009). Otherwise, it becomes a case of governing without "general consent" (Qudrat-I Elahi, 2009: 1169). Therefore, the concepts of "governance" and "good governance" require an in-depth analysis (Qudrat-I Elahi, 2009: 1168). This is more important in relation to the stakeholders in balancing conservation and development which interconnects the State, private sector, and civil society in Africa. The two, conservation and development, are inseparable from human development or needs/well-being of soceity. Caring of the well-being of society means "leading long and healthy life, being knowledgeable and having a decent standard of living" (Qudrat-I Elahi, 2009: 1169). This means political, economic, and social opportunities have to be provided to the beneficiaries of this process (Qudrat-I Elahi, 2009: 1169). There has to be empowerment, cooperation, equity, sustainability, and security of beneficiaries and role players in the process (Qudrat-I Elahi, 2009: 1169). All this requires one to delve a little bit into stakeholder management theories.

Stakeholder Management Theory

An effective governance framework is required to respond to the varying needs and complex social and environmental challenges of stakeholders (Louw & Venter, 2013). Effective governance framework is anchored in stakeholder management theory with its two variants: *accountability* being the "duty to provide and account one is held responsible" and *organizational theory* being "the relationship between corporation and its stakeholders (Orij, 2010: 870). Both are underpinned by organizational management and ethics (Phillips *et al.*, 2003). Furthermore, the theory "integrates the resource-based view, market-based view and socio-political perspectives" in order to address the "morals and values of organizational behaviours toward its stakeholders" (YuLong & Hunter, 2015: 250). It is concerned with who has "an input in decision-making" and who "benefits from the outcome of such decisions" (Phillips *et al.*, 2003: 487). In this context, stakeholders are defined as any person or group who can affect or is affected by the activities or achievements of an organization, including decisions, policies and practices, or goals of the organization (Bryson, 2004; Duckworth & Moore, 2010; Freeman, 1984, 1994). Collaboration among these stakeholders brings together different kinds of resources and competencies that may assist in implementing programmes and activities (Louw & Venter, 2013; Phillips *et al.*, 2003).

However, stakeholder theory has been criticized for assuming that the varying interests of stakeholders can be balanced against each other (Blattberg, 2004; Bryson, 2004; Hummels, 1998), and for remaining characterized by hierarchical participation levels (Arnstein, 1969). Theoretically, organizations should be "managed for the benefit of all individuals or groups who have a stake in or a claim" in them (Hummels, 1998: 1404). This makes stakeholder theory indispensable from the strategic management of organizations (Freeman & McVea, 2001).

Governance systems of organizations are continuously challenged to become better and more relevant to stakeholders in Africa. This is against the long-established opinion that "an organization has power, relative to an element of its task environment, to the extent that the organization has the capacity to satisfy needs of that element and to the extent that the organization monopolizes that capacity" (Rhodes, 2007; Thompson, *et al.*, 2017: 30–31). The situation in which the "dominant coalition retains some discretion" and the power to determine "which relationships are seen as a problem and which resources will be sought" is now being challenged from grassroots levels (Rhodes, 2007: 1246). Furthermore, previously ignored stakeholders are challenging the "rules of the game to regulate the process of exchange" in organizations (Rhodes, 2007: 1246). The governance of heritage in Africa, being a coalition of national and international governance systems, is not spared from these challenges. Stakeholders at various levels are advocating for an inclusive and integrated decision-making system to replace the top-bottom approach enjoyed by SBMS and WHC by a bottom-up approach. Regardless of which approach is taken, both need an inclusive and transparent approach for stakeholders. These challenges posed by stakeholders point out that governance is an evolving concept which needs to benefit from constant reviews based on empirical experiences. Governance remains an important process as it facilitates decision-making concerning public life and economic and social development of any nation (Peters and Pierre, 1998).

Governance Principles

Stakeholder management is underpinned by governance principles. The "traditions and institutions by which authority in a country is exercised" defines the term governance (Kaufmann *et al.*, 1999: 1). Others define governance as "the action or manner of governing – of directing, guiding, or regulating individuals, organizations, or nations in conduct or action" (Kaufmann *et al.*, 1999: 1). This includes the "capacity of the government to effectively formulate and implement sound policies" for running the country (Kaufmann *et al.*, 1999: 1). However, governance is "broader than government, covering non-state actors" such as individuals, public, private, and voluntary sectors (Lynn, 2012; Rhodes, 2007: 1246). These are stakeholders. Overall, governance "comprises mechanisms, processes and institutions, through which citizens and groups articulate their interests, exercise their legal rights, meet their obligations and mediate their differences" (Qudrat-I Elahi, 2009: 1169). Governance should "secure the civic rights and freedoms of all people, including the poorest of the poor and marginalized groups" (Qudrat-I Elahi, 2009: 1169; UNDP, 2009).

Governance is as good as the structure and processes that support its application in particular organizational context. Historically, such structures "range from hierarchical decision-making to a variety of decentralized forms of coordination including norms and institutions" (Schneider, 2012: 135). These structures and processes determine who is engaged, how they are engaged, and how it is accounted for (Qudrat-I Elahi, 2009). It is simply about how societies express their interests to decision-makers, including how they become part of the decision-making process itself (Qudrat-I Elahi, 2009). The key actors (Figure 3.1) in any governance structure and processes include the state, public, and private sector (Qudrat-I Elahi, 2009; UNDP, 1997).

The State is the principal actor of government that facilitates participation and provides an enabling environment to other elements of the society (UNDP, 1997). It creates "an enabling environment for sustainable human development" by "establishing and maintaining stable, effective and fair legal-regulatory frameworks for public and private activities" (Qudrat-I Elahi, 2009: 1170). Furthermore, the State provides resources, such as information, technical expertise, research and development programs, physical infrastructure, grants-in-aid, or incentive schemes (Qudrat-I Elahi, 2009; UNDP, 1997). Additionally,

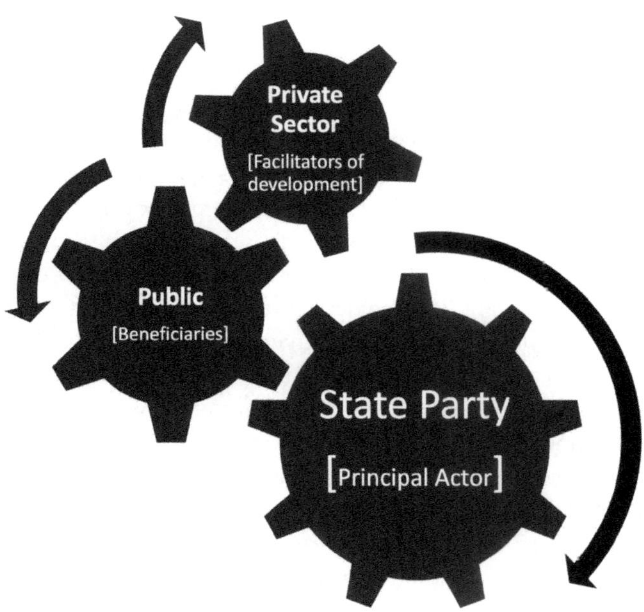

FIGURE 3.1 Key actors in governance structures and processes (Source: Qudrat-I Elahi, 2009)

it promotes stability and equity in the marketplace, conflict resolution mechanisms, encourages cultural diversity and social integration, as well as effective and accountable public services (Qudrat-I Elahi, 2009). On the other hand, the private sector is the facilitator or the engine delivering sustainable human development for the benefit of the society (Qudrat-I Elahi, 2009: 1169; UNDP, 1997, 2009). Private sector is responsible for generating jobs in support of government targets and related initiatives (Qudrat-I Elahi, 2009). This is done through varying and multiple development initiatives, some of which are located at WHS. The State has to establish and maintain a conducive environment as an enabler for the private sector to deliver (Qudrat-I Elahi, 2009).

The Civil society consists of the complex of citizens, individuals, and groups "interacting among themselves socially, politically and economically" (Qudrat-I Elahi, 2009: 1171). It operates between the State and private sector (Qudrat-I Elahi, 2009). These include but are not limited to cultural and religious groups, cooperatives and community development organizations, academic institutions, environmental groups, business associations, political parties, charity organizations, trade unions, non-governmental organizations, learners, and professional associations (Taruvinga, 2019; UNDP, 1997). These are broadly classified as stakeholders. Civil society has an obligation to protect the "rights of all citizens" (Qudrat-I Elahi, 2009: 1171). This is done through mobilizing society to participate in the decision-making processes on issues of sustainable development (SD) or access government social services, including providing their own services in areas unreachable by the public bureaucracy (Qudrat-I Elahi, 2009). Society influences public policies (including access thereof) and also provides "checks and balances on government power" (Qudrat-I Elahi, 2009: 1171). All this leads to the institutionalization of social interaction, thereby defining a nation's social capital (Qudrat-I Elahi, 2009).

Overall, governance is supposed to connect stakeholders in participatory, transparent, and accountable manner. This entails upholding the rule of law (fair and impartially enforcement of legal frameworks), promoting transparency (free flow and access of information directly to those concerned), encourage responsiveness (serving the concerned stakeholders), and push for consensus-oriented decisions among different stakeholders (Qudrat-I Elahi, 2009). Governance equally promotes equity and ensures multivocality (Qudrat-I Elahi, 2009: 1169). However, multivocality may be impractical and constrained by politics, time, and resources. Additionally, governance is about effectiveness and efficiency in managing stakeholders (Qudrat-I Elahi, 2009; UNDP, 1997). Finally, governance promotes accountability by decision-makers to their stakeholders, shareholder, or public (Qudrat-I Elahi, 2009). However, accountability, transparency, and the rule of law are intertwined and inseparable in application (UNDP, 1997). Overall, good governance should be responsive to the present and future needs of the society. It should also help us to understand how power, legitimacy and urgency identifies with stakeholders.

Power, Legitimacy, and Urgency

Stakeholder identification is currently based on three key attributes: power, legitimacy, and urgency (Sachs & Rühli, 2011: 38). *Power* is how shareholders control resources critical to the organization (Orij, 2010; Ullmann, 1985). Shareholders are also stakeholders of any entity. That power is their ability to influence the use of resources towards a desired outcome. This power may reside in voting, politics, decision-making, legislation, economics, and information. *Urgency* is a scenario where "stakeholders want their wishes to be fulfilled quickly", while *legitimacy* is where "certain actions fit with the expectations and demands of the other party, manager, or stakeholder. In the context of legitimacy, *action* is what is "reasonable within a subsystem" (Orij, 2010: 871).

The three attributes (power, legitimacy, and urgency) define how the roles of stakeholders of any institution are defined (Orij, 2010: 871). Practically, organizations tend to involve those that have the "power to influence the processes or outcomes of the decision", followed by those with "legitimate authority to enable the execution and urgency to utilize their influence (Eden & Ackermann, 1998; YuLong & Hunter, 2015: 252). However, Donaldson and Preston (1995) argue that stakeholders are involved for different reasons and in different ways. Stakeholders have other reasons beyond the three as a rationale for their involvement. Therefore, firms or corporations should be responsive to the interests of stakeholder demands (Gray, 1988; Healy & Palepu, 2001; Orij, 2010; Robertson, 1992).

Stakeholder Participation Models

Stakeholder participation models are defined by the unilateral (dyadic) and multilateral (network) modes of interaction (Sachs & Rühli, 2011: 8). The *unilateral mode* of interaction is interest-guided (Sachs & Rühli, 2011). In this mode, stakeholders interact when either they "see their interest jeopardized or when they intend to identify with other stakeholders" or when this is assumed to be based on a rational behaviour (Sachs & Ruhli, 2011: 39). However, "individuals are not completely rational, and have multiple interests and different identities" (Sachs & Ruhli, 2011: 39). Therefore, stakeholder interactions and identities are important in this mode. Organizational identity is defined as the assumed relationship between a company and its stakeholders (Sachs & Rühli, 2011: 39). Theoretically, there is always a conflictual and collaborative relationship between the firm and its stakeholders (Butterfield *et al.*, 2004; Freeman, 1984; Sachs & Rühli, 2011). Both conflictive and collaborative relationships require one to understand the dialogue forms used for interaction or communication (Sachs & Rühli, 2011). Dialogue form will determine stakeholders' influence on business decisions from inception to implementation and monitoring (Sachs & Rühli, 2011). For instance, roundtables comprising of all stakeholders, or their representatives, is a dialogue form that will result in mutual decisions on business and stakeholder perspective (Sachs & Rühli, 2011; Freeman, 1984). Reciprocal engagements and mutual understanding of stakeholders ensure that "decision making authority lies not necessarily exclusively with the company" (Sachs & Rühli, 2011: 40).

On the other hand, *multilateral (network) mode* of interaction places emphasis on multi-stakeholder settings (Sachs & Rühli, 2011). In these settings, firms or institutions are "structurally embedded in networks of relationships with stakeholders, who are also tied to each other within networks" (Sachs & Rühli, 2011: 40). Theoretically, institutions are the centre of a network as long as they have stakeholder networks, and at the same time they are part of it (Sachs & Rühli, 2011). Bilateral dialogues are needed to maintain value creation out of relational embeddedness and mutual multi-lateral processes for any institution (Sachs & Rühli, 2011). Therefore, management policies and practices have to take into account the multiple stakeholder interests, expectations, and needs (Post *et al.*, 2002). Managing the links with stakeholders proactively or interactively results in benefits arising out of competitive advantage (Porter & Kramer, 2002, 2006, 2011; Porter & Reinhardt, 2007; Porter & Van der Linde, 1995). However, it is difficult to sufficiently show what happens in multi-stakeholder settings due to unpredictable behaviours of stakeholders (Sachs & Rühli, 2011). Moving from unilateral to multilateral modes of interaction as the complexity and interconnectivity of business activity grows has become a trend nowadays (Sachs & Rühli, 2011).

Furthermore, the models for involving stakeholders are normative, instrumental, and descriptive (Phillips *et al.*, 2003; YuLong & Hunter, 2015), as referred to in Figure 3.2. The *normative model* is for stakeholders to "whom an organization has a direct moral to attend to their well-being" (Phillips *et al.*, 2003: 489). Their interactions are based on moral guidelines, and these are informed by economic and social/philosophical foundations (Boatright, 1994; Bosse *et al.*, 2009; Donaldson & Preston, 1995; Etzioni, 1998; Goodpaster, 1991; Hendry, 2001; Hill & Jones, 1992). The normative model views such players as primary stakeholders, whose continuing participation is vital to the company (Louw & Venter, 2013). In heritage governance, these would be stakeholders critical for retaining the significance of the sites. In other contexts, these are referred to as critical stakeholders, which a company cannot do without (Taruvinga, 2014; YuLong & Hunter, 2015).

The *Instrumental model* is for stakeholders who are involved on the "basis of whether/how much benefit such involvement would bring to the organization's business performance" (YuLong & Hunter, 2015: 252). Often referred to as stakeholders with "strategic dimensions and critical to the survival of an organization", these employ strategies based on their access to resources (Sachs & Rühli, 2011: 37). They influence the firm's behaviour through coalitions with other stakeholders (Sachs & Rühli, 2011). They have the power, influence, and ability to affect the organizational strategy and are equally defined by their legitimate or moral claims, risks, and contracts (Sachs & Rühli, 2011). These stakeholders

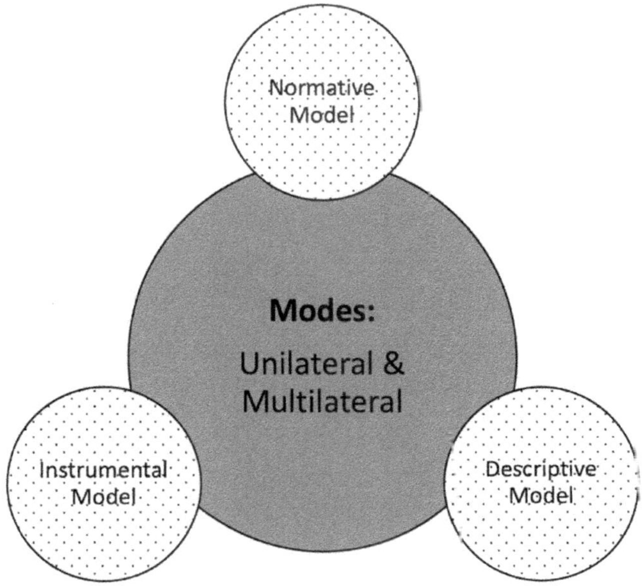

FIGURE 3.2 Stakeholders modes and models (Source: YuLong & Hunter, 2015)

have a contractual relationship with the organization, which should result in the achievement of set business goals, increased profitability, growth, and sustainability (Wood and Jones, 1995).

The *descriptive model* is for stakeholders whose motivations are based on "social capital theory", in particular "how organization views and presents itself to the community to which it belongs to" (YuLong & Hunter, 2015: 253). Individual companies are considered building blocks of the greater community (YuLong & Hunter, 2015: 253). The motivation in this case is to develop a "functional organizational identity that is conducive to its community environment" (YuLong & Hunter, 2015: 253). This approach identifies relevant stakeholder groups, their claims and concerns and subsequently identifies social initiatives to mitigate them and ensures an organization's corporate mission acknowledges these concerns (YuLong & Hunter, 2015). Descriptive motivation underscores painstaking iterations of contests and negotiations between stakeholders and site managers. This can be viewed as a "constellation of co-operative and competitive interests possessing intrinsic value" (Donaldson & Preston, 1995: 66).

The normative, instrumental, and descriptive dimensions are not mutually exclusive but complementary to each other. They collectively promote "active communication, building cooperative, and trusting relationships" among stakeholders, while guaranteeing "stakeholders that their interests are being fairly handled" (YuLong & Hunter, 2015: 253). Organizations have a "multi-fiduciary" ethical responsibility to their shareholders and stakeholders (Goodpaster, 1991). Though stakeholder partnerships exist at different scales and takes different forms, they are important in establishing strategic alliances (Berman *et al.*, 1999; Hendry, 2001; Louw & Venter, 2013). When stakeholders engage, they create a complex web of relationships between and among themselves, and this requires good governance approaches to manage (Louw & Venter, 2013). Governance should be compatible with "typical approaches to democracy and social justice, in which interests of the nominally powerless must be given weight" (Boyte & Kari, 1996; Bryson, 2004; Lebacqz, 1986; Lewis, 1991).

Global-Local Nexus Concept: Stakeholder Interactions

To further understand stakeholder interactions, the stakeholder theory is linked to the global-local nexus concept (Deegan, 2012; Robertson, 1992). The concept, developed by Robertson (1992), focuses on global-local relationships, as well as between the universal and the particular interactions, thereby presenting the world as a single place, a process commonly known as globalization (Deegan, 2012). Globalization "refers both to the compression of the world and the intensification of consciousness of the world as a whole", thereby promoting "concrete global interdependence and in consciousness of the global as a whole" (Robertson, 1992: 8). It is an arena where globalization and localization tensions as well as trends manifest themselves (Deegan, 2012). Robertson (1992) assumed that the world should be understood and interpreted in reference to the dynamics of this world as a whole. Therefore, distinguishing the micro from macro approaches is misleading (Robertson, 1992). It is further argued that "individuals are as much a part of the globalization process as any other basic category of socio-theoretical discourse" (Robertson, 1992: 8). Globalization dissolves the autonomy of actors and practices into contemporary world order. Robertson (1992: 175) argues that globalization captures "*the form* in terms of which the world has moved towards "unicity" and does not refer to a multitude of historical processes of this transition".

Globalization is a twofold process: the interpretation of the universalization of particularism and the particularization of the universalism (Robertson, 1992). This assumes that globalism leads to the construction of the local. "*Universalism*" follows principles, laws, and enforce a rule-based approach, whereas "*particularism*" keeps a relationship rather than upholding a universal principle in every situation (Trompenaars, 1993). When the forces of globalization and localization, "*universalism*" and "*particularism*", clash and collide with each other, they create the notion of "*glocalize*" (Robertson, 1992). In this process, the global and *local* blend, while the universal values are picked up in the local, but at the same time, local values are brought up to the global arena (Robertson, 1992). Therefore, globalization involves the construction of locality from a western perspective, and there seems to be no good reason to see globalization as promoting homogenization (Neu, 2009). Global ideas and processes are, however, contested and their inherent dynamics provoke a reaction or resistance at some level. To provoke a positive reaction, it is important to link universal ideas, policies, and principles with local circumstances through bridging guidelines (Trompenaars, 1993). Spreading global concepts requires that numerous actors be involved and contribute to the interpretation of such a concept which defies the homogenous approach (Neu, 2009). This implies that universal ideas and processes are interpreted and absorbed differently according to the vantage point and history of particular groups in a given context. Globalization thus creates a single arena in which all stakeholders pursue their goals by deliberate comparison with others or using common standards as yardsticks (Robertson, 1992).

The nexus concept argues that "homogenization" goes hand in hand with, or intertwines with, "heterogenization" (Neu, 2009; Robertson, 1992). Robertson's "glocalization" approach is important in that it helps in understanding how global concepts are perceived, negotiated, and become meaningful in the local arena (Neu, 2009). Various perceptions and interpretations of a global concept are developed in the local (Neu, 2009). It is in this context that a local-global nexus model is proposed central to understanding the MMSF. This nexus should encourage global processes to negotiate and interface with local context of the site rather than imposing themselves at the local. The proposed local-global nexus turns the current top-down approach into a bottom-up decision-making process at WHS as will be demonstrated in the MMSF being used in this discussion. Managing stakeholders using MMSF is "akin to conducting an orchestra - one is neither the composer, nor the skilled player of any one instrument - instead one's role is to help coordinate the different voices, to create a space that gives weight and shape to them all at that level" (Prieto *et al.*, 2014). In moving from a "cacophony to a symphony, the conductor needs the skill and indeed humility to recognize and give weight to these different voices" (Prieto *et al.*, 2014). Boom, a world winning song is produced! However, the choice of which stakeholders should be involved is inherently political in nature, involves judgement, and has ethical consequences (Bryson, 2004; Cooper, 1998; Lewis, 1991). The MMSF, discussed below, should provide a rational approach in a heavily contested stakeholder terrain at WHS in Africa.

Towards MMSF

Acknowledging the growing governance complexities of WHS, the MMSF is proposed as an inclusive and dynamic governance approach and structure to bring together stakeholders in handling conservation and development at WHS in Africa. Responding to complex social and environmental contexts requires partnership with various stakeholders (Hummels, 1998; Louw & Venter, 2013; Sachs & Rühli, 2011).

"Multiple" and "Multi-Layered" Perspective

MMSF brings together the views, perceptions, power matrix, roles, responsibility, relations, and ability of the multiple and multi-layered stakeholders at WHS.

MMSF defines the "multiple" and "multi-layered" stakeholders at WHS. The term "*multiple*" is defined as the engagement and involvement of more than one individual or group of stakeholders, while "*multi-layered*" refers to the hierarchical levels of participation by stakeholders at WHS. Stakeholders operate at local, regional, national and international levels (nexus concept) as interested and affected players at WHS. Both "multiple" and "multi-layered" stakeholders have, theoretically, the ability to operate at any of these levels. Also, their interests and expectations are assumed and, theoretically, acceptable at all hierarchical levels of governance. Therefore, "*multiple and multi-layered stakeholders*" refers to the many and overlaid stakeholders with multiple interests at WHS and are operating at different levels of governance structures. Furthermore, "*multiple*" and "*multi-layered*" are used in their combinative nature to denote the complexity of WHS governance in balancing conservation and development.

Localization and Globalization Perspectives of MMSF

The local-global nexus proposed by MMSF is an arena where vertical-horizontal synergies and conflicts between localization and globalization, as well as within each force are integrated to develop an inclusive stakeholder governance framework for WHS. Overall, MMSF is an arena where tensions between the trends of localization and globalization manifest themselves through the roles of multiple and multi-layered stakeholders at various hierarchical levels of participation at the site. However, the power and impact of localization on globalization has not been effectively embraced in the governance of WHS. MMSF argues that if local perspectives are ignored or replaced by global dimensions, this has far-reaching consequences on the future of WHS in developing nations which are seized with accelerated developmental programmes.

The identification and classification of stakeholders proposed by the MMSF takes into consideration power-interest matrix, legitimacy, influence capability, and the participation-planning matrix (Bryson, 2004; Charron, 2007; Eden & Ackermann, 1998), as well as value-based approaches (Pikirayi, 2016). The power-interest matrix approach determines the interest and power "bases" that must be taken into consideration; the "*players*" - those with interest and significant powers, "*subjects*" - those who have an interest but little power, "*context setters*" - those who have power but little interest and, the "*crowd*" - the stakeholders with little interest or power in solving a particular problem (Bryson, 2004: 31). Power defines the direction of interests of a particular stakeholder, while the base of power identifies the source of power which is needed to direct these interest (Bryson, 2004; Eden & Ackermann, 1998). Power can come from being in control of legislation, resources, and access to support mechanisms (Charron, 2007). While winning coalitions are viable political strategies for achieving better outcomes, these are influenced heavily by bases of power. It is thus important to determine which stakeholder is critical to strategic management or a strategy change effort (Bryson, 2004).

Identifying Stakeholders Using MMSF

MMSF governance framework and structure theoretically identifies stakeholders of WHS as follows:

i. **Stakeholders with "universalized" interests, power, and means to impose decisions**. These stakeholders do not own the sites but own the international processes that give them power and intergovernmental statutory means to implement their mandates as enshrined in the ratified 1972 World Heritage Convention and other related instruments through -SBMS. UNESCO is one such example.
ii. **Stakeholders with "nationalized" interest, power, and means to make double-barrelled decisions**. These are nationalized stakeholders playing multiple and conflicting roles such as policing legal instruments, determining resource allocation, making decisions and controlling other stakeholder participation at Mapungubwe Cultural Landscape World Heritage Site (MCLWHS). These roles are inseparable from politics and governmental approach of the SBMS that benefit from taxpayers. These stakeholders include national and provincial governments, districts and local municipalities, and their associated State-funded agencies. These stakeholders are forever fighting to protect their legal authority and political directives. Furthermore, their governance approach is more of compliance with standing protocols as opposed to being part of the solution to broader community challenges. In addition, they are characterized by bureaucratic rigidity on matters beyond their competence.
iii. **Stakeholders with a "localized" interest but without power to make decisions and implement them**. These stakeholders assume they have a right to use their heritage or any other resource near the site in a manner they desire to meet their daily livelihoods but are expected to ask for permission from the SBMS. They make a contribution by providing their points of view on being consulted but have no power to ensure their opinions are included in the formal decisions at heritage sites. They usually resign themselves to their fate of being secondary players at their own heritage sites. These also include community-based development trusts or associations, which only have grassroots power, and beyond that require other stakeholders to survive.
iv. **Stakeholders with interest and means to exploit renewable and non-renewable resources at Heritage Sites.** These are stakeholders with the means and power to exploit both renewable and non-renewable resources at heritage sites. They come with political, financial, bilateral, and multilateral support to undertake socio-economic developments in support of national development targets. They are the bridge between the development aspirations of communities and the inability of SBMS in fully meeting these aspirations. However, any investment and partnerships for these

stakeholders are calculated on the basis of maximizing profitability and sustainability of their operations (Taruvinga, 2010).

v. **Stakeholders with "non-legalized interests", but that have a voice and means to influence conservation and socio-economic development decisions at WHS.** These stakeholders have a direct interest on heritage matters as independent policy watchdogs effecting good citizenship stewardship on heritage matters (Brown & Hay-Edie, 2014). These include self-regulating professional associations with binding constitutions governing their professional and ethical conduct. They usually align themselves to heritage compliance authorities as guardians and providers of independent, balanced, and credible opinions on heritage matters. These could be double-barrelled, depending on how they are benefiting from the whole process and they easily oscillate among the spectrum of multiple and multi-layered stakeholders giving advice or pushing their own agenda.

Failure to attend to any of these categorized stakeholders and their related interests can lead to disasters (Bryson, 2004). World Heritage and related institutions should allow the once perceived "enemies" to become "partners" in conservation and socio-economic development at WHS. This is important in decolonizing the notions of conservation using dynamics at localities where heritage is one of the many resources available to the society. The MMSF proposes that local perspectives of heritage, including its perceived role and function in socio-economic development, are powerful forces that should influence governance of WHS. Local perspectives on conservation and use of heritage are an irreversible impetus for the local-global nexus, which should define such governance framework.

Conclusion: Understanding Stakeholder Perspectives

With the above understanding of MMSF, it is clear that engaging and interpreting stakeholder perspectives at WHS have become more critical than before. Central to this process is how to identify, engage, and interpret them using available quantitative and qualitative analytical tools, but without reducing their opinions to a mere statistical matter. Qualitative research is defined as "non-experimental research that focuses on the occurrence of social, cultural and political human behaviour" traits (Glen, 2009; Salkind, 2012: 13). It seeks to reveal important and underlying factors and deals with subjective data that is produced by respondents (Salkind, 2012; Welman *et al.*, 2005). On the other hand, quantitative approach is defined as a "dynamic form of analysis of verbal and visual data that is oriented toward summarizing the informational contents of that data" (Sandelowski, 2000: 338). Their application together in interpreting stakeholders leads to mixed research approach. Mixed research methods have the advantage of not restricting or constraining the researcher's choices (Altheide, 1987; Black, 1999; Creswell & Plano Clark, 2007; Glen, 2009).

The identification of stakeholders is informed by multiple factors and sources beyond legal frameworks and policies of SBMS and UNESCO. The sources available include on primary source documents, secondary and archival materials, attendance registers, minutes of meetings, background and concept documents on programmes, brochures, newspapers, maps, planning and policy documents, memorandums, annual reports, publications, and archived materials. This also includes identifying development industries, non-governmental organizations, heritage professional associations, foundations, and any other entity with an interest or affected by decisions at WHS. Therefore, stakeholder identification is never limited by the geographic location and boundary delineation of a WHS. These multiple sources assist in eliciting "meaning, gaining understanding and developing empirical knowledge" on varied positions of stakeholders (Bowen, 2009: 27). However, it is important to look at the documents subjected to archival analysis in a critical and objective way (Bowen, 2009; Ferriday & Muir-Cochrane, 2006).

The engagement of stakeholders at heritage takes various forms and approaches. Whatever form and approach is adopted, good governance should prevail in order to give equal opportunities to stakeholders for being part of the solution arising from shared values and mutually-beneficial partnerships connecting stakeholders, SPs and UNESCO. For instance, digital platforms are becoming popular, but caution has to be taken as not all stakeholders have access to and resources for such platforms. This includes domesticating engagement tools into local languages preferred by stakeholders as opposed to using universal languages. With the understanding developed in this chapter on MMSF, the main case study of this discussion, MCLWHS is introduced in the next chapter, followed by the views of multiple and multi-layered stakeholders.

4

HISTORY, VALUES, SIGNIFICANCE, AND MANAGEMENT OF THE CASE STUDY

Mapungubwe Cultural Landscape World Heritage Site

Introduction: The Case Study

This chapter presents Mapungubwe Cultural Landscape World Heritage Site (MCLWHS) as the primary case study for this discussion (Plate 4.1). MCLWHS is a national park, national heritage, and World Heritage Site (WHS), as well as part of the Greater Mapungubwe Transfrontier Conservation Area (GMTFCA) connecting Botswana, South Africa, and Zimbabwe. The significance and outstanding universal value (OUV), management, and the emerging conservation, socio-economic, and stakeholder issues at MCLWHS are discussed. The "local outstanding universal values" anchored on the spirituality, and intangible values of Indigenous and Descendant Communities (IDCs) are also discussed. Interfacing WHS and local value propositions at MCLWHS, broadens our understanding of the localness notion of heritage, and equally, gives the local and regional context of the site. This helps in identifying all the significances, related stakeholders and their aspirations at MCLWHS. In this context, issues discussed include accessibility and beneficiation, implications of spirituality and intangible values, and the multiple and cross-cutting legislations (at national, interstate, and international levels). Other issues discussed relate to stakeholder management framework, diversity of land uses, and the emerging prioritization of cultural heritage in a sea of biodiversity at Mapungubwe National Park (MNP). Overall, the chapter synthesizes the opportunities and challenges emanating from the evolving relationship between conservation and development at MCLWHS. Interestingly, MCLWHS itself is a manifestation of socio-economic development process, making it a rich heritage-laboratory for providing futuristic perspectives on the relationship between conservation and development.

DOI: 10.4324/9781003149361-4

PLATE 4.1 Mapungubwe Hill, Mapungubwe National Park, South Africa (credit: McEdwards Murimbika@2021)

Historical Context: Mapungubwe Cultural Landscape, South Africa

MCLWHS, the main case study of this discussion on the relationship between conservation and development, is located inside the Mapungubwe National Park (MNP) at the confluence of *Shashe* and *Limpopo* rivers in Limpopo Province, South Africa (Plate 4.2). The southern boundary of the site cuts across geometrical citrus farms, while the northern boundary is the Limpopo River, which forms the frontier between South Africa and the neigbhouring states of Botswana and Zimbabwe (Figure 4.1). The *Shashe-Limpopo* confluence is central to the history and heritage of the area, with the archaeological site of Mapungubwe hill considered very important (Apley, 2000; Berry & Cadman, 2007; Calabrese, 2000, 2005; Huffman, 2000, 2004, 2005; Smith, 2006; Steyn & Nienaber, 2000; Van Schalkwyk & Hanisch, 2002). Therefore, the broader territorial landscape of the site is home to both culture and nature values, of which the former is recognized as World Heritage. Of importance also, from a socio-economic context, is that Limpopo Province is one of the least developed regions of South Africa.

The site has a four-tier status: national park, national heritage, WHS, and GMTFCA connecting Botswana, South Africa, and Zimbabwe. The significance of the MCLWHS is traced from its broader and regional context, including the management framework taking into consideration this four-tier status. The OUV of the site is key to understanding the manifestation of World Heritage

PLATE 4.2 Shashe Limpopo Confluence Area (credit: Janette Deacon)

and its interface with the local and regional context of the site. This creates an opportunity to understand the evolving relationship between conservation and socio-economic developments at the site and the role of stakeholders in the process.

MNP is intricately linked with this history of establishing "gardens of Eden" in Africa (Nelson, 2003), in which nature reserves and parks, also referred to as Protected Areas (PA), were created for biodiversity and wildlife conservation (Ditchkoff *et al.*, 2006). All this was done at the back of forced removals of rural communities off their land, and them being deprived of access to it and its resources (Munthali, 2007; Ndoro & Wijesuriya, 2014). They were pushed to marginal parts of the Limpopo area. The history of Mapungubwe is intertwined with the process of establishing Dongola Botanical Reserve in 1922 (Berry & Cadman, 2007). Nine farms were set aside to create the reserve to advance nature conservation. This marks the beginning of bitter and contentious political debate of the 1940s on the relationship between conservation and other land uses in the area (Berry & Cadman, 2007). Political debates for and against nature conservation emerged. This was led by General Jan Smuts (Prime Minister of South Africa; 1919–1924, 1939, and 1948), Dr Illtyd Buller, Pole Evans (a botanist), Andrew Conroy (Minister of Lands during the second term of General Jan Smuts), and Dr Bernard Price, an influential engineer, businessman, and philanthropist (Berry & Cadman, 2007). Jan Smuts and Dr Evans used their political and environmental knowledge to advance nature conservation, resulting in the proclaiming of Dongola Wildlife Sanctuary in 1947 (Apley, 2000). This reinforced the 1922 conservation model in the Mapungubwe area.

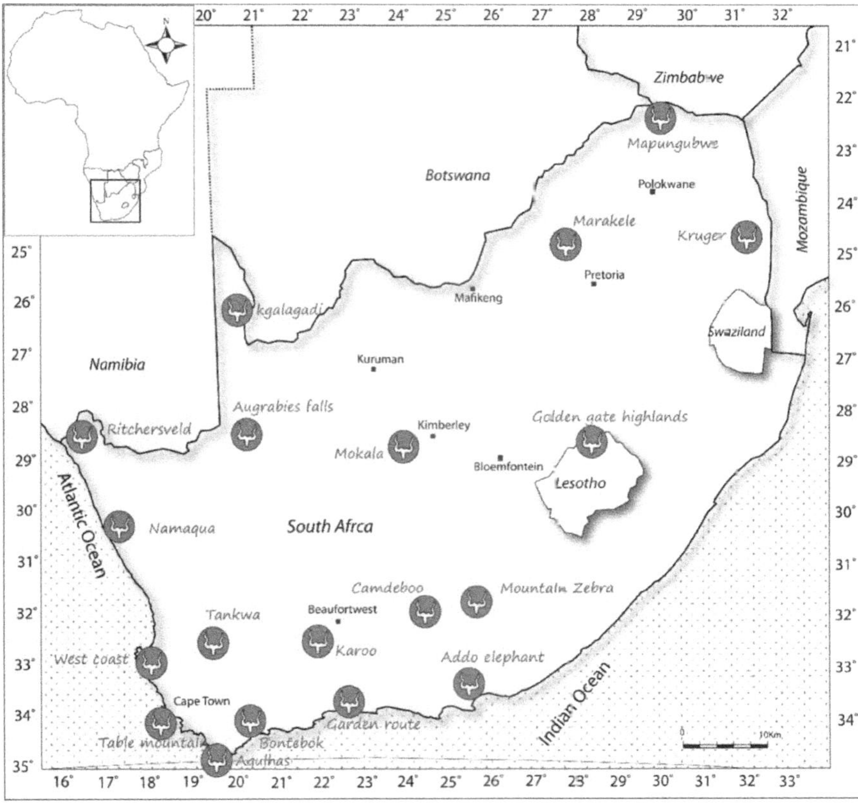

FIGURE 4.1 Location of MCLWHS in relation to other National Parks, South Africa

Regrettably, the Dongola Wildlife Sanctuary was deproclaimed when General Smuts lost his political mandate to the opposition, the National Party (Apley, 2000). The National Party had opposed the concept from the beginning as they argued that it interfered with the farming livelihoods of Afrikaners in the area. The party considered farming to be a more important land use in Mapungubwe area compared to nature conservation. This Afrikaner perspective never prioritized the needs of IDCs of the area, as forced relocations were used to displace them to other areas to pave way for the new farmers. An attempt to revive the Dongola Wildlife Sanctuary in 1968 failed (Berry & Cadman, 2007). The racial segregation policies (apartheid) of South Africa were in action as a tool for alienating the majority in the process, including the 1913 Land Act that appropriated land in favour of white settlers arriving in South Africa.

The next push for advancing nature conservation, and its exclusionary approach, is linked to the Vhembe Game Reserve, first established at a provincial level and subsequently proclaimed as the Vhembe-Dongola National Park (VDNP) in 1998 (Berry & Cadman, 2007). This transition from provincial to national park demonstrates the entrenchment of nature conservation in the area.

VDNP become a precursor to the MNP proclaimed in 2004, but only became operational in 2006 (DEA, 2014b; SANParks, 2013b). Parallel to this, the discovery of kimberlite diamonds on Venetia farm by De Beers in the 1980s also set in motion the "revival" of the nature conservation initiative started by Dr Evans in Limpopo area. De Beers then curved the Venetia Limpopo Nature Reserve (VLNR) out of sixteen farms in 1990 (Carruthers, 2006). From all these successive developments advancing nature conservation, coupled with the regional importance of the culture and biodiversity of Mapungubwe, a grand conservation idea emerged: the GMTFCA established in 2006. This is supported by South Africa, Zimbabwe, and Botswana. In order to understand how MCLWHS is being managed, the evolving significance (cultural, natural, and regional perspectives) and the emerging management framework is presented from historic times to the present.

Significance of Mapungubwe Cultural Landscape

MCLWHS is a cultural landscape endowed with both cultural and natural values that contribute to its national and international significance. However, it is the cultural values that gives it a World Heritage Status under the provisions of the Convention. However, some of these cultural values are only of national and local significance.

Cultural Values and Significance

MCLWHS reflects on the society that utilized it from the past to the present times (Carruthers, 2006; Lowenthal, 2005; Pikirayi, 2016). It is an important cultural landscape at national, regional, and international levels, but its significance initially ignored the spirituality associated with IDCs of the site. As pointed earlier, the IDCs were relocated outside the area during the colonial period, but some are slowly coming back to the site through the land claims on the South Africa side (Taruvinga, 2019). Mapungubwe Hill was declared a National Heritage Site in 2002 under the National Heritage Resources Act (25 of 1999) and in 2003 became a WHS as a cultural landscape under criteria (ii), (iii), (iv), and (v), a status equally recognized by South African World Heritage Convention Act 49 of 1999 (SANParks, 2013a). MCLWHS is a cultural site located inside a PA, typical of many other sites in Africa such as the Rock Art and archaeological sites enclosed in the Matobo National Park in southwestern Zimbabwe.

The cultural significance and evolving land uses at MCLWHS can be traced back to its layered history and successive occupation from pre-colonial times to the present (Calabrese, 2000, 2005; Huffman, 2000, 2004, 2005; Smith, 2006; Steyn & Nienaber, 2000; Van Schalkwyk & Hanisch, 2002). The Mapungubwe archaeological site, when connected to similar sites in Botswana, Zimbabwe, and Mozambique, collectively define the earliest civilizations in ancient southern Africa (Pikirayi, 2016). A farmer and prospector, E.S.J. van Graan, "re-discovered" Mapungubwe on Greefswald farm in 1932 with the assistance of a local informant named Mowena (Berry & Cadman, 2007; Pikirayi, 2016). However, it is also argued that a "bush-wacker" named Francois Lotriet was the first European to see the site, a discovery shared with Mowena (Berry & Cadman, 2007: 48). This assertion is absurd to believe because Mowena as a local and the locals would have known about this site before the arrival of Europeans in the area, similarly, like at Victoria Falls (Zambia and Zimbabwe),where David Livingstone is credited with discovering the site. This is typical exploration and discovery accounts relating to Africa when an explorer like David Livingstone is erroneously credited with discovering the Mosi oa Tunya (Victoria Falls). The point is that local communities know their heritage, including locations of the various attributes, as part of their creation and knowledge processes transmitted from one generation to another.

Despite the above minor discussion on exploration and discoveries, MCLWHS is testament to the rise and fall of one of the first indigenous kingdoms in Southern Africa between 900 AD and 1300 AD (Hall & Smith, 2000). Human presence in the Mapungubwe area stretches over at least a million years covering over 400 recorded archaeological sites (Hall & Smith, 2000). These include Stone Age, Rock Art, Iron Age, and recent historical sites (Hall & Smith, 2000; SANParks, 2010; Van Doornum, 2005). The confluence of the *Limpopo* and *Shashe* rivers has served as an interconnector of culture and nature of the area from the past

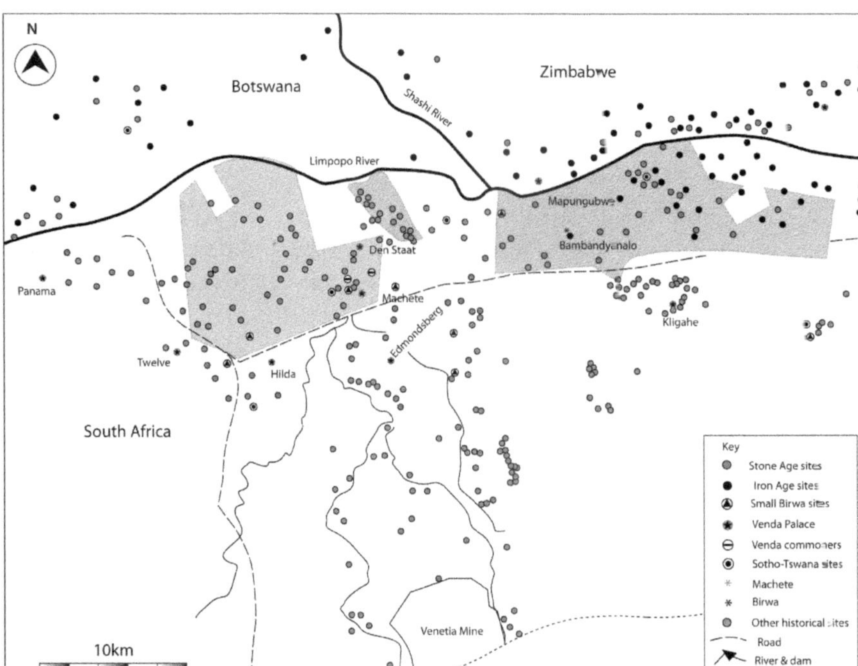

FIGURE 4.2 Cultural heritage sites in the Greater Mapungubwe Area

to present times as illustrated on the map below showing distribution of cultural sites in the area (Hall & Smith, 2000; Van Doornum, 2005).

Stone Age Period

Stone Age sites in the MCLWHS constitute the earliest archaeological evidence of human occupation in the area (Kuman *et al.*, 2004: 2009; Pollarolo & Kuman, 2009). Stone Age communities were hunter-gatherers residing in caves, rock shelters, and open areas during the Early, Middle, and Late Stone Age periods (Deacon, 2011; Hall & Smith, 2000; Van Doornum, 2005). Testimonial to this are attributes such as rock paintings, stone tools, and the cultural practices among the living descendant of San communities now scattered in the Kalahari Desert and many parts of Southern Africa (Davidson-Hunt *et al.*, 2016; Eastwood, 2003, 2006; Eastwood & Blundell, 1999). These attributes define the nature and character of the attendant societies. Hunter-gatherer presence in the MCLWHS continued during and post the Mapungubwe days (see Hall & Smith, 2000; Manyanga, 2007; Chirikure *et al.*, 2013; Van Doornum, 2005, 2007). Based on excavations, Van Doornum (2005) argues that the San people occupied Mapungubwe for nearly 13,000 years, after which they were supplanted by the incoming Iron Age farmers, who occupied the area between 900 and 1300 AD (Hall & Smith, 2000).

Through San (Hunter-gatherer) ethnography, it is broadly accepted that rock art is a record of beliefs and experiences of the people who created it (Deacon, 1993, 2005, 2011; Eastwood & Blundell, 1999; Lewis-Williams, 1994). The neighbouring landscapes of Botswana and Zimbabwe in the Shashe-Limpopo confluence are also dotted with many rock art sites linked to the same traditions and beliefs. This attests to the fact that it was an expansive landscape which transcended modern political boundaries. Current geopolitical boundaries in Africa are a result of colonialism which disregarded pre-colonial inter- and intra-community connections through intermarriages, cultural affiliations, and being members of the same polities spread over vast landscapes divided by rivers and other geographical phenomenon (Pikirayi, 2016). This extant territorial dimension defines the multiplicity and complexity of communities associated within this expansive cultural landscape and who should be considered in the management of broader territorial issues at MCLWHS.

Iron Age Period: 900–1300 AD

The Iron Age period of MCLWHS, which can be divided into three distinct but overlapping phases, is a record of the Early Iron Age farming migrants from West Africa between 350 AD and 600 AD (Huffman, 2005). As they migrated southwards from West Africa, they met San hunter-gatherers (Huffman, 2005). The period 900–1020 AD is associated with *Zhizho* settlements in the broader *Shashi-Limpopo* Basin, with the main settlement located on Schroda Farm (Calabrese,

2000, 2007; Hanisch, 1980). The next significant phase is the Leopard's Kopje period, 1020–1220 AD, with the settlement at *Bambandyanalo*, while class distinction (when Mapungubwe Hill became significant) is thought to have become pronounced in 1220–1290 AD (Carruthers, 2006; Huffman, 2005; Maggs, 2000). The three phases represent a growing and dense settlement associated with social classification system at Mapungubwe (Carruthers, 2006). Thus, Mapungubwe is a pre-colonial state representing the first stage in a development that would lead to the rise of Great Zimbabwe, Khami, and many other stone-walled sites in Southern Africa (Hall *et al.*, 2006; Huffman, 2000, 2007). The pre-colonial state of Mapungubwe lasted about 80 years between 1220 AD and 1300 AD (Huffman, 2000, 2007). At its peak, Mapungubwe had 5000–9000 residents, but the identities of the inhabitants remain contested by the IDCs (Carruthers, 2006; Huffman, 2005; Smith, 2006). However, and based on decolonized ethnographic approaches, it is now argued that the site had a population of less than 5000 inhabitants (Chirikure et al., 2010; Ndoro, 2017). This revised population is calculated based on the traditional family set-up and space configuration derived from ethnographies of IDCs (Chirikure et al., 2010; Ndoro, 2017).

The Iron Age period of Mapungubwe has been researched from the 1932 (see Calabrese, 2000, 2005; Fouché, 1937; Galloway, 1959; Gardner, 1963; Hanisch, 1980, 1981; Huffman, 2000, 2004, 2005; Pikirayi, 2016; Smith, 2006; Steyn, 1994; Steyn & Nienaber, 2000; Van Schalkwyk & Hanisch, 2002; Voigt, 1984). These researches point to a powerful Mapungubwe kingdom, characterized by extensive wealth-smithing and social stratification or complex social system (Apley, 2000; Huffman, 2005; Pikirayi, 2016). The regular flooding of the Limpopo River provided silt and water for crops, thereby sustaining the Kingdom (Huffman, 2005; Manyanga, 2007). The accumulation of wealth in the form of gold, ivory, glass beads, and cotton cloth gave birth to a social classification system which witnessed the elite and the sacred leader residing at the top of Mapungubwe Hill, while commoners lived in the valley below the hill (Huffman, 2005; Maggs, 2000). This is evident from the distinctive graves, associated grave goods (such as a golden rhinoceros), and an elaborate settlement on the hilltop (Huffman, 2005). The social, economic and ideological values relating to state formation made Mapungubwe flourish (Huffman, 2005; Pikirayi, 2016).

Furthermore, participating in the Indian Ocean trade network along the east coast of Africa also made Mapungubwe an influential kingdom. Their access to rich natural resources and climatic conditions that allowed sustainable agriculture defined the economic character of this trading kingdom (Calabrese, 2005; Carruthers, 2006; Huffman, 2000, 2005). Mapungubwe had long-distance trade connections with the Chinese, Arab, and Indian merchants from as long as 4000 years ago (Calabrese, 2000; Huffman, 2004; Steyn & Nienaber, 2000; Van Schalkwyk & Hanisch, 2002). The location of Mapungubwe at the crossings of the north/south and east/west routes in southern Africa enabled it to control trade through the East African ports to India and China and throughout southern Africa. From its hinterland, it harvested gold and ivory – commodities in scarce supply elsewhere – which brought it some form of wealth displayed in such imports as Chinese porcelain and Persian glass beads (Mapungubwe Nomination Dossier, 2003).

In addition, research has pointed to the wider regional importance of this state formation system (Carruthers, 2005). Recent research see the development of complexity as taking place at different places, with Mapungubwe being one of them (Chirikure, 2013; Chirikure & Taruvinga, 2017; Huffman, 2007; Manyanga, 2000, 2007; Mothulatshipi, 2008; Pikirayi, 2016). Research at Mapela Hill (Zimbabwe) is challenging past interpretations on Mapungubwe as the single centre of development and it being the earliest site of this typology (Chirikure et al., 2010; Ndoro, 2017). It is increasingly being argued that the "development of socio-political complexity and urbanism was much broader than currently thought" (Chirikure et al., 2010: 586). It is now argued that pre-colonial southern Africa populations had a "permanent interest and possibly additional dwellings in the countryside where important subsistence activities took place" (Chirikure et al., 2010: 585).

According to Chirikure and Taruvinga (2017: 1), "the demography of prehistoric and contemporary populations occupies an important role in interdisciplinary and multi-scalar discourses on sustainability". This is important in understanding the "ecological and sustainability implications of various population estimate on key resources such as land for agriculture and animal husbandry" in the area (Ndoro, 2017: 11). For example, the previously suggested population of 20,000 people at Great Zimbabwe was based on wrong ethnographic assumptions (Ndoro, 2017). Such a population would have made it unsustainable from an ecological perspective (Ndoro, 2017). Furthermore, Chirikure and Taruvinga (2017: 14) argues that "low populations are essential to achieve ecological sustainability". This implies that any study at MCLWHS should be considered within the dynamics of the broader regional context of the site. This regional context points to the presence of multiple resources resulting in production specializations that benefitted the society (Chirikure et al., 2010). Production extended beyond subsistence levels as evidenced by craft specialists, smiths and smelters, traders, and merchants (Chirikure et al., 2010). This resulted in population agglomeration in the area (Chirikure et al., 2010).

Another dimension emerging from recent research is that neither "K2 nor Mapungubwe qualify to be the region's incipient Zimbabwe culture centres" based on the fact that "Mapela had a fully developed Zimbabwe culture much earlier than Mapungubwe" (Chirikure *et al.*, 2014: 7). It is further argued that "various communities exhibiting identical cultural practices all contributed and participated in early state formation" in the region (Chirikure *et al.*, 2014: 17). Therefore, new archaeological evidence and the associated radiocarbon dating from recovered artefacts suggest that early state formation started with the many Leopard's Kopje sites in south-western Zimbabwe and adjacent regions (Chirikure *et al.*, 2014: 17). The cultural practices of the present-day Shona and Venda peoples originated during this historical process, and this should be taken

PLATE 4.3 Archaeological sites in Mapungubwe area (credit: Janette Deacon)

into consideration in redefining the place of the site in the broader regional context of this tradition (Ralushai, 2003).

The collapse of Mapungubwe as a state is attributed to several factors, among them the onset of the Little Ice Age at the end of the 13th century AD and perhaps a strong El-Nino (Carruthers, 2006; O'Connor & Kiker, 2004). This points to the ability of indigenous communities to interpret and respond to impact of climate change, yet these experiences are ignored in current climate change discussions. The decline is also attributed to the centre of the regional power shifting to Great Zimbabwe (1290–1450 AD) and later to Khami, owned by Rozvi people, in 1450–1820 AD (Carruthers, 2006). The factors made it impossible to sustain the growing population and the power base for such a classified society (Carruthers, 2006). While new evidence may point to revised chronologies of these interconnected sites, it is not a matter for discussion in this research (Taruvinga, 2019). These emerging interpretations are discussed to provide a revised context towards understanding socio-economic issues and histories of IDCs at the site (Taruvinga, 2019). What remains clear is that Mapungubwe remains interwoven into the regional story of how social complexity emerged in Southern Africa (Plate 4.3).

Spirituality and Mapungubwe: Intangible Dimensions

Mapungubwe has always been associated with spirituality from the pre-colonial period to the present times and is largely associated with the alienated IDCs of Mapungubwe (Taruvinga, 2019). The IDCs of Mapungubwe include the *Venda*,

Lemba, *Leshiba*, *Machete*, *Tshivhula*, and *Vhangona* communities (Huffman, 2014; Pikirayi, 2016). These IDCs are the traditional custodians of sites and the associated intangible values of Mapungubwe (Nelson, 2003; Ndoro *et al.*, 2018; Pwiti & Ndoro, 1999). However, these IDCs contest about the ownership of the site. According to oral traditions of IDCs, the name *Mapungubwe* means "place or hill of Jackals", or alternatively, "place where Jackals eat" (Fouche, 1937: 1). From various ethnic and linguistic groups in the region, including the Pedi, the Swazi, and Sotho, the name also means "place of wisdom" and "the place where the rock turns into liquid" (Hall *et al.*, 2006). Also, from a Shona/Venda perspective, Mapungubwe means the "rocks of the bateleur eagles" as the word *mapungu* refers to eagle in the plural while the singular noun is "*chapungu*". The word *ibwe* means stone or rock (Hall *et al.*, 2006). Therefore, "Mapungubwe" as one word denotes the presence of bateleur eagles in that area (Hall *et al.*, 2006). These eagles are believed to have the same spiritual symbolism as the soapstone birds recovered at Great Zimbabwe World Heritage Site in Zimbabwe (Ndoro, 2001). The context of the name, Mapungubwe, denotes strong spiritual affinities between the place and beliefs of IDCs, for instance, the eagle is considered a messenger of God and the ancestors (*Vadzimu*).

The spirituality of the MCLWHS revolves around the varying traditions and practices of the mentioned IDCs, including all of them attributing their identities to the site (Taruvinga, 2019). For the Lemba clan, MCLWHS is their trade centre from pre-colonial times hence their name *Mushavi* "Trader" (Chirikure & Taruvinga, 2017). They were potters and beads manufacturers, black smiths and worked with diamond, gold, copper and silver (Chirikure & Taruvinga, 2017). As for the Leshiba clan, they practiced rain calling ceremonies and celebration of new fruits among many other practices (Chirikure & Taruvinga, 2017). In addition to this, the Machete clan, had spiritual practices supported by *dzin'anga* being persons who were the eyes and ears to see things from far on behalf of the King, (Chirikure & Taruvinga, 2017). There are three prominent pools for them in the Limpopo River, namely, *masivhunde, masibulele, mbangwe* (Chirikure & Taruvinga, 2017). The Tshivhula clan view themselves as the direct descendants of Mapungubwe as supported by similar ceremonies (Chirikure & Taruvinga, 2017). Also, the Vhangona clan consider Mapungubwe as their judicial and economic capital, with cotton and gold mining have been their preoccupation (Chirikure & Taruvinga, 2017).

The connecting thread among these IDCs is the mentioned religions and spiritual practices. It is the reason why they request free access to the site as part of their spiritual protocols (Chirikure & Taruvinga, 2017). Therefore, the continued identification of past and contemporary communities based on a ceramic industries fixation is a disservice to the complex and intricate history of IDCs at MCLWHS (Pikirayi, 2016). There are many facets of cultural traditions that need to be interrogated beyond such narrow archaeological approaches to understand the link between the past and present communities from an intangible perspective (Pikirayi, 2016). The archaeology of MCLWHS cannot be separated

TABLE 4.1 Description of the OUV of MCLWHS (Source: Mapungubwe Nomination Dossier, 2003)

Criteria	Description of outstanding universal values
(ii)	An important interchange of human values that led to far-reaching cultural and social changes in Southern Africa between AD 900 and 1300.
(iii)	A testimony to the growth and subsequent decline of the Mapungubwe state which at its peak was the largest kingdom in the African subcontinent.
(iv)	A powerful state trading through the East African ports with Arabia and India was a significant stage in the history of the African sub-continent.
(v)	Illustrates the impact of climate change and record the growth and then decline of the kingdom of Mapungubwe as a clear record of a culture that became vulnerable to irreversible change.

from the oral histories and social memory of IDCs (Hanisch, 2008). It cannot also be separated from the associated land uses relating to intangible or spiritual dimensions of IDCs at MCLWHS. Understanding the complex and intricate history of IDCs, their social memory, and vested livelihood interests requires understanding their spirituality or intangible dimensions.

Outstanding Universal Value: World Heritage Site

Mapungubwe was inscribed as a WHS in 2003 using criteria (ii), (iii), (iv), and (v). The OUV of the site illustrates a complex and hierarchical/socially stratified society in Southern Africa, in particular the Iron Age period;

At the time of inscription, the authenticity and integrity of MCLWHS was largely considered not to have been subjected to any destructive form of human intervention since it was abandoned (Mapungubwe Nomination Dossier, 1999). Even then, agricultural activities in the area were also considered as not having an impact on the cultural landscape (Mapungubwe Nomination Dossier, 1999). However, the inscription emphasized the monumentality of the landscape, without much reference to the spirituality or intangible dimensions of IDCs of the MCLWHS. As such, the intangible elements of this site are considered important and could point to an associative landscape, given their connection with ancestral and historical Venda (Pikirayi, 2016).

The delineation of the boundaries of MCLWHS was unclear from the beginning, yet the World Heritage Committee proceeded on listing the site. The core area of the MCLWHS had twenty-two farms, measuring approximately 28,168.66 hectares and this area falls under the South African National Parks (SANParks). The buffer zone of MCLWHS has been a contentious matter since 2009, resulting in it being reduced from 265,900 to 133,600 hectares to accommodate the authorized extractive process (coal mining) on the northern side of

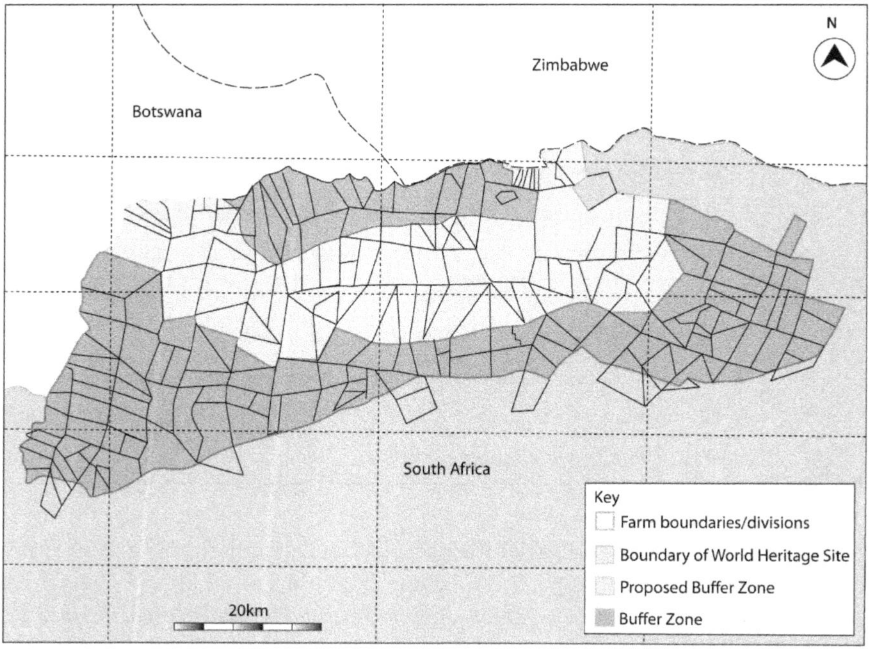

FIGURE 4.3 Core area and buffer zone of MCLWHS

the greater Mapungubwe area. During the amendment of the buffer zone, the World Heritage Committee imposed a reactive monitoring process at the site, and this created tension with the parties involved. The buffer zone is supposed to offer extra protection to the core area. The IDCs were not involved in the decision to amend the buffer zone as this became a technocratic process led by the SP, however, with litigations from other stakeholders strategy to resolve the impasse at local levels with the State (Taruvinga, 2019).

Nature Values and Significance

MNP, inside which MCLWHS is located, is one of the nineteen Public Parks of South Africa under the management of South African National Parks (SANParks). MNP is home to scenic landscapes, unique ecological systems, and biodiversity of significance that warrant protection (Berry & Cadman, 2007). The unique ecological system of MNP is supported by sandstone formations, riverine forests, woodlands, and the famous baobab trees (Figure 4.4). MNP has twenty-four acacia and eight *Commiphora* species, a fairly dense population of shrubby mopane trees, and other trees such as fever, ana, leadwood and fig (Berry & Cadman, 2007). There are about 387 bird species, which include the black eagle, thick-billed cuckoos, Kori bustards, grey crowned cranes, great white pelicans, southern pied babbler, crimson-breasted shrike, and black-faced waxbill, among many others in MNP. The mammals in MNP include many African game species that

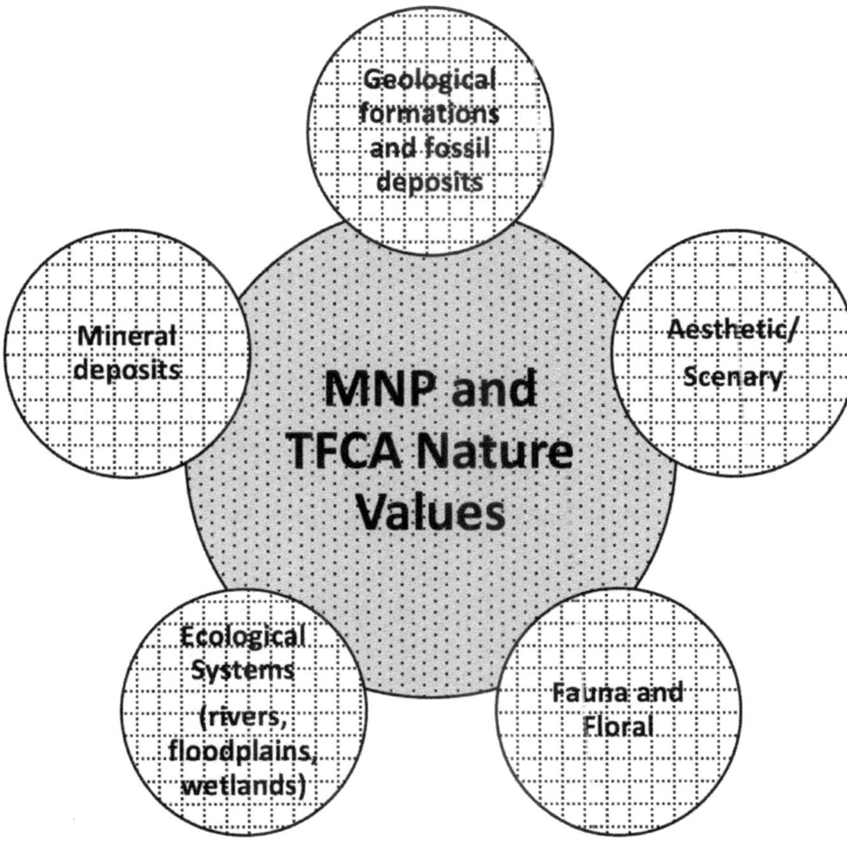

FIGURE 4.4 Summary of the nature values and significances of MNP

roam freely with the Greater Limpopo-Shashe Transfrontier Conservation Area. These include zebra, African bush elephant, white rhino, leopard, Transvaal lion, eland, bushbuck, waterbuck, impala, klipspringer, baboon, pigs, and giraffe. Reptiles in MNP include the giant plated lizards, ground *agama*, speckled thick-toed geckos, different snake species (python, black mamba, puff-adders, and snouted black cobra), and the Nile crocodiles in the Shashe and Limpopo rivers (Bhatasara *et al.*, 2013). Despite this biodiversity richness, MNP has threats that include mining, habitat loss, agriculture, alien species invasion, climate change, unsustainable harvesting of natural resources, and poor land use practices (DEA, 2014a). This also includes stakeholder-related issues and tensions among IDCs desiring to access benefits coming with these natural resources (DEA, 2014a).

Transfrontier Conservation Values

The Greater Mapungubwe area is a shared heritage resource (with both culture and nature values) for Botswana, South Africa, and Zimbabwe. These shared heritage values led to the creation of the GMTFCA between the three State Parties (SPs). The idea was formulated in colonial times as an Interstate Park; however, due to lack of financial support, it failed to materialize (Berry & Cadman, 2007). This idea was revived only 50 years later by the Peace Park Foundation (PPF), which was established in 1997 to support the creation of Transfrontier Parks in Southern Africa. In June 2006, Botswana, South Africa, and Zimbabwe signed a memorandum of understanding (MoU) establishing a 4800 km^2 Trans-Frontier Conservation Area constituting the GMTFCA. The GMTFCA covers a complex mosaic of communal areas, National Parks, and private conservancies in Botswana, South Africa, and Zimbabwe (Berry & Cadman, 2007). In Botswana, the TFCA covers Northern Thuli National Park and adjacent private conservancies, while in South Africa, it includes private land, state-owned land and National Parks areas. In Zimbabwe, it includes portions of Maramani communal lands and some privately owned commercial farms. The establishment of the GMTFCA is based on a combination of catchment, cadastral, biodiversity, and cultural values of the transient area covering Botswana, South Africa, and Zimbabwe (DEA, 2014).

TFCAs are usually etched out of National Parks, nature reserves, and conservancies as a bioregion; however, in the case of GMTFCA, communal and private land was involved (Sinthumule, 2014). This brings about land ownership and use dynamics that require effective management. The GMTFCA provides insight into the ecological, environmental, and economic dimensions of the Greater Mapungubwe Area (DEA, 2014b). The Minister of Environment and Tourism from Botswana, Kisto Mokaila, pointed out that the GMTFCA had a dual role of promoting both conservation and socio-economic development in the area (Berry & Cadman, 2007). Therefore, conservation and socio-economic development are equally expected to drive each other in the GMTFCA.

The three SPs are already signatories to relevant international conventions governing such transfrontier parks. Botswana, South Africa, and Zimbabwe are signatories to the Convention on Biological Diversity (Rio de Janeiro, 1992), the United Nations Convention to Combat Desertification (Paris, 1994), the 1972 World Heritage Convention, the 1973 Convention on International Trade in Endangered Species of Wild Flora and Fauna (CITES), and many other relevant agreements. At a regional level, these countries are signatories to the 1992 Treaty of the Southern African Development Community (SADC), the Southern African Development Community Wildlife Policy (1997), and the Southern African Development Community Protocol on Wildlife Conservation and Law Enforcement (Maputo, 1999) and enjoy cordial political relations through bilateral agreements as SPs. While it is a noble thing, international instruments tend to override domestic matters of African SPs. But at least the GMTFCA, by theoretically removing political boundaries, brings nostalgia about pre-colonial communities, who enjoyed fluid social boundaries in the area. The only challenge is that local communities are not part of the GMTFCA as will be discussed in the next chapter.

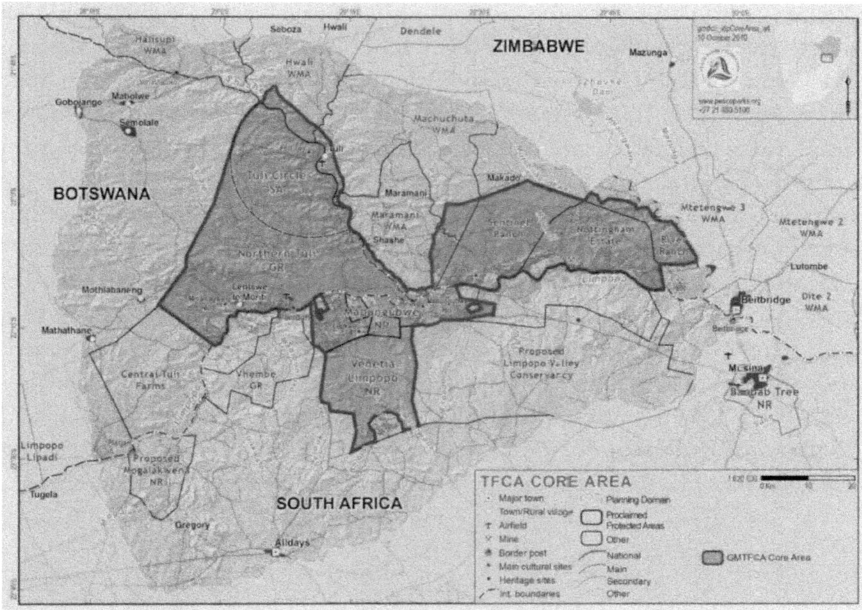

FIGURE 4.5 Core area and planning domain of the GMTFCA (Source: DEA, 2011b)

Management of Mapungubwe Cultural Landscape

The management framework of MCLWHS is inseparable from its four-tier status and is primarily informed by the relevant suite of national and international heritage laws, policies and guidelines promulgated from time to time. Therefore, the cross-cutting legal matrixes of MCLWHS are expressed in the Integrated Management Plan (IMP) of the site. However, the management of the cultural heritage, in its diversity as outlined in the White Paper on Arts, Heritage and Culture of 1998, falls under the Department of Sport, Arts and Culture (DSAC). The DSAC is thus responsible for policy and strategic guidance on management of cultural heritage (tangible and intangible, moveable and immovable) in South Africa. This function is devolved to the South African Heritage Resources Agency (SAHRA) established in terms of section 11 of the National Heritage Resources Act (NHRA) of 1999. While SAHRA administers the NHRA, the DSAC appoints public entities (for example, Robben Island Museum) to manage specific heritage sites in compliance with NHRA provisions. However, DSAC has no direct mandate to oversee management of WHS as this mandated is with the Department of Fisheries, Forestry and Environment (DFFE). DEFF is the focal point for South Africa on the WHC. This denotes the administrative complexity of managing the site with overlays of reporting and parallel lines of accountability. This is similar and contrary at the same time to arrangements in other countries, for instance, the National Museums and Monuments of Kenya (NMK) in Kenya and the National Museums and Monuments of Zimbabwe (NMMZ) in

Zimbabwe who are the legal custodians and Management Authorities (MAs) of heritage sites, including WHS. However, for NMMZ, cultural sites in Matobos and Victoria Falls Monument are both inside PAs managed by another State entity. Since 2018, DSAC has been reviewing the White Paper on Arts, Culture and Heritage in order to align heritage management to evolving good practices and socio-political conditions of the nation. These arrangements have varying management and governance implications; however, it is not the subject of this discussion, except to mention that all of the proposals are still and systematically sidelining local communities. In relation to national development priorities of South Africa, DSAC is aligned to Chapter 15 of the National Development Plan that promotes social cohesion and nation building as a strategy for development in South Africa. Department of Sport, Arts and Culture (DSAC) collaborates with government departments such as the Department of Women, Children and People with Disabilities, National Department of Tourism, and others in creating jobs and economic development opportunities using arts, culture, and heritage.

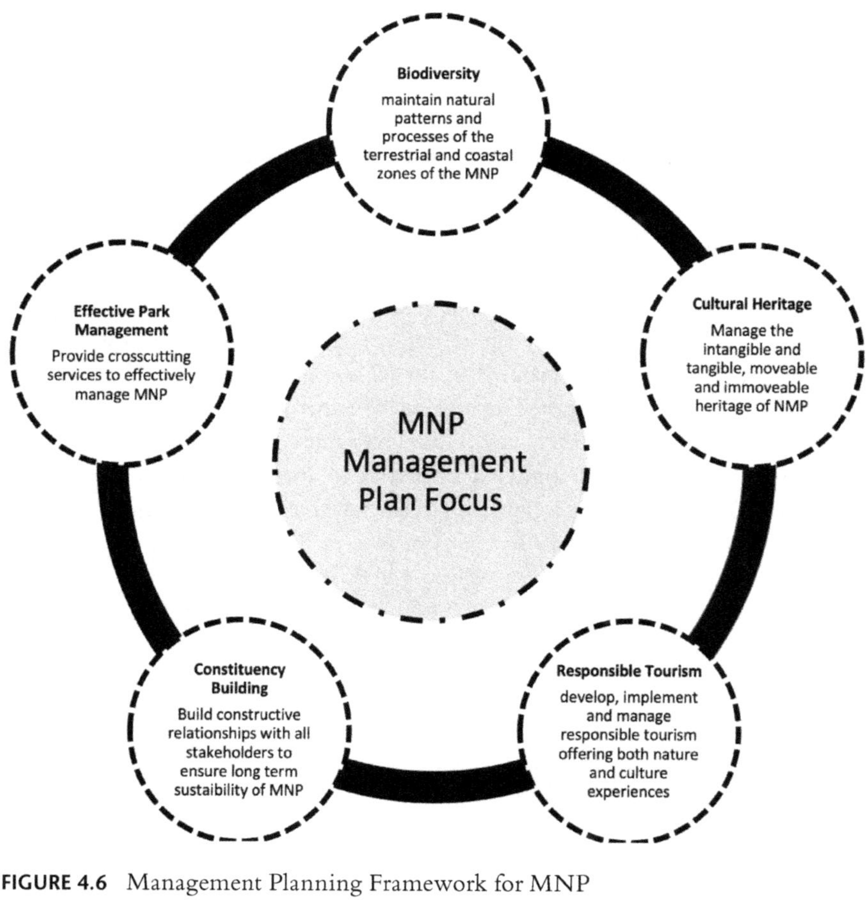

FIGURE 4.6 Management Planning Framework for MNP

Furthermore, in 2012, DSAC adopted *Mzansi Golden Economy Strategy* to strategically position arts and culture sector in job creation and promoting economic development in South Africa. The strategy also promoted quality education and rural development through the establishment of libraries, monuments, and community centres for the benefit of the broader society (DAC, 2012). The strategy was expected to create 5 million jobs by 2022 (DAC, 2012). This remains a pipeline dream, including being seriously impacted by the COVID-19 pandemic which has induced closure of some public and private heritage institutions in South Africa. The Arts, Culture, and Heritage (ACH) is continuously being positioned as an economic growth sector worth domestic investment (DAC, 2012). This requires dialogue with stakeholders to encourage growth in employment-creating activities (DAC, 2012). Despite all this visionary approach, DSAC and related entities suffer from under-funding, poor strategic stakeholder engagements, and the inability to retain partnerships (Taruvinga, 2014, 2019). Therefore, management of heritage is increasingly becoming inseparable from stakeholders and socio-economic developments.

Management as a National Heritage Site

As a National Heritage Site, MCLWHS is managed by SANParks as part of the MNP. However, SAHRA monitors compliance of SANParks with provisions of the NHRA of 1999. NHRA is a sequential improvement of heritage laws promulgation dating back to the apartheid era of South Africa, starting with the Bushman Relics Protection Act of 1911, followed by the Natural and Historical Monuments of 1923, the Relics and Antiques Act of 1934, and the National Monuments Act of 1969. It is the National Monuments Act of 1969 which was subsequently repelled by the NHRA of 1999. In its formulation, NHRA borrowed some aspects from other national heritage legislations, for instance from Canada, Australia, and New Zealand, further reinforcing a colonial mentality that such Acts are best examples without fully domesticating them to evolvining local needs. Also, international documents such as the Burra Charter and the 1972 World Heritage Convention were used in shaping the NHRA. This further points to a process that largely ignores the African experiences, including presenting abstracted heritage concepts and practices devoid of local perspective. The 'so-called' legal consultations are just a public relations exercise in many African countries, where State-Based Management Systems (SBMS) yield the power to push such processes with Parliament. While the NHRA perpetuates Western approaches to heritage management, it also enhances some policy elements at domestic levels (Deacon, 2015). Despite the contradictory nature of external influences, the guiding principles in the drafting of the NHRA were driven by the governance philosophy of the new democratic South Africa, equally linking the governance philosophy presented in the White Paper on Arts and Culture, for example, transparency, inclusivity, consultation, and engagement (Deacon, 2015). It also includes additional or broader categories of heritage resources not previously protected, such as intangible heritage or living heritage, historical graves, graves of "victims of conflict", underwater archaeology, and heritage objects (Deacon, 2015).

NHRA provides for an integrated management system for national heritage resources in South Africa. This covers the identification, conservation, protection, and promotion of heritage resources for the present and future generations through a three-tier management system: Grade 1 (national estate), Grade II (provincial heritage), and Grade III (local heritage). Grade I sites have exceptional qualities qualifying them for declaration as of national significance, while Grade II sites have special and significant qualities worthy of a provincial status. Grade III sites cover all other heritage resources worthy of conservation at local levels. National heritage is managed by appointed MAs (such as SANParks, Robben Island Museum, etc.) who comply with NHRA, while provincial heritage sites are managed by the Provincial Heritage Resources Authority (PHRA). Local heritage is also managed by PHRAs via the departments of culture, or custodial local municipalities such as City of Johannesburg in Gauteng Province of South Africa.

While SANParks has management responsibility of MCLWHS, SAHRA has an oversight compliance monitoring function at the site. SAHRA provides the oversight regulatory and management framework, including guidelines on maintaining inventories of heritage resources and undertaking Heritage Impact Assessments (HIA). In Limpopo Province, the Limpopo Heritage Authority (LIHRA) has struggled with obtaining the approval of SAHRA due to "lack of competence, financial resources, skills and equipment" (CALS, 2016: 42). This is compromising all other provincial and local sites in the province. However, SAHRA does not provide any funding to any MAs such as SANParks for the administration of the site, meaning that their role is limited to compliance monitoring only. All other cultural sites not recognized as Grade 1 in the broader area of this Mapungubwe are under the management of the dysfunctional Limpopo Heritage Resources Authority (LIHRA).

Theoretically, NHRA empowers the public and the long-alienated indigenous communities by making provisions for them to be consulted (Ndoro & Chirikure, 2009). The permitting process of NHRA for such developments has provisions for consultation with the public before authorization is given (Ndlovu, 2011). This also includes promoting good management practices, encouraging communities to nurture and conserve their own legacy as a bequeath to future generations (Taruvinga, 2019). Furthermore, NHRA defines cultural identity as foundational to the spiritual well-being of local communities. Additionally, NHRA also affirms diverse cultures and cultural heritage manifestations as a way of redressing past inequities, including previously neglected research into rich oral traditions and customs of local communities in South Africa (Taruvinga, 2019).

However, the ideology of NHRA serves in the reproduction rather than the transformation of the social order with its known inequalities, being a strategy of containment and social closure (Deacon, 2015). Ideology guides the legal system by formulating rules for behaviour and inspiring ideas that benefit society (Deacon, 2015). An ideology can only be successful if it persuades the majority of people to observe such rules (Deacon, 2015). From this perspective, NHRA has achieved limited success owing to the piecemeal implementation of these principles and underfunding of SAHRA (Ndoro & Chirikure, 2009). SAHRA struggles to keep pace with monitoring developments around heritage sites as illustrated by the delays in issuing permits, including the increasing number of buildings becoming historic buildings as the NHRA automatically protects any building which is 60 years old. Neither has SAHRA properly costed its Act to ensure sustainability of operations and their mandate. Therefore, the success of a law is not about the number of convictions for breaking it, but about persuading the general public to value their heritage (Deacon, 2015). In this context, legislation punishes people only after they have broken the rules (Deacon, 2015). This is often too late because the resources are already destroyed by the time the matter is taken to court (Deacon, 2015). Progressive legislation should be proactive with society.

Management as a National Park

MNP is under the administration of SANParks, an entity under the Department of Environment, Fisheries and Forestry (DEFF). DEFF provides regulatory framework in the management of the environment and the related biodiversity in South Africa through the National Environmental Management Act (NEMA) of 2004. The Act defines environment as referring to the "conditions and influences under which any individual or thing exists, lives or develops" ("White Paper on Environmental Management Policy"). These conditions and influences include renewable and non-renewable natural resources such as air, water, land, and all forms of life; the socio-economic and political, cultural, working, and other factors; and the culture, socio-economic, politics, and value systems that determine the interaction between people and the environment. They also include the use of natural resources, and the values and meanings that people attach to life forms, ecological systems, physical and cultural landscapes, and places. Furthermore, the Act requires PA to submit IMPs to ensure the protection, conservation, and management of their heritage resources. These IMPs are approved by the Minister of DFFE, demonstrating political commitment to conservation and development in PAs.

In this context, DFFE strives to balance conservation and sustainable development, while providing equitable distribution of the benefits derived from natural resources to society (Section 24 of the Constitution). DFFE provides regulatory processes in the form of Environmental Impact Assessments (EIA), Strategic Environmental Impact Assessments (SEA), and where culture is concerned, HIA regulated by SAHRA. In the context of WHS, the application of the International Council on Monuments and Sites (ICOMOS) Guidelines on HIA for Cultural World Heritage Properties is enforced by Department of

Fisheries, Forestry and Environment (DFFE) as a representative of the SP to UNESCO on the 1972 World Heritage Convention.

SANParks, established in terms of the National Environmental Management Protected Areas Act 57 of 2003, undertakes the daily management of all National Parks, including MNP. Other parks include Kruger, Golden Gate, Table Mountain, and the Mountain Zebra National Parks among many more others. SANParks conserves, protects, controls, and manages the culture, environment, and biodiversity values in national parks. SANParks also manages Transfrontier initiatives involving any of their National Parks, such as GMTFCA. All National Parks are conservation and tourism magnets in South Africa. In order to fulfil this cultural mandate, SANParks established a Heritage Department under People and Conservation Division. The division has a mandate to oversee the management of cultural components in all their parks, yet the majority of the staff in this division are nature and community specialists with a skeletal team focusing on cultural heritage. The placement of the Heritage Department under People and Conservation is also questionable, given that nature conservation is under the Scientific Services Division of SANParks (Taruvinga, 2019). Nature and Culture are equally important values of heritage in South Africa, yet they are treated differently by SANParks. This portrays the colonial mentality of prioritizing nature conservation over cultural values. Even at the Ngorongoro Conservation Area (NCA) in Tanzania, cultural values were subservient to nature until recently when NCA Management Authority placed emphasis on it after the site become a mixed property after the recognition of Laeotoli as WHS in 2010. NCA even absorbed the mandate of cultural heritage, through an arrangement with the Department of Antiquities. Management of cultural sites in Matobo Hills World Heritage Site suffered the same fate until its recognition by UNESCO, but serious tensions between NMMZ and the Zimbabwe Parks and Wildlife Management Authority have continued over the years.

SANParks desires to "develop, expand, manage and promote a system of sustainable national parks that represents biodiversity and heritage assets, through innovation and best practices for the just and equitable benefit of current and future generations" (SANParks website). Programmes promoting access, benefit sharing, socio-economic development, and improved well-being of local communities living adjacent to parks are thus implemented (SANParks, 2013a). Socio-economic development is implemented through two vehicles, the Expanded Public Works Programme (EPWP) and the Social Investment Programme (SIP). EPWP is a government-expanded Public Works programme that creates employment and transfers incomes to poor households on a temporary and seasonal basis inside parks (SANParks, 2013b). The programme supports environment and culture programmes inside parks. On the other hand, SIP promotes legacy development projects within rural communities bordering parks (SANParks, 2013a). These legacy projects include constructing school facilities and establishing computer centres. The SIP is supported through a community fund, derived from a levy charged on all accommodation services in national parks (SANParks, 2013a).

Despite having these beneficiation mechanisms in place, MNP faces the challenge of finding sustainable solutions to the growing gap between socio-economic needs of neighbouring communities and the long-term conservation goals. There is need to create shared values among stakeholders of the MNP. At MNP, this is happening within the fast-changing and dynamic socio-political environment of the Limpopo Province. While SANParks has been implementing good practices in wildlife conservation, they have been criticized for ethcs-fortressing conservation, thereby continuing to marginalize the needs of local communities (Carruthers, 2006). In response, SANParks has developed policies on heritage management, community engagement, and socio-economic development. Their "Policy on conservation, management and promotion of cultural heritage resources in SANParks" came into effect in 2011. However, funding for cultural heritage by SANParks remains inadequate (Taruvinga, 2019).

Interstate Governance of the GMTFCA

The GMTFCA is managed by the memorandum of agreement (MoA) signed by the SPs of Botswana, South Africa, and Zimbabwe in 2006. The MoA promotes ecosystem integrity, biodiversity conservation, cultural heritage resources management, and sustainable socio-economic development across international boundaries. Respective institutions in the parties to the MoA constitute the management structure and collectively, they policy the implementation of the MoA in addition to applicable legislations in each country. The structures of the GMTFCA include the Trilateral Ministerial Committee (TMC), Trilateral Technical Committee (TTC), and the Trilateral Advisory Committee (TAC). These structures bring together regional role players from the public, private, and communal sector with vested interests in both cultural and natural heritage. Representation from communities in the GMTFCA is limited to Zimbabwe, the Maramani Community, as they are more of organized private conservancies with interest on tourism. They do not represent the previously marginalised local communities of the area. The successful land claims in the area on the South African side may add to the number of communities involved in the GMTFCA. Respective government departments provide guidance and political support to their entities in this management matrix. For instance, in South Africa, DFFE has the mandate for the overall coordination of South African component of TFCAs, interdepartmental and inter-governmental communication, assisting with marketing and the formalization of policies, and monitoring tools for use in the GMTFCA. It also acts as the lead agent in negotiations and entering into international agreements with different role players among many other administrative responsibilities. To guide the role of these players, Article 4 of the MoA articulates the management imperatives summarized in the diagram below.

At country level, each signatory to the MoA has a TFCA National Steering Committee (NSC) supported by a Management Committee (MC), and these would feed into the TTC organized as four working groups on Conservation,

FIGURE 4.7 Management imperatives outlined in the MoA of the GMTFCA

Community Development, Safety and Security, and Tourism as reflected in the diagram below. Each SP to the GMTFCA was expected to develop a Management Plan for its respective area, which would lead to a Joint Management Plan for the GMTFCA (Taruvinga, 2019). This Joint Management Plan was expected to address conservation, cultural, tourism and revenue matters, and other issues of common interest and mutual impact within the GMTFCA. The finalization and implementation of the Joint Management Plan has not happened as expected, and each SP is in its own corner due to limited funding and different national priorities (DEA, 2011; Taruvinga, 2019).

The land reform process in Zimbabwe also had far-reaching implications on the reality of the GMTFCA as land changed hands abruptly and in a chaotic and violent manner. Also, Zimbabwe was expected to relocate some cattle from

Maramani to relieve grazing area outside the core of the GMTFCA, and this has not happened (Sinthumule, 2014). The land claims by IDCs in the Mapungubwe area is perceived as a threat to the operationalization of the GMTFCA. Another dimension of both GMTFCA and WHS, not yet fully explored and implemented, is the extension of the MCLWHS to Botswana and Zimbabwe given their cultural connection and presence of similar material culture (Taruvinga, 2019).

Management as WHS

As a WHS, Mapungubwe is managed through the provisions of the 1972 World Heritage Convention, which provides for inscription, conservation, and promoting of sites such as MCLWHS. In addition, the Operational Guidelines on the Implementation of the 1972 World Heritage Convention provide further extended framework for managing such sites. The Convention is coordinated by the World Heritage Committee supported by the ABs. Compliance with the Convention includes submitting state of conservation reports, updated management plans, agreeing to reactive monitoring missions, and, where necessary, accepting to be placed on the List of World Heritage in Danger in order to mitigate aggravating conservation issues, as well as be delisted in case of loosing the OUV, authenticity, and integrity. This constitutes both statutory and non-statutory compliance by the SP. The Convention also provides International Assistance under the World Heritage Fund, as well as mobilization of national and international support in favour of the 1972 World Heritage Convention.

As for South Africa, DFEF has promulgated the South African World Heritage Convention Act (WHCA) No. 49 of 1999 as a way of domesticating the Convention. South Africa is the first sub-Saharan country to do so. DFFE is the SP representative to the Convention and thus coordinates World Heritage management in South Africa. To this end, DFFE has developed binding national guidelines and structures for implementing the Convention in South Africa. In terms of the South WHCA, No. 49 of 1999, in particular sections 8, 9, and 25 (1), DFFE appoints MAs of WHS in South Africa. The MAs are expected to develop and implement IMPs, as well as comply with the Environmental Regulatory framework governing PA. Also, the performance of the MAs is assessed by the Minister in terms of section 12 (2) of the same Act based on audited annual reports via the Biodiversity and Conservation Directorate supported by the Protected Areas Systems Management and the World Heritage Departments as illustrated in Figure 4.8. The South African World Heritage Convention Act has the following provisions for, but not limited to, monitoring the state of conservation; cultural and environmental protection and sustainable development; being sensitive to the people and their needs, including their physical, psychological, developmental, cultural, and social interests; engaging and involving all interested and affected parties in the governance of sites; promoting intergovernmental coordination and harmonization of policies; and compliance with cross-cutting legislations and plans affecting WHS in South Africa. Overall, the

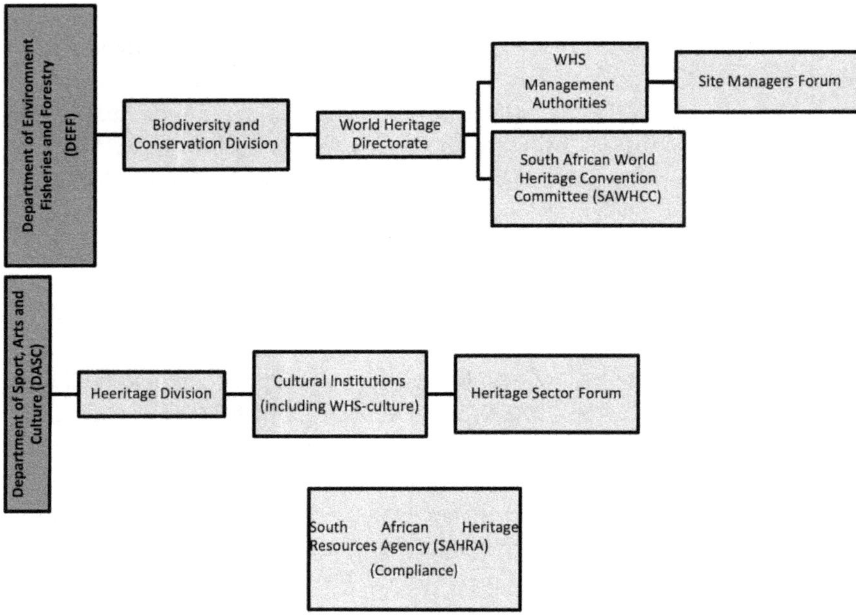

FIGURE 4.8 World Heritage management structure aligned to respective Mandates of Government Departments in South Africa

monitoring and evaluation framework of MCLWHS is in line with the World Heritage Reporting cycles as outlined in the Operational Guidelines and the Convention.

The WHCA, No. 49 of 1999, also provides governing structures linked to the management of WHS, and these include the South African World Heritage Convention Committee (SAWHCC) established in 1997, Site Managers Forum (SMF), and Mas as already explained. The SAWHCC members are appointed by DFFE for a 5-year period and it advises the Minister of DFFE on all matters pertaining to the Convention, Operational Guidelines on the Implementation of the Convention and the WHCA. The SAWHCC recommends to the DFFE properties which meet the criteria for inclusion on the South African Tentative List, examines the state of conservation reports for inscribed properties, and makes recommendations on nomination files to be submitted to UNESCO among many of its functioning mandate. It was also empowered with powers to develop and, where necessary, review the national guidelines on World Heritage management, as well as seek innovative ways to assist potential as well as existing WHS with financial and technical assistance. The Committee also reviews and provides advice on the IMPs of WHS, as well as advise the Minister on World Heritage matters under discussion during World Heritage Committee sessions. In addition, the Committee is tasked with identifying training and capacity-building needs at national level, including raising public awareness of WHS (DEA, 2013).

The SMF is constituted by the ten site Managers of existing World Heritage properties in South Africa. SMF shares information and experiences managing World Heritage and supports SAWHCC. The SMF further advocates for the engagement and involvement of local communities, accompanied by visible socio-economic opportunities at WHS. Some of the challenges that face SAWHCC and SMF include poor attendance by its members, inability to attract funding, and lack of consistency in decision-making around World Heritage matters, especially around impacts of socio-economic developments at sites, in particular the contentious extractive industries in South Africa (DEA, 2013). For example, financial limitations do not permit all SAWHCC and SMF members to attend the World Heritage Committee sessions as part of the State Delegation.

Cross-cutting Management Dimensions of MCLWHS

MCLWHS in its four-tier status interfaces with cross-cutting governance system and interprovincial laws implemented by both the National and Provincial Governments, as well as applicable regional instruments. The Limpopo Department of Economic Development, Environment and Tourism, the Provincial Heritage Resources Agency, the Vhembe and Capricorn District Municipalities, as well as the Musina and Blouberg Local Municipalities are important role players in the Mapungubwe area. As a catchment area, Mapungubwe area has been a source of livelihoods for many centuries, of which water abstraction needs to be regulated to avoid catastrophic consequences to the ecological systems of the area (Taruvinga, 2019). This is done through the National Water Act of 1998 administered by the National Department of Water Affairs and Sanitation. Also, MCLWHS is situated in an area with both renewable and non-renewable resources, which includes minerals. The National Department of Mineral Resources (DMR), through the Minerals and Petroleum Resources Development Act (MRPRDA) of 2002, provides equitable access and sustainable development of the nation's mineral and petroleum resources in South Africa. MRPRDA recognizes that these resources contribute to national development (Taruvinga, 2019). The MRPRDA further recognizes the need to protect the environment for the benefit of present and future generations, ensure ecologically sustainable development of mineral and petroleum resources in order to promote economic and social development and local and rural development, as well as promote social upliftment of communities affected by extractive industries (Taruvinga, 2019). At the same time, it also affirms the State's commitment to guaranteeing security of tenure in respect of prospecting and mining operations through an internationally competitive and efficient administrative system (Taruvinga, 2019).

Another cross-cutting dimension at Greater Mapungubwe Area is the promotion of sustainable development and protection of the environment by the Limpopo Department of Economic Development, Environment and Tourism (LEDET). LEDET is expected to contribute to the growth of the economy and job

TABLE 4.2 Cross-cutting legal frameworks for MCLWHS

Level of Protection	Legal Instrument (s)
National	National Parks Act (No.57 of 1976)
	National Monuments Act (Act No. 28 of 1969): National Heritage Resources Act (Act No. 25 of 1999)
	South African World Heritage Convention Act (Act No. 49 of 1999)
	Environment Conservation Act (Act No. 73 of 1989)
	National Environmental Management Act (Act No. 107 of 1998)
	Minerals and Petroleum Resources Development Act (MRPRDA) of 2002
	National Water Act of 1998
	National Environmental Management: Biodiversity Act (Act No. 10 of 2004)
	National Environmental Management: Protected Areas Act (Act No. 57 of 2003)
	Finance Management Act (Act No. 1 of 1999)
	Conservation of Agricultural Resources Act (Act No. 43 of 1983)
	Protected Areas Act, 2005 (Act No. 57 of 2003)
	South African Heritage Resources Act, 1999 (Act No. 25 of 1999)
	Biodiversity Act, 2004 (Act No. 10 of 2004)
	A Framework for Development and Implementing Management Plans for South African national Parks, 2008
	Agreement for the Establishment of the Vhembe/Dongola National Park, 1996
	Tourism Act No. 3 of 2014
	Disaster Management Act No 57 of 2002
Regional	1992 Treaty of the Southern African Development Community
	2003 Protocol on Shared Watercourses (Revised)
	1999 SADC Protocol on Wildlife Conservation and Law Enforcement
	2002 Protocol on Forestry
	1992 Treaty of the Southern African Development Community
	African (Banjul) Charter on Human and People's Rights
	2001 New Partnership for Africa's Development (NEPAD)
	1968 African Nature Convention
	2003 African Convention on the Conservation of Nature and Natural Resources (Revised)
	1991 Treaty establishing the African Economic Community
	African (Banjul) Charter on Human and People's Rights
International	1972 World Heritage Convention
	Convention on International Trade in Endangered Species of Wildlife Fauna and Flora (CITES),1975
	Ramsar Convention on Wetlands, 1971
	1997 Watercourses Convention
	Bonn Convention on Migratory Species of Wildlife Animals

creation through targeted interventions. Their mandate also includes the protection and enhancement of environmental assets and natural resources. LEDET also has the responsibility of strategically positioning Limpopo Province as a competitive and preferred tourist destination within the South African Development Community (SADC). This includes the role of municipalities such Musina, Thulamela, Mkhado, and Collins Chibane in such initiatives over and above their primary responsibility of meeting the needs of residents (Taruvinga, 2019).

The Integrated Management Planning Framework of MCLWHS

The integrated management of MCLWHS in its four-tier status is coordinated through the MNP and World Heritage Site Management Plan. The plan promotes both the conservation of culture and nature (SANParks, 2013b). SANParks applies Strategic Adaptive Management (SAM) as the foundation of all the planning processes as parks cannot function in isolation from their broader context (Herman, 2013b; SANParks, 2013b). SAM is "strategic (facilitate action with foresight and purpose), adaptive (facilitate learning whilst we are doing), and participatory (facilitate engagement and co-learning with stakeholders)", (Grant *et al.*, 2008). Interactions between and among the socio-economic and ecological components, as well as sub-systems in the broader context are unpredictable, hence the SAM approach. Furthermore, the suite and complexity of stakeholders having varied or even conflicting expectations, worldviews, and values demands such any approach. Overall, SAM "integrates research, planning, management and monitoring in repeated cycles of learning how to better define and achieve goals" in dealing with emerging complexities while providing optimal solutions and desired management outcomes (SANParks, 2019: 9).

Furthermore, parks in South Africa use a zoning approach on land uses. These zones indicate what activities may take place in different sections of the park and the conservation objectives of those sections. The zoning of MNP "was based on an analysis and mapping of the sensitivity and value of the park's biophysical, heritage and scenic resources" (SANParks, 2005; SANParks, 2019: 45). MNP land uses include primitive, low-intensity, and high-intensity leisure zones as illustrated in Figure 4.9 (SANParks, 2019). Cultural heritage is protected and covered under special management overlays, "which designate specific areas of the park that require special management interventions" (SANParks, 2019: 51).

The zoning adopts a deliberate spatial planning framework to make provisions for conservation, tourism, and visitor experience initiatives (SANParks, 2013b). Also, the above zones that meet the requirements of the park's Conservation Development Framework (CDF) are reflected in the diagram below.

The integrated management framework of MNP demands engaging and consulting interested and affected stakeholders, including local communities. MNP does this through Park Forum (governed through the Park Forum Charter adopted in 2011). Other stakeholders are engaged as prescribed by governing laws. Even the NHRA (of 1999) requires stakeholders to be consulted in all these processes. In order to improve implementation of set targets for National Parks, and now equally applicable to all WHS in South Africa, DFFE introduced the Management Effectiveness Tracking Tool (METT). METT assesses the effectiveness of management strategies by paying attention to the context, planning, inputs, processes, outputs, and outcomes of the strategies as illustrated below (Figure 4.11). From 2014/15, MCLWHS has scored METT assessments averaging 75–81% compared to all other WHS in South Africa. This demonstrates consistency and effectiveness of the management system used at the site by SANParks.

Zone	Description	Zone as a % of the total size of the MNP	% of highly sensitive areas in the zone
Primitive	• Conservation of natural values, in particular remote and wilderness areas of MNP as sensitive environments • Emphasis is on preserving the solitude, remoteness, serenity and wilderness qualities, with controlled access in relation to the number, frequency and size of groups that visit such areas	64.18	56.89
Low Intensity Leisure	• Tourism activities, in particular for day and overnight tourists • Conservation mandate is to mitigate impacts of a high level of tourism and infrastructure development through careful planning and active management. • Cultural heritage sites, picnic sites, and game-viewing areas are parts of this zone.	35.62	43.11
High Intensity Leisure	• Tourism activities, in particular concentration and containment of commercial, tourism, managerial, operational and industrial park activities within a restricted and designated area, which is robust enough to tolerate development. • Designed to retain a level of ecological integrity consistent with that of a protected area, though it has the highest level of deviation from the natural pristine state allowed in the zone. • This are the areas with high density of tourism and infrastructural development, with shops, restaurants, and interpretive centre	0.20	0.00

FIGURE 4.9 MNP zones and descriptions as of 2019 (Source: SANParks, 2019)

However, METT has consistently pointed out deficiencies at most WHS in South Africa such as lack of capacity/resources to enforce regulations and by-laws due, limited budgets, low staffing levels, limited number of socio-economic development projects that benefit local stakeholders, lack of training and equipped staff, as well as effective community liaison structures at sites.

Conclusion: Systemic Exclusionary or Inclusive Management Approaches?

The advancing of nature conservation, and how this was largely supported by the apartheid government in the Mapungubwe area, including how it continues today under the new democracy, denotes "power-politics" in deciding land uses, by who and for what benefit. The historical context of Mapungubwe gives birth to a complex management system of the site characterized by a systemic exclusionary

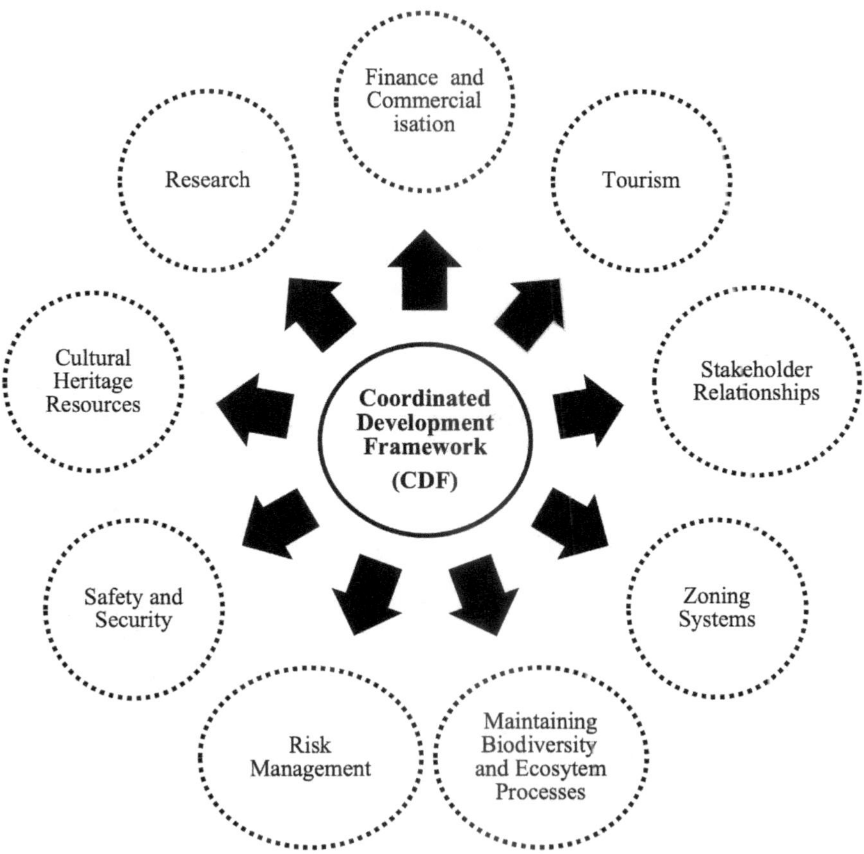

FIGURE 4.10 Principles of the coordinated development framework applicable to National Parks in South Africa (CDF)

engagement approach, which made it impossible for local communities to be involved, yet it is their undeniable cultural footprint exhibited in the Mapungubwe area. This systemic exclusionary engagement approach has been demonstrated in this discussion by the dominance of SBMS supported by UNESCO's systems through the Convention. Furthermore, this approach manifests at most heritage sites in Africa owing to the colonial blueprint which has largely remained as an active but mistakenly considered dormant force in managing heritage in Africa. When MCLWHS is considered in its four-tier status as a National Heritage site, WHS, PA, and GMTFCA, its management framework is supported by multiple and cross-cutting legislations (national and interstate) advocating for this exclusion of other stakeholders. However, this systemic exclusionary approach is now being challenged by stakeholders, in particular, local communities who are advocating for an "inclusionary engagement" approach. But the complexity of IDCs, including their differences, is an inherent factor militating against their inclusion as will be demonstrated in the next chapter in relation to MCLWHS and other sites.

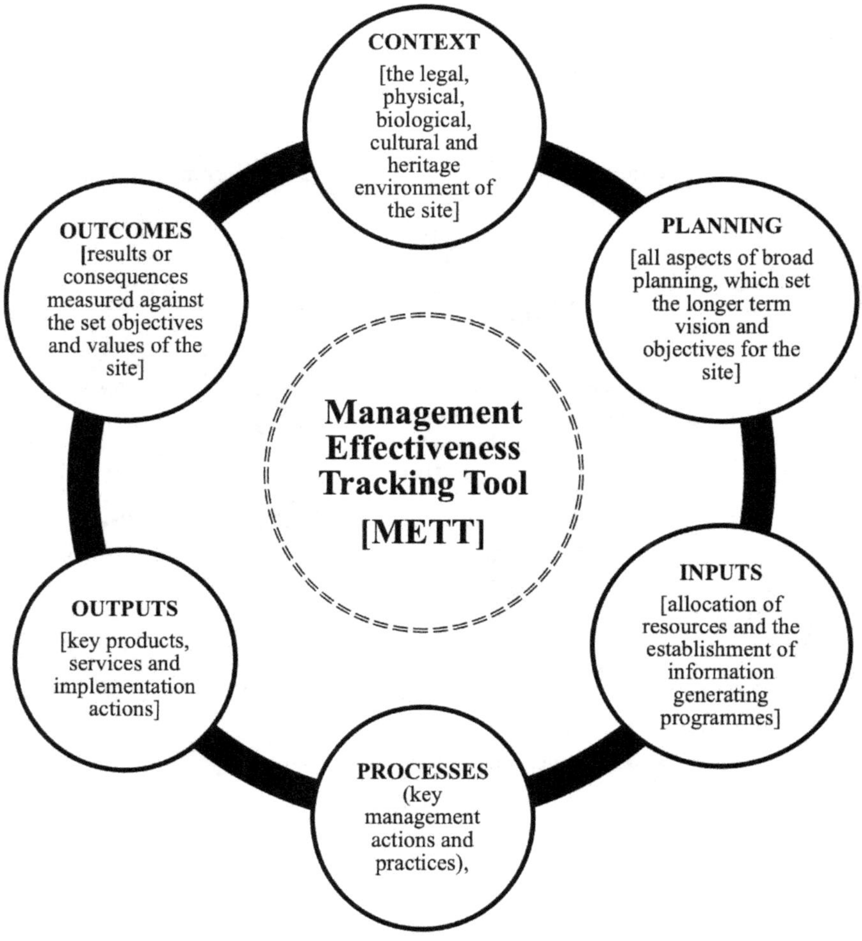

FIGURE 4.11 Management Effectiveness Tracking Tool applied at National Parks and WHS in South Africa

While the Traditional Management Systems could provide more democratic and communally driven management approaches as presented in Chapter 3, this was overridden by both apartheid and post-apartheid approaches emphasizing the role of the SBMS. The net effect is that the evolving complex management system of MCLWHS has become impenetrable for stakeholders without detailed managerial and scientific knowledge as competences highly sought by SBMS such as SANParks. This complex managerial phenomenon in Africa, fully supported by legal or formalized systems, needs to be challenged, if other stakeholders, including local communities, are to exercise their grassroots power and experiences they had in the pre-colonial periods. The sustenance of systemic exclusionary approach in Africa should not be allowed to continue in the post-democracy period. A deeper systemic critique of this approach is long overdue and needs to be replaced by a more inclusive or grassroots approach.

Related to this exclusionary approach is how nature has remained more prioritized compared to culture in National Parks across Africa. This prioritization has and continues to exclude local communities in the process, except for the peripheral initiatives such as CAMPFIRE. At MCLWHS, cultural aspects have been under-funded over the years, while lack of capacity/experts in the area has become acceptable, yet rangers are roaming national parks in the name of biodiversity and anti-poaching activities. In response to this, SANParks appointed a General Manager for Heritage Management in 2017 and is moving towards capacitating the Heritage Division which has the huge mandate of looking after all cultural heritage in national parks under SANParks. It remains to be seen whether this division will embark on an aggressive transformatory approach to have culture recognized as a driver and enabler for sustainable socio -economic development WHS located inside PA. This includes capacitating the Department with adequate resources (human, technical and financial). Only time will tell as political support remains aligned to saving the big five in National Parks. This bias further demoralizes local communities, whose perception of heritage combines both cultural and nature as inseparable values of their cosmological and beliefs in any landscape. MCLWHS is not an exception to this dichotomy and contradiction engineered by the exclusionary management biased towards biodiversity. This also relates to how land re-possessed by IDCs through legal processes shall be forever zoned for conservation purposes, including their futile and prolonged fight for increased access into the parks for their spirituality and well-being as a society.

PLATE 4.4 Mapungubwe Hill perimeter dry stone wall (credit: Eric Mathoho@2021)

5

UNMASKING STAKEHOLDERS AT WORLD HERITAGE SITES

Introduction

This chapter systematically identifies, categorizes, and profiles stakeholders at Mapungubwe Cultural Landscape World Heritage Site (MCLWHS) using the historical context, heritage values (significance), legislative environment, interests and relationships among stakeholders, evolving land uses, emerging opportunities, and challenges of managing conservation and development. All these variables were analyzed using historiographical approach and consulting primary and secondary documents to identify stakeholders at MCLWHS. The identified stakeholders include Indigenous and Descendant Communities (IDCs), farmers, extractive industries, hoteliers, hunters, academics, politicians, learners, lawyers, and multiple government departments with varying influence and decision-making powers. By reconstructing stakeholder profiles from the past to the present at MCLWHS, an empirical perspective on their role is brought to the fore. From management and governance perspectives, this brings out the opportunities, conflicts, tensions, and rivalries among stakeholders, claims, and counter-claims over the ownership, land uses and access to resources, power matrixes, and stakeholder engagement approaches at MCLWHS. These opportunities, tensions, and contradictions characterize stakeholder governance typical of MCLWHS and many other World Heritage Sites (WHS) in Africa. This also points to how the State-Based Management Systems (SBMS) and World Heritage Convention (WHC) have betrayed the pioneering spirit, engagement, and decision-making approaches of traditional communities by maintaining exclusionary approach as opposed to inclusivity. Where possible, parallels are drawn from the experiences of other WHS.

DOI: 10.4324/9781003149361-5

Stakeholders: From Passive to Active "Social Power"

Stakeholders have largely been masked in general terms and processes, without intentionally identifying and profiling them and their needs at heritage sites. Identifying and profiling stakeholders, including their interests, is an evolving, involving and complex process. Stakeholders may be identified or reconstructed using value-based, historiographical, and archival approaches, among many other approaches. The process of identifying them from the past to the present awakens otherwise considered passive stakeholders, who become active social power and alternative voices at WHS. At MCLWHS, a historiographical analysis of the past to the present identified as many stakeholders as possible (Taruvinga, 2019). The sources analysed are the history, significance, evolving land uses, and state of conservation reports. Overall, MCLWHS, like other WHS, illustrates evolving multiple and multi-layered stakeholders that invariably have and continue to use, exploit, and maximize on "natural resources to reinforce social, political and economic values" (Carruthers, 2006: 10). The successive communities of MCLWHS have, in one way or another, been involved in economic activities of the landscape. These include "archaeological communities", IDCs, farmers, extractive industries, and many others (Pikirayi, 2011, 2016). Despite this history of successive occupations by different groups, there is a paucity of data for some periods as to who were the occupants of the site (Pikirayi, 2016). As an approach, it was thus important to situate the stakeholder analysis into the chronology of occupation and uses of Mapungubwe using a combination of historiography and archival analysis.

Historiography uncovers the meaning of cyclical and linear time in relation to the experience of the past and the future expectation of concerned people (Rüsen, 1996). It has the ability to answer different questions and within independent intellectual views (Goodman, 2005; Iggers, 2005; Ralushai, 2003; Rüsen, 1996). Historiography sources for MCLWHS refer to historical narratives published over the years, archaeological materials, primary records of the site, and other published materials. This also includes newspaper archives, consultation documents, minutes of meetings, oral traditions, commission reports, and any other documents that can give insights into stakeholder identities, interests, and their relations with others. Overall, archival research overcomes exclusions and silences in other dominant accounts of authorized heritage discourse (Flinn, 2011). It also addresses "colonial categories of knowledge" that "flattened the multi-sided experiences of people in colonies" (Cooper, 1994: 1517). It should be understood that archives "are not passive storehouses of old stuff, but active sites where social power is negotiated, contested, confirmed" (Schwartz & Cook, 2002: 1). However, archival practice is also "political, loaded with meaning, pressures, and consequences" (Flinn, 2007: 1; Schwartz & Cook, 2002). Hence, there is need to have a balanced and analytical approach (Flinn, 2007, 2008).

However, history-making and archiving are never neutral or disinterested activities (Flinn, 2008). Using archival approaches without other complementary methods such as value-based approach used at WHS, creates challenges. Value-based approach identifies stakeholders on the basis of the values and significances they attach to a WHS through time and space. "Through time" creates a links to the archival sources. Also, universally accepted and conventional approaches of identifying stakeholders, such as interest and needs analysis matrix, legislative mandates analysis, and decision-making power levels are integrated within this value-based approach, historiography, and archival analysis at WHS. This creates a holistic and inclusive stakeholder identification approach at WHS. This inclusive approach was used at MCLWHS. All these approaches helped in tracing and reconstructing the emerging stakeholders of MCLWHS, which include the IDCs, farmers, extractive industries, tourism actors and hoteliers, hunters, academics, politicians, learners, lawyers, civil society, and multiple government departments. These stakeholders have varying influences, interests and decision-making powers at MCLWHS. Also, these stakeholders can be analyzed in relation to conservation and developments at MCLWHS. Understanding how the systemic exclusionary approach of SBMS has faced multiple governance challenges emanating from these stakeholders is important in this process. Stakeholder interests, views or perceptions, and contributions are important in understanding their positions on conservation and sustainable development (SD) at WHS in Africa.

"Dormant" Giants Awakened at WHS

The identification and profiling of views, interests, contestations, conflicts, and dynamics of the multiple and multi-layered stakeholders at MCLWHS are herein presented.

IDCs: Spirituality, Intangibles, and Contested Narratives

The IDCs of MCLWHS are constituted by local communities that once lived in the Mapungubwe area from pre-colonial to recent historical periods until they were forcefully moved to pave way for the park. This also includes those being resettled in the area following successful land restitution processes. IDCs are intrinsically linked to the cultural history of the site in its broader territorial definition. Their history and spirituality, cultural practices, intertwining intangible processes, and tangible places in the landscape cannot be delinked from the Mapungubwe landscape (Murimbika, 2006; Taruvinga, 2019). These IDCs include the *Venda, Lemba, Leshiba, Machete*, and *Tshivhula*, all connected in their complex histories as Venda communities (Pikirayi, 2016). Despite the historical, social, or spiritual connections, these Venda communities have always had socio-economic interests in the area since pre-colonial times (Taruvinga, 2019). The historiography and archival records show that they have only been recorded since the 1940s (Esterhuysen & Smith, 2007; Fouche, 1937; Van Warmelo, 1940). IDCs' connection with heritage sites extends to recent times, even though they were resettled outside Protected Areas (PA) during colonial times (Huffman, 2014; Pikirayi, 2016). Deliberate documentation of IDCs intensified in the post-apartheid era with a view of redefining their identities and promoting social cohesion, typical of many similar experiences in Africa. While multiple IDCs are connected with MCLWHS, their respective narratives are contested, especially around the ownership and use of the site. However, their group identity as "*vaVenda*" is more important in identifying them as IDCs of MCLWHS (Esterhuysen & Smith, 2007; Huffman, 2014).

The group identity, "*vaVenda*", is viewed as a present-day variation of the Zimbabwe culture of Kalanga, Karanga, and Sotho-Tswana groups and languages (Huffman, 2014; Lestrade, 1927; Wentzel, 1983). They are considered as an "outgrowth of a merger between Shona (*Khami*) and Sotho-Tswana (*Icon*) ceramic styles" (Huffman, 2012). The *Icon* pottery is found at over 65 sites in Mapungubwe, and the Twamamba, who made Khami pottery, moved into the Limpopo valley (Huffman, 2012). It is argued that the *vaVenda* consolidated their transformation before the arrival of Singo, who moved across the Limpopo in the 1690s (Beach, 1980). Singo, related to Changamire Rozwi, conquered the independent chiefdoms and consolidated *vaVenda* as a nation (Beach, 1980). Singo ruled from Dzata in Nzhelele valley (Stayt, 1931). Historical records and oral traditions indicate that the *Tshivula* dynasty of Twamamba intermarried with the Singo people (Huffman, 2014; Van Warmelo, 1940). Later, and due to internal disagreements, Tshivula moved to Mavhambo, east of the Saltpan. Also, the first Machete residing in Mavhambo was dispatched to the North (Huffman, 2014). On the other hand, the Birwa dynasty (North Sotho group) and the Sotho-Tswana groups lived around Leokwe Hill (Van Warmelo, 1953). Birwa probably occupied the western side of Leokwe hill, which was associated with the ceremonies (Calabrese, 2007). The pottery recovered from Tauyatswala, and associated cattle kraals and stone walling, is similar to that of southwest Zimbabwe (Van Schalkwyk, 1994). Birwa settlements, especially the walling, have a link with Bambandynalo, which is presumed to be the headquarters of Thaha, a Kalanga Chief (Huffman, 2014). Scholars argue that Birwa, who occupied Letshidzhili, which was not far from Leokwe Hill, experienced some tensions with Bambandynalo (Van Warmelo, 1940). Leokwe Hill was abandoned in the mid-1860s because of these tensions (Huffman, 2014). Dzata is the only paramount Chief Capital in Venda, and the throne was abandoned with the rise of the legendary leader Thohoyandou. As such, Dzata was divided into three competing contemporary dynasties of Mphephu, Tsuvhase, and Mphaphuli (Huffman, 2014). The multi-ethnic intermarriages between Birwa, Sotho-Tswana, and Venda contributed to the Sotho-ization of Machete and the Tshivula (Beuster, 1879; Hofmeyr, 1890; Huffman, 2014).

The North and Sotho Tswana groups later spread to the new Bobirwa (west of Blouberg, South Africa) and Gwanda areas (Matebeland south, Zimbabwe), while some Birwa moved to south of Limpopo due to the Ngwato expansion from central Botswana (Beach, 1974). The connection between the Khami phase (Zimbabwe) and Venda broadly signifies the regional connection between Mapungubwe and similar sites in the region, hence the close linkages between lineages in South Africa and Zimbabwe. Sotho-Tswana people intermarried with the Birwa. Also, another Tswana group known as Kwena, under the leadership of Moseri, moved into the area about the same time (Van Warmelo, 1940). Moseri was the grandfather of Tshiwana and Mavhina, the two brothers that are recorded in relation with Mapungubwe (Esterhuysen & Smith, 2007; Fouche, 1937; Ralushai, 2002). There is speculation that Moseri may have moved into Mapungubwe and married the sister of Machete 1 in the 1840s (Fouche, 1937). The Tshivula dynasty also intermarried with Birwa. These intermarriages imply that a multi-ethnic community co-existed and shared the Limpopo valley during Machete I's era (Bonner & Carruthers, 2003).

The Tshivula are considered the most senior dynasty of the Twamamba, dating back to the 15th century, while the Machete lineage is viewed as a minor division of the Tshivula (Van Warmelo, 1940). The Tshivula dynasty could have used Machemma as their early headquarters, which dates to the 15th century according to archaeological excavations (Huffman, 2014; Ralushai, 2002). From an archaeological perspective, this Khami Phase in Mapungubwe marks the arrival of the Tshivula dynasty (Huffman, 2014). It is also argued that Raletaupe, son of Chief Tshivula, was named as "Machete" meaning "Mr Be Quiet" (Van Warmelo, 1940). At that time, Leokwe Hill was under Chief Thaha, who was a Kalanga (Huffman, 2014). Also present at this time was the Sotho Tswana (Huffman, 2005, 2014). When Chief Machete died, his son Rantshana took over but was assassinated shortly after through witchcraft, and this led to the abandonment of Leokwe Hills (Huffman, 2014). The reigning Machete was probably living on the flat alluvial terraces near Limpopo until the 18th century (Holmgren *et al.*, 2003; Trevor & Mellor, 1908). The end of the 18th century also coincided with frequent droughts which made subsistence agriculture less successful in the area. The Machete chieftainship at Mapungubwe eventually ended with the arrival of the Dutch and Europeans in the Limpopo basin as apartheid began taking shape advanced by the former (Huffman, 2014).

Two other palaces that appear to be connected to the Machete dynasty are Edmondsburg ridge south of Leokwe Hill and the Kilsyth area that overlooks the large vlei on Den Staat (Huffman, 2014). Den Staat could have become the capital after the unusual death of Machete II due to witchcraft. A small palace located on Hilda range, which is opposite to Kolope, and Edmondsburg ridge have single palace walls in a defensive position (Huffman, 2014). These two sites could possibly mark the gradual decline of the Venda political power in the area. The Limpopo River played an important role in separating chiefdoms. One of the best-known Venda *Misanda* or Chiefdom is located on the Mmamagwa Hill near the confluence of Motloutse and Limpopo rivers. Another *misanda* stands on a ridge on the Ratho farm, while another one overlooks a stream on Parma (Huffman, 2014). It appears Mmamagwa was the capital of the dynasty separate from Parma and Ratho, and all were separate from Machete (Huffman, 2014). Other *misanda* are found across the Limpopo in Zimbabwe. What is clear is that the Khami and Venda *misanda* are identical thereby signalling the close relationship between the communities of the two neighbouring countries (Huffman, 2014).

The Tsolwe hill, which is located on south-western edge on the Mapungubwe National Park (MNP), is associated with Chief Lesiba (*Leshiba, Lishivha*). Lesiba was another senior Tshivula chief, who usurped power from Mankadiko (brother of Machete I of Leokwe Hill). In order to control Lesiba, who had become a threat, Field Cornet J.G. Duvenage took his guns (Huffman, 2014; Archives of the State Secretary, Incoming documents for 1856, 11: 402). A Mamadi Chief, known as Madidimalo, attacked Lesiba and forced him to flee to Tsolwe (Huffman, 2014). As a result, Mankadiko's chieftainship was restored as they were connected to Madidimalo through intermarriages. Intermarriages promoted heterogeneity in the Limpopo Valley (Huffman, 2014). The death of Machete II marked the decline of the Chieftaincy in the 1930s. The decline may be attributed to the fact that Mapungubwe was no longer as productive as it was in the 19th century (Khami phase). Also, the exhaustion of resources due to the large number of Venda capitals in a very short period contributed as well.

The Venda sites in Mapungubwe confirm that agricultural lands were limited due to poor climate, and this contributed to possible low population numbers and poverty among local communities (Huffman, 2014). Therefore, issues of addressing socio-economic issues among IDCs can be traced back to this period (Taruvinga, 2019). Also, the rinderpest epidemic of 1896–1898 decimated cattle herds that existed in the area and the advent of firepower with the arrival of Boers in the area were contributing factors to the challenges IDCs faced (Huffman, 2014). Ivory trade was used to expand Boer influence in the area, and the acquisition of farms also contributed to the decline of the Machete Chieftainship (Huffman, 2014). A few stonewalls on isolated hilltops on Machete testify to Boer military activities in the area (Huffman, 2014). Boers began to own land in the Limpopo, and a similar process was taking place in the then Rhodesia, now Zimbabwe. After the Boer wars, Europeans become interested in minerals, leading to the establishment of mines such as Seta Mine (Trevor & Mellor, 1908). The net effect was IDCs being pushed out of the Mapungubwe area and are currently occupying the north and south sides of the Soutpansberg Mountains, with others residing much further away from the Mapungubwe.

Contextualizing the narratives of IDCs into the contemporary period of MCLWHS is important. In 2007, IDCs of MCLWHS claimed and reburied about 143 remains originally excavated and housed at the University of Pretoria (UP) (Nienaber *et al.*, 2008; Pikirayi, 2016). The systematic research at the site by UP from 1933 resulted in a huge archaeological collection, which includes human remains. The reburial of these human remains at MCLWHS was led by the Mapungubwe Steering Committee established by the President of South Africa (Nienaber *et al.*, 2008). The Committee included claimant groups (IDCs), Department of Fisheries, Forestry and Environment, (DFFE), Department of Sport, Arts and Culture (DSAC), South African National Parks (SANParks),

Limpopo Provincial Government, South African Heritage Resources Agency (SAHRA), Ditsong National Cultural History Museum, UP, and University of Witwatersrand (Wits). The claimant groups were constituted by the Vhangona Cultural Movement, Lemba Cultural Association, and Tshivula Royal Family. Despite this inclusive composition of IDCs, their consistent representation in Committee meetings was challenged as other participants felt it was not reflective of the inclusive stakeholdership of the IDCs at Mapungubwe (Nienaber *et al.*, 2008). Venda groups were considered overrepresented in the Committee at the expense of other cultural groups (Nienaber *et al.*, 2008). For instance, the San community, who are well organized and had expressed their willingness to be involved in the reburial process, were only invited in the last meeting when everything else was already set for implementation (Nienaber *et al.*, 2008). The question is how far can inclusivity be undertaken given the inherent and underlying contestations among IDCs themselves and the ensuing land contestations to legitimize respective importance? For instance, the San/Hunter-Gatherer communities are now calling themselves the first citizens or nation of South Africa. This is now creating tensions with other citizens.

During the reburial exercise, political agendas and interdepartmental rivalries, especially between DFFE and DSAC, surfaced, thereby derailing processes (Nienaber *et al.*, 2008). This made it difficult to build relationships beyond the reburial process among stakeholders at MCLWHS (Nienaber *et al.*, 2008). Even the funding for the reburial process become a burden for the UP as claimants could not provide any resources as agreed in the Reburial Policy signed by all concerned parties (Nienaber *et al.*, 2008). From the beginning, the reburial process was characterized by claims, counter-claims, negotiations, and compromises by IDCs (Nienaber *et al.*, 2008: 64). This also included renegading on previously agreed positions. However, IDCs greatly assisted in establishing the reburial protocols, and created a platform to correct past mistakes in handling human remains, including scientific interpretations devoid of the spiritual understanding of IDCs on such matters (Pikirayi, 2016). The reburial processes promoted ethical dimensions and traditional protocols in dealing with human remains (Nienaber *et al.*, 2008). The reburial exercise proved that IDCs can "conserve, deploy, and remake the meanings they link to archaeological sites" through their present social conditions and are able to "generate futures using these sites" (Pikirayi, 2011: 120).

Reburial of human remains is a phenomenon associated with many sites worldwide (Legassick & Rassool, 2000; Pikirayi, 2016). Global debates on ethical treatment and respect of human remains by scientists and the African campaign on the need to repatriate these back to their communities were an impetus for South Africa at MCLWHS. However, some scholars called for a negotiated approach on dealing with such sensitive matters (Steyn & Nienaber, 2005). In the process, IDCs demanded that the interpretation of the site should reflect local community histories, while research agendas should highlight their "memory and tradition to counter monolithic, authoritative, dominant, and highly contested narratives about the pasts of these communities" (Pikirayi, 2016: 121). IDCs argued that the research agendas should take into account their own interests (Pikirayi, 2016), including them being viewed as knowledge producers as opposed to being research targets (Taruvinga, 2020).

On the negative side, the *vaVenda* challenged the repatriation process as a fixated approach of past human group identities by SBMS, including universities (Pikirayi, 2016). This did not resonate with their diversity and the growing social complexity of the local histories of the MCLWHS characterized and driven by residential, non-descent, and non-local descent communities of the area (Pikirayi, 2016). The collective voices of the IDCs highlight their spiritual expectations and contemporary needs at the site (Pikirayi, 2016). The reburial ceremony was viewed as a strategy of addressing the long known impact of illegal excavation of human remains at the site by academics (Pikirayi, 2016). While IDCs were involved in the reburial, they were not in full control of the processes as would have been expected of traditional custodians. As a remedy, heritage management should empower local communities to engage their past in a manner that benefits their present lives too (Pearson & Sullivan, 1995). Furthermore, IDCs demanded access into the site for their spirituality, which SANParks still has to formally resolve (Pikirayi, 2016).

The reburial ceremony came with additional challenges associated with the process and materials used for the funeral rites against what is provided for by the spirituality of IDCs. The human remains were reburied in non-degradable plastic materials (high density polyurethane, which can last 400 years if not exposed to thermal or chemical influences) as opposed to degradable traditional materials which the IDCs may have preferred in line with the concept of allowing buried remains to become part of the soil. Also, an access manhole was included in the structures of the burial chambers to allow future access. The involvement and systematic control of the process by academics during the exercise defied the traditional protocols shrouding the rituals, secrecy, and practices of such ceremonies in Africa. Only "kin and kith" are supposed to witness such processes. The use of non-degradable plastic materials further reinforced a long suspected but undeclared hope that with passage of time, academics could access the remains for further research using new technologies (Nienaber *et al.*, 2008). Harbouring future scientific hopes in the process defied the traditional approach to burials where degradable materials are used. This violated traditional rites and the rights of IDCs. While it is common knowledge that funerals bring people together in the community irrespective of their connection with the deceased, it is only the descendants of the deceased that lead the burial processes and not strangers like academics and SBMS. By controlling the process, Academics, who are supposed to observers, were indirectly saying the IDCs were not knowledgeable about their own funeral rites. Of course, politics and high-ranking politicians played a role in the process. The top-down approach of scholars in the repatriation exercise and their continued "*archaeologicalization*" of communities requires a mindset shift in order to redefine the rules of engagement and practices (Pikirayi, 2016: 122).

The question now is, when shall these ancestors be allowed to rest and, how should they be wrestled from the hands, eyes, and evolving advanced research tools of these academics? However, interfacing of IDCs with academics during this exercise represented a long-overdue dialogue on these sensitive and emotive issues in Africa. Once again, the site was re-consecrated as a sacred burial landscape, of which IDCs demanded unconditional access for these rites on a regular basis in the future. Though it is traditionally genuine to care for your ancestors, it could also be interpreted as a process of legitimizing land claims in the area in the future. The reburial exercise provides an interesting dimension relating to IDCs, which had never been properly documented in local histories until they emerged to stake their claims at the site (Pikirayi, 2011). While this reburial exercise in some ways authenticated the ownership of the land by IDCs, it also demonstrated that no modern community occupies the MCLWHS or can lay claim to an organic association with it (Carruthers, 2006). Recently in Ghana (2020), UNESCO was consulted on whether a deceased Royal could or could not be buried at the Asante Buildings WHS, and regrettably, the recommendation was that this person be buried outside the site to avoid it becoming a grave site. The question is whose decision is this supposed to be, UNESCO, State Party (SP), or IDCs? The usurping of local powers remains reinforced at WHS in Africa. The thinking that MCLWHS was "spiritually and culturally uncontested landscape" is now a myth due to new information generated by research in the area (Carruthers, 2006: 3). The current counter-claims among IDCs who have lodged a number of land restitution claims in the area are testament to this history of contestations through time at the site.

Academics: Own Interest as Professionals Representing Others!

Academics and universities have interacted with MCLWHS since 1933 until now, making them the "unofficial representatives" of the "archaeological community" and IDCs of MCLWHS. This is typical at many heritage sites in Africa. The archaeological communities cover the Stone Age and Iron Age periods of MCLWHS. MCLWHS is widely understood today due to the extensive research undertaken by academics. They have advanced science until recently when Traditional Management Systems (TMS) begun to be considered as well. The UP dominates, and still wields the academic authority it established in 1933 at Mapungubwe. UP signed an agreement with Mapungubwe site soon after it was "discovered" by Van Graan in 1933 (Apley, 2000). UP took ownership of the site, the research process, and the excavation extensively (Apley, 2000). In the early years of research, UP successfully blocked mineral prospecting or mining at the site in order to pave way for research (Apley, 2000). This marked the beginning of the academic stand-off against extractive process at the site, a process that continues today by many academics. UP research, including its stance on the mining at that time, increased awareness on the significance of the site worldwide. However, extractive industries were beginning to emerge at a large scale in the area (Apley, 2000). As a pre-occupation of colonial administrations in Africa, mining and agriculture took centre stage while heritage sites were destroyed, and local communities were dispossessed access to such resources and ancestral lands.

Other universities and scholars have also undertaken interdisciplinary researches in the Mapungubwe area. These include University of Cape Town (UCT), Wits, University of Johannesburg (UJ), Northwest University of South Africa, University of South Africa (UNISA), (all universities are in South Africa), University of Botswana (UB), Botswana, University of Zimbabwe (UZ), Zimbabwe, and many other research institutions, including partner universities from outside Africa. Multiple themes are being researched: wildlife management, biodiversity, geology, archaeology, heritage management, TMS, history, tourism, and hospitality aspects. Universities such as Wits, UCT, and University of Venda (UNIVEN) have intensified their research in the area over the last two decades. UZ and UB are collaborating with South African universities to further research in the area as well. The is helping in developing holistic, inclusive, and decolonized stories of MCLWHS (Ndoro, 2008; SANParks, 2013). The interest of academics and universities at MCLWHS is advancing professional growth, knowledge production and conservation. Both the academics and academic institutions have contradictory relationships with the land uses at MCLWHS from many viewpoints.

From a positive side, and firstly, academics have benefitted through research since 1933, with some of them carving careers and high-level academic qualifications from MCLWHS. Second, MCLWHS has become an academic laboratory for interdisciplinary research, furthering the understanding of the the diverse heritage of the area. This benefits various academics, social scientists, public, and community organizations. Third, this academic research has benefitted from multiple funding grants coming through Environmental Impact Assessments (EIA), Heritage Impact Assessments (HIA), and commissioned specialist research by land users. This happens as part of compliance with heritage regulatory frameworks of SPs. For instance, De Beers and Coal of Africa (CoAL) have funded conservation and management planning projects as part of mitigating negative impacts of their extractive processes, which the Management Authority would not have been able to fund due to financial limitations.

However, and contrary to the above positives, most academics are anti-development and systematically embark on campaigns pushing a conservative conservation agenda. This is contradictory in that while they understand the needs of other stakeholders, in particular developers and local communities, they prefer to side with strict conservation approaches. They are contradicting the very nature of the archaeological communities who practised agriculture, mining, and other extractive processes for their own livelihoods in such landscapes. These communities were architects of economic processes in the area. The anti-development campaigns are pushed through Professional Associations and legal litigation (CALS, 2016). They derive their power from heritage regulatory frameworks, while at the same time using funding from land users or developers to conduct impact assessments which empower them with information for litigation and for their own academic growth. With the exception of a few, who holistically look at the bigger picture in relation to the socio-economic conditions at the local levels of WHS, the rest usually offer hard-line positions without room for interrogating alternative strategies favouring both conservation and development. Others even provide inconclusive EIA or HIA reports, like what happened at MCLWHS in relation to coal extraction by CoAL, which was only resolved through subsequent assessments by a different team. This induced a second HIA conducted by different professionals, which was then used by the SP to get a favourable decision from the WHC.

At MCLWHS, academics are a double-edged sword as they benefit on one hand, and on the other, they choose to remain oblivious to the socio-economic needs of communities, who are the knowledge producers. This is an ethical matter for the profession itself. Academics are increasingly being seen as "scientific obstacles" to the growing needs of stakeholders at heritage sites. Also, the fact that academics are themselves also stakeholders is often not considered, and they are erroneously considered spokespersons of SBMS and WHC. They often act as spokespersons of fossilized "archaeological" communities. They represent their own interest in the name of others!

Farmers: Crops, Livestock, Fruits, and Wildlife

Colonialism ushered new farmers and almost relegated subsistence farming to the periphery of agricultural and economic production systems in Africa. Local communities become slaves (labourers) to their new masters. Prior to the arrival of the Dutch, the British had invaded South Africa in a wave that eventually spread to Zimbabwe, Malawi, and Zambia. Later, the Dutch colonial frontier opened up the Limpopo area to frontiersmen for leisure hunting, mining, and farming (Berry & Cadman, 2007; Carruthers, 2006). The European hunters arriving in the Limpopo area in the mid-1800s included Fredrick Courtney Selous, Henry Cotton Oswell, and Roualeyn Gordon Cumming (Berry & Cadman, 2007). The Buys people arrived from Eastern Cape, followed by the Voortrekkers in the late 1830s (Carruthers, 2006). The Anglo-Boer war (1899–1902) marked the establishment of land settlement schemes in the Limpopo area (Carruthers, 2006: 6). It is the only recorded war to have taken place in this area (Berry & Cadman, 2007). This disrupted subsistence farming of IDCs, including triggering their relocations to marginal areas.

The extension of the railway line and the "burgeoning copper industry" during apartheid brought about development in the Musina area (Carruthers, 2006: 6). However, the drought and depression of the 1930s halted this development (Carruthers, 2006). By the end of World War 1, the apartheid government was advocating for every available farm be settled and the farmers be given support for crop or animal husbandry (Berry & Cadman, 2007). This included supporting "speculator" farmers who had the practice of moving cattle onto the land for fattening before selling them off to buyers (Berry & Cadman, 2007). However, the carrying capacity of the land, including its suitability for such farming options, had not been assessed (Berry & Cadman, 2007). Furthermore, cheap land was also given to about forty employees of the Messina Copper Mine as an incentive to keep them working at the white-owned mine (Berry & Cadman, 2007). From this, it is clear that the land uses at Mapungubwe were beginning to dramatically change. Also, land ownership changed hands, with the incoming Boers becoming legal landlords while IDCs were pushed to Venda communal lands, far away from MCLWHS.

In the last century, commercial farming, with extensive irrigation systems, has become more pronounced in the Mapungubwe area. Since the 1950s, citrus, tomatoes, oranges, and other crops have been produced, including rearing cattle alongside wildlife management (Carruthers, 2006). The 2013 land use audit shows hectares allocated for various agricultural activities: natural grazing (2751 hectares), citrus (715 hectares), maize (523 hectares), tomato (351 hectares), and cotton (46 hectares) (DEA, 2013a). Though cattle ranching was intensified in the 1970s, it was gradually replaced by conservation-friendly game ranching (DEA, 2014b). Agriculture contributes to the economy of the greater Musina Local Municipality (DEA, 2014a).

Private Sector: Extractive Industries, Profits, Unfulfilled Promises, and Social Responsibility

MCLWHS has witnessed private sector players coming through various approaches as either purely private initiatives, public-private partnerships, or Corporate Social Responsibility (CSR). The CSR approach supports SANParks in educational programmes raising awareness on the significance of MCLWHS. One such example was the memorandum of agreement (MoA) signed between SANParks and *Pick n Pay* in 2004 supporting the "Kids in the Parks" (KiP) educational programme. *Pick n Pay* is a large chain supermarket in South Africa (DEA, 2014a). The KiP, designed to introduce 5000 children (Grades 5–7) and 50 teachers from previously disadvantaged communities to importance of national parks, was supported by DFFE and the Department of Education. *Pick n Pay* provided a bus and other goodies, while SANParks offered accommodation, expertise, and facilities for the educational tours to MCLWHS. During the tours, children and teachers were exposed to environmental education and the cultural significance of MNP. This programme surpassed its target by catering for 7069 children and 430 teachers. This programme imparted writing, observation, and experiential skills to learners and teachers. It also strengthened school-based environmental management practices, conservation ethics in local communities, environmental educational policies, and efforts in preserving and sustaining heritage in its diversity (DEA, 2014a). However, this kind of SCR has been slow to manifest at most heritage sites due to lack of immediate benefits for the supporting companies. For the private sector, all they need is to be guaranteed mileage in terms of publicity, which then translates to their market share or becoming the brand of choice by public. The only element of the KiP programme which was not well considered was the budget for follow-ups with the schools, which forms a valuable part of the sustainability of the programme (DEA, 2014a).

Issues relating to the incompatibility of resource utilization and conservation have dominated heritage discourse on the continent for a long time (Meskell, 2011). These are driven largely by the private sector. At MCLWHS, mining has been in existence from the past, and continues to be an economic activity. MCLWHS is located in an area very rich in mineral deposits (CALS, 2016). De Beers established the Venetia Diamond Mine and begun to operate inside the 26000-hectare Venetia Limpopo Nature Reserve in the early 1990s. Another mining company, CoAL, granted a licence in 2010, was expected to extract an estimated 441 million tonnes of coal from a site located south of MCLWHS for 29 years (DEA, 2014b). Stakeholders associated with this activity include but are not limited to geologists, explorers, mining companies, workers' union, employees, government departments, and individuals interested in exploitation of mineral resources in the area.

The extraction of coal at Mapungubwe in 2010 is a good example of how different stakeholders respond to developmental priorities at WHS. The open cast mining methodology at the Vele Colliery attracted the attention and varied responses of multiple stakeholders, among them UNESCO, heritage experts, professional associations, environmental institutions, IDCs, and the public at large. On the receiving end of these varied responses was the developer and government departments responsible for EIA, HIA, and other permitting processes. These included DEFF, Department of Mineral Resources (DMR), Department of Water Affairs and Sanitation, SANParks, and SAHRA. Underlying these interactions among stakeholders was the dilemma between conservation and development as MCLWHS "may have one mining stain on its veneer but its spiritual and natural specialness runs much deeper than that" (CALS, 2016: 13). Despite this, it was evident and is still today that mining contributes to the economy of the area (DEA, 2014b). Mining potential remains higher in the Mapungubwe area (Plate 5.1) due to the existence of rich mineral deposits (DEA, 2014). For instance, Riedel farm in the East was not included in the MCLWHS as there is potential to yield high profits from mining (DEA, 2014b). Coalfields of Mapungubwe also contain methane gas, which is not yet being exploited (DEA, 2014b).

Mining-related stakeholders at MCLWHS stretch from the pre-colonial period to the present as presented in Chapter 4. What has changed is the magnitude and scale of extractive industries in the area, a process which intensified in the colonial and post-colonial period. This is equally applicable in the African region. Archaeological communities extracted minerals at Mapungubwe, a process which continued with IDCs, and was commercialized by colonial authorities as the scramble for Africa heightened. The British South Africa Company, led by imperialist John Cecil Rhodes, explored, extracted, and looted minerals across Southern Africa. Compliance with EIA and HIA laws appears to have

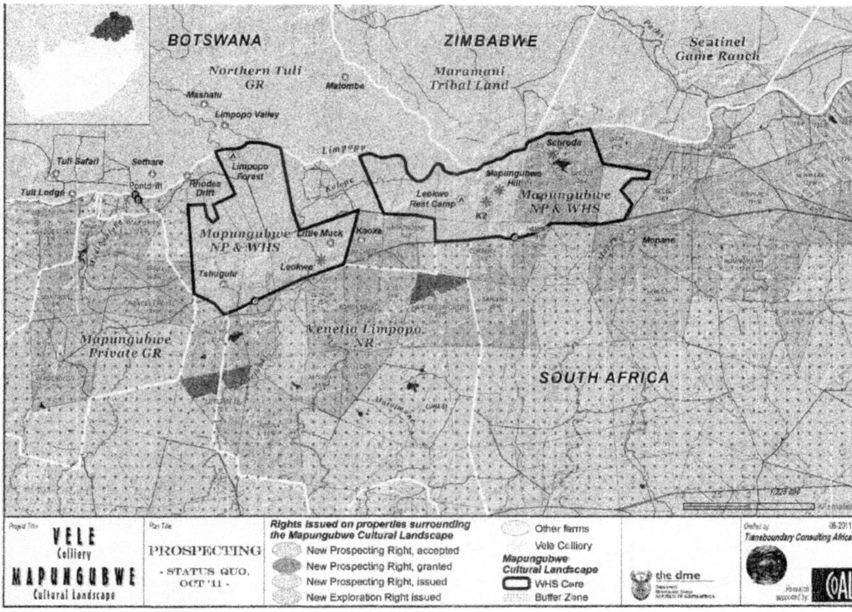

PLATE 5.1 Mining exploration and MCLWHS (Source: Siyathembana, 2012)

been weaker in colonial times than today, though challenges are still rife in this area. EIA only become intensified in the post-colonial era as a way of stopping the hemorrhaging of resources by the private sector without respecting the environment and cultural resources. Monitoring of such extractives has become an intense process worldwide, including the emergence of "Green Corps" among many others. During colonial period, Mines and Minerals legislations were given too much power over all other legislations, a situation that still prevails in post-colonial Africa, for example in Kahuzi-Biega National Park in eastern Democratic Republic of the Congo.

Extractive industries at MCLWHS have attracted the attention of the following stakeholders:

- **Environmental Justice/Legal Stakeholders**
  The Centre for Applied Legal Studies (CALS) at the Wits applied various legal and social interventions promoting environmental justice in view of the coal extraction project at Mapungubwe (CALS, 2016). Environmental justice is concerned with implementing "social justice in relation to the environment" (CALS, 2016: 45). Social justice processes demand that "affected communities equally participate in the decision making process regarding impacts of development on their lived environment" (CALS, 2016: 45; Stacey, 1999). Such participation is expected to lead to "equitable sharing of benefits and burdens of development" (CALS, 2016: 45). Social justice respects, protects, promotes, and fulfills human rights of the affected

societies (CALS, 2016; Stacey, 1999). In this process, SPs should be the "primary bearer and protector of the rights of individuals" as international laws do not "bind corporations" yet they have "influence over people's lives" (CALS, 2016: 46). Mechanisms of accountability need to be supported by effective governance systems (CALS, 2016).

At MCLWHS, CALS argued that while Section 24 of the South African Constitution gives rights to the environment, it should be acknowledged that human rights to an environment are "not harmful to health and well-being" of the society (CALS, 2016: 45). CALS argued that CoAL operations at MCLWHS would deprive people of their rights to have a non-harmful environment and posed threats to water, food security, and livelihoods due to environmental degradation (CALS, 2016). In order to retain the rights of communities, CALS used various approaches: litigation, negotiations, and collaborative compliance monitoring against CoAL (CALS, 2016). A Community Engagement Policy (CEP) was developed and defined the principles and engagement rules with local communities to ensure mining respected heritage in totality (CALS, 2016). They further interrogated the governance of MCLWHS and the Greater Mapungubwe Transfrontier Conservation Area (GMTFCA) to ensure rights of communities are protected.

- **From "Mapungubwe Action Group to Save Mapungubwe Coalition"**
  The Mapungubwe Action Group (MAG) compromises commercial farmers, IDCs, private conservancies, and lodger owners. MAG has threefold objectives: (i) protect the integrity of MCLWHS, (ii) ensure that development benefits inhabitants and interested and affected parties of Greater Mapungubwe Area, and (iii) promote ecologically SD in the area (CALS, 2016). Emerging from the MAG, the "Save Mapungubwe Coalition" (SMC) was subsequently formed and supported by professional associations, among them Association for Southern African Professional Archaeologists (ASAPA), Birdlife South Africa, Wilderness Foundation South Africa, World Wide Fund for Nature South Africa, and Peace Parks Foundation (CALS, 2016). The SMC became a localized platform for sharing knowledge and expertise, pooling of limited resources, and combining different forms of legal expertise in the environmental sector (CALS, 2016: 27).
- **Decision-makers and Extractive Industries at MCLWHS**
  Decision-makers at MCLWHS include DFFE, SANParks, DSAC, DWAS, DMR, Limpopo Department of Economic Development, Environment and Tourism (LEDET), SAHRA, Vhembe and Capricorn District Municipalities, PHRAs, and UNESCO. All these decision-makers, except for UNESCO, are backed by South Africa legislation. The Convention is the backbone of UNESCO. CALS (2016) points out the different stances taken by DFFE, DMR, and DWAS on the coal extraction matter. DEFF viewed the mining operation as a threat to the PA as a whole (but at the same time had an undeclared soft approach for its contribution to national development priorities),

TABLE 5.1 Roles and responsibilities of SMC members

SMC member organization	Area of expertise and influence
Association of Southern African Professional Archaeologists (ASAPA)	• Upholding integrity and ethics of professionals • Heritage impact assessments • Significance assessment • Research into critical heritage thematic areas • Lead discussions with DEA on buffer zone modifications
Birdlife South Africa (BL)	• Conservation of birds in South Africa • Protect important and threatened bird areas • Facilitate stakeholder collaboration in conservation projects
Wilderness Foundation South Africa (WF)	• Protect and sustain African wilderness and wildlands • Educational programmes on biodiversity
World Wide Fund for Nature South Africa (WWFNSA)	• Inspire South Africans to live in harmony with nature • Protect the integrity of natural ecosystems as sources of sustainable livelihoods • Protect water systems under threat from developments
Peace Parks Foundation (PPF)	• Establish Transfrontier conservation areas (TFCAs) in South Africa, Mozambique, Swaziland and Zimbabwe • Facilitate funding for the establishment of TFCAs
Endangered Wildlife Trust (EWT)	• Protect the integrity of threatened species and ecosystems • Align ecosystems to broader social and economic imperatives • Multi-stakeholder governance

while DWAS had a softer approach (CALS, 2016). The issuing of a mining licence to CoAL by DMR was ample evidence of an unbalanced approach to SD in the Mapungubwe area by government decision-makers (CALS, 2016). For DMR, it is difficult to push an environmental agenda and at the same time support national development targets by permitting mineral extraction. DSAC, another government department, is unable to enforce her mandate given the financial limitations and the remote-control approach on cultural heritage through a Western Cape-based agency, SAHRA (CALS, 2016). At Provincial level, LEDET has the difficult responsibility of balancing economic development, environment, and tourism in the greater Limpopo area (CALS, 2016). The same applies to District and Local Municipalities, which MCLWHS falls under. This paints a picture of national and provincial State players fighting for their own specific mandates but accountable to the same government (Taruvinga, 2019). What is different is the political backing of these structures as South Africa is polarized in parliament due to existence of a considerable number of opposition parties. On the other hand, and as previously discussed, UNESCO wielded its power through Committee decisions, hence enforced Reactive Monitoring Missions to the site. UNESCO did not support this coal extraction initiative at all and was pushing very hard for a moratorium approach.

- **ASAPA and Extractive Industries at MCLWHS**
  ASAPA is a professional association of heritage and related experts for Southern Africa, whose main objective is to promote research and conservation and upholding ethics or professionalism among its members. ASAPA is a watchdog association for heritage ethics and practices in Southern Africa. At MCLWHS, and as part of the MAG and SMC, ASAPA provided technical advice on cultural heritage matters. They consistently raised flags against coal extraction, including undertaking litigation measures against the mining company and DFFE (CALS, 2016). However, the heritage values of MCLWHS put forward by ASAPA were still embedded in the archaeological and World Heritage framework of the site without in-depth reference to the intangible values of IDCs who are supposed to be beneficiaries (CALS, 2016). This paints a picture that the decolonization of heritage is just a process justified for academic reasons without it practically benefiting the broader public as a strategic tool for negotiating beyond the realms of authorized heritage discourse and science. IDCs are active knowledge producers, rather than subjects of research, and as such their voice matters in the process. Their heritage resides in the spirituality of their cultures, which can only be accessed through them (Chirikure & Taruvinga, 2017). Their history and culture should have framed how environmental justice would benefit them as opposed to their history being used to justify the need to uphold conservation only (Taruvinga, 2019). SBMS, professional associations, and academics have, regrettably so, retained the monopoly of representing the IDCs in Africa. This trend needs to be dismantled as IDCs have their own voices, like how the Maasai have done so in East Africa. Sadly, IDCs have no financial capacity to push their own interests or present acceptable credentials to become members of professional associations. The question is who should represent IDCs in professional association meetings deciding matters that affect them?

Land Restitution: Restorative Aspirations, Judiciary Players and the Reality!

South Africa, just like all previously colonized African countries, is not an exception to land restitution processes of the post-colonial period. Land is a resource for socio-economic development, well-being of society and a facilitator of nation building (RDLR, 2017). This process is initiated by IDCs as claimants. Other role players include lawyers, land commissioners, magistrates, community advocacy officers, provincial and district officers, and NGOs representing and facilitating community restorative aspirations. This includes surveying Departments and Community Property Associations. These stakeholders are either validators, current custodians, or facilitators of processes relating to land claims.

Land claims by IDCs at MCLWHS are supported by the Restitution of Land Rights Act of 1994 (amended in 2014) and its judiciary mechanisms. The act provides for different resolution options for restitution claims: receiving land, cash compensation for land, and any other resolution as agreed in the arbitration process. DFFE and Lands Affairs have signed a MoA that provides a framework for dealing with land claims in PA. As of January 2008, 74,698 out of 79,696 claims had been settled in South Africa (DEA, 2008). Of all PA in South Africa, Limpopo Province had 26 claims compared to Mpumalanga with 30 (Figure 5.1). These land claims can be interpreted as "taking back" IDCs into their ancestral areas where they were once evicted from. This is the restorative aspiration of IDCs in Africa. The claims have no respect to other land uses (i.e., farming, hoteliers, mining, etc.); all they want is their land back. Land restitution has witnessed protracted negotiations between current land owners and claimants as facilitated by applicable legal systems and instruments. The Limpopo Land Claims Commission is the driver of this restitution process in the Mapungubwe Area.

Land restitution has resulted in tensions and accusations among the players involved at MCLWHS. In 2013, the Leshiba and Machete families accused DMR of issuing exploration licences to mining companies on land they had laid claims without their consent. In response, both DMR and Department of Rural Development and Land Restitution (DRDLR) disputed this interpretation (IOL, 2013). They rightly argued that the law does not prohibit the granting of a right over land where there is a claim, hence their reason for issuing the licences (IOL, 2013). Claimants can only have a say once the land is ceded back to them as the rightful owners. DMR further explained that it is mining

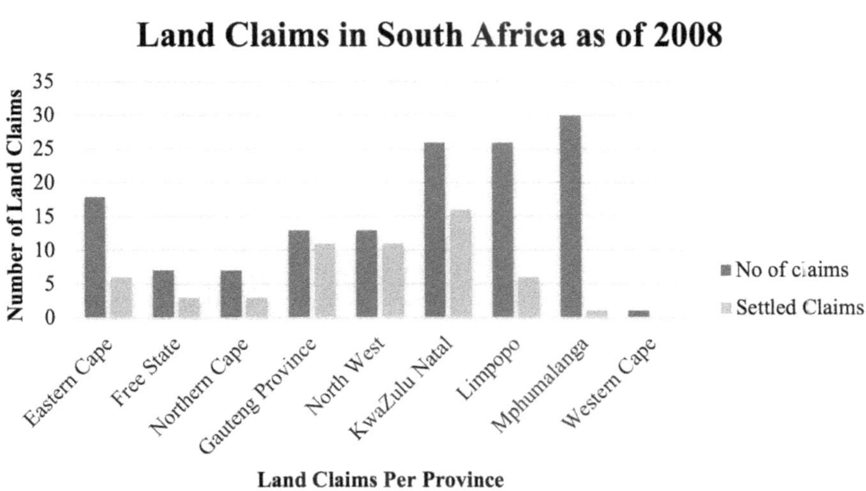

FIGURE 5.1 Number of lodged and settled land claims in PA of South Africa

(Source of statistics: DEA now DEFF)

companies, in this case Vele Colliery, who are supposed to consult communities in the process (IOL, 2013).

Land claims pose another challenge in that once a decision is made to award the land back to the concerned IDC claimants, internal fights regarding ownership and use of the land become rife among the claimants themselves. This was the case of Machete and Tshivuli families' conflict based on internal disagreements on the ownership of the restituted land. This defeated the united front they portrayed during the land claiming process. Additionally, resettling IDCs in their former areas is an involving and costly process with unintended consequences, such as change of ownership and compensation for person handing over the land. Furthermore, land restitution is a complex legal process involving detailed research to validate claims, ensuring transparency from a legal perspective, and determining the cash compensation for the IDCs where moving back into the area is now impossible. MCLWHS is riddled with the outstanding land claims by IDCs, which may in the long term have undeterminable impacts on both conservation and development.

GMTFCA Stakeholders: "Mixed Grill"

The GMTFCA covers a complex mosaic of communal areas, national parks, private conservancies, tourism players, mining and public facilities in Botswana, South Africa, and Zimbabwe (Berry & Cadman, 2007). These different land uses automatically generate multiple stakeholders with different aspirations, hence a "mixed grill". Socio-economic development is a major issue in the area, with tourism, mining, and agriculture dominating. Since the 1940s, conservation has dominated existing land uses in the area (Hoogendoorn *et al.*, 2019; Sinthumule, 2016). The GMTFCA approach demands more land for conservation, which increases human-wildlife conflicts (Sinthumule, 2016). It is argued that "conservation and socio-economic objectives are not mutually reinforcing, rather they are in conflict" in the area (Sinthumule, 2016: 113). This "creates a gap between conservation agencies, local communities and irrigation farmers" in the area (Sinthumule, 2016: 113). Also, human-wildlife conflicts are an issue in the area, especially in the Maramani Communal Area (Zimbabwe). However, the "integration of various land-use practices within one area benefits conservation goals at the expense of local communities and irrigation farmers" (Sinthumule, 2016: 103). Addressing local community livelihoods and development remains a challenge in the GMTFCA (Sinthumule, 2016).

Military Stakeholders: Peace, Security, and Border Patrols

Military outfits have and continue to play a role in Africa as enablers of peace, security, and protecting nations, though sometimes they are used for the wrong reasons! The struggle for freedom in South African begun with wars of resistance against colonial domination waged by traditional chiefs, followed by a well-coordinated national struggle against apartheid (Kros, 2007). In the process, national movements such as the Pan African Congress, Black Consciousness Movement, African National Congress, among many others, emerged (Delius & Cope, 2007; NHC, 2011). This was in response to the creation of the Union of South Africa in 1910 by the colonial government, and subsequently the apartheid (racial) governance system. The majority (black people) became the minority in their own mother country. The Apartheid era (1948–1994) witnessed gross human rights violations against blacks throughout South Africa. The UN eventually declared apartheid as a crime against humanity (NHC, 2011). During the 1970–1980s, the apartheid army used the Mapungubwe area as a defence line against terrorists destabilizing the country from the neigbhouring Zimbabwe (Carruthers, 2006). The Kruger National Park was used in a similar way to create a security buffer countering guerrilla movement to and from Mozambique. But sometimes the military use of PA would be rejected by the army, as was the case with the Rustenburg Game Reserve in South Africa (Berry & Cadman, 2007). At present, military units are functioning as borderline control resources. In line with this, Mapungubwe confluence area is heavily controlled by the army and is inaccessible to local communities who are forced to use formal entry points such as Musina and Ponte Drift Border Posts. This maintains security of the nation, including reducing incidents of illegal border crossings from neighbouring countries. This has intensified with the COVID-19 pandemic as boat and helicopter patrols have been reinforced to assist with the integrated management of the confluence area. The presence of military personnel and base stations is an extra layer of security for MCLWHS. It also assists in combating illegal extraction of resources from such areas. However, corruption has been recorded too within this sector with regard to facilitating illegal crossings to and from South Africa. However, border controls remain critical to controlling the hegemony of any State, and South Africa is no exception. The army or military personnel, as well as security cluster operating in the area stakeholders of MCLWHS.

Tourists: Domestic, Regional, and International Arrivals

MCLWHS, in all its tier status, is a tourism destination and attracts domestic, regional, and international visitors. Some are repeat visitors, especially domestic ones. The significance and three-tier status encourages tourism and prompts national pride and social cohesion (Marschall, 2005). Tourists visit the site for multiple reasons, among them to explore a new destination, relax, family recreation, and educational reasons (Herman, 2013b; Van der Merwe *et al.*, 2009). In terms of what attracts or motivates visitors at MCLWHS, heritage attributes are ranked third highest with a mean of 3.47 out of 5 attributes (Herman, 2013a). The other top four motivators are: park attributes (3.15), escape attributes (3 64), and natural attributes with a rating of 3.87 (Herman, 2013a: 943). From a heritage perspective, visitors are interested in learning about culture and history and experience what a WHS is all about (Herman, 2013a). The increasing number of visitors to MNP "generates market growth for the park" as a tourist destination (Herman, 2013a: 943). Overall, tourism activities include bird watching, heritage tours, game drives, eco trails, confluence lookout, guided and hiking trails, and mountain biking. MCLWHS elevates the tourism profile of Limpopo Province (Carruthers, 2006). Research on visitor profiles of the site shows that these cut across all age groups and different source markets (Herman, 2013a).

The Mapungubwe Tourism Initiative and the Annual Mapungubwe Arts Festival have become cornerstones of tourism experience in Limpopo (Carruthers, 2006). Also, the Culture Week, introduced by SANParks since 2018, adds a flare to tourism by attracting ordinary South Africans to the site every year for almost free. This has been diversified now to include the Mapungubwe Annual Lecture Series. This is designed to ensure that visitor experiences are locally grounded and exposed to the broader and untold historical significance of the site (Chirikure & Taruvinga, 2017). The Inaugural programme in 2018 had fourfold objectives: (i) a seminar initiating discussion between IDCs and scientists on the process of decolonizing knowledge, (ii) interpretation and presentation of MCLWHS, (iii) cultural performances to celebrate the Mapungubwe Kingdom as an African achievement with tentacles extending into Botswana and Zimbabwe, and (iv) the Mapungubwe Annual Lecture Series (Chirikure & Taruvinga, 2017).

Visitorship trend for MCLWHS shows a gradual increase over the years (SANParks, 2010b, 2012). The current interpretation centre, whose Architect, Peter Rich, won the "World Building of the Year" during the World Architectural Awards in 2009, has played a role in increasing visitors to MCLWHS. However, there are limitations for MCLWHS as a tourist destination: being remotely located on the northern edge of South Africa, the absence of public transport between MCLWHS and Musina town, absence of substantial interpretation of cultural values, limited shops and alternative restaurants, harsh weather conditions (very hot), and threat of malaria (DEAT, 2002b; Norton, 2000; SiVEST Consortium, 2002). In addition, the MCLWHS has the lowest average occupancy rate of 25% compared to other National Parks in South Africa (SANParks, 2010a). Despite this, MCLWHS provides accommodation and interpretive facilities inside the park. Accommodation facilities include the Tshugulu Lodge, Leokwe Rest, Vhembe Bush, Mazhou, and the Limpopo Forest Tent Camps. However, outside the park, private accommodation and restaurant facilities exist. These accommodation facilities include Dongola Ranch, Kaoxa Bush Camp, Chinaka Bodge, Mopane Bush Lodge, Iphofolo Lodge, and many others. Roads connecting the main gate and areas of interest in the MNP have been developed and continue to be maintained to avoid degradation and erosion resulting from continued use by tourists, while roads outside the park are maintained by Provincial Government Authorities.

The COVID-19 pandemic negatively impacted visitorship to MCLWHS, but is also providing opportunities to do things differently, including advancing digitization and digitalization approach, as well as use of drone technologies for monitoring activities at MCLWHS. These internal and external factors may

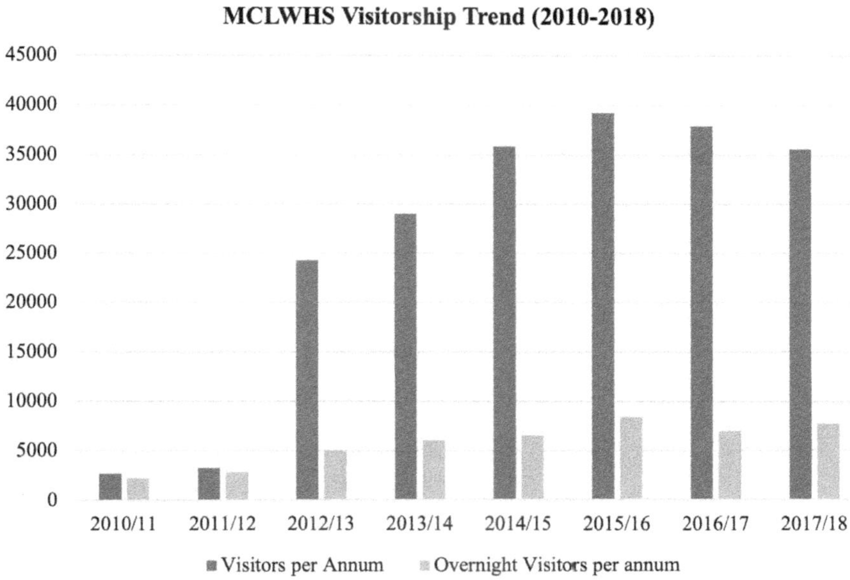

FIGURE 5.2 Visitorship trends at MCLWHS (Source: MCLWHS, 2018)

threaten the future and sustainability of tourism at MCLWHS (Herman, 2013b). It is also argued that the adaptive management strategy of SANParks does not address the relationship between tourism, world heritage issues, mining, TFCA, and private ownership of land in MCLWHS (Herman, 2013b). This matter has been left to individual proprietors in the area.

Politicians and MCLWHS

The South African World Heritage Convention Committee (SAWHCC), established by DFFE, advises the Minister on World Heritage issues in South Africa. While SAWHCC is driven by good practices and appointed experts, its functioning cannot be delinked from aspirations of politicians beyond WHS. At MCLWHS, SAWHCC was kept abreast on the coal extraction matter through technical briefings by the World Heritage Unit (DFFE) and SANParks. Having considered all submissions made regarding the matter, SAWHCC recommended DFFE develop guidelines for future decision-making processes (DEA, 2013b). Furthermore, SAWHCC recommended an Environmental Management Framework to mitigate the impacts of mining and also endorsed the offset negotiations with CoAL (DEA, 2010). It also recommended the elevation of the matter to the Presidency via Cabinet in view of the status of MCLWHS (DEA, 2013a). The Committee warned the Cabinet that mining could result in MCLWHS being placed on the List of World Heritage in Danger by UNESCO (DEA, 2011a). Additionally, the Committee alluded to the negative impacts on the GMTFCA (DEA, 2011b: 1), including the extension of MCLWHS into Zimbabwe and

Botswana (Taruvinga, 2019). The Cabinet memorandum also acknowledged the political complexity of this matter given that many government departments were invariably affected by this development, and all had varied positions on the matter. Hence, Cabinet was requested to provide political guidance towards obtaining a collective and balanced position for the SP (DEA, 2011a).

Complicating this collective position is that DFFE was expected to defend both mining and conservation at MCLWHS (DEA, 2011a). Contrary to this, DMR continued issuing prospecting licenses in the area without consulting DFFE. This was against the spirit embodied in a moratorium announced by DFFE on such activities at MCLWHS. This demonstrates the difficulties and contradictions among government departments on managing conservation and development at heritage sites due to the views of presiding politicians. While Cabinet was processing the matter, the Parliament of South Africa, in its oversight capacity, demanded answers from DFFE in view of the successive UNESCO decisions on the matter (DEA, 2011a). In response, DFFE indicated that SANParks and the mining company were implementing the UNESCO decision in preparation for further reporting in 2012 (DEA, 2011a; ICOMOS, 2011). Regarding inadequacy on public consultations, DEFF confirmed that all relevant documents were made public via the website except for the Reactive Monitoring Mission reports which were available to the public on request (DEA, 2011b). Making documents available on request is against the notion of transparency in good governance. Such information should be made available for the public when they need it. Transparency and access to information is critical ingredient for good governance in any setup. SBMS are still entrenched in information filtering, unless a court order instructs them to release it.

Emerging Stakeholders at MCLWHS

The majority of stakeholders at MCLWHS cannot be disassociated from the history, heritage values, land use, exploitation, and maximization of both renewable and non-renewable resources in the area (Carruthers, 2006). This continues in the present times as MCLWHS acts as an economic and tourism magnet (Carruthers, 2006). While the voices of IDCs dominated in the pre-colonial period, SBMS and UNESCO are domineering in this space now. This is symptomatic of Africa at large and works against transparency and open and reciprocated stakeholder management. For instance, IDCs' participation in Forum meetings comes at a cost and loss of livelihoods as they sacrifice time and personal resources. However, dealing with IDCs who are "unidirectional, relying on a number of oral traditions and material culture" to justify their involvement is an equally complex matter (Huffman, 1996). However, better understanding of IDCs and their connection with the landscape is being gained from decolonized research (Pikirayi, 2016; Ralushai, 2002; Ralushai & Gray, 1977). IDCs are always competing with commercial players, public entities, extractive industries, farmers, hoteliers, and tour operators to get the attention of SBMS at MCLWHS

(DEA, 2014a). All these stakeholders have varying interests and mandates in the area. This is further complicated by land ownership and associated power. Land ownership includes contractual partners, private conservancy owners, landowners, land claimants, private tourism operators, commercial farmers, and mining concessions (DEA, 2013a).

Furthermore, IDCs are strategically used as an "entry point" in heritage activism by heritage professionals to fight against development. However, they are quickly sidelined as agenda setting shifts. While SMC member organizations acknowledged the centrality of IDCs, they did not even explain their role and value in the coalition matrix. Is this a case where IDCs were name-dropped to justify environmental justice litigation? If environmental justice is meant to protect the rights of the people, whose rights were being protected given that the IDCs are not involved at the battlefront? Neither are they meaningfully benefiting from the exploitation of resources at the site. Experts still consider themselves the mouthpiece of IDCs, yet IDCs can better express their own views. The voices of IDCs remain muffled and obscured by those with power and resources at MCLWHS. At the UNESCO level, all MCLWHS stakeholders are represented by DFFE, with no opportunity of being funded for the same. If they wish to participate at UNESCO meetings, they have to draw on their own resources and attend as observers without voting powers, unless they lobby a 'party' with such rights to advance their agenda.

Conclusion

At MCLWHS, the emerging stakeholders have varying interests, perceptions and views on conservation and development issues, and this determines relations among them. From a governance perspective, there are constant conflicts due to tensions over land ownership, use and access to resources, counter-claims, intense negotiations, and rivalries among stakeholders. However, in this conflicted situation, SP, DFFE, SANParks and UNESCO still wield more power at MCLWHS. The mining industry has resources to take a legal route in conflicts, while IDCs are dependent on goodwill of other stakeholders such as academics, professional associations and Nong-governmental organisations. The power pendulum always tilts towards SBMS, with UNESCO in the shadows. This points to lack of an inclusive governance framework that capitalizes on the diversity and differences of stakeholders for their own mutual benefit. The views and opinions of these stakeholders on conservation and development must be solicited at WHS as an ongoing process.

6

EMPIRICAL STAKEHOLDER PERSPECTIVES

Emerging Profiles, Views, and Perceptions

Introduction

Based on a mixed research methodology applied during fieldwork at Mapungubwe Cultural Landscape World Heritage Site (MCLWHS) between 2015 and 2019, and the subsequent quantitative and qualitative analysis of the data, this chapter presents the empirical perspective of stakeholders. Field methodology included archival and primary sources analysis, conducting specialist seminars and interviews, questionnaire surveys, and attending stakeholders' meetings at MCLWHS and during World Heritage Convention (WHC) sessions. All these field methods were woven together using the applied iterative principles of the Delphi technique to develop areas of common agreement and those of variance among stakeholders. The field results demonstrate stakeholder aspirations, perceptions, needs, and their role in the decision-making at MCLWHS, typical of the cry at other World Heritage Sites (WHS) in Africa. The empirical perspectives presented include, but are not limited to, their (stakeholders) knowledge on the significance of the site, roles and responsibilities in the governance processes, beneficiation frameworks, and relationship between conservation and development. This summarizes their perceptions, power-interest matrix, and influence capabilities at MCLWHS. This brings to the fore empirical perspectives on stakeholders, conservation, and development at MCLWHS. The field results are illustrated using graphic illustrations cascading from broad thematic variables to specific sub-themes. This sets the foundation for the discussion in Chapter 7, which equally gives an opportunity to relate outcomes to other WHS selected as secondary case studies.

DOI: 10.4324/9781003149361-6

Field Approach: Stakeholders Perspectives

Building on stakeholder profiles developed in the preceding chapters, a mixed field methodology combining desktop surveys, seminars, and fieldwork was conducted to solicit their views on conservation and sustainable development (SD) at MCLWHS between 2015 and 2019 in South Africa. Desktop surveys focused on the state of conservation (SOC), UNESCO, Department of Fisheries, Forestry and Environment, (DFFE), and South African National Parks (SANParks) reports, including administrative records or documentation relating to MCLWHS. Consultative seminars held included the Association for Southern African Professional Archaeologists (ASAPA) Conference (Harare, Zimbabwe) attended by the 20 experts, the Mapungubwe Kingdom Seminar (MCLWHS, South Africa) attended by 135 participants (including Indigenous and Descendant Communities, IDCs), and the Robben Island Seminar on World Heritage and SD (Cape Town, South Africa) attended by 25 experts from African region and Europe. The fieldwork revolved around a questionnaire distributed through various platforms, among them emails, workshops, and conferences in South Africa. While the questionnaires targeted 300 stakeholders, only 145 questionnaires were received back. Also, fieldwork included repeat field visits to MCLWHS. All this data was analyzed using both qualitative and non-qualitative analytical methods and the emerging views are presented below.

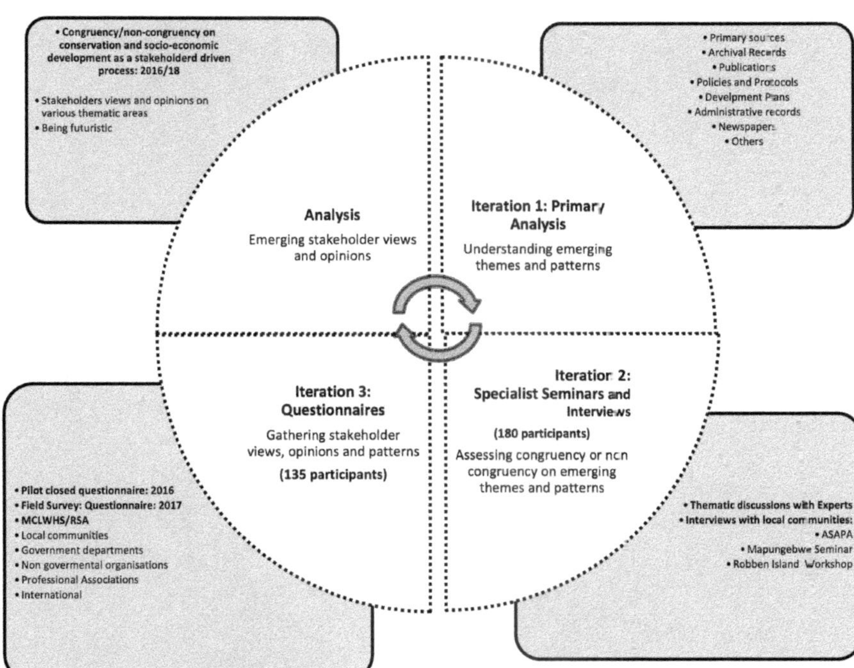

FIGURE 6.1 Mixed research methodology applied at MCLWHS

State of Conservation: MCLWHS

The conservation challenges at MCLWHS were analyzed using the 13 primary threat typologies as defined by UNESCO, which are illustrated in Figure 6.2. Management and institutional threats are cross-cutting among these typologies identified and developed at WHS over the years.

Conservation, agriculture, and mining issues at MCLWHS were identified at the time of inscription, and this was expected to be addressed by the State Parties (SPs) (Table 6.1). MCLWHS has consistently submitted SOC reports to UNESCO in line with WHC decisions: 36 COM 7B.48 (36th Session, Saint-Petersburg in 2012), 37 COM 7B.43 (37th Session, Phnom Penh, 2013), and 38 COM 8B.48 (38th Session, Doha, 2014). However, MCLWHS is not an exception as many WHS across the globe face similar challenges pitying conservation and development, for example, the uranium extraction at Selous Game Reserve (Tanzania), current mining explorations at Matobo Hills WHS (Zimbabwe), and

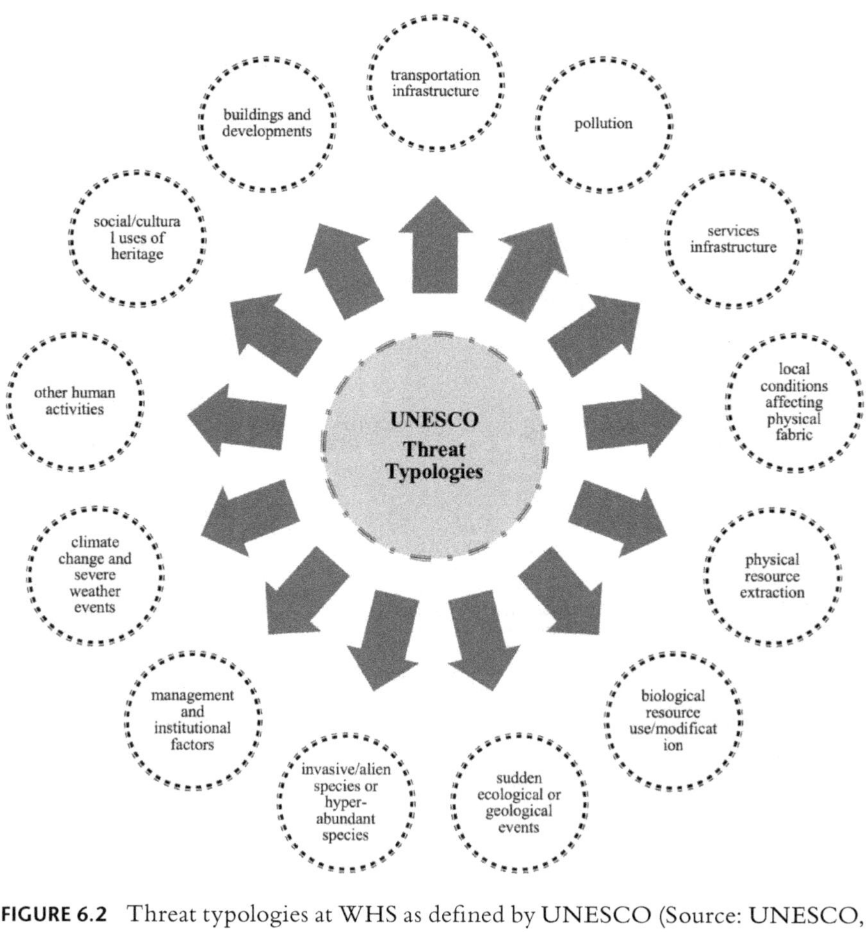

FIGURE 6.2 Threat typologies at WHS as defined by UNESCO (Source: UNESCO, http://whc.unesco.org/en/soc/)

TABLE 6.1 Conservation issues identified at the time of inscription at MCLWHS (Nomination Dossier, 1999)

Factor	Commitment	Status as of 2018
Conservation	Properties in the core area shall be acquired in order to avoid competing land uses in MCLWHS.	• A number of Private Nature Reserves and privately owned farms are in the process of inclusion into the MNP. • Portion 27/1MS of Den Staat Farm is part of the Land Claim & is being acquired to be part of the MNP. • The -IMP 2012-18 has made provisions for MNP Consolidation Programme as part of resolving this issue (page 41).
Agriculture	Farming in the core area shall be reduced or phased out over the next 5 years.	Nothing has been achieved as Farmers within the core are not keen to sell their farms. Farmers have several reasons why they are not keen and among them is that; they do not want to lose farm incomes, it is part of a speculative holding to land tenancy for commercial benefit, and that MCLWHS should protect their land instead of acquiring it.
Mining	Impact of mining shall be closely monitored and footprints shall not be allowed to expand.	While mitigations have been implemented, this matter remains a thorny issue in the management of MCLWHS as mining operations continue. This will be discussed below in detail.

the hotel construction at the forts and castles (Ghana). MCLWHS is thus a microcosm of how conservation and development are manifesting at WHS in Africa.

When MCLWHS is contextualized in the current state of threats at WHS across the world, the graph below (Figure 6.2) shows that these threats are increasing. More than 574 properties have been subjected to SOC reporting as of March 2021, culminating in 3796 individual reports. Africa accounts for 13% of these SOC reports covering 75 WHS in 31 SPs as of March 2021. In South Africa, eight out of the ten WHS have been subjected to SOC reporting. Cultural sites such as MCLWHS, Robben Island World Heritage Site (RIWHS), and Cradle of Humankind dominate SOC reporting, accounting for 50% as of March 2021. At a regional level, MCLWHS is one of the many cultural properties which have the highest frequency of SOC reporting in Africa.

The emerging pattern in South Africa reflects that SOC issues are becoming more pronounced at cultural sites both regional and worldwide. In Africa, the ranking is as follows: culture (48%), nature (45%), and mixed (nature and culture) properties (7%) as of March 2021. The submission of SOC reports has been increasing from 2003, with a peak submission in 2014 (http://whc.unesco.org/en/soc/). This demonstrates increasing challenges at WHS and requires

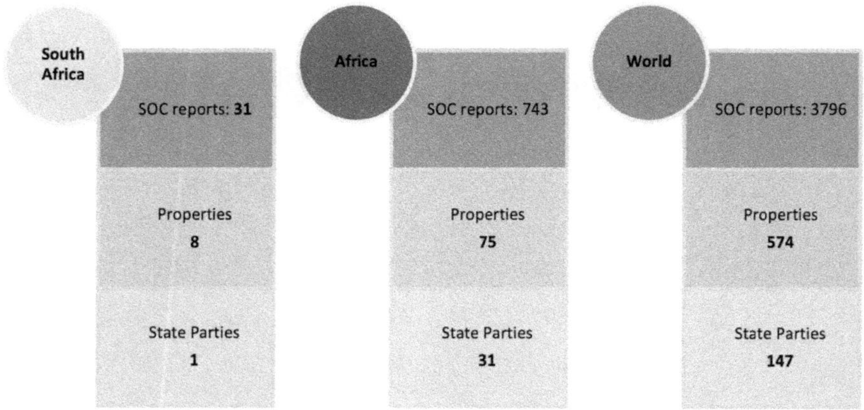

FIGURE 6.3 SOC reporting patterns: South Africa versus regional and international as of January 2021 (Source: http://whc.unesco.org/en/soc/)

stakeholder-driven solutions to augment scientific approach of State-Based Management Systems (SBMS).

The threats bedeviling MCLWHS are multiple and cross-cutting (Figure 6.4). These threats include physical resource extraction (mining), management and institutional factors (legal frameworks, financial resources, management activities, management systems and plans), biological resource use/modification (land conversion), and buildings and development (housing, major visitor accommodation and associated infrastructure, and industrial areas).

Mining Impacts at MCLWHS

Mining contributes 28–30% to the economy of the greater Musina Local Municipality due to the existence of rich mineral deposits in the area (DEA,

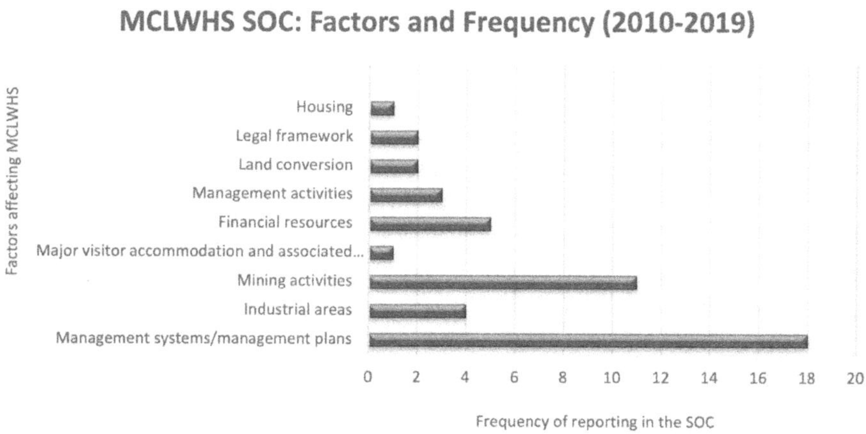

FIGURE 6.4 Conservation threats at MCLWHS

2014a; LEDET, 2012; Limpopo Provincial Treasury, 2017). Minerals in the Mapungubwe area include high- and middle-grade coking coal, diamonds, platinum group metals, iron ore, chromium, diamonds, antimony, phosphate, gold, emeralds, scheelite, magnetite, vermiculite, silicon, mica, and copper (CIC Energy, 2012). IDCs were involved in artisanal extraction of minerals as previously presented. Venetia Diamond Mine started operating in Venetia Limpopo Nature Reserve in the early 1990s (Carruthers, 2006; DEA, 2014a). Riedel farm (east of Mapungubwe) was excluded from the MCLWHS boundaries as there was potential for mining (DEA, 2014a). Also, coalfields in the area, though containing methane gas, are yet to be exploited (DEA, 2014a). Though extraction of minerals is beneficial to some stakeholders, this process comes with multiple threats to MCLWHS as discussed below.

The decision by the SP to grant a coal "*mining right permit*" to the Limpopo Coal Company (Pty) Ltd, later called CoAL, in March 2010 created tensions among stakeholders at MCLWHS. The permit was issued in terms of the Mineral and Petroleum Resources Development Act (MPRDA) No. 28 of 2002 by the Department of Mineral Resources (DMR). The issuance of exploration licenses at WHS in Africa remains a challenge, yet a necessary step in development goals and targets for SP. The "*mining right permit*" covered a total surface area of 8663 hectares covering the following farms: Bergen Op Zoom 124 MS, Semple 155 MS (Consolidation of Almond 120 and Semple 119 MS), Portion 3, 4, 5, 6, 13, 14 and Remainder Extent of the farm Over Vlakte 125 MS and Voorspoed 836 MS (Consolidation on Remaining Extent of the farm Newmark 121 MS), as well as Portion 1 of Bergen Op Zoom 124 MS (SANParks, 2011a). The mining venture entailed construction of roads, plant area, and supportive infrastructure (SANParks, 2011a).

Furthermore, the location of the mining venture close to MCLWHS created apprehension given the widely known negative impacts of coal extraction on landscapes. Subsequently, the SP requested a boundary modification in line with the provisions of the Operational Guidelines on the Implementation of the World Heritage Convention (OGWHC) in order to accommodate the mining venture, which was expected to yield 441 million tonnes of coal for 29 years (DEA, 2014). The SP considered the buffer zone outlined at the time of inscription as sufficient enough to protect the outstanding universal value (OUV) (SANParks, 2011). However, based on the documents submitted at the time of inscription, UNESCO indicated that the mining area was inside the original buffer zone of the site (WHC, 2011). The "original" buffer zone covered Venetia-Limpopo Nature, Vhembe Nature, and Limpopo Valley Game Reserves (Nomination Dossier, 2003; SANParks, 2011a). This had been delineated based on land that was available and whose owners were willing to be part of MCLWHS (SANParks, 2013a). Additionally, the WHC noted that the buffer zone had not been formally promulgated as requested at the time of inscription (WHC, 2011). However, both parties could not agree with each other on this boundary issue. As such, UNESCO recommended that the mining be stopped but the SP insisted on continuing with this venture on the basis of mitigation measures proposed by the developer via an approved Environmental Impact Assessment (EIA). These issues and inherent discrepancies picked up on this matter triggered a Joint World Heritage Centre/International Council on Monuments and Sites (ICOMOS) Reactive Monitoring Mission to MCLWHS (SANParks, 2011a).

The Joint Mission consistently raised the boundary demarcation issues, including the negative impacts of the mining venture on the OUV, in particular the effects of opencast versus underground mining operations at the site (WHC, 2011). However, the Joint Mission equally made some erroneous observations, for instance regarding the height of infrastructure at the mine, which they pegged at 40 metres, yet it was actually below 20 metres. It could have been a miscalculation made from a helicopter fly-past around MCLWHS, yet this became a factual discussion of WHC. The responsibility is on both the SP and Joint Mission to verify facts beyond reasonable doubts. The SP eventually re-submitted the amended buffer zone maps to UNESCO, which were subsequently approved, paving way for the mining venture to proceed (SANParks, 2013).

These opposing positions of the SP and UNESCO regarding the coal venture at MCLWHS led to "agonizing and painful" negotiations over the years and at some point, led to the halting of the mining operation in 2010 and 2011 to allow for further consultations and the conducting of an additional OUV-Based Impact Assessment (as provided by ICOMOS guidelines). This included closing knowledge gaps in the original EIA commissioned by the mining company itself. The halting process also ensured that the mining company obtains the necessary authorizations required before resuming mining activities (WHC, 2012). However, and parallel to the halting, CoAL was issued with penalties by DFFE for having implemented some infrastructure projects before obtaining all environmental authorizations (WHC, 2012). Mining only resumed towards the end of 2011 after issuance of authorization permits by DFFE and DMR, including the submission of the second Heritage Impact Assessment (HIA) as requested by WHC. According to the SP, the postponement of a Joint Mission to MCLWHS the same year did not warrant continued upholding of the halting of mining as it would be unfair to the mining company which had fully complied with both national and international compliance protocols (SANParks, 2012a). The Joint Mission further encouraged CoAL to join the International Council of Mining and Metals (ICMM) as part of enforcing the 2000 Cairns UNESCO Decision that encouraged mining companies to uphold "WHS as no-going zone for mining".

The mining venture of CoAL created tensions among government departments in South Africa, in particular DFFE and DMR because of their varying mandates in the area. Their visible tensions and contradictory positions on the issuance of permits and prospecting licences outplayed themselves at both national and international levels. This eventually forced DFFE and DMR to sign an agreement in 2008 adopting one system regulating all environmental matters (SANParks, 2011). Through this agreement, DFFE was given full powers to control environmental functions while DMR retained exclusive responsibility of issuing both prospecting and mining rights in Protected Areas (PA). For instance, the Terms of Reference (ToRs) for the second HIA commissioned within the framework of this agreement included assessing the impact on the heritage beyond the boundaries of the MCLWHS, thereby covering the broader cultural landscape (SANParks, 2012a).

The second HIA recommended mitigations measures for heritage sites in the area being mined by CoAL, including a research to link sites identified in the process with those found in the core area of the MCLWHS (Siyathembana, 2012). However, the HIA concluded that there was no stakeholder "consensus regarding the meaning, purpose, nature and extent of the buffer zone of MCLWHS" (SANParks, 2012a: 4). Stakeholders always have different conceptualizations of what constitute such a buffer zone based on their perspectives (Siyathembana, 2012). This created confusion and tensions among stakeholders, and the SP worsened this by presenting contradictory perspectives on this matter. UNESCO recommended that the SP develop guidelines and regulations to govern the activities earmarked within such a zone (SANParks, 2012a). In response, the SP conducted workshops on buffer zones in 2014 and 2016, resulting in policy recommendations on Buffering Mechanisms that are being finalized by DFFE. The HIA also noted that mining activities are either taking place or have already been licensed in Botswana and Zimbabwe, thereby further threatening the objectives of the Greater Mapungubwe Transfrontier Conservation Area (GMTFCA) (Siyathembana, 2012).

Overall, and consistent with the SP's position to UNESCO, the second HIA concluded that the mining activities by CoAL had no direct impact on the OUV, core area, and buffer zone of MCLWHS (Siyathembana, 2012). The HIA noted that the impact on other sites located seven kilometers away from the mining area would be minimal and appropriate mitigation measures were recommended (Siyathembana, 2012). Regarding stakeholders who raised concerns around the mining initiative, the HIA recommended a broad-based and continuous process of stakeholder engagement facilitated by the SP at MCLWHS (Siyathembana, 2012). However, UNESCO expressed dissatisfaction with the stakeholder consultation processes during the second HIA. In response, the SP requested that UNESCO separate stakeholder dissatisfaction from issues of socio-economic benefits from mining versus dissatisfaction with the HIA process (SANParks, 2012a). The SP cited compliance with national laws on stakeholder consultations and the existence of the Park Management Forum constituted by representatives of stakeholders to MCLWHS (SANParks, 2012a). In addition, the SP highlighted that a special Environmental Management Committee had been established to advise on environmental management issues and also to monitor the environmental impact of the mine. This was in line with provisions of the Section 24G Environmental Authorization, paragraph 17 under Management of the Activity (SANParks, 2012a). On this basis, the SP refused to implement the recommended new "Mapungubwe Management Committee" citing that it would be an overload on their governance system (SANParks, 2012a).

Given the varying interpretations of the HIA results and process, UNESCO encouraged the SP to take measures for the protection, conservation, and consolidation of the archaeological evidence at MCLWHS, in particular the Leopard's Kopje (K2) site, which the Joint Mission reported to be seriously deteriorated (WHC, 2012). Regarding the extraction of resources in the Mapungubwe Area, the SP, through SANParks and South African Heritage Resources Agency (SAHRA), continued to monitor the implementation of the HIA and EIA relating to all mining operations, in particular the De Beers Consolidated Mines Pty Ltd at Venetia. The company is undertaking an underground expansion project. In addition, SANParks is working closely with the company to monitor impact of water abstraction which may affect the riverine forest (SANParks, 2017).

The Joint Mission also recommended that the entire landscape covering the Mapungubwe Kingdom be considered for inclusion into the site, of which the SP argued that such an approach would not be practical (SANParks, 2012a; WHC, 2012). Such an approach would have naturally intensified the perception that UNESCO uses the WHC to advance conservation ahead of other national aspirations. However, had the SP also agreed to this approach, this would mean including places of spirituality as heeded by IDCs, thereby intensifying calls for inclusive governance at MCLWHS. The default position of SANParks was not to accept this UNESCO suggestion, thereby advancing the need to balance conservation and SD at MCLWHS. Also, including the entire landscape covering Mapungubwe Kingdom would be synonymous with the GMTFCA, which comes with serious transboundary governance challenges. This would have equally pointed to extending the MCLWHS as a transboundary WHS covering Botswana and Zimbabwe. It would further reinforce the notion of conservation first and the needs of society as secondary as has always been with the establishment of Mapungubwe National Park (MNP). The question now is whether UNESCO was recommending this to further conservation as prescribed by the Convention or support the call of IDCs. In my opinion, it would have been the former as this is in tandem with calls to extend the site to Botswana and Zimbabwe. This would be contrary to IDCs' economic aspirations at MCLWHS as exclusion would follow.

These contestations characterize the emotive discussions between UNESCO and SP on such issues. They also indicate the limitations with Reactive Monitoring Missions by UNESCO within a very short space of time and usually using experts who are not familiar with the local dynamics and context of the site (Taruvinga, 2019). There have been increasing calls for UNESCO and Advisory Bodies (ABs) to involve more African experts during their missions in Africa, including sending them to other continents for similar activities as part of implementing the "universal notion" of the Convention. While African World Heritage Fund (AWHF) is pushing for such capacity building in Africa with the support of ABs, the visible practical results are still a long way from manifesting. For example, while ICOMOS has been advocating more representation from Africa on its Board, the election system is still reinforcing the status quo, and now there is a deliberate process to appoint Africa and Arab region representatives on the Board outside the long-established voting procedures. However, African countries also have to reorganize themselves beyond divisive lingual lines and rally behind candidates to the Board as one region, including establishing vibrant forum for Site Managers of WHS in Africa.

Management and Institutional Factors at MCLWHS

The management and institutional factors at MCLWHS revolve around the management system, inadequate financial resources, and legal frameworks. The management system issue emanated from the absence of an Integrated Management Plan (IMP) in 2011, but this was subsequently addressed in 2013 with the development of such a plan (SANParks, 2011b), and the latest plan being approved in 2020. The IMP has gradually given attention to cultural heritage at MCLWHS. The inadequate financial resources issue relates to how funding for cultural heritage is always secondary to that of nature values. Culture is less prioritized from a budgeting perspective as illustrated in Chapter 4 (SANParks, 2013a; 2013b), and this is symptomatic of PA in Africa. However, at Ngorongoro Conservation Area (NCA) (Tanzania), cultural heritage has been elevated alongside the nature values. NCA has made considerable investment in interpretation infrastructure, conservation, and capacity building in this area. This was possible after the Department of Antiquities (Tanzania) agreed to hand over the culture component to NCA as mixed WHS since 2010.

Biological Resource Use/Modification (Land Conversion)

Given the historical and current land uses at MCLWHS, including the associated land claims, WHC requested the SP to finalize the transfer of targeted land to SANParks, as well as conclude the establishment of the Trans-Frontier Conservation Area (SANParks, 2012d). Resolving land ownership, contractual, and acquisition issues remains a critical process for the successful land conversions at MCLWHS (Sinthumule, 2014). For GMTFCA, communal and private land was involved (Sinthumule, 2014). It remains important to acquire outstanding land for the GMTFCA if it is to become a reality (Sinthumule, 2014). This is over and above negotiating and buying private land in the core area of the MCLWHS as part of consolidating the conservation strategy (Sinthumule, 2014: 64). In this mix, MCLWHS still faces land claims from IDCs (SANParks, 2013a). However, successful claimants are forced to become part of the "*conservation train*".

In order to resolve this land issue at MCLWHS, SANParks partnered with Peace Parks Foundation (PPF) to facilitate negotiations and acquisitions of land for either contractual agreements or outright ownership by SANParks to support conservation (Sinthumule, 2014). PPF partnered with Rupert Family Foundation, De Rothschild Foundation, De Beers, National Parks Trust (NTP), and Wildlife Fund for Nature in South Africa to assist SANParks with these processes. On the South African side, MNP (including leased freehold land not owned by SANParks) and Venetia National Park (VNP) were included in the proposed GMTFCA (Sinthumule, 2014). With regard to contractual agreements with land owners, SANParks becomes the authority for conservation, while the land owner retains the ownership of the land (Sinthumule, 2014). Land under this arrangement includes Welton (1997), Rhodes drift (1998), Little Muck, Armenia, and Mona (2002), and Riedel (2011), all on a 99-year lease (Sinthumule, 2014). These leases are renewable for a further 25 years (Sinthumule, 2014). Outright purchases of land by SANParks secured ownership of Hamilton and Janberry (2005), Samaria (2005), and Balerno (2006). For UNESCO, these acquisitions and contractual arrangements are important in retaining the area as a conservation zone. Therefore, the reluctance by private owners to sell their land to SANParks remains a stumbling block to the consolidation of the long-drawn conservation programme and the GMTFCA (Sinthumule, 2014). At the same time, it points to the increasing pressure for conservation to coexist alongside development.

Resolving land uses, ownership, and tenures is not an overnight process and cannot be done at the instruction and time schedule of UNESCO as there are local stakeholders and related dynamics with nothing to do with WHC processes. Land issues are emotive, contested, negotiated, and protracted legal processes given the historical context of MCLWHS, including the narratives of dispossessed IDCs. Additionally, consolidation of land uses through contractual agreements creates a problem in that while conservation system is under a single entity, the land involved is under a multiplicity of ownership and conservation models (Sinthumule, 2014). Land can change hands anytime, and this has its own inherent risks. The conservation model at MCLWHS is still in its infancy and needs to be nurtured while continuously addressing inherent and unforeseen stakeholder challenges. Its effectiveness is a function of time and continuous stakeholder involvement. SANParks continues with negotiating for more contractual agreements despite land claims by IDCs at MCLWHS (Sinthumule, 2014).

While the GMTFCA was envisioned to enhance conservation, it has created other unintended impacts, for instance, the non-maintenance of fences along some sections of the borders by SANParks is now affecting farmers, who have to deal with elephants (Sinthumule, 2014). There are also illegal immigrants from Zimbabwe through the porous border. The illegal immigrants are 'escaping' from the economic and political situation in Zimbabwe. On the other hand, commercial farmers have a perception that SANParks is "de-fencing" borders to create more problems for them and push the conservation agenda (Sinthumule, 2014). Farmers also think that the land consolidation and GMTFCA initiative will stop commercial farming as a land use though it has coexisted with conservation in the area since the 1980s (Sinthumule, 2014). From a conservation perspective, commercial farming is a challenge to convert the whole area into a bioregion (Sinthumule, 2014). Similarly, in NCA, the nomadic lifestyle of the Maasai, including their gradual intensifying "*sedentary life*" in the site associated with increased farming and mushrooming structures, is viewed as "anti-conservation" by the Ngorongoro Conservation Area Authority (NCAA), though the two have coexisted as a great experiment since the 1960s. In order to make NCA a 100% conservation zone, NCAA undertook feasibility and benchmarking studies to resettle the Maasai into permanent sedentary lifestyles outside NCA. The implementation was met with limited success despite the "*basket of benefits*" offered to volunteering Maasai families. This experimentation deteriorated relations between the NCAA and the Maasai, including allegations raised by the Maasai that the International Union on Conservation of Nature (IUCN) and later the ICOMOS had both instigated this move against them. ABs are erroneously politicized in the role while acting on behalf of the WHC. In response to this evolving complex situation, UNESCO has been enforcing monitoring missions to NCA.

Recently (in 2020), Tanzania has muted the idea of permanently settling communities in PA where they are currently living, such as NCA, which could have solved the simmering tensions between them and PA. However, this was not implemented, which has helped trigger widespread protests (since 2021) by the Maasai in NCA. They are demonstrating in Tanzania and Europe demanding that the forced relocations be stopped as it is violating their rights and access to land like any citizen of Tanzania. While both NCAA and the Maasai agree on benefits of conservation and tourism, both desire more from both. The financial allocations to Maasai by NCAA for infrastructural development, education bursaries, and health services are testimonial to this and inspire them. They cannot afford to walk away from their landscape and become another experiment once again. For NCCA, conservation and tourism are inseparable in their quest to achieve financial sustainability through tourism for the sake of NCA. The bottom line is that the "1960 experiment" should be fair and beneficial to all founding parties; however, the SD approach is needed to balance conservation and development.

Buildings and Development

In 2010, when the ASAPA lodged an appeal against the decision to grant the mining rights to CoAL in the MCLWHS, they raised a concern over construction of buildings and their potential negative impact on the entire landscape (http://whc.unesco.org/en/soc/554/). The appeal cited the inadequacy of the EIA and related HIA aspects, including its failure to consider the wider landscape of the site. The ASAPA argued that this venture would encourage infrastructure developments that would negatively impact the integrity of the cultural landscape (http://whc.unesco.org/en/soc/554/). Such infrastructure includes houses, roads, shelters, sanitation facilities, and administrative offices. This would result in vegetation clearance in the area. This included the non-quantified negative impacts on archaeological sites and places of spirituality at MCLWHS. Overall, the ASAPA concluded that the entire Vele Basin area is of great importance, and any development, even on a small scale, would be detrimental to the generation of knowledge from the rich and diversified cultural resources of the area.

However, some housing, major visitor accommodation, and associated infrastructure at MCLWHS were already in existence before inscription, equally demonstrating balance between conservation and tourism. These include staff houses, camping grounds, lodges, research facilities, administrative offices, resting places, and infrastructure supporting various land uses. One latest and interesting additional development is the SANParks Multipurpose Facility Centre constructed outside the main gate of MCLWHS, which did not seek approval and permits like any other development from SAHRA. In response, SAHRA demanded that SANParks complies with all permitting requirements. This points to three things: Management Authorities (MAs) at WHS tend to deviate from compliance and permitting conditions for developments initiated by themselves, a process which they normally enforce on other institution desiring to undertake such developments in their areas. Second, this may point to how PA are jealously guarded compared to adjacent areas where such developments can be imposed. If it was inside the PA, maybe the response of SANParks and conservation watchdogs would have been different. Third and last, the Heritage division at SANParks, though under-capacitated, was quiet while this construction was taking shape without raising an alarm on non-compliance.

The size of the construction footprint would have naturally triggered section 10 of the National Heritage Resources Act of 1999 (NHRA). The case of MCLWHS may begin to paint a picture that MA-driven developments generally find themselves not having to fully comply with heritage laws. However, one cannot make an overarching conclusion on how rampant this problem is at WHS in Africa without a forensic study. In this process, UNESCO becomes the much-needed watchdog to ensure compliance in the process of balancing conservation and developments at WHS. However, a site such as RIWHS (South Africa) enforces full compliance through appointment of independent consultancies, which then process permit applications with SAHRA and DFFE as an independent process on behalf of RIWHS. Additionally, internal impact assessment tools exist at RIWHS, which are used to determine whether a development or any event has impact on the site, and what would be the course of action thereof from a compliance perspective. This helps in separating powers, responsibilities, conflict of interest, and ethical issues by MAs of WHS, especially in implementing internally driven developments projects.

MCLWHS in the Eyes of Experts

The Experts Seminar during the 2015 ASAPA Conference brought together over 20 heritage experts to discuss WHS and stakeholders in Africa. The seminar was themed "Multiple and multi-layered stakeholder of heritage sites" and was attended by site managers and academics from African countries. The seminar had an eight-member panel giving individual interventions on broad questions around the theme followed by a plenary discussion with other participants. The perceptions and views of these experts are discussed below.

Stakeholder Theory in Heritage Contexts

Experts pointed out that stakeholder theory is interpreted and invariably applied at WHS in Africa. The term "stakeholder" is interchangeably used with "local communities" at WHS in Africa. Whichever terminology is used, they are all interest and affected groups. These groups are complex, conflicted, and have fluid boundaries and memberships or networks. Interestingly, experts argued that local communities and UNESCO are also stakeholders too. While stakeholders are those with "stakes", experts argued that they are not equal in reality because of their unequal powers in the governance of heritage. Experts acknowledged constant challenges in dealing with the stakeholders at WHS. Overall, WHS fail to clearly define the stake and the respective stakeholders, including their decision-making roles. This includes dealing with overlapping members and groups within groupings of stakeholders at WHS, which may induce confusing role plays. While unpacking stakeholders, it is important to establish their common ground, the inherent or underlying politics and power relations.

Identifying Multiple and Multi-layered Stakeholders at Heritage Site

Panelists argued that identifying and defining stakeholders in a community setup is challenging from multiple perspectives. The process should be a broad-based approach given that this is beyond conservation. There is need to include other broader territorial aspects and related stakeholders in the process. Apart from significance or value-based approach, socio-economic expectations and land uses, all other issues should be addressed in the process. Though this is ideal, it will never be an inclusive and democratic process given the legalistic approach of SBMS and politics of participation in Africa. SBMS always create their "governance comfort zone". Experts pointed out that those stakeholders and their roles are site specific, and equally, WHS are peculiar and specific places with own historical and current contexts. On the relationship between identification of stakeholders and the delineation of WHS boundaries, experts concluded that the latter is a scientific process led by themselves and the supportive SBMS. This makes experts the jury and judge in both processes, which is not good governance. However, panelists implored heritage practitioners to live in the "present", in which they have to acknowledge current challenges of the society rather than reinforcing past exclusionary approaches.

Power, Legitimacy, Influence, Capability, and Role of Stakeholders

From a stakeholder governance perspective, panelists raised the disjuncture between national and international legislation provisions and aspirations of stakeholders at WHS in Africa. Experts argued that, for instance, local communities have little appreciation of heritage legislations as they perceive them as tools used by SBMS to suffocate them and their aspirations. Furthermore, governance of stakeholders at WHS was viewed as a "one-way" street with the SP and UNESCO listening to each other more often at the expense of multi-vocality at WHS. SP take advantage of their structured and bureaucratic governance structures while sidelining all other stakeholders. Panelists further argued that having an inclusive stakeholder approach at WHS remains a fallacy and a wishful dream in Africa. Furthermore, experts noted that inequality is a part of humanity even if there is a desire to be inclusive. The notion of a one-size-fits-all does not apply given the varying interests of stakeholders, but rather the issue should be how to manage and give stakeholders fair roles at WHS.

With specific reference to local communities, experts argued that engaging and involving them should form the foundation of resolving conflicts and tensions associated with balancing conservation and developments at WHS. Ideally, inputs from local communities should find expression in the decision-making framework of WHS, rather than being archived in administrative and compliance templates of SBMS. In reality, heritage institutions always prevail over local communities, hence management plans always have a strong bias towards conservation. One panelist argued that at NCA, it is clear that heritage experts have preconceived ideas or positions even before they consult the Maasai on conservation and development issues. Local communities should be consulted with an open mind and without pre-conceived notions. The mindset of superiority by SP and experts should not continue to dominate at WHS at the expense of other stakeholders. Stakeholders have different levels of power, legitimacy, influence, and capabilities, and these should be used to "curve" their tailor-made roles at WHS.

Panelists also raised the issue of "site management committees" established at WHS, which places stakeholders in the lowest decision-making position of SBMS. Higher responsibility in such committees remain the privy of SBMS. At Matobo Hills World Heritage Site (Zimbabwe), such a Committee was interchangeably chaired by National Parks & Wildlife Management Authority and National Museum and Monuments of Zimbabwe, with other stakeholders playing a supportive role. This approach is symptomatic of a widespread practice in which SP in Africa and related entities retain governance dominance at WHS. Even at UNESCO levels, designated State Representatives are appointed without input from both local communities and professionals, further reinforcing the exclusionary approach of SBMS. SP inputs at UNESCO sessions are supposed to represent all stakeholders, including local communities, yet this is often not true.

Furthermore, panelists argued that engaging and involving stakeholders should be a governance, moral, and ethical issue for SP. It should be viewed as a localized process rather than a compliance approach. This means attendance registers are not sufficient to demonstrate their involvement. The SP should have the responsibility to submit detailed minutes or accounts of such consultations. This makes issues raised by stakeholders available for independent analysis. The effort by UNESCO in making SP more accountable in this area should be supported by

SP. However, panelists noted with concern that while UNESCO acknowledges the role of local communities and stakeholders, World Heritage governance does not give them a decision-making voice. They are not officially represented at Committee sessions. A strong political commitment is required to implement stakeholder approach at WHS in Africa by both SBMS and UNESCO.

Stakeholders and Benefits from Development at WHS

The expert discussions on benefits accruing to stakeholders from development at WHS showed interesting patterns. Experts argued that benefits accruing to stakeholders are a function of local planning rather than international processes. Often heritage institutions themselves are being brought on board at a later stage in broader territorial development. This is through enforcing EIA and HIA as a compliance rather than a planning process. Heritage institutions, as stakeholders to other economic spheres, want to play the overall deciding role on whether development can or cannot happen, yet they are not involved from conception stages. Continuing with such a path is a progressing risk in that heritage institutions are being pushed to the periphery of decision-making on development planning levels. If this is to be avoided, heritage institutions should register on specialist databases to become visible on consultations relating to broader development issues in Africa. While experts argued that development is not a function or competence of heritage institutions, interaction with stakeholders competent in these areas would be beneficiary to heritage governance. The development sector should equally and proactively involve heritage institutions from the beginning of their planning process.

Furthermore, panelists argued that heritage is a resource, which should be sensitive to the existence of other resources alongside it. Heritage should be sensitive to renewable and non-renewable resources at WHS. Conservation and development should equally plan for each other. In this context, one has to deliberately know the aspirations of the associated stakeholders at WHS. Therefore, heritage institutions cannot use a "*conservation blanket*" to derail development perceived to be a threat to heritage, including in the buffer zones of sites, without considering developmental aspirations. To this end, experts argued that the stakeholder approach may be used to facilitate conservation and development as trade-off processes at WHS without underplaying either one of them. According to experts, stakeholder approach may be a governance vehicle for negotiating alternative solutions at WHS.

In regard to how stakeholders are benefiting from WHS, the experts agreed that this is a contentious area which needs empirical studies to ascertain what is actually happening. Stakeholders, for instance local communities, still do not see the benefits of being involved at WHS. Experts argued that WHS do not give back in a meaningful and measurable way to local communities. This makes them question the World Heritage concept. However, the scale and magnitude at which different stakeholders are accruing benefits from WHS remain difficult to quantify in the absence of predetermined, partial, objective, relational, and statistical indicators. As an example, and from a relational perspective, one cannot compare the benefit of harvesting thatching grass and selling curios against extraction of minerals from the same environment. Additionally, and while mining is at the heart of developing countries, local communities have no decision-making power on these issues, except being peripheral beneficiaries of the processes. The mining agendas around WHS are largely driven by private and profit-driven entities. If the World Heritage concept is to remain relevant to stakeholders, these inequalities have to be addressed through an inclusive stakeholder approach.

In assessing socio-economic benefits accruing from development at WHS, experts argued that the intrinsic aspects be considered as wells. For instance, depriving IDCs access to their culture at WHS is detrimental to the future of the Convention in Africa. Furthermore, promoting World Heritage interpretation without local community inputs is perpetuation of a scientific approach devoid of a community-driven expression of what the site means to them. Narratives are still heavily influenced by archaeological precepts and conservation mindsets. For instance, IDCs as knowledge producers should be actively involved in the interpretation of their own narratives at WHS. At RIWHS, former political prisoners (affectionately known as Ex Political Prisoners, (EPPs)) are involved through various strategies in developing the memorialization and interpretation materials; however, this remains a contestation between and among individual and collective experiences of EPPs, intra-politics associated with different political associations, and present-day challenges of socio-economic beneficiation, which has been slow and skewed for EPPs. To mitigate these challenges, RIWHS has established an Ex Political Prisoners Advisory Forum (EPPAF), which is the link between the site and over 3600 EPPs as well as a conduit through which heritage programming at the site is discussed with them and the Robben Island Museum Council, being the highest governance level of RIWHS.

Overarching Views of Experts

Firstly, the experts pointed out the dilemma in terminologies used to refer to role players at WHS. The term "local community" does necessarily refer to stakeholders, but they agreed that the latter includes the former. Experts noted that identifying and involving stakeholders should be an inclusive process and a broader approach based on geographical location, aspirations, and socio-economic context of the WHS. They argued that most stakeholders live outside the WHS, yet they are supposed to benefit from areas inside or closer to the WHS. Second, experts argued that trusting SP in representing other stakeholders in World Heritage governance processes is a fallacy given that they do not necessarily represent stakeholders, unless there is political pressure or it is unavoidable. Experts lamented the tendency at WHS, where SP initiate socio-economic developments such as tourism and related downstream industries, but these are not necessarily benefiting intended beneficiaries such as local communities. These initiatives are designed to augment the inadequate grants received by Heritage institutions from national governments. This is for their own sustainability, not that of the others. The Experts Seminar concluded that SP are reluctant to implement a stakeholder approach given the accountability and transparency that comes with it. Third, experts noted that SP, in particular institutions responsible for heritage, have the tendency of not actively participating in broader development planning processes, unless compliance with heritage protocols is an issue, which then attracts their attention. Being primarily concerned with enforcing heritage compliance framework is marginalizing the relevance of heritage in contributing to innovative and creative approaches in resolving societal issues such as socio-economic development at WHS. Fourth, experts concluded that there is a need to define and quantify what are the socio-economic benefits at WHS, including developing indicators for measuring them. Fifth, experts repeatedly highlighted that heritage landscapes are different and ever-changing, implying that one-size-fits-all governance approach is not practical and effective, especially in view of the different socio-economic aspirations of stakeholders. And last but not the least, the experts concluded that benefiting from a WHS does not mean that there are no conflicts among stakeholders. World Heritage governance should have conflict resolution mechanisms and tools.

"Kingdom, Spirituality and Access": IDCs and MCLWHS

During the Mapungubwe Festival and the Mapungubwe Kingdom Seminar held in 2016 and 2017 (respectively) at MCLWHS, the IDCs (see chapter 4) were engaged on the subject matter. The 2016 consultations aimed at understanding the IDCs landscape, while the 2017 seminar was towards "Revisiting Mapungubwe". Both events were jointly hosted by SANParks and the National Arts Council (NAC) of South Africa. The 2017 seminar was attended by over 150 participants from government departments, IDCs, academics, representative of tourism industries, and creative artists. During the seminar, SANParks emphasized the importance of involving IDCs in the management, interpretation, and use of MCLWHS. This theoretically denotes a shift from a strict conservationist approach at MCLWHS. Related to this, SANParks acknowledged the outstanding and unresolved land claims at the site, which are subject to legal processes beyond its jurisdiction.

Our Kingdom, Our Spirituality

During the seminar, IDCs made profound submissions based on their aspirations in relation to various dimensions of MCLWHS. IDCs requested SANParks to (i) partner with them in the management and development of the site, (ii) grant them access and infrastructural support for spiritual ceremonies at the site, (iii) decolonize the narrative being used at the site, including them being allowed to train tour guides on their history, and (iv) involve them in research projects as the creators, custodians, and authentic voices of their own landscape. Furthermore, individual communities raised other pertinent issues. The Machete and Leshiba families entreated SANParks and the government to finalize the land claim issue, which could provide socio-economic opportunities to them as "restored" landowners. The Tshivhula Royal Council raised the issue of traditional artefacts in their custody and the need to come up with a collections management strategy for them at MCLWHS, including how they can be used to decolonize the narrative. Furthermore, the Vhangona community demanded that their ancestral possessions in the custody of universities should be returned to MCLWHS without conditions as a consolidation of the reburial exercise in 2008. The reburial of human remains at MCLWHS remains incomplete and desecrated as all associative cultural objects are still outside the landscape. From Traditional Management Systems (TMS), spirits of the deceased cannot rest until all these outstanding processes are complied. This begs a question as to when the spirits venerated by IDCs at MCLWHS are going to rest given the reburial approach used or it shall become an intensified localized call from them legitimizing their access to such landscapes? The point is that repatriation policies and processes have remained largely oblivious to TMS and their foundational ethics and principles in handling such sensitive cultural materials. Taking TMS into consideration avoids pure "*academising*" and underpinning local solutions with "*science*", which has increased tensions between the local and global perspectives at WHS in Africa.

Paradigm Shift: From Exclusion to Inclusion

The "Inaugural lecture on MCLWHS", delivered by Dr Ndoro (Director General of ICCROM) during the seminar, argued for a paradigm shift from the notion of PA to an inclusive approach giving an opportunity to stakeholders, including local communities, in deciding on conservation and development at WHS. Furthermore, the decolonization agenda across the varying thematic issues and strategic functions of the site was reiterated. Academics, who attended the seminar of 2017, equally highlighted the same issues, however, with emphasis on spirituality and research, ownership of the site, and socio-economic matters relating to MCLWHS. In response, IDCs expressed concerns with being sidelined in the development of a consolidated history of MCLWHS by academics. Furthermore, IDCs lamented on how their ancestors and associative burial objects continue to be treated in the landscape without an appropriate collections management strategy, equally informed by their Indigenous Knowledge Systems. All these issues would require innovative and creative strategies to be developed together with the IDCs going forward.

The notion of inclusive governance structures at MCLWHS that involves IDCs was echoed repeatedly by speakers during the 2017 seminar. The IDCs themselves have always lamented about the skewed governance of MCLWHS, which is always in favour of SBMS. While IDCs acknowledged their presence in the Park Forum at MCLWHS, they considered it mainly motivated by the interests of SANParks and other powerful stakeholders. During the academic seminar, IDCs pointed out that they are no longer fighting for ownership of MCLWHS as it "naturally" belongs to them but are rather concerned with how they are going to be involved in the conservation and development of the site. Stakeholders at the local levels of WHS agree that the governance status quo dominated by SBMS and UNESCO can no longer continue. There has to be a deliberate and inclusive approach to involve others in a meaningful way going forward.

Long-known Practised Relationship: Conservation and Development

In relation to balancing conservation and development in MCLWHS, 2017 seminar participants noted that this is not a new phenomenon as extractive industries have been operating in the area for many decades, starting with pre-colonial institutions, at Mapungubwe. The concern of IDCs is over the issuance of mining permits without their involvement at WHS. Both IDCs and academics acknowledged the opportunities and challenges associated with mining that could potentially affect the site in different ways. This requires moving towards common decision-making processes and regional integration with other local communities in Botswana and Zimbabwe on these sensitive socio-economic issues. Overall, the seminars recommended partnership with IDCs as equal players to become a priority by SANParks and other development stakeholders. Above all, access for IDCs to the site for their cultural and spiritual activities was considered important going forward. In as much as SBMS are looking for sustainability for the operations, IDCs are looking for the same for their livelihoods. Equally, other stakeholders are looking for same as well for their own mandates.

Emerging Stakeholders Perspectives at MCLWHS

Based on fieldwork conducted between 2015 and 2018 at MCLWHS, opinions of the multiple and multi-layered stakeholders on conservation and development at MCLWHS were gathered. A questionnaire survey was administered but was not restricted to MCLWHS itself because of its status. The assumption was that MCLWHS is known and is a benefit to the public, especially in South Africa. The survey target areas were in and around MCLWHS. The "target" stakeholders included academics, universities, civilians, IDCs, tourists, school children, students, farmers, government employees, and many other diverse people. The field survey targeted 300 stakeholders; however, only 140 responses (46% of the target sample) were received back. Only 123 responses (41% of the targeted sample) were analyzed as they were fully completed while the remaining 17 responses were considered incomplete. This was due to either perceived complexity of the questionnaire or lack of time to respond and interest in the subject matter. The main questions and sub-questions of the questionnaire were derived from thematic and empirical issues identified through desktop surveys, specialist seminars during the ASAPA Conference, and pre-survey consultations with local communities at MCLWHS. The five cascading thematic areas of the questionnaire were – identifying stakeholder profiles, understanding stakeholder knowledge on World Heritage, roles and responsibilities of stakeholders, benefits of protecting MCLWHS, and relationship between conservation and socio-economic development at WHS. The emerging stakeholder perceptions presented in this section follow this thematic line of the questionnaire.

Regarding the stakeholder profiles, all the 123 stakeholders were from South Africa, and at gender profile level, 73% were female, while 27% were male. The dominance of females was considered priceless as gender balance is a priority in heritage issues worldwide. From a stakeholder category perspective, the represented categories were dominated by academia (28%), local communities (22%), and heritage institutions (16%).

The dominance of academia is not shocking given their history and relationship with heritage at MCLWHS, while the local community representation represents increased awareness on heritage issues and shifting patterns in stakeholder engagement at MCLWHS. This pattern is further confirmed by the dominance of stakeholder interests in "*education*" (63%), "*socio-economic development*" (37%), "*conservation*" (33%), "*social interests*" (28%), and "*others*" (7%) as illustrated in Figure 6.6. The dominance of education by a wide margin is testimony to the level of awareness raised by SANParks regarding MCLWHS.

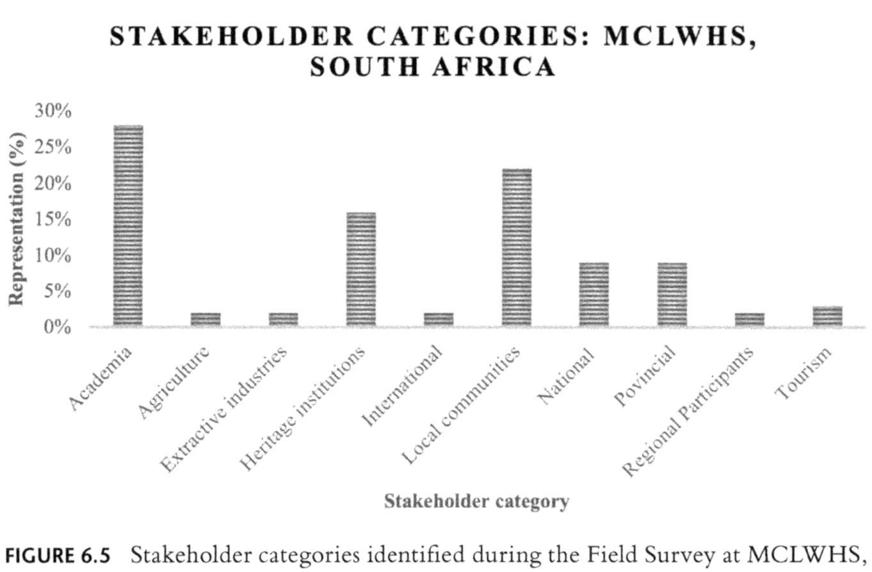

FIGURE 6.5 Stakeholder categories identified during the Field Survey at MCLWHS, South Africa

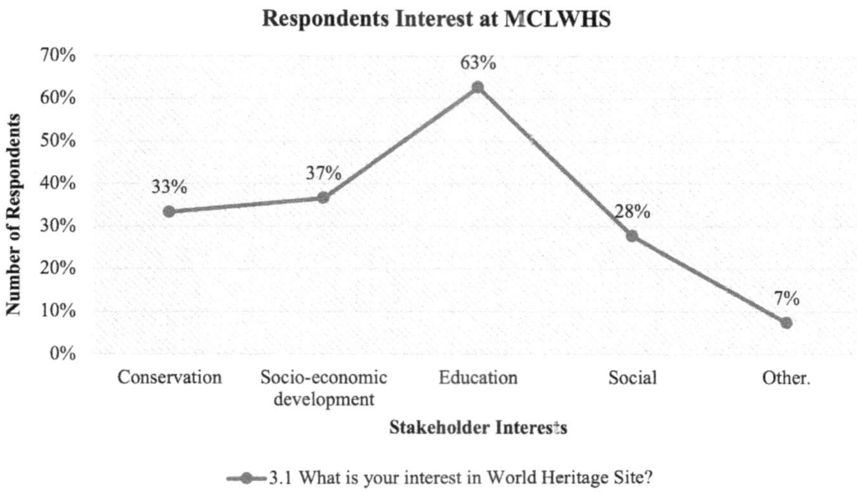

FIGURE 6.6 Interests of stakeholders interviewed in South Africa in 2017/18

Stakeholders and Their Knowledge on MCLWHS

The knowledge of stakeholders on the values and attributes (significance) of MCLWHS identifies the following values as important: cultural values (65%), education (52%), social (44%), tourism (37%), and biodiversity (28%), as illustrated in Figure 6.7. On the lower end of prioritized values by stakeholders were agriculture (11%), infrastructure (10%), hunting (7%), and mining (5%). Overall,

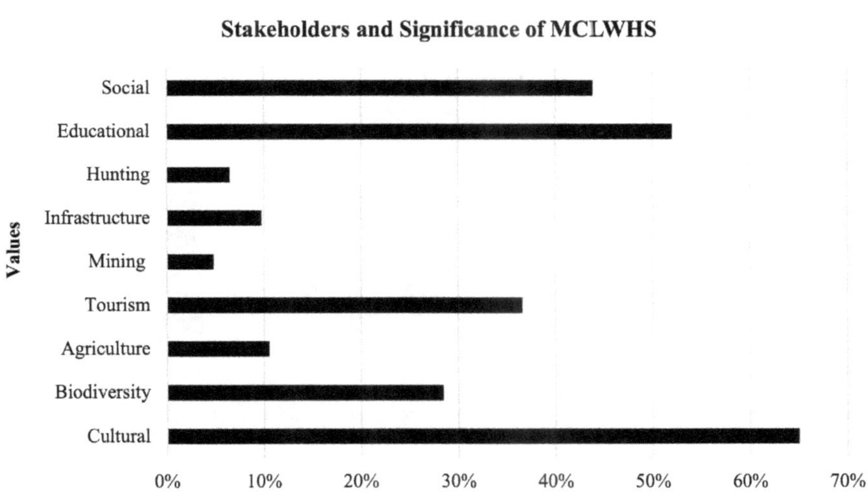

	Cultural	Biodiversity	Agriculture	Tourism	Mining	Infrastructure	Hunting	Educational	Social
■ Series1	65%	28%	11%	37%	5%	10%	7%	52%	44%

FIGURE 6.7 Significance of WHS to stakeholders

culture is the dominant value despite the fact that biodiversity values have been prioritized for many decades at MCLWHS. This denotes the increasing efforts by SANParks in profiling cultural heritage in PA.

When it came to stakeholder knowledge on the protection and management systems at MCLWHS, the majority of stakeholders (42%) said they were not familiar with conservation plans, while 46% of stakeholders stated that they were responsible for the protection of MCLWHS. The latter is interesting in the sense that while SANParks is in control at MCLWHS, there are other stakeholders with land holdings and having decision-making power in the process. However, this could also point to increased engagement of other stakeholders in the management of the site. Regarding the effectiveness of conservation plans at MCLWHS, 50% of the stakeholders did not know whether conservation plans are effective or not, while 40% confirmed that conservation plans are effective. While the majority of stakeholders stated they were responsible for the protection of the site in some way and are familiar with conservation plans, they do not know the effectiveness of these plans. This may demonstrate their non-involvement in the governance of the site, especially in the implementation, evaluation, and monitoring of the management plans, which is predominately a performance management process in the ambit of SANParks. Possibly, some stakeholders may have come to know about management plans from a consultation process. Overall, the knowledge of stakeholders on the protection and management systems at MCLWHS is predominantly positive and affirming increasing stakeholder engagement and involvement.

Stakeholder awareness on the socio-economic developments at MCLWHS ranked tourism (78%), infrastructure (32%), agriculture (24%), and mining (11%) as the dominant ones. This provides some insightful perspectives on MCLWHS

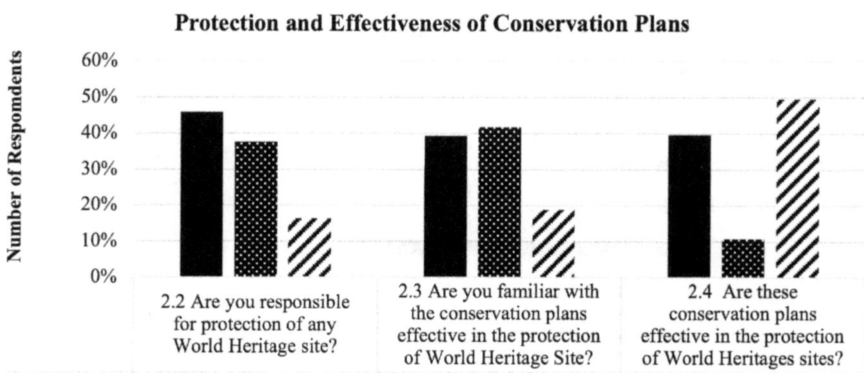

	2.2 Are you responsible for protection of any World Heritage site?	2.3 Are you familiar with the conservation plans effective in the protection of World Heritage Site?	2.4 Are these conservation plans effective in the protection of World Heritages sites?
Yes	46%	39%	40%
No	38%	42%	11%
Don't Know	16%	19%	50%

Variable Questions/Response Options

FIGURE 6.8 Stakeholder knowledge on protection, conservation, and effectiveness of conservation plans at MCLWHS

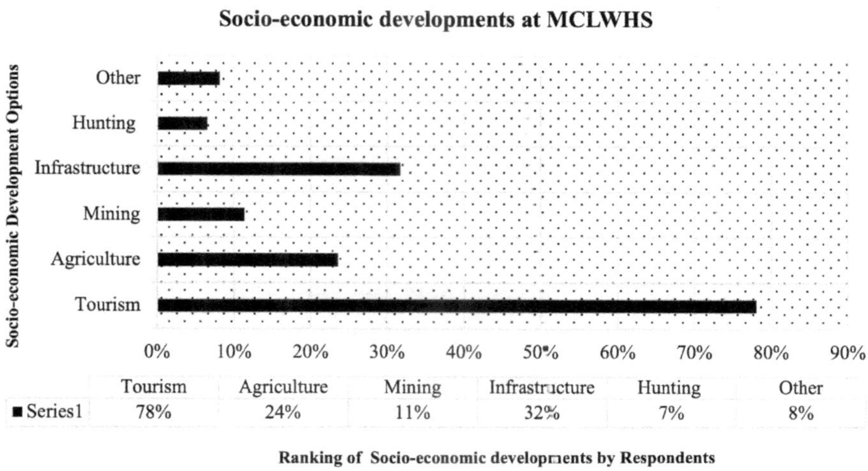

FIGURE 6.9 Stakeholders ranking of socio-economic developments at MCLWHS

in meeting the well-being of the society. Overall, stakeholders at MCLWHS identify more with tourism than with mining, which happens to be one of the major activities at MCLWHS. This is because MCLWHS is part of the national and international tourism itinerary. It is also the pinnacle of attractions in the Limpopo Province (South Africa).

Roles and Responsibilities of Stakeholders at MCLWHS

The field survey showed that 47% of the stakeholders are involved in the protection of MCLWHS, while 37% are not involved, and the remaining 16% did not know whether they are involved or not in the protection of the site. However, when it came to whether as stakeholders they provide resources in the protection of MCLWHS, the majority of stakeholders indicated that they provide "nothing" (25%), and at least 21% of stakeholders were not sure, while and in variable but lower percentages, some stakeholders provide resources for "social corporate responsibility", "technical resources", "infrastructure support", "financial resources", and "human capital" at the site (Figure 6.10). Related to the above is how all the stakeholders involve local communities in their programmes at MCLWHS. Fifty percent of stakeholders said "*yes*" they do involve local communities, while 28% said it was "*not applicable*" to them, and the remaining 22% said they don't involve local communities.

On how important it was to have stakeholders involved in the decision-making process on conservation and development at MCLWHS, the majority of the stakeholders (38%) said it was "*very important*", while 25% said it was "*important*", with only 17% considering it "*extremely important*" (Figure 6.11). Therefore, most stakeholders at MCLWHS would like to be involved in the decision-making process at MCLWHS.

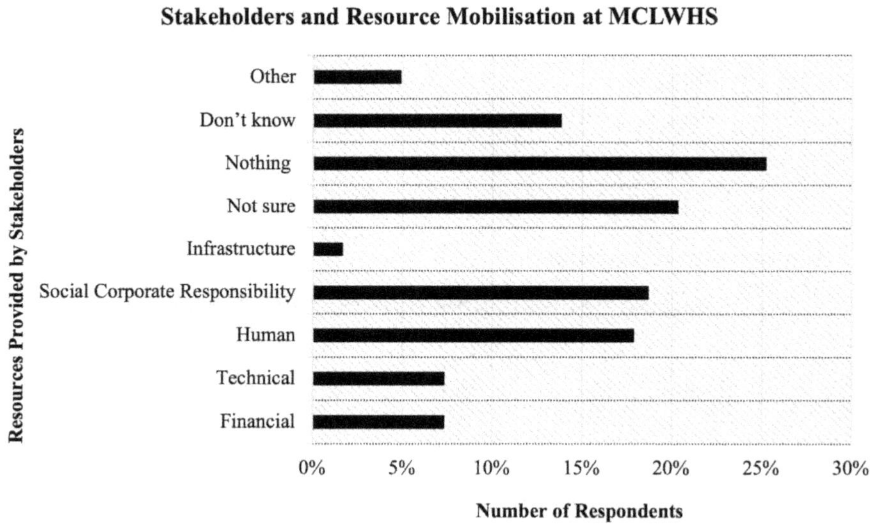

FIGURE 6.10 Stakeholders and resources they provide for the protection of MCLWHS

Stakeholder motivations (reasons) for desiring to be involved in the decision-making processes at MCLWHS are dominated by "commitment" (38%) and "professional" (28%), while 18% "did not know" their motivations. Therefore, stakeholders at MCLWHS are mainly driven by their "commitment" (Figure 6.12).

When it comes to stakeholders' level of influence on conservation decisions, it is evident that cumulatively, stakeholders have an "*average*" (29%) to "*low*" (19%) influence on conservation at MCLWHS (Figure 6.13). When compared to their level of influence on development decisions at MCLWHS, the results show that stakeholders rated their influence as "very low" on the following socio-economic

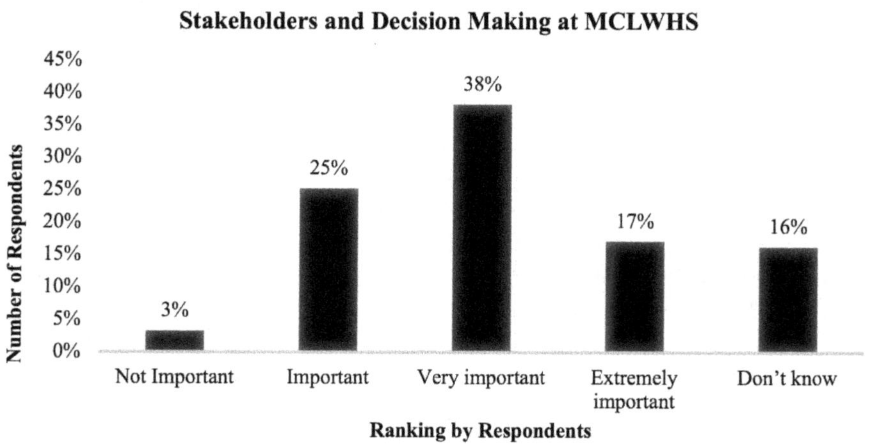

FIGURE 6.11 Stakeholders' perspective and experiences in the decision-making process at MCLWHS

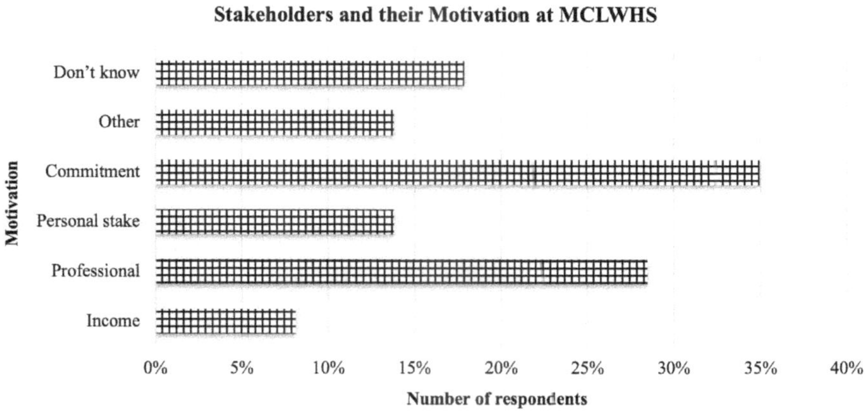

FIGURE 6.12 Stakeholders and their motivation to participate in the management of MCLWHS

activities: hunting (38%), others (31%), extractive industries (30%), infrastructure (27%), agriculture (26%), and tourism (19%), as illustrated in Figure 6.14. This is followed by average and low influences on decisions around socio-economic developments. In the "*very high*" level of influence on decision-making relating to socio-economic activities, the following ranking emerged: tourism (11%), infrastructure (6%), extractive industries (4%), hunting (3%), agriculture (3%), and others (2%), as illustrated in Figure 6.14. Tourism is a critical socio-economic aspect of the MCLWHS when compared to all others.

For the stakeholders who considered their influence as "*high*" in decision making on socio-economic development, they selected tourism (17%) and

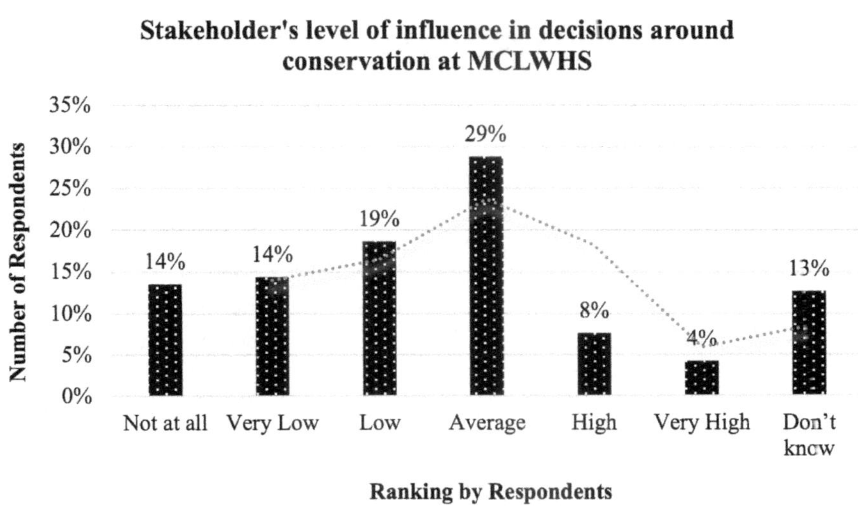

FIGURE 6.13 Stakeholders level of influence in decision-making on conservation issues at MCLWHS

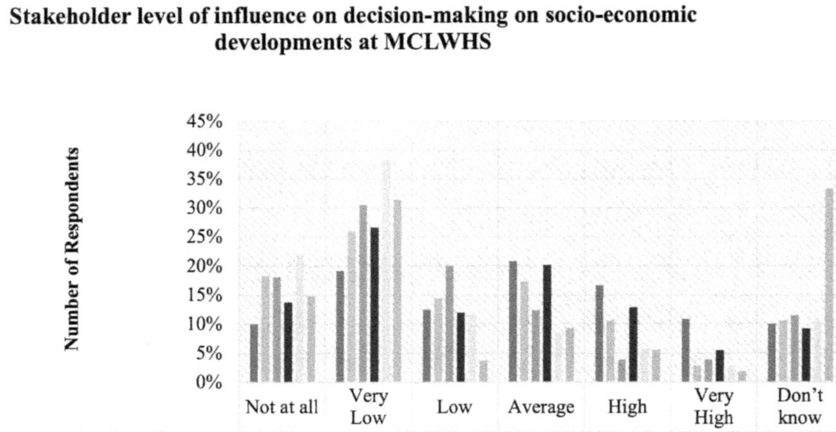

	Not at all	Very Low	Low	Average	High	Very High	Don't know
Tourism	10%	19%	13%	21%	17%	11%	10%
Agriculture/ farming	18%	26%	14%	17%	11%	3%	11%
Extractive industries(mining)	18%	30%	20%	12%	4%	4%	11%
Infrastructure development	14%	27%	12%	20%	13%	6%	9%
Hunting	22%	38%	12%	9%	6%	3%	11%
Other	15%	31%	4%	9%	6%	2%	33%

Ranking by Respondents

FIGURE 6.14 Stakeholders level of influence in decision-making on socio-economic issues at MCLWHS

infrastructure (13%) as their highest-ranking options (Figure 6.14). For those who considered their influence socio-economic activities as "*average*", they ranked tourism (21%) again as the highest followed by infrastructure (20%). However, some stakeholders considered their influence as "*low*" and extractive industries (20%) dominates, followed by agriculture (14%). Those who considered their influence on socio-economic development aspects to be "*very low*", they ranked hunting (38%) as the highest under this category (see Figure 6.14).

Furthermore, some stakeholders revealed that they do not have any influence (*not at all*) in the decision-making process around socio-economic initiatives especially in hunting (22%) and extractive industries (18%). Others indicated that they "*didn't know*" whether they have or do not have influence in the decision-making process relating to socio-economic development aspects and "*others*" (33%) dominate under this category. Overall, the highest ranking per each socio-economic aspect is "*very low*" except for tourism, which is ranked "*average*" at MCLWHS (Figure 6.15).

An analysis of stakeholder impact on other interested and affected stakeholders at MCLWHS revealed the following rankings (Figure 6.16): "*very low*" impact on national government (25%) and international community (24%), "*average*" impact for provincial government (25%) and learners (23%), "*high*" for local communities (22%) and learners (19%), "*very high*" for learners (19%) and tourists (15%), while "*low*" is reflected for tourists (12%) and local communities (12%).

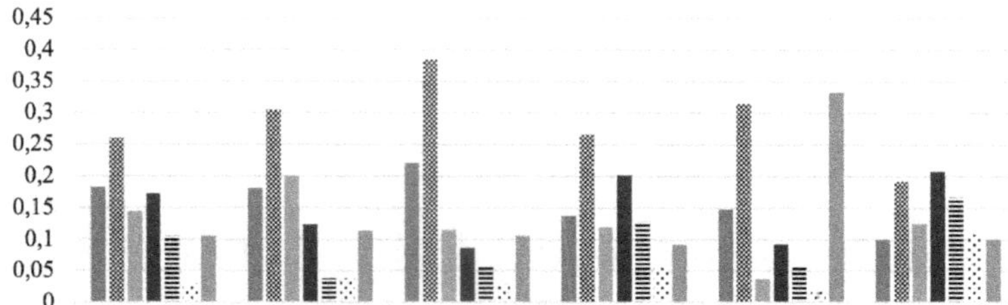

	Agriculture/ farming	Extractive industries(mining)	Hunting	Infrastructure development	Other	Tourism
Sum of Not at all	0,182692308	0,180952381	0,221153846	0,137614679	0,148148148	0,1
Sum of Very Low	0,259615385	0,304761905	0,384615385	0,266055046	0,314814815	0,191666667
Sum of Low	0,144230769	0,2	0,115384615	0,119266055	0,037037037	0,125
Sum of Average	0,173076923	0,123809524	0,086538462	0,201834862	0,092592593	0,208333333
Sum of High	0,105769231	0,038095238	0,057692308	0,128440367	0,055555556	0,166666667
Sum of Very High	0,028846154	0,038095238	0,028846154	0,055045872	0,018518519	0,108333333
Sum of Don't know	0,105769231	0,114285714	0,105769231	0,091743119	0,333333333	0,1

FIGURE 6.15 Pivot chart showing highest ranking level of influence per each socio-economic aspect

However, some stakeholders revealed that they have no impact at all (*not at all*) on other stakeholders, and this is dominated by international community (19%) and national government (18%), while those who "*don't know*" whether they impact or do not impact other stakeholders was dominated by international community (16%) and national government (15%). When all these responses (Figure 6.16) are considered in the collective to find unique values per each stakeholder category against impact ranking options, the pivot chart below (Figure 6.17) shows that the highest ranked impact option is "*average*" for provincial government (0.25), followed by "*low*" (0.24) for international community, "*very low*" (0.24) for national government, and "*average*" (0.23) for learners and local

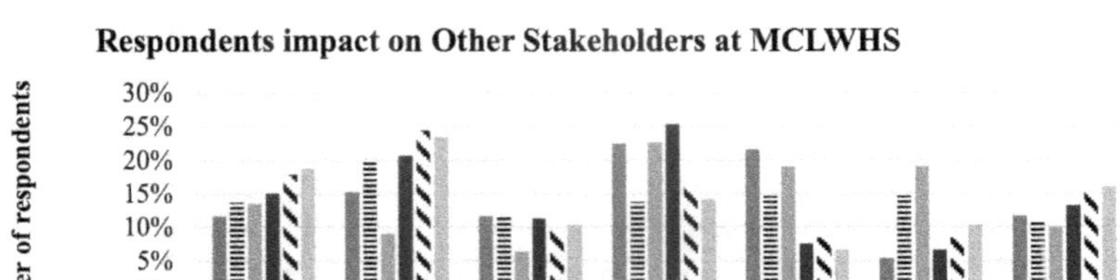

	Not at all	Very Low	Low	Average	High	Very High	Don't know
Local communities	12%	15%	12%	23%	22%	5%	12%
tourists	14%	20%	12%	14%	15%	15%	11%
Learners	14%	9%	6%	23%	19%	19%	10%
Provincial Goverments	15%	21%	11%	25%	8%	7%	13%
National Government	18%	25%	9%	16%	8%	8%	15%
International Community	19%	24%	10%	14%	7%	10%	16%

Ranking by Respondents

FIGURE 6.16 Stakeholders and their impact on other stakeholders at MCLWHS

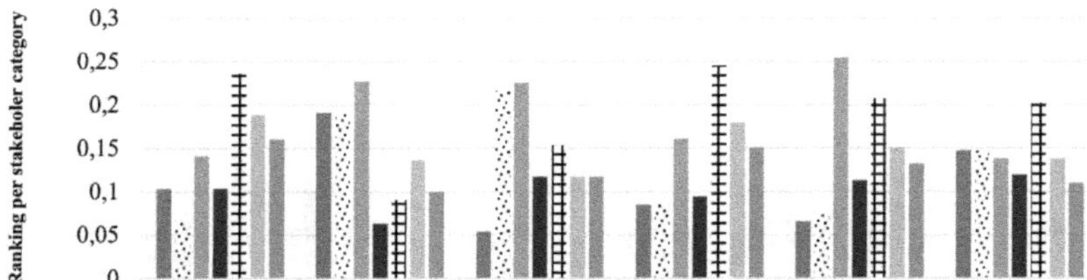

	International Community	Learners	Local communities	National Government	Provincial Goverments	tourists
Sum of Very High	0,103773585	0,190909091	0,054054054	0,08490566	0,066037736	0,146788991
Sum of High	0,066037736	0,190909091	0,216216216	0,08490566	0,075471698	0,146788991
Sum of Average	0,141509434	0,227272727	0,225225225	0,160377358	0,254716981	0,137614679
Sum of Low	0,103773585	0,063636364	0,117117117	0,094339623	0,113207547	0,119266055
Sum of Very Low	0,235849057	0,090909091	0,153153153	0,245283019	0,20754717	0,201834862
Sum of Not at all	0,188679245	0,136363636	0,117117117	0,179245283	0,150943396	0,137614679
Sum of Don't know	0,160377358	0,1	0,117117117	0,150943396	0,132075472	0,110091743

Stakeholder categories

FIGURE 6.17 Pivot chart showing the highest ranked impact option per stakeholder category

communities. Thus, at MCLWHS, the predominantly impacted stakeholder category is provincial government (0.25) given its proximity to the site.

With regard to stakeholder consultations and their involvement in the implementation of national and international legislation at MCLWHS, the majority of stakeholders (83%) were not consulted in the establishment of any national park in South Africa. This is a true reflection of how this system of national parks evolved in the colonial period as a closed process engineered and pushed by scientists only. Also, this points to intergeneration transition in which the elderly people have passed on and the new generation, born post these processes, is unaware of these issues. Furthermore, the majority of stakeholders (79%) were also not consulted during the inscription of Mapungubwe as a National Heritage site. The same pattern emerges in relation to the inscription of MCLWHS, with 81% of stakeholders stating that they were not consulted. Also, the majority of stakeholders confirmed that they were not consulted in the demarcation of the core area (82%) and buffer zone (87%) of MCLWHS. The same pattern repeats itself on consultations relating to socio-economic developments at MCLWHS, where 79% of the stakeholders were not consulted. It is no different in respect to the management of MCLWHS, where only 20% were consulted. This pattern denotes how PA and WHS in Africa have always strategically consulted stakeholders as a compliance process rather than as a good governance approach underpinned by principles of accountability, transparency, equity and inclusiveness, empowerment, and broad-based engagement and involvement.

Predominantly, the above analysis reveals that the majority of stakeholders are not consulted in the heritage processes at MCLWHS. Demarcation of the buffer zone boundary stands out as the area where most stakeholders are not consulted, yet this is where most of the developments at MCLWHS are situated.

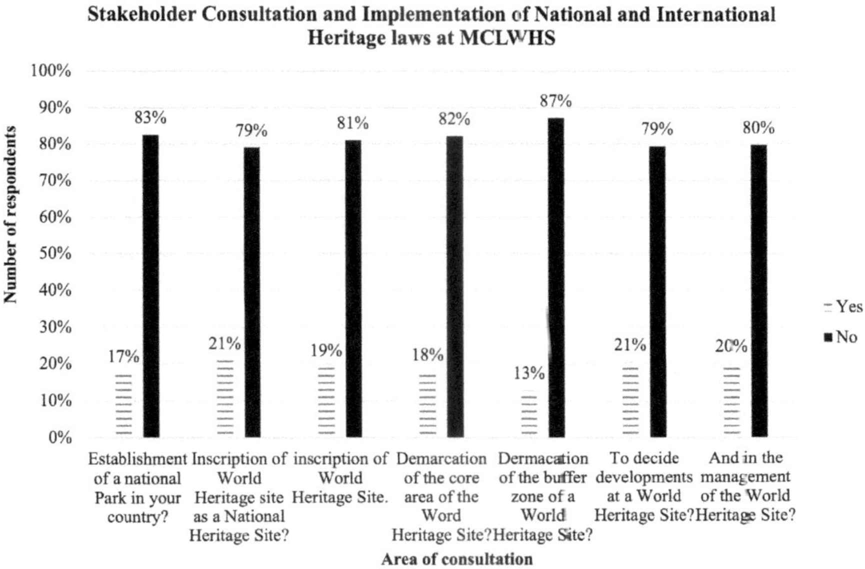

FIGURE 6.18 Stakeholder consultations in implementing national and international legislation

This may be explained by analyzing challenges of stakeholders at MCLWHS. Regarding issues and challenges affecting stakeholders at MCLWHS, the following ranking emerged: lack of financial resources (77%), communication at national levels (68%), communication at provincial levels (67%), politics (67%), time constraints in attending meetings (64%), communication at local levels (61%), and communication at international levels (55%). However, some of the stakeholders pointed out that there are no issues and challenges affecting them as stakeholders at MCLWHS (Figure 6.19). For IDCs this could be a geographical, time, and resources limitation, while for the development sector, it is each man for himself approach, and for both, the continued dominance of SBMS approach at MCLWHS.

Overall, there are more cross-cutting issues and challenges affecting stakeholders at MCLWHS. The majority of stakeholders are affected by "*lack of financial resources*" followed by "*lack of communication*" at national and provincial levels, "*politics*" and "*time constraints in attending meetings*" at the site. This equally explains the emerging pattern on consultation and involvement in the implementation of national and international laws at MCLWHS.

Stakeholder Views on the Benefits of Protecting MCLWHS

Stakeholders always have expectations of benefitting from processes and activities they are engaged and involved in, including those opportunistic moments that are presented to them by other stakeholders. It is human nature to be expectant

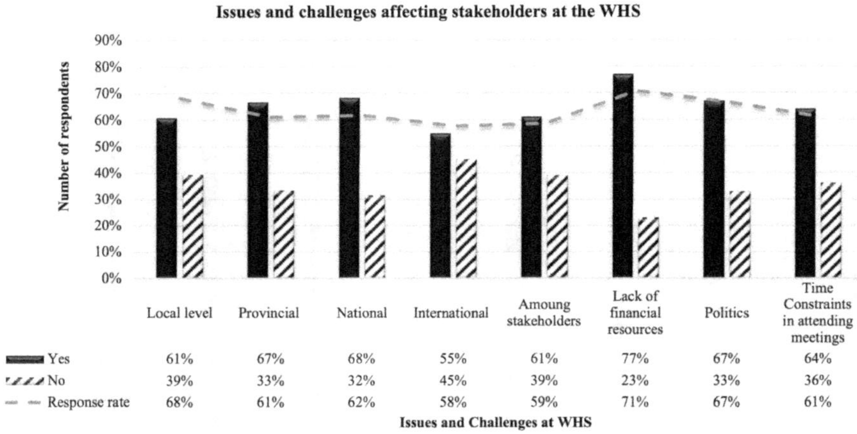

FIGURE 6.19 Issues and challenges affecting stakeholders at the MCLWHS

as part of realizing this well-being. They have expectations for both conservation and development. Stakeholders expressed their opinions on benefits from (i) derived from extracted resources, (ii) socio-economic developments implemented, (iii) the contribution of conservation and socio-economic development to each other, and (iv) the level of importance between development and conservation at MCLWHS.

The majority of stakeholders (78%) do not extract any resources from MCLWHS, while 22% do so (Figure 6.20). This confirms existence of stringent controls associated with PA and WHS in Africa prioritizing conservation over other resources. This has made WHS become conservation fortresses against

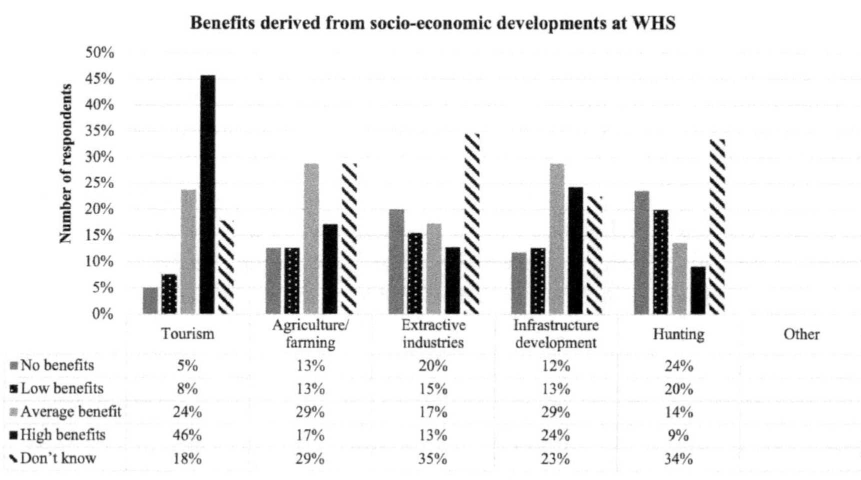

FIGURE 6.20 Stakeholder opinions on benefits of socio-economic development at MCLWHS

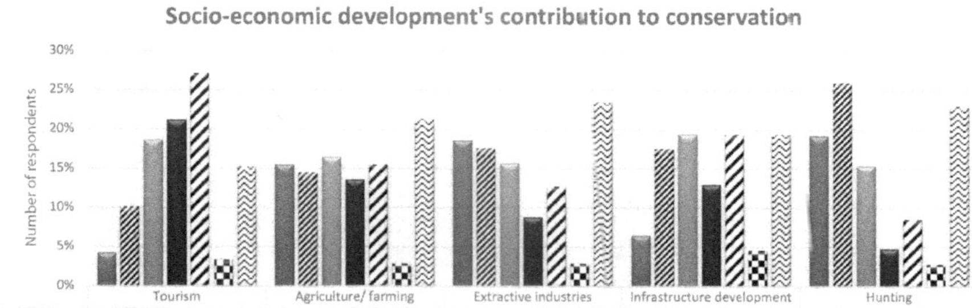

	Tourism	Agriculture/farming	Extractive industries	Infrastructure development	Hunting
Not at all	4%	16%	19%	6%	19%
Very Low	10%	15%	18%	18%	26%
Low	19%	17%	16%	19%	15%
Average	21%	14%	9%	13%	5%
High	27%	16%	13%	19%	9%
Very High	3%	3%	3%	5%	3%
Don't know	15%	21%	24%	19%	23%

Socio economic aspects and contribution to conservation

FIGURE 6.21 Stakeholder opinions on contribution of socio-economic development to the protection and conservation of MCLWHS

development. However, stakeholders are challenging this and advocating for increased resource utilization at WHS for meeting their livelihoods. Regarding the level of benefits derived from the different aspects of socio-economic developments at MCLWHS, stakeholder ranking was as follows (Figure 6.21): tourism as the option with "*high benefits*" (46%) and infrastructure development (24%). This signifies that tourism is more important compared to commercial agriculture (ranked 17% in this category) in the area. This could be linked to job opportunities and sustainable livelihoods for a long term. Agriculture's highest ranking is "*average benefits*" (29%), while the highest ranking for extractive industries and hunting was "*don't know*" (35% and 34%, respectively). Overall, the majority of stakeholders confirmed that they "*do not know*" anything about deriving such benefits from the site.

Stakeholder views on how socio-economic aspects impact protection and conservation of MCLWHS revealed an interesting pattern, often not analyzed in many studies in Africa. The overall ranking per each socio-economic aspect and its contribution to conservation (Figure 6.21) shows the following ranking: tourism, "*high*" (27%), infrastructure development (19%), and agriculture (19%). On the "very high" level, infrastructure development (5%) is above all other aspects (all ranked at 3%). And 21% of stakeholders "*don't know*" the contribution of development to conservation. Extractive industries are largely considered not contributing to conservation as its ranks: "*don't know*" (24%), "*not at all*" (19%), and "very low" (18%). Hunting is ranked as having "*very low*" (26%) contribution to conservation at MCLWHS. On the other hand, stakeholders "*don't know*" what agriculture (21%) and extractive industries (24%) are contributing in this process.

Overall, and based on the pivotal analysis below (Figure 6.22), the majority of the stakeholders expressed that socio-economic activities do not contribute to the protection and conservation of MCLWHS: "*not at all*".

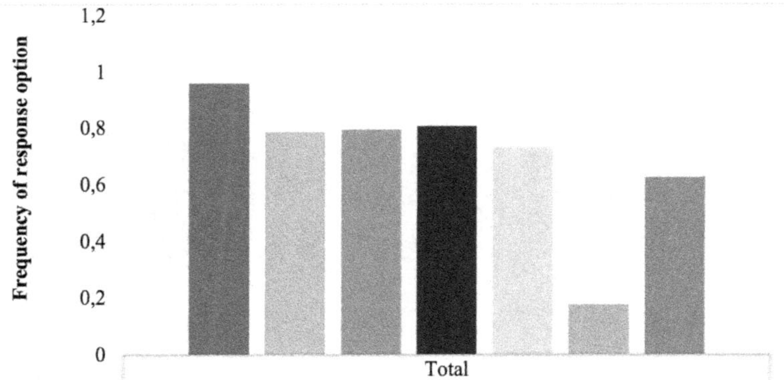

	Total
Sum of Not at all	0,9602626
Sum of Very Low	0,787713216
Sum of Low	0,79605188
Sum of Average	0,809840139
Sum of High	0,734212458
Sum of Very High	0,178337612
Sum of Don't know	0,628318936

FIGURE 6.22 Pivotal analysis of the overall responses on the contribution of socio-economic development to the protection and conservation of MCLWHS

Stakeholder comparative analysis on the importance of benefits between conservation and development at MCLWHS (Figure 6.23) showed conservation as "*very important*" (58%) compared to socio-economic development (29%). The latter is considered "*important*" (48%) by stakeholders. Considering both

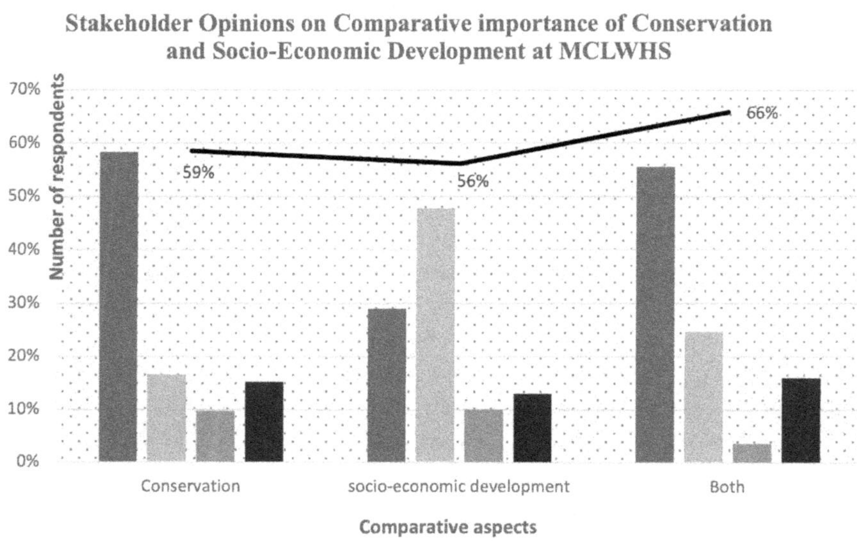

FIGURE 6.23 Stakeholder opinions on comparative importance of conservation and socio-economic development at MCLWHS

conservation and development, stakeholders ranked them as "*very important*" (56%), signifying the growing relationship between the two over the years. Overall, stakeholders agree that conservation is more important than development at MCLWHS.

Relationship between Conservation and Socio-economic Development

The ongoing discourse on conservation and development at WHS makes it important to solicit the views of stakeholders on the relationship between the two. Field assessments focused on understanding (i) how the two contribute to each other, (ii) whether they should coexist or not, (iii) kinds of developments that could be permitted, and (iv) how the socio-economic value of heritage should be measured at MCLWHS. All these are issues pertinent in current discussions on how to localize SD at WHS. On how conservation and development contribute to each other at WHS (Figure 6.24), the former contributes more (80%) compared to the latter in a significant way than the latter does to the former (71%). On a comparative basis of both, heritage contributes more to socio-economic development than the latter contributes to it. On whether conservation and development should co-exist or not at MCLWHS, the majority of stakeholders (71%) said yes, further confirming calls made at various discussion sessions wherever these matters were discussed.

Regarding the "kinds" of developments that may be allowed at WHS, stakeholders made the following rankings (Figure 6.25): tourism (71%), infrastructure development (41%), agriculture (34%), human settlement (28%), extractive industries (18%), dam constructions (17%), all (15%), and others (6%). The term "all" refers to any development options that stakeholders may put on the table at a WHS. Overall, stakeholders said development should be allowed at MCLWHS. It is not surprising that tourism was considered the most appropriate development compared to all other land use options. This is a result of the sustained

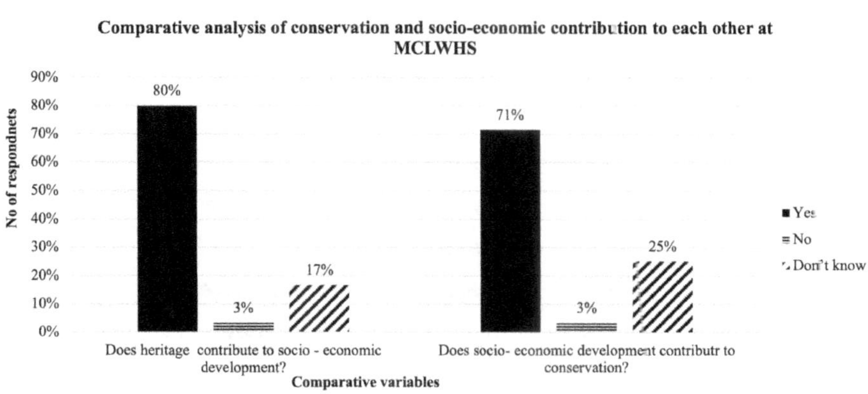

FIGURE 6.24 Comparative analysis of the contribution of heritage and socio-economic development to each other

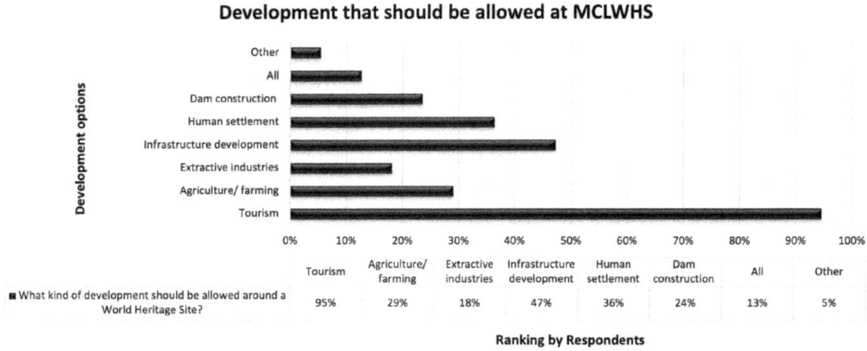

FIGURE 6.25 Types of socio-economic developments that should be allowed at MCLWHS

link between WHS and tourism opportunities, the advancing of sustainable and responsible tourism, though negative impacts of tourism are often underplayed.

Regarding how to measure the contribution of heritage to SD targets, stakeholders at MCLWHS suggested multiple measuring indicators (Figure 6.26) and these include: "*employment/jobs created*" (58%), "*level of conservation*" (50%), "*level of community development*" (49%), "*number of tourists*" (43%), "*contribution to gross domestic product*" (40%), "*number of business/investors*" (34%), "*infrastructure development*" (30%), "*return on investment*" (26%), and "*others*" (6%). Emphasis on job creation, conservation, community development, and tourist indicators point to community aspirations. However, it may be easier to measure conservation indicators at WHS (as these are outlined in the desired SOC statement) than SD

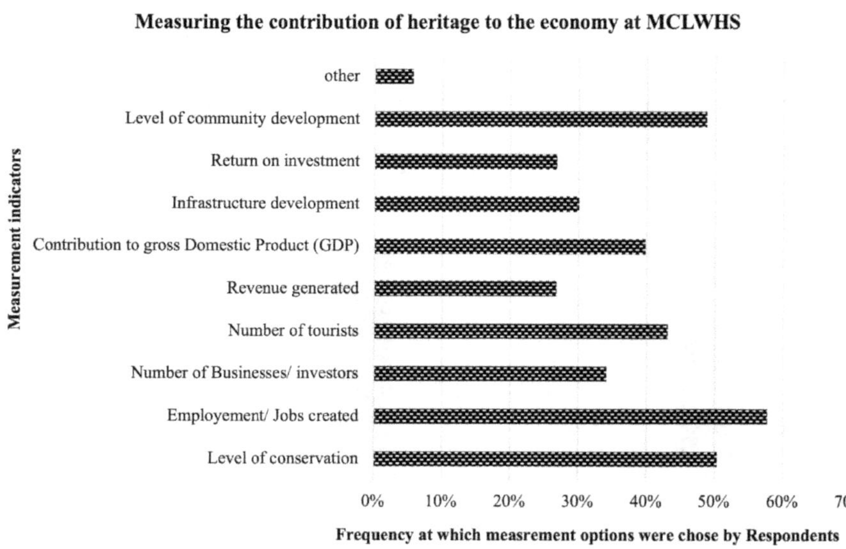

FIGURE 6.26 How economic value of heritage should be measured at MCLWHS

indicators, which are linked to national statistics. Heritage component is often subsumed under tourism statistics, hence the call for culture statistics, and this matter will be discussed in the next chapter.

Conclusion

The perspectives and patterns emerging from an inclusive analysis of all stakeholders bring out perceptions, aspirations, power-interest matrix, challenges, and opportunities and influence capabilities on conservation and development at MCLWHS. The perspectives and patterns form the basis of the discussion on stakeholder-driven conservation and SD at MCLWHS in the next chapter. That discussion draws empirical patterns and experiences at other selected WHS in Africa.

7

STAKEHOLDERS, CONSERVATION, AND DEVELOPMENT

Introduction: Crossing and Integrating Frontiers

This chapter provides an analytical context of emerging stakeholders, conservation, and development issues at World Heritage Sites (WHS) in Africa. These are frontiers that need to engage, negotiate, and integrate into each other at WHS. Using Mapungubwe Cultural Landscape World Heritage Site (MCLWHS), the discussion places emphasis on major issues around stakeholder profiling and their level of awareness on World Heritage, their roles, and their futuristic perspectives on integrating conservation and development at WHS. Parallels are drawn from the experiences of other selected WHS in Africa in the process of integrating stakeholders, conservation, and development frontiers. However, this analytical context and empirical evidence from WHS cannot be divorced from the historical evolution of each WHS through time and space. Neither can it be divorced from evolving and recorded multiple land uses at each WHS. These land uses amplify multiple voices advocating for conservation and development to co-exist at WHS. This analysis reinforces the fact that the land uses at WHS in Africa, including MCLWHS, are still heavily influenced by the conservation thinking advanced by State-Based Management Systems (SBMS) and Conventions. It also argues that there are less benefits accruing from conservation compared to those coming from broader development initiatives at MCLWHS. The chapter argues that it is not enough just to identify and acknowledge stakeholders without giving them a decision-making role at WHS. Advancing multivocality at WHS does not mean that stakeholders are oblivious to threats on the site, especially those coming from developments; it is just that they want to sustain their livelihoods too. In this discussion, both Multiple and Multi-layered Stakeholder Theory (MMST) and sustainable development (SD) principles are recommended as catalytic processes of an adaptive management

DOI: 10.4324/9781003149361-7

approach balancing conservation and development at WHS in Africa. From a futuristic perspective, the discussion notes that an adaptive management approach facilitates crossing and integrating the stakeholders, conservation, and development frontiers at WHS in Africa. It nurtures the emerging relationship among the frontiers. The localization of Sustainable Development Goals (SDGs) at the site lacks an adaptive management approach geared towards providing local solutions to local challenges.

Multiple and Multi-Layered Stakeholders at WHS

The concept of stakeholders is gradually gaining momentum in WHS governance in Africa. Good governance demands that stakeholders be involved in the decision-making process of an institution. Dominance of one group at the exclusion of others is not implementing good governance. For example, at most WHS, the multiple and multi-layered stakeholders are often left out by SBMS, UNESCO, and their selected academics or experts (Table 7.1). Their dominance surfaces in meetings, interpretation, and conservation programmes at MCLWHS

TABLE 7.1 Historical and emerging stakeholders at MCLWHS

Pre-colonial Period 900-1380AD	Colonial Period 1300AD-1993	Post-Apartheid Period 1994-present
• Stone Age Communities • Iron Age Communities • Traders from Asia and East Africa Coast, • Regional Polities in Botswana and Zimbabwe	• Indigenous and descendant communities, • Commercial farmers • National and Provincial governments • Hunters • Associations • Non-Governmental Organisations • Tour Operators • Mining companies • Academics/Universities • Learners	• Indigenous and descendant communities, • Commercial farmers • National and Provincial governments • Hunters • Professional Associations • Non-Governmental Organisations • Tour Operators • Extractive Industries • Academics/Universities • International Communities *UNESCO • Interstate Parties (Zimbabwe and Botswana) • Learners

(Apley, 2000; Carruthers, 2006; Nienaber *et al.*, 2008; Pikirayi, 2016). As such, the archaeological mindset and narratives dominate interpretation at the expense of spirituality at WHS (Chirikure & Taruvinga, 2017; Ndoro, 2017). Indigenous and Descendent Communities (IDCs) have always complained about this dominance and approach (Chirikure & Taruvinga, 2017). WHS governance has to adopt open and transparent ways of involving stakeholders in its processes.

What is of Value to Stakeholders at WHS?

Stakeholders always place diverse values and significances on WHS, and this is critical in understanding them. However, these values are dynamic and society-driven despite legal prescription of what is of significance or not. Culture is not static, so is the creation of new values, which may change the configuration of stakeholders at WHS. At MCLWHS, cultural heritage (65%) and educational (52%) values are of significance to stakeholders compared to all other values. This is typical of WHS in Africa. This reflects on how cultural heritage and educational awareness is being prioritized in Protected Areas (PA) by South African National Parks (SANParks). For IDCs, intangible values and spirituality constitute the cultural heritage values of MCLWHS. There has been more broad-based research in the area profiling the cultural heritage values. Similar developments have happened at other WHS, such as Ngorongoro Conservation Area (NCA), where cultural values have received increased attention and prioritization from a budgeting perspective. Historically, nature values had been prioritized at NCA, while the Department of Antiquities (DoA) was remotely managing the archaeological sites. The handover of the cultural values to NCA through a memorandum of agreement by DoA was a bold decision towards a single management approach at NCA. Both the nature and culture values of NCA have resulted in the site becoming a mixed WHS in 2010.

The dominance of the educational values at MCLWHS confirms the success, consistency, and resilience of heritage educational and outreach programmes, the people, and park initiatives, including the culture week programmes being implemented by SANParks. This is replicating itself at many WHS in Africa. Robben Island World Heritage Site (RIWHS) has vibrant heritage education programmes, on and off-site. These have raised awareness on RIWHS as a cultural landscape with its multiple and multi-layered values. Also, at MCLWHS, social (44%), tourism (34%), and biodiversity (28%) values are also considered important by stakeholders. These are linked to the tourism perspectives of WHS and the need to promote social cohesion and nation-building in respective State Parties (SPs). For MCLWHS, the Greater Mapungubwe Transfrontier Conservation Area (GMTFCA) is an impetus to these additional values. However, the low ranking of agriculture, infrastructure, and hunting values possibly signifies how these influence livelihoods at local levels compared to stakeholders far away from MCLWHS. For example, with agriculture, most employees are from Zimbabwe and Botswana, which are sources of seasonal and cheap labour.

Despite the long history of extractive industries and the perceived economic benefits at MCLWHS, these industries were ranked low by stakeholders. The reason could be that stakeholders are not benefitting from their proceeds. IDCs argue that beneficiaries of extractive industries are SP and the private sector (Chirikure & Taruvinga, 2017). Also, the increasing negative impacts of extractive industries could be another reason. However, given the African Union (AU) Agenda 2063, aiming for self-sustenance in Africa, equally supported by National Development Plans (NDP), more extractive processes may happen at WHS in Africa. Despite negative dimensions, extractive processes indeed support SD in developing nations.

MMST and WHS in Africa

The net influence of stakeholders at WHS in Africa is embedded in their legitimacy, interests, power, and political relations as outlined by MMST. Stakeholder influence is a force that needs to find space in the governance of WHS. As argued in Chapters 2 and 3, MMST can facilitate good governance at WHS, including gradually changing the current top-down ("with SP and UNESCO as kings") approach to a bottom-up approach ("with stakeholders as kings"). The bottom-up approach gives birth to the "local-global nexus" of MMST, in which the local influences the global at various operating levels of the WHS (Figure 7.1).

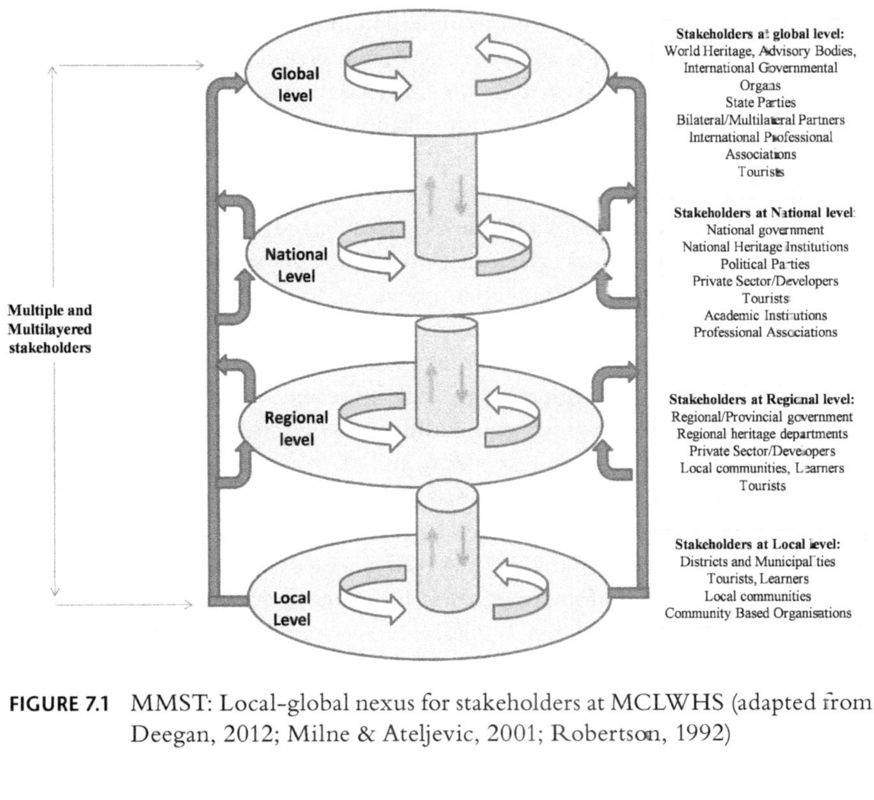

FIGURE 7.1 MMST: Local-global nexus for stakeholders at MCLWHS (adapted from Deegan, 2012; Milne & Ateljevic, 2001; Robertson, 1992)

This engagement interconnects and permits interpenetration among stakeholders. The paradox is that global processes can no longer be conceived without a local context, while local context cannot escape the global ideas (Levitt, 1983). However, this nexus is in a constant mode of transition as stakeholder composition, legitimacy, interests, political, and power relations change through time.

Using MCLWHS profile, the confirmed emerging stakeholder categories at WHS in Africa are:

- **Stakeholders with "universalized" interests, power, and means to impose decisions at WHS**: These stakeholders operate at a global level supported by intergovernmental organizations policing the UNESCO Conventions. These include UNESCO, World Heritage Convention (WHC), and Advisory Bodies (ABs) [International Council on Monuments and Sites (ICOMOS), International Union on Conservation of Nature (IUCN), and International Centre for the Study of the Preservation and Restoration of Cultural Property (ICCROM)]. Their primary goal is to protect and maintain the spirit of Conventions and related protocols. However, some experts sent by ABs on WHS missions are not familiar with the local context and dynamics of a WHS they are supposed to be advising on. The link between these stakeholders and institutions of higher learning in Africa is still very weak compared to Europe or Asia, yet this could deepen the understanding of local dynamics and create access to data for decision-making purposes at WHS. Also, understanding the diverging local forces of stakeholders takes time and effort beyond a "five-day" mission to a WHS. This is further compounded by the financial costs involved, which does not allow for broad consultations with stakeholders at the local levels.

  One dimension with the role of these stakeholders is the interpretation of their objectives, for instance, with Reactive Monitoring Missions (RMM) to WHS in Africa. The WHC uses RMM a lot and this is supported by ABs. The mindset of RMM experts is gathering or investigating facts on direct and indirect, actual and potential threats, understanding the broader management issues, governance structures, and their effectiveness at the WHS. This is an analysis beyond mere observations at WHS in order to establish root causes and mitigation thereof for recommendation to the SP. An independent eye always sees things differently! For the SP receiving RMM, the mindset is conflicted: responding and reacting to perceived and alleged ineffective management, which may lead to placement on endangered list; opportunity to be assisted and embarrassment of being discussed by other SP; and interpreting the Terms of Reference (TORs) for RMM in order to provide factual information towards a solution versus prevailing politics beyond WHS issues. For stakeholders, it is always the opportunity to share their own perspective on the matter at hand.

  As a result of these mindsets, and while the TORs for RMM are statutory decisions of the WHC, they are often interpreted by SP as "fault finding" leading to "unfavourable measures" (Figure 7.2). This emanates from the

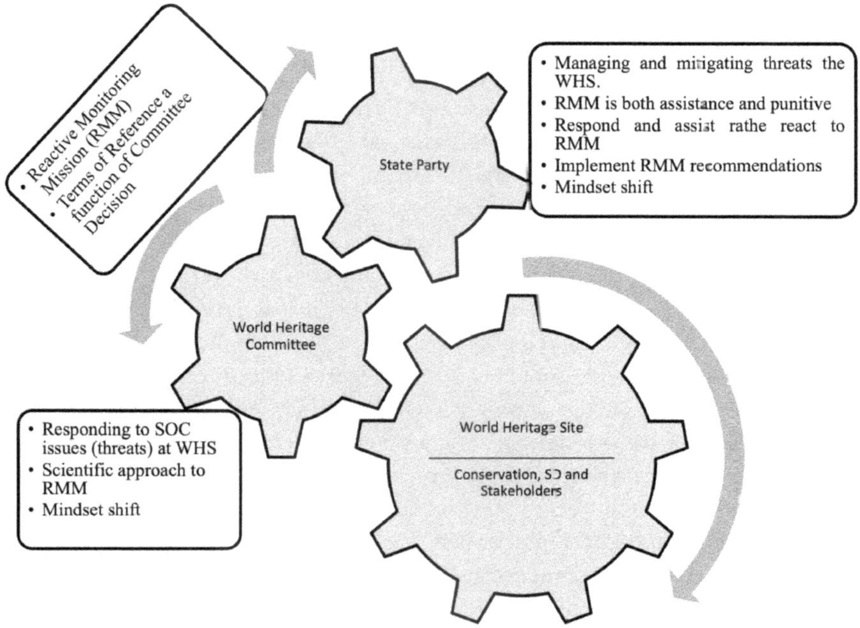

FIGURE 7.2 Reactive monitoring mission dynamics at WHS

perception of SP that inscription is an end in itself and for tourism, yet management of the WHS becomes a priority and a going concern. This includes the contradictory perception of RMM and SP on the role of stakeholders in the process. For the WHC, stakeholders are sources of information not necessarily obtainable from SP, while for the SP, stakeholders are "sell-outs" as they provide negative information to the WHC due to local dynamics beyond the comprehension of RMM. The irony with both WHC and SP is that the very same stakeholders are observers at WHC sessions, yet they are directly consulted at WHS level. They can give information but cannot be decision-makers in the process. Therefore, stakeholder consultation during RMM is often indirectly controlled by SP. Resultant RMM reports are often contested by SP despite their positive side of guiding protection measures, unlocking political support for Management Authorities (MAs), and systematically introducing conservation tools and policies, previously lacking at most WHS.

The frustration of SP with RMM is that prioritized recommendations from the WHC are not always in line with their capacity and resources, and are not sensitive to national planning cycles, which influences time frames for implementing and reporting back to WHC. The WHC meets annually in June, yet most SP financial years run from January to December, meaning decisions are made halfway into the financial year when budgets are already fixed, approved, and being spent. Of course, the assumption is that SP should have already prioritized WHS maintenance. For WHC, the frustration is the absence of baseline indicators for monitoring state of conservation, especially for cultural heritage when compared to nature sites in Africa. It is work in progress and evolving at most WHS in Africa.

- **Stakeholders with "nationalized" interest, power, and means to make double-barrelled decisions at WHS**: These stakeholders are mainly national and provincial governments with their respective departments, heritage institutions, and agencies with power and resources to implement decisions at WHS. As argued in previous chapters, these are the SBMS, who are the official and legal custodians of WHS, except where IDCs are in charge. Even though this may be the case, the SP is still involved in the process. As argued before, these stakeholders represent all other local stakeholders at international levels. Such representation stifles multivocality on conservation and development issues at WHS in Africa. Often UNESCO receives submissions from stakeholders contradicting the position painted by the SP at Committee sessions. As argued before, the WHC pushes this back to the SP for resolution as they are more familiar with local dynamics on conservation and development issues at sites. These issues can be classified as manifestation of dissonance at WHS in Africa. Rather, dissonance should be viewed as a seed for multivocality being embraced at WHS. However, dissonance is also a double-edged sword depending on the motives of the involved stakeholders, context, and the mediator in the process. In the process, the SP has a bigger responsibility to address these issues in a transparent and inclusive manner.

In this process, the SP has dual voice and power to either align or not align with other stakeholders on heritage matters depending on whether the decision to be taken is in their favour or not. Additionally, legislative-based stakeholders such as SANParks, South African Heritage Resources Agency (SAHRA), Department of Sport, Arts and Culture (DSAC), Department of Forestry, Fisheries and Environment (DFFE), and Department of Mineral Resources (DMR) have varying and sometimes conflicting mandates at WHS. For instance, DFFE and DMR contradicted each other publicly on the mining activities at MCLWHS. This demonstrates that governance of public entities is a show of legal and administrative powers but still serving the same society. These stakeholders have a dual mandate to balance conservation and SD in their territories, whether UNESCO is there or not.

Equally, SP are increasingly becoming very political at WHC sessions. This is leveraged through bilateral, multilateral, and economic blocks beyond the scope of the Convention (Meskell, 2011). This reinforces a sense of national sovereignty, often expressed in policy and strategic announcements and state of nation addresses by SP. For instance, the economic philosophy of South Africa, largely characterized by a mixed economy (capitalism and socialism), influences the economic, investment, and ethical behaviour of the State Party (Louw & Venter, 2013). Another example was the "brewing" threat of possible withdrawal from the Convention by

Tanzania during the Selous Game Reserve WHS issue in which the SP was requesting permission for a minor modification of the boundary to facilitate uranium extraction as part of eradicating observable poverty in the country. It was only through political pressure and technicalities on boundary modifications that favourable decisions were granted to SP. Even the invoking of the IUCN-brokered principle of "no go area zone" for PA failed with this SP.

- **Stakeholders with a "localized" interest but without power to take decisions and implement them at WHS**: These stakeholders are mainly constituted by local communities or IDCs in their multiple manifestations, tradition-based organizations, and influential individuals at WHS in Africa. These stakeholders are broadened to include ordinary citizens of the SP, who may have remote connections with the WHS. Often, they do not bring meaningful resources on board, except their databank of intangible knowledge and social memory. As argued before, these stakeholders operate at local and regional levels, with limited national and international influences at WHS, unless consulted on a specific matter by the SP. This creates tensions between them and the SP, especially when they have differing positions on such matters. While IDCs have been marginalized over the decades at MCLWHS, they are now on the verge of wielding power vested through ownership of successfully reclaimed land. To them, this is their capital investment at MCLWHS, which should create access and benefits for their well-being. This spells birth of localized "political muscles" to influence conservation and development at WHS (Taruvinga, 2010).
- **Stakeholders with interest, power, and means to exploit renewable and non-renewable resources at WHS as localities of development:** These stakeholders are predominantly in the private sector: extractive industries, commercial farmers, tourism operators, bilateral and multilateral development partners. Their investment is based on principles of profitability and sustainability, with conservation being secondary in the process (Taruvinga, 2010). This is why some developments have destroyed heritage sites in Africa for a long time. Ironically, and if they are to assist SP in meeting SDG targets, they have to proceed with developments under the guidance of compliance authorities. They contribute to job creation, poverty alleviation, and infrastructure development among many others. The use of social corporate responsibility programmes by these stakeholders is a strategy to give back to the community and also create buy-in among other stakeholders. However, the materiality of benefits accruing to other stakeholders from developers varies from one initiative to another. Corruption has ensured that the benefits end up in the hands of a few politically connected individuals rather than accruing at grassroots level. This has given birth to the concept of "*#StateCapture*" in South Africa by rich and politically connected individuals, as well as the powerful companies with little respect to good governance.

While this stakeholder category upholds SD principles, Heritage Impact Assessment (HIA) and Environmental Impact Assessment (EIA), they often implement these half-heartedly in their quest to have a higher return on their investment within the shortest timeframe. The acid drainage from past mines affecting Cradle of Humankind World Heritage Site (COHWHS) in South Africa is one such example. It points to unfulfilled promises of an environmental rehabilitation plan that was developed as part of the EIA for mining. Overall, these stakeholders "borrowed power" from the SP through permits and licences as development partners. They also have "financial capacity" to challenge legal decisions opposed to their interests at WHS, such as what happened with CoAL at MCLWHS.

- **Stakeholders with "non-legalized interests", but who have influential voice and means to influence conservation and development decisions at WHS**: This stakeholder category is constituted by lobby groups, professional associations, academics and non-governmental organizations with interest in conservation and development matters at WHS. At MCLWHS, these include the Association of Southern African Professional Archaeologists (ASAPA), the Mapungubwe Coalition Group (MCG), and Environmental Lobbying groups. They are mainly driven by conservation perspectives, while their appetite on development is defined by its nature and potential impact on WHS. They are effective at aggressive and intensive lobbying for conservation and the upholding of SD principles. Their voices and litigation approach usually forces SP to be responsive. However, their litigations often induce delays in the decision to either continue or not continue with development until all their concerns are addressed. This frustrates both the private sector and SP as it sometimes leads to financial losses as was the case with CoAL at MCLWHS. IDCs often support these lobbying groups when it comes to the protection of their heritage. Professional associations have become many and more diversified in their focus on the African continent.

Stakeholder Management: Dynamic and Continuous Process

The stakeholder categories presented above are continuously emerging through time at WHS in Africa. Therefore, stakeholder management is a dynamic and continuous process, with in-built periodic assessments in order to align with the evolving needs, interests, and views of stakeholders. Relationships between and among stakeholders are largely influenced by how they relate to heritage values and conservation and construct development opportunities at WHS. Periodic assessment is a strategic fit-element of the advocated adaptive management approach at WHS. While the MMST creates a "responsive" local-global nexus at WHS and furthermore acknowledges the complexity and intricate relationships among the stakeholders, conflicts among them are inevitable. Periodic assessments should pick current and potential conflicts among stakeholders. As such, conflict resolution should be a in-built mechanism in the adaptive governance of WHS. Conflict resolution mechanisms give an equal and fair opportunity to interested and affected stakeholders at WHS, which may lead to stakeholder-driven binding decisions, including taking disagreeing positions on conservation and development at WHS. The benefit of the latter is that innovative mechanisms outside the comfort zone may emerge and become instrumental in building resilient bridges for conservation and development. In this process, SP should play a catalytic role as opposed to defending their legalistic and compliance approaches when dealing with stakeholders with different opinions. This is critical in guaranteeing the future of WHS in an ever-changing socio-economic and operational context with competing ideas and needs.

As witnessed at MCLWHS, NCA, and RIWHS, stakeholder multivocality is now challenging the wholesale concept of conservation and the "conservative" approach to development at WHS. These stakeholders desire both conservation and development. Development, or commercialization of WHS, is a necessary evil for the well-being of the society. Additionally, SP are not adequately funded and have to rely on alternative revenue sources such as tourism. Both SP and society can use the same resource for different outcomes but informed by the same principle: sustainability. What needs to be addressed is the quality of conservation and development, definition of measurable benefits, and consistent application of SD as a catalytic process. As such, the future of WHS lies in the concerted effort of stakeholders (Chirikure, 2013; Munjeri, 2004a). Heritage institutions should improve their legal, ethical, and philanthropic responsibilities towards stakeholders. They should also guarantee power and legitimacy to stakeholders in the governance process at WHS. However, an African adage says, "when two elephants fight, it's the grass that suffers". The elephants at WHS are the SP and UNESCO, while the grass refers to stakeholders without power and legitimacy in the governance process at MCLWHS. As such, stakeholders become losers at most WHS in Africa, unless they invoke political support. Empowerment linked to good governance may resolve this situation. Empowerment is "a shift in balance between the powerful and the powerless, between the dominant and the dependent" (Cole, 2006: 4; Maton, 2008; YuLong & Hunter, 2015). In the process, stakeholders "acquire valued resources and basic rights, and achieve greater control over their lives and environment" (Cole, 2006; Maton, 2008; YuLong & Hunter, 2015: 254).

The multivocality of stakeholders at WHS has an inherent weakness embedded in internal contestations, counterclaims, and changing narratives among them, especially IDCs. Without being united and focused, IDCs stand to lose ground in their negotiations at WHS. This is often used by SP to play them against each other and prolong negotiations while business as usual continues. However, contestations and counterclaims are an opportunity for further investigation to correct distorted histories and narratives of IDCs. If conflict resolution mechanisms are built into adaptive management, it has the potential to "build unity in IDCs diversity" due to the commonality of their experiences and interests. Also, contestations are opportunities for improving dialogue and communication among stakeholders. Involving stakeholders at WHS is "time and aspirations sensitive" in an ever-changing context of the WHS.

Stakeholders Aspirations in World Heritage Governance

Identification of stakeholders at WHS is not an end in itself. There is a need to define their roles and responsibilities within the governance process, regardless of the dominance of SP and WHC in the process. Emphasis should be placed on making them decision-makers and resource mobilizers and defining the benefits of their involvement in WHS governance. These are considerations towards promoting the bottom-up approach outlined in the MMST. This is not an overnight process but rather a mindset shift in SBMS and WHS governance approaches.

WHS Governance and Stakeholders

At MCLWHS, our main case study, the sum-average pivotal analysis (Figure 7.3) of all potential responses on whether stakeholders are involved or not involved in the management of MCLWHS, ranks response options as: yes (1.2), no (0.9), and don't know (0.8). This seemingly contradictory and confusing scenario could be interpreted from multiple viewpoints. First, being consulted by SANParks may be confused by stakeholders with being involved in the governance of the site, signalling a lack of understanding on what governance is. Second, stakeholders

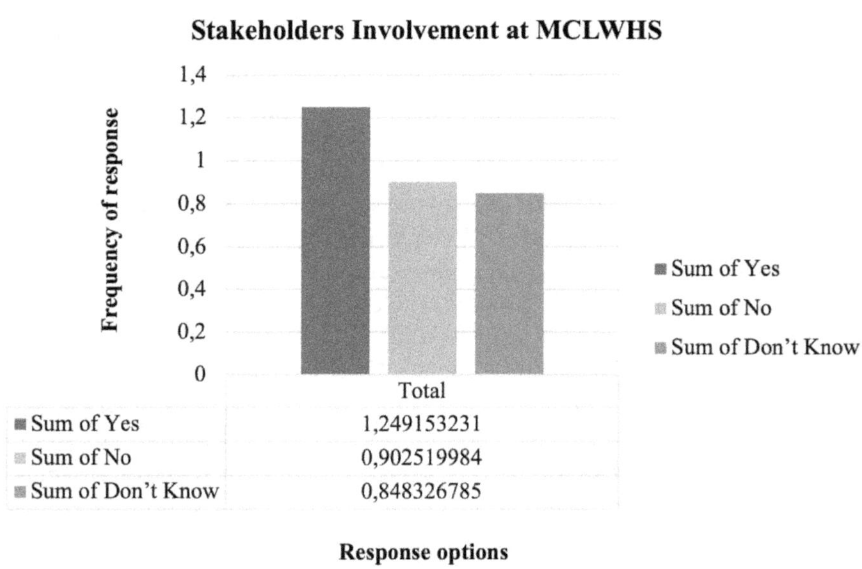

FIGURE 7.3 Sum average response of stakeholders on their role in management of MCLWHS

such as extractive industries, commercial farmers, and the tourism sector at MCLWHS are in constant dialogue with SANParks given their cross-cutting interests and existing contractual land agreements. Third, given the long history of academics at MCLWHS and how they have advanced the conservation agenda over the years, it could have influenced these responses on governance. Understandably, there are stakeholder engagements through periodic consultations and limited involvement in MCLWHS governance processes via the Park Forum. Site committees are popular at WHS; however, the final decision-maker is the MA of a WHS. This explains the "middle road" understanding by stakeholders on the effectiveness or non-effectiveness of management plans at MCLWHS.

Empowering stakeholders entails availing governance opportunities without SBMS and UNESCO fearing losing their statutory power at WHS. It is better to turn contestants into decision-makers than keep them at arm's length. SP should explore emerging approaches among them but not limited to People-Centred Approaches (PCA), Rights-Based Approaches (RBA), and others. These promote recognition, transparency, and multivocality yet retaining diversity among stakeholders. MCLWHS has an opportunity of further building on the People and Conservation Programme to offer real governance opportunity to stakeholders, including intensifying planning of the socio-economic aspirations as a component of management plans, rather than creating a buffer between them and the wildlife. There is a need to dilute the governance rigidity of PA and SBMS in Africa (Nicholson, 2002). While online consultations are becoming popular due to the impact of COVID-19, they are equally inaccessible due to data costs associated with downloading such voluminous documents. The consultative approaches used by SP and UNESCO at WHS do not necessarily foster ownership for stakeholders on management plans. Stakeholders cannot assess the effectiveness of what they do not own or know how it is being implemented. These processes are a privy of SBMS at WHS in Africa with a strong and remote influence of UNESCO at WHS (Donnachie 2010; Meskell, 2011, 2012).

Additionally, the World Heritage processes remain a function of SP as the ratifying organ. In the process, UNESCO assumes good governance is taking place at SP levels, implying that stakeholders are being involved. However, UNESCO regularly receives objections to its processes from stakeholders who are indirectly represented by SP in World Heritage governance (Meskell, 2012). At WHC level, stakeholders are only afforded an observer status, thereby reducing them to lobby or pressure groups. The question is: do the SP and UNESCO fully understand and represent the aspirations of all these stakeholders? Statutory obligations of both SP and WHC should be gliding towards good governance where the local and universal engage each other without preconceived positions on matters of conservation and development. Furthermore, SP and WHC should not erroneously consider consultation to be synonymous with consent when dealing with stakeholders. Consultation is a process while consent is the outcome of this process. Consent can be affirmative or not affirmative.

The dilemma between local and universal dimensions at WHS is that if a proposed development is approved by the SP, the concerned stakeholders support the SP to defend their ventures, like what happened with uranium extraction in Tanzania at Selous Game Reserve. Side events were organized in partnership with the developers at WHC sessions to support the development. UNESCO is then left to defend the conservation agenda by the SP. However, if SP says no to a proposed development, the opposite is true, and both SP and UNESCO become a combined force against development. This pendulum is dependent on SP's decision on a development. Sometimes, stakeholders demand audience with the WHC if they are not satisfied with the position of the SP (Meskell, 2013b). On a progressive note, UNESCO now requests stakeholder consultation minutes, but the weakness still remains in how UNESCO pushes contentious issues back to SP who already have predefined positions. This creates tensions and conflict among stakeholders at WHS, including UNESCO being accused of failing to make SP resolve such issues. Governance of WHS can no longer remain the privy of SP and UNESCO. Legal prescripts and financial limitations are not limitations to radically reform World Heritage governance at all levels. World Heritage governance should carry values of equitable representation, inclusivity, and transparency to the benefit of all stakeholders. Solutions to local challenges at WHS in Africa are negotiated settlements supported by good governance.

The above confirms the need to look deeply into the relationship between the WHS and surrounding communities (DEA, 2013a). This is important in bridging the two-way corridor of conservation and development. Conservation corridor appears as a blocked space by SP, WHC, and experts, while the development corridor represents the space with all other stakeholders with "desires" that are at variance with those stakeholders in the conservation corridor. The 2015 Policy on World Heritage and SD is now trying to permeate and bridge the two corridors. Theoretically, what separates the two corridors is the current top-down governance model used by SP and WHC. This model needs to be collapsed in order for stakeholders to "relate on mutually beneficial terms". This can only be negotiated through a radically transformed WHS governance framework.

Stakeholder Perspectives on Conservation at WHS

Stakeholder influence on conservation decisions at WHS remains peripheral to that of SP, WHC, and experts, even in cases where intangible heritage exists, which should be a privy of IDCs as knowledge producers and custodians. Democracy in Africa does not necessarily translate into good governance being established across various sectors under the control of the SP. As argued before, heritage governance remains exclusionary, however, under new democratic administrators in Africa. The few stakeholders involved at WHS are motivated by their own commitment and professional stakes. The "average-low" influence on conservation by stakeholders at MCLWHS has multiple reasons.

First, experts, SP, and WHC have become more "superior" than other stakeholders, and they "know it all". Experts are the embodiment of conservation at WHS and the advisers to SP and WHC. Second, conservation matters are first resolved in the governance boardrooms of heritage institutions without other stakeholders before being "pushed down in their throats for rubber stamping" and implemented at the WHS. Third, IDCs, who were long relocated to other areas, are "distance-managed" and thus have limited influence opportunities in WHS governance. SP and heritage institutions take advantage of this to retain conservation approach in their realm of governance. However, IDCs still have a major role to play at WHS, like in Mali (Dogons), Uganda (Buganda tombs), and Kenya (Mijikenda Forests) where they are actively contributing to conservation solutions. At RIWHS, Ex Political Prisoners (EPPs) are guiding the restoration of the Blue Stone Quarry (BSQ) Site using their social memory to ensure the process respects the authenticity of the site. The BSQ was used in the 1960s to break EPPs mentally, physically and spiritually by forcing them to drain sea water from the quarry site and quarried stones on a daily basis. The EPP Reference Group on BSQ provides the much-needed guidance.

The Oranjemund shipwreck discovery by NAMDEB (Namibia) provides another example where a mining stakeholder balanced conservation and development (Chirikure *et al.*, 2010). NAMDEB, a mining giant in Namibia, discovered the shipwreck during mining explorations in Oranjemund (Namibia). As a result, diamond extraction was halted to allow for the proper archaeological rescue of the shipwreck, of which the recovered collections will inform and enhance exhibitions for a planned interpretation centre (Chirikure *et al.*, 2010).

PLATE 7.1 Blue Stone Quarry Site (RIWHS)

NAMDEB supported the documentation, rescue excavations, and relocation exercises of this archaeological operation (Chirikure *et al.*, 2010). Parallel to this is the technical work initiated to conserve the recovered archaeological artefacts. This support should not be treated as incidental as many developers have played similar roles at different sites in Africa. This is a major contribution to heritage databanks and conservation processes at WHS in Africa. This demonstrates that both conservation and development can become equal winners at the end of the day. Extractive industries avail resources that are usually scarce from SP, heritage institutions, and WHC. Therefore, focusing only on the negative aspects of extractive industries without equally quantifying their contribution to conservation is injustice to this erroneously defined partnership at heritage sites in Africa. WHS in Africa have to ascertain and quantify these benefits beyond the known and perceived negative impacts of development. The involvement of IDCs and/or local communities by other stakeholders remains an issue of concern at WHS in Africa. However, this should be interpreted in the sphere of influence for each stakeholder at a WHS rather than being a matter of World Heritage governance.

Having formal governance arrangements with IDCs at heritage sites averts unfortunate developments like what happened at Domboshava Rock Art site (Zimbabwe), where extensive vandalism took place, and twice, the site museums owned and operated by National Museums and Monuments of Zimbabwe (NMMZ) were burnt. The reason for this was NMMZ had barred local communities from undertaking rainmaking ceremonies at the site for years. In a protest, firstly, the site museum was subjected to arson attacks followed by the defacing of paintings with enamel paint, thereby threatening the significance of the site. The defacing was a direct challenge to the ossifying monumentalism approach prioritizing archaeological deposits and paintings at the expense of the intangible dimensions (rainmaking ceremony) venerated by local communities. Second, the local communities, with the support of other stakeholders, constructed a bottle store in a forest considered sacred and central to the rainmaking ceremony for income generation, capitalizing on tourist arrivals at the site and the mushrooming local population around the site due to its proximity to Harare, the capital city of Zimbabwe. The local population is an answer to challenges of tourism seasonality, which affects many heritage sites in Africa. This builds domestic resilience based on the 'appetite' and 'resilience' of local instead of international arrivals at heritage sites in Africa, especially in the post pandemic period.

Previously, NMMZ had persuaded the local communities to construct an amphitheatre outside the boundaries of the site, but this project was never completed due to a combination of compliance issues, financial constraints, and its inappropriateness to a community not excited with theatre. Hence, the bold decision to construct a restaurant inside the site that could benefit them as a place of unwinding in a sacred forest. The "*Ndambakurimwa*" (the uncultivable sacred forest) is now home to thriving downstream tourism activities, with both the restaurant and the monument benefiting from the tourist flow to the area. Even if NMMZ wanted to institute a legal process against this development, chances are very high that the Goromonzi District Council responsible for community development would have politicized the matter given that the site is a magnet for local tourism and a source for tax revenue for them. The selling of beers at the restaurant is not far-fetched as the oral traditions at Domboshava confirm that beer was brewed and served in the same forest during the rainmaking ceremony led by a diviner. The traditionally thatched structures characterizing the restaurants further creates a sense of belonging in a local area that has become a sea of urban development with a high number of modern and expensive houses mushrooming. The expansive and architecturally awesome houses near the site illustrate the lifestyles of the 'filthy rich' young people in Zimbabwe, popularly known as "*mbingas*". The Domboshava area is now highly urbanized given its proximity to Harare, and an opportunity missed by the NMMZ was to partner with this vibrant community initiative.

The Domboshava Rock Art site restaurant is the only place that resembles a communal experience within a rural setup being "urbanized". In the long term, this could actually contribute to the extended preservation of the forests, as they give a context to this creative and innovative initiative by the local community. Only the owners and creators of this sacred heritage can make such bold decisions. Beneficiation mechanisms for IDCs have to be built around such creative and innovative processes that appeal to the sense of recognition and worthiness at the site. Satisfaction of IDCs through active involvement at the site may assist in overshadowing the demand for material benefits as a primary benefit. Material benefits should be enshrined as secondary spin-offs derived from the active involvement of IDCs in the interpretation of WHS. There is a need for formal governance arrangement with IDCs at WHS as this could assist in dealing with issues of common interest and mistrust and unlock potential resources for both conservation and the well-being of the society.

Resource Mobilization at WHS

Resources (in particular financial ones) to support conservation are shrinking at both national and international levels due to unprecedented cycles of economic depressions, politics, pandemics, and conflicts, among many other factors. At the same time, conservation remains capital-intensive as a non-profit-making venture, yet heritage management can be viewed as a business without losing its integrity and traditional mandate. For those that mobilize some resources for use at WHS, they cater to different dimensions, including conservation, capacity building, and management planning. At MCLWHS, this caters to social corporate responsibility (19%), human resources (18%), technical and financial resources (7% each) with infrastructural resources accounting for only 2%. Contextualizing resource mobilization into the profile of stakeholders interviewed at MCLWHS allows for an interesting analysis. The majority of stakeholders interviewed were young (below 18 years) and very old (above 60 years) people. Both age categories cannot provide resources as they fall into what one can call "beneficiary category" of any WHS. Young people are "dependencies" of their guardians while the older people are "pensioners" either about to retire, retired, or semi-retired and are more concerned with their personal needs or well-being in their 'golden years'.

Equally, the IDCs seem not able to mobilize resources capitalizing on how they are classified by other stakeholders as impoverished, yet they are prosperous in their own right. Sense of entitlement regarding accruing benefits from their historic landscapes may be at play too on their side. IDCs argue that their cultural resources and benefits thereof are being appropriated by SBMS and experts. Despite this situation, they too want to benefit from this appropriation. However, if IDCs were to assume some responsibilities, including mobilizing resources, they can strategically position themselves in World Heritage governance. By doing so, they would give weight to their own voices in the governance process at WHS. In a capitalist world, the resources you bring to the table give you power and influence. Within the wide range of existing and potential resources, stakeholders should find something to bring on board at MCLWHS. Stakeholders cannot meaningfully benefit from processes they have not invested in one way or another.

At the same time, creativity and innovation by WHS MAs remains very low in this area. They recycle traditional resource mobilization approaches, which yield the same results biased towards conservation only. At Great Zimbabwe World Heritage Site (GZWHS), lodges and rondavels, owned and operated by NMMZ, lack the flare of the attributes that define the architectural significance of the site, yet these could persuade clients to use them, thereby generating additional revenue for the site (Taruvinga, 2017). These have not been optimally used while the appropriation of these attributes by the private sector, for instance, Great Zimbabwe Hotel and Great Zimbabwe Inn, makes them preferred accommodation options as clients marvel at their link with the site. If the NMMZ rondavels and lodges are renovated to have a flare of the site, they will give an exceptional and locational advantage to clients in appreciating the site. Anyone could pay a premium to sleep and wake up inside GZWHS. Creative marketing is a competence outside heritage capacity, thus demanding interdisciplinary approach. Additionally, the staff village located close to the hotel and inside the site is better placed for commercial accommodation given its strategic position for aerial and sky-view of GZWHS (Taruvinga, 2017). This could give birth to upper-class overflow facilities linked to existing similar facilities around GZWHS. WHS sites in Africa have to become "Smart WHS Hubs" following the concept of "Smart Cities" wherein innovation, creativity and multi-dimensions and disciplines (including ICT, infrastructure, alternative energy sources, planning and rural development) are infused for the benefit of the site and well-being of society.

Resource mobilization gives power to SBMS operating at WHS in Africa. For instance, SANParks, just like many heritage institutions in Africa, has power derived from SP grant investment into conservation at MCLWS. The grant is made possible due to income tax revenue levied from the public via treasury strategies, which is then disbursed to government entities. However, this investment remains insufficient, forcing MAs of WHS to explore additional and alternative revenues sources to make good on all of their obligations. This compels heritage institutions to implement development assumed to be in tandem with conservation. This is either self-driven or through partnerships with contractual arrangements that guarantee minimum revenue accruing to them as they are risk averse. Self-driven development, mainly tourism, puts heritage institutions in direct competition with other stakeholders investing in similar activities. Heritage institutions suffer from lack of quality and business acumen compared to the private sector. GZWHS is a classic experience with its internally driven tourism and accommodation which has failed to match the quality of similar establishments around the site. This is where heritage institutions need partnerships with private sector, or alternatively establishing commercial divisions with competent and skilled personnel, as opposed to making heritage practitioners a "one-stop shop" for functions even beyond their competence. This is the space for shared values and outcomes using mutually-beneficial partnerships, further affirming the muted "Smart WHS Hubs" in Africa.

Stakeholders Perspectives on SDGs

Understanding how stakeholders construct socio-economic values at heritage sites is important for planning purposes, including determining how conservation embraces SDGs. Most WHS in Africa are locations endowed with renewable and non-renewable resources. Using MCLWHS as the primary case study, this section analyses stakeholder awareness levels on developments, their ideal roles in development decisions versus their actual influence on developments at WHS.

Level of Stakeholder Awareness on Developments at WHS

Regarding stakeholder awareness levels on existing developments at MCLWHS, tourism emerged as the dominant one, followed by infrastructure, agriculture, and mining. This confirms the reality of the nature of largely accepted SD initiatives at WHS in Africa. Tourism and accompanying infrastructure are dominant. At MCLWHS, RIWHS, and NCA, cultural tours gradually became popular with the WHS status, while nature tours have long established a strong clientele base. At MCLWHS, the dominance of nature tours is further reinforced by the GMTFCA dimension. School tours pay lower tariffs as part of making heritage accessible to them. This concessionary rate is a widespread approach for leaners in many African countries, which explains why educational interest is relatively ranked higher at WHS. Educational tour concessions are an empowerment vehicle for heritage education in Africa towards building social cohesion, nation building, and reconciliation in previously colonized and racially divided nations.

The cultural tours at MCLWHS have also attracted criticism from IDCs, just like at most WHS in Africa, as they view them as archaeologically biased, and they ignore the spirituality and sacredness associated with this appropriated landscape. IDCs highlighted that the tour guiding narrative is largely constructed around archaeological debates and colonial narratives. From this criticism, an opportunity exists for SANParks to creatively involve IDCs in reimagining the interpretation and presentation of MCLWHS to the public. Reimagining demands decolonized and community-driven research to augment deficient and existing information databanks. These databanks lack detailed information on their spirituality and the sacredness of the landscapes associated with WHS. The results thereof could be used in reviewing the interpretation and guiding framework at MCLWHS. Spirituality, sacredness, indigenous knowledge on plants and animals, and their histories could diversify the narrative offered to public.

In the process, members of IDCs could either become trainers of guides, sources of information or guides at MCLWHS, like at RIWHS where EPPs are prison guides and now have an Intergenerational Learning Unit to train successor generation of guides. EPPs have always been appointed as Prison Guides at RIWHS since its establishment. They provide authentic narratives and offer a lifetime engagement opportunity to visitors. Additionally, they evoke the emotional atmosphere related to their experiences as prisoners. The succession plan at RIWHS is geared towards ensuring that they transfer their social memory and narratives to a new generation of guides towards establishing systems similar to those operating at slave sites such as Forts and Castles in Ghana. Such sites have trained guides around slavery to the extent that when they offer their services, one may think they actually experienced slavery itself, yet it is their long-gone ancestors who were slaves. Succession is also a process of retaining the authenticity of their social memory into the future of the site. The RIWHS model could be considered and varied where applicable at MCLWHS. There is nothing stopping SANParks from accommodating IDCs as tour guides.

While this is the case at most WHS in Africa, other types of developments have slowly found expression at WHS worldwide. This includes mining, agriculture, and infrastructural developments. As pointed in earlier chapters, developments have both known positive and negative impacts, and these have been recorded at sites (Briedenhann & Wickens, 2004; Deacon, 2006; Salazar, 2010a, 2010b; Scheyvens, 2003). The point is that development is both a going and growing concern for society, and therefore their formulation and implementation is equally the same. WHS have to be on the lookout for these developments in order to find points of synergies rather than focusing on compliance issues only. The visibility of agriculture and mining at MCLWHS confirms their contribution to the GDP of Limpopo in South Africa, approximately 2% and 28% respectively (Limpopo Provincial Treasury, 2017). These activities attract both seasonal and permanent work forces, induce infrastructural developments (accommodation, roads, and processing facilities), and equally feed the tourism industry supported by heritage attributes inside their properties. This is the pinnacle of development for Musina Town, now also bourgeoning because of cross-border business trading with Zimbabwe and the rest of Southern African Development Community (SADC) countries. Also, the peculiar and unique cultural connections of MCLWHS should be leveraged to introduce competitive GMTFCA products and tariffs towards creating a localized niche for sustainable tourism given seasonal challenges associated with the tourism industries in Africa. MCLWHS needs to find a tariff model that balances accessibility at GMTFCA, national, and regional levels. Understandably, this may make heritage destinations expensive for international communities, but international arrivals have more buying power due to their strong economies. Maintaining single tariffs is no longer acceptable in the African tourism sector, and perpetuating it as well condemns nationals of SP in Africa to life-time inaccessibility of their own heritage due to poorly performing economies.

Stakeholder Influence on Socio-Economic Decisions

Having understood the stakeholder awareness levels on developments, it is important to investigate their influence in the decision-making process on developments at WHS. At MCLWHS, the highest overall ranking of stakeholder influence on decision-making on development is *very low* (1.72), with *very high* (0.27) influence being the least ranked (Figure 7.3). Though this is the case, some stakeholders have average influences on decision-making around tourism, infrastructure, and agriculture initiatives at MCLWHS. Overall, the power resides with the owner and initiators of development at WHS. The rest of the stakeholders, including SP and heritage institutions, are only informed through Park Forum meetings or when EIA and HIA compliance is undertaken. This pattern confirms that other stakeholders are not necessarily consulted by the developers. Developers directly deal with heritage institutions, SP, and compliance entities, while all other stakeholders become collateral players when the need arises. Therefore, good governance is not just a matter of concern to SP, heritage institutions, and UNESCO, developers have even more closed governance systems that keep all other stakeholders in the dark regarding their operations. Also, the sandwiching of these developments in between public parks, private conservancies, mining concessions, commercial farms, and tourism establishments leaves very little room for any other stakeholder category outside this circle to influence decision around any of these developments. The ownership of such development initiatives, for instance, mining and agriculture, is predominantly in the hands of the private sector. The question that arises is whose development is it for and who are the beneficiaries, if the intended beneficiaries not involved from the beginning to the end?

The bottom line is that decision-making on development issues resides with those with power, capital and the ability to implement them, similar to conservation issues. Often, heritage institutions come at the tail end of launching these investments, wherein they enforce HIA and EIA protocols. This creates tensions

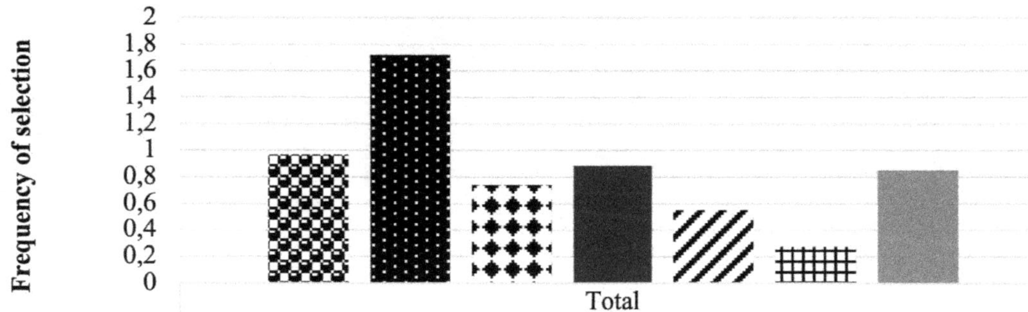

	Total
Sum of Not at all	0,970561362
Sum of Very Low	1,721529201
Sum of Low	0,740918477
Sum of Average	0,886185697
Sum of High	0,552219366
Sum of Very High	0,277685269
Sum of Don't know	0,850900628

Multiple Response options for Respondents Influence on Socio-economic decisions at MCLWHS

FIGURE 7.4 Ranking of overall stakeholder influence on development aspects at the MCLWHS

with the developers as the heritage institutions sometimes insist on retrospective compliance, which derails implementation plans and sometimes induces losses for the developers. Sadly, and in the process, other stakeholders, especially IDCs, are often sandwiched between the promised benefits and conservation aspirations pushed by heritage institutions. These dynamics are common at WHS and have become the "shifting tensions" that can only be dealt through stakeholder management approach. This demands an inclusive and good governance system which creates a conducive environment to address broader territorial development issues alongside conservation at WHS in Africa.

Both heritage institutions and supposed beneficiaries of developments have to be engaged from the conceptualization stage of development throughout its life cycle. RIWHS offers an illustrative example how EPPs are contributing to decision-making processes on developments at the site. RIWHS muted the installation of photovoltaic solar panels in order to reduce dependents on diesel-generated power, thereby reducing carbon emissions at the site, as well as "greening" the island as tourist destination. Three location options were considered: the agriculture precinct (hydroponics area), the cricket pitch, and the area close to the village precinct on the island. It is important to note that RIWHS as an island has limitations in that any possible development can only be in the core area as the buffer zone is part of the Atlantic Ocean. The three proposed areas were all subjected to a preliminary HIA impact assessment, with the preferred location being subjected to a full EIA. The area close to the village precinct was considered not suitable due to its limited size, proximity to tourism route, and constant colonization by breeding birds. The agricultural precinct was equally not suitable because it is one of the attributes illustrating imprisonment and is also close to the Limestone Quarry site and within the perimeter of the Maximum-Security Prison. More importantly, EPPs worked in this precinct during their incarceration at Robben Island. The cricket pitch became the only possible location for the installation. This pitch is located behind the administration office of the island.

What guided the decision to choose the cricket pitch was the social memory of EPPs. During EIA consultations, the EPPS eliminated all other options on the basis of their connection with these spaces as political prisoners. As prisoners, they had never interacted with the cricket pitch. The EPPs confirmed that the pitch was used by prison warders. As such, EPPs had no difficulties in endorsing the cricket pitch as a preferred project site. In other words, while the entire landscape is important to them, they have no particular emotional or memory attachment and connection with this particular space compared to the quarry sites and the agriculture precinct where they used to work. To the EPPs, the cricket pitch could be "sacrificed" for the benefit and sustenance of the island. They even went further and suggested that RIWHS should use the savings to be made by installing solar energy to support underfunded heritage programmes. While one could look at this decision as obliteration of hurtful landscapes by EPPs, the bottom line is that the consultation and the subsequent memory trade-off was a game changer in balancing conservation and development. Also, the project itself did not have the potential to dramatically alter the landscape given it was camouflaged by a number of buildings close to it, unless one is flying past the island (visual impact). Additionally, a buffering layer of indigenous trees was planted to camouflage it from approaching and departing passenger ferries to the island. Should the need arise, the installation plant can be removed with minimum effort and impact, and the space can be easily rehabilitated back to the pre-construction era.

Benefits of Development at WHS

Development cannot be without benefits to stakeholders or part of stakeholders at WHS. Hence it is important to discuss the level of benefits derived from development by stakeholders at WHS. This defines the materiality of development benefits in meeting the well-being of the society. At MCLWHS, the majority of stakeholders indicated that they "do not know anything" (1.37) about such benefits accruing to them, followed by average (1.12) and high benefits (1.09) as illustrated in Figure 7.4. The fact is the majority of interviewed stakeholders are geographically far away from MCLWHS; they may not be benefitting that much, except those in the localized environment of the site, such as seasonal and permanent work forces supporting various development activities at MCLWHS. This is confirmed by the selection of socio-economic activities from which these benefits are coming from: high benefits are derived from tourism (46%), infrastructure (24%), agriculture (17%), extractive industries (13%), and hunting (9%).

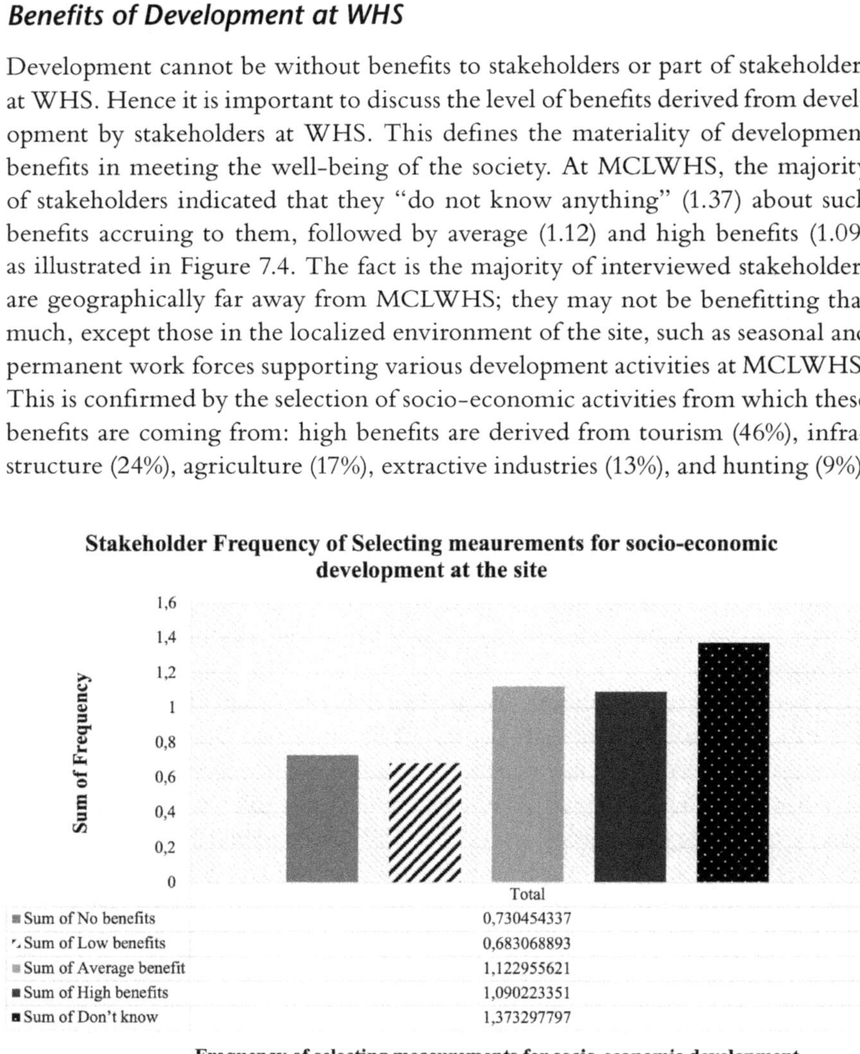

FIGURE 7.5 Pivot chart showing the highest ranked level of benefit accruing from development at MCLWHS.

However, these benefits have a tendency of spilling over to extended families living beyond the boundaries of MCLWHS. This was not, however, quantified in this study.

While tourism benefits are dominant among stakeholders at WHS, the IDCs argue that they are not benefitting at all from MCLWHS. They do not have access to tourism opportunities, and neither are they allowed to run craft centres at MCLWHS in order to connect with flowing visitors. Their presence as part of the touristic experience in one way or another may be a vital connection at MCLWHS. However, their request to access MCLWHS in order to carry out ceremonies has not been formally resolved, yet this could be capitalized on for a niche cultural experience packaged like the annual *Kuomboka* Festival of the Barotse Floodplain in western Zambia (Flint, 2007). The festival is hosted by *Lozi* peoples and rides on their history and heritage as a strong symbol of identification for African cultures and practices that have survived the test of time (Flint, 2007). The site is being considered for World Heritage Listing. However, ownership of the heritage has a bearing on its use and the beneficiation distribution to contesting IDCs (Flint, 2007). This study acknowledges that dealing with scattered IDCs in MCLWHS remains a complex matter compared to those of the Barotse Cultural Landscape. While the study also acknowledges the existing contestations and complex dynamics among IDCs, they still have a role to play at MCLWHS. Establishing community museums in their current geographic locations far away from the site and creating connecting tourism nodes with MCLWHS may be uneconomical given the current experience with a plethora of similar white elephant museums and underfunded community, private, and public museums in Africa. COVID-19 has forced many of these institutions to either scale down or permanently close operations.

Alternative spaces to accommodate community-driven development initiatives have to be identified at WHS. At MCLWHS, the administrative facility constructed outside the main entrance gate could have been used to implement creative and innovative projects that are mutually beneficial to SANParks and IDCs. For instance, the current range of curios at MCLWHS is limited and tantamount to an academic book shop. Therefore, a robust and community-driven craft centre could be established by SANParks. This administration centre would be a strategic location at MCLWHS in order to take advantage of the tourist trails and main entrance. The centre could have traditional curios and meals associated with IDCs, which may give life and diversity to the current restaurant operated by SANParks through a contractual arrangement with a service provider. Many IDCs have cultural objects in their possessions that could be replicated for commercial purpose at MCLWHS. However, IDCs do not trust SANParks with these collections given its history and inability to recover cultural objects appropriated by the University of Pretoria (UP) over the years. UP has promised to honour collections repatriation once SANParks has established secure facilities and is implementing good collections management practices. IDCs are not in the picture of this scientific dialogue. SANParks is now strengthening this aspect throughout parks. If SANParks is serious about resolving IDCs' concerns, it has to make conditions and infrastructure conducive for the repatriation of these objects from Pretoria, including those that are in the custody of the IDCs themselves.

From a governance perspective, the community craft centre can be managed by trained IDC entrepreneurs either as a trust or joint venture with SANParks or whatever form it may take. Golden Gate National Park has a similar arrangement which can be replicated in other National Parks in South Africa. This would be part of building self-worth among IDCs, allowing their reconnection with the landscape in a semi-commercialized sense and building their sustainable involvement at MCLWHS. This IDC-driven approach for tourism could also give birth to a process where facilities within MCLWHS could be renamed using indigenous names and attributes important to the IDCs; a renaming approach (decolonizing interpretation and spaces) which is consistent with post-colonial processes in democratic nations. This approach will reduce material beneficiation to a secondary matter as IDCs are looking for official recognition in the first place. All this will strategically position IDCs to be part of the socio-economic dimensions of MCLWHS in a broader and sustainable way.

However, dealing with community-driven developments at WHS has both positives and negatives dimensions as illustrated by a curio market located outside GZWHS. The selling of curios (including carved replicas of the famous soap stone birds) is now in the hands of local communities. While this downstream industry is functioning, more can be done to create better relations among them, heritage institutions, and consumers. The current location of the curio market has its own limitations: absence of social amenities, limited and risky parking space, dislocation from the site, security concerns, and absence of value-add processes to create a competitive marketing edge while creating and curating lasting memories about GZWHS. As an alternative, the former golf course inside GZWHS could have been used as it is already archaeologically disturbed and could be managed differently. Such proximity to GZWHS would entice visitors to "consume" both the narratives and curios under a community socialized ambience setup at GZWHS. Also, value-added processes to the curio products, including communities acquiring entrepreneurship skills and product management, are important. For instance, curios could be packaged in materials that carry indigenous taste, terminology, and manifestation, all being the cornerstones of decolonizing the interpretation of GZWHS. At Goree Island (Senegal), local communities have free trading opportunities, and hence the island is littered with curio shops with diverse products. This may appear chaotic to a first-time visitor to Goree Island, including on the ferries to Goree island, but the liberal approach breeds competition, which benefits consumers, improves quality and range of products, and instills a sense of ownership which helps in protecting the site as the chicken that lays eggs for the community on the island. This is win-win situation between communities and WHS as both are capitalizing and optimizing on tourist arrivals. Curio industries running as a family business have a higher tendency of sustainability and continuity compared to cooperatives that are fraught with infighting and corruption, and eventually suffer from the tragedy of commoners (no one taking responsibility leading to their total collapse).

At MCLWHS, and peculiar to SANParks, there is a 1% levy charged on all park accommodation and activity reservations. Through this levy, SANParks supports sustainable and strategic initiatives benefitting communities around parks such as clinics, schools, water, and sanitation projects (SANParks, 2018). It is an opportunity for philanthropic approach in a heritage context, and this can be replicated at other WHS in Africa. This SANParks initiative, coupled with other infrastructural developments in the area, may be the reason why stakeholders ranked infrastructure as the second dominant development at MCLWHS. Therefore, WHS could adapt this levy concept to generate additional revenue in order to support community development at WHS. However, this requires negotiating with public tourism boards who are already collecting such levies in the name of promoting tourism, which includes these WHS. It is important that WHS have their share of these levies as the "chicken that lays the eggs" being consumed by tourists. That "chicken" belongs to the community.

Benefits accruing from agriculture have cross-cutting impacts beyond MCLWHS itself. At the local level, Musina area has many households involved in agriculture such as citrus, cattle, and game farming. Agriculture is a labour-intensive activity (GMTFCA TTC, 2010; Limpopo Provincial Treasury, 2017). In 2011, approximately 2,261 households were involved in agriculture (STATS S.A., 2011). Despite this, the soaring unemployment rate has risen from 19.3% (2016) to 19.6% (2017) in the Limpopo Province (Limpopo Provincial Treasury, 2017). Additionally, the COVID-19 pandemic has induced massive job losses due to the underperforming economy. Most productive sectors are closing or scaling down operations. Despite this, agriculture at MCLWHS still offers permanent and seasonal opportunities to farm workers. Agriculture also employs 10% of the population and contributes about 2% of the GVA-R Sectoral composition of Limpopo Province (Limpopo Provincial Treasury, 2017). However, agriculture poses challenges to conservation in the greater Mapungubwe area due to continuous water abstraction from the rivers and underground water sources (Berry & Cadman, 2007). Also, private ownership of the agricultural land infringes on free movement of tourists and local communities in the confluence area (Berry & Cadman, 2007).

Regarding extractive industries at MCLWHS, these are known to provide both short- and long-term benefits to the society. The benefits, which have impact beyond WHS, include employment opportunities, career awareness and orientation campaigns, internship programmes, infrastructural development, contribution to GDP, and support to manufacturing industries. At MCLWHS, extractive industries employ about 6% of the population and contribute 28% to the overall economy of Limpopo Province (Limpopo Provincial Treasury, 2017). The contribution of this industry to SDGs agenda of South Africa is immense and is recognized at political levels (Creamer, 2013). Extractive industries are expected to benefit both mining communities and labour-sending areas, while upholding sustainable principles (Creamer, 2013). This underlines the political rhetoric associated with extractive industries in Africa: profit maximization yet benefits accruing to labour-sending areas have remained low. The latter overshadows the positive benefits of these extractive industries in other areas. This is also due to the fact that extractive industries have negative impacts on the environment and society and, furthermore, are not fulfilling the promised benefits to communities (Chirikure *et al.*, 2010). The same applies to hunting, which provides some employment benefits through guiding services and assisting professional hunters, but trophy benefits do not automatically reach local communities. These are disbursed through some structured mechanism that always disadvantages local communities while conservancy owners get the lion's share.

Relating the above to other sites can be summarized as follows. Examples of successful development include the high employment levels, curio centre and conferencing facilities achieved by the COHWHS in Gauteng Province of South Africa, the major re-positioning for tourism at Tarragona World Heritage site (Spain) and Goree Island (Senegal), and the blossoming craft and hotel industry around the Mombasa and Lamu Islands WHS in Kenya. However, in Tanzania, the proposed re-opening of the Laetoli footprint trackway for tourism attracted the attention of UNESCO, resulting in numerous reactive monitoring missions to the site and scientific discussions to find a balance between conservation and the tourism appetite for hominid footprints trackway to be opened up to visitors (Taruvinga, 2012). The point of the matter was that the SP of Tanzania saw an opportunity to diversify visitors experience using Laetoli and increase visitor numbers to the archaeological sites in the Ngorongoro Conservation Area. Currently, visitorship in Ngorongoro Conservation Area is concentrated in the crater where it is easier to see wildlife and the bomas, where Maasai communities dwell.

Discussing how stakeholders are accruing benefits from development at WHS cannot be fully understood without discussing how they impact each other from a socio-economic perspective. This matter, rarely discussed, is equitable to the spirit of *Ubuntu*, an ancient African concept, that explains "how people centeredness approach, permeable walls, partisanship and progeny all interact together as stakeholders take collective decisions on conservation and SD at WHS" (Michael, 2007; Taruvinga, 2017). Ubuntu is a Nguni proverb that states "*umuntu ngumuntu ngabantu*" – "a person is a person through other persons" (Gade, 2011: 303; Michael, 2007). People centredness is the role of people in making conscious decisions that affect them as a whole, while "permeable walls" is communication that is not restricted by hierarchy, political position, educational background, and position in society (Gade, 2011; Taruvinga, 2017). Partisanship is how loyalty is fostered among individual members of the community while "progeny" refers to how the individual members of the stakeholders promote collective decision-making which affects them as whole (Gade, 2011; Taruvinga, 2017). Summarised, Ubuntu promotes the common good of society and "humanness as an essential element of human growth" (Venter,

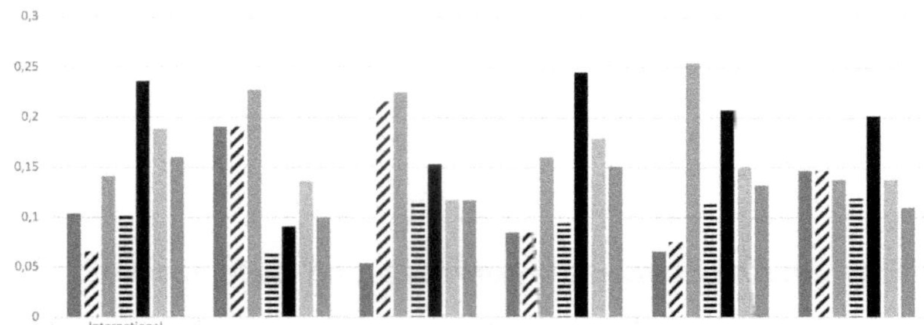

	International Community	Learners	Local communities	National Government	Provincial Goverments	tourists
Sum of Very High	0,103773585	0,190909091	0,054054054	0,08490566	0,066037736	0,146788991
Sum of High	0,066037736	0,190909091	0,216216216	0,08490566	0,075471698	0,146788991
Sum of Average	0,141509434	0,227272727	0,225225225	0,160377358	0,254716981	0,137614679
Sum of Low	0,103773585	0,063636364	0,117117117	0,054339623	0,113207547	0,119266055
Sum of Very Low	0,235849057	0,090909091	0,153153153	0,245283019	0,20754717	0,201834862
Sum of Not at all	0,188679245	0,136363636	0,117117117	0,179245283	0,150943396	0,137614679
Sum of Don't know	0,160377358	0,1	0,117117117	0,150943396	0,132075472	0,110091743

FIGURE 7.6 Pivot chart showing impact of stakeholders on each other based on their socio-economic activities.

2004: 149). At MCLWHS, this humanness (impact on other stakeholders from a socio-economic perspective) is mainly "average" (1.15) followed by "very low impact" (Figure 7.6).

The "average impact" at MCLWHS demonstrate stakeholder relations woven with different, rather than common interests, business goals, and expectations. It is more of power, legitimacy, and urgency determining the "vertical and horizontal relations among stakeholders" (Sachs & Rühli, 2011: 38). Ubuntu is at the periphery of this business-relationship among stakeholders at MCLWHS. Also, the average impact also explains the long-known tensions and weaker impacts among stakeholders at MCLWHS, similarly at other WHS in Africa. However, "very high" impact is on learners, although it is equally ranked under "very low" impact. Learners are cross-cutting beneficiaries of the stakeholders at most WHS in Africa. The high impact on learners at MCLWHS emanates from career guidance and educational programmes offered by SP, heritage institutions, and the private sector. Also, in their own perspective, learners do not have the means to induce impact on other stakeholders. Under "high" impact ranking, local communities come out top, but are also ranked "very low" impact. This points to oscillating impacts on local communities, however, dependent on the goodwill of other stakeholders. Both IDCs and learners are unlikely to benefit due to their geographical disconnection with MCLWHS.

Also, notably, provincial government received the "highest average impact", while tourism comes last in this same ranking, with local communities in between the two at MCLWHS. Interestingly, the "very low impact" is on international communities, national government, and tourists. The remoteness of MCLWHS could explain this, being far from major cities and airports. National government's low ranking could be resulting from its contradictory role as an initiator of conservation and SDGs, as well as the watchdog on compliance across policies covering both nature and culture values. The average impact ranking among stakeholders is an opportunity to domesticate an adapted social corporate responsibility approach in heritage governance, possibly underpinned by *Ubuntu* principles. WHS in Africa are devoid of such approaches, implying they are failing to learn from other sectors and disciplines.

Camaraderie between Conservation and Development

The comparative analysis of conservation and SD at WHS leads to understanding their contribution to each other, including how to measure their individual and collective contribution in meeting SDG indicators in Africa. Due to conservation and development having co-existed from time immemorial (see Chapter 2), they have developed some camaraderie over the years, but this has remained unexplored at WHS until recently with the adoption of the 2015 World Heritage Policy on SD. At MCLWHS, conservation appears to contribute more to development than the latter contributes to the former. Conservation provides opportunities for sustainable tourism, controlled infrastructural developments, and many other downstream livelihood industries as guided through HIA and EIA processes. Culture has become a catalyst for development. On what is more important between conservation and development, stakeholder responses at MCLWHS showed that former is "very important" compared to the latter considered as "important". However, stakeholders acceded that both are very important and mutually beneficial to each other at MCLWHS. The history and experience at MCLWHS show that, while conservation is more on the compliance path, it has been deliberately building more "positive bridges" with development since precolonial times (see Chapters 3 and 4). Therefore, camaraderie between conservation and SD symbolizes progressive, inescapable, and sometimes conflicting synergies. This area needs more research in Africa to paint a clearer picture on this evolving beneficial relationship.

Development Benefitting Conservation at WHS

The benefits of development to conservation are important in understanding the nature of the symbiotic relationship between the two. It is the ground for moving towards scientific and social trade-off towards their co-existence at WHS. At MCLWHS, most stakeholders "*don't know*" (1.03) how development contributes to conservation (Figure 7.6). This could be explained by the fact that heritage institutions always have biased reporting towards conservation indicators with little inference to the contribution of development at WHS. Equally glaring is the absence of and inaccessible public and private datasets on how development contributes to conservation at WHS in Africa. Private companies jealously guard their datasets and only disclose such information in their statutory annual audited reports. Stakeholders at WHS in Africa have not created buy-in from each other on these matters. Silo planning of heritage institutions may explain this situation

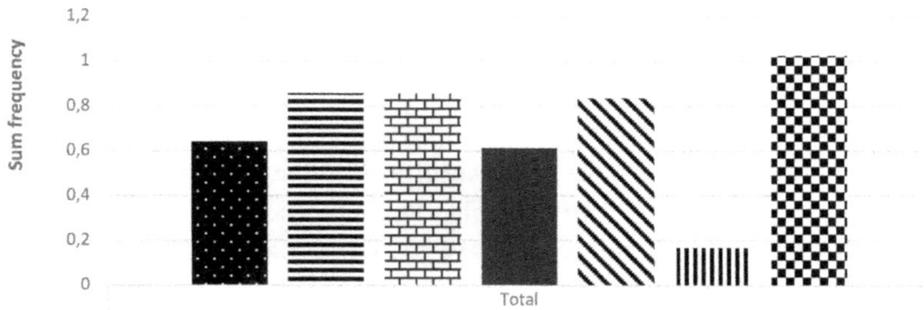

	Total
Sum of Not at all	0,641109704
Sum of Very Low	0,859337832
Sum of Low	0,856642565
Sum of Average	0,613728564
Sum of High	0,834960133
Sum of Very High	0,167578734
Sum of Don't know	1,026642399

Frequency of selection by Participants

FIGURE 7.7 Pivot chart showing the highest ranked contribution response option.

at WHS in Africa. Equally, this may explain why conservation and development are always perceived as conflicting at WHS.

Despite the above quandary, there is a "high" level of dominance of tourism (27%), infrastructure development (19%), and agriculture (16%) contributing to conservation at MCLWHS. While tourism largely complies with the conservation protocols at most WHS, there are recorded negative impacts. Both tourism and agriculture seem to have gradually developed a symbiotic relationship with conservation at MCLWHS. Regarding the contribution of infrastructure developments to conservation, only qualitative inferences can be made, such as creating access to the site (roads), provision of sanitary and ablution facilities, collections and conservation facilities among others. Depending on the size of the infrastructure footprint, negative impacts may be recorded at WHS. Measuring the contribution of all these dimensions to conservation at MCLWHS remains difficult in the absence of statistical datasets. This equally applies to most WHS in Africa.

While stakeholders (19%) indicated that extractive industries are not contributing to conservation at MCLWHS, archival analysis demonstrates the opposite. Stakeholder views are affected by absence of statistical datasets on the matter, including the inability to access such information from private sector. From archival analysis, activities and research leading to the discovery and recognition of MCLWHS were partly financed by extractive industries. Explorers, mines, and farmers settling in the Mapungubwe area invested resources in heritage management one way or another. They raised awareness on Mapungubwe, which in turn attracted funding for research and conservation activities. Regrettably, cultural objects were looted and appropriated in the process from Africa, but this cannot overshadow how explorers, commissioned researchers, and mining companies variedly and unintentionally contributed to knowledge production and conservation in Africa. One can argue this knowledge had colonial context, but this has become the rational for decolonizing heritage knowledge and spaces in Africa. For example, many scholars have conducted EIA and HIA fully paid by developers at many WHS in Africa. This research has resulted in the creation of massive heritage databanks in the custody of universities, research institutions, academics, and museums. For example, the University of Pretoria (South Africa) holds priceless collections from MCLWHS, which have now become the subject of local repatriation discussion. IDCs want the collections at MCLWHS while the university has no faith in SANParks' capacity to look after this collection. Also, other universities in South Africa have benefitted from scholarships/grants awarded by extractive industries, thereby furthering academic and scholarly works. Additionally, EIA and HIA have developed and implemented developer-funded mitigation plans, thereby supporting conservation at WHS. For example, De Beers developed a heritage management plan for the property where they are extracting minerals at MCLWHS.

The sharing of information between heritage institutions and developers may further synergies between conservation and development at WHS in Africa. For instance, without SANParks sharing their conservation needs and challenges, and equally, developers sharing their needs and challenges, both parties would not know how to assist each other at MCLWHS. Without developers proactively engaging WHS on conservation issues, they are not able to assist, yet this may favour their social corporate responsibility programmes. Heritage as a shared value for society should be a natural fit for social corporate responsibility programmes. The partnership between African World Heritage Fund (AWHF) and NAMDEB in 2012 to rescue the Oranjemund Shipwreck underwater site is testimonial to how conservation and development can benefit each other. AWHF awarded a recognition certificate to NAMDEB for their role in the process. In the midst of tensions between conservation and development, stakeholders should find beneficial opportunities, which may not have been conceivable and feasible before due to biased focuses on each other and in the absence of constructive dialogue among them. While the SP is the beneficiary of corporate taxes and most partnerships, these hardly filter to heritage compared to security, health, education, mining, and agriculture, which are prioritized in national budget allocations. Constructive confrontation and pro-activeness are steps towards resolving conservation and development tensions at WHS in Africa.

Measuring Heritage Contribution to Socio-Economic Indicators at WHS

Measuring the contribution of heritage to SDGs (culture statistics) at WHS remains a matter of concern at WHS in Africa. Though generally known but cannot be proven by solid statistical datasets (culture statistics), heritage as a resource provides opportunities for "job creation, infrastructure development, and educational opportunities" and "revenue alike" (Carruthers, 2006; Lowenthal, 2005: 28; Ndoro, 2015: 393). While culture statistics are largely non-existent at heritage sites in Africa, some can be extracted, such as the number of visitors and jobs created at WHS. A stakeholder assessment on how these statistics should be measured at MCLWHS was undertaken. Stakeholders indicated potential and multiple measuring indicators, among them jobs created, state of conservation, level of community development, number of tourists, contribution to GDP, number of business/investors, infrastructure development, and return on investment at WHS. The indicators could be both direct or indirect, as well as quantitative and qualitative.

Direct indicators would be those that are directly measured by heritage institutions, while indirect ones would be inferred from measurements of other stakeholders, such as tourism and national statistic offices. Quantitative measurements refer to a number of phenomenon presented as output while qualitative measures the quality of the phenomenon; in this case, it would be the well-being of the society. The absence of culture statistics makes WHS peripheral to development issues. This also creates a flaw in aligning heritage to SD indicators. This affects budgetary allocations to conservation and further influences the legal classification of heritage institutions as non-profit, thereby making them non-bankable businesses, yet they are magnets of tourism worldwide. Sites such as RIWHS, COHWHS in South Africa, and NCA (Tanzania) generate over 40% of their total annual operational budget, although currently impacted by the COVID-19 pandemic in 2020/21.

The proposed culture statistics indicators could benefit from broader SDG indicators which are already measured in different countries and regions. Applying these indicators at WHS is not a far-fetched matter. For example, SANParks has already conceptualized "development as being one of the three core pillars of the organisation of conservation and tourism" (Mketeni, 2016: 2). In this conceptualization, SANParks argues that PA are strategically placed to become drivers of local economic development that should benefit neighbouring local communities (Mketeni, 2016). Therefore, SANParks continuously looks for sustainable opportunities and strategies that uplift the livelihoods of the local communities around PA (Mketeni, 2016). WHS should adopt this conceptualization for local economic development and become more relevant to grassroots issues of society. This points to adapted social responsibility strategies contributing towards "national identity, national pride, national and transformational economy and politics" at MCLWHS (Carruthers, 2006: 11). This way, cultural statistics will become a decision-making tool at WHS on balancing conservation and SD. In this process, various government departments, development monitoring agencies and universities need to conduct baseline studies to establish culture statistics towards closing this gap at WHS in Africa. This would assist in unearthing the hidden and the assumed "very insignificant" contribution of heritage to the GDP of African countries. Such baseline indicators will help WHS understand how far they are aligned to SDG and related indicators (Hart, 1996;

Miller, 2001). What cannot be measured may not exist (Daly & Cobb, 1990) and may be difficult to argue for. Some of the existing SDG indicators can be adapted and localized at WHS in Africa.

Camaraderie and Co-existence as a Reality!

The camaraderie and co-existence between conservation and SD is a reality at WHS in Africa. Regional meetings and reports, including UNESCO itself via the 2015 World Heritage Policy of SD, confirm this reality. For example, the 2015 Policy aims to "ensure that the conservation and management of World Heritage properties are aligned with broader sustainable development objectives namely environmental sustainability, inclusive social development, inclusive economic development, as well as fostering peace and security" (Labadi, *et al.*, 2020: 7). In the same breadth, MCLWHS stakeholders suggested different types of developments that should be allowed at WHS in the present and the future. Such suggestions include already known developments, however classified by UNESCO as threats to WHS. However, for all these developments to happen at WHS, the localization of SDGs is needed in Africa, in particular borrowing from the pioneering spirit and experiences of IDCs, involving other stakeholders and ensuring the 2015 Policy is aligned to local context of the WHS.

A regional workshop on the 2015 Policy held at RIWHS in 2019 and attended by 30 participants from 18 countries from the Africa region recommended many practical approaches. Participants included site managers, decision-makers, universities, private sector, UNESCO, ABs, and research institutions. The main objective was to discuss how best to implement practices and principles on conservation and SD at WHS in Africa. Localization was defined as connecting and empowering the full range of actors and stakeholders, from policy makers (policy level) to communities and individuals at grassroots level (Labadi *et al.*, 2020). Localization has the potential to "fulfil the African Union aspirations of "Agenda 2063; The Africa We Want" (African Union, 2015), the 2016 Ngorongoro Declaration on Safeguarding African World Heritage as a Driver of Sustainable Development, Recommendations of the 2018 Position Paper on World Heritage and Sustainable Development in Africa, and the 2012 Recommendations of the 40th Anniversary Conference on the World Heritage Convention themed "Living with World Heritage in Africa" (Labadi, *et al.*, 2020: 7). The RIWHS meeting adopted an action plan for 3–4 years, and this will be coordinated by AWHF and African SPs. Another outcome is the publication of case studies from Africa demonstrating various levels of localization of SDGs and associated challenges. Implementing SDGs is a stakeholder-driven process that capitalizes on their different competencies, experiences, and wisdom towards achieving pragmatic, practical, and inclusive approach in balancing conservation and development at WHS. For example, meaningful community participation together with public sector support present opportunities for the development of small-scale indigenous tourism projects in less developed areas (Briedenhann & Wickens, 2004; Brown & Hay-Edie, 2014).

An emerging concept in South Africa is the application of the nodal approach/crowdfunding in promoting sustainable tourism and increased benefits to stakeholders. The nodal approach prioritizes destinations rather than individual tourism projects/products/activities (Manetsi, 2017). This approach brings together various efforts, funds, and influence of stakeholders in a locality of development (Manetsi, 2017). This improves relationships with other departments or entities, and among stakeholders towards strengthening development linkages with other sectors (Manetsi, 2017). The National Department of Tourism (NDT) is applying this nodal approach at WHS under the Destination Development Initiative (DDI) in South Africa. The DDI nodal approach promotes the identification and prioritization of clusters/nodes in particular locality of development. NDT defines a "tourism node" as a defined geographic area with a concentration of clearly defined tourism potential or attractiveness, with the following attributes: collection of tourist attractions in close proximity, accessible or potential to enable accessibility, having basic infrastructure or potential for development and enhancements thereof, having supporting tourism facilities and services or potential thereof, and potential for tourism and commercial value chain opportunities (Manetsi, 2017).

In selecting the node for clustering of resources and stakeholders input, NDT considers existing investment commitments, relevance to jobs, poverty alleviation, and sustainability of the projects (Manetsi, 2017). It also considers whether the node addresses tourism infrastructure gaps at high usage tourism destinations and iconic attractions, as well as underperforming tourism destinations and national attractions (Manetsi, 2017). Supporting tourism infrastructure development at new priority tourism destinations is also an important criterion considered by NDT (Manetsi, 2017). Other criteria applied include clear destination development goals; existing or potential tourism drivers; avoiding stand-alone products but fostering significant clustering and compactness; and significant level of population concentration. NDT also considers the destination for potential upgrades and linkages with existing attractions, destinations, precincts, and routes (Manetsi, 2017). Nodal approach can be applied across various socio-economic initiatives at localities of development, but this requires greater stakeholder cooperation, which is where MMST plays an integrative role as discussed before.

Stakeholder Opportunities and Challenges at WHS

Analyzing the opportunities and challenges of stakeholders is a step towards understanding and improving engagements among them, as well as mitigating their conflicts at WHS. However, it appears there are more challenges than opportunities for stakeholders at WHS in Africa.

Opportunities for Stakeholders at MCLWHS

Stakeholders, regardless of discipline and sector, are always looking for opportunities at WHS. Opportunity is universally understood to be a time or set of circumstances that makes it possible to do something. Stakeholders at MCLWHS ranked potential opportunities between 32% and 45%, with communication at international and local levels being considered important. Communication or dialogue is central to balancing conservation and SD at WHS. At MCLWHS, 36% of stakeholders are willing to attend meetings and another 23% have financial resources to do so. Therefore, there is a need to capitalize on the growing communication opportunity at local and national levels at WHS. Heritage institutions and stakeholders should not be surprising each other by taking decisions without discussing them together in order to establish areas of common agreement and variance, including mitigating the latter. Stakeholders and heritage institutions should demonstrate how their respective mandates can benefit each other. Lessons learnt from the past should be part of shaping this dialogue at WHS with a view of riding on the positives while minimizing the negatives. Without communicating with each other, conservation and development run the risk of self-engineered and sidelining solutions detached from broader discussions and processes in Africa seeking the integration of the two. The effectiveness of the Park Forum (MCLWHS), EPPAC (RIWHS), Ngorongoro Pastoral Council (NCA, Tanzania), and many other stakeholder management approaches at WHS in Africa should be interrogated towards enhancing and improving governance approach beyond legislative consultations.

Challenges for Stakeholders at MCLWHS

Stakeholder management at WHS is often fraught with multiple and unresolved challenges, of which some are historical while others are evolving in the present. At MCLWHS, these challenges were identified and ranked as follows: lack of financial resources, communication challenges at national and provincial levels, politics, time constraints in attending meetings, and generally, communication among stakeholders at various levels. The lack of financial resources, communication challenges, and politics at MCLWHS point more towards localized challenges for most stakeholders at WHS in Africa. Financial resourcing affects their active and consistent participation in decision-making processes at WHS. The learners, impoverished IDCs and older generations, may not have the financial capacity to participate in WHS governance. Therefore, decisions are taken in their absence, and they are only informed of these decisions through public platforms such as newspapers and websites, or through Park Forum meetings like at MCLWHS. Heritage institutions and stakeholders should leverage each other, including considering strategies for implementing good governance at WHS. Though this sounds idealistic, if done, governance will not remain dominated by those with power and resources, and neither will it remain an emotive process that induces a crisis management mode at WHS. That crisis mode further induces retrospective engagements with stakeholders, rather than proactively dialoguing before a crisis emerges at WHS.

The communication challenge identified at national and provincial levels at MCLWHS is typical of most of the WHS in Africa. Communication remains key to any dialogue at WHS. Having communication challenges at WHS could be a sign of mistrust, lack of transparency, and absence of defined communication channels among stakeholders (Aas *et al.*, 2015). Such challenges can lead to unnecessary tensions and legal battles bordering on counter accusations and non-disclosure of information among stakeholders. As an example, site managers are known to resist "supplying pertinent information to interest groups" (Pedersen, 2002: 38). This only breeds distrust, thereby limiting the site manager's ability to deal with the public in a transparent and honest manner (Keith, 1994; Pedersen, 2002: 38). Hierarchical communication structures employed by heritage institutions like SANParks are fraught with bureaucracy, confidentiality, and secrecy that inhibits effective communication with stakeholders (Aas *et al.*, 2015). Stakeholders are always seeking information in order to make informed decisions, yet this is what is denied to them.

Even among stakeholders, the same happens, like the lack of trust between the Mapungubwe Coalition Group and CoAL, which led to litigation after the former accused the latter of not disclosing critical information. The same mistrust also ensued during the trade-off negotiations between Department of Environmental Affairs (now DFFE) and CoAL, even in the hands of a broker between the parties. Certain information is often considered sensitive, yet it is critical to solicit honest and informed opinions from interested and affected stakeholders at WHS. This equally applies to how other stakeholders jealously guard sensitive information not meant for public consumption, unless demanded through legal processes. Even though formal structures of organizations are necessary, they sometimes, and when coupled with cultural norms, make it "difficult to elicit the opinions of certain stakeholder" at WHS (Pedersen, 2002: 38). On the other side, open and transparent communication can be a threat to the powers of management authorities at WHS (Pedersen, 2002: 38). Integrity and credibility are issues that all stakeholders operating at WHS should improve in order to create a conducive environment for mitigating contentious issues beyond legislative limitations. Collective participation may improve communication among stakeholders at WHS (Hillman & Michael, 1999; Keith, 1994; Morrel, 2007; Pedersen, 2002; Ury *et al.*, 1983).

Regarding political challenges, these are widespread at WHS across the world, and neither are new as the past has always been politicized and contested. The contestation has intensified with contemporary politics filtering into governance levels of WHS (Meskell, 2011). Politics happen at local, national, regional, and international levels. At WHC level, the Committee has become politically polarized with increased participation of politically appointed permanent delegations to UNESCO, Ambassadors, multilateral and bilateral representatives, and in some cases, Ministers deployed by SP during World Heritage Committees. This also includes economic blocs such as the Economic Community of West African States (ECOWAS), Southern Africa Development Community (SADC),

AU, Common Market for East and Southern Africa (COMESA), International Monetary Fund (IMF), World Bank, the coalition of Brazil, Russia, India, China, and South Africa (BRICS) and many others lobbying for favourable decisions supporting their economic partners through various platforms and channels.

At MCLWHS, political contestations are rife, including discussions on the role of the site in establishing national identity in a once racially fractured nation (Pikirayi, 2016). Nation-building, identity construction, and social cohesion are tools used to politicize heritage as a unifying factor in Africa. At MCLWHS, IDCs are divided along political lines characterizing the democratic South Africa. Also, based on archival analysis at MCLWHS, it is clear that politics is as old as the site itself, and has filtered into the land restitution, conservation, developments, and the establishment of the GMTFCA. In essence and either way, UNESCO has at some point bowed down to national political dynamics and localized aspirations at WHS. Also, in South Africa, a subtle political tension between DFFE and DMR extended to a recent inscription, Barberton Mountains WHS (Mpumalanga, South Africa), where mining is a threat to attributes of the site. The inscription hanged in balance as political principals had to make the final decision on the matter, even though credibility of the Outstanding Universal Value (OUV) had been favourably reviewed by IUCN in its advice to the WHC. Such situations leave other stakeholders in the "limbo" and unsure of the actual position of the SP. Political principals (ministers) sit in the same cabinet meetings where such contradictory decisions are taken, yet they can resolve matters at that level. This tension may be interpreted as "egos" of political principals playing out at heritage sites in Africa. Mining has always received an upper hand over conservation in Africa. These political dimensions force stakeholders to align themselves with the "camp" that can make their own dreams a reality.

Another example of a political decision that achieved a positive result for the SP and its mining stakeholders, and the nation at large, relates to when UNESCO conceded to uranium extraction at Selous Game Reserve WHS (Tanzania). This World Heritage Committee decision was granted against heavy political lobbying, which at some point witnessed over 30 officials from Tanzania attending a side event in 2012 at a WHC session in Paris (France) to lobby other SP on the matter. At a political level, the request for a boundary modification by the SP was premised on mitigating poverty affecting her citizens, while at the same time retaining the OUV of the site. At a technical level, and in terms of the Operational Guidelines on the Implementation of the World Heritage Convention (OGWHC), while the request for a minor boundary modification is permissible, this was going against the commonly agreed principle of "no go zone" in protected areas signed between IUCN and the mining community barring any form of mining in natural WHS. The SP emphasized the provisions of the OPWHC as subsuming this IUCN and mining community agreement. On the home front, the SP was consistent with its position that WHC should not be an obstacle to development in Tanzania. These political dynamics almost witnessed the government considering withdrawing from the WHC, and also the SP stopped developing further nomination files for potential sites on the Tentative List. Politics is thus a double-edged sword that can be used by stakeholders to their own advantage.

Conclusion

The above analysis reinforces that the management of WHS in Africa is still driven by conservation thinking as advanced by SBMS and supported by UNESCO Conventions, with limited engagement and involvement of stakeholders, who may be solution bearers to their own local challenges. Equally important is the need to understand stakeholder perspectives on the relationship between conservation and development at WHS in Africa as an ongoing process. This assists in unpacking the relationship developing between the two in Africa. The opportunities and challenges bedeviling stakeholders at WHS are equally important in unpacking past mistakes, strengths and forging the future. This cannot be divorced from assessing the elasticity and relevance of heritage management in embracing stakeholder management at WHS. Applying good governance approaches in stakeholder management is important (King Code IV, 2016). This includes adopting "a stakeholder-inclusive approach that balances the needs, interests, and expectations of material stakeholders in the best interest of the organisation over time" (King Code IV, 2016). This is equally important in the localization of SDGs alongside conservation as a stakeholder-driven process in Africa.

Furthermore, this discussion confirms that both conservation and SD are grassroots stakeholder-driven issues, which are equally vulnerable to the ever-changing local and global contexts. From pre-colonial times, these issues have been dealt with by communities, thereby confirming that heritage and use of resources around it has always been a local issue requiring learning by implementing. Grassroots issues require localized solutions. These solutions can only be unlocked through *psychological empowerment* emanating from self-esteem and pride of cultural practices/traditions which are embedded in the values of the site; *social empowerment*, derived from increased community cohesion when they are brought together; *economic empowerment*, by maintaining ownership/control of resources; and through *political empowerment*, being participation in decisions, development of leadership, and increased local governance over resources (Scheyvens, 2003; Sofield, 2003; YuLong & Hunter, 2015). Stakeholder empowerment is more "action driven" and aims at "changing power structures to remove barriers that prevent people from participation" (YuLong & Hunter, 2015: 254). WHS governance must negotiate with other competing socio-economic needs in the same locality and how to do it is the focus of the last chapter.

8

CONCLUSION

From Local to Global Perspectives

Introduction: Abandoning the Conservation Comfort Zone

This concluding chapter recommends the process of deconstructing stakeholders and their profiles, localization of Sustainable Development Goals (SDGs), implementing Adaptive Management Approach (AMA) to improve the effectiveness of State-Based Management Systems (SBMS) and World Heritage Site (WHS) governance, converting challenges into opportunities, creativity, and innovation as a function of multidisciplinary mindsets at WHS. Furthermore, it argues that conservation must abandon its comfort zone by embracing development value propositions at WHS. From a futuristic perspective, WHS governance should be a continuous process of managing continuity and change as conservation and development are going to have a life-long relationship. As stakeholder-driven processes, conservation and development must find each other, otherwise delisting of WHS shall become a norm, a threat looming at Selous Game Reserve WHS (Tanzania) and a reality for Liverpool (United Kingdom), which was delisted from World Heritage List in July 2021. These are the binary choices between heritage and development challenging World Heritage across the globe. This challenges WHS to do things differently to get different results fully knowing that failure to do so is risking the future of conservation in Africa.

"Inseparable and Inexorable Relations" between Conservation and Sustainable Development?

Conservation and sustainable development (SD) are "inseparable" (indissoluble) and "inexorable" (unavoidable) stakeholder-driven processes at WHS. This relationship has its roots in pre-colonial Africa. Currently, competing demands are being placed on both processes. Due to a lack of statistical data, it is challenging

DOI: 10.4324/9781003149361-8

to confirm how they contribute to each other in reality. It appears that conservation contributes more to SD than what the latter is contributing to conservation. However, more research is needed as there seems to be some empirical evidence that shows that development has and continues to contribute to conservation through identification, documentation, and management mitigation offered through Environmental Impact Assessment (EIA) and Heritage Impact Assessment (HIA), which are developer-funded. Equally, and of further concern, demonstrating how heritage contributes to achieving development indicators also suffers from paucity of data. Therefore, developing baseline cultural statistics towards monitoring the alignment of conservation to development goals is a priority in Africa.

The evolving "inseparable" and "inexorable" relationship between conservation and SD is no longer a debatable issue; the issue now is rather, how to practically manage their co-existence as mutually beneficial and stakeholder-driven processes at WHS (Taruvinga, 2019). There is a lack of practical integration at WHS level, hence the need for integrated management approach at WHS. Heritage should be managed beyond conservation mandate, implying it has to become integral to the needs of the society. Heritage integration into broader territorial processes at WHS remains a negotiated process at different levels that includes integrating (i) legislation and policies applicable, (ii) conservation and SD value proposition, (iii) stakeholders, and (iv) reviewing governance framework (Figure 8.1). Practically integrating conservation and SD is now an unavoidable choice for WHS in Africa. The slow pace at which the Convention has aligned itself to SD is viewed as a burden that hinders development (Labadi, 2018). WHS have to "accept compromises with development to secure a more positive heritage future" (Chirikure, 2014: 242).

Conservation, stakeholder, and development needs are mutually non-exclusive at heritage sites since pre-colonial Africa. Earlier it was a communally driven process but it now involves multiple stakeholders. Therefore, localizing SDGs at WHS is an inevitable stakeholder-driven process. However, this requires a conducive governance framework to be in place at WHS. Such a governance framework links conservation to broader territorial planning processes at WHS. In the same vein, AMA should be a catalyst in linking the experiences of conservation, development, and stakeholders at WHS. Adaptive management demands learning from doing (experience) and is also effective when it considers a multidisciplinary approach. Multidisciplinary (also interdisciplinary) approaches draw and benefit from multiple disciplines beyond conservation. It assists in creating solutions based on understanding the complex and conflicting situations as opportunities for innovative solutions (Duval *et al.*, 2019). It is impossible to resolve the complex situation of conservation and SD without looking at other disciplines. Heritage governance should not remain in the conservation compliance enforcement mode. Furthermore, heritage experts should become part of the solution rather than being obstacles to the integration of conservation, stakeholders, and SD at WHS. More importantly, inclusive stakeholder approaches

Level of Integration	Areas of integration	Integration Outcome
1. Integrating legislation and policies applicable at WHS.	• Creating an Integrated Policy Framework at WHS: giving each legislation equal opportunity for the protection of the WHS. • Developing an Integrated Management System (for instance combining modern and traditional practices at WHS, or whichever systems are operating at the WHS). • Adequate resourcing of integrated approach is needed at WHS.	**All applicable legislations have a role and have to negotiate point of congruency on conservation and SD.**
2. Integrating Conservation and SD at WHS.	• Integrating the caring of heritage and the well-being of society at WHS. • Moving conservation planning into broader territorial planning and development. • Adequate resourcing of this balancing act between conservation and SD.	**Localising the 2015 World Heritage Policy on SD, AU2063 Agenda and National Development Plans.** -
3. Integrating stakeholders or role players at WHS.	• Identifying and recognising all Stakeholders, interested and affected. • Establishing Stakeholder Management Framework, including "mutual rules of engagement". • Defining roles and responsibilities for stakeholders. • Stakeholder customised policy intervention.	**Stakeholders have a responsibility to play at WHS.**
4. Governance level	• Adapted Governance frameworks • Decision Making involving stakeholders • Good governance practices	**Inclusive and transparent decision making at WHS**

FIGURE 8.1 Levels of integrated management approach at WHS

such as Multiple and Multi-layered Stakeholders Theory (MMST) should be adopted as preferred governance frameworks to facilitate this transition at WHS.

Overall, the nature and local relevance of the WHS value proposition to stakeholders should be the reason why they support or do not support its continued implementation in Africa. This value proposition should be aligned to the development value proposition of the society, otherwise conservation value proposition becomes a risk to itself. The development value proposition, if not facilitated to materialize, will degenerate into a political issue, which may unintentionally reverse the gains of conservation at WHS. The "pristine approach" of conservation should be dismantled at WHS in Africa (Nelson, 2003; Taruvinga, 2020). The contribution of the WHS value proposition to development indicators should be demonstrated by measurable quantitative indicators (culture statistics). The existing SDG- and stakeholder-proposed indicators are important in establishing WHS-specific cultural statistics. Sustaining poverty in the name of conservation is no longer tenable and threatens the future of WHS in Africa.

Deconstructing, Constructing, and Reconstructing Stakeholder Profiles at WHS

Engaging and involving stakeholders is inseparable from legislative context, evolving land uses, development needs, politics, social environment, and heritage value layers at WHS. Neither can it be considered a one-off process because stakeholders are continuously evolving and reimagining ways of being relevant at WHS. Therefore, deconstructing, constructing, and reconstructing stakeholders is a dynamic and site-specific process. This discussion has provided alternative methods of doing this, including how to categorize stakeholders using Multiple and Multi-layered Stakeholder Framework (MMSF) (Figure 8.2).

FIGURE 8.2 Categories of stakeholders identified at WHS through MMSF

MMSF attests that the local perspectives of stakeholders cannot be replaced by universal dimensions of the World Heritage Convention (WHC). Ignoring stakeholders makes WHS more vulnerable to destruction in the long term. Though conflicts may arise between and among stakeholders at WHS, these can be mediated through conflict resolution mechanisms. The future of WHS lies in the hands of stakeholders, often overlooked, but are a powerful force with alternative voices and adaptive capabilities built on their past and present experiences. For instance, Indigenous and Descendent Communities (IDCs) have wielded these capabilities since pre-colonial times, but they were largely untapped by SBMS and WHC. While MMST recognizes that there are more benefits realizable from negotiated agreements with stakeholders at WHS apart from enforcing legislative frameworks, this is beyond enforcing the conservation mantra at WHS in Africa. Stakeholders are co-creators of knowledge, conservation, and development practices in various ways, as well as in making decisions relating to their own well-being. Heritage should be about co-creating and engagement with stakeholders as an adaptive process of an inclusive governance framework. Heritage management should not be a unilateral process reserved only for authorized role players. It is an inclusive process that should involve other stakeholders as well.

Stakeholders and the Development Value Proposition at WHS

The "construction" of development value proposition, including its benefits at WHS, is a stakeholder function negotiating within the rigidity of WHS value proposition (conservation). Stakeholders decide on the nature and type of developments that solve their problems regardless of what conservation can say. They envision benefits for themselves in order to meet their needs. While stakeholders are aware of developments at WHS, the sad part is that the majority of them are not aware of any benefits accruing to them. This points to those with power and resources to initiate developments, including determining their nature, type, and the quantum of benefits that should accrue to other stakeholders. Hence the very low stakeholder influence on decision-making process related to developments at WHS, except for the instigators themselves, and IDCs when they are fully or partial managing a site. Furthermore, the observable dysfunctional connection among and between stakeholders at WHS is equally contributing to this skewed pattern favouring those with power, legitimacy, and resources. Meanwhile, WHS are always watching for opportunities to enforce compliance favouring conservation. The challenge with this approach is that WHS value proposition does not operate on an island; it's implemented in the midst of other competing needs. Only formal and inclusive governance structures, as opposed to goodwill approach, can improve this observed skewed pattern. Heritage values are not static but rather dynamic, so is their use by the society.

Localization of SDGs at WHS

While WHC is a universal concept, heritage itself is a function and a manifestation of society-driven processes at the local levels of WHS. This means, first and foremost, heritage is local before it is universalized. So is its use and application in development issues. The same applies to development needs, which are primarily local needs of the society. Therefore, conservation decisions should consider these local needs before trying to align with the exogenous or universal concepts that have little sympathy with local needs. If conservation cannot do so, it is likely not going to be supported by stakeholders, nor is it going to negotiate with development as a learning by doing process at WHS. What needs to be localized is not the conservation and development practices as these are evident from pre-colonial times. Rather, localization should be about how to persuade this localized experience with that of the new and exogenous perspectives. Localization is not an imposing process; it is rather an engagement of good practices and innovation from different players without making any proposal subservient to another. It is negotiating in good faith for a commonly agreed cause or outcome. It is better to support rather than appropriate local aspirations of stakeholders at WHS.

The above engaging localization approach lacks a mindset shift at global levels as local input is often not viewed and considered favourably. This induces local resistance from stakeholders. For instance, pushing the adoption of the 2015 Policy on World Heritage and SD without this mindset is a futile exercise in Africa. Conservation and development should practically facilitate their integration while mitigating the actual and perceived negative impacts from both. When conservation stops development, this negatively impacts society's well-being aspirations. Equally, when development threatens or destroys heritage, this also negatively impacts conservation aspirations. The complexity here is that all stakeholders are custodians of either or both conservation and SD processes, however, with varying responsibilities, powers, and expected outcomes. In the process, both conservation and development should bring positive and measurable benefits to the society. These are two necessary processes at WHS. The cautionary approach of conservation in this process stems from its inability to (i) embracing creative and innovative development approaches and techniques and (ii) accommodate stakeholders (Nelson, 2003). The localization of a common understanding on how conservation and SD can pragmatically engage at WHS is important. This is because conservation and development have never been mutually exclusive as local processes since pre-colonial times. Also, both have a commitment and aspirations to the past, present, and future stakeholders. Their co-existence or integration should be responsive to the past and present experiences of stakeholders while upholding common commitment to the future of WHS.

Furthermore, conservation needs to align with national, regional, and the African Union (AU) 2063 development plans in order to establish sustainable partnerships at the local levels of WHS. This helps in building relationships and liaising with stakeholders beyond the heritage discipline (ADCOM, 2017/2). This also paves the way for building bridges between heritage and development sectors. Equally, this will distil some sanity given the "self-view" of heritage practitioners as being the "all-encompassing experts" at WHS. They need behavioural, psychological, technical, and mindset changes in order to appreciate other disciplines which are beneficial to them and their functions at WHS. Ability to listen to other sectors and experts, collaborating and dissenting voices in areas where one is not competent is wisdom and power in any decision-making process for any sector. While the localization of SDGs is assumed to guarantee conservation at WHS, not all developments are sustainable in nature (Taruvinga, 2020; UNESCO, 2010), hence the risk of unintended threats being created in the process, which needs the experience of other sectors and experts who are competent in a particular area. This contradictory position needs to be addressed in an inclusive and holistic manner as stakeholder-driven process at WHS. In this process, considering the human, social, economic, and political dimensions is equally important. Given the broader geographic context of WHS, innovative and creative approaches, including in the area of governance, are required at WHS (UNESCO, 2010). Also, and while rigid and stricter scientific process are essential in managing conservation and development, stakeholder-negotiated outcomes should be given an opportunity at WHS. Balancing the two is no longer a scientific process only, but a broad-based social approach that requires an inclusive governance approach at WHS.

Governance and Adaptive Management at WHS

As noted from empirical perspectives and discussions in this book, most stakeholders have an average role in WHS governance in Africa. Equally, they do not provide meaningful resources in the process, and neither are they consulted as an ongoing process, nor are they involved in the decision-making processes at WHS. SBMS are still in charge because of their legal power and the resources at their disposal. Despite the advent of democracy at political level, heritage governance has remained in the realm of colonial philosophy, thereby reinforcing alienation and exclusionary perspectives. Despite this, stakeholders still have governance aspirations at WHS. On the other hand, WHS governance comes with responsibilities that these stakeholders should be willing to assume if they are to be part of this process. What needs to be explored is how WHS governance can be adapted to effectively involve other stakeholders in the management of WHS.

As discussed earlier, WHS governance represents an "extremely complex" system supposedly "involving multiple stakeholders", yet empirically it is marginalizing them (Molated, *et al.*, 2017). This governance system is defined by power, legitimacy, and resources channelled towards conservation. However, this system is becoming insufficient in WHS value proposition given the intensifying development value proposition at WHS. In this case, WHS governance cannot remain in the realm of the all-knowing experts, Advisory Bodies (ABs) and State Parties (SP), yet it operates at the local levels of WHS. The net effect is that managing WHS in relation to other land uses within its proximity now requires a lot of conscious effort from the SP, SBMS, UNESCO, and stakeholders (Selman, 2004, 2007). Thus, WHS governance needs moral, ethical, constructive, and inclusive dimensions that benefit interested and affected stakeholders. WHS governance cannot afford to remain in a conservation mode that is supported by top-down management approaches in an ever-changing operational environment. This demands that a bottom-up management approach be considered at WHS in order to replace "the usual sector or one-dimensional approaches with new transversal or multidimensional ones" (Ripp, 2018: 2). By doing so, WHS governance will be "taking into account the role of each part in the whole structure" of stakeholders (Ripp, 2018: 2; European Union, 2010). This further supports the AU 2063 Agenda, which is advocating for a fully participatory and bottom-up approach in all spheres of the economy in Africa. It is important to involve society in the "ongoing process of defining and managing cultural heritage" (Council of Europe, 2005: Faro Convention, Article 1c).

There is a popular adage that says, "nothing about me without me" (a phrase initially coined by Valerie Billingham during the Salzburg Global Seminar presentation "Through the Patient's Eyes" in 1998) that is applicable to both conservation and development. This means neither conservation nor development can be planned from the beginning to the end without involving the intended beneficiaries. Neither can the two be planned without considering each other at WHS anymore. The overall implication of this saying is WHS have to effectively involve stakeholders in their planning process, while developers have to equally involve conservation in their planning. The bottom-line is that conservation, stakeholders, and development should involve each other as stakeholder-driven processes at WHS. As such, WHS governance has become a social and political construct reflecting the interests and needs of the society in its diversity (Chirikure *et al.*, 2016; Labadi & Logan, 2015; Leick & Lang, 2018; Pike *et al.*, 2017; Tomlinson & Branston, 2017). In the above context, WHS governance demands an appetite for AMA, which builds on learning by doing as shared experiences among stakeholders in their areas of operation at WHS. This allows for multivocality at WHS.

AMA is defined as the continual learning process which can be used for site-specific management that is operating in ever-changing circumstances (Walters, 1986). The approach is not reactive but rather responsive in nature. It enhances "information flow among policy actors and provides opportunities for creating shared understandings" in the decision-making processes (McLain & Lee, 1996: 437). This approach incorporates "knowledge from multiple sources" and "use of multiple systems models" while supporting "new forms of cooperation among stakeholders" (McLain & Lee, 1996: 437). AMA borrows from adaptive control process theory, which provides a framework for constructing decision-making controls (Bellman, 1961). Such controls are capable of learning from experiences (Bellman, 1961). Central to AMA is consultation, collaboration, monitoring, quantitative modelling, and simulations (Schreiber *et al.*, 2004). AMA anticipates how to deal with any unforeseen surprises, which are inevitable when dealing with multiple stakeholders, contradictory interests and needs, as well as changing operating environments (Timmerman, 1986). It also has the ability to deal with broad impact of policies (Timmerman, 1986).

Interestingly, AMA has been practised by many IDCs for centuries through Traditional Management Systems (TMS) and equally applied in Protected Areas management in Africa. Being proactive and creative is necessary in developing and fostering effective stakeholder collaborations (Louw & Venter, 2013). More important, participation of stakeholders can never be quantified as either too little or too much (Bryson, 2004). Neither can one say they have fully exhausted alternative approaches to get the best from stakeholders or a situation. AMA is an evolving process and has attained a degree of success in Australia (e.g., Kakadu and other Australian Northern Territory areas), where different local communities have power in the decision-making structures of WHS. They have entered into agreements with mining companies and derive royalties from them. In other instances, they have decided against mining, which makes them less affluent to those deriving royalties. These are choices underlying the dynamics of society. Similarly, the agreements reached with the ǂKhomani and South African National Parks (SANParks) in the Kgalagadi National Park in South Africa are emerging examples of "learning by doing" that still needs detailed analysis as to their impact. Adaptive management is context-sensitive, which makes solutions very localized to each WHS.

If AMA is adopted as part of WHS governance, maintaining the conservation mantra becomes increasingly difficult as it creates a sense that no generation is allowed to enjoy the benefits of this "past in the present and the future". This renders preserving WHS for future generations meaningless. Whoever has been viewed as the future generation in the "past" should be given reasonable use rights in the present as this is their time. Future generations cannot remain in perpetuity. By advocating for AMA, the point being made is that heritage should learn from implementing rather than stopping development without trying and mitigating impacts along the way together with stakeholders. Simply put, this should trigger change management towards the systematic transformation of heritage governance in Africa. It has to become responsive rather than being a legal reaction to broader planning processes of the society.

WHS governance system should be underpinned by an open, transparent, inclusive, and accountable framework (Philips, 2003: 16). This is where MMST governance framework comes into play as it promotes effective participation of stakeholders, in particular those at grassroots level who are often voiceless, powerless, resource-impoverished, and lacking opportunities, unless and until they yield to the powerful SBMS, UNESCO, and other stakeholders with privileges to act as they deem fit. MMST provides governance flexibility which encourages varied opinions, policy decisions, and local discussions to inform global processes. Global processes are often oblivious to the gravity of local needs of stakeholders at WHS. Stakeholders are intricately linked with each other in achieving their aspirations, resolving conflicts, encouraging holistic views, and pursuing common solutions in a culturally diverse environment (Jamal & Getz, 1995; YuLong & Hunter, 2015). Overall, MMST caters for the normative, instrumental, and descriptive stakeholders operating at different levels of WHS (YuLong & Hunter, 2015). These stakeholders have a "shared value" that "looks for opportunities to solve social problems" (Bouwer, 2018: 75). Africa has to develop governance models it desires rather than continuing with those that are not effectively working for them. Borrowing and learning from others should be contextualized rather than applied as an abstract truth. It is naïve for one to think that conservation and SD will always deliver on their full promises to stakeholders without a deliberate effort to do things differently at WHS. There are inherent public relations promises in both dimensions that have to be "sanitized" with the support of stakeholders through an AMA.

Converting Challenges into Opportunities: Reimagining the Role Stakeholders

This discussion observed that there are more challenges with stakeholders than opportunities at WHS in Africa. Stakeholders are seen as opposed to conservation in favour of development, yet they are equally concerned with both. Stakeholders argue that they are disadvantaged because of limited financial resources, ineffective communication, politics, time constraints, conflicts and tensions, inaccessibility, growing intergenerational gaps, and bureaucratic governance structures of SBMS and UNESCO. While these challenges appear insurmountable to SP and UNESCO, they are actually opportunities for developing a responsive and inclusive governance framework at WHS, just like how democratic governance systems have been curved out of and under challenging conditions in Africa. Applying SD principles only in this process is not sufficient. There is a need to embrace solutions offered by stakeholders themselves given their different spheres of influence. Furthermore, there shall never be equal roles and benefits for stakeholders at WHS, but what is needed is turning contestants into solution bearers to their own challenges while optimizing the magnitude of the benefits accruing to them and conservation.

However, and despite the above opportunity, it is unfortunate that IDCs remain subdued in WHS governance. IDCs are "dispossessed former authorities" of pre-colonial Africa. They made sound decisions that improved livelihoods and shaped regional economic development, including influencing modern-day economic blocks, however, with varied structural arrangements due to changing times. Giving them a governance role now is like "re-empowering them"; however, this is curtailed by positions leveraged by current legal and bureaucratic governance systems. While some traditional leaders play a role in the current political structures of African SP, their involvement in heritage management remains at arm's length, with the SBMS still wielding "dominant" authority. Also, participation of traditional leaders in civic functions is ceremonial in Africa except in a few circumstances such as the Ingwenyama Trust in the KwaZulu-Natal (KZN) province of South Africa. This trust owns large tracts of land in KZN administered through the Zulu King. They have powers to make decisions regarding the land and associated developments. What is not clear in this emerging pattern is the actual power and benefits accruing to such stakeholders in capitalistic economies of Africa riddled with corruption and notions of #State Capture in the case of South Africa.

Overall, IDCs are limited by legislation, lack of clarity, limited resources, and the long-overdue land redistribution. Their role, and equally that of other stakeholders, has to be reimagined as a local process in heritage management in Africa. It has to be reimagined beyond the political rhetoric of civic functions and that of WHS governance. The experiences of IDCs embedded in their pioneering spirit and practices of balancing conservation and use of resources have remained largely untapped and often neglected in favour of new approaches and terminologies, yet the very same exist in their own local vocabulary and experiences. Reimagining the role of stakeholders in WHS governance cannot be done without them, neither can it be left to chance, including World Heritage not becoming a silent observor on the unfolding land issue in Africa, which equally affects its processes". World Heritage has to become part of the solution while balancing both the caring of heritage and needs of local communities. It has to be a deliberately engaging process among SBMS, WHS governance, and stakeholders.

Beyond the Conservation Comfort Zone: Multi- and Inter-Disciplinary Approach

Reimagining the role of stakeholders and implementing AMA at WHS demands doing things differently in order to get the different results compared to the present experience. There is a need to explore different strategies and techniques that facilitate moving towards mutually beneficially outcomes beyond conservation at WHS. In the process, different experts and disciplines have a role to play. Generally, heritage management has not effectively harnessed benefits from the unlimited innovation and creativity that comes with disciplines other than conservation itself. There is too much "inward thinking and in-breeding" within heritage management systems, yet the opportunity of multidisciplinary approach exists. Conservation always defaults to its traditional dimension when faced with challenges beyond its practice such as development. Therefore, there is a need to harness innovation and creativity embedded in multidisciplinary approaches for the purposes of solving the varying and emerging multiple challenges at WHS. The varying geographic, social, political, and economic contexts of heritage sites is a conducive environment for multi-disciplinary approach to aid adaptive response beyond conservation benefits.

When this is contextualized in the relationship between WHS and academic institutions in Africa, it is unfortunate that the latter still struggles with detaching itself from historical and colonially choreographed course outlines designed to instill scientific facts and philosophies rather than adapting and being responsive to the emerging challenges and needs of WHS, including embracing the much-needed Indigenous Knowledge Systems (IKS) dimensions (Taruvinga, 2018). The question is how relevant are these historical facts to the emerging challenges of society and prospective employers of heritage trained graduates? University curricula should not be fixated on mastering traditional courses but rather explore new multidisciplinary thematic areas, which would make curricula dynamic and relevant to contemporary needs. Heritage curricula should promote interdisciplinary mindsets by linking heritage to other disciplines such as tourism, development, entrepreneurship, business management, economics, financial management, disaster management, climate change, engineering, and many others relevant to broader territorial planning processes in which WHS are located. This demands cross-cutting inter- and intra-faculty courses being made available to heritage students. Expecting a heritage expert to be a one-stop shop at WHS is a huge mistake and a risk to the future of heritage. Furthermore, heritage training programmes in Africa should go beyond the conservation mantra and take a cue from technical and artisanal curricula which links students with current practical developments in their industry, including emerging technologies. There is need to link academic institutions with both World Heritage and broader socio-economic disciplines as part of changing mindsets at training level, which will then improve practices at WHS. Heritage is never in a social vacuum in any given context, neither is it the only discipline that can offer solutions to society.

A multi-disciplinary approach will prepare future heritage practitioners to become innovative and creative in interfacing with evolving conservation and development needs of society. This requires synergies among universities themselves, and between universities and heritage institutions, as well as between stakeholders and experts operating at WHS (Taruvinga, 2018). This transformation is not possible if it is not seen and implemented through the "eyes or lenses" of the affected/interested stakeholders and challenges that need to be solved (Taruvinga, 2018). Without this, solutions offered shall remain theoretical, temporary, and consultant-dependent without changing mindsets and transferring skills among the intended beneficiaries, including the stakeholders themselves (Taruvinga, 2018). Therefore, interdisciplinary approach will ensure the transfer of the highly sought after multifaceted skills and interdisciplinary mindsets at WHS in Africa.

Futuristic Perspectives on Conservation and SD

Solutions to local challenges at WHS cannot only be provided through enforcing laws and conventions. Stakeholders are always naturally looking for solutions to their own challenges, regardless of what these laws or conventions permit, prohibit, or do not provide for. For example, IDCs want the historical land issue resolved, which is inadvertently linked to places now curved as WHS. Repossessing that land carries the hope of a better life for them. As such, there is a need to understand the impact of land restitution on conservation and SD in the context of the envisioned compensatory land uses by successful claimants at WHS in Africa. Land reform is a moral, cultural, social, political, and economic imperative for the society, and this affects WHS in many ways in Africa. Failing to resolve the land issue at WHS means WHS governance is not part of the solution to grassroots challenges. In South Africa, the San communities have repeatedly pronounced their position as the "first citizens" of the nation, including demanding their land back. Petitions after petitions, demonstrations after demonstrations on these issues are happening in South Africa and other Africa countries. Resolving these land issues requires strategies beyond the conservation approach. WHS have to become part of the solution to this politically charged issue in Africa. Lessons have to be learnt from cases where land has been returned to rightful custodians, in particular, how the land is being used and what is nature and impact on both conservation and development.

The land use issue relating to Maasai communities in Ngorongoro Conservation Area (NCA) (Tanzania) is not new and neither is it their own making at all, but it is a classical illustration of dynamics associated with the great experiment on the co-existence of people and wildlife dating back to 1965. The nomadic lifestyle of Maasai has always been a "living dimension" of many landscapes in Tanzania for times immemorial. With NCA being established, their lifestyle was gradually "frozen" through subtle enforcement of restrictions advancing or leveraging conservation of wildlife. This is affecting their traditional lifestyle and seasonal movements, access to water sources and salt pans, as well as their shifting cultivation system. This has led to tensions, largely because the 1965 vision of coexistence between the Maasai and wildlife has placed emphasis on the latter in recent years, instead of balancing both. In mitigation, the NCA has diverted substantial financial resources for their development as a community, but the restrictions on their lifestyle and associated land uses remain as matters of primary concern to them. They are strategically and intentionally being indirectly pushed into sedentary lifestyle inside and outside the NCA in exchange for these financial and technical resources from NCA. At the same time, the Maasai community still views the NCA as their "place" and continues to argue that their lifestyle should be permitted while contributing to conservation as they have always had a relationship with the landscape. This is a clear cry for culture to be allowed to be the determinant of change for them rather than a conservation decision. This challenges NCA as a WHS, thereby pitying the SP and UNESCO against the communities.

Pursuant to the above, there is an ongoing discussion in Tanzania on how such communities should be permanently settled in the parks where they are currently residing. This is not far off from a long-term local solution to addressing a century-old crisis issue of land alienation in Africa. While parks are meant for conservation and tourism, that should not be used to subvert solving local land uses and ownership issues. This includes national heritage in private hands in many parts of Africa. Settling the Maasai communities permanently in the

NCA should be seen as an opportunity of developing an adaptive conservation and development approach at the WHS. This should, however, take the experiences of both over the years without making either of them subservient to the other, nor usurping local dimensions at the property in the name of universal conservation aspirations. This requires dialogue among stakeholders, including developing an integrated conservation framework for the site.

Another emerging classical example is the Barotse Plains Cultural Landscape (BPCL) in Western Zambia proposed for inscription as WHS. The nomination process of BPCL is riddled with unresolved land uses, socio-economic, political, and other pertinent local issues that stretch for decades. As a result of this, an official objection against the inscription by some members of the community has been lodged with UNESCO, African World Heritage Fund (AWHF), and the National Heritage Conservation Commission (NHCC) of Zambia in September 2020. One can only imagine the magnitude of conflicts and tension that will ensue if the inscription proceeds without resolving these matters. The proposed Outstanding Universal Value (OUV) of the BPCL is not questionable as it illustrates an exceptional interaction of human society and settlement over time under the influence of the physical constraints and opportunities presented by the natural environment (massive floodplain supported by river system). This interaction reflects the complex and long-standing traditions of the Lozi people and their social and economic organization, which is inseparable from origin, conservation, and ecological functioning of the Barotse Plains. The ongoing interactions are illustrated by dynamic and living cultural tradition of water and land management involving transhumance as well as land drainage system based on an elaborate and robust TMS. This system and accompanying traditions have sustained subsistence agriculture, navigation systems, settlement mounds, as well as the social, political, and cultural structures that have existed for over 400 years.

The objection against the proposed inscription was lodged by the Barotse National Freedom Alliance (BNFA)-Namungelele. BNFA was launched in 2012 as a political organization bringing together "all activists" groups. The predecessor to BNFA is the Barotse Cultural Association (BCA) established in 1992. The activists groups constituting BNFA are the Linyungandambo, Barotse Freedom Movement, Movement for the Restoration of Barotseland, Barotse Imilemas, Barotse Youth league, and others. The purpose of BNFA is to implement the sovereign resolutions of March 2012 Barotse National Council in a "pacific manner and protecting the human rights of the peoples of Barotseland" (BNFA, 2020). According to the BNFA, the history of BPCL is linked to the Barotseland Agreement of 1964, which in an intricately crafty manner brought together Barotseland and Northern Rhodesia as an independent Unitary State called Zambia, with the former retaining her governance system (BNFA, 2020). Furthermore, as per the Zambia Local Government Act No. 69 of 1965, the powers and privileges of the Litunga of Barotseland were taken away, including the abolition of the Barotse National Council and seizing of their budgets, thereby precipitating the current tensions simmering between Zambia (BNFA, 2020).

All land and the resources of the Barotseland become subject to national controls of the Zambia government (BNFA, 2020).

Informed by all these perceived losses, BNFA lodged a complaint with the African Commission on Human & People's Rights in Banjul, Gabon – Communication 429/12 titled the "Ngambela of Barotseland and Others vs The Republic of Zambia over Zambia government occupation of Barotseland". Furthermore, BNFA invited the Government of Zambia to sign the Permanent Court of Arbitration (PCA) submission agreement to allow peaceful resolution of the matter, and this was supported by over 10,000 signatures (BNFA, 2020). This political issue remains unresolved, and has precipitated the objection to WHS inscription process by Zambia. On the other hand, the Government of Zambia remains committed to maintaining a unitary Statehood, including maintaining peace and order to ensure that the nation continues to develop in every sphere of the economy. The tension between BNFA and the SP of Zambia reflects a deep-seated local political issue preceding the birth of the Convention. The BNFA objection against the proposed inscription of BPCL is premised on the following.

TABLE 8.1 Reasons cited by BNFA for objecting to the inscription of Barotse Cultural Landscape, Zambia (Source: BNFA Letter to National Heritage Conservation Commission, Zambia, 2020)

BNFA REASONS FOR OBJECTING THE INSCRIPTION OF BAROTSE CULTURAL LANDSCAPE AS A WORLD HERITAGE SITE	
(i)	Barotseland no longer a territory under the jurisdiction of the Zambian government and therefore, Zambia is not the appropriate SP applicant to UNESCO;
(ii)	Zambian government working towards in erasing the national, cultural, and traditional identity of Barotseland and her people;
(iii)	Zambian government failing to bring development to Barotseland before and after the unilateral repudiation of the Barotseland Agreement 1964, and finally that;
(iv)	Barotse cultural landscape is owned by and home of the Barotse people, as the economic, social and political hub of Barotseland.

The BNFA objection letter (copied to the United Nations Secretary General and the Director, UNESCO World Heritage Centre) states their willingness to amplify the objection to the inscription application at an appropriate forum (BNFA, 2020). The question is then, what is an appropriate forum being demanded by BNFA? Is this mediation at local, national, or international level?

If it is at the national level, it is localized dialogue to find each other beyond WHS issues. Equally, if it is at the international level, then UNESCO has to broker discussions on a local matter, however, being sensitive to the sovereignty of the SP. And how does the site managers of the BPCL, as scientific technocrats on culture and nature conservation, feature in this politically sensitive matter at the local level? Are they strategically positioned to address these interconnecting political and economic issues at BPCL? Is it a political or stakeholder engagement forum capitalizing on the proposed inscription for political mileage to score victory on decades-long issues? It is clear that the proposed inscription is not the issue at stake, but it is being used to attract attention to long-known political and socio-economic issues relating to BPCL. This demonstrates that stakeholder issues at WHS are not restricted to conservation matters only. There are other issues that have to be resolved to avoid WHC becoming a victim of these dynamics.

As part of stakeholder engagement at WHS, there is a need sometimes to exclude UNESCO from addressing local issues. The SP needs to work towards reaching amicable solutions with stakeholders before they become issues of concern to UNSECO. Though this may not be consensual due to varied and historical positions, local dialogue is likely to yield more practical solutions than "international arbitration". Instead of dealing with the localness of these issues, SBMS often use UNESCO to subvert providing solutions to long-standing issues that are equally beyond the scope of the Convention. Stakeholders should be afforded attention in their own sphere of operation and influence in addressing local issues at WHS. Stakeholders are themselves solution bearers. Conservation solutions cannot solve political issues at WHS, neither are heritage managers trained as politicians and in conflict resolution. Political issues require political solutions, while threats to WHS require conducive socio-economic and political context in order to be mitigated. Balancing the competing needs of BPCL requires a scientific, political, economic, and socially negotiated process. It cannot be solved by simply enforcing legislations and defensive positions of SBMS. Additionally, solving local issues at WHS is linked to the emerging application of People Centered Approaches in heritage management, in particular for protected areas traditionally preserved as enclaves of biodiversity (Thompson & Wijesuriya, 2018). This paradigm shift promotes both the caring for heritage and the well-being of the society (Thompson & Wijesuriya, 2018). New issues emerging include conflicts, civil wars, rights-based and knowledge issues, as well as pandemics such as COVID-19 that require mitigation at WHS.

Conclusion: Local Challenges, Local Solutions

The relationship between conservation and SD is as old as humanity itself and thus remains evolving. The two are localized and stakeholder-driven issues requiring grassroots solutions. Currently these solutions are engineered from the exclusionary governance systems of SBMS and WHC. The weaknesses of these governance systems in Africa are that they remain heavily leaning towards legal powers and conservation mantra and have been slow in embracing development and accommodating stakeholders at WHS. As much as conservation and development are not mutually exclusive to each other, so is heritage governance and stakeholder involvement in post-colonial Africa. Neither is the past, present, and the future mutually exclusive to each other, because there is adaptation and continuity through time and space. However, adapted heritage governance has the potential to offer practical solutions to challenges at WHS. Heritage governance needs to integrate itself in broader territorial planning and stakeholder management processes. This will strategically position conservation to provide for the well-being of the society. Development is inevitable and is here to stay, and WHS value proposition may become a risk to itself if it maintains the current approach and pace in embracing development. An emerging case of the Liverpool Waterfront WHS (United Kingdom) testifies to this. The SP has developed skyscrapers considered interfering with the integrity of the WHS, hence forcing UNESCO to recommend delisting. What more in developing nations desiring an improved state of life and well-being, as well as regenerating left-behind nations and communities in line with SDGs? It is becoming a reality that UNESCO cannot continue preserving everything in a pristine way. It appears that more SPs will rather opt to have their sites delisted in the future in order to solve local challenges.

While localization of SDGs may be viewed as the best strategy to integrate conservation and development dimensions at WHS, this should be a stakeholder-driven process supported by AMAs. Stakeholders have an equal opportunity of playing an increased role at WHS given their varying past and present experiences. This will equally benefit in shifting the mindsets of the hard-liner conservationist approach in Africa. Equally important, and as part of decolonizing long-entrenched heritage governance systems in Africa, legislation reviews and accompanying guidelines that formally embrace stakeholders are urgently needed. Moving towards inclusive stakeholder management approaches is a good step in WHS governance worldwide. Such an approach will encourage an inclusive stakeholder-driven decision-making process on conservation and development at WHS.

The WHS value proposition to its stakeholders should be the compelling reason why it must remain in existence as an international concept in a localized environment. Conservation means nothing if it cannot give meaningful and measurable expression to socio-economic needs of stakeholders at the local levels of the site. The reality is that stakeholders at grassroots levels want more than conservation at WHS. The future of WHS is now dependent on how fast adaptive management is adopted by SBMS and UNESCO to mitigate the impact of their top-down approach in heritage governance and, equally, embrace development as a lifelong relationship with conservation. WHS experts can no longer impose solutions on stakeholders. Rather, stakeholders, together with these experts, should provide solutions to their own challenges. Effective change at WHS will only happen as long as it is supported from the bottom rather than imposed from the top. The localness of heritage and how it should be used shall remain on the radar of WHS governance and stakeholders for a long time. Engaging and involving stakeholders in practically balancing conservation and development at WHS in Africa is now a priority. Either WHC is with them, or they are without it: denoting how binary choices between heritage and development are becoming a reality!

REFERENCES

Aas, C., Ladkin, A. & Fletcher, J. 2015. Stakeholder collaboration and heritage management. *Annals of Tourism Research* 32(1): 28–48.

Abungu, O.K. 2019. *Personal Comment during the SANParks Capacity Building Workshop*. Kruger National Park, South Africa.

Abungu, O.K. 2016. Introduction. In Abungu, O.K., Ndukuyakhe, N., Thiombiano, E.F.I., Zulu, J. & Varrissou, S. (eds.). *Traditional Management Systems at Heritage Sites in Africa*. Midrand, South Africa: 8–10.

Abu-Zeid, M.A. & El-Shibini, F.Z. 2010. Egypt's high Aswan Dam. *International Journal of Water Resources Development* 13: 209–218. https://doi.org/10.1080/07900629749836

ADCOM. 2017. Document on World Heritage and Sustainable Development. ICOMOS General Assembly, New Delhi, India. *Unpublished*.

Affolder, N.A. 2009. The private life of environmental treaties. *Journal of International Law* 10: 510–525.

African Union. 2006. *Charter for African Cultural Renaissance*. Addis Ababa, Ethiopia: African Union.

African Union. 2013. *African Union Agenda 2063. "A Shared Strategic Framework for Inclusive Growth and Sustainable Development". Background Note*. 21st Ordinary Session of the Assembly of Heads of State and Government of the African Union. Addis Ababa, Ethiopia: African Union. *Unpublished*.

African Union. 2015. *Agenda 2063: The Africa We Want*. Addis Ababa: Ethiopia: African Union.

AKTC. 2005. *"Al-Azhar Park, Cairo and Revitalisation of Darb Al-Ahmar Project Brief"*. www.akdn.org/hcp/egypt.asp [Accessed 10 November 2014].

Albert, M.T. 2012. Perspectives of World Heritage: Towards future-oriented strategies with the five 'Cs'. In Albert, M.T., Richon, M., Viñals, M.J. & Witcomb, A. (eds.). *Community Development through World Heritage, World Heritage Papers 31*. Paris, France: UNESCO: 32–38.

Albert, M.T., Richon, M., Vinals, M.J. & Witcomb, A. (eds.). 2012. *Community Development Through World Heritage, World Heritage Papers* 31. Paris, France: UNESCO

Alberts, H.C. & Hazen, H.D. 2010. Maintaining authenticity and integrity at cultural World Heritage sites. *The Geographical Review* 100(1): 56–73.

Altheide, D.L. 1987. Reflections: Ethnographic content analysis. *Qualitative Sociology* 10(1): 65–77. https://doi.org/10.1007/BF00988269

Andrias, L.M., Lekgowa, D., Satau, B. & Satau, G.W. 2017. *World Heritage Sites in Botswana-An Indigenous Perspective*. Indigenous Peoples of Africa Co-ordinating Committee (IPACC). Maun, Botswana: Open Society Initiative for Southern Africa.

Archives of the State Secretary, Incoming documents for 1856, 11: 402. Pretoria: National Archives of South Africa.

Arnstein, S.R. 1969. A ladder of citizen participation. *Journal of the American Institute of Planners* 35(4): 216–224. https://doi.org/10.1080/01944366908977225

Artur, D. 2003. *Projecto Transfronteiriço De Chimanimeni - Papel Da Cultura Nativa Na Preservação Do Meio Ambiente: o Caso De Nhakayedo*. Chimoio: ARPAC.

Apley, A.G., 2000. The cultural management of upward mobility in Botswana. New York University. *A dissertation submitted in partial fulfillment of the requirements for the degree of Doctor of Philosophy*. Department of Anthropology, New York University, USA.

Askew, M. 2010. The magic list of global status: UNESCO, world heritage and the agendas of States. In Labadi, S. & Long, C. (eds.). *Heritage and Globalization*. London: Routledge: 19–94.

Ayittey, G.B. 1992. *Africa Betrayed*. New York: St. Martin's Press.

Baker, W.G. 2003. *Arabs, Islam, and the Middle East*. Dallas: Brown Books Publishing Group.

Barton, H. 2000. *Sustainable Communities. The Potential for Eco-Neighbourhoods*. London: Earthscan.

Basu, P. & Modest, W. (eds.). 2015. *Museums, Heritage, and International Development (Routledge Studies in Culture and Development)*. New York and London: Routledge.

Beach, D.N. 1974. Ndebele raiders and Shona power. *Journal of African History* 15(4): 633–651.

Beach, D.N. 1980. *The Shona and Zimbabwe 900–1850*. Gweru: Mambo Press.

Beel, D., Jones, M. & Jones, I.R. 2016. Regulation, governance and agglomeration: Making links in city-region research. *Regional Studies, Regional Science* 3(1): 509–530.

Bellman, R.E. 1961. *Adaptive Control Processes: A Guided Tour*. Princeton, New Jersey: Princeton University Press.

Berkes, F. & Davidson-Hunt, I.J. 2006. Biodiversity, traditional management systems, and cultural landscapes: Examples from the boreal forest of Canada. *International Social Science Journal* 187: 35–47.

Berkes, F., Colding, J. & Folke, C. 2000. Rediscovery of traditional ecological knowledge as adaptive management. *Ecological Applications* 10(5): 1251–1262.

Berman, S.L., Wicks, A.C., Kotha, S. & Jones, T.M. 1999. Does stakeholder orientation matter? The relationship between stakeholder management models and firm financial performance. *The Academy of Management Journal* 42(5), Special Research Forum on Stakeholders, Social Responsibility, and Performance: 488–506. http://www.jstor.org/stable/256972 [Accessed 20 October 2015].

Berry, M. & Cadman, M. 2007. *Dongola to Mapungubwe: The 80-Year Battle to Conserve the Limpopo Valley*. Swartwater: Mmabolela Press: 46.

Bertacchini, E., Liuzza, C., Meskell, L. and Saccone, D. 2016. The politicization of UNESCO World Heritage decision making. Public Choice, 167(1): 95–129.

Beuster, C.L. 1879. Das volk der vavenda. *Zeitschrift Fiir Geschichte Erdkunde*. Berlin. 14: 236–24A.

Bhatasara, S., Nyamwanza, A.M. & Kujinga, K. 2013. Transfrontier Parks and development in southern Africa: The case of the Great Limpopo Transfrontier Park. *Development Southern Africa* 30(4–5): 629–639. https://doi.org/10.1080/0376835X.2013.837377

Bhatnagar, B. & Aubrey, C. 1992. Participatory development and the World Bank: Potential directions for change. *World Bank Discussion Paper.* Washington, DC: The World Bank.

Blacik, L. 2007. A critical assessment of the impact of World Heritage sites designation in sub-Saharan Africa, Carleton College, Sociology and Anthropology, Switzerland. *Unpublished paper.*

Black, T.R. 1999. *Doing Quantitative Research in the Social Sciences: An Integrated Approach to Research Design, Measurement and Statistics.* London: Sage Publications.

Blanchard, L. & Trotter, R.T. 2001. What is community? An evidence-based definition for participatory public health. *American Journal of Public Health* 91: 1929–1938.

Blattberg, C. 2004. Welfare: Towards the patriotic corporation. *From Pluralist to Patriotic Politics: Putting Practice First.* New York: Oxford University Press: 172–184.

Bloemers, T. 2010. The cultural landscape and heritage paradox: Protection and development of the Dutch archaeological-historical landscape and its European dimension. In Bloemers, T. (ed.). *The Cultural Landscape & Heritage Paradox: Protection and Development of the Dutch Archaeological-Historical Landscape and Its European Dimension.* Amsterdam: Amsterdam Press: 3–16.

Boatright, J.R. 1994. Fiduciary duties and the shareholder-management relation: Or, what's so special about shareholders? *Business Ethics Quarterly* 4(4): 393–407. https://doi.org/10.2307/3857339

Boccardi, G. 2015. From mitigation to adaptation. In Albert, M. (ed.). *Perceptions of Sustainability in the Heritage Studies.* Berlin: de Gruyter: 87–97.

Bokova, I. 2013. *Keynote Address at the United Nations General Assembly. Thematic Debate on Culture and Development.* 67th Session of the United Nations General Assembly, United Nations Headquarters, New York, Paris, France: UNESCO.

Bonner, P. & Carruthers, E.J. 2003. The recent history of the Mapungubwe area. Mapungubwe Cultural Heritage Resources Survey. *Unpublished report commissioned by the Department of Environmental Affairs and Tourism*; funded by the Norwegian Government.

Borona, G.K. 2015. *Comments during a seminar discussion on local communities and conservation.* University of British Columbia, Canada. *Unpublished.*

Bosse, D.A., Phillips, R.A. & Harrison, J.S. 2009. Stakeholders, reciprocity, and firm performance. *Strategic Management Journal* 30(4): 447–456.

Bouwer, C. 2018. Inclusive Growth for Africa. *Sawubona.* April 2018: 75.

Bowen, G.A. 2009. Document analysis as a qualitative research method. *Qualitative Research Journal* 9(2): 27–40. https://doi.org/10.3316/QRJ0902027

Boyte, H. & Kari, N. 1996. *Building America: The Democratic Promise of Public Work.* Philadelphia: Temple University Press.

Brace, C. 2003. Envisioning England: The visual in countryside writing in the 1930s and 1940s. *Landscape Research* 28: 383–404.

Brida, G.J., Chirikure, L. & Faccioli, M. 2011. Residents' perception and attitudes towards tourism impacts: A case study of the small rural community of Folgaria (Trentino – Italy). *Benchmarking: An International Journal* 18(3): 359–385.

Briedenhann, J. & Wickens, E. 2004. Tourism routes as a tool for the economic development of rural areas: Vibrant hope or impossible dream? *Tourism Management* 25(1): 71–79.

Brown, J. & Mitchell, B. 2000. The stewardship approach and its relevance for protected landscapes. *The George Wright Forum* 17: 70–79.

Brown, J. & Hay-Edie, T. 2014. *Engaging Local Communities in Stewardship of World Heritage.* Paris: UNESCO World Heritage Centre. World Heritage Papers 40.

Brumann, C. 2009. Outside the glass case: The social life of urban heritage in Kyoto. *American Ethnologist* 36(2): 276–299. https://doi.org/10.1111/j.1548-1425.2009-01135.x

Brumann, C. & Berliner, D. (eds.). 2016. *World Heritage on the Ground: Ethnographic Perspectives.* New York and Oxford: Berghahn Publishers.

Brundtland Report. 1987. *Our Common Future: Report of the World Commission on Environment and Development (WCED).* Oxford: Oxford University Press.

Bryson, J.M. 2004. What to do when stakeholders matter: Stakeholder identification and analysis techniques. *Public Management Review* 6(1): 21–53. https://doi.org/10.1080/14719030410001675722.

Buckley, P.J., Clegg, L.J., Cross, A., Liu, X., Voss, H. & Zheng, P. 2010. The determinants of Chinese outward foreign direct investment. In Buckley. P (ed.). *Foreign Direct Investment, China and the World Economy.* London: Palgrave Macmillan.

Bujra, A. 2002. *Africa: From the OAU to the African Union.* Tripoli, Libya: African Centre for Research and Training in Social Development: 117–124.

Butterfield, K.D., Reed, R. & Lemak, D.J. 2004. An inductive model of collaboration from the stakeholder's perspective. *Business & Society* 43(2): 162–195. https://doi.org/10.1177/0007650304265956

Byrne, D. 1991. Western hegemony in archaeological heritage management. *History and Anthropology* 5(2): 269–276.

BNFA. 2020. *Barotse National Freedom Alliance-Namungelele.* https://bnfa.info/48-2/ [Accessed 18 September 2020].

Calabrese, J.A. 2000. Metals, ideology and power: The manufacture and control of materialised ideology in the area of the Limpopo-Shashe confluence. *South African Archaeological Society Goodwin Series* 8: 91–99.

Calabrese, J.A. 2005. Ethnicity, class and polity: The emergence of social and political complexity in the Shashe-Limpopo valley of southern Africa, AD 900–1300. *Unpublished PhD thesis.* University of Witwatersrand.

Calabrese, J.A. 2007. The emergence of social and political complexity in the Shashi-Limpopo valley of Southern Africa, AD 900 to 1300. *Ethnicity, Class, and Polity. BAR International Series 1617.* Cambridge Monographs in African Archaeology 69. Oxford: Archaeopress.

CALS. 2016. *The Mapungubwe Story: A Campaign for Change.* Johannesburg: South Africa: Centre for Applied Legal Studies, University of Witwatersrand.

Cameron, C. 2014. Entre chien et loup: World Heritage Cultural Landscapes on the Fortieth Anniversary of the World Heritage Convention. In Conserving Cultural Landscapes: Routledge: 79–92.

Carruthers, J. 2006. Mapungubwe: An historical and contemporary analysis of a World Heritage cultural landscape. *Koedoe* 49(1): 1–13.

Carruthers, J. 2012. *Evolution and Innovation in Wildlife Conservation.* New York: Routledge.

Cassel, S.H. & Pashkevich, A. 2013. World Heritage and tourism innovation: Institutional frameworks and local adaptation. *European Planning Studies*, April 8: 1–16.

Charron, D.C. 2007. Stockholders and stakeholders: The battle for control of the corporation. *Cato Journal* 27(1): 1–22.

Chijoriga, M.M. 2012. New Areas for Enhancement of Revenue Collection and Broadening Tax Base of the Country. Paper Presented at the Consultative Forum Organised by the CAG, 2 March. nao.go.tz/?wpfb_dl=48 [Retrieved 15 August 2014].

Chimakonam, J.O., Agu, S.N. & Agbo, J.N. 2014. The nemesis of individualistic ontology in globalization and the practice of liberal democracy in post-colonial Africa. *The Journal of Pan African Studies* 7(4): 137–155.

Chirenje, L.I., Giliba, R.A. and Musamba, E.B., 2013. Local communities' participation in decision-making processes through planning and budgeting in African countries. *Chinese journal of population resources and environment, 11*(1): 10–16.

Chirikure, S., Mukwende, T. and Taruvinga, P., 2016. Post-colonial heritage conservation in Africa: Perspectives from drystone wall restorations at Khami World Heritage site, Zimbabwe. *International Journal of Heritage Studies, 22*(2):65–178.

Chirikure, S. & Taruvinga, P. 2017. Report on the Inaugural Lecture Series on the Mapungubwe Kingdom Whose? Revisiting the Mapungubwe Kingdom, 31 August–2 September 2017, Mapungubwe Cultural Landscape World Heritage Site, Limpopo, South Africa. *Unpublished report.*

Chirikure, S., Manyanga, M., Pikirayi, I. and Pollard, M., 2013. New pathways of socio-political complexity in Southern Africa. *African Archaeological Review, 30*(4): 339–366.

Chirikure, S. 2013. Heritage conservation in Africa: The good, the bad, and the challenges. *South African Journal of Science* 109(1/2), Art #a003: 3. https://doi.org/10.1590/sajs.2013/a003

Chirikure, S. 2014. Power imbalance and unequal benefit at UNESCO World Heritage site. *Africa Review of Books/Revue Africaine Des Livres* 10(1).

Chirikure, S. & Pwiti, G. 2008. Community involvement in archaeology and cultural heritage management. *Current Anthropology* 49(3): 467–485.

Chirikure, S., Manyanga, M., Ndoro, W. & Pwiti, G. 2010. Unfulfilled promises? Heritage management and community participation at some of Africa's cultural heritage sites. *International Journal of Heritage Studies* 16(1–2): 30–44. https://doi.org/10.1080/13527250903441739

Chirikure, S., Manyanga, M., Pollard, A.M., Bandama, F., Mahachi, G. & Pikirayi, I. 2014. Zimbabwe culture before Mapungubwe: New evidence from Mapela Hill, South-Western Zimbabwe. *PLoS ONE* 9(10): e111224.https://doi.org/10.1371/journal.pone.0111224

Chirikure, S., Mukwende, T. & Taruvinga, P. 2015. Post-colonial heritage conservation in Africa: Perspectives from drystone wall restorations at Khami World Heritage Site, Zimbabwe. *International Journal of Heritage Studies* 22(2): 165–178. https://doi.org/10.1080/13527258.2015.1103300.

Chirikure, S., Sinamai, A., Goagoses, E. & Ndoro, W. 2010. Maritime archaeology and trans-oceanic trade: A case study of the Oranjemund shipwreck cargo, Namibia. *Journal of Maritime* 5: 37–55. https://doi.org/10.1007/s11457-010-9059-9

CIC Energy. 2012. *Mmamabula Coalfield.* https://en.wikipedia.org/wiki/Mmamabula [Retrieved 21 September 2012]

Claudi, I.B. 2011. The New Kids on the Block: BRICS in the World Heritage Committee. Department of Political Science, University of Oslo, Norway. *Unpublished MA thesis.*

Cock, J. & Fig, D. 2000. From colonial to community-based conservation: Environmental justice and the National Parks of South Africa. *Society in Transition* 31: 22–35. https://doi.org/10.1080/21528586.2000.10419008

Cole, S. 2006. Information and empowerment: The keys to achieving sustainable tourism. *Journal of Sustainable Tourism* 14(6): 629–644.

Cooper, C., Fletcher, J., Fyall, A., Gilbert, D. & Wanhill, S. 2008. *Tourism: Principles and Practice.* Harlow: Prentice Hall Financial Times.

Conrad, D. 2005. *Empires of Medieval West Africa: Ghana, Mali, and Songhay.* New York: Facts on File, Inc.

Cooper, F. 1994. Conflict and connection: Rethinking colonial African history. *The American Historical Review* 99(5): 1516–1545. https://doi.org/10.2307/2168387.

Council of Europe. 2005. *Council of Europe Framework 1972 World Heritage Convention on the Value of Cultural Heritage for Society*. Strasbourg: Council of Europe.

Craig, A.M., Graham, W.A., Kagan, D., Ozment, S. & Turner, F.M. 2012. *The Heritage of World Civilizations, Combined Volume*, Brief 5th Edition. Peachpit Press. Berkeley, California, United States.

Creamer, M. 2013. Anglo top brass give De Beers Venetia diamond mine a new mid-century horizon. *Mining Weekly*, 1 November 2013. http://www.miningweekly.com/article/anglo-top-brass-give-de-beers-venetia-diamond-mine-a-new-mid-century-horizon-2013-11-01/rep_id: 3650 [Accessed 12 June 2018, Cape Town, South Africa].

Creswell, J.W. & Plano Clark, V.L. 2007. *Designing and Conducting Mixed Methods Research*. Thousand Oaks, CA: Sage.

DAC & AWHF. 2014. *Africa celebrating 40th Anniversary of the 1972 World Heritage Convention. Conference on Living with World Heritage in Africa*, 26–29 September 2012, African World Heritage Fund and Department of Arts and Culture.

DAC. 2012. *Mzansi Golden Economy*. Department of Arts and Culture, South Africa.

Daly, H.E., 1992. Allocation, distribution, and scale: towards an economics that is efficient, just, and sustainable. *Ecological economics*, *6*(3):185–193.

Dalkey, N.C. 1969. The Delphi Method: An Experimental Study of Group Opinion. *Futures* 1(5): 408–426.

Dalkey, N.C., Rourke, D.L., Lewis, R. & Snyder, D. 1972. *Studies in the Quality of Life: Delphi and Decision Making*. US Rand Corporation, Lexington Books. Lanham, MD United States.

Dalkey, N. & Helmer, O. 1963. An experimental application of the Delphi Method to the use of experts (online publication). *Management Science* 9(3). https://doi.org/10.1287/mnsc.9.3.458

Daly, H.E. & Cobb, J.B. 1990. *For the Common Good: Redirecting the Economy Towards Community, the Environment and a Sustainable Future*. London: Green Print.

Davidson-Hunt, I.J., Asselin, H., Berkes, F., Brown, K., Idrobo, C.J., Jones, M. & Rodriguez, M. 2016. The use of biodiversity for responding to globalised change. In Davidson-Hunt, I.J., Suich, H., Meijer, S.S. & Olsen, N. (eds.). *People in Nature: Valuing the Diversity of Interrelationships between People and Nature*. Gland, Switzerland: IUC: 19–31.

DEA 2008. *Conservation for the People with People*. Pretoria, South Africa: Department of Environmental Affairs.

DEA 2010. *Minutes of the SAWHCC Meeting*. Pretoria, South Africa: Department of Environmental Affairs. *Unpublished minutes*.

DEA. 2011a. Greater Mapungubwe Transfrontier Conservation Area Integrated Development Plan. Presentation made to Stakeholders in MCLWHS in September 2011. Pretoria, South Africa. *Unpublished Plan*.

DEA. 2011b. Memorandum to the Minister of Arts and Culture on the MCLWHS. Department of Environmental Affairs, Pretoria, South Africa. *Unpublished memorandum*.

DEA. 2013a. Audit Report on Implementation of South African 1972 World Heritage Convention Committee Decisions. Department of Environmental Affairs, Pretoria, South Africa. *Unpublished report*.

DEA. 2013b. *Minutes of the Greater Mapungubwe Transfrontier Conservation Area Tri-Lateral Technical Committee Meeting Held in Pretoria*. Pretoria, South Africa: Department of Environmental Affairs. *Unpublished minutes*.

DEA. 2014a. Report on the South African World Heritage Buffer Zone Expert Workshop, Maloti - Drakensberg Park World Heritage Site (South Africa). Department of Environmental Affairs, Pretoria, South Africa. *Unpublished report.*

DEA. 2014b. *Conservation for the People with the People.* Pretoria, South Africa: Department of Environmental Affairs.

DEA. 2015. Mapungubwe Nomination. Presentation made to stakeholders by DEA. February 2015. Pretoria, South Africa.

DEA. 2018. South African World Heritage Site Managers Forum (SMF). Department of Environmental Affairs, Pretoria, South Africa. *Unpublished Terms of Reference.*

Deacon, J. 1993. *Management Guidelines for Rock Art Sites. In Two Wilderness Areas in the Western Cape.* Project Funded by the Department of Environmental Affairs and Tourism. Pretoria, South Africa, Department of Environmental Affairs and Tourism.

Deacon, J. 2005. Management strategies for African rock art. In Deacon, J. (ed.). *The Future of Africa's Past: Proceedings of the 2004 TARA Rock Art Conference Nairobi.* Nairobi: Trust for African Rock Art: 90–96.

Deacon, J. 2006. Rock art conservation and tourism. *Journal of Archaeological Method and Theory* 13(4): 379–399.

Deacon, J. 2011. Rock art and hunter-gatherer communities in relation to World Heritage. In Sanz, N. & Keenan, E. (eds.). *Human Evolution: Adaptations, Dispersals and Social Developments (HEADS): World Heritage Thematic Programme.* Paris: UNESCO World Heritage Centre: 145–171.

Deacon, J. 2015. Great expectations: Archaeological sites and the National Heritage Resources Act (Act 25 of 1999). *South African Archaeological Bulletin* 70(202): 220–223.

DEAT. 2002a. *Mapungubwe World Heritage Site Nomination.* Department of Environmental Affairs and Tourism. Pretoria, South Africa: Government Printer.

DEAT. 2002b. *National Responsible Tourism Guidelines.* Department of Environmental Affairs and Tourism, Pretoria, South Africa.

DEAT. 2003. Mapungubwe Nomination Dossier. Department of Environment and Tourism, Pretoria, South Africa. *Unpublished.*

Deegan, N. 2012. The local-global nexus in the politics of World Heritage: Space for community development. In Albert, M.T., Richon, M., Vinals, M.J. & Witcomb, A. (eds.). *Community Development Through World Heritage, World Heritage Papers* 31. Paris, France:UNESCO: 77–83.

Delius, P. & Cope, R. 2007. Hard-fought frontiers: 1845-1883. In Delius, P. (eds). *Mpumalanga History and Heritage.* University of KwaZulu-Natal Press, KwaZulu Natal, South Africa: 137–199.

Dena, G. 1992. Public sphere and private life: Toward a synthesis of current historiographical approaches to the old regime. *History and Theory* 31(1): 1–20.

Desai, M.D. & Lee-Thorp, J.N. 2000. Indigenous gold mining in Southern Africa: A review. *South African Archaeological Society Goodwin Series* 8: 91–99.

Di Giovine, M.A. 2008. *The Heritage-Scape: UNESCO, World Heritage, and Tourism.* New York: Lexington Books.

Ditchkoff, S.S., Saalfeld, S.T. & Gibson, C.J. 2006. Animal behaviour in urban ecosystems: Modifications due to human-induced stress. *Urban Ecosystems* 9: 5–12.

Donaldson, T. & Preston, L. 1995. The stakeholder theory of the corporation: Concepts, evidence and implications. *The Academy of Management Review* 20(1): 65–69.

Donnachie, I. 2010. World Heritage. In Harrison, R. (ed.). *Understanding Global Heritage. Understanding the Politics of Heritage.* Manchester: United Press: 115–153.

Du Plessis, C. 2000. Cities and sustainability: Sustaining our cultural heritage. In Brandon, P., Lombardi, P. & Perera, S. (eds.). *Cities and Sustainability: Sustaining Our Cultural Heritage.* Conference Proceedings, Kandalama: Sri Lanka.

Duckworth, H.A. & Moore, R.A. 2010. *Social Responsibility: Failure Mode Effects and Analysis* CRC Press (Taylor and Francis Group), London, New York.

Duval, M., Smith, B., Hœrlé, S., Bovet, L., Khumalo, N. & Lwazi Ehengu, L. 2019. Towards a holistic approach to heritage values: A multidisciplinary and cosmopolitan approach. *International Journal of Heritage Studies* 25(12): 1279–1301. https://doi.org/10.1080/13527258.2019.1590449

Eagles, P.F.J., McCooL, S.F. & Haynes, A.P. 2002. *Sustainable Tourism in Protected Areas: Guiding for Planning and Management.* Gland, Switzerland: IUCN Programme on Protected Areas: 120–122.

Eastwood, E.B. 2006. Animals behaving like people: San Rock paintings of Kudu in the Central Limpopo Basin, Southern Africa. *South African Archaeological Bulletin* 61(183): 26–39.

Eastwood, E.B. & Blundell, G. 1999. Rediscovering the rock art of the Limpopo–Shashi confluence area, Southern Africa. *Southern African Field Archaeology* 8: 17–27.

Eastwood, E.B. 2003. A cross-cultural motif in San, Khoekhoe and Northern Sotho rock paintings of the Central Limpopo Basin, Southern Africa. *South African Archaeological Bulletin* 58: 14–26.

Eboreime, J. 2009. Challenges of heritage management in Africa. In *Cultural Heritage and the Law: Protecting Immovable Heritage in English-Speaking Countries of Sub-Saharan Africa*: 1–5.

Eboreime, O.J. 2008. Understanding cultural sites and the preparation of Nomination Dossier for World Heritage listing. In Ali Biu, Z. (ed.). *Conservation and Management of Heritage Sites in Nigeria.* Abuja: UNESCO: 22–27.

Eden, C. & Ackermann, F. 1998. *Making Strategy: The Journey of Strategic Management.* London: Sage Publications.

Edroma, E.L. 2004. Linking universal and local values for the sustainable management of World Heritage sites. In de Merode, E., Smeets, R. & Westrik, C. (eds.). *Linking Universal and Local Values: Managing a Sustainable Future for World Heritage.* Paris: UNESCO World Heritage Centre: 36.

Enders, J.C. & Remig, M. 2014. *Theories of Sustainable Development: An Introduction.* New York: Routledge.

Englebert, P. 2009. *Africa, Unity, Sovereignty and Sorrow.* Boulder, Colorado: Lynne Rienner Publishers, Inc.

Esposito, A. & Gaulis, I. 2010. The cultural heritage of Asia and Europe: Global challenges and local initiatives. *Unpublished Background Document to the Roundtable in Preparation for the Fourth ASEM Culture Ministers' Meeting 2010.* Leiden: International Institute of Asian Studies.

Esterhuysen, A. & Smith, J. 2007. Stories in stone. In Delius, P. (ed.). *Mpumalanga History and Heritage.* University of KwaZulu-Natal Press, KwaZulu Natal, South Africa: 41–68.

Etzioni, A. 1998. A communitarian note on stakeholder theory. *Business Ethics Quarterly* 8(4): 679–691.

European Union. 2010. *Toledo Declaration on Urban Development.* Brussels: European Union.

Evans, P. 1995. *Embedded Autonomy: States and Industrial Transformation.* Princeton, NJ: Princeton University Press.

Eyong, C.T. & Foy, I.I. 2006. Towards alternative strategies for sustainable development in Africa. *International Journal of Sustainable Development* 1(2): 133–156.

Falola, T. & Aderinto, S. 2010. *Nigeria, Nationalism, and Writing History.* Rochester: University of Rochester Press.

Ferriday, J. & Muir-Cochrane, E. 2006. Demonstrating rigor using thematic analysis: A hybrid approach of inductive and deductive coding and their development. *International Journal of Qualitative Methods* 5(1): 80–92.

Fleminger, D. 2006. *World Heritage Sites of South Africa: Mapungubwe Cultural Landscape.* Johannesburg: 30° South Publishers: 72.

Flinn, A. 2007. Community histories, community archives: Some opportunities and challenges. *Journal of the Society of Archivists* 28(2): 151–176.

Flinn, A. 2008. Migrations, disputed heritages and multicultured identities: Archives in a post-colonial society. *Archives and Manuscripts* 36(2): 54–75.

Flinn, A. 2011. Archival activism: Independent and community-led archives, radical public history and the heritage professions. *InterActions: UCLA Journal of Education and Information Studies* 7 (2).

Flint, L. 2007. Contradictions and challenges in representing the past: The Kuomboka Festival of Western Zambia. *Journal of Southern African Studies* 32(4): 701–717. https://doi.org/10.1080/03057070600995483

Freeman, R. E. 1984, *Strategic management: A stakeholder approach.* Pittman Publishing, Marshfield, MA

Forssman, T. 2014. The spaces between places: A landscape study of foragers on the Greater Mapungubwe landscape, Southern Africa. *Azania: Archaeological Research in Africa* 49(2): 282. https://doi.org/10.1080/0067270X.2014.891874

Fouche, L. 1937. *Mapungubwe: Ancient Bantu Civilization on the Limpopo.* Cambridge: Cambridge University Press.

Fowler, P. 2002. World Heritage cultural landscapes, 1992–2002: A review and prospect. In UNESCO (ed.). *Cultural Landscapes: The Challenges of Conservation, World Heritage 2002 Shared Legacy, Common Responsibility Associated Workshops; World Heritage Papers.* Ferrara, Italy: 16–33.

Fowler, P. 2003. *World Heritage Cultural Landscapes 1992–2002.* Paris: UNESCO. *World Heritage Papers 6.*

Fredheim, H.L. & Manal Khalaf, M. 2016. The significance of values: Heritage value typologies re-examined. *International Journal of Heritage Studies* 22(6): 466–481.

Freeman, R.E. 1984. *Strategic Management: A Stakeholder Approach.* Boston: Pitman.

Freeman, R.E. & Harrison, J.S. (eds.). 2001. *The Blackwell Handbook of Strategic Management.* Oxford: Blackwell.

Freeman, R.E. & McVea, J. 2001. A stakeholder approach to strategic management. In Hitt, M.A., Freeman, R.E. & Harrison, J.S. (eds.). *The Blackwell Handbook of Strategic Management.* Oxford: Blackwell: 189–207.

Fries, J. & Heemans, J. 1992. Natural forest management in semi-arid Africa: Status and research needs. *Unasulva* 43: 9–15.

Gade, C. 2011. The historical development of the written discourses on ubuntu. *South African Journal of Philosophy* 30(3): 303–329.

Galla, A. (ed.). 2012. *World Heritage: Benefits Beyond Borders.* Cambridge: Cambridge University Press.

Galloway, A. 1959. *The Skeletal Remains of Bambandyanalo.* Johannesburg: Witwatersrand University Press.

Gardner, G.A. 1995. Mapungubwe 1935-1940. *The South African Archaeological Bulletin* 10(39): 73–77.

Gardner, G.A., 1963. *Mapungubwe. v. 2, Report on excavations at Mapungubwe and Bambandyanalo in Northern Transvaal from 1935 to 1940.* Pretoria: Van Schaik, South Africa.

Getty. 2007. Manual and notes for the South African rock art initiative: A program of the Getty Conservation Institute in collaboration with the Southern African Rock Art Project (SARAP), SANParks and SAHRA. Rock art interpretation and presentation workshop, MCLWHS, Limpopo. *Unpublished manual.*

Giddings, B., Hopwood, B. & O'Brien, G. 2002. Environment, economy and society: Fitting them together into sustainable development. *Sustainable Development* 10(4): 187–234. https://doi.org/10.1002/sd.199

Glen, B. 2009. Document analysis as a qualitative research. *Qualitative Research Journal* 9(2): 27–40.

GMTFCA TTC. 2010: Greater Mapungubwe Transfrontier Conservation Area Integrated Development Plan, 1st Edition. Department of Environmental Affairs, Pretoria, South Africa.

Goh, D.P.S. 2008. Postcolonial disorientations: Colonial ethnography and the vectors of the Philippine nation in the imperial frontier. *Postcolonial Studies* 11(3): 259–276.

Goodman, J. 2005. Fictional history. *Rethinking History* 9(2–3): 237–253 https://doi.org/10.1080/1362520500149178

Goodpaster, K. 1991. Business ethics and stakeholder analysis. *Business Ethics Quarterly* 1(1): 53–73. https://doi.org/10.2307/3857592

Gray, S.J. 1988. Towards a theory of cultural influence on the development of accounting systems internationally. *Abacus* 24(1): 1–15.

Grant, R., Sherwill, T., Rogers, K., Biggs, H., Freitag, S , Hofmeyr, M., *et al.*, 2008, *A framework for developing and implementing management plans for South African National Parks*, South African National Parks, Pretoria, South Africa.

Grogan, K., Birch-Thomsen, T. & Lyimo, J. 2013. Transition of shifting cultivation and its impact on people's livelihoods in the Miombo Woodlands of Northern Zambia and South-Western Tanzania. *Human Ecology* 41: 77–92.

Harper, C., Marcus, R. and Moore, K., 2003. Enduring poverty and the conditions of childhood: lifecourse and intergenerational poverty transmissions. *World development*, *31*(3): 535–554.

Healy, P.M. & Palepu, K.G. 2001. Information asymmetry, corporate disclosure, and capital markets. A review of the disclosure empirical literature. *Journal of Accounting and Economics* 31(1–3): 405–440.

Halbert, C. 1993. How adaptive is adaptive management? Implementing adaptive management in Washington State and British Columbia. *Reviews in Fisheries Science* 1(3): 261–283.

Hall, R.N. & Neal, W.G. 1902. *The Ancient Ruins of Rhodesia (Monomatapae Imperium).* London: Methuen and Co.

Hall, S. & Smith, B. 2000. Empowering places: Rock shelters and ritual control in farmer–forager interactions in the Northern Province, South Africa. *South African Archaeological Society*, Goodwin Series 8: 30–46.

Hall, S., Miller, D., Anderson, M. & Boeyens, J. 2006. An exploratory study of copper and iron production at Marothodi, an early 19th-century Tswana town, Rustenburg district, South Africa. *Journal of African Archaeology* 4(1): 3–35.

Hanisch, E. 2008. Re-interpreting the origins of Dzata. In Swanepoel, N., Esterhuysen, A. & Bonner, P. (eds.). *Five Hundred Years Rediscovered: Southern African Precedents and Prospects.* Johannesburg: Witwatersrand University Press: 119–134.

Hanisch, E. 1981. Schroda: A Zhizo site in the Northern Transvaal. In Voigt, E. (ed.). *Guide to Archaeological Sites in the Northern and Eastern Transvaal.* Pretoria: South African Association of Archaeologists: 37–53.

Hanisch, E.O.M. 1980 An archaeological interpretation of certain Iron Age sites in the Limpopo/Shashi Valley. *MA dissertation*, University of Pretoria.

Hardi, P. & Zdan, T. 1997. *Assessing Sustainable Development.* Winnipeg: International Institute for Sustainable Development.

Harrison, R. 2010b. Introduction. In Harrison, R. (ed.). *Understanding Global Heritage. Understanding the Politics of Heritage.* Manchester: United Press: 1–4.

Harrison, R. 2010a. The politics of heritage. In Harrison, R., (ed.). *Understanding Global Heritage. Understanding the Politics of Heritage.* Manchester: United Press: 154–196.

Hart, M. 1996. *Characteristics of effective indicators.* http://www.subjectmatters.com/indicators.

Haughton, G. 1999. Environmental justice and the sustainable city. *Journal of Planning Education and Research* 18: 233–243.

Hellriegel, D., Jackson, S.E. & Slocum, J.W. 1999. *Management.* Cincinnati: South-Western.

Hendry, J. 2001. Missing the target: Normative stakeholder theory and the corporate governance debate. *Business Ethics Quarterly* 11(1): 159–176. https://doi.org/10.2307/3857875

Herman, U.P. 2013a. A visitor profile of Mapungubwe National Park and World Heritage site. *SAIMS 2013 Conference Proceedings.* Southern African Institute for Management Scientists: 935–945.

Herman, U.P. 2013b. Development of a tourism management framework for Mapungubwe National Park. *Unpublished PhD dissertation,* Potchefstroom Campus, North-West University.

Hill, C.W. & Jones, T.M. 1992. Stakeholder-agency theory. *Journal of Management Studies* 29: 131–154. https://doi.org/10.1111/j.1467-6486-992-tb00657.x

Hillman, A.J. & Michael, A.H. 1999. Corporate political strategy formulation: A model of approach, participation, and strategy decisions (online publication). *Academy of Management Review* 24(4). https://doi.org/10.5465/amr.1999.2553256

Hirsch, P. & Warren, C. 1998. *The Politics of Environment in Southeast Asia: Resources and Resistance.* New York: Routledge.

Hofmeyr, I. 1890. *Twintig Jaar in Die Zoutpansberg.* Cape Town: Rose.

Hoggart, R. 2011. *An Idea and Its Servants: UNESCO from Within.* Piscataway, NJ: Transaction.

Holden, P. & Mathabatha, S. 2007. The politics of resistance: 1948-1990. In Delius, P. (ed.). *Mpumalanga History and Heritage.* University of KwaZulu-Natal Press: 393–461.

Holmgren, K., Lee-Thorp, J.A., Cooper, G.R., Lundblad, K., Partridge, T.C., Scott, L., Sithaldeen, R., Talma, A.S. & Tyson, P.D. 2003. Persistent millennial-scale climatic variability over the last 25,000 years in southern Africa. *Quaternary Science Reviews* 22: 2311–2326.

Hopwood, B., Mellor, M. & O'Brien, G. 2005. Sustainable development: Mapping different approaches. *Sustainable Development* 13: 38–52.

Hoogendoorn, G., Kelso, C. & Sinthumule, I. 2019. Tourism in the Great Limpopo Transfrontier Park: A review. *African Journal of Hospitality, Tourism and Leisure* 8(5): 1–15. https://doi.org/10.4314/sajpem.v30i3.69578

Hsu, C. & Sandford, B. 2007. Minimizing non-response in the Delphi process: How to respond to non-response. *Practical Assessment, Research & Evaluation* 12(17): 62–78.

http://en-unesco.org/about-us/introducing-unesco [Accessed 23 August 2015].

http://localizingthesdgs.org/library/tools/monitoring-and-evaluation [Accessed 21 December 2017, Cape Town, South Africa].

http://sociology.emory.edu/faculty/globalization/theories03.html [Accessed 1 April 2015].

http://www.businessdictionary.com/definition/stakeholder.html#ixzz3E2tUM5Rd

https://ourworldindata.org/world-population-growth#population-size-vs-population-growth-rate

https://www.jstor.org/stable/24562597

Huffman, T.N. 1996. *Snakes & Crocodiles: Power and Symbolism in Ancient Zimbabwe.* Johannesburg: Witwatersrand University Press.

Huffman, T.N. 2000. Mapungubwe and the origins of the Zimbabwe culture. *Goodwin Series, Vol. 8, African Naissance: The Limpopo Valley 1000 Years Ago*. South African Archaeological Society: 14–29. http://www.jstor.org/stable/3858043 [Accessed 21 April 2018].

Huffman, T.N. 2004. Beyond data: The aim and practice of archaeology. *South African Archaeological Bulletin* 59: 66–69.

Huffman, T.N. 2005. *Mapungubwe: Ancient African Civilization on the Limpopo*. Johannesburg: Wits University Press.

Huffman, T.N. 2007. *Handbook to the Iron Age: The Archaeology of pre-Colonial Farming Societies in Southern Africa*. Scottsville: University of KwaZulu-Natal Press.

Huffman, T.N., 2012. Debating the 500 Year Initiative: history, anthropology or both?. *South African Archaeological Bulletin*, *67*(196): 231–243.

Huffman, T.N. 2009. Mapungubwe and Great Zimbabwe: The origins and spread of social complexity in Southern Africa. *Journal of Anthropological Archaeology* 28: 37–54.

Huffman, T.N. 2014. Social complexity in southern Africa. *African Archaeological Review* 32(1). https://doi.org/10.1007/s10437-014-9166-3

Hughes, L. & Harrison, R. 2010. Heritage, colonialism, and postcolonialism. In Harrison, R. (ed.). *Understanding the Politics of Heritage*. Manchester, NH: Manchester University Press: 234–269.

Hulme, D. & Shepherd, A. 2003. Conceptualizing chronic poverty. *World Development* 31(3): 403–423.

Hummels, H., 1998. Organizing ethics: A stakeholder debate. *Journal of Business Ethics*, *17*(13): 1403–1419.

ICMM. 2008. A sustained commitment to improved industry performance, sustainable development framework: 17. International Council on Metals and Mining (ICMM).

ICOMOS. 1965. *Venice Charter*. Paris: ICOMOS.

ICOMOS. 2011. *Guidance on Heritage Impact Assessments for Cultural World Heritage Properties*. Paris: ICOMOS.

ICOMOS. 2017. *ICOMOS Action Plan: Cultural Heritage and Localizing the SDGs*. Paris: ICOMOS.

Igboin, B.O. 2011. Colonialism and African cultural values. *African Journal of History and Culture* 3(6): 96–103. http://www.academicjournals.org/ajhc

Iggers, G.G. 2005. Historiography in the twentieth century. *History and Theory* 44: 469–476. https://doi.org/10.1111/j.1468-2303.2005.00337.x

IMF. 2018. *International Monetary Fund Annual Report*. https://www.imf.org/external/pubs/ft/ar/2018/eng/spotlight/making-the-system-work-better/index.htm

IOL. 2013. Land restitution stand-off over mining rights. www.IOL.co.za/business-report/companies/land-restitution-stand off-over-mining-rights-1588027

Istasse, M. 2016. Affects and senses in a World Heritage site: The sensual and emotional side of Fez Medina. In Brumann, C. & Berliner, D. (eds.). *World Heritage on the Ground: Ethnographic Perspectives*. Oxford: Berghahn: 37–59.

Itani, J. 2007. Effects of socio-economic changes on cultivation systems under customary land tenure in MBozi District, Southerzania. *African Study Monographs* 34: 57–74. http://dx.doi.org/10.14989/68483

Jamal, T.B. & Getz, D. 1995. Collaboration theory and community tourism planning. *Annals of Tourism Research* 22(1): 186–204.

Johannesburg Declaration. 2002. Report of the World Summit on Sustainable Development, 26 August–4 September 2002, Johannesburg, South Africa, A/CONF.199/20 and Key Outcomes of the Summit, UN.

Johannesburg Resolutions. 2013. Recommendations of the International Conference on "Living with World Heritage in Africa", 26–29 September 2012, Benon, South Africa. *Unpublished Resolutions.*

Johnson, G.S.K. 2002. *Exploring Corporate Strategy*, 6th edition. Harlow, England: Pearson Education.

Johnson, J.T., Howitt, R., Cajete, G., Berkes, F., Louis, R.P. and Kliskey, A., 2016. Weaving Indigenous and sustainability sciences to diversify our methods. *Sustainability Science*, 11(1), pp. 1–11.

Jokiletho, J. & Cameron, C. 2008. *The World Heritage List: What is OUV? Defining the Outstanding Universal Value of Cultural World Heritage Properties*. Berlin: Bassler Verlag.

Jokiletho, J. 2011. World Heritage: Observations on decisions related to cultural heritage. *Journal of Cultural Heritage Management and Sustainable Development* 1(1): 61–74.

Jones, O. & Cloke, P. 2002. *Tree Cultures: The Place of Trees and Trees in Their Place*. Oxford: Berg.

Jones, R. & Shaw, B. 2012. Thinking locally, acting globally? Stakeholder conflicts over UNESCO World Heritage inscription in Western Australia. *Journal of Heritage Tourism* 7(1): 83–96. https://doi.org/10.1080/1743873X.2011.632482

Jopela, A. 2006. "Custódia Tradicional do Património Arqueológico na Província de Manica: Experiência e práticas sobre as pinturas rupestres no Distrito de Manica, 1943-2005". *Dissertação De Licenciatura*. Maputo: UEM/FLCS.

Jopela, A. 2010. Traditional custodianship of rock art sites in Central Mozambique: A case study from Manica district. *Studies in African Past* 8: 161–177.

Jopela, A. 2011. Traditional custodianship: A useful framework for heritage management in southern Africa? *Conservation and Management of Archaeological Sites* 13(2–3): 103–122.

Jopela, A. 2016a. Conserving a World Heritage site in Mozambique. Entanglements between politics, poverty, development and governance of the island of Mozambique. In Labadi, S. & Logan, W. (eds.). *Urban Heritage, Development and Sustainability: International Frameworks, National and Local Governance*. London: Routledge: 37–56.

Jopela, A. 2016b. The traditional custodianship system in Southern Africa. In Abungu, O.K., Ndukuyakhe, N., Thiombiano, E.F.I., Zulu, J. and Varrissou, S. (eds.). *Traditional Management Systems at Heritage Sites in Africa*. Midrand, South Africa: 8–10.

Joy, C. 2016. UNESCO is what? World heritage, militant Islam and the search for a common humanity in Mali. In Brumann, C. & Berliner, D. (eds.). *World Heritage on the Ground: Ethnographic Perspectives*. Oxford: Berghahn: 60–77.

Judd, R.C. 1972. Use of Delphi methods in higher education. *Technological Forecasting and Social Change* 4(2): 173–186.

Kalman, H. 2014. *Heritage Planning*. New York: Routledge.

Kashef, A.I. 1981. Technical and ecological impacts of the High Aswan Dam. *Journal of Hydrology* 53(1–2): 73–84. https://doi.org/10.1016/0022-1694(81)90037-8

Kaufmann, D., Kraay, A. & Zoido-Lobatón, P. 1999. Governance matters. *World Bank Policy Research Department Working Paper* No. 2196.

Kauzya, J.M. and Balogun, J., 2005. Governance and public administration reforms and innovations in African countries: A focus on achievements, setbacks and future directions. https://citeseerx.ist.psu.edu/viewdoc/download?doi=10.1.1.602.4144&rep=rep1&type=pdf. *Downloaded in August 2019*

Kayizzi-Mugerwa, S. 1999. *The African Economy: Policy, Institutions and the Future*. Routledge.

Keith, K.P. 1994. *Team Decision-Making Techniques*. Irvine, California: Richard Chang Associates, Inc., Publications Division.

Keitumetse, S.O. 2007. Celebrating or marketing the indigenous? International rights organizations, national governments, and tourism creation. In Burns, P. and Novelli, M. (eds.). *Tourism and Politics: Global Frameworks and Local Realities. Series: Advances in Tourism Research*. Elsevier, London: 109–122.

Keitumetse, S.O. 2009. The eco-tourism of cultural heritage management (ECT-CHM): Linking heritage and 'environment' in the Okavango Delta Regions of Botswana. *International Journal of Heritage Studies* 15(2–3): 223–244. https://doi.org/10.1080/13527250902890811

Keitumetse, S.O. 2011. Sustainable development and cultural heritage management in Botswana: Towards sustainable communities. *Sustainable Development* 19: 49–59. https://doi.org/10.1002/sd.419

Kigongo, R. & Reid, R. 2007. Local communities, politics and the management of the Kasubi Tombs, Uganda. *World Archaeology* 39(3): 371–384. https://doi.org/10.1080/00438240701563094

King Code IV. 2016. http://www.iodsa.co.za/? pageAboutKingIV [Assessed May 2016, Cape Town, South Africa].

Koch, E. 1998. Ecofile. *Out There* (March).

Kretzmann, J.P. & McKnight, J.L. 1993. *Building Communities from the Inside Out: A Path Toward Finding and Mobilising a Community's Assets*. Chicago: ACTA Publications.

Kros, C. 2007. Mpumalanga's heritage. In Delius, P. (ed.). *Mpumalanga History and Heritage*. University of KwaZulu-Natal Press: 463–501.

Kuman, K., Le Baron, J.C. & Gibbon, R.J. 2004. Earlier Stone Age archaeology of the Vhembe-Dongola National Park (South Africa) and vicinity. *Quaternary International* 129: 23–31.

Labadi, S. 2013. *UNESCO, Cultural Heritage and Outstanding Universal Value*. Walnut Creek, CA: AltaMira Press.

Labadi, S. & Logan, W.S. 2015. *Urban Heritage, Development and Sustainability: International Frameworks, National and Local Governance*. London/New York: Routledge.

Labadi, S. 2018. Historical, theoretical and international considerations on culture, heritage and (sustainable) development. In Larsen, P.B. and Logan, W (eds.). *World Heritage and Sustainable Development*; Routledge: New York, NY, USA: 37–49.

Labadi, S., Giliberto, F., Taruvinga, P. and Jopela, A. (eds.), 2020. *World heritage and sustainable development in Africa: Implementing the 2015 policy*. African World Heritage Fund (AWHF), Midrand, South Africa: https://awhf.net/wp-content/uploads/2020/12/PUBLICATION-World-Heritage-and-Sustainable-Development-in-Africa-1.pdf

Landes, D.S. 1998. *The Knowledge Economy*. London: Butterworth–Heinemann.

Lane, J.E. & Ersson, S. 2016. *Culture and Politics*. London: Routledge.

Lane, M.B. 2003. Participation, decentralization, and civil society: Indigenous rights and democracy in environmental planning. *Journal of Planning Education and Research* 22(4): 360–373. https://doi.org/10.1177/0739456X03022004003

Larsen, P.B. & Wijesuriya, G. 2015. Nature-culture interlinkages in World Heritage: Bridging the gap. *World Heritage Review* (75): 4–15.

Lebacqz, K. 1986. *Six Theories of Justice*. Minneapolis: Augsburg Publishing House.

LEDET. 2012. Budget speech of Limpopo Department of Economic Development and Environment and Tourism. [Accessed 9 July 2016]. *Unpublished*.

Legaspi, P.E. 2005. *Overview of Governance Framework. A Working Draft on a Handbook on LGU-SPA Partnership*. UP NCPAG.

Legassick, M. & Rassool, C. 2000. *Skeletons in the Cupboard: South African Museums and the Trade in Human Remains, 1907–1917*. Kimberley: McGregor Museum.

Leick, B. & Lang, T. 2018. Re-thinking non-core regions: Planning strategies and practices beyond growth. *European Planning Studies* 26(2): 213–228.

Lestrade, G.P. 1927. Some notes on the ethnic history of the BaVenda and their Rhodesian affinities. *South African Journal of Science* 24: 486–495.

Levitt, T. 1983. The globalization of markets. *Harvard Business Review* 61: 92–102.

Lewis, C.W. 1991. *The Ethics Challenge in Public Service: A Problem-Solving Guide*. San Francisco: Jossey-Bass.

Lewis-Williams, J.D. 1994. *Contested Images: Diversity in Southern African Rock Art*. Johannesburg: Witwatersrand University Press.

Lewis-Williams, D. 2004. *Discovering Southern African Rock Art*. Cape Town: David Phillip.

Limpopo Provincial Treasury. 2017. *Limpopo Socio-Economic Review and Outlook: 2018/19*. South Africa: Limpopo Provincial Government.

Lindeman, C. 1981. *Priorities within the Health Care System: A Delphi Survey*. Kansas City, MO: American Nurses' Association.

Linstone, H. & Turloff, M. 1975. *The Delphi Method: Techniques and Applications*. New York: Sage Publications.

Logan, W. & Reeves, K. 2009. *Places of Pain and Shame: Dealing with Difficult Heritage*. London: Routledge.

Loubser, J.H.N. & Laurens, G. 1994. Depictions of domestic ungulates and shields: Hunter-gatherer and agro pastoralists in the Caledon River Valley area. In Dowson, T.A. & Lewis-Williams, J.D. (eds.). *Contested Images: Diversity in Southern African Rock Art*. Johannesburg: Witwatersrand University Press: 83–118.

Louw, L. & Venter, P. 2013. *Strategic Management: Developing Sustainability in Southern Africa*, 3rd Edition. Cape Town: Oxford University Press.

Lowenthal, D. 2005. *The Past is a Foreign Country*. London: Cambridge University Press.

Ludwig, B. 1997. Predicting the future: Have you considered using the Delphi methodology? *Journal of Extension* 35(5): 1–4. http://www.joe.org/joe/1997october/tt2.html

Lynam, T., De Jong, W., Sheil, D., Kusumanto, T. and Evans, K. 2007. A review of tools for incorporating community knowledge, preferences, and values into decision making in natural resources management. Ecology and society, 12(1).

Lynn, L.E. 2012. The many faces of governance: Adaptation? Transformation? Both? Neither? In Levi-Faur, D. (ed.). *The Oxford Handbook of Governance*. Oxford: Oxford University Press: 49–64.

MacEachern, S. 2010. Seeing like an oil company's CHM programme: Exxon and archaeology on the Chad Export Project. *Journal of Social Archaeology* 10(3): 347–366.

MCLWHS, 2018. Mapungubwe National Park Annual Report. SANparks, Pretoria, South Africa. *Online*.

MacEachern, S., Archer, D.J.W. & Garvin, R.D. 1989. Households and communities. *Proceedings of the Twenty-First Annual Conference of the Archaeological Association of the University of Calgary*. University of Calgary, Archaeological Association.

Maddison, A. 2001. *The World Economy: A Millennial Perspective*. Paris: OECD.

Maggs, T. 2000. African Naissance: An introduction. In Leslie, M. & Maggs, T. (eds.). *Goodwin Series, Vol. 8, African Naissance: The Limpopo Valley 1 000 Years Ago*. Cape Town: The South African Archaeological Society: 1–4.

Mahachi, G. & Kamuhangeri, E. 2008. Administrative arrangements for heritage resources management in sub-Saharan Africa. In Ndoro, W., Mumma, A. & Abungu, G. (eds.). *Cultural Heritage and the Law. Protecting Immovable Heritage in English Speaking Countries of Southern Africa. ICCROM Conservation Studies 8*. Rome: ICCROM: 43–51.

Mahachi, G. and Ndoro, W., 1997. The socio-political context of southern African Iron Age studies with special reference to Great Zimbabwe. *Caves, monuments and texts: Zimbabwean archaeology today:* Studies in Archaeology, 4. Uppsala: Uppsala University, pp. 89–108.

Makuvaza, S. 2016. Local Indigenous Communities, Traditional Conservation Practices and the Management of the Matobo Hills in South Western Zimbabwe. Archaeological Studies. Leiden University: Leiden University Press. *Unpublished thesis.*

Maluleke, L. & Steenkamp, C. 1998. *The Makuleke Experience. Workshop on Land Claims on Conservation Land: 3–4 September 1998.* Pretoria, South Africa: IUCN South Africa Country Office. *Unpublished.*

Manetsi, T. 2017. Concept document on tourism nodes. National Department of Tourism, Pretoria, South Africa. *Unpublished report.*

Manyanga, M. 2007. *Resilient Landscapes: Socio-Environmental Dynamics in the Shashi-Limpopo Basin, Southern Zimbabwe c.AD 800 to the Present.* Uppsala: Department of Archaeology and Ancient History, Uppsala University.

Mapungubwe Nomination Dossier. 1999. South African National Parks, Pretoria, South Africa.

Maradze, J. 2003. Back to the Old School? Revival of traditional management systems in Zimbabwe. Papers from the 14th General Assembly and Scientific Symposium of ICOMOS, October 2003, Victoria Falls. Available at: http://www.internationalicomos.org/victoriafalls2003 [Accessed February 2005]. *Unpublished.*

Marschall, S. 2005. Making money with memories: The fusion of heritage, tourism and identity formation in South Africa. *Historia* 50(1): 103–122.

Martino, J.P. 1983. *Technological Forecasting for Decision Making.* New York: Elsevier.

Matless, D. 1998. *Landscape and Englishness.* London: Reaktion.

Maton, K.I. 2008. Empowering community settings: Agents of individual development, community betterment, and positive social change. *American Journal of Psychology* 41(1): 4–21.

Matthias, R. 2017. Community involvement as a key strategy for heritage-based urban development, City of Regensburg. In Monika, G. & Matthias, R. (eds.). *Community Involvement in Heritage Management.* Joint Project of European Union/Council of Europe CUMUS and EUROCITIES: 66–70.

Marton-Lefe`vre J. 2008. Intervention at World Heritage Committee. Quebec International Union for Conservation of Nature, Gland, Switzerland.

Mazibuko, N.P. 2000. *Community Participation in Tourism Development at KwaNgcolosi.* KwaZulu-Natal: University of Zululand. *Unpublished MRT dissertation.*

McCartney, M. 2009. Living with dams: Managing the environmental impacts. *Water Policy* 11(S1): 121–139. https://doi.org/10.2166/wp.2009.108

McLain, R.J. & Lee, R.G. 1996. Adaptive management: Promises and pitfalls. *Environmental Management* 20(4): 437–448.

Meskell, L., 2009. *Cosmopolitan Archeologies.* Duke University Press. Durham and London

Meskell, L.M. 2011. *The Nature of Heritage: The New South Africa.* UK: John Wiley & Sons.

Meskell, L.M. 2011. From Paris to Pontdrift: UNESCO meetings, Mapungubwe and mining. *South African Archaeological Bulletin* 66(194): 149–156.

Meskell, L.M. 2012. The rush to inscribe: Reflections on the 35th session of the World Heritage Committee, UNESCO Paris, 2011. *Journal of Field Archaeology* 37(2): 145–151.

Meskell, L.M. 2013a. UNESCO and the fate of the World Heritage Indigenous Peoples Council of Experts (WHIPCOE). *International Journal of Cultural Property* 20(2): 155–174.

Meskell, L.M. 2013b. UNESCO's World Heritage conservation at 40: Challenging the economic and political order of international heritage conservation. *Current Anthropology* 54(4): 483–494.

Meskell, L.M. 2014. States of conservation: Protection, politics, and pacting within UNESCO's World Heritage Committee. *Anthropological Quarterly* 87(1): 217–243.

Meskell, L. & Brumann, C. 2015. UNESCO and new world orders. In Meskell, L. (ed.). *Global Heritage: A Reader,* 1st Edition. John Wiley & Sons, Inc. Oxford, UK: 22–42.

Michael, H. & Negri, A. 2000. *Empire.* Cambridge: Harvard University Press.

Michael, B. 2007. *Reconciliation: The Ubuntu Theology of Desmond Tutu.* Pilgrim Press, Cleveland, Ohio, USA.

Michalopoulos, S. & Papaioannou, E. 2012. Pre-Colonial Ethnic Institutions and Contemporary African Development. Working Paper 18224, National Bureau of Economic Research, Cambridge, MA. http://www.nber.org/papers/w18224 [Accessed 20 July 2018]

Miller, D., Desai, N. & Lee-Thorp, J. 2000. A review: Indigenous gold mining in Southern Africa. *Goodwin Series, Vol. 8, African Naissance: The Limpopo Valley 1000 Years Ago.* Cape Town: South African Archaeological Society: 91–99.

Miller, G. 2001. The development of indicators for sustainable tourism: Results of a Delphi survey of tourism researcher. *Tourism Management* 22: 351–362.

Milne, S. & Ateljevic, I. 2001. Tourism, economic development and the global-local nexus: Theory embracing complexity. *Tourism Geographies* 3(4): 369–393. https://doi.org/10.1080/146166800110070478

Mkapa, B.W., 2010. Leadership for growth, development, and poverty reduction: An African viewpoint and experience. *Leadership and growth* 19–79.

Mketeni, F. 2016. From the CEO's desk. *SANParks Time,* June 2016.

Moukala, E. 2019. *The State of the World Heritage Convention in Africa. A Position Paper Prepared for World Heritage Discussion on the Implementation of Sustainable Development at World Heritage Sites.* Paris: UNESCO. *Unpublished.*

Molatedi, L., Lekgowa, D., Satau, B. & Satau, G.W. 2017. *World Heritage Sites in Botswana: Indigenous Perspective.*: Indigenous Peoples of Africa Coordinating Committee (IPACC). Website: www.ipacc.org.za, Email: ipacc@iafrica.com. Open Society Initiative for Southern Africa (OSISA).

Mothulatshipi, S.M. 2008. Landscape archaeology of the Later Farming Communities of the Shashe-Limpopo Basin, eastern Botswana: Land use diversity and human behaviour. University of Edinburgh. *Unpublished PhD thesis.*

Morrel, W.E.I. 2007. Participatory planning in Indonesia, seeking a path to democracy. *Policy Studies* 28(1): 1–15.

Moyo, D. 2009. Why foreign aid is hurting Africa. *The Wall Street Journal.*

Müller, M. 1998. Cultural heritage protection: Legitimacy, property, and functionalism. *International Journal of Cultural Property* 7(2): 396–409.

Mumma, A. 2000. Legal aspects of cultural landscape protection in Africa. *The 1972 World Heritage Convention and Cultural Landscapes in Africa.* Paris:UNESCO.

Mumma, A. 2002. Legal aspects of cultural landscape protection in Africa. In *Cultural Landscapes: The Challenges of Conservation, Shared Legacy.* Common Responsibility Associated Workshops, World Heritage Papers: 156–159.

Mumma, A. 2003. Community-based legal systems and the management of World Heritage sites. In de Merode, E., Smeets, R. & Westrik, W. (eds.). *Linking Universal and Local Values: Managing a Sustainable Future for World Heritage.* Paris: UNESCO World Heritage Centre: 43–44.

Mumma, A. 2005. The link between traditional and formal systems. *Legal Frameworks for the Protection of Immovable Cultural Heritage in Africa.* ICCROM, Rome Italy: 22–24.

Munjeri, D., 2002. Smart Partnership: Cultural Landscapes in Africa. *World Heritage Papers,* 7: 134–142. UNESCO, Paris, France.

Munjeri, D. 2004a. Anchoring African cultural and natural heritage: The significance of local community awareness in the context of capacity-building. In de Merode, E., Smeets, R. & Westrik, W. (eds.). *Linking Universal and Local Values: Managing a Sustainable Future for World Heritage*. Paris: UNESCO World Heritage Centre: 75–80.

Munjeri, D. 2004b. Tangible and intangible heritage: From difference to convergence. *Museum International* 56(1–2): 12–20.

Munthali, S.M. 2007. Transfrontier conservation areas: Integrating biodiversity and poverty alleviation in Southern Africa. *Natural Resources Forum* 31: 51–60. https://doi.org/10.1111/j.1477-8947.2007.00130

Murimbika, M.T. 2006. Sacred powers and rituals of transformation: An ethnoarchaeological study of rainmaking rituals and agricultural productivity during the evolution of the Mapungubwe state, AD 1000 to AD 1300. University of the Witwatersrand. *PhD thesis.*

Muringaniza, J. 1998. Community participation in archaeological heritage management in Zimbabwe: The case of Old Bulawayo. University of Cambridge. *Unpublished MPhil thesis.*

Mwanja, R. 2014. Credibility of Sources in Determining Authenticity: The Case of Kasubi Tombs of Uganda. Paper presented during the NARA+20 Conference, Aichi, Japan. *Unpublished.*

National Development Planning Commission. 2016. *National Development 2030.* South Africa: Government of South Africa.

Ndlovu, N. 2011. Legislation as an instrument in South African heritage management: Is it effective? *Conservation and Management of Archaeological Sites* 13(1): 31–57. https://doi.org/10.1179/175355211X13097877338932

Ndoro, W. 2001. *Your Monument, Our Shrine: The Preservation of Great Zimbabwe.* Department of Archaeology and Ancient History. Studies in African Archaeology 19. Uppsala: Uppsala University.

Ndoro, W. 2003. Traditional and customary heritage systems: Nostalgia or reality? The implications of managing heritage sites in Africa. In de Merode, E., Smeets, R. & Westrik, W. (eds.). *Linking Universal and Local Values: Managing a Sustainable Future for World Heritage.* Paris: UNESCO World Heritage Centre: 81–84.

Ndoro, W. 2005. *Your Monument, Our Shrine: The Preservation of Great Zimbabwe.* Rome: ICCROM.

Ndoro, W. 2006. Building the capacity to protect rock art heritage in rural communities. In Agnew, N. & Bridgland, J. (eds.). *Of the Past to the Future: Integrating Archaeology and Conservation. Proceedings of the Conservation Theme of the 5th World Archaeological Congress Washington D.C., 22-26 June 2003.* Los Angeles: The Getty Conservation Institute: 336–339.

Ndoro, W. 2008. Legal definitions of heritage. In Ndoro, W., Mumma, A. and Abungu, G. (eds.). *Cultural Heritage and the Law. Protecting Immovable Heritage in English-Speaking Countries of Sub-Saharan Africa.* Rome: ICCROM: 25–36.

Ndoro, W. 2015. World Heritage sites in Africa. What are the benefits of nomination and inscription? In: *A Companion in Heritage Studies.* John Wiley & Sons, Inc, UK: 392–409.

Ndoro, W. 2017. Inaugural lecture on Mapungubwe Kingdom. Lecture presented during the launch of the MCLWHS lecture series, Limpopo, South Africa. *Unpublished presentation.*

Ndoro, W. & Chirikure, S. 2009. The practice of heritage management in sub-Saharan Africa. In Stanley-Price, N. & King, J. (eds.). *Conserving Authenticity.* Rome: ICCROM: 136–145.

Ndoro, W., Chirikure, S. & Deacon, J. 2018. *Managing Heritage in Africa: Who Cares?* London: Routledge.

Ndoro, W., Mumma, A. & Abungu, G. (eds.). 2008. *Cultural Heritage and the Law. Protecting Immovable Heritage in English-Speaking Countries of Southern Africa*. Rome: ICCROM Conservation Studies, 8.

Ndoro, W. & Wijesuriya, G. 2014. Heritage management and conservation: From colonization to globalization. In Meskell. L. (ed.). *Global Heritage: A Reader*. Oxford and West Sussex: John Wiley & Sons Ltd. 12: 131–149.

Nelson, R.H. 2003. Environmental colonialism: "Saving" Africa from Africans. *The Independent Review* 8(1): 65–86.

Neu, C. 2009. Daseinsvorsorge – Eine Einführung. In Neu, C. (ed.). *Daseinsvorsorge. Eine Gesellschaftswisserschaftliche Annaherung*. Wiesbaden: 9–19.

Nhamo, A., Saetersdal, T. & Walderhaug, E. 2007. Ancestral landscapes: Reporting on rock art in the border regions of Zimbabwe and Mozambique. In Pwiti, G. & Katsamudanda, S. (eds.) *The Ancestral Landscapes of Manyikaland. Zimbabwea*, NMMZ, Harare, Zimbabwe: 9: 43–62.

NHC, 2011. National Liberation Heritage Route (NLHR) Business Case. National Heritage Council of South Africa, Pretoria, South Africa. *Unpblished*.

Nicholas, L., Thapa, B. & Pennington-Gray, L. 2009. Public sector perspectives and policy implications for the Pitons Management Area World Heritage Site, St Lucia. *International Journal of Sustainable Development & World Ecology* 16(3): 205–216. https://doi.org/10.1080/13504500902919730

Nicholson, G. 2002. The common heritage of mankind and mining: An analysis of the law as to the high seas, outer space, the Antarctic and World Heritage. *New Zealand Journal of Environmental Law* 6: 177–198.

Nienaber, W.C., Keough, N., Steyn, M. & Meiring, J.H. 2008. Reburial of the Mapungubwe human remains: An overview of process and procedure. *The South African Archaeological Bulletin* 63(188): 164–169. http://www.jstor.org/stable/20475012

Nkoana-Mashabane, M. 2013. The 'S' in BRICS: An African Perspective. In Ngongula, G. (ed.). *Inspire South Africa*. Cape Town: All Media Africa International: 18–19.

Norton, P. 2000. *An Overview of Tourism Development Potential in the Mapungubwe Area*. Johannesburg: DBSA.

Operational Guidelines. 2019. Operational Guidelines on the Implementation of the 1972 World Heritage Convention. UNESCO, Paris, France. https://whc.unesco.org/en/guidelines/; *online guidelines*.

O'Connor, T.G. & Kiker, G.A.K. 2004. Collapse of the Mapungubwe society: Vulnerability of pastoralism to increasing aridity. *Climatic Change* 66(1/2): 49–66.

O'Neil, J. 2001. *Building Better Global Economic BRICs*. New York: The Goldman Sachs Group.

Okoduwa, A.I. 2008. Corruption: A quagmire from colonial rule in Nigeria, c1900-1960 A.D. A.A.U. *African Studies Review* 7.

Orij, R. 2010. Corporate social disclosures in the context of national cultures and stakeholder theory. *Accounting, Auditing & Accountability Journal* 23(7): 868–889. https://doi.org/10.1108/09513571011080162

Paul, S. 1987. *Community Participation in Development Projects. The World Bank Experience*. Washington DC: World Bank.

Pearson, M. & Sullivan, S. 1995. *Looking after Heritage Places: The Basics of Heritage Planning for Managers, Landowners and Administrators*. Melbourne: Melbourne University Press.

Pedersen, A. 2002. *World Heritage Sites: A Practical Manual for World Heritage Site Managers*. Paris: UNESCO World Heritage Centre.

Perbi, A. 2001. *Slavery and Slave Trade in Pre-colonial Africa*. www.latinamericanstudies.org/slavery/perbi.pdf [Retrieved 20 August, 2014]

Peterson, D., Gavua, K. & Rassool, C. (eds.). 2015. *Heritage Politics in Africa*. Cambridge: Cambridge UP.

Silverman, R. (ed.). 2014. *Museum as Process: Translating Local and Global Knowledges*. London: Routledge.

Stromgard, 1989. Adaptive Strategies in the Breakdown of Shift- ing Cultivation: The Case of Mambwe, Lamba, and Lala of Northern Zambia. *Human Ecology* 17(4): 428–444.

Peters, B.G. and Pierre, J., 1998. Governance without government? Rethinking public administration. *Journal of public administration research and theory*, *8*(2):223–243.

Philips, R. 2003. *Stakeholder Theory and Organizational Ethics*. San Francisco: Berret-Koehler Publishers.

Phillips, R., Freeman, E. & Wicks, A.C.W. 2003. What stakeholder theory is not. *Business Ethics Quarterly* 13(4): 479–502.

Pike, A., Rodríguez-Pose, A. & Tomaney, J. 2017. Shifting horizons in local and regional development. *Regional Studies* 51(1): 46–57. https://doi.org/10.1080/00343404.2016.1158802

Pikirayi, I. 2011. *Tradition, Archaeological Heritage Protection and Communities in the Limpopo Province of South Africa*. Ethiopia: OSSREA, African Books Collective.

Pikirayi, I. 2016. Archaeology, local knowledge, and tradition. The quest for relevant approaches to the study and use of the past in Southern Africa. In Schmidt, P. & Pikirayi, I. (eds.). *Community Archaeology and Heritage in Africa: Decolonizing Practice*. London: Routledge: 113–134.

Pikirayi, I., Sulas, F., Musindo, T.T., Chimwanda, A., Chikumbirike, J., Mtetwa, E., Nxumalo, B. & Sagiya, M.E. 2016. Great Zimbabwe's water. *Wiley Interdisciplinary Reviews: Water* 3(2): 195–210. https://doi.org/10.1002/wat2.1133.

Pollarolo, L. & Kuman, K. 2009. Field and technical report: Excavation at Kudu Koppie Site, Limpopo Province, South Africa. *South African Archaeological Bulletin* 64(189): 69–74.

Poncian, J. & Mgaya, E.S. 2015. Africa's leadership challenge in the 21st century: What can leaders learn from Africa's pre-colonial leadership and governance? *International Journal of Social Science Studies* 3(3): 106–115.

Porter, M.E. & Kramer, M.R. 2006. Strategy and society: The link between competitive advantage and corporate social responsibility. *Harvard Business Review* 84(12): 78–92.

Porter, M.E. & Kramer, M.R. 2011. Creating shared value. *Harvard Business Review* 89(1/2): 62–77. http://hbr.org/2011/01/the-big-idea-creating-shared-value

Porter, M.E. & Kramer, M.R. 2002. The competitive advantage of corporate philanthropy. *Harvard Business Review* 80(12): 56–69.

Porter, M.E. & Van der Linde, C. 1995. Green and competitive: Ending the stalemate. *Harvard Business Review* 73(5): 120–133.

Porter, M.E. & Reinhardt, F.L. 2007. A strategic approach to climate. *Harvard Business Review*: 1–4.

Post, J.E., Preston, L.E., Sauter-Sachs, S. & Sachs, S. 2002. *Redefining the Corporation: Stakeholder Management and Organizational Wealth*. Stanford: Stanford University Press.

Preston, L.E. & Sapienza, H.J. 1990. Stakeholder management and corporate performance. *Journal of Behavioural Economics* 19(4): 361–371.

Prieto, M.B., Clark, K., Munjeri, D., Nordenflycht, J. & Turner, M. 2014. From cacophony to symphony - The challenge of authenticity in voice when dealing with multiple heritage stakeholders. A discussion paper presented during the NARA+20 Conference, Aichi, Japan. *Unpublished paper.*

Pwiti, G. & Ndoro, W. 1999. The legacy of colonialism: Perceptions of the cultural heritage in Southern Africa with specific reference to Zimbabwe. *African Archaeological Review* 16(3): 143–153.

Pwiti, G. & Ndoro, W. 2001. Heritage management in Southern Africa: Local, national and international discourse. *Public Archaeology* 2(1): 21–34.

Qudrat-I Elahi, K. 2009. UNDP on good governance. *International Journal of Social Economics* 36(12): 1167–1180. https://doi.org/10.1108/03068290910996981

Rahman, A. 2005. *People's Budgeting at the Local Government Level in Bangladesh.* New York: United Nations.

Ralushai, N.M.N. & Gray, J.R. 1977. Ruins and traditions of the Ngona and the Mbedzi among the Venda of the Northern Transvaal. *Rhodesian History* 8: 1–11.

Ralushai, N.M. 2002. *A Preliminary Report on the Oral History of Mapungubwe.* Sibasa: Department of Environmental Affairs & Tourism.

Ralushai, N.M. 2003. *Mapungubwe Oral History Project. Additional Information on the Oral History of Mapungubwe.* Johannesburg: Department of Tourism and Environment. *Unpublished.*

Rees, W., Wackernagel, M. and Testemale, P., 1996. *Our ecological footprint: reducing human impact on the Earth* (pp. 3–12). Gabriola Island, BC: New Society Publishers.

Rhodes, R.A.W. 2007. Understanding governance: Ten years on. *Organization Studies* 28: 1243–1264.

Richard, A.J., Bell, D.C. & Carlson, J.W. 2000. Individual religiosity, moral community, and drug user treatment. *Journal for the Scientific Study of Religion* 39(2): 240–246. https://doi.org/10.1111/0021-8294.00019

Rico, T. 2008. Negative heritage: The place of conflict in World Heritage. *Conservation and Management of Archeological Sites* 10(4): 344–352.

Ripp, M. 2018. Heritage as a system and process that belongs to local communities: Reframing the role of local communities and stakeholders. *PhD project: Heritage-based urban development.* Senftenberg: Brandenburg University of Technology Cottbus. https://www.researchgate.net/institution/

Robertson, R. 1992. *Globalization - Social Theory and Global Culture.* London: Sage.

Rodney, W. 1973. *How Europe Underdeveloped Africa.* Dar es Salaam: Tanzania Publishing House.

Rodney, W. 1982. *How Europe Underdeveloped Africa.* Washington: Howard UP.

Rotberg, R.I., 2004. Strengthening governance: Ranking countries would help. *The Washington Quarterly, 28*(1): 71–81.

Rossler, M. 2006. World heritage cultural landscapes: A UNESCO flagship programme 1992 – 2006. *Landscape Research* 31(4): 333–353.

Rossler, M. 2010. World Heritage and cultural heritage management: New conservation challenges. Paper presented at XVth International Summer University at ISES, Koszeg, Hungary. *Unpublished.*

RDLR, 2017: *Rural Development and Land Reform (RDLR) Annual Report 2016/17.* Department of Rural Development and Land Reform, Pretoria, South Africa

Rüsen, J. 1996. Some theoretical approaches to intercultural comparative historiography. *History and Theory* 35(4): 5–22. https://doi.org/10.2307/2505441.

Sachs, J.D. 2012. From millennium development goals to sustainable development goals. *The Lancet* 379(9832): 2206–2211.

Sachs, S. & Rühli, E. 2011. *Stakeholders Matter: A New Paradigm for Strategy in Society.* Cambridge: Cambridge University Press.

Saetersdal, T. 2004. Places, People and Ancestors: Archaeology and Society in Manica, Mozambique. Bergen: University of Bergen. *Unpublished PhD thesis.*

Salazar, N.B. 2010a. *Envisioning Eden: Mobilizing Imaginaries in Tourism and Beyond.* London: Berghahn Books.

Salazar, N.B. 2010b. The globalization of heritage through tourism. In Labadi, S. & Long, C. (eds.). *Heritage and Globalization.* London: Routledge: 131–145.

Salkind, N.J. 2012. *Exploring Research*, 8th edition. University of Kansas: Pearson.

Samantha, M. 2012. Stakeholders: Essentially contested or just confused? *Journal of Business Ethics* 108(3): 285–298. https://doi.org/10.1007/s10551-011-1090-8.

Sandelowski, M. 2000. Combining qualitative and quantitative sampling, data collection, and analysis techniques in mixed-method studies. *Research in Nursing & Health* 23(3): 246–255. https://doi.org/10.1002/1098-240x(200006)23:3<246:aid-nur9>3.0.co;2-h

SANParks. 2005. *Mapungubwe National Park: Park Management Plan.* South African National Parks, Pretoria, South Africa.

SANParks. 2006. *Inaugural National Parks Week. Submission Made to the SANParks Executive Management by the Communications Department.* Pretoria, South Africa: SANParks. *Unpublished internal submission.*

SANParks. 2010a. *Greater Mapungubwe Transfrontier Conservation Area Integrated Development Plan.* Pretoria, South Africa: Greater Mapungubwe Transfrontier Conservation Area Trilateral Technical Committee.

SANParks. 2010b. *Mapungubwe National Park: Park Management Plan.* Pretoria, South Africa: South African National Parks: 9–70.

SANParks. 2011a. *About SANParks: Profile.* http://www.sanparks.org/about/default.php [Accessed 8 February 2018].

SANParks. 2011b. *State of Conservation Report for the Mapungubwe Cultural Landscape World Heritage Site: Department of Environmental Affairs.* Pretoria, South Africa: South African National Parks. *Unpublished.*

SANParks. 2012a. *Mapungubwe National Park Management Plan 2013.* Pretoria, South Africa: South African National Parks.

SANParks. 2012b. State of conservation report for the Mapungubwe Cultural Landscape World Heritage Site. Unpublished SOC Report Submitted to World Heritage Committee by Government of South Africa. Department of Environmental Affairs, South Africa.

SANParks. 2012c. *Parks A to Z: Kruger National Park.* http://www.sanparks.mobi/parks/kruger [Accessed 1 October 2012].

SANParks. 2012d. *Mapungubwe National Park and World Heritage Site Management Plan 2012.* Pretoria, South Africa: South African National Parks: 21.

SANParks. 2012e. *South African National Parks Strategic Plan for Fiscal 2012/13-2016/17.* Pretoria, South Africa: South African National Parks.

SANParks. 2013a. Mapungubwe National Park and World Heritage Site Management Plan (2013-2018). Department of Environmental Affairs, South Africa. *Unpublished.*

SANParks. 2013b. State of conservation report for the Mapungubwe Cultural Landscape World Heritage Site. Unpublished SOC Report Submitted to World Heritage Committee by Government of South Africa. Department of Environmental Affairs, South Africa.

SANParks. 2015. Mapungubwe World Heritage Site Environmental Management Framework. Pretoria, South Africa: South African National Parks. *Unpublished framework.*

SANParks. 2017. State of conservation report for the Mapungubwe Cultural Landscape World Heritage Site. Unpublished SOC Report Submitted to World Heritage Committee by Government of South Africa. Department of Environmental Affairs, South Africa.

SANParks. 2018. https://www.sanparks.org/about/news/default.php?id=1682 [Accessed May 2018].

SANParks. 2019. *Mapungubwe National Park and World Heritage Site Management Plan: 2019-2029.* Pretoria, South Africa: South African National Parks.

Schneider, V. 2012. Governance and complexity. In Levi-Faur, D. (ed.). *The Oxford Handbook of Governance.* Oxford: Oxford University Press: 129–142.

Scheyvens, R. 2003. *Tourism for Development, Empowering Communities.* New Jersey: Prentice Hall.

Schreiber, E.S.G., Bearlin, A.R., Nicol, S.J. & Todd, C.R. 2004. Adaptive management: A synthesis of current understanding and effective application. *Ecological Management & Restoration. Linking Science and Practice* 5(3): 177–182. https://doi.org/10.1111/j.1442-8903.2004.00206.x

Schwartz, J. & Cook, T. 2002. Archives, records and power: The making of modern memory. *Archival Science* 2: 1–19.

Selman, P. 2004. Community participation in the planning and management of cultural landscapes. *Journal of Environmental Planning and Management* 47(3): 365–392. https://doi.org/10.1080/0964056042000216519.

Selman, P. 2007. Landscape sustainability at the national and regional scales. In Benson, J.F. & Roe, M. (eds.). *Landscape and Sustainability*, 2nd Edition. London: Routledge: 104–117.

Shackelford, S.J. 2008. The tragedy of the common heritage of mankind. *Stanford Environmental Law Journal* 27: 101–120. http://papers.ssrn.com/sol3/papers.cfm?abstract_id=1407332

Sheridan, M. & Nyamweru, C. 2008. *African Sacred Groves: Ecological Dynamics and Social Change.* Oxford, Athens, & Pretoria: James Currey, Ohio University Press, and UNISA Press.

Simmons, D. 1994. Community participation in tourism planning. *Tourism Management* 15(2): 98–108.

Sinamai, A. 2003. Contested heritage: The socio-political contexts of Zimbabwe sites in Zimbabwe, South Africa and Botswana. South Africa: University of the Western Cape. *Unpublished Masters Thesis.*

Sinamai, A. 2019. *Memory and Cultural Landscape at the Khami World Heritage Site, Zimbabwe.* London: Routledge. https://doi.org/10.4324/9781351022026

Sinthumule, N.I. 2014. Land use change and bordering in the Greater Mapungubwe Transfrontier Conservation Area. University of Cape Town. *PhD thesis.*

Sinthumule, N. 2016. Multiple-land use practices in transfrontier conservation areas: The case of Greater Mapungubwe straddling parts of Botswana, South Africa and Zimbabwe. *Bulletin of Geography. Socio-Economic Series, Sciendo* 34(34): 103–115.

Sinthumule, N.I. 2017. Unfulfilled promises: An exposition of conservation in the Greater Mapungubwe Transfrontier. *Africa Today*, 64(1): 55–73. https://doi.org/10.2979/africatoday.64.1.03

Siravo, F. 2014. Planning and managing historic urban landscapes. In Bandarin, F. & van Oers, R. (eds.). *Reconnecting the City: The Historic Urban Landscape Approach and the Future of Urban Heritage.* 161–175.

Sivest Consortium. 2002. Mapungubwe Tourism Initiative Conceptual Tourism Destination Plan. *Unpublished plan.*

Siyathembana. 2012. Plan relating to the establishment of the Vele Colliery near Mapungubwe World Heritage Site, Musina, Limpopo Province: South Africa. Report prepared for Limpopo Coal Company (Pty) Ltd for submission to the Department of Environmental Affairs. *Unpublished report.*

Smith, A. 2010. Queer theory and native studies: The heteronormativity of settler colonialism. *GLQ* 16(1–2): 41–68. https://doi.org/10.1215/10642684-2009-012

Smith, L. 2004. *Archaeological Theory and the Politics of Cultural Heritage.* Oxford: Routledge.

Smith, L. 2006. *Uses of Heritage.* London/New York: Routledge, Taylor & Francis Group.

Smith, B.C. 2007. *Good Governance and Development.* New York: Palgrave Macmillan.

1972 World Heritage Convention Concerning the Protection of the World Cultural and Natural Heritage. Paris: World Heritage Center.

Ok Sofield, T. 2003. *Empowerment for Sustainable Tourism Development.* Oxford: Pergamon.

Songorwa, A.N. 1999. Community-based wildlife management (CWM) in Tanzania: Are the communities interested? *World Development* 27(12): 2061–2079.

Sparks, D., Madhlopa, A., Keen, S., Moorlach, M., Dane, A., Krog, P. & Dlamini, T. 2014. Renewable energy choices and their water requirements in South Africa. *Journal of Energy in Southern Africa* 25(4): 80–92.

Stacey, H. 1999. Environmental justice and transformative law in South African and some cross-jurisdictional notes about Australia, the United States and Canada. *Acta Juridica* 36: 36–70.

STATS, S.A. 2011. http://www.statssa.gov.za/?page_id=993&id=musina-municipality [Accessed June 2018, Cape Town, South Africa].

Stayt, H.A. 1931. *The Bavenda.* London: Oxford University Press for the International African Institute.

Steyn, M. 1994. An assessment of the health status and physical characteristics of the prehistoric populations from Mapungubwe. University of the Witwatersrand, Johannesburg. *PhD thesis.*

Steyn, M. & Nienaber, W.C. 2005. Repatriation of human remains in South African. In Strkalj, G., Pather, N. & Kramer, B. (eds.). *Voyages in Science - Essays by South African Anatomists in Honour of Phillip V. Tobias's 80th Birthday.* Pretoria: Content Solutions, South Africa: 159–177.

Steyn, M. & Nienaber, W. 2000. Iron Age human skeletal remains from the Limpopo Valley and Soutpansberg area. *Goodwin Series* 8: 112–116. South African Archaeological Society, South Africa. https://doi.org/10.2307/3858052

Stromgard, P. 1989. Adaptation strategies in the breakdown of shifting cultivation: The case of Mambwe, Lamba, and Lala of Northern Zambia. *Human Ecology* 17(4): 427–444.

Stovel, H. 2007. Effective use of authenticity and integrity as World Heritage qualifying conditions (online publication). *City & Time* 2(3): 3. http://www.ct.ceci-br.org

Strzepek, K.M., Gary, W.Y., Richard, S.J.T. & Mark, W.R. 2008 The value of the high Aswan Dam to the Egyptian economy. *Ecological Economics* 66(1): 117–126.

Sullivan, S. 2004. Local involvement and traditional practices in the World Heritage system. In de Merode, E., Smeets, R. & Westrik, C. (eds.). *Linking Universal and Local Values: Managing a Sustainable Future for World Heritage.* Paris: UNESCO World Heritage Centre: 49–55.

Swan, L. 1994. Early gold mining on the Zimbabwean Plateau. *Studies in African Archaeology* 9. Uppsala: Societas Archaeologica Upsaliensis.

Swan, L. 2007. Minerals and managers: Production contexts as evidence for social organization in Zimbabwean prehistory. Studies in Global Archaeology Series 12. PhD dissertation, Archaeology and Ancient History, University of Uppsala, Sweden

Szirmai, A. 2005. *The Dynamics of Socio-Economic Development: An Introduction*. Cambridge: Cambridge University Press.

Tapela, B.N. & Omara-Ojungu, P.H. 2012. Towards bridging the gap between wildlife conservation and rural development in post-apartheid South Africa: The case of the Makuleke community and the Kruger National Park. *South African Geographical Journal* 81(3): 148–155. https://doi.org/10.1080/03736245.1999.9713675

Taruvinga, P. and Ndoro, W., 2003. The vandalism of the Domboshava rock painting site, Zimbabwe: Some reflections on approaches to heritage management. *Conservation and management of archaeological sites*, *6*(1): 3–10.

Taruvinga, P. 2007. Community participation and rock art management in Zimbabwe. In Deacon, J. (ed.). *African Rock Art: The Future of Africa's Past*: TARA, Nairobi, Kenya: 39–48.

Taruvinga, P. 2008. Comments. In Chirikure, S. and Pwiti, G.,(authors). Community involvement in archaeology and cultural heritage management: An assessment from case studies in Southern Africa and elsewhere. *Current Anthropology*, *49*(3):467–485

Taruvinga, P. 2010. Community Participation and Heritage Management in Africa. Paper Presented at the Rock Art Workshop, Morocco. *Unpublished.*

Taruvinga, P. 2012. *AWHF Mission to Nigeria; World Heritage Nomination Training Programme*. Pretoria, South Africa. *Unpublished.*

Taruvinga, P. 2013. Heritage Cluster Report to RIM Council, Cape Town, South Africa. *Unpublished.*

Taruvinga, P. 2014. Conservation, tourism and social responsibility: Challenges of developing Robben Island World Heritage Site as a sustainable 'hybrid' institution. Paper Prepared for a Workshop on Heritage and Sustainable Development, Morocco. *Unpublished.*

Taruvinga, P. 2017a. Conservation, stakeholders and local politics: The management of the Matobo Hills World Heritage Site, South Western Zimbabwe. In Makuvaza, S. (ed.). *Aspects of Management Planning for Cultural World Heritage Sites*. Springer, UK: 147–161.

Taruvinga, P. 2017b. Living Heritage, Contestations and Conservation: The Case of Robben Island World Heritage Site, South Africa, Cape Town, South Africa. *Unpublished paper.*

Taruvinga, P. 2017c. Building robust heritage industries: Making heritage responsive to socio-economic needs of African State Parties. A paper presented during the *International Conference on African Cultures (ICAC 2017)*, National Art Gallery of Zimbabwe, 11–13 September 2017, Harare, Zimbabwe. *Unpublished.*

Taruvinga, P. 2018. Building bridges between theory and practice; aligning university curriculums to the interdisciplinary needs of World Heritage sites: Case of Robben Island World Heritage site. Paper presented at the UNESCO Workshop on "World Heritage & Educational Institutions in Africa", Great Zimbabwe, Zimbabwe, 19–21 April 2018. *Unpublished.*

Taruvinga, P. 2019. Stakeholders, conservation and socio-economic development: the case of Mapungubwe Cultural Landscape World Heritage Site, South Africa. PhD dissertation. Department of Archaeology, University of Cape Town, Cape Town, South Africa. *Unpublished.*

Taruvinga, P. 2020a. Betrayal of the pioneering spirit: Domesticating international commitments on sustainable development at World Heritage sites in Zimbabwe. ROSA, Harare. *In print.*

Taruvinga, P. 2020b. World Heritage, sustainable development and Africa. *Oxford Research Encyclopedia of Anthropology*. Oxford: Oxford University Press. https://doi.org/10.1093/acrefore/9780190854584.013.240

Taruvinga, P. & Ndoro, W. 2003. The vandalism of the Domboshava rock painting site, Zimbabwe: Some reflections on approaches to heritage management. *Conservation and Management of Archaeological Sites* 6(1): 3–10. https://doi.org/10.1179/135050303793137983

Taruvinga, P. & Varissou, S. 2013. Conservation practices at African World Heritage sites: The case of Kasubi Tombs and Great Zimbabwe. Midrand, South Africa: African World Heritage Fund. *Unpublished paper.*

Taylor, M. 2006. CBNRM and pastoral development in Botswana: Implications for San land rights. International Land Coalition, Rome. *Paper for Presentation at Workshop on Environment, Identity and Community Based Natural Resource Management: Experiences of the San in Southern Africa*, Oxford University African Studies Centre.

Taylor, R.E. & Judd, L.L. 1989. Delphi method applied to tourism. In Witt, S.F. & Moutinho, L. (eds.). *Tourism Marketing and Management Handbook*. UK:Prentice Hall: 95–98.

Thomas, D. & Preston, L.E. 1995. The stakeholder theory of the corporation: Concepts, evidence and implications. *Academy of Management Review* 20(1): 65–91.

Thompson, J.D., Zald, M.N. and Scott, W.R., 2017. *Organizations in action: Social science bases of administrative theory*. Routledge, UK.

Thompson, J. & Wijesuriya, G. 2018. From "sustaining heritage" to "heritage sustaining broader societal wellbeing and benefits": An ICCROM perspective. In Larsen, P.B. & Logan, W. (eds.). *World Heritage and Sustainable Development: New Directions in World Heritage Management*. London:Routledge.

Tilley, H. (ed). 2007. *Ordering Africa*. Manchester: Manchester UP.

Timmerman, P. 1986. Mythology and surprise in the sustainable development of the biosphere. In Clark, W.C. & Munn, R.E. (eds.). *Sustainable Development of the Biosphere*. New York: Cambridge University Press: 435–453.

Tomlinson, J.P.R. & Branston, R. 2017. Firms, governance and development in industrial districts. *Regional Studies* 52(10): 1410–1422.

Tosun, C. & Timothy, D. 2003. Arguments for community participation in tourism development. *Journal of Tourism Studies* 14(2): 2–11.

Trevor, T.G. & Mellor, E.T. 1908. *Report on the Reconnaissance of the North-Western Zoutpansberg District*. Transvaal Mines Department. Pretoria: Government Printer.

Trompenaars, F. 1993. *Riding the Waves of Culture: Understanding Cultural Diversity in Business*. London: Nicholas Brealey: 6–10.

Tunbridge, J.E. & Ashworth, G.J. 1996. *Dissonant Heritage: The Management of the Past as a Resource in Conflict*. UK: John Wiley & Sons.

Turner, R.K. 1988. *Sustainable Environmental Management*. New York: Belhaven Press.

Turner, S.D., 2012. World Heritage Sites and the extractive industries. *Independent study commissioned by IUCN in conjunction with the UNESCO World Heritage Centre, ICMM and Shell.[online] Available at:* ⟨ http://www. icmm. com/document/3787⟩(accessed 30 July 2015).

Turoff, M. & Hiltz, S.R. 1995. Computer-based Delphi processes. In Adler, M. & Ziglio, E. (eds.). *Gazing into the Oracle: The Delphi Method and Its Application to Social Policy and Public Health*. London: Kingsley Publishers: 56–88.

Turtinen, J. 2000. *Globalising Heritage: On UNESCO and Transnational Construction of a World Heritage*. Stockholm: Stockholm Center for Organizational Research.

Ullmann, A.A. 1985. Data in search of a theory: A critical examination of the relationships among social performance, social disclosure, and economic performance of US firms. *Academy of Management Review* 10(3): 540–557.

UNCSD. 2012. *Rio 012 issues briefs*. Produced by the UNCSD Secretariat, No. 6.

UNESCO. 2020. List of endangered World Heritage sites. UNESCO, Paris, France. http://www.whc.unesco.org/end/list/stat; *online data.*

UNESCO. 2014. How the 1972 World Heritage Convention contributes to sustainable development? Paris. http://www.whc.unesco.org [Accessed 26 July 2014].

UNESCO. 2011. *World Heritage in Africa Region, Main Results: Second Cycle Periodic Reporting.* Paris: UNESCO World Heritage Centre.

UNESCO. 1972. *The 1972 World Heritage Convention Concerning the Protection of the World Cultural and Natural Heritage.* Paris: UNESCO.

UNESCO. 2010. *WHC-10/34.COM/5D. 5D. 1972 World Heritage Convention and Sustainable Development.* Brasilia, Brazil: World Heritage 34 COM. *Unpublished.*

United Nations. 2013. *A New Global Partnership: Eradicate Poverty and Transform Economies Through Sustainable Development. The Report of the High-Level Panel of Eminent Persons on the Post-2015 Development Agenda.* New York: United Nations Publication.

UNDP. 1997. *Governance & Sustainable Human Development. A UNDP Policy Document.* New York: United Nations Development Program.

UNDP. 2009. *Fast Facts, United Nations Development Programme.* www.undp. org/governance/docs/Overview_pub_FastFactsDemGov.pdf [Accessed 1 June 2009].

UN. 2020. *World Population Prospects 2019.* Department of Economic and Social Affairs Population Dynamics. Population Division, United Nations, New York.

UN. 1992. United Nations Conference on Environment and Development, Rio de Janeiro, Brazil, 3-14 June 1992. https://www.un.org/en/conferences/environment/rio1992. Accessed 10 June 2018.

UNDP. 2020. *Human Development Report.* United Nations Development Programme, New York.

Ury, F., Ury, R., Ury, W. & Patton, B. 1983. *Getting to Yes: Negotiating Agreement Without Giving In.* London: Penguin Books.

Van Blarcom, B.L. & Kayahan, C. 2011. Assessing the economic impact of a UNESCO World Heritage designation. *Journal of Heritage Tourism* 6(2): 143–164. https://doi.org/10.1080/1743873X.2011.561858

Van der Merwe, P., Saayman, M., Du Plessis, L. & Kruger, M. 2009. A marketing analysis of overnight visitors to Mapungubwe National Park. Potchefstroom: Northwest University. *Unpublished research report.*

Van Doornum, B.L. 2005. Changing Places, Spaces and Identity in the Shashe Limpopo Region of Limpopo Province, South Africa. University of the Witwatersrand, Johannesburg, South Africa. *PhD dissertation.*

Van Schalkwyk, J.A. & Hanisch, E.O.M. 2002. *Sculpted in Clay: Iron Age Figurines from Schroda, Limpopo Province, South Africa.* Pretoria: National Cultural History Museum.

Van Schalkwyk, J.A. 1994. Randjies: Report on excavations of a Late Iron Age site on the Mogalakwena River, North Western Transvaal. *Research by the National Cultural History Museum* 3: 100–18.

Van Warmelo, N.J. 1940. The copper miners of Musina and the early history of the Zoutpansberg. *Ethnological Publication 8.* Pretoria: Government Printer.

Van Warmelo, N.J. 1953. Die tlokwa en birwa van Noord-transvaal. *Ethnological Publication 29.* Pretoria: Government Printer.

Venter, E. 2004. The notion of Ubuntu and communalism in African educational discourse. *Studies in Philosophy and Education* 23: 149–160. https://doi.org/10.1023/B:SPED.0000024428.29295.03

Viñals, M.J. & Maryland, M. 2012. Heritage, tourism and local community interactions within the framework of site management. In Albert, M.-T., Richon, M., Viñals, M.J. & Witcomb, A. (eds.). *Community Development Through World Heritage.* Paris: UNESCO: 40–47.

Voigt, E. 1984. Mapungubwe: An archaeological interpretation of an Iron Age community appendix III. *Transvaal Museum Monographs* 1(1). https://hdl.handle.net/10520/AJA090799001_130

Von Droste, B. 2012a. World Heritage and globalization: UNESCO's contribution to the development of global ethics. In Albert, M.T., Richon, M., Vinals, M.J. & Witcomb, A. (eds.). *Community Development Through World Heritage, World Heritage Papers (31).* Paris: UNESCO: 10.

Von Droste, B. 2012b. Empowerment of the Convention: From theory to practical application (1981–2000). *World Heritage Review* 64: 12–19.

Walters, C. 1986. *Adaptive Management of Renewable Resources.* New York: MacMillan.

Waterton, E. 2010. *Politics, Policy and the Discourses of Heritage in Britain.* UK: Palgrave Macmillan.

Waterton, E. & Watson, S. (eds.). 2015. *The Palgrave Handbook of Contemporary Heritage Research.* Palgrave and Macmillan.

WCED. 1987. *Our Common Future.* World Commission on Environment and Development (WCED). New York: Oxford University Press.

Welman, J.C., Kruger, F. & Mitchell, B. 2005. *Research Methodology*, 3rd edition. Cape Town: Oxford University Press Southern Africa.

Wentzel, P.J. 1983. *The Relationship between Venda and Western Shona, Vol. 3.* Pretoria: University of South Africa.

WHC. 2010. *World Heritage and Sustainable Development: World Heritage Committee Thirty-Four Session*, Brasilia, Brazil, 25 July–3 August 2010. Paris: UNESCO.

WHC. 2011. Mapungubwe Cultural Landscape (South Africa) (C 1099bis). World Heritage Centre, UNESCO, Paris. http://whc.unesco.org/en/soc/385/

WHC. 2012. Reactive Monitoring Mission to Mapungubwe World Heritage Site. World Heritage Centre, Paris, France.

World Commission on Culture and Sustainable Development Conference. 1995. UNESCO, Paris, France. https://unesdoc.unesco.org/ark:/48223/pf0000104333. *online*

Wood, D.J. & Jones, R.E. 1995. Stakeholder mismatching: A theoretical problem in empirical research on corporate social performance. *The International Journal of Organizational Analysis* 3(3): 229–267. https://doi.org/10.1108/eb028831

Wood, M. 2000. Making connections: Relationships between international trade and glass beads from the Shashe-Limpopo area. *Goodwin Series, Vol. 8, African Naissance: The Limpopo Valley 1000 Years Ago.* South African Archaeological Society, South Africa: 78–90.

Yager, T.R., Bermudez-Lugo, O., Mobbs, P.M., Newman, H.R., Taib, M., Wallace, G.J. & Wilburn, D.R. 2010. *The Mineral Industries of Africa. Minerals Yearbook, 3.*

Yousuf, M.I. 2007. Using experts' opinions through Delphi technique. *Practical Assessment, Research and Evaluation* 12(4). http://pareonline.net/getvn.asp?v=12&n=4

YuLong, L. & Hunter, C.Y.L. 2015. Community involvement for sustainable heritage tourism: A conceptual model. *Journal of Cultural Heritage Management and Sustainable Development* 5(3): 248–262.

INDEX

Note: Page references in *italics* denote figures and in **bold** tables.

Lightning Source UK Ltd.
Milton Keynes UK
UKHW051955070822
406931UK00005BA/21

9 780367 694807